# Application of Innovative Power Electronic Technologies

# Application of Innovative Power Electronic Technologies

Editors

**Ching-Ming Lai**
**Yitao Liu**

MDPI

Basel • Beijing • Wuhan • Barcelona • Belgrade • Novi Sad • Cluj • Manchester

*Editors*

Ching-Ming Lai
Department of Electrical Engineering, National Chung Hsing University
Taichung, Taiwan

Yitao Liu
College of Mechatronics and Control Engineering,
Shenzhen University
Shenzhen, China

*Editorial Office*
MDPI
St. Alban-Anlage 66
4052 Basel, Switzerland

This is a reprint of articles from the Topic published online in the open access journals *Energies* (ISSN 1996-1073), *Electricity* (ISSN 2673-4826), and *Electronics* (ISSN 2079-9292) (available at: https://www.mdpi.com/topics/power_electronic).

For citation purposes, cite each article independently as indicated on the article page online and as indicated below:

| |
|---|
| Lastname, A.A.; Lastname, B.B. Article Title. *Journal Name* **Year**, *Volume Number*, Page Range. |

**ISBN 978-3-0365-9634-1 (Hbk)**
**ISBN 978-3-0365-9635-8 (PDF)**
**doi.org/10.3390/books978-3-0365-9635-8**

# Contents

energies

MDPI

Article

# Starter for Voltage Boost Converter to Harvest Thermoelectric Energy for Body-Worn Sensors

**Grzegorz Blakiewicz, Jacek Jakusz and Waldemar Jendernalik ***

Faculty of Electronics, Telecommunications and Informatics, Gdańsk University of Technology, 80-233 Gdańsk, Poland; grzegorz.blakiewicz@pg.edu.pl (G.B.); jacek.jakusz@pg.edu.pl (J.J.)
* Correspondence: waldemar.jendernalik@pg.edu.pl

**Abstract:** This paper examines the suitability of selected configurations of ultra-low voltage (ULV) oscillators as starters for a voltage boost converter to harvest energy from a thermoelectric generator (TEG). Important properties of particularly promising configurations, suitable for on-chip implementation are compared. On this basis, an improved oscillator with a low startup voltage and a high output voltage swing is proposed. The applicability of n-channel native MOS transistors with negative or near-zero threshold voltage in ULV oscillators is analyzed. The results demonstrate that a near-zero threshold voltage transistor operating in the weak inversion region is most advantageous for the considered application. The obtained results were used as a reference for design of a boost converter starter intended for integration in 180-nm CMOS X-FAB technology. In the selected technology, the most suitable transistor available with a negative threshold voltage was used. Despite using a transistor with a negative threshold voltage, a low startup voltage of 29 mV, a power consumption of 70 μW, and power conversion efficiency of about 1.5% were achieved. A great advantage of the proposed starter is that it eliminates a multistage charge pump necessary to obtain a voltage of sufficient value to supply the boost converter control circuit.

**Keywords:** CMOS; low voltage; low power; starter; boost DC-DC converter; energy harvesting; Colpitts oscillator

**Citation:** Blakiewicz, G.; Jakusz, J.; Jendernalik, W. Starter for Voltage Boost Converter to Harvest Thermoelectric Energy for Body-Worn Sensors. *Energies* **2021**, *14*, 4092. https://doi.org/10.3390/en14144092

Academic Editor: Ching-Ming Lai

Received: 1 June 2021
Accepted: 4 July 2021
Published: 6 July 2021



## 1. Introduction

Rapid development of body-worn sensors systems integrated on a chip imposes a significant demand on small and efficient power supply circuits. Batteries are avoided in these types of applications because of their relatively large size and limited life span. Among alternative energy sources such as photovoltaic cells [1], vibration transducers [2], or the thermoelectric generators (TEGs) [3–5], the latter ones are widely used due to their small size, portability and suitability for on-body applications. TEGs exploit temperature gradient between the human skin and the ambient environment in practical situations. Small TEG size and low temperature difference (typically 1–2 °C) limit the upper voltage bound below 100 mV. Such a low voltage can be applied to slow circuits based on MOS transistors operating in the subthreshold region. However, many systems-on-chip implementing sophisticated functions require MOS transistors operating in the strong inversion region, which involves the use of supply voltages of around 1 V. Although boost converters can be used to raise voltages as low as 20–40 mV [6–11] to the required supply level, they also need a voltage within 1 V for efficient energy conversion. For this reason, an additional starter is used to allow "cold start" of the converter at low voltages. Such a circuit is turned on for a short time needed to start the converter, and then turned off once the converter supplies sufficient voltage to its own control circuits. In practical implementations of ULV boost converters, very significant difficulties are associated with realization of their starters.

Starter solutions designed to initialize the operation of on-chip boost converters are described in the literature [7–10,12–17]. The solutions can be classified into two major groups: starters based on classic ring oscillators using only transistors [12,16,17], and starters based on oscillators with inductors or transformers [7,9,10,12–14]. In ring oscillators, each stage must have a gain greater than one in order to start and sustain oscillations. Satisfying this condition at very low supply voltages is very difficult, where all transistors have relatively low transconductance and drain-source resistance [18–21]. The lowest supply voltage reported for this type of starter is 60 mV [13]. This level was achieved by applying a mechanism for technology corner detection and automatic reconfiguration of the ring oscillator to achieve almost optimal operating conditions. The ring oscillator using only transistors generates output voltages that are limited by the supply voltage, therefore, they need an additional multistage (in some solutions up to 40 stages [13]) charge pump to obtain a sufficiently high voltage. As a result, ring-oscillator-based starters are fairly complicated and have a relatively low efficiency. On the other hand, they can be fully integrated on a chip.

Inductor or transformer-based starters support much lower supply voltages. In these circuits, a transformer or inductor acts as a load for a MOS transistor, allowing for voltage swing greater than the supply voltage and eliminating the inevitable voltage drop across the active load as in the case of a transistor-only ring oscillator. Consequently, they are capable of self-starting at supply voltages as low as 25–40 mV [7,9]. Moreover, an output voltage within 1 V can be achieved without additional charge-pump-based voltage multipliers owing to boosting voltage in a transformer or a LC resonant tank. In this regard, such starters are preferable over ones based on ring oscillators owing to their reliability, even with the supply voltages of only 25–40 mV. The disadvantage of these solutions is the need for inductors, which often have to be implemented as off-chip components to meet the high quality factor requirement.

In terms of the feasibility of reducing the startup voltage, the starters based on the Colpitts oscillator are particularly promising [12,14,18,19]. The paper [19] gives an example implementation of a low-voltage Colpitts oscillator that operates even at 3.5 mV supply voltage. This circuit has indeed a very low inrush voltage but reveals a limitation of the output voltage swing. As a result, the circuit requires an additional multi-stage voltage multiplier to achieve a voltage close to 1 V.

This paper describes a low voltage starter based on an improved Colpitts oscillator featuring a high output voltage swing, which allows using only a few stages of a charge pump.

## 2. Ultra-Low-Voltage Oscillators

Three configurations of ULV oscillator potentially suitable as a boost converter starter are depicted in Figure 1.

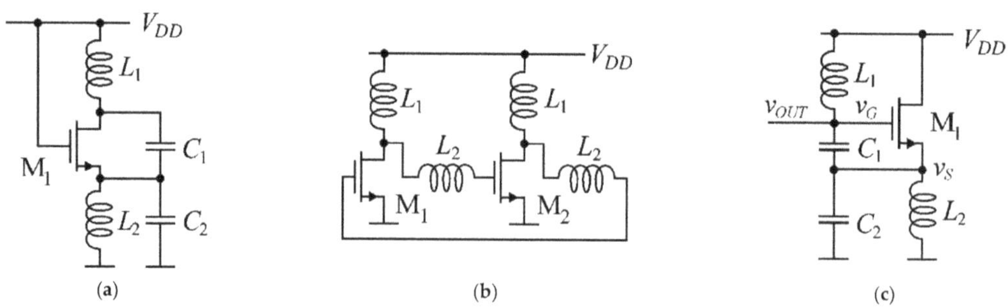


**Figure 1.** Ultra-low-voltage oscillators: (**a**) ESCO; (**b**) ESILRO; (**c**) proposed HSCO.

The oscillators in Figure 1a,b are ULV solutions discussed in [18,19], referred to as enhanced swing Colpitts oscillator (ESCO) and inductive-load enhanced swing ring oscillator (ESILRO). The circuit in Figure 1c is a proposed high swing Colpitts oscillator (HSCO). The common feature of all the ULV circuits is the application of the full supply voltage ($V_{DD}$) to bias the transistor gate and drain. In this way, a sufficiently high intrinsic gain of the transistor can be achieved at a relatively low supply voltage. For this reason, two inductors are necessary to avoid voltage drop between the $V_{DD}$ rail, drain and gate of the transistor. An unavoidable disadvantage of ESCO and ESILRO is limited voltage swing at the drains of the transistors. The reason for this effect is the pn junction between the drain and the substrate, which is forward biased during the negative halves of the generated periodic voltage waveform.

This undesirable effect is avoided in HSCO by connecting the resonant tank with the inductor $L_1$ to the transistor gate and design the feedback loop ($C_1$, $C_2$) so that the voltage scaling factor $v_G/v_S$ is much larger than unity. Under such conditions, the voltage swing on the resonant tank is virtually unlimited. At the same time, voltage across the pn junction formed between the source and the substrate is highly reduced, which prevents it from forward bias. In the small-signal model of the circuit, shown in Figure 2, the biasing inductor $L_2$ is represented by its dynamic resistance $R_{L2}$ at oscillation frequency $\omega_0$.

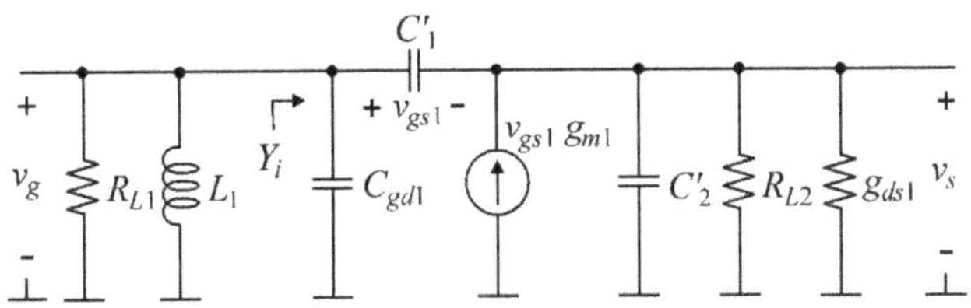


**Figure 2.** Small-signal model of HSCO shown in Figure 1c.

In Figure 2, the transistor $M_1$, represented by $g_{m1}$ and $g_{ds1}$, together with the capacitors $C'_1$, $C'_2$ and the dynamic resistance $R_{L2}$ form a circuit that implements the admittance $Y_i$ with negative real part (conductance). The negative conductance allows compensation for the losses of inductor $L_1$, represented by $R_{L1}$. The real and imaginary parts of the admittance $Y_i(\omega)$ seen from the gate of $M_1$ are

$$\mathrm{Re}\{Y_i(\omega)\} = \frac{\omega^2 C'^2_1 R_0 \left[1 - g_{m1} R_0 \frac{C'_2}{C'_1}\right]}{(1 + g_{m1}R_0)^2 + \omega^2 (C'_1 + C'_2)^2 R_0^2} \tag{1}$$

$$\mathrm{Im}\{Y_i(\omega)\} = \omega C_{gd1} + \frac{\omega C'_1 \left[1 + g_{m1}R_0 + \omega^2 R_0^2 C'_2 \left(C'_1 + C'_2\right)\right]}{(1 + g_{m1}R_0)^2 + \omega^2 \left(C'_1 + C'_2\right)^2 R_0^2} \tag{2}$$

where: $C'_1 = C_1 + C_{gs1}$, $C'_2 = C_2 + C_{sb1} - 1/(\omega^2 L_2)$, $R_0 = 1/(g_{ds1} + 1/R_{L2})$. The negative real part of the admittance $Y_i(\omega)$ can be obtained when the following condition is satisfied

$$\frac{C'_2}{C'_1} > \frac{1}{R_0 g_{m1}} \tag{3}$$

For a transistor operating at low supply voltage, this condition also means that $C'_2 >> C'_1$, due to the fact that $1/R_0 >> g_{m1}$. Therefore, further considerations are restricted to such a case. The $L_1$ inductor losses will be compensated and oscillations with frequency $\omega_0$ will start if the following conditions are met

$$\mathrm{Re}\{Y_i(\omega_0)\} < -1/R_{L1} \tag{4}$$

$$\mathrm{Im}\{Y_i(\omega_0)\} = -\frac{1}{\omega_0 L_1} \tag{5}$$

With the condition $C'_2 >> C'_1$, the Equation (1) can be simplified to

$$\text{Re}\{Y_i(\omega)\} \cong \frac{\omega^2 C'^2_1 R_0 \left(1 - g_m R_0 \frac{C'_2}{C'_1}\right)}{1 + \omega^2 C'^2_2 R_0^2} \tag{6}$$

The simplified Equation (6) was derived under assumption that $g_{ds1} >> g_{m1}$ which is satisfied when $M_1$ operates at very low supply voltages. To satisfy the Equation (4), transconductance of $M_1$ must fulfill

$$g_{m1} > \left(\frac{C'_1}{C'_2}\right)\frac{1}{R_0} + \left(\frac{C'_2}{C'_1}\right)\frac{1}{R_{L1}} + \frac{1}{\omega_0^2 C'_1 C'_2 R_0^2 R_{L1}} \tag{7}$$

The absolute minimum transconductance $g_{m1}$ can be achieved when both inductors are lossless ($R_{L1}, R_{L2} \to \infty$). In such a case, oscillations will start when the transconductance of $M_1$ meets the condition

$$g_{m1} > \left(\frac{C'_1}{C'_2}\right)\frac{1}{R_0} \overset{g_{ds1} \gg 1/R_{L2}}{\to} \left(\frac{C'_1}{C'_2}\right) g_{ds1} \tag{8}$$

It is worth noting that in theory by minimization of $C'_1/C'_2$ ratio an arbitrarily small value of $g_{m1}$ can be achieved, and thus an arbitrarily low supply voltage can be obtained. In practice the inductors are lossy, therefore $g_{m1}$ can be reduced by minimizing the last component of Equation (7) and by selecting the optimal value of the $C'_2/C'_1$ ratio. Reduction of the last component in Equation (7) requires that

$$\omega_0^2 C'_1 C'_2 R_0^2 \overset{g_{ds1} \gg 1/R_{L2}}{\to} \frac{\omega_0^2 C'_1 C'_2}{g_{ds1}^2} >> 1 \tag{9}$$

Under the assumption that Equation (9) is fulfilled, the optimal value of the capacitance ratio is given as

$$\frac{C'_2}{C'_1} = \sqrt{\frac{R_{L1}}{R_0}} \overset{g_{ds1} \gg 1/R_{L2}}{\to} \sqrt{g_{ds1} R_{L1}} \tag{10}$$

and the minimum transconductance required to start the oscillations is

$$g_{m1,min} > \frac{2}{\sqrt{R_{L1} R_0}} > 2\sqrt{\frac{1 + g_{ds1} R_{L2}}{R_{L1} R_{L2}}} \overset{g_{ds1} \gg 1/R_{L2}}{\to} 2\sqrt{\frac{g_{ds1}}{R_{L1}}} \tag{11}$$

Note that according to Equation (9), it is advantageous to choose the highest possible oscillation frequency ($\omega_0$) and the largest capacitances $C'_1$, $C'_2$. Thus, in order to minimize the transconductance required to start oscillations in Equation (7), small inductances $L_1$ and $L_2$ and large capacitances $C'_1$, $C'_2$ are advantageous, which is also very beneficial for integration of the oscillator.

## 3. Minimum Supply Voltage Required to Start ULV Oscillators

Low supply voltage and high output swing are important criteria for choosing an oscillator as a starter for ULV boost converter. To identify which of the configurations shown in Figure 1 is best suited as a starter, the minimum supply voltages of these circuits were determined and compared. In further considerations ESILRO is omitted due to the need for four inductors, which is difficult to integrate. N-channel MOS transistors with near-zero threshold voltage as well as negative threshold voltage are good candidates for implementing ULV oscillators because they provide relatively high drain current at supply voltages below a few tens of mV [18,19,21]. Therefore, the minimum supply voltages of ESCO and HSCO were determined for the transistor operating in the weak and strong inversion regions.

For the sake of simplicity, in the compared oscillators it was assumed that both inductances as well as their losses are identical ($L_1 = L_2$, $G_{L1} = G_{L2}$). According to the analysis presented in [19], oscillations will start in ESCO if the following condition holds

$$\frac{g_{ms1}}{g_{md1}} > a + \frac{a^2+1}{a-1}\frac{G_{L1}}{g_{md1}} \tag{12}$$

where

$$a = 1 + \sqrt{\frac{2G_{L1}/g_{md1}}{G_{L1}/g_{md1}+1}} \tag{13}$$

where $g_{ms1}$ and $g_{md1}$ are the source and drain transconductances [19,20]. The ratio of these conductances for strong and weak inversion regions can be approximated [19,20] by

$$\frac{g_{ms1}}{g_{md1}} = \frac{V_{DS}V_{T0}}{V_{DS}(1-n) - V_{T0}} \tag{14}$$

$$\frac{g_{ms1}}{g_{md1}} = \exp\left(\frac{V_{DS}}{U_T}\right) \tag{15}$$

where $V_{T0}$ is the threshold voltage, $n$ is the slope factor of the current-voltage characteristic in the weak inversion region, $U_T$ is the thermal voltage, and $V_{DS}$ is the drain-source voltage.

The minimum supply voltage necessary to start oscillations in ESCO can be determined from Equations (12)–(15) for the strong and weak inversion regions, respectively [19]

$$V_{DD,min} > \frac{\left(1-\frac{G_{L1}}{g_{md1}}\right)V_{T0}}{1-(1-n)\frac{G_{L1}}{g_{md1}}} \tag{16}$$

$$V_{DD,min} > U_T \ln\left(a + \frac{a^2+1}{a-1}\frac{G_{L1}}{g_{md1}}\right) \tag{17}$$

In HSCO, oscillations start when Equation (11) is satisfied. Based on this equation, the minimum ratio of the gate transconductance ($g_{m1}$) to the output conductance ($g_{ds1}$) of the transistor can be determined as

$$\frac{g_{m1,min}}{g_{ds1}} > 2\sqrt{\frac{G_{L1}}{g_{ds1}}} \tag{18}$$

where $G_{L1}/g_{ds1}$ represents the ratio of $L_1$ losses, modeled by $G_{L1} = 1/R_{L1}$, to the transistor output conductance. The ratio of the transistor transconductance to its output conductance for the strong and weak inversion regions can be defined [19,20] as

$$\frac{g_{m1}}{g_{ds1}} = \frac{nV_{DS}}{V_{DS}(1-n) - V_{T0}} \tag{19}$$

$$\frac{g_{m1}}{g_{ds1}} = \frac{1}{n}\left(\exp\left(\frac{V_{DS}}{U_T}\right) - 1\right) \tag{20}$$

The minimum supply voltage for HSCO, $V_{DD,min}$, determined based on Equations (18)–(20), is for the strong and weak inversion regions, respectively

$$V_{DD,min} > \frac{-2V_{T0}\sqrt{G_{L1}/g_{ds1}}}{n-2(1-n)\sqrt{G_{L1}/g_{ds1}}} \tag{21}$$

$$V_{DD,min} > U_T \ln\left(1 + n\sqrt{G_{L1}/g_{ds1}}\right) \tag{22}$$

The comparison of the minimum supply voltage, $V_{DD,min}$, defined by Equations (16), (17), (21) and (22), for ESCO and HSCO is plotted in Figure 3a,b, for the strong and weak inversion regions, respectively. The calculations for the strong inversion region are based on extracted values of $V_{T0}$ and $n$ for particular $V_{GS} = V_{DS} = V_{DD}$ voltage values for a native n-channel MOS transistor (W/L = 2500 μm/1 μm) in X-FAB 180-nm CMOS technology. The plots for the weak inversion were made for $U_T$ = 26 mV and similar values of the parameter $n$.

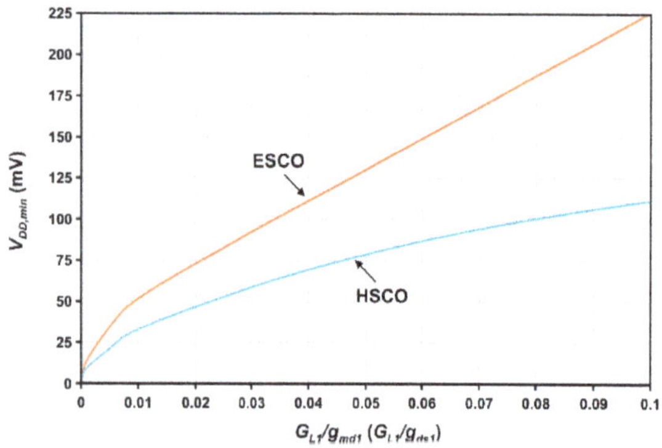


(a)

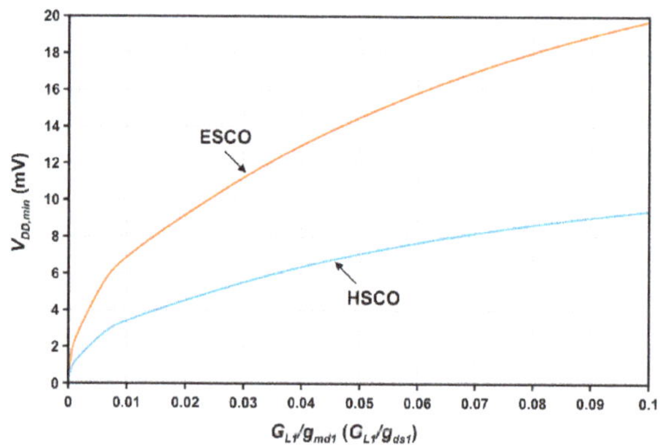


(b)

**Figure 3.** The minimum supply voltage, $V_{DD,min}$, required to start oscillations in ESCO and HSCO as a function of $G_{L1}/g_{md1}$ ($G_{L1}/g_{ds1}$), for the transistor operating in: (**a**) the strong inversion region; (**b**) the weak inversion region.

The results in Figure 3a,b show that HSCO requires significantly lower supply voltage compared to ESCO. This feature is particularly evident for high values of $G_{L1}/g_{ds1}$, which corresponds to use of low quality factor inductors. The general Equations (16), (17), (21) and (22), defining the minimum startup voltage, were derived based on a simplified linear model. To provide more in-depth investigation of the considered oscillators properties, number of HSCO and ESCO oscillator designs using native n-channel MOS transistors with negative threshold voltage were prepared for a 180-nm X-FAB technology. For each oscillator design, the component parameters were optimized to obtain the minimum startup voltage, $V_{DD,min}$, for selected values of: $G_{L1}/g_{md1}$ ($G_{L1}/g_{ds1}$), inductances $L_1 = L_2$, and inductor quality factors $Q_L$. A summary of the oscillators parameters is shown in Table 1.

**Table 1.** Summary of parameters of ULV oscillators.

| $G_{L1}/g_{ds1}$ | $V_{DD,min}$ | $C_1$ | $C_2$ | $L_1$ | $L_2$ | $Q_L$ | $f_0 = \omega_0/(2\pi)$ |
|---|---|---|---|---|---|---|---|
| | | | | **HSCO** | | | |
| 0.001 | 10 mV | 3.8 nF | 120 nF | 10 μH | 10 μH | 255 | 820 kHz |
| 0.005 | 24 mV | 4 nF | 60 nF | 10 μH | 10 μH | 51 | 826 kHz |
| 0.01 | 37 mV | 4.2 nF | 50 nF | 10 μH | 10 μH | 26 | 806 kHz |
| | | | | **ESCO** | | | |
| 0.001 | 20 mV | 73 nF | 9 nF | 10 μH | 10 μH | 275 | 770 kHz |
| 0.005 | 37 mV | 45 nF | 9 nF | 10 μH | 10 μH | 55 | 758 kHz |
| 0.01 | 54 mV | 45 nF | 11 nF | 10 μH | 10 μH | 29 | 715 kHz |

Based on the data from Table 1, one can see that HSCO oscillator, compared to the ESCO, exhibits lower inrush voltage for similar values of $G_{L1}/g_{ds1}$ and inductor quality factor $Q_L$. For the variants $G_{L1}/g_{ds1}$ = 0.001, $Q_L$ = 255, and $G_{L1}/g_{ds1}$ = 0.005, $Q_L$ = 51 a series of simulations for HSCO and ESCO were performed showing the peak-to-peak value of the output voltage, $v_{OUT\text{-}PP}$, as a function of the supply voltage $V_{DD}$ (Figure 4). For the high inductor quality factor case, a noticeable limitation of the output voltage is observed as the ESCO supply voltage increases. This effect is caused by forward biasing of the pn junction between the transistor drain and substrate.

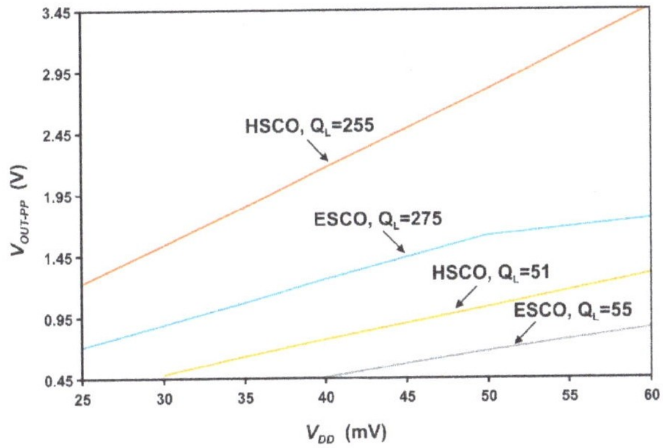


**Figure 4.** Peak-to-peak values of the output voltage, $v_{OUT\text{-}PP}$, as a function of the supply voltage, $V_{DD}$, for HSCO and ESCO.

For all the considered ULV oscillators, the main factor limiting amplitude of the output voltage is a relatively large drain to source conductance of a transistor. Due to this fact, the transistor internal gain is relatively small, especially in HSCO. The gain in this configuration can be increased by increasing the supply voltage $V_{DD}$, but this is unfavorable because ULV oscillators should operate for the lowest possible supply voltage. To overcome this limitation, an improved version of HSCO was developed, as shown in Figure 5.

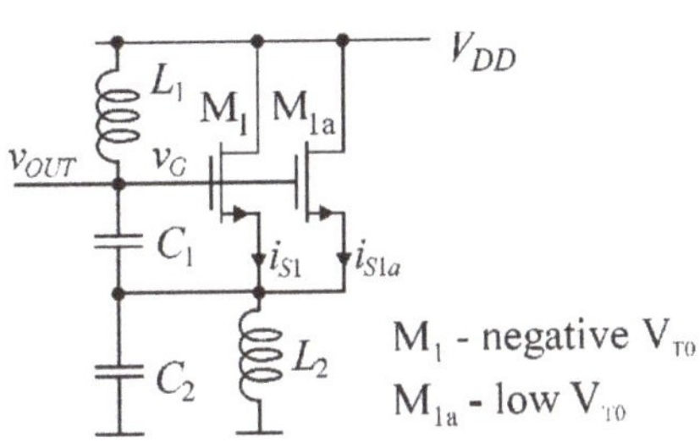


**Figure 5.** Improved version of HSCO.

In this oscillator, two transistors connected in parallel are used. $M_1$ is a transistor with negative or near-zero threshold voltage, whereas $M_{1a}$ is a low-threshold-voltage transistor. The transistor $M_1$ plays the same role as in the circuit of Figure 1c and serves mainly to initiate the oscillation, while $M_{1a}$ acts as an additional booster which is activated when the output voltage $v_{OUT}$ reaches a sufficient amplitude. Notice that introduction of $M_{1a}$ does not lead to noticeable increase of the output conductance of the composite transistor ($M_1$ + $M_{1a}$), because it is switched off for most of the time of the periodic waveform. This transistor is only turned on for short periods of time when the voltage at its gate exceeds a threshold value. Time waveforms illustrating operation of the circuit are shown in Figure 6.

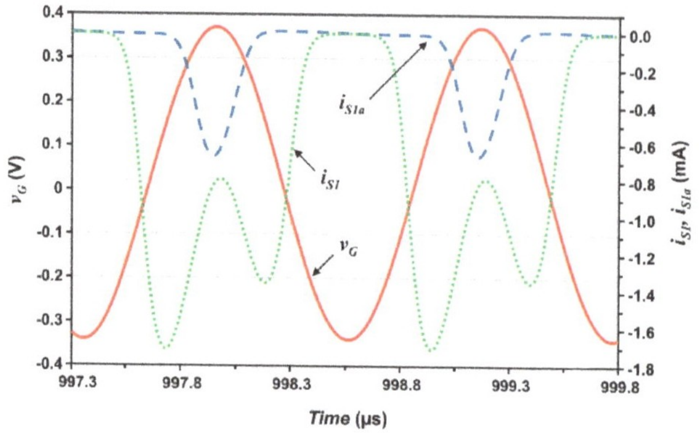


**Figure 6.** Time waveforms of the signals in the improved HSCO.

To demonstrate the advantage of the improved oscillator, the peak-to-peak values of the output voltage as a function of the supply voltage for ESCO, HSCO and improved HSCO are plotted in Figure 7.

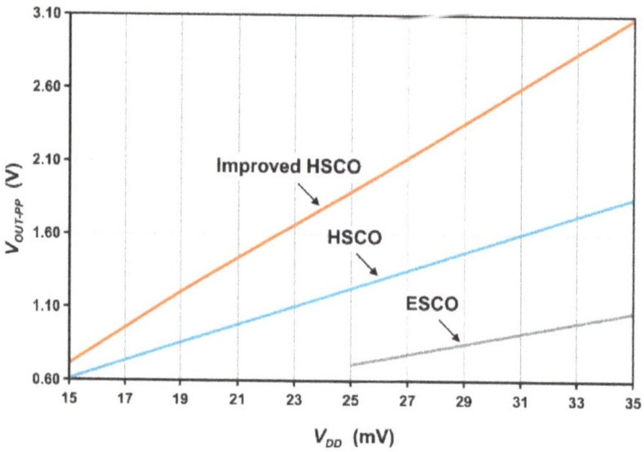


**Figure 7.** Peak-to peak values of the output voltage, $v_{OUT\text{-}PP}$, as a function of the supply voltage, $V_{DD}$, for ESCO, HSCO and improved HSCO ($G_{L1}/g_{ds1} = G_{L1}/g_{dm1} = 0.0025$, $Q_L = 100$).

## 4. Starter Based on the Improved HSCO

The key component of the proposed starter is the improved HSCO oscillator (Figure 5) which is capable of generating the highest output voltage at low supply voltages (Figure 7). As the plots in Figure 3 show, the most advantageous is to use a transistor with a near-zero threshold voltage that operates in the weak inversion region. In the chosen X-FAB 180 nm CMOS technology a zero-threshold-voltage transistor is not available. Thus, the most suitable transistor available was used, namely a n-channel native MOS transistor with a negative threshold voltage of approximately −180 mV. The starter consists of the improved HSCO and a 3-stage voltage multiplier, as shown in Figure 8.

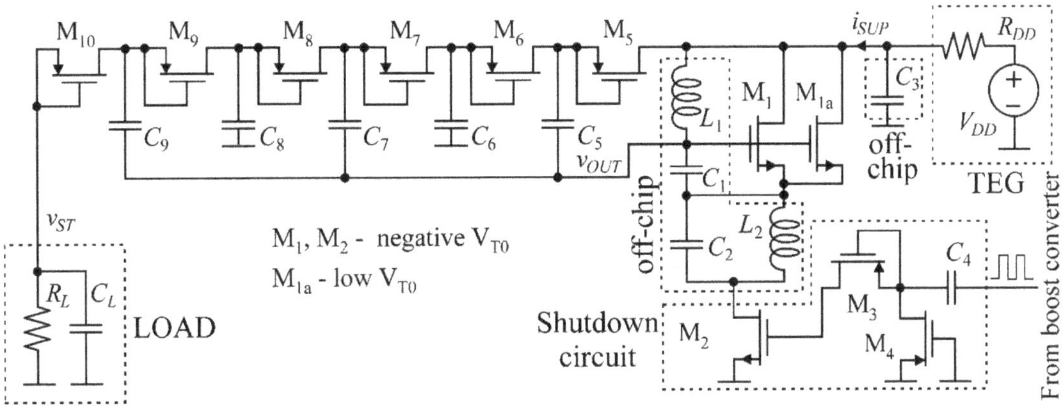


**Figure 8.** Starter based on the improved HSCO.

This solution is additionally equipped with a shutdown circuit which allows its disabling when the main boost converter starts running. The shutdown circuit, consisting of $M_2$, $M_3$, $M_4$, and $C_4$ is placed in the main current path and allows for complete cutoff of the supplying current. With this solution, when the main boost converter is running, the starter does not increase power consumed from the supply source. $M_2$ in the shutdown circuit requires a negative voltage to be turned off. This voltage is obtained through rectification (by means of $M_3$, $M_4$, and $C_4$) the square wave voltage applied to control the boost converter switches. The parameters of the starter components are given in Table 2.

**Table 2.** Parameters of the starter components.

| $M_1$ | $M_{1a}$ | $M_2$ | $M_3$, $M_4$ | $M_5$–$M_{10}$ |
|---|---|---|---|---|
| 1500 µm/1 µm | 75 mm/0.22 µm | 30 mm/1 µm | 0.22 µm/0.22 µm | 50 µm/0.22 µm |
| $C_1$<br>2.9 nF | $C_2$<br>100 nF | $C_3$<br>10 µF | $C_4$<br>0.5 pF | $C_5$–$C_9$<br>15 pF |
| $C_L$<br>25 pF | $R_L$<br>1 MΩ | $L_1$, $L_2$<br>10 µH | $Q_{L1}$, $Q_{L2}$<br>100 | $R_{DD}$<br>5 Ω |

The HSCO startup process is illustrated in Figure 9 showing the oscillator output voltage, $v_{OUT}$, waveform. Two phases of startup can be clearly observed in this plot. The first one, covering the time interval up to about 1.4 ms, is mainly related to operation of the transistor $M_1$. The second phase begins when the voltage amplitude exceeds the threshold voltage of $M_{1a}$. During this time interval, additional current pulses are generated to further increase the amplitude of $v_{OUT}$. The plot also illustrates the shutdown moment, which occurs at 2 ms.

The voltage waveform at the output of the starter, $v_{ST}$, is shown in Figure 10. It can be seen that obtaining $v_{ST}$ above 1 V became possible only after boosting the oscillator by the transistor $M_{1a}$. The effectiveness of the shutdown circuit can be determined based on the waveform of the current $i_{SUP}$ sourced from the supply $V_{DD}$. In shutdown state, $i_{SUP}$ decreases to a few tens of nA.

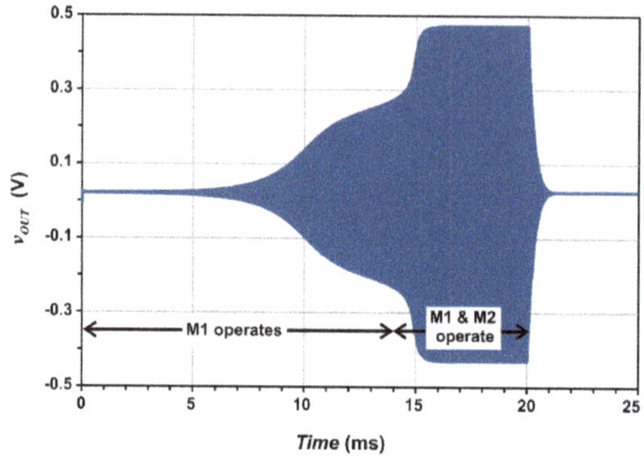


**Figure 9.** The output voltage, $v_{OUT}$, waveform in HSCO illustrating the startup process.

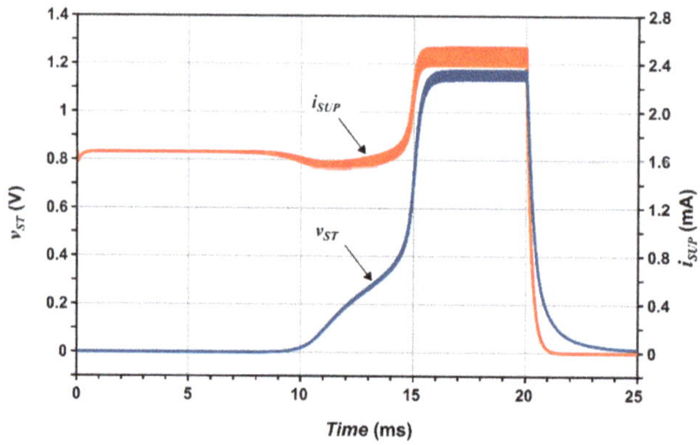


**Figure 10.** Waveforms of the output voltage ($v_{ST}$) and current ($i_{SUP}$) sourced from supply in the starter.

The effect of temperature changes and variations of technology parameters, represented by corners, on the startup voltage is shown in Figure 11. The worst case startup voltage is 29 mV at 50 °C under worst-speed (WS—slow NMOS, slow PMOS) and worst zero (WZ—slow PMOS, fast PMOS) corners. Therefore, the proposed starter enables reliable start at supply voltages greater than 29 mV.

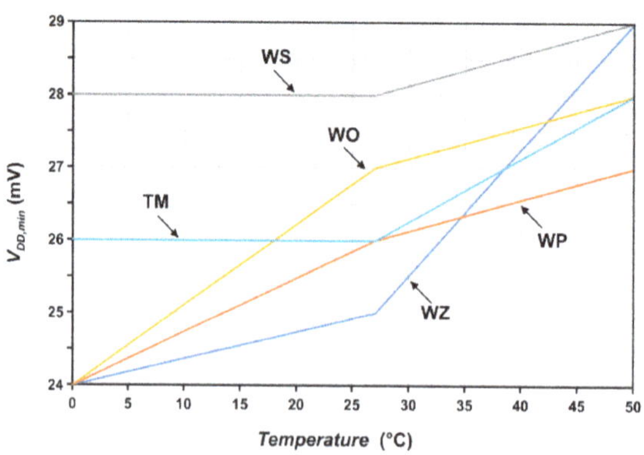


**Figure 11.** Minimal startup voltage, $V_{DD,min}$, as a function of temperature and variations of technology parameters, represented by corners.

A comparison of the main features of the developed starter with solutions reported in the literature is given in Table 3. The developed starter has a startup voltage which is within the lower range of voltages generated by TEGs. It should be emphasized that the startup voltage can be further reduced for CMOS technologies where zero threshold voltage transistors are available. The developed starter needs two external inductors ($L_1$, $L_2$), where $L_2$ can be of low quality, and two external capacitors ($C_1$, $C_2$). In the presented solution, the oscillation frequency is about 1 MHz, but can be increased to more than 10 MHz, allowing further reduction of external inductors and capacitors leading to a higher degree of miniaturization. The power efficiency of the proposed starter is approximately 1.5%. However, it should be emphasized that such low efficiency is not important since the starter operates only in the initial phase of the converter start-up and then is immediately turned off.

**Table 3.** Comparison of starters.

| Parameters | [7] | [9] | [10] | [12] | This Work [1] |
|---|---|---|---|---|---|
| CMOS technology | 130 nm | 130 nm | 65 nm | 130 nm | 180 nm |
| Minimal startup voltage | 21 mV | 40 mV | 50 mV | 11 mV | 29 mV |
| Output voltage | 1 V | 2 V | 1.2 V | 1 V | >1 V |
| Power consumption/conversion efficiency | N/A<br>N/A | N/A<br>N/A | N/A<br>N/A | N/A<br>N/A | 70 µW<br>1.5% |
| External components | Transformer<br>2 capacitors<br>1 diode | Transformer<br>3 capacitors<br>1 diode | 2 inductors<br>2 capacitors | 4 inductors<br>4 capacitors | 2 inductors<br>3 capacitors |

[1] Results of simulations.

## 5. Conclusions

A comparison of ULV oscillators, suitable for on-chip implementation, in terms of critical features for their use as starters in boost converters was presented. Based on the analysis, the improved oscillator configuration featuring a small startup voltage and a large output voltage swing was proposed. These features allow for simplification of the starter design. With the developed oscillator, a starter for on-chip implementation was designed that meets the requirements for using TEG as a power source. Compared to known solutions, the proposed starter offers greater miniaturization owing to the possibility of using smaller inductors with lower quality.

**Author Contributions:** Conceptualization, G.B.; methodology, G.B.; formal analysis, G.B.; investigation, G.B., J.J. and W.J.; writing—original draft preparation, G.B., J.J. and W.J.; writing—review and editing, G.B., J.J. and W.J. All authors have read and agreed to the published version of the manuscript.

**Funding:** This research was funded in part by National Science Centre of Poland under the grant 2016/23/B/ST7/03733.

**Institutional Review Board Statement:** Not applicable.

**Informed Consent Statement:** Not applicable.

**Data Availability Statement:** This study did not report any data.

**Conflicts of Interest:** The authors declare no conflict of interest. The funders had no role in the design of the study; in the collection, analyses, or interpretation of data; in the writing of the manuscript, or in the decision to publish the results.

## References

1. Prabha, R.D.; Rincon-Mora, G.A. CMOS photovoltaic-cell layout configurations for harvesting microsystems. In Proceedings of the IEEE International Midwest Symposium on Circuits and Systems (MWSCAS), Columbus, OH, USA, 4–7 August 2013; pp. 368–371. [CrossRef]
2. Ramadass, Y.; Chandrakasan, A. An efficient piezoelectric energy harvesting interface circuit using a bias-flip rectifier and shared inductor. *IEEE J. Solid State Circ.* **2010**, *45*, 189–204. [CrossRef]

3. Leonov, V.; Torfs, T.; Fiorini, P.; Van Hoof, C. Thermoelectric converters of human warmth for self-powered wireless sensor nodes. *IEEE Sens. J.* **2007**, *5*, 650–657. [CrossRef]
4. Yang, Y.; Wei, X.-J.; Liu, J. Suitability of a thermoelectric power generator for implantable medical electronic devices. *J. Phys. D Appl. Phys.* **2007**, *40*, 5790–5800. [CrossRef]
5. Morais, F.; Carvalhaes-Dias, P.; Duarte, L.; Spengler, A.; de Paiva, K.; Martins, T.; Cabot, A.; Siqueira Dias, J. Optimization of the TEGs Configuration (Series/Parallel) in Energy Harvesting Systems with Low-Voltage Thermoelectric Generators Connected to Ultra-Low Voltage DC–DC Converters. *Energies* **2020**, *13*, 2297. [CrossRef]
6. Carlson, E.; Strunz, K.; Otis, B. A 20 mV input boost converter with efficient digital control for thermoelectric energy harvesting. *IEEE J. Solid State Circ.* **2010**, *45*, 741–750. [CrossRef]
7. The, Y.-K.; Mok, P.K.T. Design of Transformer-Based Boost Converter for High Internal Resistance Energy Harvesting Sources with 21 mV Self-Startup Voltage and 74% Power Efficiency. *IEEE J. Solid State Circ.* **2014**, *49*, 2694–2704. [CrossRef]
8. Ramadass, Y.; Chandrakasan, A. A battery-less thermoelectric energy harvesting interface circuit with 35 mV startup voltage. *IEEE J. Solid State Circ.* **2011**, *46*, 333–341. [CrossRef]
9. Im, J.-P.; Wang, S.-W.; Ryu, S.-T.; Cho, G.-H. A 40 mV Transformer-Reuse Self-Startup Boost Converter with MPPT Control for Thermoelectric Energy Harvesting. *IEEE J. Solid State Circ.* **2012**, *47*, 3055–3067. [CrossRef]
10. Weng, P.-S.; Tang, H.-Y.; Ku, P.-C.; Lu, L.-H. 50 mV-input batteryless boost converter for thermal energy harvesting. *IEEE J. Solid State Circ.* **2013**, *48*, 1031–1041. [CrossRef]
11. Katic, J.; Rodriguez, S.; Rusu, A. A Dual-Output Thermoelectric Energy Harvesting Interface with 86.6% Peak Efficiency at 30 μW and Total Control Power of 160 nW. *IEEE J. Solid State Circ.* **2016**, *51*, 189–204. [CrossRef]
12. Radin, R.L.; Sawan, M.; Galup-Montoro, C.; Cherem Schneider, M. A 7.5-mV-Input Boost Converter for Thermal Energy Harvesting with 11-mV Self-Startup. *IEEE Trans. Circuits Syst. II Exp. Briefs* **2020**, *67*, 1379–1383. [CrossRef]
13. Dezyani, M.; Ghafoorifard, H.; Sheikhaei, S.; Serdijn, W.A. A 60 mV Input Voltage, Process Tolerant Start-Up System for Thermoelectric Energy Harvesting. *IEEE Trans. Circuits Syst. I Reg. Pap.* **2018**, *65*, 3568–3577. [CrossRef]
14. Lim, B.-M.; Seo, J.-I.; Lee, S.-G. A Colpitts Oscillator-Based Self-Starting Boost Converter for Thermoelectric Energy Harvesting with 40-mV Startup Voltage and 75% Maximum Efficiency. *IEEE J. Solid State Circ.* **2018**, *53*, 3293–3302. [CrossRef]
15. Blakiewicz, G.; Jakusz, J.; Kłosowski, M.; Jendernalik, W.; Szczepański, S. Light-Powered Starter for Micro Power Boost DC-DC Converter for CMOS Image Sensors. *Circuits Syst. Signal Process.* **2020**, *39*, 1195–1212. [CrossRef]
16. Goeppert, J.; Manoli, Y. Fully Integrated Startup at 70 mV of Boost Converters for Thermoelectric Energy Harvesting. *IEEE J. Solid State Circ.* **2016**, *51*, 1716–1726. [CrossRef]
17. Yi, H.; Yin, J.; Mak, P.-I.; Martins, R.P. A 0.032-mm$^2$ 0.15-V Three-Stage Charge-Pump Scheme Using a Differential Bootstrapped Ring-VCO for Energy-Harvesting Applications. *IEEE Trans. Circuits Syst. II Exp. Briefs* **2018**, *65*, 146–150. [CrossRef]
18. Galup-Montoro, C.; Schneider, M.C.; Machado, M.B. Ultra-Low-Voltage Operation of CMOS Analog Circuits: Amplifiers, Oscillators, and Rectifiers. *IEEE Trans. Circuits Syst. II Exp. Briefs* **2012**, *59*, 932–936. [CrossRef]
19. Machado, M.B.; Schneider, M.C.; Galup-Montoro, C. On the Minimum Supply Voltage for MOSFET Oscillators. *IEEE Trans. Circuits Syst. I Reg. Pap.* **2014**, *61*, 347–357. [CrossRef]
20. Enz, C.; Krummenacher, F.; Vittoz, E.A. An analytical MOS transistor model valid in all regions of operation and dedicated to low-voltage and low-current applications. *J. Anal. Integr. Circuits Signal Process.* **1995**, *8*, 83–114. [CrossRef]
21. Alioto, M. Ultra-Low Power VLSI Circuit Design Demystified and Explained: A Tutorial. *IEEE Trans. Circuits Syst. I Reg. Pap.* **2012**, *59*, 3–29. [CrossRef]

*energies*

*Article*

# Do Perceived Risk, Perception of Self-Efficacy, and Openness to Technology Matter for Solar PV Adoption? An Application of the Extended Theory of Planned Behavior

**Arifa Tanveer [1], Shihong Zeng [1,2,3,*], Muhammad Irfan [4,5,*] and Rui Peng [1,2,*]**

1 Applied Economics Department, College of Economics and Management, Beijing University of Technology, Beijing 100124, China; arifa_tanveerbjut@yahoo.com
2 Finance and Economics Development Research Center, College of Economics and Management, Beijing University of Technology, Beijing 100124, China
3 Beijing Modern Manufacturing Development Research Base of Beijing Philosophy and Social Sciences, Beijing University of Technology, Beijing 100124, China
4 School of Management and Economics, Beijing Institute of Technology, Beijing 100081, China
5 Center for Energy and Environmental Policy Research, Beijing Institute of Technology, Beijing 100081, China
* Correspondence: zengshihong@bjut.edu.cn (S.Z.); irfansahar@bit.edu.cn (M.I.); pengrui1988@bjut.edu.cn (R.P.); Tel.: +86-15652070825 (M.I.)

**Abstract:** Solar PV (photovoltaic) technology has gained considerable attention worldwide, as it can help reduce the adverse effects of $CO_2$ emissions. Though the government of Pakistan is adopting solar PV technology due to its environmental friendliness nature, studies focusing on consumer's acceptance of solar PV are limited in the country. This research aims to close this knowledge gap by looking into the various considerations that may influence consumers' willingness to adopt (WTA) solar PV for household purposes. The study further contributes by expanding the conceptual framework of the theory of planned behavior by including three novel factors (perceived risk, perception of self-efficacy, and openness to technology). The analysis is based on questionnaire data collected from 683 households in Pakistan's provincial capitals, including Lahore, Peshawar, Quetta, Gilgit, and Karachi. The proposed hypotheses are investigated using the state-of-the-art structural equation modeling approach. The empirical results reveal that social norms, perception of self-efficacy, and belief about solar PV benefits positively influence consumers' WTA solar PV. On the contrary, the perceived risk and solar PV cost have negative effects. Notably, the openness to technology has an insignificant effect. This study can help government officials and policymakers explore cost-effective, risk-free technologies to lessen the environmental burden and make the country more sustainable. Based on research results, study limitations, as well as prospective research directions, are also addressed.

**Keywords:** solar PV technology; consumers; willingness to adopt; theory of planned behavior; structural equation modeling; Pakistan

**Citation:** Tanveer, A.; Zeng, S.; Irfan, M.; Peng, R. Do Perceived Risk, Perception of Self-Efficacy, and Openness to Technology Matter for Solar PV Adoption? An Application of the Extended Theory of Planned Behavior. *Energies* **2021**, *14*, 5008. https://doi.org/10.3390/en14165008

Academic Editor: Carlos Henggeler Antunes

Received: 16 July 2021
Accepted: 11 August 2021
Published: 15 August 2021



## 1. Introduction

Climate change is a global environmental challenge, and to minimize its negative impacts, governments are implementing environmentally friendly technologies worldwide [1]. The persistent advancement in the population and globalization development has increased the energy demand [2,3]. Although being more technologically evolved than ever before, most countries still depend purely on crude oil to generate electricity [4]. About two-thirds of the world's increasing energy demand is owing to fossil fuels [5]. The continuous use of traditional sources in electricity generation is the primary cause of global climate change [6]. As a result, to combat global warming, the world must transition to clean energy sources [7,8]. Solar PV, a form of clean energy, has become more common in recent times and reached a global installed capacity of 303 GW, with a healthy annual

33-percent growth rate [9]. By 2025, solar PV is projected to meet 4% of the global electricity demand [10].

Similarly to other developing countries, Pakistan is facing environmental degradation and the worst energy crisis [11]. Due to Pakistan's massive population and economic growth, the energy demand has increased over the last two decades [12]. Pakistan's existing energy structure is heavily dependent on costly imported fuels such as oil and gas. Fuel imports from other countries are the only choice because Pakistan's domestic fuel supplies are inadequate to meet the country's energy demands. The country spends nearly 60% of its foreign exchange on fossil fuel imports [13]. An enormous dependence on traditional electricity generation methods is the main cause of environmental problems and a burden on economic development [14]. Solar PV is being introduced in Pakistan to address climate change issues and provide a long-term strategy for resolving the country's energy crisis [15].

Solar PV, a novel energy green technology, effectively decreases the cost of imported oil and minimizes $CO_2$ emissions [16]. Different countries have taken steps to raise the proportion of solar energy in their portfolio structure [17,18]. According to the sustainable global progress report 2020, solar PV rose 12% and generated electricity of 115 GW in 2019. Until 2019, the estimated worldwide solar PV output reached 627 GW [19].

Solar PV projects are believed to boost the quality of life for residents in numerous ways, such as providing job opportunities for people [15], they can help to reduce $CO_2$ emission [20], and it is the cheapest source of renewable energy and would be helpful to sustain the prices of electricity [21]. Electricity is considered a vital component in economic growth [22]. Solar PV technology helps to generate electricity at the household level and minimize carbon emissions [23]. The value of solar PV is highlighted by the fact that home appliances are one of the largest sources of $CO_2$, accounting for 70% of global emissions [24]. Fossil fuels (coal, oil, and gas) are the major energy sources of Pakistan [25]. The country spends USD 7 M annually on importing such costly energy sources [26]. According to the Pakistan Economic Survey, households are one of the largest energy consumers in Pakistan, with a share of 49.1% [27]. Other industries and sectors, including construction, transportation, and agriculture, also have a major share in energy consumption and carbon emission. As the country's energy structure is primarily based on fossil fuels, environmentally and economically friendly technologies are needed to minimize carbon emissions, on the one hand, and reduce the import of costly fossil fuels, on the other hand [28].

Solar PV is the most sustainable energy alternative for meeting the rising energy demand while also preserving the climate. Furthermore, the importance of solar PV adoption in Pakistan is demonstrated by [29], who claim that if 20% of the land in Baluchistan were covered in solar panels, then Pakistan's total electricity demand could be met. This demonstrates the considerable capability of solar PV technology to meet Pakistan's electricity needs. However, persistent electricity shortages can be addressed at the household level by using solar PV. Despite its enormous potential, solar PV has a low acceptance rate in Pakistan.

Several researchers have examined the dwellers' willingness to adopt solar PV in developed countries. Previous studies were conducted in economies where the governments already have defined strategic policies to achieve zero carbon emissions. Polo Lopez, Lucchi, and Franco [30] examined the potential, barriers, assessment criteria, and acceptance of building integrated photovoltaic in heritage buildings and landscapes. The authors concluded that it is necessary to protect the compatibility requirements for architecture and landscape. The possibilities of emerging solar products, which are currently not well-introduced in the market, are endless, thanks to advanced customization capabilities that enable a better integration into contexts of special heritage-protected buildings, thereby preserving their cultural and essential values. In another study, [31] established a conceptual framework for integrating solar energy systems into heritage sites and buildings to preserve their cultural and natural values, while lowering primary energy consumption, increasing comfort levels, mitigating environmental impacts, and enhancing the technical quality and economic outlays. The authors of [32] found that Australia attains 6% of its necessities from renewable resources. The authors concluded that Australia has diligently used renewable resources in recent years, which has had a positive environmental impact. According to the report of [33], the share of solar PV to generate electricity globally has increased by 28.3%. Irfan et al. [22] examined consumers' acceptance of solar energy by employing the theory of planned behavior from a Chinese perspective and discovered that consumers are positively influenced by awareness, self-efficacy, and the belief in the benefits of solar energy. Consumers' willingness to adopt solar PV is considered as the crucial factor to determine social acceptance [34]. Subsequent studies also identify the factors that may affect consumer's acceptance. Likewise, [35] investigated that acceptance is high when consumers believe that solar PV could mitigate carbon mission, benefit society, and increase job opportunities for people. Their study further revealed that descriptive factors such as age, income, education, and location also affect social acceptance [36].

A few scholars have conducted studies in developing countries as well to analyze the consumers' willingness to adopt solar PV. Alrashoud and Tokimatsu [37] examined considerations that may either empower or dissuade Saudi Arabian people from purchasing solar photovoltaic (PV) systems. The research found that education had the greatest positive effect, while the installation cost was the greatest barrier to adoption. Another study conducted by [38] in Ethiopia found that consumers have a high willingness to pay for solar energy to generate electricity, and the tendency to pay is positively influenced by economic variables such as age, income, and education. Likewise, [39] examined the solar PV social acceptation in six major states of India. They found that villagers show concern about the cost of solar PV, and they consider that solar PV is the best alternative, but it is more expensive in villages than in urban cities.

Along these lines, former researchers considered solar PV adoption from the following standpoints: (i) economic sustainability [40], (ii) social and environmental factors [41–43], (iii) barriers and drivers [29], (iv) the moderating role of policy and propaganda [24], (v) and financial incentives and subsidies [44]. Despite former researchers' long-standing interest, the inclination to find consumers' risk perceptions, self-efficacy, and openness to technology regarding solar PV adoption has been largely ignored. This research gap prompted us to add to the existing literature by evaluating the influence of these novel factors that could shape households' willingness to adopt solar PV. The study makes three major contributions in this regard. First, we considered all possible factors that may affect consumers' willingness to adopt (WTA) solar PV. Second, we added three new factors to the theory of planned behavior (TPB), including the perceived risk, perception of self-efficacy, and openness to technology. Previous studies have never taken these considerations into account in any context before, which is another contribution of this study. Third, unlike previous studies, the current research findings went beyond the previous research findings. For instance, the perceived risk appeared to be a significant factor in the acceptance of solar PV. Similarly, self-efficacy perception remains a vital component of TPB's theoretical structure. On the other hand, openness to technology plays a minor role in solar PV acceptance.

This study examines the willingness of Pakistani consumers to adopt solar PV using both existing and proposed novel factors. Pakistan has plenty of solar power resources, and if they are used properly, they can meet all of the country's current and potential energy needs [45]. However, as a developing country, Pakistan faces significant challenges in developing solar energy, including technological constraints, developers' reluctance to invest in solar PV generation, policies, and economic woes [46]. In addition, our research provides a robust conceptual framework by extending the TPB model and adding novel factors to better understand consumers' acceptance of solar PV.

The rest of the research is organized as follows: The theoretical framework is explained in Section 2. The formulation of hypotheses is shown in Section 3. Research methods are depicted in Section 4. The study's findings and implications are mentioned in Section 5. A discussion of research results is included in Section 6. Section 7 concludes the study, offers valuable policy guidance and discusses study limitations.

## 2. Theoretical Framework

To investigate the buying behavior of consumers, various theoretical models have been used in the literature, such as TPB, the theory of reasoned action (TRA), and theory of self-efficacy (TSE) [47,48]. However, TPB is a commonly used model that predicts and identifies consumer behavior [49]. Professor Ajzen was the first one to study behavioral intention, notably with Fishbein. The TPB model suggests that a person's behavioral intentions control his or her actions [50]. TPB explains that individuals' behavior is shaped by their notable beliefs and the subsequent evaluations of a particular action. Several researchers have employed this theory to evaluate consumer behavior in different fields and contexts (see Table 1).

**Table 1.** Usage of TPB in different fields and contexts.

| Theoretical Model | Country | Industry | Proposed Factors | Author |
|---|---|---|---|---|
| **TPB** | Portugal | Travel | Perceived behavioral control, subjective norm, and attitude | [51] |
| **TPB** | Denmark | Food | Perceived behavioral control, subjective norm, and attitude | [52] |
| **TPB** | Australia | Health | Perceived behavioral control, subjective norm, and attitude | [53] |
| **TPB** | Australia | Education | Attitude, social norm, and behavioral intentions | [54] |
| **TPB** | Lithuania | Solar | Attitude, environmental concern, and subjective norm | [55] |
| **TPB** | Malaysia | Solar | Attitude, subjective norm, and behavior | [56] |
| **TPB** | The Netherlands | Agriculture | Behavior, normative, and control belief | [57] |
| **TPB** | China | Solar | Perception about self-effectiveness, belief of solar energy benefits, and perception of neighbors' participation | [58] |
| **TPB** | Pakistan | Health | Risk perceptions of the pandemic, perceived benefits of face masks, and unavailability of face masks | [59] |
| **TPB** | Pakistan | Health | Self-efficacy, perceived risk, pandemic knowledge, and ease of pandemic prevention adoption | [60] |

As previously stated, TPB's model can identify and forecast consumer behavior adoption. Several researchers have added additional variables to the model to enhance behavior prediction and clarify why certain people find it hard to put their good intentions into effect [61–63]. The TPB model can incorporate other critical variables that specifically affect behavior and intention, in addition to the factors that make up the theory itself. Based on the justification from the literature, we added three additional factors (perceived risk, perception of self-efficacy, and openness to technology) with existing factors (belief about solar PV benefits, solar PV cost, and social norms) to the TPB model to investigate consumers' WTA solar PV. The study's framework is portrayed in Figure 1. Consumers' adoption was negatively influenced by perceived risk. Perceived risk was the customers' assessment of the probability of safety and security incidents and the corresponding consequences. Reasonably, perceived risk may negatively influence the dwellers' willingness. Perception of self-efficacy indicated "a person's assessment of how easy or difficult it is to conduct a specific action". Individuals believed that they had the requisite expertise, resources, or opportunities to adopt new technology successfully. The perception of self-efficacy plays a crucial role and positively influenced consumers' adoption. The third factor was openness to technology that was defined as "whether consumers try new technologies or stick with existing ones". We examined households' opinions by determining their willingness to adopt solar PV (see Figure 1).

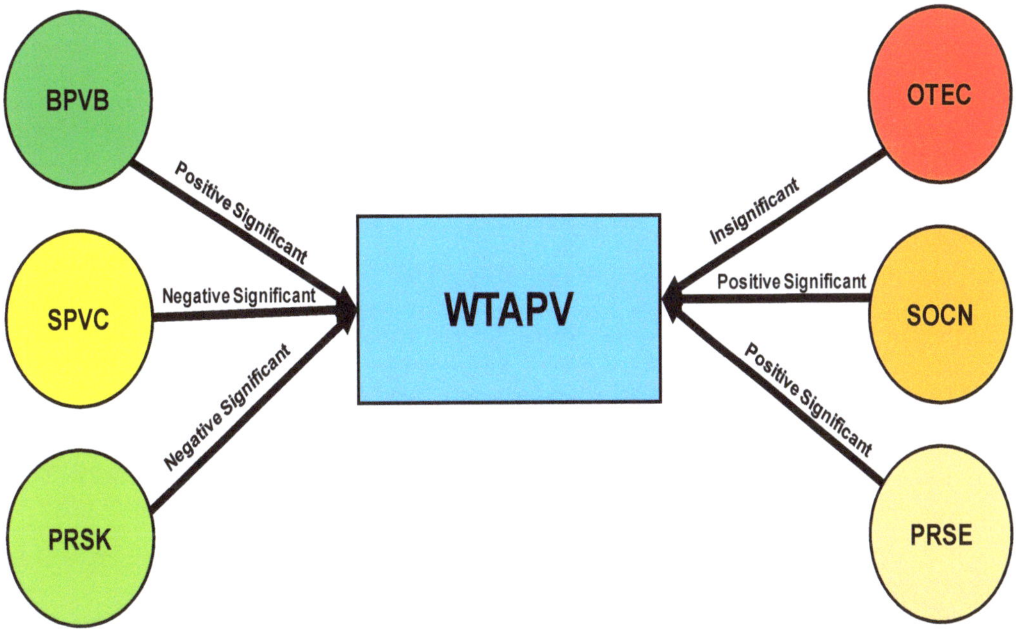


**Figure 1.** Study framework depicting the adoption of solar PV. Notes—BPVB: belief about solar PV benefits; SPVC: solar PV cost; PRSK: perceived risk; OTEC: openness to technology; SOCN: social norms; PRSE: perception of self-efficacy; WTAPV: willingness to adopt solar PV.

## 3. Development of Research Hypotheses

### *3.1. Belief about Solar PV Benefits (BPVB)*

Belief about solar PV benefits (BPVB) was defined as consumers' perceptions about solar PV that it has various benefits such as energy protection measures, combating climate change, and energy-saving [64]. The study found that farmers' assumptions about perceived benefits had a major impact on their adoption in India [65]. People distinguished between traditional and renewable energy sources, making purchasing decisions based on their socio-economic status [66]. As a result, initiatives are required to increase residents' understanding of the advantages of solar power use, such as improved air quality and lower carbon emissions, as well as to educate them about the negative consequences of dependence on energy that is based on fossil fuel [67]. Further research carried out by [68] in Sweden found that folks who are constantly aware of the benefits of solar PV and how it helps to alleviate the burden of electricity are highly motivated to adopt solar PV. In light of these considerations, we formulated the following hypothesis:

**Hypothesis 1 (H1).** *Consumers' willingness to adopt solar PV is positively influenced by BSPVB.*

### *3.2. Solar PV Cost (SPVC)*

As per the results of previous studies, there is a negative correlation between the cost of solar energy and its public acceptance. For instance, [69] concluded that solar PV was still more costly than traditional electricity. During purchasing circumstances, consumers also think about cost details to address monetary deficits [70]. Solar energy is relatively expensive since solar PV projects necessitate a large initial investment [15]. According to [38], rising prices are discouraging customers from adopting solar PV. Rising prices are the major deterrent to solar PV deployment [71]. Another study was conducted to determine consumers' desire to adopt solar home systems and solar PVs. The research results indicate that respondents are discouraged by high costs [72]. As a result of these findings, we came up with the following hypothesis:

**Hypothesis 2 (H2).** *Consumers' willingness to adopt solar PV is negatively impacted by SPVC.*

### *3.3. Perceived Risk (PRSK)*

In this context, perceived risk refers to a customer's belief that solar PV technology is risk-oriented. Consumers utilizing a new service or product were concerned about safety risks. This has a significant effect on consumers' interest in that particular product or service and, thus, on the acceptance of that technology [73]. A study by [74] reported that consumers should feel comfortable when interacting with technologies to increase adoption. Furthermore, results revealed that customers were hesitant to implement internet of things (IoT) services due to a lack of protection. Lee [75] discovered that security risk negatively affects consumers' attitudes toward online banking. In terms of customer confidence in adoption, the security risk is frequently cited as a major concern. Similarly, [76] also found that attitudes toward smart meter adoption were negatively influenced by perceived danger, to which privacy and safety issues are important antecedents. The study suggests the following hypotheses based on the preceding literature review:

**Hypothesis 3 (H3).** *Consumers' willingness to adopt solar PV is negatively influenced by PRSK.*

### *3.4. Openness to Technology (OTEC)*

Individual judgments about the usefulness of technology would be influenced by their willingness to try new things [77]. In recent years, research has increasingly focused on the impact of transparency on people's interactions with technology. Openness to technology has been linked to an inquiring mind, intelligence, and intellectual interests because it reflects the responsiveness of an individual to new perspectives [78]. High openness to technology is described as a desire to learn and understand things that are new to them, and embrace innovative ways to solve problems and adopt new technology [79]. Individuals vary in their level of openness towards technology due to their diverse backgrounds and life experiences [80]. Watjatrakul [81] found that students eager to learn new things want to do so in a constantly changing world. The practical importance of online learning is positively influenced by openness to technology. Students who can take risks and try new things are more concerned about the content of online learning. Based on these study findings, we formulated the hypothesis as follows:

**Hypothesis 4 (H4).** *Consumers' willingness to adopt solar PV is positively impacted by OTEC.*

### *3.5. Social Norms (SOCN)*

Social norms refer to the social influence on consumers' WTA solar PV [82]. Every country has its specific economic characteristics and individual culture [83]. Researchers observed that farmers inspired other farmers to engage in conservation initiatives by recommending them to attend agro-environmental programs [84]. It means that social pressure is often present during the execution of a specific action [85]. Another study conducted by [86] in three villages of South Korea's Geumsan county examined how residents who engage in the rural landscape development program have a substantial effect and motivation on other residents' perceptions. The majority of participants demonstrated an interest in and involvement in group activities. Lopes et al. [87] found a significant link between SOCN and residents' energy conservation behavior. According to [88], the actions of society have a huge influence on residents' intentions to purchase solar PV. Subsequently, it is essential to look at the impact of this crucial factor on WTA solar PV in Pakistan. Thus, we formulated the hypothesis as follows:

**Hypothesis 5 (H5).** *Consumers' willingness to adopt solar PV is positively impacted by SOCN.*

*3.6. Perception of Self-Efficacy (PRSE)*

In the context of solar PV, understanding regarding self-efficacy refers to how convenient or difficult it is for consumers to adopt solar PV [89]. PRSE has been shown in many studies to have a substantial impact on consumers' decisions. According to a study, PRSE has a favorable impact on customer perceptions of solar PV adoption [90]. Talpur et al. [91] revealed that PRSE was discovered to be a significant factor in the acceptance of solar PV. Another study conducted by [92] provides support for applying the theory of planned behavior in the United States on adolescents' perceptions of competence to protect themselves in their future occupational workforce. Similarly, PRSE influences customers' decision making, as per [93]. According to the analysis, PRSE plays a significant role in buyers' decisions to support solar PV. Thus, we formulated the hypothesis as follows:

**Hypothesis 6 (H6).** *The willingness of consumers to adopt solar PV is positively influenced by PRSE.*

## 4. Research Methods

*4.1. Target Population: Provincial Capitals of Pakistan*

The study's target population included residents of all five provincial capitals (Lahore, Peshawar, Quetta, Gilgit, and Karachi). As the Pakistani government intends to develop provincial capitals and make them the mainstream regions of the country, the urbanization trend has been steadily growing. The majority of the population is migrating from rural areas to the country's capital centers, searching for better jobs, education, and healthcare opportunities. In view of the economic, energy, and resource structure, these provincial capitals represent the country's unique characteristics, and the energy demand is increasing day by day in these capitals [94]. For instance, Lahore, the provincial capital of Punjab, is usually recognized as the cultural capital of Pakistan. This is Pakistan's second-largest city, with a population of 12,642,000 in 2020 [95]. Due to the rapid increase in population, the city faces a huge electricity shortage. Peshawar is the provincial capital of KPK. It is located next to the eastern terminus of the historic Khyber Pass near to the Afghan border. The city covers an area of 1257 $km^2$ and has a population of about 2 M [96].

On the other hand, Baluchistan is Pakistan's largest province by area, with Quetta as the provincial capital and the province's most urbanized city, hosting 29% of the province's total urban population. Provincial capital Karachi is the largest city in Sindh province and is known as the business center of Pakistan and the world's second-largest Islamic city with a population of about 24 M [97]. Gilgit is the fifth provincial capital of Pakistan. It possesses an enormous economic potential through tourism, tremendous renewable energy resources, minerals, and precious stones, and its strategic location that facilitates Pakistan's only road-to-road trade link with China: the CPEC's linchpin [98].

The survey description is provided in Table 2. The following three criteria were considered while conducting the questionnaire survey. (i) Due to COVID-19's second wave, Pakistan's government imposed a smart lockdown in all of the country's cities at the end of December 2020. Therefore, the authors used an online survey to collect data. (ii) We chose 800 participants using a convenient random sampling methodology [99], and a total of 683 responses were received. (iii) We followed the Comfrey and Lee's scale to determine the adequacy of sample size. For instance, [100] recommended the following scale: (very poor—50), (poor—100), (fair—300), (very good—500), (excellent—1000 or more). According to this scale, our study sample size (683 respondents) falls under the "excellent" category, ensuring that the sample size represents this research and supports its findings.

**Table 2.** Survey description.

| Parameters | Value |
|---|---|
| **Time frame** | January, February, March (2021) |
| **Location of the survey** | Provincial capitals of Pakistan |
| **Size of the sample** | 800 |
| **Valid responses** | 683 |
| **Response rate** | 85.4% |

### 4.2. *Questionnaire Development*

The questionnaire's items were mainly adopted from past studies that had been checked through a thorough review. The sources of data used to calculate each item and the development of the questionnaire are mentioned in Table 3. The self-administered questionnaire consisted of two parts: one for profiling and another for the elements that were used to assess each construct. In Section A, there were six questions about gender, age, marital status, as well as education, household income, and occupation. Section B had 39 questions, including seven questions for belief about solar PV benefits, social norm, and perception of self-efficacy—five questions were about solar PV cost and perceived risk. Four questions were asked about openness to technology and willingness to adopt solar PV. Section B items were rated on a seven-point Likert scale ranging from 1 (strongly disagree) to 7 (strongly agree). The detailed questionnaire is reported in Appendix A.

**Table 3.** Measurement source.

| Constructs | Items | Source |
|---|---|---|
| **BPVB** | 7 | [101] |
| **SPVC** | 5 | [101] |
| **PRSE** | 7 | [101] |
| **PRSK** | 5 | [102] |
| **OTEC** | 4 | [103] |
| **SOCN** | 7 | [104] |
| **WTAPV** | 4 | [105] |

## 5. Data Analysis and Empirical Results

To investigate the research hypotheses and model that were considered, SEM was used by employing SPSS and AMOS version 26. SEM is a realistic method that provides reliable and concrete results when determining the relationship between various variables [106]. The method has several advantages over conventional techniques. A precise estimation of measurement errors and observed variables is used to estimate latent factors and validate the model for pattern evaluation and execution [107]. Furthermore, the majority of multivariate methods implicitly ignore the measurement error. On the other hand, the SEM measures all dependent and independent variables by taking the measurement error into account [108]. The technique produces reliable and eloquent results due to its robustness and reliability [109]. Considering the benefits of SEM, we used it in our research because it was the most efficient way to evaluate the relationship between all of the variables under consideration. The descriptive statistics of the data are presented in Table 4. The descriptive statistics were scrutinized by means and standard deviations.

**Table 4.** Descriptive statistics of the data.

| Variables | Observations | Items | Mean | Std. Dev | Coefficient of Variation (CV) |
|---|---|---|---|---|---|
| **BPVB** | 683 | 7 | 3.630 | 0.590 | 0.162 |
| **SPVC** | 683 | 5 | 2.811 | 1.509 | 0.536 |
| **PRSK** | 683 | 5 | 3.324 | 0.154 | 0.046 |
| **OTEC** | 683 | 4 | 3.919 | 0.574 | 0.146 |
| **SOCN** | 683 | 7 | 2.603 | 0.661 | 0.253 |
| **PRSE** | 683 | 7 | 2.906 | 1.563 | 0.537 |
| **WTAPV** | 683 | 4 | 2.472 | 0.367 | 0.148 |

Notes: Dependent variable—WTAPV.

### *5.1. Respondents' Profile*

The survey peculiarities of the 683 respondents are summarized in Table 5. Male respondents outnumbered female respondents, with 398 male respondents accounting for 58.27% of the overall sample and 285 female respondents accounting for 41.72% of the total sample. In addition, the vast majority (459, 67.20%) were married, followed by unmarried respondents (224, 32.79 percent). Nearly 42.75 percent of respondents were between the ages of 26 and 35, followed by 36 to 45-year-olds (19.91 percent), 46 to 55 (11.71 percent), up to age 25 (16.39 percent), and those aged 56 and more (9.2 percent). The majority of respondents (45%) had an MS/MPhil degree, followed by a Ph.D. (16%), master's (29%), bachelor's degree (8%), and people with intermediate or below qualifications (4 percent). Furthermore, 287 (42.02 percent) respondents had a monthly household income of PKR 26,000–45,000, and 326 (47.73 percent) had their own business. See Table 5.

**Table 5.** Respondents profile N = 683.

| Characteristics | Option | Frequency | Percentage |
|---|---|---|---|
| **Gender** | | | |
| | Male | 398 | 58.27% |
| | Female | 114 | 41.72% |
| **Age** | | | |
| | Up to 25 | 112 | 16.39% |
| | 26–35 | 292 | 42.75% |
| | 36–45 | 136 | 19.91% |
| | 46–55 | 80 | 11.71% |
| | 56 and above | 63 | 9.22% |
| **Marital Status** | | | |
| | Single | 224 | 32.79% |
| | Married | 459 | 67.20% |
| **Education** | | | |
| | PhD | 112 | 16% |
| | MS/MPHIL | 295 | 45% |
| | Master's | 195 | 29% |
| | Bachelor's | 55 | 8% |
| | Intermediate or below | 26 | 4% |
| **Income (PKR)** | | | |
| | Up to 25,999 | 82 | 12% |
| | 26,000–45,999 | 287 | 42.02% |
| | 46,000–65,999 | 156 | 22.84% |
| | 66,000–85,999 | 102 | 14.93% |
| | Above 86,000 | 56 | 8.19% |
| **Occupation** | | | |
| | Government employee | 112 | 16.39% |
| | Private Job | 156 | 22.84% |
| | Own business | 326 | 47.73% |
| | Farmer | 89 | 13.03% |

## 5.2. Measurement Model Testing

The measurement model was evaluated using a validity and reliability test. There were two types of validity tests, one was convergent validity, as determined by average variance extracted (AVE), and discriminant validity calculated by Heterotrait–Monotrait ratios (HTMT) [110]. Outer loadings were used to calculate reliability, which included items reliability.

Table 6 presents the discriminant and convergent validity test. Discriminant validity determined that construct measures should not be highly correlated with one another theoretically and were not found to be highly correlated. HTMT is the most robust approach for determining discriminant validity. As a result, HTMT criteria were used to assess the discriminant validity [111] assert that if the values of (HTMT) were greater than 0.90, there were some validity problems. Kline [112] argues that there is discriminant validity if values are greater than 0.85. Table 6 indicates that all values were within the recommended criteria, measuring constructs were not overlapping each other and justified the discriminant validity test. Convergent validity refers to the degree to which two measures of constructs that should be related theoretically are actually related. The convergent validity test acceptability criteria are that AVE's value is well above the suggested value of 0.5 [113]. In Table 6, all measuring constructs had values greater than 0.5, indicating that they were highly related.

**Table 6.** Correlation, discriminant validity, and convergent validity analysis.

| Factors | OTEC | PRSE | SOCN | BPVB | PRSK | SPVC | WTAPV | AVE | MSV |
|---|---|---|---|---|---|---|---|---|---|
| **OTEC** | (0.708) | | | | | | | 0.501 | 0.126 |
| **PRSE** | 0.268 | (0.822) | | | | | | 0.676 | 0.274 |
| **SOCN** | 0.327 | 0.491 | (0.823) | | | | | 0.678 | 0.327 |
| **BPVB** | 0.355 | 0.523 | 0.376 | (0.751) | | | | 0.564 | 0.274 |
| **PRSK** | 0.175 | 0.419 | 0.545 | 0.305 | (0.776) | | | 0.603 | 0.524 |
| **SPVC** | 0.341 | 0.187 | 0.258 | 0.329 | 0.229 | (0.832) | | 0.693 | 0.116 |
| **WTAPV** | 0.299 | 0.507 | 0.572 | 0.418 | 0.724 | 0.244 | (0.737) | 0.544 | 0.524 |

Notes: Diagonal values in parentheses represent the root square of AVEs.

Furthermore, the factor loadings and composite reliability (CR) test was performed to assess the consistency and stability of all variable elements. Table 7 presents the analysis of outer loadings and composite reliability. The recommended value of 0.5 was met by all outer loadings [114]. Composite reliability was used to ensure that the scale item was internally consistent. The criteria to check the reliability analysis were acceptable and have been used in numerous research [115]. Furthermore, Table 7 demonstrates that CR and Cronbach $\alpha$ values were far above the acceptable limit of 0.7, indicating that all scale items were internally consistent [113]. The outcomes of the measurement model demonstrated that the data were reliable and valid.

An Exploratory Factor Analysis (EFA) was used to find the detrimental conceptual model. Before performing EFA, the Kaiser–Meyer–Olkin (KMO) and Bartlett's sphericity tests (BTS) were used to assess the data fit [116]. The KMO test was used to determine the proportion of variance shared by variables. If the proportion was lower, more data were suitable for the factor analysis. The value should have been as close to 1.0 as possible. If the value was less than 0.50, it indicated that data were unsuitable for factor analysis. Table 8 evidences that the KMO value was 0.918, suggesting that a factor analysis could be performed [117]. Similarly, BTS generated a significant value of 9985.49 [118], satisfying the EFA requirement (see Table 8). The data were then scrutinized using the confirmatory factor analysis (CFA) to ensure that it was suitable for the proposed research context. The measurement model's content validity was verified since all items were substantially loaded on their respective constructs (see Figure 2).

**Table 7.** Factor loadings and results of reliability analysis.

| Factors | Items | Standard Loadings | CR | Cronbach-α |
|---|---|---|---|---|
| **Belief about solar PV benefits** | | | 0.900 | 0.902 |
| | BPVB1 | 0.563 | | |
| | BPVB2 | 0.834 | | |
| | BPVB3 | 0.722 | | |
| | BPVB4 | 0.661 | | |
| | BPVB5 | 0.896 | | |
| | BPVB6 | 0.899 | | |
| | BPVB7 | 0.613 | | |
| **Solar PV cost** | | | 0.918 | 0.916 |
| | SPVC1 | 0.726 | | |
| | SPVC2 | 0.776 | | |
| | SPVC3 | 0.902 | | |
| | SPVC4 | 0.865 | | |
| | SPVC5 | 0.823 | | |
| **Perception of self-efficacy** | | | 0.935 | 0.937 |
| | PRSE1 | 0.638 | | |
| | PRSE2 | 0.834 | | |
| | PRSE3 | 0.797 | | |
| | PRSE4 | 0.857 | | |
| | PRSE5 | 0.855 | | |
| | PRSE6 | 0.673 | | |
| | PRSE7 | 0.726 | | |
| **Perceived risk** | | | 0.883 | 0.890 |
| | PRSK1 | 0.887 | | |
| | PRSK2 | 0.973 | | |
| | PRSK3 | 0.675 | | |
| | PRSK4 | 0.664 | | |
| | PRSK5 | 0.567 | | |
| **Openness to technology** | | | 0.800 | 0.873 |
| | OTEC1 | 0.728 | | |
| | OTEC2 | 0.740 | | |
| | OTEC3 | 0.673 | | |
| | OTEC4 | 0.671 | | |
| **Social norms** | | | 0.936 | 0.928 |
| | SOCN1 | 0.774 | | |
| | SOCN2 | 0.800 | | |
| | SOCN3 | 0.939 | | |
| | SOCN4 | 0.967 | | |
| | SOCN5 | 0.829 | | |
| | SOCN6 | 0.728 | | |
| | SOCN7 | 0.740 | | |
| **Willingness to adopt SPV** | | | 0.826 | 0.818 |
| | WTAPV1 | 0.655 | | |
| | WTAPV2 | 0.665 | | |
| | WTAPV3 | 0.661 | | |
| | WTAPV4 | 0.616 | | |

Notes: extraction method–maximum likelihood; rotation method–Promax with Kaiser normalization.

**Table 8.** KMO and Bartlett's test.

| **Kaiser–Meyer–Olkin Measure of Sampling Adequacy** | | **0.918** |
|---|---|---|
| **Bartlett's Test of Sphericity** | Approx. Chi-Square | 9985.49 |
| | df | 741 |
| | Sig. | 0.000 |

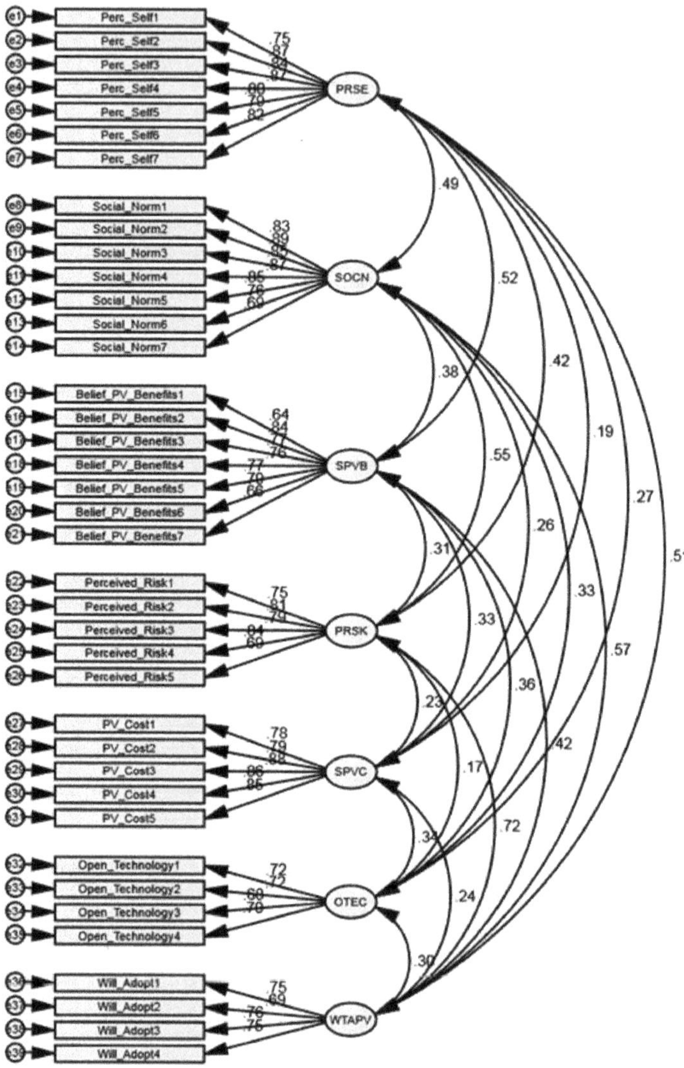


**Figure 2.** Confirmatory factor analysis, representing that all items substantially loaded on the respective constructs.

### 5.3. Structural Model Testing

Upon reviewing the measurement model, we determined the structural model and evaluated the hypotheses. The collinearity diagnostic is shown in Table 9. Collinearity is a predictor–criterion concept that can be used to determine whether or not proposed variables calculated the same constructs. The collinearity test was recommended as a viable alternative for detecting common method bias. If the variance inflation factors (VIFs) are less than 10, the model is stated to be free of common method bias [119]. As a result, every latent variable had its VIFs generated. The results showed that the VIFs of all latent variables were less than 10. Our model was found to be free of common method bias. Table 10 illustrates the communalities finding. Osborne, Costello, and Kellow [120] define that communalities greater than 0.4 are acceptable. To determine the number of variables, only those with an eigenvalue greater than one were considered (see Table 11).

**Table 9.** Collinearity diagnostic test.

| Variables | Collinearity Statistics | |
|---|---|---|
| | Tolerance | VIF |
| **BPVB** | 0.619 | 1.615 |
| **SPVC** | 0.782 | 1.279 |
| **PRSK** | 0.565 | 1.770 |
| **OTEC** | 0.723 | 1.382 |
| **SOCN** | 0.536 | 1.865 |
| **PRSE** | 0.565 | 1.769 |

Notes: Dependent variable—WTAPV.

**Table 10.** Communalities findings.

| Variables | Communalities | |
|---|---|---|
| | Initial | Extraction |
| **BPVB** | 1.000 | 0.501 |
| **SPVC** | 1.000 | 0.640 |
| **PRSK** | 1.000 | 0.871 |
| **OTEC** | 1.000 | 0.510 |
| **SOCN** | 1.000 | 0.516 |
| **PRSE** | 1.000 | 0.664 |
| **WTAPV** | 1.000 | 0.803 |

Notes: extraction method–maximum likelihood.

**Table 11.** Eigenvalues and cumulative variance.

| Variables | Initial Eigenvalues | | | Extraction Sums of Squared Loadings | | |
|---|---|---|---|---|---|---|
| | Total | % of Variance | Cumulative % | Total | % of Variance | Cumulative % |
| 1 | 12.353 | 31.673 | 31.673 | 11.838 | 30.354 | 30.354 |
| 2 | 3.879 | 9.946 | 41.620 | 3.459 | 8.869 | 39.224 |
| 3 | 3.472 | 8.903 | 50.522 | 3.185 | 8.166 | 47.390 |
| 4 | 2.413 | 6.187 | 56.710 | 2.067 | 5.301 | 52.690 |
| 5 | 2.142 | 5.493 | 62.203 | 1.833 | 4.701 | 57.392 |
| 6 | 1.932 | 4.955 | 67.157 | 1.550 | 3.975 | 61.367 |
| 7 | 1.147 | 2.941 | 70.098 | 0.793 | 2.032 | 63.399 |

Notes: rotation method—Promax with Kaiser normalization; cumulative variance—63.399%.

The Path diagram of the structural model is shown in Figure 3. There were three different levels of significance considered: (***) indicates significance at the 0.1% ($p$ = 0.001), (**) signifies significance at the 1% level ($p$ = 0.01), whereas (*) signifies significance at the 5% level ($p$ = 0.05). Insignificant paths are denoted by dotted lines, while constant lines denote significant paths. The belief about solar PV benefits H1 ($\beta$ = 0.09, $p$ = 0.01) and perception of self-efficacy H6 ($\beta$ = 0.10, $p$ = 0.01) were statistically significant at 5% and had a positive influence on dwellers' adoption of solar PV, according to the path diagram. Thus, we accepted the *H1* and *H6* hypotheses. Consumers' willingness to adopt solar PV was negatively influenced by perceived risk H3 ($\beta$ = −0.4, $p$ = 0.05) and positively influenced by social norms H5 ($\beta$ = 0.12, $p$ = 0.05). On the other hand, solar PV cost H2 ($\beta$ = −0.01, $p$ = 0.001) substantially influenced customer willingness at the 1% significance level. The path coefficient did not validate the hypothesis H4 (β = 0.80) because the variable "openness to technology" did not substantially impact dwellers' willingness to adopt, and it was refuted. The path of structural models and hypothesis significance are depicted in Table 12. The $R^2$ value was determined to be 0.73, indicating a significant interpretation because it exceeded the 0.35 value suggested by (Cohen, 2013). Various fitness measurements were also used to see whether the data were well-fitting for the proposed model. All fit index values were in line with the prescribed requirements [121], according to the findings reported in Table 13. The Heckman test was used to analyze endogeneity to preserve that the findings were reliable [122]. Our findings showed that there was no endogeneity bias in our findings (see Table 14).

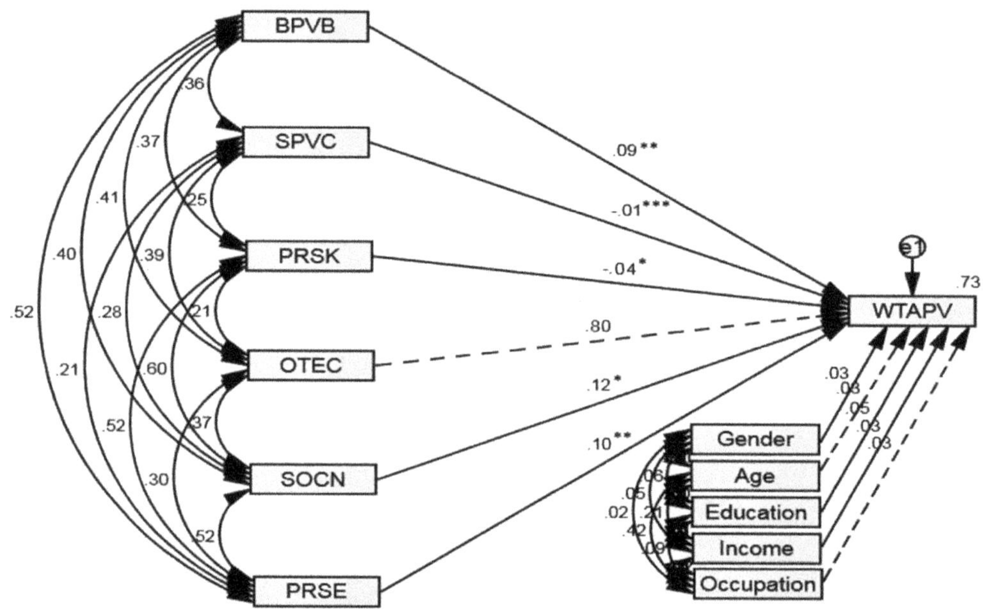


**Figure 3.** Diagram of structural model's path. Notes: *** $p < 0.00$, ** $p < 0.01$, * $p < 0.05$.

**Table 12.** Results of hypotheses.

| Hypotheses | Structural Paths | | | β-Value | f-Value | Result | $R^2$ |
|---|---|---|---|---|---|---|---|
| **H1** | BPVB | → | WTAPV | 0.09 ** | 227.4 *** | Accepted | 0.73 |
| **H2** | SPVC | → | WTAPV | −0.01 *** | 178.6 *** | Accepted | |
| **H3** | PRSK | → | WTAPV | −0.04 * | 131.4 *** | Accepted | |
| **H4** | OTEC | → | WTAPV | 0.80 | 235.9 *** | Rejected | |
| **H5** | SOCN | → | WTAPV | 0.12 * | 130.5 *** | Accepted | |
| **H6** | PRSE | → | WTAPV | 0.10 ** | 149.7 *** | Accepted | |

Notes: *** $p < 0.00$, ** $p < 0.01$, * $p < 0.05$.

**Table 13.** Goodness-of-fit indices values of the measurement and structural model.

| Fit Index | Description | Recommended Criterion | Values Based on a Structural Model |
|---|---|---|---|
| **CFI** | Comparative fit index | >0.9 good fit | 0.994 |
| **NFI** | Normed fit index | >0.9 good fit | 0.969 |
| **IFI** | Incremental fit index | >0.9 good fit | 0.973 |
| **TLI** | Tucker–Lewis index | >0.9 good fit | 0.986 |
| **GFI** | Goodness of fit | >0.9 good fit | 0.983 |
| **RMSEA** | Root mean squared error of approximation | <0.08 good fit | 0.026 |
| **$X^2$/df** | Chi-square | <3 good fit | 1.245 |
| **SRMR** | Standardized root mean squared residual | <0.09 good fit | 0.020 |

**Table 14.** Endogeneity test.

| Hypotheses | Structural Paths | | | β-Value | t-Statistics | Description |
|---|---|---|---|---|---|---|
| **H1** | BPVB | → | WTAPV | 0.052 ** | 2.955 | Not dissimilar |
| **H2** | SPVC | → | WTAPV | −0.376 *** | −8.702 | Not dissimilar |
| **H3** | PRSK | → | WTAPV | −0.693 * | −1.471 | Not dissimilar |
| **H4** | OTEC | → | WTAPV | 0.305 *** | 3.487 | Not dissimilar |
| **H5** | SOCN | → | WTAPV | 0.083 * | 6.761 | Not dissimilar |
| **H6** | PRSE | → | WTAPV | 0.671 ** | 5.613 | Not dissimilar |

Notes: *** $p < 0.00$, ** $p < 0.01$, * $p < 0.0$.

## 6. Discussion and Implications

### *6.1. Belief about Solar PV Benefits and Willingness to Adopt Solar PV*

The analysis results validated that consumers' WTA solar PV the is significantly influenced by their beliefs about its benefits. The findings of [66] also demonstrated that buyers' buying decisions were established on a solid faith in the advantages of the specific thing they needed to buy. Likewise, another recent study in Zambia factors influencing households' intention to adopt solar energy solutions revealed that due to its numerous advantages and the region's vast solar energy generation potential, solar energy solutions have become an attractive alternative to grid-based electricity, and many households are influencing their adoption [123]. Consumers can adopt solar PV if they can see the real advantages of using it over non-renewable resources. Similarly, a study conducted in India found that consumers are more receptive to adopting new technologies if they believe in the innovative benefits of technology [124]. Thus, in the early stages of its implementation, the Pakistani government must emphasize the advantages of solar PV, and consumers' confidence in solar PV will grow over time.

### *6.2. Solar PV Cost and Willingness to Adopt Solar PV*

The study's findings added to the body of knowledge by indicating that SPVC negatively influenced dwellers' acceptance. Users' intentions to install a solar PV system were mostly influenced by price, and previous studies back up these findings [125,126]. Due to the higher cost of rooftop solar PV, a recent study in the United States discovered that low- and moderate-income (LMI) households are less likely to install it. The findings indicate that when financial incentives, PV leasing, and property-assessed financing are used, PV adoption among LMI households in established markets increases and solar installation expansion is facilitated [127]. As a response, policymakers should consider price value when developing and implementing energy policies in Pakistan. Household tax relief and other incentive programs are needed. In addition, solar PV companies should spend more on technology and research to tackle the technology's perceived price issue.

### *6.3. Perceived Risk and Willingness to Adopt Solar PV*

The results confirmed that consumers' perceptions of risk had an adverse influence on their adoption of solar PV. Past research also supports this result [73,74]. Physical and health protection has indeed become a global regulatory concern. As a result, Pakistan's government should empower solar PV companies to manufacture risk-free technology and employ personnel with experience and a willingness to assist and support their clients, even after they have purchased their products. Simultaneously, marketers can provide consumers with sufficient knowledge and advice on operating a solar PV system. The removal of these risk beliefs gives people more interest in adopting new technologies, which can be a key factor in the future.

### *6.4. Openness to Technology and Willingness to Adopt Solar PV*

Openness is a personality trait that indicates that individuals with a high openness level are intellectually curious and receptive to unique ideas and pro-environmental practices. These individuals are more apt to take risks. Additionally, they are more susceptible to adopting new technologies quickly and easily. As per results, hypothesis four had no substantial effect on willingness to adopt solar PV. This result differed from previous research findings, which showed that customers with a high degree of openness were enthusiastic about emerging technology, and their acceptability was high [128,129]. Previous studies conducted in developed countries show that the adoption of new emerging technologies is common there. However, in Pakistan, literacy rates are relatively low, and people are more comfortable using technology. Thus, in Pakistan, purchasing decisions are not affected directly by the openness to technology.

### 6.5. *Social Norms and Willingness to Adopt Solar PV*

As per findings, SOCN had an important influence on dwellers' willingness to adopt solar PV. Research findings supported previous research by those who found that SOCN has a major influence on dwellers' intentions. Another study by [130] found that dwellers' behavior is directly affected by social norms. According to [131], the result of a consumer's product intake is predetermined by social norms. When people have a negative perception of a product, their intake of that product drops dramatically [132]. Nevertheless, the results showed that promoting conditions were determined by the outcome of social norms. Individuals who portray solar energy as having a positive environmental effect would be more likely to accept it [133]. Pakistan has a socially constructed structure, and the behavior of society and neighbors has a huge effect on people's minds. The previous impression of peers to the use of solar PV may influence dwellers' actions in such a way that a positive experience allows for solar PV acceptance. On the contrary, negative experiences have a different effect. Each country has its distinct social characteristics. As a result, companies use corporate governance as a social tool. If community members believe that the norms are fair or in everyone's best interest, they will share and adhere to them, as well as adopt those technologies. Similarly, researcher explore the corporate governance in the Romanian economy using the Bucharest Stock Exchange's corporate governance code. Corporate governance refers to the procedures and policies that a business uses to accomplish its stated goals. In conclusion, implementing a corporate governance code in the Romanian energy system has increased companies' overall liquidity, which contributes to an increase in the overall performance [134]. As a result, SOCN plays a key role in the decision-making framework.

### 6.6. *Perception of Self-Efficacy and Willingness to Adopt Solar PV*

The findings revealed that consumers' perceptions of self-efficacy had a positive impact on their willingness. This conclusion is backed up by previous research [135–137]. Another study conducted in Telangana revealed that ease of use plays a vital role in customers' attitude towards solar energy [138]. The energy efficiency, energy savings, and environmentally friendly solar PVs inherent characteristics are all factors that may encourage residents to use it. PVs for domestic use are also handy and simple to install, with such a long lifespan. Dwellers' assessments of technology and understanding about how to use it would boost dweller's interest in solar PV as vital developments in the near future.

## 7. Conclusions and Policy Recommendations

Being an environmentally friendly technology, solar PV has the potential of reducing carbon emissions; however, studies examining consumers' WTA solar PV technology are limited. This study focused on this research gap by incorporating novel factors (perceived risk, self-efficacy, and openness to technology) with existing factors (social norm, solar PV cost, and belief of solar PV benefits) in the conceptual framework of TPB to comprehensively investigate the consumer's adoption mechanism. Data were gathered from Pakistan's provincial capitals and analyzed using SEM. The research findings revealed that social norms, perception of self-efficacy, and belief about solar PV benefits positively affected consumer willingness to adopt solar PV. On the other hand, perceived risk and solar PV cost negatively affected consumers' willingness. Interestingly, openness to technology had an insignificant impact.

Pakistan, such as other developing countries, is facing environmental degradation issues. Global recognition has been accorded to sustainable energy technologies. Additionally, a number of countries are putting in significant amounts of investment in these technologies. This study's conclusions have significant ramifications for the government and solar PV vendors' efforts to promote solar PV deployment in Pakistan. An alternative energy board and the Ministry of climate change should create awareness programs for the public and promote the benefits of solar PV. They should reassure the public that solar PV provides sustainable solutions for combating climate change and the planet's ongoing degradation. As a consequence, dwellers will consider solar PV more beneficial and will feel more confident installing and using it.

The study's findings are instructive for policymakers and developers. Cost was a significant concern for many users; therefore, the Pakistani government should provide subsidies and sufficient financial assistance in collaboration with solar PV companies. Make sure the availability of financial incentives, such as grants, assist people install home solar PV systems. Solar PV companies must develop risk-free technologies and recruit a knowledgeable workforce which could provide sufficient technical support to their customers. Furthermore, in order to attract multinational businesses, the government should provide tax relief; this would increase market pressure, causing companies to boost product quality. When consumers are weighing the pros and cons of different solar PV options, such initiatives will help them decide to adopt solar PV. The importance of perceived risk, self-efficacy, and the price were highlighted in the conceptual context for this analysis, and the results were robust for Pakistan, indicating that the findings may be transferable to other countries.

There were a few limitations to this study that should be considered in future works. First, the analysis only considered provincial capitals, where the infrastructure and quality of living are much superior to other cities. This flaw can be addressed in future research by incorporating consumer input from other cities. Second, we approached the analysis from the consumer's perspective; future research can explore the supply side. Third, due to the country-wide lockdown situation during COVID-19, conducting a large-scale questionnaire survey was impossible. In this respect, a sample of 683 respondents was insufficient for a country with a 37 M population. Subsequent researchers should extend the sample size to overcome this constraint. Finally, it would be interesting to perform a cost–benefit analysis and determine the Levelized cost of electricity. This critical feature may be addressed in future investigations in order to add to the existing body of knowledge.

**Author Contributions:** Conceptualization, M.I. and A.T.; methodology, M.I.; software, M.I.; validation, M.I.; formal analysis, M.I.; investigation, M.I.; resources, S.Z.; data curation, M.I.; writing—original draft preparation, A.T.; writing—review and editing, M.I., A.T., S.Z., R.P.; visualization, M.I.; supervision, M.I., S.Z.; project administration, S.Z.; funding acquisition, S.Z. All authors have read and agreed to the published version of the manuscript.

**Funding:** This research was funded by Beijing Nova Program of Science & Technology under grant Z191100001119100 for Rui Peng.

**Institutional Review Board Statement:** Not applicable.

**Informed Consent Statement:** Informed consent was obtained from all subjects involved in the study.

**Data Availability Statement:** The data will be made a vailable on reasonable request from the corresponding author.

**Conflicts of Interest:** The authors declare no conflict of interest.

## Acronyms and Abbreviations

| | |
|---|---|
| BPVB | Belief about solar PV benefits |
| OTEC | Openness to technology |
| PRSE | Perception of self-efficacy |
| PRSK | Perceived risk |
| SEM | Structural equation modeling |
| SOCN | Social norms |
| Solar PV | Solar photovoltaic |
| SPVC | Solar PV cost |
| TPB | Theory of planned behavior |
| WTAPV | Willingness to adopt solar PV |

## Appendix A. Questionnaire Survey

| Part A: Demographics | Options | |
|---|---|---|
| Gender | Male | Female |
| Marital status | Single | Married |
| Age | Up to 25 | 26–35 |
| | 36–45 | 46–55 |
| | 56 and above | |
| Qualification | PhD | MS/MPhil |
| | Master;s | Bachelor's |
| | Intermediate or below | |
| Income (PKR) | Up to 25,999 | 26,000–45,999 |
| | 46,000–65,999 | 66,000–85,999 |
| | Above 86,000 | |
| Occupation | Government Employee | Private Job |
| | Own Business | Farmer |

| Part B. Factors Influencing Consumers' Willingness to Adopt Solar PV | | | | | | | |
|---|---|---|---|---|---|---|---|
| **Perceived Risk** | **1** | **2** | **3** | **4** | **5** | **6** | **7** |
| Solar PV adoption would come with a lot of safety risks | | | | | | | |
| Adopting solar PV energy at home is risky, as it may not provide the optimal energy solution | | | | | | | |
| Due to malfunctions or misuse, solar PV may cause some unpleasant physical side effects | | | | | | | |
| There is a risk that electricity generation will decrease as a result of a lack of sunlight or snow covering solar panels for extended periods. | | | | | | | |
| There is a risk that components will generate less electricity than expected over time | | | | | | | |
| Solar PV cost | | | | | | | |
| Solar PV charges extra costs for production | | | | | | | |
| Solar PV installations require a high up-front cost. | | | | | | | |
| It is expensive to generate electricity from solar PV. | | | | | | | |
| It is not economically friendly | | | | | | | |
| The cost of solar PV is very high and I cannot afford it | | | | | | | |
| Openness to technology | | | | | | | |
| I am a friendly person, and I welcome new technologies with open arms | | | | | | | |
| I think deeply and I am constantly coming up with new ideas | | | | | | | |
| I am enthusiastic about new technologies and I prefer to use them | | | | | | | |
| I have a favorable attitude toward new technology that has the potential to benefit the environment | | | | | | | |
| Perception of self-efficacy | | | | | | | |
| I have the necessary information to adopt solar PV | | | | | | | |
| I have full power to adopt solar PV | | | | | | | |

| |
|---|
| I have power over all resources to adopt solar PV |
| I have power over the adoption of solar PV services |
| My home has sufficient space for solar PV |
| Solar PV energy is not difficult to use; therefore, I am capable of adopting it |
| I would not have any difficulty adopting solar PV if I wanted to |
| Social norms |
| The majority of people who are important to me encourage me to use solar PV |
| Those who have my best interests at heart would prefer that I purchase solar PV |
| When it comes to energy-saving behavior, most of the people who matter to me believe that I should buy solar PV |
| People close to me believe that solar PV should play a significant role in reducing greenhouse gas emissions |
| People I care about adopt solar PV |
| Solar PV adoption is increasing among those in my close surroundings |
| Most of the people in my social network advise me to use solar PV. |
| Solar PV benefits |
| Adopting solar PV will help to reduce carbon emission. |
| Adopting solar PV will help to improve the clean environment.. |
| The development of solar PV projects will aid in the reduction in greenhouse gas emissions and the creation of new job opportunities. |
| Solar PV adoption will aid in the improvement of the energy structure. |
| Solar PV is the most efficient source of energy. |
| Generally, solar PV systems do not require much maintenance. |
| Solar PV energy will assist in lowering electricity bills. |
| Willingness to adopt solar PV |
| I am willing to pay more for solar PV because it saves energy. |
| I am willing to adopt solar PV because I can afford it. |
| I strongly encourage others to adopt solar PV because it contributes to a clean environment |
| Solar PV appeals to me because it is eco-friendly |

Notes: 1—strongly disagree; 2—disagree; 3—slightly disagree 4—neutral; 5—slightly agree; 6—agree, 7—strongly agree.

## References


1. Irfan, M.; Zhao, Z.Y.; Ahmad, M.; Batool, K.; Jan, A.; Mukeshimana, M.C. Competitive assessment of Indian wind power industry: A five forces model. *J. Renew. Sustain. Energy* **2019**, *11*, 063301. [CrossRef]
2. Nasir, M.; Khan, H.A.; Khan, I.; Hassan, N.U.; Zaffar, N.A.; Mehmood, A.; Sauter, T.; Muyeen, S.M. Grid load reduction through optimized PV power utilization in intermittent grids using a low-cost hardware platform. *Energies* **2019**, *12*, 764. [CrossRef]
3. Li, Y.; Yang, X.; Ran, Q.; Wu, H.; Irfan, M.; Ahmad, M. Energy structure, digital economy, and carbon emissions: Evidence from China. *Environ. Sci. Pollut. Res.* **2021**, 1–24. [CrossRef]
4. Irfan, M.; Zhao, Z.Y.; Panjwani, M.K.; Mangi, F.H.; Li, H.; Jan, A.; Ahmad, M.; Rehman, A. Assessing the energy dynamics of Pakistan: Prospects of biomass energy. *Energy Rep.* **2020**, *6*, 80–93. [CrossRef]
5. IEA Tracking Transport. 2020. Available online: https://www.iea.org/reports/tracking-transport-2020 (accessed on 3 August 2021).
6. Ahmad, M.; Zhao, Z.Y.; Irfan, M.; Mukeshimana, M.C.; Rehman, A.; Jabeen, G.; Li, H. Modeling heterogeneous dynamic interactions among energy investment, SO2 emissions and economic performance in regional China. *Environ. Sci. Pollut. Res.* **2020**, *27*, 2730–2744. [CrossRef]
7. Hussain, A.; Oad, A.; Ahmad, M.; Irfan, M. Do financial development and economic openness matter for economic progress in an emerging country ? Seeking a sustainable development path. *J. Od. Risk Financ. Manag.* **2021**, *14*, 237. [CrossRef]
8. Ahmad, M.; Jabeen, G.; Irfan, M.; Işık, C.; Rehman, A. Do inward foreign direct investment and economic development improve local environmental quality: Aggregation bias puzzle. *Environ. Sci. Pollut. Res.* **2021**, *1–21*, 34676–34696. [CrossRef] [PubMed]
9. Pan, Y.; Yao, X.; Wang, X.; Zhu, L. Policy uncertainties: What investment choice for solar panel producers? *Energy Econ.* **2019**, *78*, 454–467. [CrossRef]
10. International Energy Agency. Renewables 2020—Analysis-IEA. Available online: https://www.iea.org/reports/renewables-2020 (accessed on 9 June 2021).
11. Irfan, M.; Hao, Y.; Panjwani, M.K.; Khan, D.; Chandio, A.A.; Li, H. Competitive assessment of South Asia's wind power industry: SWOT analysis and value chain combined model. *Energy Strateg. Rev.* **2020**, *32*, 100540. [CrossRef]
12. Ur Rehman, S.A.; Cai, Y.; Siyal, Z.A.; Mirjat, N.H.; Fazal, R.; Kashif, S.U.R. Cleaner and Sustainable Energy Production in Pakistan: Lessons Learnt from the Pak-TIMES Model. *Energies* **2020**, *13*, 108. [CrossRef]

13. Ur Rehman, S.A.; Cai, Y.; Mirjat, N.H.; Walasai, G.D.; Shah, I.A.; Ali, S. The future of sustainable energy production in Pakistan: A system dynamics-based approach for estimating hubbert peaks. *Energies* **2017**, *10*, 1858. [CrossRef]
14. Kavari, G.; Tahani, M.; Mirhosseini, M. Wind shear effect on aerodynamic performance and energy production of horizontal axis wind turbines with developing blade element momentum theory. *J. Clean. Prod.* **2019**, *219*, 368–376. [CrossRef]
15. Irfan, M.; Zhao, Z.Y.; Ahmad, M.; Rehman, A. A techno-economic analysis of off-grid solar PV system: A case study for Punjab Province in Pakistan. *Processes* **2019**, *7*, 708. [CrossRef]
16. Rezaee, M.J.; Yousefi, S.; Hayati, J. Root barriers management in development of renewable energy resources in Iran: An interpretative structural modeling approach. *Energy Policy* **2019**, *129*, 292–306. [CrossRef]
17. Merino, I.; Herrera, I.; Valdés, H. Environmental Assessment of energy scenarios for a low-carbon electrical network in Chile. *Sustainability* **2019**, *11*, 5066. [CrossRef]
18. Valdés, H.; Leon, G. Cogeneration process technical viability for an apartment building: Case study in Mexico. *Processes* **2019**, *7*, 93. [CrossRef]
19. REN21. Renewables 2021 Global Status Report. 2021. Available online: https://www.ren21.net/wp-content/uploads/2019/05/GSR2021_Full_Report.pdf (accessed on 12 July 2021).
20. Sweerts, B.; Pfenninger, S.; Yang, S.; Folini, D.; van der Zwaan, B.; Wild, M. Estimation of losses in solar energy production from air pollution in China since 1960 using surface radiation data. *Nat. Energy* **2019**, *4*, 657–663. [CrossRef]
21. Kabir, E.; Kumar, P.; Kumar, S.; Adelodun, A.A.; Kim, K.H. Solar energy: Potential and future prospects. *Renew. Sustain. Energy Rev.* **2018**, *82*, 894–900. [CrossRef]
22. Irfan, M.; Hao, Y.; Ikram, M.; Wu, H.; Akram, R.; Rauf, A. Assessment of the public acceptance and utilization of renewable energy in Pakistan. *Sustain. Prod. Consum.* **2021**, *27*, 312–324. [CrossRef]
23. Mahmudul, H.M.; Rasul, M.G.; Akbar, D.; Narayanan, R.; Mofijur, M. A comprehensive review of the recent development and challenges of a solar-assisted biodigester system. *Sci. Total Environ.* **2021**, *753*, 141920. [CrossRef] [PubMed]
24. Ali, S.; Poulova, P.; Akbar, A.; Javed, H.M.U.; Danish, M. Determining the Influencing Factors in the Adoption of Solar Photovoltaic Technology in Pakistan: A Decomposed Technology Acceptance Model Approach. *Economies* **2020**, *8*, 108. [CrossRef]
25. Irfan, M.; Zhao, Z.-Y.; Ahmad, M.; Mukeshimana, M. Solar energy development in Pakistan: Barriers and policy recommendations. *Sustainability* **2019**, *11*, 1206. [CrossRef]
26. Wang, Z.; Ali, S.; Akbar, A.; Rasool, F. Determining the influencing factors of biogas technology adoption intention in Pakistan: The moderating role of social media. *Int. J. Environ. Res. Public Health* **2020**, *17*, 2311. [CrossRef]
27. Pakistan Economic Survey Energy. Islamabad. Available online: http://www.finance.gov.pk/survey/chapters_19/Economic_Survey_2018_19.pdf (accessed on 17 July 2021).
28. Rehman, A.; Ma, H.; Ozturk, I.; Ahmad, M.; Rauf, A.; Irfan, M. Another outlook to sector-level energy consumption in Pakistan from dominant energy sources and correlation with economic growth. *Environ. Sci. Pollut. Res.* **2021**, *28*, 33735–33750. [CrossRef]
29. Qureshi, T.M.; Ullah, K.; Arentsen, M.J. Factors responsible for solar PV adoption at household level: A case of Lahore, Pakistan. *Renew. Sustain. Energy Rev.* **2017**, *78*, 754–763. [CrossRef]
30. Polo Lopez, C.S.; Lucchi, E.; Franco, G. Acceptance of building integrated photovoltaic (Bipv) in heritage buildings and landscapes: Potentials, barrier and assessment criteria. In Proceedings of the Rehabend, Granada, Spain, 28–30 September 2020; pp. 1636–1644.
31. Lucchi, E.; Polo Lopez, C.S.; Franco, G. A conceptual framework on the integration of solar energy systems in heritage sites and buildings. *IOP Conf. Ser. Mater. Sci. Eng.* **2020**, *949*, 012113. [CrossRef]
32. Li, H.X.; Edwards, D.J.; Hosseini, M.R.; Costin, G.P. A review on renewable energy transition in Australia: An updated depiction. *J. Clean. Prod.* **2020**, *242*, 118475. [CrossRef]
33. IRENA. Renewable Capacity Statistics. 2020. Available online: https://irena.org/publications/2020/Mar/Renewable-Capacity-Statistics-2020 (accessed on 13 July 2021).
34. Paravantis, J.A.; Stigka, E.; Mihalakakou, G.; Michalena, E.; Hills, J.M.; Dourmas, V. Social acceptance of renewable energy projects: A contingent valuation investigation in Western Greece. *Renew. Energy* **2018**, *123*, 639–651. [CrossRef]
35. Simpson, G.; Clifton, J. Subsidies for residential solar photovoltaic energy systems in Western Australia: Distributional, procedural and outcome justice. *Renew. Sustain. Energy Rev.* **2016**, *65*, 262–273. [CrossRef]
36. Yuan, X.; Zuo, J.; Ma, C. Social acceptance of solar energy technologies in China-End users' perspective. *Energy Policy* **2011**, *39*, 1031–1036. [CrossRef]
37. Alrashoud, K.; Tokimatsu, K. Factors influencing social perception of residential solar photovoltaic systems in Saudi Arabia. *Sustainability* **2019**, *11*, 5259. [CrossRef]
38. Entele, B.R. Analysis of households' willingness to pay for a renewable source of electricity service connection: Evidence from a double-bounded dichotomous choice survey in rural Ethiopia. *Heliyon* **2020**, *6*, e03332. [CrossRef] [PubMed]
39. Aklin, M.; Cheng, C.Y.; Urpelainen, J. Social acceptance of new energy technology in developing countries: A framing experiment in rural India. *Energy Policy* **2018**, *113*, 466–477. [CrossRef]
40. Khalid, A.; Junaidi, H. Study of economic viability of photovoltaic electric power for Quetta—Pakistan. *Renew. Energy* **2013**, *50*, 253–258. [CrossRef]
41. Irfan, M.; Zhao, Z.Y.; Li, H.; Rehman, A. The influence of consumers' intention factors on willingness to pay for renewable energy: A structural equation modeling approach. *Environ. Sci. Pollut. Res.* **2020**, *27*, 21747–21761. [CrossRef]

42. Irfan, M.; Zhao, Z.Y.; Rehman, A.; Ozturk, I.; Li, H. Consumers' intention-based influence factors of renewable energy adoption in Pakistan: A structural equation modeling approach. *Environ. Sci. Pollut. Res.* **2020**, *28*, 432–445. [CrossRef] [PubMed]
43. Jabeen, G.; Ahmad, M.; Zhang, Q. Factors influencing consumers' willingness to buy green energy technologies in a green perceived value framework. *Energy Sources Part B Econ. Plan. Policy* **2021**, 1–17. [CrossRef]
44. Kamran, M.; Mudassar, M.; Ahmed, S.R.; Mehmood, H.S.; Joyia, F.M.; Hussain, M.; Ahmad, H.; Saeed, S. Socio-economic acceptance for stand-alone Solar PV systems: Survey evidence from Southern Punjab, Pakistan. *Int. J. Renew. Energy Res.* **2019**, *9*, 301–308.
45. Tahir, Z.R.; Asim, M. Surface measured solar radiation data and solar energy resource assessment of Pakistan: A review. *Renew. Sustain. Energy Rev.* **2018**, *81*, 2839–2861. [CrossRef]
46. Akhtar, S.; Hashmi, M.K.; Ahmad, I.; Raza, R. Advances and significance of solar reflectors in solar energy technology in Pakistan. *Energy Environ.* **2018**, *29*, 435–455. [CrossRef]
47. Gong, X.; Ye, Z.; Liu, K.; Wu, N. The effects of live platform exterior design on sustainable impulse buying: Exploring the mechanisms of self-efficacy and psychological ownership. *Sustainability* **2020**, *12*, 2406. [CrossRef]
48. Procter, L.; Angus, D.J.; Blaszczynski, A.; Gainsbury, S.M. Understanding use of consumer protection tools among Internet gambling customers: Utility of the Theory of Planned Behavior and Theory of Reasoned Action. *Addict. Behav.* **2019**, *99*, 106050. [CrossRef]
49. Ajzen, I. From Intentions to Actions: A Theory of Planned Behavior. In *Action Control*; Springer: Berlin/Heidelberg, Germany, 1985; pp. 11–39.
50. Ajzen, I.; Fishbein, M. A Bayesian analysis of attribution processes. *Psychol. Bull.* **1975**, *82*, 261–277. [CrossRef]
51. Japutra, A.; Loureiro, S.M.C.; Molinillo, S.; Ekinci, Y. Travellers' mindsets and theory of planned behaviour. *Tour. Manag. Perspect.* **2019**, *30*, 193–196. [CrossRef]
52. Pandey, S.; Ritz, C.; Perez-Cueto, F.J.A. An application of the theory of planned behaviour to predict intention to consume plant-based yogurt alternatives. *Foods* **2021**, *10*, 148. [CrossRef]
53. Liddelow, C.; Mullan, B.; Novoradovskaya, E. Exploring Medication Adherence Amongst Australian Adults Using an Extended Theory of Planned Behaviour. *Int. J. Behav. Med.* **2020**, *27*, 389–399. [CrossRef] [PubMed]
54. Opoku, M.P.; Cuskelly, M.; Pedersen, S.J.; Rayner, C.S. Attitudes and self-efficacy as significant predictors of intention of secondary school teachers towards the implementation of inclusive education in Ghana. *Eur. J. Psychol. Educ.* **2020**, 1–19. [CrossRef]
55. Liobikienė, G.; Dagiliūtė, R.; Juknys, R. The determinants of renewable energy usage intentions using theory of planned behaviour approach. *Renew. Energy* **2021**, *170*, 587–594. [CrossRef]
56. Alam, S.S.; Ahmad, M.; Othman, A.S.; Shaari, Z.B.H.; Masukujjaman, M. Factors affecting photovoltaic solar technology usage intention among households in Malaysia: Model integration and empirical validation. *Sustainability* **2021**, *13*, 1773. [CrossRef]
57. Sok, J.; Borges, J.R.; Schmidt, P.; Ajzen, I. Farmer Behaviour as Reasoned Action: A Critical Review of Research with the Theory of Planned Behaviour. *J. Agric. Econ.* **2020**, *72*, 388–412. [CrossRef]
58. Irfan, M.; Elavarasan, R.M.; Hao, Y.; Feng, M.; Sailan, D. An assessment of consumers' willingness to utilize solar energy in china: End-users' perspective. *J. Clean. Prod.* **2021**, *292*, 126008. [CrossRef]
59. Irfan, M.; Akhtar, N.; Ahmad, M.; Shahzad, F.; Elavarasan, R.M.; Wu, H.; Yang, C. Assessing public willingness to wear face masks during the COVID-19 pandemic: Fresh insights from the theory of planned behavior. *Int. J. Environ. Res. Public Health* **2021**, *18*, 4577. [CrossRef] [PubMed]
60. Ahmad, M.; Akhtar, N.; Jabeen, G.; Irfan, M.; Anser, M.K.; Wu, H.; Isek, C. Intention-based critical factors affecting willingness to adopt Novel Coronavirus prevention in Pakistan: Implications for future pandemics. *Int. J. Env. Res. Public Health* **2021**, *18*, 6167. [CrossRef] [PubMed]
61. Abreu, J.; Wingartz, N.; Hardy, N. New trends in solar: A comparative study assessing the attitudes towards the adoption of rooftop PV. *Energy Policy* **2019**, *128*, 347–363. [CrossRef]
62. Walters, J.; Kaminsky, J.; Gottschamer, L. A systems analysis of factors influencing household solar PV adoption in Santiago, Chile. *Sustainability* **2018**, *10*, 1257. [CrossRef]
63. Bashiri, A.; Alizadeh, S.H. The analysis of demographics, environmental and knowledge factors affecting prospective residential PV system adoption: A study in Tehran. *Renew. Sustain. Energy Rev.* **2018**, *81*, 3131–3139. [CrossRef]
64. Komendantova, N.; Yazdanpanah, M. Impacts of Human Factors on Willingness to Use Renewable Energy Sources in Iran and Morocco. *Environ. Energy Econ. Res.* **2017**, *1*, 141–152. [CrossRef]
65. Kumar, V.; Syan, A.S.; Kaur, A.; Hundal, B.S. Determinants of farmers' decision to adopt solar powered pumps. *Int. J. Energy Sect. Manag.* **2020**, *14*, 707–727. [CrossRef]
66. Colmenares-Quintero, R.F.; Rojas, N.; Kerr, S.; Caicedo-Concha, D.M. Industry and academia partnership for aquatic renewable energy development in Colombia: A knowledge-education transfer model from the United Kingdom to Colombia. *Cogent Eng.* **2020**, *7*, 1829805. [CrossRef]
67. Madurai Elavarasan, R.; Pugazhendhi, R. Restructured society and environment: A review on potential technological strategies to control the COVID-19 pandemic. *Sci. Total Environ.* **2020**, *725*, 138858. [CrossRef]
68. Palm, J.; Eidenskog, M.; Luthander, R. Sufficiency, change, and flexibility: Critically examining the energy consumption profiles of solar PV prosumers in Sweden. *Energy Res. Soc. Sci.* **2018**, *39*, 12–18. [CrossRef]

69. Fadlallah, S.O.; Benhadji Serradj, D.E. Determination of the optimal solar photovoltaic (PV) system for Sudan. *Sol. Energy* **2020**, *208*, 800–813. [CrossRef]
70. Bayulgen, O.; Benegal, S. Green Priorities: How economic frames affect perceptions of renewable energy in the United States. *Energy Res. Soc. Sci.* **2019**, *47*, 28–36. [CrossRef]
71. Rigo, P.D.; Siluk, J.C.M.; Lacerda, D.P.; Rosa, C.B.; Rediske, G. Is the success of small-scale photovoltaic solar energy generation achievable in Brazil? *J. Clean. Prod.* **2019**, *240*, 118243. [CrossRef]
72. Shakeel, S.R.; Rajala, A. Factors Influencing Households' Intention to Adopt Solar PV: A Systematic Review. In Proceedings of the Advances in Intelligent Systems and Computing; Springer: Cham, Switzerland, 2020; Volume 1209, pp. 282–289. [CrossRef]
73. Yildirim, H.; Ali-Eldin, A.M.T. A model for predicting user intention to use wearable IoT devices at the workplace. *J. King Saud Univ. Comput. Inf. Sci.* **2019**, *31*, 497–505. [CrossRef]
74. Falcone, R.; Sapienza, A. On the users' acceptance of IoT systems: A theoretical approach. *Information* **2018**, *9*, 53. [CrossRef]
75. Lee, M.C. Factors influencing the adoption of internet banking: An integration of TAM and TPB with perceived risk and perceived benefit. *Electron. Commer. Res. Appl.* **2009**, *8*, 130–141. [CrossRef]
76. Choi, B.; Lee, I. Trust in open versus closed social media: The relative influence of user- and marketer-generated content in social network services on customer trust. *Telemat. Inform.* **2017**, *34*, 550–559. [CrossRef]
77. Devaraj, U.S.; Easley, R.F.; Michael Crant, J. How does personality matter? Relating the five-factor model to technology acceptance and use. *Inf. Syst. Res.* **2008**, *19*, 93–105. [CrossRef]
78. Christensen, A.P.; Cotter, K.N.; Silvia, P.J. Reopening Openness to Experience: A Network Analysis of Four Openness to Experience Inventories. *J. Pers. Assess.* **2019**, *101*, 574–588. [CrossRef]
79. Hsu, H.Y.; Liu, F.H.; Tsou, H.T.; Chen, L.J. Openness of technology adoption, top management support and service innovation: A social innovation perspective. *J. Bus. Ind. Mark.* **2019**, *34*, 575–590. [CrossRef]
80. Saef, R.; Woo, S.E.; Carpenter, J.; Tay, L. Fostering socio-informational behaviors online: The interactive effect of openness to experience and extraversion. *Pers. Individ. Dif.* **2018**, *122*, 93–98. [CrossRef]
81. Watjatrakul, B. Online learning adoption: Effects of neuroticism, openness to experience, and perceived values. *Interact. Technol. Smart Educ.* **2016**, *13*, 229–243. [CrossRef]
82. Daxini, A.; Ryan, M.; O'Donoghue, C.; Barnes, A.P. Understanding farmers' intentions to follow a nutrient management plan using the theory of planned behaviour. *Land Use Policy* **2019**, *85*, 428–437. [CrossRef]
83. Dan, R.L.; Buglea, A.; Anis, C. The main models of corporate governance. In Proceedings of the International Academic Conferences, (No. 3105360); International Institute of Social and Economic Sciences: Prague, Czech Republic, 2015.
84. Villamayor-Tomas, S.; Sagebiel, J.; Olschewski, R. Bringing the neighbors in: A choice experiment on the influence of coordination and social norms on farmers' willingness to accept agro-environmental schemes across Europe. *Land Use Policy* **2019**, *84*, 200–215. [CrossRef]
85. Shakeel, S.R.; Rahman, S. Ur Towards the establishment of renewable energy technologies' market: An assessment of public acceptance and use in Pakistan. *J. Renew. Sustain. Energy* **2018**, *10*, 045907. [CrossRef]
86. Park, E.-Y.; Lee, H.-S. Residents' Perception and Participation Satisfaction on Rural Landscape Improvement Projects—Focused on the Rural Revitalization Project in Geumsan. *J. Korean Soc. Rural Plan.* **2019**, *25*, 1–10. [CrossRef]
87. Lopes, J.R.N.; Kalid, R.d.A.; Rodríguez, J.L.M.; Ávila Filho, S. A new model for assessing industrial worker behavior regarding energy saving considering the theory of planned behavior, norm activation model and human reliability. *Resour. Conserv. Recycl.* **2019**, *145*, 268–278. [CrossRef]
88. Jayaraman, K.; Paramasivan, L.; Kiumarsi, S. Reasons for low penetration on the purchase of photovoltaic (PV) panel system among Malaysian landed property owners. *Renew. Sustain. Energy Rev.* **2017**, *80*, 562–571. [CrossRef]
89. Ajzen, I. The theory of planned behavior: Frequently asked questions. *Hum. Behav. Emerg. Technol.* **2020**, *2*, 314–324. [CrossRef]
90. Aggarwal, A.K.; Syed, A.A.; Garg, S. Factors driving Indian consumer's purchase intention of roof top solar. *Int. J. Energy Sect. Manag.* **2019**, *13*, 539–555. [CrossRef]
91. Talpur, M.A.H.; Chandio, I.A.; Baig, F.; Shaikh, F.A.; Napiah, M. Energy Crisis and Household's Perception about Solar Energy Acceptance: District Hyderabad, Pakistan. *Sindh Univ. Res. J. Sci. Ser.* **2017**, *49*, 601–604. [CrossRef]
92. Guerin, R.J.; Toland, M.D. An application of a modified theory of planned behavior model to investigate adolescents' job safety knowledge, norms, attitude and intention to enact workplace safety and health skills. *J. Saf. Res.* **2020**, *72*, 189–198. [CrossRef]
93. Weng, G.S.; Zailani, S.; Iranmanesh, M.; Hyun, S.S. Mobile taxi booking application service's continuance usage intention by users. *Transp. Res. Part D Transp. Environ.* **2017**, *57*, 207–216. [CrossRef]
94. Solangi, Y.A.; Shah, S.A.A.; Zameer, H.; Ikram, M.; Saracoglu, B.O. Assessing the solar PV power project site selection in Pakistan: Based on AHP-fuzzy VIKOR approach. *Environ. Sci. Pollut. Res.* **2019**, *26*, 30286–30302. [CrossRef] [PubMed]
95. Azam, M.; Jahromy, S.S.; Raza, W.; Raza, N.; Lee, S.S.; Kim, K.H.; Winter, F. Status, characterization, and potential utilization of municipal solid waste as renewable energy source: Lahore case study in Pakistan. *Environ. Int.* **2020**, *134*, 105291. [CrossRef]
96. Ahmad, M.; Jamal, A.; Tang, X.W.; Al-Sughaiyer, M.A.; Al-Ahmadi, H.M.; Ahmad, F. Assessing potable water quality and identifying areas of waterborne diarrheal and fluorosis health risks using spatial interpolation in Peshawar, Pakistan. *Water* **2020**, *12*, 2163. [CrossRef]
97. Memon, I.A.; Kalwar, S.; Sahito, N.; Napiah, M.B. Factors that Influence Travelers'willingness to Use or Not Park-and-Ride Service in Putrajaya And Karachi Cbd. *PalArch's J. Archaeol. Egypt/Egyptol.* **2021**, *18*, 720–734.

98. Ali, M.; Wazir, R.; Imran, K.; Ullah, K.; Janjua, A.K.; Ulasyar, A.; Khattak, A.; Guerrero, J.M. Techno-economic assessment and sustainability impact of hybrid energy systems in Gilgit-Baltistan, Pakistan. *Energy Rep.* **2021**, *7*, 2546–2562. [CrossRef]
99. Elfil, M.; Negida, A. Sampling methods in clinical research; an educational review. *Arch. Acad. Emerg. Med.* **2019**, *7*, 52. [CrossRef]
100. Comrey, A.L.; Lee, H.B. *A First Course in Factor Analysis*, 2nd ed.; Lawrence Erlbaum Associates, Inc.: Hlilsdale, NJ, USA, 1992; ISBN 0805810625.
101. Liu, W.; Wang, C.; Mol, A.P.J. Rural public acceptance of renewable energy deployment: The case of Shandong in China. *Appl. Energy* **2013**, *102*, 1187–1196. [CrossRef]
102. Parasuraman, A. Technology Readiness Index (Tri): A Multiple-Item Scale to Measure Readiness to Embrace New Technologies. *J. Serv. Res.* **2000**, *2*, 307–320. [CrossRef]
103. Molla, A.; Abareshi, A. Organizational green motivations for information technology: Empirical study. *J. Comput. Inf. Syst.* **2012**, *52*, 92–102. [CrossRef]
104. Fang, W.T.; Ng, E.; Wang, C.M.; Hsu, M.L. Normative beliefs, attitudes, and social norms: People reduce waste as an index of social relationships when spending leisure time. *Sustainability* **2017**, *9*, 1696. [CrossRef]
105. Chou, J.S.; Kim, C.; Ung, T.K.; Yutami, I.G.A.N.; Lin, G.T.; Son, H. Cross-country review of smart grid adoption in residential buildings. *Renew. Sustain. Energy Rev.* **2015**, *48*, 192–213. [CrossRef]
106. Steenkamp, J.B.E.M.; Baumgartner, H. On the use of structural equation models for marketing modeling. *Int. J. Res. Mark.* **2000**, *17*, 195–202. [CrossRef]
107. Novikova, S.I.; Richman, D.M.; Supekar, K.; Barnard-Brak, L.; Hall, D. NDAR: A Model Federal System for Secondary Analysis in Developmental Disabilities Research. In *International Review of Research in Developmental Disabilities*; Academic Press Inc.: London, UK, 2013; Volume 45, pp. 123–153.
108. Sardeshmukh, S.R.; Vandenberg, R.J. Integrating Moderation and Mediation: A Structural Equation Modeling Approach. *Organ. Res. Methods* **2017**, *20*, 721–745. [CrossRef]
109. Belaïd, F. Untangling the complexity of the direct and indirect determinants of the residential energy consumption in France: Quantitative analysis using a structural equation modeling approach. *Energy Policy* **2017**, *110*, 246–256. [CrossRef]
110. Akbar, A.; Ali, S.; Ahmad, M.A.; Akbar, M.; Danish, M. Understanding the antecedents of organic food consumption in pakistan: Moderating role of food neophobia. *Int. J. Environ. Res. Public Health* **2019**, *16*, 4043. [CrossRef]
111. Gold, A.H.; Malhotra, A.; Segars, A.H. Knowledge management: An organizational capabilities perspective. *J. Manag. Inf. Syst.* **2001**, *18*, 185–214. [CrossRef]
112. Kline, R.B. *TXTBK Principles and Practices of Structural Equation Modelling*, 4th ed.; Guilford Publications Inc.: New York, NY, USA, 2015; ISBN 9781609182304.
113. Anderson, J.C.; Gerbing, D.W. Structural equation modeling in practice: A review and recommended two-step approach. *Psychol. Bull.* **1988**, *103*, 411. [CrossRef]
114. Hair, J.F.; Sarstedt, M.; Hopkins, L.; Kuppelwieser, V.G. Partial least squares structural equation modeling (PLS-SEM): An emerging tool in business research. *Eur. Bus. Rev.* **2014**, *26*, 106–121. [CrossRef]
115. Naqshbandi, M.M.; Kaur, S.; Ma, P. What organizational culture types enable and retard open innovation? *Qual. Quant.* **2015**, *49*, 2123–2144. [CrossRef]
116. Mondiana, Y.Q.; Pramoedyo, H.; Sumarminingsih, E. Structural Equation Modeling on Likert Scale Data With Transformation by Successive Interval Method and With No Transformation. *Int. J. Sci. Res. Publ.* **2018**, *8*, 398–405. [CrossRef]
117. Kaiser, H.F. An index of factorial simplicity. *Psychometrika* **1974**, *39*, 31–36. [CrossRef]
118. Liza Ghazali, P.; Arni Syed Jaaffar, S.; Hazimi Mohammed Foziah, N.; Malek Tambi, A.A.; Amalina Md Nawi, F.; Mamat, M.; Bahri Mohammad, S.; Mohd Nazri Wan Daud, W.; Sadad Mahmud, M. The Construction of A New Mathematical Model for Islamic Home Financing. *Asian Acad. Manag. J.* **2019**, *24*, 33–41. [CrossRef]
119. Field, A. *Discovering Statistics Using IBM SPSS Statistics*; Sage: Riverside, CA, USA, 2013; ISBN 1446274586.
120. Osborne, J.W.; Costello, A.B.; Kellow, J.T. *Best Practices in Exploratory Factor Analysis*; CreateSpace Independent Publisher: Scotts Vally, CA, USA, 2011; ISBN 9781500594343.
121. Ikram, M.; Zhou, P.; Shah, S.A.A.; Liu, G.Q. Do environmental management systems help improve corporate sustainable development? Evidence from manufacturing companies in Pakistan. *J. Clean. Prod.* **2019**, *226*, 628–641. [CrossRef]
122. Huit, G.T.M.; Hair, J.F.; Proksch, D.; Sarstedt, M.; Pinkwart, A.; Ringle, C.M. Addressing endogeneity in international marketing applications of partial least squares structural equation modeling. *J. Int. Mark.* **2018**, *26*, 1–21. [CrossRef]
123. Zulu, S.; Zulu, E.; Chabala, M. Factors influencing households' intention to adopt solar energy solutions in Zambia: Insights from the theory of planned behaviour. *Smart Sustain. Built Environ.* **2021**. Vol. ahead-of-print No. ahead of print. [CrossRef]
124. Roy, S.; Mohapatra, S. Problems of adoption of solar power and subsequent switching behavior: An exploration in India. *Int. J. Energy Sect. Manag.* **2021**. Vol. ahead-of-print. [CrossRef]
125. Kardooni, R.; Yusoff, S.B.; Kari, F.B. Renewable energy technology acceptance in Peninsular Malaysia. *Energy Policy* **2016**, *88*, 1–10. [CrossRef]
126. Luthra, S.; Kumar, S.; Garg, D.; Haleem, A. Barriers to renewable/sustainable energy technologies adoption: Indian perspective. *Renew. Sustain. Energy Rev.* **2015**, *41*, 762–776. [CrossRef]
127. O'Shaughnessy, E.; Barbose, G.; Wiser, R.; Forrester, S.; Darghouth, N. The impact of policies and business models on income equity in rooftop solar adoption. *Nat. Energy* **2020**, *6*, 84–91. [CrossRef]

128. Ashkanasy, N.; Bowen, P.L.; Rohde, F.H.; Wu, C.Y.A. The Effects of User Characteristics on Query Performance in the Presence of Information Request Ambiguity. *J. Inf. Syst.* **2007**, *21*, 53–82. [CrossRef]
129. Korukonda, A.R. Differences that do matter: A dialectic analysis of individual characteristics and personality dimensions contributing to computer anxiety. *Comput. Hum. Behav.* **2007**, *23*, 1921–1942. [CrossRef]
130. Pérez-Sánchez, M.d.L.Á.; Tian, Z.; Barrientos-Báez, A.; Gómez-Galán, J.; Li, H. Blockchain Technology for Winning Consumer Loyalty: Social Norm Analysis Using Structural Equation Modeling. *Mathematics* **2021**, *9*, 532. [CrossRef]
131. Ge, W.; Sheng, G.; Zhang, H. How to Solve the Social Norm Conflict Dilemma of Green Consumption: The Moderating Effect of Self-Affirmation. *Front. Psychol.* **2020**, *11*, 24. [CrossRef]
132. Bach, L.; Hopkins, D.; Stephenson, J. Solar electricity cultures: Household adoption dynamics and energy policy in Switzerland. *Energy Res. Soc. Sci.* **2020**, *63*, 101395. [CrossRef]
133. Mundaca, L.; Samahita, M. What drives home solar PV uptake? Subsidies, peer effects and visibility in Sweden. *Energy Res. Soc. Sci.* **2020**, *60*, 101319. [CrossRef]
134. Avram, R.L.; Buglea, A.; Avram, A. The Impact of Corporate Governance on the Company? S Performance. In *Proceedings of International Academic Conferences*; International Institute of Social and Economic Sciences: Prague, Czech Republic, 2017; Volume 32, pp. 29–47. [CrossRef]
135. Ortenburger, D.; Wąsik, J.; Mosler, D. Perception of Self-Efficacy and Health-Related Behavior in Context of Taekwon-Do Sport Camps. *Sustainability* **2021**, *13*, 4645. [CrossRef]
136. Gielnik, M.M.; Bledow, R.; Stark, M.S.; Gielnik, M.M.; Bledow, R. *A Dynamic Account of Self-Efficacy in Entrepreneurship*; American Psychological Association Inc.: Washington, DC, USA, 2020; Volume 105.
137. Melchionda, M.M.; Aletti, G.; Mauri, P.A. Validation of a self-efficacy survey for Italian midwifery students with regard to breastfeeding support. *Nurse Educ. Pract.* **2019**, *37*, 9–14. [CrossRef] [PubMed]
138. Verma, S.; Sreeramulu, D. An Empirical Study to Analyze Challenges Associated with Solar Products Purchase Intention for Domestic Usages in Telangana. *Rev. Geintec Gest. Inov. E Tecnol.* **2021**, *11*, 3048–3060. [CrossRef]

energies

MDPI

Article

# A Modified Grid-Connected Inverter Topology for Power Oscillation Suppression under Unbalanced Grid Voltage Faults

Cheng Luo [1], Xikui Ma [1], Lihui Yang [1,*], Yongming Li [1], Xiaoping Yang [2], Junhui Ren [2] and Yanmei Zhang [2]

[1] State Key Laboratory of Electrical Insulation and Power Equipment, School of Electrical Engineering, Xi'an Jiaotong University, Xi'an 710049, China; lc19950522@stu.xjtu.edu.cn (C.L.); maxikui@xjtu.edu.cn (X.M.); yongmingli@stu.xjtu.edu.cn (Y.L.)
[2] Xi'an XD Power System Co., Ltd., Xi'an 710075, China; yangxp@xdps.com.cn (X.Y.); renjh@xdps.com.cn (J.R.); zhangym@xdps.com.cn (Y.Z.)
* Correspondence: lihui.yang@mail.xjtu.edu.cn; Tel.: +86-029-8266-8630 (ext. 201)

**Abstract:** Under unbalanced grid voltage faults, the output power oscillation of a grid-connected inverter is an urgent problem to be solved. In the traditional topology of inverters, it is impossible to eliminate power oscillation and simultaneously maintain balanced output current waveform. In this paper, considering the solvability of reference current matrix equation, the inherent mechanism of inverter output power oscillation is analyzed, and a modified topology with auxiliary modules inserted in series between the inverter output filter and the point of common coupling (PCC) is proposed. Due to the extra controllable freedoms provided by auxiliary modules, the inverter could generate extra voltage to correct PCC voltage while keeping balance of output current, so as to eliminate the oscillation of output power. Simulation and experimental results verify the effectiveness of the proposed topology.

**Keywords:** modified gird-connected inverter topology; negative-sequence component; power oscillation; unbalanced grid voltage fault

**Citation:** Luo, C.; Ma, X.; Yang, L.; Li, Y.; Yang, X.; Ren, J.; Zhang, Y. A Modified Grid-Connected Inverter Topology for Power Oscillation Suppression under Unbalanced Grid Voltage Faults. *Energies* **2021**, *14*, 5057. https://doi.org/10.3390/en14165057

Academic Editor: Mohamed Benbouzid

Received: 18 July 2021
Accepted: 15 August 2021
Published: 17 August 2021



## 1. Introduction

The grid-connected inverter is the vital interface module for distributed generation (DG) systems, including wind power generation, photovoltaic power generation, to be connected to the grid. It can directly determine the value and direction of current and power and is crucial for the safe operation of the grid [1,2]. Small and medium-sized DG systems are often connected to the grid through the power distribution network. However, due to the abnormal weather conditions, large load switching on–off, insulation failure, human error and so on, voltage faults often occur in the power distribution network. Most of voltage faults can be attributed to the asymmetric faults of grid voltage, for example, the unbalanced voltage fault caused by single-phase grounding is a representative type [3]. In the case of asymmetric faults, the grid-connected inverter is required to have the low voltage ride through (LVRT) capability so as to avoid the chain reaction of DGs disconnection with the grid [4,5]. When the voltage at the point of common coupling (PCC) drops, grid code requires the inverter to keep connecting with the grid for a certain period of time [3]. At the same time, the grid-connected inverter should have the power transmission capacity under abnormal voltage at PCC [1,6]. Under unbalanced PCC voltage, the traditional current closed-loop control strategy which only controls positive-sequence current in essence will cause output power oscillation so as to enlarge the voltage ripple of DC side bus and damage the output current quality [6–9]. To enhance the performance of inverters, it is necessary to eliminate the output power oscillation of grid-connected inverters under unbalanced PCC voltage.

At present, the main methods to eliminate the output power oscillation of grid-connected inverter under unbalanced grid voltage can be divided into two categories: the first type is to improve the control strategy; the second one is to change the topology of the inverter. In [6,10,11], different proportions of negative-sequence current is added to the current reference to decrease power oscillation. The main problem is that the output current has difficulty meeting the requirements of grid codes, and there may be risk of current overrun. In [12,13], the power reference is modified by considering the upper current limit value of switch tube to guarantee that the output current will not exceed the maximum allowable value. However, the waveform of the output current is seriously distorted. A positive and negative-sequence conductance and susceptance control scheme is proposed in [14]. By optimizing the value of negative-sequence conductance, the peak value of the output current can be controlled, but the power oscillation is not effectively reduced. The second type of strategy is to eliminate the power oscillation by changing the topology of the inverter. In [15], a three-phase four-wire system with zero-sequence current channel is proposed. The introduction of zero-sequence current increases the controllable quantity of the system, which is conducive to eliminating power oscillation and improving current quality. Nevertheless, the power oscillation still exists when adopting the current closed-loop control strategy. In one word, the reference current matrix equation of grid-connected inverter in the topologies mentioned above cannot meet the solvability condition, which is the essential reason that the power oscillation could not be eliminated while the current waveform is balanced.

In this paper, a modified grid-connected inverter topology with auxiliary modules inserted in series between PCC and the output filter of each phase is proposed, which could increase the controllable freedoms of reference current equation of inverter so as to make the reference current equation of the inverter meet the solvability condition. Then, the oscillation of the output power is eliminated, and the negative-sequence current is avoided to be injected into power grid simultaneously. The simulation and experimental results are presented to verify the effectiveness of the modified topology.

## 2. Relationship between Power Oscillation and Grid Voltage in Traditional Topology

Typical grid voltage faults can be divided into seven categories [16], most of them are asymmetric faults. Unbalanced voltage fault caused by single-phase grounding is a representative one [3]. The unbalanced voltage at PCC will cause the oscillation of output power and the distortion of output current, which will affect the safe operation of the grid-connected inverters [11].

### 2.1. Relationship among Output Current, Power Oscillation and Unbalanced Voltage

The traditional control strategy of grid-connected inverters under unbalanced grid voltage can be summarized as a unified control strategy [10,11,17], which uses different values of adjustment coefficient ($-1 \leq k \leq 1$) to reflect different control strategies, as shown in Table 1. No matter what control strategy is used, it is always difficult to achieve the optimal output current and power at the same time. A quantitative numerical analysis is given as follows.

**Table 1.** Output effect under different adjustment coefficient $k$.

| $k$ | Control Strategy | Characteristics | | |
|---|---|---|---|---|
| | | Oscillation Cancellation | | Current Quality |
| | | Active Power | Reactive Power | |
| 1 | Average Active Reactive Control | × | √ | × |
| 0 | Balanced Positive Sequence Control | × | × | √ |
| −1 | Positive Negative Sequence Control | √ | × | × |
| Other value | Trade-off between power oscillation cancellation and current quality | | | |

(1). Numerical analysis of output current

Since there is no zero-sequence current component channel in the three-phase three-wire power distribution network, the voltage at PCC and output current can be expressed in Equation (1):

$$\begin{bmatrix} u_a \\ u_b \\ u_c \end{bmatrix} = U^+ \begin{bmatrix} \sin(\omega t + \theta_{vp}) \\ \sin(\omega t + \theta_{vp} - 120^\circ) \\ \sin(\omega t + \theta_{vp} + 120^\circ) \end{bmatrix} + U^- \begin{bmatrix} \sin(\omega t + \theta_{vn}) \\ \sin(\omega t + \theta_{vn} + 120^\circ) \\ \sin(\omega t + \theta_{vn} - 120^\circ) \end{bmatrix} + U^0 \begin{bmatrix} \sin(\omega t + \theta_0) \\ \sin(\omega t + \theta_0) \\ \sin(\omega t + \theta_0) \end{bmatrix} \tag{1}$$

$$\begin{bmatrix} i_a \\ i_b \\ i_c \end{bmatrix} = I^+ \begin{bmatrix} \sin(\omega t + \theta_{ip}) \\ \sin(\omega t + \theta_{ip} - 120^\circ) \\ \sin(\omega t + \theta_{ip} + 120^\circ) \end{bmatrix} + I^- \begin{bmatrix} \sin(\omega t + \theta_{in}) \\ \sin(\omega t + \theta_{in} + 120^\circ) \\ \sin(\omega t + \theta_{in} - 120^\circ) \end{bmatrix} \tag{2}$$

where $U^+$, $U^-$ and $U^0$ are the amplitudes of positive, negative and zero-sequence components of the voltage, $I^+$ and $I^-$ are the amplitudes of positive and negative-sequence components of output current, $\theta_{vp}$, $\theta_{vn}$ and $\theta_0$ are the initial phase angles of positive, negative and zero-seqence components of the voltage, $\theta_{ip}$ and $\theta_{in}$ are the initial phase angles of positive, negative-sequence component of output current and $\omega$ is the angular frequency of the voltage, respectively.

When the three-phase inverter is connected to the grid, the current reference value is shown in Equations (3) and (4) [12].

$$\begin{aligned} I_\alpha^* &= \tfrac{2}{3}\tfrac{P^*}{(U^+)^2+k(U^-)^2}\left[U^+\sin(\omega t + \theta_{vp}) + kU^-\sin(\omega t + \theta_{vn})\right] \\ &+ \tfrac{2}{3}\tfrac{Q^*}{(U^+)^2+k(U^-)^2}\left[-U^+\cos(\omega t + \theta_{vp}) - kU^-\cos(\omega t + \theta_{vn})\right] \end{aligned} \tag{3}$$

$$\begin{aligned} I_\beta^* &= \tfrac{2}{3}\tfrac{P^*}{(U^+)^2+k(U^-)^2}\left[-U^+\cos(\omega t + \theta_{vp}) + kU^-\cos(\omega t + \theta_{vn})\right] \\ &+ \tfrac{2}{3}\tfrac{Q^*}{(U^+)^2-k(U^-)^2}\left[-U^+\sin(\omega t + \theta_{vp}) + kU^-\sin(\omega t + \theta_{vn})\right] \end{aligned} \tag{4}$$

where $P^*$ and $Q^*$ are the reference values of active power and reactive power, respectively.

After transformation to *abc* coordinate system, Equations (5)–(7) are obtained,

$$I_a^* = \frac{2}{3}\sqrt{A_1^2 + (kA_2)^2 + 2kA_1A_2\cos(\theta_{vp} - \theta_{vn} - 2\delta)} \cdot \sin(\omega t + \psi_a) \tag{5}$$

$$I_b^* = \frac{2}{3}\sqrt{A_1^2 + (kA_2)^2 + 2kA_1A_2\cos(\theta_{vp} - \theta_{vn} - 2\delta - 240^\circ)} \cdot \sin(\omega t + \psi_b) \tag{6}$$

$$I_c^* = \frac{2}{3}\sqrt{A_1^2 + (kA_2)^2 + 2kA_1A_2\cos(\theta_{vp} - \theta_{vn} - 2\delta + 240^\circ)} \cdot \sin(\omega t + \psi_c) \tag{7}$$

where,

$$A_1 = \sqrt{\left[\frac{P^*U^+}{(U^+)^2 + k(U^-)^2}\right]^2 + \left[\frac{Q^*U^+}{(U^+)^2 - k(U^-)^2}\right]^2}$$

$$\delta = \arctan\frac{Q^*\left[(U^+)^2 + k(U^-)^2\right]}{P^*\left[(U^+)^2 - k(U^-)^2\right]}$$

$$A_2 = \sqrt{\left[\frac{P^*U^-}{(U^+)^2 + k(U^-)^2}\right]^2 + \left[\frac{Q^*U^-}{(U^+)^2 - k(U^-)^2}\right]^2}$$

$$\psi_a = \arctan\frac{A_1\sin(\theta_{vp} - \delta) + kA_2\sin(\theta_{vn} - \delta)}{A_1\cos(\theta_{vp} - \delta) + kA_2\cos(\theta_{vn} - \delta)}$$

$$\psi_b = \arctan\frac{A_1\sin(\theta_{vp} - \delta - 120^\circ) + kA_2\sin(\theta_{vn} - \delta + 120^\circ)}{A_1\cos(\theta_{vp} - \delta - 120^\circ) + kA_2\cos(\theta_{vn} - \delta + 120^\circ)}$$

$$\psi_c = \arctan\frac{A_1\sin(\theta_{vp} - \delta + 120^\circ) + kA_2\sin(\theta_{vn} - \delta - 120^\circ)}{A_1\cos(\theta_{vp} - \delta + 120^\circ) + kA_2\cos(\theta_{vn} - \delta - 120^\circ)}.$$

When three-phase inverter is in grid-connected operation state, the current closed-loop control strategy is often used [8], so the output current could accurately track the reference value. The maximum value of output current is given as Equation (8):

$$\begin{aligned} I_{\max} = \quad & \max\Big\{ \tfrac{2}{3}\sqrt{A_1^2 + (kA_2)^2 + 2kA_1A_2\cos(\theta_p - \theta_n - 2\delta)}, \\ & \tfrac{2}{3}\sqrt{A_1^2 + (kA_2)^2 + 2kA_1A_2\cos(\theta_p - \theta_n - 2\delta - 240^\circ)}, \\ & \tfrac{2}{3}\sqrt{A_1^2 + (kA_2)^2 + 2kA_1A_2\cos(\theta_p - \theta_n - 2\delta + 240^\circ)} \Big\}. \end{aligned} \tag{8}$$

It can be seen from Equations (5)–(7) that the amplitudes and phase angles of three-phase currents are related to the amplitudes of positive and negative-sequence voltages, initial phase angles of positive and negative-sequence voltages, power reference value and adjustment coefficient $k$. To ensure that the output current of grid-connected inverters meets the grid codes [18,19], the only way is to set the adjustment coefficient $k$ as 0, which means to use the balanced positive sequence control (BPSC) strategy [20].

According to Equation (8), when the parameters of Table 2 are adopted, the relationship between the maximum value of output current amplitude and the adjustment coefficient $k$ is shown in Figure 1. When $k$ changes from $-1$ to 1, the maximum value of output current amplitude decreases firstly and then increases, and the BPSC control method, which means $k$ as 0, can ensure that the current stress of the inverter is minimum under the same power.

(2). Analysis of power oscillation

According to the instantaneous power theory, the active power and reactive power can be expressed as Equations (9) and (10):

$$\begin{aligned} p \ &= \tfrac{3}{2}u_\alpha\left[A_1\sin(\omega t + \theta_{vp} - \delta) + kA_2\sin(\omega t + \theta_{vn} - \delta)\right] \\ &+ \tfrac{3}{2}u_\beta\left[-A_1\cos(\omega t + \theta_{vp} - \delta) + kA_2\cos(\omega t + \theta_{vn} - \delta)\right] \end{aligned} \tag{9}$$

$$\begin{aligned} q \ &= -\tfrac{3}{2}u_\alpha\left[-A_1\cos(\omega t + \theta_{vp} - \delta) + kA_2\cos(\omega t + \theta_{vn} - \delta)\right] \\ &+ \tfrac{3}{2}u_\beta\left[A_1\sin(\omega t + \theta_{vp} - \delta) + kA_2\sin(\omega t + \theta_{vn} - \delta)\right] \end{aligned} \tag{10}$$

where $u_\alpha$ and $u_\beta$ are the $\alpha\beta$ axis components of the PCC voltage, respectively.

When expanding Equations (9) and (10), the quadratic term is the fluctuating power, as shown in Equations (11) and (12):

$$\begin{aligned} \Delta p \ &= -(1+k)\tfrac{P^*}{(U^+)^2 + k(U^-)^2}U^+U^-\cos(2\omega t + \theta_{vp} + \theta_{vn}) \\ &-(1+k)\tfrac{Q^*}{(U^+)^2 - k(U^-)^2}U^+U^-\sin(2\omega t + \theta_{vp} + \theta_{vn}) \end{aligned} \tag{11}$$

$$\begin{aligned} \Delta q \ &= (1-k)\tfrac{P^*}{(U^+)^2 + k(U^-)^2}U^+U^-\sin(2\omega t + \theta_{vp} + \theta_{vn}) \\ &-(1-k)\tfrac{Q^*}{(U^+)^2 - k(U^-)^2}U^+U^-\cos(2\omega t + \theta_{vp} + \theta_{vn}). \end{aligned} \tag{12}$$

When the parameters shown in Table 2 are adopted, the relationship between the power oscillation value and the adjustment coefficient $k$ is shown in Figure 2. When $k$ changes from $-1$ to 1, the oscillation value of output power can reach 0.4 p.u. at most. The change trend of active power oscillation value is just opposite to that of reactive power oscillation value. It is impossible to make active power oscillation value and reactive power oscillation value minimal at the same time by changing $k$. When BPSC is used, the oscillation value of active power and reactive power can reach 0.2 p.u., which means that the oscillation of output power is not eliminated.

**Table 2.** Operating parameters of main circuit used in simulations.

| Symbol | Description | Value (p.u.) |
|---|---|---|
| $V_a$ | Amplitude value of A-phase voltage | $220\sqrt{2}$ V (1 p.u.) |
| $V_b$ | Amplitude value of B-phase voltage | $110\sqrt{2}$ V (0.5 p.u.) |
| $V_c$ | Amplitude value of C-phase voltage | $220\sqrt{2}$ V (1 p.u.) |
| $P_0$ | Output power | 10 kW (1 p.u.) |
| $f_0$ | Fundamental frequency | 50 Hz |
| $f_{sw}$ | Operating frequency | 10 kHz |
| $L_f$ | Output inductor | 0.044 p.u. |
| $V_{dc}$ | DC voltage | 700 V |
| $k_p$ | Proportional coefficient | 2.0 |
| $k_i$ | Integral coefficient | 1.0 |

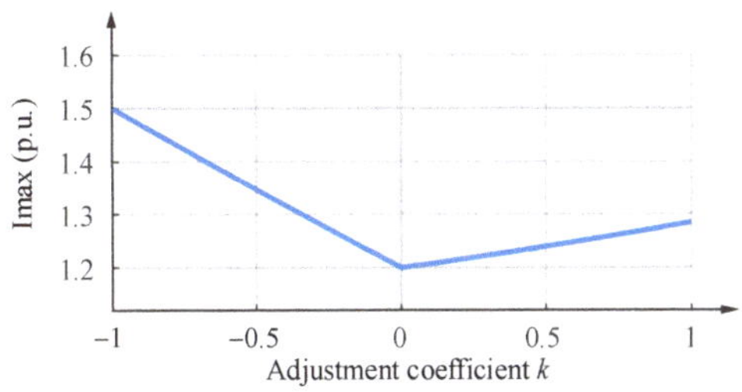


**Figure 1.** Maximum value of output current amplitude varying with the adjustment coefficient.

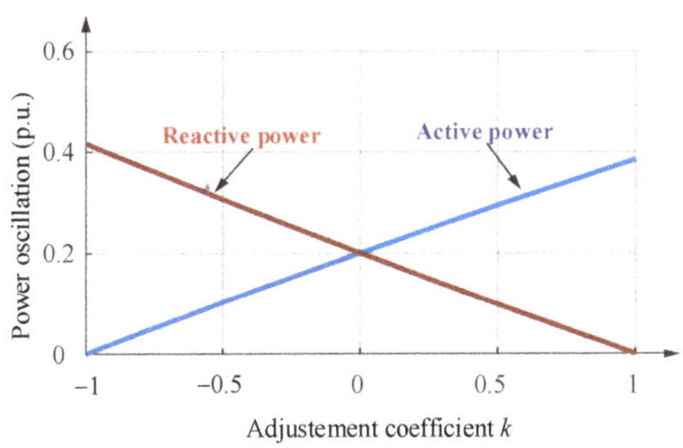


**Figure 2.** Oscillation value of output power varying with the adjustment coefficient.

The variation of $k$ represents different control strategies. The power oscillation cannot be eliminated by just changing the control strategy when the topology is not optimized. The next section will give a strict theoretical proof.

*2.2. Deficiency of Traditional Inverter Topology*

After Clark transformation, Equation (1) can be written as Equation (13):

$$\begin{bmatrix} u_\alpha \\ u_\beta \end{bmatrix} = \frac{2}{3}\begin{bmatrix} 1 & -\frac{1}{2} & -\frac{1}{2} \\ 0 & \frac{\sqrt{3}}{2} & -\frac{\sqrt{3}}{2} \end{bmatrix}\begin{bmatrix} u_a \\ u_b \\ u_c \end{bmatrix} = \begin{bmatrix} u_\alpha^+ + u_\alpha^- \\ u_\beta^+ + u_\beta^- \end{bmatrix} \quad (13)$$

where

$$\begin{bmatrix} u_\alpha^+ \\ u_\beta^+ \end{bmatrix} = \begin{bmatrix} U^+ \sin(\omega t + \theta_{vp}) \\ -U^+ \cos(\omega t + \theta_{vp}) \end{bmatrix} \begin{bmatrix} u_\alpha^- \\ u_\beta^- \end{bmatrix} = \begin{bmatrix} U^- \sin(\omega t + \theta_{vn}) \\ U^- \cos(\omega t + \theta_{vn}) \end{bmatrix}$$

Three-phase output currents can be written as Equation (14):

$$\begin{bmatrix} i_\alpha \\ i_\beta \end{bmatrix} = \frac{2}{3}\begin{bmatrix} 1 & -\frac{1}{2} & -\frac{1}{2} \\ 0 & \frac{\sqrt{3}}{2} & -\frac{\sqrt{3}}{2} \end{bmatrix}\begin{bmatrix} i_a \\ i_b \\ i_c \end{bmatrix} = \begin{bmatrix} i_\alpha^+ + i_\alpha^- \\ i_\beta^+ + i_\beta^- \end{bmatrix} \tag{14}$$

where

$$\begin{bmatrix} i_\alpha^+ \\ i_\beta^+ \end{bmatrix} = \begin{bmatrix} I^+\sin(\omega t + \theta_{ip}) \\ -I^+\cos(\omega t + \theta_{ip}) \end{bmatrix} \begin{bmatrix} i_\alpha^- \\ i_\beta^- \end{bmatrix} = \begin{bmatrix} I^-\sin(\omega t + \theta_{in}) \\ I^-\cos(\omega t + \theta_{in}) \end{bmatrix}$$

The instantaneous output power is given as Equation (15):

$$\begin{bmatrix} p \\ q \end{bmatrix} = \begin{bmatrix} v_\alpha i_\alpha + v_\beta i_\beta \\ v_\alpha i_\beta - v_\beta i_\alpha \end{bmatrix} = \begin{bmatrix} P_0 + P_{c2}\cos(2\omega t) + P_{s2}\sin(2\omega t) \\ Q_0 + Q_{c2}\cos(2\omega t) + Q_{s2}\sin(2\omega t) \end{bmatrix} \tag{15}$$

where $P_0$ is the DC component value of active power, $P_{c2}$ and $P_{s2}$ are the coefficients of cosine and sinusoidal terms for the second oscillation value of active power, $Q_0$ is the DC component value of reactive power, $Q_{c2}$ and $Q_{s2}$ are the coefficients of cosine and sinusoidal terms for the second oscillation value of reactive power.

$P_0$, $P_{c2}$, $P_{s2}$, $Q_0$, $Q_{c2}$ and $Q_{s2}$ can be expressed by matrix as Equation (16):

$$\begin{bmatrix} P_0 \\ Q_0 \\ P_{c2} \\ P_{s2} \\ Q_{c2} \\ Q_{s2} \end{bmatrix} = \begin{bmatrix} v_d^+ & v_q^+ & v_d^- & v_q^- \\ v_q^+ & -v_d^+ & v_q^- & -v_d^- \\ -v_d^- & -v_q^- & -v_d^+ & -v_q^+ \\ -v_q^- & v_d^- & v_q^+ & -v_d^+ \\ -v_q^- & v_d^- & -v_q^+ & v_d^+ \\ v_d^- & v_q^- & -v_d^+ & -v_q^+ \end{bmatrix}\begin{bmatrix} i_d^+ \\ i_q^+ \\ i_d^- \\ i_q^- \end{bmatrix} \tag{16}$$

where $v_d^+$, $v_q^+$ are the $dq$ axis components of $u_\alpha^+$, $u_\beta^+$; $v_d^-$, $v_q^-$ are the $dq$ axis components of $u_\alpha^-$, $u_\beta^-$; $i_d^+$, $i_q^+$ are the $dq$ axis components of $i_\alpha^+$, $i_\beta^+$; $i_d^-$, $i_q^-$ are the $dq$ axis components of $i_\alpha^-$, $i_\beta^-$ after Park transformation, respectively.

When the output current meets the grid codes [18,19], the negative-sequence current should be set as zero, then Equation (16) will be changed into Equation (17):

$$\begin{bmatrix} P_0 \\ Q_0 \\ P_{c2} \\ P_{s2} \\ Q_{c2} \\ Q_{s2} \end{bmatrix} = \begin{bmatrix} v_d^+ & v_q^+ \\ v_q^+ & -v_d^+ \\ -v_d^- & -v_q^- \\ -v_q^- & v_d^- \\ -v_q^- & v_d^- \\ v_d^- & v_q^- \end{bmatrix}\begin{bmatrix} i_d^+ \\ i_q^+ \end{bmatrix}. \tag{17}$$

In Equation (16), the rank of the coefficient matrix is 4, the rank of the augmented matrix is 5. The rank of the coefficient matrix is not equal to the rank of the augmented matrix, so Equation (16) has no solution. This is the reason why the control strategy in [10–14] has difficulty eliminating the power oscillation.

In Equation (17), the rank of the coefficient matrix is 2, the rank of the augmented matrix is 3. The rank of the coefficient matrix is not equal to the rank of the augmented matrix, so Equation (17) has no solution. This is the reason why the BPSC control strategy has difficulty eliminating power oscillation for the traditional topology.

In the same way, it can be found that after adding a zero-sequence component in [15], Equation (17) becomes:

$$\begin{bmatrix} P_0 \\ Q_0 \\ P_{c2} \\ P_{s2} \\ Q_{c2} \\ Q_{s2} \end{bmatrix} = \begin{bmatrix} v_d^+ & v_q^+ & v_{\mathrm{Re}}^0 & v_{\mathrm{Im}}^0 \\ v_q^+ & -v_d^+ & 0 & 0 \\ -v_d^- & -v_q^- & v_{\mathrm{Re}}^0 & -v_{\mathrm{Im}}^0 \\ -v_q^- & v_d^- & -v_{\mathrm{Im}}^0 & -v_{\mathrm{Re}}^0 \\ -v_q^- & v_d^- & 0 & 0 \\ v_d^- & v_q^- & 0 & 0 \end{bmatrix} \begin{bmatrix} i_d^+ \\ i_q^+ \\ i_{\mathrm{Re}}^0 \\ i_{\mathrm{Im}}^0 \end{bmatrix} \tag{18}$$

where $v_{\mathrm{Re}}^0$, $v_{\mathrm{Im}}^0$ are the real and imaginary components of zero-sequence voltage at PCC, $i_{\mathrm{Re}}^0$, $i_{\mathrm{Im}}^0$ are the real and imaginary components of zero-sequence output current, the detailed definition refers to [15].

In (18), the rank of the coefficient matrix is 4, the rank of the augmented matrix is 5. The rank of the coefficient matrix is not equal to the rank of the augmented matrix, so Equation (18) has no solution. This is the reason why the BPSC control strategy has difficulty eliminating power oscillation for three-phase four-wire inverter in [15].

## 3. Principle and Advantages of Modified Topology for Grid-Connected Inverter

According to the discussion in Section 2, the reason why the traditional topology is incapable of eliminating the power oscillation is that the number of controllable free variables is small, which leads to the mismatch between the order of the current reference coefficient matrix and the order of the augmented matrix. From the point of view on hardware, the number of controllable variables in the current reference equation can be increased by changing the topology structure, so that the equation can meet the solvability conditions.

### 3.1. Modified Topology of Grid-Connected Inverters

The proposed topology of grid-connected inverter is shown in Figure 3. The auxiliary modules are inserted in series between the output filter of inverter (the points $a_1$, $b_1$ and $c_1$ in Figure 3a) and PCC (the points $a_2$, $b_2$ and $c_2$ in Figure 3a).

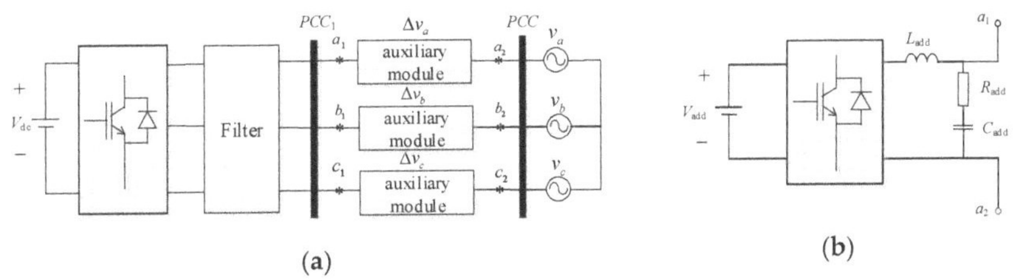


**Figure 3.** Structure block diagram of the modified topology of grid-connected inverter: (**a**) general system diagram. (**b**) Auxiliary module block diagram.

The three-phase auxiliary modules are independent of each other and can be separately controlled. Figure 3b shows the detailed internal block diagram of the auxiliary module which is essentially a single-phase inverter. $L_{\mathrm{add}}$ and $C_{\mathrm{add}}$ constitute the output filter circuit of the auxiliary module to filter the switching subharmonics in the circuit. $R_{\mathrm{add}}$ is the damping resistance of the output filter circuit to prevent the possible oscillation of the LC filter circuit. For the voltages at $\mathrm{PCC}_1$, Equation (19) could be obtained as:

$$v_{a1} = v_a + \Delta v_a v_{b1} = v_b + \Delta v_b v_{c1} = v_c + \Delta v_c \tag{19}$$

where $\Delta v_a$, $\Delta v_b$ and $\Delta v_c$ are the output voltages of auxiliary modules in phases A, B and C, respectively.

Accordingly, Equation (17) can be changed into Equation (20):

$$\begin{bmatrix} P_0 \\ Q_0 \\ P_{c2} \\ P_{s2} \\ Q_{c2} \\ Q_{s2} \end{bmatrix} = \begin{bmatrix} v_d^+ + \Delta v_d^+ & v_q^+ + \Delta v_q^+ \\ v_q^+ + \Delta v_q^+ & -v_d^+ - \Delta v_d^+ \\ -v_d^- - \Delta v_d^- & -v_q^- - \Delta v_q^- \\ -v_q^- - \Delta v_q^- & v_d^- + \Delta v_d^- \\ -v_q^- - \Delta v_q^- & v_d^- + \Delta v_d^- \\ v_d^- + \Delta v_d^- & v_q^- + \Delta v_q^- \end{bmatrix} \begin{bmatrix} i_d^+ \\ i_q^+ \end{bmatrix} \tag{20}$$

where $\Delta v_d^+$, $\Delta v_q^+$ are the positive-sequence $dq$ axis components of $\Delta v_a$, $\Delta v_b$ and $\Delta v_c$; $\Delta v_d^-$, $\Delta v_q^-$ are the negative sequence $dq$ axis components of $\Delta v_a$, $\Delta v_b$ and $\Delta v_c$ after positive and negative-sequence separation and Park transformation.

When the output of the auxiliary module is set to satisfy Equation (21):

$$\begin{aligned} \Delta v_d^- &= -v_d^- \\ \Delta v_q^- &= -v_q^- \\ \Delta v_d^+ &= 0 \\ \Delta v_q^+ &= 0, \end{aligned} \tag{21}$$

a new equation, Equation (22) can be obtained from Equation (20):

$$\begin{bmatrix} P_0 \\ Q_0 \\ P_{c2} \\ P_{s2} \\ Q_{c2} \\ Q_{s2} \end{bmatrix} = \begin{bmatrix} v_d^+ & v_q^+ \\ v_q^+ & -v_d^+ \\ 0 & 0 \\ 0 & 0 \\ 0 & 0 \\ 0 & 0 \end{bmatrix} \begin{bmatrix} i_d^+ \\ i_q^+ \end{bmatrix}. \tag{22}$$

It can be further simplified as Equation (23):

$$\begin{bmatrix} v_d^+ & v_q^+ \\ v_q^+ & -v_d^+ \end{bmatrix} \begin{bmatrix} i_d^+ \\ i_q^+ \end{bmatrix} = \begin{bmatrix} P^* \\ Q^* \end{bmatrix}. \tag{23}$$

The coefficient matrix rank of Equation (23) is 2, the rank of the augmented matrix is 2. The number of equations is equal to the number of variables, so that Equation (23) has a unique solution. The corresponding current reference value can be solved as Equation (24):

$$\begin{bmatrix} i_d^* \\ i_q^* \end{bmatrix} = \begin{bmatrix} i_d^+ \\ i_q^+ \end{bmatrix} = \begin{bmatrix} v_d^+ & v_q^+ \\ v_q^+ & -v_d^+ \end{bmatrix}^{-1} \begin{bmatrix} P^* \\ Q^* \end{bmatrix}. \tag{24}$$

When it is needed to limit the current, the upper limit of the reference current can be adjusted by changing the value of $\Delta v_d^+$ and $\Delta v_q^+$ in Equation (21).

The overall control structure of the system is shown in Figure 4. After positive and negative sequence separation module, positive-sequence components $v_a^+$, $v_b^+$ and $v_c^+$, negative-sequence components $v_a^-$, $v_b^-$ and $v_c^-$, and zero-sequence components $v_a^0$, $v_b^0$ and $v_c^0$ are derived from PCC voltages $v_a$, $v_b$ and $v_c$. Following that, $v_d^+$ and $v_q^+$ are generated from $v_a^+$, $v_b^+$, $v_c^+$ through $abc/dq$ transformation. Then, the current reference values $i_d^*$ and $i_q^*$ are given from current reference generator module according to Equation (24). After that, the current reference values $i_d^*$ and $i_q^*$ generate the PWM waves that control the main circuit through the PI module as well as processing the decoupling components $\omega_0 L_f i_d$ and $\omega_0 L_f i_q$. The negative-sequence voltage components $v_a^-$, $v_b^-$, $v_c^-$ and the zero-sequence voltage components $v_a^0$, $v_b^0$, $v_c^0$ are added and then create reference values $\Delta v_a^*$, $\Delta v_b^*$, $\Delta v_c^*$ of the auxiliary module output voltage by multiplying by minus one. The three reference values $\Delta v_a^*$, $\Delta v_b^*$ and $\Delta v_c^*$, respectively, generate PWM waves for three auxiliary module circuits through the PR controller.

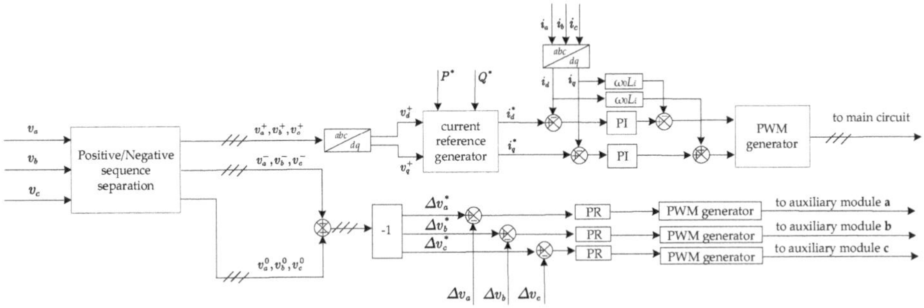


**Figure 4.** Overall control structure diagram.

The Decoupled Double Synchronous Reference Frame Phase-Locked Loop (DDSRF-PLL), which has better performance under unbalanced voltage, is adopted to achieve phase detection [21]. The structure of DDSRF-PLL is shown in Figure 5. After positive and negative sequence abc/dq transformation, $v_d^+$, $v_q^+$, $v_d^-$ and $v_q^-$ are derived from the grid voltage $v_a$, $v_b$ and $v_c$. Then, $v_d^{+*}$, $v_q^{+*}$ are generated from $v_d^+$, $v_q^+$ after a decoupling network. Similar to the phase detection principle of the Synchronous Reference Frame Phase-Locked Loop (SRF-PLL) [21], the phase of the system $\hat{\theta}$ could be obtained by making $v_q^{+*}$ approach zero through the PI controller. The detailed structure of decoupling network is expressed as in [21].

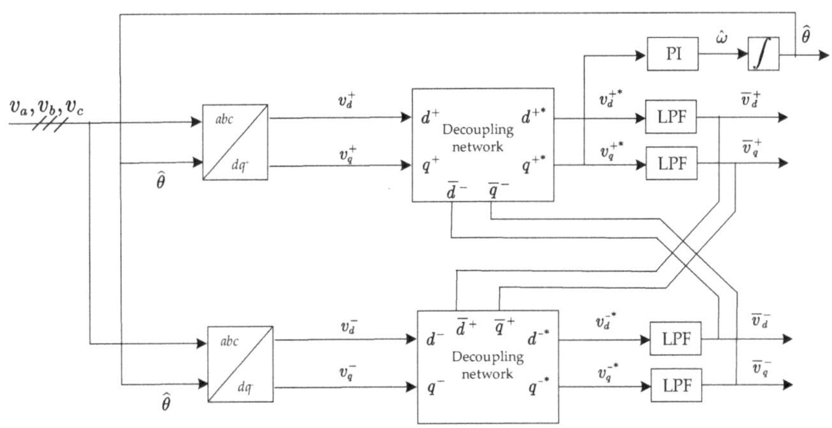


**Figure 5.** The structure of DDSRF-PLL.

### 3.2. *Advantages of Modified Topology*

The voltage and current vector trajectories of the traditional topology and the modified topology are shown in Figure 6 [9], respectively. When single-phase voltage fault occurs, the trajectory of voltage vector will be distorted. When the voltage vector trajectory shows a spindle-shaped variation, as the trajectory of $U_{tra}$ in Figure 6, in order to keep the output power constant, the trajectory of the current amplitude becomes an ellipse, as the trajectory of $I_{tra}$ in Figure 6. Where, $I_{max}$ is the maximum value of the allowable current amplitude of the system. It can be seen that the amplitude of $I_{tra}$ has exceeded the value of $I_{max}$ near the long axis of the actual current track, as shown in the shaded area, which may give rise to over-current fault. When the auxiliary modules are inserted into the system, the unbalanced voltage is corrected to a standard circle, as the trajectory of $U_{mod}$ in Figure 6, and the corresponding current amplitude trajectory is also a standard circle, as the trajectory of $I_{mod}$ in Figure 6. Since the amplitude of $U_{mod}$ can be adjusted by changing the value of $\Delta v_d^+$ and $\Delta v_q^+$, so that the radius of the trajectory of $U_{mod}$ can be ensured not to be too small, then the corresponding trajectory of $I_{mod}$ can be guaranteed to be included in the circular trajectory of $I_{max}$, so as to avoid the risk of overcurrent fault.

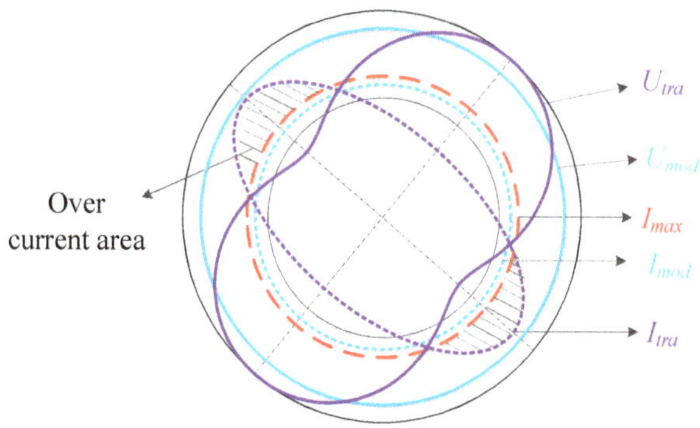


**Figure 6.** Change of voltage and current amplitude trajectories of the traditional and the modified topologies.

### 3.3. Capacity Design of Auxiliary Modules

Supposing that the fault occurs in phase B, and the voltage amplitude of phase B is $\beta$ times of its original normal value, then the vector format of three-phase voltage is described as Equation (25):

$$\boldsymbol{V}_a = V\angle\varphi \quad \boldsymbol{V}_b = \beta V\angle(\varphi - 120^\circ) \quad \boldsymbol{V}_c = V\angle(\varphi + 120^\circ) \tag{25}$$

The positive, negative and zero-sequence voltage components of phase A are obtained as Equation (26), [21].

$$\begin{bmatrix} \boldsymbol{V}_{a(1)} \\ \boldsymbol{V}_{a(2)} \\ \boldsymbol{V}_{a(0)} \end{bmatrix} = \frac{1}{3}\begin{bmatrix} 1 & h & h^2 \\ 1 & h^2 & h \\ 1 & 1 & 1 \end{bmatrix}\begin{bmatrix} \boldsymbol{V}_a \\ \boldsymbol{V}_b \\ \boldsymbol{V}_c \end{bmatrix} = \begin{bmatrix} \frac{2+\beta}{3}V\angle\varphi \\ \frac{1-\beta}{3}V\angle(\varphi - 60^\circ) \\ \frac{1-\beta}{3}V\angle(\varphi + 60^\circ) \end{bmatrix} \tag{26}$$

where

$$h = e^{j120^\circ} = -\frac{1}{2} + j\frac{\sqrt{3}}{2} h^2 = e^{j240^\circ} = -\frac{1}{2} - j\frac{\sqrt{3}}{2}$$

The positive, negative and zero-sequence voltage components of phase B and phase C are obtained as Equations (27) and (28):

$$\begin{bmatrix} \boldsymbol{V}_{b(1)} \\ \boldsymbol{V}_{b(2)} \\ \boldsymbol{V}_{b(0)} \end{bmatrix} = \begin{bmatrix} \frac{2+\beta}{3}V\angle(\varphi - 120^\circ) \\ \frac{1-\beta}{3}V\angle(\varphi + 60^\circ) \\ \frac{1-\beta}{3}V\angle(\varphi + 60^\circ) \end{bmatrix} \tag{27}$$

$$\begin{bmatrix} \boldsymbol{V}_{c(1)} \\ \boldsymbol{V}_{c(2)} \\ \boldsymbol{V}_{c(0)} \end{bmatrix} = \begin{bmatrix} \frac{2+\beta}{3}V\angle(\varphi + 120^\circ) \\ \frac{1-\beta}{3}V\angle(\varphi + 180^\circ) \\ \frac{1-\beta}{3}V\angle(\varphi + 60^\circ) \end{bmatrix}. \tag{28}$$

In the modified topology, the average powers of the three auxiliary modules are shown in Equation (29):

$$P_a = \frac{1-\beta}{6}VI \, P_b = -\frac{1-\beta}{3}VI \, P_c = \frac{1-\beta}{6}VI. \tag{29}$$

When $\beta$ changes between [0, 1], the power curve of the auxiliary module is shown in Figure 7. Auxiliary module of the fault phase-phase B essentially injects power into the circuit. When the output power of the grid-connected inverter is used as the power base value, which is $P_{base} = (3/2)VI$, the maximum value of the injected power is 0.22 p.u. ($\beta$ = 0). Phases A and C are non-fault phases. Their auxiliary modules absorb power from the circuit, and the maximum absorbed power is 0.11 p.u. ($\beta$ = 0). Therefore, for the single-phase voltage fault, the power capacity of the auxiliary module designed above 0.22 p.u. can meet the requirements no matter what the drop depth is and no matter which phase the fault occurs in.

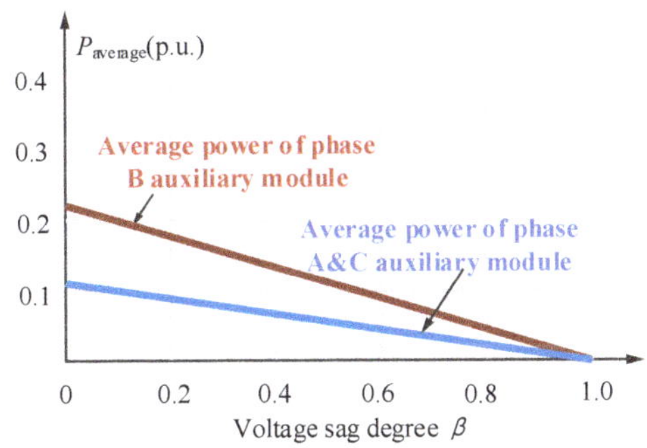


**Figure 7.** Output power of the auxiliary module.

According to the clearing time of inverters under voltage faults shown in Figure 8, in case of voltage drop, the grid-connected inverter only needs to maintain the connection time of 2 s at most. Therefore, the energy needed to be absorbed or released by the auxiliary module will not be too large, and the operation of the inverter during the LVRT period will not cause damage to the energy storage devices in the auxiliary module.

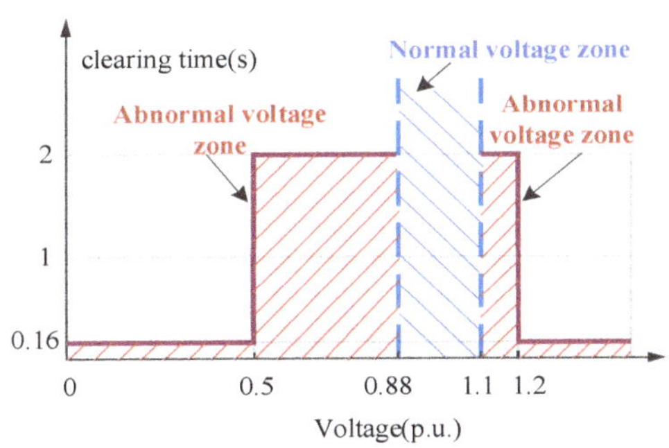


**Figure 8.** Clearing time of inverters under different voltage faults.

## 3.4. Stability Analysis of the Modified Topology

Since the parameters of the three branches of the inverter are identical, according to [7,22], the whole system can be simplified shown in Figure 9.

In Figure 9, $v^*$ is the output voltage reference of the auxiliary module, $G_{add}(s)$ is the transfer function of the auxiliary module voltage in open mode. $L_{add}$, $C_{add}$ and $R_{add}$ are inductance, capacitor and resistor of the auxiliary module, respectively. $G_{cli,m}(s)$ denotes the current reference to output transfer function of grid-connected inverter, $Y_{oi,m}(s)$ represents the equivalent output admittance of the system, $i^*_{gm}(s)$ and $i_{gm}(s)$ are the reference current and output current of system, respectively, and their detailed form are expressed as [22].

When we adopt the parameters in Tables 3–5, the pole-zero maps of the current closed transfer functions for the traditional topology and the modified topology are shown in Figure 10, respectively. In Figure 10, $p_1$ and $p_2$ are the dominant closed-loop poles of the current transfer function for the traditional topology and the modified topology, respectively. Although $p_2$ is closer to the imaginary axis than $p_1$, they are still very close to each other and far away from the imaginary axis. Therefore, compared with the traditional topology, the output stability of the modified topology is reduced, but all poles of the modified topology system are still in the left half plane and the whole system is still stable.

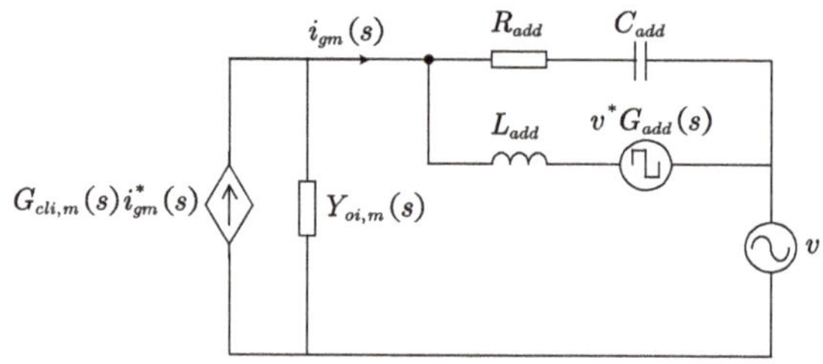


**Figure 9.** Equivalent structure diagram of the modified topology.

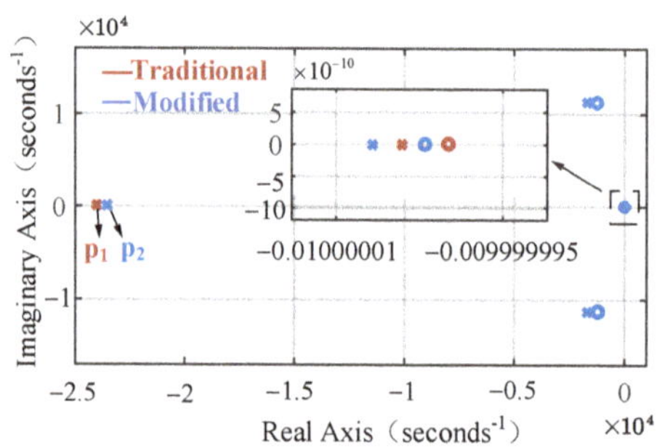


**Figure 10.** Pole-zero maps of current closed transfer functions for the traditional topology and the modified topology.

**Table 3.** Operating parameters of main circuit used in experiments.

| Symbol | Parameter | Value (p.u.) |
|---|---|---|
| $V_a$ | Amplitude value of A-phase voltage | 50 V (1 p.u.) |
| $V_b$ | Amplitude value of B-phase voltage | 25 V (0.5 p.u.) |
| $V_c$ | Amplitude value of C-phase voltage | 50 V (1 p.u.) |
| $P_0$ | Output power | 105 W (1 p.u.) |
| $f_0$ | Fundamental frequency | 50 Hz |
| $f_{sw}$ | Operating frequency | 10 kHz |
| $L_f$ | Output inductor | 0.044 p.u. |
| $V_{dc}$ | DC voltage | 120 V |
| $k_p$ | Proportional coefficient | 2.0 |
| $k_i$ | Integral coefficient | 2.0 |

**Table 4.** Operating parameters of auxiliary module used in experiments.

| Symbol | Parameter | Value |
|---|---|---|
| $V_{add}$ | DC voltage of auxiliary module | 36 V |
| $f_{add}$ | Switching frequency of auxiliary module | 10 kHz |
| $C_{add}$ | Filter capacitor of auxiliary module | 15 μF |
| $L_{add}$ | Filter inductor of auxiliary module | 800 μH |
| $R_{add}$ | Damping resistance of auxiliary module | 2.0 Ω |
| $k_p$ | Proportional coefficient | 0.01 |
| $k_r$ | Resonant coefficient | 4.0 |

**Table 5.** Parameters of auxiliary module used in simulations.

| Symbol | Parameter | Value |
|---|---|---|
| $V_{add}$ | DC voltage of auxiliary module | 150 V |
| $f_{add}$ | Switching frequency of auxiliary module | 10 kHz |
| $C_{add}$ | Filter capacitor of auxiliary module | 15 μF |
| $L_{add}$ | Filter inductor of auxiliary module | 800 μH |
| $R_{add}$ | Damping resistance of auxiliary module | 2.0 Ω |
| $k_p$ | Proportional coefficient | 0.02 |
| $k_r$ | Resonant coefficient | 16.91 |

## 4. Simulation Results

With the detailed discussion of the modified topology, the MATLAB/Simulink was performed to achieve the verification. In the simulation, the fault type is that the voltage of phase B drops to 0.5 p.u. The parameters of other main circuits are shown in Table 2, and the parameters of the auxiliary modules are shown in Table 5. The results of the traditional topology and the modified topology are shown in Figures 11 and 12, respectively.

Comparing the results in Figures 11 and 12, when the modified topology is adopted, the unbalanced voltage at $PCC_1$ can be corrected, and the inverter output current keeps being balanced. The oscillation value of the output active power is reduced from 0.49 to 0.17 p.u., and the output reactive power oscillation value is reduced from 0.46 to 0.15 p.u. The oscillation value of active power for the modified topology is 33.88% of that of the traditional topology and the reactive power oscillation value is reduced to 32.17% of the value of traditional topology. The simulation results show that the modified topology can correct the unbalanced voltage at the output of the inverter and reduce the oscillation of the output active power and reactive power.

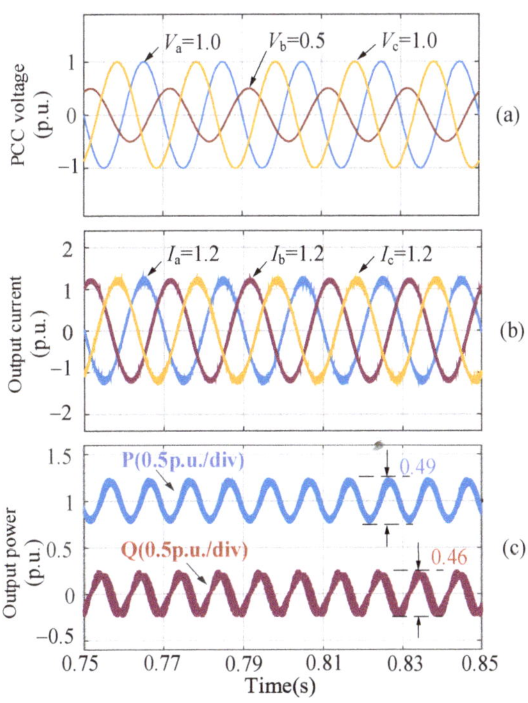


**Figure 11.** Simulation results of the traditional topology, when $V_b$ drops to 0.5 p.u. (**a**) Output voltage at $PCC_1$; (**b**) output current; (**c**) output active power and reactive power.

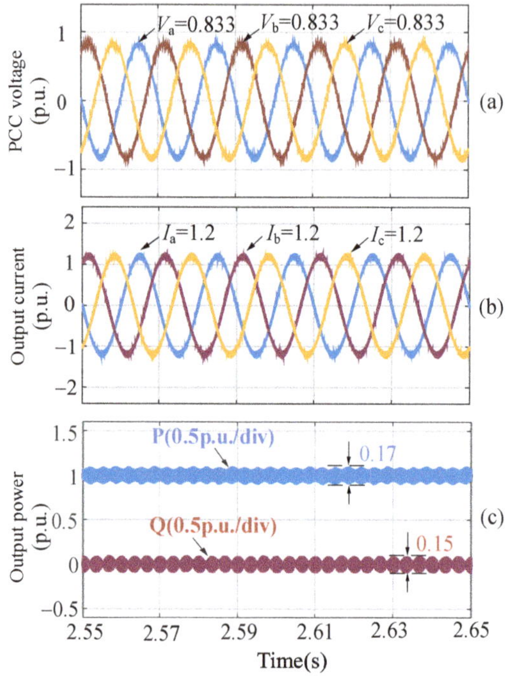


**Figure 12.** Simulation results of the modified topology, when $V_b$ drops to 0.5 p.u. (**a**) Output voltage at $PCC_1$; (**b**) output current; (**c**) output active power and reactive power.

## 5. Experimental Results

In order to confirm the simulation results, the three-phase converter was verified by downscaling the levels of voltage and power. The overall configuration of the experimental setup is shown in Figure 13. The core control algorithm is implemented on TMS320F28335, while the intelligent power module (IPM) is PM150RLA120. IT6516C DC source is used to generate DC voltage and Chroma 61702 AC source is implemented to generate AC voltage to simulate grid voltage. The auxiliary module consists of three identical single-phase inverters, all of which adopt the filter composed of $L_{add}$, $R_{add}$ and $C_{add}$. $L_f$ is the filter inductance of grid-connected inverter output circuit. In the experiment, the fault type is that the voltage of phase B drops to 0.5 p.u. The parameters of other main circuits are shown in Table 3, and the parameters of auxiliary modules are shown in Table 4. The experimental results of the traditional topology and the modified topology are shown in Figures 14 and 15, respectively.

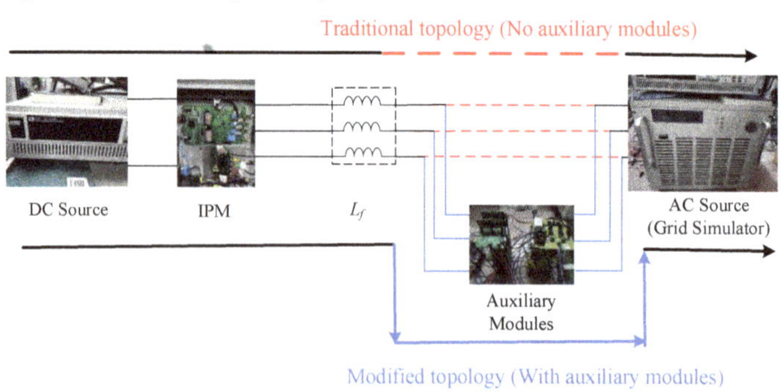


**Figure 13.** Configuration of the experimental setup.

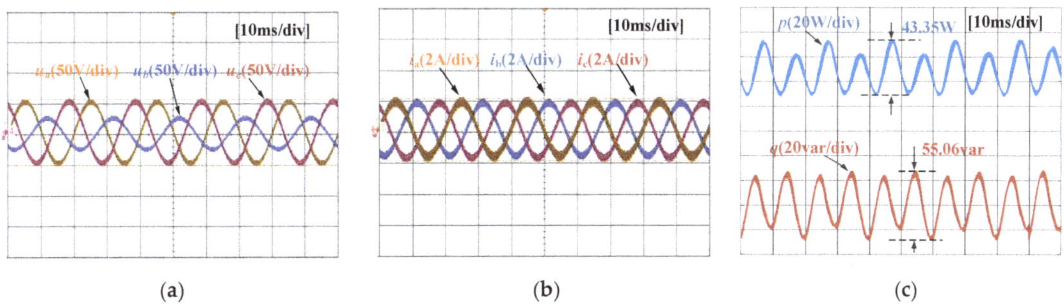


**Figure 14.** Experimental results of the traditional topology, when $V_b$ drops to 0.5 p.u. (**a**) Output voltage at $PCC_1$; (**b**) output current; (**c**) output active power and reactive power.

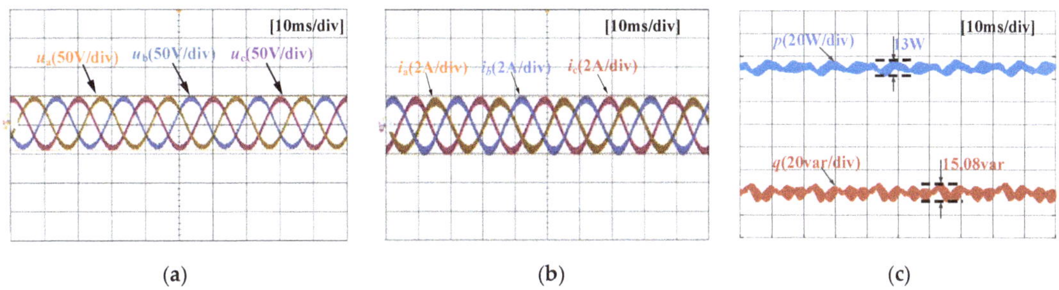


**Figure 15.** Experimental results of the modified topology, when $V_b$ drops to 0.5 p.u. (**a**) Output voltage at $PCC_1$; (**b**) output current; (**c**) output active power and reactive power.

Comparing the results of the traditional topology and the modified topology in Figures 14 and 15, the unbalanced voltage at $PCC_1$ can be corrected under the modified topology, and the inverter output current keeps balanced. The oscillation value of the output active power is reduced from 43.35 to 13 W, and the output reactive power oscillation value is reduced from 55.06 to 15.08 var. The oscillation value of active power in the modified topology is 29.99% of that in traditional topology and the reactive power oscillation value is reduced to 27.39% of the value of the traditional topology. The experimental results show that the modified topology can correct the unbalanced voltage at the output of the inverter and reduce the oscillation of the output active power and reactive power.

## 6. Conclusions

The grid-connected inverter with modified topology could compensate the negative-sequence and the zero-sequence components of the output voltage, so that the reference current equation of the inverter meets the solvability condition, thus eliminating the output power oscillation while the output current waveform still meets the requirements of grid codes. The simulation and experimental results show that the modified topology can effectively correct the unbalanced voltage and reduce the output active power oscillation and reactive power oscillation.

When the grid voltage is balanced, grid voltage does not contain negative-sequence components and zero-sequence components. Therefore, the output voltage reference value of the auxiliary module is zero at this time, and the auxiliary module will not output voltage, nor will it absorb power from the circuit or inject power to circuit. The auxiliary module is always connected to the circuit regardless of the unbalanced grid voltage fault or normal grid voltage, but when the grid voltage is balanced, the auxiliary module will not play a role.

The auxiliary voltage source can be a DC voltage source, a large capacitor, a super capacitor, or other energy storage units. This article aims to illustrate the function of the auxiliary module, so in the experiment, only the DC voltage source is used. In the future, the methods of using large capacitors or other energy storage structures to replace the additional DC power source in the auxiliary module and also designing the effective charging and discharging topologies of auxiliary modules as well as reducing the cost of hardware will be explored.

**Author Contributions:** Conceptualization, C.L., X.M. and L.Y.; methodology, C.L., X.M. and L.Y.; software, C.L. and Y.L.; validation, C.L., X.Y. and J.R.; data curation, Y.Z.; writing—original draft preparation, C.L. and X.M.; supervision, X.M.; project administration, X.M. All authors have read and agreed to the published version of the manuscript.

**Funding:** This work was supported by China Scholarship Council under Grant 202006280458.

**Institutional Review Board Statement:** Not applicable.

**Informed Consent Statement:** Not applicable.

**Data Availability Statement:** Not applicable.

**Conflicts of Interest:** The authors declare no conflict of interest.

## References

1. Blaabjerg, F.; Liserre, M.; Ma, K. Power electronics converters for wind turbine systems. *IEEE Trans. Ind. Appl.* **2012**, *48*, 708–719. [CrossRef]
2. Jia, J.; Yang, G.; Nielsen, A.H. Investigation of grid-connected voltage source converter performance under unbalanced faults. In Proceedings of the 2016 IEEE PES Asia-Pacific Power and Energy Engineering Conference (APPEEC), Xi'an, China, 25–28 October 2016; pp. 609–613.
3. Tleis, N. *Power Systems Modeling and Fault Analysis: Theory and Practice*, 1st ed.; Elservier: Amsterdam, The Netherlands, 2008.
4. Nian, H.; Cheng, P.; Zhu, Z.Q. Independent operation of DFIG-based WECS using resonant feedback compensators under unbalanced grid voltage conditions. *IEEE Trans. Power Electron.* **2015**, *30*, 3650–3661. [CrossRef]
5. Taul, M.G.; Wang, X.; Davari, P.; Blaabjerg, F. An overview of assessment methods for synchronization stability of grid-connected converters under severe symmetrical grid faults. *IEEE Trans. Power Electron.* **2019**, *34*, 9655–9670. [CrossRef]
6. Taul, M.G.; Wang, X.; Davari, P.; Blaabjerg, F. Current reference generation based on next generation grid code requirements of grid-tied converters during asymmetrical faults. *IEEE J. Emerging Sel. Top. Power Electron.* **2020**, *8*, 3784–3797. [CrossRef]
7. Wang, X.; Blaabjerg, F.; Chen, Z.; Wu, W. Resonance analysis in parallel voltage-controlled distributed generation inverters. In Proceedings of the 2013 Twenty-Eighth Annual IEEE Applied Power Electronics Conference and Exposition (APEC), Long Beach, CA, USA, 17–21 March 2013; pp. 2977–2983.
8. Rocabert, J.; Luna, A.; Blaabjerg, F.; Rodríguez, P. Control of power converters in AC microgrids. *IEEE Trans. Power Electron.* **2012**, *27*, 4734–4749. [CrossRef]
9. Ng, C.H.; Ran, L.; Bumby, J. Unbalanced-grid-fault ride-through control for a wind turbine inverter. *IEEE Trans. Ind. Appl.* **2008**, *44*, 845–885. [CrossRef]
10. Rodriguez, P.; Timbus, A.V.; Teodorescu, R.; Liserre, M.; Blaabjerg, F. Flexible active power control of distributed power generation systems during grid faults. *IEEE Trans. Ind. Electron.* **2007**, *54*, 2583–2592. [CrossRef]
11. Guo, X.; Liu, W.; Zhang, X.; Sun, X.; Lu, Z.; Guerrero, J. Flexible control strategy for grid-connected inverter under unbalanced grid faults without PLL. *IEEE Trans Power Electron.* **2015**, *30*, 1773–1778. [CrossRef]
12. Guo, X.; Liu, W.; Lu, Z. Flexible power regulation and current-limited control of grid-connected inverter under unbalanced grid voltage faults. *IEEE Trans. Ind. Electron.* **2017**, *64*, 7425–7432. [CrossRef]
13. Camacho, A.; Castilla, M.; Miret, J.; Borrell, A.; de Vicuña, L.G. Active and reactive power strategies with peak current limitation for distributed generation inverters during unbalanced grid faults. *IEEE Trans. Ind. Electron.* **2015**, *62*, 1515–1525. [CrossRef]
14. Du, X.; Wu, Y.; Gu, S.; Tai, H.; Sun, P.; Ji, Y. Power oscillation analysis and control of three-phase grid-connected voltage source converters under unbalanced grid faults. *IET Power Electron.* **2016**, *9*, 2162–2173. [CrossRef]
15. Ma, K.; Chen, W.; Liserre, M.; Blaabjerg, F. Power controllability of a three-phase converter with an unbalanced AC source. *IEEE Trans. Power Electron.* **2015**, *30*, 1591–1604. [CrossRef]
16. Saccomando, G.; Svensson, J.; Sannino, A. Improving voltage disturbance rejection for variable-speed wind turbines. *IEEE Trans. Energy Conv.* **2002**, *17*, 422–428. [CrossRef]
17. Jia, J.; Yang, G.; Nielsen, A.H. A review on grid-connected converter control for short-circuit power provision under grid unbalanced faults. *IEEE Trans. Power Deliv.* **2018**, *33*, 649–661. [CrossRef]
18. IEEE Standard 929. *IEEE Recommended Practice for Utility Interface of Photovoltaic (PV) Systems*; IEEE: Piscataway, NJ, USA, 2000.

19. IEEE Standard 1547. *IEEE Standard for Interconnecting Distributed Resources with Electric Power Systems*; IEEE: Piscataway, NJ, USA, 2003.
20. Mortazavian, S.; Shabestary, M.M.; Mohamed, Y.A.I. Analysis and dynamic performance improvement of grid-connected voltage-source converters under unbalanced network conditions. *IEEE Trans. Power Electron.* **2017**, *32*, 8134–8149. [CrossRef]
21. Teodorescu, R.; Liserre, M.; Rodriguez, P. *Grid Converters for Photovoltaic and Wind Power Systems*, 1st ed.; Wiley: Hoboken, NJ, USA, 2011.
22. Wang, X.; Blaabjerg, F.; Wu, W. Modeling and analysis of harmonic stability in an AC power-electronics-based power system. *IEEE Trans. Power Electron.* **2014**, *29*, 6421–6432. [CrossRef]

energies

MDPI

Article

# An Investigation into Conversion of a Fleet of Plug-in-Electric Golf Carts into Solar Powered Vehicles Using Fuzzy Logic Control

Wafaa Saleh [1,2,*,†], Shekaina Justin [1], Ghada Alsawah [1,3], Areej Malibari [4] and Maha M A Lashin [1,5]

1 College of Engineering, Princess Nourah Bint Abdulrahman University, Riyadh 84428, Saudi Arabia; SJThambi@pnu.edu.sa (S.J.); gaalsawah@pnu.edu.sa (G.A.); mmlashin@pnu.edu.sa (M.M.A.L.)
2 School of Engineering and The Built Environment, Edinburgh Napier University, Edinburgh EH10 5D, UK
3 Mechanical Engineering Department, Higher Technological Institute, 10th of Ramadan City 44629, Egypt
4 Department of Computer Science, Faculty of Computing and IT, King Abdulaziz University (KAU), Jeddah 21589, Saudi Arabia; aamalibari1@kau.edu.sa
5 Mechanical Department, Faculty of Engineering-Shoubra, Banha University, Banha 13518, Egypt
* Correspondence: wsshoukry@pnu.edu.sa or w.saleh@napier.ac.uk
† Visiting Professor.

**Abstract:** This paper presents an investigation factors that need to be considered in the design and selection of components for the conversion of a fleet of plug-in electric golf carts at Princess Nourah Bint Abdelrahman University, (PNU), Riyadh, Kingdom of Saudi Arabia (KSA), into solar power energy. Currently, the plug-in electric golf carts are powered by a set of deep-cycle lead-acid battery packs consisting of six units. Solar energy systems (photovoltaics and solar thermal) provide significant environmental benefits and opportunities over the traditional and conventional sources. Therefore, they can contribute positively to many aspects of the built environment and societies. There are many factors that affect the energy generated from the solar panel system. These include type and dimension of the solar panels, weight, speed, acceleration, and other characteristics of the used golf carts, and the energy efficiency of the solar energy system, as main factors that affect the green energy generated to operate the carts. The energy values needed to power the electric cart were calculated and optimized using traction energy calculation and optimized using a fuzzy logic analysis. The fuzzy logic system was developed to assess the impacts of varying dimensions of solar panel, vehicle speed, and weight on the energy generation. Initial calculations show that the replacement cost of the batteries can be up to approximately 75 percent of the operating cost. Together with the indirect cost benefits of achieving zero tail-pipe emission and the comfort of silent operation, the cost of operation using solar energy can be significant when compared with the cost of battery replacement. In order to achieve better efficiency, supercapacitors can be investigated to replace the conventional batteries. The use of fuzzy logic successfully facilitated the optimization of system operation conditions for best performance. In this study, fuzzy logic and calculated data were used as an optimization tool. Future work may be able to use fuzzy logic with experimental data to demonstrate feasibility of utilizing fuzzy logic systems to assess energy generation processes. Future investigations could also include investigation of other factors and methodologies, such as various types of batteries, supercapacitors, solar panels, and types of golf carts, together with different techniques of artificial intelligence to assess the optimum system specifications.

**Keywords:** golf carts; electric carts; solar energy; conversion of electric carts

**Citation:** Saleh, W.; Justin, S.; Alsawah, G.; Malibari, A.; Lashin, M.M.A. An Investigation into Conversion of a Fleet of Plug-in-Electric Golf Carts into Solar Powered Vehicles Using Fuzzy Logic Control. *Energies* **2021**, *14*, 5536. https://doi.org/10.3390/en14175536

Academic Editors: K.T. Chau and Tek Tjing Lie

Received: 15 June 2021
Accepted: 25 August 2021
Published: 4 September 2021



## 1. Introduction and Literature Review

Electric vehicles have been recognized as a key technology in the mobility sector that are associated with reducing emissions, greenhouse gases, and fuel consumption. Therefore, electric carts are good candidates to demonstrate and begin the journey for achieving a much more sustainable mobility future. Small electric vehicles seem to be excellent options to achieve adaptation to solar energy initiatives in the transportation sector [1,2]. In order for this technology to achieve its target, it has to be powered by a source of renewable energy such as solar, wind, or natural gas. Solar energy technologies are fundamental components of a sustainable energy future since they provide clean, renewable, and a home energy source. In comparison to traditional energy sources, solar energy systems have significant environmental benefits. The development and operation of solar systems have considerable environmental consequences that have been demonstrated in literature [3].

Electromobility can contribute to more ecologically responsible, climate-friendly mobility. However, in terms of electric vehicles, it has not yet reached its full potential. There are two significant impediments to purchasing an electric automobile from the user's perspective: it is more expensive than a car with an internal combustion engine and it has a shorter range than a normal vehicle. Intermodal concepts based on electric shared vehicles can be a potential option because the total cost of ownership of the electric vehicles may be split among many users, and as part of the public transportation system, it can also serve as a supplement for local and long-distance travel [4].

The golf automobiles, in contrast, are created to be durable, dependable, and safe when used for long periods of time. There are many forms available for such cars (see the PNU golf cart picture, Figure 1). Solar panels, batteries, and electronics that can be used to power these types of golf carts have recently become more affordable, functional, and reliable, allowing for new and inventive uses. Various forms of solar-panel-powered electric carts have been seen in various studies. The solar powered golf cart can be used in many purposes including short distance travel where there is solar incidence, within campuses travel, and outside superstores to transport customers with mobility issues, elderly, and children [5].

**Figure 1.** Princess Nourah Bint Abdelrahman University's golf cart.

In comparison to traditional energy sources, solar energy systems (photovoltaics and solar thermal) also provide significant environmental benefits, contributing to the long-term development of human evolution and well-being [6]. Solar energy is an abundant energy resource and the fastest growing renewable energy source in the world. In photovoltaic (PV) technology, many categories of solar panels have been produced and currently are used. Further development of similar products is also progressing in addition to the optimization techniques that contribute to cost reduction and energy efficiency [7].

The solar cell's efficiency is stated as the ratio of the calculated energy generation, the besieged direct current (DC) yield, and the actual alternating current (AC) yield generated [8]. This efficiency is often affected by factors such as dirt, sand, and snow together with other obstacles such as shade, etc. Electrical efficiency of the PV module (in this case, the higher the temperature of the panel, the lower the performance of the modules), also affects the system's overall efficiency [9].

In recent years, the market for solar energy has greatly expanded and further expansion is expected. This is due to the global interest in more renewable sources and because of the overall goal of achieving more sustainable societies.

The case of light electrical vehicles has been broadly studied by Keshri, Bertoluzzo, and Buja [10]. De Pinto et al. [11] investigated the allowance of the range of the EV via PV panels [11]. Kim et al. [12] investigated issues related to battery savings of the electrical vehicles while utilizing the vehicle's surface areas in order to optimize charging time [12], which is a new area of research [13]. The autonomy of the car showed an increase in both studies. Okedu [14] investigated the use of solar panels in hydrogen cars [14]. Other studies have focused on the PV integration and their applications [15]. Direct applications in buildings [16], PV parks [17] or on renewable-energy-based EV charging stations, specifically wind-based [18] and solar-based [19]. The available research is very scarce in innovative methods of solar integration, such as powering buggies, golf cars, and other options that can be used for sports or local transporting functions. This paper's objective is to explore whether or not such an investment in the PNU University's Golf carts would show potential saving in energy and provision of renewable energy.

Solar golf carts are mostly powered by a photovoltaic (PV) or thin film panel mounted on top of the existing roof, or using a PV panel as the roof itself (see Figure 1). A controller converts the sun's energy to charge the golf cart's supercapacitor bank. In terms of solar energy supply, the solar panels may well increase the distance the cart can travel on a single charge and hence can provide potential increased usage of these golf carts. In this paper, a case study is presented where we investigate the potential benefits of powering a university's fleet of battery-powered golf carts into solar energy power. The aim of the paper also is to briefly investigate the operation cost and its implications of adopting green energy to operate golf carts at Princess Nourah Bint Abdulrahman University (PNU) campus in Riyadh, Kingdom of Saudi Arabia, for transportation within the university, instead of using electric batteries. The paper is structured as follows: Section 1 provides an introduction and literature review of the study. The case study is presented in Section 2. System specification is overviewed in Section 3 with a discussion of the general specifications of the proposed main system and the operating characteristics. The methodology is presented in Section 4 where the calculations of energy needed to power the electric cart are presented. The fuzzy logic control system, which is used to assess optimum energy generation relative to the system specifications, is discussed in Section 5. Section 6 concludes the findings of the study and discusses future work.

## 2. Case Study

Princess Nourah Bint Abdelrahman University (PNU) in Riyadh, KSA, is the first all-female university in Riyadh that was established in 2006 with the aim of serving the development and progress of the Kingdom. Riyadh is 587 m above sea level. Riyadh has a desert climate. During the year, there is virtually no rainfall. According to Köppen and Geiger, this climate is classified as BWh. The average annual temperature is 26.2 °C | 79.2 °F in Riyadh. Precipitation averages about 66 mm | 2.6 inch per year. The driest month is June, with 0 mm | 0.0 inch of rainfall. Most of the precipitation here falls in April, averaging 14 mm | 0.6 inch. The university's mission is distinguished by its leadership in education and scientific research. It contributes to establishing a knowledge-based economy with societal and global partnerships. The campus is 13 million square meters in size, with a maximum capacity of 60,000 students. The campus has 600 high-tech smart buildings, large-capacity student residencies, various models of faculty residence units, and three spacious, state-of-the-art recreation centers. In addition, it has preschools, primary schools, intermediate, and secondary schools. It also has an elegant central library, research centers, a university hospital, student support centers, student-accessible sport facilities, a convention center, and an automated metro system. Due to the large size of the campus, efficient means of transporting systems are in operation. These include an

11.5 km automated guideway metro system together with a fleet of 150 golf cars—100 plug-in-electric and 50 petrol engined that are used as a multipurpose utility vehicles.

The plug-in-electrics typically include 6–8 batteries of 8–6 volts/each. The recharging is usually carried out overnight at different centers located inside the university campus. Each car is charged every 2 to 3 days depending on the usage and the condition of its batteries. Batteries are replaced every 2 to 3 years. Therefore, the operating efficiency of the golf cars is rather limited in addition to the high cost associated with the operation of these cars. There are also operating cost implications regarding the golf cars used. In terms of the cost involved in operating the batteries, this is relatively high, as each unit costs about 850 to 900 SAR (226.7 to 240 USD) costing an average of around 4534–4800 USD on an annualized basis for the fleet. Assuming that each cart is operated on an average of about 2000 km annually, the cost of the battery works out to be about 0.12 USD per km of operation. Even after considering the indirect cost benefits accrued on account of the zero tail-pipe emission and the comfort of silent operation, the cost of operation is quite undesirable. Moreover, the cost of operating and maintaining the charging centers are also involved, though they are not as high when compared with the cost of battery replacement.

## 3. System Specifications

The annual solar incidence at PNU location is about 2350 kWh, translating into a daily mean solar incidence of 6.44 kWh [20]. There are only a few days in the year when the solar incidence is low due to cloud cover or when the angle of incidence is relatively positive for energy generation.

The PNU golf cart roof has a dimension of 2 m × 1 m. The cart roof is the location where harvesting of solar energy is undertaken. When the surface area is increased, the generated energy is increased. Table 1 shows the estimated energy generated by different dimensions and conversion efficiency ranges as found in the market, which have been based on published findings [20–23].

**Table 1.** The energy generated by different dimensions and conversion efficiency ranges as found in the market.

| Solar Panel Dimension | Energy Generated (kWh) | | | |
|---|---|---|---|---|
| | $\eta = 15\%$ | $\eta = 18\%$ | $\eta = 21\%$ | $\eta = 24\%$ |
| 1.8 m × 0.9 m | 1.56 | 1.88 | 2.19 | 2.50 |
| 2.0 m × 1.0 m | 1.9 | 2.32 | 2.70 | 3.09 |
| 2.5 m × 1.2 m | 2.9 | 3.48 | 4.06 | 4.64 |

For example, with a 2.0 m × 1.0 m area of the roof covered with solar panel, the energy generated by a panel with 24% efficiency should work out to be 2.7 kWh per day on an average, which is more than sufficient to meet the daily usage for the golf cart, as illustrated later.

When the area of the roof is increased, thereby increasing the panel area, and when the conversion efficiency of the solar panel used is high, the supercapacitor capacity can be decreased. This will help in reducing the cost and mass of the supercapacitor module. The increase in mass of solar panel is relatively low when compared with increase in mass of the supercapacitor on account of higher capacity.

We need to compare this energy with the energy produced from the battery pack fitted in the golf cart, which consists of either 6 units of 8 V-170 Ah or 8 units of 6 V-170 Ah deep-cycle lead-acid batteries connected in series totaling an energy capacity of 8.16 kWh. The net energy available to be discharged from the battery pack is about 4 kWh. Based on the available data, as 4 kWh energy is utilized over a period of 2 to 3 days, the mean daily distance covered is in the range of about 10 to 15 km. These data indicate the optimum capacity needed for the supercapacitor bank.

As the charging would be done with solar PV panel installed on the roof of the cart, and as the supercapacitor could be fully discharged, a supercapacitor bank with a capacity of 1.5 to 2 kWh is considered optimal to meet the current usage. In any unlikely event of a mismatch between solar incidence and demand for transport, for example, recharging 1 kWh using a 100 kWh charger would take no more than 50 s.

The parameters that can be considered for selecting the solar panels are mean efficiency, efficiency variance with respect to temperature, type of the solar panels and their connections to withstand stress and vibrations, mass per unit area of the panel, cost of the panel per unit area for panels fabricated as per specified dimensions, cost per unit area of readily available panels, and flexibility of the panel [20–23]. Panels designed exclusively for transport applications or panels fabricated with similar design should be considered. The solar panels suitable for this application are thin-film rolled panels, perovskite single junction and multijunction panels, thick-film panels and semiflexible silicon panels [24,25]. It should be noted, however, that the commercial availability of some of these panels may not be fully guaranteed. Since there are many factors involved, the investigation in this paper focuses on the solar panel types (or dimension, in this case), speed of the golf cars, and energy efficiency as main factors that affect the green energy generated to operate the carts. The energy values needed to power the electric cart was calculated and assessed using traction energy calculation and optimized using fuzzy logic analysis. The fuzzy logic system was developed to assess the impacts of varying dimensions of solar panel, speed of vehicle, and weight on the energy generation.

## 4. Methodology Traction Energy Calculation

Vehicle acceleration, maximum velocity, and range are the main parameters that affect traction energy calculations. In order to consider these forces, it is therefore required to include the force due to rolling resistance, the aerodynamic force, the acceleration force, and the rotational acceleration force [26]. The impact of driver behavior is represented in the acceleration force while the impact of the passengers should be included in the total weight of the vehicle. For this analysis and for simplification, the energy required for operating the golf cart is assumed to be the sum of two components only, as given below:

1. Rolling resistance force
2. Aerodynamic drag force

Assuming that the mean speed of the golf cart is 25 km/h, mass is 500 kg, air density 1.2 kg/m$^3$, and the energy conversion efficiency of the cart $\eta$ = 0.75, the electricity consumption is calculated as follows (Equation (1)):

$$E_{25} = \frac{v\left(c \cdot m \cdot g + 0.5 c_d \cdot p \cdot A \cdot v^2\right)}{\eta} \tag{1}$$

where $c$ = 0.06 (rolling resistance coefficient)
$m$ = 500 kg (mass of the cart)
$g$ = 9.81 m s$^{-2}$ (gravitational acceleration)
$c_d$ = 0.7 (aerodynamic drag coefficient)
$p$ = 1.2 kg m$^{-3}$ (density of air)
$v$ = 6.94 m s$^{-1}$ (velocity of cart)
$\eta$ = 0.75 (energy conversion efficient of the cart)
$A$ = 1.5 m$^2$ (frontal area)
$E_{25}$ = 3238.67 Wh

This is equal to 3.238 kWh. On the other hand, the energy generation obtained in Section 3 above is 2.7 kWh. Given that the speed of the golf cart is much lower than 6.94 ms$^{-1}$, the energy generated should be sufficient to run the cart. For comparison, Table 2 shows the level of energy need, energy consumption, and energy generation from solar panels. From the table, it appears that while the theoretical value for energy requirement per day is 3.238 kWh, the actual value needed to run the cart is slightly lower; a value of 2.66 kWh seems sufficient.

**Table 2.** Summary of golf cart energy requirement and energy generation from batteries and solar panels.

| Energy Need kWh | Energy Generation from Battery Set kWh (over 2–3 Days) | Energy Generation from Solar Panel System kWh |
|---|---|---|
| 3.238 | 4.0 (1.33/day) | 2.7 |

Other than that, the factors that can be investigated in this case that impact electricity consumption are vehicle speed, solar panel dimensions, and energy efficiency factors. In the section below, a fuzzy logic control system is developed to investigate impacts of various values of these factors on the energy production. Fuzzy logic controllers (FLCs) have the following advantages over the conventional controllers: they are easily developed and processed using very simple software, they can cover a broader range of operating conditions, and they are more readily adaptable in natural language terms. It should be noted, however, that it is possible to use other conventional methods.

## 5. Fuzzy Logic Control System

In managing complicated situations, fuzzy logic systems are effective prediction tools. These models or groups are mathematical representations of the ambiguity and hazy information. They are capable of distinguishing, expressing, explaining, and operating this ambiguous and imprecise information. They can be programmed to calculate a difference between data sets by emulating human thinking processes [27]. Because there is no "standard" mathematical model available or one that can be derived from physical laws for effectively modeling pedestrians' crossing behavior, fuzzy logic is clearly effective in this type of inquiry. Furthermore, because fuzzy systems were first offered to represent human decision-making by weighing rules, it appears sensible to utilize them to simulate human behavior. Fuzzy logic has a wide range of applications, including transportation (for example, [28–31]). In comparison to other industries, however, transportation has a considerable dearth of fuzzy logic system applications. The membership functions are an important aspect of fuzzy logic analysis. The membership functions can take a variety of shapes including triangular, trapezoidal, and Gaussian [32,33] for assessing the efficiency of estimation of various forms. Because of their simplicity and ease of computation, the triangular and trapezoidal membership functions are frequently used in various applications. As a result, adopting a triangle form as a first step in any research is reasonable. Other forms can be utilized if the system's performance is not adequate.

A nonlinear translation of input data sets to scalar output data is a fuzzy logic system [27]. As shown in Figure 2, the fundamental mechanisms of a fuzzy system are a fuzzifier and a defuzzifier, if–then rules, and an inference engine. The use of a fuzzy logic system has the advantage of being able to manage challenges involving imprecise and partial data. They can also simulate arbitrary or sophisticated nonlinear functions.

### 5.1. Fuzzy Logic Analysis

The values of electricity required for driving golf cars used as transportation inside PNU are affected by cart speed and solar panel area. The mathematical equation for calculating the electricity needed to power the golf carts and to confirm a fuzzy logic control system was designed and implemented for the car working variables (cart speed and roof area), which enabled a prediction of the electrical power required for driving the car.

The fuzzy logic system was developed using the traction energy calculations presented earlier in the paper. The membership function of fuzzy logic is a fundamental component of a fuzzy logic system and the most crucial stage in fuzzy set construction. The membership function controls the fuzziness feature of a fuzzy set. The only requirement for a membership function is that it must fluctuate between 0 and 1. Because of their simplicity and ease of calculation, triangle and trapezoidal membership functions are frequently used in various applications, as opposed to other function types such as bell, sigmoidal, asymmetric, LR, and Gaussian, which may need more sophisticated mathematical procedures. In addition, defining a triangular and trapezoidal membership function requires fewer data. In our instance, it might be most appropriate to utilize a triangle shape.

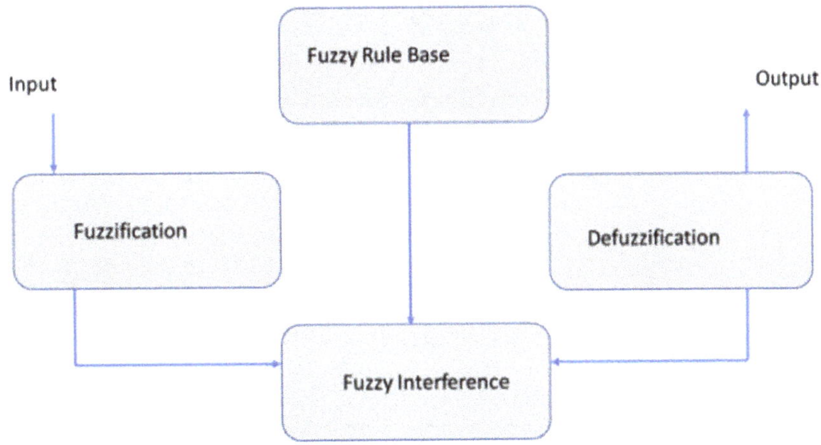


**Figure 2.** Architecture of the fuzzy system.

Table 3 lists the names and input and output variables that correspond to the parameter ranges with the triangular membership functions. Three parameters were included to define the input variables for building and implementing fuzzy logic systems (low, medium, and high).

**Table 3.** Levels and name of inputs (triangular membership function) used and output variables and range values.

| Variables | Inputs | | |
|---|---|---|---|
| | Low | Medium | High |
| Cart velocity km/h | 22.5–27.5 | 27.5–32.5 | 32.5–37.5 |
| Solar panel dimension m$^2$ | 1.5–1.74 | 1.74–2.26 | 2.26–4.26 |
| Energy conversion efficiency% | 14–16 | 16–20 | 20–24 |

*5.2. Detection of Electrical Power Using the Fuzzy System*

The fuzzy system procedure that is used in this study follows Mendel's (1995) methodology, which consists of three steps:

1. Fuzzification process;
2. Inference process based on if–then rules;
3. Defuzzification process.

The first step is to fuzzify the data, which can be done by using fuzzy linguistic variables and membership functions to convert the input data to a fuzzy set. The inference procedure is then carried out using if–then rules. Finally, the defuzzification stage is accomplished, and fuzzy outputs are crisped using membership functions (Figure 3). Each of these steps is further illustrated below.

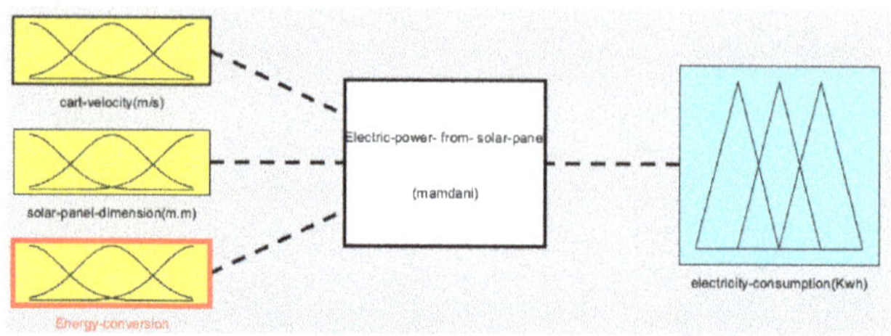


**Figure 3.** Fuzzy logic system for electric power from solar panel.

### 5.3. Fuzzification Process

The fuzzy logic system's first stage is the fuzzification process. Fuzzy linguistic variables and membership functions are used to turn the input data into a fuzzy set at this stage. Three to seven language variables are commonly used to muddle the input and output components. It would be inefficient to use less than three values since it would be difficult to discriminate the output after fuzzy inference [29–34]. The number of fuzzy rules employed would become more complicated if there were more than seven.

The membership function in the fuzzification process is built in this case utilizing three parameters a, b, c for each input and output variable, representing low, medium, and high values of each of the inputs, with variable ranges shown in Table 3. The estimating toolkit that comes with MATLAB was used. The z coordinates of the three vertices in a fuzzy set L are represented by the triangle membership function, which is determined by the three parameters a, b, and c. (a and c are the lower and upper boundaries, where membership degree is 0; b is the center, where membership degree is 1).

$$\mu_A(z) = \begin{cases} 0 \ \ if\, z \leq a \\ \frac{z-a}{b-a}\, if\, a \leq z \leq b \\ \frac{c-z}{c-b}\, if\, b \leq z \leq c \\ 0 \ \ if\, z \geq c \end{cases} \tag{2}$$

The real-time data can then be matched with the fuzzy model established, and a mathematical explanation for selecting a determined value for a rule, such as sum-of-squared-error minimization or any other applicable rule, can be provided using this toolkit. For operating a golf automobile, a fuzzy system is employed to determine the values of electrical power obtained from a solar panel. Low, medium, or high membership functions are available for all variables. This was done for all input and output variables in the same way (Table 3). The values and structure of inputs, outputs, and critical gap values for the three adopted ranges are shown in Table 1 and Figure 4.

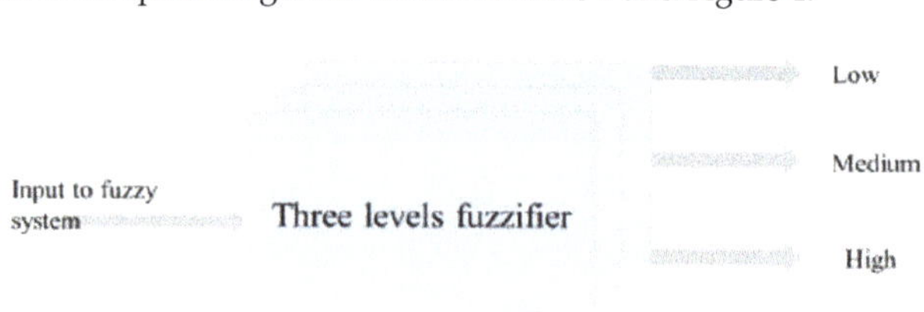


**Figure 4.** Fuzzification process of electric power from the solar panel fuzzy logic system.

Figure 5a–d depicts the fuzzy logic system's three levels of inputs and outputs, along with the membership function ranges for the inputs and outputs, respectively. The membership functions used are of the triangular type, as seen in the fuzzy system's inputs and outputs. As previously stated, triangle membership functions are one of several options for determining the best values for the introduced criteria. While the simplicity of this division is apparent, it does not exclude the construction of additional partitions that are optimal in terms of the specified requirements for the input and output interface while also performing better in terms of processing block optimization.

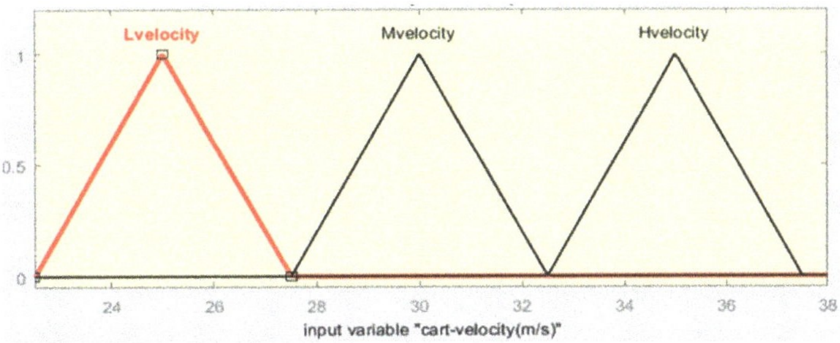


(a)

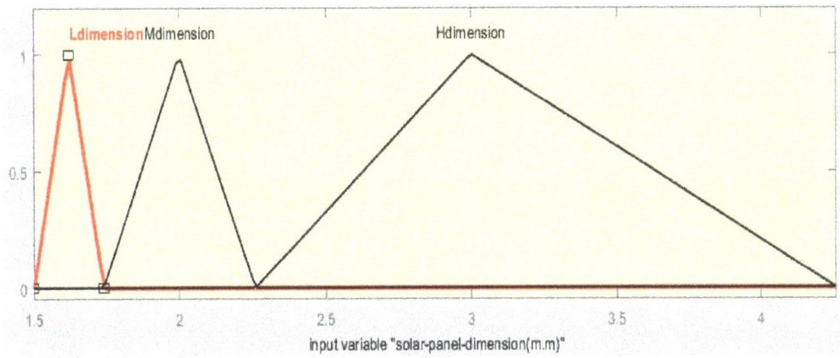


(b)

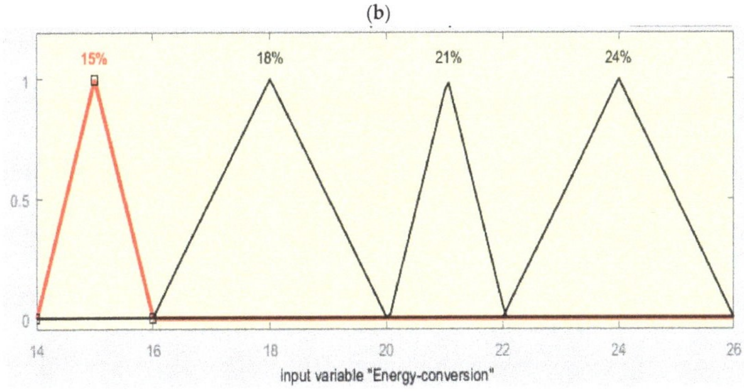


(c)

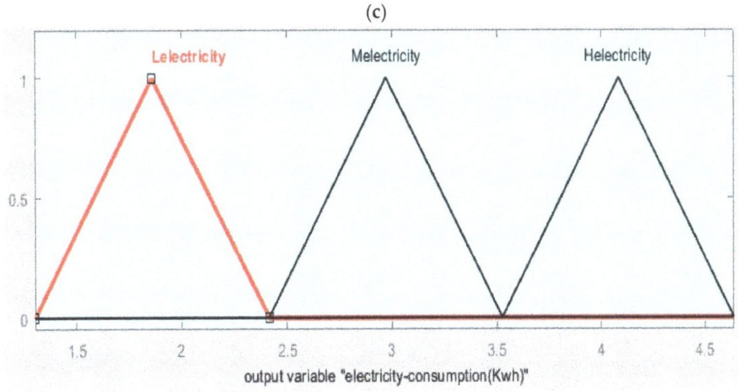


(d)

**Figure 5.** Low, medium, and high values for inputs (**a**–**c**) and output (**d**).

### 5.4. *Inference Process Based on If–Then Rules*

Inference operations are carried out utilizing linguistic if–then rules to determine the relationships between the input and output variables, which are based on a wide range of if–then rules. The linguistic fuzzy logic control system that is a part of the suggested fuzzy control or fuzzy decision step, generates the if–then rules. Table 4 lists some of the if–then rules that were employed in this inquiry. To cover all possible outcomes, fifteen if–then rules were employed in this research.

**Table 4.** A few examples of if–then rules that were used as part of the proposed fuzzy decision.

| Fuzzy Rules (If–Then Rules)—Relationship between Inputs and Outputs |
| --- |
| If (cart-velocity (m/s) is Lvelocity) and (solar-panel-dimension (m.m) is Ldimension) and (energy-conversion-efficiency-of-the-cart is 15%) then (electricity-consumption (kWh) is Lelectricity) |
| If (cart-velocity (m/s) is Lvelocity) and (solar-panel-dimension (m.m) is Mdimension) and (energy-conversion-efficiency-of-the-cart is 21%) then (electricity-consumption (kWh) is Helectricity) |
| If (cart-velocity (m/s) is Lvelocity) and (solar-panel-dimension (m.m) is Hdimension) and (energy-conversion-efficiency-of-the-cart is 24%) then (electricity-consumption (kWh) is Helectricity) |

### 5.5. *Defuzzification Step*

The process of defuzzification is the final step in the optimization process using a fuzzy control system. The defuzzification procedure is based on each person's membership function and the computation of their decisions. If the performance does not meet the assessment criteria, some fuzzy rules will need to be adjusted and the test performed again. Using the same inputs (Table 3), a graphical illustration of the defuzzification process is shown in Figures 4 and 6 and Table 5, which illustrate the outcomes of the fuzzy system.

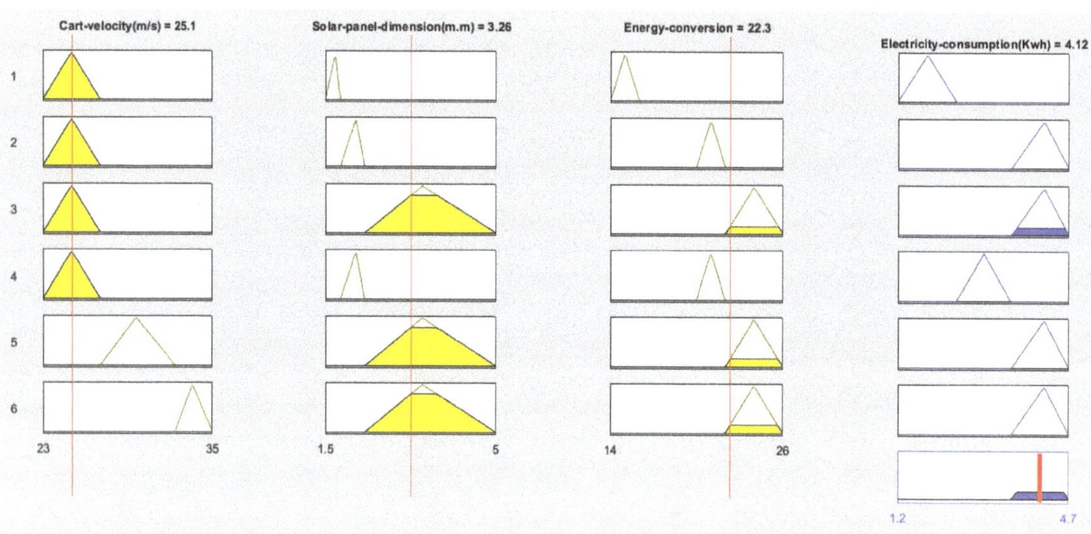


**Figure 6.** Results of electric power from the solar panel fuzzy logic system.

**Table 5.** Levels and name of outputs obtained from the fuzzy logic system and input values.

| **Variables** | **Inputs** | | |
|---|---|---|---|
| | **Low** | **Medium** | **High** |
| Cart velocity km/h | 22.5–27.5 | 27.5–32.5 | 32.5–37.5 |
| Solar panel dimension $m^2$ | 1.5–1.74 | 1.74–2.26 | 2.26–4.26 |
| Energy conversion efficiency % | 14–16 | 16–20 | 20–24 |
| Variables | **Output** | | |
| | Low | Medium | High |
| **PV Electricity production kWh** | **1.3–2.413** | **2.42–3.53** | **3.52–4.64** |

The results from fuzzy analysis (Figure 5 and Table 6) show that the minimum PV electrical energy production is achieved at lower values of cart speed and energy conversion efficiency factors. The maximum PV electricity energy production obtained is 4.17 kWh and this was acquired with maximum input parameter values.

**Table 6.** Output results from the proposed fuzzy system.

| **Values** | **Inputs** | | | **Output** |
|---|---|---|---|---|
| | **Cart Velocity** | **Solar Panel Dimension** | **Energy Conversion** | **PV Electricity Production** |
| Minimum value | 23.1 | 2.04 | 14.8 | 1.86 |
| Maximum value | 25.2 | 3.28 | 21.2 | 4.17 |

## 6. Conclusions

The aim of this paper is to investigate an optimum energy generation methodology that can be implemented using a renewable energy source. The energy generation is needed to power a set of golf cars at PNU campus in Riyadh, KSA. It is proposed therefore, to adopt a green energy system rather than using electricity from the grid to charge the batteries that are used to power the golf carts. As Riyadh, KSA, consistently receives very good solar radiation, calculations were carried out to assess the energy requirement to power these vehicles and to determine if the area of the roof is sufficient to meet mean daily energy requirements.

Based on the mean utilization derived from the charging interval, initial investigations show that the replacement cost of the battery is about 75 percent of the operating cost.

Then, based on the theoretical energy production for traction and the charging interval, the mean distance covered daily was calculated. Assuming that each cart is making an average of about 2000 km annually, the cost of the battery is assumed to be 0.12 United States Dollar (USD) per km of operation.

If the proposed adoption of green energy is implemented, the annual expenditure incurred in the replacement of batteries would be totally considered as initial and running cost savings. In addition, the energy consumed in charging the batteries is saved, which results in green-energy environmental benefits leading to sustainability of resources and lower emissions resulting from less conventional electricity production. In addition, there will be savings on consumables such as acid and distilled water. Moreover, there will be savings in labor spent on recharging, maintaining, and replacing vehicles. Using fuzzy logic facilitated the optimization of system operation conditions for best performance, taking into account three variables: cart velocity, solar panel dimensions, and energy conversion efficiency. Other variables are also possible. Future investigations could include investigating various types of batteries, capacitors, solar panels, and types of golf carts, together with using different techniques of analytical and artificial intelligence to assess the optimum system specifications. In this case, fuzzy logic facilitated the optimization of system operation conditions for best performance based on the three variables identified. These could be changed, and the optimization conditions could also be altered to include different conditions.

It is further suggested that research should examine the reduction of selected pollutant emissions, lower traditional fuel consumption, and extent of decreased operating cost when PV panels are used to power golf carts. Further research into alternative fuels and appropriateness for their use in this case is also recommended.

**Author Contributions:** Conceptualization, W.S. and S.J.; methodology, all; software, M.M.A.L. and A.M.; validation, W.S., S.J. and G.A.; formal analysis, S.J. and M.M.A.L.; investigation, W.S. and G.A. resources, A.M. and W.S.; data curation, M.M.A.L. and S.J. writing—original draft preparation, W.S. and S.J.; writing—review and editing, all; visualization, all; supervision, W.S.; project administration, W.S.; funding acquisition College of Engineering, PNU. All authors have read and agreed to the published version of the manuscript.

**Funding:** The APC was funded by DSR and PNU.

**Institutional Review Board Statement:** Not applicable.

**Acknowledgments:** This research project was funded by the Deanship of Scientific Research, Princess Nourah bint Abdulrahman University, through the Program of Research Project Funding after Publication, grant No (41-PRFA-P-39).

**Conflicts of Interest:** The authors declare no conflict of interest.

## References

1. Helmers, E.; Marx, P. Electric cars: Technical characteristics and environmental impacts. *Environ. Sci. Eur.* **2012**, *24*, 1–15. Available online: http://www.enveurope.com/content/24/1/14 (accessed on 25 August 2021). [CrossRef]
2. Kobayashi, S.; Plotkin, S.; Ribeiro, S.K. Energy efficiency technologies for road vehicles. *Energy Effic.* **2009**, *2*, 125–137. [CrossRef]
3. Gunerhan, H.; Hepbasli, A.; Giresunlu, U. Environmental Impacts from the Solar Energy Systems. *Energy Sources Part A Recovery Util. Environ. Eff.* **2008**, *31*, 131–138. [CrossRef]
4. Wappelhorst, S.; Sauer, M.; Hinkeldein, D.; Bocherding, A.; Glaß, T. Potential of electric carsharing in urban and rural areas. *Transp. Res. Procedia* **2014**, *4*, 374–386. [CrossRef]
5. Milani, R.; Couto, L.C.; Soria, R.; Szklo, A.; Lucena, A.F. Promoting social development in developing countries through solar thermal power plants. *J. Clean. Prod.* **2020**, *246*, 119072. [CrossRef]
6. Tsoutsos, T.; Frantzeskaki, N.; Gekas, V. Environmental Impacts from the Solar Energy Technologies. *Energy Policy* **2005**, *33*, 289–296. [CrossRef]
7. Nižetić, S.; Papadopoulos, A.M.; Giama, E. Comprehensive analysis and general economic-environmental evaluation of cooling techniques for photovoltaic panels, part I: Passive cooling techniques. *Energy Convers. Manag.* **2017**, *149*, 334–354. [CrossRef]
8. Peters, I.M.; Liu, H.; Reindl, T.; Buonassisi, T. An energy yield calculator for solar cells worldwide. In Proceedings of the 2018 IEEE 7th World Conference on Photovoltaic Energy Conversion, WCPEC 2018—A Joint Conference of 45th IEEE PVSC, 28th PVSEC and 34th EU PVSEC, Berlin, Germany, 10–15 June 2018; pp. 453–454.
9. Zhang, H.L.; Van Gerven, T.; Baeyens, J.; Degrève, J. Photovoltaics: Reviewing the European feed-in-tariffs and changing PV efficiencies and costs. *Sci. World J.* **2014**, *2014*, 404913. [CrossRef] [PubMed]
10. Keshri, R.; Bertoluzzo, M.; Buja, G. Integration of a photovoltaic panel with an electric city car. *Electr. Power Compon. Syst.* **2014**, *42*, 481–495. [CrossRef]
11. De Pinto, S.; Lu, Q.; Camocardi, P.; Chatzikomis, C.; Sorniotti, A.; Ragonese, D.; Iuzzolino, G.; Perlo, P.; Lekakou, C. Electric vehicle driving range extension using photovoltaic panels. In Proceedings of the 2016 IEEE Vehicle Power and Propulsion Conference, VPPC, Hangzhou, China, 19 December 2016; pp. 1–6.
12. Kim, J.; Wang, Y.; Pedram, M.; Chang, N. Fast photovoltaic array reconfiguration for partial solar powered vehicles. In Proceedings of the 2014 International Symposium on Low Power Electronics and Design, ISLPED '14, La Jolla, CA, USA, 11 August 2014; pp. 357–362.
13. Gupta, K.; Achathuparambil Narayanankutty, R.; Sundaramoorthy, K.; Sankar, A. Optimal location identification for aggregated charging of electric vehicles in solar photovoltaic powered microgrids with reduced distribution losses. *Energy Sources Part A Recovery Util. Environ. Eff.* **2020**, 1–16. [CrossRef]
14. Okedu, K.E. Integrated renewable energy system: Intelligent adjustable solar panel and hydrogen cars. *Renew. Energy Sources Appl. Emerg. Technol.* **2016**, *17*, 131–158.
15. Koroneos, C.; Nanaki, E.; Xydis, G. *Air Conditioning Systems and Their Applicability—An Exergy Approach, Resources, Conservation and Recycling*; Elsevier: Amsterdam, The Netherlands, 2010; Volume 55, pp. 74–82. [CrossRef]

16. Saber, E.M.; Lee, S.E.; Manthapuri, S.; Yi, W.; Deb, C. PV (photovoltaics) performance evaluation and simulation-based energy yield prediction for tropical buildings. *Energy* **2014**, *71*, 588–595. [CrossRef]
17. Xydis, G. A techno-economic and spatial analysis for the optimal planning of wind energy in Kythira Island, Greece. *Int. J. Prod. Econ.* **2013**, *146*, 440–452. [CrossRef]
18. Xydis, G.; Nanaki, E. Wind energy based electric vehicle charging stations sitting. A GIS/wind resource assessment approach. *Challenges* **2015**, *6*, 258–270. [CrossRef]
19. Mouli, G.R.C.; Leendertse, M.; Prasanth, V.; Bauer, P.; Silvester, S.; van de Geer, S.; Zeman, M. Economic and $CO_2$ emission benefits of a solar powered electric vehicle charging station for workplaces in the Netherlands. In Proceedings of the 2016 IEEE Transportation Electrification Conference and Expo(ITEC), Dearborn, MI, USA, 27–29 June 2016.
20. Zell, E.; Gasim, S.; Wilcox, S.; Katamoura, S.; Stoffel, T.; Shibli, H.; Engel-Cox, J.; Al Subie, M. Assessment of solar radiation resources in Saudi Arabia. *Solar Energy* **2015**, *119*, 422–438. [CrossRef]
21. Li, X.; Li, P.; Wu, Z.; Luo, D.; Yu, H.Y.; Lu, Z.H. *Review and Perspective of Materials for Flexible Solar Cells, Materials Reports: Energy*; Elsevier: London, UK, 2020.
22. Zinaddinov, M.; Milshtein, S. Design of light-weight solar panels. In Proceedings of the 2019 IEEE 46th Photovoltaic Specialists Conference (PVSC), Chicago, IL, USA, 16–21 June 2019.
23. IRENA. *Future of Solar Photovoltaic: Deployment, Investment, Technology, Grid Integration and Socio-Economic Aspects (A Global Energy Transformation: Paper)*; International Renewable Energy Agency: Abu Dhabi, United Arab Emirates, 2019.
24. Abderrezek, M.; Djeghlal, M.E. Comparative study on $Cu_2ZnSn(S,Se)_4$ based thin film solar cell performances by adding various back surface field (BSF) layers. *Chin. J. Phys.* **2020**, *63*, 231–239.
25. Rokonuzzaman, M.; Hossam-E-Haider, M. Design and Implementation of Maximum Power Point Tracking Solar Charge Controller. Presented at the 2016 3rd International Conference on Electrical Engineering and Information Communication Technology (ICEEICT), Dhaka, Bangladesh, 22–24 September 2016.
26. Shaikh, M. Design of an Electric Golf Cart with Batteries & Solar Panel for 6 Seat Capacity. *Int. Res. J. Eng. Technol.* **2019**, *6*, 2103–2122.
27. Ansari, A. *The Basics of Fuzzy Logic: A Tutorial Review, Computer Education*; Computer Education Group: Stamford, UK, 1998; Volume 88, pp. 5–9.
28. Chai, C.; Shi, X.; Wong, Y.D.; Er, M.J.; Gwee, E.T.M. Fuzzy logic-based observation and evaluation of pedestrians' behavioral patterns by age and gender. *Transp. Res. Part F Traffic Psychol. Behav.* **2016**, *40*, 104–118. [CrossRef]
29. Islam, N.; Haider, M.; Uddin, M.B. Fuzzy logic enhanced speed control system of a VSI-fed three phase induction motor. In Proceedings of the 2nd International Conference on Electrical and Electronics Engineering, Rome, Italy, 22–23 July 2019; pp. 296–301.
30. Demiroz, Y.I.; Onelcin, P.; Alver, Y. Illegal road crossing behavior of pedestrians at overpass locations: Factors affecting gap acceptance, crossing times and overpass use. *Accid. Anal. Prev.* **2015**, *80*, 220–228. [CrossRef] [PubMed]
31. Saady, G.E.; Sharaf, A.M.; Makky, A.; Sherbiny, M.K.; Mohamed, G. Ahigh performance induction motor drive system using fuzzy logic controller. In Proceedings of the 7th Mediterranean Electrotechnical Conference, Antalya, Turkey, 12–14 April 1994; Volume 3, pp. 1058–1061.
32. Wu, B.; Yan, X.; Wang, Y.; Soares, C.G. Selection of maritime safety control options for NUC ships using a hybrid group decision-making approach. *Saf. Sci.* **2016**, *88*, 108–122. [CrossRef]
33. Hsien-Chung, W. The optimality conditions for optimisation problems with fuzzy-values objective functions. *J. Optim.* **2008**, *57*, 473–489.
34. Goedtel, A.; da Silva, I.N.; Serni, P.J.A. Load torque identification in induction motor using neural networks technique. *Electr. Power Syst. Res.* **2007**, *77*, 35–45. [CrossRef]

energies

Article

# Evaluation of Electrical Performance and Life Estimation of PPs for HVDC Power Cable

Seung-Won Lee [1], Hae-Jong Kim [1], Ik-Su Kwon [1] and Jang-Seob Lim [2,*]

1 Superconductivity Research Center, Korea Electrotechnology Research Institute, Seongju-dong, Changwon-si 51543, Korea; 815lsw@keri.re.kr (S.-W.L.); hjkim@keri.re.kr (H.-J.K.); iskwon@keri.re.kr (I.-S.K.)
2 Division of Marine Mechatronics, Mokpo National Maritime University, Mokpo-si 58654, Korea
* Correspondence: Janylim@mmu.ac.kr; Tel.: +82-055-280-2535

**Abstract:** Demand and need for the application of high voltage direct current (HVDC) are increasing because of high capacity and long-distance transmission. Research on polypropylene (PP) that can increase the operation temperature compared to existing insulation is constantly being considered. This study aimed to evaluate the electrical performance and estimate the life of HVDC application of PP. In this study, a DC V-t characteristic tests were conducted on three types of PP sheets at a temperature of 110 °C. In addition, a life estimation formula based on the electrical stress was derived and the electrical performances were evaluated. The experimental results show that the life exponent of material mixed with block copolymer, homo polymer and high density polyethylene (HDPE) was 23 and the electrical performance was 17% better than block copolymer, thereby demonstrating the reliability and electrical performance for application of HVDC.

**Keywords:** HVDC; polypropylene; life exponent; V-t characteristic test; long term reliability; life estimation

**Citation:** Lee, S.-W.; Kim, H.-J.; Kwon, I.-S.; Lim, J.-S. Evaluation of Electrical Performance and Life Estimation of PPs for HVDC Power Cable. *Energies* **2021**, *14*, 5673. https://doi.org/10.3390/en14185673

Academic Editors: Mario Marchesoni, Ching-Ming Lai and Yitao Liu

Received: 6 August 2021
Accepted: 8 September 2021
Published: 9 September 2021



## 1. Introduction

Demand for high voltage direct current (HVDC) is increasing due to higher voltage of power cables and long-distance transmission. Classically, the insulations for HVDC uses mass oil, which has recently been converted to polymers due to environmental, production, and transmission capacity issues [1–3]. Among them, polypropylene (PP) has an operating temperature of 110 °C, has a higher transmission capacity than existing mass impregnated (MI), oil filled (OF) and cross linked polyethylene (XLPE), and can be recycled. Currently in terms of cable insulation, PP has a lot of interest [4,5].

In general, the design life of HVDC MI, OF and XLPE power cables is 30 to 40 years. Since HVDC cables are operated underground or on the seabed for a long period of time, the evaluation of life and long-term reliability is considered as an important factor for the development of DC insulating materials [6,7].

This study aimed to evaluate the electrical performance and long-term reliability of PPs for HVDC application.

Several researchers are using the conductivity, space charge, dielectric strength, V-t characteristic test, partial discharge (PD) and dissipation factor to evaluate reliability of materials [8–15]. Among them, the V-t characteristic test is believed to be major factor contributing toward the evaluation of long term reliability using life exponent [13,15,16]. Other researchers used the life exponent that is a simple numerical value to assess the reliability of XLPE, epoxy, polypropylene laminated paper (PPLP) and low density polyethylene (LDPE) materials. The life exponent of materials ranged from 15 to 30 based on additive type and AC and DC application [13–19]. However, as the design life of power cables was not considered when evaluating the long-term reliability, it is impossible to determine the performance of materials during actual operation. Above all, the reliability evaluation and life estimation of PPs for HVDC application has not been considered yet.

In general, the breakdown voltage of an insulating material depends on the temperature and the pressure applied to the test sample [20,21]. However, during the V-t characteristic test, the temperature of the test object could not be known in real time because the dummy sample was not installed, and there was no description of a method that could equalize the measuring force of the samples.

Unlike previous studies, we calculated the life exponent of PPs that can increase the transmission capacity and demonstrated the HVDC applicability by comparing life exponent of other materials. Furthermore, the electrical performance of PPs with a design life of 40 years for power cables was evaluated and priorities were established. In addition, the failure data was analyzed using the Weibull distribution, the value of 63.2% of the Weibull distribution function of the data of time to breakdown was applied to derive the lifetime estimation according to the electrical stress. Temperature sensors were installed in the dummy sample to improve the reliability of the V-t characteristic test of the PPs, the electrodes and micrometer were combined so that the uniform measuring force was applied to the sheets.

As for the test contents, a short-term ascent test and a V-t characteristic test were conducted at 110 °C on three types of PP sheets in which block copolymer, homo polymer and high density polyethylene (HDPE) were mixed.

This paper contributes to long-term reliability and life evaluation during research and development of HVDC insulating materials, and the calculated life exponent should be used as a reference for calculating the deterioration coefficient with the type and prequalification (PQ) test of PP cables for application of HVDC [22].

The remainder of this paper is organized as follows. Section 2 presents the discussion of the theory for calculating the life exponent and life estimation formula and Weibull distribution. In addition, we propose a circuit for V-t test. In Section 3, we demonstrate how to conduct the V-t characteristic test at 110 °C and Section 4 presents the evaluation of electrical performance.

## 2. Theory

### 2.1. Inverse Power Law

Insulating materials for power cables are exposed to various stresses. As the operating time of the cable increases, aging progresses and the insulation performance degrade. V-t test is an experimental method based on electrical stress of aging factors. As a result, the life exponent n and the formula of life estimation due to electrical stress can be calculated, and it is mainly used in the long-term reliability evaluation method. The equation of V-t is as follows [23,24].

$$V^n \times t = k \tag{1}$$

The life equation using Equation (1) is as follows.

$$L(V) = kV^{-n} \tag{2}$$

where L is the time-to-breakdown in hours, usually it is the Weibull probability of 63.2%, V is the applied voltage to materials, and n, k, are constants to be determined for the specific tested material or device. The inverse power law is considered valid, if the data being plotted on log-log graph fits a straight line [25].

In Equation (2), n is called the life exponent and is used to evaluate the long-term reliability of insulating materials and calculate the aging factor [22]. The output formula of the n is as follows. Figure 1 shows the V-t characteristic curve.

$$V_1^n \times t_1 = V_2^n \times t_2 \tag{3}$$

$$n = \frac{\log\left(\frac{t_2}{t_1}\right)}{\log\left(\frac{V_1}{V_2}\right)} \tag{4}$$

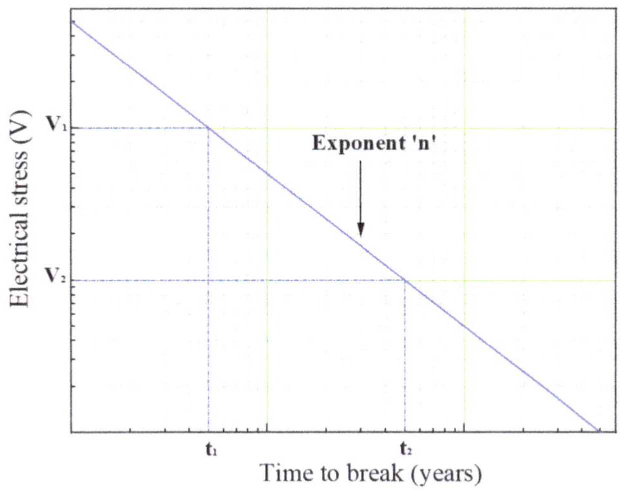


**Figure 1.** The V-t characteristic curve.

### *2.2. Weibull Distribution*

Statistical methods such as normal distribution, exponential distribution, logarithmic distribution and Weibull distribution are used to analyze and evaluate the failure data of materials. Among them, the Weibull distribution function is widely used for the reliability analysis of products associated with failures because it is possible to determine the failure modes in the early, constant, and wear out failures by using the scale parameter and shape parameter [26,27]. The following is the formula for the two-parameter Weibull distribution function [28].

$$F(t) = 1 - \exp\left[-\left(\frac{t}{\alpha}\right)^{\beta}\right] \quad (5)$$

In the equation, t represents the breakdown time, F (t) is the probability of failure, alpha is a scale parameter greater than 0, and means the 63.2% value of the failure probability. In the case of beta, it is a shape parameter, which means the type of data.

## 3. Experimental

### *3.1. Test Sample*

In general, PPs are divided into two types, homo polymer and copolymer, depending on the use of co-monomer. Homo polymer refers to polypropylene obtained by polymerizing only propylene monomer without using co-monomer. Further, as a general copolymer, there is a block copolymer polymerized by adding ethylene to propylene. Homo polymer has the advantages of heat resistance, mechanical properties, chemical resistance, and recycling, but is vulnerable to impact resistance at low temperatures. In the case of block copolymer, the impact resistance was improved.

The materials were blended with different proportions of block copolymer, homo polymer and HDPE. Table 1 shows three types of the test samples. The block polymer was selected because it best suits the properties required for the cable, and when the homo polymer and block polymer are mixed in half, it was selected to confirm the recyclability of the homo polymer. Furthermore, the quantity of ethylene included in the block copolymer of this paper was about 15%, and the reason why HDPE was mixed with the mixture of homo polymer and block copolymer was because of corrosion resistance, and the reason why homo polymer and block polymer had a larger ratio than HDPE was to emphasize the heat resistance of PP. In general, submarine and underground cables are corroded from various sources. HDPE has better corrosion resistance than LDPE, XLPE and other materials and has the advantage of melting.

**Table 1.** The test samples for V-t characteristic test.

| Description of the Test Samples |
|---|
| Block copolymer = 100% |
| Block copolymer : Homo polymer = 50% : 50% |
| Block copolymer : Homo polymer : HDPE = 30% : 30% : 40% |

Prior to sheet production, the hot press was heated to 200 °C for 5 min. Then, compound supplied by the cable manufacture was added to the molding flask, and the mixture was pressurized at 10 MPa for 5 min, and rapidly cooled for 2 min using a cooling press of 10 °C. Figure 2 shows the equipment for manufacturing the PP sheets and the completed test sample.

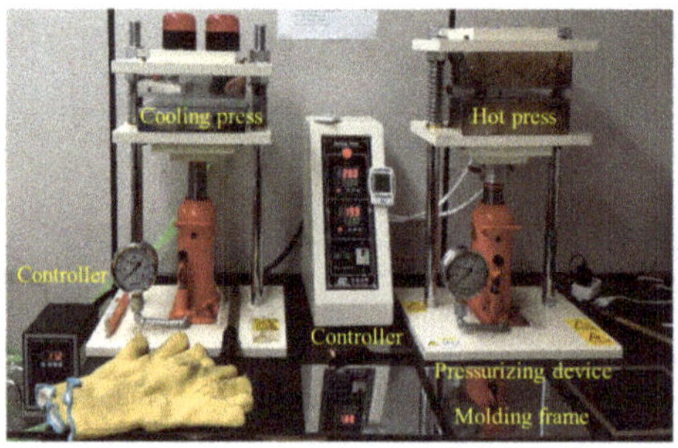


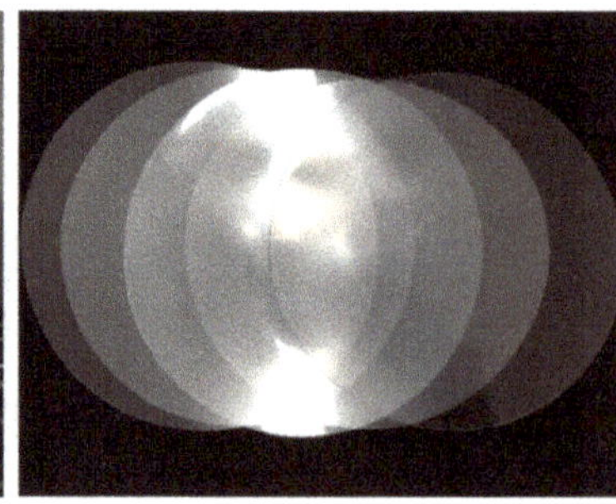

**Figure 2.** The equipment for manufacturing test samples.

The outer diameter of the PP sheet was 22 cm, the thickness was 0.2 mm, and the error range of the thickness was within 5%.

With reference to the IEC 62431 standard, the number of specimens for the ramping test was set to 10 [29]. In addition, the V-t characteristic test was performed 20 times, and 5 specimens per test voltage were used.

### 3.2. Experimental Setup

HVDC generator, voltage dividers, oven, oil bath, data acquisition system (DAS) and PC for temperature measurement were used as equipment for the breakdown test of PP materials.

The HVDC voltage source has a maximum voltage and current of 120 kV and 30 mA, and is configured as a high-voltage transformer, a clamp circuit and peak detector that combines a capacitor and a diode, a protection resistor and a voltage divider. HVAC is generated from the transformer, DC voltage is superimposed via the clamp circuit, and HVDC is generated through the peak detector. The DC voltage from the power source is connected in series with the protection resistor and then applied in parallel to the voltage divider and the test sample. The applied HVAC transformer is 220 V input and 60 kV output, and the diode is for rectifying the AC voltage. It is combined with a diode and a capacitor to the voltage doubler [30]. Capacitors and diodes were used that could withstand a HVDC voltage of 100 kV or higher, and a limited resistor of 120 kΩ was also configured to protect the circuit from overvoltage and overcurrent when the test sample was breakdown. In addition, the voltage divider that connected 300 MΩ and 30 kΩ in series was used to measure the DC voltage applied to the test sample in real time. Figure 3 shows a circuit diagram for the HVDC generator. In general, in order to perform a DC breakdown test, the ripple factor of the HVDC generator is important, and its value must be 3% or less. Note that if the value is exceeded, DC voltage characteristics are not reflected in the test.

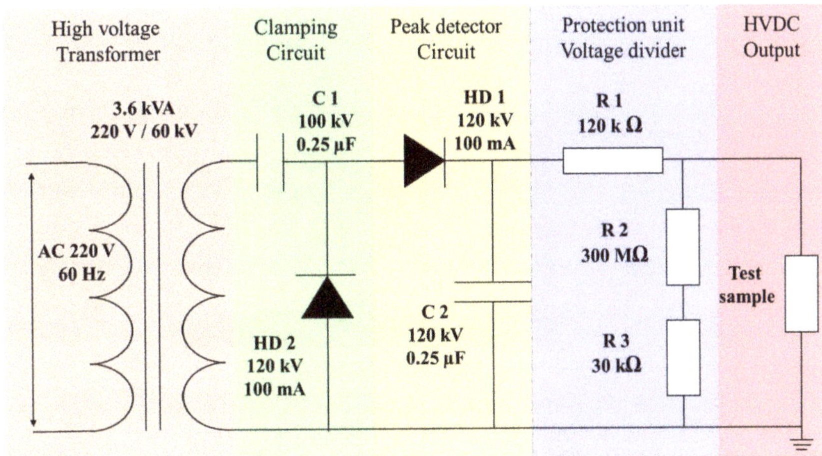


**Figure 3.** The circuit for breakdown voltage test.

Figure 4 shows the schematic representation of ramping and the V-t characteristic test. An oven was used to maintain a constant temperature of 110 °C for the test sample, and a temperature sensor was constructed to the dummy sample, and the temperature of the PPs was measured in real time. The transformer oil was put into an oil bath made of stainless steel to prevent flashover, and the oil bath was grounded. For the supporting structure of the breakdown test a polytetrafluoroethylene (PTFE) that was not deformed even at a temperature of 110 °C or higher was applied. A heat-resistant silicon-insulated cable was used to apply DC voltage in the air and oil inside the oven.

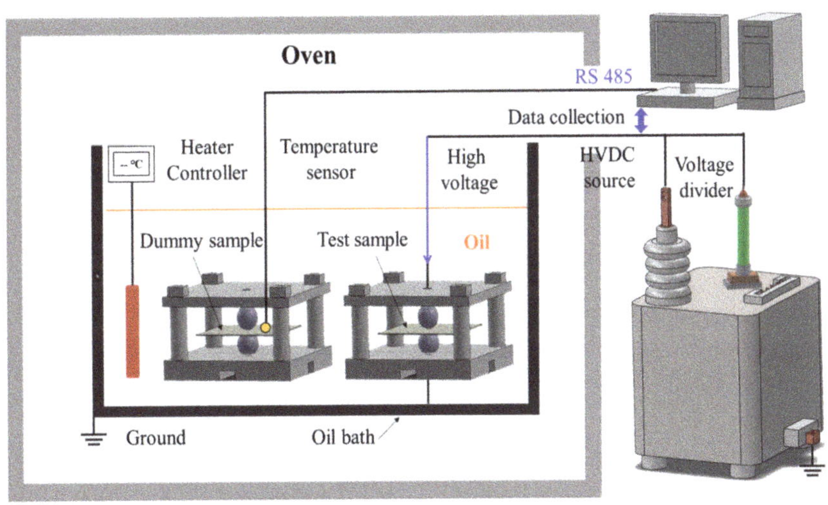


**Figure 4.** Schematic representation of the experimental setup for V-t characteristic test.

Commercialized micrometers and electrodes were combined for the constant measuring force applied to the test sample. The test electrodes were designed so that the constant force was applied to PP sheet during the V-t characteristic test. The electrode was made of a stainless steel with an outer diameter of 20 mm referring to the standard of IEC 62431. The temperature and voltage applied to the test sample were saved in real time using RS 485 communication. Figure 5 is a picture of the test electrode and the V-t characteristic test inside the oven.

(a) (b)

**Figure 5.** The pictures of V-t test: (**a**) electrode with micrometer and PTFE structure; (**b**) dummy sample inside oil for measuring temperature.

### 3.3. Test Conditions

The ramping test was performed under a step-up condition at 2 kV per second in accordance with the IEC standard [29]. The DC V-t characteristic test was performed with voltage until it broke. The test was carried out while maintaining the constant operating temperature of 110 °C for the PPs.

## 4. Results

### 4.1. Ramping Test

In order to determine the approximate test voltage of the V-t characteristic test, a ramping test was performed on three types of PPs, and the Weibull distribution 63.2% of breakdown data was 40 kV or more. Figure 6 shows the breakdown data after the ramping test of the test samples as a Weibull distribution function. Table 2 specifies the scale and shape parameters of three types of PPs after ramping test.

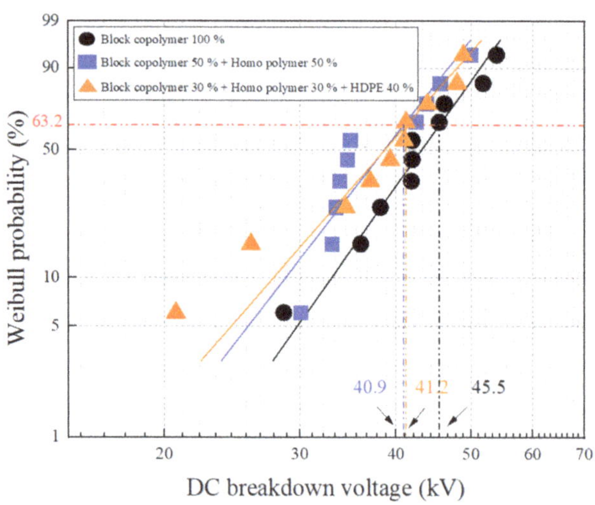


**Figure 6.** The Weibull distribution after ramping test of the test samples.

**Table 2.** Specification of the scale and shape parameters of three types of test samples.

| | Block Copolymer = 100% | Block Copolymer : Homo Polymer = 50% : 50% | Block Copolymer : Homo Polymer : HDPE = 30% : 30% : 40% |
|---|---|---|---|
| Scale parameter | 45.5 | 40.9 | 41.2 |
| Shape parameter | 6.9 | 6.3 | 5.6 |

### 4.2. DC V-t Test

After the ramping test, a DC V-t characteristic test was performed based on a breakdown voltage of 63.2%. For the test voltage, four voltage levels were applied between a maximum of 47 and 36 kV. The breakdown data according to time was plotted using the Weibull distribution function, and a V-t characteristic curve was made using the value of the breakdown time according to the applied voltage. The life exponent was calculated through the slope of the graph, and the life estimation equation of the PPs was derived through the relationship between voltage and time.

In Figures 7a, 8a and 9a the analyses are shown of the breakdown data using the Weibull distribution function, and Figures 7b, 8b and 9b display the figure for calculating the life exponent and life estimation equation. In Figures 7b, 8b and 9b, the red data is the Weibull 63.2% of the breakdown time according to the applied voltage, and the life estimation was derived from the four red data. In the figure, the fastest breakdown data in Figure 9 was 23 s when 47 kV was applied, and the longest lasting data in Figure 8 was 87,138 s when the voltage 36 kV was applied.

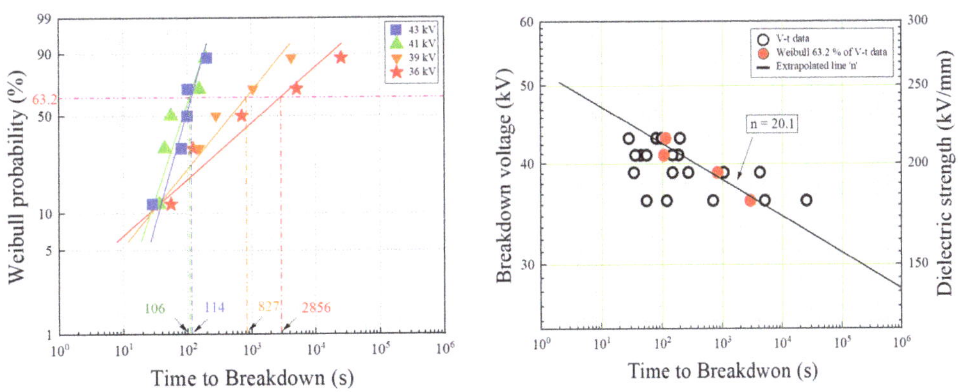


(a) (b)

**Figure 7.** Breakdown data analysis on block copolymer: (**a**) Weibull distribution according to time to breakdown; (**b**) V-t graph to obtain life exponent and life estimation formula.

### 4.3. Life Exponent and Estimation

The V-t characteristic test was carried out on three types of PPs, and the V-t characteristic curve was drawn according to electrical stress. Using Equations (2) and (4), the life exponent n was calculated, and the life estimation formula was also derived.

In general, the life exponent is used in a way that makes it easy to evaluate the long-term reliability of insulating materials. It is judged that the larger the life exponent of the insulating materials, the better the life characteristics. Each life exponent is as follows: the block copolymer was 20.1, when the block copolymer and homo polymer PP were mixed, it was 21.1, and when the block copolymer PP, homo polymer and HDPE were mixed, it was 23.2. It was shown that as the mixture was added to the block copolymer, the long-term reliability properties improved.

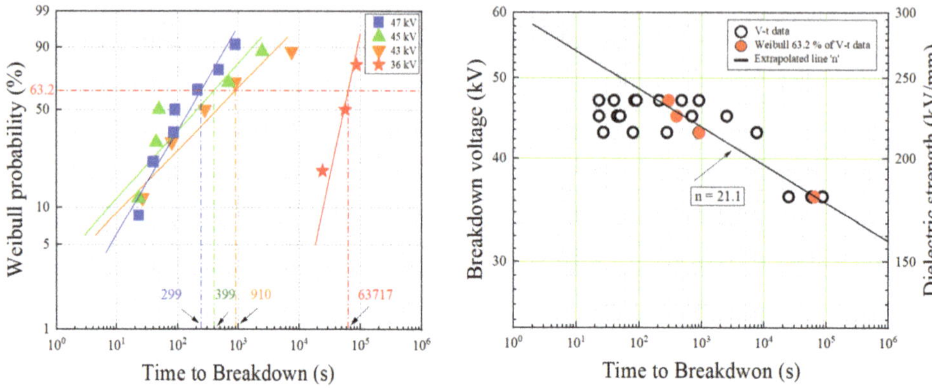


(a) (b)

**Figure 8.** Breakdown data analysis on block copolymer and homo polymer: (**a**) Weibull distribution according to time to breakdown; (**b**) V-t graph to obtain life exponent and life estimation formula.

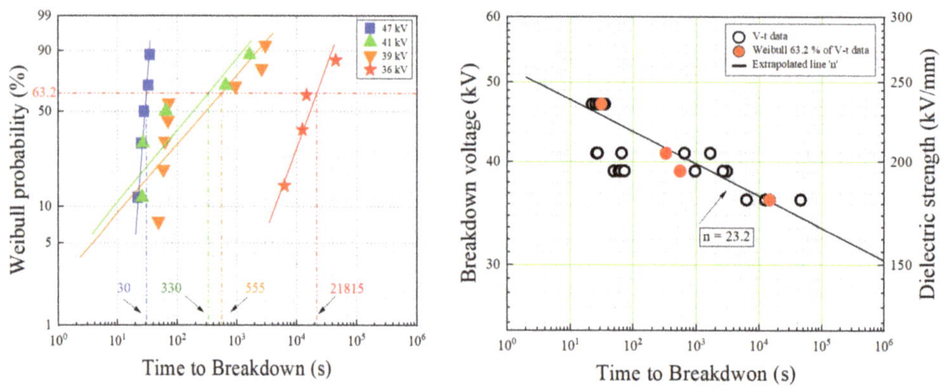


(a) (b)

**Figure 9.** Breakdown data analysis on block copolymer, homo polymer and HDPE: (**a**) Weibull distribution according to time to breakdown; (**b**) V-t graph to obtain life exponent and life estimation formula.

Since the design life of a power cable is usually 40 years, an equation that can calculate the long-term operating life due to electrical stress was needed. Table 3 shows the life exponent and life estimation formula of other PPs with the mixing ratio of additives.

**Table 3.** The life exponent and life estimation formula according to PPs.

| Description of the Test Samples | Life Exponent | Life Estimation Formula |
|---|---|---|
| Block copolymer = 100% | 20.1 | $L(V) = 6 \times 10^{34} \times V^{-20.13}$ |
| Block copolymer : Homo polymer = 50% : 50% | 21.1 | $L(V) = 4 \times 10^{37} \times V^{-21.09}$ |
| Block copolymer : Homo polymer : HDPE = 30% : 30% : 40% | 23.2 | $L(V) = 10^{40} \times V^{-23.23}$ |

### *4.4. Evaluation of Electrical Performance*

The use of life exponent in assessing long-term reliability and performance is very relative. So, we required a way to express the electrical performance as a percentage.

Figure 10 shows the electrical stress of 100 years of PP insulating material using the lifetime equation and when comparing the three materials, the larger breakdown voltage value according to life is more economical because it can reduce the insulation thickness applied to the cable. Electrical performance was evaluated by electrical stress tolerance based on a block copolymer with a design life of 40 years as a standard. The PP with HDPE was 17%, and the electrical performance was excellent. Table 4 shows the performance evaluation of dielectric strength of PPs in 40 years.

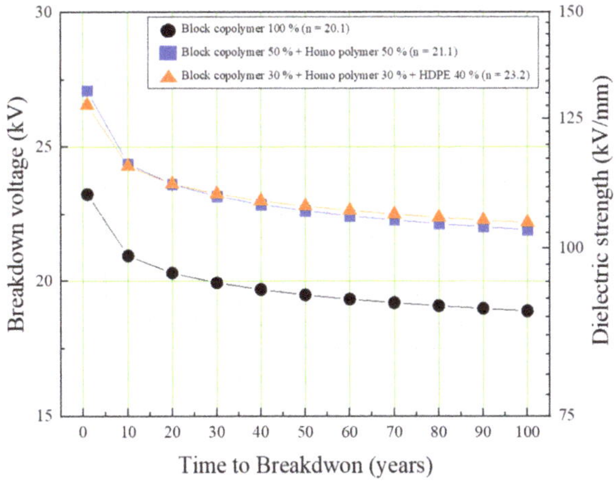


**Figure 10.** V-t curve considering the long term operation of power cables.

**Table 4.** Performance evaluation of dielectric strength of PPs in 40 years.

| Description of the Test Samples | Breakdown Voltage | Performance |
|---|---|---|
| Block copolymer = 100 % | 19.6 kV | Benchmark |
| Block copolymer : Homo polymer = 50% : 50% | 22.8 kV | 16% |
| Block copolymer : Homo polymer : HDPE = 30% : 30% : 40% | 23 kV | 17% |

## 5. Discussion

Many researchers have used the life exponent as a method for assessing long-term reliability, but some have not considered it in experiments and analyses.

Initially, the experimental part did not use a dummy sample during the V-t test and did not present a method that could provide a uniform pressure on the test samples.

Second, the long-term reliability of the materials was evaluated only by the life exponent, which is a relative numerical value in the analysis. Based on the life equation, the evaluation of electrical performance at the design life of the cable should be performed.

In this study, dummy samples were used to measure temperature in real time, and a commercially available micrometer was used to present a method for a constant measuring force of samples. The V-t characteristics test of the test samples was conducted at 110 °C. As shown in Figures 7–9, the life exponent according to the breakdown voltage and time to breakdown is quantitatively calculated. The block copolymer had a value of 20.1, the case block copolymer and homo polymer were mixed, with a value of 21.1, and the block copolymer, homo polymer and HDPE were mixed, with a value of 23.2. We compared the electrical performance between the materials mixed block copolymer, homo polymer and HDPE to prove the long-term reliability and applicability of HVDC. In addition, we evaluated the electrical performance which is the cable design life of 40 years. The electrical performance of the insulation mixed block copolymer, homo polymer and HDPE is up to 17% higher than block copolymer. Additionally, it was found that the mixed material without HDPE was also about 16% higher than the block copolymer, and the influence of HDPE seems to be inadequate. However, HDPE has a corrosion resistance to water, which will prevent the aging and failure due to moisture penetration of the submarine and underground cable and will ensure long-term reliability. Based on this study, evaluation of electrical performance with PPs when exposed to water is also essential, and further research on the long-term reliability of the block copolymer and HDPE mixed materials are not mentioned in this paper.

There are some points and limitations that need to be improved in this study. Since the content of this study is related to the life due to electrical stress, additional studies according to the thermal and mechanical properties are also required. In addition, the reliability of this model depends on the number of breakdown data. Because the Weibull distribution function is a statistic, the reliability of the analysis increases with more data. Conversely, this model is vulnerable and can fail if the amount of data is small. We test to meet the minimum number provided by the IEC and require additional testing to obtain more reliable results. Further study of the related attributes of conductivity and space charge for a DC application is also required. It is also necessary to evaluate the long-term reliability of cables to which PP insulating materials are applied. In this paper, we can contribute as a reference for HVDC application of PP and insulation, and contribute to the life and long-term reliability evaluation of insulating materials through the DC V-t.

In addition, as described in this paper, Figures 1 and 6, Figures 7–10 use a software program called Origin.

## 6. Conclusions

In this study, electrical performance was evaluated for HVDC application of PP. DC V-t characteristics tests were performed on three types of s, the life exponent was calculated, and the life estimation equation of the PP insulating material due to electrical stress was derived.

The implications of this study can be summarized as follows. First, the life exponent of PPs is 20 or more, which is equivalent to or higher than that of conventional materials, demonstrating the applicability of HVDC. Second, the long-term reliability was evaluated by deriving the life estimation equation of PP insulating material due to electrical stress. Third, based on the third life estimation equation, the PP with HDPE had a 17% better electrical performance than the block polymer based on the design life of the cable of 40 years.

**Author Contributions:** This paper is a result of the collaboration of all co-authors. S.-W.L. and I.-S.K. conceived and designed the study. H.-J.K. and S.-W.L. established the model and drafted the manuscript. J.-S.L. refined the language and provided statistical information. H.-J.K. helped with the corrections. S.-W.L. designed and performed the experiments. All authors have read and agreed to the published version of the manuscript.

**Funding:** This research was funded by the Ministry of Science and ICT (MSIT) (No. 21A01058).

**Institutional Review Board Statement:** Not applicable.

**Informed Consent Statement:** Not applicable.

**Data Availability Statement:** Not applicable.

**Acknowledgments:** This research was supported by Korea Electrotechnology Research Institute (KERI) Primary research program through the National Research Council of Science &Technology (NST) funded by the Ministry of Science and ICT (MSIT) (No. 21A01058).

**Conflicts of Interest:** The authors declare no conflict of interest.

## References


1. Mazzanti, G.; Marzinotto, M. *Extruded Cables for High-Voltage Direct-Current Transmission*; WIley-IEEE Press: Hoboken, NJ, USA, 2013; pp. 99–106.
2. Kazuki, T.; Sukuki, H.; Hara, M.; Watanabe, K. Research and development of /spl plusmn/250 kV DC XLPE cables. *IEEE Trans. Power Deliv.* **1998**, *13*, 7–16. [CrossRef]
3. Wang, H.; Cao, J.; He, Z.; Yang, J.; Han, Z.; Chen, G. Research on Overvoltage for XLPE Cable in a Modular Multilevel Converter HVDC Transmission System. *IEEE Trans. Power Deliv.* **2015**, *31*, 683–692. [CrossRef]
4. Huang, X.; Fan, Y.; Zhang, J.; Jiang, P. Polypropylene based thermoplastic polymers for potential recyclable HVDC cable insulation applications. *IEEE Trans. Dielectr. Electr. Insul.* **2017**, *24*, 1446–1456. [CrossRef]
5. Yoshino, K.; Ueda, A.; Demura, T.; Miyashita, Y.; Kurahashi, K.; Matsuda, Y. Property of Syndiotactic Polypropylene and Its Application to Insulation of Electric Power Cable -Property, Manufacturing and Characteristics. In Proceedings of the 7th International Conference on Properties and Applications of Dielectric Materials (ICPADM), Nagoya, Japan, 1–5 June 2003.
6. Rong, X.; Shaoxin, M.; Yuli, W. Long Term Reliability Tests of 500 kV XLPE Cable System. In Proceedings of the 2014 International Conference on Power System Technology, Chengdu, China, 20–22 October 2014. [CrossRef]
7. Nakiri, T.; Maekawa, Y.; Takashima, H.; Imajo, T.; Hirota, H.; Ishii, K.; Nakao, Y.; Shimada, M. Long-term reliability testing of 500 kV DC PPLP-insulated oil-filled cable and accessories. *IEEE Trans. Power Deliv.* **1999**, *14*, 319–326. [CrossRef]
8. Morette, N.; Ditchi, T.; Oussar, Y. Partial Discharges Measurements and Analysis as an Evaluation Tool for the Reliability of Polymeric-Insulated Cables used under HVDC Conditions. In Proceedings of the 2018 IEEE 2nd International Conference on Dielectrics (ICD), Budapest, Hungary, 1–5 July 2018. [CrossRef]
9. Wang, S.; Yu, S.; Xiang, J.; Li, J.; Li, S. DC breakdown strength of crosslinked polyethylene based nanocomposites at different temperatures. *IEEE Trans. Dielectr. Electr. Insul.* **2020**, *27*, 482–488. [CrossRef]
10. Zhang, C.C.; Li, Y.F.; Hu, M.Y.; Ma, F.L.; Zhao, H.; Han, B.Z. Conductivity Properties of XLPE Insulation Used for HVDC Cable after Accelerated Thermal Ageing. In Proceedings of the 2018 12th International Conference on the Properties and Applications of Dielectric Materials (ICPADM), Xi'an, China, 20–24 May 2018.
11. Dakka, M.A.; Bulinski, A.; Bamji, S. Correlation Between Space Charge Development and Breakdown in Polymeric Insulation under dc Field. In Proceedings of the 2004 IEEE International Conference on Solid Dielectrics, Toulouse, France, 5–9 July 2004.
12. Li, W.; Zhang, X.; Zhao, A.; Wu, L.; Ren, Z.; Deng, J.; Zhang, G. Studies of the Polarization/Depolarization Current Characteristics of XLPE Cable. In Proceedings of the 2016 IEEE PES Asia-Pacific Power and Energy Engineering Conference (APPEEC), Xi'an, China, 25–28 October 2016.
13. Shirasaka, Y.; Murase, H.; Okabe, S.; Okubo, H. Cross-sectional comparison of insulation degradation mechanisms and lifetime evaluation of power transmission equipment. *IEEE Trans. Dielectr. Electr. Insul.* **2009**, *16*, 560–573. [CrossRef]
14. Yoon, J.; Kwon, J.; Ryu, J.; Bae, C.; Choi, J.; Kim, Y. A Study of Relationship between V-t and Tanδ Characteristic on Epoxy Resin. In Proceedings of the 2019 IEEE Electrical Insulation Conference (EIC), Calgary, AB, Canada, 16–19 June 2019.
15. Lalam, F.B.; Djemmal, F. DC Electrical Ageing of XLPE under Hydrostatic Pressure. *Adv. Mater. Sci. Eng.* **2017**, *2017*, 3936163. [CrossRef]
16. Murata, Y.; Sakamaki, M.; Abe, K.; Inoue, Y.; Mashio, S.; Kashiyama, S.; Matsunaga, O.; Igi, T.; Watanabe, M.; Asai, S.; et al. Development of High Voltage DC-XLPE Cable System. *SEI Tech. Rev.* **2013**, *101*, 55–62.
17. Tiwana, A.P.S.; Reddy, C.C. Life Estimation of LDPE Film under Stepped-Stress Voltage Application. In Proceedings of the 2016 IEEE Conference on Electrical Insulation and Dielectric Phenomena (CEIDP), Toronto, ON, Canada, 16–19 October 2016.
18. Fujimori, A.; Tanaka, T.; Takashima, H.; Imajo, T.; Hata, R.; Tanabe, T.; Yoshida, S.; Kakihana, T. Development of 500 kV DC PPLP-insulated oil-filled submarine cable. *IEEE Trans. Power Deliv.* **1996**, *11*, 43–50. [CrossRef]
19. Wang, Y.; Lv, Z.; Wang, X.; Wu, K.; Zhang, C.; Li, W.; Dissado, L. Estimating the inverse power law aging exponent for the DC aging of XLPE and its nanocomposites at different temperatures. *IEEE Trans. Dielectr. Electr. Insul.* **2016**, *23*, 3504–3513. [CrossRef]
20. Xia, L.; Qinxue, Y.; Minghao, L.; Kai, Z.; Lisheng, Z. Temperature Dependence of AC / DC Breakdown Strength of HVDC XLPE Cable Slices. In Proceedings of the 2016 International Conference on Condition Monitoring and Diagnosis (CMD), Xi'an, China, 25–28 September 2016.
21. Hu, H.; Jia, Z.; Wang, X.; Shi, Z.; Zhang, C. Simulation of Mechanical Pressure of XLPE-SiR Interface in Cable Joints Based on Operating Temperature and Interference Parameters. In Proceedings of the 2020 IEEE International Conference on High Voltage Engineering and Application (ICHVE), Beijing, China, 6–10 September 2020.
22. Brochure CIGRE 496. *Recommendations for Testing DC Extruded Cable Systems for Power Transmission at a Rated Voltage up to 500 kV*; CIGRE Working Group B1.32; CIGRE: Paris, France, 2012.
23. Feilat, E.A.; Grzybowski, S.; Knight, P.J. Electrical Aging Models For Fine Gauge Magnet Wire Enamel of Flyback Transformer. In Proceedings of the IEEE SoutheastCon 2000, 'Preparing for The New Millennium' (Cat. No. 00CH37105), Nashville, TN, USA, 9 April 2000.
24. Matsumoto, S.; Sato, T.; Honda, H.; Yokoyama, S. Approximate Formula for Surface V-t Characteristics of 6.6-kV Distribution Insulator. In Proceedings of the 2012 International Conference on Lightning Protection (ICLP), Vienna, Austria, 2–7 September 2012.

25. Ying, L.; Hui, B.; Yu, S.; Xiaolong, C. The Measurement of Voltage Endurance Coefficient by Electrical Treeing Test for XLPE Cable Insulation. In Proceedings of the 2013 IEEE International Conference on Solid Dielectrics (ICSD), Bologna, Italy, 30 June–4 July 2013.
26. Su, P.; Yin, Y.; Zheng, X.; Xuan, Y.; Wu, J. Investigation of the Thickness Effect on DC Breakdown Strength for HVDC Flexible Cable Insulation Associated with Space Charge. In Proceedings of the 2019 IEEE Electrical Insulation Conference (EIC), Calgary, AB, Canada, 16–19 June 2019.
27. Yuan, W.; Wang, T.; Ni, H.; Gao, M.; Ding, Y.; Li, Y.; Zhao, Y.; Zhang, Q. Weibull Statistical Analysis of Size Effects on the Impulse Breakdown Strength in Transformer Oil. In Proceedings of the 2017 IEEE 19th International Conference on Dielectric Liquids (ICDL), Manchester, UK, 25–29 June 2017.
28. Zhang, L.; Zhou, Y.; Cui, X.; Sha, Y.; Le, T.H.; Ye, Q.; Tian, J. Effect of nanoparticle surface modification on breakdown and space charge behavior of XLPE/$SiO_2$ nanocomposites. *IEEE Trans. Dielectr. Electr. Insul.* **2014**, *21*, 1554–1564. [CrossRef]
29. IEC 62431-1. Electric Strength of Insulating Materials—Test Methods—Part 1: Tests at Power Frequencies. 2013. Available online: https://webstore.iec.ch/publication/1101 (accessed on 6 September 2021).
30. Kwon, J.-M.; Kwon, B.-K. High Step-Up Active-Clamp Converter with Input-Current Doubler and Out-put-Voltage Doubler for Fuel Cell Power Systems. *IEEE Trans. Power Electron.* **2009**, *24*, 108–115. [CrossRef]

*energies*

MDPI

*Article*

# A New Arrangement of Active Coils for Wireless Charging of UAV

**Zhengwang He [1], Zhiyong Li [1,*], Ruoyue Wang [1], Ying Fan [2] and Minqian Xu [1]**

[1] School of Automation, Central South University, Changsha 410083, China; 8207181024@csu.edu.cn (Z.H.); 0908170424@csu.edu.cn (R.W.); 1406160104@csu.edu.cn (M.X.)
[2] School of Electric Engineering, Guangdong Mechanical & Electronical College of Technology, Guangzhou 510515, China; 2014010008@gdmec.edu.cn
* Correspondence: lizy@mail.csu.edu.cn; Tel.: +86-138-7312-7689

**Abstract:** This paper presents the design and optimization of a wireless power transfer (WPT) charging system based on magnetically coupled resonant technology, applied to an Unmanned Aerial Vehicle (UAV). In this paper, a charging system, including dual active transmitter coils and a single receiver coil, is proposed. The dual transmitting coils adopt a coaxial structure with different radii. This structure simplifies the calculation of the complex mutual inductance between the coils to a function of mutual inductance only related to the value of the radial misalignment. Aiming toward a constant charging power, the optimal transmission efficiency of electric energy is achieved by controlling the input voltages of the active coils, which are solved via a set of equations defined as Lagrange multipliers. The simulation results of the 570 V and 85,000 Hz system verified the validity of the proposed wireless UAV charging scheme.

**Keywords:** Lagrange multiplier; magnetically coupled resonant technology; power compensation; radial misalignment

**Citation:** He, Z.; Li, Z.; Wang, R.; Fan, Y.; Xu, M. A New Arrangement of Active Coils for Wireless Charging of UAV. *Energies* **2021**, *14*, 5754. https://doi.org/10.3390/en14185754

Academic Editor: José Gabriel Oliveira Pinto

Received: 21 July 2021
Accepted: 9 September 2021
Published: 13 September 2021



## 1. Introduction

UAVs undoubtedly have high research potential and application value, but the flight time of UAVs restricts their development. This is because a high-energy density lithium battery, which is adopted to the UAV, only permits a flight time of about 20–40 min [1].

Thus, researchers began to research the application of magnetically coupled resonant technology to charge UAVs for solving this limitation [2–5]. This method creates a base station where the batteries can be automatically charged after landing [3]. The proposed magnetically coupled resonant system in [2] is mainly composed of two coils, namely, a single transmitting coil and a single receiving coil. The power transmission efficiency between the two coils is a function of the vertical distance between the two coils and the power supply. When the two coils are coaxial, the power transmission efficiency of the coil is the highest, and when the distance between the two coils is increased, the power transmission efficiency decreases. In the practical charging process of UAVs, the coaxial position of the receiving coil and the transmitting coil cannot be guaranteed, namely, radial misalignment, which will lead to a reduction in output power and charging time delay [6]. Therefore, it is necessary to research the power compensation problem of misaligned charging.

Scholars have adopted various methods to solve this problem [7–18]. In [7–11], a reconfigurable resonator and numerous transmission coil arrays with different structures have been used to improve transmission efficiency. A relatively complete mutual inductance model for analyzing magnetically coupled resonant technology was established. A planar transmitting resonator with an array of $2 \times 2$ using a single feeding loop was presented in [12]. This structure could maintain high transfer efficiency in a plane. However, the power transfer efficiency function determined by this structure is still two-dimensional in essence. This leads to the need for complex control strategies to achieve the goal of constant output power, which is not worth it in practical application. In [13], two circular planar spiral coils were applied to both receive and transmit circuits of a WPT system. The main disadvantage of this configuration is that it reduces the load capacity of the UAV. In [14], an automatic landing procedure for UAV was proposed, which fundamentally reduces the possibility of radial misalignment. A WPT charging system with a movable transmitting coil was introduced in [15]. When the UAV lands, the mobile transmitting coil can automatically align with the receiving coil. Of course, it requires a high-precision positioning system. A lightweight wireless charging system was presented in [17]. This design can improve the load capacity of UAVs without affecting the operation of various equipment carried by UAVs. In [18], Noriaki Oodachi et al. have proposed a wireless power transmission system to solve this problem by using phase weights of transmission coil arrays. When the receiving coil is not coaxial with the transmitting coil, according to the direction between the transmission coil and the receiving coil, the transmission coil of the coil array is excited by the transmission circuit and the corresponding phase weight to achieve the purpose of power compensation. However, this method also requires a large number of coils, and the control algorithm and implementation are complex, which increase the costs.

The system proposed in this paper is simple and effective. In order to compensate the power reduction caused by radial misalignment, two coaxial active coils with different radii are used as the transmitting coils, and the compensation is realized by controlling the different input power of the two coils. When the output power of one transmitting coil decreases due to radial misalignment, the other transmitting coil with different radii that are still coaxial is applied to compensate the output power, so as to maximize the transmission efficiency of the system, which satisfies the Lagrange multiplier. In addition, by controlling the input voltage of the two active coils of the system, a power supply with arbitrary output power can be designed. This design simplifies the efficiency function from a function with two-dimensional variables to a function with only a radial variable, so there is no need for a complex control algorithm, which makes it more practical.

## 2. Proposed Wireless Charging of UAV System

This paper presents the design of a UAV charging system with magnetically coupled resonant technology, as shown in Figure 1. The transmitting coil uses double coaxial active coils, with the receiving coil and transmitting coils coupled by air. The vertical distance between the UAV and the charging pile is 10 cm. Both the double transmitting circuit and the single receiving circuit adopt the topology of inductance and capacitor in series. The whole system realizes power transmission through the full resonance of coil and capacitor.

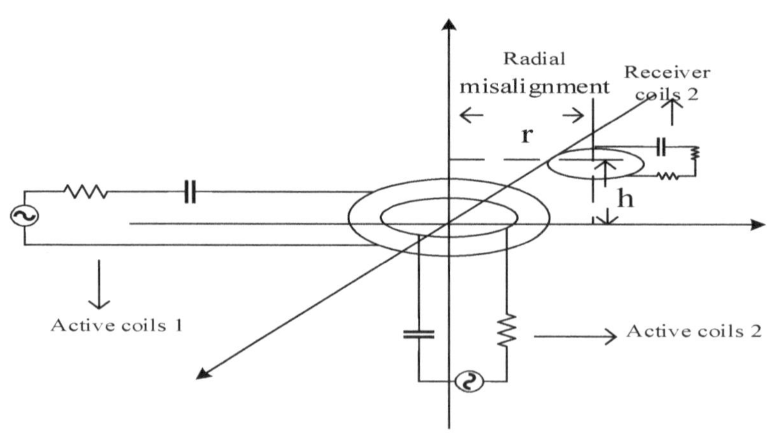


**Figure 1.** Proposed system for UAV charging.

In order to research the specific internal strategy of the new charging pile, two mathematical models, including the power model and mutual inductance model of the system, are established according to the actual charging situation of the UAV, and the theoretical analysis is carried out. The power model analyzes the relationship between the output power of the system and the input voltage of the two active coils; the transmission efficiency is found when the mutual inductance and output power are constant. The mutual inductance model analyzes the relationship between the mutual inductance of the system and the radial misalignment caused by the landing error when the UAV is charging, and provides a scheme to compensate for the system's output power reduction that is simple and effective. A 570 V, 85,000 Hz system is built in MATLAB/Simulink, and the output power of the system is obtained. The transmission efficiency of the system can reach 82%. At the same time, it is confirmed that the double transmitting coils have a good compensation effect on the power reduction caused by the radial misalignment of the system; the correctness of the control strategy is also verified.

### *2.1. Model of System Mutual Inductance and Radial Misalignment Distance*

In most of the experimental studies, the transmitting coil and receiving coil are placed in parallel, and their central axis position is the same. However, when the UAV actually stops, it cannot be guaranteed that it will stop in the axial direction of the transmitting coil. When the position of the coil deviates, the mutual inductance M will change. When the system frequency is fixed and the coil resistance is constant, the change in mutual inductance M is the most important factor affecting system performance. Therefore, when the receiving coil has radial misalignment, the mutual inductance between the receiving coil and the transmitting coil changes, which leads to a change in system power and efficiency.

Only the mutual inductance mode of the single transmitting and single receiving coils is considered, as shown in Figure 2. The center of the transmitting coil is placed in the coordinate system (0, 0, 0), and the center of the receiving coil is placed in the coordinate system (0, t, h).

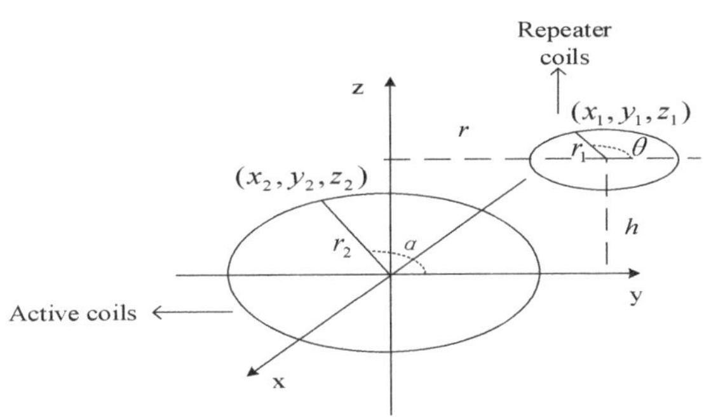


**Figure 2.** One active to receiver charging mode.

The parameter equations of the transmitting and receiving coils are listed as follows:

$$\begin{cases} x_1 = r_1 \times \cos\theta \\ y_1 = r_1 \times \sin\theta + r \\ z_1 = h \\ x_2 = r_2 \times \cos\alpha \\ y_2 = r_2 \times \sin\alpha \end{cases} \tag{1}$$

Therefore:

$$\begin{cases} dl_1 = (r_1 \times \cos\theta - r_1 \times \sin\theta)d\theta \\ dl_2 = (r_2 \times \cos\alpha - r_2 \times \sin\alpha)d\alpha \\ dl_1 \times dl_2 = r_1 \times r_2 \cos(\theta - \alpha)d\theta d\alpha \end{cases} \tag{2}$$

According to the Neumann formula, when the number of turns of the receiving coil is $N_1$ and the number of turns of the transmitting coil is $N_2$, the mutual inductance formula between them is as follows:

$$M = \frac{N_1 \times N_2}{4\pi} \times \frac{dl_1 \times dl_2}{R} = \frac{N_1 \times N_2}{4\pi} \int_0^{2\pi}\int_0^{2\pi} \frac{r_1 \times r_2 \cos(\theta - \alpha) d\theta d\alpha}{D} \tag{3}$$

In the above formula:

$$D = \sqrt{(r_1 \times \cos\theta - r_2 \times \cos\alpha)^2 + (r_1 \times \sin\theta + r - r_2 \times \sin\alpha)^2 + h^2}$$

Analysis of Equation (3) shows that the mutual inductance of the system decreases with the increase in radial misalignment, and the output power decreases with the decrease in mutual inductance. Therefore, it is necessary to compensate for the mutual inductance reduction caused by the radial misalignment of the single transmitter and single receiver system.

Considering double transmitting and single receiving coils, the two transmitting coils are coaxial with different radii. With the increase in radial misalignment, the mutual inductance between the two transmitting coils and the receiving coil changes according to Equation (3). However, for the receiving coil, the total mutual inductance is the sum of mutual inductance between the two transmitting coils and the receiving coil, which compensates for the reduction in mutual inductance; Maxwell simulation verifies this conclusion.

### 2.2. Model of Output Power and Input Voltage

We define the mutual inductance between coil 1 and coil 2 as $M_{12}$, and define $M_{13}$ and $M_{23}$ in the same way. When the UAV lands on the charging pile, the relative position between the receiving coil and the two transmitting coils is fixed, so the mutual inductance between the two coils is fixed, i.e., $M_{13}$, $M_{23}$ and $M_{12}$ are constant. The output power of the system is only related to the input voltage of the active coil, as shown in Figure 3. The detailed model analysis is as follows:

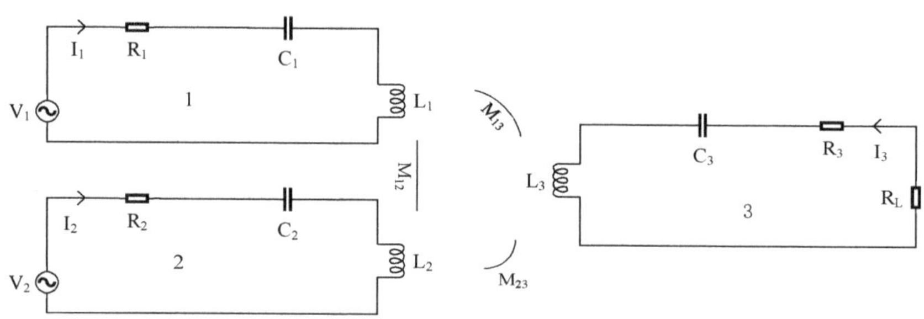


**Figure 3.** Two active to receiver charging modes.

In order to simplify the model, the parameters of the circuit model are symmetrical, namely, $R_1 = R_2 = R_3 = R$. The transmitter and receiver of the system will work at the same frequency, which is defined as:

$$\omega = 2\pi f = \frac{1}{\sqrt{C_1 L_1}} = \frac{1}{\sqrt{C_2 L_2}} = \frac{1}{\sqrt{C_3 L_3}} \tag{4}$$

where f is the fundamental frequency of the power supply.

By listing the voltage equation of each circuit, the following voltage and current (Equation (5)) can be obtained:

$$\begin{pmatrix} V_1 \\ V_2 \\ 0 \end{pmatrix} = \begin{pmatrix} R_1 + j\omega L_1 + \frac{1}{j\omega C_1} & -j\omega M_{12} & -j\omega M_{13} \\ -j\omega M_{12} & R_2 + j\omega L_2 + \frac{1}{j\omega C_2} & -j\omega M_{23} \\ -j\omega M_{13} & -j\omega M_{23} & R_3 + R_l + j\omega L_3 + \frac{1}{j\omega C_3} \end{pmatrix} \begin{pmatrix} I_1 \\ I_2 \\ I3 \end{pmatrix} \quad (5)$$

The input impedance of the system is solved. According to the defined system operating frequency, order $V_2 = 0$, the following equation (Equation (6)) can be obtained:

$$Z_{\text{sent1}} = R_1 + \frac{(\omega M_{12})^2(R_3 + R_l) - j\omega^3 M_{12}M_{13}M_{23}}{R_2(R_3 + R_l) + (\omega M_{23})^2} + \frac{(\omega M_{13})^2 R_2 + j\omega^3 M_{12}M_{13}M_{23}}{R_2(R_3 + R_l) + (\omega M_{23})^2} = R_1 + Y_2 + Y_3 \quad (6)$$

$Y_2$ and $Y_3$ in Equation (6), respectively, indicate the mutual inductance influence of loop 2 and loop 3 on loop 1:

$$\begin{cases} Y_2 = \frac{(\omega M_{12})^2(R_3+R_l) - j\omega^3 M_{12}M_{13}M_{23}}{R_2(R_3+R_l)+(\omega M_{23})^2} \\ Y_3 = \frac{(\omega M_{13})^2 R_2 + j\omega^3 M_{12}M_{13}M_{23}}{R_2(R_3+R_l)+(\omega M_{23})^2} \end{cases} \quad (7)$$

Therefore, the input current $I_1$ of active coil 1 can be expressed as:

$$I_1 = \frac{V_1}{Z_{\text{sent1}}} = \frac{V_1}{R_1 + Y_2 + Y_3} \quad (8)$$

Aligned, the input current $I_2$ of active coil 2 can be expressed as:

$$I_2 = \frac{V_2}{Z_{\text{sent2}}} = \frac{V_2}{R_2 + Y_2' + Y_3'} \quad (9)$$

Among:

$$\begin{cases} Y_2' = \frac{(\omega M_{12})^2(R_3+R_l) - j\omega^3 M_{12}M_{13}M_{23}}{R_1(R_3+R_l)+(\omega M_{13})^2} \\ Y_3' = \frac{(\omega M_{23})^2 R_1 + j\omega^3 M_{12}M_{13}M_{23}}{R_1(R_3+R_l)+(\omega M_{13})^2} \end{cases} \quad (10)$$

Considering two active coils, the output current $I_3$ of the receiver can be expressed as:

$$I_3 = \frac{j\omega M_{13} \times I_1 + j\omega M_{23} \times I_2}{R_3 + R_l} = \frac{j\omega M_{13}}{R_3 + R_l} \times \frac{V_1}{R_1 + Y_2 + Y_3} + \frac{j\omega M_{23}}{R_3 + R_l} \times \frac{V_2}{R_2 + Y_2' + Y_3'} \quad (11)$$

Among $Z_{sent2} = R_1 + Y_2' + Y_3'$. Therefore, the input power $P_{\text{in}}$ and output power $P_{\text{out}}$ of the system can be expressed as:

$$P_{\text{in}} = \text{Re}(V_1 \times I_1^*) + \text{Re}(V_2 \times I_2^*) = \frac{(V_1)^2}{R_1 + Y_2 + Y_3} + \frac{(V_2)^2}{R_2 + Y_2' + Y_3'} \quad (12)$$

$$P_{\text{out}} = I_3^2 \times R_l = \left(\frac{j\omega}{(R_3 + R_l)}\right)^2 \times \left(\frac{M_{13}V_1}{R_1 + Y_2 + Y_3} + \frac{M_{23}V_2}{R_2 + Y_2' + Y_3'}\right)^2 \times R_l \quad (13)$$

Therefore, the efficiency of the system $\eta$ can be calculated as follows:

$$\eta = \frac{P_{\text{out}}}{P_{\text{in}}} = \left(\frac{j\omega}{R_3 + R_l}\right)^2 \times \frac{Z_{\text{sent1}} \times Z_{\text{sent2}} \times \left(\frac{M_{13}V_1}{Z_{\text{sent1}}} + \frac{M_{23}V_2}{Z_{\text{sent2}}}\right)^2}{(V_1)^2 \times Z_{\text{sent2}} + (V_2)^2 \times Z_{\text{sent1}}} \times R_l \quad (14)$$

Thus, the value of $\eta$ is related to the two input voltages of the system and the mutual inductance between the two coils. Additionally, the observation Equations (13) and (14) show that when the output power and radial misalignment are fixed, there must be the maximum transmission efficiency.

The output power $P_{out}$ is set as constant at 640 W. The goal is to maximize transmission efficiency. The optimal problem formulation is Equation (15):

$$\begin{cases} Max: \left(\eta\eta = \frac{P_{out}}{P_{in}} = \left(\frac{j\omega}{R_3+R_l}\right)^2 \times \frac{Z_{sent1}\times Z_{sent2}\times\left(\frac{M_{13}V_1}{Z_{sent1}}+\frac{M_{23}V_2}{Z_{sent2}}\right)^2}{(V_1)^2\times Z_{sent2}+(V_2)^2\times Z_{sent1}} \times R_l\right) \\ P_{out}-640=0 \end{cases} \tag{15}$$

Lagrange multiplier $\lambda$ is introduced and a Lagrangian function is constructed:

$$L = \eta - \lambda \times (P_{out} - 640) \tag{16}$$

When the partial derivative of Equation (16) is found, they can be set to zero:

$$\begin{cases} \frac{\partial L}{\partial V_1} = \frac{\partial \eta}{\partial V_1} - \lambda \times \frac{\partial P_{out}}{\partial V_1} = 0 \\ \frac{\partial L}{\partial V_2} = \frac{\partial \eta}{\partial V_2} - \lambda \times \frac{\partial P_{out}}{\partial V_2} = 0 \\ P_{out} = 640 \end{cases} \tag{17}$$

The corresponding $V_1$ and $V_2$ are obtained when the efficiency is maximum, as long as $V_1$ and $V_2$ are satisfied (Equation (17)).

The flowchart of the methodology mentioned is introduced in Figure 4.

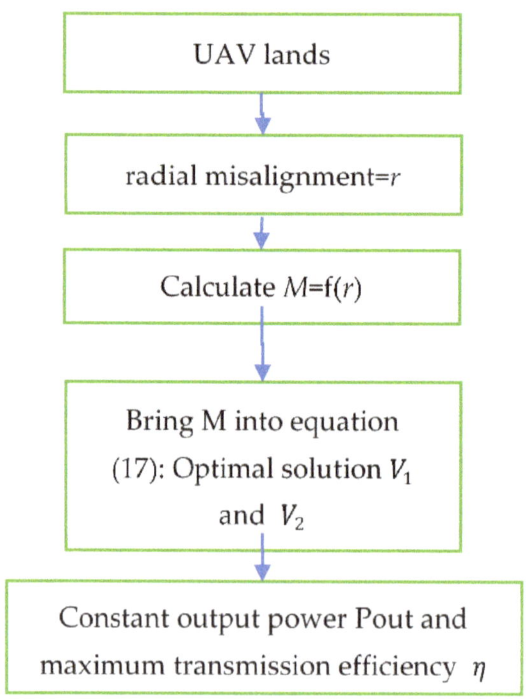


**Figure 4.** The flowchart of the methodology.

## 3. Simulation

The proposed 570 V and 85,000 Hz power model is simulated in MATLAB/Simulink, and the rationality of the model is verified. The parameters of the system are listed in Table 1. The mutual inductance parameters of the coil are simulated by the finite element method in Maxwell.

**Table 1.** Parameters of the system.

| Parameter | Value | Parameter | Value |
|---|---|---|---|
| $L_1$ & $L_3$ | 500.0 μH | Peak amplitude of $V_1$ & $V_2$ | 570 V |
| $C_1$ & $C_3$ | 0.0072 μF | f | 85,000 Hz |
| $R_1$ & $R_2$ & $R_3$ | 0.2 Ω | $M_{13}$ | 175 μH |
| $R_L$ | 20 Ω | $M_{23}$ | 300 μH |
| $M_{12}$ | 400 μH | $C_2$ | 0.0036 μF |
| $L_2$ | 1000 μH | | |

### 3.1. *Simulation of Mutual Inductance Parameters*

When the coil has radial misalignment, double transmitting and single receiving coils are considered. The radii of the two transmitting coils are 26 cm and 36 cm, respectively, and the radius of the receiving coil is 26 cm. The vertical distance between the transmitting coil and the receiving coil is kept at 10 cm, and the medium is air. The finite element simulation, as shown in Figure 5a, is carried out in Maxwell. The relationship between mutual inductance and radial distance in the two transmitting coils and receiving coils can be obtained, as shown in Figure 5b,c. It is found that the mutual inductance between the two transmitting coils and the radial distance decreases with the increase in radial misalignment. Combined with the power model, the power of the system also decreased. For the receiving coil, mutual inductance is the sum of the two, which compensates for the reduction in mutual inductance and power of the single transmitting coil. At the same time, the self-inductance of the 36 cm coil is twice that of the other two 26 cm coils, and the self-inductance is only related to the material and radius of the coil.

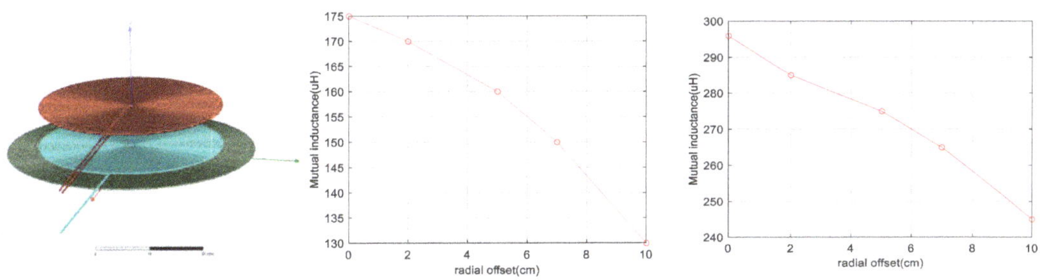


(**a**) Maxwell simulation diagram

(**b**) Relationship between mutual inductance and radial distance of 26 cm transmitting coil

(**c**) Relationship between mutual inductance and radial distance of 36 cm transmitting coil

**Figure 5.** Maxwell simulation. (**a**) Snapshot of Maxwell simulation diagram (**b**) Simulation result of 26 cm transmitting coil. (**c**) Simulation result of 36 cm transmitting coi.l.

### 3.2. *Power*

According to the coil parameters calculated in model A, the simulation output waveform of the system simulated by Simulink is shown in Figure 6. A snapshot of the simulation file is shown in Figure 6a. We use a transformer to replace the resonant coils and their mutual inductance. As for the high-frequency AC power required by the system, we directly use AC power to generate it. Both the double transmitting circuit and single receiving circuit adopt the topology of inductance and capacitor in series. In other figures, $U_{out}$ is the output voltage waveform. $U_{in}$ is the input voltage of the active coil. $I_{in}$ is the input current waveform of the active coil. $I_{out}$ is the output current waveform. Output power is the output active power of the system and input power1 and input power2 are the input active power of the system. According to the simulated voltage and current, the transmission efficiency of the system can be calculated, which reaches 94%. Comparing the simulation results with the theoretical results, the output power error is less than 5%.

At the same time, the simulation results show that the output current and voltage waveform is sine wave, but there are phase differences, which is caused by mutual inductance resonance not being complete, and therefore, there is phase lag.

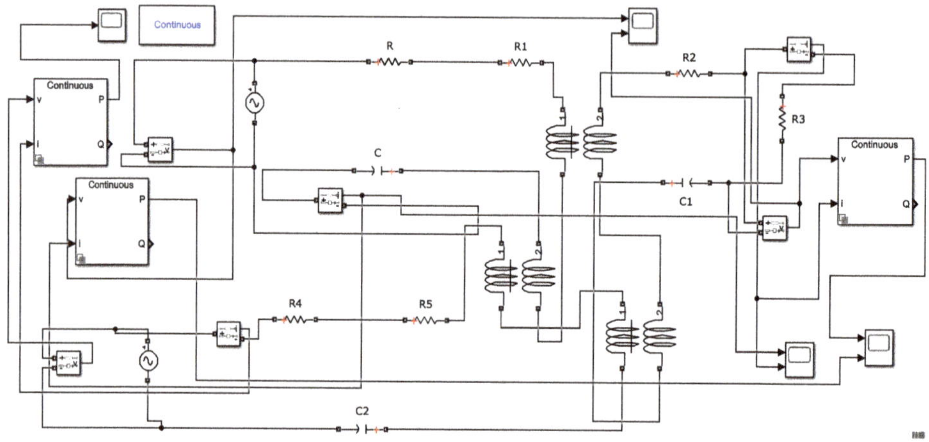


(**a**) One snapshot of the simulation file

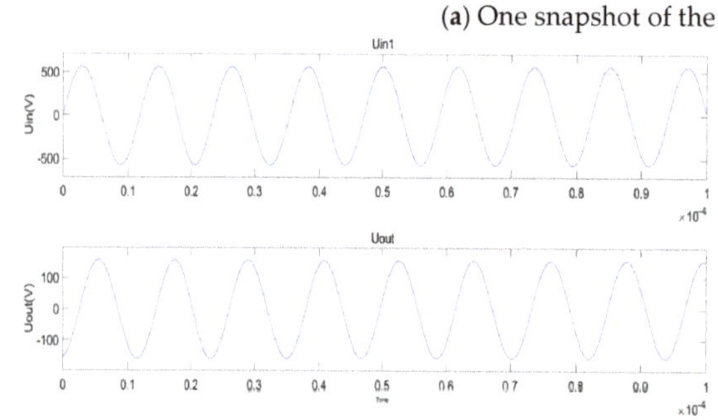


(**b**) Two to one system: the input voltage and the load voltage

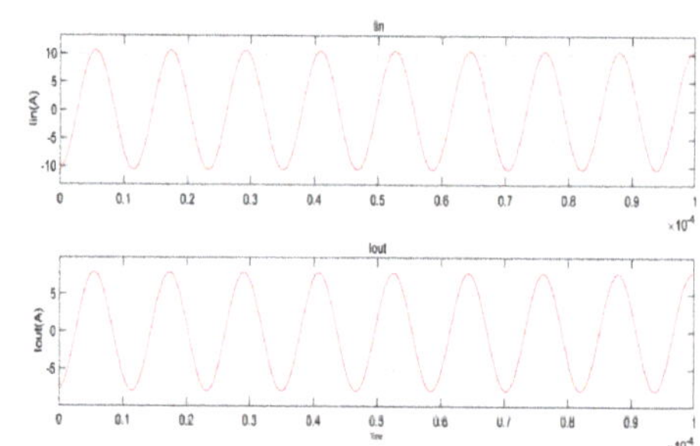


(**c**) The input current and the load current

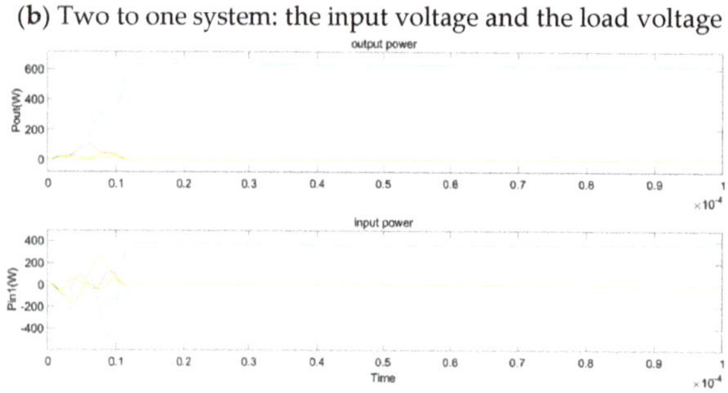


(**d**) The output power and the input power1

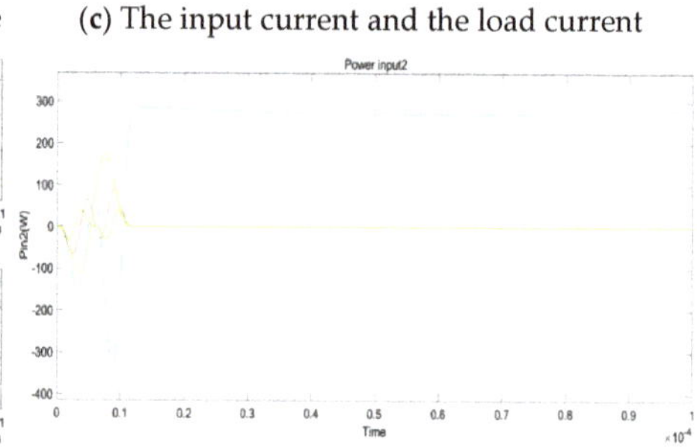


(**e**) The input power2

**Figure 6.** Waveforms of the transmitter currents and the load voltage. (**a**) Snapshot of Matlab simulation (**b**) Voltage (**c**) Current (**d**) Output power and input power 1 (**e**) input power 2.

By changing the value of the input voltage, an arbitrary power output can be achieved. Table 2 shows the simulated output power under different voltage peaks, which is close to the theoretically derived value.

In addition, different radial misalignment will cause a decrease in the mutual inductance between the receiving coil and the transmitting coil, resulting in a decrease in power. As shown in Figure 7, when the peak input voltages of the two active coils are both 570 V, the output power is negatively correlated with the radial misalignment.

Finally, when the radial misalignment is constant with the input voltage of one active coil, the simulation between the output power and the input voltage of the other active coil verifies the correctness of the control strategy.

The verification result is shown in Figure 8. At this time, the radial misalignment of the system is 5 cm, and the input voltage of an active coil is 570 V. The output power of the system increases as the input voltage of the active coil increases.

**Table 2.** Comparison of simulation and theoretical values of output power.

| | Value | Value | Value | Value | Value |
|---|---|---|---|---|---|
| Peak Amplitude of Input Voltage ($V_1 = V_2$) (V) | 300 | 400 | 500 | 570 | 600 |
| Simulation (W) | 180 | 310 | 500 | 640 | 710 |
| Theoretical (W) | 186 | 320 | 525 | 662 | 739 |

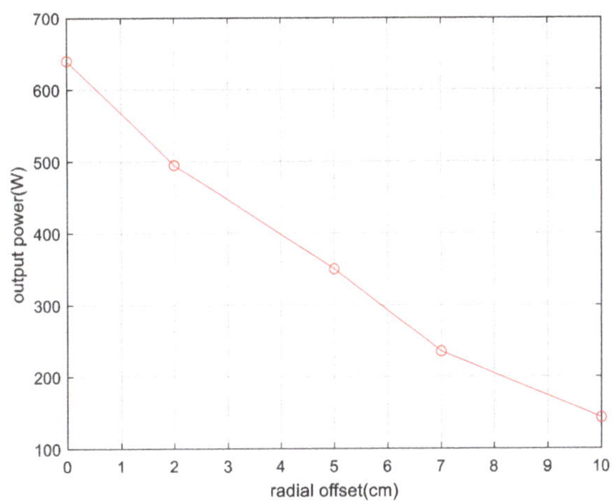


**Figure 7.** The output power and the radial misalignment.

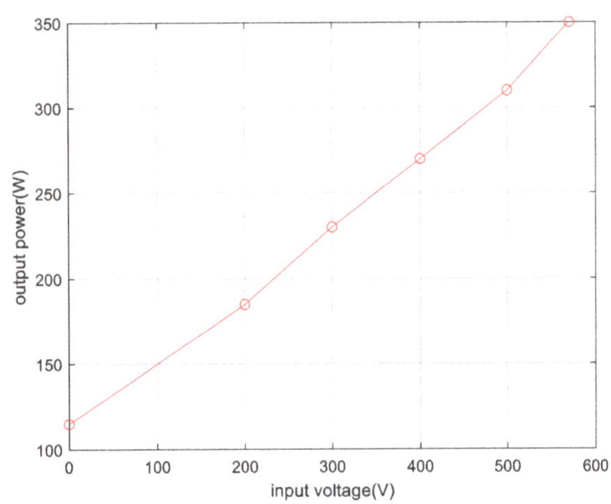


**Figure 8.** The output power and the input voltage.

## 4. Discussion

This paper presents a new wireless power transmission charging system for UAVs, and establishes two mathematical models to perfect the theoretical basis of the system. We also determine the control strategy between output power and input voltage when there is radial misalignment. The structure of a double transmitting coil and single receiving coil can better compensate for the mutual inductance and output power drop caused by radial misalignment. At the same time, the output power of the system can be changed by controlling the input voltage of the system to realize a power supply of any output power, which is simple and easy to operate. When the output power is constant, maximum power transfer efficiency is obtained by controlling the input voltage of the two coils. Numerical and simulation results have been presented in terms of electrical performance to demonstrate the validity of the proposed structure.

**Author Contributions:** Conceptualization, Z.H. and Z.L.; methodology, Z.H.; software, Z.H.; validation, M.X., Y.F. and R.W.; formal analysis, Z.H.; investigation, M.X.; resources, Z.L.; data curation, R.W.; writing—original draft preparation, Z.H.; writing—review and editing, Z.H.; visualization, R.W.; supervision, Z.L.; project administration, Z.H.; funding acquisition, Z.L. All authors have read and agreed to the published version of the manuscript.

**Funding:** This research received no external funding.

**Data Availability Statement:** The data presented in this study are available on request from the corresponding author.

**Acknowledgments:** Thanks are due to Wenjing Xiong and Zixi Liu, of Central South University, for their help in the testing.

**Conflicts of Interest:** The authors declare no conflict of interest.

## References

1. Lee, B.; Kwon, S.; Park, P.; Kim, K. Active power management system for an unmanned aerial vehicle powered by solar cells, a fuel cell, and batteries. *IEEE Trans. Aerosp. Electron. Syst.* **2014**, *50*, 2253–2268. [CrossRef]
2. Kurs, A.; Karalis, A.; Moffatt, R.; Joannopoulos, J.D.; Fisher, P.; Soljačic, M. Wireless Power Transfer via Strongly Coupled Magnetic Resonances. *Science* **2007**, *317*, 83–86. [CrossRef] [PubMed]
3. Lee, D.; Zhou, J.; Lin, W.T. Autonomous Battery Swapping System for Quadcopter. In Proceedings of the 2015 International Conference on Unmanned Aircraft Systems (ICUAS), Denver, CO, USA, 9–12 June 2015; pp. 118–124.
4. Tak, Y.; Park, J.; Nam, S. Mode-Based Analysis of Resonant Characteristics for Near-Field Coupled Small Antennas. *IEEE Antennas Wirel. Propag. Lett.* **2009**, *8*, 1238–1241.
5. Ishizaki, T.; Komori, T.; Ishida, T.; Awai, I. Comparative study of coil resonators for wireless power transfer system in terms of transfer loss. *IEICE Electron. Express* **2010**, *7*, 785–790. [CrossRef]
6. Lee, S.G.; Hoang, H.; Choi, Y.H.; Bien, F. Efficiency improvement for magnetic resonance based wireless power transfer with axial-misalignment. *Electron. Lett.* **2012**, *48*, 339–340. [CrossRef]
7. Lee, H.H.; Kang, S.H.; Jung, C.W. MR-WPT with reconfigurable resonator and ground for laptop application. *IEEE Microw. Wirel. Compon. Lett.* **2018**, *28*, 269–271. [CrossRef]
8. Zhengmin, Z.; Zhiyong, L. *Research on the Double Load Performance of Magnetic Coupled Resonant Wireless Power Transfer*; Central South University: Changsha, China, 2020.
9. Alshhawy, S.; Barakat, A.; Yoshitomi, K.; Pokharel, R.K. Separation-Misalignment Insensitive WPT System Using Two-Plane Printed Inductors. *IEEE Microw. Wirel. Compon. Lett.* **2019**, *29*, 683–686. [CrossRef]
10. Jeon, S.-J.; Seo, D.W. Maximum Output Power Improvement Using Negative Coil in Over-Coupled WPT System. *IEEE Microw. Wirel. Compon. Lett.* **2020**, *30*, 810–813. [CrossRef]
11. Kang, S.H.; Choi, J.H.; Harackiewicz, F.J.; Jung, C.W. Magnetic Resonant Three-Coil WPT System Between Off/In-Body for Remote Energy Harvest. *IEEE Microw. Wirel. Compon. Lett.* **2016**, *26*, 741–743. [CrossRef]
12. Jolani, F.; Yu, Y.; Chen, Z.D. Electromagnetic Modeling and Optimization of Magnetic Resonant Coupling Wireless Power Transfer Using Coil Array. In Proceedings of the IEEE MTT-S International Conference on Numerical Electromagnetic and Multiphysics Modeling and Optimization (NEMO), Ottawa, ON, Canada, 11–14 August 2015.
13. Campi, T.; Dionisi, F.; Cruciani, S.; De Santis, V.; Feliziani, M.; Maradei, F. Magnetic Field Levels in Drones Equipped with Wireless Power Transfer Technology. In Proceedings of the 2016 Asia-Pacific International Symposium on Electromagnetic Compatibility (APEMC), Shenzhen, China, 17–21 May 2016; pp. 544–547.
14. Junaid, A.B.; Konoiko, A.; Zweiri, Y.; Sahinkaya, M.N.; Seneviratne, L. Autonomous Wireless Self-Charging for Multi-Rotor Unmanned Aerial V ehicles. *Energies* **2017**, *10*, 803. [CrossRef]

15. Choi, C.H.; Jang, H.J.; Lim, S.G.; Lim, H.C.; Cho, S.H.; Gaponov, I. Automatic Wireless Drone Charging Station Creating Essential Environment for Continuous Drone Operation. In Proceedings of the 2016 International Conference on Control, Automation and Information Sciences (ICCAIS), Ansan, Korea, 26–29 October 2016; pp. 132–136.
16. He, X.; Bito, J.; Tentzeris, M.M. A Drone-Based Wireless Power Transfer and Communications Platform. In Proceedings of the 2017 IEEE Wireless Power T ransfer Conference (WPTC), Taipei, Taiwan, 10–12 May 2017; pp. 1–4.
17. Campi, T.; Cruciani, S.; Feliziani, M. Wireless Power Transfer Technology Applied to an Autonomous Electric UAV with a Small Secondary Coil. *Energies* **2018**, *11*, 352. [CrossRef]
18. Oodachi, N.; Ogawa, K.; Kudo, H.; Shoki, H.; Obayashi, S.; Morooka, T. Efficiency Improvement of Wireless Power Transfer Via Magnetic Resonance Using Transmission Coil Array. In Proceedings of the IEEE International Symposium on Antennas and Propagation (APSURSI), Spokane, WA, USA, 3–8 July 2011; pp. 1707–1710.

energies

MDPI

Review

# Multilevel Inverter: A Survey on Classical and Advanced Topologies, Control Schemes, Applications to Power System and Future Prospects

**Subhashree Choudhury [1], Mohit Bajaj [2], Taraprasanna Dash [3], Salah Kamel [4] and Francisco Jurado [5,*]**

[1] Department of EEE, Siksha 'O' Anusandhan (Deemed to be University), Bhubaneswar 751030, India; subhashreechoudhury@soa.ac.in
[2] National Institute of Technology, Delhi 110040, India; mohitbajaj@nitdelhi.ac.in
[3] Department of ECE, Siksha 'O' Anusandhan (Deemed to be University), Bhubaneswar 751030, India; taradash@soa.ac.in
[4] Department of Electrical Engineering, Faculty of Engineering, Aswan University, Aswan 81542, Egypt; skamel@aswu.edu.eg
[5] Department of Electrical Engineering, University of Jaén, EPS Linares, 23700 Jaén, Spain
* Correspondence: fjurado@ujaen.es

**Abstract:** In recent years, multilevel inverters (MLIs) have emerged to be the most empowered power transformation technology for numerous operations such as renewable energy resources (RERs), flexible AC transmission systems (FACTS), electric motor drives, etc. MLI has gained popularity in medium- to high-power operations because of numerous merits such as minimum harmonic contents, less dissipation of power from power electronic switches, and less electromagnetic interference (EMI) at the receiving end. The MLI possesses many essential advantages in comparison to a conventional two-level inverter, such as voltage profile enhancement, increased efficiency of the overall system, the capability of high-quality output generation with the reduced switching frequency, decreased total harmonic distortions (THD) without reducing the power of the inverter and use of very low ratings of the device. Although classical MLIs find their use in various vital key areas, newer MLI configurations have an expanding concern to the limited count of power electronic devices, gate drivers, and isolated DC sources. In this review article, an attempt has been made to focus on various aspects of MLIs such as different configurations, modulation techniques, the concept of new reduced switch count MLI topologies, applications regarding interface with renewable energy, motor drives, and FACTS controller. Further, deep insights for future prospective towards hassle-free addition of MLI technology towards more enhanced application for various fields of the power system have also been discussed. This article is believed to be extremely helpful for academics, researchers, and industrialists working in the direction of MLI technology.

**Keywords:** flexible AC transmission systems (FACTS); multilevel inverter (MLI); renewable energy resource (RER); selective harmonic elimination (SHE); space vector control (SVC); total harmonic distortions (THD)

**Citation:** Choudhury, S.; Bajaj, M.; Dash, T.; Kamel, S.; Jurado, F. Multilevel Inverter: A Survey on Classical and Advanced Topologies, Control Schemes, Applications to Power System and Future Prospects. *Energies* **2021**, *14*, 5773. https://doi.org/10.3390/en14185773

Academic Editor: Ching-Ming Lai

Received: 19 July 2021
Accepted: 7 September 2021
Published: 13 September 2021



## 1. Introduction

Recently, the MLI has been regarded as state of the art technology for power conversion from DC to AC in the fields of generation, transmission, distribution, and employment of electrical energy. They have been innovated as an competent solution for wide range of power applications such as RER [1–6], FACTS [7–10], high-voltage direct current (HVDC) [11–14], static compensators (STATCOMs) [15–18], unified power flow controllers (UPFCs) [19–21], dynamic voltage restorers (DVRs) [22,23], active filters (AFs) [24,25], motor drives used for traction/transportation [26–33], marine propulsion [34–38], conveyors [39,40], mine hoists [41], magnetic resonance imaging system (MRI) [42] and induction heating power supply [43]. The conventional two-level inverters fail to operate in the medium voltage range due to the semiconductor's blockage voltage limitation. However, MLIs play a key role in medium-voltage and high-power operations.

Further, for the same power ratings, MLIs have merit over a two-level inverter in terms of reduced harmonic contents of line-to-line voltage that are fed to load with respect to its level of switching frequencies [44]. The primary reasons behind the MLI serving as a vital revolution in the era of industrialization for potent performance are as follows: (i) compatible in design; (ii) capability of operating at higher current and voltage due to its modular structure; (iii) lesser voltage derivatives on power electronic switches as the voltage stress gets divided across the switches at various levels; (iv) easy interface with RER and motor drives where the load sharing is brought about by DC link; (v) enhanced power quality performance with low harmonic content and better electromagnetic compatibility; (vi) produces lower common-mode voltage; (vii) ability of transformer-less operation; (viii) enhanced efficiency level due to switching at fundamental frequency thus decreasing losses due to conduction and switching; (ix) draws input current with fewer distortions and (x) uses various control approaches and reduced switching states for achieving fault-tolerant operation. Table 1 highlights the primary differences between a traditional two-level inverter and the MLI.

**Table 1.** Comparison between traditional two-level inverter and MLI.

| Properties | Two-Level Inverter | MLI |
|---|---|---|
| Structure | Complicated | Modular |
| Operation at high voltage and current | Can operate (for parallelized structures) | Can operate |
| stress on power electronic switches | More | Less |
| Application | Low voltage | High voltage |
| Power quality performance | Low | High |
| Harmonic content | Low | High |
| Electromagnetic interference (EMI) Immunity/susceptibility | Less | More |
| Production of common-mode voltage | Higher | Lower |
| Ability of transformer-less operation | No | Yes |
| Efficiency | Low | High |
| Switching losses | High | Low |
| Operation at the fundamental frequency | Fails | Can operate |
| Input current distortions | High | Low |
| Fault tolerant operation | Impossible | Possible |
| Rate of change of voltage | High | Low |
| Ability to operate at low/high/ fundamental frequency | More | Less |
| Production of multiple voltage level | Not possible | Possible |
| Electromagnetic interference (EMI) generation | High | Low |

The basic principle behind MLI is that it consists of one or more DC sources and an array of low-rated power semiconductor switches for generating an output voltage with a stepped voltage waveform to achieve higher power levels [45,46]. The main objective of MLI is to synthesize an approximately sine wave of voltage with various steps by using the appropriate switching signal of the power electronic switches with the help of different direct current voltage sources such as batteries, supercapacitors, fuel cells, solar panels, etc. [47]. The number of levels of the output waveform can be increased for attaining pure sinusoidal voltage without the use of heavy transformers and passive filters [47,48]. The most traditional MLI topologies, according to their structure, are classified into three types (i) neutral point clamped (NPC); (ii) flying capacitor (FC), and (iii) cascaded H-bridge (CHB) [49]. The three-level NPC type of MLI is widely used in the application of motor drive. The FC type of MLI uses balancing capacitors on phase buses to generate multilevel output voltage waves clamped by capacitors instead of diodes. The CHB is extensively used as it's more flexible due to its modular configuration and uses the least number of semiconductor switches for a particular level of operation. It generally comprises an array of H-bridge cells for synthesizing a required voltage from numerous isolated DC sources [50]. However, the classical MLIs undergo a crucial shortcoming: the increase in the count of driving circuits and devices in proportion to the number of levels makes the control circuit very complicated. The increased number of device components, in turn, reduces the overall stability of the system. Consequently, many authors have investigated new configurations intending to optimize the utilization of components and enhance the output voltage waveform [51].

Amongst the different topologies developed are the cascaded H-bridge, hybrid series and parallel sources (HSPSs), criss-cross, packed-U cells, and cascaded-bipolar switched cells have been reported by many researchers and have acquired worldwide consideration. Authors in [49] have researched the classical Voltage Source Inverter (VSI) development for the MLI application. Consequently, the research has been expanded to investigate the performance of numerous control approaches to improve the operation of MLIs and optimize their performance with various topologies. The modulation methods in MLIs have been endorsed by their range of switching frequencies. Nevertheless, the high-frequency modulation methods are restricted in the medium voltage range because the losses incurred during switching reduce the inverter component values and the system's overall efficiency.

The SPWM approach is implemented in industrial applications to lower the harmonic content on the output voltage waveform. Space vector modulation (SVM) technique possesses remarkable performance in 3-level MLI topologies. Other techniques involving modulation methods at a low switching frequency that have attained more demand in a broader field of function are Staircase modulation, space vector control (SVC), and selective harmonic elimination (SHE) [39]. A detailed study of these techniques has been discussed further in this article. Figure 1 illustrates the summary of the review methodology adopted in the present research work.

The significant contributions of this research paper include:

(1) More than 260 recent research articles have been critically reviewed, and a detailed literature summary about the evolution, classification of the MLI topologies, various modulation techniques, and application have been presented.
(2) A detailed discussion concerning the various types of conventional MLI techniques has been carried out.
(3) An extensive valuation of a wide range of possible new reduced switch count MLI topologies has been explored and presented in-depth.
(4) A clear idea of the different modulation techniques required for MLI has been apprehended.
(5) Further, this research paper also highlights a thorough knowledge of MLI novel applications in various fields of power system networks such as renewable energy interface, FACTS controller, and motor drives based on a comprehensive research survey.
(6) Finally, this article also elaborately discusses the issues faced by the present MLI technology and suggests some more in-depth vital points to be adopted in the future for further better application in the electric power grid network.
(7) This comprehensive article is alleged to serve as a good guidance for enhancing the understanding of readers, academics, and industrialists for researching in the area of MLI concerning the proper choice of topology for definite application, accurate selection of parameters, switching, control schemes, and application in other power system fields.

The entire research article is structured into seven major sections. Section 2 addresses the development of MLI Topologies. A thorough classification of MLI topologies in terms of classical versus new reduced switch count topology and single versus multiple DC sources is projected in Section 3. Section 4 represents the detailed categorization of various modulation schemes depending upon the switching frequency. The novel applications of MLI after a rigorous survey of literature on MLI are highlighted in Section 5. The key points to be dealt with seriously for further enhancement in MLI technology in the future have been meticulously discussed in the Section 6. Finally, the conclusion from the entire study is presented in Section 7.

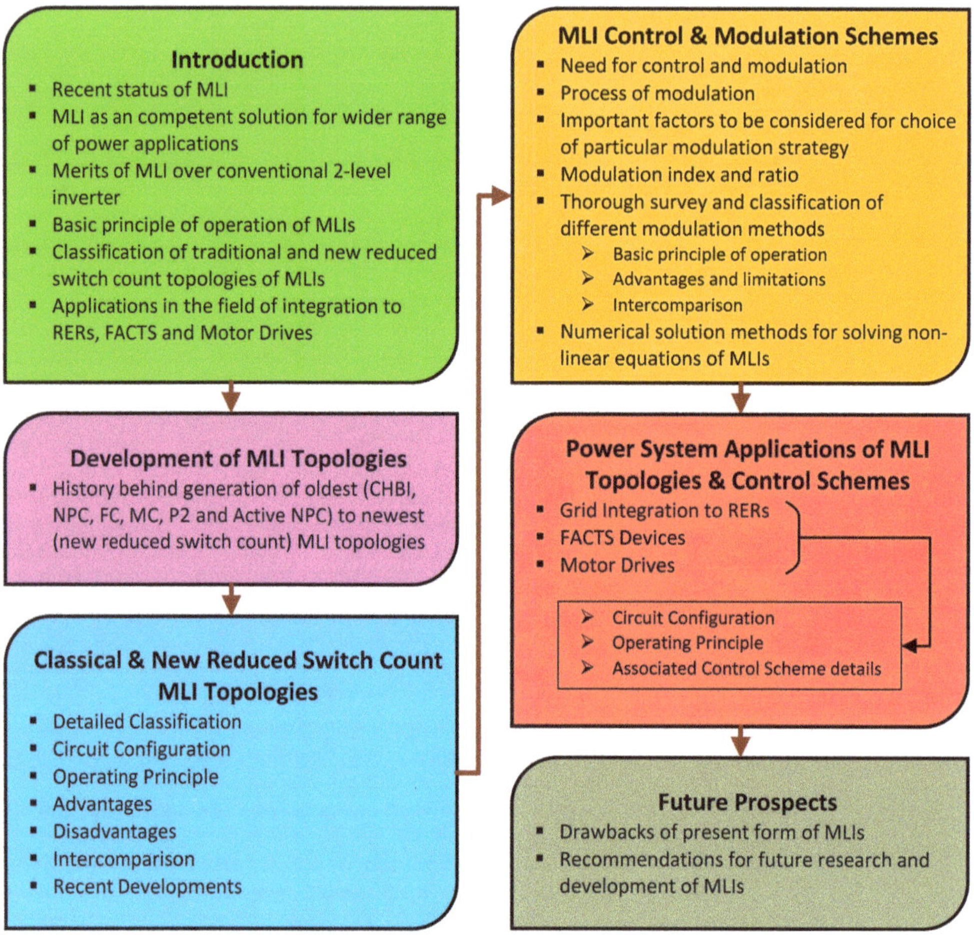


**Figure 1.** A summary of the review methodology adopted in the present research work.

## 2. Development of MLI Topologies

An outlook on the development of various topologies of MLI is depicted in Figure 2. A thorough study of this evolution has been discussed in this section. CHB was first developed by Baker and Bannister [52] in the 1970s, which possessed the capability of generating multilevel voltage using different DC source voltage [53–55]. The next advancement in MLI topology was the NPC inverter, first evolved by Nabae et al. [56] in the 1980s [57–59]. In the 1990s, Meynard and Foch [60] and Lavieville et al. [61] developed the FC type of MLI. These three classical types of MLI-based topologies were regarded as the foundation of most MLI inverters presented in the modern era. The next invention was the modular multilevel converter (MMC), primarily applied in industrialization [62,63]. In 2000 [64], a generalized MLI named P2 was introduced. Authors in [65] have proposed active neutral point clamped (ANPC) topology. Many other new MLI configurations with application-based approaches have been proposed and conferred in the recent past. Authors have also focused on developing a new reduced switch count topology of MLI [66,67]. The new topology of

MLI can be categorized as (i) symmetric type based on the use of equal DC sources [68–80]; (ii) asymmetric type based on the use of unequal DC sources [69,70,81–85]; (iii) having inherent negative level [68–70,81–85]; (iv) without inherent negative level [72–80]; (v) use of capacitors links with single sources [81–86]; (vi) regenerative configuration for operation as both an inverter and a rectifier [73–75] and (vii) hybrid approach of NPC, FC and CHB [84–87].

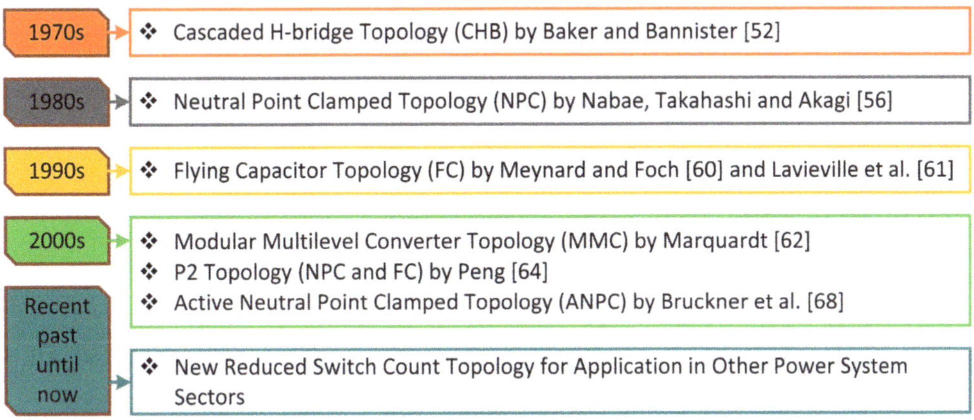


**Figure 2.** An outlook on the development of various MLI topologies.

## 3. Classical and New Reduced Switch Count MLI Topologies

The concept of MLIs was first introduced in 1975 [52] and then pursued by combative work to develop numerous topologies based on modifying the configuration regarding arrangements of power electronic semiconductor switches and DC sources. Research was also conducted to achieve greater power by the series combination of switches using DC sources with a small voltage range to convert power and synthesize voltage. The most frequently used voltage sources which could be configured in one (single) and more (multiple) units are RER, batteries, and capacitors. The broad categorization of MLI on several DC sources used and their structure is given in Figure 3. In addition, a comprehensive survey of the literature yields various classifications of MLI based on different strategies as suggested by many researchers.

Nevertheless, this research article is an attempt to broadly classify all possible types of MLIs depending upon the structure of DC source (either single or multiple units) used. The multiple uses of capacitors can produce multiple voltage levels. Table 2 summarizes the significant advantages and disadvantages of various MLI configurations.

### *3.1. Single DC Source*

This type of topology is extensively endorsed in the industrial field, considering its simple structure, high efficiency, and power rendering [88]. The different categories of MLIs using a single DC source are discussed below [89].

#### 3.1.1. Neutral Point Clamped Multilevel Inverter (NPC-MLI)

Nabae, Takashi, and Akagi in 1981 proposed the diode clamped multilevel inverter (DC-MLI), also termed NPC-MLI [56]. These inverters have been broadly adopted on account of their immense proficiency in high-power and medium-voltage operations with comparatively high efficiency. It is a three-level structure with two diodes that clamp the switching voltage to half the supply voltage magnitude. Further, it guarantees equal sharing of the supply voltage among the two halves of the switches being held at these points consisting of a neutral point between them. The middle voltage level is termed the neutral point. This topology employs multiple capacitor banks in series for providing multiple DC voltage levels, as illustrated in Figure 4.

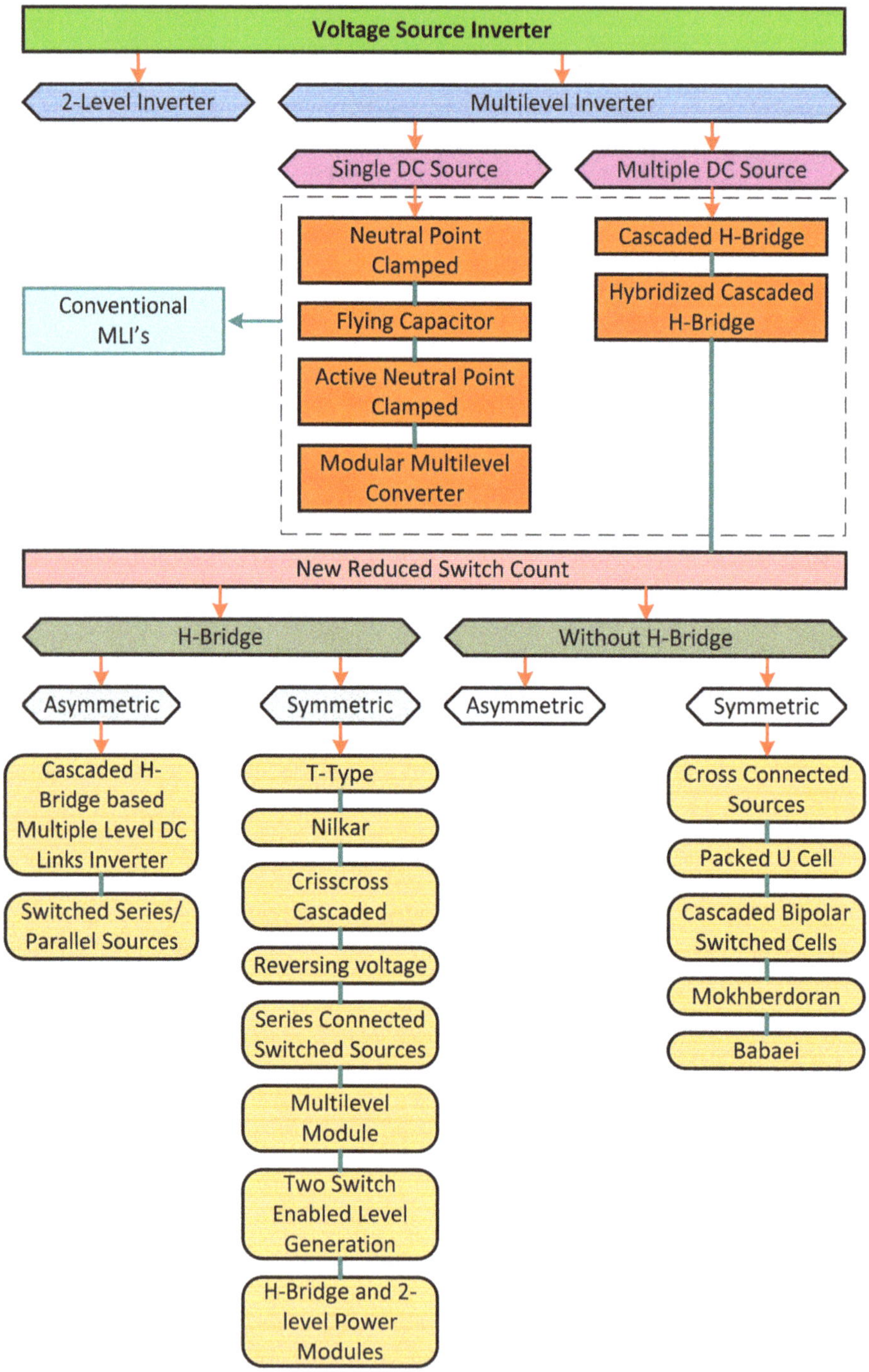


**Figure 3.** Overall classification of MLI.

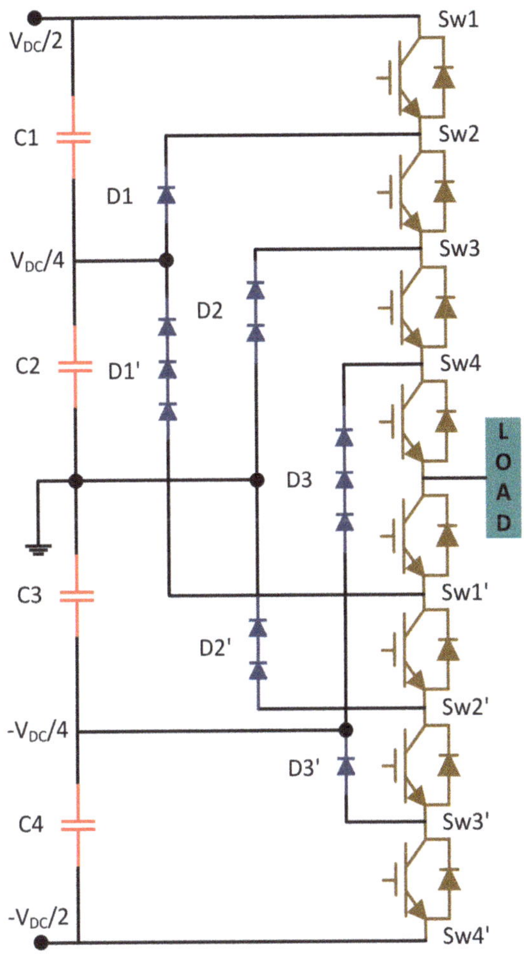


**Figure 4.** Topology of neutral point clamped multilevel inverter.

The major benefits of this topology are: (i) the capacity demand of the converter is minimized as all of the phases share a common DC bus; (ii) enhanced capability of reducing the THD; (iii) the capability of the capacitors to be pre-charged as a group; (iv) provides higher efficiency and (v) the number of switches, clamping diodes and capacitors can be enhanced to obtain higher voltage levels. However, this topology faces some demerits such as: (i) implementation cost can be increased as additional reactors are needed for mitigation of elevated voltage levels and THD of current as this topology uses fundamental frequency for switching; (ii) although voltage level can be increased by enhancing the number of capacitors and clamping diodes it increases the complexity of the overall configuration; (iii) due to deviation in the switching characteristics, the voltage at the neutral point fluctuates which creates difficulty in the static and dynamic sharing of the voltage across the switching units and (iv) due to imbalance in capacitor voltage and over-voltage issues, the actual application of this topology are confined to three-level only. NPC-MLI finds its application in the following ways: static VAR compensation (SVC), variable-speed motor drives, high voltage system interconnections, and be assimilated with high voltage AC/DC transmission system [90–95].

3.1.2. Flying Capacitor Multilevel Inverter (FC-MLI)

In 1992, Meynard and Foch [60] and Lavieville et al. [61] proposed the FC-MLI topology to mitigate the issue of static and dynamic sharing of the voltage across semiconductor switches as built in the NPC-MLI topology. The main architecture of FC-MLI is identical to NPC-MLI. Nevertheless, here the clamping diodes are replaced by capacitors, as illustrated in Figure 5. This topology is termed FC. In FC, capacitors that replaced the diodes of the DC-MLI are autonomous (flying) as compared to the other capacitors present in the total configuration. The voltage on every capacitor varies from that of the adjacent capacitor. This MLI topology possesses numerous advantages as compared to NPC-MLI, such as: (i) improved stability of the FC is attained by the redundancy of switching within the phase; (ii) capability in controlling both the active and reactive power; (iii) it has transformerless working and (iv) more flexible in synthesizing voltage by using capacitors instead of clamping diodes.

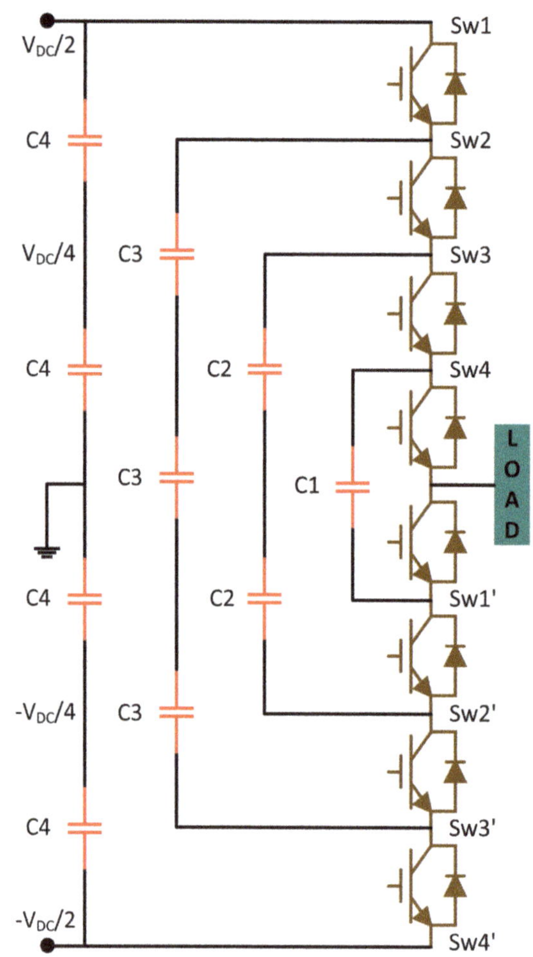


**Figure 5.** Topology of flying capacitor multilevel inverter.

Although this topology marks the issues of the NPC, it also has some disadvantages such as: (i) the proper charging and discharging control of capacitors is restricted due to an increase in the voltage levels; (ii) the increment in the number of the capacitor leads to increased cost, reduced lifetime and system becomes bulky; (iii) increase in the number of levels leads to increase the number of capacitors thus limiting its operation to maximum 3–5 level; (iv) complexity in start-up and same level pre-charging of all the capacitors; (v) high switching losses due to operation at high frequency and (vi) large numbers of levels creates packaging issues. The major attractive applications of FC-MLI includes [39,60,94]: (i) static var generation; (ii) AC motor drives; (iii) active filter operations; (iv) switched converters; (v) sinusoidal current rectifier and (vi) converters with THD-reducing capacities.

### 3.1.3. Active Neutral Point Clamped Multilevel Inverter (ANPC-MLI)

Bruckner et al. [68,96] have proposed the ANPC-MLI topology. This inverter helps to overcome the insufficient and uneven losses which are shared between the outer and inner switches. It was possible by placing power switches rather than normal diodes [97]. Figure 6 demonstrates the nine-level ANPC inverter with the combination of NPC and FC-based MLI topologies [97]. In this configuration, the number of the two-level inverter can be found out by $(n - 1)/2$, where 'n' is the number of output levels of the inverter. So, four number of two-level inverters were cascaded to obtain a nine-level inverter. In Figure 6, switches 'Sw1' to 'Sw8' and capacitors 'C1' to 'C3' belong to the first part, whereas the switches 'Sw17' to 'Sw24' and capacitors 'C7' to 'C9' belong to the second part of the MLI. Switches 'Sw9' to 'Sw16' and capacitors 'C4' to 'C6' belong to part three, used for connecting the inverter to the load.

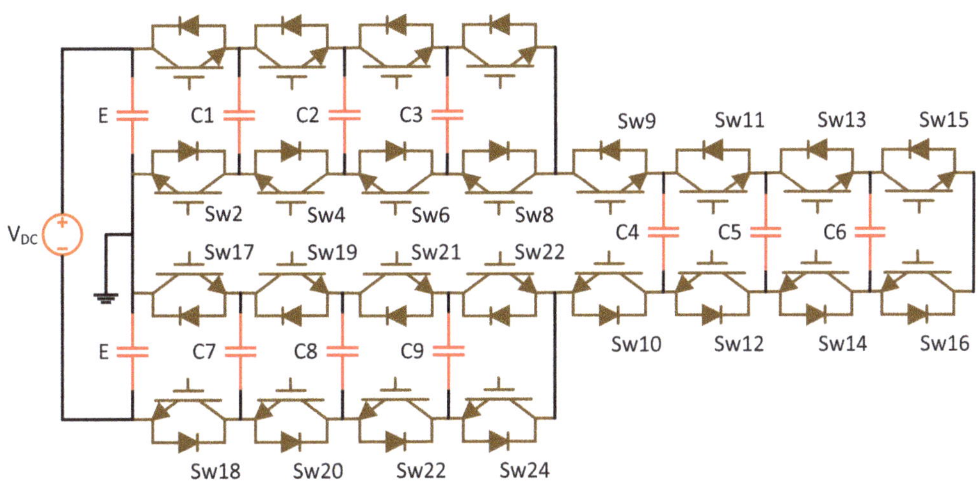


**Figure 6.** Topology of active neutral point clamped multilevel inverter.

### 3.1.4. Modular Multilevel Converter (MMC)

The MMC was first proposed in 2001 by Lesnicar and Marquardt in a German patent [62], and later in 2002, it was employed for a wide-scale power range [98]. MMC offers numerous merits in comparison to other available topologies such as [99,100]: (1) large range of voltage operation by cascading cells; (2) independent PQ control; (3) negligible losses; (4) high modularity; (5) low switching frequency; (6) output almost sinusoidal so does not need AC filters; (7) mechanical structure is simple; (8) voltage and current quality generated is high and (9) high reliability, availability, and efficiency. Therefore, they are often preferred for medium to high voltage applications because of their high-quality output and modular structure.

Over recent years, MMC has been successfully operated as an efficient power converter in numerous power system applications such as HVDC system [101–104], FACTS devices [105–107], energy storage devices [108–110], electric vehicles [111,112], motor drives [113–115], active power filters [116,117] and renewable energy [118–121]. Presently, the primary vital concerns of MMC include capacitor voltage balancing (CVB) and circulating current suppression (CCS). Authors in [122] have reported an efficient CVB method for MMC using the carrier-based phase shift pulse width modulation method for designing a flexible mission profile emulator for the test of MMC under various working conditions. In [123], a step up non-isolated DC-DC MMC having self-voltage balancing and soft-switching has been discussed. Researchers in [124] and [125] have proposed a sensorless switch clamp MMC for voltage balancing and inter-cluster voltage balancing control for MMC during unbalanced grid voltage. CVB control strategy for MMC has been reported in [126,127]. A novel voltage balancing control with dv/dt reduction for 10-kV SiC MOSFET-based medium voltage MMC has been proposed in [128]. Authors have also carried out hardware implementation based on the peak current mode switching cycle control for CVB of MMCs [129]. Numerous techniques have been designed and implemented for the voltage balancing of MMCs, as reported in the literature [130–134]. The CCS in MMC has been carried out efficiently through various techniques as proposed by several researchers [135–137]. In [138], authors have analyzed effective ways of CVB and CCS for a three-phase four-wire split capacitor DSTATCOM. Authors have applied various methods such as fuzzy logic controller [139,140], predictive control method [141], dead beat control [142], sliding mode approach [143], and frequency adaptive spatial repetitive [144] for suppressing the circulating current harmonics in an MMC. Many other methods have been studied and incorporated for efficient CCS of MMC [145–149].

The basic circuit of the three-phase MMC is illustrated in Figure 7, which consists of a DC voltage source, three phases (legs) with two arms per leg (upper and lower arm), each arm consisting of series connections of several (N) sub modules (SM) producing a multilevel voltage signal at its output terminal and a series inductor for smoothening and filtering the circuits. Different types of SM topologies are reported in the literature [99,100]. Recently, authors have proposed MMC basing on interleaved half-bridge submodules [150]. Some of the basic configurations are illustrated in Figure 8. Half-bridge SM, full-bridge SM, single clamped, and double clamped and shown in Figure 8A, Figure 8B, Figure 8C, and Figure 8D, respectively.

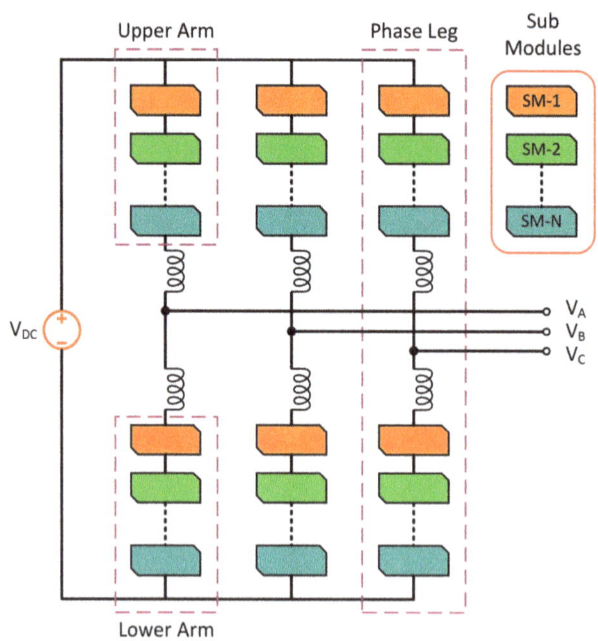


**Figure 7.** Topology of modular multilevel inverter.

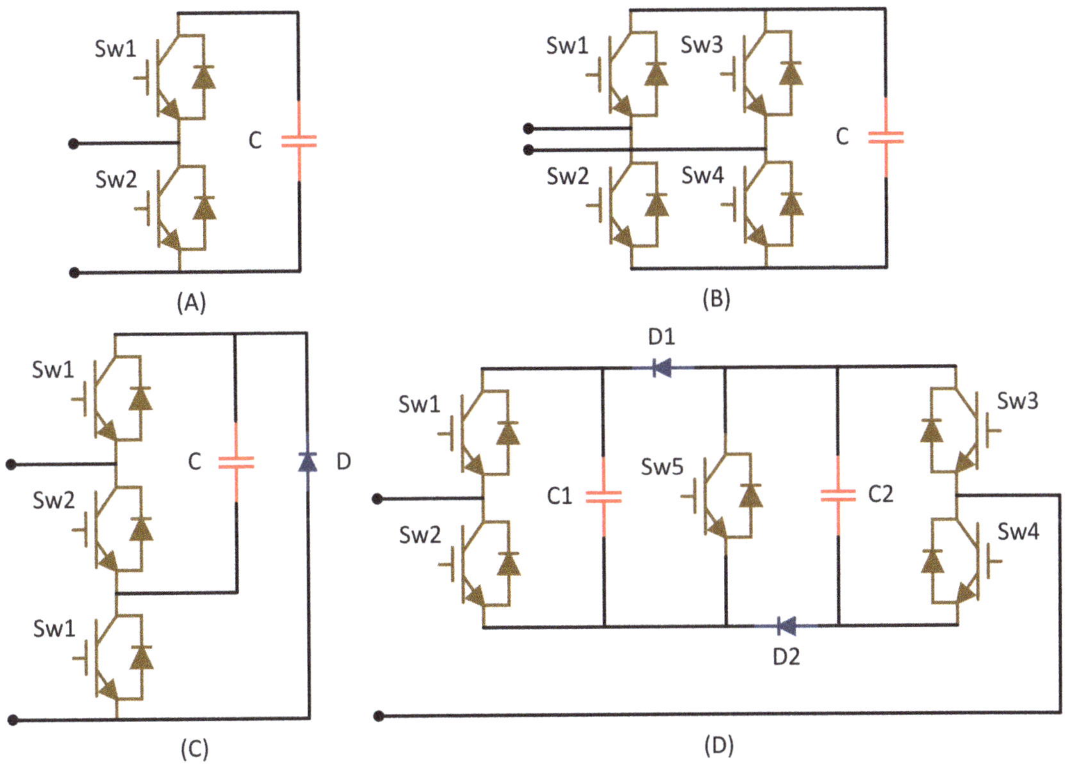


**Figure 8.** Topologies of sub modules (**A**) half-bridge, (**B**) full-bridge, (**C**) single clamped, and (**D**) double clamped.

### 3.2. Multiple DC Source

Authors in [33,90,93] have proposed MLI topologies with multiple sources (DC) as the use of a single source (DC) is limited for achieving greater voltage levels. In this section, various topologies of MLI with multiple DC sources are discussed in detail.

#### 3.2.1. Basic Multiple DC Source Topology

Cascaded H-bridge Multilevel Inverter (CHB-MLI)

Baker and Bannister [52] proposed the first patent on this topology which was considered to be a viable substitute to previously described topologies as it requires a significantly fewer number of power devices. This topology was termed CHB-MLI, which constitutes the series connection of H-bridges with separate DC sources. Numerous series-connected H-bridge structures generate the multilevel stepped waveform. By cascading the general H-bridge cell, the resultant CHB-MLI can form an unlimited number of levels theoretically. This property of CHB-MLI allows modulation. The advancement of the technique for eminence follows the trade-off for high power medium voltage operations, which have now reached the level of megawatt by industries. It is an effective solution to voltage imbalance found in the NPC and FC configurations due to its modular structure. CHB generally comprises power conversion cells, each of which is supplied by an isolated DC source on the DC side, obtained from batteries, ultracapacitors, fuel cells, and series connected on the AC side [151]. A schematic diagram of the CHB topology has been shown in Figure 9.

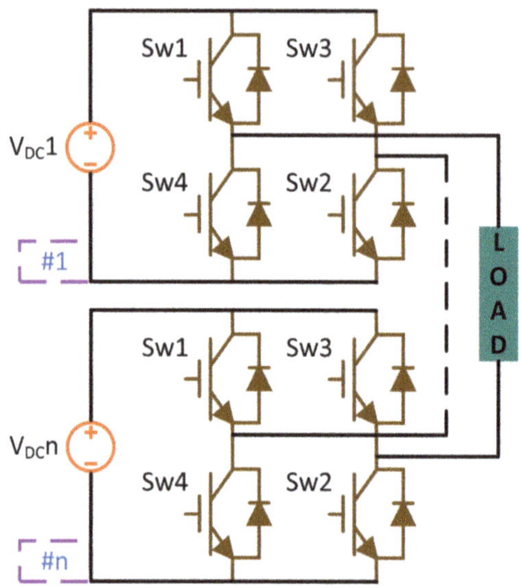


**Figure 9.** Topology of cascaded H-bridge multilevel inverter.

The advantages of CHB-MLI topology are as follows: (i) easy modulation, control, protection, and maintenance during failure due to compatible structure; (ii) capable of handling higher voltages and absence of voltage imbalance; (iii) ability to eliminate common-mode voltages by proper selection of modulation scheme; (iv) generates almost sinusoidal output and hence almost requires no output filter; (v) less component requirement being equated to other topologies; (vi) no requirement of flying capacitors or clamping diodes and (vii) uniform distribution of load power amidst all switching devices [152]. However, despite several merits, this topology faces some serious drawbacks such as: (i) requirement of numerous separate DC sources and ii) need of many DC link voltage controllers. Various fields of application of this topology include RER interface, motor drives, electric vehicle drives, laminators, blowers, fans, conveyors, DC power source utilization, frequency link systems, and power factor compensators [39,56,153–155].

Hybridized Cascaded H-bridge Multilevel Inverter (HCHB-MLI)

HCHB-MLI was first introduced by Odeh and Nnadi [156], as illustrated in Figure 10. HCHB-MLI signifies that the inverter is employed with: (i) various semiconductor device technology; (ii) various amplitude and characteristics of DC sources; and (iii) combination of many modulation strategies [157]. Figure 10 depicts a nine-level HCHB-MLI, which consists of two DC sources. Two five-level hybrid inverters are interconnected for providing nine-level of voltages per cycle. This topology is quite identical to the earlier discussed CHB-MLI topology. However, the significant difference between HCHB and CHB topology are: (i) in HCHB, each H-bridge cell is added with an auxiliary switch for the harmonic profile improvement of output waveforms; (ii) the number of devices required in this HCHB topology is comparatively significantly less in comparison to the CHB topology for the same level of output voltage waveform [158] and (iii) with the operational and switching activities, HCHB topology possesses double RMS output voltage, voltage steps quantity and reduced number of DC sources. However, the major demerit is that it cannot be employed for high voltage applications.

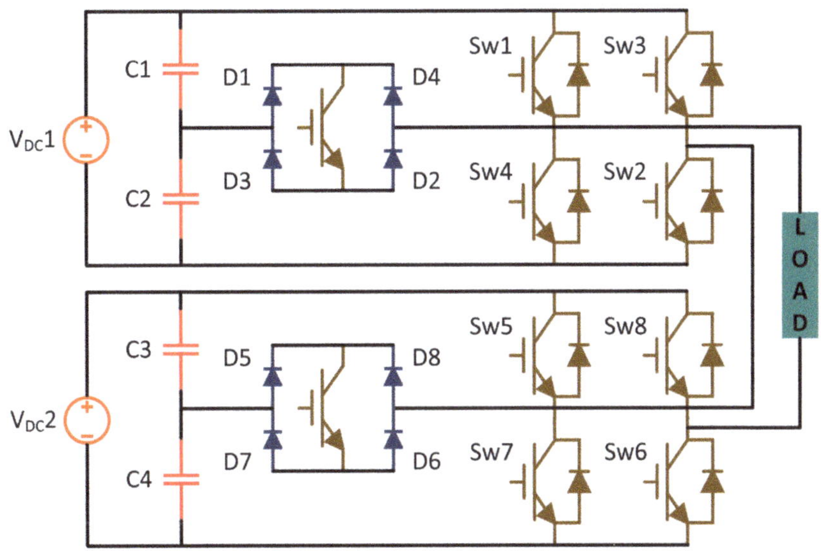


**Figure 10.** Topology of hybridized cascaded H-bridge multilevel inverter.

### 3.2.2. New Reduced Switch Count Topology

Topologies with H-bridge broadly consist of asymmetric and symmetric types.

H-bridge Topologies

(i) Asymmetric H-bridge Topology

Asymmetric MLI is usually CHB-MLI type in which the value of the voltage of any one DC source varies dynamically [159,160]. The main merits include: (i) generation of output voltage waveform with minimum THDs [39,161,162]; (ii) requires reduced number of semiconductor devices in comparison to symmetric topology [163]; (iii) needs only 12 switching units to attain seven, nine, fifteen and twenty-one levels [164]; (iv) operates with reduced dimension and cost of the inverter [165] and (v) enhanced reliability due to operation with fewer semiconductor switches and capacitors [166]. Asymmetric topologies with H-bridge have been further classified into two types, and the details are highlighted below [167].

A. Cascaded H-bridge based Multiple Level DC Links Inverter

Gui Jia Su et al. [72] proposed this topology which comprises the CHB with multiple level DC links (MLDCL). Figure 11 represents an MLDCL inverter consisting of 2 input DC sources. The circuit comprises of 'n' number of half-bridge units in cascade, and each unit has one DC source with two switches connected in series. A stepped DC voltage waveform is produced with the help of these cascaded units, which are also known as level-generation parts. A full multilevel AC waveform is generated using the H-bridge by changing the polarity of the output voltage. MLDCL topology utilizes only fewer semiconductor switches to generate the same output voltage level [39]. Major applications of this configuration include: (i) permanent magnet (PM) motor drives with low range of power (<100 kW); (ii) metal oxide semiconductor field effect transistors (MOSFETs); (iii) insulated gate bipolar transistors (IGBTs) and (iv) solar and fuel cell integration [168,169].

**Table 2.** A comprehensive summary of merits and demerits of various MLI configurations.

| Configuration | Merits | Demerits |
| --- | --- | --- |
| NPC [56,90–95] | ➢ Modular in design<br>➢ Better dynamic response<br>➢ Cost effective (only for three-level structure)<br>➢ Less number of DC sources needed for operation<br>➢ Good fault-tolerant characteristics<br>➢ Absence of floating capacitors<br>➢ Back-to-back inverters are employed<br>➢ Maximum efficiency when operated at fundamental frequency<br>➢ Plays a vital role for application in industries | ➢ DC level may discharge due to improper control and monitor<br>➢ An increase in voltage level results in an increase in the number of clamping diodes<br>➢ Uneven distribution of losses across inner and outer switches<br>➢ Costlier and less reliable due to an increase in number of diodes |
| FC [39,60,61, 94] | ➢ Can be employed for fault-tolerant operation<br>➢ Uses less number of DC sources<br>➢ Cost effective if used for high-level structures<br>➢ Does not require clamping diodes<br>➢ Ability to lessen voltage derivation stress on the power electronic components | ➢ Capacitor requirements are greated, which is more expensive than diodes<br>➢ Voltage balancing unit is complex<br>➢ Switching efficiency is minimum<br>➢ Complication in overall system control<br>➢ Complex structure<br>➢ Less reliable<br>➢ Use of more voltage sensors<br>➢ Gives rise to ripples at low switching frequencies |
| ANPC [68,96,97] | ➢ Losses across each switch are uniform<br>➢ Dynamic response is good<br>➢ Simple in structure<br>➢ Cost effective, compact in size, and requires no floating capacitor for a three-level structure<br>➢ Uses power switches than normal diodes | ➢ For levels greater than 3, more numbers of floating capacitors are required<br>➢ Control circuit is complex in comparison to NPC<br>➢ When operated at higher voltage levels, power switches of various voltage ratings are needed |
| MMC [62,98–121] | ➢ Possesses voltage scalability by cascading identical cells<br>➢ Capable of independent PQ control<br>➢ Losses are negligible<br>➢ Level of modularity at the cell level is maximum<br>➢ Switching frequency is low<br>➢ Does not require AC filters as output is purely sinusoidal<br>➢ Simple in structure<br>➢ Quality of voltages and currents is high<br>➢ High availability, reliability, and efficiency | ➢ Need of more power electronic interfaces<br>➢ Uses more number of capacitors and cells<br>➢ Control unit for circulating current and voltage balancing of cell is needed |

**Table 2.** *Cont.*

| Configuration | Merits | Demerits |
|---|---|---|
| CHB [39,52,56, 151–155] | ➢ Structure is simple and modular<br>➢ Ease of extending to higher levels<br>➢ Simplicity in storing and packaging<br>➢ Voltage derivative stress is lowered due to the production of common-mode voltage<br>➢ Reliability is higher<br>➢ Control is simple<br>➢ Absence of floating capacitors<br>➢ Minimum harmonics in the input current<br>➢ Ability to operate both at fundamental and switching frequencies<br>➢ Output signal has less distortion without any filter unit<br>➢ Best suited for applications of fault tolerance<br>➢ Needs only unidirectional switches<br>➢ It can employ asymmetric source topology<br>➢ Can operate as single DC source unit | ➢ Requirement of more power switches<br>➢ The level of output voltage is comparatively less<br>➢ Cost of implementation is large<br>➢ Voltage rating across switches varies for asymmetric configuration<br>➢ Requires a greater number of DC sources<br>➢ Requires more complex controller unit |
| HCHB [156–158] | ➢ Presence of an auxiliary switch for harmonic profile improvement of output waveforms<br>➢ Requires a smaller number of switching units and devices<br>➢ Utilizes reduced number of DC sources | ➢ Cannot be employed with high voltage application<br>➢ Costly for implementation |
| MLDLC [39,72,168, 169] | ➢ Uses less number of semiconductor switches<br>➢ Circuit layout is optimized<br>➢ No requirement of extra clamping diodes or capacitors | ➢ Costly as it involves capacitors for storage<br>➢ With the increase in number of DC sources, output level increases<br>➢ Power rating of device increases due to reduced device count which leads to damage of the device by causing its operating temperature to exceed safe levels as well increases its overall cost |
| SSPS [168,170–172] | ➢ Ability to generate more levels of output voltage using very few numbers of switches<br>➢ Simple structure<br>➢ Uniform sharing of load is possible<br>➢ Needs less number of gate driver circuit<br>➢ Can operate as single DC source configuration | ➢ Cannot be applied for fault-tolerant operations<br>➢ Switches with the highest voltage rating fail to operate at fundamental frequency<br>➢ Cannot be operated as asymmetric configuration<br>➢ Due to reduced switch count, the overall power rating is higher leading to threat in damage to the entire system |

**Table 2.** *Cont.*

| Configuration | Merits | Demerits |
|---|---|---|
| T-type [80–82,173] | ➢ Control structure is simple<br>➢ Does not require floating capacitors and diodes | ➢ Voltage derivative stress is higher on switches<br>➢ Switching losses are high<br>➢ Operation at high-frequency results in low efficiency<br>➢ Cannot operate for high voltage and power applications |
| Nilkar [174] | ➢ Can operate with less number of switches<br>➢ The net harmonic content is less<br>➢ Employs batteries and capacitors as the DC voltage source | ➢ Complicated structure<br>➢ Costly as power rating of reduced number of switches is more<br>➢ Not feasible for fault-tolerant operation<br>➢ Operation with less number of switches increases power rating and in turn makes its operating temperature exceed safe levels |
| CCHB [51,175] | ➢ Peak inverse voltage of the system is low<br>➢ Both symmetric and asymmetric topology are possible<br>➢ Ability to operate both with positive & negative voltage | ➢ Switches with bidirectional operation are required<br>➢ Needs isolated input DC link for operation |
| RV [78,176] | ➢ Three-phase operation can be carried out by single DC link<br>➢ Non-isolated type of DC links are operated<br>➢ Rated switches are operated at peak voltage and at switching frequency | ➢ Load sharing is not uniform<br>➢ Asymmetric configuration is not possible |
| SCSS [76,77] | ➢ Modular structure<br>➢ Rated switches can operate at peak voltage and switching frequency | ➢ System requires to operate at only symmetric configuration<br>➢ Rating of voltage varies from one switch to another<br>➢ Load sharing is not uniform |
| MLM [79] | ➢ Can operate with reduced number of DC voltage sources<br>➢ Requires fewer semiconductor switches, transistors, and power diodes | ➢ Fails for application in asymmetric configuration<br>➢ Requires isolated DC sources<br>➢ Power rating of device increases due to reduced device count which leads to damage of the device by causing its operating temperature to exceed safe levels as well increases its overall cost |
| 2SELG [83,168] | ➢ Needs minimum number of switches<br>➢ Structure is simple | ➢ Fails to operate at fundamental switching frequency<br>➢ Complex control<br>➢ Requires isolated DC sources<br>➢ Use of minimum number of switches leads to increased cost and temperature level higher |
| HBTPM [177] | ➢ Simple structure<br>➢ Appropriate for high voltage applications | ➢ Fails to synthesize various levels of voltage waveform at bus end<br>➢ All the individual levels accessible by sources could not be achieved<br>➢ Fails to operate in asymmetric configuration |

**Table 2.** *Cont.*

| Configuration | Merits | Demerits |
|---|---|---|
| CCS [51,69] | ➢ Ability to produce possible values of minimum step voltage<br>➢ Requires less number of basic sub-inverter cells and switching devices<br>➢ Possesses minimum blocking voltage for a particular level<br>➢ Modular in structure | ➢ Operates only with isolated DC sources<br>➢ Requires on-state switches<br>➢ Not so cost effective<br>➢ Operation with less number of switches increases power rating and in turn makes its operating temperature exceed safe levels |
| PUC [168,178–182] | ➢ Structure is simple<br>➢ Possibility of adding more crossover switches | ➢ Not modular<br>➢ Operation with asymmetric configuration is mandatory<br>➢ Different switches have different voltage ratings<br>➢ Fault-tolerant applications are impossible<br>➢ Cost of implementation is more |
| CBSC [71] | ➢ Reduced operational cost<br>➢ Simple circuit | ➢ Cannot work with asymmetric topology |
| Mokhberdoran [51,183] | ➢ High level of voltage can be achieved by cascading basic units in series<br>➢ Enhanced system efficiency<br>➢ Modular structure<br>➢ Cost effective | ➢ Requirement of switches and gate driver circuit is more<br>➢ Individual DC sources are required for operation |
| Babaei [51,70,184] | ➢ Simple and modular structure<br>➢ Power is equally shared among all cells<br>➢ Operation for symmetric and asymmetric configuration is possible | ➢ Requires DC sources that are isolated<br>➢ Needs switches for various ratings of voltage |

B. Switched Series/Parallel Sources based Multilevel Inverter (SSPS-MLI)

Hinago and Koizumi [170] proposed the SSPS-MLI topology. It is comprised of two major units; the "level-generation" unit, where a staircase voltage waveform with positive polarity is generated, and the "polarity-generation" unit, where the staircase DC voltage waveform gets converted to AC voltage as depicted in Figure 12. This configuration possesses the primary merit of generating more output voltage levels using very few numbers of switches in contrast to other conventional MLI topologies. Therefore, SSPS-MLI can be applied in the areas of: (i) electric vehicle where a single unit of the battery can be composed of several series-connected battery cells [171]; (ii) vehicle drive system to meet voltage or power need by a possible combination of two or more sources either in series or in parallel and (iii) traction purposes by possible joining of multiple sources in various flexible configuration [168,172].

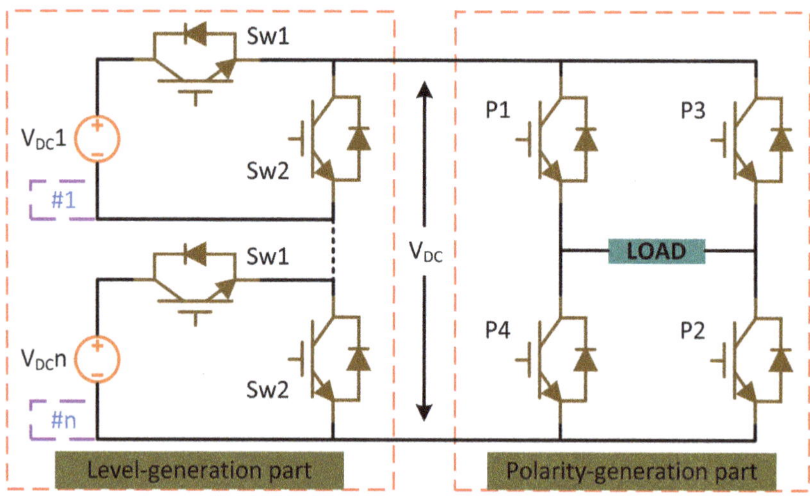


**Figure 11.** Topology of multiple level dc links inverter.

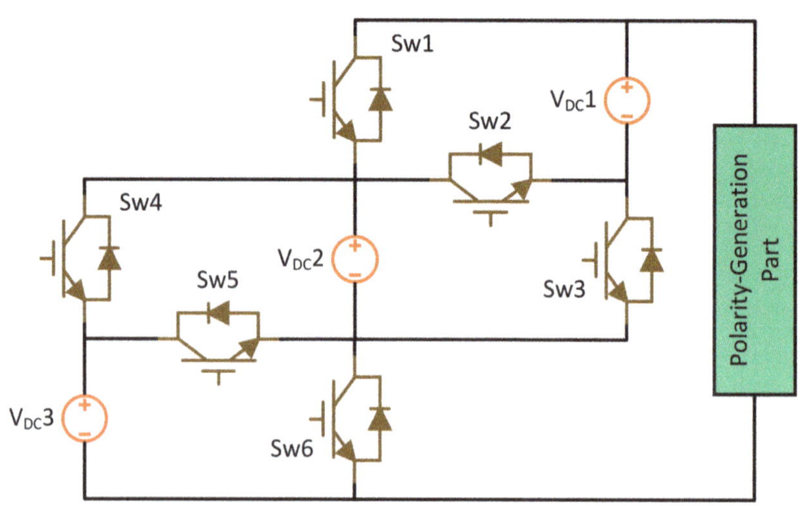


**Figure 12.** Topology of switched series/parallel sources based multilevel inverter.

(ii) Symmetric H-bridge Topology

This family of MLI usually consists of inverters where the magnitude of all the isolated supply DC sources to each of the H-bridge cells is identical. The symmetric H-bridge topology is further classified into different types. This section throws light on each of the Symmetric H-bridge topologies in specific.

A. T-type Multilevel Inverter

A novel T-type configuration with a five-level single phase inverter was first proposed by Gerardo Ceglia et al. [80–82]. The primary benefit of it is that the designed configuration reduces the use of more switches. A T-type inverter topology is depicted in Figure 13 below. Compared with other traditional topologies, T-type topology delivers an extraordinary improvement regarding the lower count of switches used and lower layout complexity. Further, almost 40–50% reduction in power switch count is achieved without diodes or capacitors [173]. This configuration has an H-bridge and an auxiliary bidirectional switch for controlling the connection of the supply from the DC sources to generate the staircase output voltage. However, this topology fails to render switching states to have all essential levels, as in the asymmetric H-bridge topologies [80].

B. Nilkar Multilevel Inverter (N-MLI)

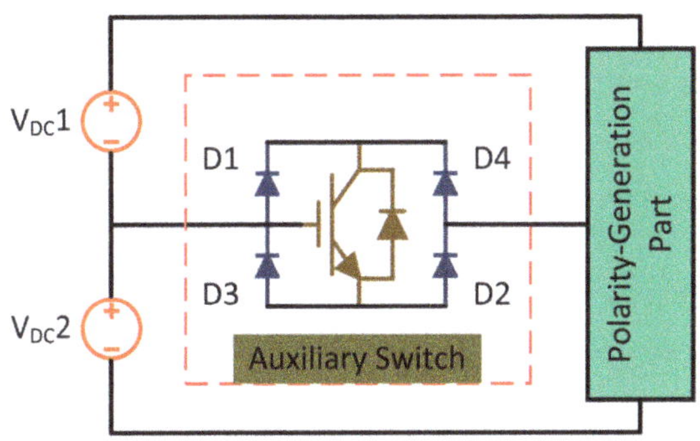


**Figure 13.** Topology of T-type symmetric H-bridge topology.

Nilkar et al. proposed the Nilkar inverter topology [174]. The basic module of N-MLI is comprised of two identical DC voltage sources with four semiconductors switching units for generating a staircase DC voltage waveform with positive polarity, which is further connected to an H-bridge consisting of 4 switching devices as demonstrated in Figure 14. The H-bridge helps in alternating the polarity of voltage for producing a complete sinusoidal alternating output signal. This topology renders many advantages, such as: (i) operates with significantly less number of switches; (ii) THD is effectively lessened in comparison to other available classical MLI topologies, and (iii) batteries and capacitors can be used as the DC voltage source. This topology can be used for various RER's (such as solar and fuel cell) interface and medium to high voltage applications in industries.

C. Crisscross Cascaded Multilevel Inverter (CCHB-MLI)

In [175], Khosroshahi proposed a CCHB-MLI topology consisting of the basic units in cascade. Figure 15 shows a CCHB inverter configuration comprising of two sources of DC voltage and a combination of one-way and two-way switches. It has two units; the first is the level generation unit, and the second is the polarity generation unit. The level-generation unit has two power switches, namely Sw2 and Sw3, which are unidirectional. The other switches Sw1 and Sw4, are bidirectional conducting and blocking switches, respectively. The four power switches P1, P2, P3, and P4 constitute the polarity-generation part. The benefits of this topology are: (i) usage of the reduced number of semiconductor switches; (ii) limited use of isolated DC voltage sources as compared to other classical topologies; and (iii) low cost and volume as compared to CHB-MLI [51].

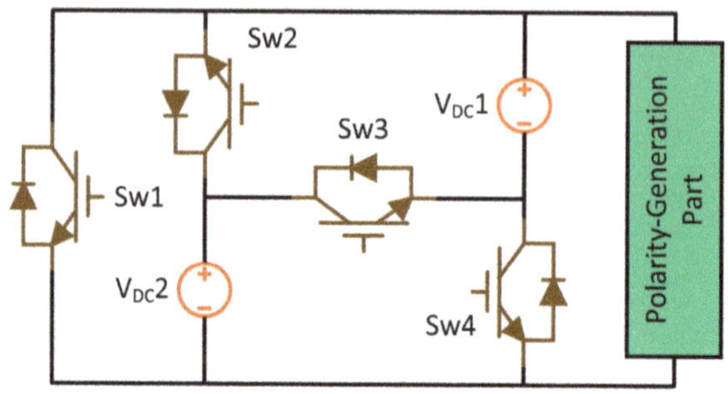


**Figure 14.** Topology of Nilkar multilevel inverter.

D. Reversing Voltage Multilevel Inverter (RV-MLI)

Reversing voltage MLI topology was first suggested by Najafi et al. in [78,176]. In this topology, the sinusoidal output voltage is produced in both level generation and polarity generation stages. The positive and negative polarity of voltages is generated in the level generation and polarity generation stage, respectively, as shown in Figure 16. By duplicating the centre stage operation with any number of levels can be possible, so application to three phases can also be extended. It has flexibility in the switching sequence and needs very few components for its work. Therefore, it can be useful in the areas of applications such as FACTS and HVDC. However, operation using different DC sources is not possible in this topology, as it is practically impossible to combine additive and subtractive DC sources.

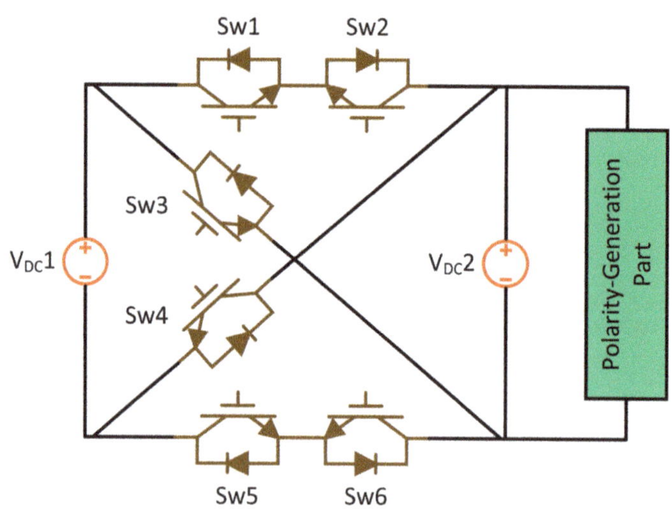


**Figure 15.** Topology of crisscross cascaded multilevel inverter.

E. Series Connected Switched Sources Multilevel Inverter (SCSS-MLI)

In this topology, the basic concept lies in the series connection of sources by the switches [76,77]. Figure 17 shows SCSS based MLI configuration. Here the poles of the voltage sources with lower magnitude are being associated with semiconductors. They are also in contact with the voltage poles with a higher magnitude of the upstream source. The link can synthesize a DC voltage with multiple levels, which take into account both the polarities with the help of the H-bridge. This structure also helps in reducing the number of switches for the symmetrical structure of the inverter. Nevertheless, the main shortcomings of this topology are: (i) power semiconductor used must be of the same rating; (ii) load sharing is impossible as input stage requires various configurations and (iii) high rated switches need to be switched with the minimum possible frequency.

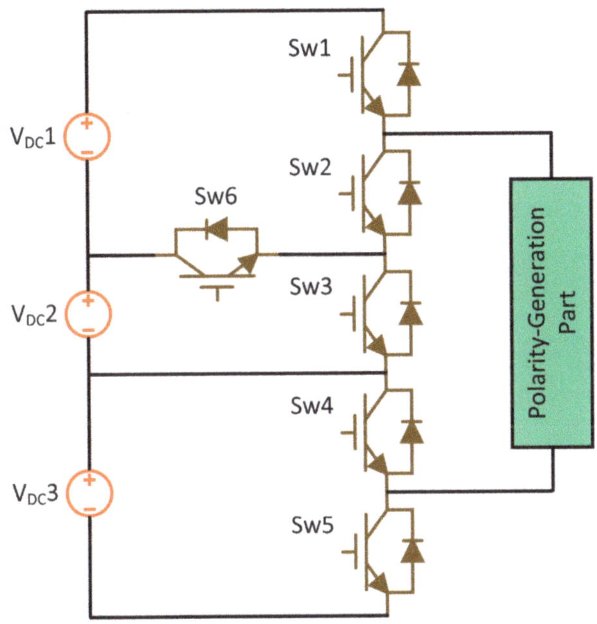


**Figure 16.** Topology of reversing voltage multilevel inverter.

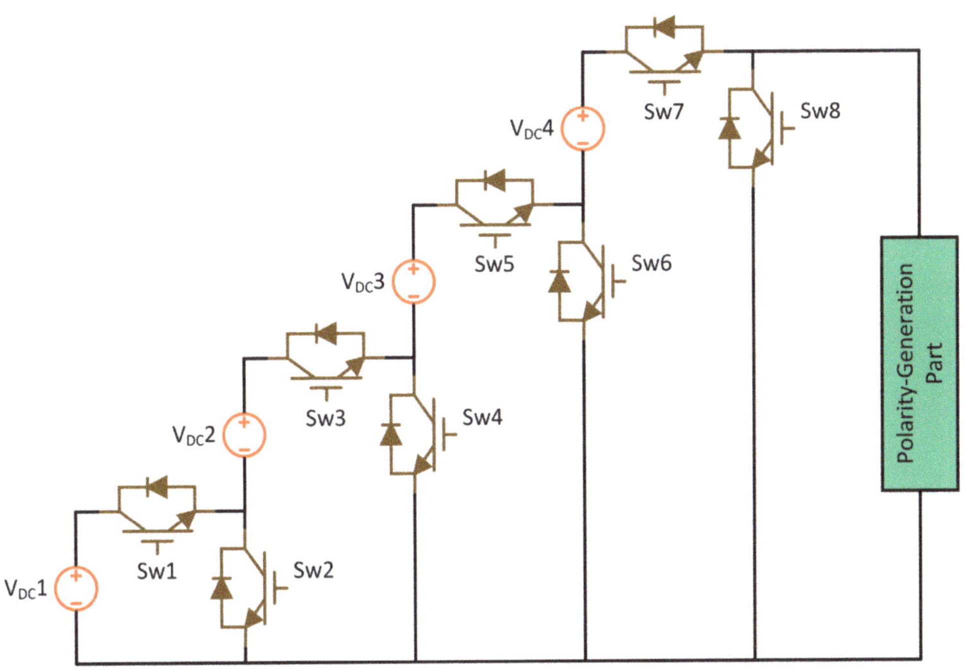


**Figure 17.** Topology of series connected switched sources multilevel inverter.

F. Multilevel Module Based Multilevel Inverter (MLM-MLI)

Babaei in [79] suggested the MLM-MLI topology, which comprises the level generation part and the polarity generation part. Figure 18 illustrates an MLM-MLI with four DC input sources. With the increase in output voltage level, this configuration can operate with the reduced number of DC voltage sources, semiconductor switches, transistors, and power diodes. The major demerit is that it fails for application in an asymmetric configuration. However, it can be used for high power quality applications that use an ample number of DC voltage sources [79].

G. Two Switch Enabled Level Generation Based Multilevel Inverter (2SELG-MLI)

Babaei in [83] discovered the level-generation based MLI Topology with two switches and seven input levels, as shown in Figure 19. This topology has two different stages, a level-generation stage and a polarity-generation stage. The name 'Two Switch' justifies that this configuration needs only two number of conducting switches in the level generation stage for synthesizing any level of voltage. However, this inverter fails to work in asymmetric topology. The major disadvantage is that as the level generation stage fails to realize the zero level of its own, the operation with a fundamental switching frequency becomes impossible [168].

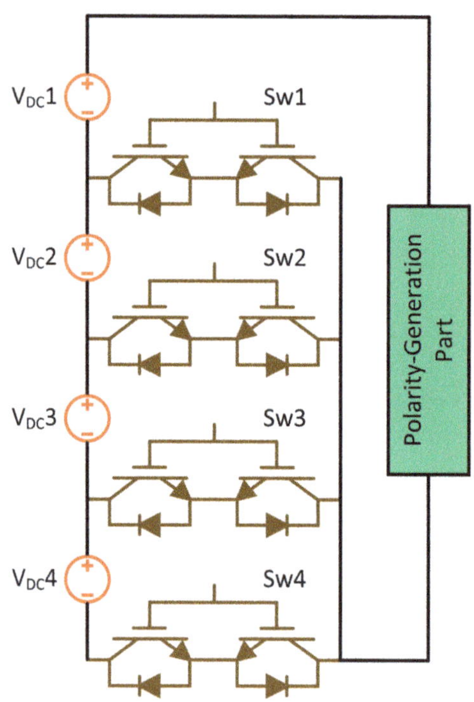


**Figure 18.** Topology of multilevel module based multilevel inverter.

H. H-bridge and two-level Power Modules Based Multilevel Inverter (HBTPM-MLI)

In [177], Suroso and Noguchi presented HBTPM-MLI. An HBTPM based inverter configuration has been illustrated in Figure 20, which consists of 4 input DC sources ($V_{DC}1$ to $V_{DC}4$). Semiconductor switches are used to interconnect the terminals with a source of low potential. Further, the proceeding source is attached to a high potential terminal using power switches. Switches P1 to P4 are for the polarity generation stage, and switches 'Sw1' to 'Sw6' form the level generation unit. The structure is simple, but it fails to synthesize various levels of the voltage waveform at the bus end. However, this topology fails to operate in asymmetric configuration for further reducing the count of the switch.

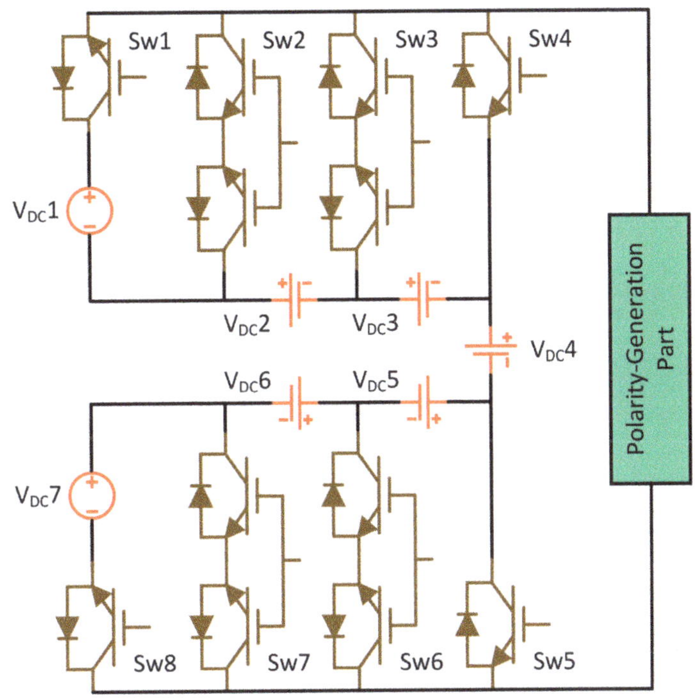


**Figure 19.** Topology of two switch enabled level generation based multilevel inverter.

Topologies without H-bridge

Topologies without H-bridge can be either asymmetric or symmetric type. However, mostly they are of a symmetric type.

(i) Asymmetric Topology without H-bridge

This category of MLI consists of numerous DC voltage sources, among which at least one differs dynamically. They do not consist of H-bridge cells or units.

(ii) Symmetric Topology without H-bridge

In this topology, the magnitude of all the isolated supply DC sources are identical, but they do not form an H-bridge configuration. In this section, different types of symmetric topology without H-bridge structure have been reviewed.

A. Cross Connected Sources Based Multilevel Inverter (CCS-MLI)

Authors in [69] have introduced the CCS-MLI topology comprising input DC sources isolated for every cell, as depicted in Figure 21. In this topology, a switch connects the two different terminals of two different sources and vice-versa. It requires a minimum number of switches for its operation and is usually active where isolated DC sources are present [51].

B. Packed U Cell Multilevel Inverter (PUC-MLI)

A novel MLI topology was offered by Youssef Ounejjar et al. and was named "PUC" [178–182]. The circuit of PUC-MLI is represented in Figure 22, which constitutes ten power semiconductor switches and four DC sources. Every individual U-cell has a single DC input level and two switching units [168]. The main advantage of this technology is that the maximum voltage-producing switch can be operated at the minimum frequency. Further, it allows easy change in the voltage level number, reduces stress on the switch, and enhances the overall converter operation.

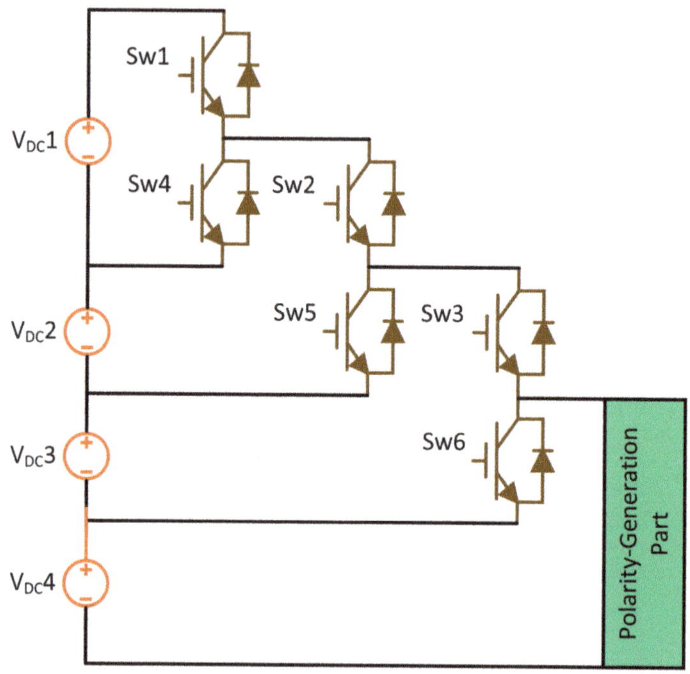


**Figure 20.** Topology of H-bridge and two-level power modules-based multilevel inverter.

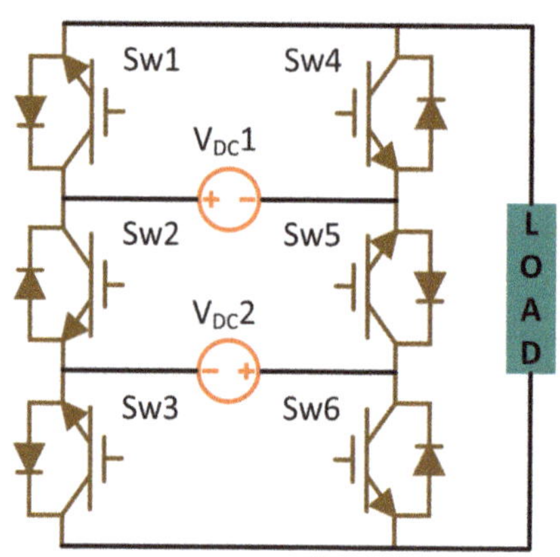


**Figure 21.** Topology of cross connected sources based multilevel inverter.

C. Cascaded Bipolar Switched Cells Multilevel Inverter (CBSC-MLI)

Babaei et al. in [71] have introduced this topology, as shown in Figure 23. The circuit comprises 4 DC sources and 10 bidirectional power semiconductor switches capable of generating voltage levels in both positive and negative polarities. Every bidirectional switch needs two IGBTs and is equal to the number of gate drive circuits. This concept helps in decreasing the operational cost and overall complexity of the circuit. The major shortcoming is that this topology is that it cannot work with an asymmetric configuration.

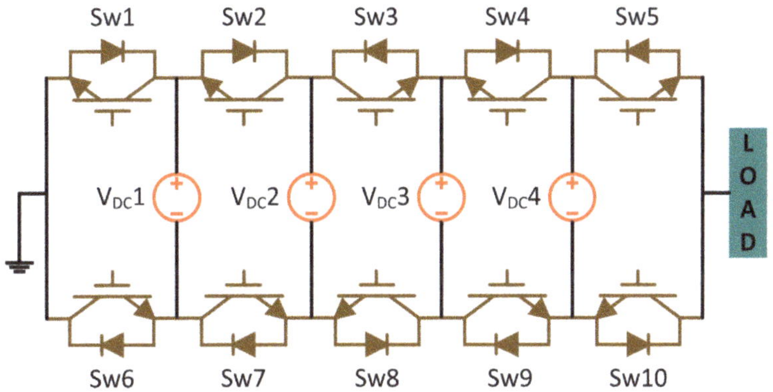


**Figure 22.** Topology of packed U-cell multilevel inverter.

D. Mokhberdoran Multilevel Inverter (M-MLI)

This topology is named after Mokhberdoran as in [183]. The basic circuit of M-MLI is illustrated in Figure 24, which comprises of two symmetric DC voltage sources, six switches, and eight diodes. Here to obtain a higher level of voltage, the basic units are cascaded in series [51]. The topology is such designed that it utilises a significantly fewer number of switching devices. The entire operation is divided into two parts depending upon frequency, such as low and high. This enhances the efficacy of the configuration. The structure is modular, and the cost of this unit is also minimum. Mokhberdoran technology usually finds its application in high power operations [183].

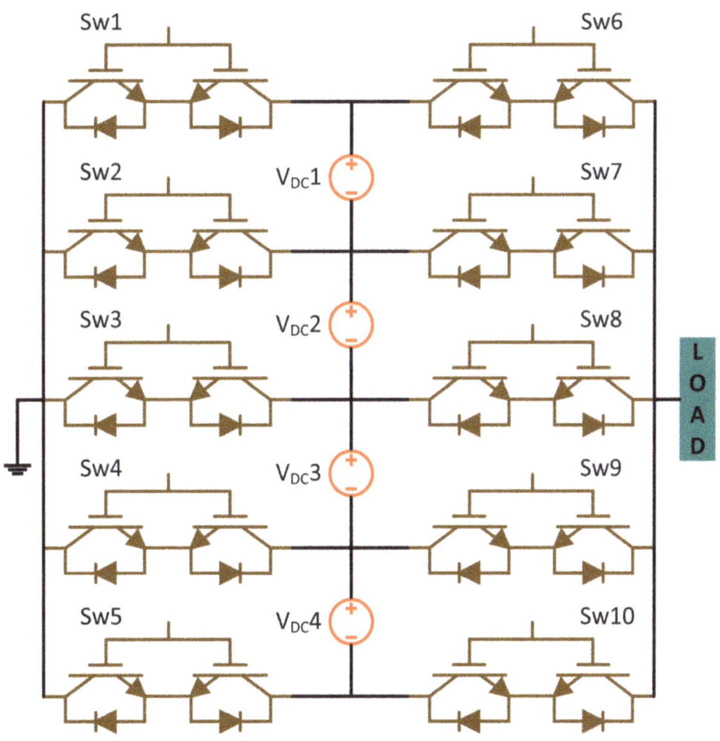


**Figure 23.** Topology of cascaded bipolar switched cells multilevel inverter.

E. Babaei Multilevel Inverter (B-MLI)

Figure 25 illustrates the B-MLI topology as proposed in [70,184]. The key elements are composed of six unidirectional switches and two symmetrical DC voltage sources. For increasing the output voltage level, the circuit developed can be reproduced by connecting in series [51].

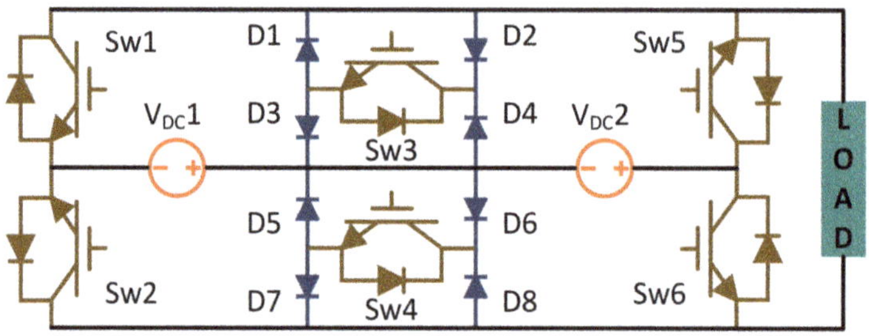


**Figure 24.** Topology of Mokhberdoran multilevel inverter.

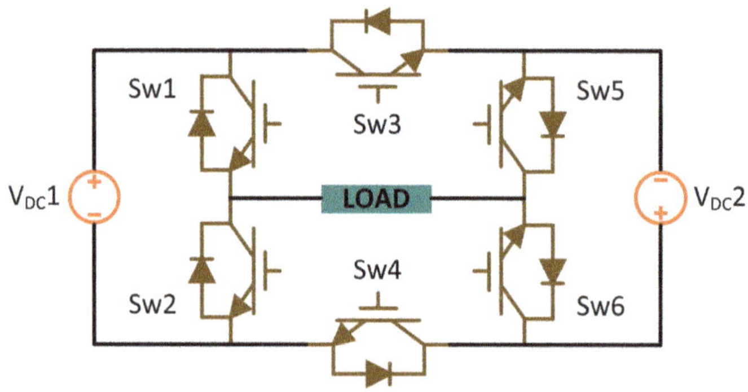


**Figure 25.** Topology of Babaei multilevel inverter.

## 4. MLI Control and Modulation Schemes

Modulation techniques play a principal role in governing the overall efficiency parameters such as harmonic reduction and switching losses used to control the inverter and turn the entire system [185]. They also have the responsibility to synthesize reference control signals to maintain all voltage sources balanced. The major purpose of modulation is to generate a staircase DC voltage signal, nearest a reference signal that is generally sinusoidal in a steady state [186]. The modulation process usually involves a variety of single or multiple attributes of a carrier signal waveform with a modulating waveform. Modulation is also referred to as a strategy to control the switching action by changing the characteristics of a particular signal (carrier signal) using another signal (reference signal). Every family of MLI has a selected appropriate modulation scheme for optimizing the circuit working and achieving the target criteria. Following are the major important factors basing on which a particular modulation technique is chosen for a particular MLI family: (i) total generated harmonics; (ii) level of distortion; (iii) frequency of switching; (iv) amount of losses and (v) response speed. The MLI modulation methods have the following needs to be fulfilled before operation: (i) quality of voltage should be high; (ii) should have modular structure; (iii) switching of multiple voltage levels simultaneously is not permitted; (iv) power devices should operate with minimum frequency; (v) load sharing should be uniform among the power modules; (vi) algorithm used for control should be easy and simple and (vii) cost of implementation must be minimum [51].

The modulation index also has a vital role in all control schemes. Modulation also depends upon modulation ratio (either over or under modulation), and the THD varies accordingly. Multiple techniques are proposed by authors in the literature depending upon the switching frequency, either fundamental or high frequency [187,188]. However, low losses are found when switched at fundamental or low frequency. A thorough survey of various modulation methods is highlighted in this section below. Figure 26 illustrates the various control and modulation schemes for MLI. Table 3 lists the merits and demerits of different modulation schemes discussed in this review article.

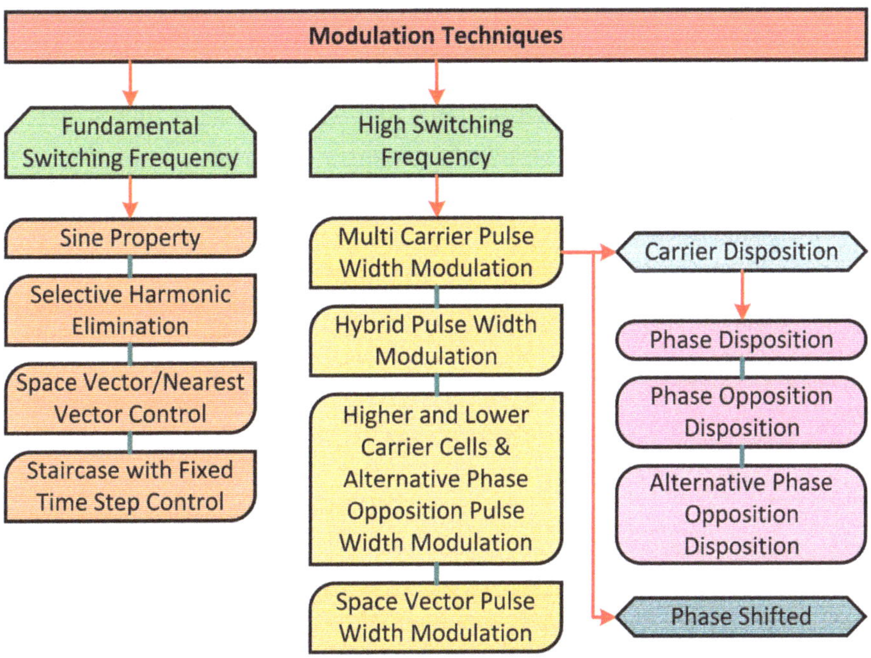


**Figure 26.** Categorization of modulation schemes for multilevel inverter.

### *4.1. Fundamental Switching Frequency Pulse Width Modulation (FSF-PWM) Techniques*

These techniques generate a staircase waveform performing single or multiple commutations of the power electronic switches [189,190]. Here, an increased number of levels is obtained by adding extra units without creating any complexity in the generation of the switching signals. Moreover, due to switching at low frequency, the losses incurred are very less. Various FSF-PWM methods have been discussed below.

#### 4.1.1. Sine Property

It is a modern method for calculating the firing angle that is to be provided to the switching units [191–193]. Calculation of firing angle is easy on adopting this technique. The firing angle is generally obtained in degrees and has the provision of appropriate conversion to any other unit of time such as 'seconds' for the easy performance of simulations.

#### 4.1.2. Selective Harmonic Elimination

Researchers in 1973 suggested a voltage control and harmonic elimination theory known as SHE. This technique is utilized for eliminating the most dominant selected lower order harmonics [194–197]. In SHE, there is a possibility of lowering the THD and the size of the output filter. As the switching angles are pre-determined off-line, it is assumed to be an open-loop modulation method [198,199]. Authors have also reported that many Fourier equations are utilized for calculating the firing angle for the switching purpose [195]. Selection of correct values of the firing angle for the Fourier series equation, the odd harmonics can be limited for any level of MLI. A microcontroller device is then used to supply these firing angles to the switches. Thus, it does not require a closed-loop controller for its implementation.

The vital functions of the SHE are: (i) maintaining the fundamental component of the waveform; (ii) harmonic reduction individually; (iii) decrease in THD, and (iv) lower switching losses. However, a major drawback of the SHE technique is that it requires the design of massive passive filters to limit the lower-order harmonics [196]. To counteract the above-cited issue of the SHE method, a novel technique known as selective harmonic mitigation (SHM) has been proposed in the literature by researchers [200]. In the SHM method, switching angles are calculated to reduce the individual harmonic distortion without the grid code limits. SHM is also an open-loop control technique and an offline procedure for calculating the switching angles and cost function that can be minimized using a search algorithm [201]. Numerous solutions can be utilized to solve the non-linear equation of MLI, such as: (i) iterative numerical methods [202]; (ii) artificial intelligence-based methods [203], and (iii) resultant theory [204]. A new technique for the SHE method named Groebner Bases Theory Method has been developed [205,206]. This method has been carried out with a three-level inverter. This method has the advantage of finding the most accurate switching angles as no initial value for iteration is required.

Recently numerous evolutionary algorithms based on SHE techniques for MLI are reported in the literature [207–218]. Authors in [185,207] have implemented a genetic algorithm (GA) to minimize low-count switch MLI distortions. In [208,209], the output voltage regulation and improvement of the harmonic profile are carried out by particle swarm optimization (PSO) technique. Additionally, PSO is used in [210,211] for the control of THD. The ant colony system (ACS) algorithm was incorporated in [212] for removing the harmonic content of a 3-phase inverter with a unipolar output voltage waveform. Researchers have used bee algorithm (BA) for ninety-seven-level CHB-MLI [213]. THD reduction is brought about by bacterial foraging algorithm (BFA) in CHB-MLI. In [215], a clonal search algorithm (CSA) has been reported for a three-phase inverter. The harmonics in the line voltage of an inductor motor are eliminated using the evolutionary programming algorithm (EPA) as in [216]. Authors in [217,218] have suggested using the differential evolution (DE) algorithm for a 3-phase inverter to enhance the level of output voltage, improve the harmonic profile for rapid convergence and better efficiency.

#### 4.1.3. Space Vector/Nearest Vector Control (SVC/NVC)

The space vector control (SVC) is also known as nearest vector control (NVC). It is reported as an alternative to SHE and can operate at a low switching frequency. Like SHE, it does not generate the average value of the required load voltage for every switching time interval. The major function of the SVC technique is to select a vector closest to the reference vector for minimizing the distance between them or the space error [187]. However, in SVC, the lower distortions produced due to switching at low frequency are usually not eliminated like in the SHE technique. The NVC technique is very simple and applicable for higher output voltage levels as the higher density of vectors can generate only small errors about the reference vector. Authors in [219] have discussed the principle of an eleven-level inverter with SVC control. Recently the SVC finds its implementation in numerous MLIs, including both classical as well as newly proposed reduced switch count technology. In [220], a five-level MLI configuration with quasi Z-source has been designed in which the NVC has been utilized as the control scheme.

#### 4.1.4. Staircase with a Fixed Time Step Control Scheme

Staircase with a fixed time step control scheme generates the output voltage signal is generated from uniform time steps within each level. The main merit is that the structure is simple and hence makes the inverter control very easy. The major drawback is that the output voltage profile consists of lower order harmonics and thus increases the THD. The waveform is divided into equal time intervals similar to the number of levels to find equal switching instants. In this control scheme, the inverter input voltage can be controlled, but the output voltage is not controllable.

### *4.2. High Switching Frequency Pulse Width Modulation (HSF-PWM) Techniques*

The HSF-PWM techniques are usually employed for high switching frequency applications in the order of kHz and consist of many commutations per cycle [39]. A detailed classification of each type of this technique has been deliberated in this section.

#### 4.2.1. Multi-Carrier Pulse Width Modulation (MC-PWM) Techniques

In this technique, only a single modulating sinusoidal signal is produced using multiple triangular carriers. Usually, the number of employed carriers is '(n-1)', where 'n' is the level of the inverter [39]. The MC-PWM techniques are further categorized into two types: (i) carrier disposition PWM and (ii) phase shifted PWM. The basic diagrams containing modulating and carrier signals for carrier disposition and phase shifted PWM methods have been reported by numerous authors in the literature [221,222].

##### Carrier Disposition Pulse Width Modulation (CD-PWM) Techniques

In this scheme, the reference waveform is produced by comparing the amplitude of the carrier waveform to a reference waveform amplitude. It is further classified into three various types [223]. The details of each are highlighted in this section.

(i) Phase Disposition (PD) Method

The important feature of the phase voltage spectrum of a PD Method is the initial distortion of the carrier. Therefore, this method generates a very good performance of the line voltage. Generally, all the carrier signals have equal amplitude and frequency and lie in one phase. PD method is usually used for asymmetric MLI, and with the increase in the number of voltage levels, the harmonic contents are decreased [224].

(ii) Phase Opposition Disposition (POD) Method

In this modulation scheme, the in-phase components are the positive carrier signals, whereas the carrier signals with negative polarity are 180° out of phase.

(iii) Alternative Phase Opposition Disposition (APOD) Method

This method involves all the carriers to be in phase opposition by 180° to the nearest carriers [224].

##### Phase-Shifted Pulse Width Modulation (PS-PWM) Technique

In the PS-PWM method, the multiple carriers are phase-shifted accordingly. It requires '(m−1)' triangular carriers for an 'm'-level inverter. This triangular carrier possesses identical frequency and amplitude; however, adjacent carriers have a definite phase shift between them [225]. The gate signals of the MLI switches are produced with the help of the on/off state of some logic circuit switches [224].

#### 4.2.2. Hybrid Pulse Width Modulation (H-PWM) Technique

The low-frequency and high-frequency modulation techniques are combined to form H-PWM. This control scheme is applied for CHB-MLI having a varying magnitude of DC sources. This method majorly aims to lower the inverter's losses by reducing the switching frequency of the higher power units. The lower power unit is controlled by using the unipolar pulse width modulation method [224].

**Table 3.** Comparative study highlighting the merits and demerits of basic modulation schemes.

| Modulation Schemes | Merits | Demerits | References |
|---|---|---|---|
| Sine Property | ➢ Easy performance of simulations<br>➢ Has constant switching frequency<br>➢ Easy thermal design | ➢ Requires more computational efforts | [191–193] |
| Selective Harmonic Elimination (SHE) | ➢ Ability to eliminate lower order harmonics<br>➢ Low harmonics<br>➢ Reduced output filter size<br>➢ Appropriate for high power application<br>➢ High efficiency<br>➢ Low losses during switching operation<br>➢ Better steady-state response | ➢ Overall dynamic response is slower<br>➢ Ineffective voltage balancing operation<br>➢ Requires massive passive filters | [194–218] |
| State Vector Control (SVC) | ➢ Can operate at low switching frequency<br>➢ Simple technique<br>➢ Low harmonics with high efficiency<br>➢ Less number of switching states<br>➢ Lower dv/dt stress<br>➢ Better dynamic response<br>➢ Does not require huge passive filters | ➢ Lower order harmonics generated are not eliminated<br>➢ Complex for structure involving more number of voltage levels | [187,219,220] |
| Phase Shifted PWM (PS-PWM) | ➢ Modular and simple structure<br>➢ Rotation of switching patterns is not required | ➢ High harmonic content<br>➢ Poor voltage balancing strategy<br>➢ Poor dynamic response | [224,225] |
| Phase Disposition PWM (PD-PWM) | ➢ Better voltage profile<br>➢ Optimal switching is achieved<br>➢ All carriers have same frequency and amplitude | ➢ Uneven power distribution<br>➢ Poor dynamic response | [224] |

#### 4.2.3. High and Low Carrier Cells and Alternative Phase Opposition Pulse Width Modulation (HLCCAPO-PWM) Technique

HLCCAPO-PWM is a modification of the PD method. Here two different carrier groups are introduced by dividing them as per each carrier period. The method renders the use of higher modulation frequency by reducing the switching losses effectively. Further, this technique also reduces the dissipation of energy as the energy shifts from the lower to higher-order harmonics. It mostly finds its application in hybrid clamped MLI.

#### 4.2.4. Space Vector Pulse Width Modulation (SV-PWM) Technique

The SV-PWM technique consists of many vector states that are utilized for modulation of the reference waveform. This method is usually based on the digital modulation method for generating PWM voltages under a known voltage [226]. Here the values of the control algorithm are directly taken from the control system [227]. However, this scheme fails to operate with a large number of levels as identifying sectors and selecting switching sequences are very crucial. Generally, for an 'n'-level inverter, '$(n-1)^2$' vector combination per sector, six sectors, and '$n^3$' switching sequences are required [228]. Reduction in the common-mode voltage, losses due to switching, and the control of the DC link voltages can be brought about by appropriate selection of modulation vector and switching combinations. Generation of a particular voltage level takes place by redundancy switching states.

Various authors have reported in the literature the application of this technique in numerous areas. In [229], a space vector hysteresis current control (SVHCC) scheme has been implemented. The SVHCC has been used with a recent MLI configuration, as suggested in [230]. The use of SVC is also projected in [231] for a T-type MLI system that implements a fault-tolerant control scheme. Amit Kumar Gupta et al. in [232] addressed a simple SV-PWM method for MLI operation in the over-modulation range. In [233], Mohan M. Renge suggested a technique for reducing common-mode voltage at the output of MLI that used the 3-D SV-PWM technique.

Better values of fundamental voltage ratio and harmonic elimination are achieved in this technique as compared to the sinusoidal PWM method. In addition, the maximum peak value of the output voltage is almost 15% more in SV-PWM than in the triangular carrier-based modulation method. Although the requirement of a look up table and identification of sectors for determining the switching intervals for all sectors make the SV-PWM technique complex, the microprocessor and digital signal processing units serve as a better solution for preparing the preparation of the process the algorithm.

## 5. Applications of MLI Topologies and Control Schemes

This section throws light on the detailed application of various MLI topologies and control schemes in different fields of power system networks, such as (i) integration of RERs to grid; (ii) FACTS devices, and (iii) electric motor drives. Table 4 provides a brief idea of the application of various MLI topologies in numerous power system domains.

**Table 4.** Application of various MLI topologies in numerous power system domains.

| MLI Topologies | Applications |
|---|---|
| NPC-MLI | ❖ High speed motor drives<br>❖ Renewable energy<br>❖ Power systems |
| FC-MLI | ❖ Renewable energy<br>❖ Motor drives |
| ANPC-MLI | ❖ Renewable energy (solar inverters)<br>❖ Active power filters |
| CHB-MLI | ❖ FACTS<br>❖ Renewable energy<br>❖ Drives |

**Table 4.** *Cont.*

| MLI Topologies | Applications |
|---|---|
| HCHB-MLI | ❖ Motor drives<br>❖ Renewable energy |
| MLDCL-MLI | ❖ PM motor drives (<100 KW)<br>❖ MOSFETs<br>❖ IGBTs<br>❖ Solar and fuel cell integration |
| SSPS-MLI | ❖ Renewable energy<br>❖ Vehicle drive system<br>❖ Traction purposes |
| T-type-MLI | ❖ AC drive system<br>❖ Renewable energy<br>❖ Power train drive |
| N-MLI | ❖ Renewable energy<br>❖ Medium/high voltage industries |
| CCHB-MLI | ❖ Motor drives<br>❖ FACTS<br>❖ Renewable energy |
| RV-MLI | ❖ FACTS<br>❖ HVDC |
| SCSS-MLI | ❖ Electric vehicles<br>❖ FACTS<br>❖ Submarine propulsion |
| MLM-MLI | ❖ Renewable energy |
| 2SELG-MLI | ❖ HVDC<br>❖ Renewable energy |
| HBTPM-MLI | ❖ Renewable energy |
| CCS-MLI | ❖ Photovoltaic system |
| PUC-MLI | ❖ Motor drives<br>❖ Renewable energy |
| CBSC-MLI | ❖ Renewable energy |
| M-MLI | ❖ HVDC<br>❖ Wind systems |
| B-MLI | ❖ HVDC |

### *5.1. Grid Integration of RERs*

The control schemes for RERs integration are categorized as (1) time and (2) frequency domain schemes. Fast Fourier Transformation is usually not employed because of more computational time and delay in computing reference signals [234]. On the other hand, p-q theory and d-q theory based on time-domain control algorithms are highly adopted due to lesser computation time in deriving the instantaneous compensating current or voltage signals. The DC quantities are assumed to be the fundamental components in the d-q algorithm. Furthermore, authors in [80] have studied the sensorless control of voltage waveform in packed U-cell topology to reduce control complexity and redundancy switching state. This operation keeps the capacitor voltage constant at half the magnitude of the DC source. Figure 27 depicts the controller for adjusting the amplitude and the phase shift of the current injected from the inverter to the grid.

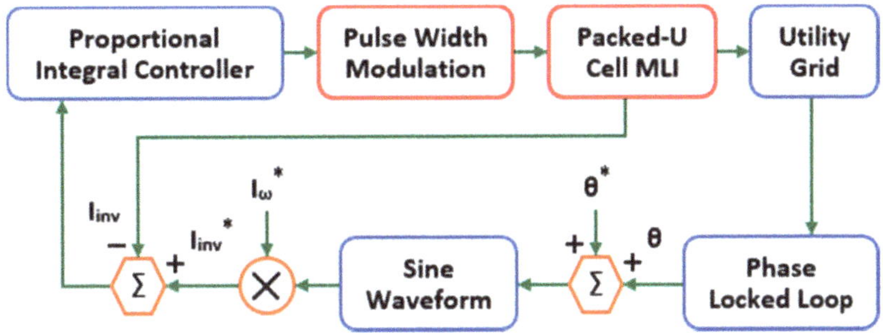


**Figure 27.** Block diagram of a single phase grid connected packed U-cell based MLI.

In [235–237], researchers have addressed the use of a solar-based MLI for enhancement in the waveform quality and reduction in issues on power quality. For a three-level NPC-MLI, instantaneous power theory control is studied for generating the perfect reference signal. Here, two control loops are defined, one for controlling the DC bus voltage and the other for controlling the current. Source active power and load reactive power are computed using the fuzzy logic controller (FLC), and in the α-β reference frame, the p-q theory was used to find the reference current [238]. Authors in [239] have proposed a digital proportional integral (PI) Controller to inject the current from the photovoltaic source to the utility grid to achieve maximum dynamic operation with minimum harmonics. Figure 28 describes the block diagram of the control part of the above study, which comprises maximum power point tracking (MPPT) control and inverter control.

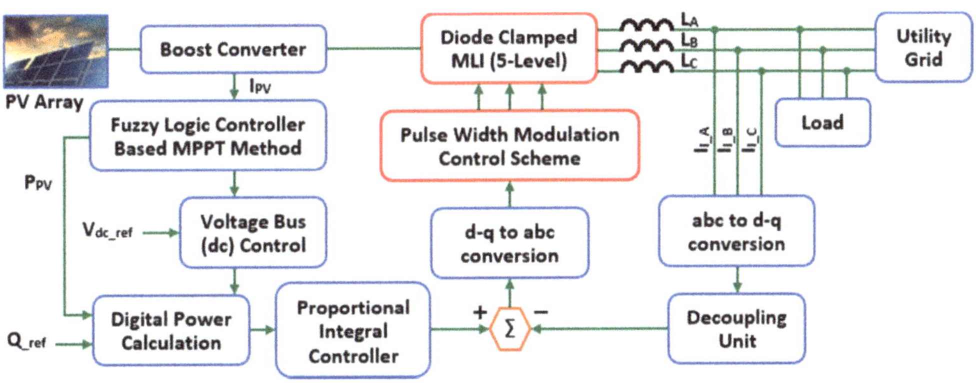


**Figure 28.** Block diagram of grid tied PV system with digital proportional integral control and fuzzy logic control.

The authors have introduced the dual-loop control method for a three-level inverter, as demonstrated in Figure 29. Here, two loops are present, the outside loop controlling the DC bus voltage and the current being controlled by the inner loop [240]. In NPC-MLI, the predictive control method is utilized to balance the DC link voltage. As suggested by authors in [241], a dual loop d-q controller is used to control the active and reactive power distribution. Different topologies of MLI in various combinations are interfaced with RERs for grid-tied applications [242–248]. In the recent past, the MLI topologies are also applied for application in the areas of marine [245] and microgrid [246,249].

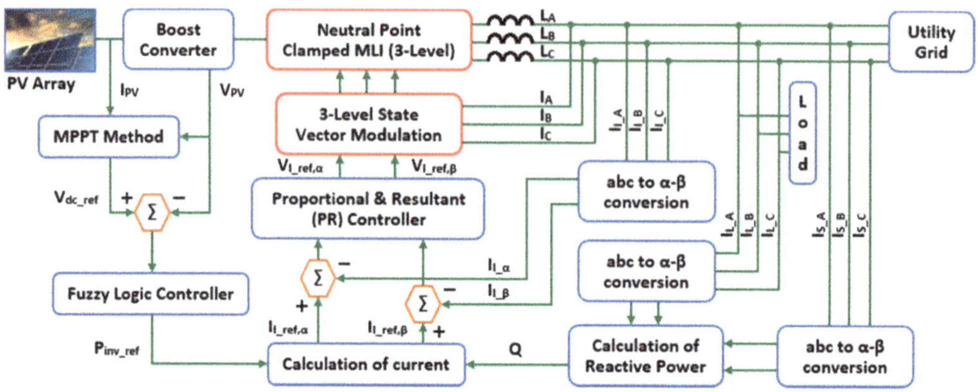


**Figure 29.** Block diagram of the dual loop control method for a three-level inverter.

### *5.2. FACTS Device*

STATCOM is considered to be a vital controller among all other types of available FACTS devices. The most appropriate topology for STATCOMs operation is the CHB-MLI for direct connection to medium voltage networks [250]. It does not need injecting active power for a normal range of operation [251]. In [17,252], authors have reported that CHB-MLI can be placed in series to achieve the operational voltage without using a transformer. A reactive current reference control scheme is proposed to enhance the transient operation of STATCOM, as depicted in Figure 30. In Figure 30, the phase locked loop (PLL) block is utilized to determine the reference phase angle of the grid voltage. The STATCOM output voltage and current are transformed into d-q reference frame vectors by adopting Park's Transformation. The feedback operation is preferred by the controller and produces the switching pulses with a proper modulation index.

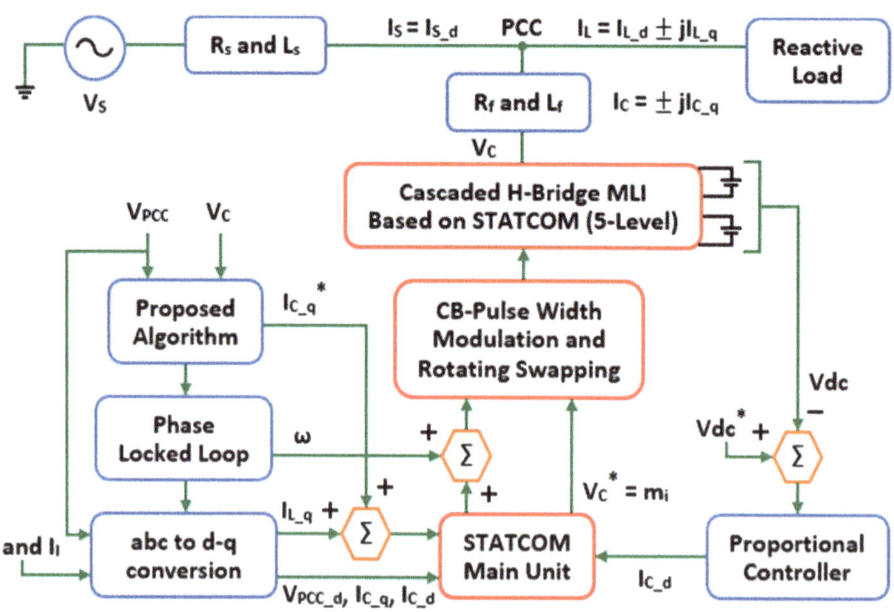


**Figure 30.** Block diagram for control technique for STATCOM application.

Researchers have suggested a novel hybrid control method for STATCOM application for delta CHB-MLI subjected to an unbalanced condition [253]. A current control method is proposed for every phase individually to control every link's active and reactive power independently. The hybrid control approach consists of four different parts as follows: (i) PLL; (ii) active current reference; (iii) reactive current reference; and (iv) instantaneous current tracking. The PLL determines the phase angle of the system. However, the compensation model is used to calculate the amplitude of the voltage, which lags the phase of reactive current by 90°. The overall value of phase current can be obtained by summing up the active and reactive current reference [254].

DVR plays a vital role in eliminating voltage-related problems like sudden rise or fall of voltage, spikes, swell, etc. It injects a voltage to prevent any disturbance in the load side voltage and is connected in series with the source side voltage [255]. The use of MLI in DVR enhances the capability of voltage injection to the maximum that can be applied for medium voltage application without using a transformer [256]. Here, the grid voltage and reference voltage comparison yield the reference value for the DVR output voltage (voltage sag value). PI controller is utilized to regulate the voltage level of the load and output voltage of the boost converter having a feedback control. The reference signal of MLI can be calculated by dividing the control output by the DC link voltage. The transformation ratio of the boost converter can be computed by its duty cycle. A crucial issue in DVR is the transformer's magnetic saturation, which causes the production of a large inrush current. Authors in [257] have suggested a novel modulation scheme called direct magnetizing flux linkage control for preventing the transformer from getting saturated, as illustrated in Figure 31. The overall modulation approach is divided into three loops depending upon their functions.

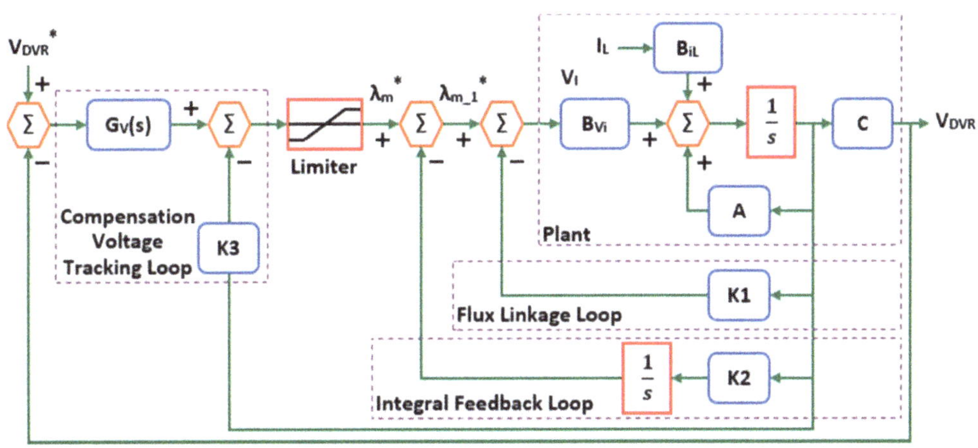


**Figure 31.** Block diagram for direct flux linkage control technique in DVR.

Loop 1 is called the flux linkage loop for tracking the flux linkage command. For attenuating the DC flux, linkage loop two, known as the integral feedback loop, is present. Loop 3 is utilized for tracking the compensation voltage command. To limit the linkage of flux, a flux linkage limiter is added between loop two and loop 3. Researchers in [258,259] have discussed the importance of MLI for unified power quality controller (UPQC) operations. A CHB-MLI is applied for DVR operation in [260].

*5.3. Motor Drives*

The basic necessity for an extraordinary function of an electric motor drive unit is its accurate torque control. A deep insight into the present literature reveals that control of field and torque for an induction machine serve as the two most remarkable control schemes [261]. In [262], authors have stated that direct control of torque can be directly switched on to an inverter without needing the regulation of stator current. However, the torque and flux generating units are separately controlled for field control. In [263], a recent control scheme is proposed for IM by CHB-MLI prone to a faulted condition. The block diagram of rotor flux linkage-oriented control as used in [263] is demonstrated in Figure 32. In [264–266], artificial neural network control [267] is used in IM to improve performance parameters. In addition to that, a reduced switch count MLI topology was executed with IM control for enhancing the quality of power, thereby lowering the THD of the output voltage.

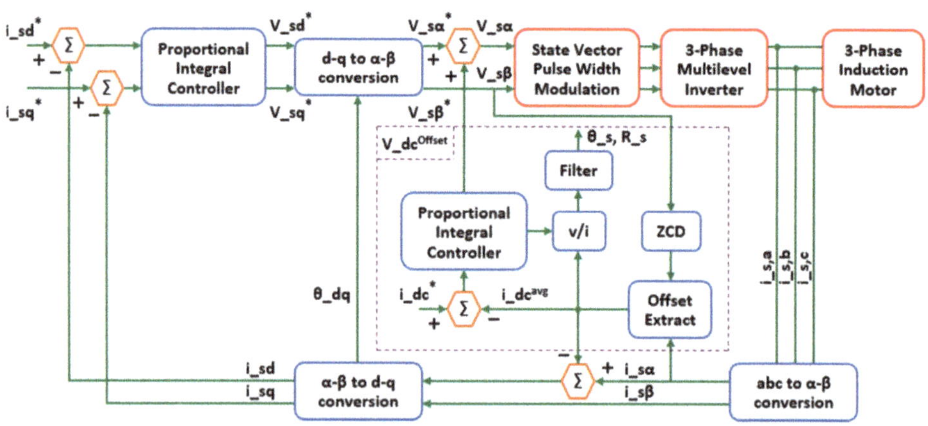


**Figure 32.** Block diagram for rotor flux linkage oriented control method of motor drive system.

## 6. Future Work

MLIs undoubtedly have been the state of art and technology for power conversion from DC to AC, electrical energy employment, and a wide range of power applications in the present era. However, some of the major shortcomings in the implementation of MLI technology can be listed: (i) cost of manufacturing enhances as the number of switching devices used is more and (ii) addition of power electronic units may lead to the incorporation of more gate driving circuits, voltage and frequency control methods and complicated switching techniques making the control unit more complicated.

In order to cope up with the above issues, researchers are opting out ways for a trade-off among the increased levels and complications in the design and working. Intending to assist the hassle-free integration of MLI technology in the present era, this comprehensive review has suggested some vital points that can be adopted in the future for further possible enhanced application of MLI technology in the various field of power systems.

(1) More research on contemporary topologies regarding the use of less power electronic interfaces, low cost, enhanced reliability, and efficiency should be developed.
(2) Studies on designing a less complex inverter should be carried out.
(3) Semiconductors with a large bandgap can be adopted for considerable increment in the switching frequency, thus facilitating the use of the smaller size of switching units.
(4) The control and modulation schemes for MLI implementation should be further enhanced with more robust, modular, and fault-tolerance capability.
(5) More in-depth work in the area of numerical methods for the solution of non-linear equations of MLI needs to be undertaken.
(6) Extensive study is also required to balance the rise in temperature of the semiconductor devices used in the MLIs.
(7) Integration of faster microprocessor units with the ability to work with high-level inverters and faster switching device applications should be adopted.

(8) Building up more robust modulation schemes to assure uniformity in the rise of temperature of all devices and reduction in the complication of the controllers.
(9) Design of reduced capacitor size by undertaking new voltage balancing methods needs to be used in MLI to enhance inverter's power density.
(10) Setting up of resonant converters basing on single DC source MLIs is recommended.

## 7. Conclusions

The upgradation and advancement of different industries and academic research globally have led to an increased demand for high energy-based efficient converters. The MLIs have attained tremendous demand due to their inherent merits and play a significant role in DC/AC conversion operations for both high/medium voltage and high power applications. In this regard, this review article attempts to critically survey the evolution of MLIs, which would serve as a prominent guideline for the researchers working in this area. This comprehensive paper throws light on the traditional MLI topologies, new reduced switch count topologies, various control approaches, applications of MLI to renewable energy interface, FACTS devices, and motor drives in detail. A thorough review of the literature reveals that the recently developed reduced switch count MLI topologies possesses many merits over the classical techniques in terms of better clarity of output waveform, low modularity, reduced number of switches, occupies minimum space, ease of control, and cost effectiveness.

Further, for easy analysis, a precise comparison among all the categories of MLI about advantages and shortcomings has been tabulated. This research article also projects a thorough idea about all conventional and newly adopted modulation techniques for different MLI topologies. This review also serves as a major objective to understand the vital role played by MLI in the areas of application for renewable energy, FACTS devices, and electric motor drives. The article's primary focus is the suggested key points to be incorporated in future research work to integrate the MLI technology more efficiently. The article is promising for obtaining maximum useful knowledge to the academicians, pursuing research in MLI field, in the fact of the suitable configuration selection for definite operation, proper schemes of switching and control, parameter selection and manifests real-time application to other sectors of the power system.

**Author Contributions:** Conceptualization, S.C and T.D; methodology, S.C.; validation, M.B., S.K. and F.J.; formal analysis, S.C and F.Z; investigation, S.C and M.B.; resources, S.K. and F.J.; writing—original draft preparation, S.C.; writing—review and editing, S.C, and T.D.; supervision, F.J.; All authors have read and agreed to the published version of the manuscript.

**Funding:** This research received no external funding.

**Institutional Review Board Statement:** NA.

**Informed Consent Statement:** NA.

**Data Availability Statement:** Authors may provide data on due request.

**Conflicts of Interest:** The authors declare no conflict of interest.

## References


1. Diaz, M.; Cardenas, R.; Espinoza, M.; Rojas, F.; Mora, A.; Clare, J.C.; Wheeler, P. Control of Wind Energy Conversion Systems Based on the Modular Multilevel Matrix Converter. *IEEE Trans. Ind. Electron.* **2017**, *64*, 8799–8810. [CrossRef]
2. Wang, M.; Hu, Y.; Zhao, W.; Wang, Y.; Chen, G. Application of modular multilevel converter in medium voltage high power permanent magnet synchronous generator wind energy conversion systems. *IET Renew. Power Gener.* **2016**, *10*, 824–833. [CrossRef]
3. Liu, Y.; Ge, B.; Abu-Rub, H.; Peng, F.Z. An Effective Control Method for Quasi-Z-Source Cascade Multilevel Inverter-Based Grid-Tie Single-Phase Photovoltaic Power System. *IEEE Trans. Ind. Inform.* **2013**, *10*, 399–407. [CrossRef]
4. Ng, C.H.; Parker, M.A.; Ran, L.; Tavner, P.J.; Bumby, J.R.; Spooner, E. A multilevel modular converter for a large, light weight wind turbine generator. *IEEE Trans. Power Electron.* **2008**, *23*, 1062–1074. [CrossRef]
5. Islam, R.; Guo, Y.; Zhu, J. A High-Frequency Link Multilevel Cascaded Medium-Voltage Converter for Direct Grid Integration of Renewable Energy Systems. *IEEE Trans. Power Electron.* **2013**, *29*, 4167–4182. [CrossRef]

6. Debnath, S.; Saeedifard, M. A New Hybrid Modular Multilevel Converter for Grid Connection of Large Wind Turbines. *IEEE Trans. Sustain. Energy* **2013**, *4*, 1051–1064. [CrossRef]
7. Chivite-Zabalza, J.; Izurza-Moreno, P.; Madariaga, D.; Calvo, G.; Rodríguez, M.A. Voltage Balancing control in 3-Level Neutral-Point Clamped Inverters Using Triangular Carrier PWM Modulation for FACTS Applications. *IEEE Trans. Power Electron.* **2013**, *28*, 4473–4484. [CrossRef]
8. Ara, A.L.; Kazemi, A.; Niaki, S.A.N. Multiobjective Optimal Location of FACTS Shunt-Series Controllers for Power System Operation Planning. *IEEE Trans. Power Deliv.* **2011**, *27*, 481–490. [CrossRef]
9. Hao, Q.; Ooi, B.-T. Tap for Classical HVDC Based on Multilevel Current-Source Inverters. *IEEE Trans. Power Deliv.* **2010**, *25*, 2626–2632. [CrossRef]
10. Soto, D.; Green, T. A comparison of high-power converter topologies for the implementation of FACTS controllers. *IEEE Trans. Ind. Electron.* **2002**, *49*, 1072–1080. [CrossRef]
11. Nami, A.; Liang, J.; Dijkhuizen, F.; Demetriades, G.D. Modular Multilevel Converters for HVDC Applications: Review on Converter Cells and Functionalities. *IEEE Trans. Power Electron.* **2014**, *30*, 18–36. [CrossRef]
12. Ghat, M.B.; Shukla, A. A new H-bridge hybrid modular converter (HBHMC) for HVDC application: Operating modes, control, and voltage balancing. *IEEE Trans. Power Electron.* **2017**, *33*, 6537–6554. [CrossRef]
13. Jung, J.-J.; Cui, S.; Lee, J.-H.; Sul, S.-K. A New Topology of Multilevel VSC Converter for a Hybrid HVDC Transmission System. *IEEE Trans. Power Electron.* **2016**, *32*, 4199–4209. [CrossRef]
14. Feldman, R.; Tomasini, M.; Amankwah, E.; Clare, J.; Wheeler, P.; Trainer, D.R.; Whitehouse, R.S. A Hybrid Modular Multilevel Voltage Source Converter for HVDC Power Transmission. *IEEE Trans. Ind. Appl.* **2013**, *49*, 1577–1588. [CrossRef]
15. Sajadi, R.; Iman-Eini, H.; Bakhshizadeh, M.K.; Neyshabouri, Y.; Farhangi, S. Selective Harmonic Elimination Technique with Control of Capacitive DC-Link Voltages in an Asymmetric Cascaded H-Bridge Inverter for STATCOM Application. *IEEE Trans. Ind. Electron.* **2018**, *65*, 8788–8796. [CrossRef]
16. Babu, N.N.V.S.; Fernandes, B.G. Cascaded Two-Level Inverter-Based Multilevel STATCOM for High-Power Applications. *IEEE Trans. Power Deliv.* **2014**, *29*, 993–1001. [CrossRef]
17. Haw, L.K.; Dahidah, M.S.A.; Almurib, H. SHE–PWM Cascaded Multilevel Inverter with Adjustable DC Voltage Levels Control for STATCOM Applications. *IEEE Trans. Power Electron.* **2014**, *29*, 6433–6444. [CrossRef]
18. Liu, Y.; Luo, F. Trinary hybrid multilevel inverter used in STATCOM with unbalanced voltages. *IEE Proc. Electr. Power Appl.* **2005**, *152*, 1203–1222. [CrossRef]
19. Liu, Y.; Yang, S.; Wang, X.; Gunasekaran, D.; Karki, U.; Peng, F.Z. Application of Transformer-Less UPFC for Interconnecting Two Synchronous AC Grids with Large Phase Difference. *IEEE Trans. Power Electron.* **2015**, *31*, 6092–6103. [CrossRef]
20. Peng, F.Z.; Liu, Y.; Yang, S.; Zhang, S.; Gunasekaran, D.; Karki, U. Transformer-Less Unified Power-Flow Controller Using the Cascade Multilevel Inverter. *IEEE Trans. Power Electron.* **2016**, *31*, 5461–5472. [CrossRef]
21. Wang, J.; Peng, F. Unified Power Flow Controller Using the Cascade Multilevel Inverter. *IEEE Trans. Power Electron.* **2004**, *19*, 1077–1084. [CrossRef]
22. Rao, G.S.; Raju, S.S. Multilevel Inverter-Based Power Quality Improvement in Grid-Connected DVR System. In *Micro-Electronics and Telecommunication Engineering*; Springer: Singapore; pp. 361–366. [CrossRef]
23. Dinesh, T. Performance Improvement of Constant Current Controller Based T-type DVR Multilevel Inverter for Solar PV Integrated with Grid. *Turk. J. Comput. Math. Educ. (TURCOMAT)* **2021**, *12*, 1465–1470.
24. Hoon, Y.; Radzi, M.A.M.; Hassan, M.K.; Mailah, N.F. Operation of Three-Level Inverter-Based Shunt Active Power Filter under Nonideal Grid Voltage Conditions with Dual Fundamental Component Extraction. *IEEE Trans. Power Electron.* **2017**, *33*, 7558–7570. [CrossRef]
25. Shu, Z.; Lin, H.; Ziwei, Z.; Yin, X.; Zhou, Q. Specific order harmonics compensation algorithm and digital implementation for multi-level active power filter. *IET Power Electron.* **2017**, *10*, 525–535. [CrossRef]
26. Quraan, M.; Tricoli, P.; Arco, S.D.; Piegari, L. Efficiency Assessment of Modular Multilevel Converters for Battery Electric Vehicles. *IEEE Trans. Power Electron.* **2016**, *32*, 2041–2051. [CrossRef]
27. Ali, M.; Mansoor, M.; Tang, H.; Rana, A. Analysis of a seven-level asymmetrical hybrid multilevel converter for traction systems. *IET Power Electron.* **2017**, *10*, 1878–1888. [CrossRef]
28. Youssef, M.Z.; Woronowicz, K.; Aditya, K.; Azeez, N.A.; Williamson, S.S. Design and Development of an Efficient Multilevel DC/AC Traction Inverter for Railway Transportation Electrification. *IEEE Trans. Power Electron.* **2015**, *31*, 3036–3042. [CrossRef]
29. Pereda, J.; Dixon, J. 23-Level Inverter for Electric Vehicles Using a Single Battery Pack and Series Active Filters. *IEEE Trans. Veh. Technol.* **2012**, *61*, 1043–1051. [CrossRef]
30. Khoucha, F.; Lagoun, S.M.; Marouani, K.; Kheloui, A.; Benbouzid, M.E.H. Hybrid Cascaded H-Bridge Multilevel-Inverter Induction-Motor-Drive Direct Torque Control for Automotive Applications. *IEEE Trans. Ind. Electron.* **2009**, *57*, 892–899. [CrossRef]
31. Du, Z.; Ozpineci, B.; Tolbert, L.; Chiasson, J.N. DC–AC Cascaded H-Bridge Multilevel Boost Inverter with No Inductors for Electric/Hybrid Electric Vehicle Applications. *IEEE Trans. Ind. Appl.* **2009**, *45*, 963–970. [CrossRef]
32. Carpita, M.; Marchesoni, M.; Pellerin, M.; Moser, D. Multilevel Converter for Traction Applications: Small-Scale Prototype Tests Results. *IEEE Trans. Ind. Electron.* **2008**, *55*, 2203–2212. [CrossRef]

33. Tolbert, L.; Peng, F.Z.; Cunnyngham, T.; Chiasson, J. Charge balance control schemes for cascade multilevel converter in hybrid electric vehicles. *IEEE Trans. Ind. Electron.* **2002**, *49*, 1058–1064. [CrossRef]
34. Bucknall, R.W.G.; Ciaramella, K.M. On the Conceptual Design and Performance of a Matrix Converter for Marine Electric Propulsion. *IEEE Trans. Power Electron.* **2009**, *25*, 1497–1508. [CrossRef]
35. Apsley, J.; Gonzalez-Villasenor, A.; Barnes, M.; Smith, A.; Williamson, S.; Schuddebeurs, J.D.; Norman, P.J.; Booth, C.; Burt, G.; McDonald, J.R. Propulsion Drive Models for Full Electric Marine Propulsion Systems. *IEEE Trans. Ind. Appl.* **2009**, *45*, 676–684. [CrossRef]
36. Gritter, D.; Kalsi, S.; Henderson, N. Variable speed electric drive options for electric ships. In Proceedings of the IEEE Electric Ship Technologies Symposium, Philadelphia, PA, USA, 27 July 2005; pp. 347–354. [CrossRef]
37. Corzine, K.A.; Lu, S. Comparison of hybrid propulsion drive schemes. In Proceedings of the IEEE Electric Ship Technologies Symposium, Philadelphia, PA, USA, 27 July 2005; pp. 355–362. [CrossRef]
38. Shuai, L.; Corzine, K. Multilevel multi-phase propulsion drives. In Proceedings of the IEEE Electric Ship Technologies Symposium, Philadelphia, PA, USA, 27 July 2005; pp. 363–370. [CrossRef]
39. Rodriguez, J.; Lai, J.; Peng, F.Z. Multilevel inverters: A survey of topologies, controls, and applications. *IEEE Trans. Ind. Electron.* **2002**, *49*, 724–738. [CrossRef]
40. Rodriguez, J.; Pontt, J.; Alzarnora, G.; Becker, N.S.; Einenkel, O.; Weinstein, A.J. Novel 20-MW downhill conveyor system using three-level converters. *IEEE Trans. Ind. Electron.* **2002**, *49*, 1093–1100. [CrossRef]
41. Kouro, S.; Rebolledo, J.; Rodríguez, J. Reduced switching-frequency-modulation algorithm for high-power multilevel inverters. *IEEE Trans. Ind. Electron.* **2007**, *54*, 2894–2901. [CrossRef]
42. Sabate, J.; Garces, L.J.; Szczesny, P.M.; Li, Q.; Wirth, W.F. High-power high-fidelity switching amplifier driving gradient coils for MRI systems. In Proceedings of the 2004 IEEE 35th Annual Power Electronics Specialists Conference, Aachen, Germany, 20–25 June 2004; Volume 1, pp. 261–266. [CrossRef]
43. Yue, Y.; Xu, Q.; Luo, A.; Guo, P.; He, Z.; Li, Y. Analysis and control of tundish induction heating power supply using modular multilevel converter. *IET Gener. Transm. Distrib.* **2018**, *12*, 3452–3460. [CrossRef]
44. Siahbalaee, J.; Sanaie, N. Comparison of conventional and new cascaded multilevel inverter topologies based on novel indices. *ISA Trans.* **2021**. [CrossRef] [PubMed]
45. Islam, M.; Mekhilef, S.; Hasan, M. Single phase transformerless inverter topologies for grid-tied photovoltaic system: A review. *Renew. Sustain. Energy Rev.* **2015**, *45*, 69–86. [CrossRef]
46. Chamarthi, P.K.; Al-Durra, A.; El-Fouly, T.H.M.; Al Jaafari, K.A. A Novel Three-Phase Transformerless Cascaded Multilevel Inverter Topology for Grid-Connected Solar PV Applications. *IEEE Trans. Ind. Appl.* **2021**, *57*, 2285–2297. [CrossRef]
47. Kazmierkowski, M.P.; Franquelo, L.G.; Rodriguez, J.; Perez, M.A.; Leon, J.I. High-performance motor drives. *IEEE Ind. Electron. Mag.* **2011**, *5*, 6–26. [CrossRef]
48. Kouro, S.; Malinowski, M.; Gopakumar, K.; Pou, J.; Franquelo, L.G.; Wu, B.; Rodriguez, J.; Perez, M.A.; Leon, J.I. Recent Advances and Industrial Applications of Multilevel Converters. *IEEE Trans. Ind. Electron.* **2010**, *57*, 2553–2580. [CrossRef]
49. Khoucha, F.; Lagoun, M.S.; Kheloui, A.; Benbouzid, M.E.H. A Comparison of Symmetrical and Asymmetrical Three-Phase H-Bridge Multilevel Inverter for DTC Induction Motor Drives. *IEEE Trans. Energy Convers.* **2010**, *26*, 64–72. [CrossRef]
50. Ali, A.I.M.; Sayed, M.A.; Takeshita, T. Isolated single-phase single-stage DC-AC cascaded transformer-based multilevel inverter for stand-alone and grid-tied applications. *Int. J. Electr. Power Energy Syst.* **2020**, *125*, 106534. [CrossRef]
51. Agrawal, R.; Jain, S. Comparison of reduced part count multilevel inverters (RPC-MLIs) for integration to the grid. *Int. J. Electr. Power Energy Syst.* **2017**, *84*, 214–224. [CrossRef]
52. Baker, R.H.; Bannister, L.H. Electric Power Converter. U.S. Patent 3,867,643, 18 February 1975.
53. Mcmurray, W. Fast Response Stepped-Wave Switching Power Converter Circuit. U.S. Patent 3,581,212, 25 May 1971.
54. Dickerson, J.A.; Ottaway, G.H. Transformerless Power Supply with Line to Load Isolation. U.S. Patent No. 3,596,369, 3 August 1971.
55. Peng, F.Z.; Lai, J.-S.; McKeever, J.; VanCoevering, J. A multilevel voltage-source inverter with separate DC sources for static VAr generation. *IEEE Trans. Ind. Appl.* **1996**, *32*, 1130–1138. [CrossRef]
56. Akira, N.; Takahashi, I.; Akagi, H. A new neutral-point-clamped PWM inverter. *IEEE Trans. Ind. Appl.* **1981**, *5*, 518–523.
57. Baker, R.H. High-Voltage Converter Circuit. U.S. Patent 4,203,151, 13 May 1980.
58. Baker, R.H. Bridge Converter Circuit. U.S. Patent 4,270,163, 26 May 1981.
59. Rodriguez, J.; Bernet, S.; Steimer, P.K.; Lizama, I.E. A survey on neutral-point-clamped inverters. *IEEE Trans. Ind. Electron.* **2009**, *57*, 2219–2230. [CrossRef]
60. Meynard, T.; Foch, H. Multi-level conversion: High voltage choppers and voltage-source inverters. In Proceedings of the PESC'92 Record. 23rd Annual IEEE Power Electronics Specialists Conference, Toledo, Spain, 29 June–3 July 1992; Volume 1, pp. 397–403. [CrossRef]
61. Lavieville, J.; Carrere, P.; Meynard, T. Electronic Circuit for Converting Electrical Energy, and a Power Supply Installation Making Use Thereof. U.S. Patent 5,668,711, 16 September 1997.
62. Marquardt, R. Stromrichterschaltungen mit Verteilten Energiespeichern. German Patent DE10103031A1, 24 January 2001.
63. Rainer, M.; Lesnicar, A.; Hildinger, J. *Modulares Stromrichterkonzept für Netzkupplungsanwendung bei Hohen Spannungen*; ETG-Fachtagung: Bad Nauheim, Germany, 2002; p. 114.

64. Peng, F.Z. A generalized multilevel inverter topology with self voltage balancing. *IEEE Trans. Ind. Appl.* **2001**, *37*, 611–618. [CrossRef]
65. Barbosa, P.; Steimer, P.; Meysenc, L.; Winkelnkemper, M.; Steinke, J.; Celanovic, N. Active neutral-point-clamped multilevel converters. In Proceedings of the IEEE 36th Power Electronics Specialists Conference, Dresden, Germany, 16 June 2005; pp. 2296–2301.
66. Omer, P.; Kumar, J.; Surjan, B.S. A Review on Reduced Switch Count Multilevel Inverter Topologies. *IEEE Access* **2020**, *8*, 22281–22302. [CrossRef]
67. Siddique, M.D.; Iqbal, A.; Memon, M.A.; Mekhilef, S. A New Configurable Topology for Multilevel Inverter with Reduced Switching Components. *IEEE Access* **2020**, *8*, 188726–188741. [CrossRef]
68. Bruckner, T.; Bernet, S.; Guldner, H. The Active NPC Converter and Its Loss-Balancing Control. *IEEE Trans. Ind. Electron.* **2005**, *52*, 855–868. [CrossRef]
69. Gupta, K.K.; Jain, S. A Novel Multilevel Inverter Based on Switched DC Sources. *IEEE Trans. Ind. Electron.* **2013**, *61*, 3269–3278. [CrossRef]
70. Babaei, E.; Laali, S.; Alilu, S. Cascaded Multilevel Inverter with Series Connection of Novel H-Bridge Basic Units. *IEEE Trans. Ind. Electron.* **2014**, *61*, 6664–6671. [CrossRef]
71. Babaei, E. A Cascade Multilevel Converter Topology with Reduced Number of Switches. *IEEE Trans. Power Electron.* **2008**, *23*, 2657–2664. [CrossRef]
72. Su, G. Multilevel DC-link inverter. *IEEE Trans. Ind. Appl.* **2005**, *41*, 848–854. [CrossRef]
73. Mudadla, D.; Rao, G.R.; Sandeep, N. Novel asymmetrical multilevel inverter topology with reduced number of switches for photovoltaic applications. In Proceedings of the 2015 International Conference on Computation of Power, Energy,, Information and Communication (ICCPEIC), Melmaruvathur, India, 22–23 April 2015; pp. 123–128. [CrossRef]
74. Gautam, S.P.; Kumar, L.; Gupta, S. A modified structure for symmetrical and asymmetrical configuration of multilevel inverter. In Proceedings of the IECON 2015-41st Annual Conference of the IEEE Industrial Electronics Society, Yokohama, Japan, 9–12 November 2015; pp. 1430–1435. [CrossRef]
75. Kashif, M.F.; Rashid, A.K. A multilevel inverter topology with reduced number of switches. In Proceedings of the 2016 International Conference on Intelligent Systems Engineering (ICISE), Islamabad, Pakistan, 15–17 January 2016; pp. 268–271.
76. Kang, S.H.; Lee, F.S. A new structure of H-bridge multilevel inverter. In Proceedings of the Annual Fall Conference of Power Electronics (KIPE Conference), Goyang, Korea, 31 October 2008; pp. 388–390.
77. Choi, W.; Kang, F. H-bridge based multilevel inverter using PWM switching function. In Proceedings of the INTELEC 2009-31st International Telecommunications Energy Conference, Incheon, Korea, 18–22 October 2009; pp. 1–5.
78. Najafi, E.; Yatim, A.H. Design and Implementation of a New Multilevel Inverter Topology. *IEEE Trans. Ind. Electron.* **2011**, *59*, 4148–4154. [CrossRef]
79. Ebrahimi, J.; Babaei, E.; Gharehpetian, G.B. A New Multilevel Converter Topology with Reduced Number of Power Electronic Components. *IEEE Trans. Ind. Electron.* **2011**, *59*, 655–667. [CrossRef]
80. Ceglia, G.; Guzman, V.; Sanchez, C.; Ibanez, F.; Walter, J.; Gimenez, M. A New Simplified Multilevel Inverter Topology for DC–AC Conversion. *IEEE Trans. Power Electron.* **2006**, *21*, 1311–1319. [CrossRef]
81. Martins, G.; Pomilio, J.A.; Buso, S.; Spiazzi, G. Three-Phase Low-Frequency Commutation Inverter for Renewable Energy Systems. *IEEE Trans. Ind. Electron.* **2006**, *53*, 1522–1528. [CrossRef]
82. Rahim, N.A.; Chaniago, K.; Selvaraj, J. Single-Phase Seven-Level Grid-Connected Inverter for Photovoltaic System. *IEEE Trans. Ind. Electron.* **2010**, *58*, 2435–2443. [CrossRef]
83. Farhadi-Kangarlu, M.; Babaei, E. A Generalized Cascaded Multilevel Inverter Using Series Connection of Submultilevel Inverters. *IEEE Trans. Power Electron.* **2012**, *28*, 625–636. [CrossRef]
84. Narimani, M.; Wu, B.; Zargari, N.R. A Novel Five-Level Voltage Source Inverter with Sinusoidal Pulse Width Modulator for Medium-Voltage Applications. *IEEE Trans. Power Electron.* **2015**, *31*, 1959–1967. [CrossRef]
85. Le, Q.A.; Lee, D.-C. A Novel Six-Level Inverter Topology for Medium-Voltage Applications. *IEEE Trans. Ind. Electron.* **2016**, *63*, 7195–7203. [CrossRef]
86. Kumari, M.; Siddique, M.D.; Sarwar, A.; Tariq, M.; Mekhilef, S.; Iqbal, A. Recent trends and review on switched-capacitor-based single-stage boost multilevel inverter. *Int. Trans. Electr. Energy Syst.* **2021**, *31*, e12730. [CrossRef]
87. Kaarthik, R.S.; Kshirsagar, A.; Gopakumar, K. Generation of Higher Number of Voltage Levels by Stacking Inverters of Lower Multilevel Structures with Low Voltage Devices for Drives. *IEEE Trans. Power Electron.* **2017**, *32*, 52–59. [CrossRef]
88. Panda, K.P.; Bana, P.R.; Panda, G. A Switched-Capacitor Self-Balanced High-Gain Multilevel Inverter Employing a Single DC Source. *IEEE Trans. Circuits Syst. II Express Briefs* **2020**, *67*, 3192–3196. [CrossRef]
89. Yeganeh, M.S.O.; Davari, P.; Chub, A.; Mijatovic, N.; Dragicevic, T.; Blaabjerg, F. A Single-Phase Reduced Component Count Asymmetrical Multilevel Inverter Topology. *IEEE J. Emerg. Sel. Top. Power Electron.* **2021**, *1*. [CrossRef]
90. Tolbert, L.; Peng, F.; Habetler, T. Multilevel inverters for electric vehicle applications. In Proceedings of the Power Electronics in Transportation (Cat. No.98TH8349), Dearborn, MI, USA, 22–23 October 1998; pp. 79–84. [CrossRef]
91. Bendre, A.; Krstic, S.; Meer, J.; Venkataramanan, G. Comparative evaluation of modulation algorithms for neutral point clamped converters. In Proceedings of the Conference Record of the 2004 IEEE Industry Applications Conference, Seattle, WA, USA, 3–7 October 2004; Volume 2, pp. 798–805. [CrossRef]

92. Joos, G.; Huang, X.; Ooi, B.-T. Direct-coupled multilevel cascaded series VAr compensators. *IEEE Trans. Ind. Appl.* **1998**, *34*, 1156–1163. [CrossRef]
93. Çolak, I.; Kabalci, E. A review on inverter topologies and developments. In Proceedings of the Electrics, Electronics and Computer Engineering Symposium, Bursa, Turkey, 26–30 November; 2008.
94. Purkait, P.; Sriramakavacham, R.S. A New Generalized Space Vector Modulation Algorithm for Neutral-point-clamped Multilevel Converters. *PIERS Online* **2006**, *2*, 330–335. [CrossRef]
95. Sadigh, A.K.; Hosseini, S.H.; Sabahi, M.; Gharehpetian, G.B. Double Flying Capacitor Multicell Converter Based on Modified Phase-Shifted Pulsewidth Modulation. *IEEE Trans. Power Electron.* **2009**, *25*, 1517–1526. [CrossRef]
96. Bruckner, T.; Bemet, S. Loss balancing in three-level voltage source inverters applying active NPC switches. In Proceedings of the IEEE 32nd Annual Power Electronics Specialists Conference (IEEE Cat. No. 01CH37230), Vancouver, Canada, 17–21 June 2001; Volume 2, pp. 1135–1140. [CrossRef]
97. Barbosa, P.; Steimer, P.; Steinke, J.; Winkelnkemper, M.; Celanovic, N. Active-neutral-point-clamped (ANPC) multilevel converter technology. In Proceedings of the 2005 European Conference on Power Electronics and Applications, Dresden, Germany, 11–14 September 2005; pp. 1–10.
98. Lesnicar, A.; Marquardt, R. An innovative modular multilevel converter topology suitable for a wide power range. In Proceedings of the 2003 IEEE Bologna Power Tech Conference Proceedings, Bologna, Italy, 23–26 June 2004; Volume 3, pp. 1–6. [CrossRef]
99. Pérez, M.; Ceballos, S.; Konstantinou, G.; Pou, J.; Aguilera, R. Modular Multilevel Converters: Recent Achievements and Challenges. *IEEE Open J. Ind. Electron. Soc.* **2021**, *2*, 224–239. [CrossRef]
100. Kurtoğlu, M.; Eroğlu, F.; Arslan, A.O.; Vural, A.M. Recent contributions and future prospects of the modular multilevel converters: A comprehensive review. *Int. Trans. Electr. Energy Syst.* **2018**, *29*, e2763. [CrossRef]
101. Vozikis, D.; Adam, G.; Rault, P.; Despouys, O.; Holliday, D. Enhanced Modular Multilevel Converter for HVdc Applications: Assessments of Dynamic and Transient Responses to ac and dc Faults. *IEEE J. Emerg. Sel. Top. Power Electron.* **2020**, 1. [CrossRef]
102. Tian, Y.; Wickramasinghe, H.R.; Pou, J.; Konstantinou, G. Loss distribution and characterization of MMC sub-modules for HVDC applications. *Int. Trans. Electr. Energy Syst.* **2021**, e13042. [CrossRef]
103. Yu, P.; Fu, W.; Wang, L.; Zhou, Z.; Wang, G.; Zhang, Z. Reliability-Centered Maintenance for Modular Multilevel Converter in HVDC Transmission Application. *IEEE J. Emerg. Sel. Top. Power Electron.* **2020**, *9*, 3166–3176. [CrossRef]
104. Martinez-Rodrigo, F.; de Pablo, S.; de Lucas, L.C.H. Current control of a modular multilevel converter for HVDC applications. *Renew. Energy* **2015**, *83*, 318–331. [CrossRef]
105. Sanz, I.; Moranchel, M.; Bueno, E.J.; Rodriguez, F.J. Analysis of medium voltage modular multilevel converters for FACTS applications. In Proceedings of the IECON 2016-42nd Annual Conference of the IEEE Industrial Electronics Society, Florence, Italy, 24–27 October 2016; pp. 6459–6464. [CrossRef]
106. Bhesaniya, M.M.; Shukla, A. Current Source Modular Multilevel Converter: Detailed Analysis and STATCOM Application. *IEEE Trans. Power Deliv.* **2015**, *31*, 323–333. [CrossRef]
107. Vural, A.M.; Wirsiy, E.N. Three-phase modular multilevel converter based unified power flow controller. *Eng. Sci. Technol. Int. J.* **2019**, *23*, 299–306. [CrossRef]
108. Liang, G.; Tafti, H.D.; Farivar, G.G.; Pou, J.; Townsend, C.D.; Konstantinou, G.; Ceballos, S. Analytical Derivation of Intersubmodule Active Power Disparity Limits in Modular Multilevel Converter-Based Battery Energy Storage Systems. *IEEE Trans. Power Electron.* **2020**, *36*, 2864–2874. [CrossRef]
109. Mo, R.; Li, H. Hybrid Energy Storage System with Active Filter Function for Shipboard MVDC System Applications Based on Isolated Modular Multilevel DC/DC Converter. *IEEE J. Emerg. Sel. Top. Power Electron.* **2016**, *5*, 79–87. [CrossRef]
110. Xu, Y.; Zhang, Z.; Wang, G.; Xu, Z. Modular Multilevel Converter with Embedded Energy Storage for Bidirectional Fault Isolation. *IEEE Trans. Power Deliv.* **2021**, 1. [CrossRef]
111. Hariri, R.; Sebaaly, F.; Kanaan, H.Y. A Review on Modular Multilevel Converters in Electric Vehicles. In Proceedings of the IECON 2020 The 46th Annual Conference of the IEEE Industrial Electronics Society, Singapore, 18–21 October 2020.
112. Quraan, M.; Abu-Khaizaran, M.; Sa'Ed, J.; Hashlamoun, W.; Tricoli, P. Design and control of battery charger for electric vehicles using modular multilevel converters. *IET Power Electron.* **2020**, *14*, 140–157. [CrossRef]
113. Zhou, S.; Li, B.; Wang, J.; Xu, D. A Modified Modular Multilevel Converter for Motor Drives Capable of High-Torque Operation at Zero/Low Motor Speeds. *IEEE Trans. Circuits Syst. II Express Briefs* **2021**, *68*, 2493–2497. [CrossRef]
114. Chakraborty, S.; Maiti, S.; Bharadwaj, C.A. A Novel AC/AC Modular Multilevel Converter for Medium Voltage Variable Frequency Vector Controlled Induction Motor Drives. In Proceedings of the 2020 IEEE International Conference on Power Electronics, Smart Grid and Renewable Energy (PESGRE2020), Cochin, India, 2–4 January 2020; pp. 1–6. [CrossRef]
115. Diab, M.; Adam, G.P.; Williams, B.W.; Massoud, A.; Ahmed, S. Quasi two-level PWM operation of a nine-arm modular multilevel converter for six-phase medium-voltage motor drives. In Proceedings of the 2018 IEEE Applied Power Electronics Conference and Exposition (APEC), San Antonio, TX, USA, 4–8 March 2018; pp. 1641–1648. [CrossRef]
116. Madi, A.I.; Hamad, M.S.; Hamdy, R.A.R.; El-Arabawy, I.F. Hybrid active power filter with modular multilevel converter. In Proceedings of the 2017 Nineteenth International Middle East Power Systems Conference (MEPCON), Cairo, Egypt, 19–21 December 2017; pp. 1100–1105.
117. Jia, G.; Chen, M.; Tang, S.; Zhang, C.; Zhao, B. A Modular Multilevel Converter with Active Power Filter for Submodule Capacitor Voltage Ripples and Power Losses Reduction. *IEEE Trans. Power Electron.* **2020**, *35*, 11401–11417. [CrossRef]

118. António-Ferreira, A.; Collados-Rodríguez, C.; Gomis-Bellmunt, O. Modulation techniques applied to medium voltage modular multilevel converters for renewable energy integration: A review. *Electr. Power Syst. Res.* **2018**, *155*, 21–39. [CrossRef]
119. Shahnazian, F.; Adabi, J.; Pouresmaeil, E.; Catalão, J.P. Interfacing modular multilevel converters for grid integration of renewable energy sources. *Electr. Power Syst. Res.* **2018**, *160*, 439–449. [CrossRef]
120. Basu, T.S.; Maiti, S. A Hybrid Modular Multilevel Converter for Solar Power Integration. *IEEE Trans. Ind. Appl.* **2019**, *55*, 5166–5177. [CrossRef]
121. Novakovic, B.; Nasiri, A. Modular Multilevel Converter for Wind Energy Storage Applications. *IEEE Trans. Ind. Electron.* **2017**, *64*, 8867–8876. [CrossRef]
122. Wang, W.; Ma, K.; Cai, X. Efficient Capacitor Voltage Balancing Method for Modular Multilevel Converter Under Carrier-Phase-Shift Pulsewidth Modulation. *IEEE Trans. Power Electron.* **2020**, *36*, 1553–1562. [CrossRef]
123. Sun, C.; Zhang, X.; Cai, X. A Step-Up Nonisolated Modular Multilevel DC–DC Converter with Self-Voltage Balancing and Soft Switching. *IEEE Trans. Power Electron.* **2020**, *35*, 13017–13030. [CrossRef]
124. Tashakor, N.; Kilictas, M.; Fang, J.; Goetz, S.M. Switch-Clamped Modular Multilevel Converters with Sensorless Voltage Balancing Control. *IEEE Trans. Ind. Electron.* **2020**, *68*, 9586–9597. [CrossRef]
125. Oghorada, O.J.K.; Zhang, L.; Han, H.; Esan, A.B.; Mao, M. Inter-cluster voltage balancing control of a delta connected modular multilevel cascaded converter under unbalanced grid voltage. *Prot. Control. Mod. Power Syst.* **2021**, *6*, 1–11. [CrossRef]
126. Luo, W.; Ma, Y.; Zheng, C. Selection-based capacitor voltage balancing control for modular multilevel converters. *J. Power Electron.* **2021**, 1–12. [CrossRef]
127. Guo, L.; Sun, Y.; Jin, N. A Capacitor Voltage Balancing Control Strategy for Modular Multilevel Converter. *Electr. Power Components Syst.* **2020**, *48*, 1410–1420. [CrossRef]
128. Ji, S.; Zhang, L.; Huang, X.; Palmer, J.; Wang, F.; Tolbert, L.M. A Novel Voltage Balancing Control with dv/dt Reduction for 10-kV SiC MOSFET-Based Medium Voltage Modular Multilevel Converter. *IEEE Trans. Power Electron.* **2020**, *35*, 12533–12543. [CrossRef]
129. Zhou, S.; Wen, B.; Wang, J.; Burgos, R.; Boroyevich, D. Design and Hardware Implementation of the Peak Current Mode Switching Cycle Control for Voltage Balancing of Modular Multilevel Converters. In Proceedings of the 2021 IEEE Applied Power Electronics Conference and Exposition (APEC), Phoenix, AZ, USA, 21–25 March 2021; pp. 1134–1139. [CrossRef]
130. Kumar, M.A.; Gopi, A.K.; Biswas, J.; Barai, M. A Voltage Balancing Scheme for Modular Multilevel Converter Based on Charge Variation in Each Cycle. *IEEE J. Emerg. Sel. Top. Ind. Electron.* **2021**, *2*, 173–183. [CrossRef]
131. Wang, C.; Hu, S.; Peng, H.; He, L.; Fan, S. Modulation Coordinated Voltage Balance Strategy for a Dual-T-Type Modular Multilevel Converter. *IEEE J. Emerg. Sel. Top. Power Electron.* **2021**, 1. [CrossRef]
132. Pourgharibshahi, H.; Jafarishiadeh, S.; Mahmoudi, H.; Zargarzadeh, H.; Ahmadi, R. Novel single-armed modular multilevel converter for reducing total converter capacitance. *IET Power Electron.* **2020**, *14*, 760–774. [CrossRef]
133. Shu, H.; Lei, S.; Tian, X. A New Topology of Modular Multilevel Converter with Voltage Self-Balancing Ability. *IEEE Access* **2019**, *7*, 184786–184796. [CrossRef]
134. Wang, Z.; Lin, H.; Ma, Y. Improved capacitor voltage balancing control for multimode operation of modular multilevel converter with integrated battery energy storage system. *IET Power Electron.* **2019**, *12*, 2751–2760. [CrossRef]
135. Kadandani, N.B.; Dahidah, M.; Ethni, S.; Muhammad, M. Lifetime and reliability improvements in modular multilevel converters using controlled circulating current. *J. Power Electron.* **2021**, 1–10. [CrossRef]
136. Kadandani, N.B.; Dahidah, M.; Ethni, S. Review of Circulating Current Control Methods in Modular Multilevel Converter. *Bayero J. Eng. Technol. (BJET)* **2021**, *16*, 62–75.
137. Huber, J.E.; Kolar, J.W. Modular Multilevel Converter Circulating Current Control with Single Active Filter Module per Phase. In Proceedings of the 2021 22nd IEEE International Conference on Industrial Technology (ICIT), Valencia, Spain, 10–12 March 2021; Volume 1, pp. 433–439. [CrossRef]
138. Lin, L.; He, J.; Xu, C. Analysis on Circulating Current and Split Capacitor Voltage Balance for Modular Multilevel Converter Based Three-phase Four-wire Split Capacitor DSTATCOM. *J. Mod. Power Syst. Clean Energy* **2021**, *9*, 657–667. [CrossRef]
139. Praveena, K.; Swarnasri, K. Circulating current harmonics suppression with fuzzy controller in modular multi-level inverter. *Mater. Today: Proc.* **2021**. [CrossRef]
140. Narayana, T.S.; Dahiya, R. DC-link voltage compensation with fuzzy-based MMC to mitigate the low-order circulating current. *Int. J. Ambient. Energy* **2021**, 1–10. [CrossRef]
141. Wang, L.; Zhang, L.; Xiong, Y.; Ma, R. Low-frequency suppression strategy based on predictive control model for modular multilevel converters. *J. Power Electron.* **2021**, 1–9. [CrossRef]
142. Chen, X.; Liu, J.; Song, S.; Ouyang, S. Circulating Harmonic Currents Suppression of Level-Increased NLM Based Modular Multilevel Converter with Deadbeat Control. *IEEE Trans. Power Electron.* **2020**, *35*, 11418–11429. [CrossRef]
143. Uddin, W.; Zeb, K.; Adil Khan, M.; Ishfaq, M.; Khan, I.; Kim, H.-J.; Park, G.S.; Lee, C. Control of Output and Circulating Current of Modular Multilevel Converter Using a Sliding Mode Approach. *Energies* **2019**, *12*, 4084. [CrossRef]
144. Kolluri, S.; Gorla, N.B.Y.; Panda, S.K. Capacitor Voltage Ripple Suppression in a Modular Multilevel Converter Using Frequency-Adaptive Spatial Repetitive-Based Circulating Current Controller. *IEEE Trans. Power Electron.* **2020**, *35*, 9839–9849. [CrossRef]
145. Chakraborty, R.; Dey, A. Circulating Current Control of Modular Multilevel Converter with Reduced Conduction Loss for Medium-Voltage Applications. *IEEE Trans. Ind. Electron.* **2020**, *68*, 9014–9023. [CrossRef]

146. Deng, F.; Heng, Q.; Liu, C.; Lyu, Y.; Wang, Q.; Liu, D.; Zhu, R. Circulating current suppression for MMC-HVDC systems with asymmetric arm impedance. *CSEE J. Power Energy Syst.* **2021**, *7*, 530–540.
147. Zhang, M.; Shen, Y.; Sun, H.; Guo, R. MMC-HVDC circulating current suppression method based on improved proportional resonance control. *Energy Rep.* **2020**, *6*, 863–871. [CrossRef]
148. Isik, S.; Alharbi, M.; Bhattacharya, S. Optimized Circulating Current Control Method based on Proportional Resonant and Proportional Integral Controllers for Modular Multi-level Converter Applications. In Proceedings of the 2020 IEEE Energy Conversion Congress and Exposition (ECCE), Detroit, MI, USA, 11–15 October 2020; pp. 5709–5715. [CrossRef]
149. Qin, F.; Gao, F.; Wang, X.; Niu, D. Circulating Current Control Method for Nine-Arm Modular Multilevel Converter. In Proceedings of the 2020 IEEE 9th International Power Electronics and Motion Control Conference (IPEMC2020-ECCE Asia), Nanjing, China, 31 May–3 June 2020; pp. 699–703.
150. Viatkin, A.; Ricco, M.; Mandrioli, R.; Kerekes, T.; Teodorescu, R.; Grandi, G. Modular Multilevel Converters Based on Interleaved Half-Bridge Submodules. In Proceedings of the 2021 22nd IEEE International Conference on Industrial Technology (ICIT), Valencia, Spain, 10–12 March 2021; Volume 1, pp. 440–445. [CrossRef]
151. Prasad, H.; Rao, R.K.; Kumar, S.K. Cascaded Operation of Hybrid Multilevel Inverter with Optimum Switching Angle Control for Power Quality Enhancement. In *Microelectronics, Electromagnetics and Telecommunications*; Springer: Singapore, 2021; pp. 21–31. [CrossRef]
152. Kalpana, K.; Nirmala, M. A Novel 31 Level Cascaded H-Bridge Inverter. In Proceedings of the 2021 International Conference on Advance Computing and Innovative Technologies in Engineering (ICACITE), Greater Noida, India, 4–5 March 2021; pp. 228–233.
153. Panagis, P.; Stergiopoulos, F.; Marabeas, P.; Manias, S. Comparison of state of the art multilevel inverters. In Proceedings of the 2008 IEEE Power Electronics Specialists Conference, Rhodes, Greece, 15–19 June 2008; pp. 4296–4301. [CrossRef]
154. Hochgraf, C.; Lasseter, R.; Divan, D.; Lipo, T. Comparison of multilevel inverters for static var compensation. In Proceedings of the 1994 IEEE Industry Applications Society Annual Meeting, Denver, CO, USA, 2–6 October 1994; Volume 2, pp. 921–928.
155. Kuhn, H.; Ruger, N.E.; Mertens, A. Control Strategy for Multilevel Inverter with Non-ideal DC Sources. In Proceedings of the 2007 IEEE Power Electronics Specialists Conference, Orlando, FL, USA, 17–21 June 2007; pp. 632–638. [CrossRef]
156. Odeh, C.I.; Nnadi, D.B.N. Single-phase 9-level hybridised cascaded multilevel inverter. *IET Power Electron.* **2013**, *6*, 468–477. [CrossRef]
157. Choi, N.S.; Cho, J.G.; Cho, G.H. A general circuit topology of multilevel inverter. In Proceedings of the PESC'91 Record 22nd Annual IEEE Power Electronics Specialists Conference, Cambridge, MA, USA, 24–27 June 1991; pp. 96–103. [CrossRef]
158. Halim, W.A.; Rahim, N.A.; Azri, M. Generalized Selective Harmonic Elimination Modulation for Transistor-Clamped H-Bridge Multilevel Inverter. *J. Power Electron.* **2015**, *15*, 964–973. [CrossRef]
159. Sarwer, Z.; Siddique, M.D.; Iqbal, A.; Sarwar, A.; Mekhilef, S. An improved asymmetrical multilevel inverter topology with reduced semiconductor device count. *Int. Trans. Electr. Energy Syst.* **2020**, *30*. [CrossRef]
160. Ponraj, R.P.; Sigamani, T.; Subramanian, V. A Developed H-Bridge Cascaded Multilevel Inverter with Reduced Switch Count. *J. Electr. Eng. Technol.* **2021**, *16*, 1445–1455. [CrossRef]
161. Jinghua, Z.; Zhengxi, L. Research on hybrid modulation strategies based on general hybrid topology of multilevel inverter. In Proceedings of the 2008 International Symposium on Power Electronics, Electrical Drives, Automation and Motion, Ischia, Italy, 11–13 June 2008; pp. 784–788. [CrossRef]
162. Manjrekar, M.D.; Lipo, T.A. A hybrid multilevel inverter topology for drive applications. In Proceedings of the APEC'98 Thirteenth Annual Applied Power Electronics Conference and Exposition, Anaheim, CA, USA, 15–19 February 1998; Volume 2, pp. 523–529.
163. Dhanamjayulu, C.; Arunkumar, G.; Pandian, B.J.; Padmanaban, S. Design and implementation of a novel asymmetrical multilevel inverter optimal hardware components. *Int. Trans. Electr. Energy Syst.* **2019**, *30*. [CrossRef]
164. Manjrekar, M.; Steimer, P.; Lipo, T. Hybrid multilevel power conversion system: A competitive solution for high-power applications. *IEEE Trans. Ind. Appl.* **2000**, *36*, 834–841. [CrossRef]
165. Bin Arif, M.S.; Sarwer, Z.; Siddique, M.D.; Ayob, S.M.; Iqbal, A.; Mekhilef, S. Asymmetrical multilevel inverter topology with low total standing voltage and reduced switches count. *Int. J. Circuit Theory Appl.* **2021**, *49*, 1757–1775. [CrossRef]
166. Colak, I.; Kabalcı, E.; Keven, G. Asymmetrical multilevel inverter topologies. In *Multilevel Inverters*; Academic Press: Cambridge, MA, USA, 2021; pp. 181–215. [CrossRef]
167. Kakar, S.; Ayob, S.B.M.; Iqbal, A.; Nordin, N.M.; Bin Arif, M.S.; Gore, S. New Asymmetrical Modular Multilevel Inverter Topology with Reduced Number of Switches. *IEEE Access* **2021**, *9*, 27627–27637. [CrossRef]
168. Gupta, K.K.; Ranjan, A.; Bhatnagar, P.; Sahu, L.K.; Jain, S. Multilevel Inverter Topologies with Reduced Device Count: A Review. *IEEE Trans. Power Electron.* **2015**, *31*, 135–151. [CrossRef]
169. Siddique, M.D.; Rawa, M.; Mekhilef, S.; Shah, N.M. A new cascaded asymmetrical multilevel inverter based on switched dc voltage sources. *Int. J. Electr. Power Energy Syst.* **2021**, *128*, 106730. [CrossRef]
170. Hinago, Y.; Koizumi, H. A Single-Phase Multilevel Inverter Using Switched Series/Parallel DC Voltage Sources. *IEEE Trans. Ind. Electron.* **2009**, *57*, 2643–2650. [CrossRef]
171. Poorfakhraei, A.; Narimani, M.; Emadi, A. A Review of Multilevel Inverter Topologies in Electric Vehicles: Current Status and Future Trends. *IEEE Open J. Power Electron.* **2021**, *2*, 155–170. [CrossRef]

172. Poorfakhraei, A.; Narimani, M.; Emadi, A. A Review of Modulation and Control Techniques for Multilevel Inverters in Traction Applications. *IEEE Access* **2021**, *9*, 24187–24204. [CrossRef]
173. Amir, A.; Amir, A.; Selvaraj, J.; Rahim, N.A. Grid-connected photovoltaic system employing a single-phase T-type cascaded H-bridge inverter. *Sol. Energy* **2020**, *199*, 645–656. [CrossRef]
174. Nilkar, M.; Babaei, E.; Sabahi, M. A new single-phase cascade multilevel inverter topology using four-level cells. In Proceedings of the 20th Iranian Conference on Electrical Engineering (ICEE2012), Tehran, Iran, 15–17 May 2012; pp. 348–353. [CrossRef]
175. Khosroshahi, M.T. Crisscross cascade multilevel inverter with reduction in number of components. *IET Power Electron.* **2014**, *7*, 2914–2924. [CrossRef]
176. Najafi, E.; Yatim, A.H.M.; Samosir, A.S. A new topology -Reversing Voltage (RV)—for multi level inverters. In Proceedings of the 2008 IEEE 2nd International Power and Energy Conference, Johor Bahru, Malaysia, 1–3 December 2008; pp. 604–608. [CrossRef]
177. Noguchi, T.; Suroso. A multilevel voltage-source inverter using H-bridge and two-level power modules with a single power source. In Proceedings of the 2011 IEEE Ninth International Conference on Power Electronics and Drive Systems, Singapore, 5–8 December 2011; pp. 262–266. [CrossRef]
178. Ounejjar, Y.; Al-Haddad, K. A novel high energetic efficiency multilevel topology with reduced impact on supply network. In Proceedings of the 2008 34th Annual Conference of IEEE Industrial Electronics, Orlando, FL, USA, 10–13 November 2008; pp. 489–494. [CrossRef]
179. Ounejjar, Y.; Al-Haddad, K. A new high power efficiency cascaded U cells multilevel converter. In Proceedings of the 2009 IEEE International Symposium on Industrial Electronics, Seoul, Korea, 5–8 July 2009; pp. 483–488. [CrossRef]
180. Ounejjar, Y.; Al-Haddad, K.; Grégoire, L. Novel three phase seven level PWM converter. In Proceedings of the 2009 IEEE Electrical Power & Energy Conference (EPEC), Montreal, QC, Canada, 22–23 October 2009; pp. 1–6.
181. Ounejjar, Y.; Al-Haddad, K. Multilevel hysteresis controller of the novel seven-level packed U cells converter. *SPEEDAM 2010* **2010**, 186–191. [CrossRef]
182. Ounejjar, Y.; Al-Haddad, K.; Grégoire, L.-A. Packed U Cells Multilevel Converter Topology: Theoretical Study and Experimental Validation. *IEEE Trans. Ind. Electron.* **2010**, *58*, 1294–1306. [CrossRef]
183. Mokhberdoran, A.; Ajami, A. Symmetric and Asymmetric Design and Implementation of New Cascaded Multilevel Inverter Topology. *IEEE Trans. Power Electron.* **2014**, *29*, 6712–6724. [CrossRef]
184. Babaei, E.; Alilu, S.; Laali, S. A New General Topology for Cascaded Multilevel Inverters with Reduced Number of Components Based on Developed H-Bridge. *IEEE Trans. Ind. Electron.* **2013**, *61*, 3932–3939. [CrossRef]
185. El-Naggar, K.; Abdelhamid, T.H. Selective harmonic elimination of new family of multilevel inverters using genetic algorithms. *Energy Convers. Manag.* **2008**, *49*, 89–95. [CrossRef]
186. Perez, M.A.; Bernet, S.; Rodriguez, J.; Kouro, S.; Lizana, R. Circuit Topologies, Modeling, Control Schemes, and Applications of Modular Multilevel Converters. *IEEE Trans. Power Electron.* **2014**, *30*, 4–17. [CrossRef]
187. Rodriguez, J.; Moran, L.; Correa, P.; Silva, C. A vector control technique for medium-voltage multilevel inverters. *IEEE Trans. Ind. Electron.* **2002**, *49*, 882–888. [CrossRef]
188. Song, B.-M.; Kim, J.; Lai, J.-S.; Seong, K.-C.; Kim, H.-J.; Park, S.-S. A multilevel soft-switching inverter with inductor coupling. *IEEE Trans. Ind. Appl.* **2001**, *37*, 628–636. [CrossRef]
189. Sirisukprasert, S.; Lai, J.; Liu, T.-H. Optimum harmonic reduction with a wide range of modulation indexes for multilevel converters. *IEEE Trans. Ind. Electron.* **2002**, *49*, 875–881. [CrossRef]
190. Rodriguez, J.; Moran, L.; Pontt, J.; Correa, P.; Silva, C. A high-performance vector control of an 11-level inverter. *IEEE Trans. Ind. Electron.* **2003**, *50*, 80–85. [CrossRef]
191. Karthikeyan, V.; Kumar, A.R.; Jamuna, V. A new multilevel inverter with BCD topology and reduction of harmonics using sine property. *Arch. Des Sci.* **2013**, *66*, 712–719.
192. Deepa, T.; Kumar, A.R. A new asymmetric multilevel inverter with reduced number of switches and reduction of harmonics using Sine Property. In Proceedings of the 2016 International Conference on Circuit, Power and Computing Technologies (ICCPCT), Nagercoil, India, 18–19 March 2016; pp. 1–8. [CrossRef]
193. Tahunguriya, S.; Kumar, A.R.; Deepa, T. Multilevel inverter with reduced number of switches and reduction of harmonics. *Middle East J. Sci. Res.* **2016**, *24*, 184–191.
194. Turnbull, F.G. Selected harmonic reduction in static DC—AC inverters. *IEEE Trans. Commun. Electron.* **1964**, *83*, 374–378. [CrossRef]
195. Patel, H.S.; Hoft, R.G. Generalized Techniques of Harmonic Elimination and Voltage Control in Thyristor Inverters: Part I—Harmonic Elimination. *IEEE Trans. Ind. Appl.* **1973**, *IA-9*, 310–317. [CrossRef]
196. Patel, H.S.; Hoft, R.G. Generalized Techniques of Harmonic Elimination and Voltage Control in Thyristor Inverters: Part II—Voltage Control Techniques. *IEEE Trans. Ind. Appl.* **1974**, *IA-10*, 666–673. [CrossRef]
197. Chenchireddy, K.; Jegathesan, V. A Review Paper on the Elimination of Low-Order Harmonics in Multilevel Inverters Using Different Modulation Techniques. In *Inventive Communication and Computational Technologies. Lecture Notes in Networks and Systems*; Ranganathan, G., Chen, J., Rocha, Á., Eds.; Springer: Singapore, 2020; Volume 145, pp. 961–971. [CrossRef]
198. Rai, M.; Tripathi, R.K. A novel multilevel inverter topology with selective harmonic elimination technique. In Proceedings of the 2014 International Conference on Power, Control and Embedded Systems (ICPCES), Allahabad, India, 26–28 December 2014; pp. 1–6. [CrossRef]

199. Alamri, B.; Sallama, A.; Darwish, M. Optimum SHE for cascaded H-bridge multilevel inverters using: NR-GA-PSO, comparative study. In Proceedings of the 11th IET International Conference on AC and DC Power Transmission, Birmingham, UK, 10–12 February 2015; pp. 1–10.
200. Siddique, M.D.; Mekhilef, S.; Padmanaban, S.; Memon, M.A.; Kumar, C. Single-Phase Step-Up Switched-Capacitor-Based Multilevel Inverter Topology with SHEPWM. *IEEE Trans. Ind. Appl.* **2020**, *57*, 3107–3119. [CrossRef]
201. Franquelo, L.G.; Napoles, J.; Guisado, R.C.P.; Leon, J.I.; Aguirre, M.A. A Flexible Selective Harmonic Mitigation Technique to Meet Grid Codes in Three-Level PWM Converters. *IEEE Trans. Ind. Electron.* **2007**, *54*, 3022–3029. [CrossRef]
202. Sun, J.; Beineke, S.; Grotstollen, H. Optimal PWM based on real-time solution of harmonic elimination equations. *IEEE Trans. Power Electron.* **1996**, *11*, 612–621. [CrossRef]
203. Kato, T. Sequential homotopy-based computation of multiple solutions for selected harmonic elimination in PWM inverters. *IEEE Trans. Circuits Syst. I Regul. Pap.* **1999**, *46*, 586–593. [CrossRef]
204. Chiasson, J.; Tolbert, L.; McKenzie, K.; Du, Z. Elimination of harmonics in a multilevel converter using the theory of symmetric polynomials and resultants. *IEEE Trans. Control. Syst. Technol.* **2005**, *13*, 216–223. [CrossRef]
205. Yang, K.; Yuan, Z.; Yuan, R.; Yu, W.; Yuan, J.; Wang, J. A Groebner Bases Theory-Based Method for Selective Harmonic Elimination. *IEEE Trans. Power Electron.* **2014**, *30*, 6581–6592. [CrossRef]
206. Yang, K.; Zhang, Q.; Yuan, R.; Yu, W.; Yuan, J.; Wang, J. Selective Harmonic Elimination with Groebner Bases and Symmetric Polynomials. *IEEE Trans. Power Electron.* **2015**, *31*, 2742–2752. [CrossRef]
207. Ponraj, R.P.; Sigamani, T. A novel design and performance improvement of symmetric multilevel inverter with reduced switches using genetic algorithm. *Soft Comput.* **2020**, *25*, 4597–4607. [CrossRef]
208. Barkat, S.; Berkouk, E.M.; Boucherit, M.S. Particle swarm optimization for harmonic elimination in multilevel inverters. *Electr. Eng.* **2009**, *91*, 221–228. [CrossRef]
209. Memon, M.A.; Siddique, M.D.; Saad, M.; Mubin, M. Asynchronous Particle Swarm Optimization-Genetic Algorithm (APSO-GA) based Selective Harmonic Elimination in a Cascaded H-Bridge Multilevel Inverter. *IEEE Trans. Ind. Electron.* **2021**. [CrossRef]
210. Taghizadeh, H.; Hagh, M.T. Harmonic Elimination of Cascade Multilevel Inverters with Nonequal DC Sources Using Particle Swarm Optimization. *IEEE Trans. Ind. Electron.* **2010**, *57*, 3678–3684. [CrossRef]
211. Sadoughi, M.; Zakerian, A.; Pourdadashnia, A.; Farhadi-Kangarlu, M. Selective Harmonic Elimination PWM for Cascaded H-bridge Multilevel Inverter with Wide Output Voltage Range Using PSO Algorithm. In Proceedings of the 2021 IEEE Texas Power and Energy Conference (TPEC), College Station, TX, USA, 2–5 February 2021; pp. 1–6. [CrossRef]
212. Sundareswaran, K.; Jayant, K.; Shanavas, T.N. Inverter Harmonic Elimination through a Colony of Continuously Exploring Ants. *IEEE Trans. Ind. Electron.* **2007**, *54*, 2558–2565. [CrossRef]
213. Kavousi, A.; Vahidi, B.; Salehi, R.; Bakhshizadeh, M.K.; Farokhnia, N.; Fathi, S.H. Application of the Bee Algorithm for Selective Harmonic Elimination Strategy in Multilevel Inverters. *IEEE Trans. Power Electron.* **2011**, *27*, 1689–1696. [CrossRef]
214. Salehi, R.; Vahidi, B.; Farokhnia, N.; Abedi, M. Harmonic Elimination and Optimization of Stepped Voltage of Multilevel Inverter by Bacterial Foraging Algorithm. *J. Electr. Eng. Technol.* **2010**, *5*, 545–551. [CrossRef]
215. Lou, H.; Mao, C.; Wang, D.; Lu, J. PWM optimisation for three-level voltage inverter based on clonal selection algorithm. *IET Electr. Power Appl.* **2007**, *1*, 870–878. [CrossRef]
216. Jegathesan, V.; Jerome, J. Elimination of lower order harmonics in Voltage Source Inverter feeding an induction motor drive using Evolutionary Algorithms. *Expert Syst. Appl.* **2011**, *38*, 692–699. [CrossRef]
217. Salam, Z.; Bahari, N. Selective harmonics elimination PWM (SHEPWM) using Differential Evolution approach. In Proceedings of the 2010 Joint International Conference on Power Electronics, Drives and Energy Systems & 2010 Power India, New Delhi, India, 20–23 December 2010; pp. 1–5. [CrossRef]
218. Hiendro, A. Multiple Switching Patterns for SHEPWM Inverters Using Differential Evolution Algorithms. *Int. J. Power Electron. Drive Syst. (IJPEDS)* **2011**, *1*, 94–103. [CrossRef]
219. Kouro, S.; Bernal, R.; Miranda, H.; Silva, C.; Rodriguez, J. High-Performance Torque and Flux Control for Multilevel Inverter Fed Induction Motors. *IEEE Trans. Power Electron.* **2007**, *22*, 2116–2123. [CrossRef]
220. Banaei, M.R.; Oskouei, A.B.; Dehghanzadeh, A. Extended switching algorithms based space vector control for five-level quasi-Z-source inverter with coupled inductors. *IET Power Electron.* **2014**, *7*, 1509–1518. [CrossRef]
221. Singh, S.; Agnihotri, A.; Bind, S.; Kumar, S. Matlab Simulation Study and Comparison of Different Multiple Carrier PWM Schemes for Multi Level CHB Inverter. In Proceedings of the 2020 IEEE First International Conference on Smart Technologies for Power, Energy and Control (STPEC), Nagpur, India, 25–26 September 2020; pp. 1–6. [CrossRef]
222. Venkatakrishna, A.; Somanatham, R.; Reddy, M.S. Phase shifted and level shifted PWM based cascaded multilevel inverter fed induction motor drive. *Int. J. Curr. Eng. Technol.* **2020**, *4*, 350–354.
223. Odeh, C.I.; Lewicki, A.; Morawiec, M. A Single-Carrier-Based Pulse-Width Modulation Template for Cascaded H-Bridge Multilevel Inverters. *IEEE Access* **2021**, *9*, 42182–42191. [CrossRef]
224. Chandwani, H.B.; Matnani, M.K. A review of modulation techniques for hybrid multilevel inverter. In Proceedings of the 2012 1st International Conference on Emerging Technology Trends in Electronics, Communication & Networking, Surat, India, 19–21 December 2012; pp. 1–7.

225. Vivek, G.; Kumar, Y.V.P. Improved Harmonic Profile of Multilevel Inverter Topology with Shifted Carrier Modulation Technique. In *Intelligent Computing in Control and Communication. Lecture Notes in Electrical Engineering*; Sekhar, G.C., Behera, H.S., Nayak, J., Naik, B., Pelusi, D., Eds.; Springer: Singapore, 2021; Volume 702, pp. 191–201. [CrossRef]
226. Silva, C.; Cordova, L.A.; Lezana, P.; Empringham, L. Implementation and Control of a Hybrid Multilevel Converter with Floating DC Links for Current Waveform Improvement. *IEEE Trans. Ind. Electron.* **2010**, *58*, 2304–2312. [CrossRef]
227. Mekhilef, S.; Kadir, M.N.A.; Salam, Z. Digital Control of Three Phase Three-Stage Hybrid Multilevel Inverter. *IEEE Trans. Ind. Inform.* **2012**, *9*, 719–727. [CrossRef]
228. Massoud, A.; Finney, S.; Williams, B. Mapped hybrid spaced vector modulation for multilevel cascaded-type voltage source inverters. *IET Power Electron.* **2008**, *1*, 318–335. [CrossRef]
229. Bharatiraja, C.; Jeevananthan, S.; Latha, R.; Mohan, V. Vector selection approach-based hexagonal hysteresis space vector current controller for a three phase diode clamped MLI with capacitor voltage balancing. *IET Power Electron.* **2016**, *9*, 1350–1361. [CrossRef]
230. Sanjeevan, A.R.; Kaarthik, R.S.; Gopakumar, K.; Rajeevan, P.; Leon, J.I.; Franquelo, L.G. Reduced common-mode voltage operation of a new seven-level hybrid multilevel inverter topology with a single DC voltage source. *IET Power Electron.* **2016**, *9*, 519–528. [CrossRef]
231. Choi, U.-M.; Blaabjerg, F.; Lee, K.-B. Reliability Improvement of a T-Type Three-Level Inverter with Fault-Tolerant Control Strategy. *IEEE Trans. Power Electron.* **2014**, *30*, 2660–2673. [CrossRef]
232. Colak, I.; Kabalci, E.; Bayindir, R. Review of multilevel voltage source inverter topologies and control schemes. *Energy Convers. Manag.* **2011**, *52*, 1114–1128. [CrossRef]
233. Renge, M.M.; Suryawanshi, H. Three-Dimensional Space-Vector Modulation to Reduce Common-Mode Voltage for Multilevel Inverter. *IEEE Trans. Ind. Electron.* **2009**, *57*, 2324–2331. [CrossRef]
234. Mondol, H.; Tur, M.R.; Biswas, S.P.; Hosain, K.; Shuvo, S.; Hossain, E. Compact Three Phase Multilevel Inverter for Low and Medium Power Photovoltaic Systems. *IEEE Access* **2020**, *8*, 60824–60837. [CrossRef]
235. Alexander, A.; Thathan, M. Modelling and analysis of modular multilevel converter for solar photovoltaic applications to improve power quality. *IET Renew. Power Gener.* **2015**, *9*, 78–88. [CrossRef]
236. Ujwala, G. Reduced Switch Multilevel Inverter Topologies and Modulation Techniques for Renewable Energy Applications. *Turk. J. Comput. Math. Educ. (TURCOMAT)* **2021**, *12*, 4659–4670.
237. Bughneda, A.; Salem, M.; Richelli, A.; Ishak, D.; Alatai, S. Review of Multilevel Inverters for PV Energy System Applications. *Energies* **2021**, *14*, 1585. [CrossRef]
238. Tsengenes, G.; Nathenas, T.; Adamidis, G. A three-level space vector modulated grid connected inverter with control scheme based on instantaneous power theory. *Simul. Model. Pr. Theory* **2012**, *25*, 134–147. [CrossRef]
239. Ravi, A.; Manoharan, P.; Anand, J.V. Modeling and simulation of three phase multilevel inverter for grid connected photovoltaic systems. *Sol. Energy* **2011**, *85*, 2811–2818. [CrossRef]
240. Tsengenes, G.; Adamidis, G. Investigation of the behavior of a three phase grid-connected photovoltaic system to control active and reactive power. *Electr. Power Syst. Res.* **2011**, *81*, 177–184. [CrossRef]
241. Liu, L.; Li, H.; Xue, Y.; Liu, W. Decoupled Active and Reactive Power Control for Large-Scale Grid-Connected Photovoltaic Systems Using Cascaded Modular Multilevel Converters. *IEEE Trans. Power Electron.* **2014**, *30*, 176–187. [CrossRef]
242. Sun, X.; Wang, B.; Zhou, Y.; Wang, W.; Du, H.; Lu, Z. A Single DC Source Cascaded Seven-Level Inverter Integrating Switched-Capacitor Techniques. *IEEE Trans. Ind. Electron.* **2016**, *63*, 7184–7194. [CrossRef]
243. Karasani, R.R.; Borghate, V.B.; Meshram, P.; Suryawanshi, H.M.; Sabyasachi, S. A Three-Phase Hybrid Cascaded Modular Multilevel Inverter for Renewable Energy Environment. *IEEE Trans. Power Electron.* **2016**, *32*, 1070–1087. [CrossRef]
244. Coppola, M.; Di Napoli, F.; Guerriero, P.; Iannuzzi, D.; Daliento, S.; Del Pizzo, A. An FPGA-Based Advanced Control Strategy of a GridTied PV CHB Inverter. *IEEE Trans. Power Electron.* **2015**, *31*, 806–816. [CrossRef]
245. Muller, N.; Kouro, S.; Malinowski, M.; Rojas, C.; Jasinski, M.T.; Estay, G. Medium-Voltage Power Converter Interface for Multigenerator Marine Energy Conversion Systems. *IEEE Trans. Ind. Electron.* **2016**, *64*, 1061–1070. [CrossRef]
246. Wang, L.; Zhang, D.; Wang, Y.; Wu, B.; Athab, H.S. Power and Voltage Balance Control of a Novel Three-Phase Solid-State Transformer Using Multilevel Cascaded H-Bridge Inverters for Microgrid Applications. *IEEE Trans. Power Electron.* **2015**, *31*, 3289–3301. [CrossRef]
247. Prabaharan, N.; Palanisamy, K. Modeling and Analysis of a Quasi-linear Multilevel Inverter for Photovoltaic Application. *Energy Procedia* **2016**, *103*, 256–261. [CrossRef]
248. Kumar, N.; Saha, T.K.; Dey, J. Sliding-Mode Control of PWM Dual Inverter-Based Grid-Connected PV System: Modeling and Performance Analysis. *IEEE J. Emerg. Sel. Top. Power Electron.* **2015**, *4*, 435–444. [CrossRef]
249. Choudhury, S. A comprehensive review on issues, investigations, control and protection trends, technical challenges and future directions for Microgrid technology. *Int. Trans. Electr. Energy Syst.* **2020**, *30*, 1–16. [CrossRef]
250. Maheswari, K.; Bharanikumar, R.; Arjun, V.; Amrish, R.; Bhuvanesh, M. A comprehensive review on cascaded H-bridge multilevel inverter for medium voltage high power applications. *Mater. Today: Proc.* **2020**, *45*, 2666–2670. [CrossRef]
251. Liang, Y.; Nwankpa, C. A new type of STATCOM based on cascading voltage-source inverters with phase-shifted unipolar SPWM. *IEEE Trans. Ind. Appl.* **1999**, *35*, 1118–1123. [CrossRef]

252. Haw, L.K.; Dahidah, M.S.A.; Almurib, H. A New Reactive Current Reference Algorithm for the STATCOM System Based on Cascaded Multilevel Inverters. *IEEE Trans. Power Electron.* **2014**, *30*, 3577–3588. [CrossRef]
253. Humayun, M.; Khan, M.M.; Hassan, M.U.; Zhang, W. Analysis of hybrid switches symmetric flying capacitor multilevel inverter based STATCOM. *Int. J. Electr. Power Energy Syst.* **2021**, *131*, 107054. [CrossRef]
254. Chunyan, Z.; Zhao, L. Advanced compensation mode for cascade multilevel static synchronous compensator under unbalanced voltage. *IET Power Electron.* **2015**, *8*, 610–617. [CrossRef]
255. Dargahi, V.; Sadigh, A.K.; Pahlavani, M.R.A.; Shoulaei, A. DC (direct current) voltage source reduction in stacked multicell converter based energy systems. *Energy* **2012**, *46*, 649–663. [CrossRef]
256. Marei, M.; Eltantawy, A.B.; El-Sattar, A.A. An energy optimized control scheme for a transformerless DVR. *Electr. Power Syst. Res.* **2012**, *83*, 110–118. [CrossRef]
257. Gao, S.; Lin, X.; Ye, S.; Lei, H.; Kang, Y. Transformer inrush mitigation for dynamic voltage restorer using direct flux linkage control. *IET Power Electron.* **2015**, *8*, 2281–2289. [CrossRef]
258. González, S.A.; Valla, M.I. UPQC implemented with cascade asymmetric multilevel converters. *Electr. Power Syst. Res.* **2015**, *124*, 144–151. [CrossRef]
259. Yang, S.; Liu, Y.; Wang, X.; Gunasekaran, D.; Karki, U.; Peng, F.Z. Modulation and Control of Transformerless UPFC. *IEEE Trans. Power Electron.* **2015**, *31*, 1050–1063. [CrossRef]
260. Galeshi, S.; Iman-Eini, H. Dynamic voltage restorer employing multilevel cascaded H-bridge inverter. *IET Power Electron.* **2016**, *9*, 2196–2204. [CrossRef]
261. Kumar, N. Bharath. Design and Analysis of Different Multi-Level Inverter Topologies For Single Phase Im Drive. *Turk. J. Comput. Math. Educ. (TURCOMAT)* **2021**, *10*, 4104–4117.
262. Holmes, D.G.; McGrath, B.; Parker, S.G. Current Regulation Strategies for Vector-Controlled Induction Motor Drives. *IEEE Trans. Ind. Electron.* **2011**, *59*, 3680–3689. [CrossRef]
263. Mora, A.; Lezana, P.; Juliet, J. Control Scheme for an Induction Motor Fed by a Cascade Multicell Converter Under Internal Fault. *IEEE Trans. Ind. Electron.* **2014**, *61*, 5948–5955. [CrossRef]
264. Zhou, D.; Zhao, J.; Liu, Y. Predictive Torque Control Scheme for Three-Phase Four-Switch Inverter-Fed Induction Motor Drives with DC-Link Voltages Offset Suppression. *IEEE Trans. Power Electron.* **2014**, *30*, 3309–3318. [CrossRef]
265. Fu, X.; Li, S. A Novel Neural Network Vector Control Technique for Induction Motor Drive. *IEEE Trans. Energy Convers.* **2015**, *30*, 1428–1437. [CrossRef]
266. Chen, Y.; Liu, T.; Hsiao, C.; Lin, C. Implementation of adaptive inverse controller for an interior permanent magnet synchronous motor adjustable speed drive system based on predictive current control. *IET Electr. Power Appl.* **2015**, *9*, 60–70. [CrossRef]
267. Liu, J.; Sun, Y.; Li, Y.; Fu, C. Theoretical harmonic analysis of cascaded H-bridge inverter under hybrid pulse width multilevel modulation. *IET Power Electron.* **2016**, *9*, 2714–2722. [CrossRef]

*energies*

MDPI

*Article*

# An Integrated Device of a Lithium-Ion Battery Combined with Silicon Solar Cells

**Hyeonsu Lim [1], Dan Na [1], Cheul-Ro Lee [1], Hyung-Kee Seo [2], O-Hyeon Kwon [3], Jae-Kwang Kim [3,*] and Inseok Seo [1,*]**

1 School of Advanced Materials Engineering, Research Center for Advanced Materials Development (RCAMD), Jeonbuk National University, Baekje-daero 567, Jeonju 54896, Korea; kady1004@naver.com (H.L.); ld3310@naver.com (D.N.); crlee7@jbnu.ac.kr (C.-R.L.)

2 Future Energy Convergence Core Center, School of Chemical Engineering, Jeonbuk National University, Baekje-daero 567, Jeonju 54896, Korea; hkseo@jbnu.ac.kr

3 Department of Solar & Energy Engineering, Cheongju University, Cheongju 360-764, Korea; ohhyeon294@naver.com

* Correspondence: jaekwang@cju.ac.kr (J.-K.M.); isseo@jbnu.ac.kr (I.S.); Tel.: +82-63-270-2295 (I.S.)

**Abstract:** This study reports an integrated device of a lithium-ion battery (LIB) connected with Si solar cells. A $Li(Ni_{0.65}Co_{0.15}Mn_{0.20})O_2$ (NCM) cathode and a graphite (G) anode were used to fabricate the lithium-ion battery (LIB). The surface and shape morphologies of NCM and graphite powder were characterized by field emission scanning electron microscopy (FE-SEM). The structural properties of NCM and graphite powder were determined by X-ray diffraction (XRD) analysis. XRD patterns of powders were well matched with those of JCPDS data. To investigate the electrochemical characteristics of NCM and graphite, cycling tests were performed after assembling the NCM-Li, the G-Li half-cell, and the NCM-G full-cell. The discharge capacity of the NCM cathode at 0.1C was 189.82 $mAh/g^{-1}$. The NCM-graphite full-cell showed 98.25% cycle retention at 1C after 50 cycles. To obtain enough charging voltage for the LIB connected with solar cells in an integrated device, eight single Si solar cells were connected in a series. The short-circuit photocurrent density for Si solar cells was 4.124 $mA/cm^2$. The fill factor and the open circuit voltage were 0.78 and 4.5 V, respectively. These Si solar cells showed a power conversion efficiency of 14.45%. The power conversion andstorage efficiency of the integrated device of the NCM battery and Si solar cells was 7.74%. Charging of the integrated device could be as effective as charging with a battery cycler.

**Keywords:** $Li(Ni_xCo_yMn_z)O_2$; graphite; lithium-ion battery; silicon solar cell; integrated device

**Citation:** Lim, H.; Na, D.; Lee, C.-R.; Seo, H.-K.; Kwon, O.-H.; Kim, J.-K.; Seo, I. An Integrated Device of a Lithium-Ion Battery Combined with Silicon Solar Cells. *Energies* **2021**, *14*, 6010. https://doi.org/10.3390/en14196010

Academic Editor: Pablo García Triviño

Received: 19 August 2021
Accepted: 18 September 2021
Published: 22 September 2021



## 1. Introduction

As the industry develops and the population grows, the world's energy demand has been increasing every year. Fossil fuels supply most of the global energy needs. In recent years, problems of depletion of fossil fuels and environmental pollution such as $SO_x$ and $NO_x$ emission problems caused by wastes of fossil fuels have emerged around the world [1]. To solve these problems, interest in renewable energy has increased. Among renewable energy sources, sunlight is the most abundant one. It is more widely used than other renewable energies such as geothermal, wind, and tidal power [2]. However, sunlight is highly dependent on weather conditions. In addition, it cannot be used at night. To use sunlight effectively, it is necessary to store solar energy in a battery or a grid [3,4].

Many research studies have reported that a converter is necessary to charge from solar energy to a battery system. However, the system that uses a converter could be more complex and expensive than a direct connection between a solar cell and a battery system [5–9]. In addition, it can be less efficient than direct charging a battery with solar energy [10]. Therefore, a direct DC current storing is in the spotlight as a promising technology using an integrated method [11].

Lithium-ion batteries (LIBs) have been applied to portable devices and electric vehicles (EV). Conventionally, LIBs are manufactured using lithium metal oxide cathode electrode materials such as $LiCoO_2$(LCO), $LiFePO_4$ (LFP), $Li(Ni_xCo_yAl_z)O_2$ (NCA), and $Li(Ni_xCo_yMn_z)O_2$ (NCM) with anode electrode materials [12,13].

Among various cathode materials, LFP has the advantages of a low price and high safety. Therefore, LFP is generally widely used in energy storage systems (ESSs). However, LFP has the disadvantages of low electronic conductivity and capacity [14,15]. NCA has a high energy density and high power. Thus, it has a higher capacity than LFP. However, NCA has a disadvantage in that its capacity retention rate is low, and its performance is degraded when stored at high temperatures [16]. Although NCM has similar properties to NCA, NCM has a higher cycle performance and C rates than those of NCA [17].

In this study, an integrated device of LIB connected Si solar cells was developed. The shape and structures of the powder of the NCM622 cathode and the graphite anode material were evaluated by field-emission scanning electron microscopy (FE-SEM) and X-ray diffraction (XRD) analysis, respectively. Electrochemical characteristics were evaluated after assembling NCM, the graphite half-cell, and the NCM-G full-cell. To obtain enough charging voltage for the battery, eight single Si solar cells in a series were connected. Electrical properties such as the fill factor (FF), the short circuit photocurrent density ($I_{SC}$), and the open circuit voltage ($V_{OC}$) were analyzed with a solar simulator. To develop an integrated device, Si solar cells and NCM-graphite LIB were combined. Photo-charging from solar cells to the NCM battery and galvanostatic charging from a cycler to the NCM battery were compared.

## 2. Materials and Methods

### 2.1. Preparation of LIB and Solar Cell

A cathode electrode slurry was prepared by adding NCM powder, polyvinylidene difluoride (PVDF), and carbon black (super P) to N-Methyl-2-pyrrolidone (NMP) solvent at a weight percent ratio of 8:1:1. The anode electrode slurry was made using graphite powder in the same way. Cathode and anode slurries were coated on Al foil and Cu foil, respectively. The coated slurry was dried and pressed to make a sheet electrode. The half-cell and the full-cell were assembled in a glove box. As an electrolyte, 1.2 M $LiPF_6$ in a 1:1 (*v/v*, %) solution of ethylene carbonate (EC) and dimethyl carbonate (DMC) was used. The half-cell was in the form of a CR 2032 coin cell. The full-cell was assembled in the form of a pouch. The pouch-type cell was 4.5 cm × 4.5 cm in size. Al and Ni were used as anode and cathode lead tabs. To obtain enough charging voltage from Si solar cells for LIB charging, a polycrystalline Si solar cell with eight single Si solar cells in a series was purchased from SAVE SOLAR Co. Ltd.

### 2.2. Characterizations

Large and small particles of NCM622 precursor were prepared using a co-precipitation method. First, large and small particles of NCM cathode were mixed at a 7:3 ratio. Graphite powder was purchased from Sigma Aldrich Co. Ltd. Surface and shape morphologies of NCM as a cathode and graphite as an anode powder were characterized by FE-SEM (model SU-70, Hitachi High-Tech Corporation, Fukuoka, Japan). The acceleration voltage for the FE-SEM measurement was 10 kV. Structures of NCM and graphite powder were analyzed with an XRD (MAX-2500, RIGAKU in Japan) with Cu-kα radiation. XRD patterns of the powder were obtained in a 2θ range of 10–80° with a scan rate of 2.0°/min and a step size of 0.05°. Diffraction intensity data for a Rietveld refinement analysis were collected in a step-scan mode with a scanning step of 0.02° and a sampling time of 2 s. Rietveld refinement was performed using a FullProf program to obtain the crystal structure parameters of NCM622. Electrochemical characteristics of NCM and the graphite half-cell and the NCM-graphite full-cell were analyzed using a battery cycler (WBCS3000S, WONATECH Co., Ltd., Seoul, Korea). Galvanostatic charge and discharge tests of the NCM-graphite full-cell were carried out over a voltage range of 2.6–4.4 V. Charge and discharge analyses were performed according to various C-rates of the NCM-graphite full-cell. Cycle retention of the NCM-graphite full-cell was examined after 50 cycles. The fill factor, the open circuit voltage, and te short circuit photocurrent density were measured by irradiating 100 mW/cm$^2$ of light on silicon solar cells with a solar simulator (PEC-L01, Peccell Technologies, Yokohama, Japan).

### 2.3. *Integrated Si Solar Cells-LIB Device*

Si solar cells and LIB were directly connected to fabricate an integrated device. The integrated sample was examined in the glovebox (WGB2200S, Woosung Hivac, Daejeon, Korea) installed in the Future Energy Convergence Core Center (FECC). The LIB was charged by irradiating light on Si solar cells with a solar simulator (PEC-L01, Peccell Technologies, Japan). Before the test, the light intensity was calibrated to 100 mW/cm$^2$. Photo-charging for the Li-ion battery using the Si solar cells was performed using a solar simulator. Discharge was then performed with a battery cycler.

## 3. Results and Discussion

Figure 1 shows FE-SEM images of NCM and graphite powder. The particle size and the shape of NCM powder are shown in Figure 1a,b. The NCM powder consisted of large and small particles with diameters of about 10 µm and 3 µm, respectively. Large particles and small particles were mixed as a bimodal. By mixing large and small particles with optimized ratios, a better energy density could be realized [18]. Park et al. [19] have reported that optimized bimodal-sized active materials can minimize voids and maximize the packing density. A bimodal structure could also improve electrochemical performance due to a large contact area between large particles and small particles [19]. The particle size and the shape of the graphite powder are shown in Figure 1c,d. FE-SEM magnitudes of Figure 1c,d were 1000 and 3000, respectively. The average size of the graphite powder was about 20 µm. As shown in Figure 1, the size of the graphite powder was greater than that of the NCM powder. Surface morphologies of the graphite powder were smoother than those of the NCM powder.

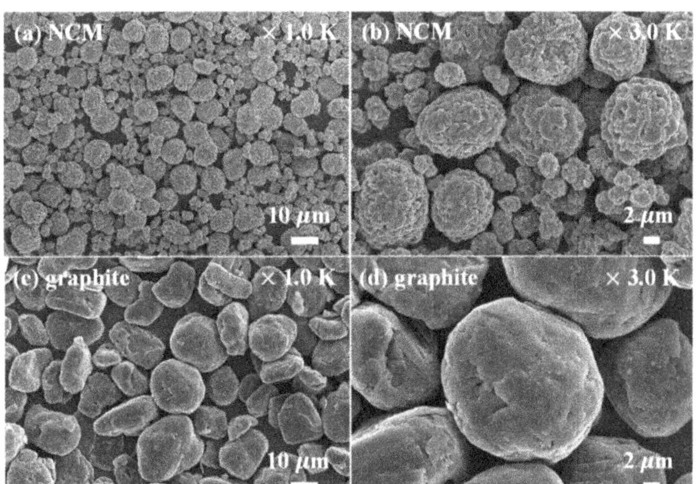


**Figure 1.** FE-SEM images of (**a**,**b**) NCM powder and (**c**,**d**) graphite powder.

XRD patterns of NCM and graphite powders are shown in Figure 2a,b. As shown in Figure 2a,b, XRD patterns of NCM and graphite powders and their joint committee on powder diffraction standard (JCPDS) data were compared. It could be noted that XRD patterns of NCM and graphite powders were very close to those of PDF data. Other XRD patterns related to impurities were not found. Results of XRD patterns confirmed that the NCM cathode powder and the graphite anode powder were successfully synthesized. For detailed structural analysis, Rietveld refinement of the NCM cathode was performed. The results are shown in Figure 2c. The structural parameters of NCM622 obtained from the Rietveld refinement are listed in Table 1. The NCM622 sample presented a hexagonal α-$NaFeO_2$ structure with an R3m space group. Structural lattice parameters of the NCM 622 cathode are also shown in Table 1. These parameters indicated that the NCM 622 cathode material was well-ordered with a layer structure. The values of $c/a$ and $I_{003}/I_{104}$ are important factors because these values indicate the degree of structural orderliness for layered structural materials. As shown in Table 1, the values of $c/a$ and $I_{003}/I_{104}$ were larger than 4.9 and 1.2, respectively. Thus, the sample had a well-ordered layered structure and a lower amount of $Li^+/Ni^{2+}$ mixing [20]. The refined atomic parameters of Ni, Co, and Mn were 0.65, 0.15, and 0.20, respectively, as shown in Table 2. While the Li1 was located in the interslab layer, the Ni1, Co1, and Mn1 were in the NCM interlayer. A slight cation mixing between Li and Ni atoms was allowed by constraining the total amount of each element to 1.00 and 0.65, respectively.

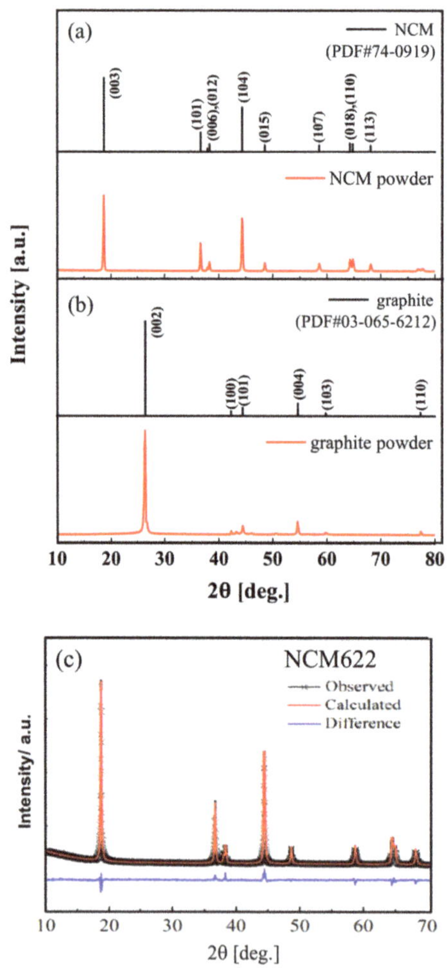


**Figure 2.** XRD patterns of (**a**) NCM (JCPDS, 74-0919) powder, (**b**) graphite (JCPDS, 03-065-6212) powder, and (**c**) Rietveld refinement of NCM622 powder.

**Table 1.** Structural parameters obtained from Rietveld refinement of XRD pattern of NCM622.

| Sample | *a* (Å) | *c* (Å) | c/a | Unit volume (Å$^3$) | $I_{(003)/(104)}$ |
|---|---|---|---|---|---|
| NCM622 | 2.910 | 14.292 | 4.911 | 101.67 | 1.575 |

**Table 2.** Atomic parameters of the pristine NMC622 material obtained from Rietveld refinement.

| Atom | X | Y | Z | Occupancy |
|---|---|---|---|---|
| Li1 | 0.00000 | 0.00000 | 0.00000 | 1.0 (fixed) |
| Ni1 | 0.00000 | 0.00000 | 0.50000 | 0.65 (fixed) |
| Co1 | 0.00000 | 0.00000 | 0.50000 | 0.15 (fixed) |
| Mn1 | 0.00000 | 0.00000 | 0.50000 | 0.20 (fixed) |
| O1 | 0.00000 | 0.00000 | 0.24224 (10) | 1.0 (fixed) |

The electrochemical results of the half-cell and the full-cell batteries are shown in Figure 3. As shown in Figure 3a, charge and discharge capacities were obtained at 0.1 C, 0.2 C, 0.5 C, and 1 C of the NCM half-cell. The NCM half-cell had a discharge capacity of 187.36 mAhg$^{-1}$ at 0.1 C. The discharge capacity was similar to that reported previously [21]. Figure 3b shows the charge and discharge capacities at 0.1 C, 0.2 C, 0.5 C, and 1 C of the graphite half-cell. The graphite half-cell had a discharge capacity of 336.82 mAhg$^{-1}$ at 0.1 C. Figure 3c shows the charge and discharge capacities at 0.1 C, 0.2 C, 0.5 C, and 1 C of the NCM-graphite full-cell. The NCM-graphite full-cell had a discharge capacity of 183.39 mAhg$^{-1}$ at 0.1 C. The discharge capacity of the full-cell was very similar to that of the NCM half-cell. Figure 3d shows the discharge capacities of the NCM-graphite battery at various C-rates. As shown in Figure 3d, the discharge capacity was almost the same at each C-rate. Thus, the NCM-graphite full-cell had a good electrochemical stability.

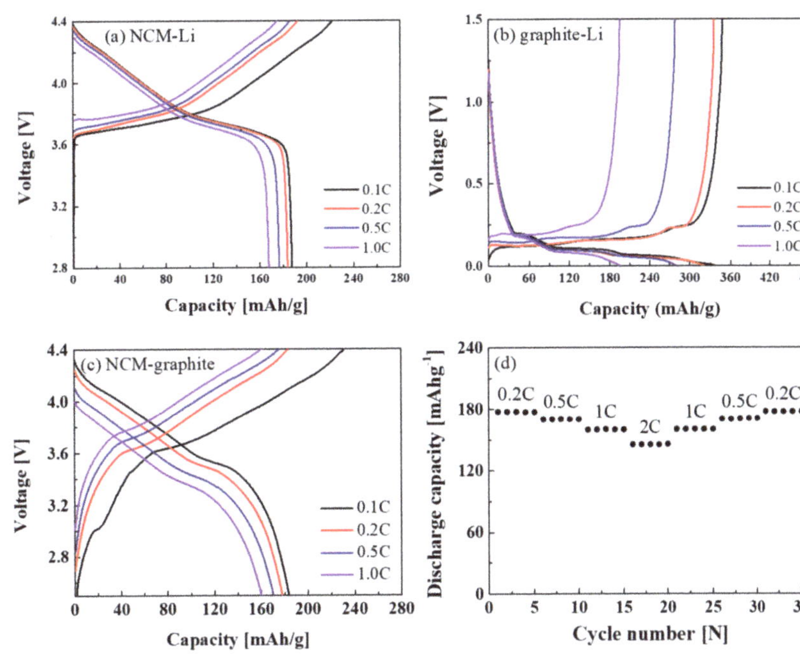


**Figure 3.** Electrochemical properties of (**a**) NCM-Li, (**b**) graphite-Li, (**c**) NCM-graphite full cell, and (**d**) rate capability of NCM-graphite full cell analyzed at various C-rates.

Charge and discharge data of the NCM-graphite full-cell until 50 cycles at the 1 C rate are shown in Figure 4a. The discharge capacity was 160.22 mAhg$^{-1}$ at the 1st cycle, 158.33 mAhg$^{-1}$ at the 5th cycle, 157.98 mAhg$^{-1}$ at the 25th cycle, and 157.41 mAhg$^{-1}$ at the 50th cycle. The NCM-graphite full-cell had a cycle retention of 98.25% after 50 cycles. This suggests that the NCM-graphite full-cell has an excellent lifecycle performance. The NCM-graphite full-cell might exhibit good performance during operation of the integrated device with Si solar cells. The discharge capacity and the Coulombic efficiency for each cycle are shown in Figure 4b. The Coulombic efficiency was 99.68% at the 1st cycle, 99.66% at the 5th cycle, 99.52% at the 25th cycle, and 99.25% at the 50th cycle. Coulombic efficiency was almost constant over 50 cycles.

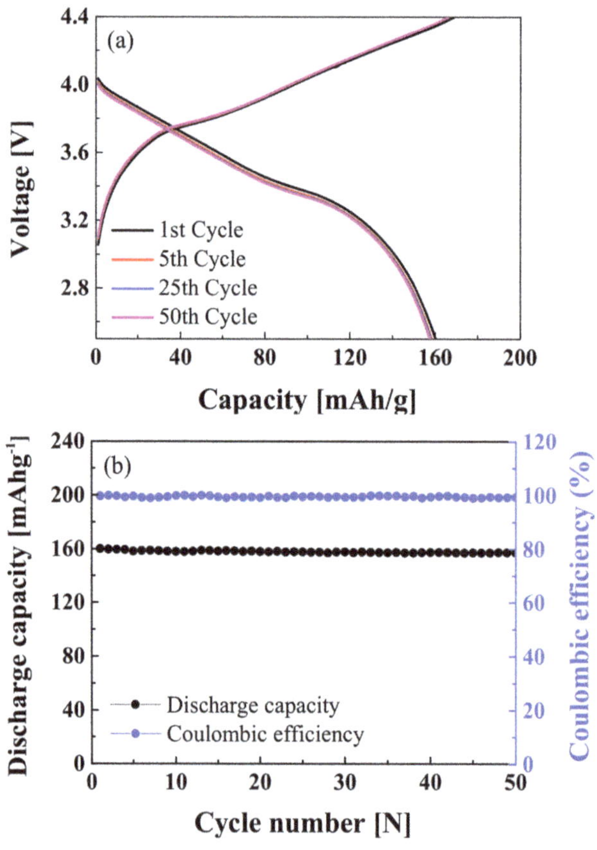


**Figure 4.** (**a**) Cycle performance of the NCM-graphite full cell over 50 cycles; (**b**) discharge capacity and Coulombic efficiency of the NCM-graphite full cell over 50 cycles.

Figure 5 shows a schematic of the integrated device using Si solar cells and the NCM-graphite battery. The voltage range of the NCM-graphite full-cell was 2.6–4.4 V. Therefore, Si solar cells were designed to obtain enough charging voltage for the battery that eight single Si solar cells (0.58 V per each single solar cell) were connected in series. Si solar cells and the NCM-graphite full-cell were directly connected without a power converter. As shown in Figure 5, the integrated Si solar cells and the battery device could be operated by self-photo-charging to the battery. The simple structure without any power converter could be an advantage.

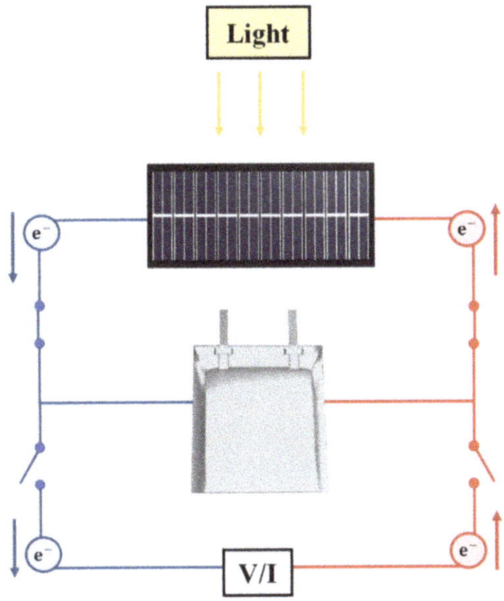


**Figure 5.** Schematic diagram of NCM-graphite integrated device with Si solar cells.

To examine the electrical property of Si solar cells, current density–voltage (J–V) curves of Si solar cells were characterized. The results are shown in Figure 6. These Si solar cells had a short-circuit photocurrent density of 4.124 mA/cm$^2$ and an open-circuit voltage of 4.5 V. These Si solar cells had enough voltage to charge the NCM-graphite battery. They had revealed a fill factor of 0.78 and a power conversion efficiency of 14.45%. Generally, efficiencies of commercialized Si solar cells are about 12–14% [22]. Thus, Si solar cells used in this study had a moderate efficiency of 14.45% based on the literature.

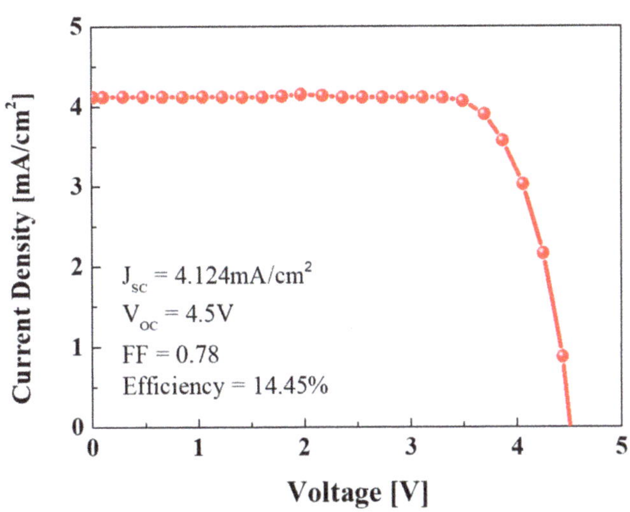


**Figure 6.** J–V curve of Si solar cells.

Figure 7 shows the results of charging and discharging tests by directly connecting Si solar cells with the NCM-graphite full-cell. The charge–discharge test was performed at a voltage range of 2.6–4.4 V. The LIB of the integrated device was charged by Si solar cells using a solar simulator. The battery was then discharged using a galvanostatic method over 10 cycles. During charge–discharge over 10 cycles, the charge–discharge behavior was almost the same. This means that the integrated device can be operated stably. The power conversion and storage efficiency of the integrated Si solar cells-NCM-graphite full-cell device was 7.74%. The integrated device showed excellent power conversion and storage efficiency at a high photo-charging voltage [23].

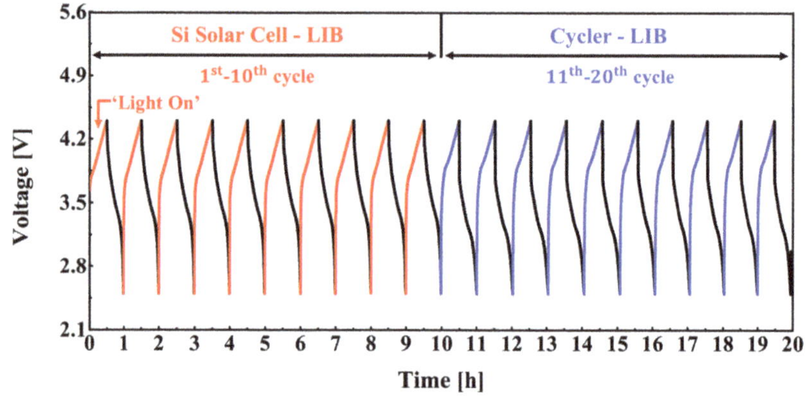


**Figure 7.** Photo-charging and discharging diagrams of the NCM-graphite full cell using Si-solar cells and charging and discharging with a battery cycler.

In general, $LiFePO_4$(LFP) cathode-based batteries (LFP-graphite, LFP-LTO, etc.) are widely used in ESS due to their advantages such as superior thermal stability and chemical stability [24]. These features provide better safety properties than other types of Li-ion batteries. LFP batteries have a low charging voltage. Thus, small amounts of single silicon solar cells were connected to match their charging voltage. LFP batteries are widely used in ESS. However, LFP batteries have lower capacities than NCM batteries [25,26]. NCM batteries have higher charging voltages than LFP batteries. The NCM-graphite full-cell in this study revealed a good cycle life of 98.25% after 50 cycles. Its capacity was higher than the capacity of LFP (~170 mAh/g). Therefore, the NCM-graphite full-cell could be a promising candidate for an integrated device. The energy storage efficiency from Si solar cells to LIB was 53.56%. After that, the integrated device was charged and discharged for 11–20 cycles with a battery cycler using a galvanostatic power supply in the voltage range of 2.6–4.4 V, similar to the charge–discharge voltage profile of the NCM-graphite battery. Photo-charging of integrated devices is an efficient way of charging using a battery cycler. Therefore, an integrated device of a LIB and a Si solar cell is a promising power system for portable devices, wearable devices, ESS, and EVs.

Figure 8a shows J–V curves of Si solar cells connected to an NCM-graphite full-cell before and after various cycle tests. The current density of Si solar cells connected to the NCM-graphite full-cell was 4.124 mA/cm$^2$ before the cycle test. As the number of cycles increased, the current density of Si solar cells connected to the NCM-graphite full-cell decreased slightly. After 1, 5, and 10 cycles, current densities were 4.122 mA/cm$^2$, 4.114 mA/cm$^2$, and 4.099 mA/cm$^2$, respectively. These Si solar cells integrated to the NCM-graphite battery revealed an open circuit voltage of 4.49 V after 10 cycles. This open circuit voltage did not change significantly over 10 cycles. Figure 8b shows the fill factor and the energy efficiency of Si solar cells integrated to the NCM-graphite full-cell before and after various cycle tests. The energy efficiency of Si solar cells connected to the NCM-graphite full-cell was 14.45% before the cycle test. As the number of cycles increased, the energy efficiency of Si solar cells connected to the NCM-graphite full-cell decreased slightly. Energy efficiencies of Si solar cells connected to the NCM-graphite full-cell after 1,

5, and 10 cycles were 14.41%, 14.32%, and 14.26%, respectively. The fill factor of the Si solar cells connected to the NCM-graphite full-cell was 0.78 before the cycle test. As the number of cycles increased, the fill factor slightly decreased. The fill factor values of the Si solar cells connected to the NCM-graphite battery after 1, 5, and 10 cycles were 0.78, 0.78, and 0.77, respectively. After 10 cycles, the energy-conversion efficiency of solar cells connected to the NCM-graphite full-cell was 98.68% of the efficiency before the cycling test. Therefore, Si solar cells were suitable devices to supply power to the NCM-graphite full-cell.

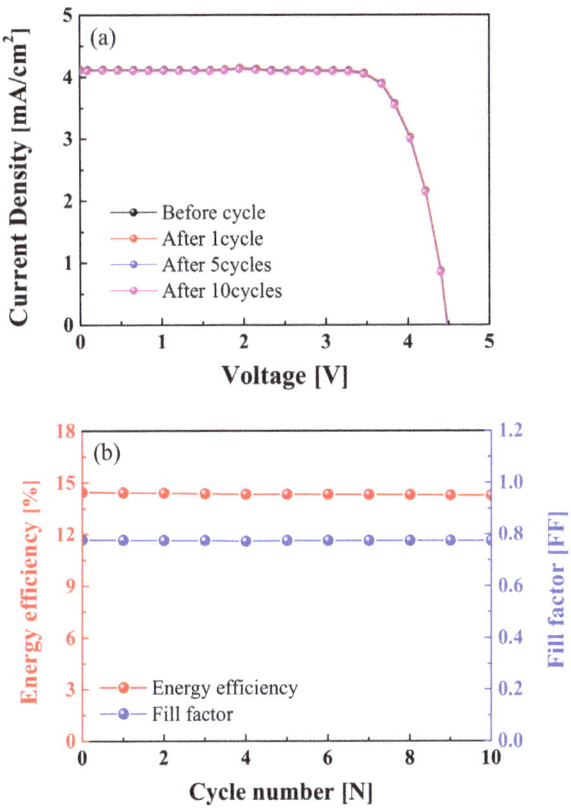


**Figure 8.** (**a**) Current density and voltage (J–V) curve; (**b**) fill factor (FF) and energy efficiency of Si-solar cells integrated to an NCM-graphite battery before and after 10 cycles.

## 4. Conclusions

In summary, we fabricated a LIB using NCM as the cathode and graphite as the anode. To obtain enough photo-charging voltage, eight single Si solar cells were connected in a series. LIB and Si solar cells were directly connected without any converter. The results of photo-charging with Si solar cells at galvanostatic discharge were very close to charging/discharging with a galvanostatic system. The power conversion efficiency of the NCM-graphite full-cell and Si solar-cells was 7.74%. The energy conversion efficiency of Si solar-cells to LIB was 53.56%, which showed a high energy conversion efficiency.

The cycle retention of the NCM-graphite battery was stable at 98.68% over 50 cycles. In addition, the NCM-graphite full-cell showed a higher capacity than reported LFP-based batteries. Charging of the integrated device could be as effective as charging with a battery cycler. Therefore, an integrated system using an NCM-graphite full-cell is a promising approach to replace an LFP-based integrated system.

**Author Contributions:** Methodology, C.-R.L.; formal analysis, D.N.; investigation, H.-K.S.; resources, O.-H.K.; data curation, H.L.; writing—original draft preparation, H.L.; writing—review and editing, I.S.; project administration, J.-K.K., I.S.; funding acquisition, I.S. All authors have read and agreed to the published version of the manuscript.

**Funding:** This work was supported by grants (2018R1A4A1025528 and 2020R1A2C2009057) of the National Research Foundation of Korea (NRF). This research was also partially supported by Cheongju University Research Scholarship Grants in 2021.

**Acknowledgments:** The authors thank Baeksang Yoon for providing technical support. The aurthors also thank anonymous reviewers for fruitful discussion and useful comments.

**Conflicts of Interest:** The authors declare no conflict of interest.

## References

1. Kannan, N.; Vakeesan, D. Solar energy for future world—A review. *Renew. Sustain. Energy Rev.* **2016**, *62*, 1092–1105. [CrossRef]
2. Nozik, A.J. Photoelectrochemistry: Applications to solar energy conversion. *Annu. Rev. Phys. Chem.* **1978**, *29*, 189–222. [CrossRef]
3. Bo, L.; Shahidehpour, M. Short-term scheduling of battery in a grid-connected PV/battery system. *IEEE Trans. Power Syst.* **2005**, *20*, 1053–1061. [CrossRef]
4. Lupangu, C.; Bansal, R.C. A review of technical issues on the development of solar photovoltaic systems. *Renew. Sustain. Energy Rev.* **2017**, *73*, 950–965. [CrossRef]
5. Licht, S. A description of energy conversion in photoelectrochemical solar cells. *Nature* **1987**, *330*, 148–151. [CrossRef]
6. Sharon, M.; Veluchamy, P.; Natarajan, C.; Kumar, D. Solar rechargeable battery-principle and materials. *Electrochim. Acta.* **1991**, *36*, 1107–1126. [CrossRef]
7. Nomiyama, T.; Kuriyaki, H.; Hirakawa, K. Photo-rechargeable battery using new layer compound $CuFeTe_2$. *Synth. Met.* **1995**, *71*, 2237–2238. [CrossRef]
8. Liao, Z.; Ruan, X. Control strategy of bi-directional DC/DC converter for a novel stand-alone photovoltaic power system. *IEEE Veh. Power Propuls. Conf.* **2008**, 1–6. [CrossRef]
9. Shiau, J.; Ma, C. Li-Ion Battery Charging with a Buck-Boost Power Converter for a Solar Powered Battery Management System. *Energies* **2013**, *6*, 1669–1699. [CrossRef]
10. Gibson, T.L.; Kelly, N.A. Solar photovoltaic charging of lithium-ion batteries. *IEEE Veh. Power Propuls. Conf.* **2009**, 310–316. [CrossRef]
11. Luo, B.; Ye, D.; Wang, L. Recent Progress on Integrated Energy Conversion and Storage Systems. *Adv. Sci.* **2017**, *4*, 1700104. [CrossRef]
12. Chu, S.; Cui, Y.; Liu, N. The path towards sustainable energy. *Nat. Mater.* **2017**, *16*, 16–22. [CrossRef]
13. Freitasgomes, I.S.; Perez, Y.; Suomalainen, E. Coupling small batteries and PV generation: A review. *Renew. Sustain. Energy Rev.* **2020**, *126*, 109835. [CrossRef]
14. Ellis, B.; Perry, L.K.; Ryan, D.H.; Nazar, L.F. Small Polaron Hopping in $Li_xFePO_4$ Solid Solutions: Coupled Lithium-Ion and Electron Mobility. *J. Am. Chem. Soc.* **2006**, *128*, 11416–11422. [CrossRef]
15. Zhao, Q.; Zhang, Y.; Meng, Y.; Wang, Y.; Ou, J.; Guo, Y.; Xiao, D. Phytic acid derived $LiFePO_4$ beyond theoretical capacity as high-energy density cathode for lithium ion battery. *Nano Energy* **2017**, *34*, 408–420. [CrossRef]
16. Watanabe, S.; Kinoshita, M.; Nakura, K. Capacity fade of $LiNi_{(1-x-y)}Co_xAl_yO_2$ cathode for lithium-ion batteries during accelerated calendar and cycle life test. I. Comparison analysis between $LiNi_{(1-x-y)}Co_xAl_yO_2$ and $LiCoO_2$ cathodes in cylindrical lithium-ion cells during long term storage test. *J. Power Sources* **2014**, *247*, 412–422. [CrossRef]
17. Chen, Y.; Song, S.; Zhang, X.; Liu, Y. The challenges, solutions and development of high energy Ni-rich NCM/NCA LiB cathode materials. *J. Phys. Conf. Ser.* **2019**, *1347*, 012012. [CrossRef]
18. Kim, K.; Park, S.; Kwon, T.; Park, J.; Ahn, H.; Lee, M. Characterization of Li-V-O nanorod phases and their effect on electrochemical properties of $Li_{1+x}V_3O_8$ cathode materials synthesized by hydrothermal reaction and subsequent heat treatment. *Electrochim. Acta* **2013**, *89*, 708–716. [CrossRef]
19. Park, C.; Lee, S.; Kim, K.; Kim, M.; Choi, S.; Shin, D. Electrochemical Properties of composite cathode using bimodal sized electrolyte for all-solid-state batteries. *J. Electrochem. Soc.* **2019**, *166*, A5318–A5322. [CrossRef]
20. Feng, Z.; Peng, Z.; Huang, X.; Rajagopalan, R.; Tang, Y.; Wang, H. Enhanced electrochemical properties of $Li(Ni_{0.8}Co_{0.1}Mn_{0.}1)O_2$ at elevated temperature by simultaneous structure and interface regulating. *J. Electrochem. Soc.* **2019**, *166*, A1439–A1448. [CrossRef]
21. Noh, H.; Youn, S.; Yoon, C.; Sun, Y. Comparison of the structural and electrochemical properties of layered $Li[Ni_xCo_yMn_z]O_2$ (x = 1/3, 0.5, 0.6, 0.7, 0.8 and 0.85) cathode material for lithium-ion batteries. *J. Power Sources* **2013**, *233*, 121–130. [CrossRef]
22. Sharma, S.; Jain, K.K.; Sharma, A. Solar cells: In research and applications—A review. *Mater. Sci. Appl.* **2015**, *6*, 1145–1155. [CrossRef]
23. Um, H.; Choi, K.; Hwang, I.; Kim, S.; Seo, K.; Lee, S. Monolithically integrated, photo-rechargeable portable power sources based on miniaturized Si solar cells and printed solid-state lithium-ion batteries. *Energy Environ. Sci.* **2017**, *10*, 931–940. [CrossRef]

24. Wang, X.; Adelmann, P.; Reindl, T. Use of $LiFePO_4$ batteries in stand-alone solar system. *Energy Proced.* **2012**, *25*, 135–140. [CrossRef]
25. Zaghib, K.; Dontigny, M.; Guerfi, A.; Charest, P.; Rodrigues, I.; Mauger, A.; Julien, C.M. Safe and fast-charging Li-ion battery with long shelf life for power applications. *J. Power Sources* **2011**, *196*, 3949–3954. [CrossRef]
26. Abe, Y.; Hori, N.; Kumagai, S. Electrochemical Impedance Spectroscopy on the Performance Degradation of $LiFePO_4$/Graphite Lithium-Ion Battery Due to Charge-Discharge Cycling under Different C-Rates. *Energies* **2019**, *12*, 4507. [CrossRef]

*energies*

MDPI

*Technical Note*

# Design and Comparison of the Performance of 12-Pulse Rectifiers for Aerospace Applications

**Fabio Corti [1,2,*], Abdelazeem Hassan Shehata [1], Antonino Laudani [3] and Ermanno Cardelli [2]**

[1] Dipartimento di Ingegneria, Università di Perugia, Via G. Duranti 67, 06125 Perugia, Italy; abdelazeemhassanshehata.atyia@studenti.unipg.it
[2] Dipartimento di Ingegneria dell'Informazione, Università di Firenze, Via di S. Marta 3, 50139 Firenze, Italy; ermanno.cardelli@unipg.it
[3] Dipartimento di Ingegneria, Università di Roma 3, Via Vito Volterra 62, 00146 Roma, Italy; antonino.laudani@uniroma3.it
* Correspondence: fabio.corti@unipg.it

**Abstract:** In this paper, a conventional 12-pulse transformer unit (CTU) and an autotransformer 12-pulse transformer unit (ATU) are compared in the view of the RTCA DO-160 standard for aircraft applications. The design of the magnetic components is proposed via a coupled FEM-circuital analysis in the time domain for an 800 Hz/2 kW system. Input AC distortion, power factor, and output DC ripple are evaluated through simulations. An accurate power loss analysis is carried out, taking into account copper losses, magnetic losses, and power losses due to power switches. The reduction in the size and weight of the ATU with respect to the CTU solution is discussed, including the need for filtering systems and the standard requirements.

**Keywords:** aerospace 12-pulse rectifier; autotransformer rectifier; magnetic component design; performance; standard requirements

**Citation:** Corti, F.; Hassan Shehata, A.; Laudani, A.; Cardelli, E. Design and Comparison of the Performance of 12-Pulse Rectifiers for Aerospace Applications. *Energies* **2021**, *14*, 6312. https://doi.org/10.3390/en14196312

Academic Editor: Javier Contreras

Received: 17 August 2021
Accepted: 28 September 2021
Published: 2 October 2021



## 1. Introduction

Aircraft with advanced electric drives represent a green solution that can reduce fuel consumption and greenhouse gas emissions, thanks to lower weight, low maintenance, and higher conversion efficiency than mechanical, hydraulic, and pneumatic systems. This transition is possible thanks to reliable and high-efficiency power converters, which allow power loss reduction, leading to higher power density, lower weight, and lower volumes. Compliance with existing international standards guarantees the adoption of power converters with acceptable values of total harmonic distortion (THD), power factor (PF), and DC output ripple, leading to increasingly reliable systems. High-harmonic components in currents and voltages and a high ripple rate in DC output voltages may result in resonances overvoltages and other problems of electromagnetic compatibility inside the onboard electrical systems [1].

The application of a CTU as an AC–DC converter represents the most used solution, even if it may not always meet the quality standard requirements in terms of input current THD and output voltage ripple [2]. Lower current harmonic content and higher power factor are achieved by using interphase transformers and impedance-matching inductors, which, in contrast, result in increased complexity, volume, weight, and cost [3]. Moreover, these components may suffer from detuning [3–9]. To improve the quality standards and reduce the overall size, AC–DC converters having a higher pulse number have been proposed [1–3]. Recently, several authors have investigated solutions based on the use of autotransformers, indicating their possible advantages and suggesting these solutions as alternatives to the use of CTUs [10–14]. In the proposed autotransformer-based solution, the windings are interconnected such that the apparent power transmitted by the actual magnetic coupling is only a portion of the total apparent power [14]. The reduced apparent power rating makes the autotransformer-based solution smaller and less costly and lets it operate at a higher efficiency than the conventional transformer-based solution [15]. This reduces the system weight and volume and seems to increase the overall reliability, capability, and maintainability; finally, it seems to provide higher durability for aircraft operations [16,17].

In this paper, an 800 Hz/2 kW 12-pulse ATU is investigated and designed, and its performance is compared with that of a CTU of the same frequency and power rate. The CTU solution is based on a Δ–ΔY transformer and two diode bridges connected in series. The ATU solution has two diode bridges in parallel.

The standards of reference used in the regulation of aircraft power supplies are CEN-EN 2282 [18], ISO 1540 [19], IEEE-1531-2020 [20], and RTCA DO-160 [21]. In this paper, the RTCA DO-160 standard has been considered, since it is the most referenced standard.

The IEEE-1531 standard recommends a limit of 8% on the THD of AC current and 5% on the THD of voltage at input AC mains [20]. The RTCA DO-160 standard indicates different maximum THD rates, depending on the device category [21].

The two configurations examined in the paper are designed for the typical avionic 115 RMS AC input voltage and 270 V output voltage. The design is made via a coupled FEM-circuital analysis. The resulting nominal parameters of the designed solutions, including their size and weight, are obtained and indicated.

As indicated by RTCA DO-160, several tests are simulated on both the CTU and the ATU to assess their compliance with this standard. In particular, four tests are simulated: (1) AC input current distortion test, (2) AC power factor test, (3) DC output voltage ripple test, and (4) phase unbalance test.

In addition to compliance with the standards, an accurate analysis of AC–DC conversion efficiency and power losses is carried out. Finally, a comparison of the CTU and ATU in terms of size, weight, and converter performance is illustrated.

The paper is organized as follows. In Section 2, the architecture and design of the CTU and ATU 12-pulse rectifier solutions are presented and discussed, including the FEM-circuital approach. In Section 3, the main features of the RTCA DO-160 standard are described. The AC current distortion, power factor, DC output voltage, and phase unbalance are evaluated through the simulation of the tests indicated by the standard in Sections 4–7, respectively. Starting from the obtained results, a comparison between the two topologies is discussed in Section 8. Finally, some concluding remarks are provided in Section 9.

## 2. Twelve-Pulse AC/DC Configurations

### *2.1. Conventional 12-Pulse Unit Design*

As shown in Figure 1a, the CTU was fed from a three-phase transformer with a Δ-connected primary winding and two secondary windings (one is Δ connected, and the other is Y connected). The magnetic core with the primary and secondary winding arrangement is shown in Figure 1b. Each secondary winding was connected to a three-phase, six-pulse diode bridge. The two bridges were connected in series to form a 12-pulse rectifier. The output voltage was usually connected to a DC load, such as batteries or a supercapacitor, which allowed increasing the power density, reducing weight [22]. As shown in Figure 1c, this configuration produced a 30° phase shift between the voltage phasors of the two bridges, and this resulted in a 12-pulse per cycle output voltage [23].

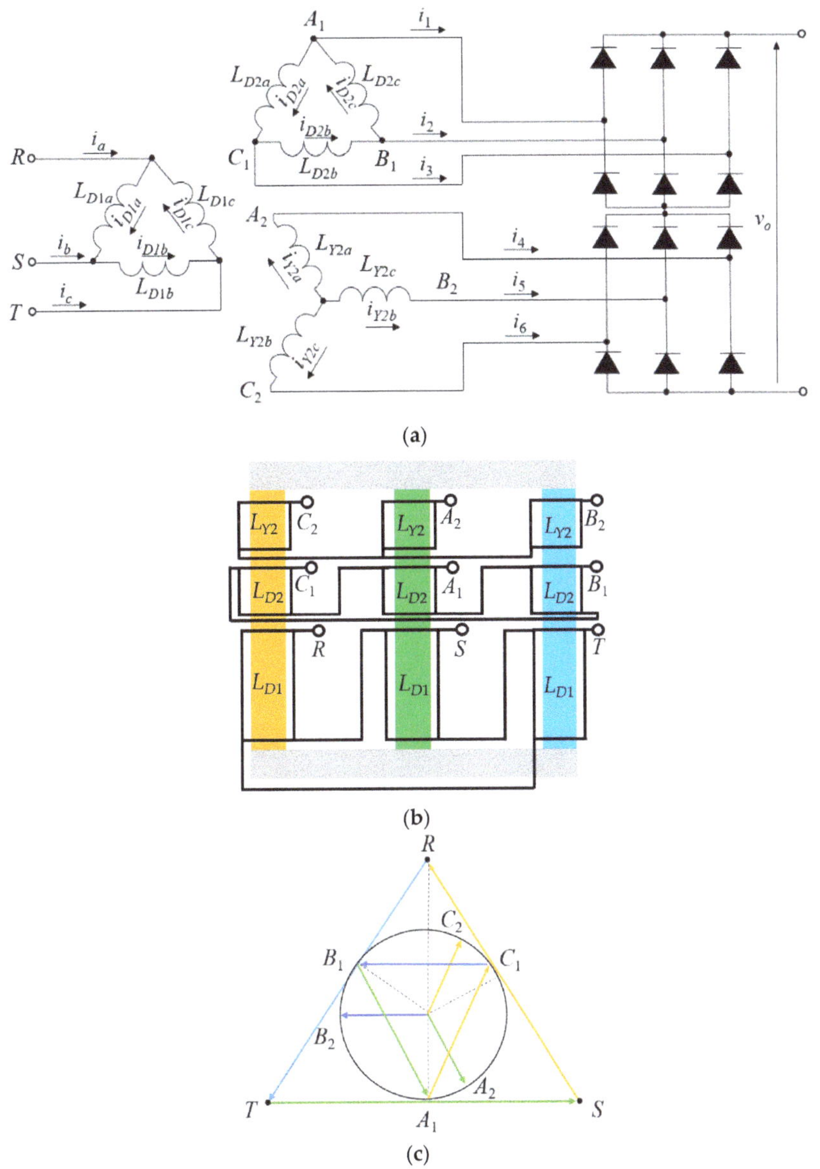


**Figure 1.** Schematic of CTU 12-pulse rectifier. (**a**) Electrical circuit. (**b**) Magnetic structure representation. (**c**) Phasor diagram representation.

The 12-pulse AC–DC converter must be suitable to operate with the design constraints shown in Table 1.

**Table 1.** Converter operating conditions.

| Parameter | Value |
|---|---|
| RMS phase-to-ground input sinusoidal voltage $V_i$ | 115 V |
| Frequency $f$ | 800 Hz |
| DC output voltage $V_o$ | 270 V |
| Nominal output power $P_o$ | 2 kW |

It was designed according to the following steps:

(1) Design of the number of turns of the windings to achieve the desired secondary no-load voltages
(2) Definition of the magnetic core cross section according to the maximum value of the working magnetic induction allowed
(3) Estimation of the maximum value of the currents in the windings
(4) Definition of the wire sections of the conductors
(5) Calculation of the resistances and the self and mutual inductances of the windings
(6) Definition of the size of the windings and the magnetic core

The design procedure was recursively run, considering at that stage a linear behavior of the magnetic material.

This first stage resulted in the output characteristics shown in Table 2.

**Table 2.** Conventional 12-pulse unit parameters.

| Parameter | Value |
|---|---|
| Primary delta turn copper wire diameter $\varnothing_{ND1}$ | 1.12 mm |
| Secondary delta turn copper wire diameter $\varnothing_{ND2}$ | 1.4 mm |
| Secondary star turn copper wire diameter $\varnothing_{NY2}$ | 1 mm |
| Number of primary delta turns $N_{D1}$ | 125 |
| Primary delta turn resistance $R_{D1}$ | 337 mΩ |
| Number of secondary delta turns $N_{D2}$ | 64 |
| Secondary delta turn resistance $R_{D2}$ | 256 mΩ |
| Number of secondary star turns $N_{Y2}$ | 37 |
| Secondary star turn resistance $R_{Y2}$ | 77 mΩ |
| Reluctance (air gap) $\mathcal{R}$ | 298,040 $H^{-1}$ |
| Primary delta inductances $L_{D1}, L_{D1a}, L_{D1b}, L_{D1c}$ | 52.4 mH |
| Secondary delta inductances $L_{D2}, L_{D2a}, L_{D2b}, L_{D2c}$ | 13.7 mH |
| Secondary star inductances $L_{Y2}, L_{Y2a}, L_{Y2b}, L_{Y2c}$ | 4.6 mH |
| Winding weight $M_{Cu}$ | 0.85 kg |
| Core weight $M_{Core}$ | 2.65 kg |
| Cross-sectional core area $S$ | $20 \times 30$ mm$^2$ |
| Core-filling factor | 0.98 |

The magnetic core material used in the design was a stacking iron-based Metglas® amorphous alloy [24]. To check the previous approximated computation of the magnetic induction and that of the magnetic inductances in the magnetic coupling coefficients between the windings, a finite element analysis (FEA) was performed by using a code developed by the authors [25–28]. In particular, a 2D analysis was performed by assuming a linear behavior of the material (constant value of the magnetic permittivity, equal to the value measured at the working magnetic induction) and a current-driven formulation of the problem. A full three-phase current system was used for the excitations, and the coupling coefficient was computed by considering the effective winding geometry. The aim of this simulation was to compute both the coupling coefficient and the maximum value of the magnetic induction in real operating conditions. The mesh discretization of the magnetic core and a close-up of the air gap region are shown in Figure 2. To compute the coupling coefficient, a postprocessing of the FEA solution was performed by evaluating the linked flux in different parts of the section surrounded by the windings. The coupling coefficient estimated was about $k = 0.98$, while the maximum value of magnetic induction (RMS value) was about $B_{RMS} = 0.85$ T. The field vectors are shown in Figure 3 at the maximum value of magnetic induction.

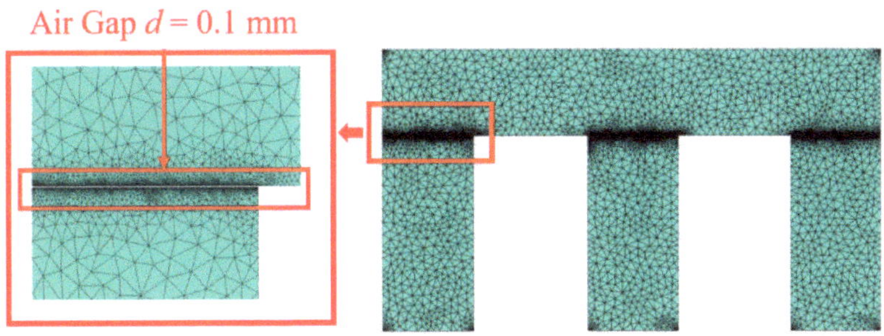


**Figure 2.** Mesh of the half section of the magnetic core used in finite element analysis.

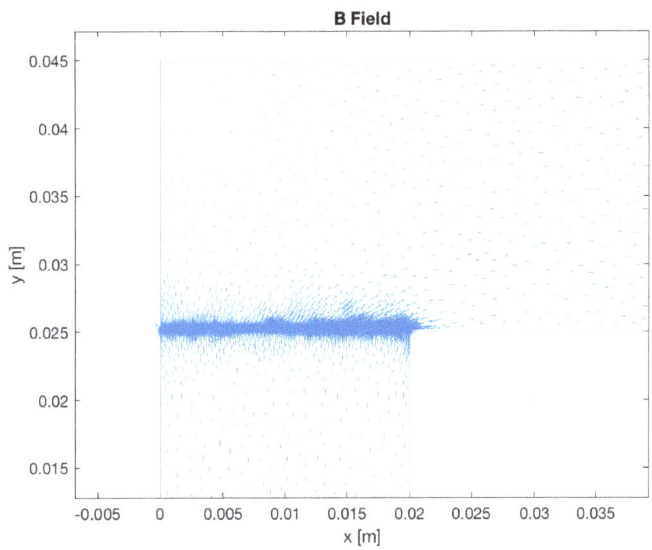


**Figure 3.** Magnetic induction field vector distribution computed by FEA in the air gap region.

*2.2. Autotransformer 12-Pulse Unit Design*

Figure 4a shows the ATU electrical circuit. It consists of a Δ-connected autotransformer with secondary windings connected to two three-phase diode bridge rectifiers; these two bridges were connected in parallel with the load. The winding representation and the voltage phasor diagram of the proposed Δ-type autotransformer are shown in Figure 4b,c, respectively.

The transformer phase-shifting angle was equal to 30°, the winding configuration of the ATU resulted in the lowest apparent power rating [3], and the size of the magnetic component reduced [10] under this condition. As described in [12,29], the proposed autotransformer achieved the simplest winding configuration when the angle was 30° (±15°). To produce a 30° phase shift between the two sets of secondary windings, the winding turn ratio should be

$$\frac{N_1}{N_2} = \frac{\sqrt{3}}{2-\sqrt{3}} \tag{1}$$

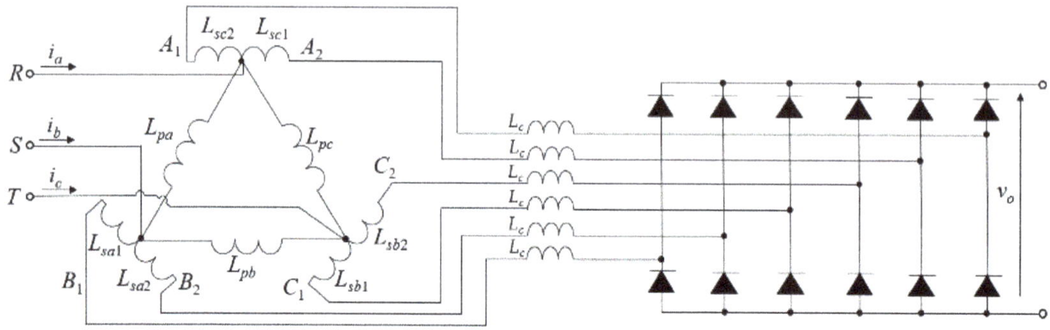


(**a**)

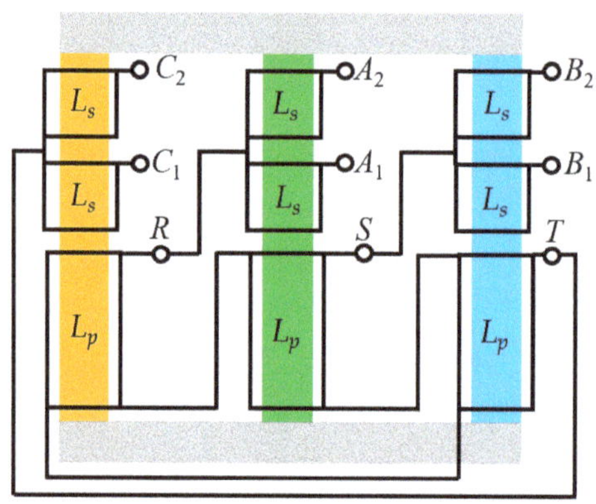


(**b**)

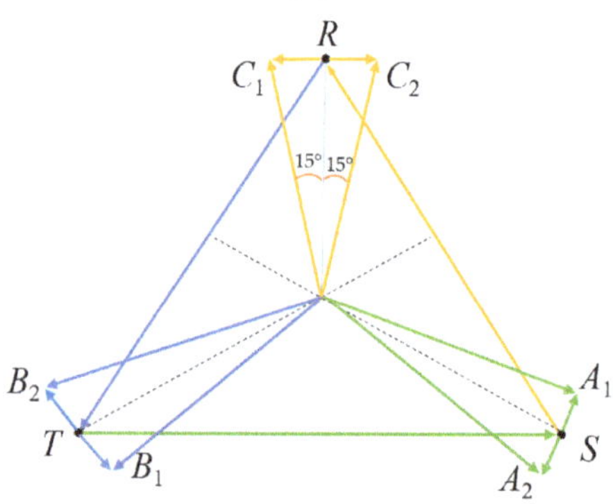


(**c**)

**Figure 4.** Schematic of an ATU. (**a**) Electrical circuit. (**b**) Structure representation. (**c**) Voltage phasor representation.

The phasors of the voltages produced by the autotransformer $V_{A1}$, $V_{B1}$, and $V_{C1}$ were at +15° with respect to the supply voltages $V_R$, $V_S$, and $V_T$, while the other set of the phasors of the voltages $V_{A2}$, $V_{B2}$, and $V_{C2}$ were at −15° with respect to supply voltages, resulting in a 12-pulse rectification [9,30,31]. The magnitudes of the secondary line voltages ($V_{A1}$, $V_{B1}$, and $V_{C1}$ and ($V_{A2}$, $V_{B2}$, and $V_{C2}$) should be equal to each other to result in a symmetrical system and to reduce the ripple in output DC voltage [11–32].

A two-step design of the ATU, requiring the same procedure described in the previous section for the CTU, was carried out. The final parameters of the designed ATU device are reported in Table 3.

**Table 3.** Autotransformer 12-pulse rectifier parameters.

| Parameter | Value |
|---|---|
| Primary turn copper wire diameter $\varnothing_{Np}$ | 0.85 mm |
| Secondary turn copper wire diameter $\varnothing_{Ns}$ | 1.25 mm |
| Number of primary turns $N_p$ | 110 |
| Primary turn resistance $R_p$ | 504 mΩ |
| Number of secondary turns $N_s$ | 17 |
| Secondary turn resistance $R_s$ | 41.87 mΩ |
| Cross-sectional core area $S$ | 20 × 20 mm$^2$ |
| Winding weight $M_{Cu}$ | 0.5 kg |
| Core weight $M_{Core}$ | 1.065 kg |
| Series inductance $L_c$ | 0.5 mH |
| Reluctance (air gap) $\mathcal{R}$ | 397,890 H$^{-1}$ |
| Primary inductance $L_p$ | 304 mH |
| Secondary inductance $L_s$ | 72.6 mH |

According to the previous FEA, a coupling coefficient of $k_{ps}$ = 0.98 between the primary and the secondary and a coupling coefficient $k_{ss}$ = 0.96 between the two secondaries were estimated.

An additional filter constituted by Δ-connected capacitors was added to the system to reduce the input current THD. The IEEE 1531-2003 standard [20] was used to size the capacitors. To determine the value of each capacitor, the required reactive power for the system $Q$ was first determined through the simulation and then the capacitance was calculated according to [33]

$$C_f = \frac{Q}{2\pi f V_i^2} \tag{2}$$

where $f$ is the fundamental frequency of the current, $V_i$ is the RMS value of the line voltage, and $Q$ is the required reactive power. By using the power measurements Matlab's block, a reactive power $Q$ of about 1700 VAR was estimated, leading to a filtering capacitor $C_f$ = 13 μF. To reduce the THD, three filtering capacitors Δ-connected with a value $C_f$ =25 μF were finally chosen.

## 3. Application of RTCA DO-160G Standard Tests

From RTCA DO-160G, the equipment intended for use on aircraft electrical systems where the primary power is supplied through an AC system with a frequency in the range of 360–800 Hz was designated as A(WF). Regarding the DC side, if the output was 270 V, the systems were designated with the letter D. Therefore, the systems studied in this paper were classified as A(WF)-D. This information is required to properly select tests that must be performed on the rectifiers to check the required compliance of the system. In addition to the tests required from RTCA DO-160G, an accurate evaluation of the conversion efficiency of the system was carried out, properly analyzing the power losses.

During the analysis, a line inductance $L_s$ = 0.1 mH was assumed to be connected in series to the three-phase voltage sources to take into account the electrical power supply system upstream of the device.

The same diode was used in the CTU and ATU rectifiers, with a forward voltage $V_F$ = 0.6 V and a conduction resistance $r_{ON}$ = 0.1 Ω. The performance of each rectifier was evaluated for three different load resistances, $R_{L1}$ = 37 Ω, $R_{L2}$ = 74 Ω, and $R_{L3}$ = 148 Ω, called here heavy load, intermediate load, and light load resistances, respectively, which represent, approximatively, a working condition at 100%, 50%, and 20% of the nominal power, respectively. These working conditions are summarized in Table 4.

**Table 4.** Converters' operating conditions.

| Parameter | Value |
|---|---|
| Diode forward voltage $V_F$ | 0.6 V |
| Diode conduction resistance $r_{ON}$ | 0.1 Ω |
| Heavy load resistance $R_{L1}$ | 37 Ω |
| Intermediate load resistance $R_{L2}$ | 74 Ω |
| Light load resistance $R_{L3}$ | 148 Ω |

As requested from the RTCA DO-160G standard, the following parameters must be evaluated.

The output voltage ripple $\Delta V_o$ was evaluated under the three different operating conditions and calculated as

$$V_{o,ripple}(\%) = \frac{V_o^{max} - V_o^{min}}{V_o^{avg}} \cdot 100 \tag{3}$$

The total harmonic distortion (THD) of the input current was calculated as

$$THD(\%) = \frac{\sqrt{I_2^2 + I_3^2 + \ldots + I_N^2}}{I_1} \cdot 100 \tag{4}$$

where $I_1$ is the amplitude of the fundamental current harmonic and $I_2, I_3, \ldots, I_N$ are the amplitudes of the second, third, ..., $N$-th current harmonic, respectively.

The power factor (PF) was also evaluated by using

$$PF = \frac{1}{\sqrt{1 + \left(\frac{THD(\%)}{100}\right)^2}} \cos(\varphi_1) \tag{5}$$

where $\varphi_1$ is the displacement angle between the fundamental components of the input current and voltage.

Another parameter of primary importance is the AC–DC conversion efficiency. As known, it can be calculated as $\eta = P_o/P_i$ with $P_o = V_o \cdot I_o$ and $P_i = P_o + P_{loss}$.

The power losses can be assumed to be constituted by two main contributions as

$$P_{loss} = P_{TRF} + P_D \tag{6}$$

where $P_D$ is the loss due to the rectifier and it is strictly related to the characteristics of the diodes. It can be calculated as

$$P_D = r_{ON} \cdot I_{D,RMS}^2 + V_F \cdot I_{D,AVG} \tag{7}$$

where $I_{D,RMS}$ and $I_{D,AVG}$ are the RSM and average current flowing through the diode, respectively. The average and RMS value of the current were calculated on Simulink, measuring the current flowing through each diode and then computing the power loss using Equation (7).

$P_{TRF}$ is the transformer power loss and can be divided into two contributions

$$P_{TRF} = P_{Cu} + P_{Core} \tag{8}$$

where $P_{Cu}$ is the ohmic loss due to the copper windings and is calculated as

$$P_{Cu} = R \cdot I_R^2 \tag{9}$$

while $P_{Core}$ is the magnetic core loss given by

$$P_{Core} = C_m \cdot f^{\alpha} \cdot B_{rms}^{\beta} \tag{10}$$

From the datasheet of the Metglas® alloy 2605SA1, the values of the parameters for this material were $\alpha$ = 1.51, $\beta$ = 1.74, and $C_m$ = 6.5 [25].

## 4. AC Current Distortion Test

In this section, the AC current distortion for both the CTU and the ATU are evaluated by using the procedure provided by the RTCA DO-160 standard.

This test must be performed for AC equipment with a maximum power consumption larger than 35 VA. The first 40th harmonics are required to satisfy the individual current harmonic limits, as shown in Table 5.

**Table 5.** Harmonic limits according to RTCA DO-160.

| Harmonic Order | Limits |
|---|---|
| 3, 5, 7 | $I_3 = I_5 = I_7 = 0.02 \cdot I_1$ * |
| Odd triplen—9,15,21,27,33,39 | $I_h = 0.1 \cdot I_1/h$ |
| 11 | $I_{11} = 0.1 \cdot I_1$ |
| 13 | $I_{13} = 0.08 \cdot I_1$ |
| Odd non-triplen—17 and 19 | $I_{17} = I_{19} = 0.04 \cdot I_1$ |
| Odd non-triplen—23 and 25 | $I_{23} = I_{25} = 0.03 \cdot I_1$ |
| Odd non-triplen—29, 31, 35, 37 | $I_h = 0.3 \cdot I_1/h$ |
| Even—2 and 4 | $I_h = 0.01 \cdot I_1/h$ |
| Even > 4 (6, 8, 10, 12, . . . , 40) | $I_h = 0.0025 \cdot I_1$ |

* As above, $I_1$ is the amplitude of the current fundamental harmonic and $Ih$ is the amplitude of the harmonic current of $h$ order.

From the RTCA DO-160G standard, the current distortion must be evaluated under two operating conditions:

A. When the circuit is supplied with a voltage waveform with $THD_V$ < 1.25%. In this case, the equipment will not demand harmonic current components above the limits shown in Table 5.

B. When the circuit is supplied with a distorted voltage waveform $THD_V$ > 1.25%. The equipment will not demand a harmonic current greater than 1.25% above the limits already specified in Table 5 for every 1% of distortion in the corresponding individual voltage harmonic.

The two waveforms used to perform test A and test B are shown in Figure 5a,b, respectively. In particular, test A was simulated assuming a three-phase voltage with $THD_V$ = 0%, while test B was simulated assuming $THD_V$ = 10% with equal RMS values of the third, fifth, and seventh harmonics.

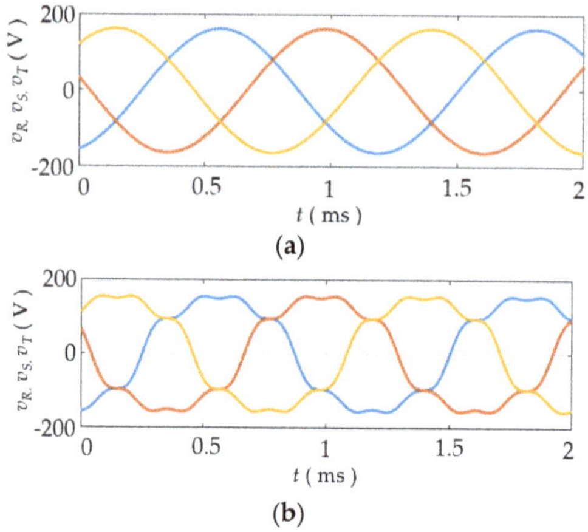


**Figure 5.** Input voltage waveforms for the simulated tests. (**a**) Phase voltage without distortion. (**b**) Distorted input voltage.

Next are summarized the results obtained for the simulated tests on both the CTU and the ATU.

### 4.1. CTU Current Distortion Evaluation

The input currents $i_a$ when the system was supplied with the voltage shown in Figure 5a,b are shown in Figure 6a,b, respectively.

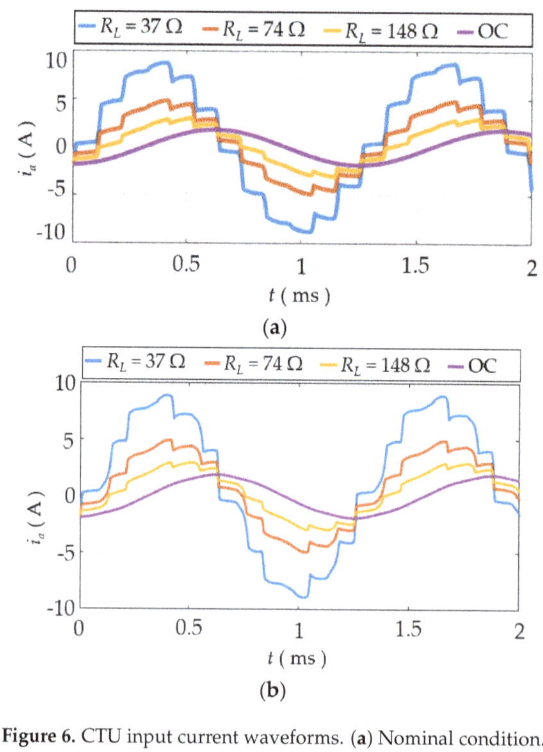


**Figure 6.** CTU input current waveforms. (**a**) Nominal condition. (**b**) Distorted input voltage.

The CTU harmonic current components under tests A and B are summarized in Table 6. The amplitude of the primary harmonic (order $H_1$ in the table) at $f$ = 800 Hz was $I_{H1}$ = 8.4 A. Each harmonic value was expressed as a percentage of the amplitude of the primary harmonic.

**Table 6.** CTU total harmonic distortion.

| Harmonic | Limit | Test A ($THD_V$ = 0%) $R_L$ = 37 Ω | $R_L$ = 74 Ω | $R_L$ = 148 Ω | Limit | Test B ($THD_V$ = 10%) $R_L$ = 37 Ω | $R_L$ = 74 Ω | $R_L$ = 148 Ω |
|---|---|---|---|---|---|---|---|---|
| $H_1$ | 100% | 100% | 100% | 100% | 100% | 100% | 100% | 100% |
| $H_2$ | 0.5% | 0.46% | 0.31% | 0.18% | 0.5% | 0.43% | 0.29% | 0.17% |
| $H_3$ | 2% | 0.49% | 0.32% | 0.19% | 6.16% | 0.52% | 0.35% | 0.2% |
| $H_4$ | 0.25% | 0.44% | 0.30% | 0.18% | 0.25% | 0.38% | 0.25% | 0.14% |
| $H_5$ | 2% | 0.41% | 0.29% | 0.17% | 6.16% | 3.52% | 2.53% | 1.09% |
| $H_6$ | 0.25% | 0.40% | 0.28% | 0.17% | 0.25% | 0.37% | 0.25% | 0.14% |
| $H_7$ | 2% | 0.36% | 0.27% | 0.17% | 6.16% | 4.39% | 3.42% | 2.03% |
| $H_8$ | 0.25% | 0.38% | 0.28% | 0.17% | 0.25% | 0.22% | 0.16% | 0.1% |
| $H_9$ | 1.1% | 0.35% | 0.28% | 0.17% | 1.1% | 0.35% | 0.26% | 0.16% |
| $H_{10}$ | 0.25% | 0.37% | 0.29% | 0.18% | 0.25% | 0.37% | 0.26% | 0.16% |
| $H_{11}$ | 10% | 6.72% | 7.44% | 6.5% | 10% | 7.04% | 7.17% | 6.27% |
| $H_{12}$ | 0.25% | 0.34% | 0.28% | 0.18% | 0.25% | 0.4% | 0.31% | 0.19% |
| $H_{13}$ | 8% | 3.5% | 4.35% | 4.18% | 8% | 4.01% | 4.4% | 4.05% |
| $H_{14}$ | 0.25% | 0.28% | 0.25% | 0.17% | 0.25% | 0.23% | 0.23% | 0.14% |
| $H_{15}$ | 0.66% | 0.26% | 0.24% | 0.16% | 0.66% | 0.34% | 0.27% | 0.17% |
| $H_{16}$ | 0.25% | 0.21% | 0.21% | 0.15% | 0.25% | 0.22% | 0.21% | 0.13% |
| $H_{17}$ | 4% | 0.17% | 0.19% | 0.14% | 4% | 3.99% | 3.37% | 2.36% |
| $H_{18}$ | 0.25% | 0.15% | 0.18% | 0.14% | 0.25% | 0.2% | 0.16% | 0.11% |
| $H_{19}$ | 4% | 0.1% | 0.15% | 0.13% | 4% | 2.84% | 2.49% | 1.78% |
| $H_{20}$ | 0.25% | 0.09% | 0.15% | 0.13% | 0.25% | 0.05% | 0.03% | 0.03% |
| $H_{21}$ | 0.47% | 0.06% | 0.13% | 0.12% | 0.47% | 0.06% | 0.06% | 0.07% |
| $H_{22}$ | 0.25% | 0.03% | 0.13% | 0.13% | 0.25% | 0.04% | 0.09% | 0.08% |
| $H_{23}$ | 3% | 1.17% | 1.82% | 2.2% | 3% | 1.1% | 1.43% | 1.69% |
| $H_{24}$ | 0.25% | 0.01% | 0.11% | 0.12% | 0.25% | 0.03% | 0.09% | 0.1% |
| $H_{25}$ | 3% | 0.71% | 1.18% | 1.55% | 3% | 0.81% | 0.91% | 1.16% |
| $H_{26}$ | 0.25% | 0.03% | 0.09% | 0.11% | 0.25% | 0.07% | 0.09% | 0.08% |
| $H_{27}$ | 0.37% | 0.04% | 0.09% | 0.11% | 0.37% | 0.1% | 0.13% | 0.11% |
| $H_{28}$ | 0.25% | 0.05% | 0.07% | 0.1% | 0.25% | 0.15% | 0.13% | 0.1% |
| $H_{29}$ | 1.03% | 0.06% | 0.06% | 0.1% | 1.03% | 2.16% | 2.22% | 1.77% |
| $H_{30}$ | 0.25% | 0.06% | 0.05% | 0.08% | 0.25% | 0.17% | 0.13% | 0.08% |
| $H_{31}$ | 0.96% | 0.07% | 0.03% | 0.07% | 0.96% | 0.87% | 1.72% | 1.45% |
| $H_{32}$ | 0.25% | 0.08% | 0.02% | 0.07% | 0.25% | 0.14% | 0.09% | 0.04% |
| $H_{33}$ | 0.3% | 0.08% | 0.02% | 0.06% | 0.3% | 0.18% | 0.11% | 0.04% |
| $H_{34}$ | 0.25% | 0.09% | 0.01% | 0.06% | 0.25% | 0.15% | 0.07% | 0.0% |
| $H_{35}$ | 0.85% | 0.5% | 0.68% | 0.87% | 0.85% | 1.32% | 0.98% | 0.61% |
| $H_{36}$ | 0.25% | 0.09% | 0.02% | 0.05% | 0.25% | 0.14% | 0.06% | 0.01% |
| $H_{37}$ | 0.81% | 0.34% | 0.4% | 0.6% | 0.81% | 0.78% | 0.87% | 0.53% |
| $H_{38}$ | 0.25% | 0.08% | 0.03% | 0.05% | 0.25% | 0.07% | 0.04% | 0.03% |
| $H_{39}$ | 0.26% | 0.08% | 0.03% | 0.04% | 0.26% | 0.06% | 0.01% | 0.04% |
| $H_{40}$ | 0.31% | 0.08% | 0.04% | 0.04% | 0.31% | 0.02% | 0.06% | 0.06% |

As can be seen from Table 6, the simulation indicated that the limits of the standards were respected for both tests A and B.

Therefore, it was indicated by the simulation results that the designed CTU complies with the current harmonic limits.

### *4.2. ATU Current Distortion Evaluation*

The ATU harmonic current components under tests A and B are summarized in Table 7. The amplitude of the primary harmonic (order $H_1$) at $f$ = 800 Hz was $I_{H1}$ = 9.6 A.

Each higher-order harmonic was expressed as a percentage of the amplitude of the primary harmonic.

**Table 7.** ATU total harmonic distortion.

| Harmonic | Test A ($THD_V$ = 0%) | | | | Test B ($THD_V$ = 10%) | | | |
|---|---|---|---|---|---|---|---|---|
| | Limit | $R_L = 37\ \Omega$ | $R_L = 74\ \Omega$ | $R_L = 148\ \Omega$ | Limit | $R_L = 37\ \Omega$ | $R_L = 74\ \Omega$ | $R_L = 148\ \Omega$ |
| $H_1$ | 100% | 100% | 100% | 100% | 100% | 100% | 100% | 100% |
| $H_2$ | 0.5% | 0.02% | 0.0% | 0.0% | 0.5% | 0.23% | 0.13% | 0.01% |
| $H_3$ | 2% | 0.66% | 0.35% | 0.18% | 6.16% | 1.52% | 2.11% | 2.38% |
| $H_4$ | 0.25% | 0.02% | 0.0% | 0.0% | 0.25% | 0.28% | 0.03% | 0.01% |
| $H_5$ | 2% | 3.65% | 2% | 1.03% | 6.16 % | 4.95% | 3.77% | 1.35% |
| $H_6$ | 0.25% | 0.02% | 0.0% | 0.0% | 0.25% | 0.22% | 0.07% | 0.01% |
| $H_7$ | 2% | 1.09% | 0.72% | 0.43% | 6.16 % | 2.9% | 2.65% | 1.78% |
| $H_8$ | 0.25% | 0.02% | 0.0% | 0.0% | 0.25% | 0.09% | 0.04% | 0.01% |
| $H_9$ | 1.1% | 0.11% | 0.08% | 0.05% | 1.1% | 0.12% | 0.04% | 0.01% |
| $H_{10}$ | 0.25% | 0.02% | 0% | 0% | 0.25% | 0.19% | 0.04% | 0.01% |
| $H_{11}$ | 10% | 0.64% | 0.51% | 0.34% | 10% | 1.58% | 1.19% | 0.94% |
| $H_{12}$ | 0.25% | 0.01% | 0% | 0% | 0.25% | 0.05% | 0.04% | 0.01% |
| $H_{13}$ | 8% | 0.36% | 0.27% | 0.19% | 8% | 0.63% | 0.51% | 0.38% |
| $H_{14}$ | 0.25% | 0.01% | 0% | 0% | 0.25% | 0.04% | 0.03% | 0.01% |
| $H_{15}$ | 0.66% | 0.11% | 0.03% | 0.02% | 0.66% | 0.05% | 0.03% | 0.01% |
| $H_{16}$ | 0.25% | 0.01% | 0% | 0% | 0.25% | 0.05% | 0.04% | 0.01% |
| $H_{17}$ | 4% | 0.28% | 0.19% | 0.15% | 4% | 1.02% | 0.5% | 0.28% |
| $H_{18}$ | 0.25% | 0.01% | 0% | 0% | 0.25% | 0.06% | 0.02% | 0.01% |
| $H_{19}$ | 4% | 0.15% | 0.13% | 0.1% | 4% | 0.42% | 0.15% | 0.16% |
| $H_{20}$ | 0.25% | 0.0% | 0% | 0% | 0.25% | 0.05% | 0.02% | 0.0% |
| $H_{21}$ | 0.47% | 0.17% | 0.03% | 0.01% | 0.47% | 0.08% | 0.03% | 0% |
| $H_{22}$ | 0.25% | 0% | 0% | 0% | 0.25% | 0.09% | 0.02% | 0% |
| $H_{23}$ | 3% | 0.12% | 0.1% | 0.07% | 3% | 0.39% | 0.39% | 0.18% |
| $H_{24}$ | 0.25% | 0% | 0% | 0% | 0.25% | 0.07% | 0.01% | 0% |
| $H_{25}$ | 3% | 0.09% | 0.06% | 0.05% | 3% | 0.22% | 0.09% | 0.07% |
| $H_{26}$ | 0.25% | 0% | 0% | 0% | 0.25% | 0.06% | 0.02% | 0% |
| $H_{27}$ | 0.37% | 0.13% | 0.06% | 0.01% | 0.37% | 0.08% | 0.02% | 0% |
| $H_{28}$ | 0.25% | 0.0% | 0% | 0% | 0.25% | 0.08% | 0.01% | 0% |
| $H_{29}$ | 1.03% | 0.04% | 0.05% | 0.04% | 1.03% | 0.21% | 0.25% | 0.17% |
| $H_{30}$ | 0.25% | 0% | 0% | 0% | 0.25% | 0.06% | 0.01% | 0% |
| $H_{31}$ | 0.96% | 0.06% | 0.04% | 0.03% | 0.96% | 0.09% | 0.1% | 0.03% |
| $H_{32}$ | 0.25% | 0% | 0% | 0% | 0.25% | 0.05% | 0.01% | 0% |
| $H_{33}$ | 0.3% | 0.06% | 0.06% | 0.01% | 0.3% | 0.06% | 0.01% | 0% |
| $H_{34}$ | 0.25% | 0% | 0% | 0% | 0.25% | 0.04% | 0.01% | 0% |
| $H_{35}$ | 0.85% | 0.05% | 0.03% | 0.03% | 0.85% | 0.2% | 0.12% | 0.13% |
| $H_{36}$ | 0.25% | 0% | 0% | 0% | 0.25% | 0.05% | 0.01% | 0% |
| $H_{37}$ | 0.81% | 0.02% | 0.03% | 0.02% | 0.81% | 0.13% | 0.07% | 0.03% |
| $H_{38}$ | 0.25% | 0% | 0% | 0% | 0.25% | 0.05% | 0.01% | 0% |
| $H_{39}$ | 0.26% | 0.04% | 0.04% | 0.02% | 0.26% | 0.05% | 0.01% | 0% |
| $H_{40}$ | 0.31% | 0.0% | 0.0% | 0.0% | 0.31% | 0.05% | 0.01% | 0% |

As can be seen from Table 6, the simulations that indicated the limits of the standards were respected for both tests A and B.

The input currents $i_a$ when the system was supplied with the voltage shown in Figure 5a,b are shown in Figure 7a,b, respectively.

In addition, for the ATU, the test simulations indicated that the input current harmonic content allows satisfying the standard limits.

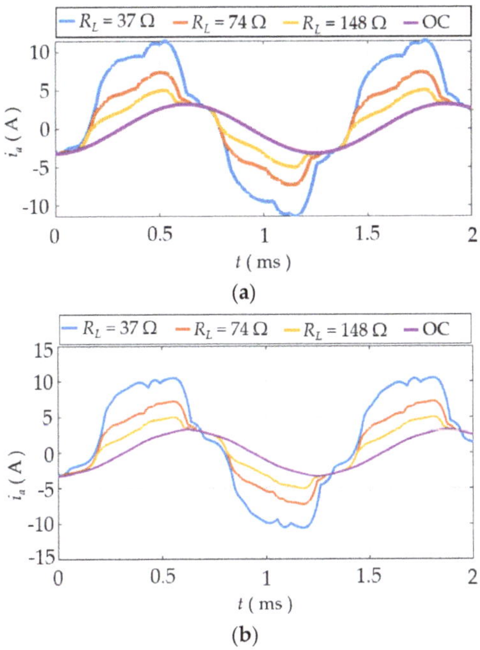


**Figure 7.** ATU input current waveforms. (**a**) Nominal condition. (**b**) Distorted input voltage.

## 5. AC Power Factor Test

Another test required by RTCA DO-160G is the AC power factor test. The power factor, defined as in Equation (5), will be equal to or higher than the values listed in Table 8.

**Table 8.** ATU total harmonic distortion and power factor limits.

| Load (kVA) | Power Factor—Leading | Power Factor—Lagging |
|---|---|---|
| <0.02 | 0.200 | 0.200 |
| 0.03 | 0.355 | 0.321 |
| 0.04 | 0.464 | 0.406 |
| 0.06 | 0.619 | 0.527 |
| 0.08 | 0.728 | 0.613 |
| 0.1 | 0.813 | 0.679 |
| 0.15 | 0.968 | 0.800 |
| ≥0.15 | 0.968 | 0.800 |

Since the proposed 12-pulse rectifiers were characterized by a 2 kW nominal load power and they were seen as ohmic-inductive loads by the source, a lagging power factor $PF > 0.8$ was required to comply with the standard requirement.

### *5.1. CTU Power Factor Test Results*

The THD of the line currents and the PF computed under different load conditions are summarized in Table 9.

The simulations indicated that the PF was always higher than the limit defined by the standard $PF > 0.8$ for all the load conditions examined. The presence of resistive-inductive loads seemed to not affect the converter behavior.

**Table 9.** CTU total harmonic distortion and power factor.

| Load | THD (%) | PF—Lagging |
|---|---|---|
| $R_{L1} = 37\ \Omega$ | 5.5 | 0.88 |
| $R_{L2} = 74\ \Omega$ | 7.4 | 0.83 |
| $R_{L3} = 148\ \Omega$ | 7.1 | 0.81 |
| $R_{L1} = 37\ \Omega$, $L_{L1} = 1.5$ mH | 5.5 | 0.88 |
| $R_{L2} = 74\ \Omega$, $L_{L2} = 1.5$ mH | 7.4 | 0.83 |
| $R_{L3} = 148\ \Omega$, $L_{L3} = 1.5$ mH | 7.1 | 0.81 |

### 5.2. ATU Power Factor Test Results

The THD of the input line current and the PF values, calculated in analogy with the case of the CTU, are summarized in Table 10.

**Table 10.** ATU total harmonic distortion and power factor.

| Load | THD (%) | PF—Lagging |
|---|---|---|
| $R_{L1} = 37\ \Omega$ | 1.89 | 0.82 |
| $R_{L2} = 74\ \Omega$ | 1.25 | 0.81 |
| $R_{L3} = 148\ \Omega$ | 0.74 | 0.80 |
| $R_{L1} = 37\ \Omega$, $L_{L1} = 1.5$ mH | 1.78 | 0.82 |
| $R_{L2} = 74\ \Omega$, $L_{L2} = 1.5$ mH | 1.25 | 0.81 |
| $R_{L3} = 148\ \Omega$, $L_{L3} = 1.5$ mH | 0.73 | 0.80 |

In addition, for the ATU case, the PF computed was always higher than the limits specified by the standard, and it can be concluded that both AC–DC converters comply with the standard requirements.

## 6. DC Current Ripple Test

The output voltage of a 12-pulse rectifier is characterized by a DC component plus a fundamental AC component with a frequency 12 times higher than that of the AC line, plus higher-order harmonics. The RTCA DO-160G standard establishes amplitude limits for different frequency contents. Thus, the harmonic content of the output voltage was analyzed to evaluate whether the two converters meet the standard limits.

### 6.1. CTU Output Voltage Ripple Results

The waveforms of the output voltage computed under nominal conditions are shown in Figure 8.

To evaluate whether the standard limits are respected, the first 20 voltage harmonics (fundamental frequency $f_0 = 800$ Hz) were evaluated at different load levels, as shown in Table 11.

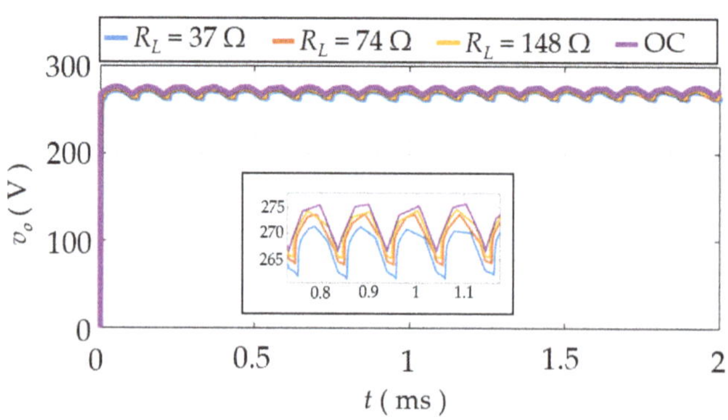


**Figure 8.** CTU output voltage waveforms.

**Table 11.** The harmonic content of CTU output voltage.

| Harmonic | DC Ripple Limit | Test A ($THD_V$ = 0%) $R_L$ = 37 Ω | $R_L$ = 74 Ω | $R_L$ = 148 Ω |
|---|---|---|---|---|
| $H_1$ | 6.4 V | 1.37 V | 0.67 V | 0.12 V |
| $H_2$ | 16 V | 1.08 V | 0.54 V | 0.10 V |
| $H_3$ | 16 V | 1.27 V | 0.62 V | 0.12 V |
| $H_4$ | 16 V | 1.22 V | 0.59 V | 0.11 V |
| $H_5$ | 16 V | 1.24 V | 0.62 V | 0.12 V |
| $H_6$ | 16 V | 1.35 V | 0.67 V | 0.13 V |
| $H_7$ | 16 V | 1.27 V | 0.65 V | 0.12 V |
| $H_8$ | 16 V | 1.32 V | 0.70 V | 0.13 V |
| $H_9$ | 16 V | 1.27 V | 0.68 V | 0.12 V |
| $H_{10}$ | 16 V | 1.27 V | 0.70 V | 0.13 V |
| $H_{11}$ | 16 V | 1.27 V | 0.70 V | 0.13 V |
| $H_{12}$ | 16 V | 4.10 V | 5.75 V | 2.09 V |
| $H_{13}$ | 16 V | 1.19 V | 0.68 V | 0.13 V |
| $H_{14}$ | 16 V | 1.13 V | 0.68 V | 0.13 V |
| $H_{15}$ | 16 V | 1.08 V | 0.67 V | 0.13 V |
| $H_{16}$ | 16 V | 1.02 V | 0.67 V | 0.13 V |
| $H_{17}$ | 16 V | 0.97 V | 0.64 V | 0.13 V |
| $H_{18}$ | 16 V | 0.91 V | 0.64 V | 0.13 V |
| $H_{19}$ | 16 V | 0.89 V | 0.63 V | 0.13 V |
| $H_{20}$ | 2.4 V | 0.83 V | 0.62 V | 0.13 V |

The average output voltage $V_o^{avg}$ and the ripple $V_o^{ripple}$ are indicated in Table 12. Simulated results indicated that the voltage amplitudes values were only marginally affected by the load, even when it was ohmic-inductive.

**Table 12.** CTU output voltage simulation results.

| Load | $V_o^{max}$ (V) | $V_o^{min}$ (V) | $V_o^{avg}$ (V) | $V_o^{ripple}$ (%) |
|---|---|---|---|---|
| $R_{L1}$ = 37 Ω | 255.9 | 244.0 | 251.7 | 4.7 |
| $R_{L2}$ = 74 Ω | 264.6 | 251.7 | 259.8 | 4.9 |
| $R_{L3}$ = 148 Ω | 269.2 | 256.1 | 262.8 | 4.9 |
| $R_{L1}$ = 37 Ω, $L_{L1}$ = 1.5 mH | 257.9 | 239.4 | 252.5 | 7.3 |
| $R_{L2}$ = 74 Ω, $L_{L2}$ = 1.5 mH | 264.2 | 250.5 | 259.1 | 5.2 |
| $R_{L3}$ = 148 Ω, $L_{L3}$ = 1.5 mH | 268.2 | 255.7 | 262.8 | 4.7 |

### *6.2. ATU Output Voltage Ripple Results*

In Figure 9, the output voltages for different loads are shown; the average output voltage $V_o^{avg}$ and the ripple $V_o^{ripple}$ are summarized in Table 13.

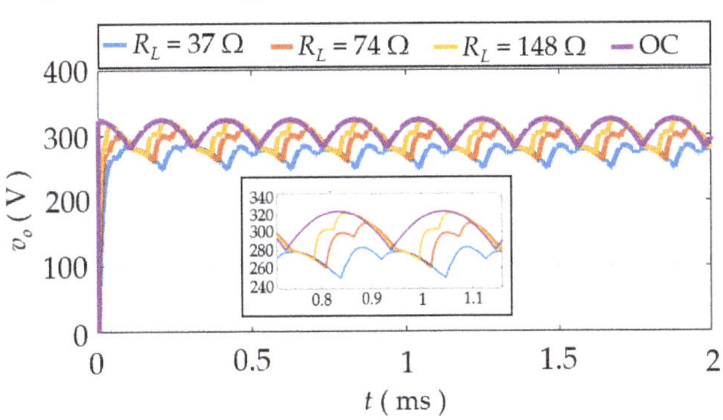


**Figure 9.** ATU output voltage waveforms.

**Table 13.** ATU with input capacitor filter—simulation results.

| Load | $V_{max}$ (V) | $V_{min}$ (V) | $V_{avg}$ (V) | $V_{ripple}$ (%) |
|---|---|---|---|---|
| $R_{L1} = 37\ \Omega$ | 283.7 | 249.8 | 272.3 | 12 |
| $R_{L2} = 74\ \Omega$ | 310.5 | 261.2 | 289.9 | 17 |
| $R_{L3} = 148\ \Omega$ | 324.4 | 270.2 | 299.6 | 18 |
| No load | 323.9 | 280.4 | 309.1 | 14 |
| $R_{L1} = 37\ \Omega$, $L_{L1} = 1.5$ mH | 286.1 | 248.3 | 272.4 | 13.8 |
| $R_{L2} = 74\ \Omega$, $L_{L2} = 1.5$ mH | 309 | 260 | 289.3 | 17 |
| $R_{L3} = 74\ \Omega$, $L_{L3} = 1.5$ mH | 324.4 | 270.2 | 299.6 | 18 |

The first 20 harmonics of the output voltage are reported in Table 14.

**Table 14.** Ripple voltage for ATU total harmonic distortion.

| Harmonic | DC Ripple Limit | Test A ($THD_V = 0\%$) | | |
|---|---|---|---|---|
| | | RL = 37 Ω | RL = 74 Ω | RL = 148 Ω |
| $H_1$ | 6.4 V | 2.24 V | 1.26 V | 0.62 V |
| $H_2$ | 16 V | 16.85 V | 16.14 V | 15.76 V |
| $H_3$ | 16 V | 2.16 V | 1.24 V | 0.62 V |
| $H_4$ | 16 V | 3.69 V | 3.92 V | 3.69 V |
| $H_5$ | 16 V | 2.02 V | 1.24 V | 0.62 V |
| $H_6$ | 16 V | 17.25 V | 19.00 V | 18.38 V |
| $H_7$ | 16 V | 1.84 V | 1.21 V | 0.62 V |
| $H_8$ | 16 V | 1.13 V | 2.24 V | 2.59 V |
| $H_9$ | 16 V | 1.65 V | 1.18 V | 0.62 V |
| $H_{10}$ | 16 V | 1.94 V | 2.62 V | 2.43 V |
| $H_{11}$ | 16 V | 1.46 V | 1.13 V | 0.62 V |
| $H_{12}$ | 16 V | 1.02 V | 2.72 V | 4.61 V |
| $H_{13}$ | 16 V | 1.29 V | 1.10 V | 0.62 V |
| $H_{14}$ | 16 V | 1.29 V | 1.67 V | 1.80 V |
| $H_{15}$ | 16 V | 1.18 V | 1.05 V | 0.59 V |
| $H_{16}$ | 16 V | 2.47 V | 1.02 V | 0.37 V |
| $H_{17}$ | 16 V | 1.11 V | 0.99 V | 0.59 V |
| $H_{18}$ | 16 V | 3.13 V | 1.27 V | 1.21 V |
| $H_{19}$ | 16 V | 1.02 V | 0.94 V | 0.59 V |
| $H_{20}$ | 2.4 V | 2.10 V | 0.99 V | 0.40 V |

The output voltage harmonic contents computed for the CTU and ATU are shown in Figure 10. Both solutions maintained the harmonic content inside the allowable standard limits.

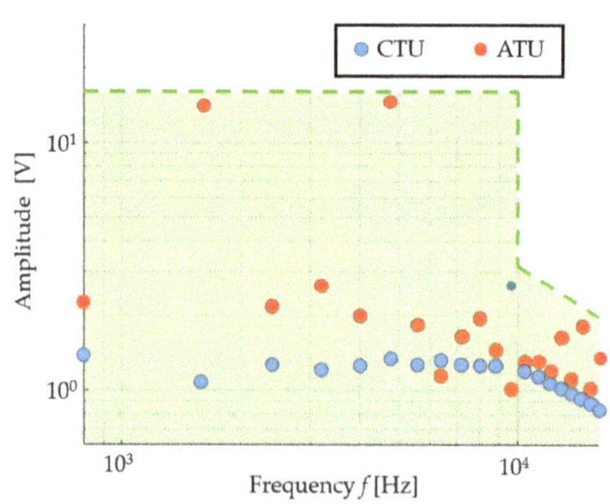


**Figure 10.** ATU input current amplitude.

## 7. Phase Unbalance Test

The RTCA DO-160G standard requires the evaluation of the effects of phase unbalances. The unbalance includes unequal voltage magnitudes at the fundamental system frequency, fundamental phase angle deviation, and unequal levels of harmonic distortion between the phases. A major cause of voltage unbalance is the asymmetry of the loads, if the loads are not uniformly shared among the three phases. The input voltage waveforms shown in Figure 11 has been used to reproduce the unbalanced operating condition.

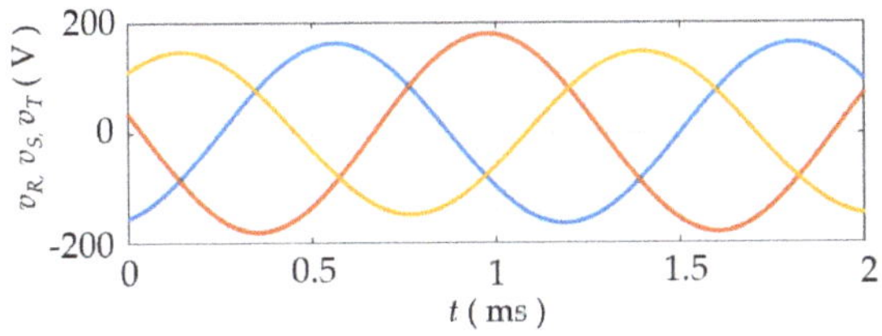


**Figure 11.** Typical input voltage waveforms under unbalanced operating conditions.

### 7.1. CTU under Unbalanced Input Voltage

In Figure 12a, the output voltages $v_o$ computed under unbalanced voltage conditions for different loads are shown. The related output voltage ripples are indicated in Table 15.

**Table 15.** CTU output voltage simulation results under unbalanced input voltage.

| Load | $V_o^{max}$ (V) | $V_o^{min}$ (V) | $V_o^{avg}$ (V) |
|---|---|---|---|
| $R_{L1} = 37\ \Omega$ | 287.6 | 247 | 267.9 |
| $R_{L2} = 74\ \Omega$ | 289.7 | 249.7 | 270.1 |
| $R_{L3} = 148\ \Omega$ | 290.8 | 251.1 | 271.8 |

The input current computed is shown in Figure 12b. The THD of the line currents and the PF for different loads are indicated in Table 16.

**Table 16.** CTU total harmonic distortion and power factor under unbalanced input voltage.

| Load | THD (%) | PF |
|---|---|---|
| $R_{L1} = 37\ \Omega$ | 12.7 | 0.9 |
| $R_{L2} = 74\ \Omega$ | 12.4 | 0.86 |
| $R_{L3} = 148\ \Omega$ | 10.4 | 0.7 |

Simulations indicated that there is an inversely proportional relationship between the THD of the input line current and the output load power, while the power factor changed from 0.86% at 10% of the load to 0.65 at full load. Finally, Figure 12c shows that independent from the load conditions, the magnetic flux density $B$(t) was sinusoidal, as expected, with an RMS value $B_{RMS} = 0.58$ T.

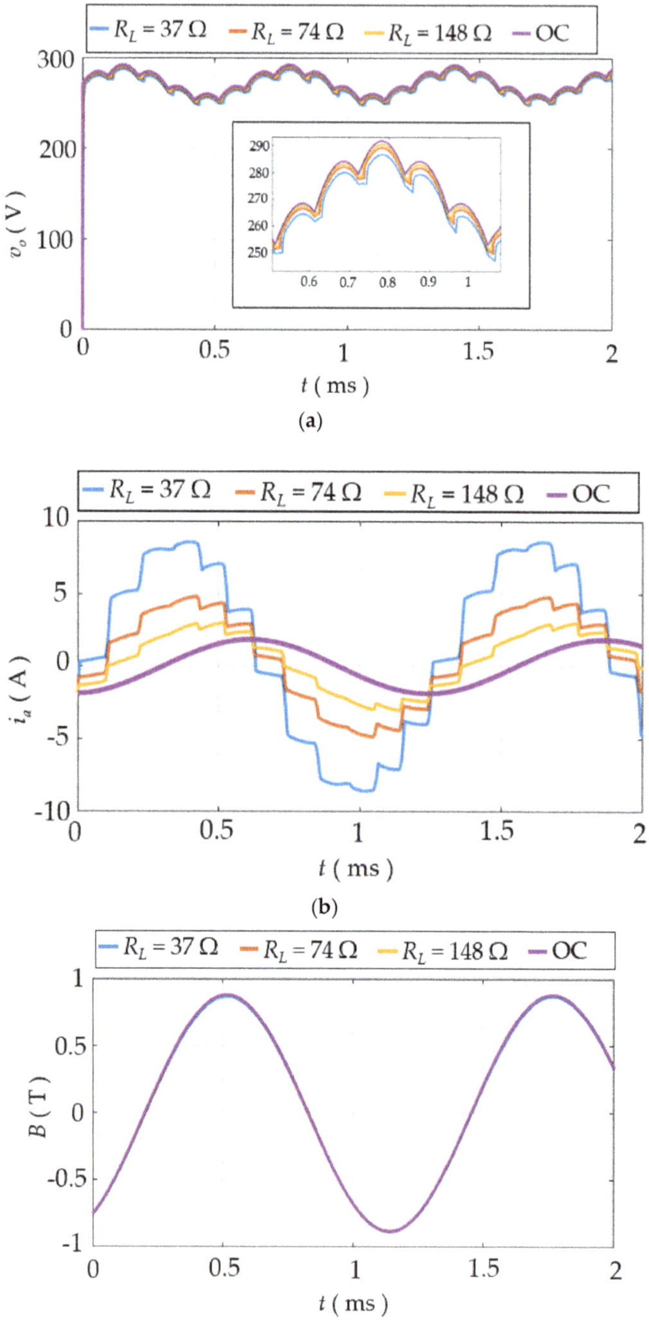


**Figure 12. CTU** Simulink/FEM simulation of a conventional 12-pulse rectifier. (**a**) Output voltage. (**b**) Input current. (**c**) Magnetic flux density.

### *7.2. ATU under Unbalanced Input Voltage*

Figure 13a shows the output voltages $v_o$ of the ATU computed for different loads. The average output voltage $V_o{}^{avg}$ and the ripple $V_o{}^{ripple}$ are summarized in Table 17.

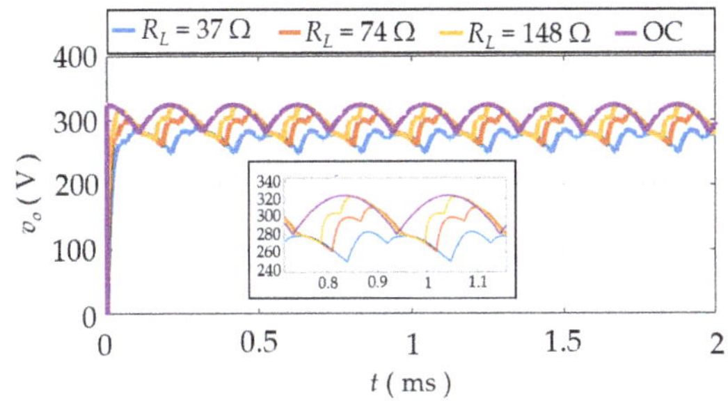


(**a**)

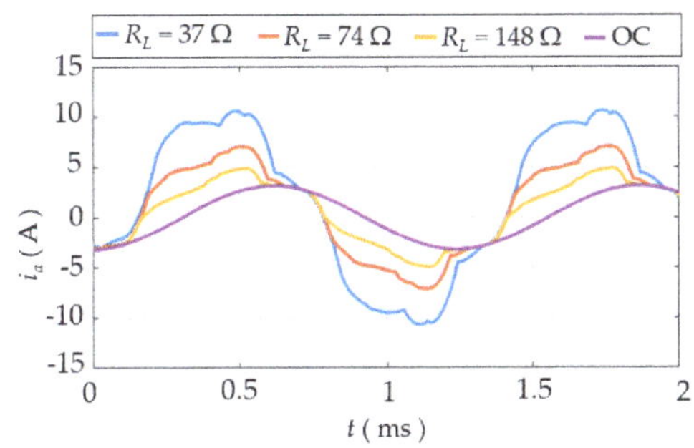


(**b**)

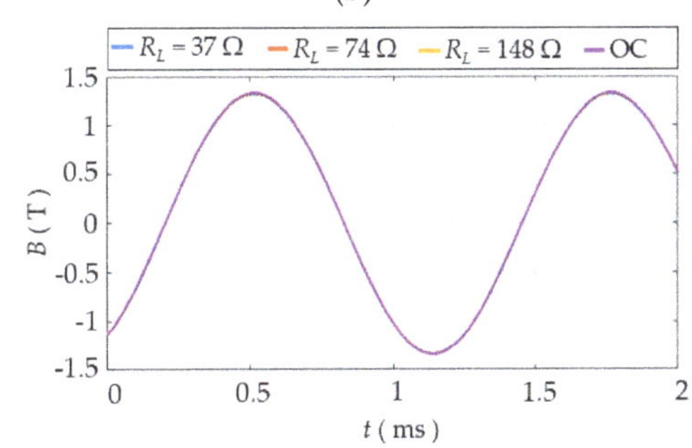


(**c**)

**Figure 13. ATU** Simulink/FEM simulation of a conventional 12-pulse rectifier. (**a**) Output voltage. (**b**) Input current. (**c**) Magnetic flux density.

**Table 17.** ATU output voltage simulation results under unbalanced input voltage.

| **Load** | $V_o{}^{max}$ **(V)** | $V_o{}^{min}$ **(V)** | $V_o{}^{avg}$ **(V)** |
|---|---|---|---|
| $R_{L1}$ = 37 Ω | 297.7 | 235.1 | 269.9 |
| $R_{L2}$ = 74 Ω | 321.8 | 248.4 | 285.5 |
| $R_{L3}$ = 148 Ω | 337.4 | 260.2 | 292.1 |

The input current is plotted in Figure 13b. The THD of the line currents and the PF for different loads are indicated in Table 18.

**Table 18.** ATU total harmonic distortion and power factor under unbalanced input voltage.

| Load | THD (%) | PF |
|---|---|---|
| $R_{L1} = 37\ \Omega$ | 7.51 | 0.75 |
| $R_{L2} = 74\ \Omega$ | 5 | 0.69 |
| $R_{L3} = 148\ \Omega$ | 2.94 | 0.56 |

As in the case of the CTU, Figure 13c indicates that independent from the load conditions, the magnetic flux density $B(t)$ was sinusoidal, as expected, with an RMS value $B_{RMS} = 0.58$ T.

## 8. Final CTU and ATU Comparison

Figure 14 shows the comparison between the estimated CTU and ATU conversion efficiency under different load conditions. The AC–DC conversion efficiency evaluated as in Section 3 was always higher than 95% under all the considered operating conditions and for both the analyzed solutions. Simulations indicated that the ATU has higher efficiency than the CTU under heavy load conditions (i.e., $R_L = 37\ \Omega$ and $R_L = 74\ \Omega$), while the ATU efficiency is comparable with the CTU's under lighter load conditions ($R_{L3} = 148\ \Omega$).

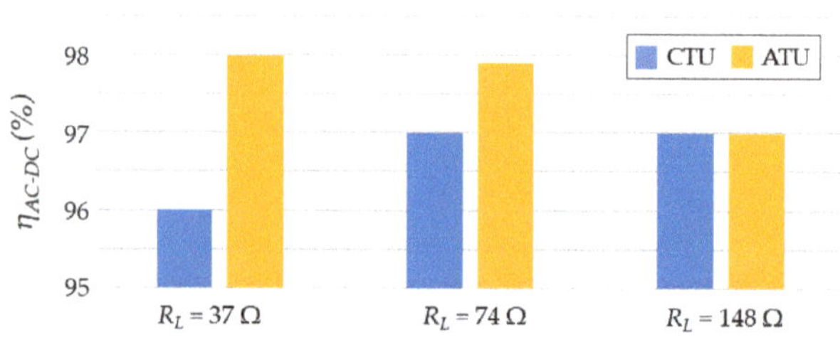


**Figure 14.** Comparison of AC–DC conversion efficiency computed under different load conditions.

The power loss contributions at different loads were analyzed for each converter, and results are summarized in Figure 15. The copper losses $P_{Cu}$, the core loss $P_{Core}$, and the diode rectifier losses $P_D$ were separately computed. The three power loss contributions for the CTU are shown in Figure 15a. The core loss was the higher contribution, independent from the operating conditions. As expected, the diode loss and the winding loss decreased under a lighter load condition. The copper loss always represents the lowest contribution.

In Figure 15b, in analogy, the different ATU-related power loss contributions are shown. In this case, the simulations indicated that the diode losses were higher than in the case of the CTU, independent from the load condition, as expected, in consideration that the two diode bridges were in series for the ATU and in parallel for the CTU. The diode losses and winding losses decreased according to the load current. The core losses, as expected, were almost constant and independent from the different load conditions.

In Figure 16, the comparison of the input current THDs predicted by the simulations by using the CTU and ATU is shown. Both the CTU and the ATU allowed meeting the standard requirements under the whole operating conditions, ensuring a THD lower than the threshold value $THD_{max} = 8\%$. Thanks to the input capacitive filter, the THD of the ATU was always lower than in the case of the CTU and <2%.

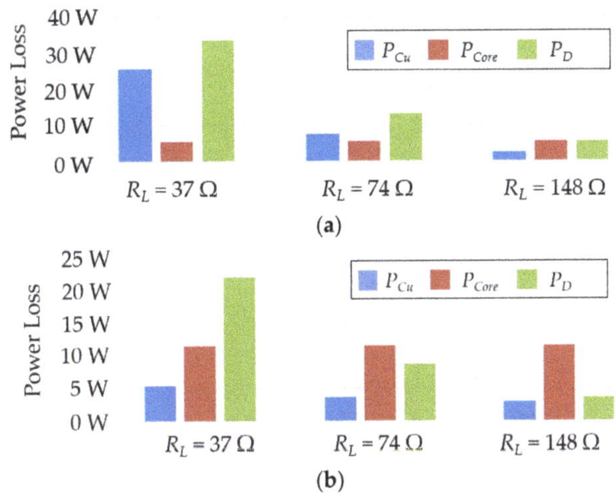


**Figure 15.** Power losses at different loads: The blue line is the winding loss, the red line is the core loss, and the green line is the diode rectifier loss. (**a**) CTU. (**b**) ATU.

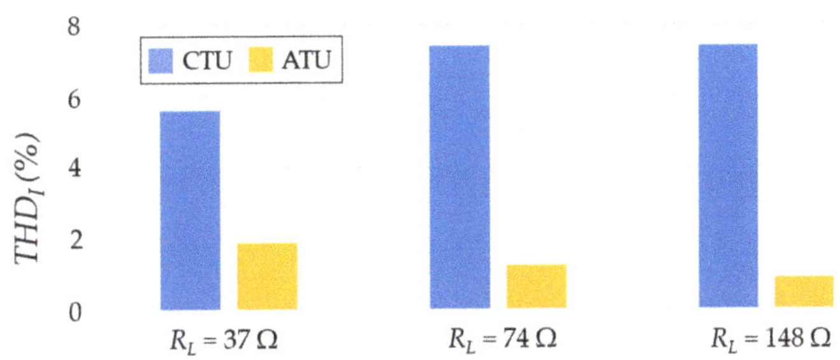


**Figure 16.** Comparison of input current THD at different loads.

The obtained results for the two topologies are summarized in Table 19.

**Table 19.** Summary of CTU and ATU characteristics.

| Characteristic | CTU | ATU |
|---|---|---|
| Higher AC–DC conversion efficiency | | X |
| Galvanic isolation | X | |
| Lower input current THD | | X |
| Lower output voltage ripple | X | |
| Good performance without additional input capacitive filter | X | |
| Good performance without additional series inductors | X | |
| Lower size and weight | | X |

## 9. Conclusions

This paper describes the design and modeling of two different AC–DC 12-pulse rectifiers, here named CTU and ATU. These devices are suitable for both terrestrial and aircraft applications. The parameters considered for the comparison of the performance of the CTU and the ATU are the conversion efficiency, output voltage, and input current THD in view of the RTCA DO-160 standard requirements. The comparison was performed by suitable numerical simulations made by a coupled FEM-circuital approach and considering variable load conditions up to the nominal power. Both pure resistive and inductive-resistive loads were considered.

Based on the simulation results, either the CTU or the ATU solutions comply with the standard requirements and limits and have a high conversion efficiency (more than 96%).

Results of the simulations indicate also that the ATU solution allows for a significant reduction in weight (more than 50% in the case study) and for an appreciable increase in efficiency (2% at nominal current) with respect to the CTU solution. However, the ATU solution needs an additional input capacitive filter, which, anyway, does not change substantially the gain in weight and size obtained and, in addition, greatly reduces the THD with respect to the case of the CTU solution and is under 2%.

Another important difference between the two topologies is that the CTU, differently from the ATU solution, allows for galvanic isolation between primary and secondary.

As future development, other 12-pulse topologies will be compared with ATU and CTU systems. The effect of the non-linearities on the magnetic core will be analyzed and techniques able to reduce copper and diode losses will be studied. Experimental validation of the obtained results will be performed.

**Author Contributions:** Conceptualization, E.C. and F.C.; methodology, E.C., F.C., and A.L.; software, F.C. and A.H.S.; validation, F.C., A.H.S., and A.L.; formal analysis, E.C., A.L., and F.C.; investigation, E.C., F.C., and A.L.; data curation, A.H.S.; writing—original draft preparation, F.C. and A.H.S.; writing—review and editing, F.C. and A.H.S.; supervision, E.C. and A.L. All authors have read and agreed to the published version of the manuscript.

**Funding:** This research received no external funding.

**Conflicts of Interest:** The authors declare no conflict of interest.

## References

1. Khan, S.; Zhang, X.; Saad, M.; Ali, H.; Khan, B.M.; Zaman, H. Comparative analysis of 18-pulse autotransformer rectifier unit topologies with intrinsic harmonic current cancellation. *Energies* **2018**, *11*, 1347. [CrossRef]
2. Singh, B.; Gairola, S.; Chandra, A.; Al-Haddad, K. Power quality improvements in isolated twelve-pulse AC-DC converters using delta/double-polygon transformer. In Proceedings of the 2007 IEEE Power Electronics Specialists Conference, Orlando, FL, USA, 17–21 June 2007; pp. 2848–2853.
3. Singh, B.; Bhuvaneswari, G.; Garg, V. Harmonic mitigation using 12-pulse AC-DC converter in vector-controlled induction motor drives. *IEEE Trans. Power Deliv.* **2006**, *21*, 1483–1492. [CrossRef]
4. Luchetta, A.; Manetti, S.; Piccirilli, M.C.; Reatti, A.; Corti, F.; Catelani, M.; Ciani, L.; Kazimierczuk, M.K. MLMVNNN for Parameter Fault Detection in PWM DC–DC Converters and Its Applications for Buck and Boost DC–DC Converters. *IEEE Trans. Instrum. Meas.* **2019**, *68*, 439–449. [CrossRef]
5. Ayachit, A.; Reatti, A.; Kazimierczuk, M.K. Magnetising inductance of multiple-output flyback dc–dc convertor for discontinuous-conduction mode. *IET Power Electron.* **2017**, *10*, 451–461. [CrossRef]
6. Saini, D.K.; Ayachit, A.; Reatti, A.; Kazimierczuk, M.K. Analysis and Design of Choke Inductors for Switched-Mode Power. *Inverters IEEE Trans. Ind. Electron.* **2017**, *65*, 2234–2244. [CrossRef]
7. Faba, A.; Rimal, H.P. Robust Lightning Indirect Effect Protection in Avionic Diagnostics: Combining Inductive Blocking Devices with Metal Oxide Varistors. *IEEE Trans. Ind. Electron.* **2018**, *65*, 6457–6467. [CrossRef]
8. Faba, A.; Gaiotto, S.; Lozito, G.M. A novel technique for online monitoring of photovoltaic devices degradation. *Sol. Energy* **2017**, *158*, 520–527. [CrossRef]
9. Mussolin, L.; Bertucci, B.; Faba, A.; Ambrosi, G.; Scolieri, G.; Tissi, F.; Gaggiotti, M.; Morelli, G.; Zhang, Z.; Koutsenko, V. Overview of the mechanical, thermal vacuum and EMI/EMC tests performed for the AMS-02 UTTPS space qualification campaign. In Proceedings of the 20th IEEE Mediterranean Electrotechnical Conference, MELECON 2020-Proceedings, Palermo, Italy, 16–18 June 2020; pp. 130–135.
10. Meng, F.; Gao, L.; Yang, S.; Yang, W. Effect of phase-shift angle on a delta-connected autotransformer applied to a 12-pulse rectifier. *IEEE Trans. Ind. Electron.* **2015**, *62*, 4678–4690. [CrossRef]
11. Lian, Y.; Yang, S.; Ben, H.; Yang, W. A 36-pulse diode rectifier with an unconventional interphase reactor. *Energies* **2019**, *12*, 820. [CrossRef]
12. Meng, F.; Yang, W.; Yang, S. Effect of voltage transformation ratio on the kilovoltampere rating of delta-connected autotransformer for 12-pulse rectifier system. *IEEE Trans. Ind. Electron.* **2012**, *60*, 3579–3588. [CrossRef]
13. Choi, S.; Lee, B.S.; Enjeti, P. New 24-pulse diode rectifier systems for utility interface of high-power AC motor drives. *IEEE Trans. Ind. Appl.* **1997**, *33*, 531–541. [CrossRef]

14. Choi, S.; Enjeti, P.N.; Pitel, I.J. Polyphase transformer arrangements with reduced kVA capacities for harmonic current reduction in rectifier-type utility interface. *IEEE Trans. Power Electron.* **1996**, *11*, 680–690. [CrossRef]
15. Yang, T.; Bozhko, S.; Wheeler, P.; Wang, S.; Wu, S. Generic functional modelling of multi-pulse auto-transformer rectifier units for more-electric aircraft applications. *Chin. J. Aeronaut.* **2018**, *31*, 883–891. [CrossRef]
16. Uan-Zo-li, A.; Burgos, R.; Wang, F.; Boroyevich, D.; Lacaux, F.; Tardy, A. Comparison of prospective topologies for aircraft autotransformer-rectifier units. In Proceedings of the IECON'03. 29th Annual Conference of the IEEE Industrial Electronics Society (IEEE Cat. No. 03CH37468), Roanoke, VA, USA, 2–6 November 2003; Volume 2, pp. 1122–1127.
17. Catelani, M.; Ciani, L.; Reatti, A. Critical components test and reliability issues for photovoltaic inverter. In Proceedings of the 20th IMEKO TC4 Symposium on Measurements of Electrical Quantities: Research on Electrical and Electronic Measurement for the Economic Upturn Together with 18th TC4 International Workshop on ADC and DCA Modeling and Testing IWADC 2014, Benevento, Italy, 15–17 September 2014; pp. 592–596.
18. CEN EN 2282:1992. Aerospace Series–Characteristics of Aircraft Electrical Supplies. Ed: European Committee for Standardization. 1992. Available online: http://store.uni.com/catalogo/en-2282-1992 (accessed on 1 October 2021).
19. International Organization for Standardization. Aerospace—Characteristics of Aircraft Electrical Systems (ISO Standard No. 1540:2006). Available online: https://www.iso.org/standard/42067.html (accessed on 1 October 2021).
20. IEEE Guide for Application and Specification of Harmonic Filters. In *IEEE Std. 1531-2020, November*; IEEE Power Engineering Society, Transmission & Distribution Committee; Available online: https://standards.ieee.org/standard/1531-2020.html (accessed on 1 October 2021).
21. RTCA-D0160D. Environmental Conditions and Test Procedures for Airborne Equipment. Ed: International Electrotechnical Commission. 2010. Available online: https://do160.org/rtca-do-160g/ (accessed on 1 October 2021).
22. Yedluri, A.K.; Anitha, T.; Kim, H.-J. Fabrication of Hierarchical $NiMoO_4/NiMoO_4$ Nanoflowers on Highly Conductive Flexible Nickel Foam Substrate as a Capacitive Electrode Material for Supercapacitors with Enhanced Electrochemical Performance. *Energies* **2019**, *12*, 1143. [CrossRef]
23. Abdollahi, R. Pulse doubling in zigzag–connected autotransformer–based 12–pulse ac–dc converter for power quality improvement. *J. Electr. Eng.* **2012**, *63*, 357–364. [CrossRef]
24. Technical bulletin for the Metglas amorphous Alloy 2605SA1. 2005. Available online: https://metglas.com (accessed on 1 October 2021).
25. Cardelli, E.; Faba, A.; Laudani, A.; Antonio, S.Q.; Fulginei, F.R.; Salvini, A. Computer Modeling of Nickel–Iron Alloy in Power Electronics Applications. *IEEE Trans. Ind. Electron.* **2017**, *64*, 2494–2501. [CrossRef]
26. Cardelli, E.; Faba, A.; Laudani, A.; Lozito, G.M.; Antonio, S.Q.; Fulginei, F.R.; Salvini, A. Implementation of the Single Hysteron Model in a Finite-Element Scheme. *IEEE Trans. Magn.* **2017**, *53*, 1–4. [CrossRef]
27. Cardelli, E.; Faba, A.; Laudani, A.; Lozito, G.M.; Fulginei, F.R.; Salvinib, A. Two-dimensional magnetic modeling of ferromagnetic materials by using a neural network based hybrid approach. *Phys. B Condens. Matter* **2016**, *486*, 106–110. [CrossRef]
28. Carcangiu, S.; Cardelli, E.; Faba, A.; Fanni, A.; Montisci, A.; Quondam, S. Moving vector hysteron model identification based on neural network inversion. In Proceedings of the 2016 IEEE 2nd International Forum on Research and Technologies for Society and Industry Leveraging a Better Tomorrow, RTSI, Bologna, Italy, 7–9 September 2016.
29. Farrugia, D.; Apap, M.; Micallef, A.; Staines, C.S. Analysis of Polygon Connected ATRU for the More-Electric Aircraft. In Proceedings of the 2020 IEEE 20th Mediterranean Electrotechnical Conference (MELECON), Palermo, Italy, 16–18 June 2020; pp. 136–140.
30. Garg, V.; Singh, B.; Bhuvaneswari, G. 24-pulse ac–dc converter for harmonic mitigation. *IET Power Electron.* **2009**, *2*, 364–374. [CrossRef]
31. Singh, B.; Bhuvaneswari, G.; Garg, V.; Gairola, S. Pulse multiplication in AC-DC converters for harmonic mitigation in vector-controlled induction motor drives. *IEEE Trans. Energy Convers.* **2006**, *21*, 342–352. [CrossRef]
32. Chen, J.; Chen, J. On reducing the shaft torque ripple of small-to-medium-scale wind energy conversion systems using multi-pulse autotransformer rectifier. *Energies* **2018**, *11*, 379. [CrossRef]
33. Das, J.C. Passive filters-potentialities and limitations. *IEEE Trans. Ind. Appl.* **2019**, *40*, 232–241. [CrossRef]

electricity

MDPI

Article

# Design, Analysis and Experimental Verification of the Self-Resonant Inverter for Induction Heating Crucible Melting Furnace Based on IGBTs Connected in Parallel

**Borislav Dimitrov [1,*], Khaled Hayatleh [2], Steve Barker [2] and Gordana Collier [2]**

1 School of Engineering, Warwick University, Coventry CV4 7AL, UK
2 School of Engineering, Computing and Mathematics, Oxford Brookes University, Oxford OX3 0BP, UK; khayatleh@brookes.ac.uk (K.H.); stevebarker@brookes.ac.uk (S.B.); gordanacollier@brookes.ac.uk (G.C.)
* Correspondence: Borislav.Dimitrov@warwick.ac.uk

**Abstract:** The object of this research was a self-resonated inverter, based on paralleled Insulated-Gate Bipolar Transistors (IGBTs), for high-frequency induction heating equipment, operating in a wide range of output powers, applicable for research and industrial purposes. For the nominal installed capacity for these types of invertors to be improved, the presented inverter with a modified circuit comprising IGBT transistors connected in parallel was explored. The suggested topology required several engineering problems to be solved: minimisation of the current mismatch amongst the paralleled transistors; a precise analysis of the dynamic and static transistors' parameters; determination of the derating and mismatch factors necessary for a reliable design; experimental verification confirming the applicability of the suggested topology in the investigated inverter. This paper presents the design and analysis of IGBT transistors based on datasheet parameters and mathematical apparatus application. The expected current mismatch and the necessary derating factor, based on the expected mismatch in transistor parameters in a production lot, were determined. The suggested design was experimentally tested and investigated using a self-resonant inverter model in a melting crucible induction laboratory furnace.

**Keywords:** induction heating; self-resonant inverter; IGBTs; IGBTs in parallel

**Citation:** Dimitrov, B.; Hayatleh, K.; Barker, S.; Collier, G. Design, Analysis and Experimental Verification of the Self-Resonant Inverter for Induction Heating Crucible Melting Furnace Based on IGBTs Connected in Parallel. *Electricity* **2021**, 2, 439–458. https://doi.org/10.3390/electricity2040026

Academic Editors: Oriol Gomis-Bellmunt and Andreas Sumper

Received: 13 July 2021
Accepted: 28 September 2021
Published: 11 October 2021



## 1. Introduction

This paper aimed to design, analyse, prototype, and experimentally test a self-resonant inverter for crucible inductance furnaces to verify the suggested circuit based on IGBTs connected in parallel. For this purpose, a methodology for analysing the transistors' characteristics and the expected mismatch in a production lot was suggested and implied as part of the design procedure. It was based on trend analysis of the transistors' datasheets, digitalising the graphical data and deriving precise polynomial equations. Such a study aimed to compensate for the lack of data on the mismatch of IGBT parameters. In this sense, the presented research's novelty consisted of modifying a well-known circuit, which was analysed with a novel methodology focused on paralleled transistors, giving the expected current mismatch and the necessary derating factor.

As a highly efficient and fast method, induction heating is applicable for numerous technological processes of heating, melting, hardening, etc. The investigated crucible furnaces are used for metal melting, such as gold, steel and copper, applied in different industries—jewellery, steel production, etc. The theoretical basis of induction systems has been described in multiple well-established literature sources [1–6], giving fundamental knowledge for the system inductor-workpiece.

A popular method for induction heating analysis is modelling and simulation with the Finite Element Method (FEM), which depicts the system's electromagnetic field inductor-workpiece. The system's thermal and electrical parameters can be analysed, as well as the required parameters of the high-frequency electronic converter to be predicted on the simulation level. Using FEM, in [7,8], an induction heating process with a moving inductor or workpiece was presented. Such systems require considerable power to be transferred for a short time for the necessary temperature to be reached at the required time. For this purpose, significant currents through the inductor, for example, 450–890 A at 36 kHz [8], must be supplied from the induction heating inverter. The optimisation of the inductor-workpiece system gave positive results [9] for the magnetic coupling but did not minimise the current provided, presented in most FEM models in the range of 106–108 $A/m^2$.

According to the research published in [10], the induction heating process with spray cooling, realised by the presented experimental equipment, can continue for a relatively long time at about 30 min. The required inductor current exceeded 400 A. The provided experiments were used for FEM model calibration. A similar study [11], based on the induction heating and FEM modelling analysis, showed a process established at a relatively low frequency of 3 kHz and 544.5 A current through the inductor. Higher frequencies, up to 20 kHz, can be applied for magnetic and nonmagnetic workpieces [12]. The presented experimental verification in [12] was completed with a low-voltage inductor, rated at 47.1 V, which gave a maximum current of 1063 A. In addition, the used capacitor tank at the resonant tank was rated at 15.41 μF or 55.7 μF, which gave a resonant frequency of between 10 kHz and 20 kHz.

Studies focused on the influence of the workpiece characteristics over the heating process [13–15] have shown that the inductor geometry characteristics have a critical impact on the electromagnetic field shape and must be considered with the specific workpiece. This phenomenon has the potential to influence the inverter characteristics influenced by the inductor length and its inductivity and, hence, the resonance frequency with the necessary capacitance tank, as well as the inverter's output current at a given power, which requires additional study.

Another group of recently published papers [16–18], focusing on the development and application of new materials, showed the induction heating application for materials' laboratory production and further investigation. The technological parameters required variations in broad ranges in output power of 10–200 kW, inductor currents up to thousands of amps, low or high inductor voltages and resonant frequencies up to 300 kHz. The exact requirements can be stated for flat-inductor systems [19] and melting induction furnaces [20]. They clearly show that with respect to the industrial equipment with specific parameters, the laboratory induction heating must be much more flexible, which would ensure its application for different laboratory tests and studies. According to these requirements, self-oscillating inverters for induction heating are good candidates for laboratory equipment as a simple structure that is highly reliability and naturally operating with inherent Zero Voltage Switching (ZVS) modulation, which gives low switching losses. In [21–23], a self-oscillating inverter with a parallel resonant tank operating at a relatively low power of 300 W was proposed. The research showed the advantages of implementing a Current Source Parallel Resonant Push-Pull Inverter (CSPRPI) and the possibilities of achieving ZVS with low switching losses. A similar topology based on a half-bridge but a series resonant tank was also proven to have a stable operation at ZVS [24]. Furthermore, the same source proved that these classes of converters can be controlled with a simple phase-locked loop based on a variable duty cycle.

A converter structure with proven application in induction heating, giving high power density and the ability to operate with soft-switching, is the full-bridge circuit with a parallel or series resonant tank, based on different modulation techniques, such as Phase-Shift Modulation (PSM), Pulse-Frequency Modulation (PFM), Pulse-Density Modulation (PDM) and Asymmetrical Duty-Cycle (ADC) [25–31]. These types of modulation offer flexible control over the wide range of output powers [25]. These inverters applied for induction heating can have a good match with a Power Factor Correction (PFC) stage [26]. An additional characteristic of the PSM is the ability of the ZVS to be achieved for the required power range under regulation with variable switching frequency [27]. It has been shown that the use of variable switching has a better efficiency than a fixed switching frequency. Despite the modulation technique, the problem with the transient overvoltages and electromagnetic interference (EMI) due to the transistor switching is mitigated with snubbers.

In contrast to the self-oscillating inverters [21–23], one of their advantages is usually based on snubber-less power stage circuits for the full-bridge PSM topology, and different types of passive snubbers such as C, RC, and RCD can be used. Their selection and design depend on the modulation technique and the match between the used IGBTs and the resonant tank parameters [28,29]. In the same class, transformer-based converters are the LLC resonant inverters. A powerful inverter of this type, rated at 25 kW with a fixed frequency of 25 kHz, is presented in [30]. Transformer-less topologies based on a full-bridge can be realised as matrix converters, whose topology requires bi-directional switches [31]. Although these converters have significant advantages, a comparative analysis shows that the self-resonant converters are based on a more straightforward structure, giving a budget-friendly induction heating system, with low EMI, without snubber circuits and low losses.

Another group of papers focused on induction heating devices for cooking [32–36], which usually have flat inductors and an installed power in the range of 1.5–2.8 kW. According to their shapes and materials, an important issue is the magnetic coupling between the inductor and the loads, i.e., pan, pot and wok [32,33]. The other parts of the control system: sensors, feedbacks, filters, etc., are also well investigated from the cooking-intended induction heaters [34–36]. The reported efficiency for cooking induction heating converters is over 95%, which can be expected as a parameter for a powerful induction heating crucible system.

The provided literature review concludes that the self-oscillating resonant converters have the simplest circuit, offering a robust and budget-friendly solution. Currently, they are applied for relatively low power in the range of 300–800 W as laboratory equipment, using cylindrical inductors for heating magnetic elements [37,38]. It can be suggested that their application can be extended to more considerable powers, over 30 kW, and a frequency range of 25–35 kHz, giving an application for fast-melting crucible inductance furnaces [39]. Such a power would require a power stage based on IGBT transistors connected in parallel to be designed and implemented. Numerous papers [40–47] present research mainly focused on IGBTs. Current equalisation with improved gate drive control implemented with Digital Signal Processors (DSPs) and precise current measurements is given in [40,41]. Strategies based on active driver circuits and current cross-reference control techniques were proposed in [42], including for Silicon Carbide (SiC) MOSFETs [43,44].

Nowadays, Revers Conducting IGBTs (RC-IGBTs), specialised for resonant inverters, i.e., applicable for induction heating systems, are offered by manufacturers [45–48]. According to the presented applications [45], the analysis of the transistors' technology and their implementation in the induction inverters [47] justifies the suggestion that paralleled RC-IGBT can operate with a good current sharing and, hence, low current mismatch. The experimental data presented with oscillograms showed inverters' soft-switching operation [45,47–50] in the induction heating resonant inverters. Furthermore, the assembling procedure applied for half-bridge resonant inverters [46] implies that current equalisation for the self-resonant induction converters based on paralleled IGBTs can be easily achievable. All of the stated suggestions require additional analytical and experimental research as such has not been found in the literature. Some statistical information about the parameters' dispersion of IGBT modules is given in [50], but no analysis or experimental data have been found for discrete transistors.

As a conclusion from the presented literature review: the self-resonant converters for induction heating [21–24] are an applicable technology giving reliable and budget-friendly solutions; induction heating inverters, with an installed power higher than 10–15 kW for industrial and laboratory research purposes [29–31], can be constructed by IGBT transistors connected in parallel, but the literature presenting such applications specifically for self-resonated inverters is insufficient; although the paralleling of transistors is an object of significant research [40–44], the application of this topology in the self-resonance architectures requires additional investigation. In general, from the literature, it is clear that the topology of self-resonant inverters for induction crucible melting furnaces has the potential to reach higher powers with paralleled transistors. However, the undesirable phenomena caused by transistors' parameters mismatches are not well described in the literature specifically for this application. Furthermore, the parallel operation in the resonant inverters can cause problems such as uneven current allocation, operation outside the ZVS region and major malfunction.

Based on the presented literature findings, the primary purpose of the suggested research was to experimentally show the possibility of resonant inverters with paralleled IGBTs operating with improved characteristics. For this purpose, a methodology for the analysis of transistors' characteristics was developed, giving a precise estimation of the paralleling IGBTs abilities. The presented approach estimated the derating factors and the expected current share through digitalising the graphical data and conducting a thorough analysis.

The paper is organised as follows: Section 2 shows an analysis of the modified self-resonated inverter based on the suggested methodology for analysis of the transistors connected in parallel; Section 3 shows an experimental verification; the conclusions are presented in Section 4.

## 2. Analysis of the Investigated Inverter, Based on RC-IGBTs in Parallel

The self-resonated converter is shown in Figure 1. The object of this research was the power stage of the inverter, which is part of the entire system presented with a block diagram. The further analysis and the experimental model presented in Section 3 were completed with discrete RC-IGBTs. In this circuit, it is necessary for the transistors' emitters to be connected at a common point (GND); off-the-shelf half-bridge IGBT modules would not be suitable. Analysing the IGBTs currently available on the market, the targeted power for the designed crucible melting induction system in a range of 20–30 kW, with a potential to reach 50 kW, could be achieved with 3 to 8 high and low side transistors connected in parallel. Such a circuit applies the necessity of a precise analysis of the IGBTs parameters to be undertaken due to the specific parasitic elements that the discrete packages, usually TO-247, TO263, etc., and their modification typically have. The gate drive circuits are based on optical drivers to supply the necessary gate drive current. An additional obstacle is the lack of information about the expected mismatch of transistors' parameters in a production lot, usually not published by the manufacturers.

The block diagram presented in Figure 1 is arranged as follows: (1) the three-phase rectifier and power factor correction (PFC) module are based on a full-bridge circuit, followed by a Boost PFC converter controlled by an ASIC (Application Special Integrated Circuit); (2) the Buck converter is a transformer-less PWM converter, but for safety purposes, a transformer-based full-bridge converter can be considered; (3) the resonant inverter was the object of this research, depicted in the power circuit. As the circuit is self-resonated, the control module operates as overcurrent and overvoltage protection, with the functionality to shut down the inverter; (4) the feedback from the inductor is performed by isolated current and voltage sensors; (5) the control and protection system is based on an STM microcontroller, including feedback networks from each module.

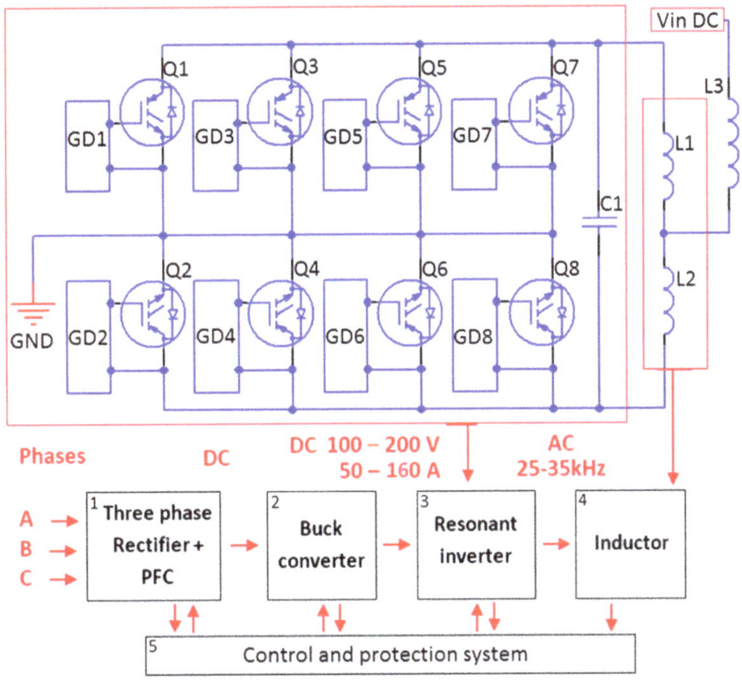


**Figure 1.** The electronic circuit of the proposed inverter for induction heating crucible melting furnace. Q1-Q8 IGBT transistors; GD1-GD8 gate drivers; C1, L1, L2 resonance tank; L3 induction heating inductor. A block diagram of the entire system for induction heating.

The steady-state and transient conditions caused, respectively, by the saturation collector-emitter voltage $V_{CE(sat)}$ and the variations in the gate-emitter threshold voltage $V_{GE(th)}$ of the IGBTs connected in parallel and, associated with them, thermal dependencies, are the most critical parameters, which must be analysed. The suggested methodology was focused on in the analysis of the variations in the transistors' parameters and the estimation of the probable current mismatch, which was designed in several steps as follows [51–55]:

Step 1. Estimating the necessary derating factor and total current.

Calculation of the derating factor Δ:

$$\Delta = 1 - \frac{I_{max}}{N_{parallel} \times I_{IGBTmax}} \tag{1}$$

where $I_{max}$ is the maximum current through all transistors connected in parallel, $N_{parallel}$ is the number of IGBT transistors in parallel, connected in the high or low side of the circuit given in Figure 1, $I_{IGBTmax}$ is the maximum current per IGBT transistor. The result from this equation under the satisfactory deration requires the number of transistors or $I_{IGBTmax}$ to be increased.

At a given derating factor, the total current in a parallel $I_{Total}$ is:

$$I_{Total} = (1 - \Delta) \times N_{parallel} \times I_{IGBTmax} \tag{2}$$

For the designed and experimentally tested inverter for induction heating, the maximum current of $I_{max} = 160\ A$ is assumed. Equation (1) is calculated for 2 to 6 transistors in parallel, and the derating factor is in the range $\Delta = 0.1 - 0.5$; the result is depicted in Figure 2A. Following the recommendations given by IGBT manufacturers and the industrial research on induction heating [45–47], as a first assumption, a derating factor of

$\Delta = 0.2$ is accepted. A suitable transistor technology for a resonant inverter can be chosen (Trenchstop, Trench gate field-stop, etc., RC-IGBT [48,49]) with Positive Thermal Coefficient (PTC). Considering the current ranges for the discrete IGBT transistors, assembled in TO-247, TO-3PN, TO-264, etc. packages, the feasible current capacity would be 50–160 A, as shown in Figure 1. With this assumption, four transistors are accepted for this design. The analysis shows that the datasheet transistor current must be selected at $I_C = 50$ A (Figure 2A), the expected current through each transistor is 40 A (Figure 2B) calculated from Equation (2), and the specific derating factor is $\Delta = 0.2$ (Figure 2C). Several IGBTs, which have been used for the analysis and experimental verification, are presented in Table A1, Appendix A.

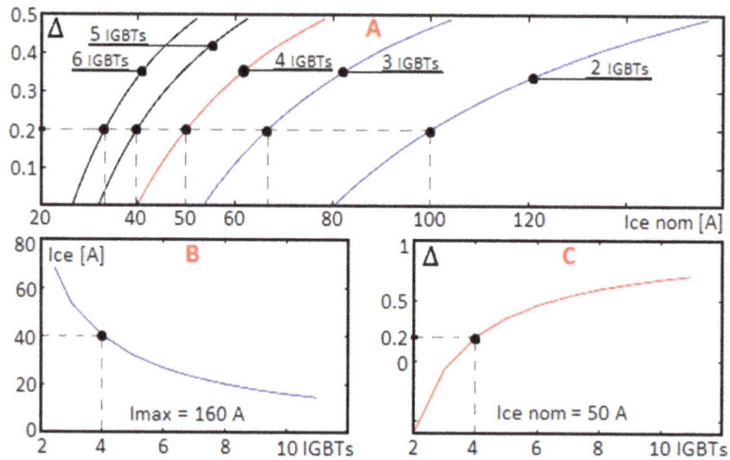


**Figure 2.** Analysis of the number of IGBTs in parallel. (**A**) Number of transistors in parallel calculated in a derating range $\Delta = 0.1 - 0.5$, giving the necessary nominal current per transistor; (**B**) expected current through each transistor at the assumed maximum current of $I_{max} = 160$ A; (**C**) the targeted derating of $\Delta = 0.2$, i.e., 20%, is feasible with $I_{CEnom} = 50$ A, 4 IGBTs in parallel.

Step 2. Selecting the IGBT, digitalising the necessary datasheet graphics and analysing the obtained data.

The critical parameters for this analysis, given in the datasheets as diagrams, are shown as examples in Figure 3 as the transconductance characteristic (Figure 3A) and the output characteristic (Figure 3B).

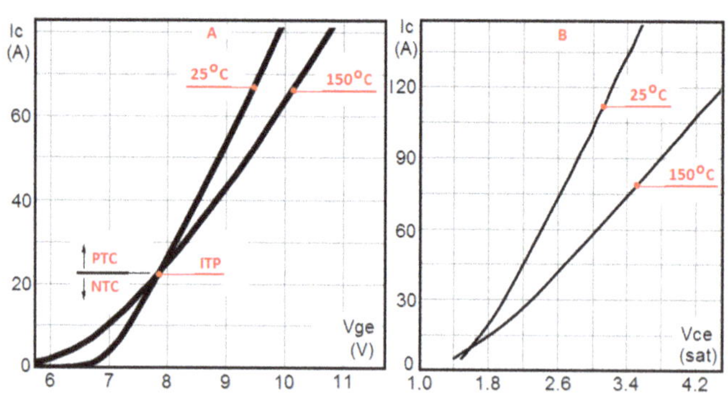


**Figure 3.** (**A**) IGBT transconductance characteristic and its dependence on the temperature; NTC—Negative Thermal Coefficient; PTC—Positive Thermal Coefficient; ITP—Isothermal Point; (**B**) IGBT output characteristics and its dependence on the temperature.

Using the IGBTs datasheet in the design procedure shows several problems that prevent a precise analysis of the paralleled transistors from being completed:

- The variation in parameters in a production lot is not apparent and cannot be analysed. Hence, the expected current variation amongst the paralleled transistors and the necessary deration factor cannot be precisely estimated.
- The precise position of the Isothermal Point (ITP, Figure 3A) and its variations, which depend on the variations in $V_{ge}$ and $I_C$, cannot be precisely estimated. This means that the designed converter would not operate under a stable Positive Thermal Coefficient (PTC) at any mode of operation, rather than a Negative Thermal Coefficient (NTC).
- There are expected variations in the Collector-Emitter voltage drop $V_{CE\ (sat)}$. Hence, the expected losses and their distribution amongst the paralleled transistors cannot be precisely estimated in a production lot.

The suggested methodology offers a solution to the stated problems as follows:

- Digitalising the graphical data available from the transistors' datasheets and converting them into equations using trend analysis. Figure 4 shows the digitalising graphics product of solving polynomial equations.
- Deriving the expected parameters variation in a production lot. A variation of ±5% can be accepted, based on the provided manufacturers' statistical [50] and experimental [51,53,54] research. The expected variation in the static characteristics is presented in Figure 4A—graphics 1–2 and graphics 3–4 parameters variations, respectively, at nominal 25 °C and maximum 150 °C temperatures. Following the same sequence, the expected variation in the dynamic characteristics is given in Figure 4B.
- Estimating the expected derating factor and current mismatch based on the expected variation. The result shows the variation in the derating factor and justifies the assumption made from Equations (1) and (2), presented in Figure 2. It also shows the PTC established at the selected $V_{ge}$ in the entire range of the static parameters.

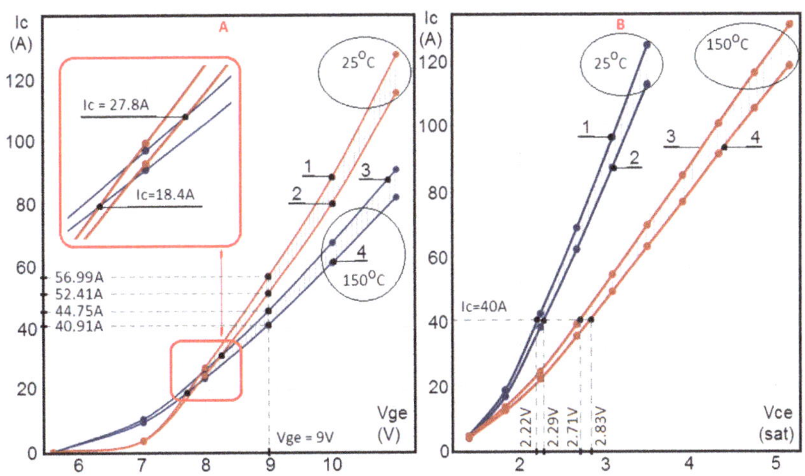


**Figure 4.** (**A**) Transconductance (dynamic) characteristics at nominal 25 °C (graphic 1 maximum and 2 minimum) and maximum 150 °C (graphic 3 maximum and 4 minimum) temperatures. (**B**) Output (static) characteristics at nominal 25 °C (graphic 1 maximum and 2 minimum) and maximum 150 °C (graphic 3 maximum and 4 minimum) temperatures.

The equations, products of the trend analysis obtained from the IGBTs datasheet (Figure 3), would be a polynomial of 4–6 degrees in order for the parameters to be described with greater precision.

Nominal transconductance characteristic, at nominal temperature 25 °C (Figure 3):

$$I_C = \left(0.0091 \times V_{GE}^5\right) + \left(-0.1009 \times V_{GE}^4\right) + \left(-3.6877 \times V_{GE}^3\right) + \left(82.845 \times V_{GE}^2\right) + \left(-571.73 \times V_{GE}\right) + 1300.2 \quad (3)$$

Maximum transconductance characteristic (Figure 4, graphic 1), i.e., +5% from Equation (3):

$$I_C = \left(0.0096 \times V_{GE}^5\right) + \left(-0.106 \times V_{GE}^4\right) + \left(-3.8721 \times V_{GE}^3\right) + \left(86.988 \times V_{GE}^2\right) + \left(-600.32 \times V_{GE}\right) + 1365.2 \quad (4)$$

Minimum transconductance characteristic (Figure 4, graphic 2), i.e., −5% from Equation (3):

$$I_C = \left(0.0087 \times V_{GE}^5\right) + \left(-0.0959 \times V_{GE}^4\right) + \left(-3.5033 \times V_{GE}^3\right) + \left(78.703 \times V_{GE}^2\right) + \left(-543.15 \times V_{GE}\right) + 1235.2 \quad (5)$$

Nominal transconductance characteristic, at high temperature 150 °C (Figure 3):

$$I_C = \left(-0.0278 \times V_{GE}^5\right) + \left(1.2169 \times V_{GE}^4\right) + \left(-21.274 \times V_{GE}^3\right) + \left(186.96 \times V_{GE}^2\right) + \left(-812.84 \times V_{GE}\right) + 1381.6 \quad (6)$$

Maximum transconductance characteristic (Figure 4, graphic 3), i.e., +5% from Equation (6):

$$I_C = \left(-0.0292 \times V_{GE}^5\right) + \left(1.2777 \times V_{GE}^4\right) + \left(-22.338 \times V_{GE}^3\right) + \left(196.31 \times V_{GE}^2\right) + \left(-853.48 \times V_{GE}\right) + 1450.6 \quad (7)$$

Maximum transconductance characteristic (Figure 4, graphic 4), i.e., −5% from Equation (6):

$$I_C = \left(-0.0264 \times V_{GE}^5\right) + \left(1.156 \times V_{GE}^4\right) + \left(-20.211 \times V_{GE}^3\right) + \left(177.61 \times V_{GE}^2\right) + \left(-772.2 \times V_{GE}\right) + 1312.5 \quad (8)$$

The derived equations are presented as matrices in Table 1. Such a format is suitable for calculation with a simple Matlab script, shown in Listing A1 the Appendix A. The designed converter is depicted in Figure 4 as $V_{ge} = 9$ V, and the expected maximum collector current due to the temperature and parameters variation would be $I_{Cmax(150^\circ C,min)} = 40.91$ A, $I_{Cmax(150^\circ C,\ max)} = 44.75$ A, $I_{Cmax(25^\circ C,min)} = 52.41$ A, and $I_{Cmax(25^\circ C,max)} = 56.99$ A. This analysis shows that the current can vary by 16 A under the worst-case conditions, i.e., if the paralleled transistors operate at the minimum and maximum temperature, assembled on different heatsinks and operating under specific cooling conditions, etc. Although such a thermal condition is less probable to be established, the correct design must consider it. If the paralleled transistors shared the same thermal conditions at the maximum temperature, the maximum current mismatch would be 4.58 A, i.e., under 20% of the maximum current. It can be expected that the accepted $\Delta = 0.2$ would be a correct assumption.

**Table 1.** Digitalisation of the transconductance characteristics, shown in Figure 4A.

| | nominal characteristic (Figure 2) | +5% data variation Figure 3A, graphic 1 | −5% data variation Figure 3A, graphic 2 |
|---|---|---|---|
| Nominal temperature 25 °C | $I_C = \begin{bmatrix} 0.0091 \times V_{GE}^5 \\ -0.1009 \times V_{GE}^4 \\ -3.6877 \times V_{GE}^3 \\ 82.845 \times V_{GE}^2 \\ -571.73 \times V_{GE} \\ 1300.2 \end{bmatrix}$ | $I_C = \begin{bmatrix} 0.0096 \times V_{GE}^5 \\ -0.106 \times V_{GE}^4 \\ -3.8721 \times V_{GE}^3 \\ 86.988 \times V_{GE}^2 \\ -600.32 \times V_{GE} \\ 1365.2 \end{bmatrix}$ | $I_C = \begin{bmatrix} 0.0087 \times V_{GE}^5 \\ -0.0959 \times V_{GE}^4 \\ -3.5033 \times V_{GE}^3 \\ 78.703 \times V_{GE}^2 \\ -543.15 \times V_{GE} \\ 1235.2 \end{bmatrix}$ |
| | Nominal characteristic (Figure 2) | +5% data variation Figure 3A, graphic 3 | −5% data variation Figure 3A, graphic 4 |
| Maximum temperature 150 °C | $I_C = \begin{bmatrix} -0.0278 \times V_{GE}^5 \\ 1.2169 \times V_{GE}^4 \\ -21.274 \times V_{GE}^3 \\ 186.96 \times V_{GE}^2 \\ -812.84 \times V_{GE} \\ 1381.6 \end{bmatrix}$ | $I_C = \begin{bmatrix} -0.0292 \times V_{GE}^5 \\ 1.2777 \times V_{GE}^4 \\ -22.338 \times V_{GE}^3 \\ 196.31 \times V_{GE}^2 \\ -853.48 \times V_{GE} \\ 1450.6 \end{bmatrix}$ | $I_C = \begin{bmatrix} -0.0264 \times V_{GE}^5 \\ 1.156 \times V_{GE}^4 \\ -20.211 \times V_{GE}^3 \\ 177.61 \times V_{GE}^2 \\ -772.2 \times V_{GE} \\ 1312.5 \end{bmatrix}$ |

Following the same data presentation format, the necessary equations for the static transistor's characteristics are given in Table 2. At the calculated points, the selected transistor would operate with PTC, considering the variations in the ITP under the worst-case conditions shown in Figure 4A. The voltage collector-emitter at the nominal expected current through each transistor of $I_C = 40$ A (Figure 4B) is $V_{CE(sat\ 25^{\circ}C\ max)} = 2.22$ V, $V_{CE(sat\ 25^{\circ}C\ min)} = 2.29$ V, $V_{CE(sat\ 150^{\circ}C\ max)} = 2.71$ V, and $V_{CE(sat\ 150^{\circ}C\ max)} = 2.83$ V. The result shows that the expected conductive losses, which are the primary losses for the analysed ZVS inverter, are feasible for the TO-273 transistor package and its modifications, as shown in Table A1, Appendix A.

**Table 2.** Digitalisation of the output characteristics shown in Figure 4B.

| | nominal characteristic (Figure 2) | +5% data variation Figure 3B, graphic 1 | −5% data variation Figure 3B, graphic 2 |
|---|---|---|---|
| Nominal temperature 25 °C | $I_C = \begin{bmatrix} -1.2487 \times V_{CE}^5 \\ 17.285 \times V_{CE}^4 \\ -96.404 \times V_{CE}^3 \\ 274.79 \times V_{CE}^2 \\ -342.33 \times V_{CE} \\ 146.59 \end{bmatrix}$ | $I_C = \begin{bmatrix} -1.3112 \times V_{CE}^5 \\ 18.149 \times V_{CE}^4 \\ -101.22 \times V_{CE}^3 \\ 288.53 \times V_{CE}^2 \\ -359.45 \times V_{CE} \\ 153.92 \end{bmatrix}$ | $I_C = \begin{bmatrix} -1.1863 \times V_{CE}^5 \\ 16.421 \times V_{CE}^4 \\ -91.583 \times V_{CE}^3 \\ 261.05 \times V_{CE}^2 \\ -325.21 \times V_{CE} \\ 139.26 \end{bmatrix}$ |
| | Nominal characteristic (Figure 2) | +5% data variation Figure 3B, graphic 3 | −5% data variation Figure 3B, graphic 4 |
| Maximum temperature 150 °C | $I_C = \begin{bmatrix} -0.0371 \times V_{CE}^5 \\ 0.7896 \times V_{CE}^4 \\ -7.2327 \times V_{CE}^3 \\ 34.392 \times V_{CE}^2 \\ -46.567 \times V_{CE} \\ 18.897 \end{bmatrix}$ | $I_C = \begin{bmatrix} -0.039 \times V_{CE}^5 \\ 0.8291 \times V_{CE}^4 \\ -7.5943 \times V_{CE}^3 \\ 36.111 \times V_{CE}^2 \\ -48.895 \times V_{CE} \\ 19.842 \end{bmatrix}$ | $I_C = \begin{bmatrix} -0.0353 \times V_{CE}^5 \\ 0.7501 \times V_{CE}^4 \\ -6.8711 \times V_{CE}^3 \\ 32.672 \times V_{CE}^2 \\ -44.238 \times V_{CE} \\ 17.952 \end{bmatrix}$ |

Step 3. Analysing the mismatch and the derating factor for the stationary and dynamic transistors' characteristics.

The mismatch factor M can be estimated from the maximum and minimum current at the given conditions from the equation:

$$M = \frac{I_{C\ max} - I_{C\ min}}{I_{C\ max}} \tag{9}$$

where $I_{C\ max}$ and $I_{C\ min}$ are the maximum and minimum current per IGBT transistor in parallel, respectively, calculated in Step 2.

Having the mismatch factor M and the accepted IGBTs in parallel, Equation (1) about the derating factor Δ can be presented as:

$$\Delta = 1 - \frac{\left(N_{parallel} - 1\right)(1 - M) + 1}{N_{parallel}} \tag{10}$$

The result from Equations (9) and (10) is depicted in Figure 5. In the range $V_{ge} = 9$–$10$ V, the derating factor varies in the range $\Delta = 0.12$–$0.21$ under the worst-case conditions, which would occur at nonthermal equivalence amongst the transistors (Figure 5A, graphic 1). The result shows that the accepted $\Delta = 0.20$ at Step 1, Equation (1), is a correct assumption. If the paralleled transistors share the same thermal equilibrium, the necessary derating factor would be in the range $\Delta = 0.034$–$0.053$, i.e., 3.4%–5.3%. Under these conditions, the system based on a 20% derating factor, matching the probable worst-case condition, would be oversized under thermal equalisation. This result is supported by an expression of the current difference in Figure 5B—graphic 1 shows the maximum current difference under the worst-case conditions and graphic 2 at thermal equalisation.

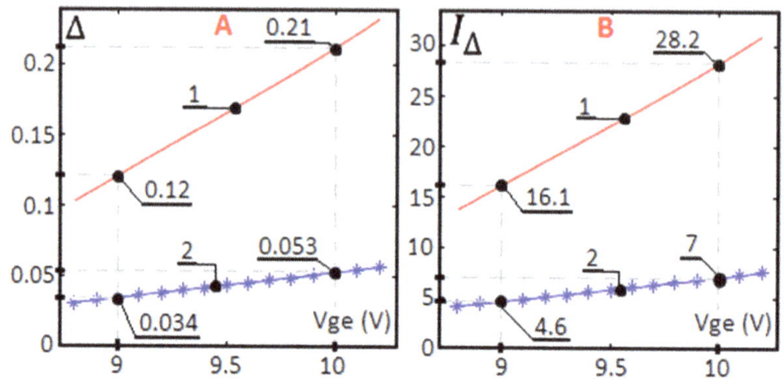


**Figure 5.** (**A**) Expected variation in the derating factor; (**B**) expected variation in the maximum current through the paralleled transistors. Graphics 1 and 2 at maximum and minimum thermal difference.

Step 4. Analysis of the single IGBT transistor parameters and the relevant selection requirements.

After the expected current through each transistor in parallel is found from the presented equations above with the necessary safety margin, the transistors can be selected according to several considerations about their basic parameters.

The continuous collector current through a single transistor must be considered according to the expected junction temperature as:

$$I_C = \frac{(T_{j\ max} - T_C)}{V_{CE} \times R_{th(J-C)}} \quad (11)$$

where $T_{j\ max}$ is the maximum junction temperature; $T_C$ is the case temperature; $V_{CE}$ is the collector-emitter saturation voltage at $I_C$; $R_{th(J-C)}$ is the junction-case thermal resistance of the transistor.

The peak collector current should be selected to be at least two times the nominal collector current for 1 ms at the maximum temperature.

$$I_{PK} = 2 \times I_C \quad (12)$$

The maximum dissipated power from a single IGBT transistor can be calculated from:

$$P_{\max IGBT} = \frac{(T_{j\ max} - T_C)}{R_{th(J-C)}} \quad (13)$$

As the analysed inverter naturally operates with ZVS and, hence, low switching losses, the selected transistors according to the necessary collector current and collector-emitter voltage usually cause an overrating in the power dissipation. This explains the relatively small-size heatsinks combined with forced convection, shown in the next part.

The switching losses during the transistors ON and OFF time depend on the amount of energy, respectively, $E_{on}$ and $E_{off}$. The analysis can be conducted according to the following equations:

$$E_{on} = \int_{T_{on1}}^{T_{on2}} V_{CE}(t) \times I_C(t)\, dt \quad (14)$$

$$E_{off} = \int_{T_{off1}}^{T_{off2}} V_{CE}(t) \times I_C(t)\, dt \quad (15)$$

where the times $T_{on1}$, $T_{on2}$, $T_{off1}$, $T_{off2}$ are, respectively, the beginning and the end of the ON and OFF periods.

As the converter operates with ZVS, the $V_{CE}$ voltages during the ON and OFF switching will be comparable with the threshold $V_{GE}$, and the times $T_{on2} - T_{on1}$ and $T_{off2} - T_{off1}$ are under 100 ns, confirmed experimentally in the next part. With this, the expected switching losses can be neglected, and the conduction losses can be accepted as dominant. Their calculation is according to the equation:

$$P_{cond} = V_{CE} \times I_C = R_{CE} \times I_C^2 (W) \tag{16}$$

For the considered packages above, TO-247, TO-3PN, and TO-264, transistors with several tens of milliohms $R_{CE}$ can be selected.

## 3. Experimental Setup

The experimental setup aimed to verify the designed inverter for induction heating based on paralleled IGBTs and the suggested methodology and to depict the current mismatch amongst the transistors experimentally. The experimental study was conducted as follows: converters were designed with 3, 4 and 5 transistors in parallel per side (Figure 1) following the presented step-by-step methodology in point 2; converters were manufactured using IGBTs listed in Table 1 Applications, using transistors from the same and different production lots; the derating factors and expected current mismatch were analysed according to the nominal current per inverter; the currents were measured through each transistor in parallel at full load during the entire melting cycle; the thermal differences were measured with an infrared camera on the heatsink surface. In addition, the layout and inductor construction improvements were implemented, leading to the minimisation of the oscillations.

Figure 6 shows an experimental model of the self-resonant converter for induction heating with its main components: 8 GIBT transistors, connected according to Figure 1; forced air cooling system with fans; resonant capacitor tank, inductor; and graphene crucible. The 4 selected transistors per side were considered according to previously presented Equations (1)–(16) with a current derating factor of 20%.

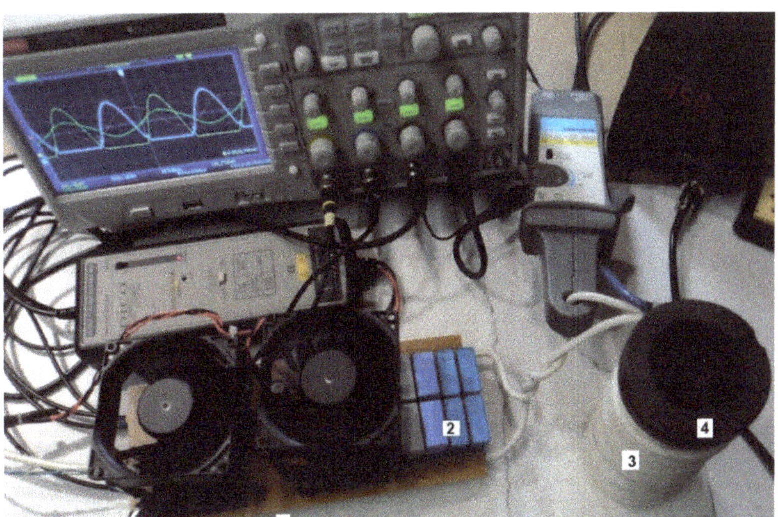


**Figure 6.** A model of self-resonant converter for induction heating: 1—IGBT transistors; 2—resonant capacitor tank; 3—inductor; 4—graphene crucible. The used measurement equipment is as follows: oscilloscope RS PRO RSDS1204CFL; current probe ELDITEST CP6220; voltage differential probe Pico.

The experimental study was conducted at the maximum converter power to depict the current mismatch under worst-case conditions. The results are illustrated with oscillograms, as follows:

- Figure 7. The oscillograms showed the expected current mismatch between transistors selected from the same production lot (1, 2) and transistors from two different lots (3, 4). Although all transistors were the same type, the first two conducted with a negligible difference, while transistor 3 showed a 25% higher current and transistor 4 switched on with a 540 ns delay (Figure 7A,B). The provided measurements clearly showed that a converter based on such a selection of IGBT transistors would be unreliable. In another selection, a different lot transistor (graphic 4, Figure 7C) could switch prematurely, conducting higher current, which is also an unacceptable condition.
- Figure 8. The oscillograms showed an acceptable current mismatch in the range of 2–5% at the peak point due to the layout issues and parasitic elements. The experiment was conducted in the following conditions: the transistors were selected from the same production lot; the resonant frequency during the entire heating process varied in the range 25–35 kHz, depending on the load conditions. The experimental result also confirmed the suggested current mismatch range of $\pm 5\%$ for the production lot. Similar results were received with the transistors shown in Table A1, Appendix A.
- Figure 9. The oscillograms presented the operation of the converter, comprised of transistors in parallel. The experiment showed a stable process in the entire frequency and power range. According to Figure 9B, the sinusoidal voltage over the inductor crossed the zero point between both sides of the voltage precisely, which experimentally proved the inverter's ability to operate in ZVS with correctly selected paralleled transistors.
- Figure 10. The proposed inverter improved from the first unit (Figure 6) by moving the capacitor tank allocated in parallel to each transistor. Although the converter layout was not an object of this research, such a layout matched better the paralleled transistors as a current mismatch amongst the IGBTs under 2% was observed.
- Figure 11. The presented thermal images provided an experimental measurement of the temperature on the heatsinks' surface. Typically, the paralleled transistors have to be assembled on a typical heat sink, sharing and equalising the temperature. In this experiment, they were intestinally separated for the thermal difference to be better depicted. The investigation showed a thermal difference between transistors in parallel with low current differences (Figure 11A,C) according to Figure 8 and high current differences (Figure 11B,D) according to Figure 7B,C.
- Figure 12. The thermal differences depicted as infrared pictures in Figure 11 are described as transient processes in an operation cycle. As Figure 12A shows, the thermal difference between transistors the Q1 and Q4 heatsink surface (Figure 7A) could reach 10 °C. The same experiment showed that the temperature difference under the worst-case conditions between transistors of the same production lot Q1 and Q2 (Figure 7A) could be minimised to 3–4 °C.
- Figure 13. The crucible reached 500 °C for 10 min; enough for melting tin with a high energy efficiency. Potentially, the entire system can be designed for higher temperatures to melt other metals, requiring higher power and, hence, powerful transistors. The system was tested with several melting cycles, giving the established transistors and crucible temperatures without overheating.
- Figure 14. The experiment showed oscillations that could potentially occur due to PCB layout issues in inverters with paralleled transistors. Usually, such oscillations occur in the gate drive circuit when transistors share the same gate drivers as the research in [51–53]. However, in this case, each transistor was controlled by an individual driver (Figure 1), and it was found that the problem originates from the conductors' length and parasitics, which occur between the PCB and the inductor.
- Figure 15 Although the above-described problems are beyond the scope of this research, a layout modification was suggested and used for all presented experimental tests. According to the circuit in Figure 15, the air-cooled inductor was divided into a

number equal to the number of paralleled transistors sections. With this, the inductors L1, L3, L5 and L7 from the high side and L2, L4, L6 and L8 from the low side correspond to inductors L1 and L2 from Figure 1. Capacitors C1-C4, in summary, represent capacitor C1 from the resonant tangent in Figure 1. Although the effect of parasitics minimisation was experimentally observed, giving the inverter's stable operation, this part requires future research and improvements.

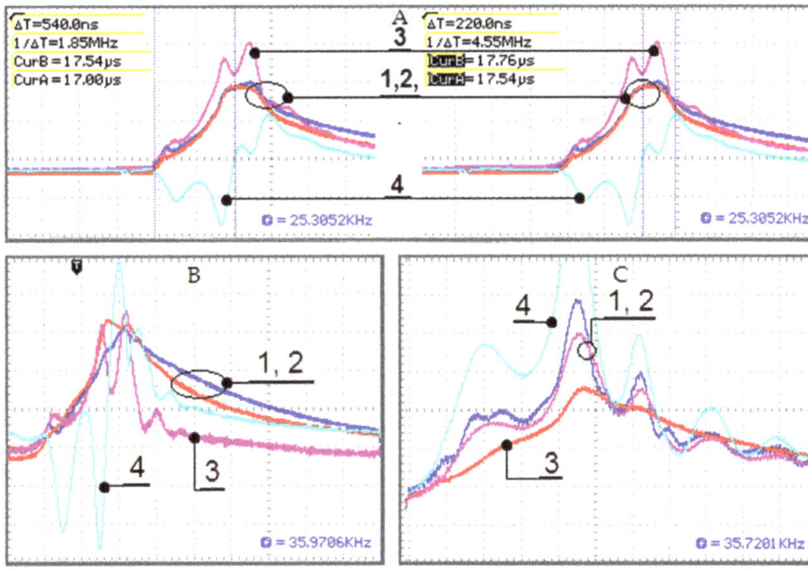


**Figure 7.** Current mismatch amongst four paralleled IGBT transistors. (**A**,**B**) switched-on with a delay of transistor 4; (**C**) switched-on prematurely of transistor 4; 1, 2—transistors with the same production lot; 3, 4—two different production lots.

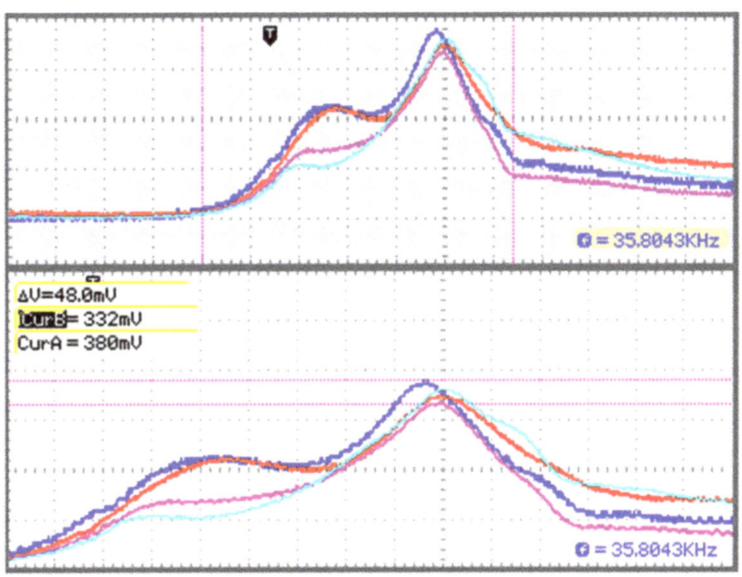


**Figure 8.** Minimum current mismatch amongst four IGBT transistors from the same production lot.

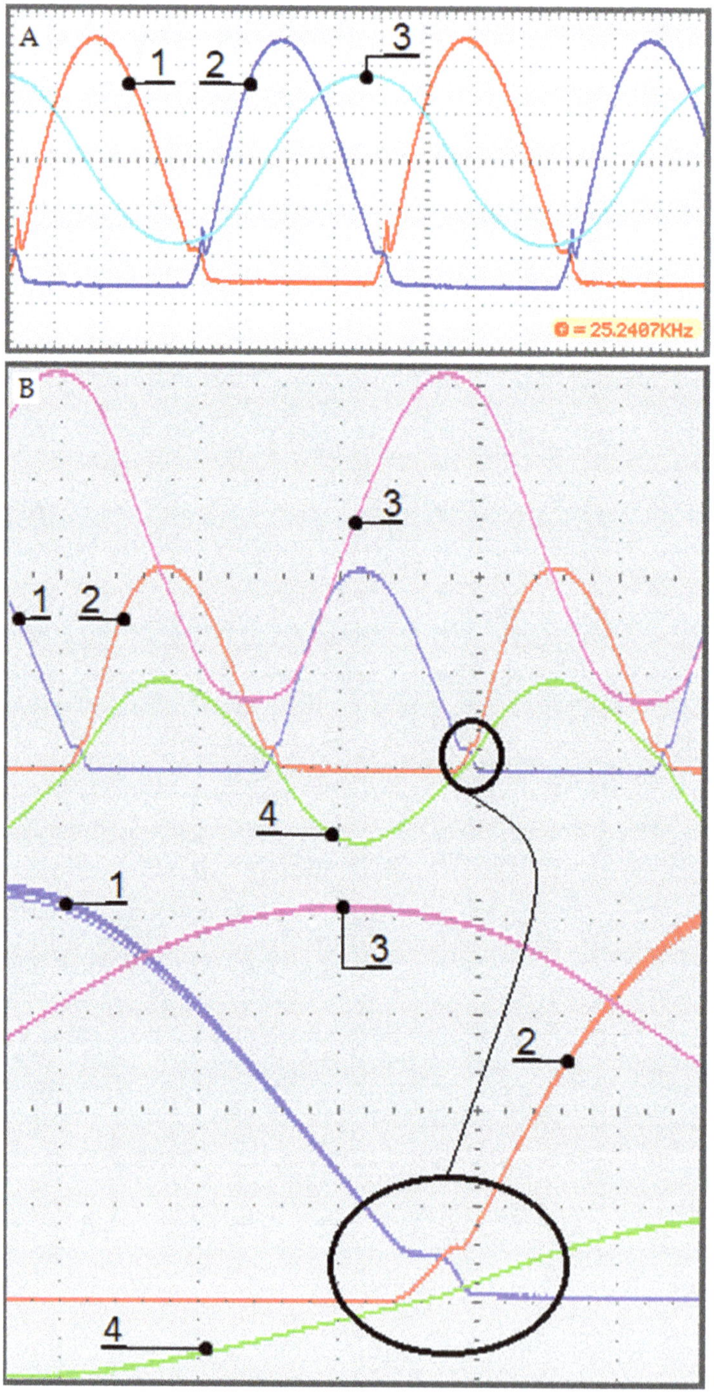


**Figure 9.** Operation of the inverter for induction heating. (**A**,**B**) 1, 2—voltages, respectively, over the paralleled high side (Q1, Q3, Q5, Q7—Figure 1) and low side (Q2, Q4, Q6, Q8—Figure 1) transistors; 3—current through the inductor; (**B**): 4—voltage crossing the zero switching point.

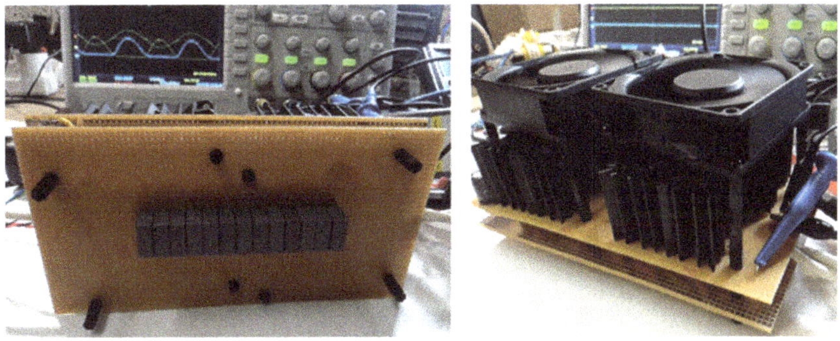

Figure 10. Improving the converter construction, given in Figure 6.

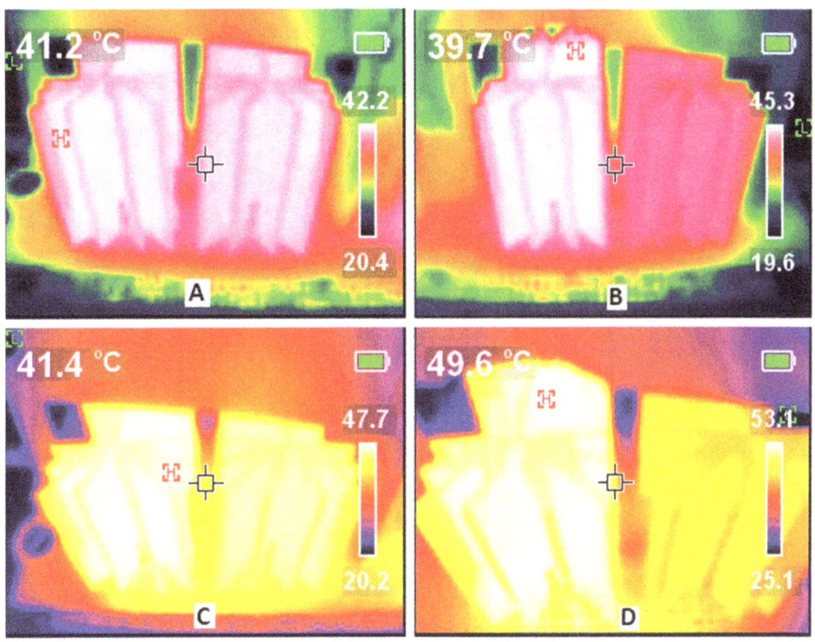


**Figure 11.** Infrared temperature measurements on the surface of the heatsink. (**A**,**C**) Appropriately selected transistors from the same production lot (Figure 7, graphics 1, 2); (**B**,**D**) transistors from different production lots (Figure 7 graphics 1, 4). The thermal measurements were performed with the infrared camera FLIR.

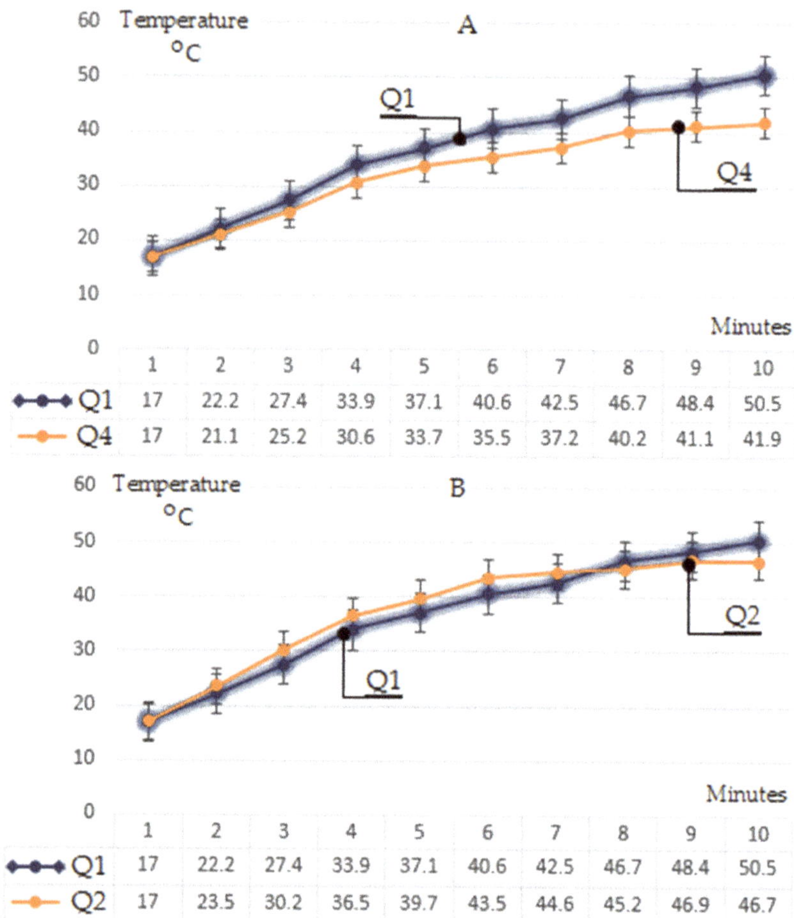


**Figure 12.** Transistors thermal difference for one melting cycle of 10 min. (**A**) Between transistors Q1 and Q4; (**B**) between transistors Q1 and Q2, both selected according to Figure 7A.

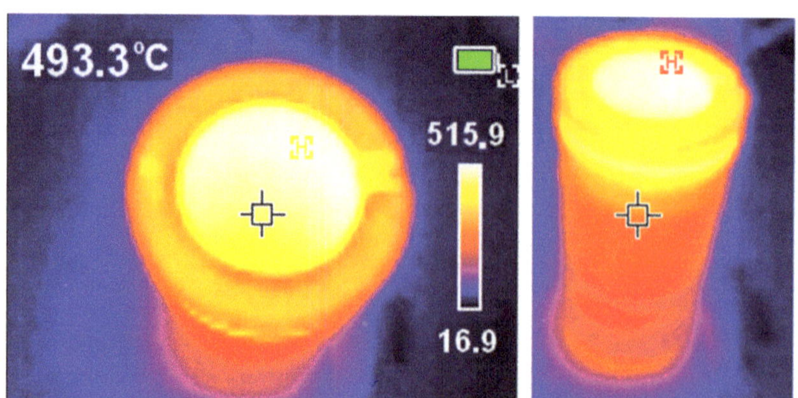


**Figure 13.** Infrared image of the graphene crucible, given in Figure 6.

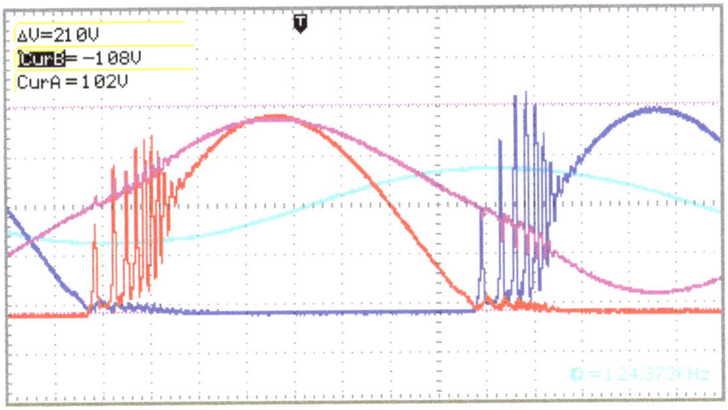


**Figure 14.** Oscillations due to PCB layout issues.

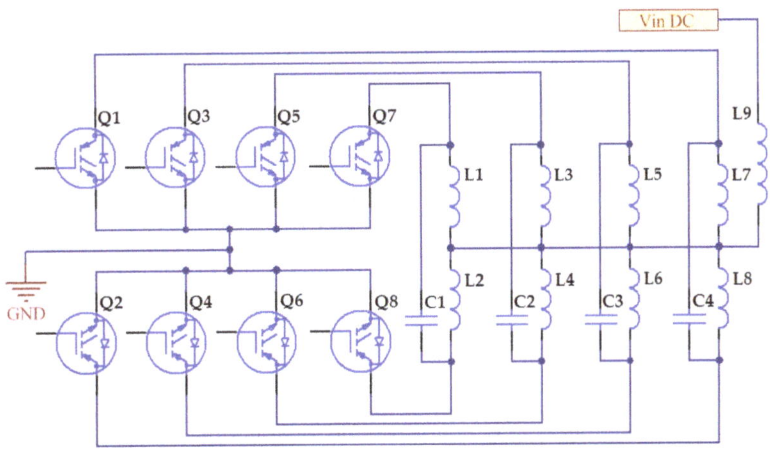


**Figure 15.** A suggested structure of connection.

## 4. Discussion

The provided experimental verification was performed with a resonant inverter for induction heating (Figure 6). As the main contribution in this research, the result showed that such a topology (Figure 1) can be completed with paralleled IGBTs, operating with low mismatch current and acceptable current difference amongst the transistors, which is supported by the presented oscillograms (Figures 7–9). The achieved current sharing amongst the paralleled IGBTs makes the ZVS mode of operation possible for a resonant inverter with paralleled transistors, which is the main advantage of the investigated topology. The presented results showed the feasibility of the investigated topology for improving the installed power of the self-resonant inverters for induction heating. As it was experimentally shown that the installed capacity can reach more than 30 kW with four IGBTs in parallel per side of the topology, it can be concluded that such a modification is applicable for small laboratory or industrial induction crucible furnaces.

The suggested methodology for analysing the paralleled IGBT transistors is based on stipulated conditions in which the transistors' static and dynamic parameters (Figure 3) are digitalised into polynomial Equations (3)–(8). The calculating procedure shows the expected parameters dispersion in a production lot (Figure 4). As a result, the necessary derating factor, the current mismatch, and the current difference amongst the paralleled transistors can be suggested (Figure 5). With this, the number of the necessary transistors in parallel can be calculated on the design level for the given application.

The temperature difference between the paralleled IGBTs, which results from the current mismatch, was in an acceptable range of several degrees Celsius. This final experiment verified the applicability of the suggested topology.

## 5. Conclusions

In this study, a methodology of the IGBTs' parallel work estimation and its application in the design procedure of a self-resonant inverter for a crucible induction system was demonstrated. The presented step-by-step approach showed satisfactory results, giving the required derating factors and current mismatch between the transistors.

The investigated self-resonant converter based on paralleled transistors was applied to an induction crucible melting system. It was shown that the presented circuit has a stable operation mode in ZVS with paralleled transistors. The conducted experiments showed a satisfactory current share and temperature equalisation. It can be concluded that the circuit has the magnitude to reach an installed power of 50–80 kW with several powerful modules in parallel, giving budget-friendly induction melting systems with inverters with high power density.

The suggested methodology was based on stipulated data, i.e., Equations (3)–(6) were derived only for the concrete transistor, and it did not have diverse characteristics. Despite that, such an approach for analysis is highly applicable for practical design and experimental verification.

**Author Contributions:** Conceptualization, B.D.; methodology, B.D., K.H., S.B., G.C.; software, B.D.; validation, B.D.; formal analysis, B.D.; investigation, B.D.; resources, G.C.; data curation, B.D., K.H., S.B.; writing—original draft preparation, B.D.; writing—review and editing, K.H., S.B.; visualization, B.D.; supervision, G.C., K.H.; project administration, G.C.; funding acquisition, N/A. All authors have read and agreed to the published version of the manuscript.

**Funding:** This research received no external funding.

**Institutional Review Board Statement:** Not applicable.

**Informed Consent Statement:** Not applicable.

**Data Availability Statement:** Not applicable.

**Conflicts of Interest:** The authors declare no conflict of interest.

## Appendix A

**Table A1.** IGBT transistors, suitable for experimental verification with resonant inverters for induction heating, operating with PTC.

| IGBT Transistor | Package | $I_{C(25^\circ C)}$(A) | $V_{CE}$(V) | $P_{W(25^\circ C)}$(W) | $V_{GE(th)}$/ $V_{CE(sat)}$ (V) |
|---|---|---|---|---|---|
| | | $I_{C(25^\circ C)}$ =30–60 A | | | |
| RGT60TS65D | TO-247N | 55 | 650 | 194 | 7.0/2.1 |
| DGTD120T25S1PT | TO-247 | 50 | 1200 | 348 | 7.0/2.4 |
| FGA50N100BNTD | TO-3P | 50 | 1000 | 156 | 7.0/1.8 |
| RGW00TS65D | TO-247N | 50 | 650 | 254 | 7.0/1.9 |
| | | $I_{C(25^\circ C)}$ =60–80 A | | | |
| NGTB40N120FL2 | TO-264 3L | 70 | 1200 | 368 | 2.6/3.7 |
| STGW39NC60VD | TO-247 | 70 | 600 | 250 | 5.75/2.4 |
| IXA45IF1200HB | TO-247 | 78 | 1200 | 325 | 6.5/2.1 |
| STGW40H65FB | TO-3PF | 80 | 650 | 283 | 6.0/2.0 |
| | | $I_{C(25^\circ C)}$ =80–120 A | | | |
| FGA60N60UFD | TO-3P | 120 | 600 | 298 | 6.5/2.4 |
| FGA60N65SMD | TO-3PN | 120 | 650 | 600 | 6.0/2.5 |
| NGTB50N120FL2 | TO-247 | 100 | 1200 | 535 | 6.5/2.4 |
| IRGP6690DPbF | TO-247AD | 140 | 600 | 483 | 6.5/1.95 |

Listing A1. Calculating the collector-emitter current (Ic) with polyval function in MATLAB:

```
clear
format short
Ic_max_hightemp = [0.0096; −0.106; −3.8721; 86.988; −600.32; 1365.2];
Vge = 9;                                          % Vge = 9 V
polyval (Ic_max_hightemp, Vge)                    % ans = 56.99 A
```

## References


1. Davies, E.J. *Conduction and Induction Heating*; The Institution of Engineering and Technology IET: London, UK, 2007; ISBN 978-0-86341-174-8.
2. Haimbaugh, R. *Induction Heat Treating*; ASM international: Materials Park, OH, USA, 2015; ISBN 978-1-62708-089-7.
3. Lupi, S.; Forzan, M.; Aliferov, A. *Induction and Direct Resistance Heating, Theory and Numerical Modeling*; Springer International Publishing: Springer Cham Heidelberg, Germany, 2015; ISBN 978-3-319-03478-2.
4. Rudnev, V.; Loveless, D.; Cook, R. *Handbook of Induction Heating*; Taylor & Francis Group: Boca Raton, FL, USA, 2017; ISBN 978-1-4665-5395-8.
5. Rapoport, E.; Pleshivtseva, Y. *Optimal Control of Induction Heating Processes*; Taylor & Francis Group: Boca Raton, FL, USA, 2007; ISBN 978-0-8493-3754-3.
6. Zinn, S.; Semiatin, S. *Elements of Induction Heating. Design, Control and Applications*; ASM International: Materials Park, OH, USA, 2007; ISBN 0-87170-308-4.
7. Shokouhmand, H.; Ghaffari, S. *Thermal Analysis of Moving Induction Heating of a Hollow Cylinder with Subsequent Spray Cooling: Effect of Velocity, Initial Position of Coil, And Geometry, Applied Mathematical Modelling 36*; Elsevier: Amsterdam, The Netherlands, 2012; pp. 4304–4323.
8. Li, F.; Ning, J.; Liang, S.Y. Analytical Modeling of the Temperature Using Uniform Moving Heat Source in Planar Induction Heating Process. *Appl. Sci.* **2019**, *9*, 1445. [CrossRef]
9. Naar, R.; Bay, F. Numerical optimisation for induction heat treatment processes. *Appl. Math. Model.* **2013**, *37*, 2074–2085. [CrossRef]
10. Zhou, X.; Thomas, B.; Hernandez, C.; Castillejos, A.; Acosta, F. *Measuring Heat Transfer During Spray Cooling Using Con-trolled Induction-Heating Experiments and Computational Models, Applied Mathematical Modelling 37*; Elsevier: Amsterdam, The Netherlands, 2013; pp. 3181–3192.
11. Djambazov, G.; Bojarevics, V.; Pericleous, K.; Croft, N. Finite volume solutions for electromagnetic induction processing. *Appl. Math. Model.* **2015**, *39*, 4733–4745. [CrossRef]
12. Jankowski, T.; Pawley, N.; Gonzales, L.; Ross, C.; Jurney, J. *Approximate Analytical Solution for Induction Heating of Solid Cylinders. Applied Mathematical Modelling 40*; Elsevier: Amsterdam, The Netherlands, 2016; pp. 2770–2782.
13. Tavakoli, M.; Karbaschi, H.; Samavat, F. Influence of Workpiece Height on the Induction Heating Process. *Math. Comput. Model.* **2011**, *45*, 50–58. [CrossRef]
14. Liu, Q.; Li, B.; Schlangen, E.; Sun, Y.; Wu, S. Research on the Mechanical, Thermal, Induction Heating and Healing Properties of Steel Slag/Steel Fibers Composite Asphalt Mixture. *Appl. Sci.* **2017**, *7*, 1088. [CrossRef]
15. Dimitrov, B.; Streblau, M.; Marinov, A. An Approach for Designing a Complex Inductor—Workpiece System for Induction Heating Detail Systems. *Technol. Educ. Manag. Inform. TEM J.* **2014**, *3*, 244–249.
16. Li, H.; Yu, J.; Wu, S.; Liu, Q.; Li, Y.; Wu, Y.; Xu, H. Investigation of the Effect of Induction Heating on Asphalt Binder Aging in Steel Fibers Modified Asphalt Concrete. *Materials* **2019**, *12*, 1067. [CrossRef]
17. Vila-Cortavitarte, M.; Jato-Espino, D.; Castro-Fresno, D.; Calzada-Perez, M. Self-Healing Capacity of Asphalt Mixtures In-cluding By-Products Both as Aggregates and Heating Inductors. *Materials* **2018**, *11*, 800. [CrossRef] [PubMed]
18. Fu, X.; Chen, Q.; Chen, X.; Zhang, L.; Yang, A.; Cui, Y.; Yuan, C.; Ge, H. A Rapid Thermal Nanoimprint Apparatus Through In-duction Heating of Nickel Mold. *Micromachines* **2019**, *10*, 334. [CrossRef]
19. Aung, M.P.; Nakamura, M.; Hirohata, M. Characteristics of Residual Stresses Generated by Induction Heating on Steel Plates. *Metals* **2018**, *8*, 25. [CrossRef]
20. Chamorro, X.; Herrero-Dorca, N.; Bernal, D.; Hurtado, I. Induction Skull Melting of Ti-6Al-4V: Process Control and Efficiency Optimisation. *Metals* **2019**, *9*, 539. [CrossRef]
21. Namadmalan, A.; Moghani, J. Self-Oscillating Switching Technique for Current Source Parallel Resonant Induction Heating Systems. *J. Power Electron.* **2012**, *12*, 851–858. [CrossRef]
22. Namadmalan, A.; Moghani, J.S.; Milimonfared, J. A Current-Fed Parallel Resonant Push-Pull Inverter with a New Cascaded Coil Flux Control for Induction Heating Applications. *J. Power Electron.* **2011**, *11*, 632–638. [CrossRef]
23. Chamsuka, P.; Srinonchat, J. Study and development Inductance melting control system using low energy. *Energy Procedia* **2013**, *34*, 945–950. [CrossRef]
24. Nagarajan, B.; Sathi, R. Phase Locked Loop based Pulse Density Modulation Scheme for the Power Control of Induction Heating Applications. *J. Power Electron.* **2015**, *15*, 65–77. [CrossRef]
25. Qi, H.; Peng, Y.; Li, Y. The Study of Phase-shift Super-Frequency Induction Heating Power Supply. *Phys. Procedia* **2012**, *24*, 101–106. [CrossRef]
26. Kawaguchi, Y.; Hiraki, E.; Tanaka, T.; Nakaoka, M. Basic Study of a Phase-Shifted Soft Switching High-Frequency Inverter with Boost PFC Converter for Induction Heating. *J. Power Electron.* **2008**, *8*, 192–199.

27. Yachiangkam, S.; Sangswang, A.; Naetiladdanon, S.; Koompai, C.; Chudjuarjeen, S. Steady-State Analysis of ZVS and NON-ZVS Full-Bridge Inverters with Asymmetrical Control for Induction Heating Applications. *J. Power Electron.* **2015**, *15*, 544–554. [CrossRef]
28. Qi, X.; Peng, Y.; Li, Y. Parameters Design of Series Resonant Inverter Circuit. *Phys. Procedia* **2012**, *24*, 133–138. [CrossRef]
29. Wang, Y.; Li, Y.; Peng, Y.; Qi, X. Research and Design on IGBT Induction Heating Power Supply. *Energy Procedia* **2012**, *16*, 1957–1963. [CrossRef]
30. Borage, M.; Tiwari, S. A 25 kW, 25 kHz Induction Heating Power Supply for MOVPE System Using L-LC Resonant Inverter, Hindawi Publishing Corporation. *Adv. Power Electron.* **2013**. [CrossRef]
31. Kumar, A.; Sadhu, P.; Mohanta, D.; Reddy, M. An Effective Switching Algorithm for Single Phase Matrix Converter in Induc-tion Heating Applications. *Electronics* **2018**, *7*, 149. [CrossRef]
32. Han, W.; Chau, K.T.; Wong, H.C.; Jiang, C.; Lam, W.H. All-In-One Induction Heating Using Dual Magnetic Couplings. *Energies* **2019**, *12*, 1772. [CrossRef]
33. Park, S.; Jang, E.; Joo, D.; Lee, B. Power Curve-Fitting Control Method with Temperature Compensation and Fast-Response for All-Metal Domestic Induction Heating Systems. *Energies* **2019**, *12*, 2915. [CrossRef]
34. Lasobras, J.; Alonso, R.; Carretero, C.; Carretero, E.; Imaz, E. Infrared Sensor-Based Temperature Control for Domestic Induc-tion Cooktops. *Sensors* **2014**, *14*, 5278–5295. [CrossRef]
35. Villa, J.; Artigas, J.; Barragan, L.; Navarro, D. An Amplifier-Less Acquisition Chain for Power Measurements in Series Reso-nant Inverters. *Sensors* **2019**, *19*, 4343. [CrossRef]
36. Oh, Y.; Yeon, J.; Kang, J.; Galkin, I.; Oh, W.; Cho, K. Sensorless Control of Voltage Peaks in Class-E Single-Ended Resonant Inverter for Induction Heating Rice Cooker. *Energies* **2021**, *14*, 4545. [CrossRef]
37. Tabaković, A.; Mohan, J.; Karač, A. Conductive Compartmented Capsules Encapsulating a Bitumen Rejuvenator. *Processes* **2021**, *9*, 1361. [CrossRef]
38. Oncu, S.; Unal, K.; Tuncer, U. Laboratory setup for teaching resonant converters and induction heating. *Eng. Sci. Technol. Int. J.* **2021**. [CrossRef]
39. Areitioaurtena, M.; Segurajauregi, U.; Urresti, I.; Fisk, M.; Ukar, E. Predicting the induction hardened case in 42CrMo4 cylin-ders. *Procedia* **2020**, *87*, 545–550.
40. Bortis, D.; Biela, J.; Kolar, J.W. Active Gate Control for Current Balancing of Parallel-Connected IGBT Modules in Solid-State Modulators. *IEEE Trans. Plasma Sci.* **2008**, *36*, 2632–2637. [CrossRef]
41. Streit, L.; Janik, D.; Talla, J. Serial-Parallel IGBT Connection Method Based on Overvoltage Measurement. *Elektron. Elektrotech.* **2016**, *22*, 53–56. [CrossRef]
42. Perez-Delgado, R.; Velasco-Quesada, G.; Roman-Lumbreras, M. Current Sharing Control Strategy for IGBTs Connected in Parallel. *J. Power Electron.* **2016**, *16*, 769–777. [CrossRef]
43. Xue, Y.; Lu, J.; Wang, Z.; Tolbert, L.; Blalock, B.; Wang, F. Active Current Balancing for Parallel-Connected Silicon Carbide MOSFETs. IEEE Energy Conversion Congress and Exposition. 2013. Available online: https://ieeexplore.ieee.org/document/6646891 (accessed on 1 April 2021).
44. Hu JAlatise, O.; Gonzalez, J.; Bonyadi, R.; Alexakis, P.; Ran, L.; Mawby, P. Robustness and Balancing of Parallel-Connected Power Devices: SiC Versus CoolMOS. *IEEE Trans. Ind. Electron.* **2016**, *63*, 2092–2102.
45. Kimmer, T. Application Note, System Application Engineering IGBT, Reverse Conduction IGBT for Inductive Cooking, In-Fineon Technologies. 2012. Available online: https://www.infineon.com (accessed on 1 April 2021).
46. Application Note, System Application IGBT, Reverse Conducting IGBT for Drives RC-Drives Cost-Optimized IGBT for Consumer Drive Application, Infineon Technologies 2010. Available online: https://www.infineon.com (accessed on 1 April 2021).
47. Application Note AN4713, Induction Cooking: IGBTs in Resonant Converters, June 2015. Available online: www.st.com (accessed on 1 April 2021).
48. Huang, X.; Chang, D.; Ling, C.; Zheng, T.Q. Research on Single-Phase PWM Converter with Reverse Conducting IGBT Based on Loss Threshold Desaturation Control. *Energies* **2017**, *10*, 1845. [CrossRef]
49. Huang, X.; Ling, C.; Chang, D.; You, X.; Zheng, T.Q. Loss Characteristics of 6.5 kV RC-IGBT Applied to a Traction Converter. *Energies* **2017**, *10*, 891. [CrossRef]
50. Application Note 5SYA 2098-00, Paralleling of IGBT modules, ABB Switzerland Ltd Semiconductors. Available online: https://library.e.abb.com/ (accessed on 1 April 2021).
51. Application Note AN 17-001, IGBT Modules in Parallel Operation with Central and Individual Driver Board, Semikron. 2017. Available online: https://www.semikron.com/ (accessed on 1 April 2021).
52. NXP UCODE G2XM SL3S1002AC2. 118. Available online: https://www.nxp.com/ (accessed on 1 April 2021).
53. AND9100/D, Parallel of IGBTs, On Semiconductor. 2014. Available online: https://www.onsemi.com/ (accessed on 1 April 2021).
54. Application Note AN 4544, IGBT Datasheet Tutorial. 2014. Available online: www.st.com (accessed on 1 April 2021).
55. Roetzer, F.; Aschauer, A.; Jadachowski, L.; Steinboeck, A.; Kugi, A. Temperature Control for Induction Heating of Thin Strips. *IFAC-PapersOnLine* **2020**, *53*, 11968–11973. [CrossRef]

energies

MDPI

Article

# Influence of Non-Linearity in Losses Estimation of Magnetic Components for DC-DC Converters

Fabio Corti [1,2,*], Alberto Reatti [2], Gabriele Maria Lozito [2], Ermanno Cardelli [1] and Antonino Laudani [3]

1 Department of Engineering, University of Perugia, Via G. Duranti 67, 06125 Perugia, Italy; ermanno.cardelli@unipg.it
2 Department of Information Engineering, University of Firenze, Via di Santa Marta 3, 50139 Firenze, Italy; alberto.reatti@unifi.it (A.R.); gabrielemaria.lozito@unifi.it (G.M.L.)
3 Department of Industrial Engineering Electronics and Mechanics, University of Roma 3, Via Vito Volterra 62b, 00146 Roma, Italy; antonino.laudani@uniroma3.it
* Correspondence: fabio.corti@unipg.it

**Abstract:** In this paper, the problem of estimating the core losses for inductive components is addressed. A novel methodology is applied to estimate the core losses of an inductor in a DC-DC converter in the time-domain. The methodology addresses both the non-linearity and dynamic behavior of the core magnetic material and the non-uniformity of the field distribution for the device geometry. The methodology is natively implemented using the LTSpice simulation environment and can be used to include an accurate behavioral model of the magnetic devices in a more complex lumped circuit. The methodology is compared against classic estimation techniques such as Steinmetz Equation and the improved Generalized Steinmetz Equation. The validation is performed on a practical DC-DC Buck converter, which was utilized to experimentally verify the results derived by a model suitable to estimate the inductor losses. Both simulation and experimental test confirm the accuracy of the proposed methodology. Thus, the proposed technique can be flexibly used both for direct core loss estimation and the realization of a subsystem able to simulate the realistic behavior of an inductor within a more complex lumped circuit.

**Keywords:** dynamic magnetic losses; ferrite core; core losses measurement; spice equivalent circuit

**Citation:** Corti, F.; Reatti, A.; Lozito, G.M.; Cardelli, E.; Laudani, A. Influence of Non-Linearity in Losses Estimation of Magnetic Components for DC-DC Converters. *Energies* **2021**, *14*, 6498. https://doi.org/10.3390/en14206498

Academic Editor: Hongwen He

Received: 12 August 2021
Accepted: 8 October 2021
Published: 11 October 2021



## 1. Introduction

DC-DC power converters are widely used in many electrical and electronic applications. The diffusion of wide-bandgap semiconductors, characterized by fast switching transients is increasing the operating frequency of DC-DC converters allowing for higher power densities [1–3].

Magnetic components are the bulkiest components of power converters, and their design must be accurate to avoid excessive weights and volumes [4]. Much effort has been spent in investigating the inductor losses generated in its winding and on its core [5]. The latter can result in both non-linear and dynamic behavior, because of the saturation and magnetic hysteresis phenomena [6]. This behavior is, in general, due to the material [7]. Considering the device geometry (i.e., the magnetic core shape), additional complexity arises due to the non-uniform distribution of the magnetic induction field across the core section, and this issue has only been partially investigated [8]. Considering these factors during the design of power converters is very difficult. For this reason, some manufacturers aim at constructive solutions that make the behavior of the magnetic component as simple as possible. In addition, they are looking for solutions with uniform distribution of magnetic field [9].

By neglecting the non-uniform magnetic induction distribution and only referring to the component data sheets, the designers are guided to choose a non-optimal magnetic component, which, under practical operating conditions, results in a worse performance than expected. Moreover, it results in oversized and overweight designs [10–13].

An optimal sizing can be achieved through calculation tools able to consider non-linearity, magnetic hysteresis, and the real non-uniform distribution of the magnetic induction in the component core, with acceptable accuracy [14–16].

The modeling of non-linearity and hysteresis in magnetic materials is achieved with different deterministic approaches, both at the micromagnetic and macromagnetic levels, but also with stochastic approaches based on specific applications of artificial intelligence [17]. The determination of distribution of magnetic field inside an inductor core is also possible by using numerical tools based on finite element methods [18,19].

Many difficulties need to be overcome when both the non-linearity, the magnetic hysteresis, and the distribution of the magnetic induction in the core must be considered in a time-domain simulation [20].

The dominant problem is that the magnetic component analysis must be set in the time-domain, and due to the non-linear nature of the equations regulating its behavior, the numerical solution (i.e., the response of the material/device) must be obtained with iterative methods. If the core geometry is considered, the discretization through the meshes of the magnetic core must be sufficiently dense [21]. Assuming this magnetic component is part of a larger time-domain simulation (e.g., a power converter), this computational process is repeated for every time-step of the simulation, resulting in a very computationally demanding scenario [22]. The computational burden of the simulation is often further increased by the very short time steps chosen to avoid numerical instability.

An interesting approach to estimate the dynamic power losses is given in [23], where the losses are calculated by a specifically designed Spice circuit, which considers the dynamic power losses. This approach is based on a model which is useful for the simulation of the behavior of DC-DC converters and allows the magnetic non-linearity to be considered along with the hysteresis and the non-uniformity of the magnetic induction. In [23], the operation of the core close to the saturation point is not considered, and this yields inaccuracies in the determination of the inductor current waveforms, which are strongly distorted in DC-DC converters. This problem has been faced in an improved model where the capability of reproducing magnetic hysteresis cycles has been introduced by referring to the non-linear behavior of the material when operated near the magnetic saturation point [24].

This paper aims to apply the approach utilized in [24], for the investigation of a DC-DC Buck converter. The proposed work focuses on the proposed model consistency when embedded in a larger design simulation. To validate this methodology, an experimental Buck converter circuit was built, and current measurements were acquired in the choke inductor. Core losses were estimated through three different techniques: the first one is based on the classic Steinmetz Equation (SE), which considers the frequency behavior of the material, yet lacks the capability of time-domain simulation under arbitrarily distorted waveforms. The second is the improved Generalized Steinmetz Equation (iGSE) technique, which introduces a methodology to apply the SE to a time-domain simulation under distorted waveform but lacks the geometric account for non-uniform field distribution in the device. The last one is the Time-Domain Core Loss with Non-Uniform Field (TDNU) technique based in the model as given in [24,25]. This technique seems also to be promising to be included in high component number circuit designs and can benefit from the optimized integration engines coupled with circuit simulation software. The paper is structured as follows. First, the two state-of-the-art methodologies used for the comparison are discussed. Then, the proposed methodology is presented, underlying the improvements introduced with respect to the classic literature approaches. In the following section, the DC-DC Buck converter design and the experimental workbench implementation are described, with particular attention to the inductor characteristics and the algorithmic approach used to estimate the losses using the three techniques. Measurements and estimated losses are then presented, along with the reconstructed hysteresis profiles, for a DC-DC Buck converter operated under different operating conditions. Conclusion and final remarks close the manuscript.

## 2. Steinmetz Equations and Improved Generalized Steinmetz Approach

Core loss estimation can be achieved through the direct application of the Steinmetz equation. This equation relates the losses to the frequency of the excitation and the intensity of the magnetic induction, and thanks to its simplicity, represents a good method to predict the core losses under sinusoidal excitation. The average power loss is given by

$$P = C_m f^{\alpha} B_m^{\beta} \tag{1}$$

where $C_m$, $\alpha$ and $\beta$ are the Steinmetz coefficients, $f$ is the frequency of the excitation and $B_m$ is the RMS value of the core magnetic flux density [26–28]. The main limit of this formula is that it results in an accurate estimation only under sinusoidal induction [26,27]. This limitation makes this formula difficult to use in time-domain simulations of non-linear devices operated at distorted currents (and, thus, H-B fields). To solve this limitation, several models were proposed. One of the most promising one is that based on the improved Generalized Steinmetz Equation (iGSE) [29,30].

Here, the average core loss is computed as

$$P = \frac{1}{T}\int_0^T |k_i| \left|\frac{dB}{dt}\right|^{\alpha} |B_m|^{\beta-\alpha} dt \tag{2}$$

where $B_m$ is the peak-to-peak flux density and

$$k_i = \frac{C_m}{(2\pi)^{\alpha-1}\int_0^{2\pi} |\cos(\vartheta)|^{\alpha} 2^{\beta-\alpha} d\vartheta} \tag{3}$$

This methodology accounts for an arbitrarily time-varying magnetic field, and for this reason, it is suitable for inclusion in time-domain simulation of magnetic materials. However, iGSE assumes a uniform distribution of the magnetic induction inside the core. It can be noted that the iGSE technique, differently from the SE approach, allows for a time-domain estimation of the core losses and can be therefore implemented in Spice environment.

## 3. Time-Domain Core Loss with Non-Uniform Field (TDNU)

A more recent method utilized to obtain a Time-Domain Core Loss estimation under non-sinusoidal excitation was presented in [23,24]. This approach allows to estimate the core power losses by using a time-domain approach rather than one based on a frequency-domain analysis. As a result, non-linearities can be considered, and estimations can be performed even with non-sinusoidal waveforms. In addition, by estimating the instantaneous power loss $p(t)$, it is possible to predict the core losses during both the transient and steady-state operation. In addition, this approach takes into account the non-uniform magnetic field distribution inside the magnetic core. The instantaneous core loss is derived as

$$p(t) = \frac{C_m}{C_{\alpha\beta}} |B_m \cos\vartheta|^{(\beta-\alpha)} \left|\frac{dB_{eff}}{dt}\right|, \tag{4}$$

where

$$C_{\alpha\beta} = (2\pi)^{\alpha} \frac{2}{\pi} \int_0^{\pi/2} (\cos\vartheta)^{\alpha} d\vartheta, \tag{5}$$

and $B_{eff}$ represents an equivalent flux density which considers the shape and the geometry of the core and approximates the effects of the non-uniformity of the magnetic field inside the core. This method consists of matching the area formed by an equivalent elliptical loop with the original hysteresis loop starting from the standard core loss coefficients. In (5), the parameter *cos* $\vartheta$ is computed as

$$\cos\vartheta(t) = \sqrt{1 - \left(\frac{B_{eff}(t) - B_{DC}}{B_m}\right)} \tag{6}$$

where $B_{DC}$ represents the DC induction bias. To compute the effective magnetic flux density $B_{eff}$, a parameter $\Delta$, called "field factor", is defined. This parameter depends on the magnetic core geometry and material, and relates the effective flux density with the current through the inductor according to

$$B_{eff}(t) = \Delta \cdot I(t). \tag{7}$$

For a toroidal core, the field factor can be calculated as in [24]

$$\Delta = \left(\frac{(\mu N)^{\beta}(2\pi)^{(1-\beta)}\frac{1}{2-\beta}\left(R_o^{(2-\beta)} - R_i^{(2-\beta)}\right)}{\pi\left(R_O^2 - R_i^2\right)}\right)^{\frac{1}{\beta}} \tag{8}$$

where $\mu = \mu_0\,\mu_r$ is the magnetic material permeability, $N$ is the number of turns, and $R_o$ and $R_i$ are the outer and inner radius of the toroidal core.

Note that if $\beta = 1$, the field factor simplifies as

$$\Delta = \frac{\mu N}{\pi\left(R_O^2 - R_i^2\right)} \tag{9}$$

Under this condition, (7) becomes

$$B_{eff}(t) = \frac{\mu N}{\pi\left(R_O^2 - R_i^2\right)} \cdot I(t). \tag{10}$$

which fully describes the case with a uniform magnetic field distribution. To properly compute the power loss $p$(t) given by (4), the actual values of $B_m$ and $B_{DC}$ must be cyclically updated. Since the proposed method works in the time-domain, the wipe out rule method is used: when the derivate of the magnetic flux density $dB_{eff}/dt$ is zero, a maximum $B_{max}$ or a minimum $B_{min}$ value of the actual hysteresis loop is reached and, therefore, the values of $B_{DC}$ and $B_m$ are updated for a correct estimation of the power loss.

This technique has an important improvement which is of fundamental importance to perform time-domain simulations: in iGSE technique the estimation of the core loss is based on the knowledge of the mean average value of the magnetic flux $B(t)$, while the method described in this section uses the effective magnetic flux $B_{eff}$ which is estimated by using (7). Moreover, this approach can be used in a lumped elements circuit and, also, it allows the SPICE subsystem modeling the inductor to be integrated in a more complex circuit, as shown in Figure 1. A detailed description of the LTSpice circuit used to compute the time-domain core power loss can be found in [24].

In the next sections, starting from the measurements of a DC-DC Buck converter prototype, a comparison between the power core losses on the inductor using the Steinmetz Equation, the iGSE and the proposed approach is presented.

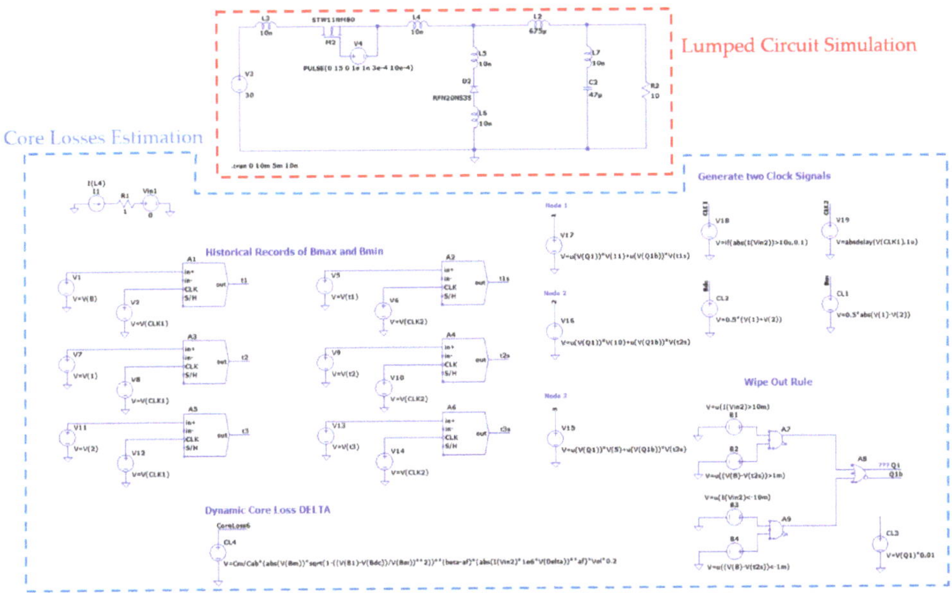


**Figure 1.** LTSpice circuit for Time-Domain Core Loss estimation based on wipe out rule [24]. The red dotted box includes the equivalent lumped circuit of the DC-DC converter. The blocks used for the real time core loss estimation are shown inside the blue box.

## 4. The Case Study: A DC-DC Buck Converter

To practically evaluate the accuracy of the three different techniques, the core loss on the inductor of a DC-DC Buck converter was analyzed. The electrical circuit of the buck converter is shown in Figure 2.

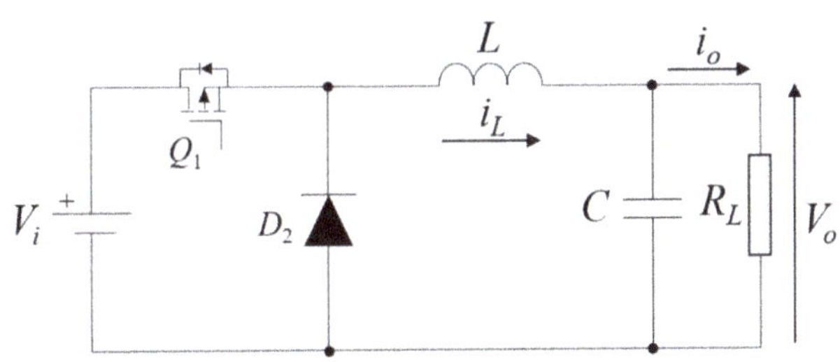


**Figure 2.** Buck converter unidirectional topology.

The KIT-CRD-3DD065P, Buck-Boost Evaluation Kit [31] was used to realize the experimental converter. The components are summarized in Table 1.

**Table 1.** Buck Converter components.

| Component | Description | Value |
|---|---|---|
| Power MOSFETs $Q_1$ | C3M0060065K | $V_{DS}{}^{max}$ = 650 V<br>$R_{DS(on)}$ = 60 mΩ |
| Body Diode $D_2$ | C3M0060065K | $V_F$ = 4.8 V |
| Output Capacitor $C$ | MAL205956479E3 | 47 µF |
| Load Resistance $R_L$ | HS100 1R J | 10 Ω |

Different operating frequencies and duty cycles were used to understand the performance of each core loss estimation technique and perform a comparison of their results.

Indeed, the proposed topology works with unidirectional behavior. This turns out in a simpler driving system but excludes the possibility to evaluate the magnetic losses of typical bidirectional topologies. However, from the point of view of the core losses, the resulting waveforms will still include both the CCM and DCM condition, resulting in a complete behavioral analysis of the phenomenon.

### 4.1. Inductor Characteristics

The inductor of Cree's KIT-CRD-3DD065P Buck-Boost Evaluation Kit [31] is based on a toroidal high temperature rated powdered core, which results in a CWS-1SN-12877 inductor; the core material is KoolMu [32]. The geometric characteristics of the core are summarized in Table 2 along with the winding number of turns.

**Table 2.** Inductor characteristics.

| Component | Description |
|---|---|
| Number of Turns $N$ | 63 |
| Inner Core Radius $R_i$ | 10.5 mm |
| Outer Core Radius $R_o$ | 20.5 mm |
| Height $H$ | 10 mm |

The anhysteretic curve of the magnetic core and the magnetic permeability are shown in Figure 3a,b, respectively.

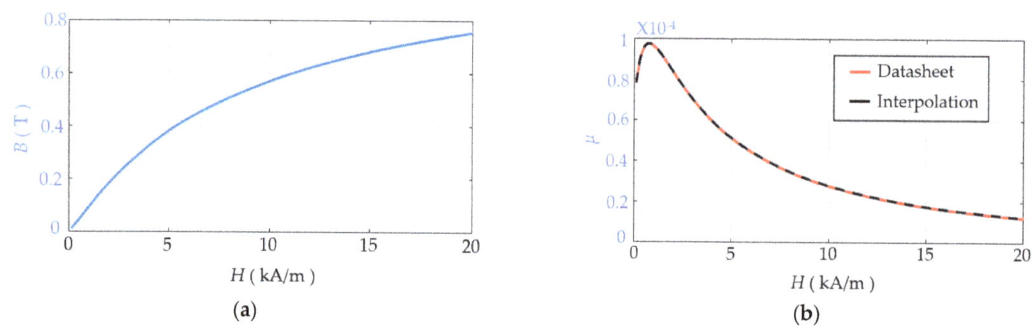


**Figure 3.** Magnetic characteristics of the toroidal core. (**a**) BH curve and permeability. (**b**) Magnetic permeability curve.

The core Steinmetz coefficient are summarized in Table 3.

**Table 3.** Steinmetz parameters from manufacturer.

| Parameter | Value |
|---|---|
| $C_m$ | 44.30 |
| $\beta$ | 1.988 |
| $\alpha$ | 1.541 |

By using in (5) the parameter values shown in Table 2, a coefficient $C_{\alpha\beta} = 8.51$ is obtained. The curve giving the magnetic core permeability $\mu$ shown in Figure 3 can be interpolated and expressed as a function of the magnetic field as follows

$$\mu = \frac{N(H)}{D(H)} = \frac{n_1 H^3 + n_2 H^2 + n_3 H + n_4}{d_1 H^3 + d_2 H^2 + d_3 H + d_4} \tag{11}$$

where, $n_1 = 1.7650 \times 10^{-17}$, $n_2 = -1.8125 \times 10^{-12}$, $n_3 = 2.551 \times 10^{-7}$, $n_4 = 6.379 \times 10^{-5}$, $d_1 = 1.4782 \times 10^{-11}$, $d_2 = 5.520 \times 10^{-7}$, $d_3 = 0.0017$, and $d_4 = 1$. As shown in Figure 3b, where the black dotted trace represents the plot of (11) and the red trace shows the real magnetic permeability the interpolation given by (7) perfectly matches the data sheet plot of the permeability.

### *4.2. Core Loss Estimation Algorithms*

The Steinmetz procedure used to compute the core loss is summarized in the block diagram shown in Figure 4. The magnetic field $H$ is deduced from the number of turns and the average axis of the toroidal core. Then, the magnetic field density $B(t)$ is computed from the BH curve, and its RMS value is used in (1) to estimate the power core loss.

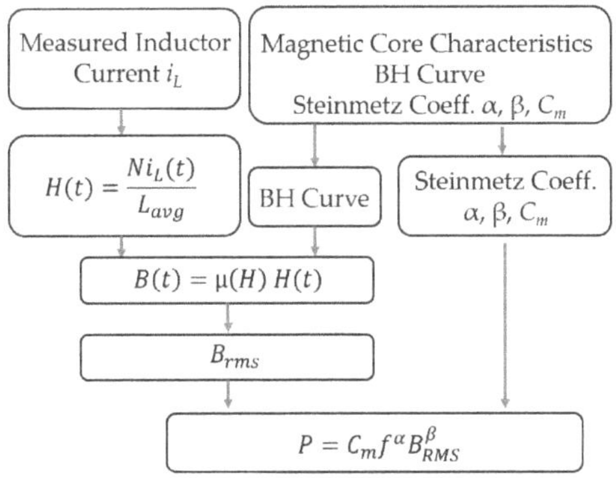


**Figure 4.** Core loss estimation using classical Steinmetz Equations.

The core loss procedure used by the iGSE technique is shown in Figure 5. The magnetic flux density and its derivative are calculated starting from the current $i_L$ and the BH curve. The Steinmetz coefficients are used to calculate the coefficient $k_i$ according to (3) and, finally, the core loss density is computed [29].

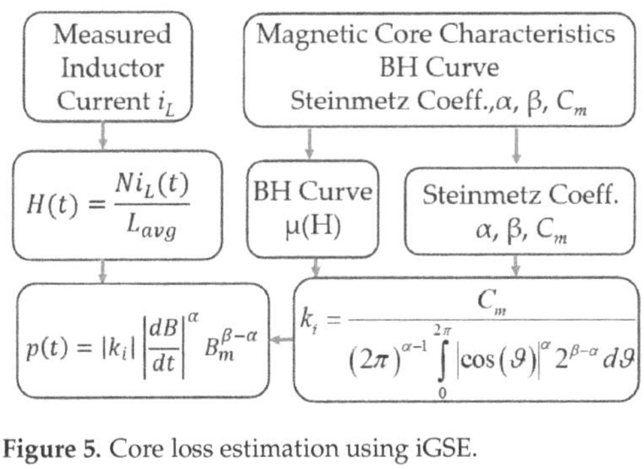


**Figure 5.** Core loss estimation using iGSE.

The procedure for the Time-Domain Core Loss density computed using the approach proposed in [23,24] is shown in Figure 6. As already discussed, the field factor is calculated from the Steinmetz coefficients and the core geometry. Then, the effective magnetic flux density is calculated from the current. Finally, the core loss density is computed by using (4).

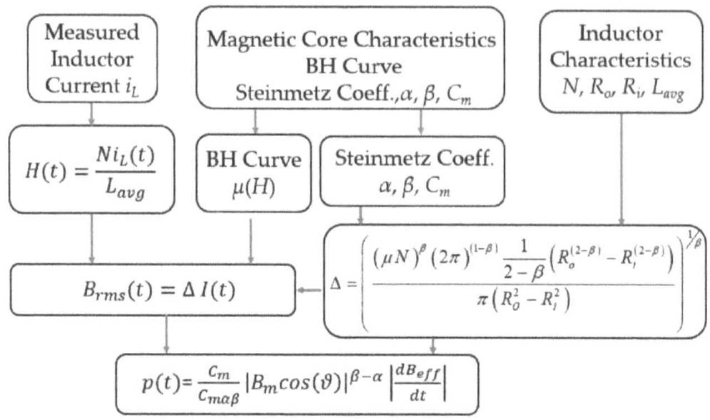


**Figure 6.** Core loss estimation using non-uniform magnetic field density.

### 4.3. *Lossy Magnetic Hysteresis Cycle Reconstruction*

The induction *B(t)* computed by iGSE and TDNU is in phase with the inductor current *I (t)*. According to Lenz law, the inductor voltage $v_L(t)$ is in quadrature with the current, resulting in a null average power loss. This means that the B-H trajectory would present no hysteresis. This is in conflict with the actual losses that are estimated by the two methodologies. To resolve this conflict, an additional artificial current term $i_{LOSS}$, in phase with the inductor voltage, must be considered. This term can be determined assuming an equivalent *R-L* parallel circuit model such as the one shown in Figure 7.

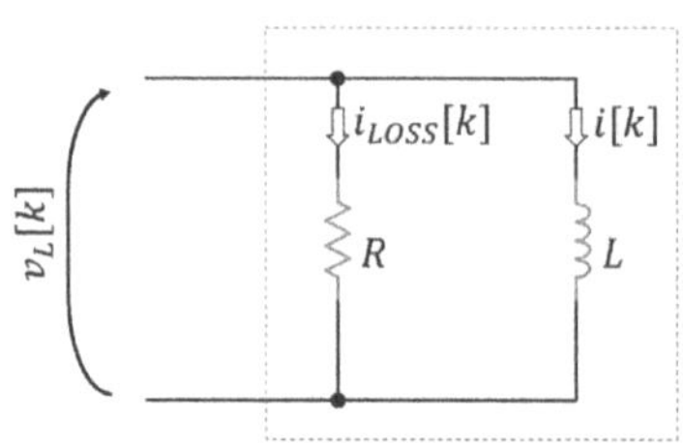


**Figure 7.** Equivalent lossy circuit model for H-B reconstruction.

In this circuit, the inductor is ideal, and the instantaneous power related to the core losses is absorbed by the resistive element. Expressing the quantities in a time-discrete domain, a loss current $i_{LOSS}[k]$ can be computed by the ratio between the computed losses $p[k]$ and the instantaneous voltage across the inductor $v_L[k]$. The latter can be computed from the numerical expression of the Lentz law. Although the magnetic field $H$ is not measured in this setup, it is possible to assume that, together with the induction $B$, it should accommodate the instantaneous value of the losses.

$$v_L[k] = NS_t \frac{B[k] - B[k-1]}{t[k] - t[k-1]} \tag{12}$$

$$i_{LOSS}[k] = \frac{p[k]}{v_L[k]} \tag{13}$$

where $S_t$ is the cross-section of the toroid. From the loss current, the loss-affected $H$ field can is derived as

$$H[k] = (i_{LOSS}[k] + i[k])\frac{N}{\lambda} \tag{14}$$

where $\lambda$ is the magnetic path length of the toroid core.

## 5. Measurements and Simulation Results

The proposed methodology is validated, both in experimental and simulated environment, through a series of different tests, aimed at assessing the consistency of the three techniques. In the first test, the inductor current used for the different loss estimation methodologies is acquired from an LTSpice simulation. In the second test, the inductor current is measured on the real DC-DC Buck converter. For both tests, the four operating conditions described in Table 4 are considered to explore different current waveforms of the inductor. These operating conditions allow to compare the core loss estimation under significative different operating conditions, considering both the CCM and DCM case. For each operating condition:

- The RMS losses are computed with three methodologies (SE, iGSE, TDNU);
- The instantaneous losses are computed with two methodologies (iGSE, TDNU);
- For the experimental data, the lossy B-H curve is reconstructed with two methodologies (iGSE, TDNU).

**Table 4.** Case studies operating conditions.

| Case | Frequency $f_s$ | Load Resistance $R_L$ | Duty Cycle $D$ |
|---|---|---|---|
| *I* | 10 kHz | 10 Ω | 0.5 |
| *II* | 1 kHz | 10 Ω | 0.3 |
| *III* | 1 kHz | 10 Ω | 0.5 |
| *IV* | 1 kHz | 10 Ω | 0.8 |

The DC-DC converter circuit model simulated by using LTSpice is shown in Figure 7. Simulations were performed as transient analysis with a minimum timestep of 10 ns to capture the high frequency non-linear dynamics of the switching components. For the same reason, parasitic inductances were added on the MOSFET, diode and output capacitor. The small timestep allowed a detailed reconstruction of the inductor current waveform, which is a critical aspect, because the current is directly related to $B$ as shown in Figure 8, and the $B$ is numerically differentiated, as shown in Figure 4, to compute the instantaneous power loss.

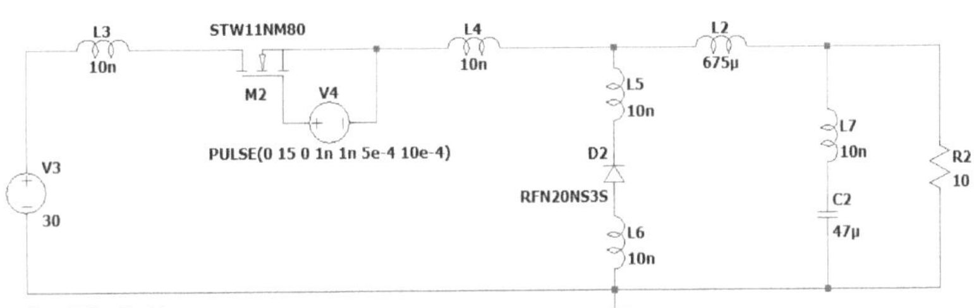


**Figure 8.** LTSpice simulation for the acquisition of the inductor current waveforms in different operating conditions of the power converter.

In Figures 9–12, the waveforms related to the Buck DC-DC converter operating in Case I, II, III and IV are shown, respectively. Each figure represents the inductor current, the instantaneous magnetic induction and the instantaneous losses.

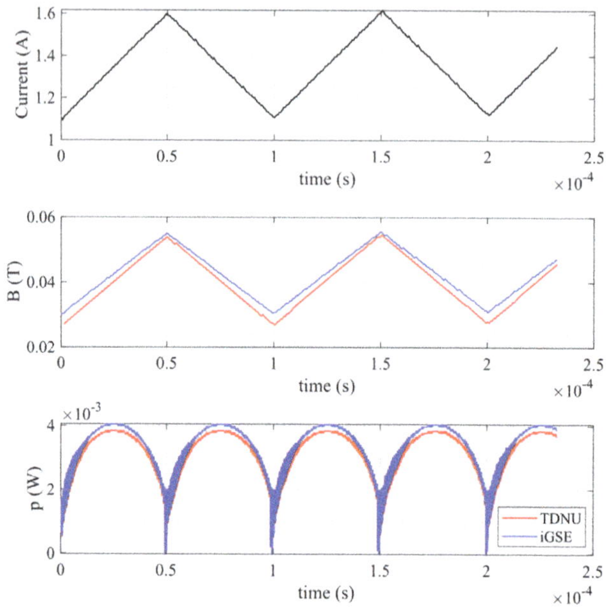


**Figure 9.** Simulated DC-DC waveforms for Case I ($f$ = 10 kHz $D$ = 0.5): Inductor current, instantaneous magnetic induction $B$, instantaneous losses $p$. Red and blue traces are the waveforms relative to the TDNU and iGSE estimation methods, respectively.

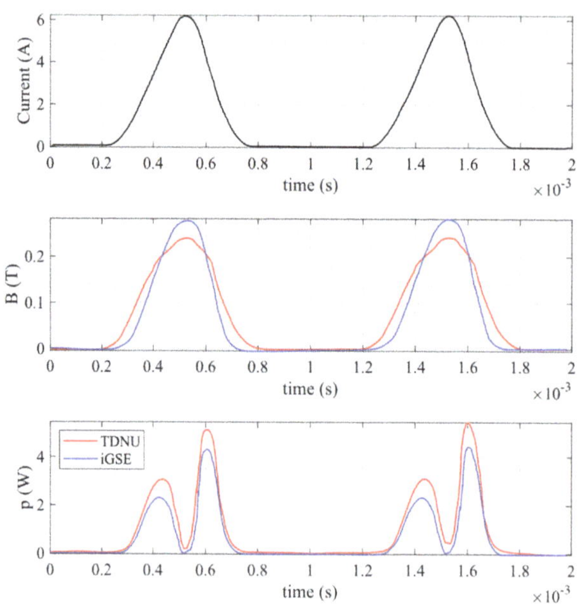


**Figure 10.** Simulated DC-DC waveforms for Case II ($f$ = 1 kHz $D$ = 0.3): Inductor current, instantaneous magnetic induction $B$, instantaneous losses $p$. Red and blue traces are the waveforms relative to the TDNU and iGSE estimation methods, respectively.

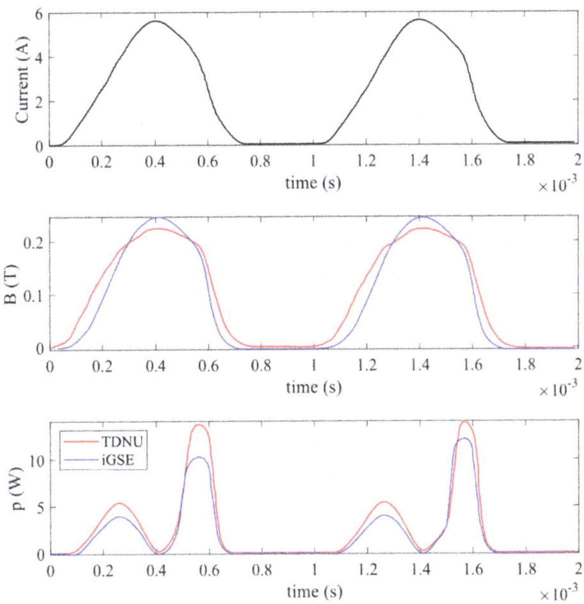


**Figure 11.** Simulated DC-DC waveforms for Case III ($f$ = 1 kHz $D$ = 0.5): Inductor current, instantaneous magnetic induction $B$, instantaneous losses $p$. Red and blue traces are the waveforms relative to the TDNU and iGSE estimation methods, respectively.

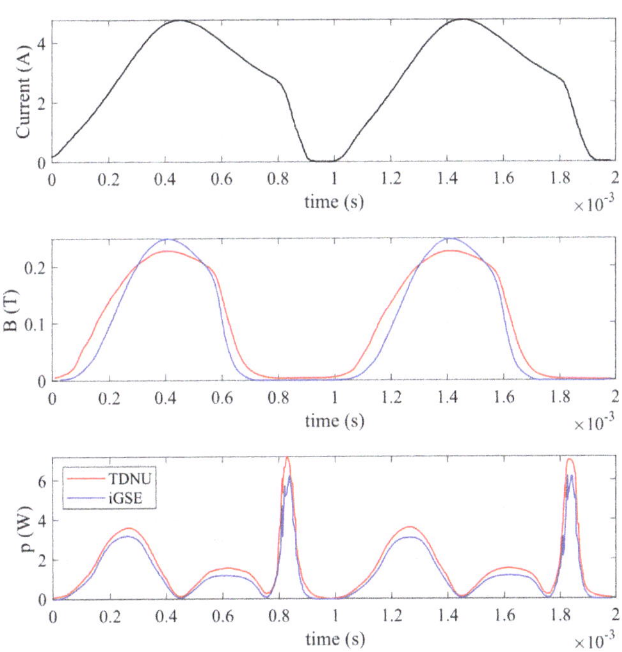


**Figure 12.** Simulated DC-DC waveforms for Case IV ($f$ = 1 kHz $D$ = 0.8): Inductor current, instantaneous magnetic induction $B$, instantaneous losses $p$. Red and blue traces are the waveforms relative to the TDNU and iGSE estimation methods, respectively.

Different measurements at different operating frequencies and duty cycles were performed and the test parameters are shown in Table 4. The converter input voltage was fixed to $V_i = 30$ V, to reproduce conditions analogous to those used in the simulation test. Figure 13 shows the experimental setup used to measure the inductor current.

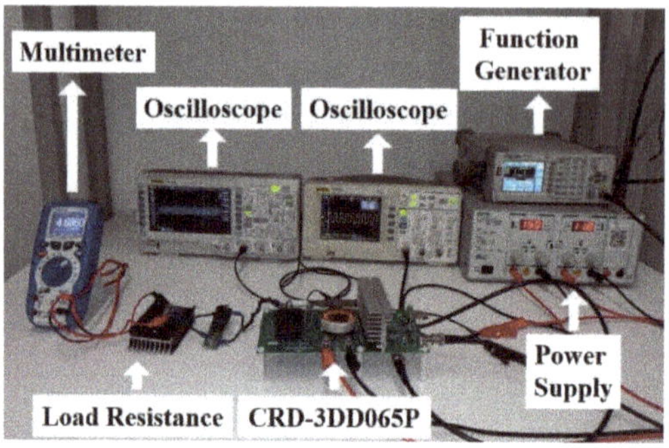


**Figure 13.** Experimental setup.

In Figures 14–17, the waveforms relative to the Buck DC-DC converter operating in Case I, II, III and IV are shown, respectively. Each figure represents the inductor current, the instantaneous magnetic induction field, the instantaneous losses, and the reconstructed lossy B-H profiles. In Table 5, the average losses for the four cases are compared between the methodologies. All the measured current waveforms are very close to the results predicted by the numerical simulations. The greatest difference between the simulated and measured data is for the Case I. This is mainly due to the highest operating frequency which increases the effect of the parasitic components. However, as shown in Table 5, the estimated core losses are consistent with those derived from the simulations. This leads to the conclusion that, even if the current waveforms might exhibit some differences, the computed core losses are not very sensible to these variations.

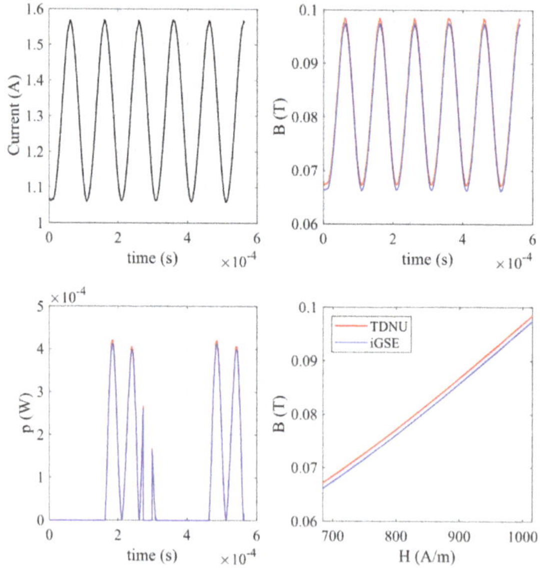


**Figure 14.** DC-DC waveforms for Case I ($f$ = 10 kHz $D$ = 0.5): Inductor current, instantaneous magnetic induction $B$, instantaneous losses $p$, reconstructed hysteresis loop. Red and blue traces are the waveforms relative to the TDNU and iGSE estimation methods, respectively.

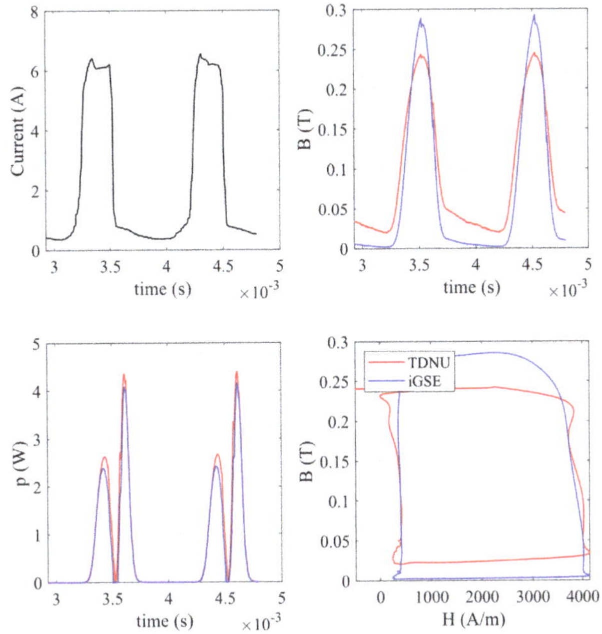


**Figure 15.** DC-DC waveforms for Case II ($f$ = 1 kHz $D$ = 0.3): Inductor current, instantaneous magnetic induction $B$, instantaneous losses $p$, reconstructed hysteresis loop. Red and blue traces are the waveforms relative to the TDNU and iGSE estimation methods, respectively.

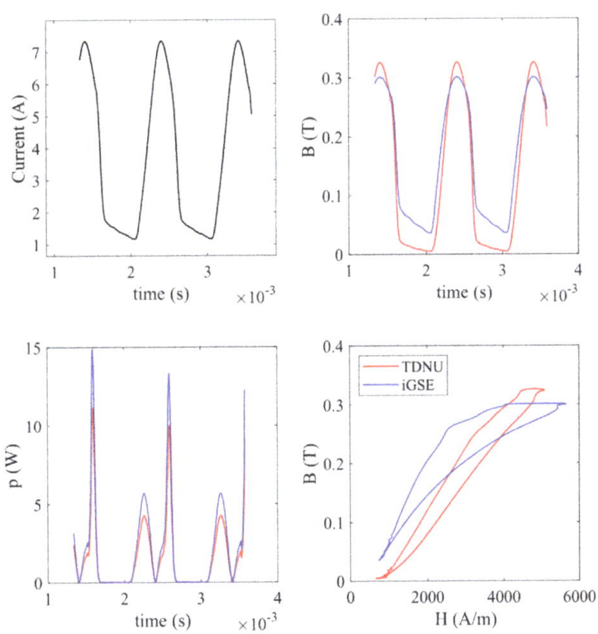


**Figure 16.** DC-DC waveforms for Case III ($f$ = 1 kHz $D$ = 0.5): Inductor current, instantaneous magnetic induction $B$, instantaneous losses $p$, reconstructed hysteresis loop. Red and blue traces are the waveforms relative to the TDNU and iGSE estimation methods, respectively.

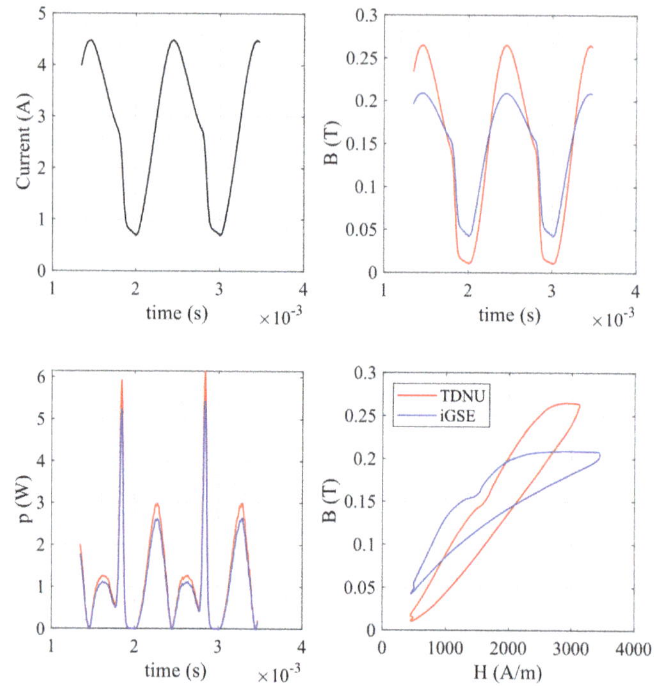


**Figure 17.** DC-DC waveforms for Case IV ($f$ = 1 kHz $D$ = 0.8): Inductor current, instantaneous magnetic induction $B$, instantaneous losses $p$, reconstructed hysteresis loop. Red and blue traces are the waveforms relative to the TDNU and iGSE estimation methods, respectively.

**Table 5.** Core losses in the four operating conditions for experimental inductor currents.

| Case | SE | iGSE | TDNU |
|---|---|---|---|
| I | 2.40 mW | 2.63 mW | 2.64 mW |
| II | 0.65 W | 0.63 W | 0.73 W |
| III | 0.58 W | 0.51 W | 0.62 W |
| IV | 2.40 W | 2.63 W | 2.64 W |

Table 6 shows the average losses determined by using the current computed through simulation for the considered cases and methodologies. Table 6 shows the core losses derived by using the experimentally measured currents. The comparison is discussed in the conclusion section.

**Table 6.** Core losses in the four operating conditions for simulated inductor currents.

| Case | SE | iGSE | TDNU |
|---|---|---|---|
| I | 2.36 mW | 2.53 mW | 2.58 mW |
| II | 0.59 W | 0.61 W | 0.71 W |
| III | 0.54 W | 0.49 W | 0.60 W |
| IV | 1.03 W | 1.06 W | 1.10 W |

## 6. Conclusions

In this paper, a novel methodology for core losses estimation was compared against two state-of-the-art approaches in the study of a DC-DC power converter. Core loss estimation in time-domain is difficult due to non-linear, dynamic and geometrical phenomena involving the magnetic material. Moreover, practical applications such as power converters usually involve non-sinusoidal excitations which further complicates the study. A unidirectional topology has been considered. The proposed methodology and the two comparison methodologies allowed the estimation of the core losses in the inductor of a DC-DC Buck converter by considering detailed magnetic behavior of the core. The obtained results showed that the TDNU methodology results in a core losses estimation comparable with the other methods, yet the estimations are usually slightly higher than others resulting from the compared methods, thanks to the ability of the TDNU to consider the field non-uniform distribution inside the core.

The proposed accurate core loss model leads to the possibility of including a real induction model inside the SPICE environment. This model is able to exhibit a consistent behavior considering core non-linearities. In fact, since the TDNU method is inherently a time-domain approach and is natively implemented in the form of a Spice circuit, it is a promising candidate to be included in larger circuit designs and can benefit from the optimized integration engines coupled with circuit simulation software.

**Author Contributions:** Conceptualization, E.C. and A.R.; methodology, F.C., A.L. and G.M.L.; software, F.C.; validation, A.R. and A.L.; investigation, F.C. and G.M.L.; data curation, F.C. and G.M.L.; writing—original draft preparation, F.C. and G.M.L.; writing—review and editing, E.C., A.L. and A.R.; supervision, E.C. and A.R. All authors have read and agreed to the published version of the manuscript.

**Funding:** This research received no external funding.

**Acknowledgments:** We would like to thank Arrow Europe for providing us the Cree CRD-3DD065P Evaluation board.

**Conflicts of Interest:** The authors declare no conflict of interest.

## References

1. Castellazzi, A.; Gurpinar, E.; Wang, Z.; Suliman Hussein, A.; Garcia Fernandez, P. Impact of Wide-Bandgap Technology on Renewable Energy and Smart-Grid Power Conversion Applications Including Storage. *Energies* **2019**, *12*, 4462. [CrossRef]
2. Millan, J.; Godignon, P.; Perpiñà, X.; Perez-Tomas, A.; Rebollo, J. A Survey of Wide Bandgap Power Semiconductor Devices. *IEEE Trans. Power Electron.* **2014**, *29*, 2155–2163. [CrossRef]
3. Loncarski, J.; Monopoli, V.G.; Leuzzi, R.; Ristic, L.; Cupertino, F. Analytical and Simulation Fair Comparison of Three Level Si IGBT Based NPC Topologies and Two Level SiC MOSFET Based Topology for High Speed Drives. *Energies* **2019**, *12*, 4571. [CrossRef]
4. Kolar, J.W.; Bortis, D.; Neumayr, D. The ideal switch is not enough. In Proceedings of the 2016 28th International Symposium on Power Semiconductor Devices and ICs (ISPSD), Prague, Czech Republic, 12–16 June 2016; pp. 15–22.
5. Detka, K.; Górecki, K.; Grzejszczak, P.; Barlik, R. Modeling and Measurements of Properties of Coupled Inductors. *Energies* **2021**, *14*, 4088. [CrossRef]
6. Musumeci, S.; Solimene, L.; Ragusa, C. Identification of DC Thermal Steady-State Differential Inductance of Ferrite Power Inductors. *Energies* **2021**, *14*, 3854. [CrossRef]
7. Salinas, G.; Serrano-Vargas, J.; Muñoz-Antón, J.; Alou, P. Thermal Resistance Matrix Extraction from Finite-Element Analysis for High-Frequency Magnetic Components. *Energies* **2021**, *14*, 3075. [CrossRef]
8. Mazgaj, W.; Sierzega, M.; Szular, Z. Approximation of Hysteresis Changes in Electrical Steel Sheets. *Energies* **2021**, *14*, 4110. [CrossRef]
9. TDK. EPCOS Data Book 2013. Ferrites and Accessories. Available online: https://www.tdk-electronics.tdk.com/download/519704/069c210d0363d7b4682d9ff22c2ba503/ferrites-and-accessories-db-130501.pdf (accessed on 9 October 2021).
10. Salas, R.A.; Pleite, J. Equivalent Electrical Model of a Ferrite Core Inductor Excited by a Square Waveform Including Saturation and Power Losses for Circuit Simulation. *IEEE Trans. Magn.* **2013**, *49*, 4257–4260. [CrossRef]
11. Corti, F.; Grasso, F.; Paolucci, L.; Pugi, L.; Luchetti, L. Circular Coil for EV Wireless Charging Design and Optimization Considering Ferrite Saturation. In Proceedings of the 5th IEEE International Forum on Research and Technologies for Society and Industry (RTSI), Florence, Italy, 9–12 September 2019; pp. 279–284. [CrossRef]
12. Locorotondo, E.; Pugi, L.; Corti, F.; Becchi, L.; Grasso, F. Analytical Model of Power MOSFET Switching Losses due to Parasitic Components. In Proceedings of the 2019 IEEE 5th International forum on Research and Technology for Society and Industry (RTSI), Florence, Italy, 9–12 September 2019; pp. 331–336.
13. Grasso, F.; Luchetta, A.; Manetti, S.; Piccirilli, M.C. Symbolic techniques in neural network based fault diagnosis of analog circuits. In Proceedings of the 2010 XIth International Workshop on Symbolic and Numerical Methods, Modeling and Applications to Circuit Design (SM2ACD), Gammarth, Tunisia, 4–6 October 2010; pp. 1–4.

14. Papamanolis, P.; Guillod, T.; Krismer, F.; Kolar, J.W. Minimum Loss Operation and Optimal Design of High-Frequency Inductors for Defined Core and Litz Wire. *IEEE Open J. Power Electron.* **2020**, *1*, 469–487. [CrossRef]
15. Guillod, T.; Papamanolis, P.; Kolar, J.W. Artificial Neural Network (ANN) Based Fast and Accurate Inductor Modeling and Design. *IEEE Open J. Power Electron.* **2020**, *1*, 284–299. [CrossRef]
16. Vidal, N.; Lopez-Villegas, J.M.; Del Alamo, J.A. Analysis and Optimization of Multi-Winding Toroidal Inductors for Use in Multilayered Technologies. *IEEE Access* **2019**, *7*, 93537–93544. [CrossRef]
17. Salvini, A.; Fulginei, F.R.; Coltelli, C. A neuro-genetic and time-frequency approach to macromodeling dynamic hysteresis in the harmonic regime. *IEEE Trans. Magn.* **2003**, *39*, 1401–1404. [CrossRef]
18. Herceg, D.; Chwastek, K.; Herceg, Đ. The Use of Hypergeometric Functions in Hysteresis Modeling. *Energies* **2020**, *13*, 6500. [CrossRef]
19. Kovacevic, I.F.; Friedli, T.; Musing, A.M.; Kolar, J.W. Full PEEC Modeling of EMI Filter Inductors in the Frequency Domain. *IEEE Trans. Magn.* **2013**, *49*, 5248–5256. [CrossRef]
20. Saeed, S.; Georgious, R.; Garcia, J. Modeling of Magnetic Elements Including Losses—Application to Variable Inductor. *Energies* **2020**, *13*, 1865. [CrossRef]
21. Mu, M.; Zheng, F.; Li, Q.; Lee, F.C. Finite element analysis of inductor core loss under DC bias conditions. *IEEE Trans. Power Electron.* **2013**, *28*, 4414–4421. [CrossRef]
22. Lin, D.; Zhou, P.; Fu, W.; Badics, Z.; Cendes, Z. A Dynamic Core Loss Model for Soft Ferromagnetic and Power Ferrite Materials in Transient Finite Element Analysis. *IEEE Trans. Magn.* **2004**, *40*, 1318–1321. [CrossRef]
23. Cui, H.; Ngo, K.D.T. Transient Core-Loss Simulation for Ferrites With Nonuniform Field in SPICE. *IEEE Trans. Power Electron.* **2018**, *34*, 659–667. [CrossRef]
24. Corti, F.; Reatti, A.; Cardelli, E.; Faba, A.; Rimal, H.P. Improved Spice Simulation of Dynamic Core Losses for Ferrites With Nonuniform Field and Its Experimental Validation. *IEEE Trans. Ind. Electron.* **2021**, *68*, 12069–12078. [CrossRef]
25. Steinmetz, C.P. On the law of hysteresis. *Proc. IEEE* **1984**, *72*, 197–221. [CrossRef]
26. Antonio, S.Q.; Faba, A.; Rimal, H.P.; Cardelli, E. On the Analysis of the Dynamic Energy Losses in NGO Electrical Steels under Non-Sinusoidal Polarization Waveforms. *IEEE Trans. Magn.* **2020**, *56*, 6300115.
27. Antonio, S.Q.; Lozito, G.M.; Ghanim, A.M.; Laudani, A.; Rimal, H.; Faba, A.; Chilosi, F.; Cardelli, E. Analytical formulation to estimate the dynamic energy loss in electrical steels: Effectiveness and limitations. *Phys. B Condens. Matter* **2020**, *579*, 411899. [CrossRef]
28. Lee, J.-I.; Shin, K.-H.; Bang, T.-K.; Kim, K.-H.; Hong, K.-Y.; Choi, J.-Y. Core-Loss Analysis of Linear Magnetic Gears Using the Analytical Method. *Energies* **2021**, *14*, 2905. [CrossRef]
29. Venkatachalam, K.; Sullivan, C.R.; Abdallah, T.; Tacca, H. Accurate prediction of ferrite core loss with nonsinusoidal waveforms using only Steinmetz parameters. In Proceedings of the 2002 IEEE Workshop on Computers in Power Electronics, Mayaguez, PR, USA, 3–4 June 2002; pp. 36–41. [CrossRef]
30. Rimal, H.P.; Ghanim, A.M.; Antonio, S.Q.; Lozito, G.M.; Faba, A.; Cardelli, E. Modelling of dynamic losses in soft ferrite cores. *Phys. B Condens. Matter* **2020**, *579*, 411811. [CrossRef]
31. Wolfspeed. Kit-CRD-3DD065P. Available online: https://www.wolfspeed.com/kit-crd-3dd065p (accessed on 9 October 2021).
32. Mag-Inc. Powder Cores. Kool-Mu Cores. Available online: https://www.mag-inc.com/Products/Powder-Cores/Kool-Mu-Cores/Kool-Mu-Material-Curves (accessed on 9 October 2021).

*energies* MDPI

*Article*

# Multicell Power Supplies for Improved Energy Efficiency in the Information and Communications Technology Infrastructures

**Michael Chrysostomou [1,*], Nicholas Christofides [1], Stelios Ioannou [2] and Alexis Polycarpou [1]**

[1] Department of Electrical Engineering, Computer Engineering and Informatics, Frederick University, Limassol 3080, Cyprus; n.christofides@frederick.ac.cy (N.C.); eng.pa@frederick.ac.cy (A.P.)
[2] Department of Electrical and Electronic Engineering, University of Central Lancashire, Preston PR1 2HE, UK; SIoannou2@uclan.ac.uk
* Correspondence: st009893@stud.fit.ac.cy; Tel.: +357-96526526

**Abstract:** The rapid growth of the Information and Communications Technology (ICT) sector requires additional infrastructure, such as more micro-datacenters and telecom stations, to support the higher internet speeds and low latency requirements of 5G networks. The increased power requirements of the new ICT technologies necessitate the proposal of new power supplies, in an attempt to support the increase in energy demand and running costs. This work provides an in-depth theoretical analysis on the losses of the individual stages of commercially available PSU and proposes a new multicell PSU, the buck PFC converter, which offers a higher overall efficiency at varying load levels. The theoretical results are verified using simulation results, via a PSIM Thermal Module, and using experimental data. The results indicate that multicell structures can improve the overall PSU efficiency by 1.2% at 50% rated power and more than 2.1% at full power. Finally, taking into consideration the economic implications of this study, it is shown that the proposed multicell structure may increase the PSU costs by 10.78%, but the payback period is in the order of just 3.3 years.

**Keywords:** micro-datacenters; Information and Communications Technology (ICT); efficiency; multicell; power supply unit

**Citation:** Chrysostomou, M.; Christofides, N.; Ioannou, S.; Polycarpou, A. Multicell Power Supplies for Improved Energy Efficiency in the Information and Communications Technology Infrastructures. *Energies* **2021**, *14*, 7038. https://doi.org/10.3390/en14217038

Academic Editors: Ching-Ming Lai and Yitao Liu

Received: 5 October 2021
Accepted: 23 October 2021
Published: 27 October 2021



## 1. Introduction

The 5G network requirements for the Information and Communication Technology (ICT) sector has caused a continuous demand for growth in micro-datacenter and Telecommunication Stations (TS) [1,2]. Datacenters accommodate the ICT data storage and processing equipment such as servers, hubs, hard discs, and ICT racks, etc. A micro-datacenter (µDC) is a datacenter of size smaller or equal to one rack. The growth of datacenters and micro-datacenters is huge nowadays and they have a great impact on global electricity consumption. In 2016, more than 1.8% of the global electrical energy was consumed by datacenters and this will keep increasing [3]. The higher speeds of radio frequency bands (such as 4G and 5G) tend to have a shorter area coverage and as a result more TS are needed. Furthermore, 5G networks have increased energy requirements compared to previous generations, and therefore several researchers have concentrated on improving the energy efficiency of communication links [4–6]. The increased number of TS and µDC will unavoidably lead to an increased energy consumption.

The impact of the power supply system to the overall energy efficiency of datacenters and TS has pushed several energy agencies to develop strict efficiency requirements, such as Energy Star and EPRI [7,8]. In the last two decades, the telecommunications industry has adopted equipment supplied by 48 V DC. In the Information Technology (IT) and Datacenter industries though, the preferred voltage system over the last few decades was 230 V AC, using though integral secondary power supplies for converting the AC 230 V to 12 V DC. With the IT and Telecom industry convergence of the last few years, both AC 230 V and DC 48 V systems were used, causing a debate as to which system would dominate the market. Over the last few years, Facebook initiated the "open compute project" (and other organizations have joined, such as Google and Microsoft), for promoting the 48 V DC distribution topology for the IT industry, because of its improved performance [9,10].

Multicell structures have been utilized in the last two decades for several applications, but the main objectives were the improvement of the voltage gain in the input or the output of a circuit and the component stress decrease. There are four different topologies of multicell structures which are:

- Input in Series Output in Series (ISOS);
- Input in Series Output in Parallel (ISOP);
- Input in Parallel Output in Series (IPOS);
- Input in Parallel Output in Parallel (IPOP).

Aiming for higher efficiency, Kasper et al. [11] suggested a 6-cell converter system, which showed the highest efficiency conversion above 1 kW, however, the suggested converter [11] has great efficiency only in specific loading conditions. Furthermore, the 6-Cell ISOP converter creates additional complexity to the system and reduced reliability, since failure of one of the cells could lead to overall system failure, especially as cells are connected in series at the input. Multicell structures have been used in various studies but not primarily for energy efficiency improvement. Arezki Fekik et al. [12] proposes a three-cell DC to DC structure for Off-Grid PV systems for improved robustness using an ISOS structure but the study does not include an experimental verification and although simulation results are promising in terms of the maximum power point tracking (MPPT) function, the efficiency is not validated. Furthermore, the reliability of the three-cell DC to DC structure is reduced since failure of one of the three cells leads to overall system failure. ISOS structures in both AC to DC and DC to DC conversion stages do not significantly improve the efficiency and have much more complicated load sharing control [13,14]. The IPOS structure which is proposed in [15] in order to improve efficiency and output voltage, reduces the overall system efficiency (maximum efficiency of 91%) and increases the complexity of the system.

The paper analyzes and compares multicell topology structures (ISOS, IPOS IPOP or ISOP) and the way that these can lead to energy efficiency improvements by investigating each stage of conversion (rectification, power factor correction and DC to DC stage through LLC and PSFB converters). The work verifies through simulations and experimental measurements that IPOP structures can lead to an improved energy efficiency conversion and proposes a new multicell buck PFC structure to increase the energy conversion efficiency. Furthermore, the work presents a technoeconomic evaluation of multicell structures compared to single structure power supply units.

The Introduction Section of the work is followed by Section 2, which presents the structure of AC/DC 48 V power supplies. The section identifies and analyzes the three individual conversion stages and specifies single-stage and two-stage conversion alternative topologies. In Section 3, the multicell structure topologies are introduced and analyzed. This analysis is subsequently followed by Section 4, the experimental setup and verification. The proposed multicell buck PFC converter is presented in Section 5 and a technoeconomic analysis follows in Section 6. The last section of the paper, Section 7, is the conclusion.

## 2. Structure of AC to DC 48 V Power Supplies

Power supply (PS) systems tend to have low efficiency under low loading conditions. It is for this reason that modular PS systems are commonly used within TS and datacenters to overcome the problem of oversized PS systems. A DC PS system may be composed of a number of power supply units (PSU), frequently called "modules", that work in parallel to feed a DC bus, creating in this way a scalable PS system. A PSU consists of three main stages: (a) The AC to DC rectification, (b) the power factor correction (PFC) stage, and (c) the DC to DC stage to step-down the voltage to the desired 48 V DC, Figure 1. A μDC should have a capacity of up to 20 kW and thus a modular μDC should have PSU modules of less than 10 kW (at least 2 modules) $P_{module} \leq (P_{rack\text{-}max}/2)$. The rectifier stage of the

PSU consists of a bridge rectifier with switches (MOSFETs, IGBT, Thyristors, etc.) or diodes. The power factor correction system is used to ensure conformity with EN61000 and that there are no excessive harmonics in the AC supply from the UPS system. The third stage of a PSU is needed to step-down the voltage from the rectification stage to the desired 48 V DC. The rectification is common in all systems but there are two options used to produce the desired 48 V DC and to satisfy the input power quality requirements. The first option is to use a two-stage process with boost PFC and DC to DC step-down converter. The second option is to use the buck PFC single-stage circuit solution (PFC and DC to DC step-down circuits combined). In a single-stage conversion circuit it is very difficult to control both the input current quality and the output voltage level at different loads, and at the same time achieve high conversion efficiency. The two-stage conversion is therefore preferred as a solution for loads above 1 kW [16–18].

$$P_{total-loss} = P_{rec} + P_{pfc-loss} + P_{dc-dc\ loss} \quad (1)$$

$$P_{rec} = 2 \cdot V \cdot f \cdot I_{in} \quad (2)$$

$$P_{pfc-loss} = I_{sw}^2 \cdot R_{ds-On\ pfc} + I_{in-rms}^2 \cdot R_{ind} + P_{c.pfc\ loss} \quad (3)$$

$$P_{dc-dc\ loss} = P_{sw-loss} + P_{trans-loss} + P_{ind-loss} + P_{cap} \quad (4)$$

$$P_{trans-loss} = P_{tr-con-loss} + P_{no\ load\ loss} \quad (5)$$

where:

$P_{rec}$: power losses from the rectification stage;
$P_{pfc-loss}$: losses of the power factor correction stage;
$P_{dc-dc\ loss}$: losses of the DC-DC converter circuit;
$P_{sw\ loss}$: MOSFET switch losses of the DC converter circuit (conduction and switching losses);
$P_{trans-loss}$: transformer losses;
$P_{ind\ loss}$: inductor losses;
$P_{cap-loss}$: capacitor losses in DC-DC converter circuit;
$P_{tr-con-loss}$: transformer load and frequency dependent losses;
$P_{no\ load\ loss}$: transformer no load loss.

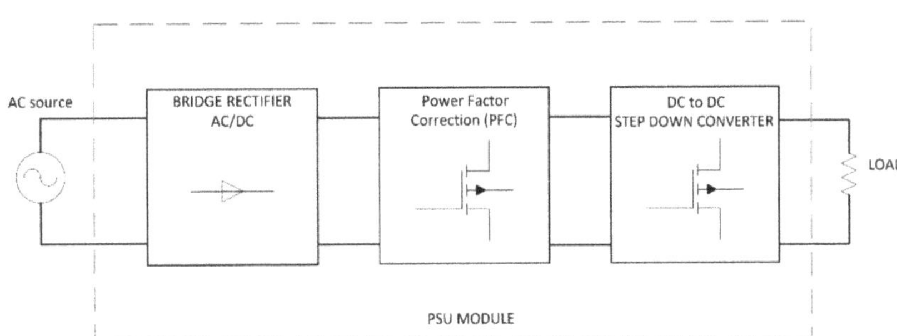


**Figure 1.** Block diagram of PSU.

## 2.1. *Single-Stage PSU*

The single-stage PSU is a buck PFC system with a bridge rectifier. Buck PFCs are used in low power applications since their structure creates dead angles at the input current waveform. Dead angles are generated from the switching of the input rectified waveform. Boost PFC circuits have the switching device in the output of the circuit and thus have higher THD, and are therefore preferred for high power applications [16,17]. On the contrary, since buck PFCs are only single-stage conversion systems, they present higher efficiency than boost converters at low power applications. Is not easy to remove dead angles in buck PFCs. Several re-searchers aimed to minimize their impact such as Saeed

Sharifi et al. [19], who introduced a new buck PFC circuit with a Z-source network that smooths out the switching effect on the input power quality, Figure 2 The Z-Source circuit was initially proposed by B. Axelrod et al. [20] with its main target to increase the voltage gain in buck converters. Fortunately, the new circuit introduces a new input inductor that not only increases the gain but also minimizes the stress on the switch and smooths out the input current dead zones.

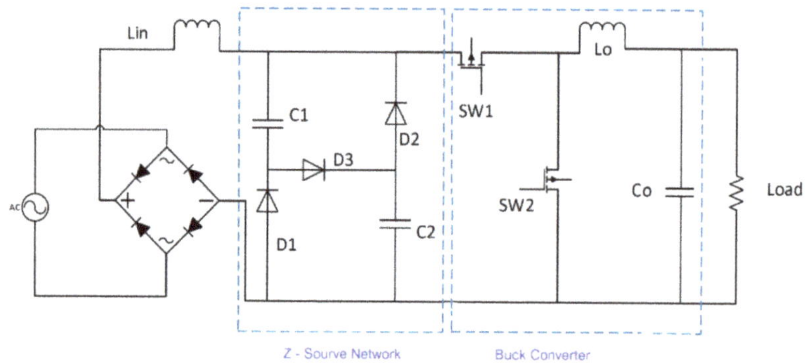


**Figure 2.** The Z source buck PFC.

### 2.2. Two-Stage PSU Solution

The two-stage solution includes the AC to DC boost PFC which converts the AC input voltage to 380–400 V DC and maintains input power quality, and a DC to DC step-down converter to step-down the voltage to the desired level. Due to imperfect switching (especially for high power applications), high switching losses occur in the converters. To step-down the voltage from 380–400 V to 48 V DC as previously described, a DC to DC step-down converter is needed. Since high energy efficiency and power density is required for such converter systems, the most commonly used low power (under 1 kW) converter circuit is the LLC half- or full-bridge converter [21–23]. These converters are resonant converters that adopt zero-voltage switching (ZVS) techniques in order to minimize switching losses, something that improves the efficiency of conversion. LLC half- or full-bridge converters, however, incorporate transformers in their structure, which creates no-load losses and therefore has low efficiency on low loads.

## 3. Multicell Structures

A multicell structure is a structure made of multiple converter systems (cells) which are put together and work in parallel or in series to form a new, larger converter system, as shown in Figure 3.

This work aims to prove that combining multiple converters (multicells) in series or in parallel can lead to overall energy efficiency improvements. Since conversion from AC 230 to DC 48 V includes more than one conversion step, multicell efficiency improvements need to be analyzed per stage (AC to DC, PFC stage and DC to DC step-down stage).

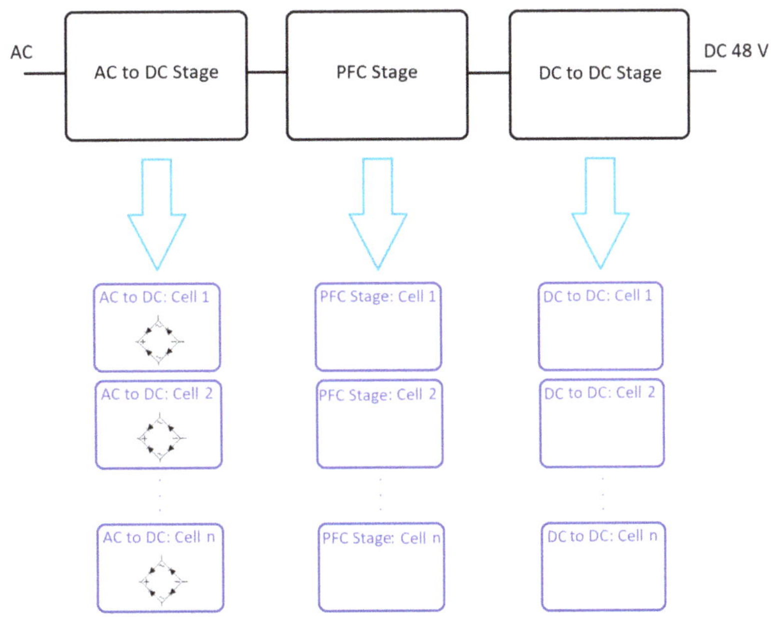


**Figure 3.** Block diagram of a multicell structure.

*3.1. The Multicell AC to DC Rectification Stage*

A multicell rectification can be constructed either with diodes or switches (MOSFETs or IGBTs, etc.), as previously mentioned. A strategy to minimize the losses in the rectification stage is to use parallel diode bridge rectifier systems [24]. Considering that each bridge circuit is a cell using multiple bridges, a multicell structure is formed. According to [24], parallel diode bridge rectifiers lead to lower losses because diodes with lower maximum current carrying capacity have less forward voltage drop, although this stands only for specific applications. The power losses are mainly generated as a result of the conduction and the switching of the diodes or the switches. Due to the low switching frequency (50 Hz or 60 Hz), the switching losses are almost negligible and the resultant losses of each of the two bridge rectifier legs arise mainly due to conduction. The use of IGBT switches would not benefit such systems since IGBT losses are similar and higher than diodes for similar power applications. GaN MOSFETs tend to have lower switching losses than silicon MOSFETs despite their higher conduction resistance (on-resistance $R_{DS-On}$). For the diode case, the conduction losses are given by:

$$P_{Dloss} = I_{rms}.V_d \tag{6}$$

where $V_d$ is the diode voltage drop and $I_{rms}$ is the average current passing through the bridge leg. For the MOSFET switch case, conduction losses are given by:

$$P_{Loss} = I_{rms}^2.R_{DS-On} \tag{7}$$

where $R_{DS-On}$ is the on-resistance of the MOSFET switch.

Connecting bridge rectifier cells in ISOS or IPOS would imply that the output voltage is a multiple of 325 V ($V_{peak}$ of AC input). This would require a very high gain to reduce it to 48 V and there would thus be higher power losses in the secondary circuits to step down the voltage. Additionally, a high voltage at the output of the bridge rectification stage would need switches that could withstand a higher voltage. Since $R_{DS-On}$ is proportional to the voltage, this would result in higher power losses. Therefore, the multicell bridge would not have better efficiency if connected in ISOS or IPOS. The ISOP structure with MOSFET switches would not improve the efficiency either, since the series switches would increase the conduction losses. Considering however that the breaking voltage is proportional to $R_{DS-On}$, switches that can withstand a lower voltage would have lower conduction losses. Replacing, therefore, one MOSFET switch with two switches of smaller $R_{DS-On}$, could in theory result in better efficiency. Examining however the characteristics of available MOSFET switches, those with breaking voltage of 50 V have an $R_{DS-On}$ as low as 3.3 mΩ (IFR3805), while switches with a breaking voltage of 200 V have an $R_{DS-On}$ as low as 4 mΩ (IXFK300N20X 3). Therefore, energy efficiency improvement could not be achieved with ISOP and MOSFETs for this type of power supply. Analyzing the circuit further, it can be seen that the ISOP structure of the bridge rectification part would not provide any energy efficiency improvement with diodes either, since in each cycle of the AC source only two switches/diodes conduct, as seen in Figure 4, as in the single cell structure, namely $D_{1a}$ and $D_{2b}$ in the positive cycle and $D_{3b}$ and $D_{4a}$ in the negative cycle. In the IPOP structure with diodes, the voltage drop increases as the number of cells increases. Efficiency is therefore slightly improved as a result of lower heat dissipation. For the case of the IPOP structure with MOSFETs, the voltage stress on the switches remains the same and the current passing through the system is divided in the parallel cell bridges, something that may lead to significant energy efficiency improvements.

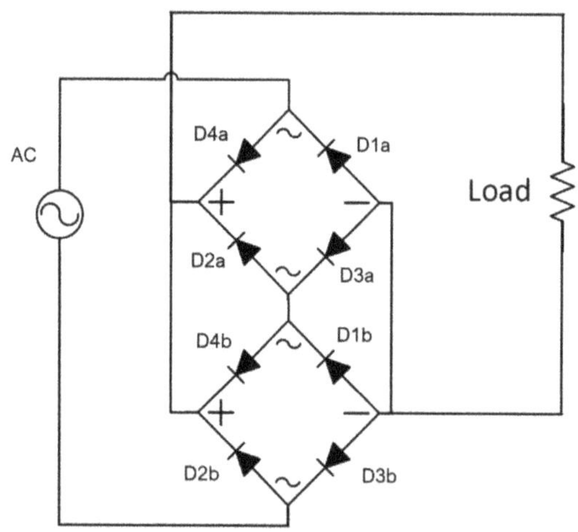


**Figure 4.** Multicell bridge ISOP circuit.

For instance, in a two-cell rectifier bridge system, the current in the two legs is halved (assuming MOSFETs with the same $R_{DS-On}$) and since $2\,(I/2)^2 < I^2$, it can be easily proven that parallel bridge structures lead to overall energy savings. For N cells, it can be shown that:

$$N\left[\frac{I}{N}\right]^2.R_{DS-On} < I^2.R_{DS-On} \tag{8}$$

## 3.2. The Multicell PFC Stage

There are different PFC structures. For single-stage converters, buck PFC can be used. For dual-stage converters, the boost PFC is widely used as well as some modified boost PFC circuits. All PFC circuits always follow the AC to DC rectification, or are incorporated into it.

### 3.2.1. Multicell Buck PFC

As discussed in the previous sections, to improve the power factor of the buck PFC circuits, the Z-source buck PFC circuit is proposed in [14] by adding a Z-source network in front of the buck converter. As seen in Figure 2, the circuit has two switches, SW1 and SW2. The losses arising in such a circuit are mainly as a result of conduction and switching, as well as due to the inductor at the input. Conduction losses are given by Equation (8) and the switching losses are given by Equation (9).

$$P_{sw} = V_{in} \cdot I_{out} \cdot f_{sw} \left( t_r + t_f \right) \tag{9}$$

where $f_{sw}$ is the switching frequency, $t_r$ is the rising time the transistor needs to switch on and $t_f$ is the falling time needed to switch off.

In a multicell parallel input configuration such as IPOP and IPOS, the input peak voltage would be 325 V DC (with ripple). For parallel cell circuits, the current in each of the parallel circuits would be split equally and the conduction losses would therefore decrease. The switching losses on the other hand cannot decrease. Considering the use of switches with a lower drain current, faster response, and thus lower $t_r$ and $t_f$, they could potentially lead to better efficiency. However, using switches with lower $t_r$ and $t_f$ and at the same lower maximum drain current capability may negatively affect the $R_{DS-On}$ (i.e., conduction losses).

Connecting the multiple cells in series would require higher gain at the output. As a result, bigger inductor and capacitor at the output would be required, which would lead to higher losses, especially on low loads. Therefore, the IPOP is the only parallel input configuration that should be examined. In a parallel configuration, the inductor winding losses are also less since the current passing through them is divided by the number of cells and, despite the fact that core losses of inductors can increase, they are negligible compared to winding losses. The inductor size in the split cells would not be affected, since the output and input voltages as well as the switching frequency remain the same. The equation for calculating the inductor size is given by:

$$L = V_{out} \cdot \frac{(V_{in} - V_{out})}{\Delta I_L} f_{sw} \cdot V_{in} \tag{10}$$

where $\Delta I_L$ is the ripple current and is proportional to the output current of each cell.

Connecting multiple buck PFC cells in series at the input (ISOP or ISOS) causes higher overall $R_{DS-On}$ resistance during the ON state. The switching losses, on the other hand, are multiplied by the number of cells but the voltage input is divided by the number of cells (assuming the case of exactly the same $R_{DS-On}$). The rising and falling times of the switches in the multicell ISOP and ISOS structures can be minimized since switches with lower voltage breaking capacity tend to have lower falling and rising times. Silicon MOSFETs can have $R_{DS-On}$ as low as 3.5 mΩ, with rising and falling times of about 170 nS (IXFN300N20X3), while GaN MOSFETs can have 12.4 ns rising and 24 ns falling time with $R_{DS-On}$ of 25mΩ (GS66516). On the other hand, inductor losses increase if the inductor size is not reduced from the single cell design. Since voltage gain is reduced in the ISOP and ISOS structures, the inductor size can also be significantly reduced. Therefore, ISOP or ISOS structures lead to reduced inductor losses but with insignificant switching loss improvement. The overall improvements are minor in relation to the complexity and increase in cost.

### 3.2.2. Multicell Boost PFC

The losses in the boost PFC circuit (as in the buck PFC) are mainly due to switching and conduction loss of the MOSFETs, the input inductor and the output capacitor. However, because of the higher voltage stress applied on the switches of the boost PFC compared with that applied on the buck PFC, higher conduction losses occur. An ISOS and ISOP structured boost PFC would lead to increased losses in MOSFET devices since they would be connected in series, even if lower $R_{DS-On}$ MOSFET switches are used. Since the inductors are in series in ISOP and ISOS structures, the overall DC resistance and hence inductor losses will increase. Considering the inductor size needed to form a multicell ISOP or ISOS structure compared to a single cell, it can be concluded that the ISOP case would lead to reduced losses compared to ISOS, but both cases would not lead to a significant reduction in the inductor's overall size and power losses (6), (7) and (8). The single cell boost PFC inductor value is given by [24]:

$$L = \frac{T}{\%Ripple} \cdot \frac{V_{ac-min}^2}{P_o} \cdot \left[1 - \frac{\sqrt{2} \cdot V_{ac-min}}{V_o}\right] \tag{11}$$

In a multicell ISOP boost PFC, the inductor value per cell is calculated by:

$$L = \frac{T}{\%Ripple} \cdot \frac{V_{ac-min}^2}{P_o \cdot N} \cdot \left[1 - \frac{\sqrt{2} \cdot V_{ac-min}}{V_o \cdot N}\right] \tag{12}$$

In a multicell ISOS boost PFC, the inductor value per cell is:

$$L = \frac{T}{\%Ripple} \cdot \frac{V_{ac-min}^2}{P_o \cdot N} \cdot \left[1 - \frac{\sqrt{2} \cdot V_{ac-min}}{V_o}\right] \tag{13}$$

For IPOS and IPOP structures as in the buck PFC case, the current is split between cells and, as previously mentioned, reduces both inductor and MOSFET losses.

### 3.3. The Multicell DC to DC Voltage Step-Down in Two-Stage Conversions

The PSUs used in the ICT industry are converting AC 230 V to DC 48 V, and therefore high gain DC to DC step-down conversion stage is needed. The most common type of circuit for high efficiency DC to DC step-down conversion is the resonant converter such as LLC [21,22] and phase shift full bridge converters (PSFB). The presence of the transformer in these circuits provides magnetic isolation to the connected load. The losses for these types of converter circuits are again down to the switches, the inductors, the capacitors and the transformers. These circuits use zero-voltage switching (ZVS) techniques to minimize the switching losses. However, these techniques do not influence the conduction losses. The losses because of the inductors and the transformers are a result of the core and $I^2$ R of windings. The core losses are given by:

$$P_{core} = P_{hysterisis} + P_{eddy} \tag{14}$$

$$P_{hysterisis} = K_h f B^n V_e \tag{15}$$

$$P_{eddy} = K_e f^2 B^2 V_e t^2 \tag{16}$$

where $K_e$ and $K_h$ are the eddy and hysteresis loss constants, $P_{eddy}$ is the eddy current power loss (W), $B$ is the flux density (Wb/m$^2$), $f$ is the frequency of magnetic reversals per second (Hz), $t$ is the core material thickness (m), $V$ is the volume of core (m$^3$) and $n$ is the Steinmetz exponent (ranging from 1.5 to 2.5 depending on the material).

Core losses are not proportional to the current passing through the coils. In inductors, the core losses are proven to be negligible compared to winding losses. In transformers, however, the larger the core size, the higher the core losses, even without any load. Paralleling transformers would reduce winding losses but would increase core losses and the overall losses would therefore also increase.

## 4. Experimental Verification of Multicell Structures

This experimental verification work validates the use of parallel converter cells within a PSU system for improved overall energy efficiency. To experimentally validate this proposal, a small-scale experimental setup was implemented with 3 buck conversion cells of a maximum power output of 18 W, input voltage of 24 V DC and output 5 V DC, connected in an IPOP structure.

Initially the setup was simulated in PSIM thermal module (version 12.0.3.) for comparison of theoretical and experimental results. The setup consists of 3 buck converter cells, a 3-pole switch for switching of the 3 parallel cells, variable loads, and two power meters measuring input and output voltage, and current and power. The buck cells used are commercially available and they employ the LM2596 controller. A photo of the experimental setup is shown in Figures 5 and 6, showing the respective schematic diagram. The input power meter used is a Lucas-Nuelle analog/digital multimeter, which is the wattmeter and power-factor meter model SO5127-1Z, which has a measurement accuracy of 2%, and a capability to work with voltages of up to 600 V and a current range of 0–20 A. The meter used in the output of the multicell conversion system is a Fluke 345 PQ meter with an accuracy in current measurements of ±0.2 A or ±1.5% and an accuracy of voltage measurements of ±1%. It is worth mentioning here that since the measurements with 3 cells, 2 cells and 1 cell are comparative, the efficiency improvement could easily be verified. As it can be seen in Figure 5, for 17 W load output, the power input with 3 buck cells is 22.8 W. For instance, at the same load (of 17 W) with 2 cells, the input power was 23.1 W. The voltage input to the 3 cells was 23.56 V and current input 0.97 A.

The efficiency of the single buck cell at different load conditions was measured, graphically analyzed and used as a baseline for energy efficiency. Following the experimental measurements with one buck converter cell, two and then subsequently three similar buck converter cells were used. Repeating the same methodology and measurement strategy, the load-efficiency curve was constructed, as seen in Figure 7. It is worth mentioning that the no-load consumption of one, two and three PFC circuits have an effect on the load-efficiency curves. As it can be seen in Figure 7, for loads corresponding to 10–20% of the full load (i.e., 1.8–3.6 W), the efficiency of one buck converter cell is higher than in the case with two and three buck converter cells. For instance, on 2 W load, the efficiency of 1 cell was 72.76% and with 3 cells it was 70.63%. Additionally, on 3.6 W load, the efficiency with 1 cell was 72.5%, with 2 cells 72.32% and with 3 cells 72.07%. This is due to the no-load losses of the PSU modules arising mainly from the inductors and the control circuitry. The no-load loss of a single buck converter cell is found to be 0.1904 W, that of two cells was equal to 0.238 W and for three cells 0.3094 W.

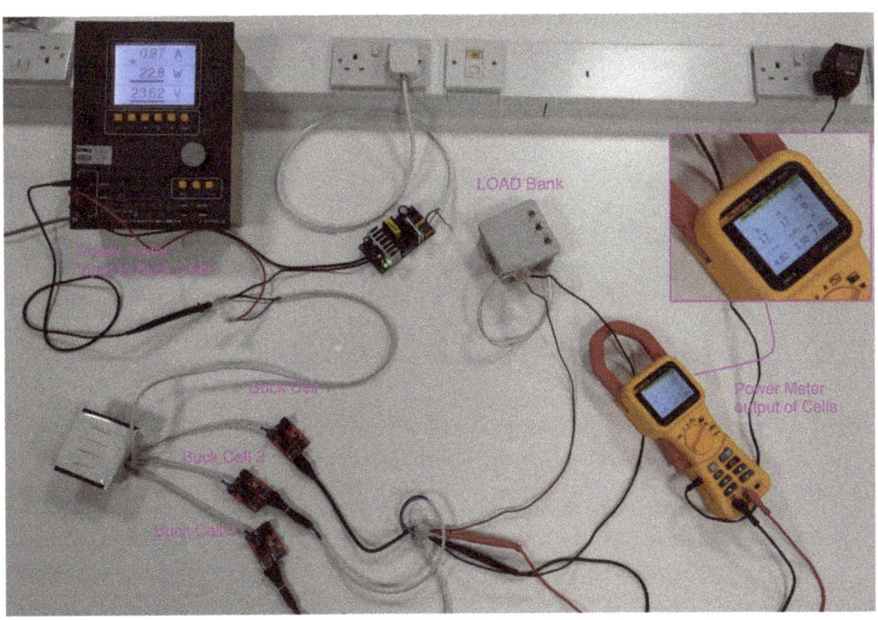


**Figure 5.** Experimental setup with parallel buck converter cells.

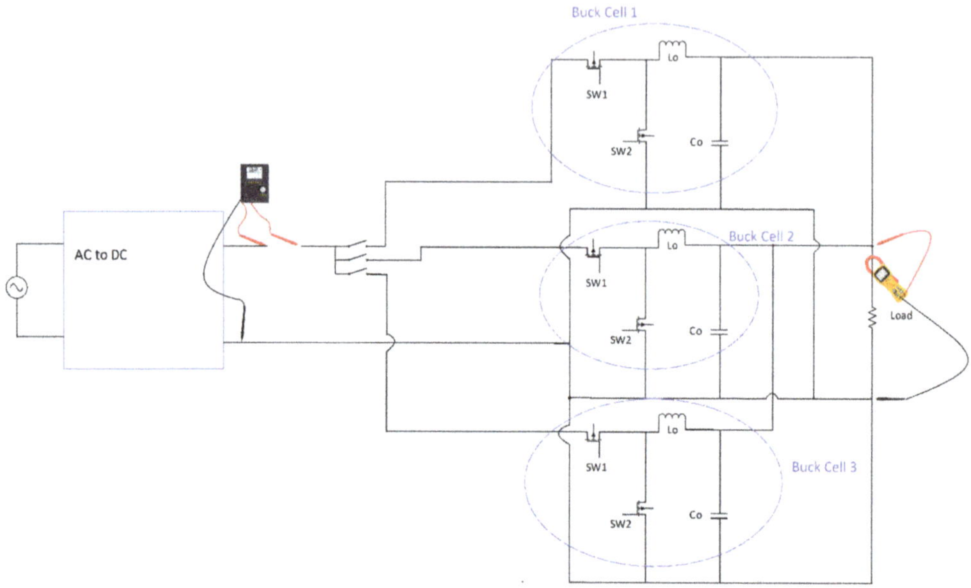


**Figure 6.** Circuit diagram of experimental setup.

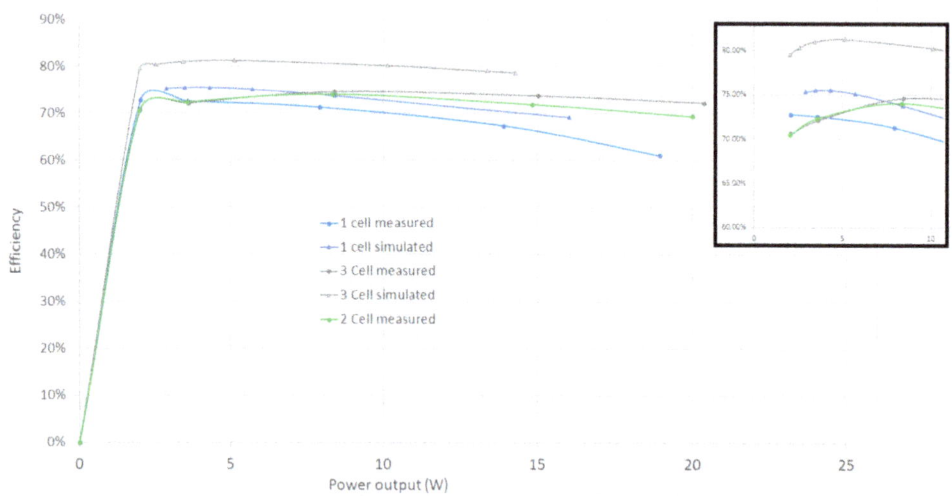


**Figure 7.** Efficiency comparison of 1-cell and 3-cell multicell buck converters.

It can be theoretically and experimentally shown that when using parallel cells of the same converter type, size and brand, the load is equally shared between the number of parallel cells. Power converters or PSUs are voltage-controlled systems in which the controller must maintain the output voltage level within certain values for different loading conditions. Either in continuous conduction mode (CCM) or discontinuous conduction mode (DCM), the small signal model of the buck converter in the secondary (output) side of the converter consists of DC sources and a series impedance/resistance. The voltage and resistance of CCM and DCM modes is of course different when analyzing the small signal model. The equivalent simplified circuit for DC power supplies proposed from this work is shown in Figure 8.

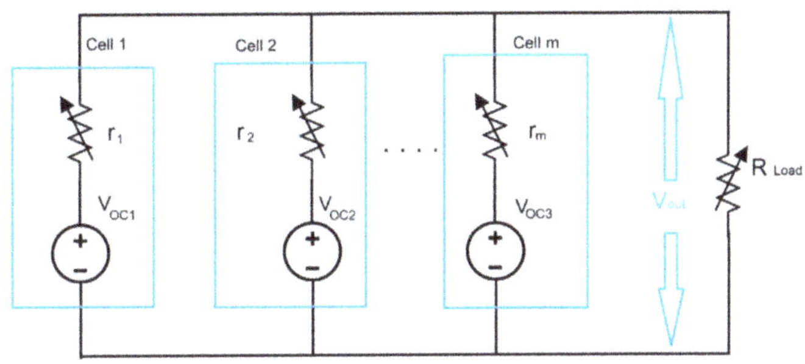


**Figure 8.** Equivalent simplified circuit of parallel DC cells connected to a common variable load.

Using the circuit proposed, the equation of output power as a function of output voltage, load and internal resistance can be derived for different cells and is given by:

$$P_{out-m} = \frac{V_{oc-m} - V_{out}}{r_m} \tag{17}$$

For cells which are multiple and dissimilar in size and type, it can be shown that connecting "w" number of cells of type A and "q" number of cells of type B in parallel supplying a common load, the voltage at the output is given by:

$$V_{out} = \frac{\left[w\,\frac{V_{oc1}}{r_1} + q\,\frac{V_{oc2}}{r_2} - P_{out}\right]}{\frac{w}{r_1} + \frac{q}{r_2}} \tag{18}$$

where: $V_{oc1}$ and $V_{oc2}$ are the cell type A and cell type B, open circuit voltage, $V_{out}$ is the output voltage of the system, $r_1$ is the internal resistance of type A buck-cell and $w$ & $q$ are the number of buck-cells of type A and B connected to the system, respectively. The power output of the system consisting of "w" cells of type A and "q" cells of type B is given by:

$$P_{out} = w.P_{out1} + q.P_{out2} \tag{19}$$

It is obvious from (12), (13), and (14), that each cell shares a portion of the output load based on the internal resistance, the open circuit voltage and the total load. For IPOP structures with the same type of converter cells, (in ideal conditions) it can be realized from equation (12) that the output power of each cell is equal. Measurements results for load sharing between cells are shown in Table 1. The current was measured using three clamp-on ammeters, which were the brand and type CEM, DT-362. The input power is measured through a Lucas-Nuelle analog/digital multimeter called SO5127-1Z. The experimental set-up is shown in Figure 9.

**Table 1.** Load sharing between three buck cells.

| Power Input Cells (W) | Input Current to Cell 1 (A) | Input Current to Cell 2 (A) | Input Current to Cell 3 (A) |
|---|---|---|---|
| 10.9 | 0.16 | 0.17 | 0.17 |
| 21 | 0.31 | 0.33 | 0.3 |
| 30.7 | 0.46 | 0.46 | 0.43 |

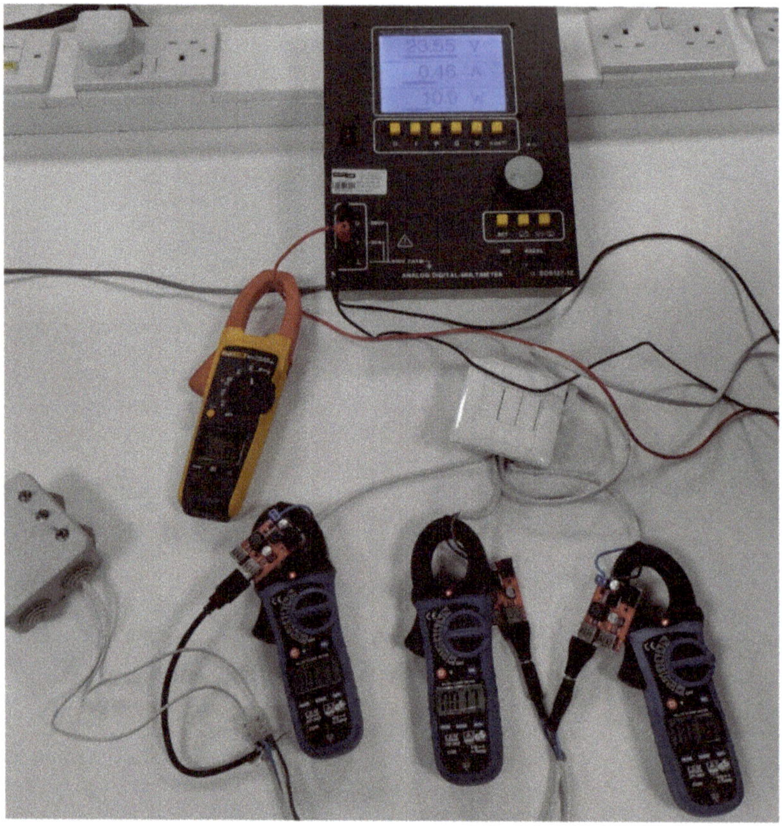

**Figure 9.** Experimental set-up of load sharing between parallel buck cells.

The no-load losses could not be measured by PSIM and this is why, in Figure 7, the simulation results show higher efficiency in all load levels compared with the respective experimental ones. Under full load, an impressive 11.18% higher efficiency was measured with three cells compared to one cell. In addition to no-load losses, PSIM is unable to measure the capacitor losses, control system consumption and other losses arising from cabling or connections, etc. Most industrial power supply units require high frequency transformers for isolation purposes that generate, however excessive, no-load losses. A single centralized transformer should be used in PSUs adopting the multicell circuitry structure. If each cell uses its own transformer, this would result in increased losses and lower overall conversion efficiency. This can be proven experimentally by combining in parallel three complete PSUs, including their individual transformer, and measuring the efficiency for the different loads. Three Eltek Valere Micro pack 250 W, 230 V AC input/48 V DC output PSUs are used to perform this experiment. The input to the three PSUs was measured using a Lucas-Nuelle analog/digital multimeter measuring the input current, applied voltage and consumed power of the system. Each of the three PSU outputs was connected to "dummy loads" of different load values whilst measuring the output current and voltage. For every different load case, input and output parameters were measured so that the efficiency could be subsequently evaluated. The plot of efficiency versus load output is shown in Figure 10, from which is it seen that the efficiency decreases as the number of PSUs connected in parallel increases. The no-load consumption of the PSUs (mainly as a result of the transformers) was measured and found to be 5 W per PSU (i.e., 2% of efficiency at full load).

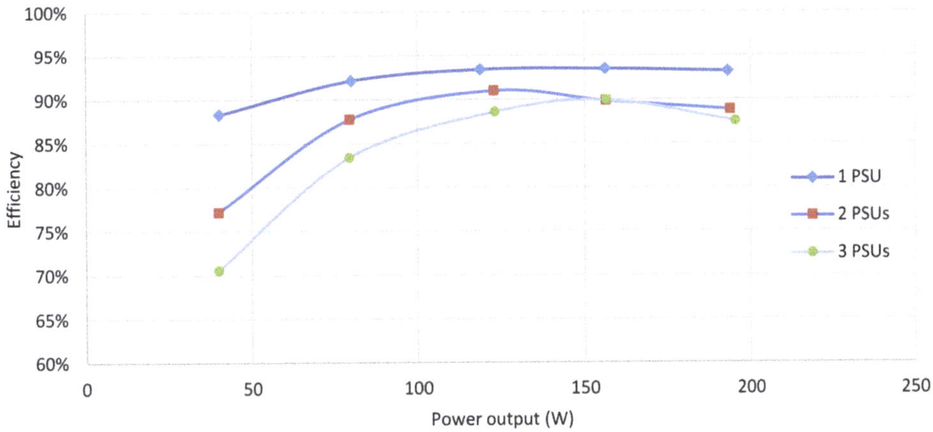


**Figure 10.** Efficiency against power output of 1, 2 and 3 PSUs working in parallel (with transformer).

## 5. Proposed Multicell Buck PFC Converter

### *5.1. Justification of AC to DC Multicell Selection*

In order to justify the selection and advantages of multicell structures, a high efficiency Z-source buck PFC with a power output of 1000 W was initially simulated. The proposed converter is an AC to DC step-down conversion circuit with improved efficiency. Despite its single-stage conversion structure, due to the Z-source filtering, it satisfies power supply regulation EN 61000-6-3. As a result, it is a transformer-less, single-stage conversion system that can be used with IPOP structures and produces an improved energy conversion efficiency.

The results with the losses at different power output levels are shown in Table 2. It should be noted that the input inductor was 3 mH and the one at the output was 92.4 μH. The output inductor design was optimized to minimize losses based on [25,26]. Table 1 shows that the total converter losses for the 200 W load output are 8.38 W and the efficiency is therefore 95.81%. For the 1000 W load case, the total losses are 55.15 W and the efficiency is 94.485%. It is interesting to note that the losses for switch 1 are lower than those of switch 2 for small loads, however, as the load increases, the losses of switch 2 become greater. Furthermore, the losses of the PFC stage are more than 40% of the total circuit loss and the output inductor losses are negligible.

**Table 2.** Simulated single cell buck PFC results.

| Power Output (W) | Total Losses (W) | Switch 1 Losses (W) | Switch 2 Losses (W) | Rectifier Losses (W) | PFC Losses (W) | L1 Loss (W) | L2 Loss (W) |
|---|---|---|---|---|---|---|---|
| 200 | 8.38 | 1.46 | 0.51 | 1.28 | 4.91 | 0.19 | 0.03 |
| 300 | 12.51 | 1.89 | 1.08 | 2.04 | 6.98 | 0.44 | 0.08 |
| 400 | 16.85 | 2.39 | 1.88 | 2.88 | 8.79 | 0.78 | 0.13 |
| 500 | 21.89 | 3.02 | 2.93 | 3.8 | 10.7 | 1.23 | 0.21 |
| 1000 | 55.15 | 8.56 | 11.07 | 8.68 | 20.93 | 5.08 | 0.83 |

Initially, the AC to DC part of the multicell rectifier was simulated with the remaining parts of the PSU in a single-cell structure. The results are shown in Table 3. The results are very promising, since the load losses at 1000 W are considerably minimized. As expected, since, for both inductors and MOSFET switches, the main losses are due to the internal resistance ($R_{indutor}$ and $R_{DS-On}$, respectively), the three-cell buck PFC has significantly lower losses (since $N(I/N)^2 \times R < I^2 \times R$).

**Table 3.** Comparison of 1-cell and 3-cell rectifier system.

| Power Output (W) | Single Cell Rectifier Losses (W) | 3 Cell Rectifier Losses (W) | Rectifier Efficiency Improvement | Overall PSU Efficiency Improvement % |
|---|---|---|---|---|
| 1000 | 8.68 | 0.00288 | >99% | 0.79% |
| 500 | 3.8 | 0.0144 | >99% | 0.70% |
| 400 | 2.88 | 0.01872 | >99% | 0.66% |
| 300 | 2.04 | 0.0261 | >98% | 0.62% |
| 200 | 1.28 | 0.03636 | >97% | 0.58% |

The significant savings affect the overall PSU energy efficiency by up to 0.79%, as demonstrated in Table 3. In the proposed multicell bridge simulation that consists of a three-cell system, it is shown that, when increasing the number of cells, the efficiency improves. The efficiency improvement is significant, especially under bigger load conditions, but as the load decreases, the efficiency gains are relatively reduced.

### *5.2. Multicell Buck PFC*

Simulating a two-cell buck PFC in PSIM's thermal module, it is revealed that inductor losses were minimized by about 50%, as shown in Table 4. The switching losses remained about the same in the case of those of a single cell system; conduction losses, however, were minimized by 52–57% for different loads, and inductor losses were improved by 50%.The overall results of the new system are shown in Figure 11. The results indicate that when increasing the number of parallel cells, the conduction and inductor losses decrease proportionally. Switching losses, on the other hand, are not affected. Therefore, using GaN switches that have low switching losses compared to silicon MOSFETs, and also minimizing conduction losses through multicells, leads to considerable efficiency improvement. The simulated system diagram is shown in Figure 12.

**Table 4.** Comparison of cost per watt for different PSU systems.

| PSU Brand | PSU Model | Power Output | PSU Price | Cost/Watt | Reference |
|---|---|---|---|---|---|
| Mean well | RCP-1000-48 | 1000 | $241.90 | $0.2419 | Jameco.com |
| Mean well | RCP-2000-48 | 2000 | $436.69 | $0.2183 | Jameco.com |
| Mean well | RST-5000-48 | 5000 | $1193.00 | $0.2386 | Jameco.com |
| Mean well | RST-10000-48 | 10,000 | $2366.00 | $0.2366 | Jameco.com |
| Jetpower | SPS60-48/CR4830 | 3000 | $360.00 | $0.1200 | Alibaba.com |
| Jetpower | SPS200-48/SR4850 | 10,000 | $1250.00 | $0.1250 | Alibaba.com |

The references "Jameco.com" and "Alibaba.com" shown in Table 4, are all accessed on 1 September 2021.

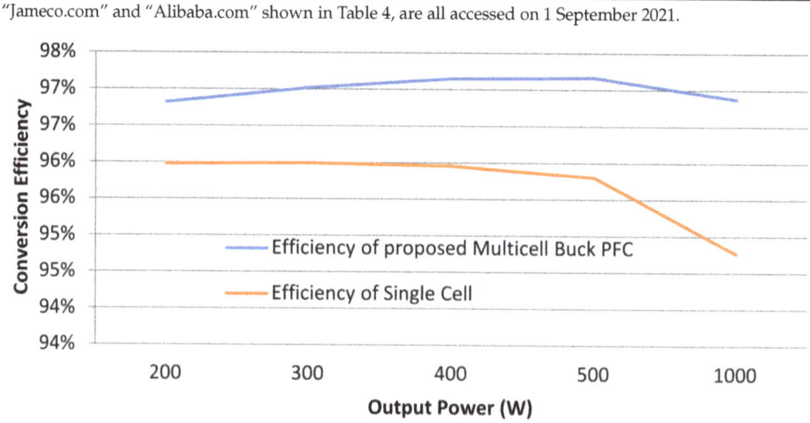


**Figure 11.** Efficiency comparison of single-cell and multicell PSU.

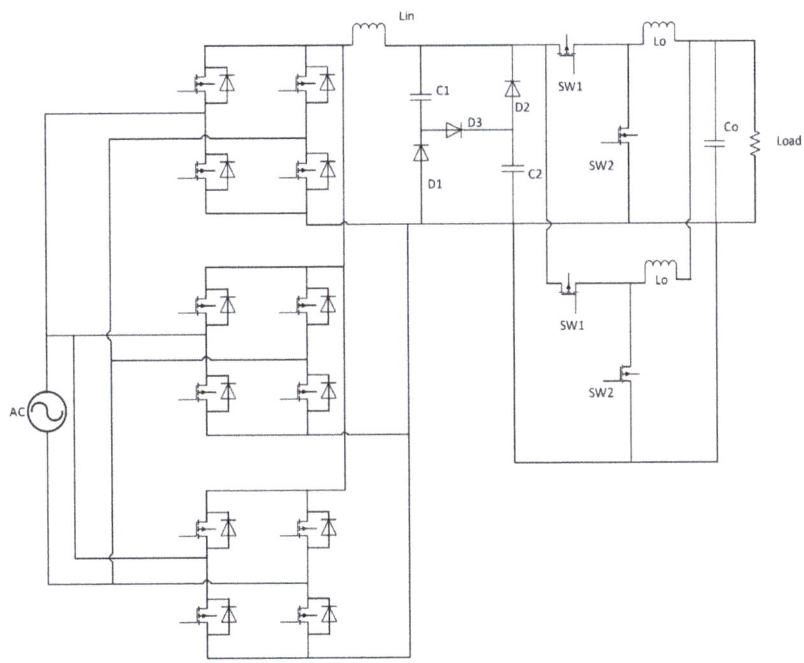


**Figure 12.** Multicell PSU circuit proposed.

In order to simulate the thermal losses of the individual components of the circuit using PSIM's thermal module, each component was modeled with its equivalent circuit, as shown in Figure 13. In the simulation, the ambient temperature and the thermal resistance of each of the components is also taken into account. The flowchart in Figure 12 illustrates the procedure used for selecting the number of cells in the multicell structure, ensuring no excessive cells are used. The first step is to identify the MOSFET switches with the lowest on-resistance ($R_{DS-On}$).

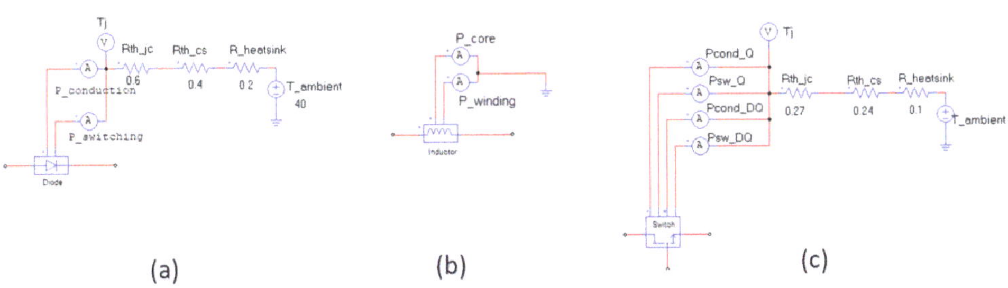


**Figure 13.** Equivalent circuits in PSIM thermal module for (**a**) diode, (**b**). inductor, (**c**). MOSFET.

All switches should have a reverse voltage breaking capacity higher than the input voltage of the source. Since the price of the switches tends to be proportional to the output current, the flowchart in Figure 14 compares the system cost with the lowest cost switch and the system cost of a multicell structure with the lowest $R_{DS-On}$ switches. The solution with the lowest system cost is selected from the product of the number of cells and the cost of the switches.

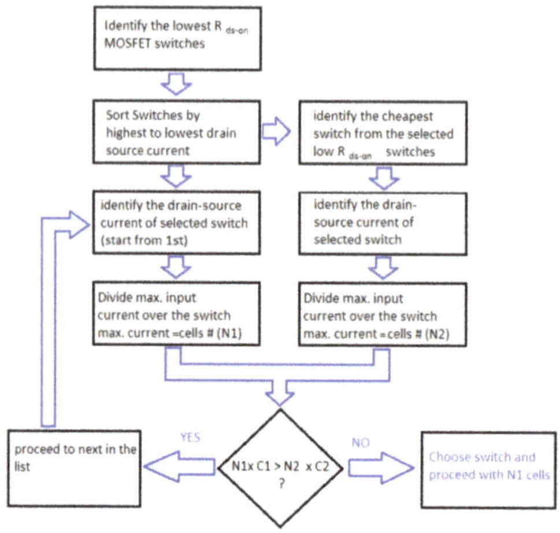


**Figure 14.** Procedure for cell/switch selection.

## 6. Discussion

The results of the research have shown that parallel cells in multicell PSU structures can lead to significant overall energy efficiency improvements in PSUs. According to [27], currently, 1.8% of global energy is supplied to the ICT industry, such as datacenters and micro-datacenters. More than 50% of this energy is supplied to the power supply units of the ICT equipment. Therefore, the impact of PSUs with improved energy efficiency could lead to significant energy savings in the ICT industry globally, and could contribute to the efforts and targets set for global energy decrease, through more efficient systems. Of course, the improvement in efficiency comes at an extra cost, as a result of the additional components used. The total manufacturing cost of a PSU system consists of the electronic components, the controller, the mounting, the enclosure, and labor costs. A PSU consisting of multicell parts would have an increased cost on components (cells) and labor costs but would not have an increased cost on the controller, the mounting and the enclosure systems. Additionally, as mentioned earlier, the transformer systems which lead to a significant cost in PSU systems should be single (centralized) and thus no additional cost should arise.

Analyzing the industry's readymade PSU prices, it can be observed that when using smaller units (as in those in multicell structures) for a 20 kW PSU (1 full rack), the cost is slightly higher. Table 4 compares the cost per watt using smaller PSUs for the same type and brand PSUs for different capacities.

The cost variation is in the order of 1–11%. For example, constructing a 20 kW power supply system with 20 pieces of RCP-1000-48 would cost 10.78% more than using 10 pieces of RCP-2000-48. Using 4 Mean Well RST-5000-48 instead of 2 RST-10000-48, would cost 0.8% more, but using Jetpower SP60-48/CR4830 instead of SPS200-48/SR4850 systems would cost less per installed Watt.

Finally, the proposed multicell IPOP structures have limited applicability for circuits that incorporate transformers. As stated in previous sections, transformers have no-load losses that do not favor decentralization (multicell structure). Centralization via the transformer part only of a multicell structure could overcome such limitations.

Further research will focus on multicell structures with planar transformers in matrix forms. Studies have shown that such transformers offer reduced no-load losses.

## 7. Conclusions

The results presented in this study demonstrate that using low-power modules such as the multicell structure for high-power PSU systems improves the energy efficiency. Using three rectifier cells and two buck PFC cells, a more efficient power supply system can be constructed for the same power output. Efficiency improvements of more than 1.2% can be achieved for 50% of the load and more than 2.1% efficiency improvement under full load. The concept of multicell efficiency improvement is experimentally proven through the testing of a three-cell, a two-cell and a single-cell buck converter system. It is shown that an efficiency improvement of up to 10% can be achieved for particular loads in the DC to DC conversion part. Energy efficiency improvements in PSUs leads directly to energy consumption minimization for the ICT industry.

The initial cost of a power supply system in a datacenter is about 5% of the total capital cost. A 10.78% increase in the power supply unit (i.e., the worst-case scenario) due to a multicell structure compared to a single cell one, corresponds to an additional 0.5% increase in relation to the total capital cost. The respective savings, however, are in the order of 2.1%, as previously mentioned. Looking ahead in the long-term, the cost benefit for this investment is viable, as with simple calculations the payback period is in the order of just 3.3 years.

**Author Contributions:** Conceptualization, M.C. and N.C.; methodology, M.C.; software, M.C.; validation, N.C., S.I. and A.P.; formal analysis, A.P.; investigation, M.C.; resources, A.P.; writing—original draft preparation, M.C.; writing—review and editing, N.C. and A.P.; visualization, S.I.; supervision, N.C.; funding acquisition, A.P. All authors have read and agreed to the published version of the manuscript.

**Funding:** This research received no external funding.

**Institutional Review Board Statement:** Not applicable.

**Informed Consent Statement:** There is no research on humans in this work.

**Conflicts of Interest:** The authors declare no conflict of interest.

## References

1. Miller, R. What 5G Will Mean to the Data Center Industry. *Datacenter Frontier*, 19 February 2019.
2. Tuffaha, N. Evolution of 5G and the Impact on Data Center Infrastructure. *Digital Reality*, 29 January 2019.
3. Shehabi, A.; Smith, S.; Sartor, D.; Brown, R.; Herrlin, M.; Koomey, J.; Masanet, E.; Horner, N.; Azevedo, I.; Lintner, W. *United States Data Center Energy Usage Report*; LBNL-1005775; Lawrence Berkeley National Laboratory: Berkeley, CA, USA, 2016.
4. Xiao, Z.; Liu, H.; Havyarimana, V.; Li, T.; Wang, D. Analytical Study on Multi-Tier 5G Heterogeneous Small Cell Networks: Coverage Performance and Energy Efficiency. *Sensors* **2016**, *16*, 1854. [CrossRef] [PubMed]
5. Zeng, F.; Li, Q.; Xiao, Z.; Havyarimana, V.; Bai, J. A Price-Based Optimization Strategy of Power Control and Resource Allocation in Full-Duplex Heterogeneous Macrocell-Femtocell Networks. *IEEE Access* **2018**, *6*, 42004–42013. [CrossRef]
6. Jiang, H.; Xiao, Z.; Li, Z.; Xu, J.; Zeng, F.; Wang, D. An Energy-Efficient Framework for Internet of Things Underlaying Heterogeneous Small Cell Networks. *IEEE Trans. Mob. Comput.* **2020**. [CrossRef]
7. Energystar. Reduce Energy Losses from Uninterruptable Power Supply (UPS) Systems. Available online: https://www.energystar.gov/products/ (accessed on 11 October 2021).
8. EPRI, 80 Plus Certified Power Supplies and Manufacturers. Available online: https://www.plugloadsolutions.com/80PlusPowerSuppliesDetail.aspx?id=0type=1 (accessed on 11 October 2021).
9. STMicroelectronics. 48 v Direct Power Conversion. 2018. Available online: https://www.st.com/en/power-management/48-v-direct-power-conversion.html (accessed on 11 October 2021).
10. Taranovich, S. Data Center Next Generation Power Supply Solutions for Improved Efficiency. *EDN Network*, 16 April 2016.
11. Kasper, M.; Bortis, D.; Deboy, G.; Kolar, J.W. Design of a highly efficient (97.7%) and very compact (2.2 kw/dm_3) isolated ac–dc telecom power supply module based on the multicell isop converter approach. *IEEE Trans. Power Electron.* **2016**, *32*, 7750–7769. [CrossRef]
12. Fekik, A.; Azar, A.T.; Kamal, N.A.; Serrano, F.E.; Hamida, M.L.; Denoun, H.; Yassa, N. Maximum Power Extraction from a Photovoltaic Panel Connected to a Multi-cell Converter. In Proceedings of the International Conference on Advanced Intelligent Systems and Informatics, Cairo, Egypt, 19–21 October 2020; pp. 873–882.
13. Yuan, J.; Chen, Y.; Yang, Y.; Blaabjerg, F.; Chen, M. High Frequency Multicell Cascaded Quasi-Square-Wave Boost Converter. In Proceedings of the 2020 IEEE 21st Workshop on Control and Modeling for Power Electronics (COMPEL), Aalborg, Denmark, 9–12 November 2020.
14. Luo, X.; Kang, L.; Lu, C.; Linghu, J.; Lin, H.; Hu, B. An Enhanced Multicell-to-Multicell Battery Equalizer Based on Bipolar-Resonant LC Converter. *Electronics* **2021**, *10*, 293. [CrossRef]
15. Gupta, R.; Kumar, A. Control of Multi-cell AC/DC and Cascaded H-bridge DC/AC-basedAC/DC/AC Converter. Taylor and Francis. *IETE J. Res.* **2021**. [CrossRef]

16. Hosseinabadi, F.; Adib, E. A soft-switching step-down pfc converter with output voltage doubler and high power factor. *IEEE Trans. Power Electron.* **2019**, *34*, 416–424. [CrossRef]
17. Rashid, M.H. *Power Electronics Handbook*; Butterworth-Heinemann: Oxford, UK, 2017.
18. Chrysostomou, M.; Christofides, N.; Ioannou, S.; Marouchos, C. Design Guidelines for Energy Efficient AC to DC Power Supplies. In Proceedings of the IEEE IECON, Toronto, ON, Canada, 13–16 October 2021.
19. Sharifi, S.; Babaei, M.; Monfared, M. A high gain buck PFC synchronous rectifier. In Proceedings of the Iranian Conference on Electrical Engineering (ICEE), Mashhad, Iran, 8–10 May 2018.
20. Axelrod, B.; Berkovich, Y.; Ioinovici, A. Switched capacitor/switched-inductor structures for getting transformerless hybrid dc–dc pwm converters. *IEEE Trans. Circuits Syst.* **2008**, *55*, 687–696. [CrossRef]
21. Fei, C.; Lee, F.C.; Li, Q. High-efficiency high-power-density llc converter with an integrated planar matrix transformer for high-output current applications. *IEEE Trans. Ind. Electron.* **2017**, *64*, 9072–9082. [CrossRef]
22. Kim, D.K.; Moon, S.; Yeon, C.O.; Moon, G.W. High-efficiency llc resonant converter with high voltage gain using an auxiliary lc resonant circuit. *IEEE Trans. Power Electron.* **2016**, *31*, 6901–6909. [CrossRef]
23. Mu, M.; Lee, F.C. Design and optimization of a 380–12 v high-frequency, high current llc converter with gan devices and planar matrix transformers. *IEEE J. Emerg. Sel. Top. Power Electron.* **2016**, *4*, 854–862.
24. Abdel-Rahman, S.; Stückler, F.; Siu, K. *Infineon Technologies: PFC Boost Converter Design Guide*; Infineon Technologies AG: Munich, Germany, 2016.
25. Ioannou, S.; Argyrou, M.C.; Marouchos, C.; Darwish, M. Efficiency investigation of a grid connected pv system with power smoothing. In Proceedings of the 54th International Universities Power Engineering Conference (UPEC), Bucharest, Romania, 3–6 September 2019; pp. 1–6.
26. Ioannou, S.; Marouchos, C.; Darwish, M.; Putrus, G. Efficiency investigation of a protection and correction solid state device for low-voltage distribution networks. In Proceedings of the 54th International Universities Power Engineering Conference (UPEC), Bucharest, Romania, 3–6 September 2019; pp. 1–6.
27. Bertoldi, P.; Avgerinou, M.; Castellazzi, L. *Trends in Data Centre Energy Consumption under the European Code of Conduct for Data Centre Energy Efficiency*; JRC Technical Report; Publications Office of the European Union: Luxembourg, 2017; ISBN 978-92-79-76445-5.

energies

MDPI

*Article*

# Demonstration of a Novel Alternating Current Hybrid Concept for a Fuel Cell–Battery Hybrid Electric Aircraft

**Debjani Ghosh , Caroline Willich *, Christiane Bauer and Josef Kallo**

Institute for Energy Conversion and Storage, Ulm University, Albert-Einstein-Allee 47, 89081 Ulm, Germany; debjani.ghosh@uni-ulm.de (D.G.); christiane.bauer@uni-ulm.de (C.B.); josef.kallo@uni-ulm.de (J.K.)
* Correspondence: caroline.willich@uni-ulm.de

**Abstract:** Hybrid electric aircraft offer the potential to decrease emissions from air travel. A new hybrid concept is proposed for a fuel cell-battery hybrid aircraft. In contrast to existing hybrids, the proposed concept puts a battery directly on the AC phases of the motor, which together with a suitable switching circuit superimposes the DC voltage from the battery on the AC voltage of the motor phase providing a voltage boost depending on the battery voltage, which can be used during a high-power demand flight phase. The system is also capable of recharging the battery during flight. The necessary switching architecture was developed and modeled in MATLAB/Simulink to verify the concept and an experimental setup was built for demonstrating the functionality. Simulation and experimental results showed a very good agreement which is very promising for the proposed new hybrid topology.

**Keywords:** hybrid system; battery; fuel cell; electric aircraft; power control

**Citation:** Ghosh, D.; Willich, C.; Bauer, C.; Kallo, J. Demonstration of a Novel Alternating Current Hybrid Concept for a Fuel Cell–Battery Hybrid Electric Aircraft. *Energies* **2021**, *14*, 7350. https://doi.org/10.3390/en14217350

Academic Editors: Felix Barreras and Mario Marchesoni

Received: 12 August 2021
Accepted: 1 November 2021
Published: 4 November 2021



## 1. Introduction

The threat of adverse climate change is becoming a prominent issue day by day. Emissions from the burning of fossil fuels in the transportation sector is one of the main contributors to the problem [1]. $CO_2$ is the largest emission component in the aviation sector, comprising 70% of the exhaust [2]. Emissions from the aviation sector also include other pollutants and greenhouse gases, making this sector one of the top 10 global emitters [3]. Although technological advances have made it possible to reduce the amount of fuel burned and emissions produced per passenger, a continued growth in air traffic outweighs the improvement. The International Civil Aviation Organization (ICAO) predicts that the international aircraft emissions can become three times higher by 2050 compared to 2015 [4]. The European Green Deal aims to achieve a 90% decrease of transportation emissions by 2050, which includes emissions from aviation sector as well [5]. Therefore, it is important to develop new technologies for aviation that will be environment friendly and help to transit from fossil fuels to green energy sources. Hybrid electric aircraft based on fuel cells (FCs) as primary energy source and batteries are a promising technology and will help to reduce $CO_2$ emission. Several such systems have already been tested successfully [6–8].

Fuel cells generate electrical power without harmful emissions and the products are only water and waste heat [9]. They also have higher efficiencies (>50%) than an internal combustion engine [10] and, in comparison to batteries, they offer the advantages of high energy densities meaning they are better suited as energy source for medium and long distance flights. Nevertheless, fuel cells have a slow dynamic response and for flight phases with high power demand, combining a fuel cell with a battery can result in an improved performance [11,12]. As determined by Lijun Gao et al., the peak power with a fuel cell-battery hybrid can be four times higher than the power fuel cell can provide alone [13]. Use of the battery as a secondary source provides a power boost when the load demand is higher, such as start or take-off. Like this, the advantage of the high-energy density of fuel cells and the high-power density of batteries are combined. There are several existing topologies for a fuel cell-battery hybrid system. These conventional hybrid systems usually connect the fuel cell and the battery in parallel connection in the DC side of the inverter.

In an active hybrid system one or more DC/DC converters are used between the fuel cell and the battery as shown in Figure 1a. In this configuration, the DC/DC converter actively controls the fuel cell and the battery to regulate the load sharing [13] and both sources can be operated in such a way that they generate to produce maximum power simultaneously. The peak power supply capability of an active hybrid is much higher than passive topology. However, this active hybrid topology has additional power losses and weight due to the DC/DC converter. According to literature, the DC/DC converter can be 3.2–65 kg for a system of approximately 100 kW range of peak load power demand [14–16].

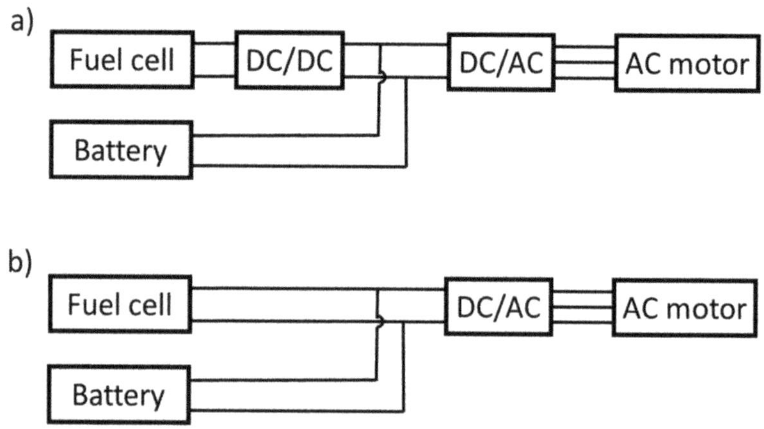


**Figure 1.** Active (**a**) and passive (**b**) fuel cell battery topology in conventional hybrid systems.

There are also existing hybrids with a passive topology as shown in Figure 1b. Here, the fuel cell and the battery are directly connected to the DC link side without any additional converters [17]. The passive topology has less power loss and cost due to the absence of DC/DC converter. Nevertheless, it needs proper coordination of the voltage-current behavior between the battery and the fuel cell and in this type of hybrid, fuel cell nominal voltage determines the battery pack design. There are also variations on the described topologies. For example, the HY4 aircraft uses a variation of a direct system with additional diodes and a separate path for recharging the battery [18,19].

This paper proposes and describes a novel hybrid concept for a fuel cell–battery drivetrain, which does not connect the fuel cell and battery in parallel on the DC side of the inverter as described above. Instead, it places three batteries directly onto the three phases of the electric motor as shown in Figure 2a. The motor phases are supplied from the inverter with AC voltage, but the batteries can only provide DC voltage. Therefore, a suitable switching circuit was developed and tested to superimpose the battery DC voltage on top of the AC voltage of the motor during each AC phase. This new hybrid configuration, with the newly developed switching circuit, makes it possible to increase the voltage on the electric motor by adding the voltage of the battery on top of the AC voltage provided by the inverter of the electric motor during each AC phase. The resulting power increase can then be used during take-off or climb if the fuel cell alone is not able to achieve the bigger load demand. The new concept excludes the necessity of a DC/DC converter between the fuel cell and the battery as found in the existing active hybrid topology while still allowing a complete control of both the fuel cell and the battery and there are no constraints regarding the voltage matching between fuel cell and battery as in the traditional passive hybrid described above. Another advantage of the newly proposed concept is that between the DC link and the AC side an inverter with a smaller power rating and lower weight can be used compared to the conventional hybrids described above since only the fuel cell voltage has to be converted by the inverter and not the battery voltage as well. The energy management between the two sources, fuel cell and battery, is crucial for an efficient operation of the hybrid drivetrain. The energy management strategy distributes the power demand of the aircraft between different power sources in such a way that each power source can be utilized optimally. There have been different techniques developed over the years for a good energy management strategy, using for example a fuzzy logic approach [20], battery involved energy managements with Deep Deterministic Policy Gradient Algorithm [21] or Soft Actor-Critic DRL Algorithm [22]. For the aircraft HY4, a Power Management Control and Delivery (PMCD) unit [19] is used to manage the power demand and control the power for different flight phases while additionally providing redundancy and more safety for the aircraft operation. In this paper, it is assumed that an existing energy management system is available on the on-board computer which will send the command to the battery unit of the AC hybrid to turn on and off as required by the mission profile.

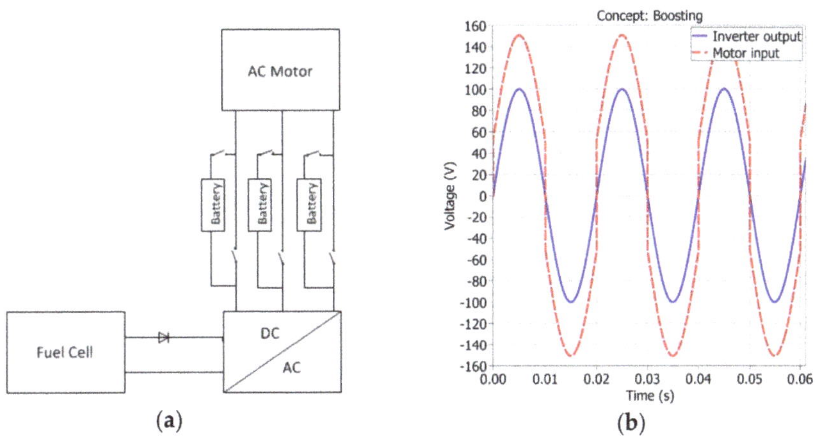


**Figure 2.** AC hybrid concept [23]. (**a**) AC hybrid topology. (**b**) Voltage boosting by AC hybrid.

## 2. Description of the Proposed AC Hybrid Concept

In the newly proposed AC hybrid concept, batteries are directly connected to the motor on each phase (Figure 2a). This configuration will superimpose the DC voltage from the battery on the AC voltage of the motor, which will provide a voltage increase directly on the motor phases, resulting in a motor voltage boost. A suitable switching circuit is required to correctly superimpose the DC voltage onto the AC voltage as illustrated in Figure 2b. The switches must be controlled in such a way that a positive DC voltage is applied to the positive half wave of the AC voltage and a negative DC voltage is added during the negative half wave of the AC voltage. Consequently, there is an increase in the peak-to-peak motor voltage. The precise control of the switching sequence at the correct phase is very important, otherwise a mismatch of the voltage polarity may cause a voltage decrease instead of increase. In Figure 2b, the output voltage of the inverter is 100 V (in blue). A 50V battery is used to superimpose the DC voltage of the battery on top of the inverter AC voltage. The result is +150 V in the positive half wave and -150V in the negative half wave. This gives an AC voltage of 150 V for the motor input, a total rise of 100 V peak to peak with a 50 V battery. With the proper switching sequence, the resulting voltage can be increased by twice the battery voltage with this new hybrid concept.

## 3. Design Requirement

The new AC hybrid concept requires an efficient high-speed circuit for connecting the battery. There are three modes of operation for this circuit:

- Voltage boost mode: The AC power is directed from the inverter to the motor via the battery. The battery is discharging.
- Direct mode: The AC power is routed directly from the inverter to the motor. The battery is disconnected.
- Recharging mode: AC power flows from the motor or from the inverter to the battery and the battery is recharged during flight.

In order to control the switches, a control circuit is also required. The control circuit receives instructions from the pilot interface (Human Machine Interface—HMI) and based on these inputs, selects the suitable mode of operation and switches accordingly. The current, power, speed, torque, etc., must also be constantly monitored to start switching at the right phase angle.

## 4. Proposed Architecture of Switching Circuit

To operate the AC hybrid, a new switching circuit was designed as seen in Figure 3. It consists of four power electronics switches, S1, S2, S3, and S4, forming a bridge architecture. These switches are used in each phase of the motor to connect the battery.

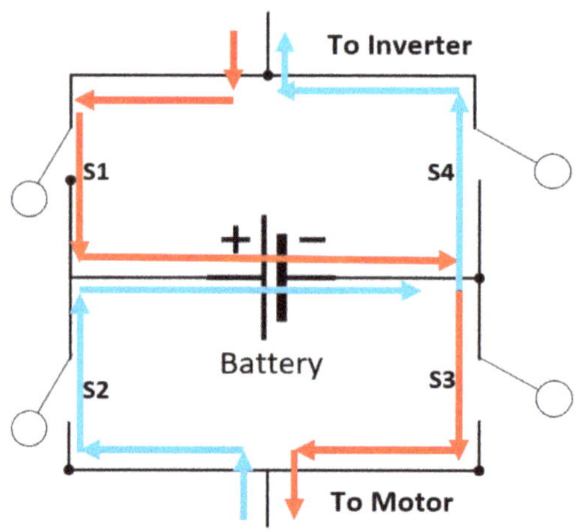


**Figure 3.** Proposed battery bridge model. Red arrows show positive half cycle switching and blue arrows show negative half cycle switching.

Table 1 shows the necessary switching configuration for the different modes. For example, during the voltage boost mode, switches S1 and S3 will be on during the positive AC half cycle in the motor voltage and current; these two switches will be turned off as soon as a negative half wave starts. Switches S2 and S4 will be turned on during the negative AC half cycle, and uninterrupted power is provided to the motor. Due to this switching arrangement, a unidirectional AC current will flow through the battery.

**Table 1.** Switching configuration.

| Mode | Positive Half Cycle | Negative Half Cycle |
|---|---|---|
| Discharging | S1, S3 on, S2, S4 off | S1, S3 off, S2, S4 on |
| Direct (FC only) | S1, S3 off, S2, S4 off | S1, S3 off, S2, S4 off |
| Recharging | S1, S3 off, S2, S4 on | S1, S3 on, S2, S4 off |

When the power demand from the motor is low, direct mode can be used in which only the fuel cell is providing power. During this mode, all above mentioned switches are off.

The recharging mode is used when the motor power demand is very low, so that the fuel cell can supply both the motor and recharge the battery. It can also be used to recuperate energy from the motor to the battery. In this mode, the switching order is the reverse of the discharge mode.

To assure the switching at the correct phase the logic shown in Table 2 is applied, where $V_{battery}$ is the battery voltage and $V_{motor}$ and $I_{motor}$ represent the motor voltage and current, respectively.

**Table 2.** Logic for the switching of the battery bridge.

| Mode | Positive Half Cycle | Negative Half Cycle |
| --- | --- | --- |
| Discharging | $V_{motor} \geq 0$ && $I_{motor} \geq 0$ | $V_{motor} \leq 0$ && $I_{motor} \leq 0$ |
| Recharging | $V_{motor} \leq V_{battery}$ && $I_{motor} \leq 0$ | $V_{motor} \leq -V_{battery}$ && $I_{motor} \leq 0$ |

For the discharging mode, zero voltage and current crossing detection is used to determine the positive and negative half cycle. Due to the inductive nature of the load, there is a phase shift between the motor voltage and current and switching for the positive or negative half cycle may only occur when the voltage and current are both positive or negative, otherwise a decrease of the motor voltage due to the mismatch of the polarity during discharge (voltage boost) mode will occur.

The recharging mode is more complex than discharging. In order to avoid an unwanted result of short discharging pulses on the battery it is necessary to insert a short time in which the battery is not charged in each switching cycle. The battery can only be charged when the motor voltage crosses the battery voltage value as the logic in Table 2 shows. This results in a short time delay between the negative and positive half cycles in which the battery is not charged, the length of which depends on the battery voltage. During this short time delay, the system is in direct mode, so that the power supply is uninterrupted.

## 5. Simulation Model

A simulation model was developed in MATLAB Simulink to verify the correct functioning of the AC hybrid concept. Figure 4 shows the simulation model of the drive with a fuel cell, the proposed battery bridge, inverter, and the motor.

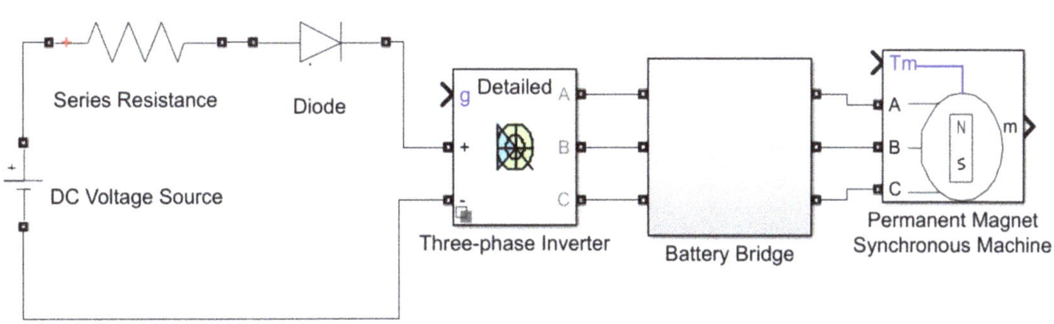


**Figure 4.** Simplified block diagram of the simulation model.

### *5.1. Fuel Cell/DC Source*

The primary power source of the drivetrain is the fuel cell. As the paper focuses on the battery bridge, a 30 V DC source with a small internal resistance in series from the Simulink library is used for this purpose. This way, it is easier to compare the hardware system as well with simulation results, as the hardware includes a motor supplied directly with 30 V DC as discussed later. A diode is connected in series to the source to prevent any current flow from inverter towards the source.

### 5.2. Inverter and Drive Control

The drive train has an inverter built with 6 IGBTs. The control for the drive consists of a proportional-integral (PI) controller in the outer control loop. The PI controller is tuned using Ziegler-Nichols method [24], which is a widely used method to tune any proportional (P), proportional-integral (PI), or proportional-integral-derivative (PID) controller. The Ziegler-Nichols method starts by setting the integral and derivative gains to zero and increases the proportional gain until it reaches the point (ultimate gain) when the output of the control loop has stable oscillations. The oscillation period and the ultimate gain value is then used to find out the tuning constant of the controller such as P, PI, or PID values.

The controller compares the setpoint speed with the current speed of the motor and generates a reference torque. The other motor parameters such as hall signals and motor currents are used with this reference torque to generate the switching signal for the gates of the IGBTs in the inverter.

### 5.3. Battery Bridge

Figure 5 shows the switching circuit inside the battery bridge block. A battery is connected to each motor phase with the help of the switching circuit. High speed switches are needed for the circuit as switching is done in each half wave of AC voltage. Hence, four MOSFET switches were selected for the purpose. The gate driver circuit of each MOSFET switch gets the gate signals from the battery bridge controller to enable switching at the correct phase and turn on or off the corresponding MOSFET switches.

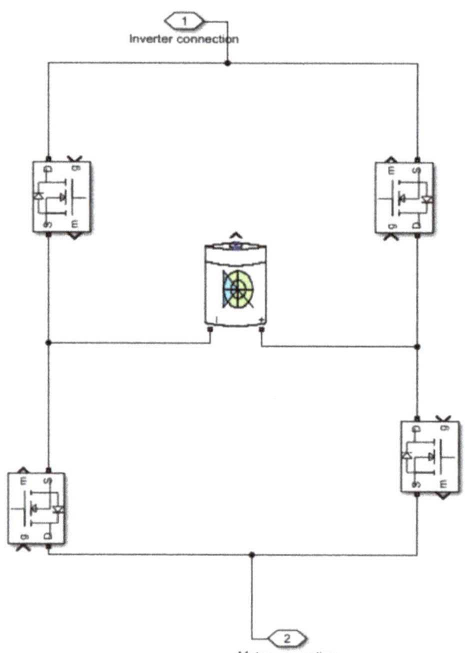


**Figure 5.** Simulink model of the battery bridge.

The model uses three battery blocks from the MATLAB/Simulink library. This battery block presents a generic dynamic model for various rechargeable batteries such as Lead-Acid, Li-ion, NiCd, and Nickel-Metal-Hydride. It is possible to simulate the temperature effect and ageing effect on the battery.

For the simulation (and for the hardware prototype), two Li-ion rechargeable cells were used in series. There are many types of Li-ion battery models available in literature with different objectives. The electrochemical model contributes to physical cell design and deals with variables such as concentration distribution and galvanostatic charge-discharge. However, this model type is unfit in this context, where only the electrical terminal behavior of the battery is required within a suitable computational time. An electrical equivalent circuit model of a battery is easy to use and suitable for circuit simulations with short computing time. The accuracy of voltage, current characteristics, state of charge, etc., of an electrical model is adequate for this particular application. Therefore, the model used in the simulation is based on an electrical equivalent circuit model of a Li-ion battery with a controlled voltage source and an internal resistance. Table 3 shows the battery parameters for a single Li-ion rechargeable cell from Samsung [25] used in the model and in the hardware.

**Table 3.** Battery parameters.

| | |
|---|---|
| Nominal Voltage (V) | 3.63 |
| Rated capacity (mAh) | 2600 (0.2C, 2.75 V discharge) |
| Initial state of charge (%) | 90 |
| Cut off voltage (V) | 2.75 |
| Maximum discharge current (mA) | 5200 |
| Maximum charge current (mA) | 2600 |

When the battery current is positive, the battery discharges and when the current is negative, it recharges. In the model it is assumed that both the discharge and recharge characteristics are the same. The model does not consider any self-discharge effect of the battery as this is not relevant for the verification of the concept.

A. Battery bridge controller

The four MOSFET switches in the battery bridge control the power flow through the battery for the different modes. These switches, in turn, are controlled by the battery bridge controller. Figure 6 presents the logical flow chart for the battery bridge controller. It is divided into four types of modes that can be selected by a user or the main computer of the aircraft.

When a particular mode is selected, the logic first checks the SOC of the battery and only then takes the necessary action. For example, if mode 2 (discharge mode) is selected, the controller will first check if the SOC level of the battery is higher or equal to 30%. Only if there is enough energy left in the battery the controller will proceed to zero detection switching according to the logic mentioned in Table 2. If the SOC is between 20% and 30%, the controller will still perform the switching and send a warning signal to the user that the battery has a low energy level. This helps the user or the pilot to determine a corrective course of action for the flight to avoid using the battery any more than necessary. If the battery has 10–20% SOC, the controller sends a critical level warning and still performs the switching. If the battery SOC is lower than 10%, the controller refuses to proceed to switching and enters the direct (FC only) mode. It is possible to override this control if a life-threatening situation occurs and the pilot must use the remaining battery energy at all costs. The different SOC levels for the warnings can of course be changed and the controller can be universally adapted according to the application and its requirements.

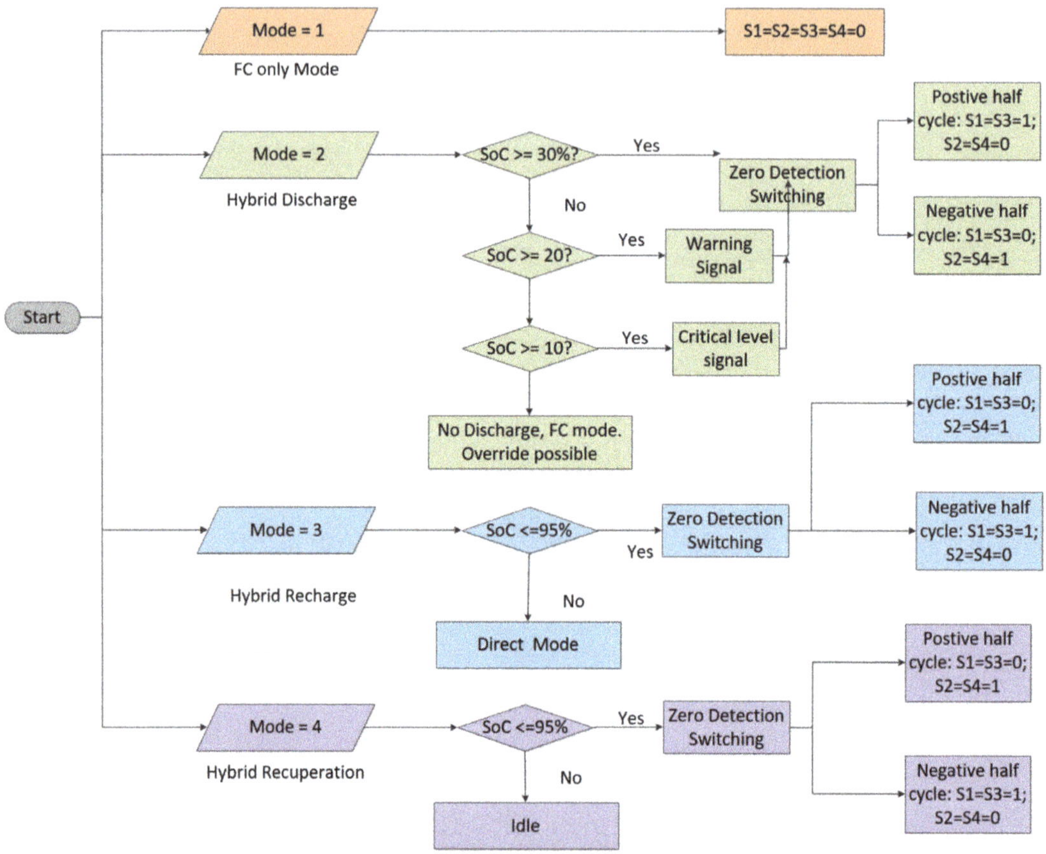


**Figure 6.** Logical flow chart for the normal operation in different modes of the battery bridge controller.

Similarly, for the recharging and recuperation modes, the controller checks if the battery is already full. It enables charging according to the logic from Table 2 if the SOC level is less or equal to 95%; otherwise, it goes to direct mode (in case of the hybrid recharging mode) or remains idle (recuperation mode).

The use of state machines for energy management is a well-known approach [26,27] and it is useful for reactive and event driven systems. Therefore, the logic chart from Figure 6 where each mode is considered as a state was programed using the MATLAB/Simulink state flow tool.

The model uses a 3-phase BLDC motor (brushless DC motor) from the Simulink library. BLDC motors are like permanent magnet synchronous motors with a trapezoidal back electromotive force (emf). It uses a closed loop speed control using a Hall sensor as mentioned before. The motor can be operated in either generator or motor mode. Therefore, it is possible to simulate the recharge/recuperation mode during flight.

The hardware prototype has a BLDC motor module from Trinamic Motion Control GmbH & Co which has a QBL5704 motor [28]. Table 4 depicts the parameters used for the simulation of the BLDC motor.

**Table 4.** Parameters of a Trinamic BLDC motor.

| | |
|---|---|
| Stator Resistance (Ω) | 0.35 |
| Stator inductance (mH) | 1 |
| Number of pole pairs | 4 |
| Torque constant (Nm/A) | 0.063 |
| Maximum peak torque (Nm) | 1.3 |
| Rotor inertia (kgm$^2$ $\times$ $10^{-6}$) | 23 |

## 6. Results

*Discharge*

A variable speed drive was simulated and Figure 7 shows the simulation results, where the desired starting speed is higher than the system can supply. It can be seen that in the first 30 s the fuel cell power is not enough to reach the desired speed of 900 RPM. Therefore, the system is switched to hybrid discharge mode at 30 s. Immediately after the battery is turned on, the motor phase voltage increases, and the motor is able to achieve the setpoint speed. The drive is switched back to direct mode once the load requirement is decreased during cruise at 200 s. With the supply of 30 V as fuel cell voltage and 7.2 V nominal battery voltage, the setup is able to provide 80–100 W of mechanical power.

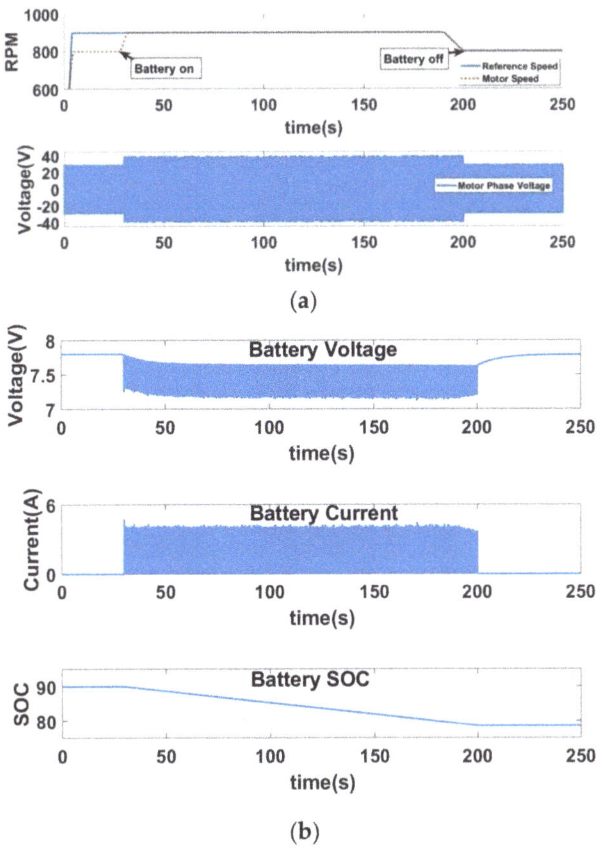


**Figure 7.** Simulation results for the battery discharge mode. (**a**) simulation results; (**b**) battery parameters.

The corresponding battery parameters (voltage, current and state of charge) during hybrid mode are also shown in Figure 7. The battery current acts like a unidirectional alternating current, varying between 0 and 5 A as expected. The battery voltage also has the shape of a unidirectional AC wave. The decrease of the battery state of charge (SOC) shows the proper discharging of the battery over time.

To better understand the battery current, it is necessary to carefully study the MOSFET current. In Figure 8, it can be seen that switches S1 and S3 carry current during the positive half wave and S2 and S4 during the negative half wave of AC. Thus, the battery only supplies a unidirectional current even though an alternating current flows through the battery bridge circuit.

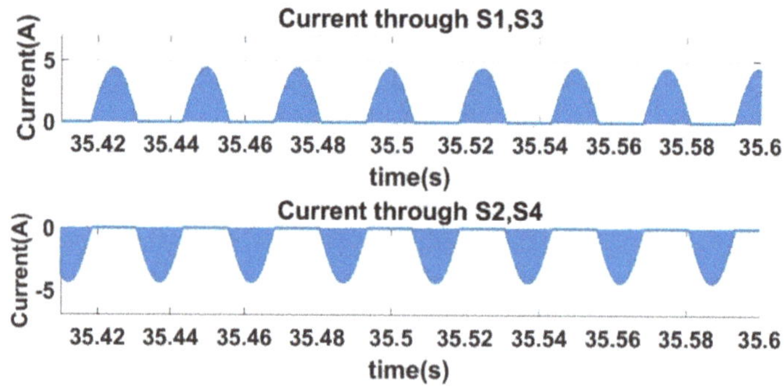


**Figure 8.** Simulated current through switches S1, S2, S3, and S4 of the battery bridge during discharge.

Figure 9 shows the results of the battery parameters during recharging. The battery recharging follows the switching sequences as described in Table 1. A dead time is required between the switching signals to prevent an unwanted short discharging of the battery. The battery has a negative current during recharging, meaning current is entering the battery. During charging the battery current shows again a unidirectional AC behavior. As expected, the SOC of the battery shows an increment over time, confirming that the battery is getting recharged successfully.

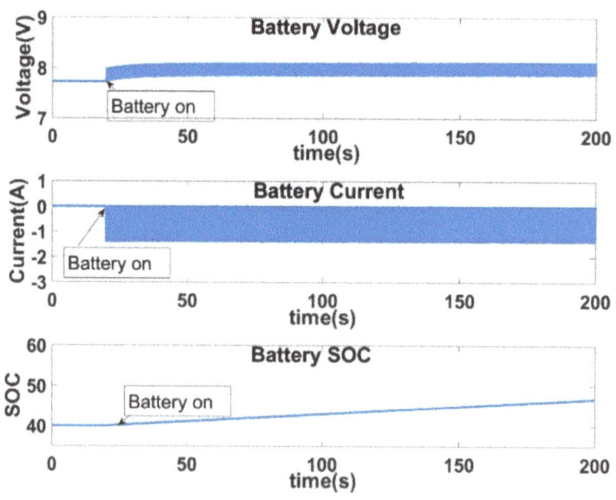


**Figure 9.** Simulated battery parameters during recharging.

## 7. Hardware Verification

A small test setup was built to demonstrate the functionality of the concept. The test set up uses a Trinamic BLDC motor along with a 3-phase inverter. For the motor control, a controller board (TMC603A) from Trinamic was used as shown in Figure 10. The board is supplied with 30V DC voltage and has six external n-MOSFETs for the inverter function. The output of the converter is usually connected to the motor phases. To test the AC hybrid, the battery bridge was inserted between the phases of the inverter and the motor as shown in Figure 10. The motor is controlled by the external microcontroller depicted in the picture. Based on the position data of the Hall sensor in the motor, the microcontroller sends pulse width modulation (PWM) signals to the inverter.

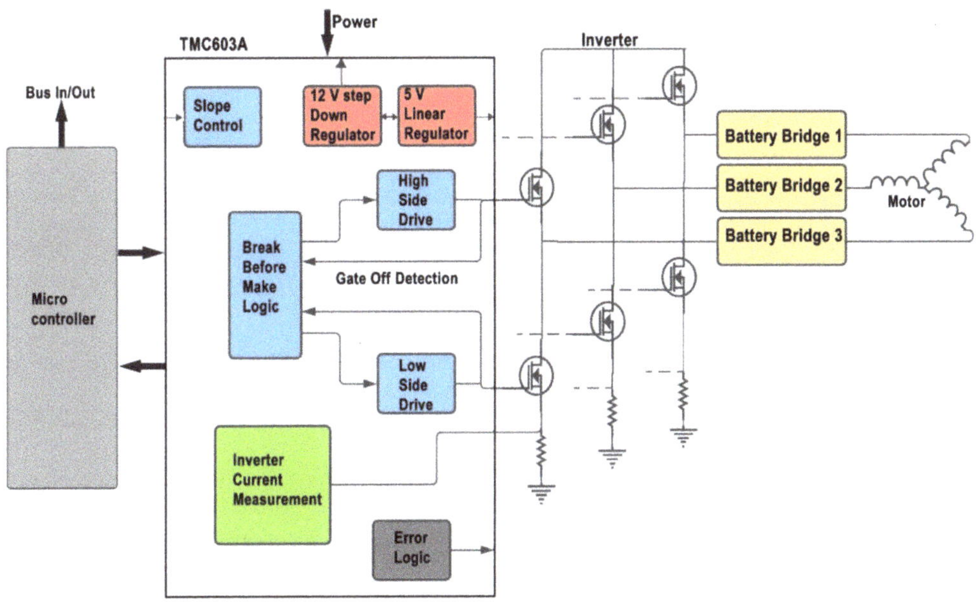


**Figure 10.** BLDC motor with the controller board and battery bridge.

The control takes place through a cascade regulation of position, speed, and current. The speed controller compares the speed set point with the motor speed measured by the sensors and calculates a set point torque which is converted into a corresponding current in the vector controller and compared with the actual motor phase current. Taking into account the measured position of the motor angle, pulses are then generated for the converter circuit. The microcontroller was programmed using the TMCL-IDE software, provided by Trinamic.

Figure 11 shows the generation concept of the gate signals for the MOSFET switches in the hardware battery bridge. It consists of a logical circuit including sensors, comparators, microcontrollers, optocoupler, resistances, and capacitors. At first, the motor phase voltage and current are measured through sensors. Next, the motor phase current is converted to a voltage signal as the comparator circuit and microcontroller can only use voltage as input for the logical process. A microcontroller (Arduino) is used to synchronously switch the battery bridge for the different operating modes. Depending on the operation modes the microcontroller sends the reference voltage to the comparator for the voltage comparison according to Table 2. In discharging mode, the reference voltage is zero and in recharging or recuperation mode the reference voltage equals the battery voltage. The comparator circuit supplies an input signal to the microcontroller (Arduino) pin every time the voltage of the motor from both positive and negative directions crosses zero voltage or the battery voltage for discharging and recharging mode, respectively. It also sends another input signal to the microcontroller every time the motor current crosses zero from either direction. Then, the algorithm shown in Figure 6 is executed in the microcontroller, two complementary PWMs are generated and fed to the gate driver circuit. The gate driver circuit provides the signals to the MOSFETs. Optocouplers are used between the microcontroller and the gate driver circuit to provide the necessary galvanic separation between the power and control circuit.

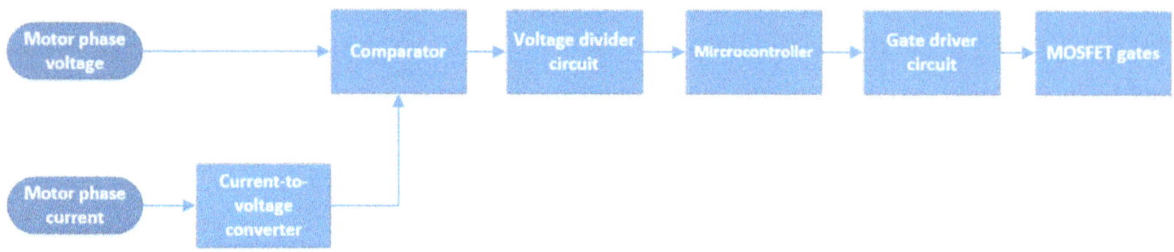


**Figure 11.** Generation concept of gate signals.

The experiment was performed under ambient lab temperature and pressure. The motor module was mounted on a stable surface and the required motor voltage of 30 V was ensured. The battery bridge included the same batteries in series as used for the simulation with a nominal voltage of 3.63 V each. It was ensured that for every motor run, the batteries have the same voltage and SOC.

Figure 12 illustrates the result of switching the battery on and off of the hardware prototype. When the battery bridge is switched on at 37.2 s, the motor speed increases from 800 RPM to about 900 RPM. This matches well with the result obtained from the simulation in Figure 7.

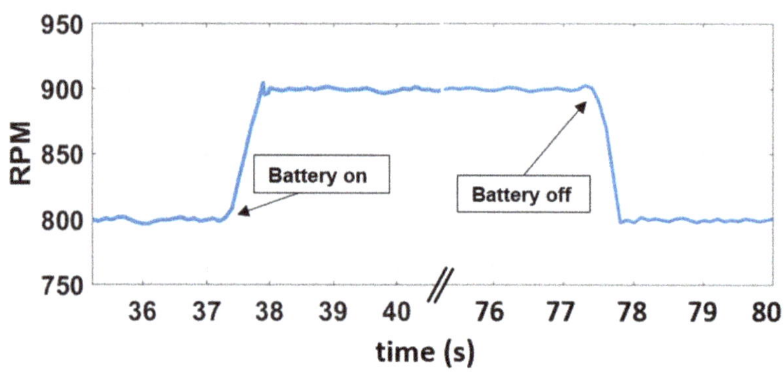


**Figure 12.** Motor speed with battery discharging.

When the battery is turned off at 77.5 s it can be observed that the motor speed also decreases indicating a reduction of motor voltage. The motor speed decreases back to 800 RPM which also corresponds well with the simulation results. It can be seen clearly that adding the battery voltage to the motor voltage boosts the motor speed.

Figure 13 shows the measured phase voltage of the Trinamic motor for the battery discharge mode. At 37.2 s, when the battery is turned on, the motor voltage increases to ±36.08 V and at 77.5 s, it reduces to 29 V as soon as the battery is disconnected. To validate the simulation data, the simulation was repeated with the same time condition as the experiment. The battery was turned on at 37.2 s and off at 77.5 s. The motor phase voltage taken as RMS value was used to compare the data between the simulation and the experiment in Figure 13. The motor waveform closely matches the one from the simulation (Figure 7). The difference between the simulation and experimental data is small, approximately 1–1.2 V for the case of the connected battery as well as the disconnected battery. This voltage drop corresponds to the losses that occur mainly in the power electronics components. Furthermore, it can be observed that the peak to peak phase voltage increases two times the battery voltage with some losses as expected from the simulation and the concept of the AC hybrid. Therefore, the hardware measurements prove the functionality of the concept.

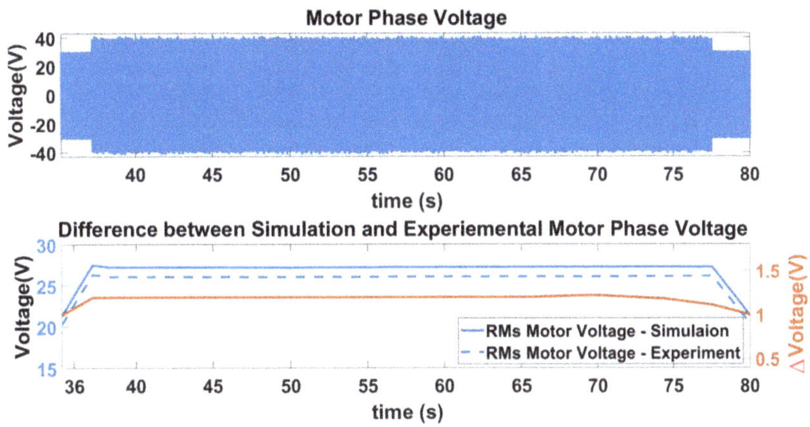


**Figure 13.** Hardware motor phase voltage and difference between simulation and experimental voltage.

Recharging of the battery was also checked with the hardware system. Figure 14 shows the open circuit voltage (OCV) of a single cell during recharging. It was determined during a cumulative motor run, meaning that the motor ran for few minutes, then the battery was taken out to measure the open circuit voltage. After that, the motor ran again for few minutes and the battery was taken again and OCV was measured. The same was performed several times recording the open circuit voltage of a cell. Since a battery OCV depends on its state of charge, it is evident from Figure 14 that the recharging mode also worked, and the battery was recharged. It can also be seen that for a higher charging current the battery is charged faster than with a lower current.

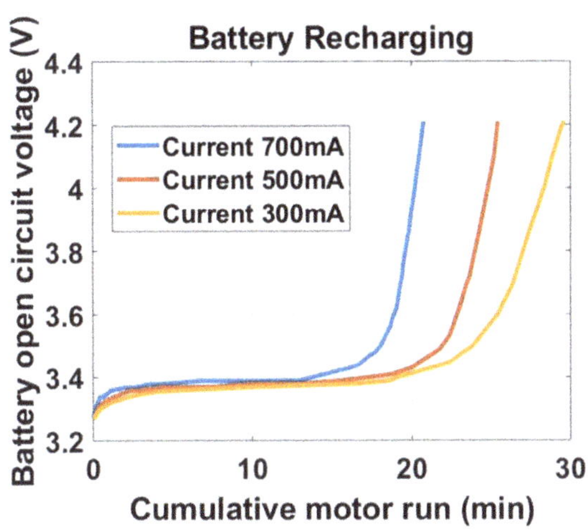


**Figure 14.** Battery open circuit voltage for one cell during recharging with different charging currents.

## 8. Conclusions

In the paper, a new AC hybrid topology for a fuel cell battery airplane was developed. In contrast to existing hybrids, the concept places the batteries directly on the phases of the motor. This proposed topology has no need for a DC/DC converter in the drivetrain architecture and at the same time provides the chance to reduce the inverter size. It therefore lowers the total weight of the drivetrain and reduces the power losses. The efficiency of the AC hybrid system is found to be 90% in simulation whereas a passive hybrid of same power range is 93%. Nevertheless, the AC hybrid provides the decoupling between two sources unlike the passive hybrid system. Therefore, it removes the constraint on the voltage-current matching as in the conventional passive hybrid systems. However, in the AC hybrid the switching must be done very precisely to avoid undesirable outcomes. Without proper synchronizing, the voltage can be decreased instead of increased. This will result in the exact opposite of the required voltage boost. Therefore, the switching of the batteries to each AC phase requires a careful synchronizing.

The necessary switching circuit was modeled and build in hardware and the different operating modes of the system (hybrid, primary energy source only, recharging of the battery via primary energy source, or recuperation) were tested to verify the concept. The results of the simulations as well as the hardware measurements confirm the functionality of the concept and there was a good agreement between the simulation and hardware results. The RPM behavior between simulation and hardware matched exactly, whereas the phase voltage of the motor showed an approximate 4% deviation that corresponds to the losses in a real circuit. The concept is therefore very promising, but further research and development is needed to scale up and optimize the hardware and assess the benefit the developed AC hybrid offers in comparison with a conventional hybrid in more detail.

**Author Contributions:** Conceptualization, D.G. and J.K.; Funding acquisition, J.K.; Investigation, D.G.; Methodology, D.G.; Project administration, C.W. and C.B.; Supervision, C.W.; Validation, D.G.; Visualization, D.G.; Writing—original draft, D.G.; Writing—review & editing, C.W. All authors have read and agreed to the published version of the manuscript.

**Funding:** This research was funded by the German Federal Ministry of Transport and Digital Infrastructure under the grant 03B10702A and 03B10702A2.

**Institutional Review Board Statement:** Not applicable.

**Informed Consent Statement:** Not applicable.

**Acknowledgments:** The presented work has received funding from the German Federal Ministry of Transport and Digital Infrastructure under the grant 03B10702A and 03B10702A2.

**Conflicts of Interest:** The authors declare no conflict of interest.

## References

1. Global Greenhouse Gas Emissions Data | Greenhouse Gas (GHG) Emissions | US EPA. Available online: https://www.epa.gov/ghgemissions/global-greenhouse-gas-emissions-data (accessed on 11 May 2021).
2. Fact Sheet | The Growth in Greenhouse Gas Emissions from Commercial Aviation | White Papers | EESI. Available online: https://www.eesi.org/papers/view/fact-sheet-the-growth-in-greenhouse-gas-emissions-from-commercial-aviation (accessed on 11 May 2021).
3. Climate Action Network and International Coalition for Sustainable Aviation joint input to the Talanoa Dialogue—Carbon Market Watch. Available online: https://carbonmarketwatch.org/publications/climate-action-network-and-international-coalition-for-sustainable-aviation-joint-input-to-the-talanoa-dialogue/ (accessed on 2 November 2021).
4. Trends in Emissions that affect Climate Change. Available online: https://www.icao.int/environmental-protection/Pages/ClimateChange_Trends.aspx (accessed on 18 June 2021).
5. Reducing emissions from aviation | Climate Action. Available online: https://ec.europa.eu/clima/policies/transport/aviation_en (accessed on 11 May 2021).
6. DLR. First flight of DLR's HY4 fuel cell light aircraft. *Fuel Cells Bull.* **2016**, *2016*, 1. [CrossRef]
7. DLR Press Release. DLR Motor Glider Antares Takes Off in Hamburg–Powered by a Fuel Cell. pp. 2–5. Available online: http://www.dlr.de/en/desktopdefault.aspx/tabid-5105/8598_read-18278/8598_page-2/ (accessed on 2 November 2021).

8. Koehler, T. Boeing Makes History with Flights of Fuel Cell Demonstrator Airplane. Boeing Frontier. 2008. Available online: https://www.boeing.com/news/frontiers/archive/2008/may/ts_sf04.pdf (accessed on 2 November 2021).
9. Thounthong, P.; Raël, S. The benefits of hybridization: An investigation of fuel cell/battery and fuel cell/supercapacitor hybrid sources for vehicle applications. *IEEE Ind. Electron. Mag.* **2009**, *3*, 25–37. [CrossRef]
10. Savvaris, A.; Xie, Y.; Malandrakis, K.; Lopez, M.; Tsourdos, A. Development of a fuel cell hybrid-powered unmanned aerial vehicle. In Proceedings of the 24th Mediterranean Conference on Control and Automation (MED), Athens, Greece, 21–24 June 2016; pp. 1242–1247. [CrossRef]
11. Chan, C.C. The state of the art of electric, hybrid, and fuel cell vehicles. *Proc. IEEE* **2007**, *95*, 704–718. [CrossRef]
12. Thounthong, P.; Raël, S.; Davat, B. Energy management of fuel cell/battery/supercapacitor hybrid power source for vehicle applications. *J. Power Sources* **2009**, *193*, 376–385. [CrossRef]
13. Gao, L.; Jiang, Z.; Dougal, R. Evaluation of active hybrid fuel cell/battery power sources. *IEEE Trans. Aerosp. Electron. Syst.* **2005**, *41*, 346–355. [CrossRef]
14. Fraunhofer. Bidirectional full SiC 200 kW DC-DC Converter for Electric, Hybrid and Fuel Cell Vehicles. 2015. Available online: https://www.iisb.fraunhofer.de/ (accessed on 2 November 2021).
15. Mon, M.M. Design of 50 kW DC-DC Converter used in Hybrid Electric Vehicle. *Int. J. Sci. Eng. Technol. Res.* **2014**, 3, 3368–3373. Available online: https://www.ijsetr.com/ (accessed on 2 November 2021).
16. PowerCell–Give Your Charger the Power it Deserves | SEMIKRON. Available online: https://www.semikron.com/products/product-lines/air-cooled-igbt-stacks/powercell.html (accessed on 30 September 2021).
17. Bernard, J.; Hofer, M.; Hannesen, U.; Toth, A.; Tsukada, A.; Büchi, F.N.; Dietrich, P. Fuel cell/battery passive hybrid powertrain with active power sharing capability. In Proceedings of the 2010 IEEE Vehicle Power and Propulsion Conference, Lille, France, 1–3 September 2010; pp. 1–5. [CrossRef]
18. Nishizawa, A.; Kallo, J.; Garrot, O.; Weiss-Ungethüm, J. Fuel cell and Li-ion battery direct hybridization system for aircraft applications. *J. Power Sources* **2013**, *222*, 294–300. [CrossRef]
19. Hoenicke, P.; Ghosh, D.; Muhandes, A.; Bhattacharya, S.; Bauer, C.; Kallo, J.; Willich, C. Power management control and delivery module for a hybrid electric aircraft using fuel cell and battery. *Energy Convers. Manag.* **2021**, *244*, 114445. [CrossRef]
20. Saib, S.; Hamouda, Z.; Marouani, K. Energy management in a fuel cell hybrid electric vehicle using a fuzzy logic approach. In Proceedings of the 2017 5th International Conference on Electrical Engineering—Boumerdes (ICEE-B), Boumerdes, Algeria, 29–31 October 2017; pp. 1–4. [CrossRef]
21. Wu, J.; Wei, Z.; Liu, K.; Quan, Z.; Li, Y. Battery-Involved Energy Management for Hybrid Electric Bus Based on Expert-Assistance Deep Deterministic Policy Gradient Algorithm. *IEEE Trans. Veh. Technol.* **2020**, *69*, 12786–12796. [CrossRef]
22. Wu, J.; Wei, Z.; Li, W.; Wang, Y.; Li, Y.; Sauer, D.U. Battery Thermal- and Health-Constrained Energy Management for Hybrid Electric Bus Based on Soft Actor-Critic DRL Algorithm. *IEEE Trans. Ind. Inform.* **2020**, *17*, 3751–3761. [CrossRef]
23. Ghosh, D.; Willich, C.; Kallo, J. Development of a novel AC hybrid concept for a fuel cell-battery hybrid electric aircraft with power electronics switches. In Proceedings of the 2018 14th International Conference on Power Electronics (CIEP), Cholula, Mexico, 24–26 October 2018; pp. 105–110. [CrossRef]
24. Widodo, A.B. PID Controllers, 2nd Edition [Åström,_Karl_J.;_Hägglund,_Tore. Available online: https://www.academia.edu/27771107/PID_Controllers_2nd_Edition_Åström_Karl_J_Hägglund_Tore_ (accessed on 30 September 2021).
25. S. SDI, Specification of product INR18650-25R. 2014. Available online: http://www.datasheet-pdf.com/datasheet/Samsung/799163/INR18650-20R.pdf.html (accessed on 2 November 2021).
26. Garcia, P.; Fernández-Ramirez, L.M.; Garcia, C.A.; Jurado, F. Energy Management System of Fuel-Cell-Battery Hybrid Tramway. *IEEE Trans. Ind. Electron.* **2009**, *57*, 4013–4023. [CrossRef]
27. Garcia, P.; Fernandez, L.M.; Garcia, C.A.; Jurado, F. Fuel cell-battery hybrid system for transport applications. In Proceedings of the 2009 International Conference on Electrical Machines and Systems, Tokyo, Japan, 15–18 November 2009. [CrossRef]
28. Motion, T.; Gmbh, C. Mot QBL5704 family Manual; 2010. Available online: https://www.trinamic.com/products/drives/bldc-motors-details/qbl5704/ (accessed on 2 November 2021).

energies

MDPI

Article

# A Compensated Peak Current Mode Control PWM for Primary-Side Controlled Flyback Converters

Chia-Hsuan Wu [1], Guan-Rong Huang [2], Cheng-Chih Chou [3,4], Ching-Ming Lai [1] and Liang-Rui Chen [3,*]

1 Department of Electrical Engineering, National Chung Hsing University, Taichung 402, Taiwan; swdemperor@hotmail.com (C.-H.W.); pecmlai@gmail.com (C.-M.L.)
2 Universal Global Scientific Industrial Co., Ltd., Nantou County 542, Taiwan; gr_huang@usiglobal.com
3 Department of Electrical Engineering, National Changhua University of Education, Changhua 50074, Taiwan; chou@itri.org.tw
4 Green Energy and Environment Research Laboratories, Industrial Technology Research Institute, Hsin-chu 31057, Taiwan
* Correspondence: lrchen@cc.ncue.edu.tw

**Abstract:** In this paper, a feedback compensator (FBC) and a Feedforward compensator (FFC) are proposed to construct a novel compensated peak current mode control pulse width modulation (CPC-PWM) for primary-side controlled flyback converters. Using the proposed FBC, the PWM duty cycle of an abnormal operating flyback converter would be descended to limit the output current for reducing power dissipation. Using the proposed FFC, the effect of delay time would be descended to reduce the over-flow current for increasing the current accuracy. In this paper, the operating principle and mathematical model are described and analyzed. Then, the component values are well designed to satisfy the electrical specifications. Finally, a prototype is designed and realized to access system performance. The experimental results show that the proposed CPC-PWM can validate in a wide input voltage range and output short conditions, which also has good current accuracy and reduces power dissipation by about 68%.

**Keywords:** feedback compensator; feedforward compensator; flyback converter

**Citation:** Wu, C.-H.; Huang, G.-R.; Chou, C.-C.; Lai, C.-M.; Chen, L.-R. A Compensated Peak Current Mode Control PWM for Primary-Side Controlled Flyback Converters. *Energies* **2021**, *14*, 7458. https://doi.org/10.3390/en14227458

Academic Editor: Federico Barrero

Received: 12 October 2021
Accepted: 1 November 2021
Published: 9 November 2021



## 1. Introduction

Because the flyback converter has a simple structure and does not require a secondary side output inductor, it has considerable advantages in size and cost. Based on the above advantages, the flyback converter is widely used in low power applications [1–16]—especially in LED lighting [8–10] and battery balancing control [11–16] to show its many advantages. With the advancement of digital technology and the cost of microcontrollers has dropped significantly. The control scheme of flyback is also moving towards digital control to obtain higher performance [17–20]. Since the flyback converter is an isolated converter, the output voltage and current are traditionally regulated by utilizing the secondary-side regulation (SSR). Usually, SSR is mainly constructed by a photocoupler. The photocoupler feedback circuit causes above 10% of the power losses at standby conditions. Recently, primary-side regulation (PSR) flyback converter (i.e., primary-side controlled flyback converter) becomes an important technology due to lower cost and standby power losses [1,17–25]. The primary-side controlled flyback converter stores energy on the transformer when the power switch is ON and releases the energy on the transformer to the load when the power switch is OFF. It is known that this topology is a constant power output. Based on the constant power output characteristics, when the output voltage of the flyback converter drops due to the load, the output current must increase. When the output voltage of the flyback converter keeps dropping, the current keeps increasing until the current is infinite, causing the problem of components burning. At the same output power, the higher input voltage results in a lower input current. On the contrary, the lower input voltage will result in a higher input current. This means that different power losses are caused by different input voltages, so a larger rated component must be selected in the circuit design to meet the range under various conditions.

As shown in Figure 1, the conventional flyback converter has primary side peak current detection overcurrent protection. Because the primary-side peak current detection protection point voltage $V_{LIMIT}$ is a fixed value, and the flyback converter working in Discontinuous Conduction Mode (DCM) has a constant power output as shown as [22]

$$P_O = \frac{1}{2} \cdot L_P \cdot f_{SW} \cdot I_P{}^2 \cdot \eta \tag{1}$$

where *Lp*, *fsw*, *Ip* and $\eta$ are the primary side inductance, switching frequency, primary side peak current, and flyback conversion efficiency, respectively.

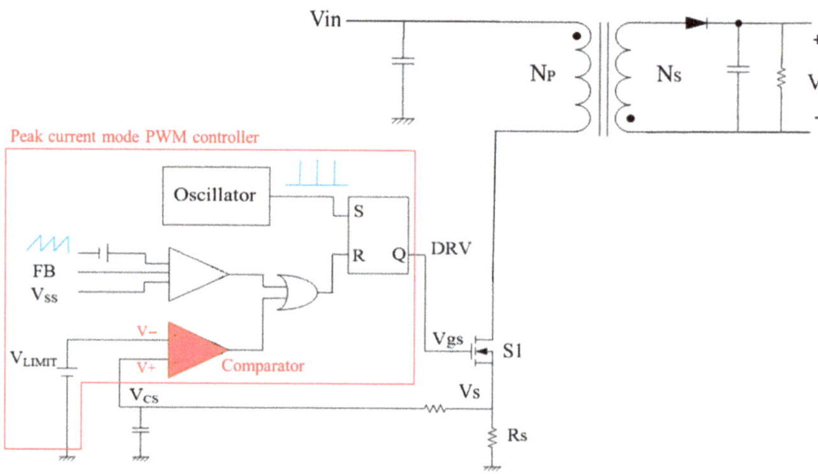


**Figure 1.** A typical flyback converter with primary side peak current mode control.

When the load on the output terminal becomes larger and larger, the output voltage becomes lower. At this time, the output current will still be larger and larger as shown in Equation (2). Therefore, the current rating of the part needs to be selected with a larger specification, just to meet the protection application.

$$I_O = \frac{1}{2} \cdot L_P \cdot f_{SW} \cdot I_P{}^2 \cdot \eta \cdot \frac{1}{V_O} \tag{2}$$

In order to overcome the above-mentioned problem, two methods were proposed and used in converters. In this first method, a current detection resistor $R_{sense}$ is connected in series with the current loop at the output end to convert the current signal into a voltage signal. Then, use the comparison circuit or the current monitor IC to judge whether the overcurrent condition is reached. If the overcurrent condition is satisfied, turn off the synchronous rectification switch. Another method is to connect a current detection resistor $R_{sense}$ in series with the ground current loop at the output end to convert the current signal into a voltage signal. Then, use the comparison circuit or the current detection IC to determine whether the overcurrent condition is occurring. If the condition is satisfied, the primary switch PWM duty is adjusted through the photocoupler control—to make the entire power supply achieve the functions of stability and overload protection. With these two methods, only a current detection resistor $R_{sense}$ and comparison control circuit or current monitor IC are needed to achieve the purpose of current detection and protection. However, the current detection resistor $R_{sense}$ is connected in series with the output main circuit. Because the current is large, in order to avoid excessive power loss, the current detection resistor is usually very small. The current detection resistance $R_{sense}$ is generally between several mΩ to tens of mΩ, depending on the output current. Therefore, the converted voltage signal is about tens of mV and hundreds of mV. The accuracy of such a small voltage signal is obviously insufficient for judgment and control. In addition, the power loss caused by the current detection resistor $R_{sense}$ under heavy load is inevitable.

In a typical primary-side controlled flyback converter, a sensing resistor $R_S$ is used to convert the current on the primary side into a voltage $V_S$ and compares it with the reference voltage $V_{LIMIT}$ of the controller. When the sensing voltage $V_S$ is greater than the reference voltage $V_{LIMIT}$, the comparator is activated to turn off the power switch. Unfortunately, in actual applications, due to the stray capacitance inside the semiconductor device, a fixed logic drive delay is caused. When the input voltage is high, the current slope $m = \frac{V_{in}}{L_P}$ on the primary side becomes steeper. Due to the delay in the response time of the comparators, logic gates, and components, the power switch cannot be turned off immediately when the sensing voltage $V_S$ is greater than $V_{LIMIT}$. As a result, the primary side current still maintains a fixed slope rise, and the overflow current $\Delta I_P$ occurs and can be shown as [22]

$$\Delta I_P = \frac{V_{in}}{L_P} \cdot t_{delay} \tag{3}$$

where $t_{delay}$ is the delay time that is the amount delay times of the comparators, logic gates, and other components.

Figure 2 shows a schematic diagram of overflow current $\Delta I_P$ changes caused by signal delay under traditional peak current mode control. It can be seen that a higher input voltage $V_{in}$ has a larger overflow current $\Delta I_P$ than a lower input voltage. This leads to a significant difference. When the input voltage is higher, the power that can be provided to the load is higher, and vice versa. This is one of the main reasons affecting line regulation of flyback converter.

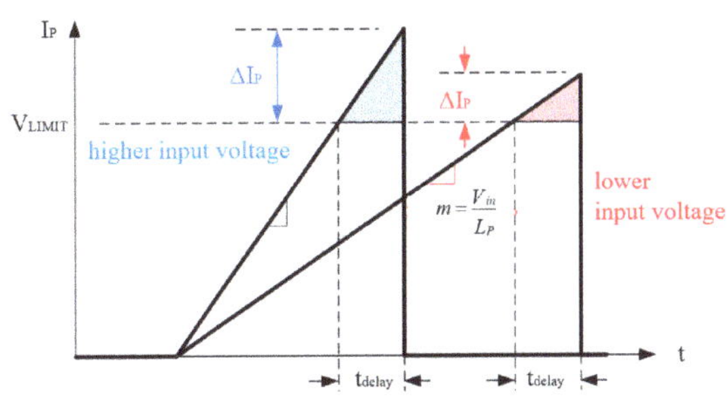


**Figure 2.** Overflow current with different input voltage.

In order to solve the above problem, many methods and techniques were proposed. The amount of overflow current $\Delta I_P$ can be estimated by using the change of current in the same $t_{dealy}$ time. According to this estimated value to modify the PWM, the influence of $t_{dealy}$ can be reduced [21]. Furthermore, the CS sample-and-hold circuit is used to measure primary side peak current $I_p$. Then, by calculating the average current, a compensation control method that is theoretically unaffected by $t_{dealy}$ can be obtained [22]. Another method is to measure the mid-point current (i.e., average current) and use the average current control method, which can effectively reduce the influence of $t_{dealy}$ and increase the accuracy of the output current [23–25]. Recently, digital technology is also used to directly measure $t_{dealy}$. Finally, $t_{dealy}$ is compensated to obtain accurate output current [17,20].

Although the above methods can be used to reduce the impact of $t_{dealy}$. However, when an overload occurs and the output voltage is reduced, the output current cannot be suppressed. Therefore, the current rating of the components needs to be selected with a larger specification, just to meet the protection application. In order to solve this problem, a Compensated Peak Current mode controlled PWM (CPC-PWM) for primary-side controlled flyback converters working in DCM is proposed in this paper. In the proposed CPC-PWM, the input voltage $V_{in}$ is sensed and can be used to compensate for the control signal delay to reduce the overflow current $\Delta I_P$ to improve line regulation. In addition, the current limiting by voltage feedback scheme [26,27] is referenced. The output voltage $V_{out}$ of flyback is sensed and be used to compensate PWM to suppress the overload current.

## 2. System Description

In order to suppress the output current $I_o$ of the flyback converter during overload protection, the feedback compensator (FBC) is proposed in this paper. The inherent characteristics of the flyback converter output voltage $V_{out}$ can be used to detect and adjust the PWM duty. Figure 3 shows the waveforms of a primary-side controlled flyback with FBC control working in DCM. As shown in Figure 3, when the output voltage $V_{out}$ drops, the power switch duty cycle needs to be reduced to suppress the output current $I_o$. This means that the PWM duty cycle $D_a$ during abnormal operation should be much smaller than the PWM duty cycle $D_n$ during normal operation. The amount of this output voltage drop $\Delta V_{out}$ is the difference between the rated output voltage $V^*$ and the output voltage $V_{out}$. Thus, the input voltage of the negative terminal of the comparator $V$- will be compensated by the amount of this output voltage drop and written as:

$$V- = V_{LIMIT} - k \cdot \Delta V_{out} = V_{LIMIT} - k(V^* - V_{out}) \tag{4}$$

where $k$ is a positive constant. A larger $k$ means a smaller duty cycle $D_a$ is obtained. Equation (4) can be rewritten as:

$$V- = k_1 \cdot V_{LIMIT} + k_2 \cdot V_{out} - k_3 \tag{5}$$

where $k_1$, $k_2$, and $k_3$ are positive constant. It can be seen from (5) that when the output voltage $V_{out}$ drops more, the voltage $V-$ becomes smaller. This reduces the PWM duty cycle and achieves the purpose of suppressing the output current $I_o$.

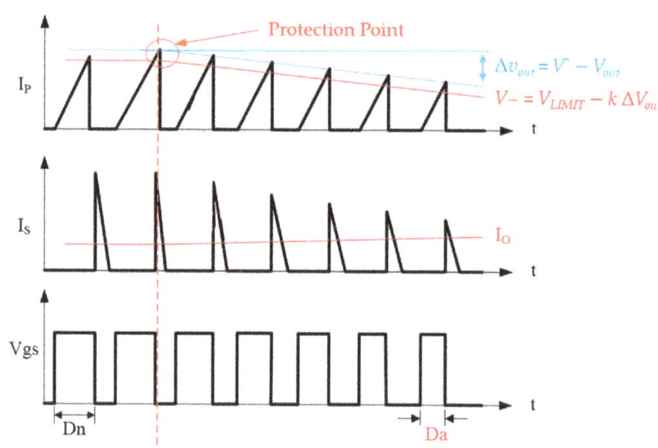


**Figure 3.** Waveform diagrams of a primary-side controlled flyback with FBC control working in DCM.

Figure 4 shows the theory expected V-I curves of a flyback with and without FBC. Comparing the V-I curves of the traditional method and the proposed method, the output current of the proposed method is much lower. It means that there is no need to choose a power component with a larger current rating. In this way, the power consumption problem caused when the output load is increased or short-circuited can be suppressed, and the components with larger specifications can be avoided.

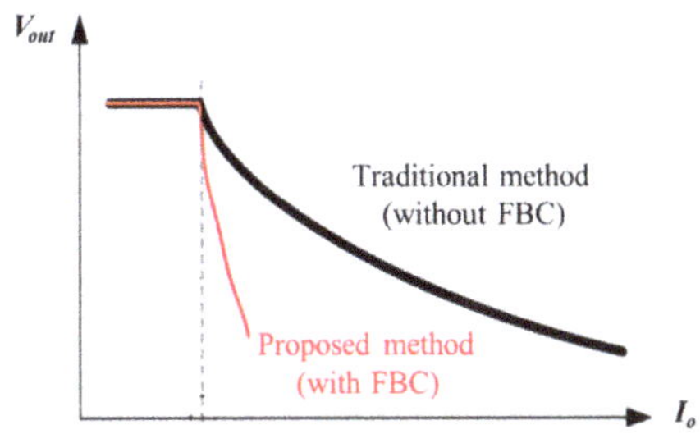


**Figure 4.** V-I curve of the traditional method and the proposed method.

In order to suppress the overflow current $\Delta I_P$ of the flyback converter, the FeedForward Compensator (FFC) is proposed in this paper. The input voltage $V_{in}$ of the flyback converter is detected and used to adjust the PWM duty, since the overflow current $\Delta I_P$ is proportional to the input voltage $V_{in}$ shown as (3). Figure 5 shows waveforms of a primary-side controlled flyback with FFC control working in DCM. As shown in Figure 5, the input voltage of the positive terminal $V+$ of the comparator is extract added by $k' V_{in}$ as written as:

$$V+ = V_S + k' \cdot V_{in} \tag{6}$$

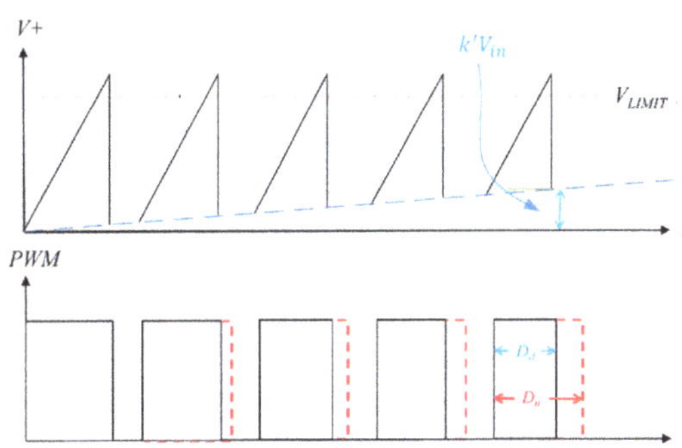


**Figure 5.** Waveforms of a primary-side controlled flyback with FFC control working in DCM.

Equation (6) can be rewritten as:

$$V+ = k_4 \cdot V_S + k_5 \cdot V_{in} \tag{7}$$

where $k_4$ and $k_5$ are positive constant. It is clear that when the input voltage $V_{in}$ increases, the voltage $V+$ becomes larger. When $V+$ becomes larger, the power switch $S1$ will be turned off early. When (8) is held, the overflow current $\Delta I_P$ caused by $t_{delay}$ is eliminated.

$$\frac{k_5}{k_4} = \frac{t_{delay} \cdot R_S}{L_P} \tag{8}$$

Based on the above discussion, the block diagram of the proposed CPC-PWM controller mentioned in this paper can be drawn as Figure 6b. Compared with the traditional peak current mode control PWM shown in Figure 6a, we can see that the FBC is added, thereby achieving the purpose of suppressing the output protection current. On the other hand, the FFC is also added, thereby suppressing the purpose of the overflow current $\Delta I_P$.

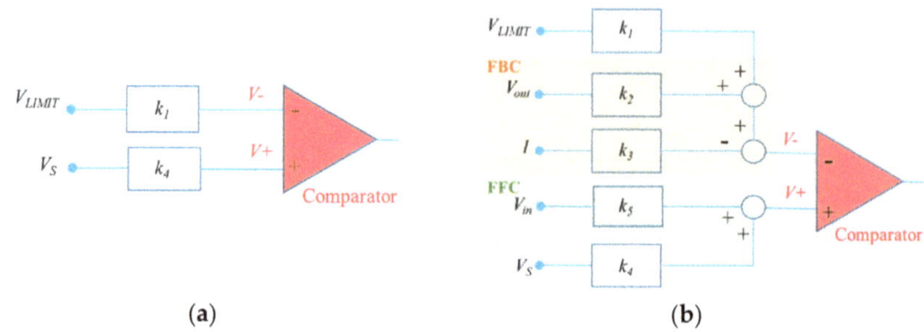


**Figure 6.** Control block diagrams of the traditional peak current mode control PWM (**a**) and the proposed CPC-PWM (**b**).

Design Example:

In order to verify the feasibility of the proposed CPC-PWM, a 5 W prototype is designed and realized. The specifications and circuit diagram of the prototype are listed and shown in Table 1 and Figure 7, respectively.

**Table 1.** Specifications of the realized prototype.

| Specification | Value |
|---|---|
| Input voltage | 17~34 V |
| Output voltage | 5 V |
| Output current | 1 A |
| Ripple voltage | 30 mV |
| Output power | 5 W |
| Overcurrent point | 1.3 A ± 2% |
| Switching frequency | 550 kHz |

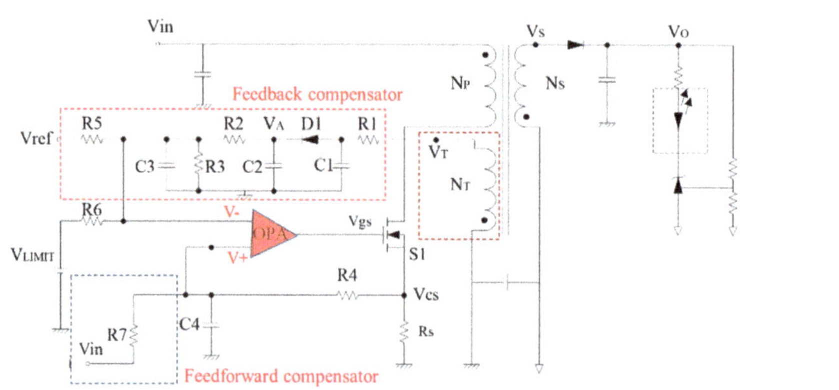


**Figure 7.** Circuit diagram of the proposed compensated peak current mode controlled PWM method.

The flyback converter proposed in this paper is mainly constructed by a transformer, an FBC, an FFB, and a power switch *S*1 as shown in Figure 7. The transformer includes a primary winding, a secondary winding, and an auxiliary winding or third winding. The output voltage of the third winding is injected into the feedback compensator circuit to adjust the PWM duty of the power switch. FBC is composed of a diode *D*1, four resistors *R*1, *R*2, *R*3, and *R*5, and three capacitors *C*1, *C*2, and *C*3, as shown as a red block in Figure 7. Due to the characteristics of the flyback converter, the energy is transferred to the secondary side when the power switch is turned off. At the same time, the third winding is used to sample the output voltage condition to obtain the voltage $V_T$. Due to the transformer leakage inductance, there will be a spike voltage and the voltage $V_A$ can be obtained only after being processed by a low-pass filter. A first-order RC filter is used and meet to (9) to filter out spike voltage

$$R1{\cdot}C1 \geq t_{spike} \tag{9}$$

where $t_{spike}$ is the period of the spike voltage.

The feedback compensator is a network circuit and can be simplified as Figure 8. The simplified circuit is composed of three voltage sources $V_A$, $V_{ref}$, and $V_{LIMIT}$, and four resistors $R2$, $R3$, $R5$, and $R6$. Voltage $V-$ can be obtained and shown as (10) by using KCL law. Compared with (5), parameters $k_1$, $k_2$, and $k_3$ can be obtained and shown as (11)–(13).

$$V- = \frac{V_{ref}{\cdot}G_5 + V_{LIMIT}{\cdot}G_6 + V_A{\cdot}G_3}{(G_5 + G_6 + G_3 + G_2)} \tag{10}$$

where $G_X$ is the reciprocal of $R_X$, that is a conductance.

$$k_1 = \frac{G_6}{(G_5 + G_6 + G_3 + G_2)} \tag{11}$$

$$k_2 = -\frac{n_3{\cdot}G_3}{n_2{\cdot}(G_5 + G_6 + G_3 + G_2)} \tag{12}$$

$$k_3 = \frac{V_{ref}{\cdot}G_5}{(G_5 + G_6 + G_3 + G_2)} \tag{13}$$

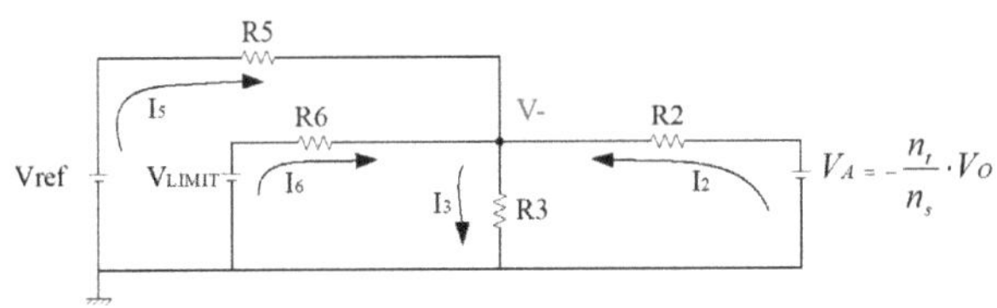


**Figure 8.** Simplified FBC circuit.

The FFC is a network circuit composed of two voltage sources $V_{in}$ and $V_s$, and two resistors $R4$ and $R7$. Voltage V+ can be obtained and shown as (14) by using KCL law. Compared with (7), parameters $k_4$ and $k_5$ can be obtained and shown as (15) and (16). In addition, the relation between resistor $R4$ and $R7$ is shown as (17).

$$V+ = \frac{V_{in}{\cdot}G_7 + V_S{\cdot}G_4}{(G_7 + G_4)} \tag{14}$$

where $G_X$ is the reciprocal of $R_X$, that is a conductance.

$$k_4 = \frac{G_4}{(G_7 + G_4)} \tag{15}$$

$$k_5 = \frac{G_7}{(G_7 + G_4)} \tag{16}$$

$$R_7 = \frac{R_4{\cdot}L_P}{t_{delay}{\cdot}R_S} - R_4 \tag{17}$$

The used components of the designed flyback converter as shown in Figure 7 are listed in Table 2.

**Table 2.** The used components and parameters in the design example.

| Parts | Value | Parts/Parameters | Value |
|---|---|---|---|
| Resistor R1 | 1.1 KΩ | Capacitor C1 | 47 pF |
| Resistor R2 | 34 KΩ | Capacitor C2 | 470 pF |
| Resistor R3 | 2.43 KΩ | Capacitor C3 | 10 nF |
| Resistor R4 | 1 KΩ | Capacitor C4 | 33 pF |
| Resistor R5 | 73.2 KΩ | Volatge Vref | 5.2 V |
| Resistor R6 | 100 KΩ | Volatge $V_{LIMIT}$ | 0.5 V |
| Resistor R7 | 290 KΩ | Volatge Vfault | 0.385 V |
| Resistor RS | 0.15 Ω | | |

## 3. Experimental Results

The feasibility and superiority of the proposed method are verified by discussing and comparing "overload current variation", "input power consumption during short circuit" and "overflow current variation".

Figure 9a–c shows the V-I curves obtained by using the traditional peak current control. Figure 9d–f shows the V-I curves obtained by using the proposed FBC method.

It can be seen from Figure 9a–c that the V-I curve has an obvious relationship with the input voltage $V_{in}$. When the input voltage $V_{in}$ is higher, the overload point will also be higher. Figure 9d–f shows the V-I curves obtained using the proposed FBC method. It can be seen that when the input voltage $V_{in}$ changes from 17 V to 34 V, there is no significant difference in the output current changes. It indicates that the proposed FBC method is significantly better than the traditional peak current control. From Figure 9d–f, we can also see that the output current during overload can be suppressed effectively by the proposed FBC as the theory predicted.

Figure 10a shows the comparison results of the measured overload point current using the traditional peak current control and the proposed FBC method. The variation rate of overload point current under different input voltages of the proposed FBC method is about 2% far better than the traditional 34.5%. In this design example, the traditional method requires at least a diode with a 1.7 A current stress. However, the proposed FBC method only needs a diode with 1.3 A current stress. Figure 10b shows the comparison results of the measured input power consumption using the traditional peak current control and the proposed FBC method under the output short circuit condition. Moreover, Figure 11b shows that the input power consumption of the proposed method under short-circuit protection is far superior to the traditional method. This means that there is no need to choose a power component with a larger current rating. In this way, the power consumption problem caused by the overload or short-circuited can be suppressed, and the components with larger specifications can be avoided.

Figure 11a–f shows the experimental waves under different input voltages of 17 V, 28 V, and 34 V when the flyback converter with/without FFC. In these figures, the positive terminal input voltage $V+$ of the comparator, the negative terminal input voltage $V-$ of the comparator, and the output voltage Vout waveforms of the flyback converter are shown. It can be seen from Figure 11a–c that without using FFC, the voltage $V-$ starts to climb from zero voltage under various input voltages. Until voltage $V-$ is greater than voltage $V+$, the power switch is turned off after the time $t_{delay}$. Observing the $V-$ waveform, it can be seen that the input voltage $V_{in}$ is larger and the overflow current $\Delta I_P$ is larger. From Figure 11d–f, can be seen that an offset voltage is injected into $V-$, so that the overflow current $\Delta I_P$ is suppressed. The injected voltage is about 51 mV, 90 mV, and 105 mV when the input voltage is 17 V, 28 V, and 34 V, respectively. According to the experimental results, the proposed method is close to the expected theoretical result. The voltage injected by different input voltages will also be different, which can reach the $V+$ voltage early to achieve the purpose of effectively reducing the overflow current $\Delta I_P$.

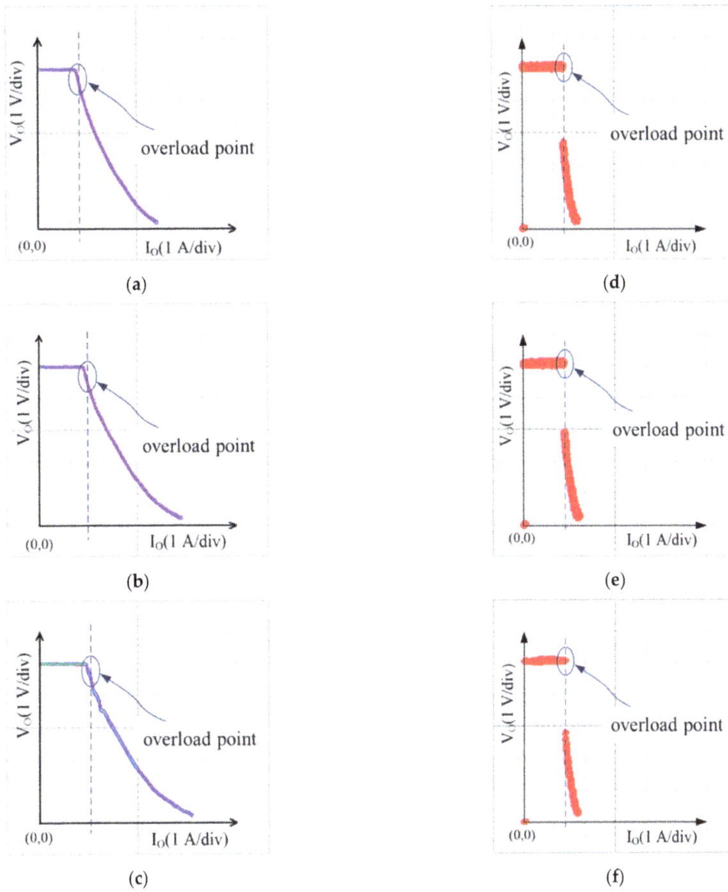


**Figure 9.** V-I curves obtained by using the traditional peak current detection circuit with input voltage (**a**) 17 V, (**b**) 18 V, and (**c**) 34 V, and by using the proposed FBC method with input voltage (**d**) 17 V, (**e**) 18 V, and (**f**) 34 V.

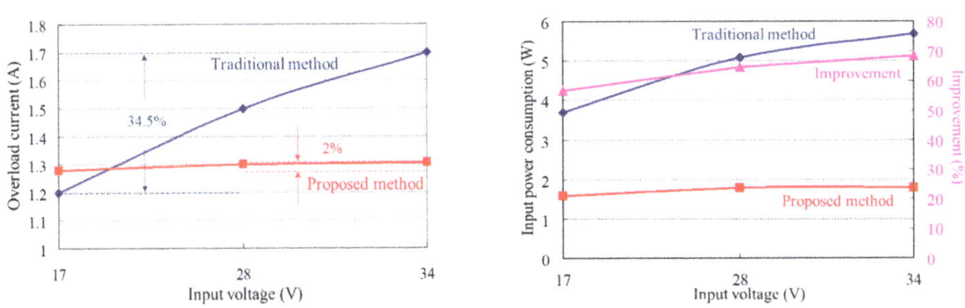


(**a**) (**b**)

**Figure 10.** Experimental results (**a**) Overload current under different input voltage, (**b**) Input power consumption under different input voltage.

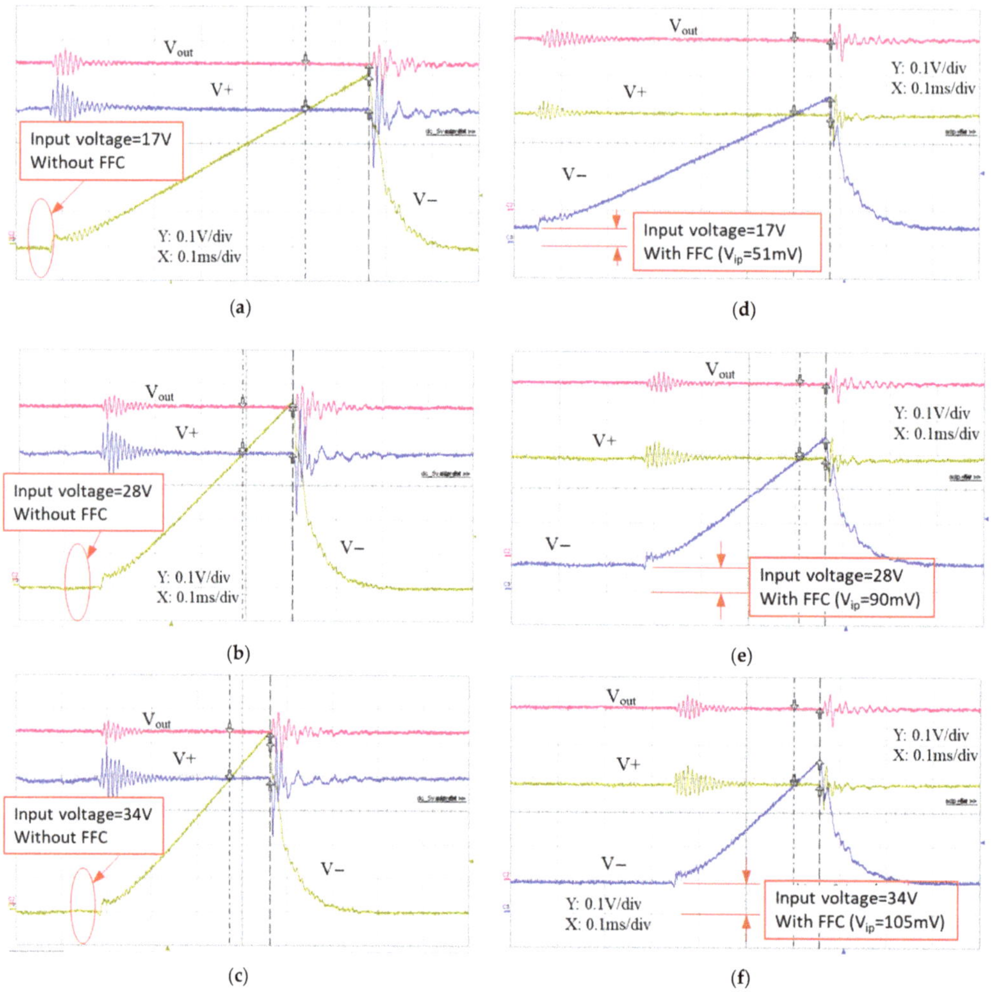


**Figure 11.** *V*−, *V*+, and Vout curves obtained by using the traditional peak current detection circuit with input voltage (**a**) 17 V, (**c**) 18 V, and (**e**) 34 V, and by using the proposed FFC with input voltage (**b**) 17 V, (**d**) 18 V, and (**f**) 34 V.

It is known that different voltage levels are required for different input voltages to be injected in the proposed FFC. From the experimental results, the accuracy of the $V_{CS}$ is still not enough. If we can analyze the characteristics of the control IC, such as the $t_{delay}$ time of different voltages, it will help the FFC to be more accurate. On the other hand, the circuits of the proposed CPC-PWM can be digitalized and simplified. For example, a small Micro-controller can be used to replace the proposed circuit, which will contribute to its simple structure, accuracy, and flexibility.

## 4. Conclusions

In this paper, a feedback compensator (FBC) and a feedforward compensator (FFC) were successfully proposed to construct a novel Compensated Peak Current mode control PWM (CPC-PWM) for primary-side controlled flyback converters. The relevant theories and methods proposed in this paper have been confirmed by circuit implementation. A 5 W prototype with a wide input range has been designed and realized to access system performance. Using the proposed FBC, the PWM duty of a flyback converter working in overload can be descended to limit the current. The power consumption problem caused by the overload or short-circuited can be suppressed, and the components with larger specifications can be avoided. Using the proposed FFC, the effect of delay time can be descended to reduce the overflow current. The experimental results show that the proposed CPC-PWM can validate in a wide input voltage range and output short conditions, which also has good current accuracy and reduces power dissipation by about 68%.

**Author Contributions:** Conceptualization, C.-H.W., G.-R.H. and L.-R.C.; methodology, C.-H.W.; software, G.-R.H.; validation, C.-H.W. and L.-R.C.; formal analysis, C.-H.W., G.-R.H. and L.-R.C.; investigation, G.-R.H. and L.-R.C.; resources, C.-C.C.; data curation, C.-H.W. and C.-C.C.; writing—original draft preparation, C.-H.W., G.-R.H. and C.-C.C.; writing—review and editing, C.-C.C., C.-M.L. and L.-R.C.; visualization, C.-C.C., C.-H.W. and C.-M.L.; supervision, L.-R.C. All authors have read and agreed to the published version of the manuscript.

**Funding:** This research received no external funding.

**Institutional Review Board Statement:** Not applicable.

**Informed Consent Statement:** Not applicable.

**Data Availability Statement:** Not applicable.

**Conflicts of Interest:** The authors declare no conflict of interest.

## References


1. Liang, T.J.; Chen, K.H.; Chen, J.F. Primary side control for flyback converter operating in DCM and CCM. *IEEE Trans. Power Electron.* **2017**, *33*, 3604–3612. [CrossRef]
2. Cheng, M.C.; Pan, C.T.; Teng, J.H.; Luan, S.W. An Input Current Ripple-Free Flyback-Type Converter With Passive Pulsating Ripple Canceling Circuit. *IEEE Trans. Ind. App.* **2017**, *53*, 1210–1218. [CrossRef]
3. Connaughton, A.; Talei, A.P.; Leong, K.K.; Krischan, K.; Muetze, A. Investigation of a soft-switching flyback converter with full secondary side-based control. *IEEE Trans. Ind. Applicat.* **2017**, *53*, 5587–5601. [CrossRef]
4. Yaqoob, S.J.; Obed, A.; Zubo, R.; Al-Yasir, Y.I.A.; Fadhel, H.; Mokryani, G.; Abd-Alhameed, R.A. Flyback Photovoltaic Micro-Inverter with a Low Cost and Simple Digital-Analog Control Scheme. *Energies* **2021**, *14*, 4239. [CrossRef]
5. Connaughton, A.; Talei, A.P.; Leong, K.K.; Krischan, K.; Muetze, A. Variable on-time control scheme for the secondary-side controlled flyback converter. *IEEE Trans. Power Electron.* **2018**, *34*, 2416–2426. [CrossRef]
6. Marroquín-Arreola, R.; Salazar-Pérez, D.; Ponce-Silva, M.; Hernández-De León, H.; Aquí-Tapia, J.A.; Lezama, J.; Saavedra-Benítez, Y.I.; Escobar-Gómez, E.N.; Lozoya-Ponce, R.E.; Mota-Grajales, R. Analysis of a DC-DC Flyback Converter Variant for Thermoelectric Generators with Partial Energy Processing. *Electronics* **2021**, *10*, 619. [CrossRef]
7. Kumar, R.; Wu, C.-C.; Liu, C.-Y.; Hsiao, Y.-L.; Chieng, W.-H.; Chang, E.-Y. Discontinuous Current Mode Modeling and Zero Current Switching of Flyback Converter. *Energies* **2021**, *14*, 5996. [CrossRef]
8. Yau, Y.-T.; Hwu, K.-I.; Liu, K.-J. AC–DC Flyback Dimmable LED Driver with Low-Frequency Current Ripple Reduced and Power Dissipation in BJT Linearly Proportional to LED Current. *Energies* **2020**, *13*, 4270. [CrossRef]
9. Adragna, C.; Gritti, G.; Raciti, A.; Rizzo, S.A.; Susinni, G. Analysis of the Input Current Distortion and Guidelines for Designing High Power Factor Quasi-Resonant Flyback LED Drivers. *Energies* **2020**, *13*, 2989. [CrossRef]
10. Ponce-Silva, M.; Salazar-Pérez, D.; Rodríguez-Benítez, O.M.; Vela-Valdés, L.G.; Claudio-Sánchez, A.; De León-Aldaco, S.E.; Cortés-García, C.; Saavedra-Benítez, Y.I.; Lozoya-Ponce, R.E.; Aquí-Tapia, J.A. Flyback Converter for Solid-State Lighting Applications with Partial Energy Processing. *Electronics* **2021**, *10*, 60. [CrossRef]
11. Hannan, M.A.; Hoque, M.M.; Ker, P.J.; Begum, R.A.; Mohamed, A. Charge Equalization Controller Algorithm for Series-Connected Lithium-Ion Battery Storage Systems: Modeling and Applications. *Energies* **2017**, *10*, 1390. [CrossRef]
12. Dimitrov, B.; Krishna, M.; Cruden, A.; Sharkh, S.; Elkhateb, A. Analysis, Design, and Experimental Validation of a Primary Side Current-Sensing Flyback Converter for Use in a Battery Management System. *Electronics* **2018**, *7*, 43. [CrossRef]
13. Zhang, X.; Min, R.; Zhang, D.; Wang, Y. An Optimized Sensorless Charge Balance Controller Based on a Damped Current Model for Flyback Converter Operating in DCM. *Energies* **2018**, *11*, 3478. [CrossRef]
14. Liu, X.; Sun, Y.; He, Y.; Zheng, X.; Zeng, G.; Zhang, J. Battery Equalization by Fly-Back Transformers with Inductance, Capacitance and Diode Absorbing Circuits. *Energies* **2017**, *10*, 1482. [CrossRef]
15. Cao, Y.; Li, K.; Lu, M. Balancing Method Based on Flyback Converter for Series-Connected Cells. *IEEE Access* **2021**, *9*, 52393–52403. [CrossRef]
16. Cao, J.; Xia, B.; Zhou, J. An Active Equalization Method for Lithium-ion Batteries Based on Flyback Transformer and Variable Step Size Generalized Predictive Control. *Energies* **2021**, *14*, 207. [CrossRef]
17. Wang, C.; Xu, S.; Lu, S.; Sun, W. A low-cost constant current control method for DCM and CCM in digitally controlled primary-side regulation flyback converter. *IEEE Trans. Emerg. Sel. Top. Power Electron.* **2017**, *6*, 1483–1494. [CrossRef]

18. Lee, Y.J.; Niou, C.P.; Chen, C.Y.; Tsai, C.H. A digital power factor controller for primary-side-regulated LED driver. *IEEE Access* **2020**, *8*, 21813–21822. [CrossRef]
19. Chen, M.; Xu, S.; Huang, L.; Sun, W.; Shi, L. A novel digital control method of primary-side regulated flyback with active clamping technique. *IEEE Trans. Circuits Syst. I Regul. Pap.* **2020**, *68*, 950–962. [CrossRef]
20. Wang, C.; Sun, D.; Zhang, X.; Hu, J.; Gu, W.; Gui, S. A Constant Current Digital Control Method for Primary-Side Regulation Active-Clamp Flyback Converter. *IEEE Trans. Power Electron.* **2020**, *36*, 7307–7318. [CrossRef]
21. Huang, P.L.; Chen, D.; Chen, C.J.; Chen, Y.M. An adaptive high-precision overpower protection scheme for primary-side controlled flyback converters. *IEEE Trans. Power Electron.* **2011**, *26*, 2817–2824. [CrossRef]
22. Zhu, Z.; Wu, Q.; Wang, Z. Self-compensating OCP control scheme for primary-side controlled flyback AC/DC converters. *IEEE Trans. Power Electron.* **2016**, *32*, 3673–3682. [CrossRef]
23. Han, Y.; Wu, Q.; Li, Y.; Zhu, Z. Average inductor current measure and control strategy for multimode primary-side flyback converters. *IEEE Trans. Power Electron.* **2020**, *35*, 13096–13103. [CrossRef]
24. Li, Y.; Zhu, Z. A constant current control scheme for primary-side controlled flyback controller operating in DCM and CCM. *IEEE Trans. Power Electron.* **2020**, *35*, 9462–9470. [CrossRef]
25. Wu, Q.; Zhu, Z. An Adaptive High-Precision OCP Control Scheme for Flyback AC/DC Converters. *IEEE Trans. Power Electron.* **2017**, *32*, 8969–8973. [CrossRef]
26. Sato, T.; Nakano, T.; Harada, K. The overload-protection characteristics of the current-mode DC-to-DC converter: Analysis and improvement. In Proceedings of the 19th Annual IEEE Power Electronics Specialists Conference, Kyoto, Japan, 11–14 April 1988; pp. 830–835.
27. *AN-6754A-Design Guideline to Replace SG6742 with FAN6754A*, Version 1.0.1; Fairchild Semiconductor: Sunnyvale, CA, USA, 2013. Available online: https://www.onsemi.com/pub/Collateral/AN-6754A.pdf (accessed on 23 November 2020).

energies

*Review*

# Power Bus Management Techniques for Space Missions in Low Earth Orbit

**Luigi Schirone [1,*], Matteo Ferrara [1], Pierpaolo Granello [1], Claudio Paris [2] and Filippo Pellitteri [3]**

1 Scuola di Ingegneria Aerospaziale, Sapienza University of Rome, 00138 Rome, Italy; matteo.ferrara@uniroma1.it (M.F.); pierpaolo.granello@gmail.com (P.G.)
2 Centro Ricerche Enrico Fermi, 00184 Rome, Italy; claudio.paris@cref.it
3 Dipartimento di Ingegneria, University of Palermo, 90128 Palermo, Italy; filippo.pellitteri@unipa.it
* Correspondence: luigi.schirone@uniroma1.it

**Abstract:** In space vehicles, the typical configurations for the Solar Array Power Regulators in charge of managing power transfer from the solar array to the power bus are quite different from the corresponding devices in use for terrestrial applications. A thorough analysis is reported for the most popular approaches, namely Sequential Switching Shunt Regulation and parallel-input Pulse Width Modulated converters with Maximum Power Point Tracking. Their performance is compared with reference to a typical mission in low Earth orbit, highlighting the respective strengths and weaknesses. A novel solar array managing technique, the Sequential Maximum Power Tracking, is also introduced in the trade-off and was demonstrated able to boost energy harvesting, especially in the presence of mismatching in the solar array. It also can achieve top levels of reliability using a rather simple control hardware. Its operation was verified both by a Matlab–Simulink model and by an experimental breadboard.

**Keywords:** satellite power sources; solar array regulation module; battery charge/discharge regulation module; maximum power point tracking (MPPT); sequential switching shunt regulation ($S^3R$); power conversion unit

**Citation:** Schirone, L.; Ferrara, M.; Granello, P.; Paris, C.; Pellitteri, F. Power Bus Management Techniques for Space Missions in Low Earth Orbit. *Energies* **2021**, *14*, 7932. https://doi.org/10.3390/en14237932

Academic Editors: Yitao Liu and Ching-Ming Lai

Received: 18 October 2021
Accepted: 23 November 2021
Published: 26 November 2021



## 1. Introduction

Electrical power systems for space applications are highly demanding in terms of reliability: the related constraints of component screening and Single Point Failure Free (SPFF) circuitry set severe limits on available design solutions. Therefore, specific circuitry and regulation architectures were developed, quite different from those normally in use for stand-alone applications in the terrestrial environment, taking into account the other unique constraints of mass budget, electromagnetic compatibility, and not-convective heat exchange.

In addition, even the operating environment is quite different from terrestrial applications [1]. Spacecraft are exposed to a severe radiative environment that affects both solar cells, via crystallographic damage mainly related to absorption of energetic particles, and electronic components, that are also affected by the secondary radiation generated as a byproduct of the interaction between cosmic rays and spacecraft materials. Sunlight is also different as, unfiltered by the atmosphere, it is richer in photons in the visible and ultraviolet (UV) range, up to the UV-B spectral region (AM0 spectrum). Solar irradiance is 1367 W/m$^2$, with $\pm$3.5% seasonal variations for eccentricity of the Earth's orbit around Sun, plus some minor random fluctuations related to solar activity. Spacecraft periodically undergo eclipses with characteristics dependent on orbit geometry: in the most common Earth orbits the time spent in darkness ranges between 30 and 70 min, and sunlight time between successive eclipses varies from 1 h to some months.

The average albedo for Earth orbits is commonly assumed to be around 30%: in reality, it varies over the sunlight time, reaching its maximum when a spacecraft flies over the icy and cloudy poles, and its minimum over the "dark" oceans.

Solar panels in a common configuration are arranged as wings, continuously rotating around their axis to ensure normal incidence of direct solar radiation. Continuity of the current paths is achieved by dedicated slip rings inside the Solar Array Drive Mechanism (SADM). Many satellites otherwise have solar cells integrated in the external surfaces of the spacecraft body. In this case, the tilt angle of solar cells may be not uniform, leading to a diminished use of their photovoltaic surfaces. On the other hand, body-mounted solar arrays benefit from reduced complexity (no SADM or slip rings are needed) and are characterized by a smaller moment of inertia, enabling a more agile attitude control of the spacecraft.

The actual solar cell temperature depends on the details of the solar panels and the specific orbit: in the case of body-mounted solar cells, their good thermal conduction keeps solar cell temperature within a few degrees with respect to the satellite body, which is typically around 0 °C, whereas in the case of solar panels arranged as solar wings the deviations can increase to several tens of degrees, due to the ineffective heat exchange with the satellite body and to the higher relevance of radiative heat exchange to the dark space and to the Earth's surface. The orbit's altitude also affects temperature via the Earth's field of view: the minimum temperatures decrease to $\approx -150$ °C passing from LEO orbits (300–1000 km altitude) to Geostationary orbits (35,786 km altitude), whereas the maximum temperatures typically vary in the range +40 °C to +80 °C. In solar wings, the operating temperature also depends on the back-surface finish of the panels. Due to the small rate of heat exchange and to the high solar cell efficiency (typically in the 28–31% range), when a solar panel is in MPPT (Maximum Power Point Tracking) the heat to be dissipated is more than 400 $W/m^2$ smaller than that observed in the open-circuit conditions, with a temperature drop that can reach a few tens of degrees.

Temperature transition rate also is a peculiar stress source for space solar arrays: in near-Earth orbits, at the end of eclipses (from eighty to six thousand times a year), the temperature of solar cells in a few minutes rises from a deeply negative value ($-30/-150$ °C, depending on the orbit altitude and on the satellite layout) to a positive value (+40 °C to +80 °C, depending on the layout of solar panels and the electrical operating point).

Mismatching is also an issue in space solar arrays. Controlled-pointing solar panels are not always used in spacecraft as sometimes the more compact approach of body-mounted solar cells is preferred. If this is the case, the different tilt angles affect the current–voltage characteristics of the single photovoltaic strings. Even in the most common case of controlled-pointing solar panels, several sources of mismatching can modify the current–voltage characteristics of a photovoltaic (PV) string in unpredictable ways. Among them, worth mentioning are temperature gradients, arising whenever the spacecraft enters eclipse. Other sources of mismatching may be partial shading, statistical dispersion of technological parameters, and solar cell failures. All of them can affect solar array uniformity by modifying the current–voltage characteristics of a string in unpredictable ways.

Since the 1970s the most widely used approach for spacecraft power conditioning has been Sequential Switching Shunt Regulation ($S^3R$) [2], because of its high power density, modularity, simplicity, and inherently high reliability. Such shunt regulators also provide high power-path efficiency [3,4], as a solar array is split into sections that can be either directly connected-to or disconnected-from the power bus, as happens in the other regulation schemes belonging to the wider family of Direct Energy Transfer (DET) regulators: the Sequential Switching Shunt Series Regulator ($S^4R$) [5], the Astrobus™ [6], the shunt module developed for the Galileo Power Conversion and Distribution Unit (PCDU) [7], and the Sequential Switching Shunt Maximum Power Point Regulator (S3MPPR) [8].

Power regulation schemes implementing Maximum Power Point Tracking (MPPT) techniques have been introduced since the 1990s [9,10]. At the expense of a larger complexity [11,12], they provide optimal energy harvesting in a wide range of operating conditions. This is critical in missions in Deep Space, such as Rosetta-Mars Express [11,13] or GOCE [14], and in certain low Earth orbits, where the solar array characteristics undergo wide variations for temperature, intensity of solar radiation, or aging. The typical MPPT Solar Array Regulators are based on classical Pulse Width Modulated (PWM) series switching converters [15]. The converter is a stepdown in most applications, such as Globalstar 1 [9], Mars Express, Rosetta [11], Venus Express [16], Goce [14], Swarm [17], Lisa Pathfinder [18], Juice, and Exomars [19]. In most cases, the configuration is the two-inductors buck configuration [20]. Step-up configurations were used for Globalstar 2 and some high-power spacecraft such as BepiColombo [13,21]. Top flexibility in the solar array configuration is achieved by using Solar Array Power Regulators (APR) comprising step-up/step-down converters [22,23], so that the solar array voltage may vary across the power bus voltage.

A first attempt to gather advantages of both approaches by introducing MPPT capability in $S^3R$ power units was proposed in [24]. It was further developed in [8,25] and then specialized for applications in spacecraft with electrical propulsion [26] or in geostationary satellites [27].

All previously introduced regulation approaches provide specific advantages and limitations so their trade-off has been widely debated [15,27]; so far, no power conditioning technique has been established as a standard. The heritage of manufacturers and agencies also affects the trade-off.

This work is intended to contribute to the debate taking into account in the trade-offs of other specific solar array management techniques capable of providing advantages with respect to all previously reported approaches.

In sectional MPPT power units [15], the solar array is split into several sections, each fitted up with a dedicated Solar Array Power Regulator (APR) with a multimode controller, capable of either independently performing MPPT or harmonically contributing to the regulation of the bus voltage or the battery current. Coordinate operation of the different APR in regulation mode is a major issue with these systems so some authors [28] developed to the purpose algorithms to be embedded in powerful supervisors.

An expedient solution appears to be provided by the Sequential Maximum Power Tracking (SMPT) bus regulation technique: the coordination problem is solved by implementing an operational sequence such that just one section at a time is kept in control mode, while the others are kept either in MPPT or in standby.

A thorough description of the main regulation techniques and SMPT operation is given in Section 2. A comparative analysis of $S^3R$ and the regulators based on MPPT or SMPT techniques is discussed in Section 3 with reference to a typical space mission in low Earth orbit, in the presence of relevant mismatching in the solar array. Section 4 is dedicated to the demonstration of SMPT operation, based on both a Matlab–Simulink model and on the experimental results provided by a breadboard built with automotive-grade components.

## 2. Solar Array Regulation Approaches

### 2.1. *Sequential Switching Shunt Regulation*

In $S^3R$ systems, the solar array is split into N sections directly connected to a power bus via blocking rectifiers. Each of them is also equipped with a shunt switch that can be tripped on in order to block power transfer to the bus (Figure 1). In order to reach a power balance between solar array generation and load absorption, the active solar array sections are set according to a pre-ordered sequence [4]: when the required solar array generation $G_{SA}$ is smaller than the power supplied by M solar array sections, but larger than the power supplied by M − 1 of them, rough regulation is provided by keeping M − 1 sections permanently enabled and the other N-(M − 1) permanently disabled. Fine regulation is achieved by hysteretic control of the duty cycle on the shunt switch of a single solar array section.

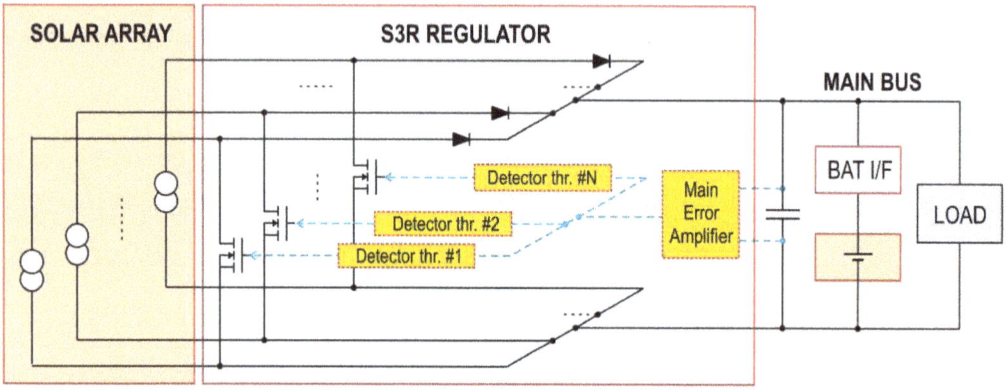


**Figure 1.** Power module with Sequential Switching Shunt Regulation.

The battery is typically interfaced to the bus via a Battery Charge Regulator (BCR) and a Battery Discharge Regulator (BDR) (full-regulated bus configuration). The BDR may be replaced by a simple rectifier (sun-regulated bus), especially in the missions involving a small Depth of Discharge (DoD).

Regulation control is performed by a Main Error Amplifier, generating both the activation sequence for the direct-current (dc) shunts and the duty cycle for the switching solar array section. It is based on a consolidated, simple, and Single Point Failure Free (SPPF) circuitry, with very high reliability.

A variant DET regulation scheme, the Sequential Switching Shunt Series Regulation ($S^4R$) has also been proposed, with an original configuration of the battery charge regulator [29] that is more efficient and decoupled from the bus regulation control. The improvement comes with a reliability issue, as in the original layout the battery-charging path is not SPFF [30].

For the sake of the analyses proposed in this work, minor differences in performance are observed among Direct Energy Transfer (DET) regulation approaches and thus only $S^3R$ will be considered.

### 2.2. *Maximum Power Point Tracking Regulation*

The photovoltaic source supplies the power bus via a series dc–dc converter enslaved to a multimode controller that can operate in either regulation or maximum power point tracking (MPPT) modes.

A survey of Maximum Power Point Tracking techniques is given in [31], whereas deeper insight about modelling and stability issues can be found in [32] and [15,33], respectively. The same algorithms used in stand-alone terrestrial applications may be used: Perturb and Observe [34] is preferred when the MPPT controller is embedded in software, whereas Incremental Conductance can be implemented by a hardware circuit, as described in [11]. The first one is more flexible and can be easily upgraded to manage quick transients or multiple maximum power points. On the other hand, some kind of digital processor is required, which can be quite expensive when carried out with space-qualified devices. The latter only requires a few discrete components but can face some stability problems in managing fast load variations [12]. The main difference with the MPPT algorithms for terrestrial applications is in the dynamical model of solar panels, which are made with solar cells based on technologies not used in terrestrial applications.

When load absorption is small, the multimode controller is in either constant-current or constant-voltage regulation mode and the operating point of the solar array typically is in the low-current region. Thus, in response to an increase of power absorption from the loads, the feedback loop increases the duty cycle, pushing the SA operating point towards higher currents [34]. When current exceeds the Maximum Power Point, further increases of duty cycle result in power reductions, and the controller switches to the MPPT mode. From there, the operating point is driven to and kept close to the global maximum power point, according to a proper tracking technique. The regulation mode starts again when the load absorption returns below the maximum available power from the solar array.

On the other hand, the system architectures are heavily affected by the requirements of reliability: several redundancies are normally introduced in the power conversion unit in order to avoid the risk of Single Point Failures. In the configuration used for many space missions [11,16], the whole solar array is split into two parts, corresponding to the spacecraft wings. In each of them (Figure 2), all photovoltaic strings are connected in parallel to supply a power module comprising three hot-redundant converters. Precisely the redundancy scheme is two-out-of-three partial redundancy [11] because each converter is sized to carry up to 50% of the overall power so that losing one of them will not affect spacecraft operation. In the space power literature, this configuration for the solar array regulator is commonly referred to as "MPPT".

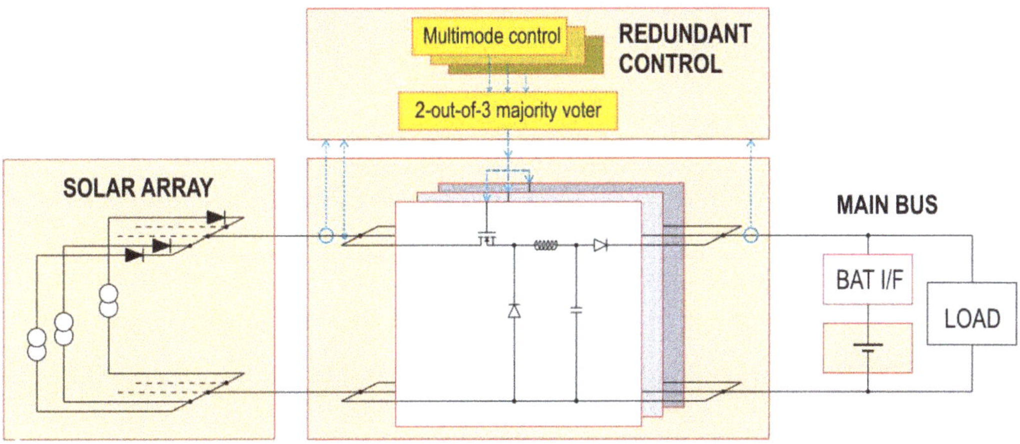


**Figure 2.** Power Module with the classic MPPT regulation for space solar arrays.

The controller also consists of three modules connected according to the majority voting redundancy scheme, allowing for the detection and dismissal of a single unmatched signal. The output bus is regulated, involving the use of both battery interface converters, with their mass and power losses.

### *2.3. Comparing MPPT and S$^3$R*

The trade-off between MPPT and S$^3$R has been widely debated in the space power community [4,11], as both provide specific advantages. This led to several proposals to combine the specific advantages of both techniques [23,24], also developing techniques tailored for specific applications [26,27].

The S$^3$R controller is based on a consolidated circuitry, simple and SPFF, that easily matches the required levels of reliability.

The MPPT-SAR scheme reported in Figure 2 involves more complex circuitry, and reliability has to be increased by the redundant connection of parallel converters. In practice, in addition to using reliable and SPFF circuitry inside the different converter blocks for their coordination, it is necessary to make them functionally independent, preventing propagation of any type of failure along any power or signal lines, and also avoiding harmful thermal couplings.

The shunt regulation approach used in S$^3$R minimizes dc losses in the power path, as just the blocking rectifiers are in series to the electrical connections. The switching losses also are small, because alternate current (ac) flows only in the section performing fine regulation, and the ac losses are substantially limited to charging/discharging of the switch capacitances, at a frequency that is typically in the 10–20 kHz range. The ripple is filtered out by a capacitor bank with no inductors. Thus, efficiency can be larger than 96%, depending on the bus voltage.

In reality, the serial dc–dc conversion used in MPPT regulators involves larger power losses [35], as the whole current undergoes switching at a frequency of a few hundreds of kHz [11], generating appreciable ac power losses both in the active devices and in the magnetic components. In space missions with few eclipses and stable operating conditions for the solar arrays, e.g., geostationary satellites, the related losses (typically a few per cent) can offset the benefits of the Maximum Power Point Tracking capability, and DET converters are preferred.

On the other hand, energy harvesting in $S^3R$ is less effective, because the solar array has the voltage tied to the bus voltage, independent of its actual current–voltage characteristics. Thus, wide margins must be taken in the solar array design in order to avoid that—in the most unfavorable conditions—parametric variations (e.g., due to heating or aging) may cause the Open Circuit Voltage of any string to fall below the bus voltage. This causes inefficient use of the solar panels, which normally are operating at voltages far from their maximum power point voltage ($V_{MP}$) even during the load peaks.

Otherwise, in the case of MPPT, especially when a converter configuration with buck-boost capabilities is chosen [22,23], all available power can be exploited independently of solar array voltage. This capability may be used for more effective sizing of solar arrays, and/or to produce extra power. It could simply be used to reduce the depth of the battery cycles, allowing for smaller batteries, with advantages also from a systems point of view. In some missions the extra available power can be used to increase, even temporarily, payload capability (e.g., in a TLC spacecraft to increase the number of transmission channels) and/or to widen the payload duty cycle.

In MPPT-APR the bus voltage is decoupled from solar array voltage, thus enabling the direct connection of the battery to the bus (unregulated bus configurations), at least in the missions involving small battery Depth of Discharge. Therefore, the mass and the power losses of both BCR and BDR may be avoided, even if they are still used in specific spacecraft, e.g., in [11].

The challenge of merging MPPT functionality into $S^3R$ systems with no reduction in reliability and a negligible increase in complexity was first faced in [24]. The proposed approach was to enslave the bus voltage in $S^3R$ to an MPP tracker [4]. This can be done by minimal modifications to the classical $S^3R$ circuitry and has a main drawback, related to the wide variations of the bus voltage (tens of volts) arising from $V_{MP}$ variations. Such variations can be managed either by widening the input range for all devices connected to the bus, at the expense of their performance, or by introducing some bound to the voltage tracking range, at the expense of the energy harvesting capability. In either case, using a full battery interface is mandatory, with the related mass and power losses.

In the systems based on the MPPT-SAR in Figure 2, as in $S^3R$, all photovoltaic strings are connected in parallel so that any source of mismatching would make the overall maximum power smaller than the sum of the maximum contributions that would be available from individual strings. Therefore, both approaches appear to be not perfectly fitted for missions in low Earth orbit, where eclipses frequently trigger temperature transitions, with the related thermal gradients inside the solar array, and sometimes partial shading.

### 2.4. Sequential Maximum Power Tracking Regulation

The effects of mismatching are reduced with the sectional MPPT configurations [15], first used in [9]: the solar array is split into N sections independently interfaced to the power bus by means of dedicated APRs (Figure 3) capable of tracking their individual MPP to provide optimal energy harvesting. This feature appears particularly interesting in spacecraft with body-mounted solar cells, where the variable angles of incidence may force photovoltaic strings to work at different operating points. Sectional MPPT may also reduce the effects of other sources of mismatching, such as temperature gradients, partial shading, statistical dispersion of technological parameters, and solar cell failures. All of them can affect solar array uniformity by modifying the current–voltage characteristics of any string in unpredictable ways.

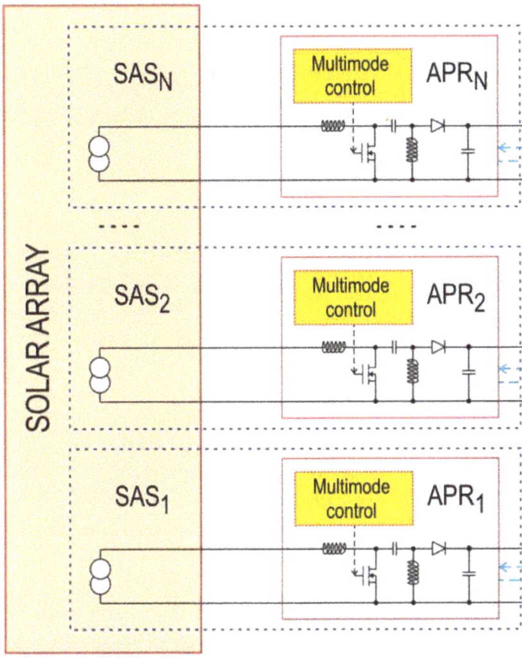


**Figure 3.** Power Module with SMPT solar array regulation.

Despite these advantages, and all other advantages of MPPT–SAR that still are effectual, sectional MPPT has seldom been used in space. An issue for its application with battery buses is the need to coordinate the different APRs during the regulation modes through effective and reliable circuitry. SMPT provides an effective solution to this issue.

Each APR can either autonomously perform current/voltage regulation based on shared feedback signals or can manage MPPT for the dedicated solar array section. At the same time, all APRs share their operational status via redundant I/O lines. Thus, the SMPT algorithm is implemented on the basis of both shared status signals and of shared feedback signals. The issue of coordinating the operation of the APRs is faced by keeping just one of them in regulation mode, with the others either in MPPT or in standby. The power-up sequence outlined in Figure 4 clarifies SMPT operation: at turn-on an APR starts operating in regulation mode, autonomously adjusting its output power to enforce the desired control law. If the feedback loop drives it in MPPT, the next APR in the sequence is turned on and takes charge of regulation. The sequence continues until the desired bus condition is achieved, e.g., with $APR_M$ in regulation mode, $APR_1$ to $APR_{M-1}$ in MPPT mode, and $APR_{M+1}$ to $APR_N$ in standby. In response to a decrease of the load absorption, the sequence is run in the opposite direction: when the feedback loop is going to quench $APR_M$, $APR_{M-1}$ exits MPPT mode to start regulation, while $APR_M$ ends in standby.

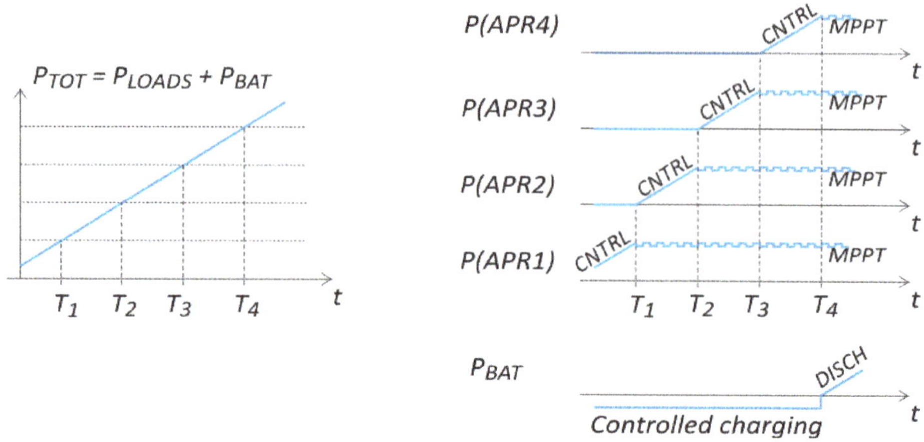


**Figure 4.** Power-up sequence in a SMPT regulator with four solar array sections. Left plot: overall power absorption vs. time. Right plot: power supplied by the single APRs and by the battery.

Thus, the operational state of all APR's is coordinated based on a small number of signals, that can be easily made SPFF.

An option for the sequence control is to use a central device, in charge of exchanging status signals and enabling signals with all APRs. The control circuitry as well as all I/O lines must be redundant and Single Point Failure Free, to avoid that a single malfunction may result in a failure for the entire power system.

In the case of low-cost missions based on Components Off The Shelf (COTS), the reliability of single components is not easily predictable, and the two-out-of-three redundancy scheme may not be sufficient to ensure required reliable operation. Thus, it could be convenient to use a distributed sequence control, where each APR autonomously manages its operational mode according to the shared status signals and a wired sequence. In order to avoid that the loss of control on a single APR may affect the target performance, it is easy to use the SMPT modularity to implement partial redundancy schemes, such as N-of-(N + 1), so that the need for a Failure Free central control unit is offset to the requirement of no-failure-propagating interfaces (signal, power and thermal) among different APRs.

In any case, SMPT also is intrinsically more reliable than the classical parallel MPPT, first because there is no input node with all PV strings connected in parallel. Moreover, the last APRs in the sequence can spend most of their time on standby and therefore undergo a reduced failure rate [36].

When SMPT regulation is compared to $S^3R$, all advantages of sectional MPPT are confirmed: it allows high flexibility in designing solar array voltage and can use MPPT to get top power harvesting even in mismatched solar arrays. It also shares some drawbacks, related to the larger complexity of the switching cells, even if this is partially compensated by the possibility of using an unregulated bus, thus avoiding the burden of the battery interface converters.

A summary of the main features of the investigated regulation techniques is provided in Table 1. The first row reports the efficiency of the conversion cell, as it represents the top efficiency theoretically reachable by the converter when neglecting all the power used to supply controllers, telemetries, redundancies, and all service circuitry. For instance, in the case of the two-out-of-three partially redundant systems reported in [11], three dc–dc converter cells are continuously running at a power level that is not exceeding 66%, until a failure event blocks one of the converter cells. Thus, the efficiency of the overall regulator is reduced due to a non-optimal power level and is also affected by the triple control circuits, so that it can hardly be evaluated, despite the fact that losses in the conversion cell are easily predicted.

**Table 1.** Summary of the main features of the investigated regulation techniques.

| | $S^3R$ | MPPT | SMPT |
|---|---|---|---|
| Conversion cell typical efficiency | 96% | 92% | 92% |
| Energy harvesting | poor | good | optimum |
| Sensitivity to mismatching | high | intermediate | none |
| Power cell complexity | low | intermediate | intermediate |
| Control complexity | low | high | intermediate |
| Power Cell Redundancy scheme | N out of N + 1 | 2 out of 3 | N out of N + 1 |
| Control redundancy | easy | complex | intermediate |
| Battery Charge Regulator | needed | optional | not needed |
| Battery Discharge Regulator | needed | optional | not needed |

## 3. Comparison among the Regulation Techniques

A trade-off analysis of the considered regulation techniques has been developed. The analyses are in reference to a specific case study, an Earth Observation Spacecraft flying in a Sun-synchronous orbit at 380 km altitude, characterized by [1], an orbital period T0 = 90 min, with 30 min eclipse duration. This trend is typical for low Earth orbits: the exact value of the irradiance duty cycle depends on orbital altitude, inclination, and eccentricity, and in a circular orbit it is lightly larger than 66% at 300 km, and smaller than 60% at 1500 km. Only in very special orbits it can be close to 100% (e.g., in down-dusk orbits, where the orbital plane is almost orthogonal to the solar vector). Due to this variability, we were forced to develop our analyses in reference to a very specific space mission, assuming that the general conclusions could be extended to similar applications, including many low Earth orbits with pulsed loads.

In order to perform a straight trade-off analysis, the solar radiation collected by the panels has been assumed to be only affected by eclipses (step transitions from darkness to sunlight, neglecting penumbra times) and by solar panel temperature. As the actual solar cell temperature depends on the details of the solar panels, in our analyses we referred to typical textbook data for low Earth orbit missions [37] (Sections 9.3–4) and assumed that temperature was in the range −60 °C (at the end of eclipse) to +40 °C (at the beginning of eclipse). Figure 5 reports the trends considered for this trade-off analysis for irradiation (orange curve, right axis) and temperature (blue curve, left axis).

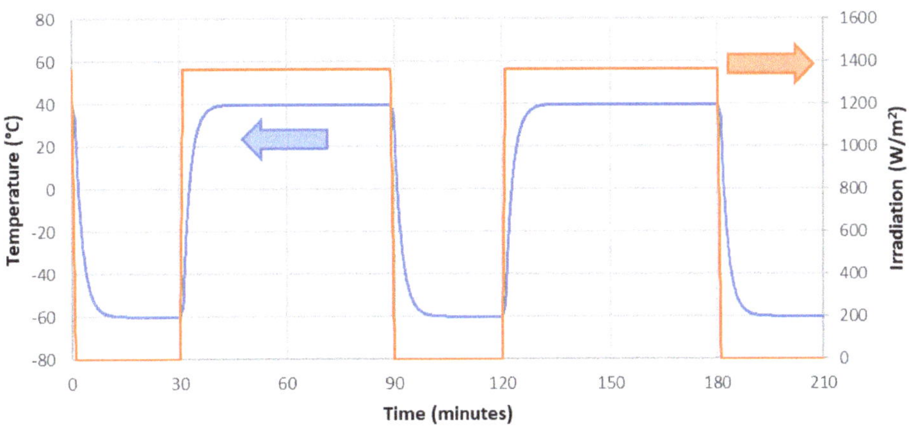


**Figure 5.** Irradiation (orange curve, right axis) and solar cell temperature (blue curve, left axis) in the considered orbit.

The solar array considered for this case study was able to provide 1.4 kW in AM0 at Beginning of Life and consisted of 1088 triple-junction solar cells, model CTJ30 by CESI, arranged in 64 strings of 17 series-connected cells, grouped as 8 independent sections. The analyses were based on their expected performance after aging by absorption of a radiation fluence of $5 \times 10^{14}$ equivalent 1 MeV electrons, corresponding to several years in a typical Sun-synchronous polar orbit.

In order to stress differences among the regulation approaches investigated in this work, mismatching among solar array sections has been taken into account. In particular, it was assumed that a section underwent a relevant reduction of the output voltage, which in real operation may be caused by damaging or shadowing a group of solar cells, with activation of the dedicated bypass diodes.

Table 2 collects the values of Open Circuit Voltage ($V_{OC}$), Short Circuit Current ($I_{SC}$) and their thermal coefficients considered for the different solar array sections.

**Table 2.** Electrical parameters of the solar array sections.

| | **Sec. 1–5** | **Sec. 6** | **Sec. 7** | **Sec. 8** |
|---|---|---|---|---|
| $V_{OC}$ (V) @AM0 | 41.8 | 44.4 | 39.2 | 31.3 |
| $I_{SC}$ (A) @AM0 | 4.2 | 4.1 | 4.3 | 4.2 |
| $\Delta V_{OC}/\Delta T$ (%/°C) | −0.217 | −0.217 | −0.217 | −0.217 |
| $\Delta I_{SC}/\Delta T$ (%/°C) | +0.0964 | +0.0964 | +0.0964 | +0.0964 |

The model for Solar Arrays accounts for temperature effects on the solar cell characteristics. Based on the thermal coefficients and other specification data published by the manufacturers [38], estimated current–voltage pairs for a representative set of temperatures were collected in a Look-Up-Table, allowing derivation by interpolating the current–voltage characteristics at any temperature in the considered range.

The available power depends on the regulation approach, as reported in Figure 6. When the loads absorb all available power from the solar array, in $S^3R$ all sections in parallel are tied to the bus voltage; in MPPT all sections in parallel are kept at the voltage corresponding to the maximum power point of the overall solar array; in SMPT optimal energy harvesting is achieved as each section independently is kept at the voltage corresponding to its Maximum Power Point.

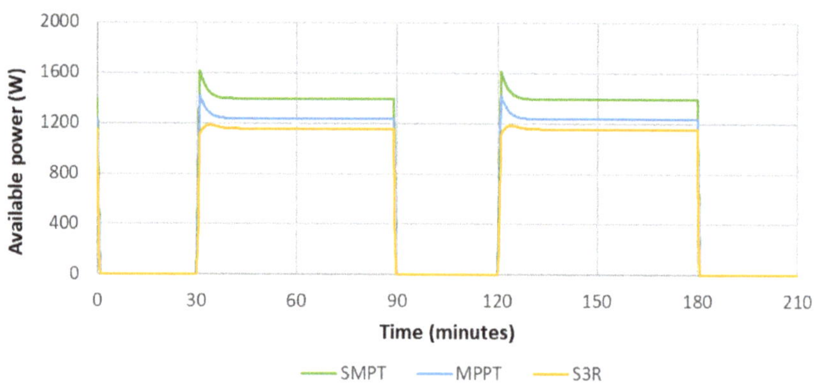


**Figure 6.** Available power from the solar array with the different regulation approaches.

The difference between the curves related to SMPT and MPPT is related to mismatching in the solar array: the common node in basic MPPT regulators keeps some sections at a voltage different from their actual $V_{MP}$, whereas in the sectional MPPT approaches, such as SMPT regulation, all solar array sections are working at their optimal voltage. Otherwise, in $S^3R$ the operating voltage for solar arrays is fixed independently of the variations of $V_{MP}$. Furthermore, due to the negative thermal coefficient for voltages, after some minutes in sunlight the sections starting with the smaller $V_{OC}$ may be forced to work in the low-current region of their characteristics, with appreciable power reductions. In Figure 7 it is clearly seen that this happens for the section 8 in the solar array, whose open-circuit voltage at 25 °C was just 3.3 V larger than the bus voltage.

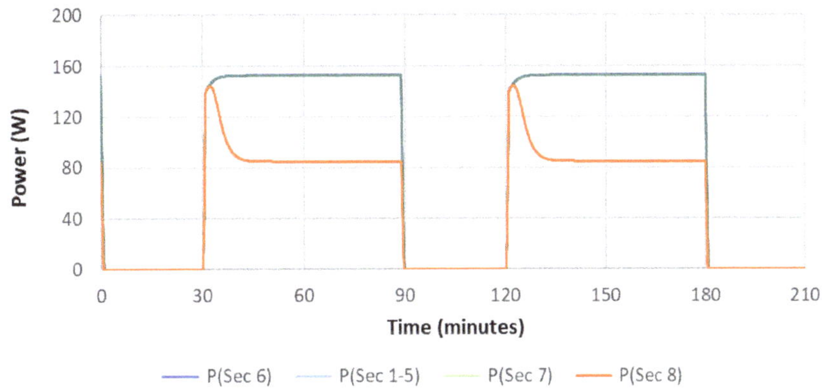


**Figure 7.** With $S^3R$ all solar array sections but section 8 supply the same output power (section 8 has a relevant decrease in Voc, whereas minor differences among Voc of sections 1–7 exist, so that the curves are superimposed).

The load profile considered for the spacecraft is reported in Figure 8. Basic activities absorb 200 W both in sunlight and in eclipse, and periodically the operations of the payload or some spacecraft subsystems result in additional peaks of 900 W for 5 min and 3.2 kW for 1 min. The conversion efficiency was assumed to be 96% for $S^3R$ and 92% for both MPPT-APRs and SMPT-APRs.

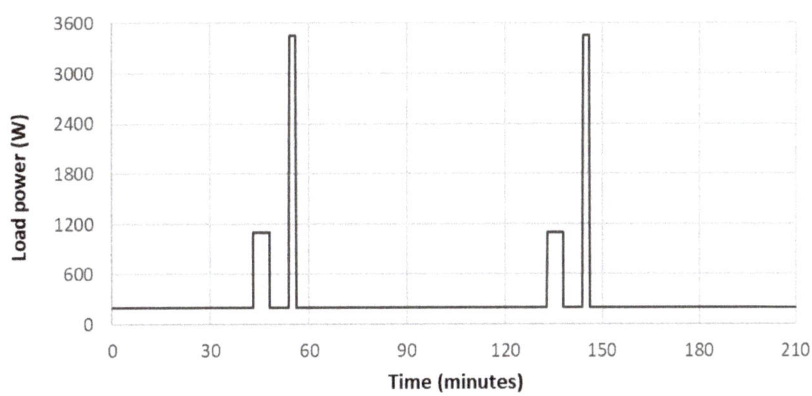


**Figure 8.** The considered load waveforms.

The battery supplies the spacecraft during eclipses and load peaks. During sunlight it absorbs 200 W for recharging, whenever enough power for the purpose is available (see Figure 9, lower curves, left axis). During the peaks of load absorption, the battery stops charging and, when needed, sustains the load with suitable discharge pulses.

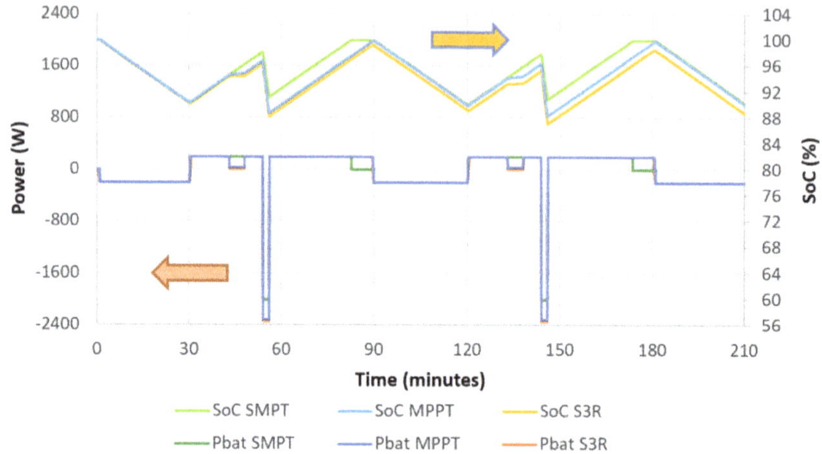


**Figure 9.** Battery State of Charge (upper curves, right axis) and Charging Power for battery (lower curves, left axis).

The trend of the battery State of Charge (SoC) (Figure 9, upper curves, right axis) shows that high load pulses may result in deep cycles in the battery: as expected, the depth of discharge is at a maximum for $S^3R$. This could affect battery duration [37] and shall be compensated by increasing the size of the battery.

For the considered case study, the overall amounts of energy per orbit delivered by MPPT and $S^3R$ are quite similar, as the larger efficiency of $S^3R$ is compensated by the poorer harvesting of the power available on the solar array. For the chosen solar array, the MPPT regulator completes battery recharging with a small margin with respect to the available sunlight time, whereas the $S^3R$ regulator is not able to fully recharge the battery recharging: larger solar arrays would be needed to enable energy balance with $S^3R$.

Otherwise, the larger amount of available power, as shown in Figure 6, enables the SMPT regulator to complete battery recharging well in advance of the next eclipse. It is clear that the case study was specifically built to highlight the strengths of the sectional MPPT approaches such as SMPT, but there is no doubt that these regulation techniques provide clear-cut advantages in terms of performance.

## 4. SMPT Simulation and Experimental Results

A model of the SMPT regulator was developed in Simulink, aiming to demonstrate the operation of the sequencing algorithm. As shown in Figure 10, the system consists of four blocks, each comprising a solar array section (in blue), an APR consisting of a dc–dc converter (in grey), and a dedicated controller in charge of managing the SMPT sequence on the basis of the status signals coming from other blocks as well as regulating the APR output current (Iout) to enforce the desired overall control law for the battery and the power bus. All converter outputs are connected in parallel and supply a battery and a variable resistive load.

The converter configuration is SEPIC [39], providing both step-up and step-down operation to allow the solar array sections to operate at voltages across the battery voltage. The circuit embedded in the submodel blocks is reported in Figure 11.

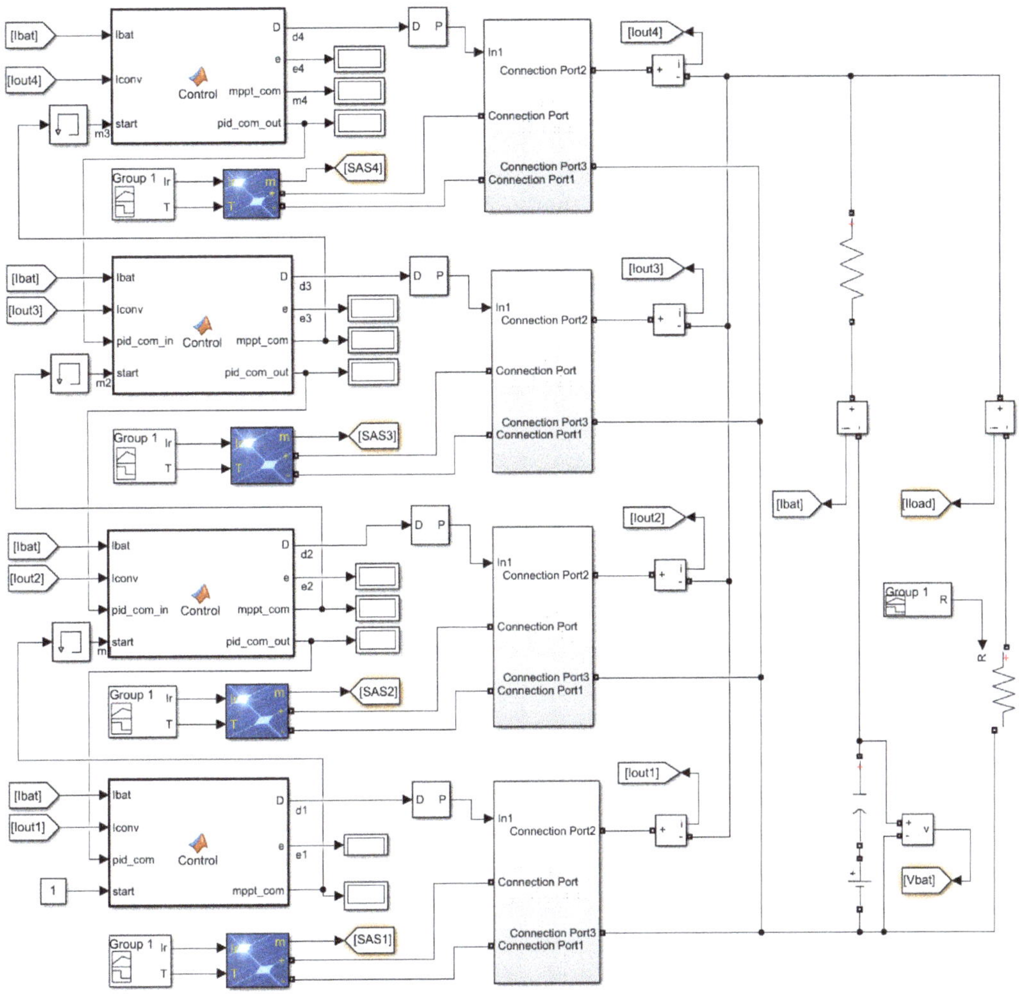


**Figure 10.** Simulink model of the SMPT technique. The subsystem blocks (in grey) model the SEPIC converters directly from their circuit schematics. The blue blocks are library blocks for photovoltaic arrays.

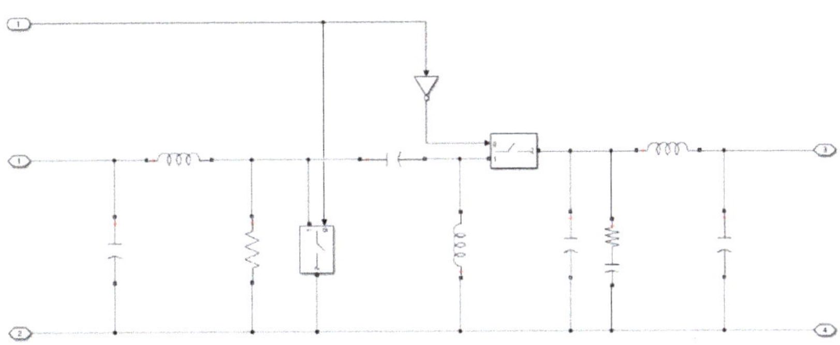

**Figure 11.** Submodel block for the SEPIC converters.

The control block of any APR sets the operating mode according to the observed battery current, to the Duty Cycle (DuC) of the corresponding converter and to the status signals received from other APRs (see Figure 12). Any single controller can work in three different modes:

- State 0: the converter is in IDLE mode, and the controller applies a default Duty Cycle (DuC) of 1, forcing the output of the photovoltaic source in short-circuit. This solution was preferred to the option of getting the idle mode with the source in open-circuit because it allows easier management of the end of eclipses.
- State 1: the controller runs the PID algorithm for control, modulating the converter DuC to stabilize the battery charging current.
- State 2: the controller is the MPPT mode, running a basic Perturb and Observe (P&O) algorithm, intended to maximize the contribution from the single block, while another APR in the sequence is in charge of fine control.

Changes of state are triggered by the signals shared with the other APRs, namely two signal lines shared with the previous APR in the start-up sequence and one signal line with the next one. More precisely, they are:

- the "START" signal. When received in input, forces an APR to exit IDLE mode and start PID control.
- the "MPPT_COM" signal, asserted when the APR is in MPPT mode. It sets the START input of the next APR to 1.
- the "PID_COM" signal, asserted when the output current of the APR drops below a specified threshold, leading the APR to enter in the IDLE mode. It forces the previous APR in the start-up sequence to stop the MPPT mode and to start the PID control.

In the Simulink model reported in Figure 10, the PID_COM_IN input and the MPPT_COM output are not provided in the last converter because there is no next converter, while the PID_COM_OUT is not provided in the first converter because there is no previous converter.

Operation is described in detail in the flow chart reported in Figure 12. An APR enters the PID control mode when it receives a positive level on the START input line. The operating mode passes to MPPT when a Maximum Power Point is detected via a change in the slope of the power-vs-duty cycle trend. At that moment, a START signal is sent to the next APR in the sequence, that enters the PID mode. Otherwise, in case the controller drives the APR output current to zero, the operating mode passes to IDLE and a signal on the PID_COM line forces the previous APR to exit MPPT and start PID mode. Adequate thresholds and hysteresis have been introduced to achieve smooth operation.

The simulation was carried out with the following parameters:

- single-section maximum power: 125 W;
- Battery start voltage: 26 V;
- Nominal charging current: 4 A;
- Battery capacity: 1 kWh;
- Load sequence (see Figure 13): 400 W to 800 W to 400 W to 140 W (1.7 Ω to 0.85 Ω to 1.7 Ω to 4.8 Ω);
- Discrete, step size $10^{-7}$ s.

The system was turned on with the bus absorption at 400 W including the load, and with the charging current for the battery set to 4 A. As depicted in Figure 14, a first APR, say $APR_1$, is activated and its output current increases attempting to achieve power balance. After a few tens of milliseconds, its output current reaches its maximum and the $APR_1$ starts operating in MPPT while the next section is activated in regulation mode. The step is repeated for the solar array sections 2 and 3 until around 0.3 s. The power balance is achieved with $APR_4$ in regulation mode.

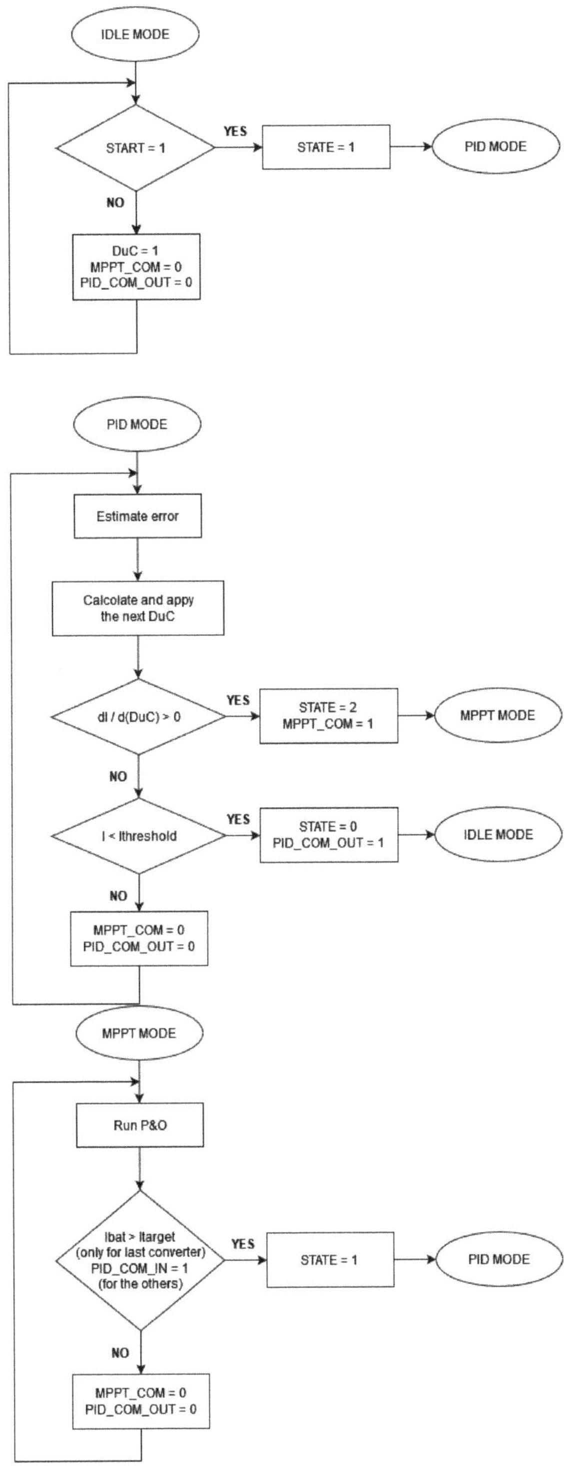


**Figure 12.** Flow chart of the control algorithm.

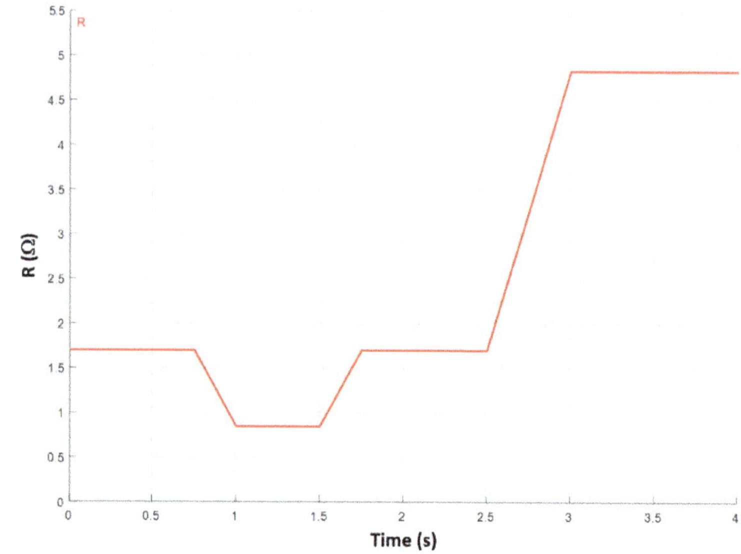


**Figure 13.** Load profile for the simulation.

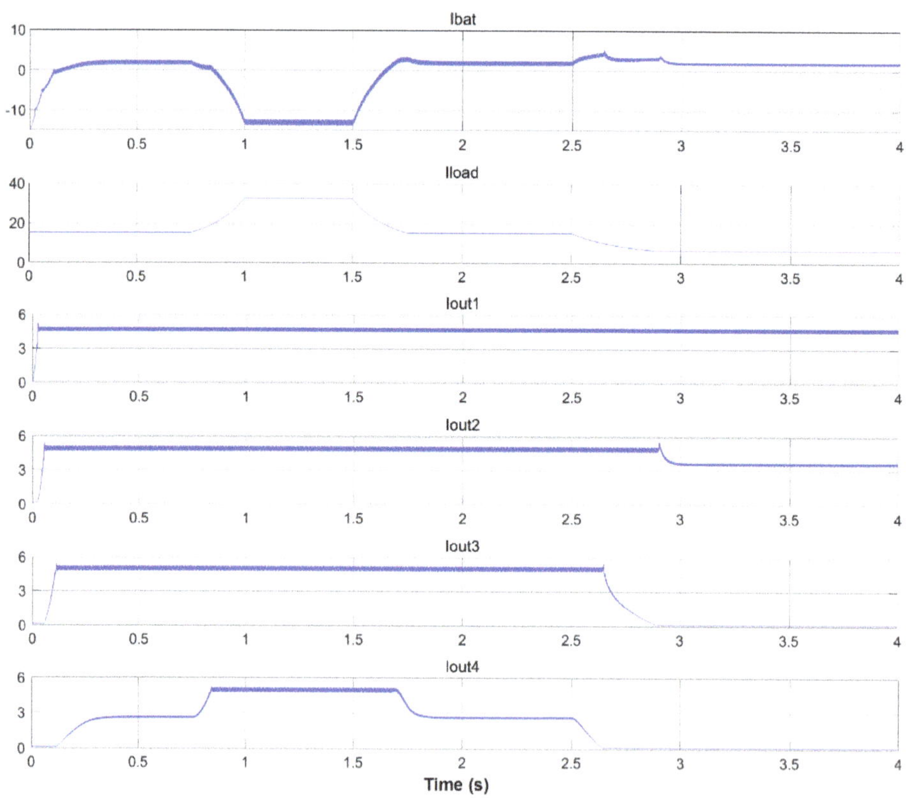


**Figure 14.** Waveforms provided by the Simulink model of the SMPT technique: Ibat: battery current; Iload: load current; Iout$n$ with $n$ = 1 to 4: output current of APR$n$. All currents are in amperes.

After 0.75 s the load absorption is further increased to 800 W. When $APR_4$ also goes into MPPT, all the available power from the solar array is delivered to the load and the battery starts discharging at nearly 12 A to balance the load demand. The load absorption returns to 400 W at 1.8 s and the previous operating point is quickly recovered. At 2.5 s a further load variation reduces the load absorption to 140 W, and the solar array sections 4 and 3 sequentially quench their outputs. The final equilibrium is found with section 2 in the regulation mode. During the transients, the control on battery current loosens and some deviations with respect to the expected trend are observed. They are particularly evident at 0.75 s, 2.5 s, and 2.7 s. An improved design of the control law will reduce this effect.

The operation of the SMPT algorithm was also experimentally demonstrated by a PCDU breadboard based on SEPIC APR converters. The regulation controller and the SMPT sequencer were both embedded in automotive microcontrollers. A detailed description of the breadboard is out of the scope of this work.

The test setup is shown in Figure 15. The photovoltaic source was provided by a Keysight E4360 two-channel Solar Array Simulator, with both channels configured as $V_{OC}$ = 44 V, $I_{SC}$ = 3.3 A, $V_{MP}$ = 39.5 V, and $I_{MP}$ = 3.2 A in order to reproduce the typical current–voltage characteristics of solar array sections consisting of 102 triple-junction solar cells in AM0 conditions [38], arranged as 6 parallel strings of 17 solar cells in series. The output bus was directly connected to a storage unit consisting of a battery bank with 36 Li-ion cells (size 26650), arranged as 4 parallel strings of 9 cells (4P9S). At the beginning of the tests, the battery was at a voltage $V_{BAT}$ = 30 V. During the tests, the load was provided by a Hewlett Packard 6060B electronic load, and the overall bus absorption was varied in the range 0–180 W in current-control mode, with the current for battery charging stabilized at 2 A.

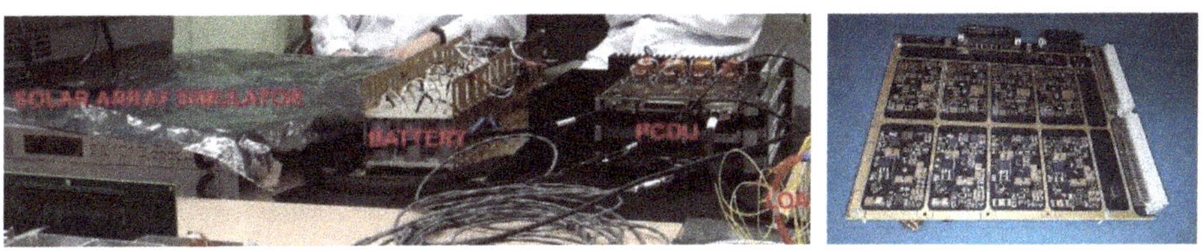


**Figure 15.** The test setup (**left**): the breadboard PCDU is supplied by a Keysight E4360 Solar Array Simulator and drives a power bus connected to a 4P9S Li-ion battery and to the loads. The converter board of the PCDU (**right**), designed to accommodate up to eight channels.

The solar array voltages observed during the tests are shown in Figure 16. As it is possible to see, the controllers used short-circuit as a starting condition so that the operating point moved in the $V < V_{MP}$ region, where the current is almost a constant. Therefore, the reported solar array voltage waveforms provide a glance at the input power of the different APRs.

At system turn-ON ($t_1$), the overall output power was $P_{OUT}$ < 120 W, resulting from battery charging at 2 A, and an additional 60 W load in parallel. The $APR_1$ (yellow trace) was controlling the charging current, while $APR_2$ (blue trace) was set to a base condition, with output power smaller than 20 W, just to keep the converter in Continuous Conduction Mode. At $t_2$ the load was increased to 120 W (overall output power Pout = 180 W), the $APR_1$ was forced into MPPT, and the control passed to $APR_2$ (blue trace). Then, at $t_3$, the load returned to 60 W: $APR_1$ first kept the MPPT mode until the output power of $APR_2$ went below base power level (at $t_4$), then $APR_1$ started again to operate in the control mode. In the meantime, $APR_2$ was set again to 20 W. The waveforms also show the system turn-off at $t_5$.

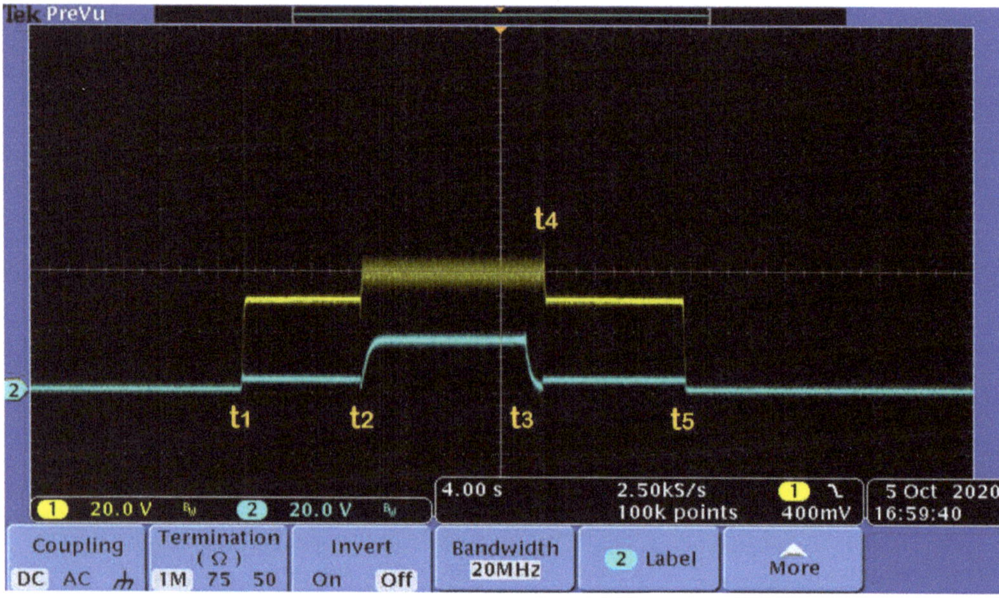


**Figure 16.** Solar array voltages during a load peak. At turn-ON the operating point is controlled by the $APR_1$ (yellow trace) with the $APR_2$ (blue trace) at a base power level. When a load peak forces $APR_1$ into MPPT, $APR_2$ takes control of the operating point. The reverse transient for a load reduction also is shown.

## 5. Conclusions

Operation of the power bus regulation techniques most popular in space applications and that of the novel SMPT bus regulation technique have been introduced and analyzed with reference to a mission in low Earth orbit. It was shown that SMPT could provide appreciable improvements: it has optimal energy harvesting in the presence of a mismatched solar array, it is suitable to drive unregulated power buses, and uses quite a simple controller that can be easily designed to be SPFF. Performance improvements are particularly evident in the presence of pulsed loads. An experimental prototype and a Simulink model allowed us to demonstrate the functionality of the SMPT technique. Future activities will be intended to develop an improved controller, capable of providing faster transient responses while remaining robust against variations of the dynamic response of both solar panels and battery. Extensive characterization of the prototype also will be carried out to assess the boundaries of the APRs' safe operating area in the perspective of future use in actual spacecraft. The research will proceed, aiming at developing new power management configurations, with enhanced reliability and larger efficiency.

**Author Contributions:** Conceptualization, methodology, investigation (equal), writing, and supervision, L.S.; software, investigation (equal), and data curation, M.F.; prototyping, investigation (support), P.G.; investigation (support), C.P. and F.P. All authors have read and agreed to the published version of the manuscript.

**Funding:** This research received no external funding.

**Institutional Review Board Statement:** Not applicable.

**Informed Consent Statement:** Not applicable.

**Data Availability Statement:** Not applicable.

**Conflicts of Interest:** The authors declare no conflict of interest.

## References


1. Pisacane, V.L. *Fundamentals of Space Systems*, 2nd ed.; Oxford University Press Inc.: New York, NY, USA, 2005.
2. O'Sullivan, D.; Weinberg, A.H. The sequential switching shunt regulator (S3R). In Proceedings of the 3rd ESTEC Spacecraft Power Conditioning Seminar, Noordwijk, The Netherlands, 21–23 September 1977.
3. Zhu, H.; Zhang, D. Design considerations of sequential switching shunt regulator for high-power applications. *IEEE Trans. Ind. Electron.* **2020**, *67*, 9358–9369. [CrossRef]
4. Blanes, J.M.; Carrasco, J.A.; Garrigós, A.; Marroquí, D.; Torres, C. Sequential Switching Shunt Regulator Parallel Power Processing Control for High Capacitance Solar Arrays. *Energies* **2021**, *14*, 429. [CrossRef]
5. Perol, P.; Olsson, D.; Haines, J.E. Device for Generating Electrical Energy for a Power Supply Bus. U.S. Patent 6181115 B1, 30 January 2001.
6. Loche, D. The Pleiades Electrical Power Subsystem. In Proceedings of the 7th European Space Power Conference, Stresa, Italy, 9–13 May 2005.
7. Jensen, H. Shunt Regulator Module. In Proceedings of the 8th European Space Power Conference, Konstanz, Germany, 14–19 September 2008.
8. Blanes, J.M.; Garrigos, A.; Carrasco, J.A.; Weinberg, A.H.; Ejea, J.B.; Sanchis, E.; Farreres, A.; Maset, E.; Soto, A.; de la Cruz, F. Sequential switching shunt maximum power point regulator (S3MPPR). In Proceedings of the 9th European Space Power Conference, Saint Raphael, France, 6–10 June 2011.
9. Denzinger, W. Electrical power subsytem of Globalstar. In Proceedings of the 4th European Space Power Conference, Poitiers, France, 4–8 September 1995; pp. 171–174.
10. Tonicello, F.; Vasquez Del Real, S. Maximum Power Point Power Tracker approach to a regulated bus. In Proceedings of the 5th European Space Power Conference, Tarragona, Spain, 21–25 September 1998.
11. Jensen, H.; Laursen, J. Power Conditioning Unit for Rosetta/Mars Express. In Proceedings of the 6th European Space Power Conference, Porto, Portugal, 6–10 May 2002.
12. Tonicello, F.; Jensen, H.; Laursen, J. Power Control Unit for Rosetta-Mars Express—Lessons Learnt. In Proceedings of the 6th European Space Power Conference, Porto, Portugal, 6–10 May 2002; pp. 269–274.
13. Lapena, E.; Rueda, P.; Garcia, O.; Cobos, J.A.; Alou, P.; Oliver, J.; del Nuevo, M.A.; de las Mulas, E.M.; Soto, A.; Rancano, J. Boost-Based MPPT For The MTM PCDU of the Bepi-Colombo Mission. In Proceedings of the 8th European Space Power Conference, Konstanz, Germany, 14–19 September 2008.
14. Caon, A.; Doerre, M.; Haines, J.E.; Landstroem, S. The GOCE electrical power system. In Proceedings of the IECEC '02. 37th Intersociety Energy Conversion Engineering Conference, Washington, DC, USA, 29–31 July 2002; pp. 87–92.
15. Tonicello, F. The control problem of maximum point power tracking in power systems. In Proceedings of the 7th European Space Power Conference, Stresa, Italy, 9–13 May 2005.
16. Loche, D. Mars Express and Venus Express Power Subsystem in Flight Behaviour. In Proceedings of the 8th European Space Power Conference, Konstanz, Germany, 14–19 September 2008.
17. Breier, N.; Kiewe, B.; Mourra, O. The Power Control and Distribution Unit for the SWARM Satellites. In Proceedings of the 8th European Space Power Conference, Konstanz, Germany, 14–19 September 2008.
18. De Luca, A.; Gray, R.; Otero, J. LISA Pathfinder Power System. In Proceedings of the 8th European Space Power Conference, Konstanz, Germany, 14–19 September 2008.
19. Croci, L.; Zanella, P.; Riva, S. The Exomars Trace Gas Orbiter electrical power subsystem. In Proceedings of the 10th European Space Power Conference, Noordwijk, The Netherlands, 13–17 April 2014.
20. White, J.L.; Muldoon, W.J. Two inductors boost and buck converter. In Proceedings of the 1987 IEEE Power Electronics Specialists Conference, Blacksburg, VA, USA, 21–26 June 1987; IEEE: Piscataway, NJ, USA, 1987; pp. 387–392.
21. Hansen, M.M. Power Conditioning Unit for BepiColombo Transfer Module. In Proceedings of the 8th European Space Power Conference, Konstanz, Germany, 14–19 September 2008.
22. Mourra, O.; Fernandez, A.; Tonicello, F. Buck Boost Regulator ($B^2R$) for Spacecraft Solar Array Power Conversion. In Proceedings of the 2010 Twenty-Fifth Annual IEEE Applied Power Electronics Conference and Exposition (APEC), Palm Springs, CA, USA, 21–25 February 2010; IEEE: Piscataway, NJ, USA, 2010; pp. 1313–1319.
23. Macellari, M.; Palmerini, G.B.; Schirone, L. On the Power System of the AMALIA moon rover. In Proceedings of the 2015 IEEE Aerospace Conference, Big Sky, Montana, USA, 7–14 March 2015; IEEE: Piscataway, NJ, USA, 2015.
24. Weinberg, A.H.; Weinberg, S.H. A new maximum power tracker topology. In Proceedings of the 6th European Space Power Conference, Porto, Portugal, 6–10 May 2002; pp. 257–262.
25. Lin, Z.; Xie, W.; Dong, M.; Lan, J. Research of S3MPR Engineering Application on Space Power Sources. In Proceedings of the 2020 IEEE 4th Information Technology, Networking, Electronic and Automation Control Conference (ITNEC), Chongqing, China, 12–14 June 2020; IEEE: Piscataway, NJ, USA, 2020; pp. 817–823.
26. Garrigos, A.; Blanes, J.M.; Carrasco, J.A.; Weinberg, A.H.; Maset, E.; Sanchis-Kilders, E.; Ejea, J.B.; Ferrreres, A. The Sequential Switching Shunt Maximum Power Regulator and its application in the electric propulsion system of a Spacecraftt. In Proceedings of the 2007 IEEE Power Electronics Specialists Conference, Orlando, FL, USA, 17–21 June 2007; IEEE: Piscataway, NJ, USA, 2007; pp. 1374–1379.

27. Garrigos, A.; Blanes, J.M.; Carrasco, J.A.; Weinberg, A.H.; Caballero, G.; Soto, A. Maximum power point tracking for GEO telecommunication satellites. In Proceedings of the 8th European Space Power Conference, Konstanz, Germany, 14–19 September 2008.
28. Bifaretti, S.; Pipolo, S.; Catalano, S.; Daprati, G.; Iacovone, V.; Scione, E. Functional study of a distributed MPPT power management system. *E3S Web Conf.* **2017**, *16*, 13008. [CrossRef]
29. Capel, A.; Perol, P. Comparative performance evaluation between the S4R and the S3R regulated bus topologies. In Proceedings of the 6th European Space Power Conference, Porto, Portugal, 6–10 May 2002.
30. Tonicello, F.; Jensen, H. Single Point Failure Free Sequential Serial and Shunt Regulator (S4R) with respect to Battery Overcharge. In Proceedings of the 8th European Space Power Conference, Konstanz, Germany, 14–19 September 2008.
31. Esram, T.; Chapman, P.L. Comparison of photovoltaic array maximum power point tracking techniques. *IEEE Trans. Energy Conver.* **2007**, *22*, 439–449. [CrossRef]
32. Femia, N.; Petrone, G.; Spagnuolo, G.; Vitelli, M. Optimization of Perturb and Observe Maximum Power Point Tracking Method. *IEEE Trans. Power Electr.* **2005**, *20*, 963–973. [CrossRef]
33. Suntio, T.; Leppäaho, J.; Huusari, J.; Nousiainen, L. Issues on solar-generator interfacing with current-fed MPP-Tracking converters. *IEEE Trans. Power Electr.* **2010**, *25*, 2409–2419. [CrossRef]
34. Brambilla, A.; Gambarara, M.; Torrente, G. Perturb and observe digital maximum power point tracker for satellites applications. In Proceedings of the 6th European Space Power Conference, Porto, Portugal, 6–10 May 2002.
35. Schirone, L.; Macellari, M. Loss Analysis of Low-Voltage TLNPC Step-Up Converters. *IEEE Trans. Ind. Electron.* **2014**, *61*, 6081–6090. [CrossRef]
36. McCluskey, F.P.; Hakim, E.B.; Fink, J.; Fowler, A.; Pecht, M.G. Reliability assessment of electronic components exposed to long-term non-operating conditions. *IEEE Trans. Compon. Pack. A* **1998**, *21*, 352–359. [CrossRef]
37. Patel, M.R. *Spacecraft Power Systems*, 1st ed.; CRC Press: Boca Raton, FL, USA, 2004.
38. Space Solar Cells—CESI. Available online: https://www.cesi.it/space-solar-cells/ (accessed on 21 June 2021).
39. Massey, R.P.; Snyder, E. High voltage single ended dc-dc converter. In Proceedings of the 1977 IEEE Power Electronics Specialists Conference, Palo Alto, CA, USA, 14–16 June 1977; pp. 156–159.

*energies*

*Article*

# Adaptive Tracking Method of Distorted Voltage Using IMM Algorithm under Grid Frequency Fluctuation Conditions

**Haoyao Nie [1,2] and Xiaohua Nie [3,*]**

1 School of Economics and Management, Nanchang University, Nanchang 330031, China; hynie003@outlook.com
2 I.H. Asper School of Business, University of Manitoba, Winnipeg, MB R3T 5V4, Canada
3 Department of Energy and Electrical Engineering, Nanchang University, Nanchang 330031, China
* Correspondence: niexiaoh@163.com

**Abstract:** This paper newly proposes an interactive multiple model (IMM) algorithm to adaptively track distorted AC voltage with the grid frequency fluctuation. The usual tracking methods are Kalman filter (KF) algorithm with a fixed frequency and KF algorithm with frequency identifier. The KF algorithm with a fixed frequency has a larger covariance parameter to guarantee the tracking robustness. However, it has a large tracking error. The KF algorithm with frequency identifier overly depends on the accuracy and stability of frequency identifier. The advantage of the proposed method is that it is decoupled from frequency detection and does not depend on frequency detection accuracy. First, the orthogonal vector dynamic (OVD) tracking model of the sine wave is established. Then, a model set covering the grid frequency fluctuation range is formed, and a new OVD-IMM tracking algorithm is proposed in combination with IMM algorithm. In the end, the effectiveness and accuracy of the proposed OVD-IMM algorithm are verified through simulations and experiments.

**Keywords:** frequency fluctuation; distorted voltage; Kalman filter; dynamic tracking model; interacting multiple model algorithm

**Citation:** Nie, H.; Nie, X. Adaptive Tracking Method of Distorted Voltage Using IMM Algorithm under Grid Frequency Fluctuation Conditions. *Energies* **2021**, *14*, 7944. https://doi.org/10.3390/en14237944

Academic Editor: Muhammad Aziz

Received: 3 November 2021
Accepted: 22 November 2021
Published: 26 November 2021



## 1. Introduction

The Kalman filter (KF), which is widely used in some power electronics fields, represents its advantages such as zero steady-state error, high real-time performance, better division of signal state and noise, and so on [1]. For example, it applies to the following fields such as harmonic estimation [2–8], travelling-wave fault location [9,10], waveform envelope detection of AC signals for power quality control equipment [11–17], power quality disturbance detection and classification [18–26], and synchronization control methods [27–33], and so on.

As we all know, as the source-grid-load changes, the power grid frequency is not fixed at 50 Hz but fluctuates with time. Frequency fluctuation obeys non-Gaussian distribution. Its fluctuation range generally reaches 49.5 Hz–50.5 Hz [34]. Some methods in the literature are to use the KF algorithm with a fixed frequency [27,28]. When tracking the AC voltage, the KF algorithm covariance parameter is 0.05. The parameter of 0.05 is larger and the frequency band is wider. It guarantees the robustness of the KF algorithm when the frequency fluctuates. However, it has a large tracking error. Another method uses KF algorithm with real-time frequency identifier, as shown in Figure 1 [29–33]. This kind of tracking method realizes adaptive tracking by feeding back the estimated value of the fundamental frequency. Different frequency detection algorithms are used in the literature, such as extended Kalman filter (EKF) combined with integrator estimation method [29], generalized averaging method [30], frequency locked loop (FLL) [31–33], etc. The disadvantage of the KF algorithm with frequency identifier is that it requires a large amount of calculation and high frequency accuracy. If the frequency identifier is inaccurate or unstable, it will affect the electrical signal tracking and parameter measurement accuracy, and even cause tracking failure.

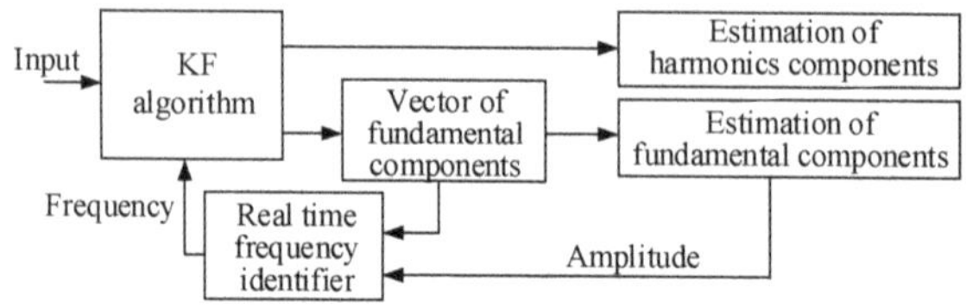


**Figure 1.** Tracking method based on KF algorithm with frequency identifier.

This paper presents a new method for tracking the distorted AC grid voltage of the frequency fluctuation using (interacting multiple models) IMM algorithm. It is shown in Figure 2. By comparing Figures 1 and 2, the advantage of IMM algorithm is that it does not require frequency detection and feedback tracking, and the tracking method is not affected by frequency detection accuracy and can achieve full self-adaptation. The IMM algorithm is an algorithm with Markov switching coefficients, in which multiple models of different frequencies work in parallel, and the tracking output is the result of the interaction of multiple filters [35–41]. The transfer between models is determined by Markov probability transfer matrix, which can effectively adjust the probability of each model. IMM estimates through the interaction between multiple models. Compared with the single-model algorithm, IMM is more adaptable to the fluctuation of the estimated object. It is used in hybrid control system tracking [38], maneuvering target tracking [39], human pose estimation [40], dynamic state estimation of distribution network [41], etc.

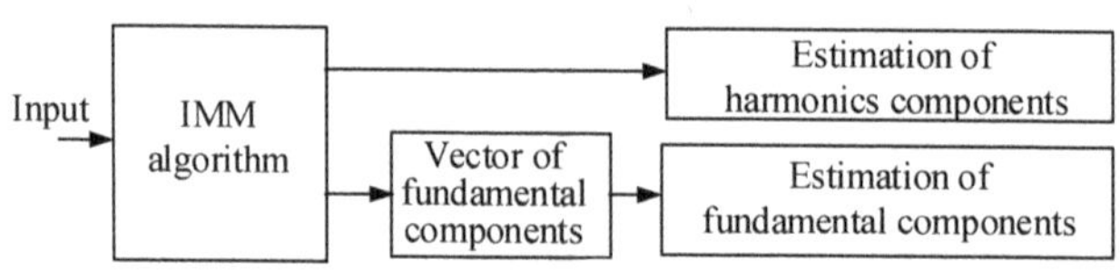


**Figure 2.** Proposed tracking method based on IMM algorithm.

The contributions of this paper are as follows: (a) the orthogonal vector dynamic (OVD) model is re-derived from the stochastic process theory [42], which made it clear that the covariance matrix is related to the sampling cycle. The orthogonal vector (OV) model used in conventional KF algorithm lacks theoretical derivation, and the covariance coefficient is only a fixed value of 0.05 [27,28]. The derivation of the OVD model lays the foundation for the application of the IMM algorithm. (b) The OVD-based IMM (OVD-IMM) algorithm of two models is proposed to adaptively track distorted AC grid voltage with the frequency fluctuation.

Section 1 introduces a new method for tracking the distorted AC grid voltage of the frequency fluctuation using (interacting multiple models) IMM algorithm. In Section 2, the OVD model is re-derived according to of stochastic process theory. The state space of the OVD is expanded. In Section 3, the OVD-IMM algorithm is presented and developed. In Sections 4 and 5, the performances of OVD-IMM algorithm are evaluated and compared by simulation and the experiment data. Section 6 makes a conclusion of the paper.

## 2. OVD Model

In Section 2, we illustrate the principle of OVD modeling. The covariance matrix of state noise for OVD model is reorganized according to the theory. The relationship between the model and the sampling cycle is clarified. We expand the state space dimension of OVD model.

### 2.1. Derivation of OVD Model

In two-dimensional space, the circular movement of the object along the midpoint is projected on the x, y axis as a pure sine and a pure cosine waveform, as shown in Figure 1. It can be expressed as

$$\begin{cases} x(t) = A_m \sin(\omega_0 t + \theta) \\ y(t) = A_m \cos(\omega_0 t + \theta) \end{cases} \tag{1}$$

where, $A_m$ is the amplitude, $\omega_0$ is the angular frequency, and $\theta$ is the phase angle. The diagram demonstrates how the object circular movement being projected in Figure 3.

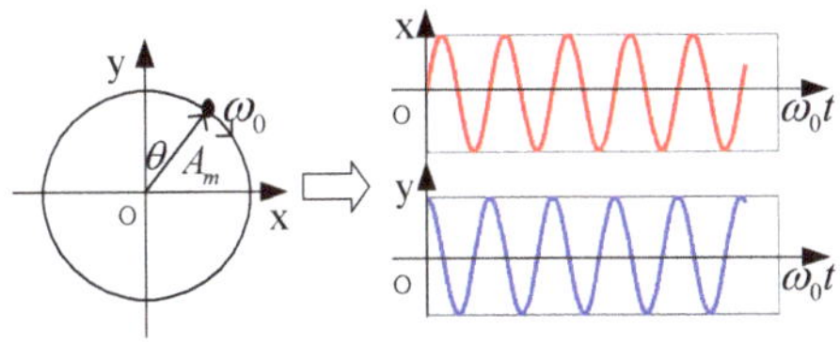


**Figure 3.** Diagram of the object circular movement being projected.

The first derivative can be expressed as

$$\begin{cases} \dot{x}(t) = \omega_0 A_m \cos(\omega_0 t + \theta) + w_x(t) = \omega_0 y(t) + w_x(t) \\ \dot{y}(t) = -\omega_0 A_m \sin(\omega_0 t + \theta) + w_y(t) = -\omega_0 x(t) + w_y(t) \end{cases} \tag{2}$$

where $w_x(t)$ and $w_y(t)$ are the input white noise, $w_x(t) \perp w_y(t)$. $w_x(t) \sim (0, \sigma_w^2)$, $w_y(t) \sim (0, \sigma_w^2)$.

The state equation can be obtained from (1).

$$\frac{d}{dt}\begin{bmatrix} x(t) \\ y(t) \end{bmatrix} = A\begin{bmatrix} x(t) \\ y(t) \end{bmatrix} + B\begin{bmatrix} w_x(t) \\ w_y(t) \end{bmatrix} \tag{3}$$

where

$$A = \begin{bmatrix} 0 & \omega_0 \\ -\omega_0 & 0 \end{bmatrix}, B = \begin{bmatrix} 1 & 0 \\ 0 & 1 \end{bmatrix}$$

Equation (3) is discretized, and we get the state equation as follows

$$\begin{bmatrix} x \\ y \end{bmatrix}_{k+1} = \Phi_k \begin{bmatrix} x \\ y \end{bmatrix}_k + W_k \tag{4}$$

where

$$\Phi_k = e^{AT} = L^{-1}\left([sI - A]^{-1}\right) = \begin{bmatrix} \cos(\omega_0 T) & \sin(\omega_0 T) \\ -\sin(\omega_0 T) & \cos(\omega_0 T) \end{bmatrix} \tag{5}$$

$$W_k = \int_{kT}^{(k+1)T} e^{A[(k+1)T - \tau]} \cdot B \cdot w(\tau) d\tau \tag{6}$$

The covariance matrix of state noise $W_k$ can be expressed as

$$Q_k = E\left[W_k W_k^T\right] = \sigma_w^2 \begin{bmatrix} T & 0 \\ 0 & T \end{bmatrix} = \sigma_w^2 T \begin{bmatrix} 1 & 0 \\ 0 & 1 \end{bmatrix} \tag{7}$$

From Equation (7), an important conclusion can be known that $Q_k$ is related to sampling cycle $T$.

### 2.2. Dimension Expansion of OVD Model

The waveform of the distorted AC signal sampled at time $t$ would be presented as

$$s(t) = \sum_{i=1}^{N} A_i(t)\sin[i\omega T + \theta_i(t)] + \sum_{i=1}^{N} w_i(t) \tag{8}$$

where $i = 1, 2, 3, \cdots, N$ represents the number of fundamental and harmonic components, $T$ is the sampling cycle. The fundamental frequency is $\omega = 100\pi \pm \mathrm{e}$, while e is the frequency fluctuation value. $\sum_{i=1}^{N} A_i(t)\sin[i\omega T + \theta_i(t)]$ and $\sum_{i=1}^{N} w_i(t)$ represent the fundamental and harmonic components, and their state noise, respectively.

For Equation (1), the dimensionality of the OVD model is expanded under the consideration of fundamental and harmonic components.

$$X(t) = \begin{bmatrix} x_1(t) & y_1(t) & x_2(t) & y_2(t) & \cdots & \mathrm{x}_n(t) & y_n(t) \end{bmatrix}^T \tag{9}$$

where $y_i(t)$ is the orthogonality of $x_i(t)$. In the condition of continuous time, the state space expression is established.

$$\frac{dX(t)}{dt} = \begin{bmatrix} M_1 & 0 & \cdots & 0 \\ 0 & M_2 & 0 & \vdots \\ \vdots & 0 & \ddots & 0 \\ 0 & \cdots & 0 & M_n \end{bmatrix} X(t) + \begin{bmatrix} w_1(t) \\ w_2(t) \\ \vdots \\ w_n(t) \end{bmatrix} \tag{10}$$

where the state noise vector is $\begin{bmatrix} w_1(t) & w_2(t) & \cdots & w_n(t) \end{bmatrix}^T$.

$$M_1 = \begin{bmatrix} 0 & \omega_m \\ -\omega_m & 0 \end{bmatrix}, M_2 = \begin{bmatrix} 0 & 3\times\omega_m \\ -3\times\omega_m & 0 \end{bmatrix}, M_n = \begin{bmatrix} 0 & \mathrm{i}\times\omega_m \\ -\mathrm{i}\times\omega_m & 0 \end{bmatrix}.$$

where $\omega_m$ represents the fixed frequency of this model, different values of $\omega_m$ represent different models.

The measurement equation is

$$z(t) = \begin{bmatrix} 1 & 0 & 1 & 0 & \cdots & 1 & 0 \end{bmatrix} X(t) + v(t) \tag{11}$$

where the measurement noise vector is $v(t)$.

We discretize Equations (10) and (11) in accordance with the principle of OVD model. Then, the state and measurement equation can be

$$\begin{cases} X_{k+1} = \begin{bmatrix} \Phi_1 & \cdots & 0 \\ \vdots & \cdots & \vdots \\ 0 & \cdots & \Phi_n \end{bmatrix} X_k + + \begin{bmatrix} W_1 \\ W_2 \\ \vdots \\ W_n \end{bmatrix} \\ z_k = H_k X_k + V_k \end{cases} \tag{12}$$

where $X_k = \begin{bmatrix} x_{1k} & y_{1k} & x_{2k} & y_{2k} & \cdots & x_{nk} & y_{nk} \end{bmatrix}^T$, $\begin{bmatrix} W_1 & W_2 & \cdots & W_n \end{bmatrix}_k^T$ is the state noise. The state transition matrixes and the measurement transfer matrix are shown below as

$$\Phi_i = \begin{bmatrix} \cos(i\omega_m T) & \sin(i\omega_m T) \\ -\sin(i\omega_m T) & \cos(i\omega_m T) \end{bmatrix} \tag{13}$$

$$H_k = \begin{bmatrix} 1 & 0 & 1 & 0 & \cdots & 1 & 0 \end{bmatrix} \tag{14}$$

The covariance matrix of state noise $W_{k_i}$ is

$$Q_{i_k} = E\left[W_{i_k} W_{i_k}{}^T\right] = \sigma_{OV}^2 T I_{2n\times 2n} \tag{15}$$

The difference from the traditional $OV$ model in the [27,28] is that $Q$ is different. The OVD model clarifies the relationship between the covariance matrix and the sampling cycle. The IMM algorithm cannot just adopt the traditional $OV$ model's covariance coefficient of 0.05. Otherwise, the IMM algorithm will bring tracking errors.

If we just choose the fixed value of 0.05, when the frequency fluctuates, the traditional $OV$ algorithm still adjusts the amplification factor K according to the covariance $Q$ of the fixed frequency. The lack of dynamic tracking performance results in the algorithm not being able to track the dynamic fluctuations in time. The IMM algorithm uses maximum likelihood estimation to determine the transition probability between models to adaptive tracking dynamic fluctuations. Therefore, the IMM algorithm has stronger and more stable adaptive tracking capability.

## 3. OVD Model-Based IMM Algorithm

In this section, the OVD Model-based IMM Algorithm (OVD-IMM) is proposed by combining OVD models and IMM algorithm. Then, the parameters of two model OVD-IMM algorithms are selected.

### 3.1. OVD-IMM Algorithm

The block diagram of IMM Algorithm is presented in Figure 4. The IMM algorithm formula is shown in Table 1.

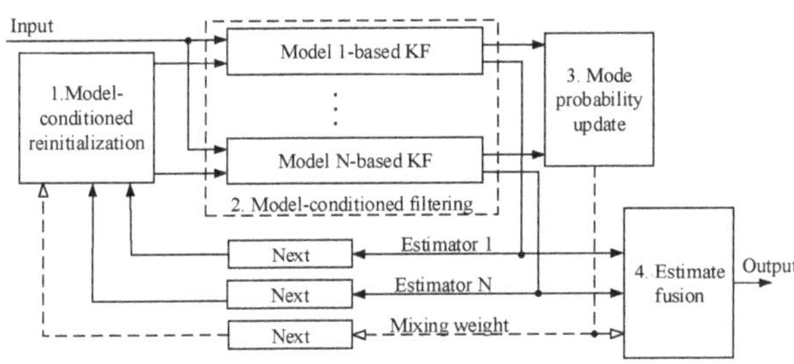


**Figure 4.** Block diagram of IMM Algorithm.

In the Table 1, $j$ represents the number of different models, $\pi_{ji}$ and $\mu_{k-1}^{(j)}$ denote the initial given values of mode probability and mixing weight, respectively. $\hat{\mathbf{X}}_{k|k}^{(i)}$ and $\mathbf{P}_{k|k}^{(i)}$ represent state estimate of filter $i$ at k and its covariance. $\overline{\mathbf{X}}_{k|k}^{(i)}$ and $\overline{\mathbf{P}}_{k|k}^{(i)}$ denote mixed conditions for filter $i$ at time $k$. $\hat{\mathbf{X}}_{k+1|k+1}$ and $\mathbf{P}_{k+1|k+1}$ indicate combined state estimate and its covariance. $\mu_k^{(i)}$ is mode probability at time $k$. $\mu_k^{j|i}$ is mixing probability at time $k$. $\mathbf{L}_k^{(i)}$ expresses the likelihood function of filter $i$.

Multiple models with different frequencies $\omega_m$ in (12) and (13) form a model set, and substituting them into formula Table 1 to form the OVD Model-based IMM (OVD-IMM) algorithm.

**Table 1.** One cycle of IMM algorithm.

| (1). Model-Conditioned Reinitialization (for i = 1,2, ... ,N): |
|---|
| Predicted mode probability: |
| Mixing weight: $\mu_k^{j\|i} \triangleq \mathbf{P}\left\{\mathbf{m}_k^{(j)} \middle\| \mathbf{m}_k^{(i)}, \mathbf{Z}^{k-1}\right\} = \pi_{ji}\mu_{k-1}^{(j)}/\mu_{k\|k-1}^{(i)}$ |
| Mixing estimate: $\overline{\mathbf{X}}_{k\|k}^{(i)} \triangleq E\left[\mathbf{X}_k \middle\| \mathbf{m}_k^{(i)}, Z^k\right] = \sum_j \hat{\mathbf{X}}_{k\|k}^{(j)}\mu_k^{j\|i}$ |
| Mixing covariance: $\overline{\mathbf{P}}_{k\|k}^{(i)} = \sum_j \left[\mathbf{P}_{k\|k}^{(j)} + (\overline{\mathbf{X}}_{k\|k}^{(i)} - \hat{\mathbf{X}}_{k\|k}^{(j)})(\overline{\mathbf{X}}_{k\|k}^{(i)} - \hat{\mathbf{X}}_{k\|k}^{(j)})'\right]\mu_k^{j\|i}$ |
| **(2). Model-Conditioned Filtering (for i = 1,2):** |
| Predicted state: $\hat{\mathbf{X}}_{k+1\|k}^{(i)} = \Phi_k^{(i)}\overline{\mathbf{X}}_{k\|k}^{(i)}$ |
| Predicted covariance: $\mathbf{P}_{k+1\|k}^{(i)} = \Phi_k^{(i)}\mathbf{P}_{k\|k}(\Phi_k^{(i)})' + \mathbf{Q}_k^{(i)}$ |
| Measurement residual: $\widetilde{\mathbf{Z}}_k^{(i)} = \mathbf{Z}_k - \mathbf{H}_k^{(i)}\hat{\mathbf{X}}_{k+1\|k}^{(i)}$ |
| Residual covariance: $\mathbf{S}_k^{(i)} = \mathbf{H}_k^{(i)}\mathbf{P}_{k+1\|k}(\mathbf{H}_k^{(i)})' + \mathbf{R}_k$ |
| Filter gain: $\mathbf{K}_{k+1}^{(i)} = \mathbf{P}_{k+1\|k}^{(i)}(\mathbf{H}_k^{(i)})'(\mathbf{S}_k^{(i)})^{-1}$ |
| Updated state: $\hat{\mathbf{X}}_{k+1\|k+1}^{(i)} = \hat{\mathbf{X}}_{k+1\|k}^{(i)} + \mathbf{K}_{k+1}^{(i)}\widetilde{\mathbf{Z}}_k^{(i)}$ |
| Updated covariance: $\mathbf{P}_{k+1\|k+1}^{(i)} = \mathbf{P}_{k+1\|k}^{(i)} - \mathbf{K}_{k+1}^{(i)}\mathbf{S}_k^{(i)}(\mathbf{K}_{k+1}^{(i)})'$ |
| **(3). Mode Probability Update (for i = 1,2, ... ,N):** |
| Model likelihood: $\mathbf{L}_k^{(i)} \triangleq p\left[\widetilde{\mathbf{Z}}_k^{(i)} \middle\| \mathbf{m}_k^{(i)}, \mathbf{Z}^k\right] \overset{assume}{=} N(\widetilde{\mathbf{Z}}_k^{(i)}; 0, \mathbf{S}_k^{(i)})$ |
| Mode probability: $\mu_k^{(i)} = \frac{\mu_{k\|k-1}^{(i)}\mathbf{L}_k^{(i)}}{\sum_j \mu_{k\|k-1}^{(j)}\mathbf{L}_k^{(j)}}$ |
| **(4). Estimate Fusion:** |
| Overall estimate: $\hat{\mathbf{X}}_{k+1\|k+1} = \sum_i \hat{\mathbf{X}}_{k+1\|k+1}^{(i)}\mu_k^{(j)}$ |
| Overall covariance: $\mathbf{P}_{k+1\|k+1} = \sum_i \left[\mathbf{P}_{k+1\|k+1}^{(i)} + (\hat{\mathbf{X}}_{k+1\|k+1} - \hat{\mathbf{X}}_{k+1\|k+1}^{(i)})(\hat{\mathbf{X}}_{k+1\|k+1} - \hat{\mathbf{X}}_{k+1\|k+1}^{(i)})'\right]\mu_k^{(i)}$ |

### *3.2. OVD-IMM Application Development*

When the IMM algorithm is applied to AC signal dynamic tracking, in order to obtain the best tracking effect, a model set that covers the frequency fluctuation mode range as much as possible must be used. However, increasing the number of models in the model set will not only increase the computational complexity, but may not necessarily improve the tracking performance. This is because an overly refined pattern space may destroy the completeness of Bayesian inference and the independence between models. Unnecessary competition from too many models will cause algorithm performance degradation.

Therefore, two frequency models covering the upper and lower limits of frequency fluctuations are adopted as a model set in this paper. The frequency of the lower limit model is $f_1 = 48.5$ Hz, the angular frequency $\omega_{m_1} = 2\pi \times f_1 = 2\pi \times 48.5$ rad/s. The frequency of the upper limit model is $f_2 = 51.5$ Hz, the angular frequency $\omega_{m_2} = 2\pi \times f_2 = 2\pi \times 51.5$ rad/s.

We calculate the amplitude and the phase angle.

$$U_{i_{k+1}} = \sqrt{\hat{x}_{i_{k+1}}^2 + \hat{y}_{i_{k+1}}^2} \tag{16}$$

$$\phi_{i_{k+1}} = \arctan\left(\frac{\hat{x}_{i_{k+1}}}{\hat{y}_{i_{k+1}}}\right) \tag{17}$$

where $i$ = 1 is the fundamental components, $i$ = 2,3,4, ... is the harmonic components. $\hat{x}_{i_{k+1}}, \hat{y}_{i_{k+1}}$ is the fundamental components of the overall estimator in Table 1(4).

The harmonics ratio for voltage (HRU) is defined as follow

$$\mathrm{HRU_n} = \frac{U_n}{U_1} \times 100\% \tag{18}$$

where $U_1$ is rms value of voltage fundamental components, $U_n$ is rms value of $n$-th voltage harmonic components.

## 4. Simulation and Evaluation

In Section 4, a fixed frequency 50 Hz model is selected to form the KF algorithm as a comparison mark to better illustrate the tracking results. We evaluate and compare OVD-IMM and KF algorithms by simulations. The HRU serves as a statistical indicator. The root means squared error (RMSE) values are adopted as statistical indicators for evaluation. At first, we set the setting of simulation. Then the tracking results of simulation data are shown in Section 4.2. Next to it, we obtain RMSE results of simulation data.

### 4.1. Simulation Setting

The distorted AC voltage signal formula with harmonics components is

$$u(t) = 220\sqrt{2} \times [1.0\sin(\omega t) + 0.05\sin(3\omega t) + 0.1\sin(5\omega t) + 0.08\sin(7\omega t)] + 3.11v(t) \tag{19}$$

The amplitude is $220\sqrt{2}$ V. The percentages of harmonic components are 5% of 3rd, 10% of 5th, and 8% of 7th. The measurement white noise is $\sigma_v = 0.01 \times 220\sqrt{2}$ V $= 3.11$ V. The sampling cycle is 0.00004 s. $\omega = 2 \times \pi \times f$ is variable. The simulation scene is divided into four situations in the condition of voltage frequency heavy fluctuations: (a) Frequency is 49.5 Hz; (b) Frequency is 50.5 Hz; (c) Frequency swell from 49.5 Hz to 50.5 Hz; (d) Frequency sag from 50.5 Hz to 49.5 Hz. The simulation voltage situations are shown in Figure 5 which includes the voltages, frequency and spectrogram of data. The simulation scene is divided into four situations in the condition of voltage frequency slight fluctuations: (a) Frequency is 49.98 Hz; (b) Frequency is 50.02 Hz; (c) Frequency swell from 49.98 Hz to 50.02 Hz; (d) Frequency sag from 50.02 Hz to 49.98 Hz. The simulation voltage situations are shown in Figure 6.

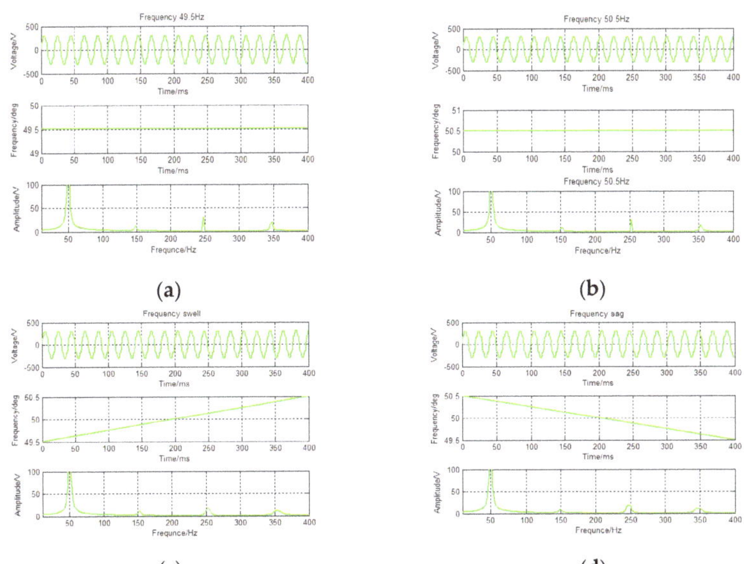


**Figure 5.** Simulation data for voltage frequency heavy fluctuations. (**a**) Frequency 49.5 Hz; (**b**) Frequency 50.5 Hz; (**c**) Frequency swell; (**d**) Frequency sag.

The parameters of the two algorithms are selected as follows. The $R$ of the two algorithms is 3.11 V. The initial covariance matrices of two algorithms are employed as $P_{0/0} = 1000 \times I_{10\times 10}$.

KF algorithms: the covariance of fundamental and harmonics components are all $Q_i = 0.05 \times I_{10\times 10}$, like [27,28]. The covariance $Q$ array of KF is larger and the frequency band is wider.

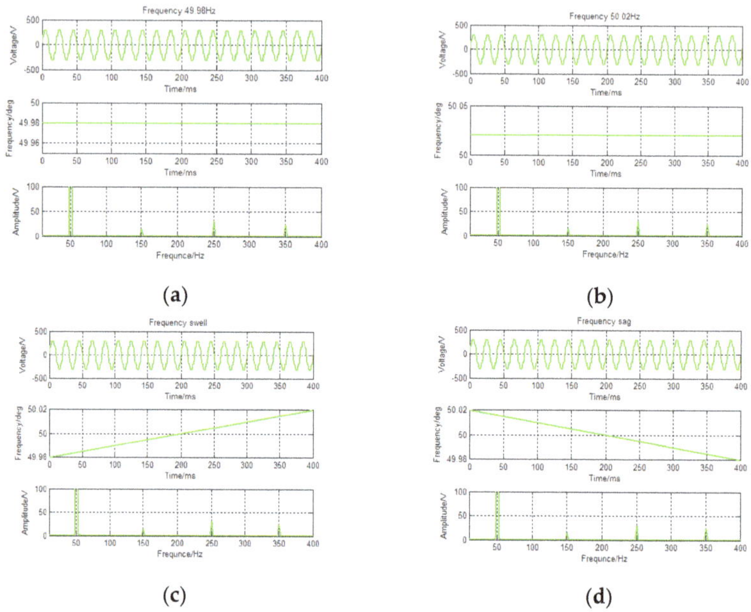


**Figure 6.** Simulation data for voltage frequency slight fluctuations. (**a**) Frequency 49.98 Hz; (**b**) Frequency 50.02 Hz; (**c**) Frequency swell; (**d**) Frequency sag.

IMM algorithms: The initial given values of mode probability and mixing weight, respectively, are: $\pi_{ji} = \begin{bmatrix} 0.98 & 0.02 \\ 0.02 & 0.98 \end{bmatrix}$, $\mu_{k-1}^{(1)} = 0.5\ \mu_{k-1}^{(2)} = 0.5$. The covariance matrix of the two models is: $Q_{i_k} = \sigma_{OV}^2 T = 10 \times T \times I_{10\times 10} = 0.0004 \times I_{10\times 10}$. In the condition of voltage frequency heavy fluctuations, the frequencies of the two models are $\omega_{m_1} = 2\pi \times 48.5$ rad/s, $\omega_{m_2} = 2\pi \times 51.5$ rad/s. And in the condition of voltage frequency slight fluctuations, the frequencies of the two models are $\omega_{m_1} = 2\pi \times 49.85$ rad/s, $\omega_{m_2} = 2\pi \times 50.15$ rad/s.

### *4.2. Tracking Results of Simulation Data*

The simulation tracking results of two algorithms for heavy and slight fluctuations are shown in Figures 7 and 8. The tracking results include fundamental and harmonics components.

It can be seen from Figures 7 and 8 that the IMM algorithm is stable and accurate in tracking no matter the fundamental wave or harmonic elements, and is not affected by frequency fluctuations. In Figures 7c,d and 8c,d, the accuracy of fundamental wave amplitude and harmonic tracking results is more obvious. Although the Q matrix of KF is large (0.05) and robust, the frequency mismatch also brings large tracking errors.

In order to quantitatively analyze the tracking results, the RMSE tracking results are evaluated at Section 4.3.

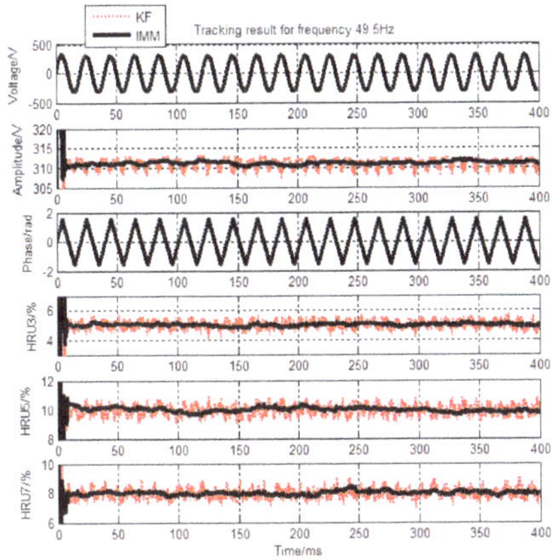


(a)

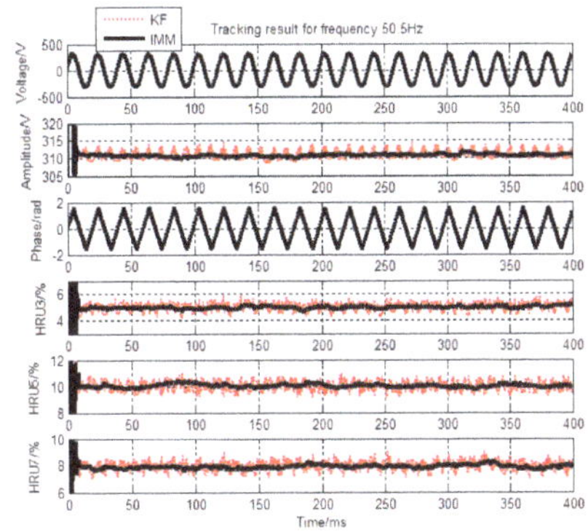


(b)

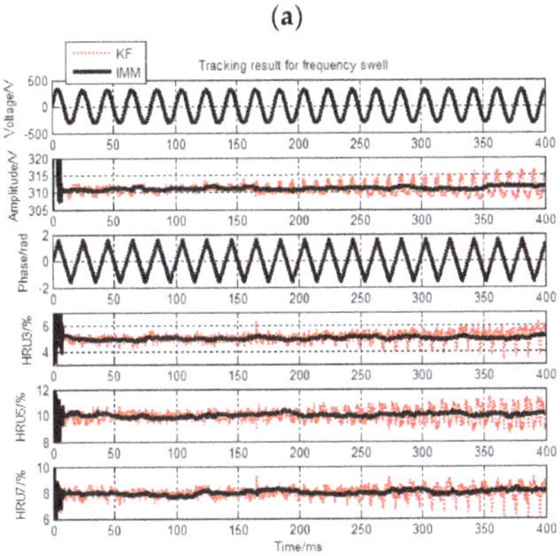


(c)

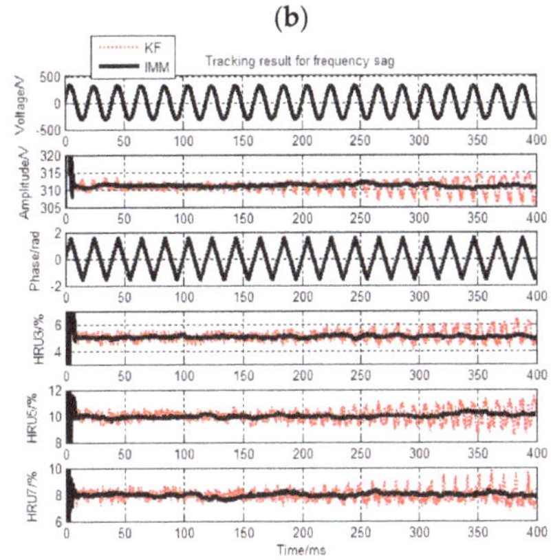


(d)

**Figure 7.** Simulation tracking results for voltage frequency heavy fluctuations. (**a**) Simulation tracking results for frequency 49.5 Hz; (**b**) Simulation tracking results for frequency 50.5 Hz; (**c**) Simulation tracking results for frequency swell; (**d**) Simulation tracking results for frequency sag.

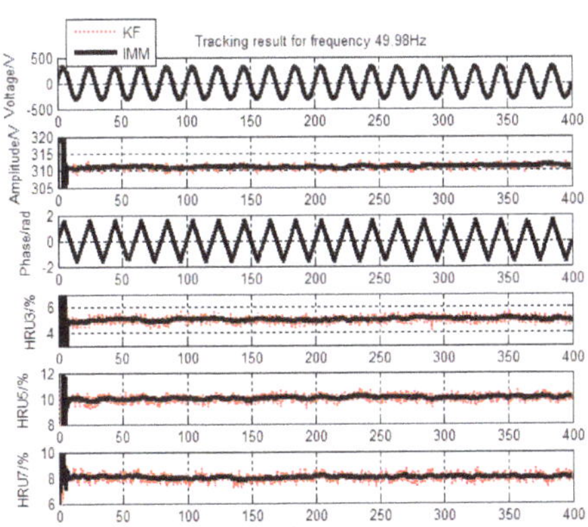


(a)

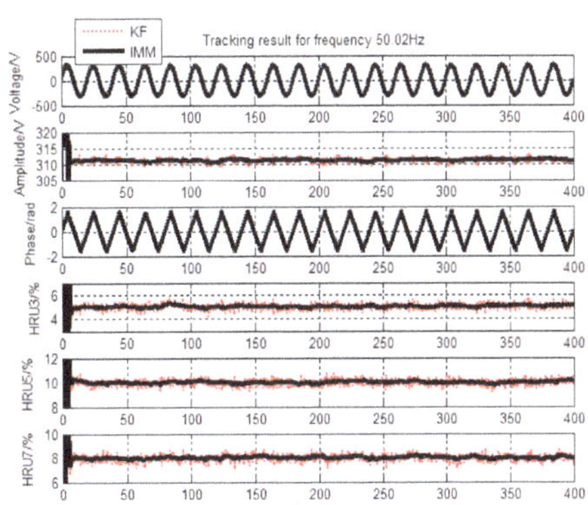


(b)

**Figure 8.** *Cont.*

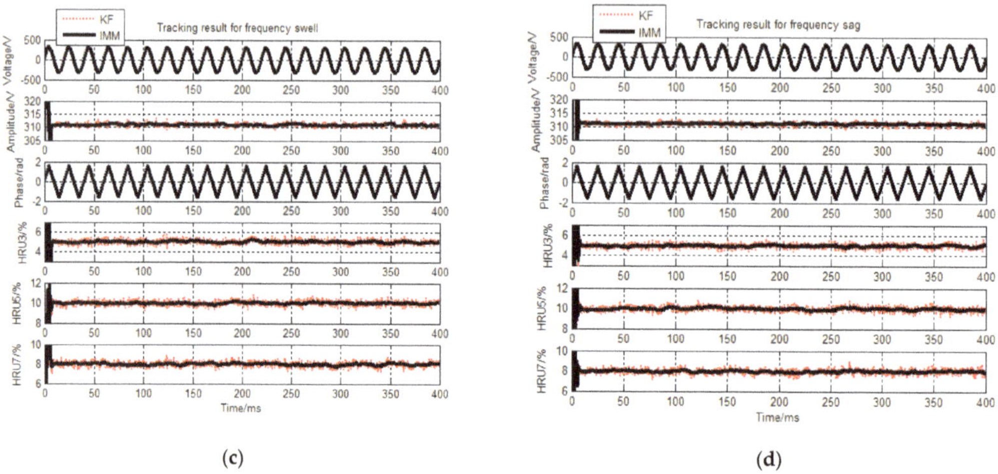


**Figure 8.** Simulation tracking results for voltage frequency slight fluctuations. (**a**) Simulation tracking results for frequency 49.98 Hz; (**b**) Simulation tracking results for frequency 50.02 Hz; (**c**) Simulation tracking results for frequency swell; (**d**) Simulation tracking results for frequency sag.

### 4.3. RMSE Results of Simulation Data

The RMSE results of two algorithms calculated by 100 times Monte Carlo simulations are shown in Figures 9 and 10. The RMSE evaluation results are shown in Tables 2 and 3.

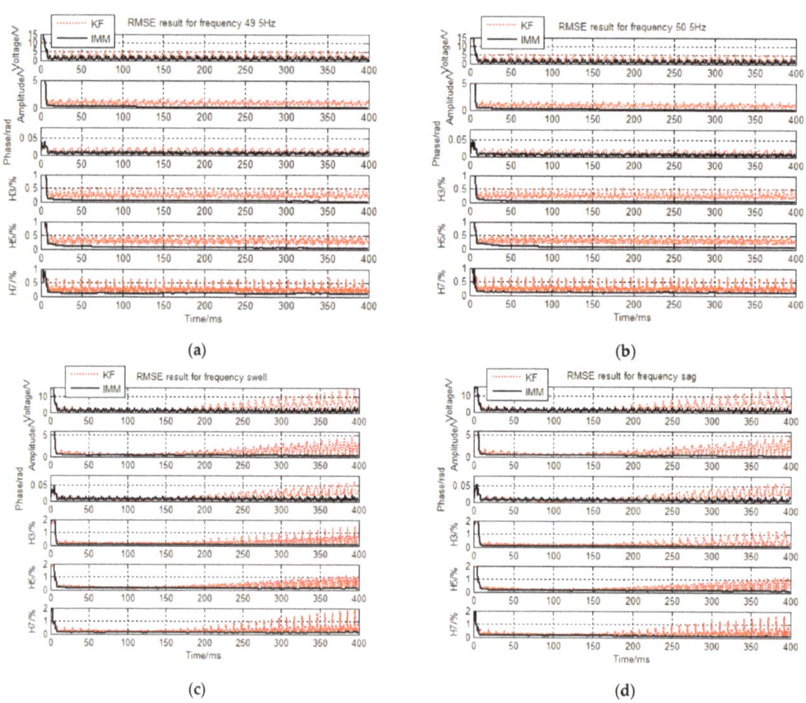


**Figure 9.** Simulation RMSE results for voltage frequency heavy fluctuations. (**a**) Simulation RMSE results for frequency 49.5 Hz; (**b**) Simulation RMSE results for frequency 50.5 Hz; (**c**) Simulation RMSE results for frequency swell; (**d**) Simulation RMSE results for frequency sag.

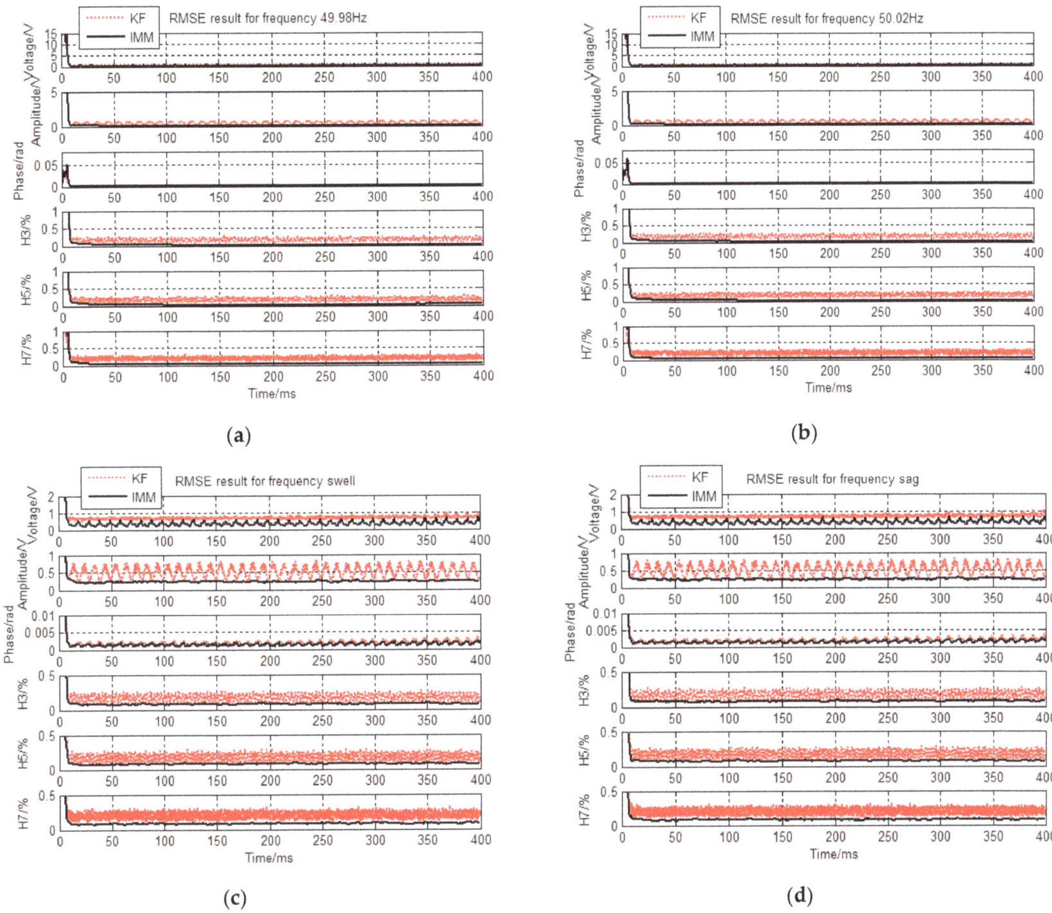


**Figure 10.** Simulation RMSE results for voltage frequency slight fluctuations. (**a**) RMSE results for frequency 49.98 Hz; (**b**) RMSE results for frequency 50.02 Hz; (**c**) RMSE results for frequency swell; (**d**) RMSE results for frequency sag.

**Table 2.** Simulation RMSE results for voltage frequency heavy fluctuations.

| RMSE Results | Algorithm | Fundament Components | | | Harmonics Percentage | | |
|---|---|---|---|---|---|---|---|
| | | Signal (V) | Amplitude (V) | Phase (rad) | H3(%) | H5(%) | H7(%) |
| For frequency 49.5 Hz | KF | 2.529610 | 0.930434 | 0.013386 | 0.262471 | 0.321099 | 0.264858 |
| | IMM | 1.115229 | 0.133475 | 0.006329 | 0.036968 | 0.070394 | 0.113563 |
| For frequency 50.5 Hz | KF | 2.543798 | 0.927858 | 0.013379 | 0.265402 | 0.322606 | 0.265622 |
| | IMM | 1.083361 | 0.151082 | 0.006080 | 0.040636 | 0.082606 | 0.108595 |
| For frequency heavy swell | KF | 3.184890 | 1.280741 | 0.016615 | 0.316734 | 0.396949 | 0.322599 |
| | IMM | 1.195442 | 0.365717 | 0.006781 | 0.092354 | 0.134780 | 0.134131 |
| For frequency heavy sag | KF | 3.152684 | 1.305744 | 0.016655 | 0.311159 | 0.400138 | 0.317231 |
| | IMM | 1.191125 | 0.358193 | 0.006817 | 0.091209 | 0.133590 | 0.115558 |

**Table 3.** Simulation RMSE results for voltage frequency slight fluctuations.

| RMSE Results | Algorithm | Fundament Components | | | Harmonics Percentage | | |
|---|---|---|---|---|---|---|---|
| | | Signal (V) | Amplitude (V) | Phase (rad) | H3(%) | H5(%) | H7(%) |
| For frequency 49.98 Hz | KF | 0.689969 | 0.509250 | 0.001745 | 0.167712 | 0.175642 | 0.199795 |
| | IMM | 0.296261 | 0.077288 | 0.001520 | 0.025432 | 0.030511 | 0.045718 |
| For frequency 50.02 Hz | KF | 0.688949 | 0.509281 | 0.001740 | 0.166633 | 0.175272 | 0.199700 |
| | IMM | 0.298582 | 0.087145 | 0.001518 | 0.025742 | 0.029914 | 0.044236 |
| For frequency slight swell | KF | 0.699127 | 0.512277 | 0.001827 | 0.167790 | 0.176890 | 0.199611 |
| | IMM | 0.350197 | 0.236394 | 0.001407 | 0.078400 | 0.077815 | 0.082226 |
| For frequency slight sag | KF | 0.701157 | 0.512460 | 0.001848 | 0.167786 | 0.176098 | 0.199778 |
| | IMM | 0.355450 | 0.237113 | 0.001442 | 0.078504 | 0.076419 | 0.083001 |

RMSE is defined as follows

$$\mathrm{RMSE} = \sqrt{\frac{1}{N}\sum_{k=1}^{N}\left(x_{ij} - \hat{x}_{ij}^{k}\right)^2} \tag{20}$$

where $N = 100$ is the number of simulations, $x_{ij}$ is j-sampled values of $x_i$ and $\hat{x}_{ij}$ j-sampled values of $\hat{x}_i$.

From Figures 9 and 10, the simulation tracking results of the IMM algorithm are stable and accurate. The KF algorithm has a large error. Especially after 200 ms, the KF algorithm has large errors in both the fundamental and the harmonics components, while the IMM algorithm is not affected.

From Table 2 to Table 3, we know that the IMM algorithm is better than the KF algorithm whether tracking results are fundamental components or harmonic components.

## 5. Experiment and Evaluation

In Section 5, we employ experiment data to evaluate and compare two algorithms. At the beginning, we introduce the experiment settings. Secondly, we compare tracking results of experiment sampling data for voltage frequency heavy and slight fluctuations.

### 5.1. Experiment Settings

The experiment platform is introduced in Figure 11a. It includes the programmable power supply and the oscilloscope. The schematic diagram of the programmable power supply is introduced in Figure 11b. Experiment data for voltage frequency heavy and slight fluctuations is conducted through a programmable power supply. The percentages of harmonic components are 5% of 3rd, 10% of 5th, and 8% of 7th. The oscilloscope of Tektronix GDS-2102 serves to sample the experiment voltage data.

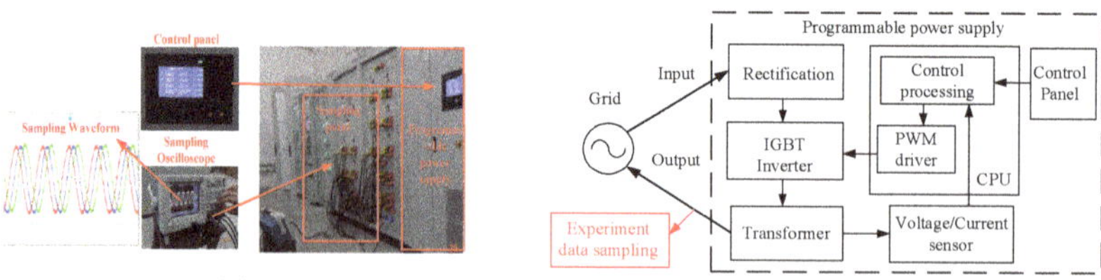


(a) (b)

**Figure 11.** Experiment platform and Schematic diagram. (**a**) Experiment platform. (**b**) Schematic diagram of the programmable power supply.

The sampling cycle is selected to be 0.00004 s. The experiment sampling voltage situations are shown in Figures 12 and 13.

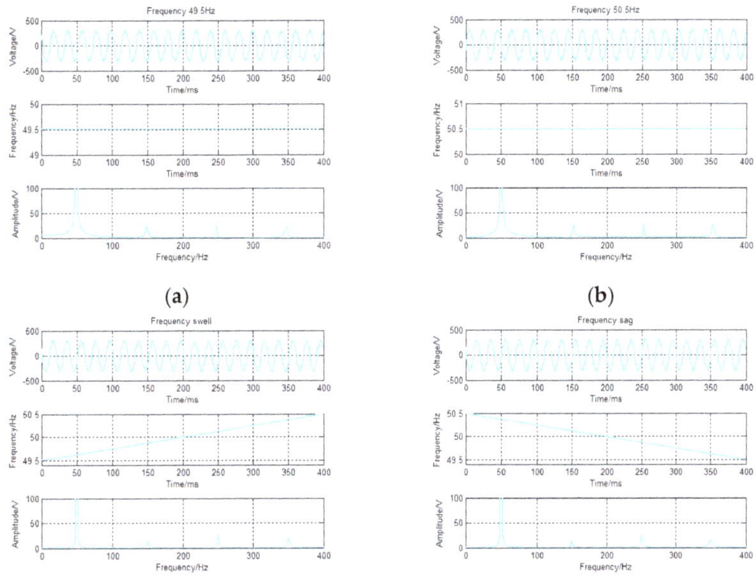


(a) (b) (c) (d)

**Figure 12.** Experiment data for voltage frequency heavy fluctuations. (**a**) Frequency 49.5 Hz; (**b**) Frequency 50.5 Hz; (**c**) Frequency swell; (**d**) Frequency sag.

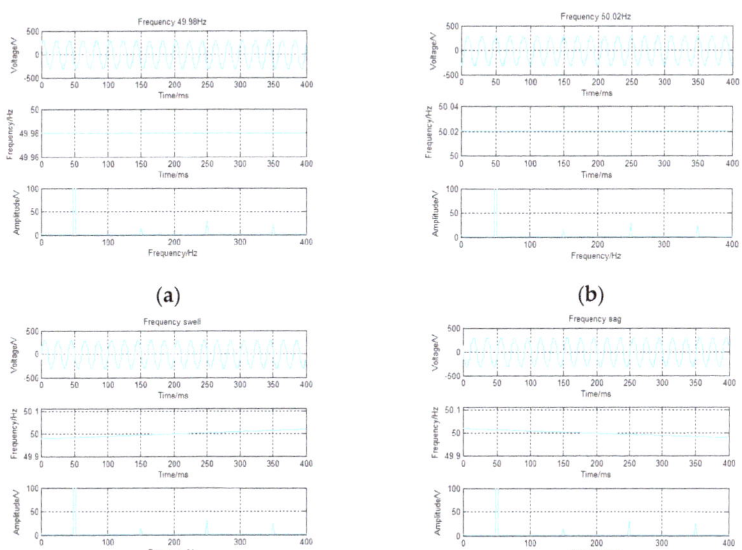


(a) (b) (c) (d)

**Figure 13.** Experiment data for voltage frequency slight fluctuations. (**a**) Frequency 49.98 Hz; (**b**) Frequency 50.02 Hz; (**c**) Frequency swell; (**d**) Frequency sag.

*5.2. Tracking Results of Experiment Sampling Data*

The model parameters of the two algorithms are the same as that of the simulation evaluation in Section 4. The experiment tracking results of voltage frequency heavy and slight fluctuations are shown in Figures 14 and 15.

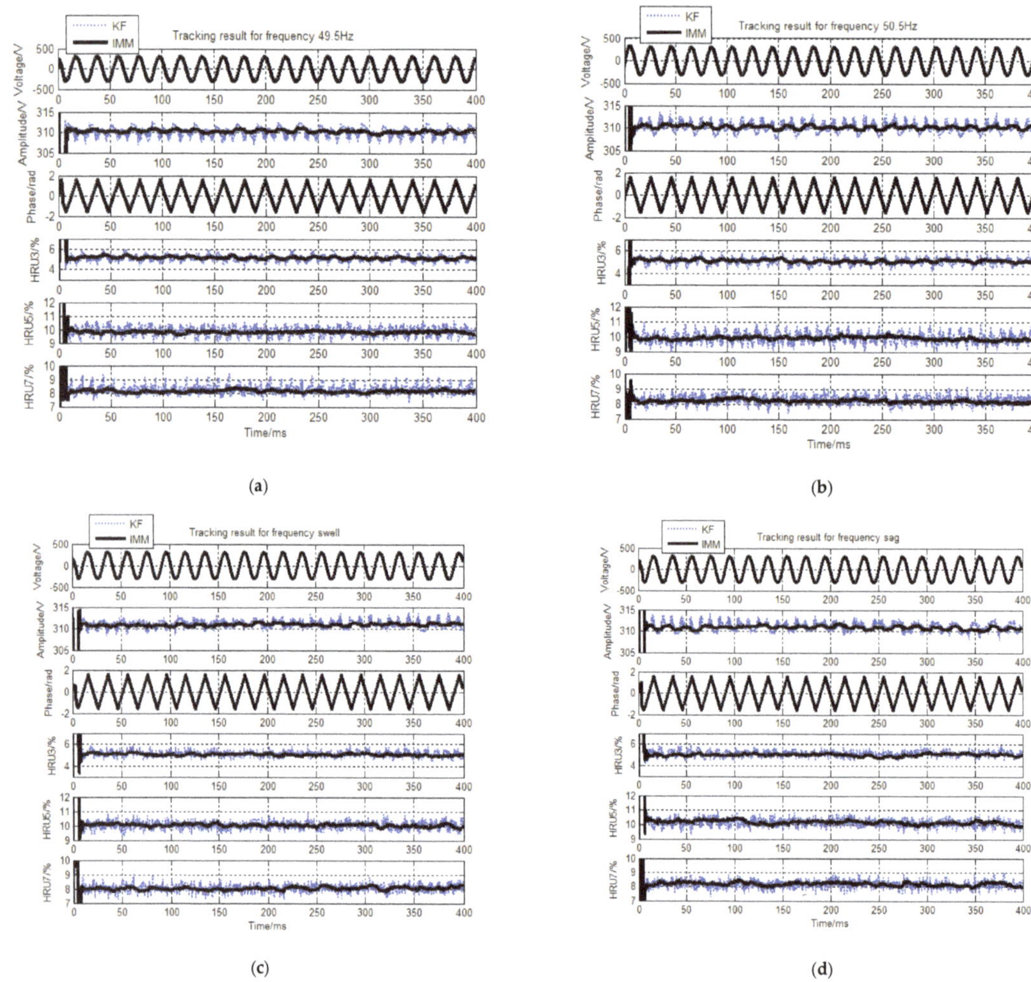


**Figure 14.** Experiment tracking results for voltage frequency heavy fluctuations. (**a**) Experiment tracking results for frequency 49.5 Hz; (**b**) Experiment tracking results for frequency 50.5 Hz; (**c**) Experiment tracking results for frequency swell; (**d**) Experiment tracking results for frequency sag.

From comparison results in Figures 14 and 15, the IMM algorithm is significantly better than the KF algorithm in the estimation. It is consistent with the simulation tracking results and conclusions.

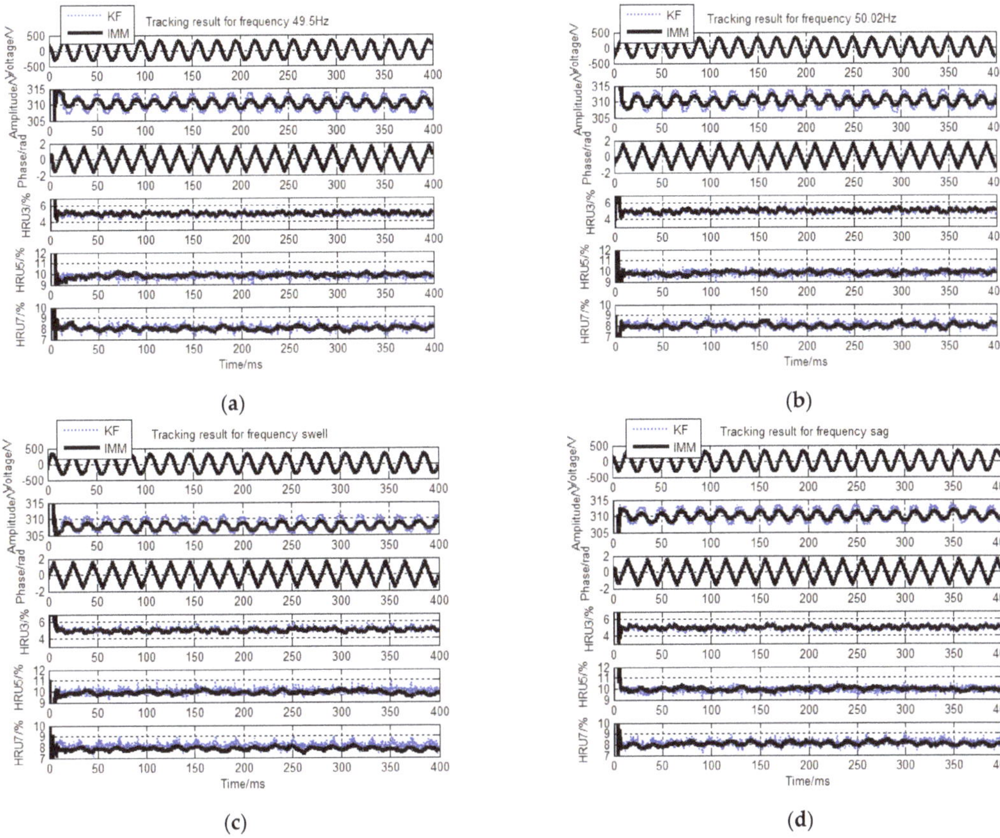


**Figure 15.** Experiment tracking results for voltage frequency slight fluctuations. (**a**) Experiment tracking results for frequency 49.98 Hz; (**b**) Experiment tracking results for frequency 50.02 Hz; (**c**) Experiment tracking results for frequency swell; (**d**) Experiment tracking results for frequency sag.

## 6. Conclusions

This paper newly presents the IMM algorithm to track the fundamental and harmonic components of the AC voltage signal under grid frequency fluctuation conditions. First of all, the orthogonal vector dynamic (OVD) model is re-derived from the stochastic process theory, which made it clear that the covariance matrix is related to the sampling cycle. Then, the IMM algorithm based on the two OVD models (OVD-IMM) is proposed. The advantage of the proposed OV-IMM algorithm is that it does not require the signal frequency detection and feedback like conventional tracking algorithms. In this paper, we simulate by two kinds of scenario data separately, which are (a) voltage frequency heavy fluctuations and (b) voltage frequency slight fluctuations. The experiment applies the same scenario data as the simulation. The results show that the fundamental and harmonic components tracking accuracy of IMM algorithm is higher than single model algorithm. Both the fundamental and harmonic components, the proposed OVD-IMM algorithm is stable and accurate, and is not affected by grid frequency fluctuations. In the future, it would be employed in some real world applications, such as synchronous control in power system [43].

**Author Contributions:** Conceptualization, H.N. and X.N.; methodology, H.N. and X.N.; software, H.N.; validation, H.N. and X.N.; formal analysis, H.N.; investigation, H.N.; resources, H.N.; data curation, H.N.; writing—original draft preparation, H.N.; writing—review and editing, H.N.; visualization, H.N.; supervision, X.N.; project administration, X.N.; funding acquisition, X.N. All authors have read and agreed to the published version of the manuscript.

**Funding:** This research was funded by National Natural Science Foundations of China, grant number 51867017 "Research on Power Quality Dynamic Tracking and Real-time Detection Method for Microgrid Complex Source Load Environment".

**Conflicts of Interest:** The authors declare no conflict of interest.

## References

1. Auger, F.; Hilairet, M.; Guerrero, J.M.; Monmassonet, E.; Orlowska-Kowalska, T.; Katsura, S. Industrial Applications of the Kalman Filter: A Review. *IEEE Trans. Ind. Electron.* **2013**, *60*, 5458–5471. [CrossRef]
2. Enayati, J.; Moravej, Z. Real-time harmonic estimation using a novel hybrid technique for embedded system implementation. *Int. Trans. Electr. Energy Syst.* **2017**, *27*, e2428. [CrossRef]
3. Enayati, J.; Moravej, Z. Real-time harmonics estimation in power systems using a novel hybrid algorithm. *IET Gener. Transm. Distrib.* **2017**, *14*, 3532–3538. [CrossRef]
4. Xu, W.; Huang, C.; Xie, X. Analysis and Application of Taylor-Kalman Filters under a Distorted Grid Condition. *IEEE Access* **2020**, *8*, 106822–106831. [CrossRef]
5. Chenchen, W.; Magaña, M.E.; Cotilla-Sánchez, E. Dynamic Frequency and Amplitude Estimation for Three-Phase Unbalanced Power Systems using the Unscented Kalman Filter. *IEEE Trans. Ins. Meas.* **2019**, *68*, 3387–3395.
6. Nie, X. Detection of Grid Voltage Fundamental and Harmonic Components Using Kalman Filter Based on Dynamic Tracking Model. *IEEE Trans. Ind. Electron.* **2020**, *67*, 1191–1200. [CrossRef]
7. Kanieski, J.M.; Tambara, R.V.; Pinheiro, H. Real-Time Detection of Interharmonics and Harmonics of AC Electric Arc Furnaces on GPU Framework. *IEEE Trans. Indu. Appl.* **2019**, *55*, 1373–1378.
8. Xi, Y.; Tang, X.; Li, Z.; Cui, Y.; Zhao, T.; Zeng, X.; Guo, J.; Duan, W. Harmonic estimation in power systems using an optimised adaptive Kalman filter based on PSO-GA. *IET Gener. Transm. Distrib.* **2019**, *13*, 3968–3979. [CrossRef]
9. Xi, Y.; Li, Z.; Zeng, X.; Tang, X.; Zhang, X.; Xiao, H. Fault location based on travelling wave identification using an adaptive extended Kalman filter. *IET Gener. Transm. Distrib.* **2018**, *12*, 1314–1322. [CrossRef]
10. Xi, Y.; Zhang, X.; Li, Z.; Zeng, X.; Tang, X.; Cui, Y.; Xiao, H. Double-ended travelling-wave fault location based on residual analysis using an adaptive EKF. *IET Signal. Process.* **2018**, *12*, 1000–1008. [CrossRef]
11. Alam, S.J.; Arya, S.R. Control of UPQC based on steady state linear Kalman filter for compensation of power quality problems. *Chin. J. Electr. Eng.* **2020**, *6*, 52–65. [CrossRef]
12. Naseri, F.; Kazemi, Z.; Farjah, E.; Ghanbari, T. Fast Detection and Compensation of Current Transformer Saturation Using Extended Kalman Filter. *IEEE Trans. Power Deliv.* **2019**, *34*, 1087–1097. [CrossRef]
13. Ahmed, H.; Biricik, S.; Benbouzid, M. Linear Kalman Filter-Based Grid Synchronization Technique: An Alternative Implementation. *IEEE Trans. Ind. Inform.* **2021**, *17*, 3847–3856. [CrossRef]
14. Samet, H.; Gashtasbi, S.; Tashakor, N.; Ghanbariet, T. Improvement of reactive power calculation in electric arc furnaces utilising Kalman filter. *IET Sci. Meas. Technol.* **2017**, *11*, 241–248. [CrossRef]
15. Zhao, S.; Huang, B.; Liu, F. Linear Optimal Unbiased Filter for Time-Variant Systems without Apriori Information on Initial Conditions. *IEEE Trans. Autom. Control* **2017**, *62*, 882–887. [CrossRef]
16. Sadigh, K.; Smedley, K.M. Fast and precise voltage sag detection method for dynamic voltage restorer (DVR) application. *Electr. Power Syst. Res.* **2016**, *130*, 192–207. [CrossRef]
17. Helena, L.; Liboni, B.; Oliveira, M.C.; Silva, I.N. On the problem of optimal estimation of balanced and symmetric three-phase signals. *Int. J. Electr. Power Energy Syst.* **2017**, *91*, 155–165.
18. Subhra, D.; Sudipta, D. Real-time cross-correlation-based technique for detection and classification of power quality disturbances. *IET Gener. Transm. Distrib.* **2018**, *12*, 1661–1671.
19. Li, J.; Qin, G.; Li, Y.; Ruan, X. Research on power quality disturbance identification and classification technology in high noise back-ground. *IET Gener. Transm. Distrib.* **2019**, *13*, 1661–1671. [CrossRef]
20. Wang, Y.; Raza, A.; Mohammed, F.P.; Ravishankar, J.; Phung, T. Detection and classification of disturbances in the islanded micro-grid by using wavelet transformation and feature extraction algorithm. *J. Eng.* **2019**, *2019*, 5284–5286. [CrossRef]
21. Chakravorti, T.; Patnaik, R.K.; Dash, P.K. Detection and classification of islanding and power quality disturbances in microgrid. *IET Signal Process* **2018**, *12*, 82–94. [CrossRef]
22. Singh, U.; Singh, S.N. Optimal Feature Selection via NSGA-II for Power Quality Disturbances Classification. *IEEE Trans. Ind. Inform.* **2018**, *14*, 2994–3002. [CrossRef]
23. Zhong, T.; Zhang, S.; Cai, G.; Huang, N. Power-quality disturbance recognition based on time-frequency analysis and decision tree. *IET Gener. Transm. Distrib.* **2018**, *12*, 4153–4162. [CrossRef]
24. Yanhui, X.; Zewen, L.; Xiangjun, Z.; Xin, T. Detection of power quality disturbance using an adaptive process noise covariance Kalman filte. *Digit. Signal Process.* **2018**, *76*, 34–49.
25. Markovska, M.; Taskovski, D.; Kokolanski, Z.; Dimchev, V.; Velkovski, B. Real-Time Implementation of Optimized Power Quality Events Classifier. *IEEE Trans. Ind. Appl.* **2020**, *56*, 3431–3442. [CrossRef]

26. Achlerkar, P.D.; Samantaray, S.R.; Sabarimalai Manikandan, M. Variational Mode Decomposition and Decision Tree Based Detection and Classification of Power Quality Disturbances in Grid-Connected Distributed Generation System. *IEEE Trans. Smart Grid.* **2018**, *9*, 3122–3132. [CrossRef]
27. Panigrahi, R.; Subudhi, B.; Panda, P.C. A Robust LQG Servo Control Strategy of Shunt-Active Power Filter for Power Quality Enhancement. *IEEE Trans. Power Electron.* **2016**, *31*, 2860–2869. [CrossRef]
28. Panigrahi, R.; Subudhi, B. Performance enhancement of shunt active power filter using a kalman filter-based H∞ control strategy. *IEEE Trans. Power Electron.* **2017**, *32*, 2622–2630. [CrossRef]
29. Ferrero, R.; Pegoraro, P.A.; Toscani, S. Synchrophasor Estimation for Three-Phase Systems Based on Taylor Extended Kalman Filtering. *Ins. Meas.* **2020**, *69*, 6723–6730. [CrossRef]
30. Golestan, S.; Guerrero, J.M.; Vasquez, J.C.; Abusorrah, A.M.; Al-Turki, Y. Single-Phase FLLs Based on Linear Kalman Filter, Limit-Cycle Oscillator, and Complex Bandpass Filter: Analysis and Comparison with a Standard FLL in Grid Applications. *IEEE Trans. Power Electron.* **2019**, *34*, 1064–1073. [CrossRef]
31. Swain, S.; Subudhi, B. Grid Synchronization of a PV System with Power Quality Disturbances using Unscented Kalman Filtering. *IEEE Trans. Sustain. Energy* **2018**, *10*, 1240–1247. [CrossRef]
32. Xu, W.; Huang, C.; Jiang, H. Analyses and Enhancement of Linear Kalman-Filter-Based Phase-Locked Loop. *IEEE Trans. Instrum. Meas.* **2021**, *70*, 6504510. [CrossRef]
33. Golestan, S.; Guerrero, J.M.; Vasquez, J.C. Steady-State Linear Kalman Filter-Based PLLs for Power Applications: A Second Look. *IEEE Trans. Ind. Electron.* **2018**, *65*, 9795–9800. [CrossRef]
34. Schäfer, B.; Beck, C.; Aihara, K.; Witthaut, D.; Timme, M. Non-Gaussian power grid frequency fluctuations characterized by Lévy-stable laws and superstatistics. *Nat. Energy* **2018**, 3, 119–126. [CrossRef]
35. Blom, H.A.P.; Bar-Shalom, Y. The interacting multiple model algorithmfor systems with Markovian switching coefficients. *IEEE Trans. Autom. Control* **1988**, *33*, 780–783. [CrossRef]
36. Hwang, I.; Seah, C.E.; Lee, S. A Study on Stability of the Interacting Multiple Model Algorithm. *IEEE Trans. Autom. Control* **2016**, *62*, 901–906. [CrossRef]
37. Lim, J.; Kim, H.-S.; Park, H.-M. Interactive-Multiple-Model Algorithm Based on Minimax Particle Filtering. *IEEE Signal Process. Lett.* **2019**, *27*, 36–40. [CrossRef]
38. Lin, H.; Lam, J.; Chen, M.Z.Q.; Shu, Z.; Wu, Z.-G. Interacting Multiple Model Estimator for Networked Control Systems: Stability, Convergence, and Performance. *IEEE Trans. Autom. Control* **2019**, *64*, 928–943. [CrossRef]
39. Ma, J.; Guo, X. Combination of IMM Algorithm and ASTRWCKF for Maneuvering Target Tracking. *IEEE Access* **2020**, *8*, 143095–143103. [CrossRef]
40. He, H.; Liu, G.; Zhu, X.; He, L.; Tian, G. Interacting Multiple Model-Based Human Pose Estimation Using a Distributed 3D Camera Network. *IEEE Sens. J.* **2019**, *19*, 10584–10590. [CrossRef]
41. Kong, X.; Zhang, X.; Zhang, X.; Wang, C.; Chiang, H.-D.; Li, P. Adaptive Dynamic State Estimation of Distribution Network Based on Interacting Multiple Model. *IEEE Trans. Sustain. Energy* **2021**, 1. [CrossRef]
42. Brown, R.G.; Hwang, P.Y.C. *Introduction to Random Signals and Applied Kalman Filtering with MATLAB Exercises*, 4th ed.; John Wiley& Sons, Inc.: Hoboken, NJ, USA, 2011.
43. Zografos, D.; Ghandhari, M. Power system inertia estimation by approaching load power change after a disturbance. In Proceedings of the 2017 IEEE Power & Energy Society General Meeting, Chicago, IL, USA, 16–20 July 2017; pp. 1–5.

energies

Article

# Research on Dynamic Modeling of KF Algorithm for Detecting Distorted AC Signal

Haoyao Nie [1,2] and Xiaohua Nie [3,*]

1 School of Economics and Management, Nanchang University, Nanchang 330031, China; hynie003@outlook.com
2 I.H. Asper School of Business, University of Manitoba, Winnipeg, MB R3T 5V4, Canada
3 Department of Energy and Electrical Engineering, Nanchang University, Nanchang 330031, China
* Correspondence: niexiaoh@163.com

**Abstract:** Kalman filter (KF) is often based on two models, which are phase angle vector (PAV) model and orthogonal vector (OV) model, in the application of distorted grid AC signal detection. However, these two models lack rigorous and detailed derivation from the principle of dynamic modeling. This paper presents a phase angle vector dynamic (PAVD) model and an orthogonal vector dynamic (OVD) model, which are combined with Kalman filter for detecting distorted grid AC signal. They reveal that the state noise covariance of the dynamic model−based KF is related to the sampling cycle, and overcome the defect of more detecting error for conventional model−based KF. Experiment and evaluation results show that the proposed KF algorithms are reasonable and effective. Therefore, this paper contributes a guiding significance for the application of KF algorithm in harmonic detection.

**Keywords:** distorted AC signal; Kalman filter; phase angle dynamic model; orthogonal vector dynamic model; state noise covariance; sampling cycle

**Citation:** Nie, H.; Nie, X. Research on Dynamic Modeling of KF Algorithm for Detecting Distorted AC Signal. *Energies* **2021**, *14*, 8175. https://doi.org/10.3390/en14238175

Academic Editor: Jaroslaw Krzywanski

Received: 5 November 2021
Accepted: 2 December 2021
Published: 6 December 2021



## 1. Introduction

The Kalman filter (KF) plays an increasingly important role in the real−time detection of grid AC signal [1]. It was first applied to the power signal processing field by Dr. A. A. Girgis in 1981 [2–4]. It mainly applies to the following fields: fundamental and harmonic components detection of grid voltage/current [5–10], phase−locked loop synchronization [11–15], power quality detection and compensation equipment [16–20], power disturbance feature extraction and machine classification [21–24], and etc. In these applications, two conventional models, phase angle vector (PAV) model and orthogonal vector (OV) model, are mainly used. The phase angles were used as vectors to establish a state-space model which could be called the phase angle vector (PAV) model through cosine expansion [3]. The orthogonal vector was used to establish a state-space model that could be called the orthogonal vector (OV) model [4]. However, the state noise covariance of the two models is only a given value based on empirical statistics of the AC signal. It is selected as a fixed value, such as $Q = 0.05 \times I$ ($I$ represents the unit matrix), when AC signal is the grid voltage [10,11]. It is selected as a fixed value, such as $Q = 0.01 \times I$, when the AC signal is the sampling current [20]. It is mainly to avoid big tracking error and enable to detect response without time delay. With the development of computer hardware and sensor technology, the sampling cycle is getting smaller, and the accuracy requirements are getting higher. The detecting accuracy of the fundamental and harmonics components for the conventional detecting model−based KF algorithm cannot meet the requirements of modern engineering applications.

The current issue is that the detecting accuracy of the fundamental and harmonics components for the conventional detecting model–based KF algorithm cannot meet the requirements of modern engineering applications. There is no fixed rule as references to find the fixed value by applying different sampling cycles. As a result, the detection accuracy cannot achieve the expected goal.

Also, the discretization model lacks of theoretical basis. In view of the fact that the previous literature did not detailedly derive the PAV and OV model from the principle of dynamic modeling. This paper proposes a phase angle vector dynamic (PAVD) model–based KF (PAVD–KF) algorithm and the orthogonal vector dynamic (OVD) model–based KF algorithm (OVD–KF) through model derivation according to the stochastic process theory [12]. It reveals that the state noise covariance of the conventional detecting model is related to the sampling cycle.

This paper contributes:

(1) We derive the discretization model from the continuous differential equation in accordance with stochastic process theory, to find the value law of fixed value of covariance under different sampling cycles.
(2) We greatly improve the detecting accuracy of fundamental and harmonics components, enabling the detection results to meet the requirements of modern engineering applications of AC current and voltage signal.

Section 1 introduces the Kalman filter mainly applies to the following fields and model–based KF existing problems. In Section 2, the related works about conventional PAV and OV model are introduced. In Section 3, the PAVD and OVD models are rederived according to of stochastic process theory. Then, the PAVD–KF and OVD–KF algorithms are proposed. In Section 4, the performances of PAVD–KF and OVD–KF algorithms are evaluated and compared by the experimental current and voltage sampling data. Section 5 concludes the paper.

## 2. Related Work

In this section, we proposed the expression of distorted AC signal first, and introduce the conventional PAV–KF and OV–KF algorithms then.

### 2.1. Distorted AC Signal

The waveform of the distorted AC signals sampled at time t can be expressed as

$$s(t) = \sum_{i=1}^{N} A_i(t)\sin[i\omega_0 T + \theta_i(t)] + \sum_{i=1}^{N} w_i(t) \tag{1}$$

where $i = 1, 2, 3, \cdots, N$, $T$ denotes the sampling cycle. For the fundamental frequency of 50 Hz, $\omega_0 = 100\pi$. $\sum\limits_{i=1}^{N} A_i(t)\sin[i\omega_0 T + \theta_i(t)]$ represents the fundamental and harmonic components. $\sum\limits_{i=1}^{N} w_i(t)$ is the state noise of fundamental and harmonic components.

### 2.2. PAV Model

The vector of fundamental and harmonic components is expressed as $X_k = \begin{bmatrix} x_1 & y_1 & \cdots & x_n & y_n \end{bmatrix}^T$, where $x_1 = A_1\cos\theta_1$, $y_1 = A_1\sin\theta_1$,$x_n = A_n\cos\theta_n$, $y_n = A_n\sin\theta_n$.

The measurement equation of model is expressed as [3]

$$z_k = H_k X_k + V_k = \begin{bmatrix} \cos(\omega_0 kT) \\ -\sin(\omega_0 kT) \\ \vdots \\ \cos(n\omega_0 kT \\ -\sin(n\omega_0 kT) \end{bmatrix}^T \begin{bmatrix} x_1 \\ y_1 \\ \vdots \\ x_n \\ y_n \end{bmatrix}_k + V_k \tag{2}$$

The state equation of model is expressed as

$$X_{k+1} = \Phi_k X_k + W_k = \begin{bmatrix} 1 & 0 & \cdots & 0 & 0 \\ 0 & 1 & \cdots & 0 & 0 \\ \vdots & \vdots & \vdots & \vdots & \vdots \\ 0 & 0 & \cdots & 1 & 0 \\ 0 & 0 & \cdots & 0 & 1 \end{bmatrix} \begin{bmatrix} x_1 \\ y_1 \\ \vdots \\ x_n \\ y_n \end{bmatrix}_k + \begin{bmatrix} W_{x_1} \\ W_{y_1} \\ \vdots \\ W_{x_n} \\ W_{y_n} \end{bmatrix}_k \quad (3)$$

The covariance matrix of state noise $W_k$ can be expressed as

$$Q_k = E\left[W_k W_k^T\right] = \sigma_{PAV}^2 I_{2n\times 2n} \quad (4)$$

*2.3. OV Model*

Considering a signal $A_k \sin(\omega_k t_k + \theta_k)$ with amplitude $A_k$, angular frequency is $\omega_k$ and phase is $\theta_k$ [4,11]. Let $x_k = A_k \sin(\omega_k t_k + \theta_k), y_k = A_k \cos(\omega_k t_k + \theta_k)$, initially, consider $A_{k+1/k} \approx A_k$, $\omega_{k+1/k} \approx \omega_k$ and $\theta_{k+1/k} \approx \theta_k$. At the time $t_{k+1} = t_k + T$, the signal $x_k$ can be expressed as

$$x_{k+1} = A_{k+1} \sin(\omega_k t_k + \omega_k T + \theta_k) = x_k \cos(\omega_k T) + y_k \sin(\omega_k T) \quad (5)$$

where, $T$ is the sampling cycle. Additionally,

$$y_{k+1} = A_{k+1} \cos(\omega_k t_k + \omega_k T + \theta_k) = -x_k \sin(\omega_k T) + y_k \cos(\omega_k T) \quad (6)$$

the state−space is extended for detection of distorted AC signal existed n frequencies harmonic components.

The vectors are expressed as $X_k = \begin{bmatrix} x_1 & y_1 & \cdots & x_n & y_n \end{bmatrix}_k^T$, where, $x_{1_k} = A_{1_k} \sin(\omega_k t_k + \theta_{1_k}), y_{1_k} = A_{1_k} \cos(\omega_k t_k + \theta_{1_k}), x_{n_k} = A_{n_k} \sin(n\omega_k t_k + \theta_{n_k})$, $y_{n_k} = A_{n_k} \cos(n\omega_k t_k + \theta_{n_k})$.

Then, the state and measurement equation is expressed as [4]

$$\begin{cases} X_{k+1} = \begin{bmatrix} \Phi_1 & \cdots & 0 \\ \vdots & \cdots & \vdots \\ 0 & \cdots & \Phi_n \end{bmatrix} X_k + W_{i_k} \\ z_k = \begin{bmatrix} 1 & 0 & \cdots & 1 & 0 \end{bmatrix} X_k + V_k \end{cases} \quad (7)$$

where, the state transition matrixes $\Phi_i$ are shown below as

$$\Phi_i = \begin{bmatrix} \cos(i\omega_k T) & \sin(i\omega_k T) \\ -\sin(i\omega_k T) & \cos(i\omega_k T) \end{bmatrix} \quad (8)$$

The covariance matrix of state noise $W_{k_i}$ can be expressed as

$$Q_{i_k} = E\left[W_{i_k} W_{i_k}^{\ T}\right] = \sigma_{OV}^2 I_{2n\times 2n} \quad (9)$$

in (4) and (9), $I_{2n\times 2n}$ is a unit matrix. $\sigma_{PAV}^2$ and $\sigma_{OV}^2$ are selected as 0.05, when AC signal is sampling voltage [10,11]. $\sigma_{PAV}^2$ and $\sigma_{OV}^2$ are selected as 0.01, when AC signal is sampling current [20].

### 2.4. PAV-KF and OV-KF Algorithm

The discrete PAV−KF and OV−KF algorithm for detecting distorted AC signal can be expressed as follow by combining PAV and OV model.

$$\begin{cases} \hat{X}_{k+1/k+1} = \Phi_k \hat{X}_{k+1/k} + K_{k+1}\left(Z_{k+1} - H_k \hat{X}_{k+1/k}\right) \\ K_{k+1} = \Phi_k P_{k+1/k} H_k^T \left(H_k P_{k+1/k} H_k^T + R_k\right)^{-1} \\ P_{k+1/k+1} = \Phi_k P_{k+1/k} \Phi_k^T - K_{k+1} H_k P_{k+1/k} \Phi_k^T + Q_k \end{cases} \tag{10}$$

The amplitude and the phase angle for the fundamental and harmonic components can be calculated as follows.

$$A_{i_{k+1}} = \sqrt{\hat{x}^2_{i_{k+1}} + \hat{y}^2_{i_{k+1}}} \tag{11}$$

$$\phi_{i_{k+1}} = \arctan\left(\frac{\hat{x}_{i_{k+1}}}{\hat{y}_{i_{k+1}}}\right) \tag{12}$$

## 3. PAVD-KF and OVD-KF Algorithm

In Section 3, we construct the AC signal continuous differential equation of phase angle vector and orthogonal vector, and derive the discretization PAVD and OVD model in accordance with stochastic process theory. Next, we combine the proposed models with KF algorithm, to obtain PAVD−KF and OVD−KF algorithm.

### 3.1. PAVFIGURE
D-KF Algorithm

The phase angle vectors of fundamental components in (2)−(3) are expressed as

$$\begin{cases} x(t) = A\cos\theta \\ y(t) = A\sin\theta \end{cases} \tag{13}$$

The first derivative can be expressed as

$$\frac{d}{dt}\begin{bmatrix} x(t) \\ y(t) \end{bmatrix} = A\begin{bmatrix} x(t) \\ y(t) \end{bmatrix} + B\begin{bmatrix} w_x(t) \\ w_y(t) \end{bmatrix} \tag{14}$$

where, $w_x(t)$ and $w_y(t)$ are the input white noise, $w_x(t) \perp w_y(t)$. $w_x(t) \sim (0, \sigma_w^2)$, $w_y(t) \sim (0, \sigma_w^2)$. Where, $A = \begin{bmatrix} 0 & 0 \\ 0 & 0 \end{bmatrix}$, $B = \begin{bmatrix} 1 & 0 \\ 0 & 1 \end{bmatrix}$.

Equation (11) is discretized, and the state equation can be obtained as follows [25].

$$\begin{bmatrix} x \\ y \end{bmatrix}_{k+1} = \Phi_k \begin{bmatrix} x \\ y \end{bmatrix}_k + W_k \tag{15}$$

where,

$$\Phi_k = e^{AT} = L^{-1}\left([sI - A]^{-1}\right) = \begin{bmatrix} 1 & 0 \\ 0 & 1 \end{bmatrix} \tag{16}$$

$$W_k = \int_{kT}^{(k+1)T} e^{A[(k+1)T - \tau]} \cdot B \cdot w(\tau) d\tau \tag{17}$$

The covariance matrix of state noise $W_k$ can be expressed as

$$Q_w = E\left[W_k W_k^T\right] = \sigma_w^2 \begin{bmatrix} T & 0 \\ 0 & T \end{bmatrix} = \sigma_w^2 T \begin{bmatrix} 1 & 0 \\ 0 & 1 \end{bmatrix} \tag{18}$$

from (18), an important conclusion can be known that $Q_w$ is related to sampling cycle $T$.

The state−space dimension of the distorted AC signal is extended under the consideration of fundamental and harmonic components. The state-space formulas of PAVD model are the same as the conventional PAV model (2)−(3). The difference is that the state noise covariance in (4). The state noise covariance of the PAVD model is

$$Q_k = \sigma_{PAVD}^2 I_{2n\times 2n} = \rho T I_{2n\times 2n} \tag{19}$$

where, $\sigma_{PAVD}^2$ is the state noise covariance of PAVD model. $\rho \leq 1$ is a coefficient. When high detection precision is required, $\rho$ is selected as a small value. When dynamic fast detecting is required, $\rho$ is selected to a large value. Then, the discrete PAVD−KF algorithm can be obtained by combining model (2), (3), (15) and KF algorithm.

### 3.2. OVD−KF Algorithm

In the two−dimensional space, the circular movement of the object along the midpoint is projected on the x, y axis as a pure sine and a pure cosine waveform. It is shown in Figure 1.

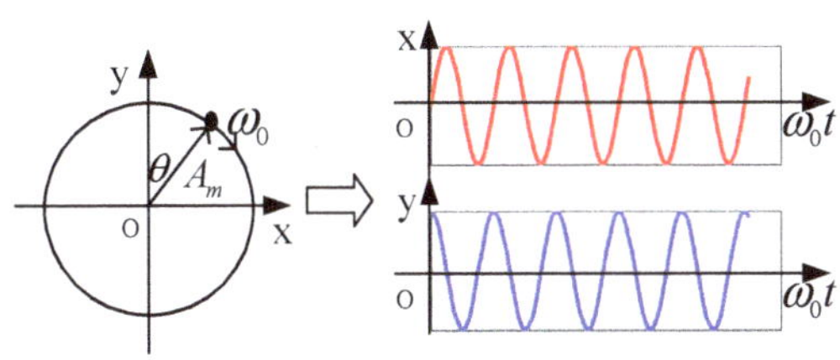


**Figure 1.** Diagram of the object circular movement being projected.

It can be expressed as

$$\begin{cases} x(t) = A_m \sin(\omega_0 t + \theta) \\ y(t) = A_m \cos(\omega_0 t + \theta) \end{cases} \tag{20}$$

where, $A_m$ is the amplitude, $\omega_0$ is the angular frequency, and $\theta$ is the phase angle.

The first derivative of (10) can be expressed as

$$\begin{cases} \dot{x}(t) = \omega_0 A_m \cos(\omega_0 t + \theta) + w_x(t) = \omega_0 y(t) + w_x(t) \\ \dot{y}(t) = -\omega_0 A_m \sin(\omega_0 t + \theta) + w_y(t) = -\omega_0 x(t) + w_y(t) \end{cases} \tag{21}$$

where, $w_x(t)$ and $w_y(t)$ are the input white noise, $w_x(t) \perp w_y(t)$. $w_x(t) \sim (0, \sigma_w^2)$, $w_y(t) \sim (0, \sigma_w^2)$.

The state equation can be obtained from (21).

$$\frac{d}{dt}\begin{bmatrix} x(t) \\ y(t) \end{bmatrix} = A\begin{bmatrix} x(t) \\ y(t) \end{bmatrix} + B\begin{bmatrix} w_x(t) \\ w_y(t) \end{bmatrix} = \begin{bmatrix} 0 & \omega_0 \\ -\omega_0 & 0 \end{bmatrix}\begin{bmatrix} x(t) \\ y(t) \end{bmatrix} + \begin{bmatrix} 1 & 0 \\ 0 & 1 \end{bmatrix}\begin{bmatrix} w_x(t) \\ w_y(t) \end{bmatrix} \tag{22}$$

Equation (22) is discretized, and the state equation can be obtained as follows [25].

$$\begin{bmatrix} x \\ y \end{bmatrix}_{k+1} = \Phi_k \begin{bmatrix} x \\ y \end{bmatrix}_k + W_k \tag{23}$$

where,

$$\Phi_k = e^{AT} = L^{-1}([sI - A]^{-1}) = \begin{bmatrix} \cos(\omega_0 T) & \sin(\omega_0 T) \\ -\sin(\omega_0 T) & \cos(\omega_0 T) \end{bmatrix} \tag{24}$$

$$W_k = \int_{kT}^{(k+1)T} e^{A[(k+1)T-\tau]} \cdot B \cdot w(\tau) d\tau \tag{25}$$

The state noise covariance matrix can be expressed as

$$Q_w = E\left[W_k W_k^T\right] = \sigma_w^2 \begin{bmatrix} T & 0 \\ 0 & T \end{bmatrix} = \sigma_w^2 T \begin{bmatrix} 1 & 0 \\ 0 & 1 \end{bmatrix} \tag{26}$$

from Equation (16), an important conclusion can be drew that $Q_w$ is related to the sampling cycle $T$.

The state-space dimension of the distorted AC signal is expanded under the consideration of fundamental and harmonic components. The state and measurement equations of OVD model are the same as the conventional OV model (4) and (5). The difference is that state noise covariance matrix of the OV model in (6). That is

$$Q_k = \sigma_{OVD}^2 I_{2n\times 2n} = \rho T I_{2n\times 2n} \tag{27}$$

where $\sigma_{OVD}^2$ is the covariance of OVD model. $\rho \leq 1$ is a coefficient. When high detection precision is required, $\rho$ is selected as a small value. When dynamic fast detecting is required, $\rho$ is selected to a large value.

Then, there are the combinations of the model (7), (8), (27), and KF algorithm. The discrete OVD−KF algorithm can be obtained.

## 4. Experiment and Evaluation

In this section, we conduct the detection of the current and the voltage with the sampling cycle of 10 µs and 40 µs. Then, comparing the detecting results for fundamental and harmonics components in PAVD−KF and PAV−KF algorithm, and in OVD−KF and OV-KF algorithm. At the last, we come to a conclusion of these results.

### *4.1. Experiment and Evaluation for AC Current Detection*

#### 4.1.1. Experiment Settings

As an experimental example, we sample the AC distorted current signal. Experiment platform consists of AC grid (380 V, 50 Hz), oscilloscope (Tektronix GDS−2102) and nonlinear load (the three−phase rectifier, R = 100 Ω and L = 1.5 mH). It is shown in Figure 2. The AC current signals are sampled by the oscilloscope. Sampling cycles are selected as 10 µs and 40 µs. The experiment sampling current data and spectrogram are shown in Figures 3 and 4. The HRI of harmonic components are measured to be approximately 24% of 5th, 10% of 7th, 10% of 11th, 5% of 13th, 5% of 17th, 4% of 19th by spectrum analysis.

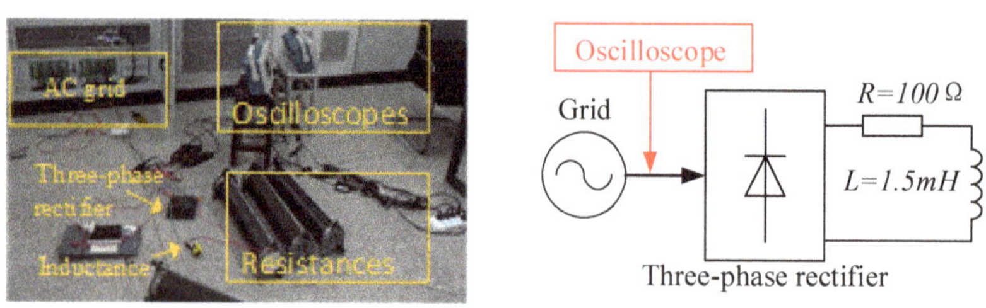


(a) (b)

**Figure 2.** Experiment platform. (**a**) Current sampling experiment platform (**b**) Principle diagram of experiment platform.

HRI is defined as follows

$$\mathrm{HRI_n} = \frac{I_n}{I_1} \times 100\% \tag{28}$$

where $I_1$ is the voltage rms value of fundamental component, $I_n$ is the voltage rms value of n−th harmonic components.

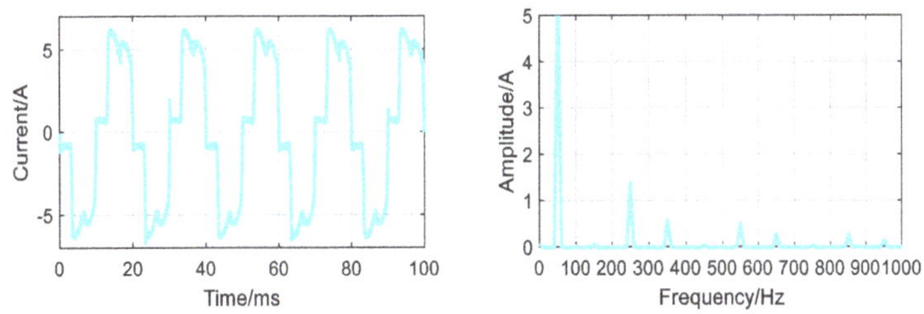


(a) (b)

**Figure 3.** Experiment current for sampling cycle 10 μs. (**a**) The sampling current data. (**b**) Spectrogram of sampling current data.

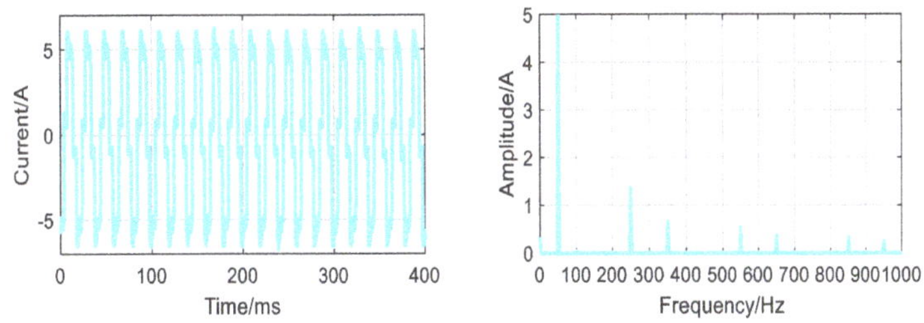


(a) (b)

**Figure 4.** Experiment current for sampling cycle 40 μs. (**a**) The sampling current data. (**b**) Spectrogram of sampling current data.

The measurement noise covariance R of the algorithms is all selected as 0.06 A, that is the measurement error of the sensor. The initial covariance matrices are selected as $P_{0/0} = 1000 \times I_{14\times14}$. The state noise covariance is selected as $Q_k = 0.01 \times I_{14\times14}$ in (4). The state noise covariance is selected as $Q_k = 0.01 \times T \times I_{14\times14}$ in (19).

### 4.1.2. Detection Results for PAVD−KF Algorithm

Two algorithms of PAV−KF and PAVD−KF are evaluated and compared. The evaluation results for sampling cycle 10 μs and 40 μs are shown in Figures 5 and 6.

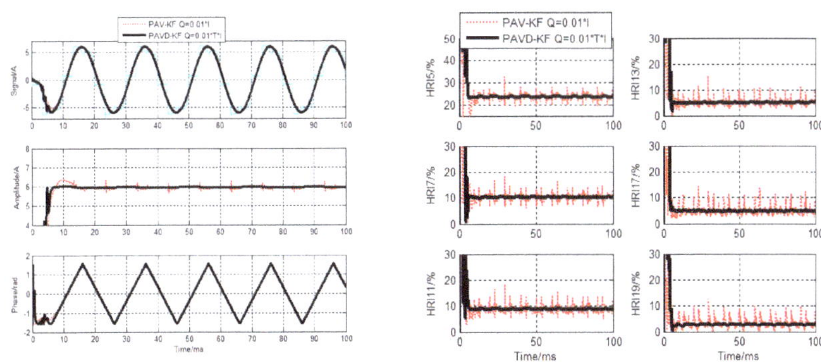


(a) (b)

**Figure 5.** PAVD−KF current detecting results for sampling cycle 10 μs. (**a**) Fundamental components detecting results (**b**) HRI detecting results.

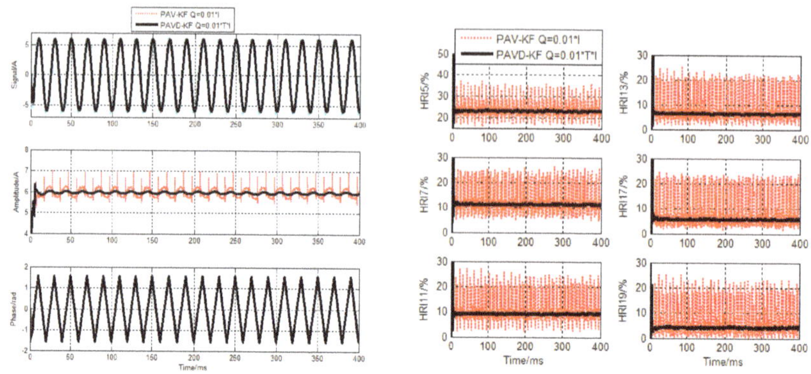


(**a**) (**b**)

**Figure 6.** PAVD−KF current detecting results for sampling cycle 40 μs. (**a**) Fundamental components detecting results (**b**) HRI detecting results.

4.1.3. Detection Results for OVD−KF Algorithm

Two algorithms of OV−KF and OVD−KF are evaluated and compared. The evaluation results for sampling cycle 10 μs and 40 μs are shown in Figures 7 and 8.

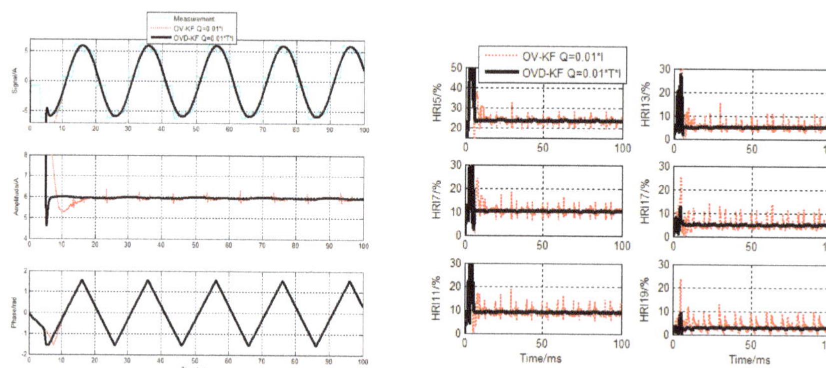


(**a**) (**b**)

**Figure 7.** OVD−KF current detecting results for sampling cycle 10 μs. (**a**) Fundamental components detecting results (**b**) HRI detecting results.

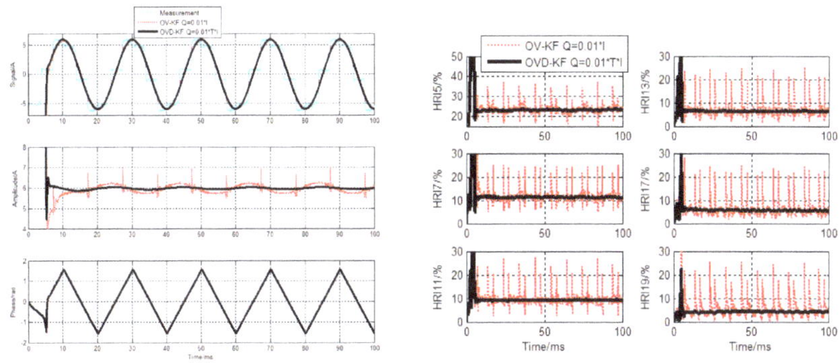


(a) (b)

**Figure 8.** OVD−KF current detecting results for sampling cycle 40 μs. (**a**) Fundamental components detecting results (**b**) HRI detecting results.

4.1.4. Summary for AC Current Detecting

It can be seen from Figures 5 and 8, when Q = 0.01 × $I$, the detecting errors of the conventional PAV−KF and OV−KF algorithms are big. In particular, the detecting errors of the conventional algorithms for sampling cycle 40 μs are rather big. The errors are too big to affect the normal operation.

Corresponding to it, when Q combines with the sampling cycle, Q = 0.01 × $T$ × $I$, regardless of the fundamental or harmonic components, the proposed detecting accuracies in PAVD−KF and OVD−KF algorithm are much higher than the conventional algorithms.

*4.2. Experiment and Evaluation for AC Voltage Detection*

4.2.1. Experiment Settings

The AC signal is grid voltage as an experimental example. Experiment platform is shown in Figure 9. The experiment of superimposing harmonics in fundamental voltage is carried out through a programmable power supply device. The AC voltage signals are sampled and the sampling cycles $T$ are selected by the oscilloscope of Tektronix GDS−2102.

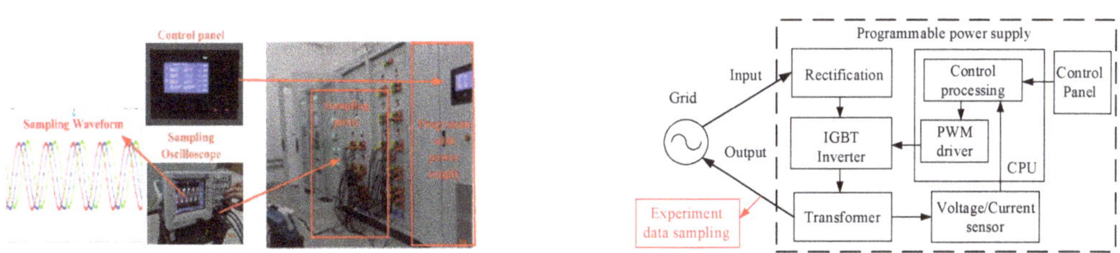


(a) (b)

**Figure 9.** Experiment platform. (**a**) Voltage sampling experiment platform. (**b**) Principle diagram of experiment platform.

The harmonics ratio for voltage (HRU) is defined as follows

$$\mathrm{HRU_n} = \frac{U_n}{U_1} \times 100\% \quad (29)$$

where $U_1$ is the voltage rms value of the fundamental component, and $U_n$ is the voltage rms value of n−th harmonic components.

Experiment voltage data for sampling cycle 10 μs is in Figure 10. Experiment voltage data for sampling cycle 40 μs is in Figure 11, and the HRU of harmonic components all are set as 5% of 3rd, 10% of 5th, and 8% of 7th.

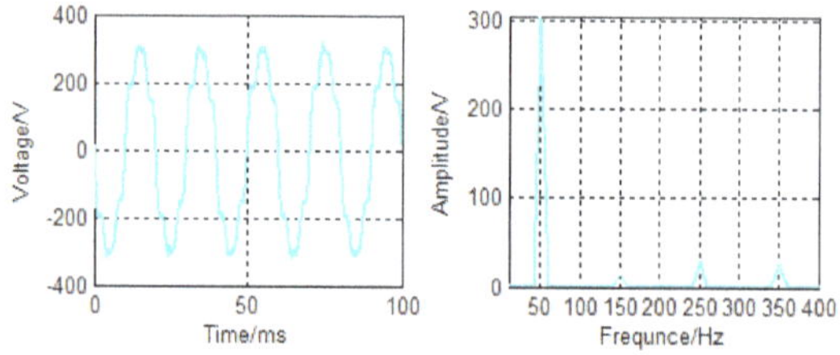


(a) (b)

**Figure 10.** Experiment voltage for sampling cycle 10 μs. (**a**) The sampling voltage data. (**b**) Spectrogram of sampling voltage data.

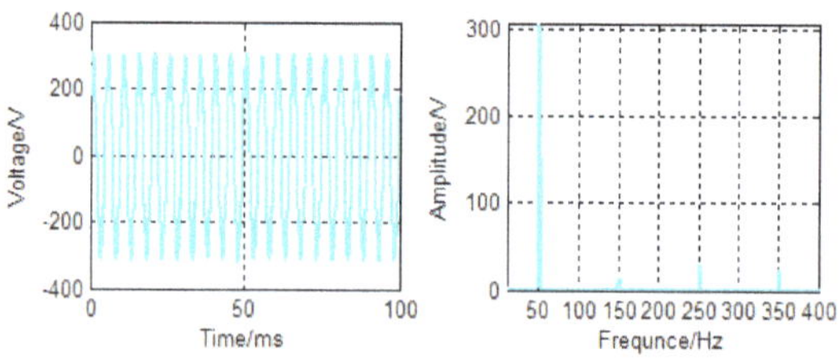


(a) (b)

**Figure 11.** Experiment voltage for sampling cycle 40 μs. (**a**) The sampling voltage data. (**b**) Spectrogram of sampling voltage data.

The R of the algorithms are all selected as 3.11 V, that is the measurement error of the sensor. The initial covariance matrices of algorithms are selected as $P_{0/0} = 1000 \times I_{10\times10}$. The state noise covariance matrix of algorithms is selected as $Q_k = 0.05 \times I_{10\times10}$. The state noise covariance matrix of algorithms is $Q_k = 0.1 \times T \times I_{10\times10}$.

#### 4.2.2. Detection Results for PAVD−KF Algorithm

PAVD−KF and PAV−KF algorithms are evaluated and compared. The detecting results of the sampling cycle 10 μs and 40 μs are shown in Figures 12 and 13.

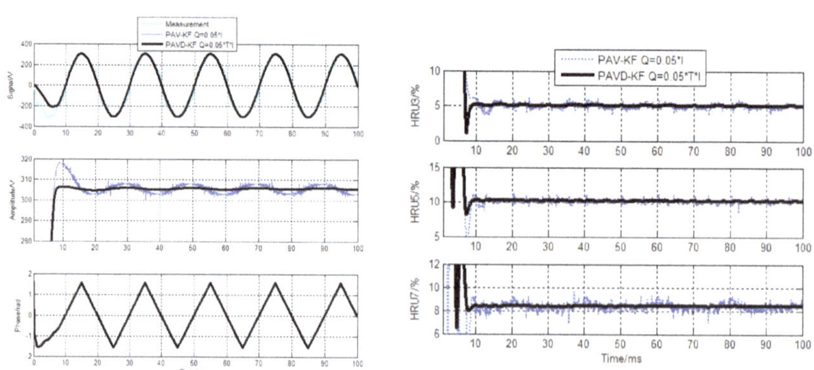


(a) (b)

**Figure 12.** PAVD−KF voltage detecting results for sampling cycle 10 μs. (**a**) Fundamental components detecting results (**b**) HRU detecting results.

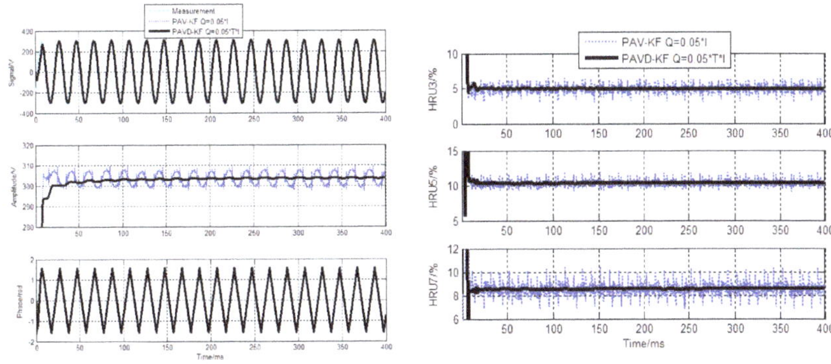


(**a**) (**b**)

**Figure 13.** PAVD−KF voltage detecting results for sampling cycle 40 μs. (**a**) Fundamental components detecting results (**b**) HRU detecting results.

4.2.3. Detection Results for OVD−KF Algorithm

OVD−KF and OV−KF algorithms are evaluated and compared. The detecting results of the sampling cycle 10 μs and 40 μs are shown in Figures 14 and 15.

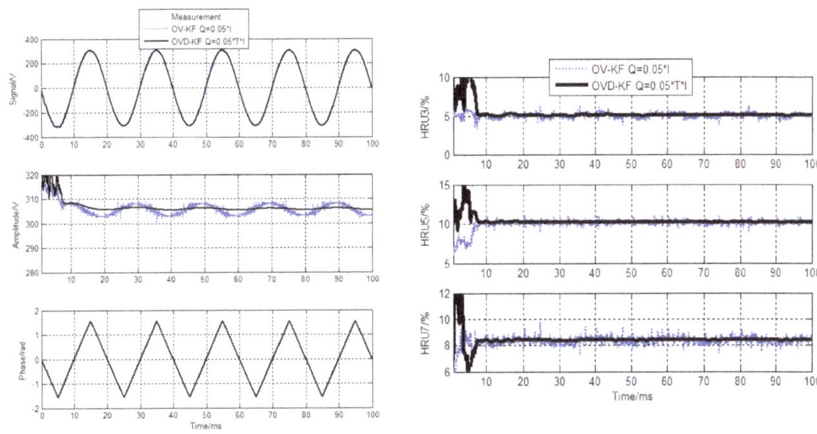


(**a**) (**b**)

**Figure 14.** OVD−KF voltage detecting results for sampling cycle 10 μs. (**a**) Fundamental components detecting results (**b**) HRU detecting results.

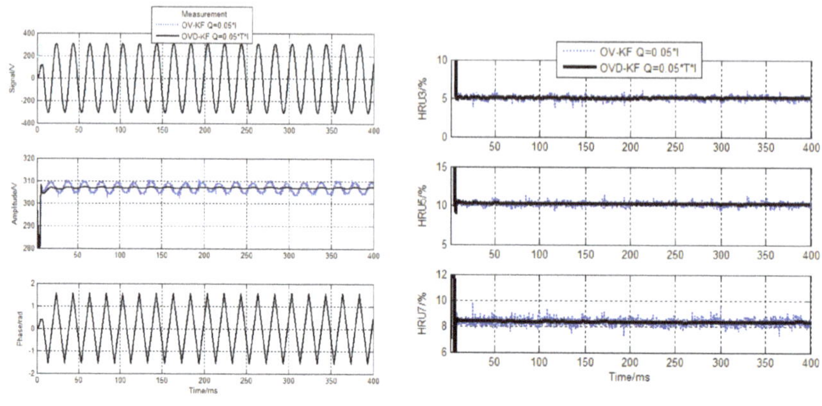


(a) (b)

**Figure 15.** OVD−KF voltage detecting results for sampling cycle 40 μs. (**a**) Fundamental components detecting results (**b**) HRU detecting results.

### 4.2.4. Summary for AC Voltage Detecting

It can be seen from Figures 12 and 13 that the amplitudes of the fundamental components of the conventional model−KF algorithm detecting are fluctuating, the stability of its detection shows poorly. Specifically, when we use the sampling cycle 40 μs, the error is too big to be ignored. From Figures 14 and 15, we know that detecting errors of the harmonics components for the conventional OV−KF algorithm are also big, though the error seems smaller compared to PAV−KF. Corresponding to it, the detecting accuracies of the fundamental and harmonic components for the OVD−KF algorithm are higher than that of the OV−KF algorithm.

Generally, when $Q = 0.05 \times I$, the conventional algorithms could basically meet the engineering requirements with some big errors under different sampling cycle conditions. When $Q = 0.05 \times T \times I$, all the novel algorithms could fulfill the requirements.

### *4.3. Summary for Experiment and Evaluation*

In general, when using the fixed value, the error of PAV model is the biggest, neither current nor voltage can meet the demand. OV model performs better, when using the sampling cycle 10 μs, the voltage can be used normally, and the current error is too big to be used. But when using the sampling cycle 40 μs, both current and voltage cannot operate normally.

The proposed algorithms in this paper, whether it is PAVD model or OVD model, it is related to the sampling cycle, so the voltage and current detecting results can meet the engineering requirement.

Take PAV model with 40 μs in current and voltage tracking for example, the conventional methods produce big errors when $Q = 0.01 \times I$ detecting the current and $Q = 0.05 \times I$ detecting the voltage. Only when $Q = 0.00001 \times I$ detecting the current, the accuracy of it would be almost close to that of our proposed algorithm, as Figure 16 shows. When $Q = 0.0005 \times I$ detecting the voltage, the accuracy would satisfy the requirement, as Figure 17 shows.

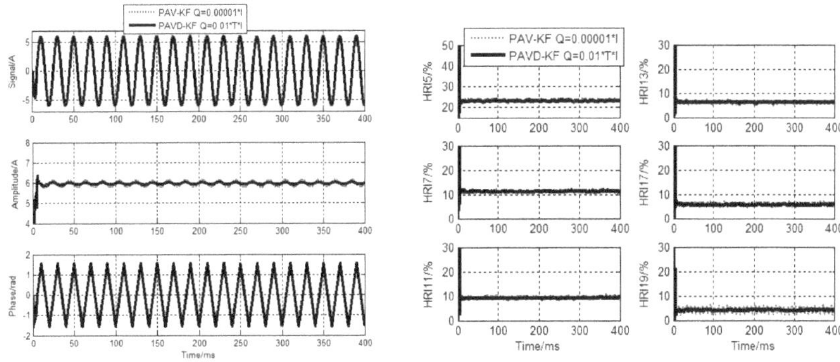


**Figure 16.** PAVD−KF current detecting results for sampling cycle 40 μs. (Q = 0.00001 × *I*).

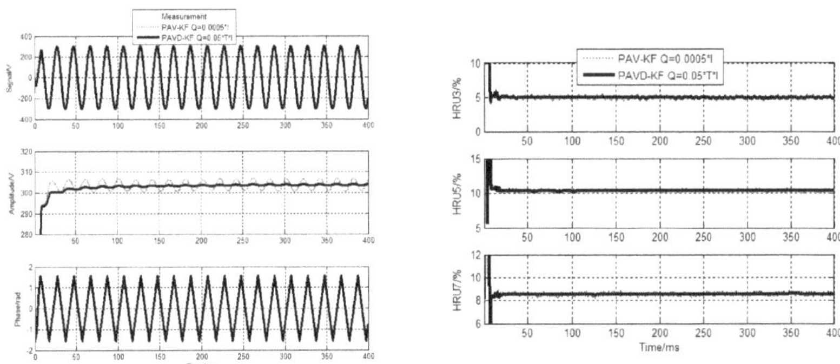


**Figure 17.** PAVD−KF voltage detecting results for sampling cycle 40 μs. (Q = 0.0005 × *I*).

## 5. Conclusions

This paper presents the PAVD−KF and OVD−KF algorithm derived from the stochastic process theory. It enhances the knowledge about the Kalman algorithms applied to the harmonic detection of the distorted AC grid. It gives a rule that the KF algorithms in the application of distorted grid AC signal detection is related to the sampling cycle and their state noise covariance cannot be simply considered as a constant.

This paper applies the actual current and voltage data of the sampling cycle 10 μs and 40 μs to detection and evaluation. The conventional algorithms have big errors in general and unable to meet engineering requirements. When using a fixed value of covariance, if we are going to obtain the ideal results, we need to spend a lot of time debugging (trying to make up) the parameters. Also, there is no theoretical source to set the fixed value of covariance. To solve it, we propose the algorithms to save much time and workload to get the ideal results. Besides, we make the algorithms more feasible and effective based on stochastic process theory. From which the results show that the proposed PAVD−KF and OVD−KF algorithms are effective and improve the dynamic detecting accuracy of grid AC signal significantly. This paper would provide a reference for the application of KF algorithms to harmonic detection, power quality control and grid synchronization in the future.

**Author Contributions:** Conceptualization, H.N. and X.N.; methodology, H.N. and X.N.; software, H.N.; validation, H.N. and X.N.; formal analysis, H.N.; investigation, H.N.; resources, H.N.; data curation, H.N.; writing—original draft preparation, H.N.; writing—review and editing, H.N.; visualization, H.N.; supervision, X.N.; project administration, X.N.; funding acquisition, X.N. All authors have read and agreed to the published version of the manuscript.

**Funding:** This research was funded by National Natural Science Foundations of China, grant number 51867017 "Research on Power Quality Dynamic Tracking and Real-time Detection Method for Micro-grid Complex Source Load Environment".

**Institutional Review Board Statement:** Not applicable.

**Informed Consent Statement:** Not applicable.

**Data Availability Statement:** Not applicable.

**Conflicts of Interest:** The authors declare no conflict of interest.

## References

1. Auger, F.; Hilairet, M.; Guerrero, J.M.; Monmasson, E.; Orlowska-Kowalska, T.; Katsura, S. Industrial Applications of the Kalman Filter: A Review. *IEEE Trans. Ind. Electron.* **2013**, *60*, 5458–5471. [CrossRef]
2. Girgis, A.A.; Brown, R.G. Application of Kalman Filtering In Computer Relaying. *IEEE Trans. Power Appar. Syst.* **1981**, *7*, 3387–3397. [CrossRef]
3. Girgis, A.A.; Hwang, T.L.D. Optimal Estimation of Voltage Phasors and Frequency Deviation. *IEEE Trans. Power Appar. Syst.* **1984**, *10*, 2943–2951. [CrossRef]
4. Girgis, A.A.; Sallam, A.A.; El-Din, A.K. Identification and tracking of harmonic sources in a power system using a Kalman filte. *IEEE Trans. Power Deliv.* **1998**, *13*, 414–420. [CrossRef]
5. Nie, X. Detection of Grid Voltage Fundamental and Harmonic Components Using Kalman Filter Based on Dynamic Tracking Model. *IEEE Trans. Ind. Electron.* **2020**, *67*, 1191–1200. [CrossRef]
6. Naseri, F.; Kazemi, Z.; Farjah, E.; Ghanbari, T. Fast Detection and Compensation of Current Transformer Saturation Using Extended Kalman Filter. *IEEE Trans. Power Deliv.* **2020**, *34*, 1087–1097. [CrossRef]
7. Swain, S.; Subudhi, B. Grid Synchronization of a PV System with Power Quality Disturbances Using Unscented Kalman Filtering. *IEEE Trans. Sustain. Energy* **2019**, *10*, 1240–1247. [CrossRef]
8. Uz-Logoglu, E.; Salor, O.; Ermis, M. Real-Time Detection of Interharmonics and Harmonics of AC Electric Arc Furnaces on GPU Framework. *IEEE Trans. Ind. Appl.* **2019**, *55*, 1373–1378.
9. Xi, Y.; Tang, X.; Li, Z.; Cui, Y.; Zhao, T.; Zeng, X.; Guo, J.; Duan, W. Harmonic estimation in power systems using an optimised adaptive Kalman filter based on PSO-GA. *IET Gener. Transm. Distrib.* **2019**, *13*, 3968–3979. [CrossRef]
10. Bagheri, A.; Mardaneh, M.; Rajaei, A.; Rahideh, A. Detection of Grid Voltage Fundamental and Harmonic Components Using Kalman Filter and Generalized Averaging Method. *IEEE Trans. Power Electron.* **2016**, *31*, 1064–1073. [CrossRef]
11. Cardoso, R.; Camargo, R.F.; Pinheiro, H.; Grundling, H.A. Kalman filter based synchronisation methods. *IET Gener. Transm. Distrib.* **2008**, *2*, 542–555. [CrossRef]
12. Swain, S.; Subudhi, B. Iterated extended Kalman filter-based grid synchronisation control of a PV system. *IET Energy Syst. Integr.* **2019**, *1*, 219–228. [CrossRef]
13. Ferrero, R.; Pegoraro, P.A.; Toscani, S. Synchrophasor Estimation for Three-Phase Systems Based on Taylor Extended Kalman Filtering. *IEEE Trans. Instrum. Meas.* **2020**, *69*, 6723–6730. [CrossRef]
14. Golestan, S.; Guerrero, J.M.; Vasquez, J.C.; Abusorrah, A.M.; Al-Turki, Y. Single-Phase FLLs Based on Linear Kalman Filter, Limit-Cycle Oscillator, and Complex Bandpass Filter: Analysis and Comparison With a Standard FLL in Grid Applications. *IEEE Trans. Power Electron.* **2019**, *34*, 1064–1073. [CrossRef]
15. Ahmed, H.; Biricik, S.; Benbouzid, M. Linear Kalman Filter-Based Grid Synchronization Technique: An Alternative Implementation. *IEEE Trans. Ind. Inform.* **2021**, *17*, 3847–3856. [CrossRef]
16. Alam, S.J.; Arya, S.R. Control of UPQC based on steady state linear Kalman filter for compensation of power quality problems. *Chin. J. Electr. Eng.* **2020**, *6*, 52–65. [CrossRef]
17. Xi, Y.; Li, Z.; Zeng, X.; Tang, X.; Liu, Q.; Xiao, H. Detection of power quality disturbance using an adaptive process noise covariance Kalman filte. *Digit. Signal Process.* **2018**, *76*, 34–49. [CrossRef]
18. Wu, C.; Magana, M.E.; Cotilla-Sanchez, E. Dynamic Frequency and Amplitude Estimation for Three-Phase Unbalanced Power Systems Using the Unscented Kalman Filter. *IEEE Trans. Instrum. Meas.* **2019**, *68*, 3387–3395. [CrossRef]
19. Panigrahi, R.; Subudhi, B.; Panda, P.C. A robust LQG servo control strategy of shunt-active power filter for power quality enhancement. *IEEE Trans. Power Electron.* **2016**, *31*, 2860–2869. [CrossRef]
20. Panigrahi, R.; Subudhi, B. Performance enhancement of shunt active power filter using a Kalman filter-based H∞ control strategy. *IEEE Trans. Power Electron.* **2017**, *32*, 2622–2630. [CrossRef]
21. Abdelsalam, A.A.; Abdelaziz, A.Y.; Kamh, M.Z. A Generalized Approach for Power Quality Disturbances Recognition Based on Kalman Filter. *IEEE Access* **2021**, *9*, 93614–93628. [CrossRef]
22. Xi, Y.; Li, Z.; Tang, X.; Zeng, X. Classification of power quality disturbances based on KF-ML-Aided S-Transform and multilayers feedforward neural networks. *IET Gener. Trans. Distrib.* **2020**, *14*, 4010–4020. [CrossRef]
23. Abdelsalam, A.A.; Eldesouky, A.A.; Sallam, A.A. Classification of Power System Disturbances using Linear Kalman Filter and Fuzzy-expert System. *Electr. Power Energy Syst.* **2012**, *43*, 688–695. [CrossRef]

24. Abdelsalam, A.A.; Eldesouky, A.A.; Sallam, A.A. Characterization of power quality disturbances using hybrid technique of linear Kalman filter and fuzzy-expert system. *Electr. Power Syst. Res.* **2012**, *83*, 41–50. [CrossRef]
25. Brown, R.G.; Hwang, P.Y.C. *Introduction to Random Signals and Applied Kalman Filtering with MATLAB Exercises*, 4th ed.; John Wiley& Sons, Inc.: Hoboken, NJ, USA, 2011.

energies

MDPI

*Article*

# One-Dimensional and Three-Dimensional Numerical Investigations of Thermal Performance of Phase Change Materials in a Lithium-Ion Battery

**Van-Tinh Huynh, Kyoungsik Chang and Sang-Wook Lee ***

Department of Mechanical Engineering, University of Ulsan, Ulsan 44610, Korea; hvtinh97@gmail.com (V.-T.H.); kschang76@ulsan.ac.kr (K.C.)
* Correspondence: leesw@ulsan.ac.kr

**Abstract:** The thermal performance of a large-format (52.3 Ah) Li-ion pouch battery with an n-octadecane PCM was investigated. A simplified 1D model was employed to estimate the transient thermal behavior. Two design parameters, the thickness and the thermal conductivity of the PCM, were considered. A 0.5 mm thick n-octadecane PCM integrated with aluminum foam reduced the battery temperature to 34.3 °C and 50.7 °C at the end stage of discharging under 3C and 5C discharge rates, respectively. The 1D results compared to the 3D results were able to predict the temperature dissipation by the PCM method at the end of discharging. The 1D approach clearly produced reliable results in predicting the thermal behavior of the PCM cooling and was superior in practical applications with its low cost and time consumption. A 3D CFD simulation was able to describe the detailed temperature uniformity in the cell, which is an important factor in the design and evaluation of a battery cooling system.

**Keywords:** lithium-ion battery; phase change material; large-format cell; electric vehicle

**Citation:** Huynh, V.-T.; Chang, K.; Lee, S.-W. One-Dimensional and Three-Dimensional Numerical Investigations of Thermal Performance of Phase Change Materials in a Lithium-Ion Battery. *Energies* **2021**, *14*, 8386. https://doi.org/10.3390/en14248386

Academic Editor: Calin Iclodean

Received: 20 October 2021
Accepted: 8 December 2021
Published: 13 December 2021



## 1. Introduction

Energy storage systems, such as fuel cells, batteries, flywheels, and ultracapacitors, play important roles in pure electric vehicles and the hybrid electric vehicle industry. There are currently three leading types of batteries in the market: Li-ion, nickel–metal hydride, and lead–acid. The Li-ion battery is a promising candidate with outstanding features, including a low self-discharge rate, high specific energy, and fast charging capacity without a memory effect [1]. However, the internal electrochemical reactions and resistances in its cells produce a considerable amount of heat. This results in fairly high and non-uniform cell temperatures when it is charged and discharged. The optimal operating temperature of a Li-ion battery ranges from 25 °C to 40 °C [2]. In addition, the maximum temperature difference within a cell should be less than 5 °C to maintain a balance between its life cycle and efficiency [3]. When the maximum temperature is higher than 80 °C, thermal runaway could happen and exothermic reactions will occur, eventually leading to catastrophic results [4].

Various thermal management strategies have been proposed, which can be classified into the active (air and liquid cooling), passive (phase change material (PCM) cooling and heat pipes), and hybrid methods. Chen et al. [5] optimized the distance between cells to improve the air cooling performance, whereby the maximum temperature of the battery pack was decreased by 4 K and the maximum temperature difference was decreased by 69%. Under uphill conditions (high heat generation and/or a high discharge rate), liquid cooling may be required to sufficiently dissipate the thermal energy. Shang et al. [6] achieved the best performance of liquid cooling when the width of the cooling plate was 70 mm, the inlet had a temperature of 18 °C, and the mass flow rate was 0.21 kg/s. In that setup, the maximum temperature was decreased by 12.61% and the temperature uniformity was increased by 20.83%.

PCMs have been shown to greatly improve the temperature uniformity of Li-ion batteries. Wang et al. [7] showed that the melting process can be sped up by using a paraffin/aluminum foam composite. Compared with pure paraffin, the heat storage time of that composite PCM was 74.4% when the heat flux was 12,000 W/m$^2$. Jilte et al. [8] proposed the best design with seven cells, seven primary containers, and one secondary container. At the environment temperature of 40 °C, the use of a nanoparticle-enhanced PCM (nePCM) can reduce the temperature of a cell to under 46 °C with the effect of natural convection.

Heat pipes are employed as effective systems to maintain homogeneity on the evaporator surface at a constant temperature while having high thermal conductivity. Putra et al. [9] investigated the cooling capacity of alcohol, acetone, and distilled water at a heat flux load of 1.61 W/cm$^2$. They demonstrated that acetone is a potential candidate with a thermal resistance of 0.22 W/°C and an evaporation temperature of 50 °C. For the hybrid method, Yang et al. [10] studied the thermal performance of a battery thermal management system (BTMS) integrated with mini-channel liquid cooling and air cooling. At 80% depth of discharge (*DOD*), the maximum temperature and the temperature difference were decreased by 11.12 K and 9.52 K, respectively, when the water flow rate increased by $2.2 \times 10^{-4}$ kg/s. If the air velocity increased from 0 to 4 m/s, the battery temperature was reduced by 2.22 K, and the temperature uniformity was decreased by 2.04 K.

Kiani et al. [11] designed a hybrid thermal management system with a nanofluid integrating metal foam and a PCM. Compared with water cooling, the nanofluid with volume fraction of 2% increased the operating time of the battery by 29% at a Reynolds number of 420. Jilte et al. [12] investigated heat dissipation in cylindrical batteries with liquid channels and a PCM. At an environment temperature of 40 °C, the surface temperature of the battery was kept under 43 °C if the PCM was applied, and under 41.2 °C if the PCM and liquid channels were used simultaneously. Depending on the desired effectiveness, the BTMS can be applied either alone or in combination, but it should meet several criteria: high reliability, easy maintenance, low power consumption, and insignificant mass.

Different numerical modeling approaches have been used to investigate the thermal behavior of Li-ion batteries. They vary from 1D thermal models without thermal interaction among the cells, to complex 3D thermal models which consider non-isothermal and heat generation rates. Jollyn et al. [13] developed a 1D transient battery model based on daily cycles, such as 12 min and 72 min. They found that a proper PCM thickness was 1.75 mm when the maximum temperature decrease was 12.9 °C. Greco et al. [14] performed simplified battery thermal management using a PCM/compressed expanded natural graphite (CENG) composite. Their 1D solutions were in good agreement with their 3D results. Hallaj et al. [15] successfully demonstrated that the maximum temperature at the core of a battery pack was 80 °C with a small temperature variation of 3 °C at a 10 A discharge rate. Furthermore, they designed a battery pack integrating a PCM matrix with a significant decrease in charge time and weight for plug-in hybrid vehicles.

In the field of Li-ion battery research, for a 1D model, Hallaj et al. [16] developed a transient-state battery prototype with a thermally homogenous domain. An example is the Sony US18650 battery, which can be scaled up to a Li-ion cell of 100 Ah capacity in safe operating conditions. Sato et al. [17] categorized the heat-generating factors of reaction in a battery as reaction heat, polarization heat, and joule heat. Yi et al. [18] estimated temperature variations over time from electrochemical reactions and ohmic heat. They found that under a constant discharge rate, the potential density on the electrodes is a function of discharge time.

Akeiber et al. [19] predicted the heat storage capacity of a PCM of paraffin (40% oil and 60% wax) by solving a 1D numerical model. Fortunato et al. [20] obtained a 2D solution for different temperature profiles and total melting times with a PCM. Samar et al. [21] investigated 2D melting procedures for paraffin wax in ANSYS Fluent software, and showed that a PCM container induced a faster melting process.

In the present study, a 1D and a 3D Li-ion battery model were evaluated and compared with experimental data under 3C and 5C discharge rates. For a 3D model, two common approaches have been developed: the equivalent circuit model and the Newman, Tiedemann, Gu, and Kim (NTGK) model; the latter was used here. The main objective was to investigate the effect of PCM thickness and thermal conductivity on thermal management performance using the CFD technique with an NTGK model.

## 2. Methodology

A Li-ion battery with a 52.3 Ah pouch cell made by SM Bexel Co., Ltd. (Gumi, Korea) was used here. A 2.3 mm thick aluminum envelope covered the active zone, and the other fundamental characteristics of the battery are listed in Table 1.

**Table 1.** Battery specifications.

| Parameter | Value | Unit |
|---|---|---|
| Width × height × thickness | 0.249 × 0.227 × 0.008 | m |
| Nominal voltage, $V_n$ | 3.75 | V |
| Nominal capacity, $Q_n$ | 52.3 | Ah |
| Electrical conductivity, $\sigma_+$ | $3.77 \times 10^7$ | S/m |
| Electrical conductivity, $\sigma_-$ | $5.96 \times 10^7$ | S/m |
| Thermal conductivity, $k_b$ | 25.5 | W/m·K |
| Specific heat, $c_b$ | 566 | J/kg·K |
| Density, $\rho_b$ | 2695 | kg/m$^3$ |
| Internal resistance, $R$ | $6.1 \times 10^{-4}$ | Ω |
| Positive electrode | Aluminum–NCM523 | |
| Negative electrode | Copper–Graphite | |
| Electrolyte | Polyethylene | |

The melting point is the primary criterion when selecting a PCM. It must be both lower than the heat source temperature and higher than the ambient environment to which the system will be subjected [22]. For systems that are designed preferentially for thermal management, the usual advice is to choose a PCM with the highest possible melting point that is considerably below the desired thermal control point (40 °C). Based on our literature review, n-octadecane was suitable for this study with its $T_{solidus}$ = 301.15 K (solidus temperature) and $T_{liquidus}$ = 303.15 K (liquidus temperature). Its main properties are summarized in Table 2.

**Table 2.** The material properties of n-octadecane.

| Parameter | Solid Phase $T<T_{solidus}$ | Mushy Zone $T_{solidus}<T<T_{liquidus}$ | Liquid Phase $T_{liquidus}$ |
|---|---|---|---|
| Density, $\rho$ (kg/m$^3$) | 814 | 769 | 724 |
| Specific heat, $c$ (J/kg·K) | 2150 | 225,000 | 2180 |
| Thermal conductivity, $k$ (W/m·K) | 0.358 | 0.255 | 0.152 |
| Pure solvent melting heat, $L$ (J/kg) | | 225,000 | |

### 2.1. One-Dimensional Mathematical Model

In this study it was assumed that heat transfer only took place through the side surfaces since the battery was thin (8 mm). In such cases, thermal behavior can be estimated in a cell without considering the thermal interaction among the cells [13]. Bernardi et al. [23] reliably predicted cell temperature and heat generation by a 1D model and proposed a simplified overpotential heat form, taking into account irreversible and reversible heat, as follows:

$$q_b = q_{irr} + q_{rev} \quad (1)$$

$$q_{irr} = RI^2 \quad (2)$$

$$q_{rev} = -IT_b \frac{\partial E}{\partial T_b} \quad (3)$$

Figures 1 and 2 show the experimental measurement of currents over time under different discharge rates.

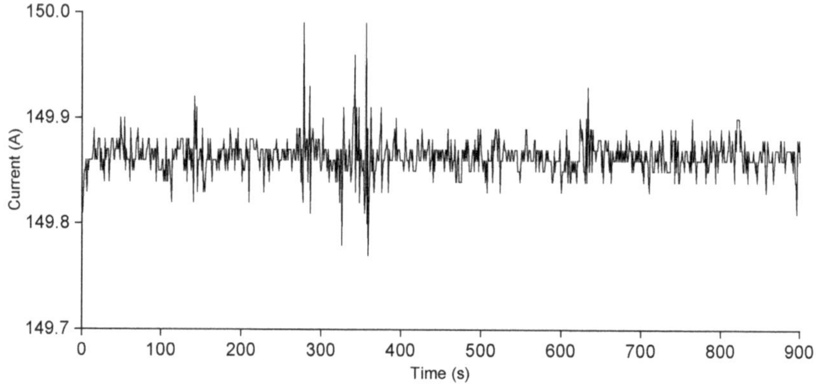


**Figure 1.** Measured electric current through the battery under 3C discharge rate.

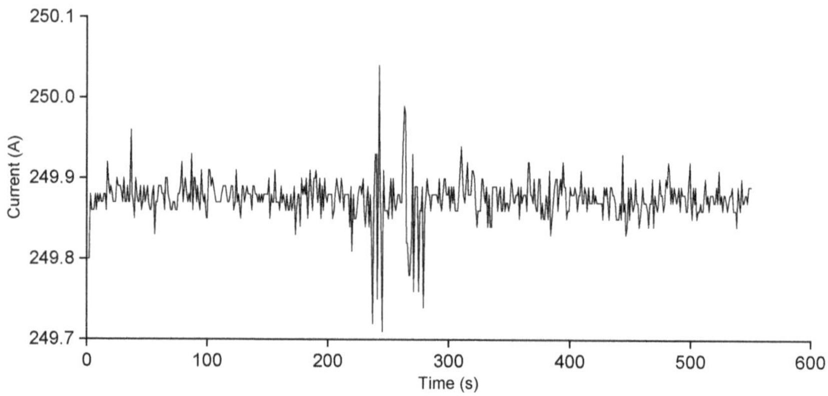


**Figure 2.** Measured electric current through the battery under 5C discharge rate.

The *DOD* is the percentage of the battery capacity that has been discharged from the fully charged battery as presented in Equation (4). An alternative form of *DOD* is the state of charge (*SOC*), wherein 0% refers to a fully discharged battery and 100% refers to a fully charged battery, as shown in Equation (5) and Figure 3a.

$$DOD = (Q/Q_0)\cdot 100 \quad (4)$$

$$SOC = 100 - DOD \quad (5)$$

The open-circuit voltage is determined from the terminal voltage by Equation (6). The entropic heat coefficient, which is estimated in Equation (7), is the derivative of the open-circuit voltage with respect to battery temperature, as shown in Figure 3b.

$$E = U + I\cdot R \quad (6)$$

$$\frac{\partial E}{\partial T_b} = \frac{E^{j+1} - E^j}{T_b^{j+1} - T_b^j} \quad (7)$$

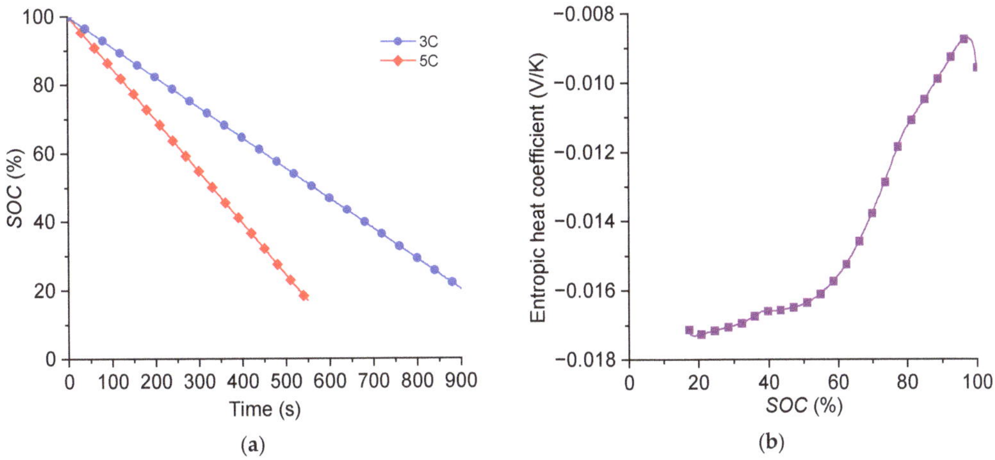


**Figure 3.** (**a**) *SOC* over time at different discharge rates. (**b**) Entropic heat coefficient as a function of *SOC*.

Reversible heat produces an entropy change that can be either negative or positive. Therefore, the entropic heat in Equation (3) is positive if endothermic, negative if exothermic during discharge. The discharge capacity and terminal voltage are also experimentally measured.

The next step is to consider the energy conservation principle. The heat generation rate equals the rate of change in the thermal energy in the cell, plus the rate of heat loss by natural convection, as represented in Equation (8). Thermal resistance for convection ($R_{conv}$ = 3.34 K/W) on a vertical surface is determined by the Nusselt number, related with the Rayleigh number and the Prandtl number, as shown in Equation (9). The Grasof number is an important standard in determining whether there is laminar flow or turbulent flow in natural convection. In this vertical plate case, the fluid flow was laminar since the Grasof number was equal to $6.11 \times 10^7$. The initial battery and environment temperature were set to 25 °C, the same as in the experimental setup.

$$q_b = m_b c_b \frac{dT_b}{dt} + \frac{2(T_b - T_a)}{R_{conv}} \tag{8}$$

$$\mathrm{Nu} = 0.68 + \frac{0.67\mathrm{Ra}^{1/4}}{\left[1 + (0.492/\mathrm{Pr})^{9/16}\right]^{4/9}} \tag{9}$$

The PCM was stored in an aluminum container, one surface of which was in contact with the battery and the other was cooled by natural convection. The dominant heat transfer mechanism was conduction in the PCM due to the thin PCM layers. Thermal energy was conducted from a battery source throughout the aluminum case and the PCM to the air at room temperature, as shown in Figure 4. In accordance with the instantaneous energy balance, heat produced by the battery could either be stored by itself or flowed into the PCM region, as follows:

$$q_b = m_b c_b \frac{dT_b}{dt} + q_{PCM} \tag{10}$$

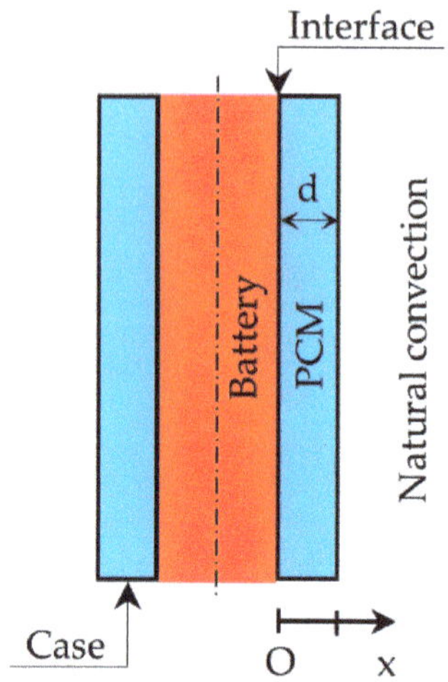


**Figure 4.** Schematic of the battery and PCM positions.

A 1D transient heat conduction equation, without heat generation, was applied to evaluate the temperature distribution in the PCM, as shown in Equation (11) [24]:

$$\frac{\partial T}{\partial t} = \alpha \frac{\partial^2 T}{\partial^2 x} \tag{11}$$

$$\alpha = \frac{k}{\rho \cdot c} \tag{12}$$

Boundary conditions were applied as follows:

$$\text{At } x = 0 \quad kA\frac{\partial T}{\partial x} = \frac{T_b - T(x)}{R_{int} + R_{cas}} = \frac{q_{PCM}}{2} \tag{13}$$

$$\text{At } x = d \quad kA\frac{\partial T}{\partial x} = \frac{T(x) - T_a}{R_{cas} + R_{conv}} \tag{14}$$

It is truly necessary to clarify two forms of heat when using a PCM. Sensible heat results in a change in material temperature without a phase change. By contrast, latent heat does not increase the temperature within the material but does cause a change of substance state [22]. Assuming that the roughness of the aluminum plates was 10 μm, thermal contact conductance was calculated to be 3640 $W/m^2 \cdot K$. Jollyn et al. [13] used a case thickness of 5 mm; accordingly, the thermal resistance for conduction was equal to $3.732 \times 10^{-5}$ K/W.

## 2.2. Three-Dimensional Mathematical Model

Numerical simulation of the heat transfer was carried out using ANSYS Fluent software. Kim et al. [25] introduced the multi-scale, multi-dimensional approach for solving interactions appearing in a vast number of length scales. This framework efficiently deals with separate solution domains at the particle ($10^{-9}$ to $10^{-8}$), electrode ($10^{-6}$ to $10^{-4}$), and cell ($10^{-2}$ to $10^{0}$) levels. At the cell scale, the current flux is governed by the following equations:

$$\nabla \cdot (\sigma_+ \nabla \varphi_+) = -j \tag{15}$$

$$\nabla \cdot (\sigma_- \nabla \varphi_-) = j \tag{16}$$

The NTGK model assumes the current flux to be a function of the potential difference between the positive and negative electrodes. Ho et al. [26] estimated two empirical fitting parameters from their experimental data. Those depend on the *DOD* of the battery and the temperature. For a 3D thermal model of Li-ion batteries, the thermal source is roughly similar to 1D modeling and the ohmic heating is also considered in the batteries, as follows:

$$q_b = j\left[E - (\varphi_+ - \varphi_-) - T\frac{dE}{dT_b}\right] + \sigma_+ \cdot \nabla^2 \varphi_+ + \sigma_- \cdot \nabla^2 \varphi_- \tag{17}$$

where $j[E - (\varphi_+ - \varphi_-)]$ is expressed as an irreversible source term and $jT(dE/dT_b)$ refers to a reversible source term. The thermal energy generated in the current collecting tab and lead wire was neglected.

Voller et al. [27] successfully solved phase change problems in the convection–diffusion–controlled mushy zone with the enthalpy–porosity technique, which relies on fixed-grid methodology. Depending upon the PCM temperature, the liquid fraction is defined as:

$$\begin{array}{ll} \beta = 0 & \text{for } T < T_{solidus} \\ \beta = 1 & \text{for } T > T_{liquidus} \\ \beta = \frac{T - T_{solidus}}{T_{liquidus} - T_{solidus}} & \text{for } T_{solidus} < T < T_{liquidus} \end{array} \tag{18}$$

Instead of tracking the melting interface, the liquid fraction referring to the liquid state in a computational domain is computed per iteration. The latent heat content is as follows:

$$\Delta H = \beta \cdot L \tag{19}$$

For melting simulations, the momentum equation is not employed. The energy equation analyzing the temperature distribution can be derived as follows:

$$\frac{\partial}{\partial t}(\rho H) + \nabla \cdot (\rho \mathbf{v} H) = \nabla \cdot (k \nabla T) + S \tag{20}$$

where $\mathbf{v}$ is the fluid velocity vector.

## 2.3. Validation

### 2.3.1. Validation of 1D Modeling

The 1D battery heat transfer model was validated against the experimental data. The error between the 1D model prediction and the experimental data was found to be within 1% at the last stage of both the 3C and 5C discharge processes. Figure 5 shows good quantitative agreement for the 3C and 5C discharge rates.

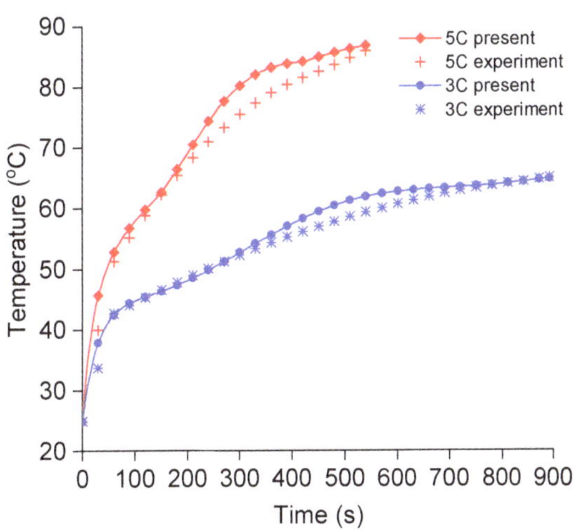


**Figure 5.** Comparison of the 1D model and the experimental maximum temperatures in the battery.

2.3.2. Validation of 3D Modeling

Ho et al. [26] conducted simulations on the same kind of Li-ion battery using ANSYS Fluent software. The NTGK model was used to predict the thermal behavior of the battery under various discharge rates: 1C, 2C, 3C, 4C, and 5C. A cell was immersed in air that had an initial temperature of 25 °C. As seen in Figure 6, there were similar trend lines between this current work and that of Ho et al. At the end of discharging, there was a 0.1% error for the 3C rate and a 0.5% error for the 5C rate. This confirmed that NTGK battery modeling was valid for various rates of discharge.

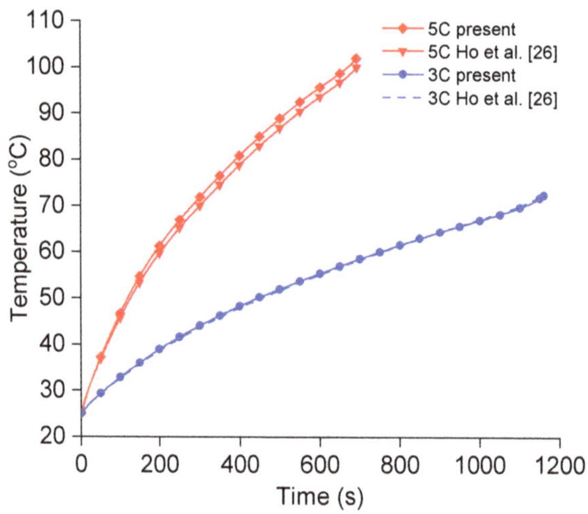


**Figure 6.** Comparison of temperatures by the NTGK model in this study and by Ho et al. [26].

A transient simulation of the thermal behavior of a Li-ion battery with a PCM was validated to the results of Javani et al. [28]. The error was found to be within an acceptable range (a maximum error of less than 1% at the end of discharging), as shown in Table 3. In order to model a cell, those researchers assumed a uniform heat generation rate within the cell domain. Due to the inconsequential thickness of the battery, this assumption was appropriate for the lumped system evaluation. We considered a transient model of a 54,374.5 W/m$^3$ heat source, especially because there was anisotropic thermal conductivity in the cell: 25 W/m·K along the cell surface and 1 W/m·K normal to the cell surface. N-octadecane, the length of which was 3 mm, was integrated around the periphery of the battery with the same thickness. It was assumed that free convection took place at all the surfaces where the battery and the PCM interfaced with the ambient air. The heat transfer coefficient was 7 W/m$^2$·K and the temperature of the environment was 294.15 K. The geometrical features of this model, which had 300,800 elements, are summarized in Table 4.

**Table 3.** Comparison of maximum temperature for the validation of the battery model at 20 min.

| Configuration | Present Study | Javani et al. [28] | Error (%) |
|---|---|---|---|
| Cell minimum temperature (K) | 303.20 | 305.05 | 0.6 |
| Cell average temperature (K) | 305.23 | 307.58 | 0.8 |
| Cell maximum temperature (K) | 306.25 | 308.43 | 0.7 |

A thorough grid independence test should be carried out to demonstrate that the element size does not affect the quantity of interest. We recorded the temperature in the cell at the end of discharging at the 5C discharge rate. As shown in Figure 7, the grid with 1,360,891 elements was sufficient to predict the temperature distribution in the computational domain.

**Table 4.** Dimensions of the battery and PCM geometry in the simulations of Javani et al. [28].

| Geometry | Length (m) | Height (m) | Thickness (m) |
|---|---|---|---|
| Cell | 0.146 | 0.194 | 0.0054 |
| Terminal | 0.035 | 0.015 | 0.0006 |
| PCM | 0.003 | 0.003 | 0.0054 |

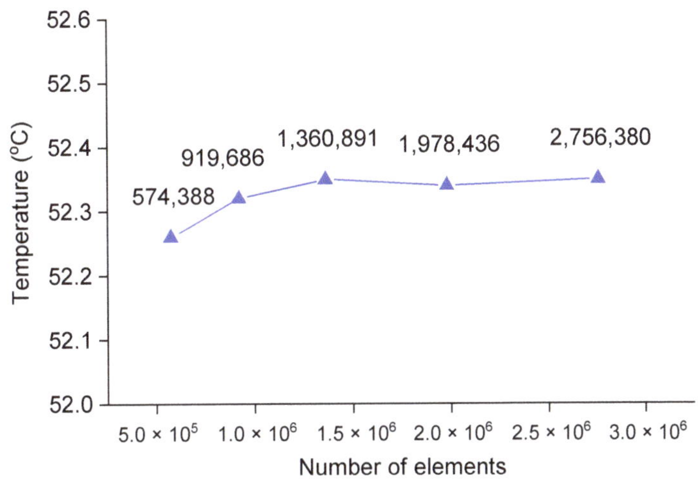


**Figure 7.** Grid independence test with different numbers of elements.

## 3. Results and Discussion

In this study, the effect of the thickness and thermal conductivity of a PCM on the cooling performance of a Li-ion battery are investigated. The PCM was of three different thicknesses, 0.5 mm, 0.55 mm, and 0.6 mm, and the thermal conductivity of the PCM was enhanced based on the practical application.

### 3.1. Effect of PCM Thickness

The major drawback with most organics, particularly the family of paraffins, is that they possess low thermal conductivity. A thin PCM easily melts fully and does not produce sufficiently effective temperature reduction. With a thick PCM there is a solid region farther from the battery and a superheated liquid region next to the battery, so instead of behaving as a thermal sink, the PCM may behave as a thermal isolator.

As shown in Figure 8, the maximum temperatures of the Li-ion batteries without a PCM were 64.86 °C and 86.83 °C at the end of the 3C and 5C discharge rates, respectively. As mentioned previously, the best temperature range is 25 °C to 40 °C and the safe temperature range is −20 °C to 60 °C for Li-ion battery operation. Moreover, the capacity fade of Li-ion cells can be 36.21% after 800 cycles when the operating temperature is around 45 °C, and the loss can even reach 70.56% after 500 cycles when the operating temperature is approximately 55 °C. At the 3C discharge rate, a PCM with a thickness of 0.6 mm or thinner can maintain the battery temperature in a safe condition. With a 0.6 mm thick PCM layer, the maximum temperature can decrease by 1.28 °C or 6.54 °C, depending on the fast or slow discharge rate.

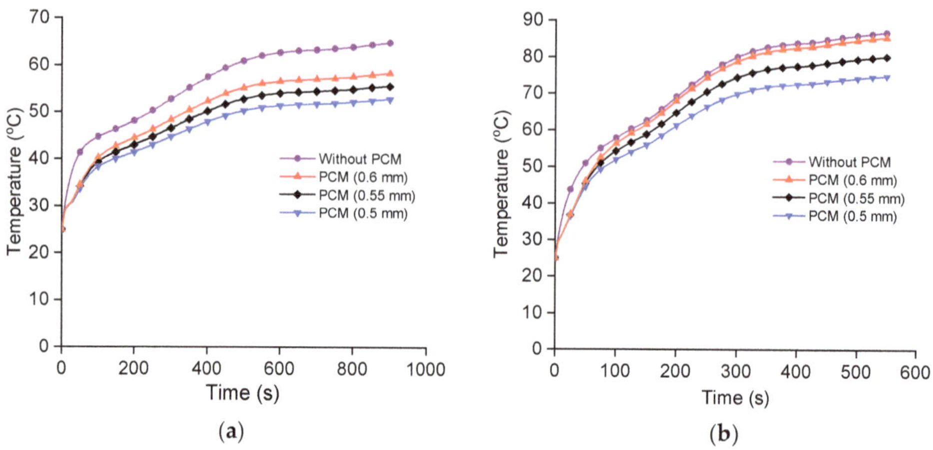


**Figure 8.** Battery temperatures for different PCM thicknesses at: (**a**) 3C discharge rate; (**b**) 5C discharge rate.

Three stages along the temperature elevation curve clearly exist in the battery with a PCM (Figure 8).

- From a starting temperature of less than 28 °C, it grows markedly. It takes about the first 3 s (5 s) to reach the limitation for the 5C (3C) discharge rate. The slope was similar to the one without PCM. The energy storage relies solely on low sensible heating (in this case, 2150 J/kg·K).
- During the melting process, the battery temperature drops into the phase change temperature range. The temperature increases slightly since thermal energy is absorbed as the heat of transformation. Under the 3C discharge rate, this process occurred in 20 s to increase the temperature to ~31 °C. The PCM started to melt from the inner to the outer layer, and then melted entirely.
- Finally, the PCM was at a liquid state with both low thermal conductivity and low specific heat. The temperature of the battery climbed considerably again but the slope was still less than the one without a PCM. The predicted temperatures had almost identical trends in their final stages.

Figure 9 is a close-up of the battery temperature change at the initial stage of the discharging process. Under the high discharge rate of 5C, it took 17.4 s to rise above 40 °C. This process could be extended to 34.8 s when the PCM thickness was 0.5 mm. Similarly, with the 3C discharge rate, it took much longer (40.3 s) to reach 40 °C. This operation could take up to 144.3 s with a 0.5 mm thick PCM.

As shown in Figure 10, at the end of the discharging process the battery temperature linearly varied with the thickness of the PCM. When the thickness decreased by 0.05 mm, the temperature also declined 5.3 °C at the 5C discharge rate. This compares with a decline of 2.7 °C at the low discharge rate of 3C.

### *3.2. Effect of PCM Thermal Conductivity*

The response time is a substantial factor in the design of a PCM system. There are several approaches to improve heat dissipation, such as enhancing the thermal conductivity, setting the ullage space far from the heat source via a potential container, and installing the PCM in the heat flow path. It is generally agreed that a material that has a high energy storage capacity will have low thermal conductivity. Therefore, much research interest has been focused on enhancing the thermal conductivity of PCMs.

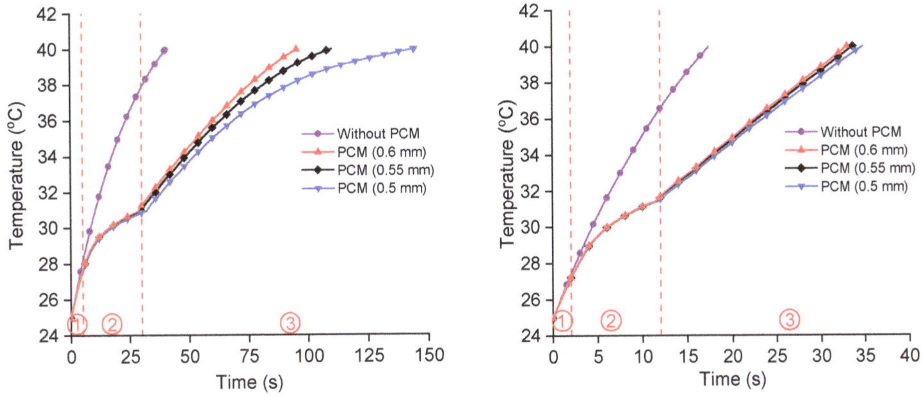


(a) (b)

**Figure 9.** Rapid increases in the temperature below 40 °C at: (**a**) the 3C discharge rate; (**b**) the 5C discharge rate (1: sensible heating of solid; 2: latent heat of fusion; 3: sensible heating of liquid).

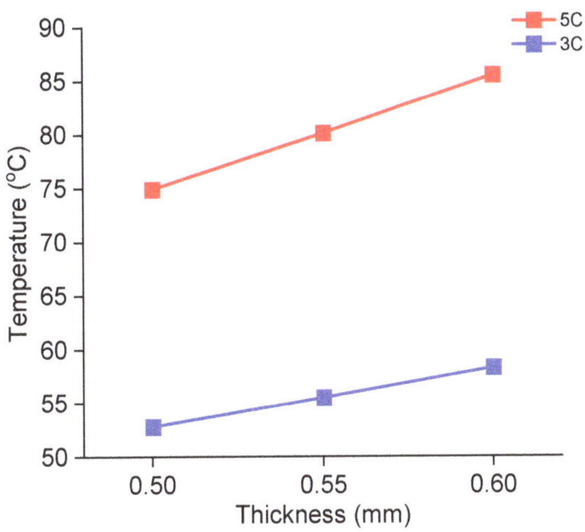


**Figure 10.** The highest battery temperature as a function of PCM thickness.

Recently, Venkateshwar et al. [29] investigated heat storage performance when embedding aluminum metal foam (MF) in n-octadecane. They used Equations (21) and (22) to determine the effective thermal conductivity of that PCM–MF composite, given as:

$$k_{s,eff} = A_1[\varepsilon \cdot k_{s,PCM} + (1-\varepsilon)k_{MF}] + \frac{1-A_1}{\frac{\varepsilon}{k_{s,PCM}} + \frac{1-\varepsilon}{k_{MF}}} \tag{21}$$

$$k_{l,eff} = A_1[\varepsilon \cdot k_{l,PCM} + (1-\varepsilon)k_{MF}] + \frac{1-A_1}{\frac{\varepsilon}{k_{l,PCM}} + \frac{1-\varepsilon}{k_{MF}}} \tag{22}$$

With the porosity ε set to 0.972, the structure with a thermal conductivity of 2.394 W/m·K conducted faster than pure n-octadecane without sacrificing the available volume.

Sari et al. [30] examined the transient thermal influence of paraffin absorbed into expanded graphite (EG). They determined that the PCM composite, with a 10% mass fraction of EG, had stable properties. The thermal conductivity of the PCM–EG linearly varied with the mass fraction of the EG. Compared with pure n-octadecane, the thermal conductivity could reach 0.828 W/m·K when integrated with the proper amount of EG.

In a macroscopic approach, the use of carbon fibers (CF) was considered by Fukai et al. [31] as an advanced technique. The effective thermal conductivity of the PCM composite was around three times higher than the pure PCM at $X_{fa} = 0.012$, as calculated in the following formula:

$$k_{eff} = \left[ \left(3.31 \times 10^{-3} + 1.69X_{fa} - 2.65 \times 10 \times X_{fa}^2\right) \cdot \left(\frac{k_f}{k_m} - 1\right)^{0.67} + 1 \right] k_m \quad (23)$$

As a result, the thermal conductivity of paraffin combined with CF was 0.678 W/m·K.

Figure 11 shows the battery temperature change during the discharging process under natural convection. The higher the thermal conductivity of the PCM composite is, the more effective the heat dissipation capacity is. Additionally, aluminum foams make the melting process more uniform. The thermal improvements of PCM composites are summarized in Table 5. At the end of the discharging process, the percentage of temperature reduction with PCM–MF was about 1.2 and 1.3 times higher than with PCM–EG and PCM–CF, respectively.

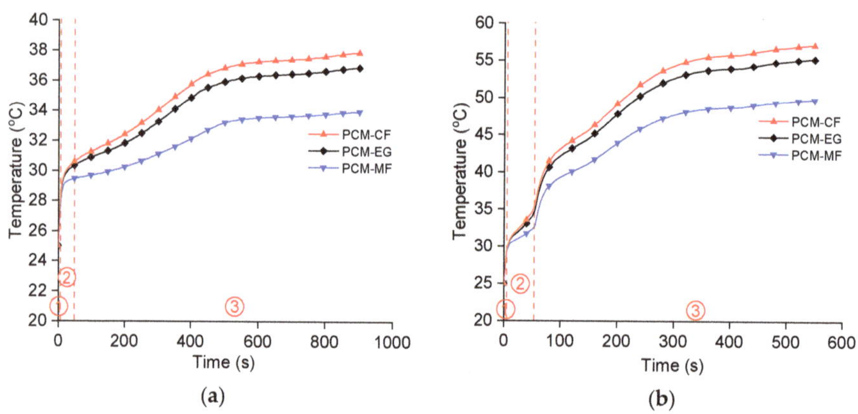


**Figure 11.** Battery temperature with various PCM composites at: (**a**) the 3C discharge rate; (**b**) the 5C discharge rate (1: sensible heating of solid; 2: latent heat of fusion; 3: sensible heating of liquid).

**Table 5.** Specifications of PCM composites and their performances.

| | Parameter | Pure PCM | PCM–MF | PCM–EG | PCM–CF |
|---|---|---|---|---|---|
| | Thermal conductivity (W/m·K) | 0.255 | 2.394 | 0.828 | 0.678 |
| 3C | Maximum temperature (°C) | 52.8 | 33.9 | 36.9 | 37.9 |
| | Percent reduction (%) | 0 | 35.8 | 30.1 | 28.2 |
| 5C | Maximum temperature (°C) | 74.9 | 49.7 | 55.2 | 57.1 |
| | Percent reduction (%) | 0 | 33.6 | 26.3 | 23.8 |

### 3.3. Comparison of 1D Calculation and 3D Simulation

We compared our 1D analysis and a 3D simulation of the 0.5 mm thick PCM–MF model. A hexahedral grid was generated for the battery and the PCM domain, while the ambient air domain was divided by polyhedral elements, as shown in Figure 12. The bottom was set to be adiabatic and the other surfaces were set to atmospheric pressure. For the pressure–velocity coupling, the ANSYS Fluent coupled algorithm was employed. A second-order interpolation scheme was used to approximate the convection terms in the momentum, energy, and potential equations, whereas a PREssure STaggering Option (PRESTO) scheme was selected for the pressure field. For the conjugate heat transfer problem, the following assumptions were applied:

- Only natural laminar convection took place in the ambient environment.
- The initial temperature of the battery and the PCM were set to the same as the ambient temperature (25 °C).
- There was no flow in the PCM liquid phase.

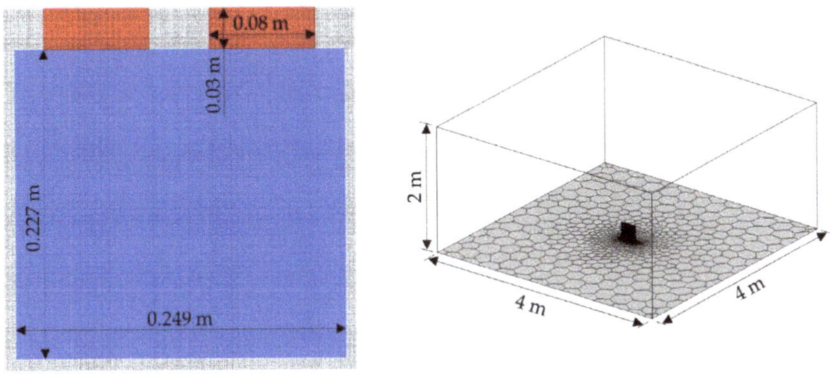


(a) (b)

**Figure 12.** (**a**) Battery and PCM immersed in air. (**b**) The computational domain.

The data from the late stage of the analysis for the 1D calculation was excluded because it was too noisy; nevertheless, we found that the 1D and 3D solutions matched well when extending the results while maintaining the curve slope (Figure 13). This agreed with the results of Greco et al. [14], who showed that cooling paths were totally dissimilar in the initial periods of the analytical and computational approaches, but they definitely matched at the end stages.

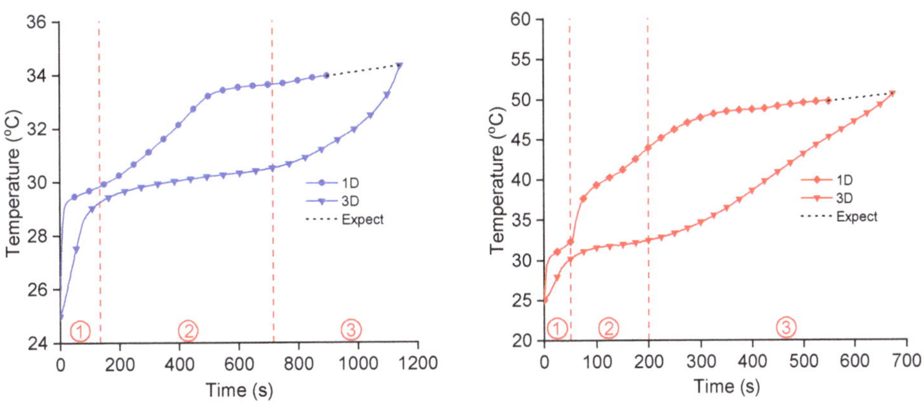


(a) (b)

**Figure 13.** Comparison of 1D and 3D simulation results for battery temperatures under: (**a**) the 3C discharge rate; (**b**) the 5C discharge rate (1: sensible heating of solid; 2: latent heat of fusion; 3: sensible heating of liquid).

Figure 14 shows the temperature uniformity in the 3D Li-ion battery model. At the low discharge rate of 3C, the tmperature disparity in the cell was maintained under 5 °C. At the higher discharge rate of 5C, it markedly escalated due to sudden heat generation.

However, the PCM exhibited outstanding improvement, with a 6.6 °C peak difference of temperature.

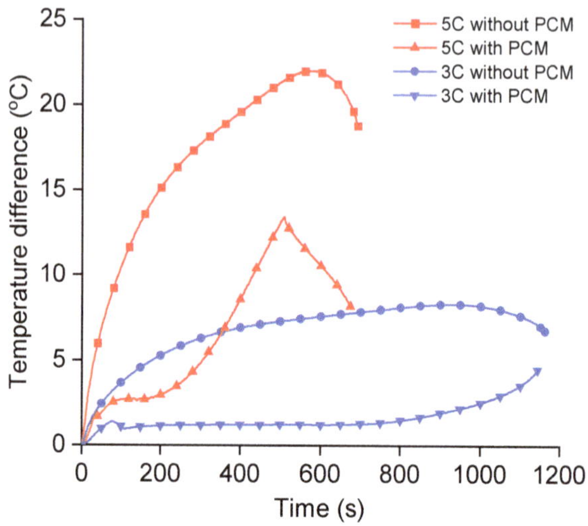


**Figure 14.** Investigation of temperature uniformity in the 3D Li-ion battery model.

At the center plane of the cell, two monitoring lines were selected to explore the temperature change at z = 0.1245 m and y = 0.1135 m, as depicted in Figure 15a. Figure 15b shows the temperature along the vertical line at the end of discharging. There was a large difference of 15.5 °C between the top and bottom positions under the 5C rate without a PCM. The use of PCM–MF caused a sudden change in temperature in the lower area of the cell at the 3C discharge rate, due to a liquid mushy PCM zone. It can be seen in Figure 15c that the temperature varied slightly in the horizontal direction and tended to be more stable when the PCM was applied. All of the differences were below 1 °C. Because the heat source was concentrated in the region close to the current collecting tabs, the temperature variation was clearly greater in the vertical direction than in the horizontal direction.

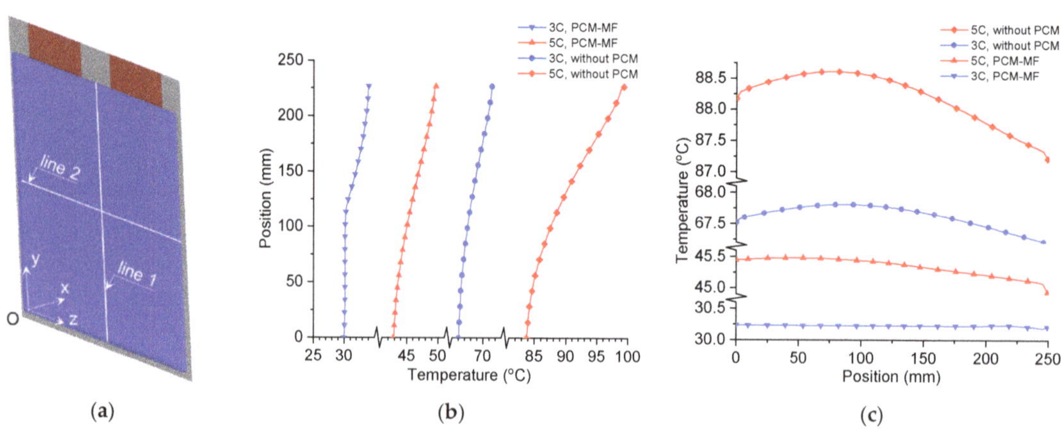


**Figure 15.** (**a**) Temperature monitoring lines. (**b**) The temperature distribution at line 1. (**c**) The temperature distribution at line 2.

Basically, the temperature distribution on the battery surface can be divided into two parts: an upper region with higher temperature and a lower region with lower temperature.

Without a PCM, the battery temperature increased more continuously and homogeneously from the bottom to the top under the 3C discharge rate. This difference was 6.8 °C. However, there was greater difference in temperature (18.8 °C) under the 5C discharge rate. The highest temperature was located near the positive and negative tabs, as shown in Figure 16.

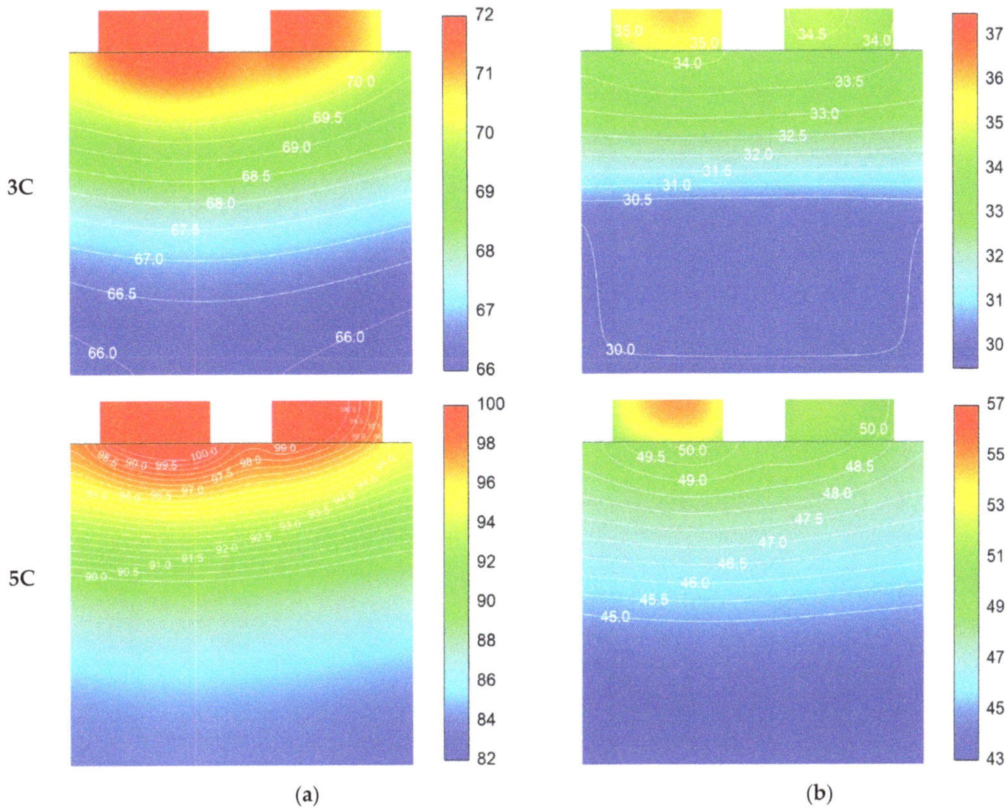


**Figure 16.** Contour of temperature on battery surfaces (°C) at the end of discharging: (**a**) without PCM; (**b**) with PCM.

Similar trends were observed in the case with the PCM. However, there was a smaller temperature difference at the 3C as well as the 5C discharge rates because thermal energy from the battery was absorbed into latent heat. The maximum temperature was located at the negative tab (left tab), which was not cooled by the PCM. For this reason, an active cooling method, such as forced convection, may be necessary. The differences in temperature were 4.5 °C and 8.2 °C on the entire battery surface under 3C and 5C discharge rates, respectively.

Depending on the amount of heat generated, the PCM can be partly or totally melted. The PCM remained in the mushy zone in the lower region under the 3C discharge rate, while it completely transformed to a liquid state under the 5C discharge rate.

## 4. Conclusions

A 1D thermal management model of a large-format Li-ion battery (52.3 Ah) with different PCM properties was evaluated. This pouch-type battery was sandwiched between two n-octadecane layers, and the heat generation agreed well with experimental data under both the 3C and 5C discharge rates. The diffusion equation that governed the thermal transfer capacity in the PCM was also solved simultaneously. By considering different

PCM thicknesses and thermal conductivities, a potential model was proposed, which was an assemblage with aluminum foam embedded in n-octadecane. The average conductivity of the resulting enhanced PCM–MF was 2.394 W/m·K. The maximum temperature of the Li-ion battery reached 86.8 °C after 550 s under the 5C discharge rate and 64.9 °C after 900 s under the 3C discharge rate without a cooling system. It was shown that the 0.5 mm thick PCM–MF composite reduced the maximum battery temperature to 49.7 °C and 33.9 °C at the 5C and 3C discharge rates, respectively.

Then, 3D transient thermodynamic simulations were carried out with an NTGK model using ANSYS Fluent software. Based on comparison with these 3D simulations, we found that the 1D analysis was capable of accurately estimating the heat dissipation by the PCM–MF at the end stage of discharging. The 1D approach clearly produced reliable results in predicting the thermal behavior of the PCM–MF cooling and was superior in practical application terms with its low cost and less time consumption.

Further study on the thermal behavior of active cooling systems with PCMs will be necessary.

**Author Contributions:** Conceptualization, V.-T.H. and K.C.; methodology, K.C.; software, V.-T.H.; resources, S.-W.L.; writing—original draft preparation, V.-T.H.; writing—review and editing, V.-T.H., K.C. and S.-W.L.; supervision, K.C. and S.-W.L.; project administration, K.C. and S.-W.L.; funding acquisition, K.C. and S.-W.L. All authors have read and agreed to the published version of the manuscript.

**Funding:** This work was supported by Upbringing Business with Innovative Urban Public Institution by the Korean Ministry of Trade, Industry, and Energy (MOTIE) (project name: Establishment of battery/ESS-based energy industry innovation ecosystem) and by the Technology Innovation Program (20006974, "Development of the multi-layered sheets and their forming/joining technologies for the fabrication of 28% lightweight battery pack/module components of electric vehicles") funded by MOTIE.

**Institutional Review Board Statement:** Not applicable.

**Informed Consent Statement:** Not applicable.

**Data Availability Statement:** Not applicable.

**Conflicts of Interest:** The authors declare no conflict of interest.

## Nomenclature

| | | | |
|---|---|---|---|
| $A$ | Surface area ($m^2$) | $t$ | Time (s) |
| $A_1$ | Correlation constant (0.35) | $T$ | PCM temperature (K) |
| $c$ | PCM specific heat (J/kg·K) | $T_a$ | Environment temperature (K) |
| $c_b$ | Battery specific capacity (J/kg·K) | $T_b$ | Battery temperature (K) |
| $d$ | PCM thickness (m) | $T_{liquidus}$ | Liquidus temperature of PCM (K) |
| $DOD$ | Depth of discharge (%) | $T_{solidus}$ | Solidus temperature of PCM (K) |
| $E$ | Open-circuit voltage (V) | $U$ | Terminal voltage (V) |
| $H$ | Enthalpy (J/kg) | $V_n$ | Nominal capacity (V) |
| $I$ | Current (A) | $v$ | Velocity (m/s) |
| $j$ | Volumetric transfer current density (A) | $x$ | Distance (m) |
| $k$ | PCM thermal conductivity (W/m·K) | $X_{fa}$ | Volume ratio of fibers to brush |
| $k_b$ | Battery thermal conductivity (W/m·K) | *Greek letters* | |
| $k_{eff}$ | Effective thermal conductivity of PCM composite (W/m·K) | $\alpha$ | Thermal diffusivity ($m^2/s$) |
| $k_f$ | Thermal conductivity of carbon (W/m·K) | $\beta$ | Liquid fraction |
| $k_{l,eff}$ | Composite thermal conductivity in liquid state (W/m·K) | $\varepsilon$ | Porosity |

| | | | |
|---|---|---|---|
| $k_{l,PCM}$ | PCM thermal conductivity in liquid state (W/m·K) | $\Delta H$ | Latent heat (J/kg) |
| $k_m$ | Thermal conductivity of paraffin (W/m·K) | $\rho$ | PCM density (kg/m$^3$) |
| $k_{s,eff}$ | Composite thermal conductivity in solid state (W/m·K) | $\rho_b$ | Battery density (kg/m$^3$) |
| $k_{s,PCM}$ | PCM thermal conductivity in solid state (W/m·K) | $\sigma_+$ | Effective electric conductivity for the positive electrode (1/Ω) |
| $k_{MF}$ | Metal foam thermal conductivity (W/m·K) | $\sigma_-$ | Effective electric conductivity for the negative electrode (1/Ω) |
| $L$ | Latent heat of material (J/kg) | $\varphi_+$ | Phase potential of the positive electrode (V) |
| $m_b$ | Battery mass (kg) | $\varphi_-$ | Phase potential of the negative electrode (V) |
| $Nu$ | Nusselt number | *Superscript* | |
| $Pr$ | Prandtl number | $j$ | Quantity of interest at the present time level |
| $q_b$ | Overpotential heat rate (J/s) | $j+1$ | Quantity of interest at the new time level |
| $q_{irr}$ | Irreversible heat rate (J/s) | *Abbreviations* | |
| $q_{PCM}$ | Heat rate through PCM (J/s) | 1D | One-dimensional |
| $q_{rev}$ | Reversible heat rate (J/s) | 3D | Three-dimensional |
| $Q$ | Discharged battery capacity (Ah) | BTMS | Battery thermal management system |
| $Q_0$ | Fully charged battery capacity (Ah) | C-rate | A measure of the rate at which a battery is discharged relative to its maximum capacity |
| $Q_n$ | Nominal capacity (Ah) | CFD | Computational fluid dynamics |
| $R$ | Internal resistance (Ω) | Li-ion | Lithium-ion |
| $R_{cas}$ | Thermal resistance of aluminum case (K/W) | NCM523 | $LiNi_{0.5}Co_{0.2}Mn_{0.3}O_2$ |
| $R_{conv}$ | Thermal resistance for convection (K/W) | NTGK | Newman, Tiedemann, Gu and Kim |
| $R_{int}$ | Thermal contact resistance (K/W) | PCM | Phase change material |
| $Ra$ | Rayleigh number | PCM–CF | Phase change material and carbon fiber |
| $S$ | Source term | PCM–EG | Phase change material and expanded graphite |
| $SOC$ | State of charge (%) | PCM–MF | Phase change material and metal foam |

## References


1. Farag, M.; Sweity, H.; Fleckenstein, M.; Habibi, S. Combined electrochemical, heat generation, and thermal model for large prismatic lithium-ion batteries in real-time applications. *J. Power Sources* **2017**, *360*, 618–633. [CrossRef]
2. Pesaran, A.A. Battery thermal models for hybrid vehicle simulations. *J. Power Sources* **2002**, *110*, 377–382. [CrossRef]
3. Zhang, H.; Li, C.; Zhang, R.; Lin, Y.; Fang, H. Thermal analysis of a 6s4p Lithium-ion battery pack cooled by cold plates based on a multi-domain modeling framework. *Appl. Therm. Eng.* **2020**, *173*, 115216. [CrossRef]
4. Hammami, A.; Raymond, N.; Armand, M. Runaway risk of forming toxic compounds. *Nature* **2003**, *424*, 635–636. [CrossRef] [PubMed]
5. Chen, K.; Song, M.; Wei, W.; Wang, S. Design of the structure of battery pack in parallel air-cooled battery thermal management system for cooling efficiency improvement. *Int. J. Heat Mass Transf.* **2019**, *132*, 309–321. [CrossRef]
6. Shang, Z.; Qi, H.; Liu, X.; Ouyang, C.; Wang, Y. Structural optimization of lithium-ion battery for improving thermal performance based on a liquid cooling system. *Int. J. Heat Mass Transf.* **2019**, *130*, 33–41. [CrossRef]
7. Wang, Z.; Zhang, Z.; Jia, L.; Yang, L. Paraffin and paraffin/aluminum foam composite phase change material heat storage experimental study based on thermal management of Li-ion battery. *Appl. Therm. Eng.* **2015**, *78*, 428–436. [CrossRef]
8. Jilte, R.; Afzal, A.; Panchal, S. A novel battery thermal management system using nano-enhanced phase change materials. *Energy* **2021**, *219*, 119564. [CrossRef]
9. Putra, N.; Ariantara, B.; Pamungkas, R.A. Experimental investigation on performance of lithium-ion battery thermal management system using flat plate loop heat pipe for electric vehicle application. *Appl. Therm. Eng.* **2016**, *99*, 784–789. [CrossRef]

10. Yang, W.; Zhou, F.; Zhou, H.; Wang, Q.; Kong, J. Thermal performance of cylindrical lithium-ion battery thermal management system integrated with mini-channel liquid cooling and air cooling. *Appl. Therm. Eng.* **2020**, *175*, 115331. [CrossRef]
11. Kiani, M.; Ansari, M.; Arshadi, A.A.; Houshfar, E.; Ashjaee, M. Hybrid thermal management of lithium-ion batteries using nanofluid, metal foam, and phase change material: An integrated numerical–experimental approach. *J. Therm. Anal. Calorim.* **2020**, *141*, 1703–1715. [CrossRef]
12. Jilte, R.; Afzal, A.; Islam, M.T.; Manokar, A.M. Hybrid cooling of cylindrical battery with liquid channels in phase change material. *Int. J. Energy Res.* **2021**, *45*, 11065–11083. [CrossRef]
13. T'Jollyn, I.; Nonneman, J.; Beyne, W.; Paepe, M.D. 1D simulations of thermally buffered prismatic betteries through the application of PCMs. In Proceedings of the 16th International Heat Transfer Conference, Beijing, China, 10–15 August 2018.
14. Greco, A.; Jiang, X.; Cao, D. An investigation of lithium-ion battery thermal management using paraffin/porous-graphite-matrix composite. *J. Power Sources* **2015**, *278*, 50–68. [CrossRef]
15. Hallaj, S.A.; Kizilel, R.; Lateef, A.; Sabbah, R.; Farid, M.; Selman, J.R. Passive thermal management using phase change material (PCM) for EV and HEV Li- ion batteries. In Proceedings of the 2005 IEEE Vehicle Power and Propulsion Conference, Chicago, IL, USA, 7–9 September 2005.
16. Hallaj, S.A.; Maleki, H.; Hong, J.S.; Selman, J.R. Thermal modeling and design considerations of lithium-ion batteries. *J. Power Sources* **1999**, *83*, 1–8. [CrossRef]
17. Sato, N. Thermal behavior analysis of lithium-ion batteries forelectric and hybrid vehicle. *J. Power Sources* **2001**, *99*, 70–77. [CrossRef]
18. Yi, J.; Kim, U.S.; Shin, C.B.; Han, T.; Park, S. Modeling the temperature dependence of the discharge behavior of a lithium-ion battery in low environmental temperature. *J. Power Sources* **2013**, *244*, 143–148. [CrossRef]
19. Akeiber, H.; Hosseini, S.; Wahid, M.; Hussen, H.; Mohammad, A. Phase Change Materials-Assisted Heat Flux Reduction: Experiment and Numerical Analysis. *Energies* **2016**, *9*, 30. [CrossRef]
20. Fortunato, B.; Camporeale, S.M.; Torresi, M.; Albano, M. Simple mathematical model of a thermal storage with PCM. In Proceedings of the AASRI Conference on Power and Energy Systems, Bari, Italy, 4–5 September 2012.
21. Ambarita, H.; Napitupulu, F.H.; Sitorus, T.B. Numerical study on melting process of phase change material as thermal energy storage. In Proceedings of the 3rd Nommensen International Conference on Technology and Engineering, Medan, Indonesia, 25–26 July 2019.
22. Fleischer, A.S. *Thermal Energy Storage Using Phase Change Materials Fundamentals and Applications*; Springer: New York, NY, USA, 2015; pp. 37–38.
23. Bernardi, D.; Pawlikowski, E.; Newman, J. A general energy balance for battery systems. *J. Electrochem. Soc.* **1985**, *132*, 5–12. [CrossRef]
24. Cengel, Y.A. *Heat Transfer: A Practical Approach*, 2nd ed.; The McGraw-Hill Companies, Inc.: New York, NY, USA, 2003; p. 75.
25. Kim, G.H.; Smith, K.; Lee, K.J.; Santhanagopalan, S.; Pesaran, A. Multi-domain modeling of lithium-ion batteries encompassing multi-physics in varied length scales. *J. Electrochem. Soc.* **2011**, *158*, A955. [CrossRef]
26. Ho, V.-T.; Chang, K.; Lee, S.W.; Kim, S.H. Transient thermal analysis of a li-ion battery module for electric cars based on various cooling fan arrangements. *Energies* **2020**, *13*, 2387. [CrossRef]
27. Voller, V.R.; Prakash, C. A fixed grid numerical modelling methodology for convection-diffusion mushy region phase-change problems. *Int. J. Heat Mass Transf.* **1987**, *30*, 1709–1719. [CrossRef]
28. Javani, N.; Dincer, I.; Naterer, G.F.; Yilbas, B.S. Heat transfer and thermal management with PCMs in a Li-ion battery cell for electric vehicles. *Int. J. Heat Mass Transf.* **2014**, *72*, 690–703. [CrossRef]
29. Venkateshwar, K.; Tasnim, S.H.; Simha, H.; Mahmud, S. Influence of metal foam morphology on phase change process under temporal thermal load. *Appl. Therm. Eng.* **2020**, *180*, 115874. [CrossRef]
30. Sari, A.; Karaipekli, A. Thermal conductivity and latent heat thermal energy storage characteristics of paraffin/expanded graphite composite as phase change material. *Appl. Therm. Eng.* **2007**, *27*, 1271–1277. [CrossRef]
31. Fukai, J.; Hamada, Y.; Morozumi, Y.; Miyatake, O. Effect of carbon-fiber brushes on conductive heat transfer in phase change materials. *Int. J. Heat Mass Transf.* **2002**, *45*, 4781–4792. [CrossRef]

energies
MDPI

Article

# Comparison of Interleaving Methods of Parallel Connected Three-Level Bi-Directional Converters

Hae-In Kim [1], Su-Hwan Kim [1], Seung-Woo Baek [1,*], Hag-Wone Kim [1,*], Kwan-Yuhl Cho [1] and Gil-Dong Kim [2]

1 Department of Control and Instrumentation Engineering, Korea University of Transportation, Chungju 27469, Korea; khi1112@ut.ac.kr (H.-I.K.); clying12@ut.ac.kr (S.-H.K.); kycho@ut.ac.kr (K.-Y.C.)
2 Propulsion System Research Department, Korea Railroad Research Institute, Uiwang 16015, Korea; gdkim@krri.re.kr
* Correspondence: bsw@ut.ac.kr (S.-W.B.); khw@ut.ac.kr (H.-W.K.); Tel.: +82-10-2296-9173 (H.-W.K.)

**Abstract:** The voltage and current ripples in the three-level bi-directional converter (TLBC) can be reduced by an interleaving technique that controls a phase difference between the modules of power converter. On the other hand, the inductor current ripple in TLBC is increased due to the circulating current between the modules. In this paper, the effects of two interleaving methods on a two-phase TLBC, Z-type and N-type, are investigated and compared. In particular, capacitor current ripple, and voltage ripple are compared by two interleaving methods verified through Powersim (PSIM) simulation.

**Keywords:** three-level bi-directional converter (TLBC); interleaving control; capacitor current ripple; parallel control

**Citation:** Kim, H.-I.; Kim, S.-H.; Baek, S.-W.; Kim, H.-W.; Cho, K.-Y.; Kim, G.-D. Comparison of Interleaving Methods of Parallel Connected Three-Level Bi-Directional Converters. *Energies* **2022**, *15*, 6. https://doi.org/10.3390/en15010006

Academic Editors: Antonio T. Alexandridis and Teuvo Suntio

Received: 27 October 2021
Accepted: 10 December 2021
Published: 21 December 2021



## 1. Introduction

In the transportation facilities including electric locomotive and electric vehicle, the kinetic energy in braking mode is converted into the electrical energy. This regenerative energy generated by the electric machine can be stored in a battery by a bidirectional power converter [1–3]. The high voltage semiconductor switches are required in such high power, high voltage applications, and this results in large losses in semiconductor switches. Therefore, research on three-level bidirectional converter (TLBC) is being actively carried out to lower the voltage stress of power semiconductors [4–7]. Figure 1 shows a TLBC circuit.

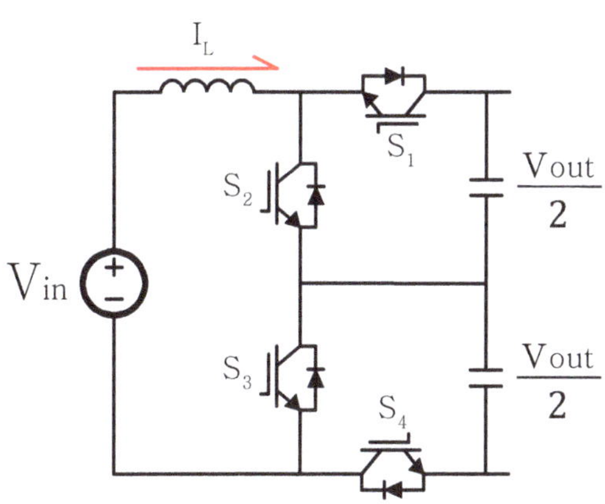


**Figure 1.** Single-module TLBC circuit diagram.

Because the TLBC in Figure 1 has a structure in which the two-level bidirectional converters are vertically symmetric, the voltage stress of the switching devices is reduced by half. This makes it feasible to reduce the loss of the elements [8]. Furthermore, the ripple of the inductor and output currents, and output voltage can be reduced by half. A decrease in the ripple of current and voltage has a positive effect on the lifespan and efficiency of passive elements [9–12].

In addition, if the power converter is configured in parallel, it can provide benefits such as the reduction of losses and heat dissipation under heavy loads, and increase in power capacity [13,14]. TLBCs that are configured in parallel generally perform interleaving operations with a phase difference on one period [5,15]. In this case, each pair of switches operate alternately so that the input current, output current and voltage ripples can be reduced. However, by the interleaving control, a circulating current is generated, which flows from the inductor of one module to that of the other module. This increases the ripple of the current flowing through the inductor [16,17]. An increase in the ripple of the inductor current results in an increase in the loss of the inductor as well as a decrease in the efficiency of the converter module. There are two interleaving methods, Z-type and N-type, in TLBCs with two-phase parallel configuration according to the order of operating switches [18]. N-type interleaving operates all the switches of one module first and then those of another. In contrast, Z-type interleaving operates the high-side switches of all the modules first and then the low-side switches. In the case of Z-type interleaving, because the phase difference between the high-side switches is smaller than that for N-type interleaving, the amount of circulating current generated between the modules of Z-type interleaving is smaller then that of N-type interleaving.

The effect of the interleaving method on the inductor current has been analyzed extensively in the reference document [18]. However, the effects on the filter capacitor are not discussed in that paper. Generally, filter capacitors can be the largest failure's sources of converter modules [19]. Therefore, in this study, the effects of interleaving methods on the filter capacitor of the current of the filter capacitor was analyzed according to the two interleaving operation methods for the two phase parallel TLBC structure. Furthermore, the control methods of two types interleaving were also analyzed. The analysis results were verified through Powersim (PSIM) simulations. This paper is organized as follows: Section 2 describes the basic operation of two-phase parallel TLBCs. Section 3 analyzes the current ripple of the filter capacitor and the control method that depend on the interleaving method. Section 4 verifies the analytical results given in Section 3 by simulations. Finally, conclusions are presented in Section 5.

## 2. Parallel Operation of Two-Phase Parallel TLBC

### 2.1. Circuit of Two-Phase Parallel TLBC

To configure TLBCs in parallel, two aspects need to be considered in connecting single TLBC. Firstly, the neutral points of the two capacitors placed between the high-side and low-side in the TLBCs in parallel must be connected. By connecting the neutral point, the current ripple of the capacitor is divided between capacitors connected in parallel, which reduces the loss of the capacitor and increases the lifespan [20]. Second, a low-side inductor must be added to prevent the short-circuiting of low-side capacitors. While interleaving the TLBC, there is a switching instance in which two low-side capacitors are short circuited. Figure 2a illustrates the low-side capacitors are short-circuited during interleaving control in TBLC.

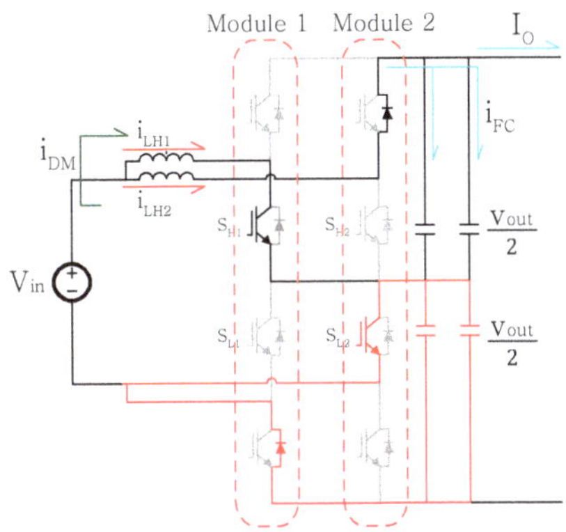


(**a**) Without low-side inductor

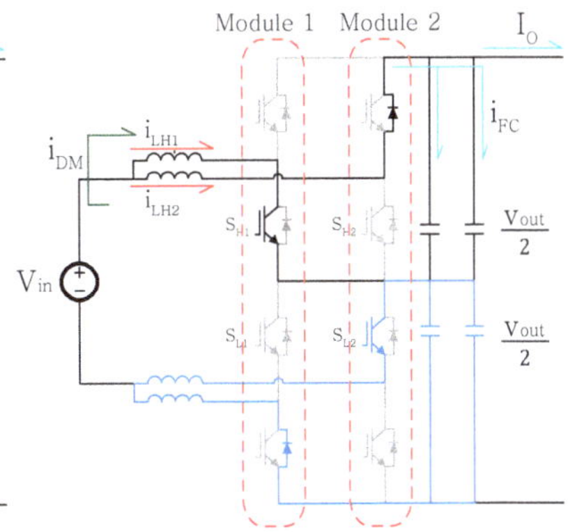


(**b**) With low-side inductor

**Figure 2.** Circuit diagram of two-phase parallel TLBC.

In such a case, the entire output voltage is applied to the high-side capacitors so that these capacitors may be destroyed by over voltage failure. Therefore, a low-side inductor should be added to prevent short circuit failure of the low-side capacitor. Figure 2b illustrates a circuit diagram of a two-phase TLBC configured in parallel with an inductor at the bottom. As described above, the two-phase parallel TLBC circuit used in this study has low-side inductors, and the neutral points of capacitors in each module are connected.

### *2.2. Interleaving Methods of 2 Parallel Three-Level Bi-Directional Converters*

In the conventional single-phase TLBC, the two switches at the high-side and low-side operate with a phase difference of 180°. However, because there are a total of four activated switches in the two-phase parallel TLBC, it must operate with a phase difference of 90° (obtained by dividing a cycle into four parts). At this time, it is categorized as N-type or Z-type interleaving according to the order of phase difference [18]. Figure 3 illustrates the switching sequence for two types of interleaving methods, N-type and Z-type. Moreover, the two filter capacitors connected in parallel are expressed equivalently as one capacitor. In N-type interleaving, the switches of first module operate in the top and bottom order, then those of second module operate in the top and bottom order—while, in Z-type interleaving, the high-side switches of two modules are turned on sequentially and then the low-side switches are turned on.

Meanwhile, a circulating current ($I_{DM}$) is generated when the parallel modules are subjected to interleaving control. This circulating current is not completely transferred to the output power, but it flows into another module. This circulating current increases the power losses as well as a current ripple of the inductor. The next section describes the circulating current for two types of interleaving methods.

### *2.3. Circulating Current Variation with an Interleaving Method*

The circulating current ($i_{DM}$) flowing between the modules owing to interleaving control and the common current ($i_{CM}$) flowing to the output terminal (without flowing to another module) are defined as in Equations (1) and (2):

$$i_{CM} = \frac{i_{LH1} + i_{LH2}}{2} \tag{1}$$

$$i_{DM} = \frac{i_{LH1} - i_{LH2}}{2} \tag{2}$$

where $i_{LH1}$and $i_{LH2}$ refer to inductor currents flowing through the high-side inductors of modules 1 and 2, respectively. The inductor current of each module can be expressed as Equations (3) and (4) using the circulation current and common current

$$i_{LH1} = i_{CM} + i_{DM} \tag{3}$$

$$i_{LH2} = i_{CM} - i_{DM} \tag{4}$$

Figure 4a shows the circuit diagram of two-phase parallel TLBC. Figure 4b shows an equivalent circuit obtained by replacing the switch and output voltage terminal of a two-phase parallel TLBC with a square wave voltage source. Point G, which is the middle point of the input voltage, is a virtual node. Point O, which is the neutral point of the capacitor, is defined as a node for interpreting the circuit by dividing it into high-side and low-side ends.

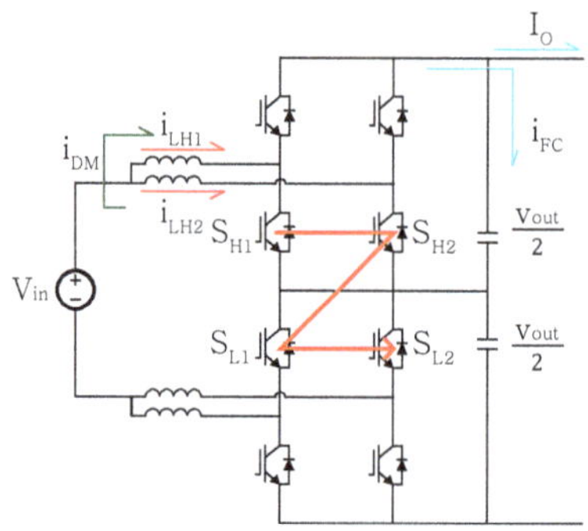


(**a**) Switching operation of Z-type interleaving

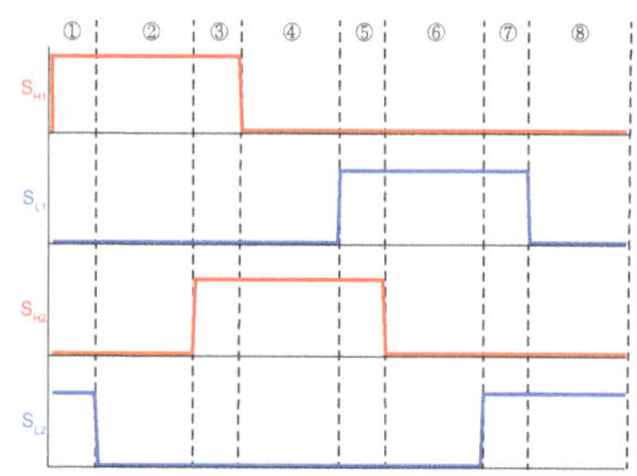

(**b**) Waveform of Z-type interleaving

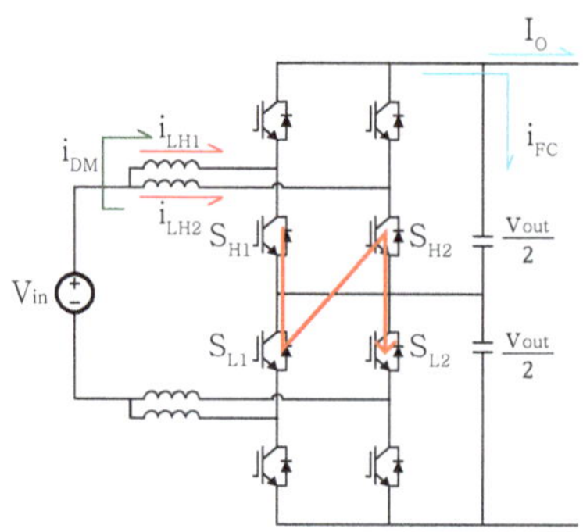


(**c**) Switching operation of N-type interleaving

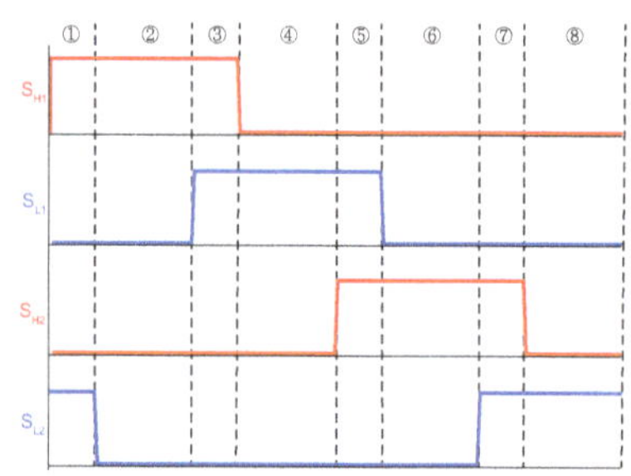

(**d**) Waveform of N-type interleaving

**Figure 3.** Switching operation according to interleaving methods.

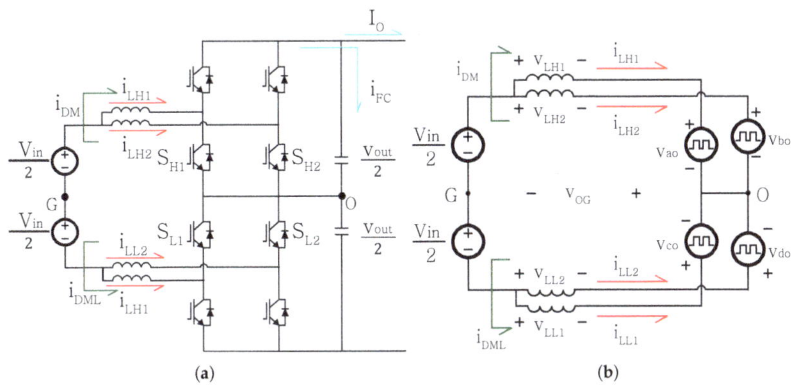


**Figure 4.** (**a**) Circuit diagram of a two-phase parallel TLBC; (**b**) equivalent circuit diagram of a TLBC with interleaving control.

The voltages applied to the inductor in Figure 4 ($v_{LH1}, v_{LH2}, v_{LL1}, v_{LL2}$) can be expressed by Equations (5)–(8):

$$v_{LH1} = \frac{v_{in}}{2} - v_{OG} - v_{ao} \tag{5}$$

$$v_{LH2} = \frac{v_{in}}{2} - v_{OG} - v_{bo} \tag{6}$$

$$v_{LL1} = -\frac{v_{in}}{2} - v_{OG} - v_{co} \tag{7}$$

$$v_{LL2} = -\frac{v_{in}}{2} - v_{OG} - v_{do} \tag{8}$$

where the square wave voltage sources $v_{ao}, v_{bo}, v_{co}$ and $v_{do}$ mean voltages corresponding to half of the output voltage according to the operation of the switch. $v_{ao}, v_{bo}, v_{co}$,and $v_{do}$ are expressed by Equations (9)–(12):

$$v_{ao} = (1 - S_{H1}) * \frac{v_{out}}{2} \tag{9}$$

$$v_{bo} = (1 - S_{H2}) * \frac{v_{out}}{2} \tag{10}$$

$$v_{co} = -(1 - S_{L1}) * \frac{v_{out}}{2} \tag{11}$$

$$v_{do} = -(1 - S_{L2}) * \frac{v_{out}}{2} \tag{12}$$

where $S_{H1}, S_{H2}, S_{L1}, S_{L2}$ represent switching functions of the high-side switch of module 1, the high-side switch of module 2, the low-side switch of module 1, and the low-side switch of module 2, respectively. Equation (13) defines a switch function. It is defined that the value of switching function is one when the switch is on and zero when it is off.

$$S_{H1} = \begin{cases} 1 & \text{Switch } S_{H1} \text{ is On} \\ 0 & \text{Switch } S_{H1} \text{ is OFF} \end{cases}, \quad S_{H2} = \begin{cases} 1 & \text{Switch } S_{H2} \text{ is On} \\ 0 & \text{Switch } S_{H2} \text{ is OFF} \end{cases}$$

$$S_{L1} = \begin{cases} 1 & \text{Switch } S_{L1} \text{ is On} \\ 0 & \text{Switch } S_{L1} \text{ is OFF} \end{cases}, \quad S_{L2} = \begin{cases} 1 & \text{Switch } S_{L2} \text{ is On} \\ 0 & \text{Switch } S_{L2} \text{ is OFF} \end{cases} \quad (13)$$

The neutral voltage of the output capacitor $v_{OG}$ can be derived by applying KCL (Kirchhoff's current law) to point O as follows:

$$\begin{aligned} & i_{LH1} + i_{LH2} + i_{LL1} + i_{LL2} = 0 \\ & \frac{v_{LH1}}{Z} + \frac{v_{LH2}}{Z} + \frac{v_{LL1}}{Z} + \frac{v_{LL2}}{Z} = 0 \\ & v_{LH1} + v_{LH2} + v_{LL1} + v_{LL2} = 0 \end{aligned} \quad (14)$$

where $Z$ denotes the impedance of each inductor (it is assumed that the inductors have equal impedance). The following equation is obtained when it is substituted into Equations (5)–(8):

$$-4v_{OG} - v_{ao} - v_{bo} - v_{co} - v_{do} = 0 \quad (15)$$

Finally, the neutral voltage $v_{OG}$ can be expressed as follows from Equations (9), (12) and (15):

$$v_{OG} = (S_{H1} + S_{H2} - S_{L1} - S_{L2}) * \frac{v_{out}}{8} \quad (16)$$

The high-side inductor voltages ($v_{LH1}, v_{LH2}$) according to the switch operation in the circuit can be rearranged by Equations (17) and (18):

$$\begin{aligned} v_{LH1} &= \frac{v_{in}}{2} - v_{OG} - v_{ao} \\ &= \frac{v_{in}}{2} + (3S_{H1} - S_{H2} + S_{L1} + S_{L2} - 4) * \frac{v_{out}}{8} \end{aligned} \quad (17)$$

$$\begin{aligned} v_{LH2} &= \frac{v_{in}}{2} - v_{OG} - v_{bo} \\ &= \frac{v_{in}}{2} + (-S_{H1} + 3S_{H2} + S_{L1} + S_{L2} - 4) * \frac{v_{out}}{8} \end{aligned} \quad (18)$$

It is difficult to distinguish the common current from the circulating current in Figure 4b; therefore, the interpretation is carried out by dividing the common mode (CM) circuit that generates a common current and the differential mode (DM) circuit that generates the circulating current. Figure 5 illustrates the two-phase parallel three-level converters divided into a CM circuit and a DM circuit. The voltages across the high side inductors in the CM circuit have the same polarity, whereas those in the DM circuit have the opposite polarity.

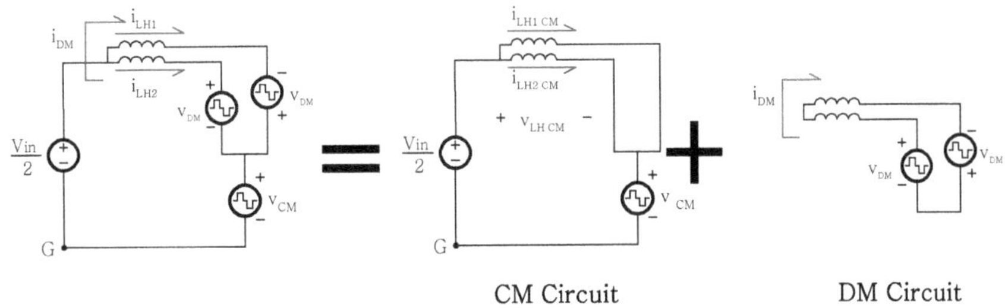


**Figure 5.** Equivalent circuit for the analysis of circulating current.

The $v_{CM}$ expressed in Figure 5 is defined by the sum of the mean of $v_{ao}$ and $v_{bo}$, and $v_{OG}$ (Equation (19)). In addition, $v_{DM}$ is defined by half of difference of $v_{ao}$ and $v_{bo}$ (Equation (21)). Furthermore, these voltages can be expressed by switching functions as given in Equations (20) and (22):

$$v_{CM} = \frac{v_{ao} + v_{bo}}{2} + v_{OG} \tag{19}$$

$$= (4 - (S_{H1} + S_{H2} + S_{L1} + S_{L2})) * \frac{v_{out}}{8} \tag{20}$$

$$v_{DM} = \frac{-v_{ao} + v_{bo}}{2} \tag{21}$$

$$= \frac{(S_{H1} - S_{H2})}{4} * v_{out} \tag{22}$$

Here, $v_{LH1}$ and $v_{LH2}$ can be expressed by Equations (23) and (24), using $v_{CM}$ and $v_{DM}$, respectively:

$$v_{LH1} = \frac{v_{in}}{2} - v_{CM} + v_{DM} \tag{23}$$

$$v_{LH2} = \frac{v_{in}}{2} - v_{CM} - v_{DM} \tag{24}$$

Moreover, the high-side inductor voltage $v_{LH\ CM}$ in the CM circuit can be expressed as Equation (25) according to the Kirchhoff voltage law of the CM circuit:

$$v_{LH\ CM} = \frac{v_{in}}{2} - v_{CM}$$

$$= \frac{v_{in}}{2} + (S_{H1} + S_{H2} + S_{L1} + S_{L2} - 4) * \frac{v_{out}}{8} \tag{25}$$

Figure 6 shows the switching waveforms ($S_{H1}, S_{H2}, S_{L1}$, and $S_{L2}$) for the Z-type and N-type interleaving method, the common mode inductor voltages ($v_{LH\ CM}$), the differential mode inductor voltages ($v_{DM}$), and the circulating currents ($i_{DM}$). The common mode voltages across high-side inductors for each module ($v_{LH\ CM}$) generate the common current, and the differential mode voltages generate a circulating current.

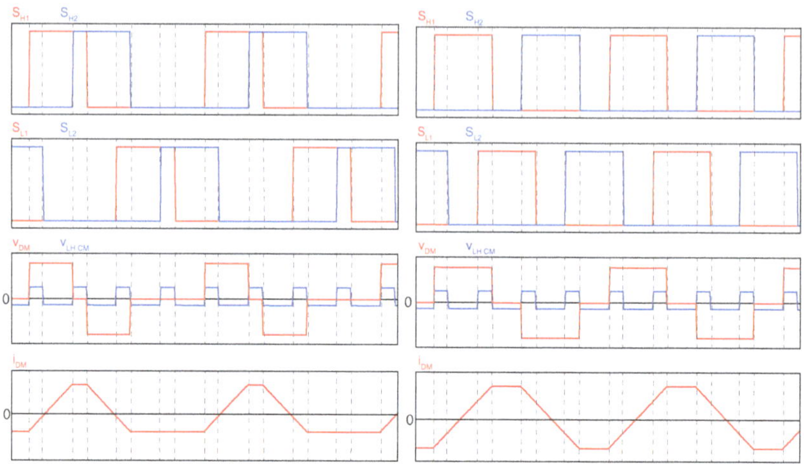


**Figure 6.** (**a**) Circulating current generated during Z-type interleaving; (**b**) circulating current generated during N-type interleaving.

As shown in Figure 6, the time in differential mode operation for Z-type interleaving is shorter than for N-type interleaving so that the circulating current for Z-type interleaving is smaller than that for N-type interleaving. The increase in the circulating current increases the ripple of the inductor current. Figure 7 illustrates the inductor current $i_L$ according to the interleaving method. The inductor current ripple in Z-type interleaving is smaller than that in N-type interleaving due to the small circulating current. This implies that the inductor losses in Z-type interleaving can be smaller than that of N-type interleaving. In the aspect of filter capacitor, however, different results may be obtained. This will be discussed in the next section.

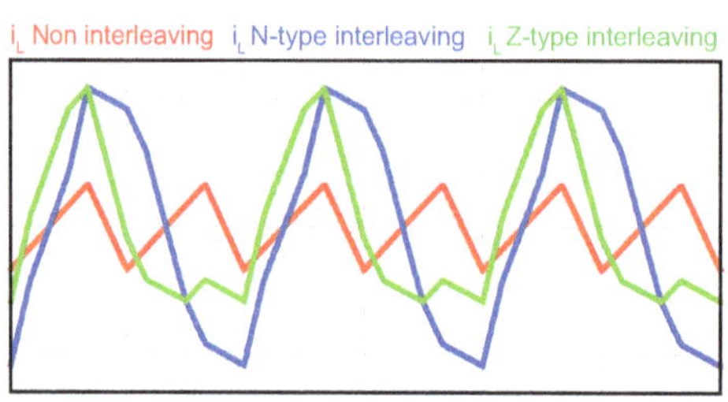


**Figure 7.** Inductor Currents for interleaving methods.

## 3. Comparison of Interleaving Methods for Capacitor's Current and Voltage

### *3.1. Impact of Interleaving on Capacitor Current Ripple of Parallel TLBC*

The effect of interleaving of the two-phase parallel TLBC on the inductor current is analyzed extensively in the reference document [18]. However, the effect on the filter capacitor is not discussed. Therefore, in this study, the effect of the interleaving method on the filter capacitor of the two-phase parallel TLBC is analyzed. The current ripple of the filter capacitor is related to the rms value of the capacitor current that is proportional to heat loss generated in the capacitor. Equation (26) represents the loss due to the equivalent series resistance (ESR) of the capacitor:

$$P_{C\ Loss} = i_{c\ rms}^2 * R_{C\ ESR} \tag{26}$$

Increased heat and loss adversely affect the capacitor lifespan and reduce the converter's efficiency. Capacitor is foremost among the causes of failure in converter modules; therefore, a capacitor's lifespan is directly related to that of the converter module [19]. The current ripple of the filter capacitor is an important indicator of the converter's lifetime. Hence, it is essential to consider the effect of the interleaving methods on the current ripple of the capacitor, which will be discussed in Sections 3.1.1 and 3.1.2.

### 3.1.1. Capacitor Current for Z-Type Interleaving

First, this section discusses the case of Z-type interleaving. Figure 3a illustrates the operating modes and switching status for Z-type interleaving method. As mentioned earlier, the high-side switches of each module operate prior to the low-side switches.

Based on the operation of the switch, a cycle of operational modes was divided into eight states. As illustrated in Figure 8, the current of the high-side and low-side capacitors according to the switching states the switch operation is expressed as Equations (27) and (28):

$$i_{FCH} = (1 - S_{H1}) * i_{LH1} + (1 - S_{H2}) * i_{LH2} - I_o \tag{27}$$

$$i_{FCL} = (1 - S_{L1}) * i_{LL1} + (1 - S_{L2}) * i_{LL2} - I_o. \tag{28}$$

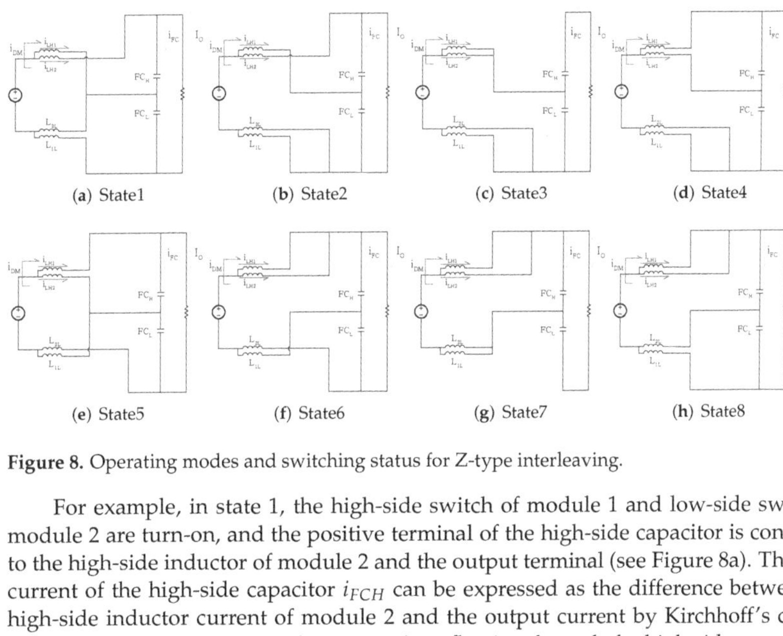

(a) State1 (b) State2 (c) State3 (d) State4

(e) State5 (f) State6 (g) State7 (h) State8

**Figure 8.** Operating modes and switching status for Z-type interleaving.

For example, in state 1, the high-side switch of module 1 and low-side switch of module 2 are turn-on, and the positive terminal of the high-side capacitor is connected to the high-side inductor of module 2 and the output terminal (see Figure 8a). Thus, the current of the high-side capacitor $i_{FCH}$ can be expressed as the difference between the high-side inductor current of module 2 and the output current by Kirchhoff's current law. Equation (29) represents the current $i_{FCH}$ flowing through the high-side capacitor at state 1, as the difference of the high-side inductor current $i_{LH2}$ of module 2 and the output current $I_o$:

$$i_{FCH} = i_{LH2} - I_o \tag{29}$$

In addition, in state 3, only the high-side switches of module 1 and 2 ($S_{H1}, S_{H2}$) are turned on, and the positive terminal of the high-side capacitor is connected only to the output terminal (see Figure 8). This implies that the inductor currents are separated from the capacitors, and the output current is supplied by only capacitors. The current flowing through the high-side capacitor $i_{FCH}$ at state 3 is expressed as Equation (30):

$$i_{FCH} = -I_o \tag{30}$$

From this analysis, the current of the high-side filter capacitor in each state can be obtained, as shown in Table 1. The current of the low-side filter capacitor can also be obtained similarly. As shown in Table 1, the high-side capacitor current $i_{FCH}$ in eight states for Z-type interleaving can be expressed by the high-side inductor currents of modules 1 and 2 ($i_{LH1}$ and $i_{LH2}$) and the output current Io.

**Table 1.** High-side capacitor current for eight operating states of a Z-type interleaving method.

| State | $i_{FCH}$ |
|---|---|
| 1 | $i_{LH2} - I_o$ |
| 2 | $i_{LH2} - I_o$ |
| 3 | $-I_o$ |
| 4 | $i_{LH1} - I_o$ |
| 5 | $i_{LH1} - I_o$ |
| 6 | $i_{LH1} + i_{LH2} - I_o$ |
| 7 | $i_{LH1} + i_{LH2} - I_o$ |
| 8 | $i_{LH1} + i_{LH2} - I_o$ |

#### 3.1.2. Capacitor Current for N-Type Interleaving

This section describes the capacitor current ripple for N-type interleaving. Figure 3b of the previous section presents a switching pattern for N-type interleaving method. Unlike Z-type interleaving (where the high-side switches of the two modules turn-on first and then the low-side switches turn-on), the high-side and low-side switches of one module turn-on first and then those of the other module turn-on in N-type interleaving.

As in the case of Z-type interleaving, a cycle was divided into eight states according to the operation of the switch. As shown in Figure 9, in state 1, a high-side switch of module 1 and a low-side switch of module 2 are turned on (this is similar to the case of Z-type interleaving), and a positive terminal of the high-side capacitor is connected to a high-side inductor of module 2 and an output terminal. Therefore, the current of the high-side capacitor $i_{FCH}$ can be expressed as the difference between the high-side inductor current of module 2 and the output current using Kirchhoff's current law. This is similar to state 1, Equation (29) of Z-type interleaving. However, states 3 and 4 alongside states 6 and 7 appear differently. For example, in state 4 of Z-type interleaving, all of the high-side switches of modules 1 and 2 are turned off. At this time, the positive terminals of the high-side capacitor are connected to both the high-side inductors and output terminals of modules 1 and 2. Therefore, the current of the top capacitor $i_{FCH}$ in the corresponding state can be expressed as the difference between the sum of the top inductor currents of modules 1 and 2 and the output current. Equation (31) represents the high-side capacitor current $i_{FCH}$ at state 4 during N-type interleaving as a function of the high-side inductor currents of modules 1 and 2 ($I_{LH1}$ and $I_{LH2}$) and the output current $I_o$:

$$i_{FCH} = i_{LH1} + i_{LH2} - I_o \tag{31}$$

This is different from $i_{LH1} - I_o$ which is $i_{FCH}$ in state 7 of N-type interleaving. At the remaining states 3, 6, and 7, the Z-type and N-type interleaving yield different results. Table 2 illustrates the current of the high-side capacitor $i_{FCH}$ in each state according to the operating modes for n-type interleaving with the high-side inductor currents of modules 1 and 2 ($i_{LH1}$ and $i_{LH2}$) and the output current $I_o$.

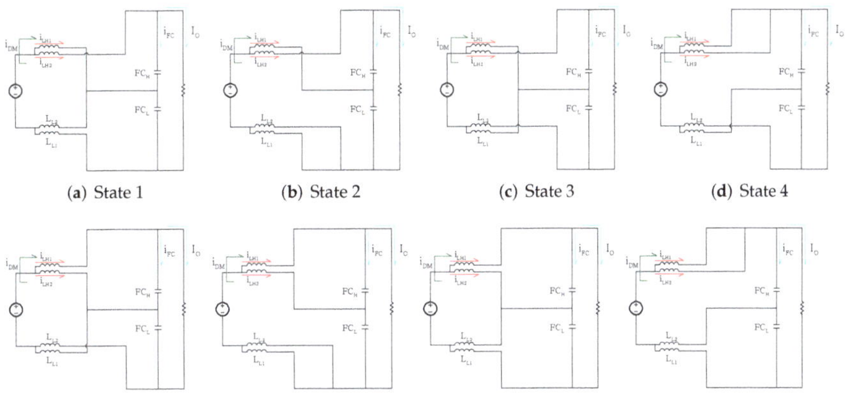

(**a**) State 1 (**b**) State 2 (**c**) State 3 (**d**) State 4

(**e**) State 5 (**f**) State 6 (**g**) State 7 (**h**) State 8

**Figure 9.** Circuit diagram according to the N-type interleaving switching operation.

**Table 2.** High-side capacitor current for eight operating states of the N-type interleaving method.

| State | $i_{FCH}$ |
|---|---|
| 1 | $i_{LH2} - I_o$ |
| 2 | $i_{LH2} - I_o$ |
| 3 | $i_{LH2} - I_o$ |
| 4 | $i_{LH1} + i_{LH2} - I_o$ |
| 5 | $i_{LH1} - I_o$ |
| 6 | $i_{LH1} - I_o$ |
| 7 | $i_{LH1} - I_o$ |
| 8 | $i_{LH1} + i_{LH2} - I_o$ |

The capacitor current was analyzed for two types of interleaving methods. The current ripple behavior for both interleaving methods appears differently due to a certain state in Z-type interleaving, which is absent in N-type interleaving. The high-side capacitor current of state 3, which exists only in the Z-type interleaving method, is equal to $-I_o$. This is an important contributing factor for the increase in the capacitor current ripple. Figure 10 illustrates the high-side capacitor current waveforms for each interleaving method. This was verified in the simulation described in Section 4.

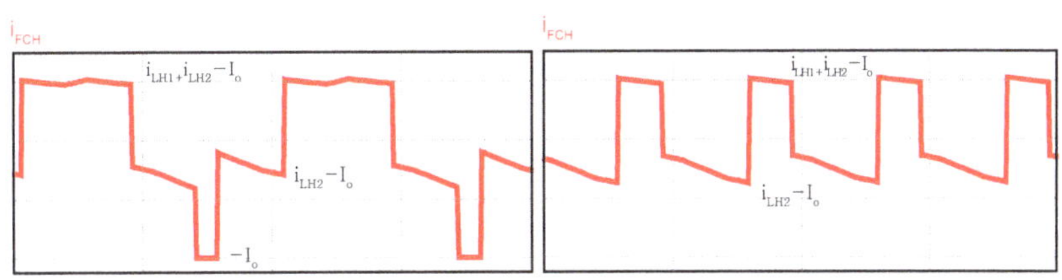


(**a**) Z-type interleaving (**b**) N-type interleaving

**Figure 10.** Capacitor current waveforms.

### *3.2. Comparative Analysis of Capacitor Voltage*

The disadvantages of Z-type interleaving control include not only an increase in loss and a decrease in the lifespan of the capacitor due to an increase in the capacitor current ripple but a negative impact on EMI. Figure 11 shows the high-side capacitor voltages according to the interleaving method.

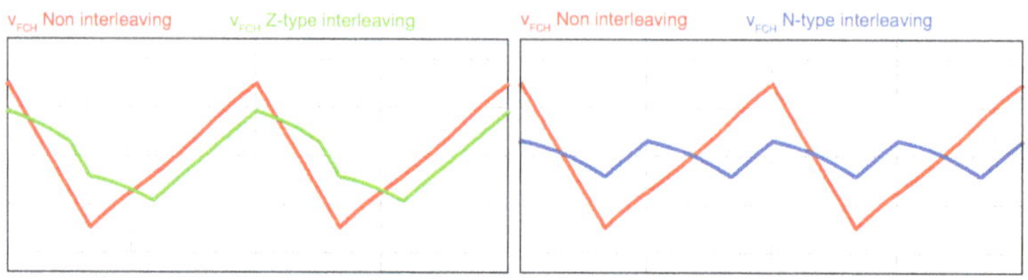


(**a**) Z-type interleaving (**b**) N-type interleaving

**Figure 11.** High-side capacitor voltage waveforms.

Figure 11 shows the high-side capacitor voltage waveforms for Z-type and N-type interleaving. The capacitor voltage of Z-type interleaving has similar waveforms to the conventional non-interleaving control, and the capacitor current is charged and discharged once per cycle. However, the high-side capacitor charges and discharges twice per cycle in N-type interleaving so that the capacitor voltage ripple is decreased. When the ripple frequency is doubled, similar to the voltage of the high-side capacitor in N-type interleaving, the cutoff frequency of the EMI filter can be increased. This enables the EMI filter size to be reduced, thereby leading to price reduction [21].

### 3.3. Current Imbalance with Z-Type Interleaving by the Capacitor Voltage

The two interleaving methods differ in terms of control as well. First, the Z-type interleaving method cannot evenly drive module currents with open loop drive. Open loop drive means that the current or voltage of the converter have no feedback signals. This implies that, in the absence of external disturbance, two module currents must be the same under same duty. However, the inductor current imbalance occurs in the Z-type interleaving control, so the inductor current is concentrated in the inductor of any one module. so that the electric power is concentrated in one module. This results in higher energy losses and heat generation. Furthermore, when the system operates at more than half the rated load, the excess power over the rated power flows on one side only. This can result in module destruction. Figure 12 illustrates the inductor current waveform for each interleaving method. In the Z-type interleaving open loop drive, the electric power is not evenly distributed to the modules, and the inductor current is concentrated in module 1. On the other hand, it can be observed that imbalance does not occur in an N-type interleaving open loop drive.

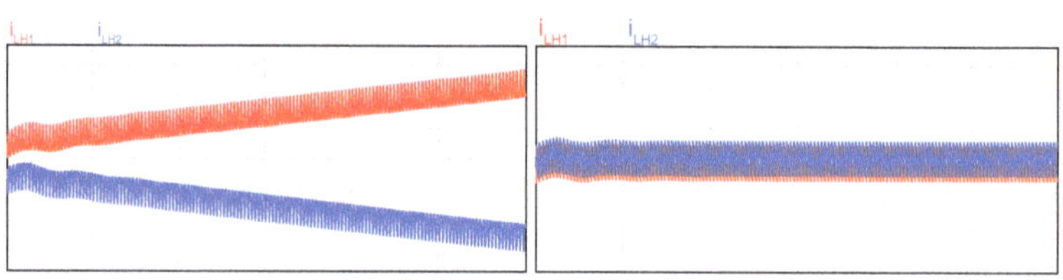


(**a**) Z-type interleaving control (**b**) N-type interleaving control

**Figure 12.** Current response in two modules for open loop drive.

The inductor current imbalance occurs due to the different voltages applied to the inductors. To describe this phenomenon, Figure 13 illustrates the high-side and low-side capacitor voltages $v_{FCH}$ and $v_{FCL}$, inductor voltages $v_{LH1}$ and $v_{LH2}$ applied to the high-side inductor of each module; and a high-side inductor current of module 1 $i_{LH1}$ with Z-type interleaving under an open loop drive.

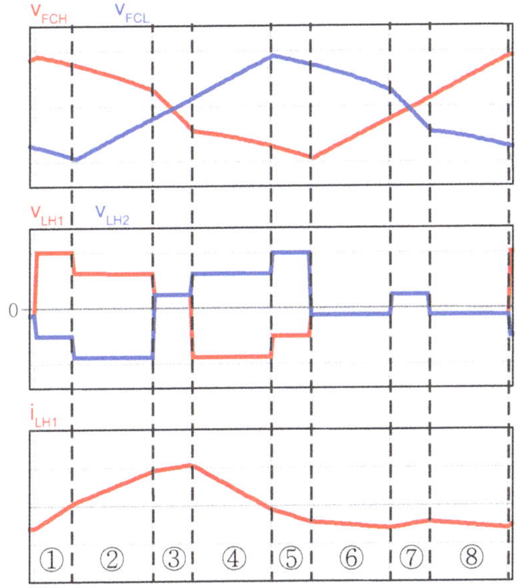


**Figure 13.** Voltage and current waveforms for Z-type interleaving with an open loop drive.

States 1–8 in Figure 13 are defined according to the switching behavior of the Z-type interleaving method. As described above, interleaving control of a module of two or more phases generates a circulating current owing to differential voltage. The differential voltages between two high-side and two low-side inductors are generated by the voltages of the high-side and low-side capacitors, respectively. However, the capacitors repeatedly charge and discharge according to the switching operation, so the potential of capacitors continuously varies. The high-side inductor voltage of module 1 is positive in states 1, 2, 3, and 7 and negative in states 4, 5, 6, and 8. During states 3, 6, 7, and 8, the current imbalance does not occur because the same voltage is applied across the high-side inductors in both modules 1 and 2. In states 1 and 2, inductor current of module 1 is more increased where the voltage of the high-side capacitor is relatively high, and for states 4 and 5, the inductor current of module 1 is less decreased where the voltage of the high-side capacitor is relatively low. As a result, the current of the high-side inductor of module 1 gradually increases as more positive and less negative voltages are repeatedly applied. This is opposite to the scenario of the high-side inductor of module 2, the inductor current of module 2 is increased by a relatively low voltage, and inductor current of module 2 is decreased by a relatively high voltage. Therefore, the current of module 2 decreases gradually with repeated charging and discharging cycles. The more the cycle is repeated, the more severe the current imbalance. In contrast, the N-type interleaving method does not cause a current imbalance. Figure 14 illustrates the states according to the switching operation in the N-type interleaving method.

In Z-type interleaving control, the voltages of the high-side capacitor in the states 1 and 2 for the module 1 inductor are different from those in states 4 and 5. In contrast, the current imbalance does not occur during N-type interleaving control because the voltages of the high-side capacitor states 1, 2, and 3 are equal to those in the states 5, 6, and 7.

As a result, in open loop (non-feedback) drive, the Z-type interleaving method leads to current imbalance. Hence, it is essential to control the closed loop with a current controller in each module with the Z-type interleaving. In contrast, with the N-type interleaving method, a relatively simple control method with low computation requirements, such as an open loop drive or closed loop current control, may be used. This was verified by the simulation described in Section 4.

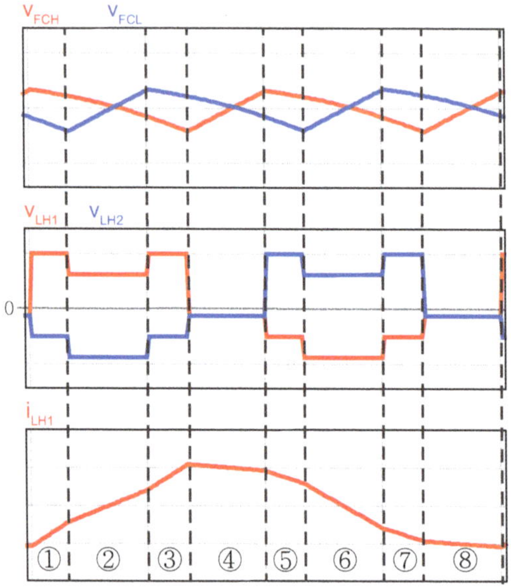


**Figure 14.** Waveforms observed when N-type interleaving is performed on a TLBC as an open loop.

## 4. Simulation

In this section, to compare and analyze the effects of the interleaving methods of the two-phase parallel TLBC, the circulating current and filter capacitor voltage and current for each interleaving method are compared. The current imbalance is also investigated for each control method. Table 3 presents the parameters used in the simulation. Two converter modules with an input voltage of 1000 V and an output voltage of 1500 V were operated under a load of 300 kW each (totally 600 kW).

**Table 3.** System parameter.

| Parameter | Value | Unit |
|---|---|---|
| Rated Power | 300 | kW |
| Input Voltage | 1000 | V |
| Output Voltage | 1500 | V |
| Load Resistor | 3.75 | Ω |
| Inductance ($L_H, L_L$) | 0.25 | mH |
| Capacitance ($C_H, C_L$) | 900 | uF |
| Switching Frequency | 5 | kHz |
| Number of Modules | 2 | - |

Figure 15 illustrates the circuit diagram used for the simulation, where the circuit consists of two-phase parallel TLBC with added low-side inductors connected to the neutral point of the filter capacitors. Closed loop control was simulated to measure and control the output voltage and inductor current, and the master–slave method was applied as parallel control. The master controller transmits phase information for interleaving operations to each slave controller and performs imbalance compensation control to match the voltages across the high-side and low-side capacitors. Figure 16 presents the circulating current $i_{DM}$ for two types of the interleaving method. As shown in Figure 16, the ripple of the circulating current for Z-type and N-type interleaving control is approximately 70 A and 100 A, respectively. This indicates that the ripple of circulating current is larger in N-type control, and, accordingly, the effective value and harmonic component are also larger in this case.

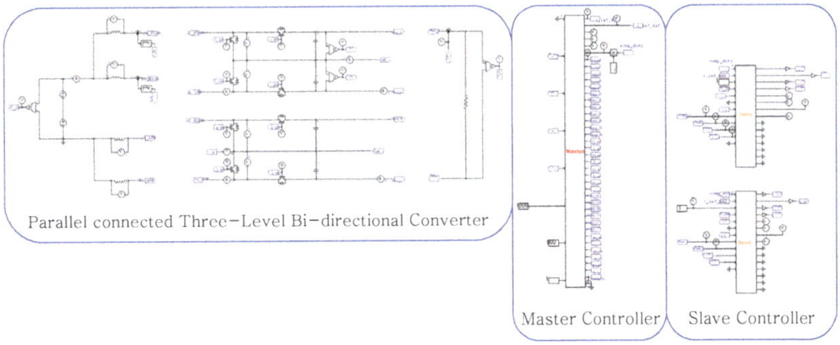


**Figure 15.** Simulation analysis circuit diagram.

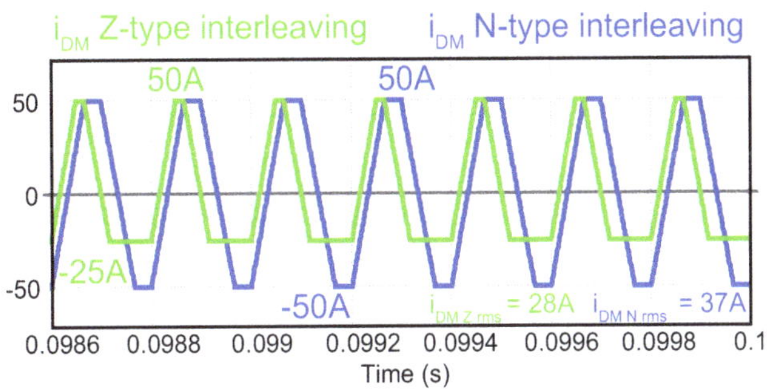


**Figure 16.** Comparison of circulating current by the interleaving methods.

Now, consider the filter capacitor voltage and current for Z-type and N-type interleaving methods. Figure 17 shows the high-side filter capacitor current $i_{FCH}$ and FFT analysis for each interleaving method. As previously discussed, the filter capacitor current ripple in Z-type interleaving is larger than that in N-type interleaving. Moreover, the rms value of the capacitor current is in Z-type interleaving is larger (93 A) than that in N-type interleaving (72 A). The results for the N-type interleaving method are affected increasing the frequency of the fundamental wave component of harmonics to a double frequency. Moreover, the magnitude of (major) harmonic component of N-type interleaving is reduced. These are significantly lower than the fundamental component without interleaving as well as the fundamental component of the Z-type interleaving method.

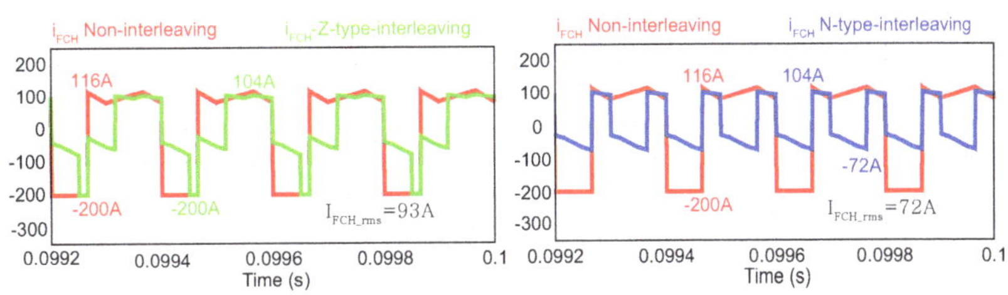


(a) (b)

**Figure 17.** (**a**) Capacitor current for Z-type interleaving; (**b**) capacitor current for N-type interleaving.

The interleaving method affects also the capacitor voltage as well as capacitor current. Figure 18 presents the capacitor voltage waveform $v_{FCH1}$ for each interleaving method.

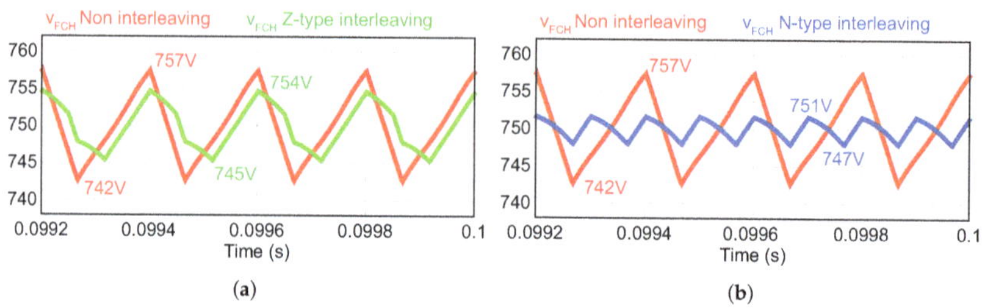


**Figure 18.** (**a**) Capacitor voltage for Z-type interleaving; (**b**) capacitor voltage for N-type interleaving.

Figure 18 reveals that the voltage ripple of a filter capacitor in Z-type interleaving control is larger than that in N-type interleaving, and that the ripple cycle is two times as long in the former than in the latter. The simulation analyses described earlier used two current controllers for each module in a voltage controller (see Figure 19a). However, for N-type interleaving, as illustrated in Figure 19b, the average current of the two modules can be controlled using a current controller. However, even in this case, Z-type interleaving causes current imbalance, as discussed in Section 3.3.

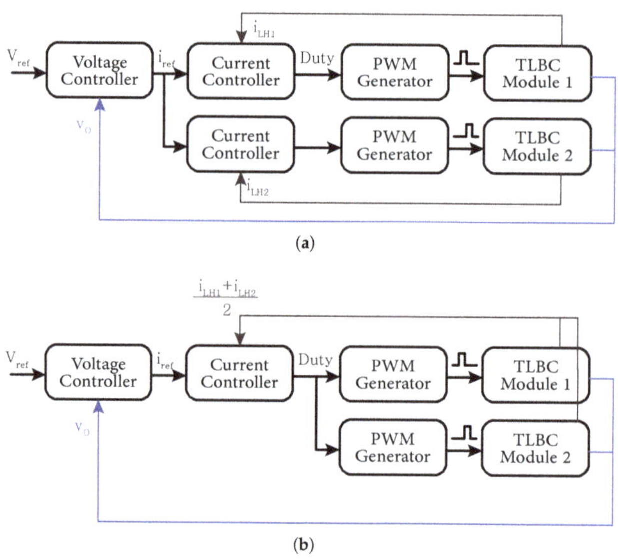


**Figure 19.** (**a**) TLBC with two current controllers; (**b**) TLBC with single current controllers.

First, as shown in Figure 19a, when single current controller is used per module, the current is effectively divided for both interleaving methods without current imbalance (Figure 20).

However, the different current response may be appeared when only one current controller is used. Figure 21 shows the current waveform for two-phase TLBC with single current controller for both interleaving methods. For the Z-type interleaving method, as illustrated in Figure 21, the inductor current is concentrated in one module, similar to the case of open loop drive. However, with the N-type interleaving method, the inductor currents are evenly distributed in each module.

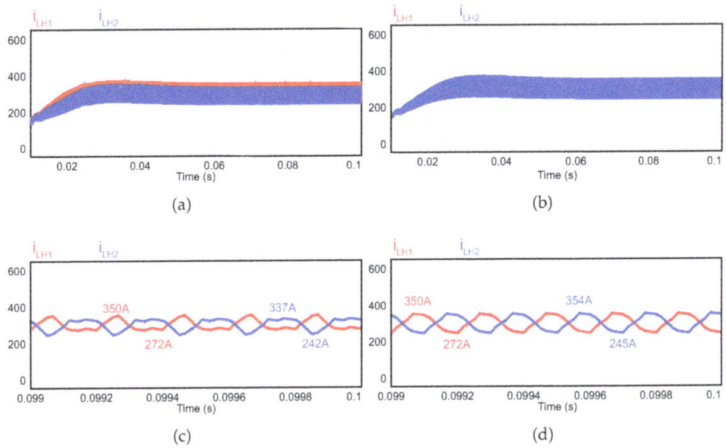


**Figure 20.** (**a**) Inductor currents for Z-type interleaving with two current controller; (**b**) inductor currents for N-type interleaving with two current controller; (**c**) enlarged (a); (**d**) enlarged (b).

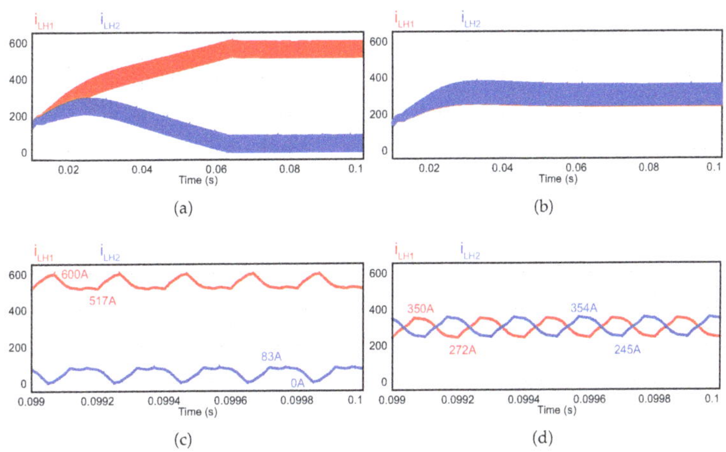


**Figure 21.** (**a**) current imbalance occurring during Z-type interleaving control in single current controller system; (**b**) current unbalance that does not occur during N-type interleaving control in a single current controller system; (**c**) enlarged (a); (**d**) enlarged (b).

The following table compares the performance of TLBC according to the interleaving method. According to each method, the rms current value of the filter capacitor $\left(i_{FCH_{rms}}\right)$, the voltage ripple of the filter capacitor and whether the open-loop control is possible and the efficiency were compared. In comparison, good things are highlighted with green, and bad things are highlighted with red.

Efficiency analysis was also carried out by PSIM simulation, and ESR of passive devices and forward voltage of IGBT were considered. Efficiency analysis was performed and considered with the same ESR for inductor and capacitor. In this analysis, copper loss was only considered for module loss. The N-type interleaving method has more inductor loss than the Z-type interleaving method, but since the capacitor loss is reduced more than that, the efficiency is good as a result. As a result, Table 4 shows that the N-type interleaving method is better.

**Table 4.** Performance comparison table according to the interleaving method.

| | Non-Interleaving | Z-type | N-type | Unit |
|---|---|---|---|---|
| | | Interleaving | Interleaving | |
| ine $i_{LH_{rms}}$ | 323 | 325 | 328 | A |
| $i_{FCH_{rms}}$ | 140 | 93 | 72 | A |
| $\Delta v_{FCH}$ | 15 | 9 | 4 | V |
| open loop drive | possible | impossible | possible | |
| Efficiency | 91.5 | 92.2 | 92.7 | % |
| ine | | | | : better |
| | | | | : worse |

## 5. Conclusions

In this study, the effect of two interleaving methods on a two-phase parallel three-level bidirectional DC/DC converter was investigated. First, the current of the filter capacitor was formulated through a switching function. Furthermore, the current and voltage of the filter capacitor were investigated by analyzing the operating modes and switching status of switches depending on the interleaving method. In the case of Z-type interleaving control, there is a state in which the capacitor bears all of the output current. Thus, the ripple of the current and voltage of the filter capacitor increases. Furthermore, the inductor current imbalance occurs between two modules due to the difference in the inductor voltages during the charging and discharging states of the high-side and low-side inductors under open loop drive. In conclusion, the N-type interleaving method has more merits to control parallel TLBC because it is superior to the Z-type interleaving method in terms of capacitor of life, EMI noise, and convenience of filter capacitor control. This has been verified by PSIM simulations, which will then be pursued through parallel control techniques for two or more phases of TLBC and other topologies.

**Author Contributions:** H.-I.K. conceptualized, wrote original draft preparation, and did simulations. S.-H.K. assisted in visualizing the data. H.-W.K. and S.-W.B. conceived the idea and wrote the paper. K.-Y.C. and G.-D.K. assisted in idea development and paper writing. All authors have read and agreed to the published version of the manuscript.

**Funding:** The research was supported by a grant from the Academic Research Program of Korea National University of Transportation in 2021. This research was supported by a grant from "Material parts technology development project" of Korea Evaluation Institute of Industrial Technology (KEIT) of the republic of Korea (No. 20001101).

**Conflicts of Interest:** The authors declare no conflict of interest.

## References

1. Chen, X.L.; Liang, D.Q.; Zhang, W.D. Braking energy recovery for electric traction based on super-capacitor and bidirectional dc-dc converter. In Proceedings of The 7th International Power Electronics and Motion Control Conference, Harbin, China, 2–5 June 2012; Volume 2, pp. 879–883. [CrossRef]
2. Lianfu, T.; Lixin, W. Research on the integrated braking energy recovery strategy based on super-capacitor energy storage. In Proceedings of the 2017 International Conference on Smart Grid and Electrical Automation (ICSGEA), Changsha, China, 27–28 May 2017; pp. 175–178. [CrossRef]
3. Wang, F.; Zhuo, F.; Wang, X.; Guo, H.; Wang, Z. Bi-directional interleaving dc/dc converter used in dual voltage system for vehicle. In Proceedings of the 2009 IEEE 6th International Power Electronics and Motion Control Conference, Wuhan, China, 17–20 May 2009; pp. 1335–1339. [CrossRef]
4. Meleshin, V.; Zhiklenkov, D.; Ganshin, A. Efficient three-level boost converter for various applications. In Proceedings of the 2012 15th International Power Electronics and Motion Control Conference (EPE/PEMC), Novi Sad, Serbia, 4–6 September 2012; pp. DS1e.9-1–DS1e.9-8. [CrossRef]
5. Kan, Z.; Li, P.; Yuan, R.; Zhang, C. Interleaved three-level bi-directional dc-dc converter and power flow control. In Proceedings of the 2018 3rd International Conference on Intelligent Green Building and Smart Grid (IGBSG), Yilan, Taiwan, 22–25 April 2018; pp. 1–4.

6. Dusmez, S.; Hasanzadeh, A.; Khaligh, A. Comparative analysis of bidirectional three-level dc–dc converter for automotive applications. *IEEE Trans. Ind. Electron.* **2015**, *62*, 3305–3315. [CrossRef]
7. Mei, Y.; Li, X.; Qi, Y. A model predictive control method for three-level bi-directional dc-dc converter in renewable generation system. In Proceedings of the 2015 18th International Conference on Electrical Machines and Systems (ICEMS), Pattaya, Thailand, 25–28 October 2015; pp. 417–421. [CrossRef]
8. Haddad, K. Three level dc-dc converters as efficient interface in two stage PV power systems. In Proceedings of the 2012 IEEE Energytech, Cleveland, OH, USA, 29–31 May 2012; pp. 1–6. [CrossRef]
9. Yao, G.; Hu, L.; Liu, Y.; Chen, A.; He, X. Interleaved three-level boost converter with zero diode reverse-recovery loss. In Proceedings of the Nineteenth Annual IEEE Applied Power Electronics Conference and Exposition (APEC '04), Anaheim, CA, USA, 22–26 February 2004; Volume 2, pp. 1090–1095. [CrossRef]
10. Bilgin, M.S.; Poyrazoglu, G.; Aktem, M.; Er, E. DC-link capacitor lifetime under various operating conditions. In Proceedings of the 2019 IEEE 13th International Conference on Compatibility, Power Electronics and Power Engineering (CPE-POWERENG), Sonderborg, Denmark, 23–25 April 2019; pp. 1–6. [CrossRef]
11. Basu, S.; Undeland, T.M. Voltage and current ripple considerations for improving lifetime of ultra-capacitors used for energy buffer applications at converter inputs. In Proceedings of the 2010 Twenty-Fifth Annual IEEE Applied Power Electronics Conference and Exposition (APEC), Palm Springs, CA, USA, 21–25 February 2010; pp. 244–247. [CrossRef]
12. Kim, Y.J.; Kim, S.M.; Lee, K.B. A switching method for improving lifetime of dc-link capacitors in hybrid ANPC inverters. In Proceedings of the 2019 IEEE Student Conference on Electric Machines and Systems (SCEMS 2019), Busan, Korea, 1–3 November 2019; pp. 1–4. [CrossRef]
13. Zeng, Y.; Li, H.; Zhang, Z.; Zheng, T.Q.; Shang, Z.; Qiu, Z.; Yuan, L.; Ding, Y. A parallel-resonant isolated bidirectional dc-dc converter with low current ripple for battery storage systems. In Proceedings of the 2019 IEEE Energy Conversion Congress and Exposition (ECCE), Baltimore, MD, USA, 29 September–3 October 2019; pp. 5548–5553. [CrossRef]
14. ElMenshawy, M.; Massoud, A. Multimodule dc-dc converters for high-voltage high-power renewable energy sources. In Proceedings of the 2019 2nd International Conference on Smart Grid and Renewable Energy (SGRE), Doha, Qatar, 19–21 November 2019; pp. 1–6. [CrossRef]
15. Shanthi, N.; Deepalaxmi, R. Performance analysis of three level interleaved boost converter with coupled inductor. In Proceedings of the 2019 Fifth International Conference on Electrical Energy Systems (ICEES), Chennai, India, 21–22 February 2019; pp. 1–6. [CrossRef]
16. Itoh, J.i.; Abe, T. Circulation current reduction for a motor simulator system using a power converter with a common mode transformer. In Proceedings of the 2009 IEEE 6th International Power Electronics and Motion Control Conference, Wuhan, China, 17–20 May 2009; pp. 1662–1667. [CrossRef]
17. Liu, L.; Liu, Z.; Chen, Q. Research and simulation of circulation current control for grid tied inverters connected in parallel. In Proceedings of the 2016 Chinese Control and Decision Conference (CCDC), Yinchuan, China, 28–30 May 2016; pp. 6036–6041. [CrossRef]
18. Lu, S.; Mu, M.; Jiao, Y.; Lee, F.C.; Zhao, Z. Coupled inductors in interleaved multiphase three-level dc-dc converter for high-power applications. *IEEE Trans. Power Electron.* **2016**, *31*, 120–134. [CrossRef]
19. Lee, J.-Y.; Kim, J.S. A study on the reliability of a hot-swap applied DVR power supply using fault tree analysis. In Proceedings of the Korean Power Electronics Society Conference, Sono Belle Byeonsan, Buan-Gun Jeollabuk-Do, Korea, 6–8 July 2021; pp. 412–413.
20. Lee, K.M.; Baek, S.W.; Kim, H.W.; Cho, K.Y.; Kang, J.W. 3-level boost converter having lower inductor for interleaving operation. *Trans. Korean Inst. Electron.* **2021**, *26*, 96–105.
21. Wang, R.R. *High Power Density and High Temperature Converter Design for Transportation Applications*; Virginia Polytechnic Institute and State University: Blacksburg, VA, USA, 2012.

*energies*

*Article*

# Operation Characteristics of Adjustable Field IPMSM Utilizing Magnetic Saturation

**Kiyohiro Iwama * and Toshihiko Noguchi**

Graduate School of Science and Technology, Shizuoka University, Hamamatsu 432-8561, Japan; noguchi.toshihiko@shizuoka.ac.jp
* Correspondence: iwama.kiyohiro.14@shizuoka.ac.jp

**Abstract:** This paper describes an interior permanent magnet synchronous motor (IPMSM) based on a new adjustable field method. The proposed PM motor achieved magnetic field control utilizing magnetic saturation. In this paper, a back electromotive force (e.m.f.) measurement test and a load test using the prototype motor were conducted to clarify if the proposed motor had a wide operation range. In the back e.m.f. measurement test, it was confirmed that the proposed motor had a wide magnetic field controllable range of 51.7%. In addition, it was revealed, through the load test, that the proposed motor had a wide operating range, including both low-speed high-torque and high-speed low-torque driving conditions. Moreover, based on electromagnetic field analysis, the magnetic field control performance of the proposed adjustable field method was compared with the conventional field weakening control and other adjustable field methods. As a comparison result, it was verified that the proposed motor had less copper loss for the magnetic field control and fewer losses in the high-speed operating range.

**Keywords:** adjustable field; IPMSM; magnetic saturation; permeability; 3D magnetic path

**Citation:** Iwama, K.; Noguchi, T. Operation Characteristics of Adjustable Field IPMSM Utilizing Magnetic Saturation. *Energies* **2022**, *15*, 52. https://doi.org/10.3390/en15010052

Academic Editors: Yitao Liu and Ching-Ming Lai

Received: 16 November 2021
Accepted: 21 December 2021
Published: 22 December 2021



## 1. Introduction

Since the Paris agreement was signed in 2015, countries around the world have been working intensively to save energy. Particularly, in the electric machine technical field, the high-efficiency motor has actively been researched. A permanent magnet synchronous motor (PMSM) using high-power density permanent magnets (PMs) has widely been used because the PMSM can make the high-efficiency and high-power density drive possible [1,2]. In addition to the efficiency and the power density, a wide driving range is one of the most important evaluation items of the motor used in automotive applications. Generally, a design that achieves both low-speed high-torque and high-speed low-torque operations is required. However, it is challenging to achieve these operations at the same time because the magnetic field of the PMSM is not adjustable. Conventionally, field weakening control has been applied to expand the high-speed operation range [3,4]. The field weakening control has traditionally employed and can undermine the magnetic field of the PMSM by using the negative $d$-axis current $i_d$. However, there is a problem whereby the copper loss increase in the high-speed operation range is not ignorable, because the field weakening control requires much current to give a counter magnetic field against the PM field. To make matters worse, the counter magnetic field generated by the negative $i_d$ may cause irreversible demagnetization of PMs in the rotor.

To solve this problem, in previous studies, an adjustable field PMSM that can control the PM-based magnetic field has been attracting attention in recent years [5–11]. The adjustable field PMSM introduced in the references [5,6] can control the magnetic field density on the rotor surface with a consequent pole structure by using a field winding. However, it has low torque density as implemented in surface permanent magnet synchronous motor (SPMSM) configuration. As known well among the researchers, the SPM configuration does not have capability to generate a reluctance torque additional to the magnet torque.

The PMs of the adjustable field SPMSM can easily demagnetize due to the long air gap length. In the reference [7], the magnetic field passing through the magnetic leakage path on the rotor core is controlled by $q$-axis magnetomotive force (m.m.f.). This adjustable field method can realize the magnetic field control with a single inverter, because the additional m.m.f. source is not required. As a result, the magnetic field is dependent on motor operation. The adjustable field principle utilizing the motor harmonics is described in the reference [8]. This method uses the 2nd order space harmonics generated by the concentrated winding structure for the magnetic field control. However, this method cannot control the magnetic field actively and arbitrarily, because the rotating speed governs the quantity of the magnetic field. Furthermore, this motor must use soft magnetic composite (SMC) material, which is difficult to produce.

The adjustable field method proposed in the reference [9] can make an expansive driving range possible, because this method can change the drive circuits that correspond to three-phase or six-phase windings, etc. However, multiple six single-phase inverters are indispensable to switch over the winding configurations. For this reason, there is a drawback that the drive system of this adjustable field PMSM can be larger than other conventional motors. Furthermore, this method has only four discrete magnetic field outputs, resulting in a discrete adjustment of the magnetic field. The IPMSM utilizing the de- and re-magnetization of the PMs can achieve wide-range control of the magnetic field and highly efficient drive [10,11]. However, the drive system for the adjustable field IPMSM based on this approach tends to be bulky, because an extremely high m.m.f. is indispensable for de- and re-magnetization. In general, several significant problems in existing adjustable field methods are the motor type, which cannot deliver the reluctance torque; detrimental influence on the motor operation and rotating speed; demagnetization of PMs; magnetic field controllability; difficulty in manufacturing; and high copper loss for magnetic field control.

The authors have been investigating a new adjustable field method focusing on the magnetic saturation of the magnetic material [12]. The magnetic saturation is a phenomenon in which magnetic permeability changes according to the intensity of the external magnetic field. Typically, this phenomenon detrimentally affects the performance of the motor output in conventional motor drives [13,14]. In the reference [13], it was confirmed, through the analysis, that d- and q-axis flux linkage and torque are affected by the magnetic saturation. In another reference about the influence of the magnetic saturation [14], the relationship between the rotor bridge and the performance of a spoke-type PMSM was mathematically derived. In addition, it was revealed that the magnetic saturation in the bridge part has a significant influence on the motor performance. As mentioned above, the magnetic saturation usually deteriorates the motor output and complicates the design and the control.

As a novelty, the authors propose a motor with an adjustable field method utilizing the magnetic saturation (the proposed motor) in this paper to overcome disadvantages of the conventional field weakening control and the previous adjustable field methods. Detailed contributions of this paper are as follows:

(1) Adjustable field capability with genuine electromagnetic operation;
(2) Capability of the reluctance torque generation;
(3) Independency of the field adjustment on the motor operation (vector control) and the rotating speed;
(4) Higher anti-demagnetization capability of PMs;
(5) Continuous magnetic field control;
(6) Better productivity of the motor hardware;
(7) Lower copper loss for magnetic field control.

Finally, the proposed motor had a wider controllable range of the magnetic field by 51.7%.

In the following section, the principle of the proposed adjustable field method is introduced and explained in detail. Back e.m.f. characteristics are explained in Section 3. The drive system of the prototype motor and its load characteristic test results are described in Sections 4 and 5, respectively. Finally, the paper concludes in Section 6 with some highlights of the research study.

## 2. Principle of Adjustable Field Method Utilizing Magnetic Saturation

As mentioned above, the proposed adjustable field method uses magnetic saturation of the magnetic material. Figure 1 shows the relationship between the magnetic flux density and the magnetic field (B–H curve) of 35JNE230, used for the prototype motor's core. The permeability is defined as a slope of the B–H characteristic. As shown in Figure 1, the permeability of the core changes depending on the external magnetic field.

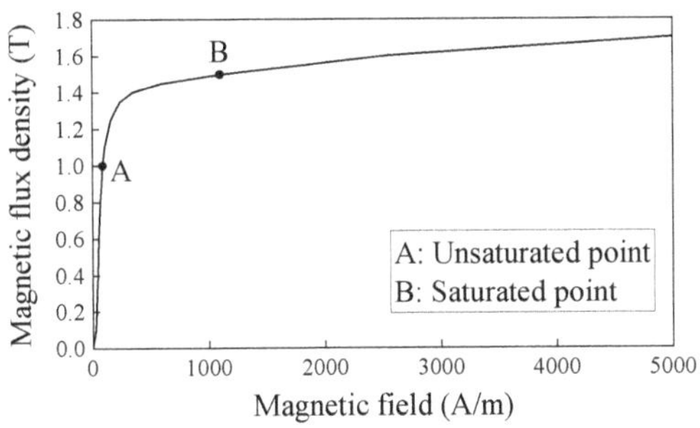


**Figure 1.** B–H curve of 35JNE230.

The magnetic circuit and specifications of the prototype motor are shown in Figure 2 and Table 1, respectively. As shown in Figure 2, the prototype motor was an IPMSM. Therefore, the prototype motor could deliver the reluctance torque. The stator and rotor cores of the prototype motor were split into two parts and an additional winding was inserted between the split stator cores. The additional winding generated a magnetic flux to cause magnetic saturation and to control the magnetic field. There were magnetic leakage paths between PMs in the rotor core in the prototype motor. The magnetic flux modulated the permeability of the magnetic leakage paths. Therefore, the additional winding and the magnetic flux were called modulation winding and modulation flux, respectively. The resistance of the modulation winding was larger than that of the armature winding. The voltage drop and inductance became large by designing the modulation winding with a large resistance value. However, the voltage drop was negligibly small compared to the phase voltage and DC-bus voltage, so the effect was small. In addition, there was no effect of the large inductance because the DC modulation current was used. S45C, a carbon steel material, was used as the rotor shaft and the stator frame to penetrate the modulation flux in the 3-dimensional (3D) magnetic path. Generally, the iron loss in the 3D magnetic path was reduced by constructing the 3D path with an SMC [15,16]. However, it costs a lot to make the SMC core, because high pressure around 1000 MPa is indispensable [17]. On the other hand, the proposed IPMSM could use a steel material as the 3D magnetic path because the modulation flux included only a DC or low-frequency component. For this reason, the 3D magnetic path could easily be realized.

**Table 1.** Specifications of prototype motor.

| Parameter | Value |
|---|---|
| Number of poles and slots | 8 poles, 48 slots |
| Armature winding | 6 turns/slot, 0.15 Ω |
| Modulation winding | 120 turns, 1.8 Ω |
| Stator diameter | $\phi$148 mm |
| Rotor diameter | $\phi$96.6 mm |
| Stack length | 63 mm |

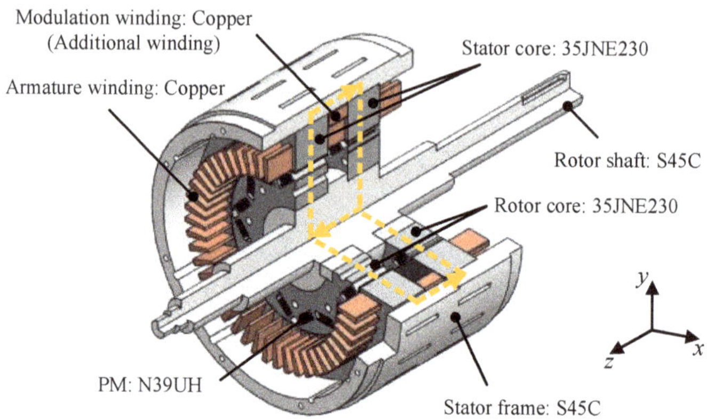


**Figure 2.** Magnetic circuit of prototype motor.

Figure 3 shows a vector plot of the modulation flux when the m.m.f. $F_m$ used for magnetic field control was given to the modulation winding. In this paper, all analysis results were obtained from the FEM using JMAG-Designer. As shown in Figure 3a, the modulation flux penetrated in the radial direction. There was a leakage flux passing through the air between the two stator cores, as shown in Figure 3b. However, the amount of the leakage flux was negligibly small because the air gap between the two stator cores was 15 mm, which is much larger than the air gap length between the stator core and the rotor core. In addition, the rotor shaft and the stator frame were magnetized by the DC modulation flux, but the residual flux was negligibly small because S45C, which is a soft magnetic material, was used as the material for the rotor shaft and the stator frame.

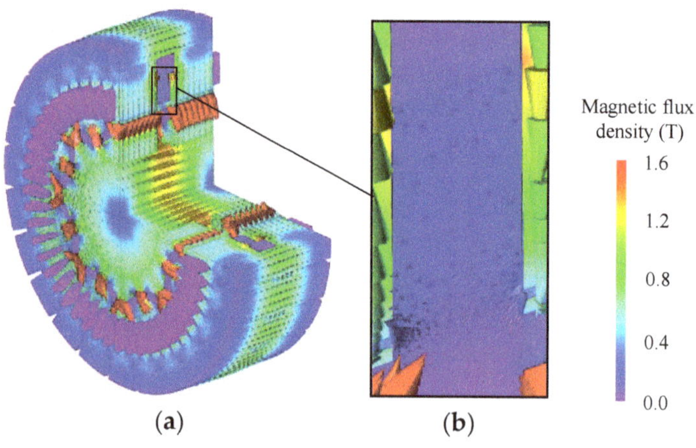


**Figure 3.** Vector plot of modulation flux with $F_m$ of 1200 AT: (**a**) modulation flux; (**b**) leakage modulation flux passing through air.

It was confirmed, from Figure 3, that the modulation flux penetrated in the radial direction. In the stator, no iron loss occurred because the modulation flux was DC flux. In addition, there was also no iron loss in the rotor due to the modulation flux. Figure 4 was prepared to examine why iron loss did not occur in the rotor. Figure 4a shows a usual DC flux and Figure 4b shows a radial DC flux similar to the modulation flux. As shown in Figure 4a, the observation point in front of the N-pole magnet was defined and the change in magnetic flux in the observation point was verified. Figure 5 shows the magnetic flux fluctuations of the $d$- and the $q$-axis components. Usually, as can be seen in Figure 5a, when the magnetic flux was transmitting, the magnetic flux component in the rotor fluctuated due to the rotation of the rotor, even if the magnetic flux was DC. On the other hand, as can be seen in Figure 5b, when the magnetic flux penetrated in all radial directions, there was no variation in the magnetic flux with respect to time. From the above results, it can be inferred that there was no effect on iron loss when the DC modulation flux was used for the adjustable field.

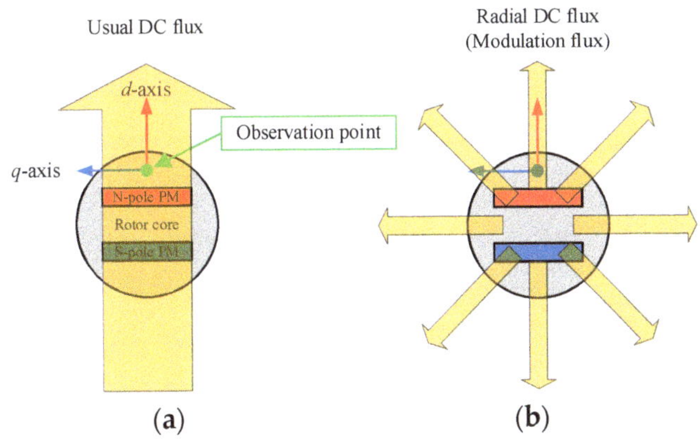


**Figure 4.** DC flux in rotor: (**a**) usual DC flux; (**b**) radial DC flux.

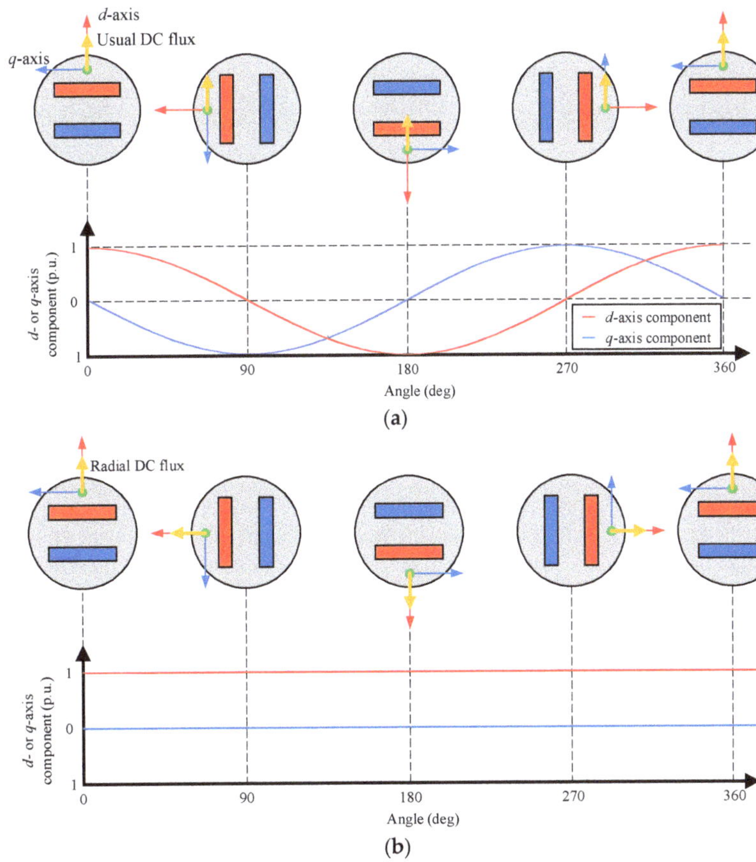


**Figure 5.** Variation in magnetic flux: (**a**) usual DC flux; (**b**) radial DC flux.

Figure 6 shows vector plots of the PM flux and the modulation flux with an $F_m$ of 0 AT, 720 AT and 1200 AT. Without the $F_m$, many PM fluxes leaked in the rotor core. In this case, the permeability of both the N- and S-pole side magnetic leakage path was 50. With an $F_m$ of 720 AT, the permeability of the N-pole side magnetic leakage path decreased to 24 by the modulation flux. Therefore, it became difficult for the N-pole PM flux to leak through the magnetic leakage path and the N-pole PM flux interlinking to the stator core increased. However, since the modulation flux penetrated in the direction weakening the N-pole magnet, it was necessary to study the de-magnetization of the N-pole PM. On the other hand, the permeability of the S-pole side magnetic leakage path increased to 3000 by the modulation flux. In this case, many modulation fluxes penetrated through the S-pole side magnetic leakage path and strengthened the magnetic field of the S-pole. With an $F_m$ of 1200 AT, the magnetic field of the proposed motor reached its limit, because the permeability of both the N- and S-pole side magnetic leakage paths decreased and the modulation flux could no longer penetrate. In this way, the proposed motor can realize magnetic field control by using magnetic saturation and modulating the permeability of the magnetic leakage paths.

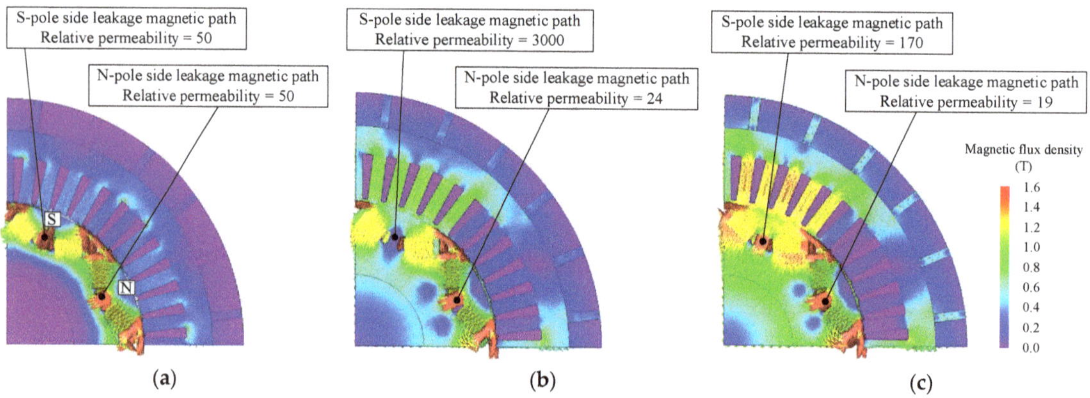


**Figure 6.** Vector plot of magnetic flux density: (**a**) PM flux without $F_m$; (**b**) PM and modulation flux with $F_m$ of 720 AT; (**c**) PM and modulation flux with $F_m$ of 1200 AT.

Figure 7 shows the magnetic flux density in the air gap between the upper stator core and rotor core with an $F_m$ of 0 AT, 720 AT and 1200 AT. By giving the $F_m$, the fundamental component increased. There are even-order components, however. Because the even-order components in the lower-side air gap were observed in the opposite phase with the even-order components in the upper-side air gap, no even-order component was generated in the back e.m.f. In addition, the DC component was also included in the gap magnetic flux density due to the radial modulation flux, but it did not affect the back e.m.f. because it did not vary with respect to time.

Figure 8 shows the rotor and the stator of the prototype motor. The rotor cores were skewed at an angle of 3.75 deg, taking advantage of the construction whereby the rotor is split into two parts. Thereby, 12th order space harmonics could be reduced because the angle between the teeth and the slots of the stator core was 3.75 deg. The gap between the stator core and the stator frame was minimized by inserting the stator core in the stator frame by shrink-fitting. Besides, slits were provided to the stator frame to suppress the iron loss. Figure 9 shows a magnetic flux density vector plot when a $q$-axis current of 80 A was supplied. However, the rotating speed was 1000 r/min. The model shown in Figure 9a had slits and the model shown in Figure 9b did not. As illustrated in Figure 9, by providing the slits, the armature flux passing in a circumferential direction could be reduced. Figure 10 shows an eddy current loss of the stator frames of the two models. The eddy current loss of the model with slits was 40.4% lower than that of the model without slits. It can be seen, from this result, that the slits were helpful in terms of iron loss.

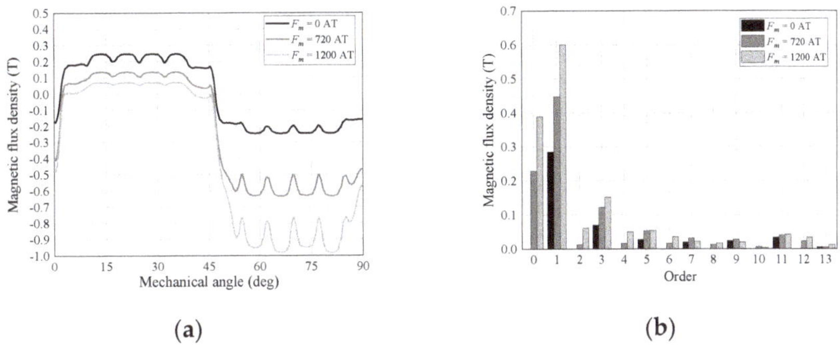


**Figure 7.** Gap magnetic flux density: (**a**) waveforms; (**b**) FFT results.

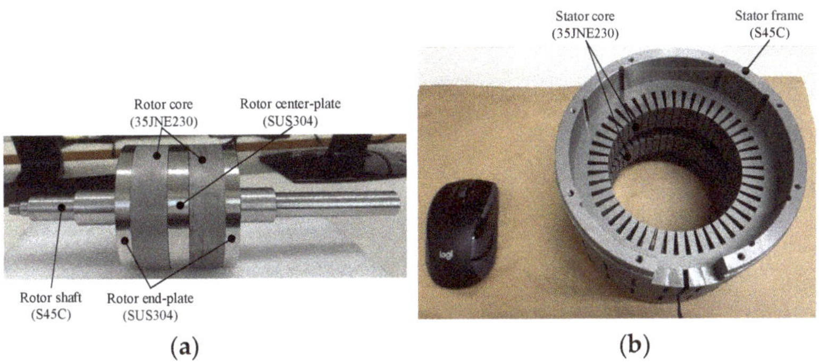


**Figure 8.** Configuration of prototype motor: (**a**) photograph of rotor; (**b**) photograph of stator.

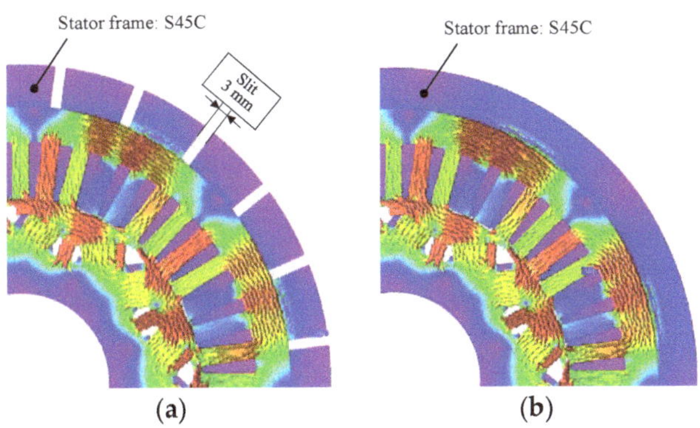


**Figure 9.** Magnetic flux density vector on $x$–$y$ cross-section with $q$-axis current of 80 A: (**a**) with slit; (**b**) without slit.

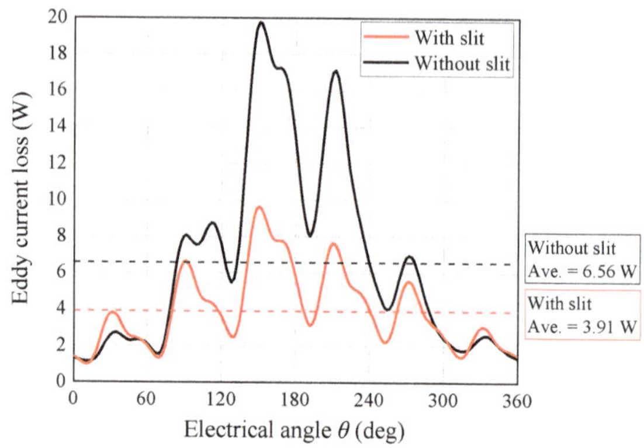


**Figure 10.** Eddy current loss of stator frame with $q$-axis current of 80 A.

## 3. Back E.m.f. Characteristic

### 3.1. Analysis of Back E.m.f.

Figure 11 shows the rotor geometry of the prototype motor model. It is crucial to design appropriately the saturation areas A and B shown in Figure 11 because the motor with the proposed adjustable field method achieves the magnetic field control by modulating the permeability of the magnetic leakage paths in the saturation areas. As shown in Figure 11, in the prototype motor, the width of the magnetic leakage paths in saturation area A was 3.2 mm and the widths of saturation area B were 1.5 mm and 0.5 mm. As shown in Table 2, the widths were parametrically changed to investigate the relationship between the width and the proposed motor performance.

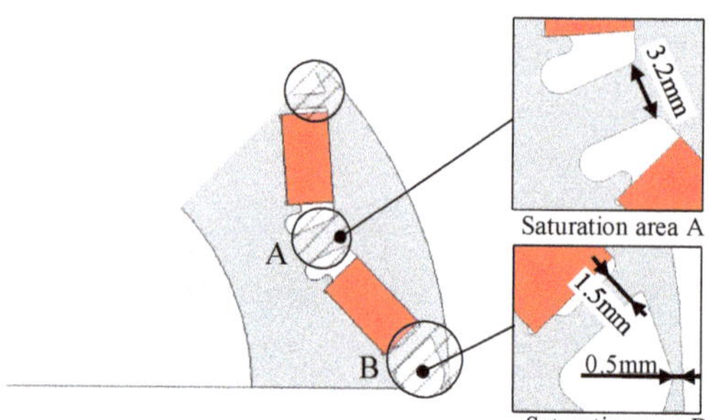


**Figure 11.** Magnetic saturation areas of prototype motor.

**Table 2.** Width of magnetic leakage paths in saturation areas A and B.

| Model | Saturation Area A | Saturation Area B |
|---|---|---|
| Model 1 | 2.1 mm | 1.00 mm, 0.5 mm |
| Model 2 | 2.7 mm | 1.25 mm, 0.5 mm |
| Model 3 | 3.7 mm | 1.75 mm, 0.5 mm |
| Model 4 | 4.3 mm | 2.00 mm, 0.5 mm |

Figure 12 shows the analysis results of the magnetic field $\Psi_a$. As shown in Figure 12, the minimum $\Psi_a$ was determined with the width of the magnetic leakage paths. On the other hand, in all models, the gradient in the $\Psi_a$ decreased when the $\Psi_a$ was about 39.6 mWb. Therefore, the maximum $\Psi_a$ was set at 39.6 mWb for all models to evaluate the performances with the same maximum fundamental component of the air gap flux. Table 3 shows the copper losses and controllable ranges of the $\Psi_a$ of each model. As shown in Table 3, the relationship between the controllable range of the $\Psi_a$ and the copper loss for the magnetic field control was a trade-off. Therefore, when the proposed motor is designed to increase the controllable range, the motor efficiency deteriorates and the motor volume is enormous.

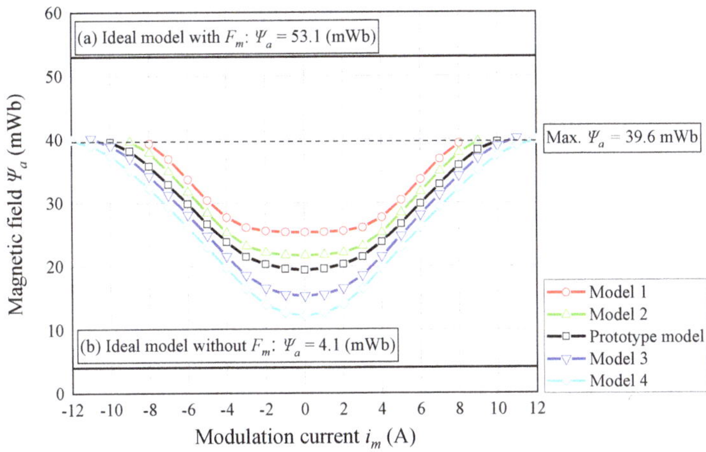


**Figure 12.** Magnetic fields of proposed motor models with different magnetic leakage path widths.

**Table 3.** Magnetic field control performances of proposed motor models.

| Parameter | | Model 1 | Model 2 | Prototype Model | Model 3 | Model 4 |
|---|---|---|---|---|---|---|
| Max. $\Psi_a$ | $i_m$ (A) | 8.1 | 8.9 | 10.0 | 10.5 | 11.7 |
| | $\Psi_a$ (mWb) | 39.6 | 39.6 | 39.6 | 39.6 | 39.6 |
| | Copper loss (W) | 118 | 143 | 180 | 198 | 246 |
| Min. $\Psi_a$ | $i_m$ (A) | 0 | 0 | 0 | 0 | 0 |
| | $\Psi_a$ (mWb) | 25.4 | 21.7 | 19.4 | 15.4 | 12.3 |
| | Copper loss (W) | 0 | 0 | 0 | 0 | 0 |
| Controllable range of $\Psi_a$ (mWb) | | 14.2 | 17.9 | 20.2 | 24.2 | 27.3 |

The ultimate design target of the proposed motor when the $F_m$ is given is to output a large $\Psi_a$ equivalent to when the saturation areas A and B shown in Figure 11 are replaced with air. In addition, the ideal design without the $F_m$ is to leak a much larger PM flux on the rotor core, equivalent to when the saturation areas A and B are replaced with 35JNE230. Therefore, the prototype motor shown in Figure 11 was compared with these two ideal designs. Figure 12 shows the magnetic field amount of the motor with saturation areas replaced with air or 35JNE230. There was a difference in the magnetic field quantity and the magnetic field quantity of the prototype motor was inferior to that of the models simulating the ideal design. Regarding the rotor geometry of the prototype motor, only the widths of the magnetic leakage paths were parametrically adjusted to make the controllable range 50% of the maximum magnetic field amount while considering the mechanical strength. In other words, there is still room to consider other elements, such as PM shape, a flux barrier position, etc. Therefore, the optimized design of the IPMSM with the proposed adjustable field method will be investigated in future works.

### *3.2. Back E.m.f. Measurement Test*

Figure 13 shows an experimental setup used in this paper. In the measurement test, the prototype motor was connected with a torque transducer TMNR-50NM, MinebeaMitsumi Inc. (Tokyo, Japan) and a load motor and the rotating speed of the prototype motor was controlled by the load motor. In this section, the back e.m.f. of the prototype motor was investigated by opening the terminal of the prototype motor and measuring the phase voltage. As mentioned above, the m.m.f. source $F_m$ for magnetic field control was required in addition to the armature m.m.f. source. The $F_m$ was supplied by the DC voltage supply.

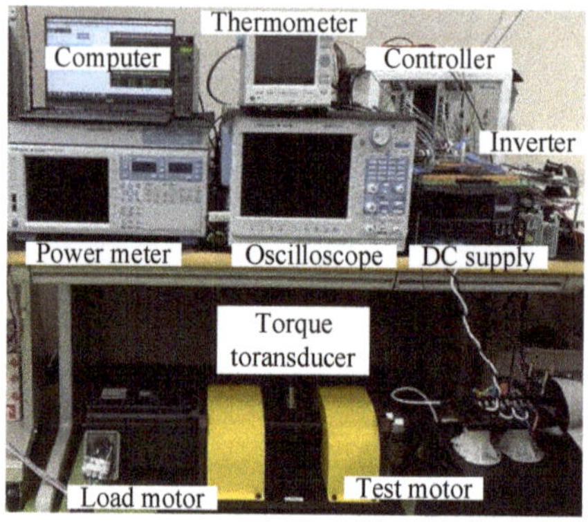


**Figure 13.** Experimental setup.

Figure 14 shows waveforms and FFT results of the back e.m.f. when an $F_m$ of 0 AT or 1200 AT was given to the modulation winding. However, the rotating speed was controlled at a constant speed of 1000 r/min by the load motor. It can be confirmed, from Figure 14, that the experimental value corresponded reasonably well with the analysis value and the fundamental component of the back e.m.f. increased by supplying the $F_m$. The fundamental components of experimental values without the $F_m$ were approximately 3% smaller than that of the analysis value. This decrement of the e.m.f. was possibly caused by the manufacturing error of the width of the magnetic leakage paths and the air gap length. In addition, the mesh used in the FEM was also a factor that may have caused the error. On the other hand, the fundamental component of the back e.m.f. with the $F_m$ of 1200 AT was 2.1 times larger than that without the $F_m$. Therefore, this result indicates that the prototype motor had a wide controllable range of the magnetic field.

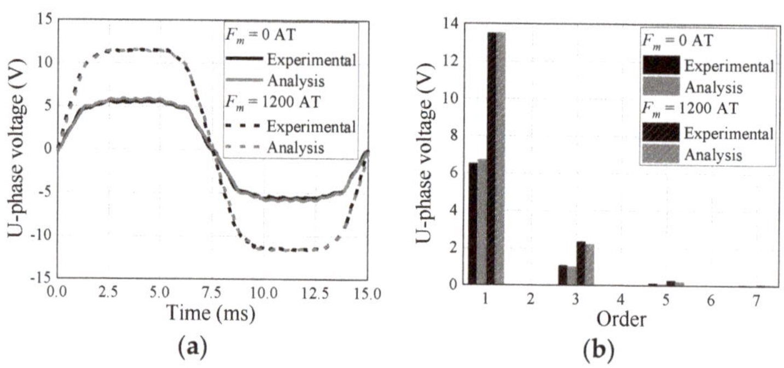


**Figure 14.** U-phase back e.m.f. at rotating speed of 1000 r/min: (**a**) waveforms; (**b**) FFT results.

Figure 15 illustrates the relationship between the magnetic field $\Psi_a$ and the modulation current $i_m$ of the prototype motor. As can be seen in the figure, the $\Psi_a$ depends on the absolute value of $i_m$ and can be approximated as

$$\Psi_a(i_m) = -1.84 \times 10^{-3}|i_m|^4 + 0.39|i_m|^2 + 18.7 \text{ (mWb)} \quad (1)$$

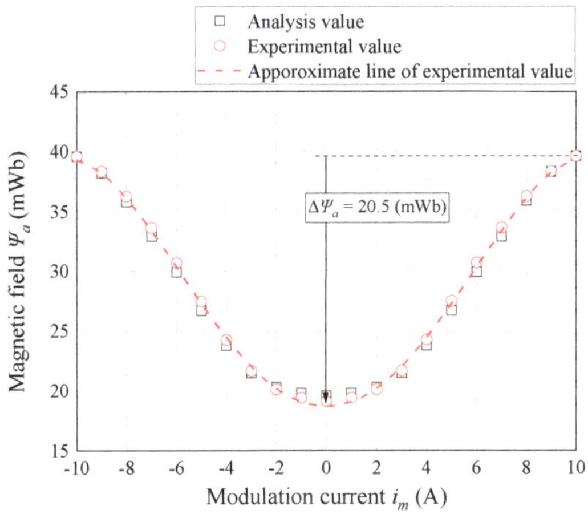


**Figure 15.** Experimental result of $\Psi_a$ and its approximate line.

### *3.3. Magnetic Field Control Performance Comparison with Other Methods*

In this section, based on the back e.m.f. measurement test results, the magnetic field control performance of the prototype motor was compared with that of the motor with other magnetic field control methods. The proposed motor was compared with three motors based on conventional techniques. The first benchmark was a conventional IPM motor. The magnetic field of the conventional IPM motor was weakened with a negative $i_d$. The second benchmark motor was a hybrid motor proposed in the reference [5,6]. The gap magnetic flux density of the hybrid motor was adjusted by the field current $i_f$. The third benchmark motor was a variable leakage flux motor introduced in the reference [7]. The magnetic leakage paths were provided in the rotor and the leakage PM flux could be adjusted by $q$-axis m.m.f. Therefore, the magnetic field of the variable leakage flux IPM motor was a function of the $q$-axis current $i_q$.

Figures 16–18 show analysis models of the conventional IPM motor, the hybrid motor and the variable leakage flux motor, respectively. In addition, Table 4 shows the specifications of the compared motors. For a fair comparison, all motors had the same material, stator geometry, core stack length, number of winding turns, PM volume and maximum current density. The rotor bridge of the conventional IPM motor was set at 1.0 mm, considering mechanical strength. Because a 3D magnetic circuit and an additional winding were not needed, the stack length and the armature winding resistance of the conventional IPM motor and the variable leakage flux motor were lower than the proposed motor and the hybrid motor. When designing an SPM motor that rotates at higher than 10,000 r/min, it is common to use a protection tube for the rotor to prevent the PM from scattering. Therefore, the air gap length is usually designed longer than the IPM motor. In this paper, the air gap length of the prototype motor, the conventional IPM motor and the variable leakage flux motor was 0.7 mm and that of the hybrid motor was 1.0 mm. It can be seen, from Table 4, that the prototype motor was inferior in terms of torque density because the motor volume, including the 3D magnetic circuit, was the most enormous and output torque was the lowest among all analysis models. The deterioration of the output torque of the prototype motor was greatly affected by the magnetic saturation of the stator teeth due to the radial modulation flux.

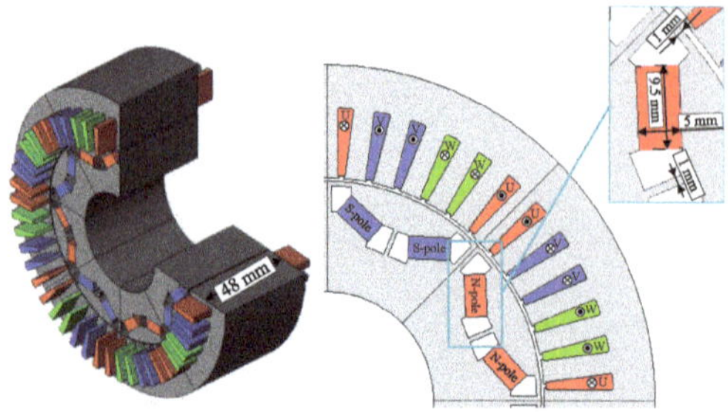


**Figure 16.** Analysis model of conventional IPM motor.

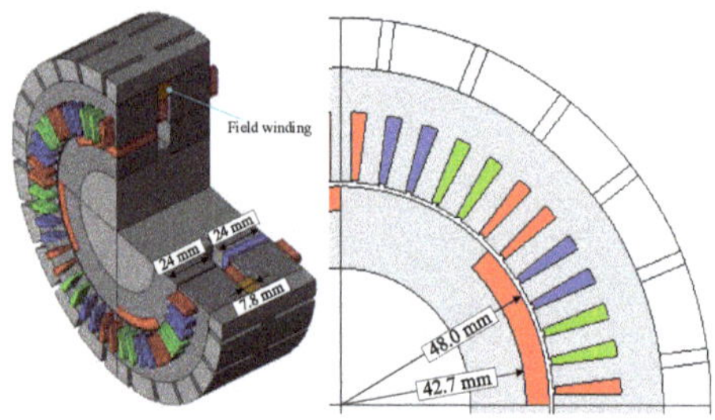


**Figure 17.** Analysis model of hybrid motor.

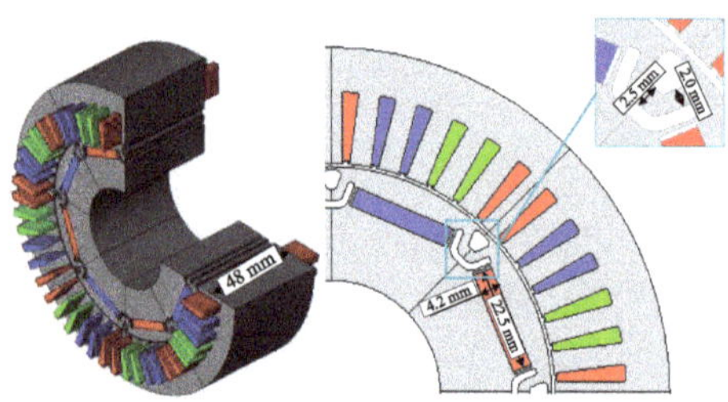


**Figure 18.** Analysis model of variable leakage flux motor.

Figure 19 shows the relationship between the $\Psi_a$ and the manipulating variable of the $\Psi_a$ of the compared motors. However, in the conventional IPM motor, the $\Psi_a$ could not be directly controlled and the $i_d$ was used to weaken the d-axis flux $\Psi_d$. As shown in Figure 15, a $\Psi_a$ of 20.5 mWb could be adjusted in the prototype motor by supplying an $i_m$ of 10 A. As shown in Figure 19a, in the conventional IPM motor, a $\Psi_d$ of 20.5 mWb was weakened with an $i_d$ of $-63$ A. In addition, from Figure 19b, in the hybrid motor, the maximum $\Psi_a$ was 53.0 mWb when an $i_f$ of 4.2 A was supplied and the minimum $\Psi_a$ was 32.3 mWb when an $i_f$ of $-4.2$ A was supplied. Thus, a $\Psi_a$ of 20.7 mWb could be controlled by an $i_f$ from $-4.2$ A to 4.2 A. On the other hand, as shown in Figure 19c, in the variable leakage flux motor, even if a maximum $i_q$ of 80 A was supplied, it was not possible to control a $\Psi_a$ of 20.5 mWb, which was roughly the same as the other motors. From this result, the variable leakage flux motor had a drawback in terms of the controllable range of the magnetic field. Table 5 shows the comparison result of the magnetic field control performance when a $\Psi_a$ of around 20.5 mWb was adjusted. However, as described above, the limit value of the magnetic field control range was 10.1 mWb in the variable leakage flux motor. It can be seen, from Table 5, that the copper loss when the $\Psi_a$ was around 20.5 mWb was 180 W for the proposed motor and 151 W for the hybrid motor, while it was 544 W for the conventional IPM motor. In other words, the proposed motor and the hybrid motor had higher magnetic field control performance.

**Table 4.** Specifications of compared motors.

| Parameter | | Prototype Motor | Conventional IPM Motor | Hybrid Motor | Variable Leakage Flux Motor |
|---|---|---|---|---|---|
| Manipulating variable for magnetic field control | | Modulation current $i_m$ | *d*-axis current $i_d$ | Field current $i_f$ | *q*-axis current $i_q$ |
| Stator core diameter (mm) | | $\phi$148 | ← | ← | ← |
| Stack length (mm) | | 63 (Core stack length: 48) | 48 | 55.8 (Core stack length: 48) | 48 |
| Number of turns | Armature winding | 6 turns/slot | ← | ← | ← |
| | Additional winding | 120 turns | - | 120 turns | - |
| Core volume ($mm^3$) | | 577,000 | 564,000 | 542,000 | 572,000 |
| Copper volume ($mm^3$) | Armature winding | 337,000 | 307,000 | 323,000 | 307,000 |
| | Additional winding | 59,000 | - | 31,000 | - |
| PM volume ($mm^3$) | | 36,000 | ← | ← | ← |
| Magnetic circuit volume ($mm^3$) | | 1,440,000 | 908,000 | 1,370,000 | 915,000 |
| Max. current density ($A_{rms}/mm^2$) | | 20 | ← | ← | ← |
| Resistance ($\Omega$) | Armature winding | 0.150 | 0.137 | 0.144 | 0.137 |
| | Additional winding | 1.80 | - | 4.29 | - |
| Max. torque (Nm) | | 12.2 ($i_m$ = 10 A, $i_q$ = 80 A) | 14.4 ($i_q$ = 80 A) | 16.6 ($i_f$ = 4.2 A, $i_q$ = 80 A) | 13.9 ($i_q$ = 80 A) |

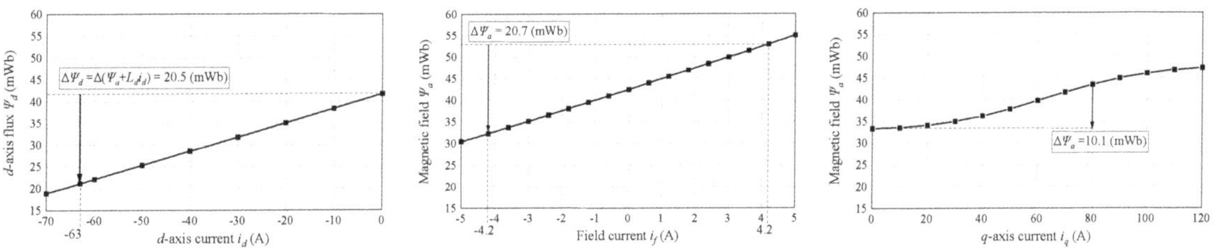


(a) (b) (c)

**Figure 19.** Relationship between magnetic field and variable for magnetic field control: (**a**) conventional IPM motor with field weakening control; (**b**) hybrid motor; (**c**) variable leakage flux motor.

**Table 5.** Comparison results of magnetic field control performance.

| Parameter | | Prototype Motor | Conventional IPM Motor | Hybrid Motor | Variable Leakage Flux Motor |
|---|---|---|---|---|---|
| Max. $\Psi_a$ or Max. $\Psi_d$ | Variable value (A) | $i_m = 10$ | $i_d = 0$ | $i_f = 4.2$ | $i_q = 80$ |
| | $\Psi_a$ or $\Psi_d$ (mWb) | $\Psi_a = 39.6$ | $\Psi_d = 41.7$ | $\Psi_a = 53.0$ | $\Psi_a = 43.5$ |
| | Copper loss (W) | 180 | 0 | 75.7 | 877 |
| Min. $\Psi_a$ or Min. $\Psi_d$ | Variable value (A) | $i_m = 0$ | $i_d = -63$ | $i_f = -4.2$ | $i_q = 0$ |
| | $\Psi_a$ or $\Psi_d$ (mWb) | $\Psi_a = 19.1$ | $\Psi_d = 21.2$ | $\Psi_a = 32.3$ | $\Psi_a = 33.3$ |
| | Copper loss (W) | 0 | 544 | 75.7 | 0 |
| Controllable range of $\Psi_a$ or $\Psi_d$ (mWb) | | 20.5 | 20.5 | 20.7 | 10.1 |

Figure 20 shows a vector plot of magnetic flux density when the rotor shafts of the prototype motor and the hybrid motor were most strongly magnetized. Although the rotor shaft diameter of the hybrid motor was larger than that of the prototype motor, the magnetic flux density of the hybrid motor rotor shaft was much higher than the prototype motor. Only the modulation flux passed through the rotor shaft in the prototype motor. On the other hand, in addition to the field flux, the PM flux penetrated to the rotor shaft of the hybrid motor. Therefore, the flux passing through the rotor shaft of the hybrid motor was higher than that of the prototype motor, resulting in a large diameter design for the rotor shaft. Figure 21 shows the B–H curve of the N39UH used for the PM and the operating points of the PM when magnetic field control was carried out in each compared model. The PM of the hybrid motor was easily demagnetized due to a low permeance coefficient caused by a wide air gap. On the other hand, it can be seen that the PM in the prototype motor was hardly demagnetized even if the same amount of the $\Psi_a$ was weakened by applying the conventional field weakening control to the IPM motor.

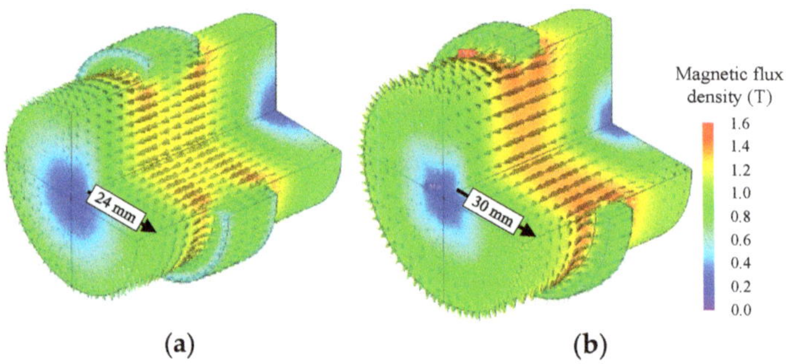


**Figure 20.** Analysis results of magnetic flux density: (**a**) prototype motor rotor shaft with $i_m$ of 10 A; (**b**) hybrid motor rotor shaft with $i_f$ of −4.2 A.

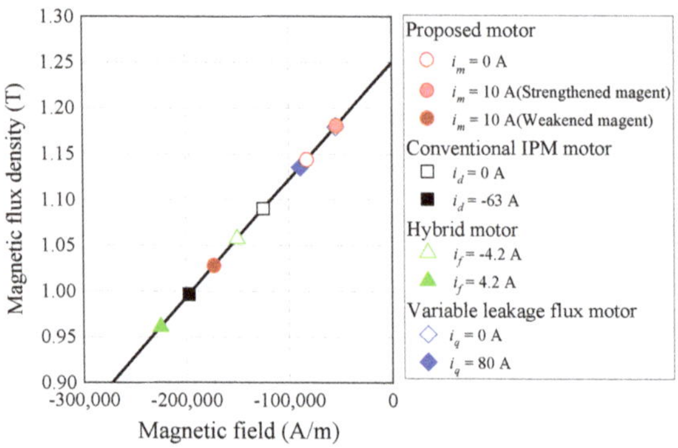


**Figure 21.** B–H characteristic of N39UH and PM operating points of each compared motor.

## 4. Drive System of Prototype Motor

From Equation (1), it can be seen that the magnetic field of the proposed motor depends on the absolute value of the $i_m$. Therefore, this paper examines magnetic field control methods by supplying the DC $i_m$.

The DC $i_m$ was controlled with the DC power supply ZX-800L, Takasago Ltd. and the armature currents were supplied by the three-phase inverter MWINV-5R022, Myway Plus Corporation. The electrical conditions and the control block diagram are shown in Table 6 and Figure 22, respectively. As shown in Table 6, the dead time was 4 μs and the error voltage of the dead time was compensated based on the reference [18,19]. In addition, as shown in Figure 22, the relationship between the $\Psi_a$ and $i_m$ expressed in Equation (1) was used for the decoupling item.

**Table 6.** Experimental conditions of current control of prototype motor.

| Parameter | Symbol | Value |
|---|---|---|
| DC-bus voltage | $V_{dc}$ | 270 V |
| Dead time | $t_d$ | 4 μs |
| Switching frequency | $f_{sw}$ | 10 kHz |
| Crossover frequency of current control | $\omega_c$ | 4000 rad/s |

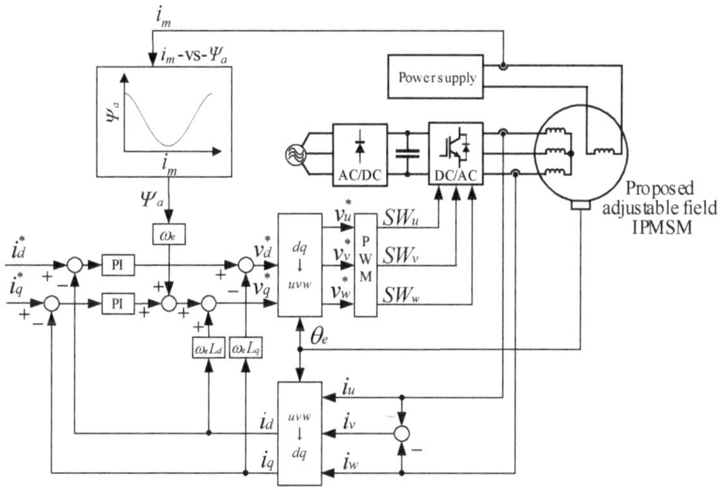


**Figure 22.** Control block diagram of current control of prototype motor when using DC $i_m$.

Figure 23 shows the experimental results at the rotating speed of 1000 r/min when an $i_m$ of 0 A or 6 A, an $i_d$ of 0 A and an $i_q$ of 20 A were given to the prototype motor. It can be confirmed from the waveforms of the three-phase line currents and $i_m$ that the line currents could be controlled regardless of the $i_m$.

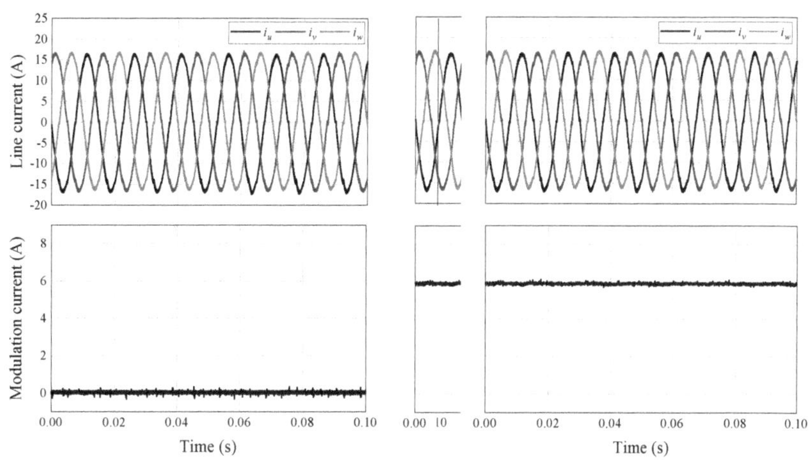


**Figure 23.** Experimental results of current control: (**a**) with $i_q$ of 20 A and $i_m$ of 0 A; (**b**) with $i_q$ of 20 A and $i_m$ of 6 A.

Table 7 shows the output torque $T_{ave}$ when a DC $i_m$ of 0 $A_{dc}$ or 6 $A_{dc}$ was given to the modulation winding. As shown in Table 7, the output torque increased from 1.49 Nm to 2.40 Nm by supplying a DC $i_m$ of 0 $A_{dc}$ or 6 $A_{dc}$. This result reveals that the prototype motor could control the $\Psi_a$ by the DC $i_m$ because the $T_{ave}$ increased with the DC $i_m$ under the same armature current condition.

**Table 7.** Comparison of output torque.

| Modulation Current | *q*-Axis Current | Output Torque |
|---|---|---|
| DC 0 $A_{dc}$ | 20 A | 1.49 Nm |
| DC 6 $A_{dc}$ | 20 A | 2.40 Nm |

In this section, based on the relationship between the $\Psi_a$ and the $i_m$, the drive system for the prototype motor is examined. The above results show that the drive system shown in Figure 22 could control the line currents independently of the DC $i_m$.

Moreover, Equation (1) means that the square wave $i_m$ can be used to output the constant $\Psi_a$ because the absolute value of the square wave is constant. The ability to control the $\Psi_a$ with the square wave $i_m$ is one of the unique points of the proposed adjustable field method. In other words, the power electronics for the proposed motor have a lot of flexibility. The proposed motor can be driven by different power electronics, optimized for magnetic field adjustment by taking advantage of the flexibility. Therefore, we will study the suitable drive system for the proposed motor in the near future. When the power electronics circuit is optimally designed for the field adjustment, the number of switching devices and the complexity of the circuit must be deeply considered because they easily deteriorate the total efficiency [20].

## 5. Load Characteristics

### *5.1. Load Analysis*

Figure 24 shows the relationships between $i_q$ and torque $T$ characteristic (I–T characteristic) of the proposed motor models with different magnetic leakage path widths shown in Figure 11 and Table 2. Figure 24a shows the I–T characteristics when the $F_m$ was not supplied. From this Figure, it can be seen that the torque constant, which is defined as the gradient of the I–T characteristic, depended on the minimum $\Psi_a$ shown in Table 3. Figure 23b shows the I–T characteristics with $F_m$. The torque constant was the same among all models because the maximum $\Psi_a$ was unified to 39.6 mWb for all models. However, the required $F_m$ was more significant for smaller minimum $\Psi_a$.

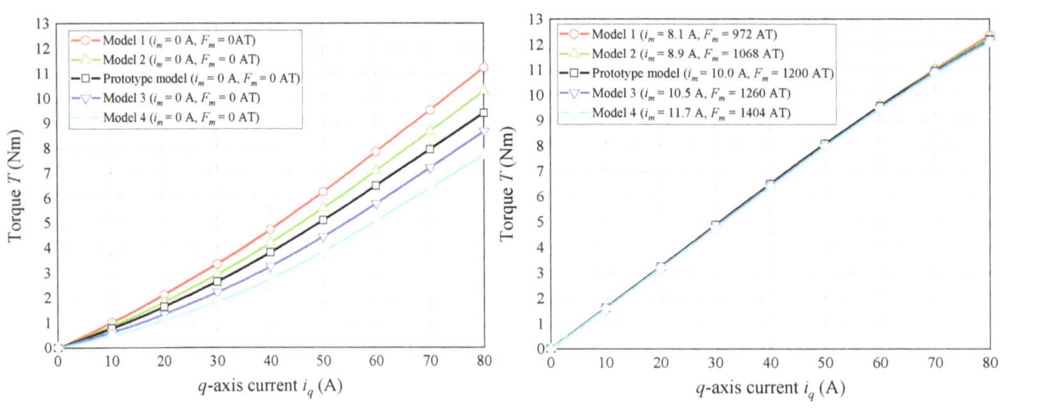


**Figure 24.** I–T characteristics of proposed motor models with different magnetic leakage path widths: (**a**) without $F_m$; (**b**) with $F_m$.

Figure 25 shows the relationships between rotating speed $N$ and $T$ characteristic (N–T characteristic) of the proposed motor models when the maximum $i_q$ of 80 A was applied. Figure 25a shows the N–T characteristics without the $F_m$. As shown in the Figure, the model with the smaller minimum $\Psi_a$ had a wider high-speed operating range. On the other hand, as can be seen in Figure 25b, for the same reason valid for the I–T characteristic, the N–T characteristics with the $F_m$ were the same for all models.

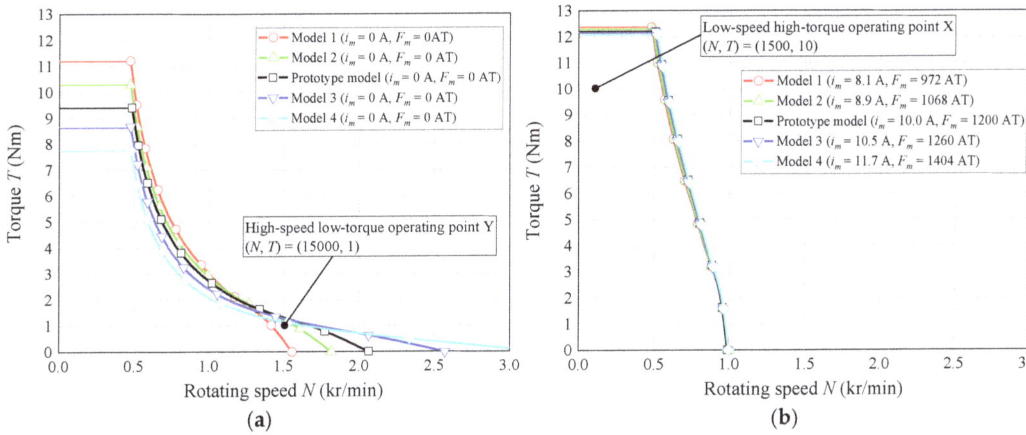


**Figure 25.** N–T characteristics of proposed motor models: (**a**) without $F_m$; (**b**) with $F_m$.

It is confirmed, from this result, that the prototype motor model could achieve the low-speed high-torque operation when the $F_m$ was given and the high-speed low-torque operation when the $F_m$ was not given. In addition, in comparing the models with the same maximum $\Psi_a$ and the different controllable range of the $\Psi_a$, it could be revealed that the copper volume and the copper loss for the magnetic field control were in a trade-off relationship with the motor operating range.

The losses at the low-speed high-torque operating point X and the high-speed low-torque operating point Y shown in Figure 25 were evaluated in comparison with the losses of other compared models shown in Figures 16–18. Table 8 shows the current conditions at each operating point. At the operating point X, the current that maximized the $\Psi_a$ of each motor was supplied and, at the operating point Y, the current that minimized the $\Psi_a$ was supplied. Figure 26 shows the loss analysis results. In this paper, the analysis method of iron loss was common to all motors. The hysteresis loss of the electromagnetic steel sheet 35JNE230 was calculated by considering its hysteresis loop. On the other hand, there are no data about the hysteresis loops of the carbon steel S45C and the PM N39UH in JMAG-Designer. Therefore, the iron loss of these parts included only the eddy current loss. Figure 26a shows the loss analysis results at the operating point X. The copper loss of the prototype motor was significant at the operating point X because the $F_m$ was necessary to cause the magnetic saturation and increase the $\Psi_a$. In addition, the copper loss of the armature winding was also the largest because the maximum $\Psi_a$ of the prototype motor was the smallest among the compared motors, as can be seen in Table 5. The modulation flux penetrated to the PM when the $F_m$ was given. However, there was almost no increase in PM eddy current loss due to the modulation flux because the modulation flux was a DC. For the same reason, additional losses of the stator frame and rotor shaft due to the modulation flux were also small. Figure 26b shows the loss analysis results at the operating point Y. Because the negative $i_d$ was supplied to the conventional motor at the operating point Y, the conventional IPM motor delivered the reluctance torque in addition to the PM torque. Therefore, the conventional IPM motor was advantageous over the other motors at the operating point Y. However, in the high-speed range, the loss of the prototype motor was the smallest among the compared motors because the additional m.m.f. was not necessary to decrease the $\Psi_a$. The PM eddy current loss of the hybrid motor was higher than the other benchmark IPM motors because of the SPM structure. Therefore, the PMs of the hybrid motor were prone to de-magnetization due to heat in addition to the temperature rise of the PMs as well as the counter magnetic field to the PMs.

**Table 8.** Current condition in load analysis.

| Operating Point | Prototype Motor | Conventional IPM Motor | Hybrid Motor | Variable Leakage Flux Motor |
|---|---|---|---|---|
| X | $i_m$ = 10 A, $i_q$ = 63 A | $i_d$ = 0.0 A, $i_q$ = 58 A | $i_f$ = 4.2 A, $i_q$ = 47 A | $i_q$ = 62 A |
| Y | $i_m$ = 0.0 A, $i_q$ = 13 A | $i_d$ = −63 A, $i_q$ = 10 A | $i_f$ = −4.2 A, $i_q$ = 7.7 A | $i_q$ = 7.5 A |

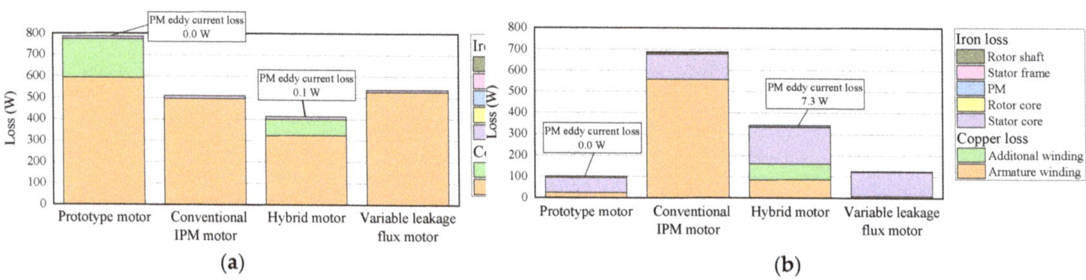


**Figure 26.** Loss analysis results: (**a**) loss at operating point X; (**b**) loss at operating point Y.

The above results show that the proposed motor could be driven at a high-speed range with less copper loss and iron loss than the field weakening control and the other adjustable field methods. However, in the analysis using FEM, there was often a difference in iron loss between the experimental and the analysis. Therefore, as a future task, it is necessary to accurately measure the drive system's efficiency and evaluate the prototype motor's performance in more detail [21].

## 5.2. Load Test

Figure 27 shows the experimental result of the I–T characteristic. In the actual load test, to evaluate the performance of the prototype motor as an adjustable field IPMSM, the $i_d$ was not included in the armature current. Besides, the maximum $i_q$ was 24 A due to the current limitation of the three-phase inverter MWINV-5R022. The prototype motor was controlled at 1000 r/min by the load motor. The $F_m$ was supplied by the DC power supply, similar to the measurement test of the back e.m.f. As shown in Figure 25, the experimental value was in good agreement with the analysis value except in the low $F_m$ condition. The torque constant could be changed according to the $F_m$. There was a difference between the experimental and analysis values when the $F_m$ was 0 AT or 360 AT. The leading cause of the difference is the decrement of $\Psi_a$ due to the production and mesh error found in the measurement test of the back e.m.f. From the above results, it is confirmed that the prototype motor could control the torque constant freely according to the $F_m$.

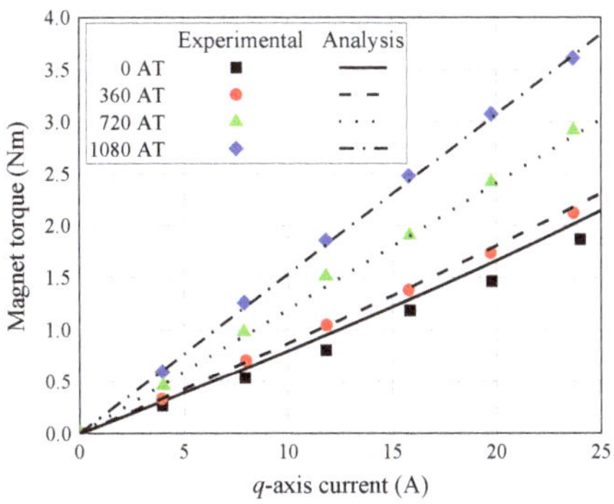


**Figure 27.** Experimental results of I–T characteristic.

Figure 28 shows the experimental result of the N–T characteristic. The phase voltage at most 30 V was supplied to the prototype motor assuming an input DC bus voltage of 60 V, because the maximum rotating speed was as low as 3000 r/min due to the speed limitation of the load system. In addition, the maximum torque was 3.7 Nm because the maximum $i_q$ was 24 A due to the current limitation of the inverter. Since the analysis did not consider iron loss and mechanical loss, the torque of the analysis result was slightly larger than that of the experimental result over the entire N–T characteristic. Especially with a large $F_m$, the harmonics were also high. Therefore, the difference between the observed and analysis values became large. In addition, when the $F_m$ was low, the measured magnet torque was slightly lower than the analyzed value because the $\Psi_a$ was smaller than expected due to the production error, similar to the I–T characteristic measurement test. However, the tendency of the N–T characteristic was in good agreement.

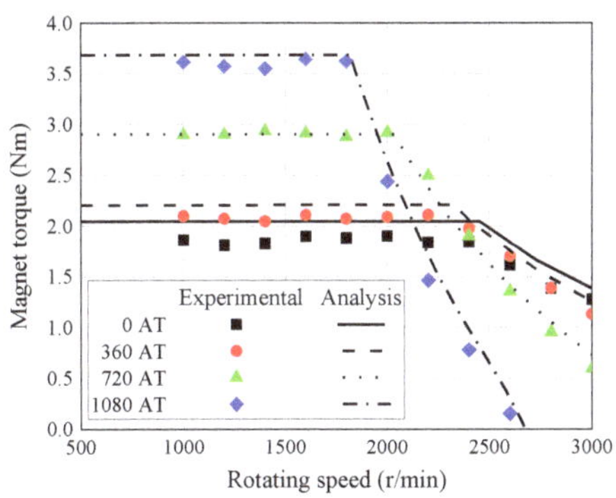


**Figure 28.** Experimental results of N–T characteristic.

Based on the above results, it can be seen that the proposed motor could achieve both the low-speed high-torque and the high-speed low-torque operation according to the $F_m$.

## 6. Conclusions

In this paper, the essential operating characteristics of the IPMSM with the proposed adjustable field method utilizing magnetic saturation were examined by conducting the measurement test of the back e.m.f. and the actual load tests.

As a result of the back e.m.f. measurement test, it was confirmed that 51.7% of the $\Psi_a$ of the prototype motor was controllable. Based on the test results, the magnetic field control performance was compared with the field weakening control and other adjustable field methods under the same amount of the magnetic field control. The proposed method has some advantages, such as a broader controllable range of magnetic field, less copper loss and anti-demagnetization capability. Detailed contributions of the proposed motor are as follows:

- Controllable range of the proposed motor magnetic field was 103% larger than that of the variable leakage flux motor, as shown in Table 5;
- Copper loss enhancement of the proposed method was up to 66.9% compared to the conventional field weakening control, as shown in Table 5;
- The proposed motor had a higher anti-demagnetization capability than the hybrid motor by 6.5%, as shown in Figures 20 and 21.

In addition, the losses of the proposed motor at two operating points, which are the low-speed high-torque point and the high-speed low-torque point, were compared with that of the benchmarks. As a result, it was seen that the copper loss of the proposed motor was the largest among the compared motors at the low-speed high-torque operating point. On the other hand, the efficiency considering the copper loss and the iron loss of the proposed motor was the highest at the high-speed low-torque operating point.

In the actual load test, the I–T characteristic and the N–T characteristic were measured. It was confirmed through the I–T characteristic measurement test that the torque constant of the prototype motor could be varied continuously according to the $F_m$. Besides, from the measurement test of the N–T characteristic, it can be seen that the prototype motor could change the operating characteristics, which are the low-speed high-torque and the high-speed low-torque, depending on the $F_m$. However, the actual load test considering the reluctance torque was not conducted. In the adjustable field PMSM, the conventional vector controls, such as maximum torque per ampere (MTPA) control, maximum torque per voltage (MTPV) control and field weakening control, considered under the constant $\Psi_a$ condition, had to be extended because the $\Psi_a$ was adjustable. Therefore, the derivation of the extended vector control algorithm for the adjustable field PMSM is an essential topic for future work.

**Author Contributions:** Conceptualization, K.I. and T.N.; methodology, K.I. and T.N.; software, K.I.; validation, K.I. and T.N.; formal analysis, K.I.; investigation, K.I.; data curation, K.I.; writing—original draft preparation, K.I.; writing—review and editing, K.I. and T.N.; visualization, K.I.; supervision, T.N.; project administration, T.N.; funding acquisition, T.N. All authors have read and agreed to the published version of the manuscript.

**Funding:** This work was supported by JSPS KAKENHI, grant number JP21K04022.

**Institutional Review Board Statement:** Not applicable.

**Informed Consent Statement:** Not applicable.

**Data Availability Statement:** Not applicable.

**Conflicts of Interest:** The authors declare no conflict of interest.

## References

1. Pellegrino, G.; Vagati, A.; Guglielmi, P.; Boazzo, B. Performance Comparison between Surface-Mounted and Interior PM Motor Drives for Electric Vehicle Application. *IEEE Trans. Ind. Appl.* **2012**, *59*, 803–811. [CrossRef]
2. Fatemi, A.; Ionel, D.M.; Popescu, M.; Chong, Y.C.; Demerdash, N.A.O. Design Optimization of a High Torque Density Spoke-Type PM Motor for a Fomula E Race Drive Cycle. *IEEE Trans. Ind. Appl.* **2018**, *54*, 4343–4354. [CrossRef]
3. Jo, C.; Seol, J.Y.; Ha, I.J. Flux-Weakening Control of IPM Motors with Significant Effect of Magnetic Saturation and Stator Resistance. *IEEE Trans. Ind. Electron.* **2008**, *55*, 1330–1340.
4. Sue, S.M.; Pan, C.T. Voltage-Constraint-Tracking-Based Field-Weakening Control of IPM Synchronous Motor Drives. *IEEE Trans. Ind. Electron.* **2008**, *55*, 340–347. [CrossRef]
5. Mizuno, T.; Nagayama, K.; Ashikaga, T.; Kobayashi, T. Basic Principles and Characteristics of Hybrid Excitation Type Synchronous Machine. *IEEJ Trans. Ind. Appl.* **1995**, *115*, 1402–1411. [CrossRef]
6. Tapia, J.A.; Leonardi, F.; Lipo, T.A. Consequent-Pole Permanent-Magnet Machine with Extended Field-Weakening Capability. *IEEE Trans. Ind. Appl.* **2003**, *39*, 1704–1709. [CrossRef]

7. Kato, T.; Hijikata, H.; Minowa, M.; Akatsu, K.; Lorenz, R.D. Design Methodology for Variable Leakage Flux IPM for Automobile Traction Drives. In Proceedings of the 2014 Energy Conversion Congress and Exposition–ECCE, Pittsburgh, PA, USA, 14–18 September 2014; pp. 3548–3555.
8. Aoyama, M.; Noguchi, T. Automatic Variable Magnetic Flux Technique in Consequent Pole Type PM-Motor Utilizing Space Harmonic. In Proceedings of the 2018 International Symposium on Power Electronics, Electrical Drives, Automation, and Motion–SPEEDAM, Amalfi, Italy, 20–22 June 2018; pp. 1315–1320.
9. Hijikata, H.; Saka, Y.; Akatsu, K.; Miyama, Y.; Arita, H.; Daikoku, A. Multi-Phase Inverter-Fed MATRIX Motor for High Efficiency Driving. *IEEJ Trans. Ind. Appl.* **2018**, *138*, 257–264. [CrossRef]
10. Sakai, K.; Hashimoto, H.; Kuramochi, S. Principle and Basic Characteristics of Variable-Magnetic-Force Memory Motors. *IEEJ Trans. Ind. Appl.* **2011**, *131*, 53–60. [CrossRef]
11. Zhu, X.; Xiang, Z.; Quan, L.; Wu, W.; Du, Y. Multimode Optimization Design Methodology for a Flux-Controllable Stator Permanent Magnet Memory Motor Considering Driving Cycles. *IEEE Trans. Ind. Electron.* **2018**, *65*, 5353–5366. [CrossRef]
12. Iwama, K.; Noguchi, T. Performance Comparison between Adjustable Field IPM Motor Based on Permeability Modulation Technique and Conventional IPM Motor. In Proceedings of the 23rd International Conference on Electrical Machines and Systems–ICEMS, Hamamatsu, Japan, 24–27 November 2020; pp. 1418–1423.
13. Chen, X.; Wang, J.; Sen, B.; Lazari, P.; Sun, T. A High-Fidelity and Computationally Efficient Model for Interior Permanent-Magnet Machines Considering the Magnetic Saturation, Spatial Harmonics, and Iron Loss Effect. *IEEE Trans. Ind. Electron.* **2015**, *62*, 4044–4055. [CrossRef]
14. Liang, P.; Chai, F.; Li, Y.; Pei, Y. Analytical prediction of magnetic field distribution in spoke-type permanent-magnet synchronous machines accounting for bridge saturation and magnet shape. *IEEE Trans. Ind. Electron.* **2017**, *64*, 3479–3488. [CrossRef]
15. Hemeida, A.; Lehikoinen, A.; Rasilo, P.; Vansompel, H.; Belahcen, A.; Arkkio, A.; Sergeant, P. A Simple and Efficient Quasi-3-D Magnetic Equivalent Circuit for Surface Axial Flux Permanent Magnet Synchronous Machines. *IEEE Trans. Ind. Electron.* **2019**, *66*, 8318–8333. [CrossRef]
16. Alipour-Sarabi, R.; Nasiri-Gheidari, Z.; Oraee, H. Development of a three-dimensional magnetic equivalent circuit model for axial flux machines. *IEEE Trans. Ind. Electron.* **2020**, *67*, 5758–5767. [CrossRef]
17. Shokrollahi, H.; Janghorban, K. Soft magnetic composite materials (SMCs). *J. Mater. Process. Technol.* **2007**, *189*, 1–12. [CrossRef]
18. Diao, N.; Sun, X.; Song, C.; Zhang, Q.; Zhang, Z. A Multi-modulation Times SVPWM for Dead Time Effect Elimination in Three-level Neutral Point Clamped Converters. *IEEE Trans. Ind. Electron.* **2021**, *68*, 5476–5485. [CrossRef]
19. Oto, Y.; Noguchi, T.; Sasaya, T.; Yamada, T.; Kazaoka, R. Space Vector Modulation of Dual-Inverter System Focusing on Improvement of Multilevel Voltage Waveforms. *IEEE Trans. Ind. Electron.* **2019**, *66*, 9139–9148. [CrossRef]
20. Mapelli, F.L.; Tarsitano, D.; Mauri, M. Plug-In Hybrid Electric Vehicle: Modeling, Prototype Realization, and Inverter Losses Reduction Analysis. *IEEE Trans. Ind. Electron.* **2009**, *57*, 598–607. [CrossRef]
21. Caruso, M.; Tommaso, A.O.D.; Lisciandrello, G.; Mastromauro, R.A.; Miceli, R.; Nevoloso, C.; Spataro, C.; Trapanese, M. A General and Accurate Measurement Procedure for the Detection of Power Losses Variations in Permanent Magnet Synchronous Motor Drives. *Energies* **2020**, *13*, 5770. [CrossRef]

*electronics*

MDPI

*Article*

# Investigation of AlGaN Channel HEMTs on β-$Ga_2O_3$ Substrate for High-Power Electronics

**A. Revathy [1,*], C. S. Boopathi [1,*], Osamah Ibrahim Khalaf [2] and Carlos Andrés Tavera Romero [3]**

[1] Department of Electrical and Electronics Engineering, SRM Institute of Science and Technology, Chennai 603203, India

[2] Al-Nahrain Nano Renewable Energy Research Center, Al-Nahrain University, Baghdad 10072, Iraq; usama81818@nahrainuniv.edu.iq

[3] COMBA R&D Laboratory, Faculty of Engineering, Universidad Santiago de Cali, Cali 76001, Colombia; carlos.tavera00@usc.edu.co

* Correspondence: arevathynew@gmail.com (A.R.); boopathc1@srmist.edu.in (C.S.B.)

**Abstract:** The wider bandgap AlGaN (Eg > 3.4 eV) channel-based high electron mobility transistors (HEMTs) are more effective for high voltage operation. High critical electric field and high saturation velocity are the major advantages of AlGaN channel HEMTs, which push the power electronics to a greater operating regime. In this article, we present the DC characteristics of 0.8 μm gate length ($L_G$) and 1 μm gate-drain distance ($L_{GD}$) AlGaN channel-based high electron mobility transistors (HEMTs) on ultra-wide bandgap β-$Ga_2O_3$ Substrate. The β-$Ga_2O_3$ substrate is cost-effective, available in large wafer size and has low lattice mismatch (0 to 2.4%) with AlGaN alloys compared to conventional SiC and Si substrates. A physics-based numerical simulation was performed to investigate the DC characteristics of the HEMTs. The proposed HEMT exhibits sheet charge density ($n_s$) of $1.05 \times 10^{13}$ $cm^{-2}$, a peak on-state drain current ($I_{DS}$) of 1.35 A/mm, DC transconductance ($g_m$) of 277 mS/mm. The ultra-wide bandgap AlGaN channel HEMT on β-$Ga_2O_3$ substrate with conventional rectangular gate structure showed 244 V off-state breakdown voltage ($V_{BR}$) and field plate gate device showed 350 V. The AlGaN channel HEMTs on β-$Ga_2O_3$ substrate showed an excellent performance in $I_{ON}/I_{OFF}$ and $V_{BR}$. The high performance of the proposed HEMTs on β-$Ga_2O_3$ substrate is suitable for future portable power converters, automotive, and avionics applications.

**Keywords:** AlGaN channel; β-$Ga_2O_3$ Substrate; breakdown voltage; high-power electronics; wide bandgap semiconductors

**Citation:** Revathy, A.; Boopathi, C.S.; Khalaf, O.I.; Romero, C.A.T. Investigation of AlGaN Channel HEMTs on β-$Ga_2O_3$ Substrate for High-Power Electronics. *Electronics* **2022**, *11*, 225. https://doi.org/10.3390/electronics11020225

Academic Editor: Amjad Anvari-Moghaddam

Received: 25 November 2021
Accepted: 6 January 2022
Published: 12 January 2022



## 1. Introduction

Group III-nitride-based heterostructure devices showed excellent performance in high-power switching electronics [1–3]. Owing to the unique material properties of GaN such as high breakdown field (3.3 MV/cm), high mobility (>1200 $cm^2$/v.s), and wide bandgap (3.4 eV), the GaN-based high electron mobility transistors (HEMTs) are exhibited high 2DEG density (two-dimensional electron gas), low on-resistance ($R_{on}$), low conduction loss, and high breakdown voltage ($V_{BR}$). Which enable the devices for high-temperature and high-power switching applications [4,5]. However, it is difficult to obtain high breakdown voltage for GaN-channel based HEMTs for smaller dimensions for portable high power switching applications like power converters, automotive electronics, and avionics. Therefore, further enhancing the power handling capability of the HEMTs, the ultra-wide bandgap (>3.4 eV) $Al_xGa_{1-x}N$ channel-based devices are widely used [6–11]. There is a wide bandgap and superior thermal stability in AlGaN material adopted as a channel for improving the high-power handling and high-temperature operations of HEMTs. Gate length 2 μm ($L_G$) AlN/$Al_{0.5}Ga_{0.5}N$ HEMT on sapphire substrate demonstrated remarkable breakdown voltage improvements [6]. $L_G$ = 0.8 μm and $L_{GD}$ = 1 μm $Al_{0.31}Ga_{0.69}N/Al_{0.1}Ga_{0.9}N$ double-channel HEMTs recorded a $V_{BR}$ of 143.5 V [8]. However, the device has shown a large negative threshold voltage ($V_{TH}$ = −11.2 V), which will increase the power loss of the device at off-state. An Al-rich AlN/$Al_{0.85}Ga_{0.15}N$ HEMTs on sapphire substrate showed 810 V of breakdown voltage ($V_{BR}$) for a gate-drain distance of ($L_{GD}$) of 10 μm [9]. Another Al-rich AlN/$Al_{0.65}Ga_{0.35}N$ HEMTs on the sapphire substrate demonstrated 770 V off-state blocking voltage for 9 μm $L_{GD}$ [10]. $L_G$ = 3 μm $Al_{0.3}Ga_{0.7}N$/$Al_{0.1}Ga_{0.9}N$ heterostructure on sapphire substrate exhibited excellent sheet carrier density and electron mobility [11], and hybrid ohmic/Schottky drain device configuration showed excellent off-state breakdown voltage. Despite high off-state blocking voltage of Al-rich AlGaN channel HEMTs [6,9,10], the current driving capability is very low due to alloy disorder scattering effects. On other hand, low Al composition AlGaN channel-based HEMTs [8,11] exhibited excellent sheet charge density and carrier mobility.

Apart from channel engineering, substrate engineering is one of the key issues in group III-nitride-based HEMTs. Sapphire (13% lattice mismatch with GaN), SiC (3.8% lattice mismatch with GaN), and Si (16% lattice mismatch with GaN) are widely used as substrate materials for GaN-based HEMTs [1–11]. Due to lattice mismatch between buffer and substrate materials, an additional AlN layer is required in between for lattice match, which involves additional costs and increases the device fabrication complexity.

Recently, β-$Ga_2O_3$ is an emerging material for power electronics applications. The wide bandgap (4.7 eV), high critical electric field (6–8 MV/cm), large-scale, high-quality bulk substrate, low defect density, and nearly lattice match (0 to 2.4% lattice mismatch for AlGaN alloys) with III-Nitride alloys make β-$Ga_2O_3$ a promising material for future high-power applications [12–18]. A numerical study on normally-off AlN/β-$Ga_2O_3$ based HEMT with GaN back barrier showed a suppressed leakage current and excellent DC characteristics [19]. In spite of high-power performance, the β-$Ga_2O_3$ channel-based HEMTs experienced low electron mobility and low sheet charge concentration [18] as a result of low current driving capability. In this work, we used β-$Ga_2O_3$ as a substrate for the following reasons. 1. Low lattice mismatch with AlGaN alloys and therefore an additional thick AlN nucleation layer is not required 2. Relatively low cost compared to GaN and SiC, and 3. Availability of large wafer size. The novelty of the work is gate field plate AlGaN channel HEMTs on β-$Ga_2O_3$ substrate.

The first $Al_{0.1}Ga_{0.9}N$ channel-based HEMTs on the β-$Ga_2O_3$ substrate is proposed in this work for simultaneous improvement in both power handling and current driving capability of HEMTs. The proposed gate field plate $Al_{0.31}Ga_{0.69}N$/$Al_{0.1}Ga_{0.9}N$ HEMTs on β-$Ga_2O_3$ substrate is investigated using ATLAS TCAD [20] and the device DC characteristics are presented. The device simulation models are validated with the experimental results. The field plate gate low Al composition (Al = 10%) AlGaN channel HEMT on β-$Ga_2O_3$ substrate showed remarkable improvement in breakdown voltage ($V_{BR}$), on-state drain current density, and low on-resistance.

## 2. Device Structure and Simulation Model

Figure 1a shows the schematic cross-section of the conventional gate $Al_{0.1}Ga_{0.9}N$ channel HEMT (Device A) on β-$Ga_2O_3$ substrate and field plate gate HEMT (Device B) is displayed in Figure 1b. Both device's epi-stack consist of 100 nm $Al_{0.1}Ga_{0.9}N$ channel, 23 nm $Al_{0.31}Ga_{0.69}N$ barrier, and 2.2 μm $Al_{0.31}Ga_{0.69}N$ buffer, 100 nm SiN passivation ($\varepsilon_r$~7). Low Al-composition $Al_{0.1}Ga_{0.9}N$ channel is used in this to improve the sheet charge density (2DEG) and enhance electron mobility (μ) [8,11]. The $Al_{0.1}Ga_{0.9}N$ layer is directly grown on the β-$Ga_2O_3$ substrate. A low lattice mismatch (<2.4%) between AlGaN buffer and β-$Ga_2O_3$ substrate alleviates the interface defects and enhanced the device performance. Moreover, the β-$Ga_2O_3$ wafer is available in a larger size and low cost compared with SiC and GaN [12–18]. Device gate length is defined as 0.8 μm ($L_G$) and asymmetrical gate to source ($L_{GS}$ = 0.8 μm) and gate to drain distance ($L_{GD}$ = 1 μm) is considered for this work in order to reduce the source resistance. To analyze the impact of filed plate structure on the breakdown performance of the AlGaN channel HEMT, a 0.75 μm field plate ($L_{FP}$) was used in Device B. Ohmic contact was realized for source and drain contact by defining the work function of the electrode as 3.4 eV and Schottky gate contact enabled by defining electrode work function as 5.2 eV. Field plate gate structure used for the proposed HEMT for improving the off-state blocking voltage of the device by re-shaping the electric field across the gate to the drain access region. The SiN passivation avoids the surface-related trapping effects and avoids the current collapse phenomena.

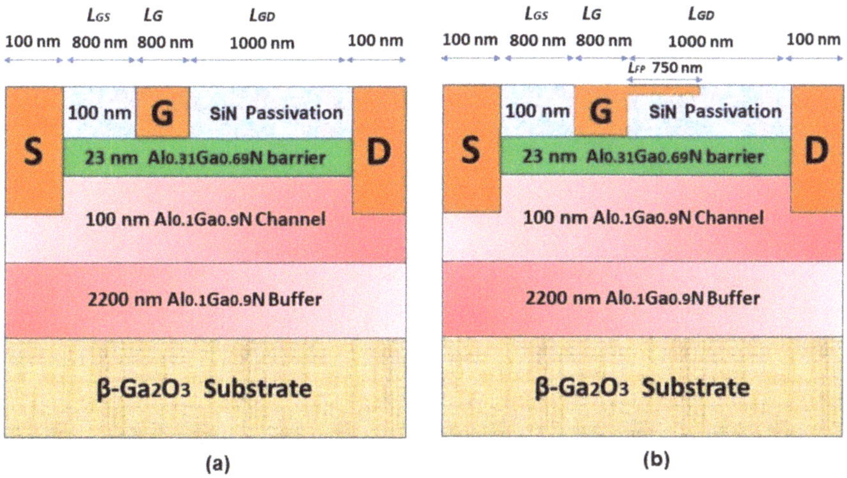


**Figure 1.** (**a**) Schematic of Conventional HEMT. (**b**) Gate field plate HEMT.

The TCAD simulation energy band diagram is plotted in Figure 2 at zero gate bias and drain bias. It shows the conduction band discontinuity at the interface of $Al_{0.31}Ga_{0.69}N/Al_{0.1}Ga_{0.9}N$ and forms a 2DEG due to spontaneous and piezoelectric polarization effects. The sheet charge density ($n_s$) of $1.05 \times 10^{13}$ cm$^{-2}$ extracted from the device simulation. The corresponding polarization charge details are displayed in Figure 3. The material parameters for the simulation are taken from [16–22] and are presented in Table 1.

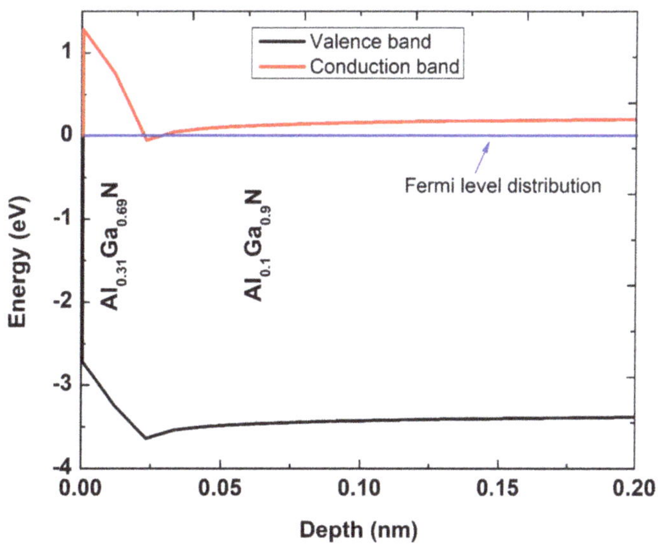


**Figure 2.** Energy band details of heterostructure at no bias conditions.

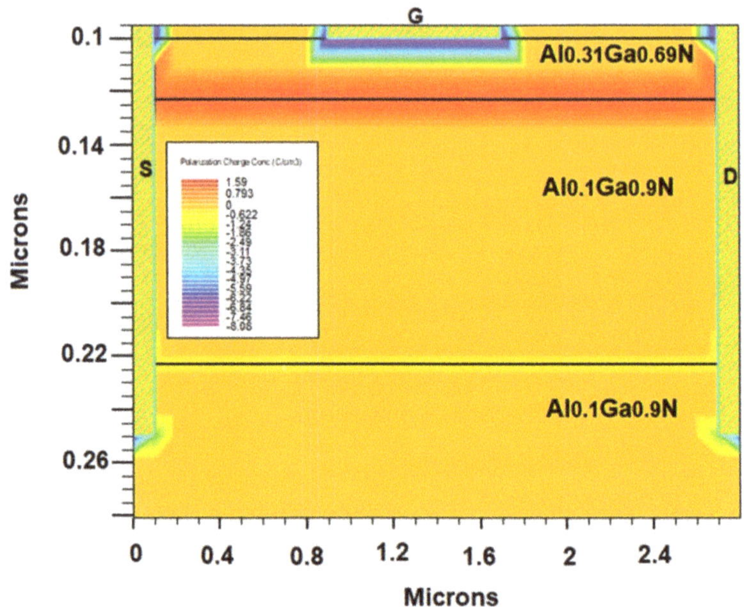


**Figure 3.** Polarization charge concentration details.

**Table 1.** Material parameters used for the TCAD simulation.

| Material Parameters | GaN | AlN | β-$Ga_2O_3$ |
|---|---|---|---|
| Bandgap, $E_g$ (eV) | 3.4 | 6.2 | 4.85 |
| Dielectric constant, ε | 9 | 8.3 | 10 |
| Electron mobility, μ ($cm^2$/V.s) | 1250 | 300 | 300 |
| Saturation velocity, v (cm/s) | 2.5 | 1.6 | 1.8 |
| Thermal conductivity, λ (W/cm.K) | 2.3 | 2.85 | 0.3 |

The basic device physics models that operate on any semiconductor devices have been derived from Schrodinger's equation, Shockley–Read–Hall (SRH) recombination model, Poisson's equations, the continuity equations, and the transport equations. The quantum electron density relies on the one-dimensional (1D) Schrodinger's equation, which relates the eigen state energies $E_{iv}(x)$ and wave function $\varphi_{iv}(x,y)$.

$$-\frac{h^2}{2}\frac{\partial}{\partial y}\left(\frac{1}{m_y^v(x,y)}\frac{\partial \varphi_{iv}}{\partial y}\right) + E_C(x,y)\varphi_{iv} = E_{iv}\varphi_{iv} \tag{1}$$

where $m_y^v(x,y)$ indicates spatial dependent effective mass and $E_C(x,y)$ represents the conduction band edge. Poisson's equations relate the electric field (E), electrostatic potential (ψ), and space charge concentration ($\rho$). Which is written as follows [20];

$$\nabla^2\psi = -\nabla E = \frac{\rho}{\varepsilon} \tag{2}$$

The continuity and transport equations describe the time-dependent electron ($n$) and hole ($p$) densities due to carrier transport, carrier generation ($G_n$ and $G_p$), and carrier recombination ($R_n$ and $R_p$) process. The continuity equations are defined by the following equations [20];

$$\frac{\partial n}{\partial t} = \frac{1}{q}\nabla J_n + G_n - R_n \tag{3}$$

$$\frac{\partial p}{\partial t} = -\frac{1}{q}\nabla J_p + G_p - R_p \tag{4}$$

The drift-diffusion carrier transport model is based on Boltzmann transport theory, which relates the current densities ($J_n$ and $J_p$) with quasi-Fermi levels ($\varnothing_n$ and $\varnothing_p$) and the quasi-Fermi levels are related with carrier concentrations and potentials through Boltzmann approximations [20].

$$J_n = -q\mu_n n\nabla\varnothing_n \tag{5}$$

$$J_p = -q\mu_p p\nabla\varnothing_p \tag{6}$$

The polarization property of group III-nitride is the major source for 2DEG, the spontaneous ($P_{SP}$) and piezoelectric polarization ($P_{PZ}$) models are used for the device simulation. The carrier mobility impacts the device on-state current density ($I_{DS}$) and transconductance ($g_m$). Therefore, the temperature, composition, and doping dependent Farahmand Modified Caughey Thomas low field mobility model (FMCT.N) and nitride specific high field mobility (GANSAT.N) models [20] are used in this work. In order to capture the trap-assisted recombination due to crystal defects and dopant, Shockley–Read–Hall (SRH) recombination model is used for the device simulation. Which is described by the following equations [20];

$$R_{net}^{SRH} = \frac{n - n_{ie}^2}{\tau_p\left[p + n_{ie}\exp\left(\frac{-E_{trap}}{KT}\right)\right] + \tau_n\left[p + n_{ie}\exp\left(\frac{-E_{trap}}{KT}\right)\right]}, \tag{7}$$

where $\tau_p$ and $\tau_n$ are electron and hole lifetime, respectively, and $E_{trap}$ defines the intrinsic Fermi energy level to trap energy level. The impact ionization model described by Selberherr used for device breakdown simulation, which relates the carrier generation ($G$) rate with electron ($\alpha_n$) and hole ionization coefficients ($\alpha_p$), and electron ($J_n$) and hole current densities ($J_p$) [20];

$$G = \alpha_p|J_p| + \alpha_n|J_n| \tag{8}$$

## 3. Results and Discussions

The current driving ($J = qvn_s$) capability of the HEMTs relies on the electron velocity ($v$), sheet charge, and 2DEG density ($n_s$). The proposed HEMT showed $1.12 \times 10^7$ cm/s of electron velocity and sheet charge density ($n_s$) of $1.05 \times 10^{13}$ cm$^{-2}$. The DC characteristics of proposed AlGaN channel HEMTs on β-$Ga_2O_3$ are presented in this section. The device output characteristics is plotted in Figure 4 for $V_{GS}$ = −4 V to 8 V and $V_{DS}$ from 0 V to 20 V. The HEMT showed a peak on-state drain current ($I_{DS}$) of 1.35 A/mm at $V_{GS}$ = 2 V and the extracted ON-resistance ($R_{ON} = \Delta V_{DS}/\Delta I_{DS}$) from the V-I curve corresponding to $V_{GS}$ = 0 V is 8.83 Ω.mm. The DC transfer characteristics of HEMT is displayed in Figure 5a at $V_{DS}$ = 10 V and $V_{GS}$ swept from −10 V to 8 V. A peak transconductance ($g_m$) of 277 mS/mm obtained at $V_{GS}$ = −1.5 V and extracted a threshold voltage of −4.31 V. The device transfer characteristics log-scale plot displayed in Figure 5b and the HEMT showed an excellent $I_{ON}/I_{OFF}$ ratio of ~$10^{14}$. The obtained $I_{DS}$ and $g_m$ are the highest results among the $Al_xGa_{1-x}N$ channel HEMTs [6–11]. The TCAD simulation is validated with the $L_G$ = 0.8 µm gate length AlGaN channel HEMT experimental results for transfer and breakdown performance [8] and depicted in Figure 6, which shows the simulation results are well correlated with experimental results.

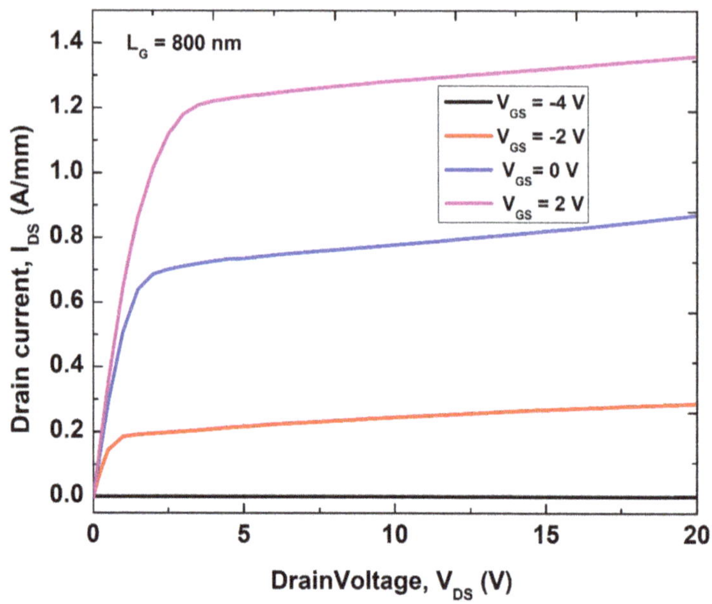


**Figure 4.** Output characteristics.

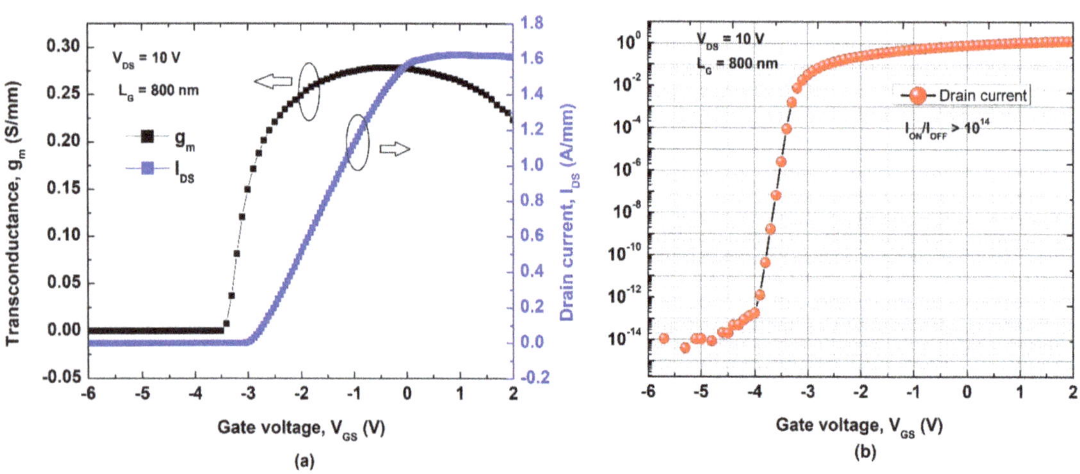


**Figure 5.** (**a**) Input characteristics. (**b**) Input characteristics in log-scale plot.

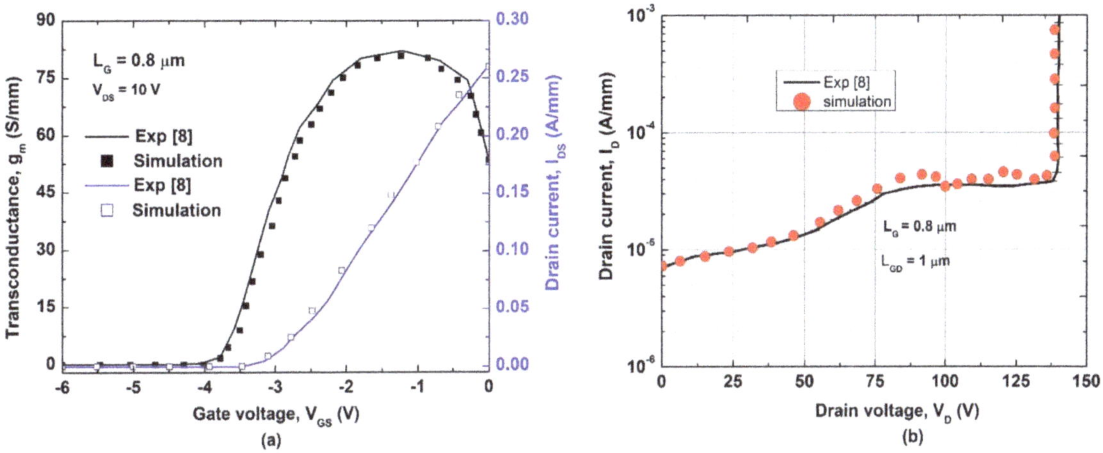


**Figure 6.** Simulation models validation with experimental results (**a**) Transfer characteristics, (**b**) Breakdown characteristics [8].

For power electronics applications, low ON-resistance ($R_{ON}$) and high breakdown voltage ($V_{BR}$) are the most critical parameters to minimize the resistive losses during high power operation of HEMTs, which determine the efficiency of the power devices. In III-nitride-based HEMTs, the 2DEG is induced by polarization, and the breakdown voltage depends on the gate-to-drain distance ($L_{GD}$). The ON-resistance of the HEMT expressed as following expressions [23];

$$R_{on} = R_{channel} + R_{drain} = \frac{L_G}{W_G} \cdot \frac{1}{q\mu n_s} + \frac{L_{GD}}{W_G} \cdot \frac{1}{q\mu n_s} \tag{9}$$

where $R_{channel}$ is the channel resistance and $R_{drain}$ is drain side access resistance. Primarily, the off-state breakdown voltage depends on $L_{GD} = \frac{V_{BR}}{E_{crit}}$ when source and drain contact resistances are neglected. In general, longer gate length and larger gate-drain spacing HEMTs typically exhibit high $V_{BR}$ but the ON-resistance also increased with $L_G$ [23]. Shorter gate length and small $L_{GD}$ HEMTs showed low breakdown voltages. Hence, the optimization of HEMTs structure is the key component for achieving high breakdown field and low $R_{ON}$ for future low loss power semiconductor devices with smaller device sizes. In this work, we used $L_G$ = 800 nm gate length and $L_{GD}$ = 1000 nm ultra-wide bandgap AlGaN channel along with gate field structure. In addition, the obtained results were compared with existing HEMTs with same $L_G$ and $L_{GD}$ [8].

The high critical electric field of AlGaN channel (~4×) HEMTs to GaN-channel HEMTs leads to high voltage operation of the device. Figure 7 shows the forward blocking characteristics of the conventional AlGaN channel and gate field plate AlGaN channel HEMTs. The off-state breakdown voltage ($V_{BR}$) is extracted from $V_D$-$I_D$ when drain leakage current reaches 1 mA/mm. As shown in Figure 7, $L_G$ = 0.8 μm conventional AlGaN channel HEMT on β-$Ga_2O_3$ substrate demonstrated 244 V of $V_{BR}$ and $L_G$ = 0.8 μm and $L_{FP}$ = 0.75 μm gate field plate HEMT on β-$Ga_2O_3$ substrate showed a $V_{BR}$ of 350 V. The field plate HEMT modulates the electric field in the device access regions (gate-drain space) and able to sustain high electric fields, which leads to enhanced breakdown voltage. Logarithmic electron concentration of the HEMTs is displayed in Figure 8 at the breakdown condition. The Depletion region of gate field plate HEMT in Figure 8b is larger than the conventional gate HEMT Figure 8a. This represents that the higher depletion width leads to high $V_{BR}$ and the smaller depletion width would decrease the $V_{BR}$ of the HEMT. Gate field plate AlGaN channel HEMT on β-$Ga_2O_3$ substrate exhibited excellent breakdown characteristics and low ON-resistance than existing AlGaN channel HEMT with identical device size [8]. This reveals the potential of AlGaN channel HEMTs on β-$Ga_2O_3$ substrate for future power switching applications with smaller device sizes. The comparison of proposed AlGaN channel HEMTs breakdown performance with previously reported AlGaN channel HEMTs [8–11,24–29] are presented in Figure 9. Our device showed excellent breakdown voltage than existing HEMTs for smaller gate length ($L_G$) and gate-drain distance ($L_{GD}$).

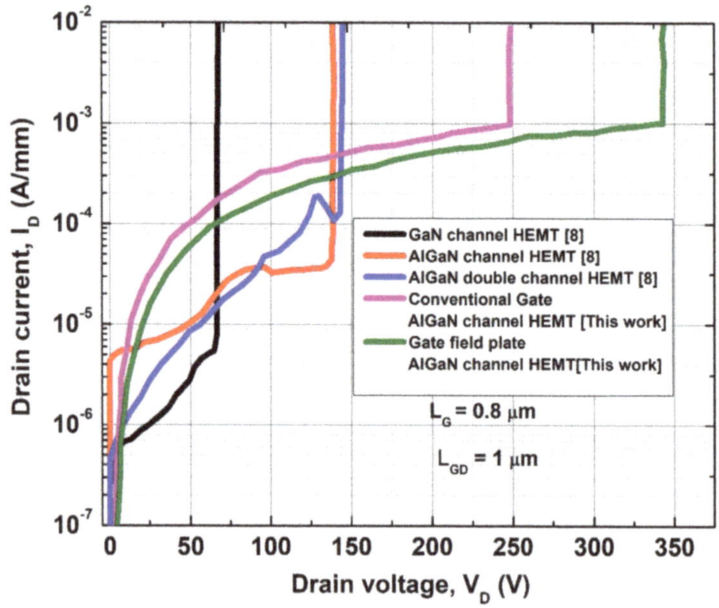


**Figure 7.** Off-state breakdown characteristics.

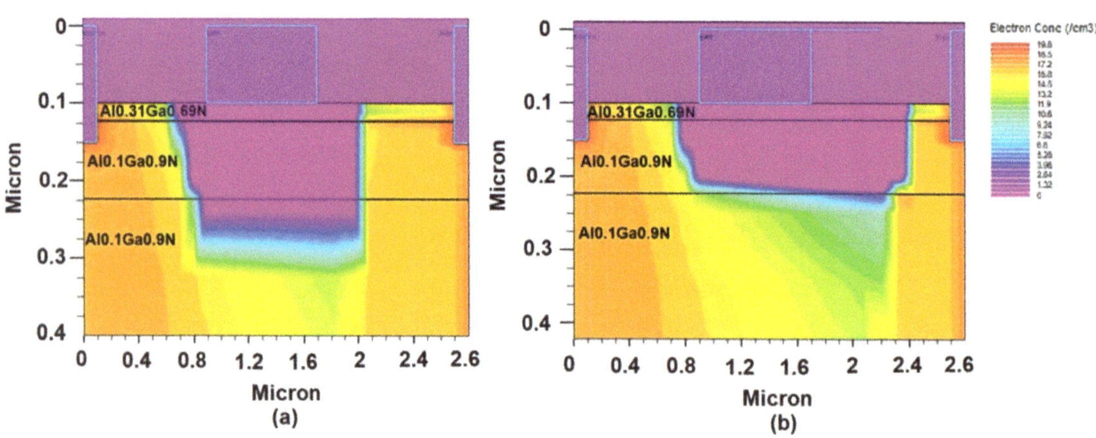


**Figure 8.** (**a**) Logarithmic electron concentration of conventional HEMT. (**b**) Logarithmic electron concentration of gate field plate HEMT.

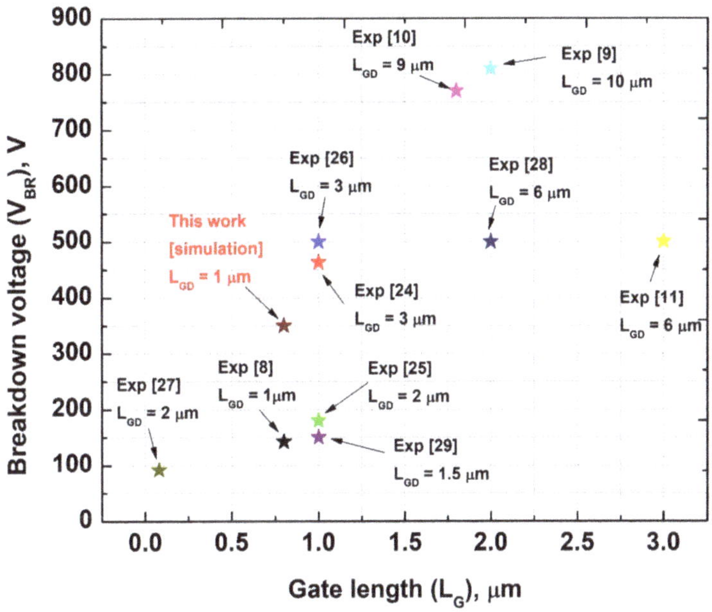


**Figure 9.** Benchmark of breakdown voltage for AlGaN channel HEMTs.

The tentative fabrication steps of the proposed HEMT are depicted in Figure 10. The epitaxial growth can be performed by the MOCVD process (Metal Organic Chemical Vapor Deposition) on the β-$Ga_2O_3$ substrate. Device mesa isolation will reduce the reducing leakage currents. This can be achieved through dry etching ICP-RIE (Inductively Coupled Plasma Reactive Ion Etching). The patterned deposition of source and drain contacts will be conducted through optical lithography and e-beam evaporation followed by high-temperature annealing. The minimum feature size used is 1 micron hence patterning is possible through optical lithography for lower feature sizes one has to depend on e-beam lithography. Then, patterning and deposition of gate metal followed by contact annealing at a lower temperature to ensure Schottky nature of the contact. Blanket deposition of silicon nitride through PECVD (Plasma Enhanced Chemical Vapor Deposition) and ALD (Atomic Layer Deposition) over the entire substrate. Selective etching of SiN for source, gate, and drain pad openings through ICP-RIE.

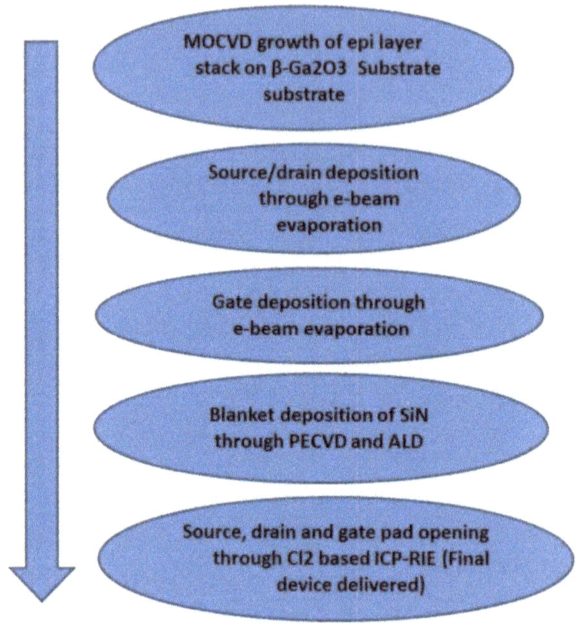


**Figure 10.** Tentative fabrication flow of AlGaN channel HEMTs.

## 4. Conclusions

AlGaN channel HEMTs on β-$Ga_2O_3$ substrate are proposed and the device DC characteristics with rectangular and gate field plate structures are analyzed. The wide-bandgap AlGaN channel along with field plate gate HEMT showed improved breakdown voltage ($V_{BR}$) of 350 V for $L_G$ = 0.8 μm and $L_{GD}$ = 1 μm, whereas the conventional gate HEMT with identical device dimensions showed 244 V of $V_{BR}$. The improved breakdown voltage from the larger depletion region at breakdown condition in field plate HEMT. The low Al composition $Al_{0.1}Ga_{0.9}N$ channel also demonstrated a peak ON-state current density of 1.35 A/mm, 8.83 Ω.mm of $R_{ON}$, and $g_m$ of 277 mS/mm. TCAD simulation and DC characteristics of the β-$Ga_2O_3$ substrate-based AlGaN channel HEMTs indicates the suitability of proposed HEMTs for future portable less weight power converters, motor driver, consumer electronics, automotive, and avionics applications. In addition, the β-$Ga_2O_3$ substrate can be utilized for all group III-nitride-based transistors for better performance.

**Author Contributions:** Conceptualization, methodology, software, validation, A.R.; formal analysis, investigation, resources, data curation, writing—original draft preparation, supervision, C.S.B.; writing—review and editing, visualization, O.I.K.; project administration, funding acquisition, C.A.T.R. All authors have read and agreed to the published version of the manuscript.

**Funding:** This research has been funded by Dirección General de Investigaciones of Universidad Santiago de Cali under call No. 01-2021.

**Institutional Review Board Statement:** Not applicable.

**Informed Consent Statement:** Not applicable.

**Data Availability Statement:** The data presented in this study will be provided based on reasonable request to corresponding authors.

**Acknowledgments:** The authors would like to thank to the SRM Institute of Science and Technology, Chennai and Universidad Santiago de Cali for providing necessary support to carry out the research work.

**Conflicts of Interest:** The authors declare no conflict of interest.

## References


1. Ma, C.T.; Gu, Z.H. Review of GaN HEMT Applications in Power Converters over 500 W. *Electronics* **2019**, *8*, 1401. [CrossRef]
2. Koksaldi, O.S.; Haller, J.; Li, H.; Romanczyk, B.; Guidry, M.; Wienecke, S.; Keller, S.; Mishra, U.K. N-Polar GaN HEMTs Exhibiting Record Breakdown Voltage Over 2000 V and Low Dynamic On-Resistance. *IEEE Electron Device Lett.* **2018**, *39*, 1014–1017. [CrossRef]
3. Huang, X.; Liu, Z.; Li, Q.; Lee, F.C. Maximizing the performance of 650-V p-GaN gate HEMTs: Dynamic $R_{ON}$ characterization and circuit design considerations. *IEEE Trans. Power Electron.* **2017**, *32*, 5539–5549.
4. Lidow, A.; Strydom, J.; de Rooij, M.; Reusch, D. *GaN Transistors for Efficient Power Conversion*; Willey: New York, NY, USA, 2015.
5. Meneghini, M.; Gaudenzio, M.; Zanoni, E. *Power GaN Devices—Materials, Applications and Reliability*; Springer: New York, NY, USA, 2017.
6. Abid, I.; Mehta, J.; Cordier, Y.; Derluyn, J.; Degroote, S.; Miyake, H.; Medjdoub, F. AlGaN Channel High Electron Mobility Transistors with Regrown Ohmic Contacts. *Electronics* **2021**, *10*, 635. [CrossRef]
7. Kaplar, R.J.; Allerman, A.A.; Armstrong, A.M.; Crawford, M.H.; Dickerson, J.R.; Fischer, A.J.; Baca, A.G.; Douglas, E.A. Review—Ultra-Wide-Bandgap AlGaN Power Electronic Devices. *ECS J. Solid State Sci. Technol.* **2017**, *6*, Q3061–Q3066. [CrossRef]
8. Zhang, Y.; Li, Y.; Wang, J.; Shen, Y.; Du, L.; Li, Y.; Wang, Z.; Xu, S.; Zhang, J.; Hao, Y. High-Performance AlGaN Double Channel HEMTs with Improved Drain Current Density and High Breakdown Voltage. *Nanoscale Res. Lett.* **2020**, *15*, 114. [CrossRef]
9. Baca, A.G.; Armstrong, A.M.; Allerman, A.A.; Douglas, E.A.; Sanchez, C.A.; King, M.P.; Coltrin, M.E.; Fortune, T.R.; Kaplar, R.J. An AlN/Al0.85Ga0.15N high electron mobility transistor. *Appl. Phys. Lett.* **2016**, *109*, 033509. [CrossRef]
10. Muhtadi, S.; Hwang, S.M.; Coleman, A.; Asif, F.; Simin, G.; Chandrashekhar, M.V.S.; Khan, A. High Electron Mobility Transistors With Al0.65Ga0.35N Channel Layers on Thick AlN/Sapphire Templates. *IEEE Electron Device Lett.* **2017**, *38*, 914–917. [CrossRef]
11. Zhang, W.; Cheng, J.; Xiao, M.; Zhang, L.; Hao, Y. High Breakdown-Voltage (>2200 V) AlGaN-Channel HEMTs with Ohmic/Schottky Hybrid Drains. *IEEE J. Electron Devices Soc.* **2018**, *6*, 931–935. [CrossRef]
12. Ito, S.; Takeda, K.; Nagata, K.; Aoshima, H.; Takehara, K.; Iwaya, M.; Takeuchi, T.; Kamiyama, S.; Akasaki, I.; Amano, H. Growth of GaN and AlGaN on (100) beta-$Ga_2O_3$ substrates. *Phys. Status Solidi (C)* **2012**, *9*, 519–522. [CrossRef]
13. Lu, Y.; Warner, J.H. Synthesis and applications of wide bandgap 2D layered semiconductors reaching the green and blue wavelengths. *ACS Appl. Electron. Mater.* **2020**, *2*, 1777–1814. [CrossRef]
14. Anju, S.; Babu, V.S.; Paul, G. Design optimization of high-frequency AlGaN/GaN HEMT on BGO substrates. *Appl. Phys. A* **2021**, *127*, 405. [CrossRef]
15. Singh, R.; Lenka, T.R.; Velpula, R.T.; Thang, B.H.Q.; Nguyen, H.P.T. Investigation of E-Mode Beta-Gallium Oxide MOSFET for Emerging Nanoelectronics. In Proceedings of the IEEE 14th Nanotechnology Materials and Devices Conference (NMDC), Stockholm, Sweden, 27–30 October 2019; pp. 1–5. [CrossRef]
16. Villora, E.G.; Shimamura, K.; Kitamura, K.; Aoki, K.; Ujiie, T. Epitaxial relationship between wurtzite GaN and $\beta$-$Ga_2O_3$. *Appl. Phys. Lett.* **2007**, *90*, 234102. [CrossRef]
17. Oshima, T.; Kato, Y.; Kawano, N.; Kuramata, A.; Yamakoshi, S.; Fujita, S.; Oishi, T.; Kasu, M. Carrier confinement observed at modulation-doped $\beta$-$(Al_xGa_{1-x})_2O_3/Ga_2O_3$ heterojunction interface. *Appl. Phys. Express* **2017**, *10*, 035701. [CrossRef]
18. Singh, R.; Lenka, T.R.; Panda, D.K.; Velpula, R.T.; Jain, B.; Bui, H.Q.T.; Nguyen, H.P.T. The dawn of $Ga_2O_3$ HEMTs for high power electronics—A review. *Mater. Sci. Semicond. Process.* **2020**, *119*, 105216. [CrossRef]
19. Song, K.; Zhang, H.; Fu, H.; Yang, C.; Singh, R.; Zhao, Y.; Sun, H.; Long, S. Normally-off AlN/$\beta$-$Ga_2O_3$ field-effect transistors using polarization-induced doping. *J. Phys. D Appl. Phys.* **2020**, *53*, 345107. [CrossRef]
20. SILVACO Int. *ATLAS User's Manual*; Device Simulation Software: Santa Clara, CA, USA, 2016. Available online: https://www.silvaco.com (accessed on 1 October 2021).
21. Farahmand, M.; Garetto, C.; Bellotti, E.; Brennan, K.F.; Goano, M.; Ghillino, E.; Ghione, G.; Albrecht, J.D.; Ruden, P.P. Monte Carlo simulation of electron transport in the III-nitride wurtzite phase materials system: Binaries and ternaries. *IEEE Trans. Electron Devices* **2001**, *48*, 535–542. [CrossRef]
22. Pearton, S.J.; Yang, J.; Cary, P.H.; Ren, F.; Kim, J.; Tadjer, M.J.; Mastro, M.A. A review of Ga2O3 materials, processing, and devices. *Appl. Phys. Rev.* **2018**, *5*, 011301. [CrossRef]
23. Meneghini, M.; De Santi, C.; Abid, I.; Buffolo, M.; Cioni, M.; Khadar, R.A.; Nela, L.; Zagni, N.; Chini, A.; Medjdoub, F.; et al. GaN-based power devices: Physics, reliability, and perspectives. *J. Appl. Phys.* **2021**, *130*, 181101. [CrossRef]
24. Nanjo, T.; Takeuchi, M.; Suita, M.; Oishi, T.; Abe, Y.; Tokuda, Y.; Aoyagi, Y. Remarkable breakdown voltage enhancement in AlGaN channel high electron mobility transistors. *Appl. Phys. Lett.* **2008**, *92*, 263502. [CrossRef]
25. Nanjo, T.; Takeuchi, M.; Imai, A.; Suita, M.; Oishi, T.; Abe, Y.; Yagyu, E.; Kurata, T.; Tokuda, Y.; Aoyagi, Y. AlGaN channel HEMTs on AlN buffer layer with sufficiently low off-state drain leakage current. *Electron. Lett.* **2009**, *45*, 1346–1348. [CrossRef]
26. Nanjo, T.; Imai, A.; Suzuki, Y.; Abe, Y.; Oishi, T.; Suita, M.; Yagyu, E.; Tokuda, Y. AlGaN Channel HEMT With Extremely High Breakdown Voltage. *IEEE Trans. Electron Devices* **2013**, *60*, 1046–1053. [CrossRef]
27. Baca, A.G.; Klein, B.A.; Wendt, J.R.; Lepkowski, S.M.; Nordquist, C.D.; Armstrong, A.M.; Allerman, A.A.; Douglas, E.A.; Kaplar, R.J. RF Performance of $Al_{0.85}Ga_{0.15}N/Al_{0.70}Ga_{0.30}N$ High Electron Mobility Transistors With 80-nm Gates. *IEEE Electron Device Lett.* **2019**, *40*, 17–20.

28. Wu, Y.; Zhang, J.; Zhao, S.; Zhang, W.; Zhang, Y.; Duan, X.; Chen, J.; Hao, Y. More Than 3000 V Reverse Blocking Schottky-Drain AlGaN-Channel HEMTs with >230 MW/cm$^2$ Power Figure-of-Merit. *IEEE Electron Device Lett.* **2019**, *40*, 1724–1727. [CrossRef]
29. Xiao, M.; Zhang, J.; Duan, X.; Zhang, W.; Shan, H.; Ning, J.; Hao, Y. High Performance $Al_{0.10}Ga_{0.90}N$ Channel HEMTs. *IEEE Electron Device Lett.* **2018**, *39*, 1149–1151. [CrossRef]

energies MDPI

*Article*

# A Bi-Directional DC Solid-State Circuit Breaker Based on Flipped Γ-Source

Jing Gao [1,*], Ziming Liu [1], Jinming Luo [1], Hongjiang Wang [1], Yiqi Liu [2] and Junyuan Zheng [2]

[1] Department of Electrical Engineering, Shenyang Institute of Engineering, Shenyang 110136, China; sielzm@sina.com (Z.L.); luojinming0830@163.com (J.L.); wanghj@sie.edu.cn (H.W.)
[2] College of Mechanical and Electrical Engineering, Northeast Forestry University, Harbin 150040, China; ee_617@nefu.edu.cn (Y.L.); zhengjunyuan0325@163.com (J.Z.)
* Correspondence: gaojing1@sie.edu.cn

**Abstract:** As the continuous consumption of non-renewable energy leads to resource shortages, distributed PV generation technology has received widespread attention. The DC microgrid is an effective way to connect distributed PV generation. However, DC microgrids have numerous challenges, such as the absence of zero current crossing point and high fault current rising rate. In the paper, a bi-directional DC solid-state circuit breaker based on flipped Γ-source is proposed to overcome some of the challenges in DC microgrid. The topology uses the mutual inductance current generated by the transformer to force the SCR naturally commutates off. Which can rapidly interrupt and isolate the faulty part, and at the same time, there will be no circulating current impact on the source side. The diode bridges allow the response to faults on either the source or load side with only a single controlled switch. Therefore, the topology is simplified while enabling it to obtain the capability of bidirectional operation protection. Finally, using the simulation software PSCAD/EMTDC and experimental platform verifies the effectiveness of the topology in this paper.

**Keywords:** PV generation; DC microgrid; solid-state DC circuit breaker; bi-directional operation protection

**Citation:** Gao, J.; Liu, Z.; Luo, J.; Wang, H.; Liu, Y.; Zheng, J. A Bi-Directional DC Solid-State Circuit Breaker Based on Flipped Γ-Source. *Energies* **2022**, *15*, 745. https://doi.org/10.3390/en15030745

Academic Editor: Enrique Romero-Cadaval

Received: 15 October 2021
Accepted: 9 January 2022
Published: 20 January 2022



## 1. Introduction

In recent years, as the consumption of non-renewable energy leads to resource shortages and the deterioration of environmental pollution, distributed renewable energy power generation technologies such as wind power generation and solar power generation have received widespread attention. DC microgrid serves as effective access to distributed power generation technologies. Compared with the AC microgrid, the DC microgrid can more efficiently and stably receive distributed renewable energy power generation systems such as wind energy and photovoltaics, energy storage devices, electric vehicles, and other DC loads. Although the DC microgrid has many advantages, the low system impedance of the DC microgrid will cause the short-circuit current to rise rapidly when the DC side fails. At the same time, the current in the DC system cannot naturally cross the zero point, and it is difficult to shut down. These problems have become the key to restricting the wide application of DC microgrids [1–4].

DC circuit breakers are divided into pure mechanical DC circuit breakers broken off by traditional mechanical switches. Solid-state circuit breakers perform interruption by power electronic devices and hybrid circuit breakers combining the above two structures [5–9]. Compared with the pure mechanical DC breakers, the solid-state circuit breakers have no mechanical switch part, and power electronic devices have good functional characteristics. The solid-state circuit breakers use the advantage of power electronic devices to realize arc-free shutdown and rapidly isolate the fault current. At the same time, its usual time is significantly increased, which is suitable for operation protection of medium and low voltage DC microgrid [10–13].

With the rapid development of power electronic devices, solid-state DC circuit breakers have also made great developments. Literature [14,15] proposed three Z-source DC circuit breaker topologies. The Z-source solid-state DC circuit breaker is relative. The traditional structure is simpler. Its special circuit structure realizes automatic commutation, which optimizes the insufficiency of forced commutation of standard DC circuit breakers. At the same time, the Z-source solid-state DC circuit breaker does not need to use additional control circuits when the short-circuit fault current rises rapidly. It can realize the natural zero-crossing point of the SCR circuit current and automatically isolate the short-circuit fault current. Utilizing the inductance current in the circuit and the characteristics of the SCR, the function of automatic detection and arc-free isolation of the fault current is realized. The problem of the complex structure of the traditional solid-state DC circuit breaker detection circuit is solved. Reference [16] proposed a novel topology of T-source DC solid-state breakers. However, the unidirectional DC circuit breaker limits its application in practical engineering. Four types of bi-directional Z-source DC circuit breakers are proposed in references [17–20]. However, these bi-directional topologies have some problems, such as an excessive number of components, a complex circuit structure, fault current having a particular impact on the source side, etc.

This paper proposed a bi-directional DC solid-state circuit breaker based on flipped Γ-source (ΓSCB). Based on keeping interrupted and isolating short circuit fault currents rapidly to reduce the cost and volume. The topology can rapidly isolate the fault current without external control, simplifying the overall circuit. There will be no circulating current impact on the source side after the fault isolation. The paper is structured as follows: Section 2 describes the overview of Z-source breakers and their operational principles. Section 3 provides the design of the ΓSCB. Section 4 analysis the fault current change rates. In Sections 5 and 6, the results of the simulation and experiment are given. Section 7 concludes the paper.

## 2. Overview of Z-Source Circuit Breakers

The DC microgrid has promoted the development of DC solid-state circuit breakers. The topology of traditional Z-source DC circuit breakers is divided into crossed topology, parallel-connected topology and series-connected topology. Table 1 explains the type of circuit breaker represented by the symbol.

**Table 1.** Nomenclature.

| Symbol | Description |
|---|---|
| ΓSCB | Γ-Source Circuit Breakers |
| ZSCB | Z-Source Circuit Breakers |
| OZSCB | O-Z-Source Circuit Breaker |
| PECB | Power Electronic Circuit Breaker |
| SSCB | Solid-State Circuit Breaker |

As shown in Figure 1, the three types of DC circuit breakers are mainly composed of SCR, capacitor, inductance, and RD buffer circuits. The red line in Figure 1 is the working state when the fault occurs. The fault current on the load side passes through the Z source capacitor and flows to the short-circuit point. Since the current in the line inductance cannot change suddenly when the reverse current in the SCR equals the steady-state load current of the line, the SCR is automatically closed, and the energy in the line decays to zero through the buffer circuit, thereby achieving the effect of quickly isolating the fault current.

Circuit structures, such as Γ-source and Z-source, are first applied to the research of inverters. With the development of PV Generation, they are applied to the field of solid-state DC circuit breakers. The crossed topology, shown in Figure 1a, fails to provide a common ground point between the source and load, so the actual application of the project is affected. The parallel-connected topology is shown in Figure 1b. The structure is transformed on the basis of the crossed type so that there is a common point between the source and the load, but when a fault occurs, the fault current will have a circulating current impact on the source side—Figure 1c shows the series-connected topology. Along with the common point, the structural characteristics of the Z-source capacitor greatly reduce the impact of the fault current on the source side, so it has become a relatively ideal solution at present. As shown in Figure 2, Three topology structures of bidirectional Z-source DC circuit are proposed by references [13,14]. Figure 2a,b shows that the characteristics of diode and SCR need to be used to achieve bidirectional operation, but the number of capacitors and inductances needs to be increased, and the current of LC branch will still have an impact on the source side. As shown in Figure 2c, the bidirectional Z-source DC circuit breaker adopting coupled inductors uses the mutual inductance current in the coupled inductors to realize automatic shutdown of the SCR, but the two sets of coupled inductors increase the volume and cost.

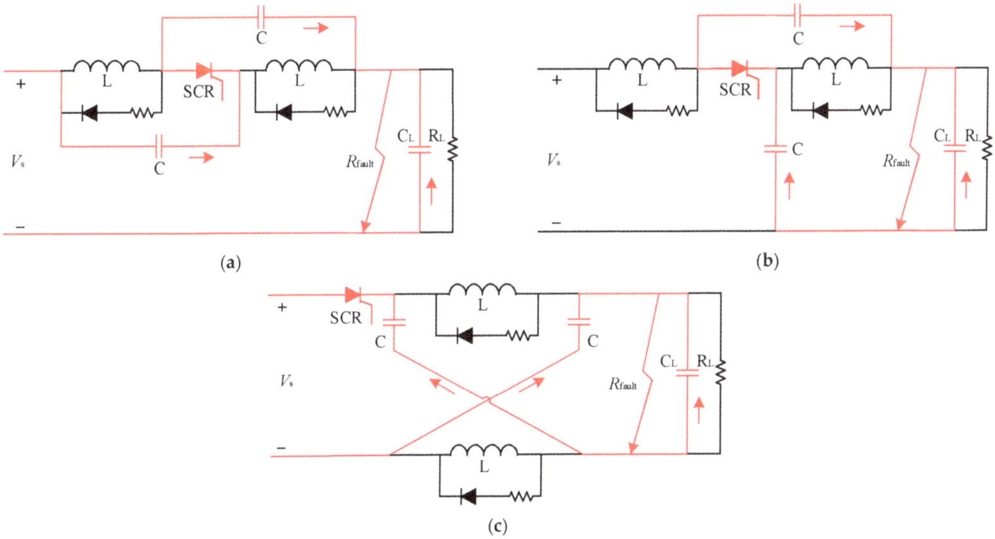


**Figure 1.** Previously proposed Z-source DC circuit breaker: (**a**) parallel-connected topology; (**b**) series-connected topology; (**c**) crossed topology.

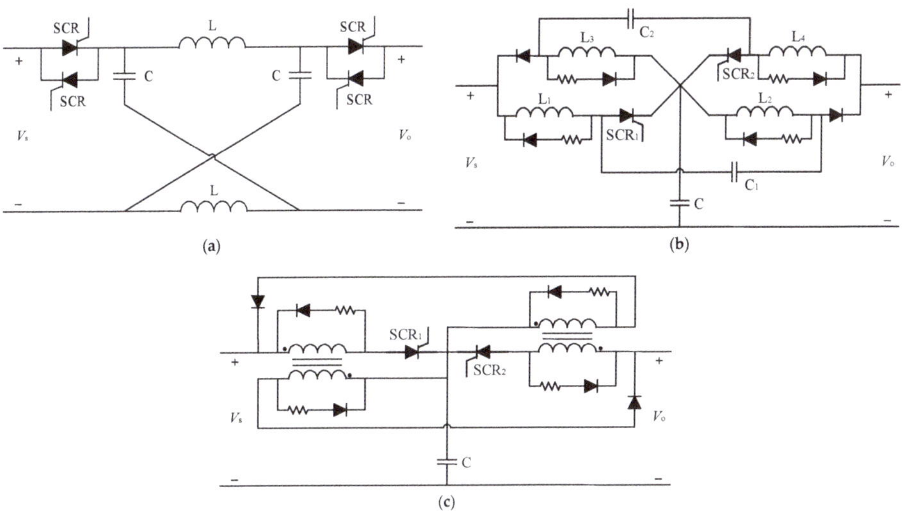


**Figure 2.** Three Bidirectional Z-source circuit breakers with coupling inductors: (**a**) Bidirectional crossed topology; (**b**) series-connected topology; (**c**) Bidirectional topology with coupled inductor.

## 3. Design of Bidirectional DC Circuit Breaker Based on Flipped Γ-Source

The bidirectional DC circuit breaker based on flipped Γ-source is shown in Figure 3, which is composed of a transformer, an SCR, capacitors, a diode bridge, a load capacitor and a load. Compared with the topology in Figure 1, the topology in this paper uses the mutual inductance current produced by the transformer to force the SCR naturally commutate off, which replaces the LC branch of the traditional Z-source topology solves the problem of circulating current impact on the source side. The bidirectional dc circuit breaker proposed in this paper has full-bridge diodes. The full-bridge arrangement of diodes allows load current to flow in either direction while retaining the same conduction path through SCR. Thus, the bidirectional dc circuit breaker limits and interrupts fault current regardless of the direction of power flow. Owing to the diode bridge circuit in the topology, only one controlled switch can realize bidirectional operation protection. Compared with the topology in Figure 2, only an SCR and a two-winding transformer are utilized to reduce the number RD buffer circuit. Thus, the topology is simplified while enabling it to have the capability of bidirectional operation protection.

To figure out the characteristics of different types of DC circuit breakers, the characteristics of various circuit breakers are analyzed in detail in Table 2. The advantages of bi-based on flipped ΓSB proposed in this paper are discussed by comparing the common ground, bidirectional operation protection capability, fault current clearing capability, and the number of SCR. Compared with traditional ZSCBs, the ΓSCB using a transformer instead of two separate inductors can offer three advantages. The first advantage is a smaller volume, the second advantage is a settable turns ratio to optimize the protection, and the third advantage is the anti-saturation characteristic. The proposed topology can achieve bidirectional energy flow and achieve bidirectional fault interruption and isolation. Obviously, this topology has a simpler circuit, fewer components, and common ground compared with the previous approach.

**Table 2.** Comparison of different DC circuit breakers.

| Category | Crosed ZSCB | Parellel ZSCB | Series ZSCB | Bi-Based on Crosed ZSCB | Bi-Based on Parellel ZSCB | Bi-Based on Coupled Inductor | Bi-Based on Flipped ΓSCB |
|---|---|---|---|---|---|---|---|
| Common Ground | No | Yes | Yes | No | Yes | Yes | Yes |
| Reflected Fault Current | Yes | No | No | Yes | No | Yes | Yes |
| Bidirectional Operation Protection | No | No | No | Yes | Yes | Yes | Yes |
| Number of SCR | 1 | 1 | 1 | 4 | 2 | 2 | 1 |

### 3.1. Working Principle

Different operating states of ΓSCB are presented in Figure 4. The forward energy flow channel consists of an SCR, the diode $D_1$ and $D_2$, capacitor C, the primary coil $L_1$ of the transformer. Similarly, the SCR, the diode $D_2$ and $D_3$, capacitor C, the primary coil $L_1$ of the transformer, constitute the channel of the reverse energy flow.

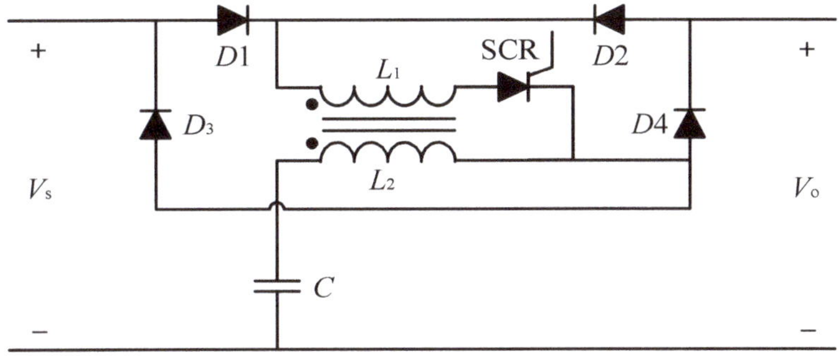


**Figure 3.** A bidirectional DC Solid-state circuit breaker based on flipped Γ-source.

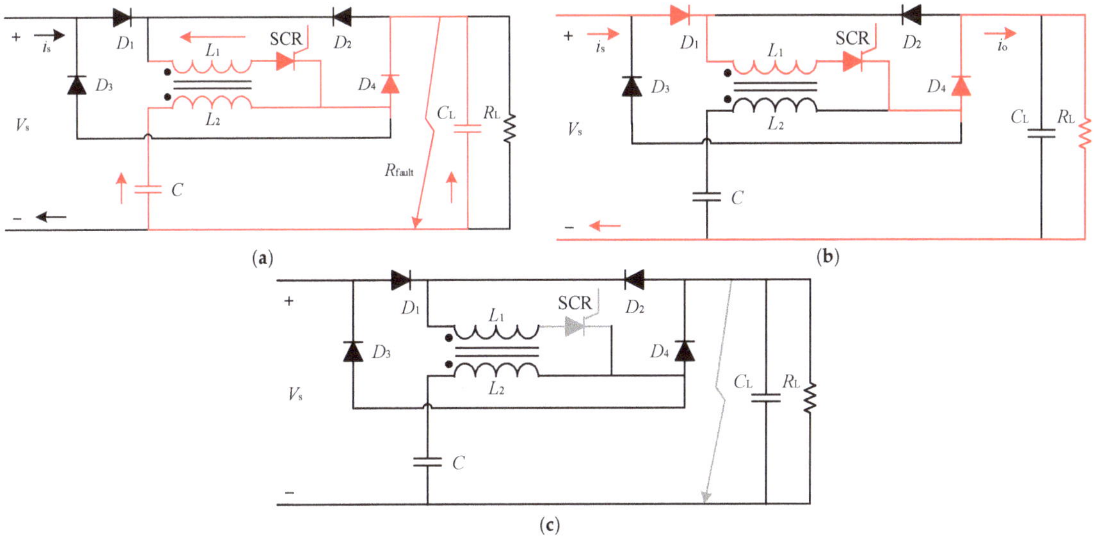


**Figure 4.** Various operating states of the flipped Γ-source DC circuit breaker: (**a**) steady state condition; (**b**) fault condition; (**c**) freewheeling state.

The normal operating states of the circuit are shown in Figure 4a. At this time, the left side is the source, and the right side is the load. In the operating state, the conduction path from the source side to the load side through $L_1$ and SCR is shown in the red line of the Figure. The voltage across the capacitor $C$ can be regarded as equal to the supply voltage under the premise of neglecting the voltage drop of SCR. When the fault occurs on the load side, the working state of ΓSCB is shown in Figure 4b. Part of the fault current generated by capacitor $C$ flows through the second coil of the transformer $L_2$ and $D_4$ to compensate for the fault current. The other part flows through load capacitor $C_L$ to compensate for the fault current. During this process, the current in the secondary coil of the transformer $L_2$ will rise rapidly. Due to the mutual inductance of the transformer, the primary coil of the transformer $L_1$ will generate an induced current opposite to the original current $i_s$, and when the two are equal, the SCR is turned off. After the SCR is turned off, the fault is completely isolated from the source side, and the current flows in the circulating path on the load side until the remaining energy decay to zero, as shown in Figure 4c. Due to the diode bridge circuit, ΓSCB does not need to add a buffer circuit, and the working principle of the forward energy flow is equal to the reverse energy flow.

### 3.2. Analysis of the Impedance Network

$k$ is the coupling coefficient of the two coupling coils, in this paper, $k = 1$. When the DC circuit breaker is in normal working condition, the mutual inductance value of the two-winding transformer is $M$.

$$M = k\sqrt{L_1 L_2} \tag{1}$$

The inductance of the primary and secondary coils of the transformer is $L_1$ and $L_2$.

According to the normal working equivalent circuit of DC circuit breakers in Figure 5, the input and output voltage transfer function is determined, and the steady-state characteristics of the circuit are analyzed. In the case of a resistive load, through theoretical derivation, the output and input voltage transfer function can be obtained as:

$$Z_{eq} = \frac{(L_1 + M)(L_2 + M)Cs^3 + R_L}{(L_2 + M)Cs^2 + R_L Cs + 1} + \frac{(L_1 + L_2 + 2M)R_L Cs^2 + (L_1 + M)s}{(L_2 + M)Cs^2 + R_L Cs + 1} \tag{2}$$

$$G_v(s) = \frac{V_o}{V_s} = \frac{R_L}{Z_{eq}} = \frac{(L_2 + M)R_L Cs^2 + R_L^2 Cs + R_L}{(L_1 + M)(L_2 + M)Cs^3 + (L_1 + L_2 + 2M)R_L Cs^2 + (L_1 + M)s + R_L} \tag{3}$$

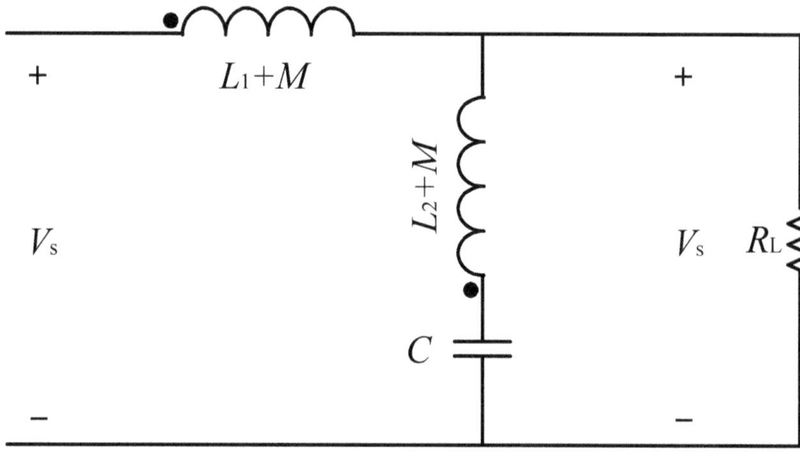


**Figure 5.** Circuit diagram when DC circuit breaker works normally.

Substituting the relevant data of the simulation parameters into Formula (3), the specific Bode diagram can be obtained, as shown in Figure 6, in the normal working state, the gain of the transfer function in the low-frequency range is approximately unity gain and the phase angle is close to zero degrees, which does not affect the normal operation of the system. In the high frequency range, it is very close to the function of the low-pass filter. Therefore, the topology structure plays a filtering role under the premise of protecting the system's normal operation and guarantees the operating efficiency of the DC microgrid.

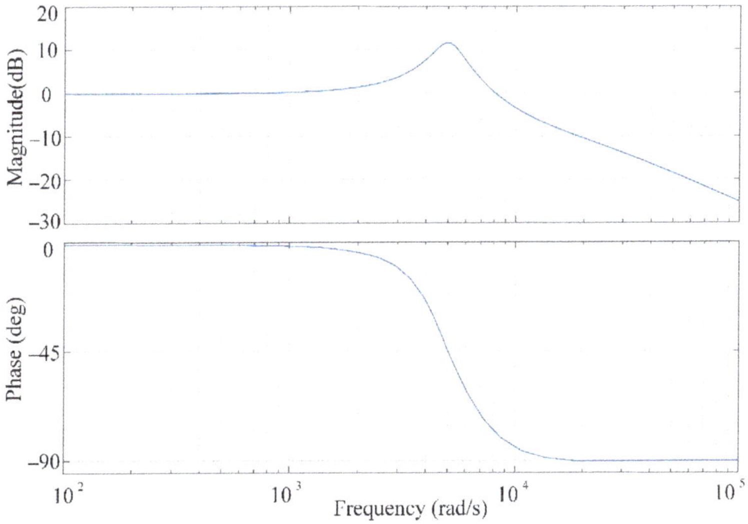


**Figure 6.** Circuit diagram when DC circuit breaker works normally.

## 4. Analysis of Fault Current Change Rate

For the parameter design of the topological structure of the reversed ΓSCB, when a fault occurs on the load side, the SCR can force automatic commutation only when the fault current rises to a critical value to achieve the purpose of fault isolation, so the rate of rising of the fault current and the sum current are important parameters that determine the automatic shutdown of the SCR. According to Figure 4b, the fault current $i_{\text{fault}}$, the secondary coil of the transformer $L_2$, capacitor $C$ and the load capacitor $C_L$ form a loop. The current path of a short-circuit fault in the DC circuit breaker shows that the fault current $i_{\text{fault}}$ can be expressed by the two capacitor line currents as:

$$i_{\text{fault}} = \frac{C + C_L}{C} i_C \tag{4}$$

$$i_{\text{fault}} = \frac{C + C_L}{C_L} i_{CL} \tag{5}$$

In the initial state of short-circuit fault, $i_{\text{fault}}$ can indicate:

$$i_{\text{fault}} = GV_o \tag{6}$$

In Formula (6), $V_o$ is the output voltage of the load, $G$ is the fault conductance. Neglecting the system line voltage drop, according to Formulas (4) and (6):

$$G_{\min} = \frac{C + C_L}{CR_L} \tag{7}$$

In Formula (7), $G_{\min}$ expresses as the smallest detectable fault conductance.

The rate of rising of the short-circuit fault current determines the detection capability of the DC circuit breaker. The following analysis assumes that the change in fault conductance with time is a linear relationship, and the fault change rate $K$ can be expressed as:

$$K = \frac{G}{\Delta t} \tag{8}$$

where $\Delta t$ is a time interval for the ramp. The fault current in terms of $K$ can be expressed as

$$i_{\text{fault}} = KV_o t \tag{9}$$

According to Formula (5), we can receive

$$KV_o t = \frac{C + C_L}{C_L}\frac{\mathrm{d}V_o}{\mathrm{d}t} \tag{10}$$

In the initial state when a fault occurs, $V_o = V_s$, from Formulas (4)–(6), we obtain

$$V_o = V_s e^{-\frac{K}{2(C+C_L)}t^2} \tag{11}$$

$V_s$ is the source input voltage, from Formulas (9) and (11), the fault current $i_{\text{fault}}$ and the current through Z-source capacitor are rewritten as

$$i_{\text{fault}} = V_s K t e^{-\frac{K}{2(C+C_L)}t^2} \tag{12}$$

$$i_C = \frac{C}{C + C_L} V_s K t e^{-\frac{K}{2(C+C_L)}t^2} \tag{13}$$

By solving Formula (13), the time at which capacitor current becomes maximum can be derived as

$$t_{\max} = \sqrt{\frac{C + C_L}{K}} \tag{14}$$

Substituting (14) into (13), maximum capacitor current can be obtained as

$$i_{C_{\max}} = V_s C \sqrt{\frac{K}{e(C + C_L)}} \tag{15}$$

N is the turn of the winding $L_1$ and $L_2$ of the transformer, respectively. The ratio of the induced current between the primary coil and the secondary coil of the two-winding transformer is $i_{L1}$:$i_{L2}$ = 1:$N$, and the induced current $i_{L2}$ on the secondary coil equals to capacitor current $i_c$, so when a fault occurs, the SCR can be automatically turned off when $\frac{1}{N} i_C \geq i_s$, $i_s$ is the steady-state load current, so it can be obtained from Formula (15):

$$\frac{1}{N} V_s C \sqrt{\frac{K}{e(C + C_L)}} \geq i_s = \frac{V_s}{R_L} \tag{16}$$

$$K \geq N^2 e \frac{C + C_L}{C^2 R_L^2} \tag{17}$$

Therefore, when the change rate of the short-circuit fault conductance satisfies Formula (17), ΓSCB can work normally. Otherwise, the DC circuit breaker will not work normally. When determining the size of the system parameters, it needs to be based on the minimum fault current. It is required to select the appropriate $C_L/C$ ratio, set the minimum fault change rate by changing the size of the capacitor value, and also consider the short-term fault current level of the SCR and the reverse recovery time to ensure the normal operation of the DC circuit breaker

In order to achieve the appropriate minimum fault ramp rate $K_{\min}$ and to ensure the safe commutation of the SCR in sufficient time, the optimal value of the inductor must be selected, and the inductor current can be obtained:

$$i_{L1} = I_L + \frac{V_s}{12(C + C_L)L_1} K t^3 \tag{18}$$

According to Formulas (13) and (18), the current flow SCR can be obtained:

$$i_{SCR} = i_{L1} + \frac{V_s C}{2(C + C_L)^2} K^2 t^3 - \frac{V_s C}{C + C_L} Kt \tag{19}$$

The inductor current should remain constant under fault conditions to minimize the detectable change rate of short-circuit faults. Therefore, the value of the inductance should be large enough so that the influence of the inductor current on the capacitor current is kept to a minimum. At the same time, the current through the SCR will not be affected.

$$L_1 = \frac{C + C_L}{6KC} \tag{20}$$

In this paper, $N = 2$, substituting (20) into (17), we obtain

$$L_1 = \frac{1}{24e} R_L^2 C \tag{21}$$

Formula (21) shows the important relationship between the load $R_L$ and the system parameters, and the result also shows that the two-winding transformer is proportional to the capacitance $C$.

## 5. Extended the ΓSCB Simulation

In order to verify the effectiveness of the DC microgrid protection design based on the bidirectional DC circuit breaker based on flipped Γ-source in this paper. A simulation model with source voltage $U_{DC}$ = 1 kV, the steady-state load current $i_s$ = 0.1 kA, and the short-circuit fault current peak value are 0.65 kA is established in the PSCAD/EMTDC software, and the working status of the DC circuit breaker when a short-circuit fault occurs on the load side is also analyzed. The two-way topology adopts two operating modes of the forward power flow and reverse power flow in the simulation. The mutual inductance value of the two-winding transformer is $M$ = 400 μH, and other relevant component parameters are shown in Table 3.

**Table 3.** The parameters of circuit breaker.

| Parameters | Values |
|---|---|
| Input voltage | 1000 V |
| Capacitor $C$ | 200 |
| Inductance of $L_1$ | 800 |
| Inductance of $L_2$ | 200 |
| Load capacitor $C_L$ | 200 uf |
| Load resistor $R_L$ | 20 Ω |
| Fault resistor | 0.2 |
| SCR off time | 50 us |

The simulation result of the topology of the ΓSCB is shown in Figure 7. When the forward energy flow channel is running, the fault occurs on the load side when the simulation is set for 0.5 s, and the capacitor $C$ and the load capacitor $C_L$ begin to discharge. The fault current rises rapidly, the magnetic flux of the secondary coil $L_2$ begins to increase, and the line current $i_{L2}$ rises.

Due to the mutual inductance between the transformer windings, the line current of the primary coil $L_1$ continues to rise, forcing the SCR current to cross zero. Due to the symmetrical structure of the diode bridge circuit, there is no resonance effect, and the fault current is cleared after 400 μs. The voltage $V_{SCR}$ on both sides of the SCR is stabilized at 1.8 kV, the voltage $V_c$ on both sides of the capacitor $C$ is stabilized at 0.75 kV, the voltages $V_{L1}$ and $V_{L2}$ on both sides of the primary and secondary windings are reduced to zero after a period of sine, and the load-side output voltage $V_o$ drops to zero. However, as the simulation time increases, the voltage on both sides of the SCR and capacitor C will slowly drop to zero after about 5.5 s.

As a result of the design of the diode bridge circuit, the ΓSCB is a symmetrical structure, so the simulation results under the reverse power flow operation of the bidirectional topology are consistent with the forward power flow operation. In the process of clearing the short-circuit fault current, it can be seen from the analysis of the simulation results that the bidirectional DC circuit breaker is designed to be suitable for the bidirectional operation of energy between different nodes and loads in the DC microgrid. Owing to the characteristics of the circuit structure, there is no need to add a buffer circuit and the ability to clear fault currents in a short time, which verifies the rationality of the topology of the ΓSCB.

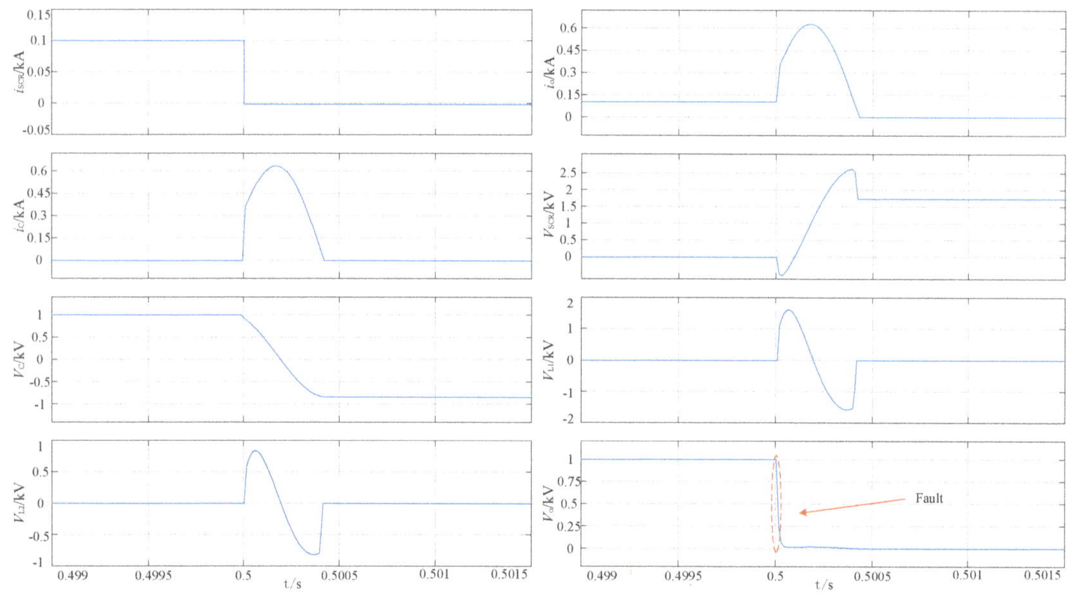


**Figure 7.** Simulation results of bidirectional DC circuit breaker based on flipped Γ-source.

The comparison simulation results of ΓSCB under different transformer parameters are shown in Figure 8. The Figure shows the change of the short-circuit fault turn-off time and the fault current peak value of ΓSCB with the capacitor C when the mutual inductance value is 400 μH and 600 μH. The circular line and the triangular line indicate the change under the condition of 400 μH, and the straight line and the diamond line indicate the change under the condition of 600 μH. Compared with the 600 μH conditions, the fault turn-off time under the 400 μH condition is short, and the fault current peak value is relatively high. With the continuous increase in the capacitor C, the difference between the fault current peak value and the turn-off time under the two values also increases. When the capacitor C is 200 μF, their gap is the smallest.

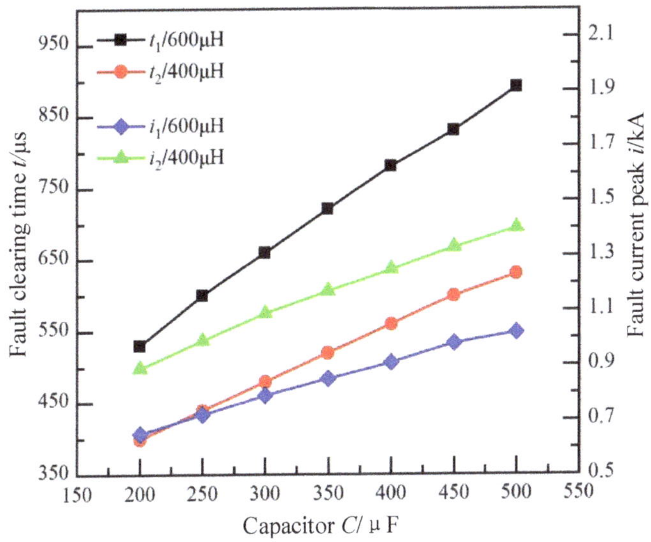


**Figure 8.** Comparison of simulation results under different parameters of ΓSCB.

## 6. Experimental Validation

In order to verify the principle of the DC microgrid protection design based on the reversed Γ source bidirectional DC solid-state circuit breaker proposed in this paper, this section builds an experimental platform with a power supply voltage of 48 V, a steady-state load current of 1.2 A. The experimental platform verifies the load-side short-circuit fault detection and breaking process. A two-way experimental circuit is built for the diode bridge structure, and the experimental circuit schematic diagram of the ΓSCB turned over is shown in Figure 9. In the diode bridge structure in Figure 9, $L_1$ and SCR are long-term current-passing branches, RL simulates load resistance, Ro simulates load-side short-circuit fault resistance, short-circuit fault occurs after the air switch $K_{SW}$ is closed, capacitor C starts to discharge, and the magnetic flux of $L_2$ rises until the SCR Automatic shutdown.

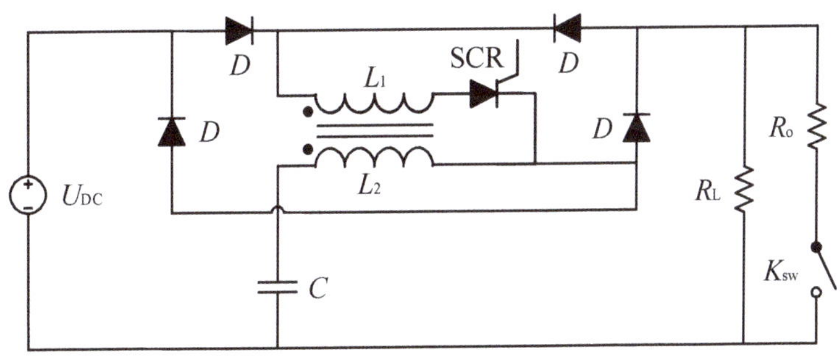


**Figure 9.** Bidirectional DC circuit breaker based on flipped Γ-source test circuit diagram.

Because the voltage level of the experiment platform is relatively low, the current-limiting module does not need to be added. Because the circuit current rises rapidly after the Ro resistance is connected, the circuit breaker is turned off according to the expected effect. In selecting experimental devices, full consideration should be given to the withstand voltage level and rated current to avoid breakdown and damage of the experimental devices and ensure the experimenters' personal safety. The physical diagram of ΓSCB turned over is shown in Figure 10.

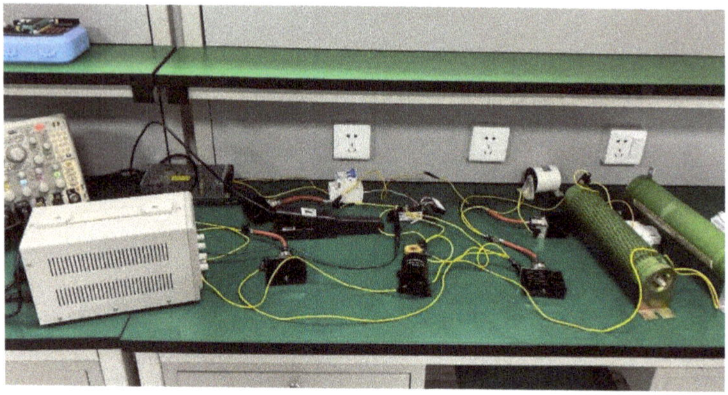

**Figure 10.** Bidirectional DC circuit breaker based on flipped Γ-source physical drawing.

After the system works normally, $K_{SW}$ is closed, Ro is connected to the circuit to simulate a short-circuit fault, and the SCR reaches the critical value to force automatic commutation, and the entire experimental operation process ends. Part of the experimental waveforms of the ΓSCB is shown in Figure 11. The experimental waveforms are SCR current waveform, output current waveform, SCR voltage waveform and output voltage waveform, and the time scale is 2 ms/div. It can be seen from the current and overvoltage waveform diagrams that under the condition of the source voltage of 48 V, the normal operating current of the entire DC system is 1.2 A, and the interruption fault current of the entire DC circuit breaker is about 400 μs. The circuit breaker topology designed in this paper reduces the number of SCR used and simplifies the overall structure on the basis of quickly shutting down the fault current. The experiment verifies the effectiveness of the topology of the flip Γ source DC circuit breaker.

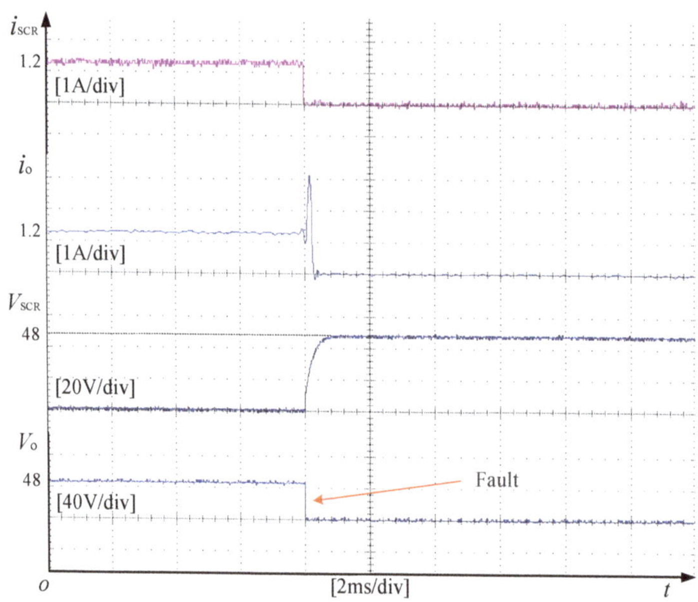


**Figure 11.** Part of bidirectional DC circuit breaker test waveform.

## 7. Conclusions

This paper proposed a DC microgrid protection design based on the flip Γ source DC circuit breaker, expounds the working principle of the flip Γ source DC circuit breaker topology and the analysis method of the failure rate. Finally, it is verified by simulation and experimental platform, and the following conclusions are obtained:

(1) When the ΓSCB works normally, by deriving the input and output voltage transfer function, it can be known that it is about unity gain in the low-frequency range and does not affect the normal operation of the system. Its effect is similar to the low pass filter and improves the operating efficiency of the DC microgrid.
(2) When a short-circuit fault occurs on the load side, this topology does not require an external control circuit, and the SCR can be quickly turned off through the mutual inductance of the transformer. At the same time, compared with the series-type Z-source DC circuit breaker using LC circuit, the source side will not produce circulating current impact after fault isolation;
(3) The topology not only has the ability to isolate currents rapidly but also meets the needs of bidirectional protection of DC microgrid. As a result of the structural characteristics of ΓSCB, only an SCR and two-wingding transformers are needed. The existence of mutual inductance of transformers also greatly reduces the demand for total inductance, thereby reducing the overall size and cost of the topology.

**Author Contributions:** J.G.: data curation, writing original draft preparation. Z.L.: project administration. J.L.: visualization. H.W.: supervision. Y.L.: software. J.Z.: validation. All authors have read and agreed to the published version of the manuscript.

**Funding:** This research was funded by Science and Technology Talents and Natural Science Foundation grant number 2021-NLTS-13-05.

**Institutional Review Board Statement:** Not applicable.

**Informed Consent Statement:** Not applicable.

**Data Availability Statement:** The study did not report any data.

**Conflicts of Interest:** The authors declare no conflict of interest.

## References

1. Aslam, M.W.; Zaffar, N.A. A novel bidirectional Z-source DC circuit breaker for DC microgrids. In Proceedings of the 2016 IEEE 7th Power India International Conference (PIICON), Bikaner, India, 25–27 November 2016; pp. 1–5.
2. Mackey, L.; Rachi, M.R.K.; Peng, C.; Husain, I. Z-Source circuit breaker utilizing Ultra-Fast Mechanical Switch for high efficiency DC circuit protection. In Proceedings of the 2017 IEEE Second International Conference on DC Microgrids (ICDCM), Nuremburg, Germany, 27–29 June 2017; pp. 452–458.
3. Corzine, K.A. Circuit breaker for DC micro grids. In Proceedings of the 2015 IEEE First International Conference on DC Microgrids (ICDCM), Atlanta, GA, USA, 7–10 June 2015; pp. 221–221c.
4. Yang, Y.; Huang, C.; Zhao, Z.; Xu, Q.; Jiang, Y. A New Bidirectional DC Circuit Breaker with Fault Decision-Making Capability for DC Microgrid. *IEEE J. Emerg. Sel. Top. Power Electron.* **2021**, *9*, 2476–2488. [CrossRef]
5. Wang, Y.; Li, W.; Wu, X.; Wu, X. A Novel Bidirectional Solid-State Circuit Breaker for DC Microgrid. *IEEE Trans. Ind. Electron.* **2019**, *66*, 5707–5714. [CrossRef]
6. Song, Y.; Yu, Y.; Wang, S.; Liu, Q.; Li, X. A Novel Efficient Bidirectional T-source Circuit Breaker for Low Voltage DC Distribution Network. In Proceedings of the 2021 IEEE 12th International Symposium on Power Electronics for Distributed Generation Systems (PEDG), Chicago, IL, USA, 28 June–1 July 2021; pp. 1–5.
7. Anselmo, I.S.; Rashid, M.H. Solid-State Circuit Breakers for D.C. Microgrid Applications. In Proceedings of the 2020 International Confer-ence on Electrical, Communication, and Computer Engineering (ICECCE), Istanbul, Turkey, 12–13 June 2020; pp. 1–4.
8. Zhang, C.; Zou, G.; Zhang, S.; Wei, X. Hybrid DC Circuit Breaker with Power Flow Control Function. In Proceedings of the 2021 6th Asia Conference on Power and Electrical Engineering (ACPEE), Chongqing, China, 8–11 April 2021; pp. 741–745.
9. Loh, P.C.; Blaabjerg, F. Magnetically Coupled Impedance-Source Inverters. *IEEE Trans. Ind. Appl.* **2013**, *49*, 2177–2187. [CrossRef]
10. Franck, C.M. HVDC Circuit Breakers: A Review Identifying Future Research Needs. *IEEE Trans. Power Deliv.* **2011**, *26*, 998–1007. [CrossRef]
11. Savaliya, S.G.; Fernandes, B.G. Analysis and Experimental Validation of Bidirectional Z-Source DC Circuit Breakers. *IEEE Trans. Ind. Electron.* **2019**, *67*, 4613–4622. [CrossRef]

12. Park, J.-D.; Candelaria, J. Fault Detection and Isolation in Low-Voltage DC-Bus Microgrid System. *IEEE Trans. Power Deliv.* **2013**, *28*, 779–787. [CrossRef]
13. Li, W.; Wang, Y.; Wu, X.; Zhang, X. A Novel Solid-State Circuit Breaker for On-Board DC Microgrid System. *IEEE Trans. Ind. Electron.* **2019**, *66*, 5715–5723. [CrossRef]
14. Corzine, K.A.; Ashton, R.W. A New Z-Source DC Circuit Breaker. *IEEE Trans. Power Electron.* **2012**, *27*, 2796–2804. [CrossRef]
15. Song, W.; An, N.; Wang, Y. A Novel Bidirectional T-source DC Circuit Breaker for DC Microgrids. In Proceedings of the 2019 14th IEEE Conference on Industrial Electronics and Applications (ICIEA), Xi'an, China, 19–21 June 2019; pp. 1540–1545.
16. Ryan, D.J.; Torresan, H.D.; Bahrani, B. A Bidirectional Series Z-Source Circuit Breaker. *IEEE Trans. Power Electron.* **2018**, *33*, 7609–7621. [CrossRef]
17. Maqsood, A.; Corzine, K. The Z-source breaker for fault protection in ship power systems. In Proceedings of the 2014 International Symposium on Power Electronics, Electrical Drives, Automation and Motion, Ischia, Italy, 18–20 June 2014; pp. 307–312.
18. Yu, J.; Kim, S.; Song, M.G.; Kim, I. A New Thyristor DC Solid-State Circuit Breaker Capable of Performing Operating Duty. In Proceedings of the 2019 22nd International Conference on Electrical Machines and Systems (ICEMS), Harbin, China, 11–14 August 2019; pp. 1–4.
19. Zhou, Z.; Jiang, J.; Ye, S.; Liu, C.; Zhang, D. A Γ-Source Circuit Breaker for DC Microgrid Protection. *IEEE Trans. Ind. Electron.* **2021**, *68*, 2310–2320. [CrossRef]
20. Qi, L.; Chen, X.; Qu, X.; Zhan, L.; Zhang, X.; Cui, X. A Novel Forced Resonant Mechanical DC Circuit Breaker by Using Auxiliary Oscillation Switch for Zero-Crossing. *IEEE Trans. Power Electron.* **2021**, *36*, 12202–12206. [CrossRef]

*energies*

MDPI

*Article*

# Design and Implementation of the Bidirectional DC-DC Converter with Rapid Energy Conversion

**Bing-Zhang Chen [1], Hsuan Liao [1,*], Linda Chen [2] and Jiann-Fuh Chen [1]**

[1] Department of Electrical and Electronics Engineering, National Cheng Kung University, Tainan 701, Taiwan; brian23330113@gmail.com (B.-Z.C.); chenjf@mail.ncku.edu.tw (J.-F.C.)

[2] Department of Electrical and Computer Engineering, University of Canterbury, Christchurch 8041, New Zealand; linday.chen@canterbury.ac.nz

* Correspondence: n28064046@gs.ncku.edu.tw

**Abstract:** The bidirectional DC-DC converters are widely used in the energy storage system (ESS) and DC distribution system. The power capacity is limited when the converter is operated with smooth power transfer. In addition, the directions of the inductor current and the capacitor voltage cannot change instantaneously. In this study, a rapid energy conversion technique for smoothing and accelerating the energy transfer under the same specification of the main components in steady state is proposed. Moreover, a bidirectional DC-DC converter with a high conversion ratio is proposed to overcome the commonly low voltage input from renewable energy sources. The operating principles of the proposed converter's step-down and step-up modes are discussed in this study. Furthermore, to achieve rapid energy conversion, digital control is a crucial component in the converter system. A digital signal processor is used as the control platform, and a control strategy is formulated to achieve rapid energy conversion. The bidirectional DC-DC prototype converter with a 24 V battery, a DC bus of 200 V, and an output power of 500 W is constructed to confirm the feasibility of rapid energy conversion. The proposed converter can be operated in CCM, BCM, and DCM conditions. The transfer period can be completed within one switching cycle when the proposed converter is operated in BCM or DCM. The energy is freewheeled before energy conversion when the proposed converter is operated in CCM condition. In the experiment, the minimum transfer period is 6.29 μs on the DCM stage.

**Keywords:** bidirectional DC-DC converter; rapid energy conversion; smooth power transfer

**Citation:** Chen, B.-Z.; Liao, H.; Chen, L.; Chen, J.-F. Design and Implementation of the Bidirectional DC-DC Converter with Rapid Energy Conversion. *Energies* **2022**, *15*, 898. https://doi.org/10.3390/en15030898

Academic Editors: Alon Kuperman, Ching-Ming Lai and Yitao Liu

Received: 27 December 2021
Accepted: 21 January 2022
Published: 26 January 2022



## 1. Introduction

In recent years, because of fast economic development, energy demands have increased rapidly, which, in consequence, caused damages to the environment and accelerated global warming during power generation. To achieve sustainable development of society and economy, environmental protection and the usage of energy have become the most important aspects [1–6]. In addition, renewable energy sources such as wind energy and solar energy are regeneratable and cause less environmental pollution when compared with conventional fossil fuel power generation. However, due to the instability in the renewable energy sources, the methods for smoothing the input energy are essential. Therefore, it is getting more common for the renewable energy system (RES) to integrate with the energy storage system (ESS) [7–10]. Figure 1 shows the RES incorporating batteries as the backup power [11]. Bryden et al. illustrated the advantages of the usage of stationary energy storage [11]. The electric vehicle (EV) charging stations can buffer the energy between the electricity grid and EVs, minimizing the maximum power required for grid connection. There are, however, some issues with the connection between the grid and converters, such as the stability of the power flow and the time required for energy transfer. The quality of the power flow and the length of the conversion time for energy transfer are the main aspects being studied in this research.

In modern days, uninterruptible power supply (UPS) systems are usually used at the utility grid as the main power source and the batteries are used as a backup power source [12,13]. Thus, for systems interfacing the ESS with the DC bus, the bidirectional DC-DC converters with power flow control have been widely discussed for effective power conversion [14–21]. Several research groups have discussed the possible methods of improving the quality of the power flow by modifying the circuit topology [14–17]. Vuyyuru et al. and Khan et al. have also illustrated different control strategies for improving the power flow quality [18–20]. In this study, we combined a control strategy designed for the proposed topologies of the DC-DC converter with a short conversion period.

The terminal voltages of renewable energy sources are usually low and with time-dependent variations. Hence, a bidirectional DC-DC converter with a high conversion ratio is commonly used to interface with the DC bus to provide stable power energy. The converter needs to reduce the load current for a smooth power transfer [21–26], which, in consequence, limits the power capacity during conversion [19,26]. Thus, this paper also analyzes the energy conversion between operating modes and proposes a technique for rapid energy conversion.

A bidirectional DC-DC converter with a short conversion period is proposed in this paper. Two different condition modes are discussed. One is the discontinuous conduction mode (DCM), or boundary conduction mode (BCM), and the other is the continuous conduction mode (CCM). In general, the inductor current decreases to zero within one switching period which allows the completion of the energy conversion in one switching cycle. However, the change of the value of the inductor current is not instantaneous. To prevent damages to components in the converter due to the high voltage spike, the inductor current should be freewheeling in the CCM condition. Therefore, the freewheeling path period needs to be considered as well.

In this paper, Section 2 discusses the operational principle of the proposed DC-DC converter with a high conversion ratio. The operational principles of the step-up mode and step-down mode are discussed. Section 3 describes the proposed control strategy of the transferring state for rapid energy conversion. A bidirectional DC-DC converter circuit is constructed with a battery voltage of 24 V and a 200 V DC bus voltage for the validation of the proposed topology design. The experimental results are presented in Section 4, followed by the conclusions and future works in the final section.

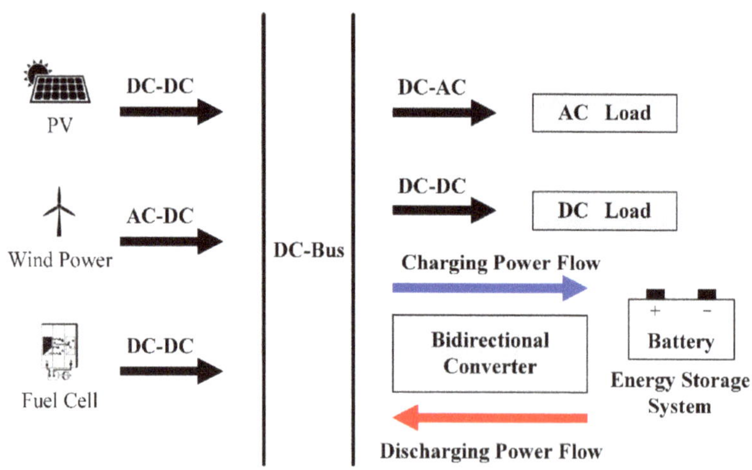


**Figure 1.** A backup battery in interfacing with various power supplies.

## 2. Operational Principle of Proposed Converter

The operational principles of the proposed bidirectional DC-DC converter are analyzed and presented in this section. Figure 2 shows the topology of the proposed bidirectional DC-DC converter, which is composed of a battery voltage $V_{bat}$, a DC bus voltage $V_{dc,bus}$, five power switches $S_1$, $S_2$, $S_3$, $S_{Aux1}$, and $S_{Aux2}$, two inductors $L_1$ and $L_2$, and two capacitors $C_{bat}$ and $C_{bus}$. Two auxiliary switches are used as the bidirectional switches for the power flow control. The following assumptions are made to simplify the circuit analysis:

a. The converter is operated at the steady state.
b. The converter is operated in CCM condition.
c. The Body diodes of the power switches $S_1$, $S_2$, $S_3$, $S_{Aux1}$, and $S_{Aux2}$ should be considered, corresponding to diodes $D_{S1}$, $D_{S2}$, $D_{S3}$, $D_{SAux1}$, and $D_{SAux2}$.
d. The capacitances of $C_{bat}$ and $C_{bus}$ are large enough to be regarded as a constant voltage source.
e. The other components are assumed with ideal conditions except for the components indicated above.

There are two operation modes when the proposed converter is analyzed in the CCM condition. The simulated waveforms of the step-up mode are shown in Figure 3 and the waveforms of the step-down mode are shown in Figure 4.

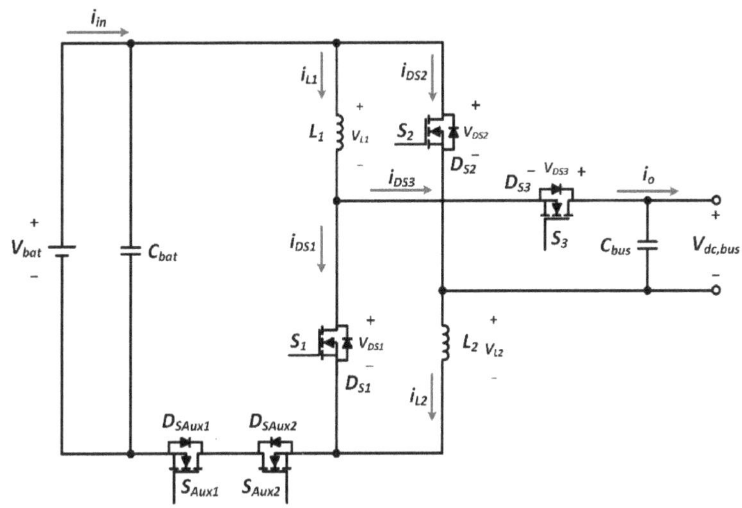


**Figure 2.** The topology of the proposed bidirectional DC-DC converter.

### 2.1. Step-Up Mode

#### 2.1.1. Mode I ($t_0$–$t_1$)

At time $t_0$, when the switches $S_1$ and $S_2$ are turned on, the switch $S_3$ is turned off. Figure 5 shows the current flowing paths of this mode. The inductors $L_1$ and $L_2$ are charged by battery voltage $V_{bat}$. This results in the inductor currents $i_{L1}$ and $i_{L2}$ to be linearly increased. The output capacitor $C_{bus}$ provides energy to load $R_{load}$. The $v_{L1}$ and $v_{L2}$ are the voltage across the inductors $L_1$ and $L_2$, which can be represented as the Equation (1). During this interval, the rate of change in $i_{L1}$ and $i_{L2}$ can be derived from both Equations (2) and (3).

$$v_{L1} = v_{L2} = V_{bat} \tag{1}$$

$$\Delta i_{L1}{}^{I} = \frac{|v_{L1}|}{L_1}(t_1 - t_0) = \frac{V_{bat}}{L} D_{dis} T_s \tag{2}$$

$$\Delta i_{L2}{}^{I} = \frac{|v_{L2}|}{L_2}(t_1 - t_0) = \frac{V_{bat}}{L} D_{dis} T_s \tag{3}$$

where $\Delta i_{L1}{}^{I}$ is the current ripple of the inductor $L_1$ during mode I, $\Delta i_{L2}{}^{I}$ is the current ripple of the inductor $L_2$ during mode I, and $D_{dis}$ is the duty ratio of step-up mode.

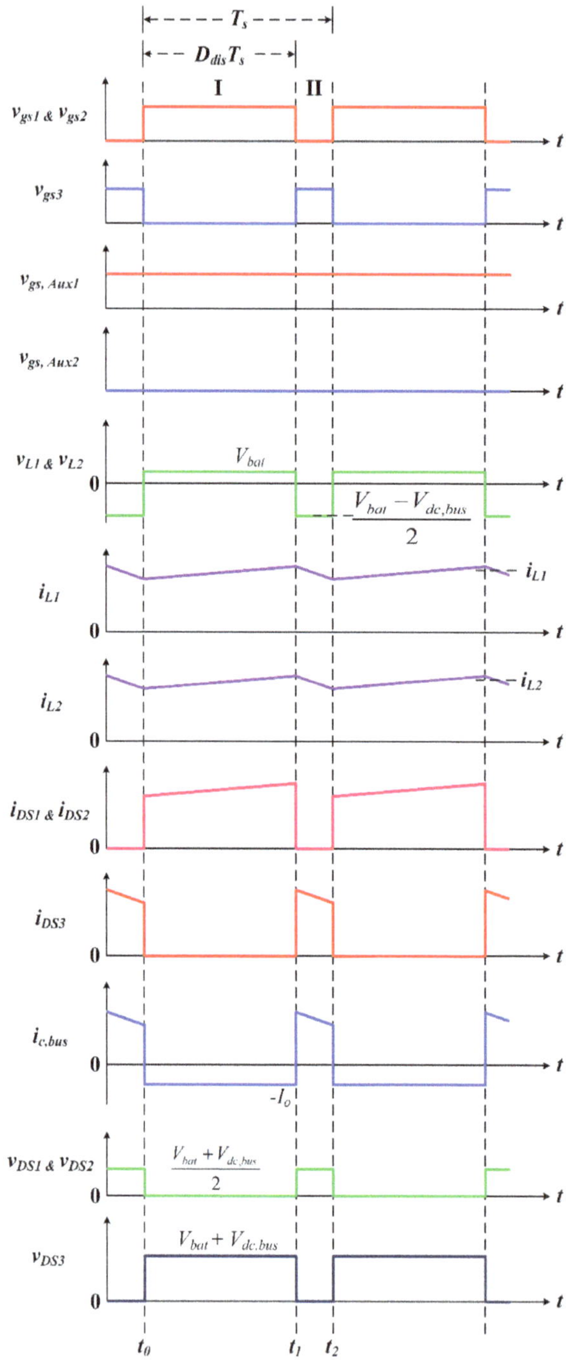


**Figure 3.** The waveforms of the step-up mode.

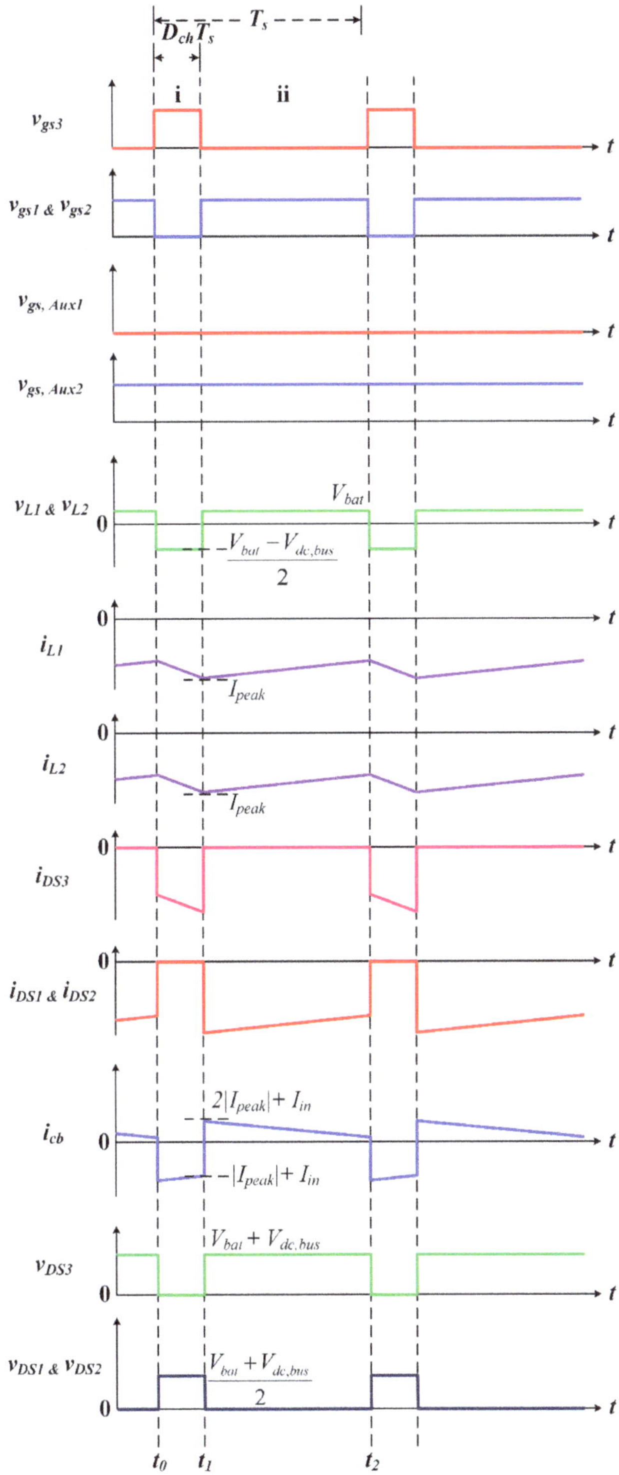


**Figure 4.** The waveforms of the step-down mode.

### 2.1.2. Mode II ($t_1$–$t_2$)

At time $t_1$, when the switch $S_3$ is turned on, the switches $S_1$ and $S_2$ are turned off. The current flowing paths of this mode are shown in Figure 6. The battery voltage $V_{bat}$ and the stored energies in the inductors $L_1$ and $L_2$ begin to charge to the output capacitor $C_{bus}$ and transfer energy to the load $R_{load}$. The inductor currents $i_{L1}$ and $i_{L2}$ are linearly decreased during this period. The voltage across the inductors $v_{L1}$ and $v_{L2}$ can be represented as Equation (4). During this interval, the rate of change in $i_{L1}$ and $i_{L2}$ can be represented as Equations (5) and (6).

$$v_{L1} = v_{L2} = \frac{V_{bat} - V_{dc,bus}}{2} \tag{4}$$

$$\Delta i_{L1}{}^{II} = \frac{|v_{L1}|}{L_1}(t_2 - t_1) = \frac{V_{dc,bus} - V_{bat}}{2L}(1 - D_{dis})T_s \tag{5}$$

$$\Delta i_{L2}{}^{II} = \frac{|v_{L2}|}{L_2}(t_2 - t_1) = \frac{V_{dc,bus} - V_{bat}}{2L}(1 - D_{dis})T_s \tag{6}$$

where $\Delta i_{L1}{}^{II}$ is the current ripple of the inductor $L_1$ during mode II, $\Delta i_{L2}{}^{II}$ is the current ripple of the inductor $L_2$ during mode II, and $D_{dis}$ is the duty ratio of step-up mode.

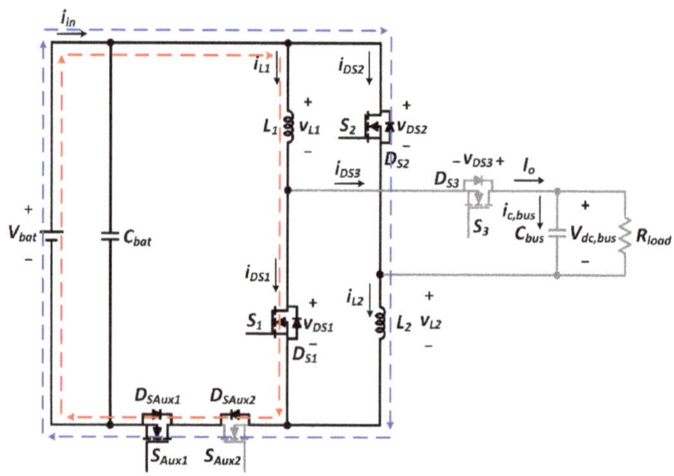


**Figure 5.** The current flow paths of step-up Mode I.

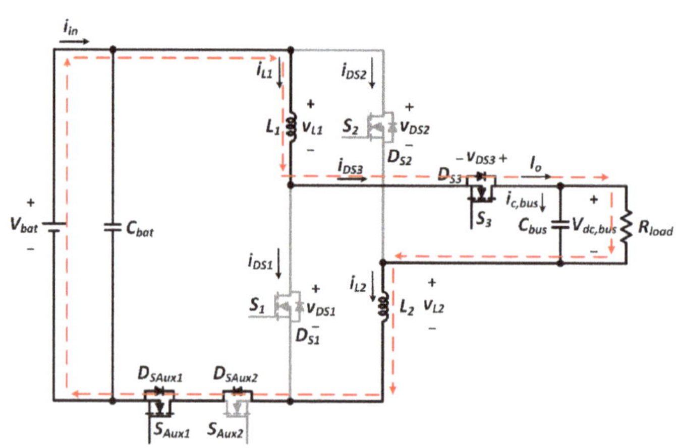


**Figure 6.** The current flow paths of step-up mode II.

### 2.2. Step-Down Mode

#### 2.2.1. Mode i ($t_0$–$t_1$)

At time $t_0$, when the switch $S_3$ is turned on, the switches $S_1$ and $S_2$ are turned off. The current flowing paths of mode i are shown in Figure 7. In this mode, the inductors $L_1$, $L_2$, and the capacitor $C_{bat}$ are charged in series by DC bus voltage $V_{dc,bus}$. Therefore, the inductor currents $i_{L1}$, $i_{L2}$ are linearly decreased.

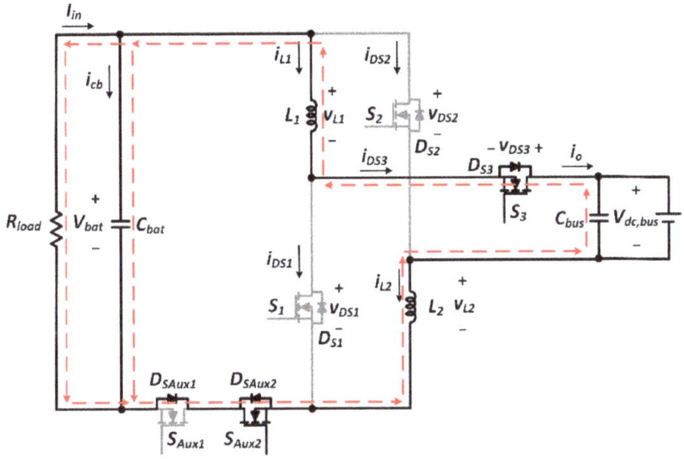


**Figure 7.** The current flow paths of step-down mode i.

#### 2.2.2. Mode ii ($t_1$–$t_2$)

At time $t_1$, when the switch $S_3$ is turned off, the switches $S_1$ and $S_2$ are turned on. The current flowing paths of mode ii are shown in Figure 8. In this mode, the inductors $L_1$, $L_2$, and the capacitor $C_{bat}$ release energies to the battery voltage $V_{bat}$. This results in the inductor currents $i_{L1}$ and $i_{L2}$ to be linearly increased.

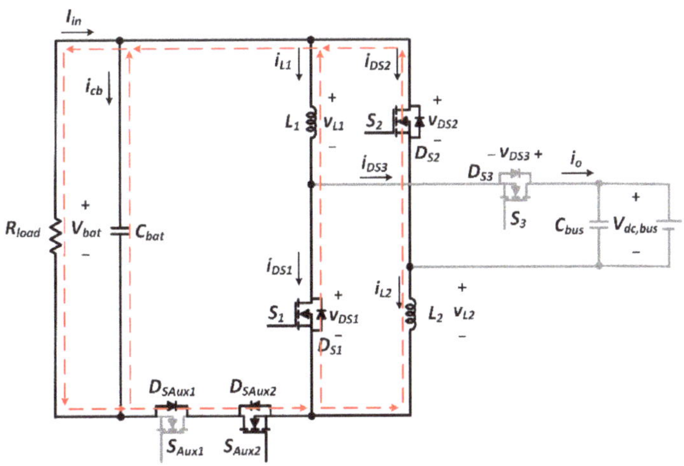


**Figure 8.** The current flow paths of step-down mode ii.

According to the voltage-second balance principle, the average voltage on inductors $L_1$ and $L_2$ in one switching cycle is zero. This can be expressed as Equations (7) and (8).

$$\frac{1}{T_s}\int_0^{T_s} v_{L1}dt = 0 \Rightarrow \int_0^{t_1} v_{L1}dt + \int_{t_1}^{t_2} v_{L1}dt = 0 \quad (7)$$

$$\frac{1}{T_s}\int_0^{T_s} v_{L2}dt = 0 \Rightarrow \int_0^{t_1} v_{L2}dt + \int_{t_1}^{t_2} v_{L2}dt = 0 \quad (8)$$

Based on Equations (1) and (4), the Equations (7) and (8) can be simplified as Equation (9):

$$V_{bat}D_{dis}T_s + \frac{V_{bat} - V_{dc,bus}}{2}(1 - D_{dis})T_s = 0, \quad (9)$$

The voltage conversion ratio of the step-up mode can be derived as Equation (10). Similarly, the voltage conversion ratio of the step-down mode can use the same approach to derive as Equation (11).

$$\frac{V_{dc,bus}}{V_{bat}} = \frac{1 + D_{dis}}{1 - D_{dis}}, \quad (10)$$

$$\frac{V_{bat}}{V_{dc,bus}} = \frac{D_{ch}}{2 - D_{ch}}. \quad (11)$$

where $D_{dis}$ is the duty ratio of the step-up mode, and $D_{ch}$ is the duty ratio of the step-down mode.

## 3. Control Strategy of Transferring State

In this section, the transferring states from step-up mode to step-down mode and vice versa are described in detail. The direction of all energy storage components in the circuit must be considered carefully because the direction of the inductor current or capacitor voltage cannot be reversed immediately.

To achieve the power flow control of the bidirectional converter, it is necessary to execute the following steps:

Step 1: Detect the status of the transfer signal and determine whether the converter changes the operating mode or not.

Step 2: The gate driving signals of the bidirectional switches $S_{Aux1}$ and $S_{Aux2}$ maintain the original operating mode until the inductor currents $i_{L1}$ and $i_{L2}$ demagnetize to zero.

Step 3: The gate driving signals of the bidirectional switches $S_{Aux1}$ and $S_{Aux2}$ change the original operating mode.

Step 4: Finally, restore the gate driving signals of the main switches $S_1$, $S_2$, and $S_3$.

### 3.1. Transferring State from Step-Up Mode to Step-Down Mode

For the converters operating in CCM condition, the driving signals of all the power switches during the transferring state are shown in Figure 9, illustrating the procedure of the transferring state, and the timing of the transferring state occurring from mode II of the step-up mode to mode i of the step-down mode. For the converters operating in BCM or DCM conditions, the transfer state occurs when the inductor current reaches zero.

During the transferring stage, the switch $S_3$ is turned on for a switching cycle to provide a path for the freewheeling of the inductor current. The inductor currents $i_{L1}$ and $i_{L2}$ are linearly decreased until the inductor currents demagnetize to zero. The original operating mode of the bidirectional switches $S_{Aux1}$ and $S_{Aux2}$ is changed. After time $t_2$, the gate driving signals of the main switches $S_1$, $S_2$, and $S_3$ are restored, and the circuit enters into the step-down mode. This results in the inductors $L_1$ and $L_2$ beginning to store the energy.

### 3.2. Transferring State from Step-Down Mode to Step-Up Mode

For converters operating in CCM condition, the gate driving signals of all the power switches during the transferring state are shown in Figure 10, illustrating the procedure of the transferring state and the timing of the transferring state occurring from mode II of the step-up mode to mode i of the step-down mode. For the converters operating in BCM or DCM conditions, the transfer state occurs when the inductor current reaches zero.

During the transferring stage, the switch $S_3$ is turned off and the bidirectional switches $S_{Aux1}$ and $S_{Aux2}$ remain in the original operating mode. The switches $S_1$ and $S_2$ are then turned on for two switching cycles to provide a path for the freewheeling of the inductor current. The inductor currents $i_{L1}$ and $i_{L2}$ are linearly increased until the inductor currents demagnetize to zero. The original operating mode of the bidirectional switches $S_{Aux1}$ and $S_{Aux2}$ is changed. After time $t_3$, the gate driving signals of the main switches $S_1$, $S_2$, and $S_3$ are restored, and the circuit enters the step-up mode. This results in the inductors $L_1$ and $L_2$ beginning to store energy.

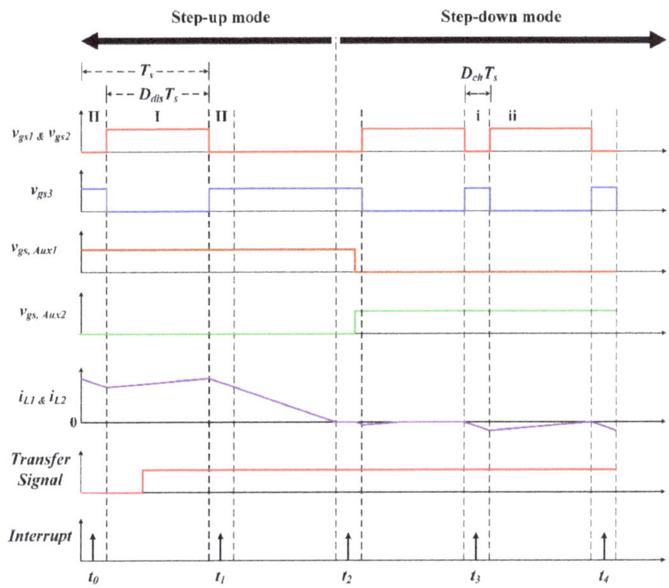


**Figure 9. The** timing of the transferring state from step-up mode to step-down mode.

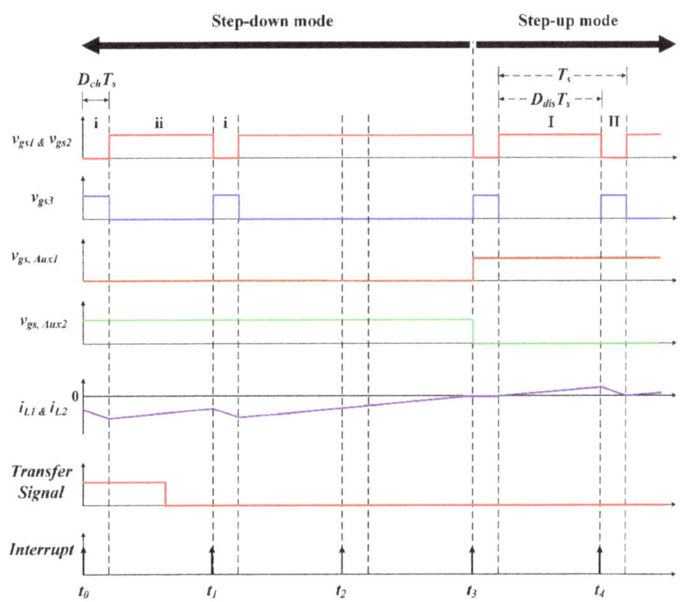


**Figure 10.** Timing of the transferring state from step-down mode to step-up mode.

### 3.3. *Flow Chart of the Digital Control System*

In general, the digital control program can be divided into the main program and the interrupt subroutine. The purpose of the main program is to configure system parameters, declare variables, and arrange the subroutines in order. In addition, the interrupt subroutine is not executed until the interrupt event occurs. The interrupt subroutine is used to evaluate events and perform complex calculations.

The flow chart of the main program is shown in Figure 11. The system parameters are first initialized at the beginning of the program and the function blocks of MCU, such as I/O pins of GPIO modules, ADC modules, and pulse width modulation (PWM) modules, are then configured. Finally, the trigger condition of the interrupt register is configured.

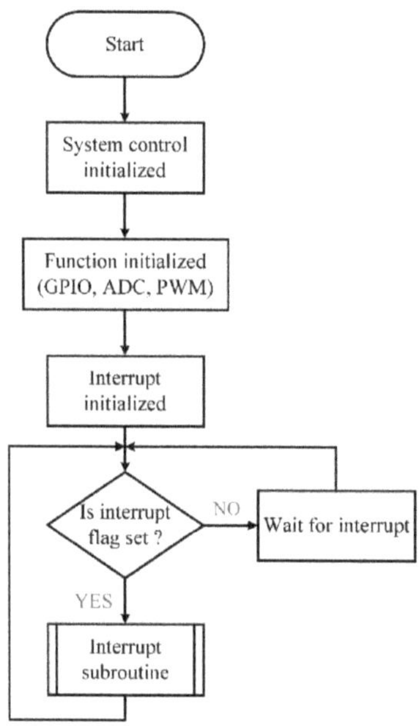


**Figure 11.** Flow chart of the main program.

The flow chart of the interrupt subroutine is shown in Figure 12. For the power flow control of the bidirectional converter to work, the following procedures should be made:

1. If the time-base counter operates in the up-count mode and the initial value of the counter is zero, the counter value increases until the PRD value is reached. The time-based counter is then reset to zero and then recount from zero to PRD value again. The cycle period of the count value is equivalent to one system cycle.
2. When the time-base counter equals zero, this event sets the flag of interrupt and enters the interrupt service routine. In addition, the ADC module is also e = set to perform the conversion.
3. After entering the interrupt service routine, the flag of interrupt must be cleared first so that the interrupt subroutine can be executed next time.
4. Finally, the state of the transfer signal is detected to determine at which mode the converter is operating. If the transfer signal is detected as 0, the converter operates in the discharging mode. On contrary, if the transfer signal is detected as 1, the converter operates in the charging mode.

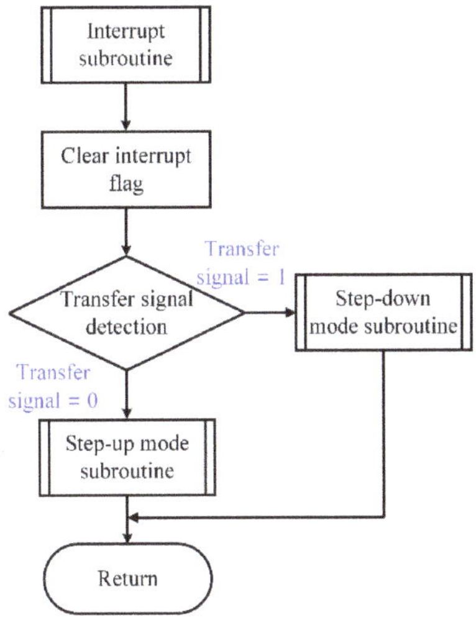


**Figure 12.** Flow chart of the interrupt subroutine.

After entering the interrupt service routine, the transfer signal is determined by the DSP. When the transfer signal is detected as 0, the converter operates in the step-up mode. On contrary, when the transfer signal is detected as 1, the converter operates in the step-down mode. In addition, the feedback circuits detect the values of the battery voltage and the DC bus voltage for voltage regulation. Finally, before leaving the interruption, the flag of interruption is cleared for the next interruption to occur. The block diagram of the control strategy is shown in Figure 13.

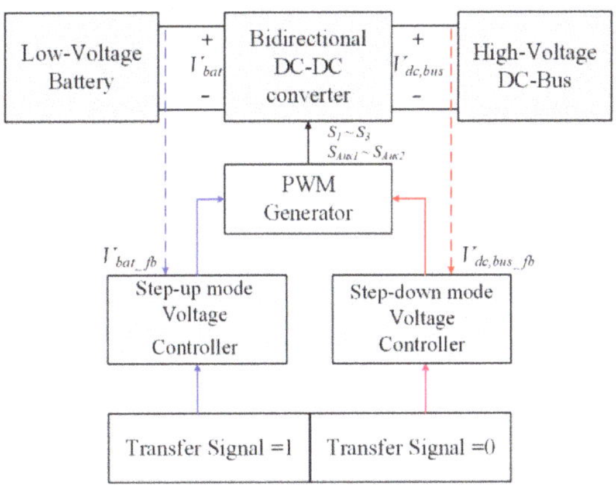


**Figure 13.** Block diagram of the control strategy.

## 4. Experimental Results

The experimental system consists of a bidirectional DC-DC converter and a DSP-based digital compensator, as shown in Figure 14. The bidirectional DC-DC converter is implemented with a battery voltage of 24 V, and a 200 V DC bus voltage. The related specifications are listed in Tables 1 and 2.

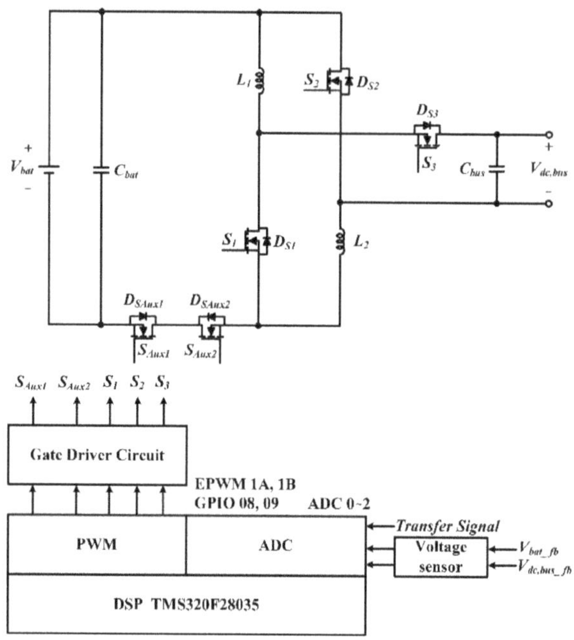


**Figure 14.** Block diagram of the experimental system for bidirectional power flow control.

**Table 1.** Specifications of the experimental system.

| Step-Up Mode | | Step-Down Mode | |
|---|---|---|---|
| Battery Voltage, $V_{bat}$ | 24 V | DC Bus Voltage, $V_{dc,bus}$ | 200 V |
| DC Bus Voltage, $V_{dc,bus}$ | 200 V | Battery Voltage, $V_{bat}$ | 24 V |
| Step-up Power, $P_{dis}$ | 500 W | Step-down Power, $P_b$ | 500 W |
| Switching Frequency, $f_{s1}$ | 50 kHz | Switching Frequency, $f_{s2}$ | 50 kHz |

**Table 2.** Parameters of the experimental system.

| Item | Value |
|---|---|
| Inductors $L_1$, $L_2$ | 185 µH |
| Capacitor $C_{bat}$ | 56 µF/35 V |
| Capacitor $C_{dc,bus}$ | 22 µF/250 V |
| Main Switches $S_1$, $S_2$, $S_3$ | IMW65R048M1H |
| Bidirectional Switches $S_{Aux1}$, $S_{Aux2}$ | FDH055N15A |

The hardware circuit of the proposed converter is shown in Figure 15, which is composed of synchronous rectification switches and bidirectional switches. The DSP is used to provide synchronous control and power flow control in the bidirectional DC-DC converter.

### *4.1. Step-Up Mode to Step-Down Mode (CCM Condition)*

The experimental waveforms with 50% load condition are shown in Figure 16. The gate driving signals $v_{gs1}$ and $v_{gs2}$ are complementary to the gate driving signal $v_{gs3}$. Moreover, the gate driving signal $v_{gs,Aux1}$ is complementary to the gate driving signal $v_{gs,Aux2}$. When the transfer signal changes from 0 to 1, the inductor currents $i_{L1}$ and $i_{L2}$ linearly demagnetize to zero, and the direction of the inductor currents immediately reverses to the opposite direction. Thus, the experimental results presented in Figure 16 validate that the proposed control strategy can effectively switch from the step-up mode to the step-down mode.

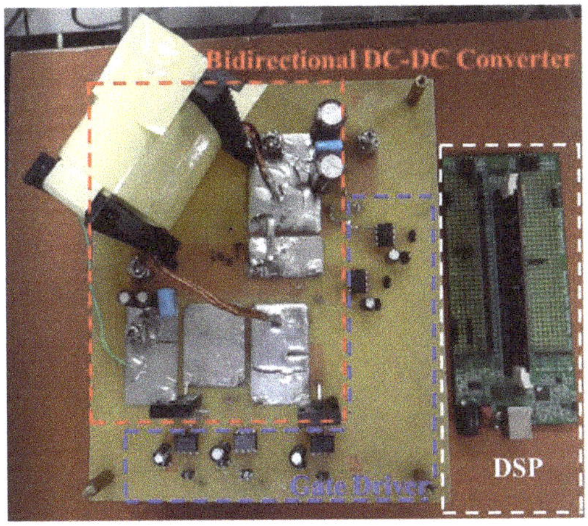


**Figure 15.** Photography of the experimental system.

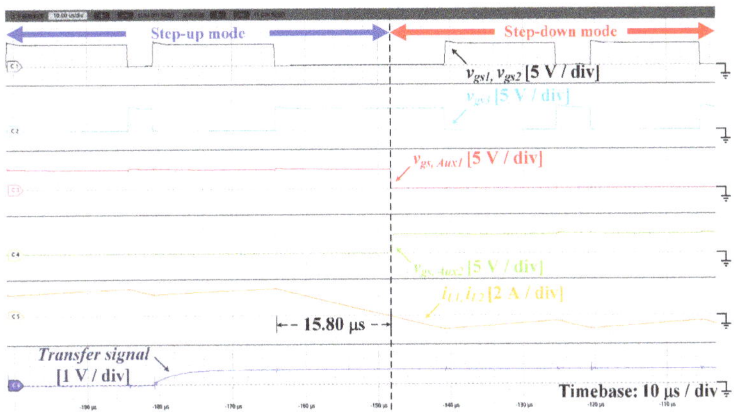


(**a**)

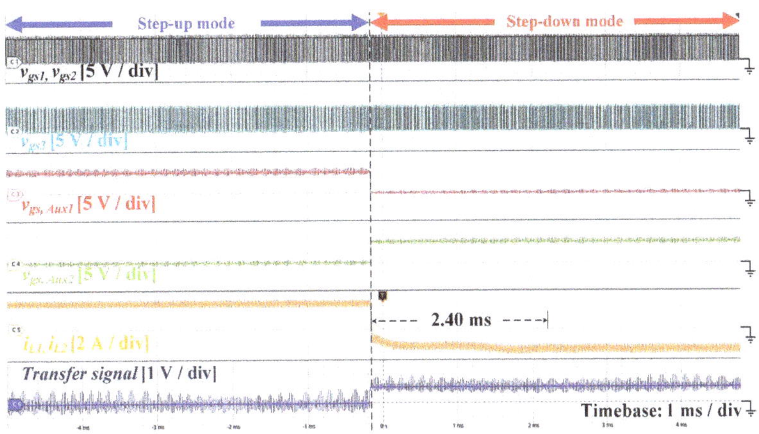


(**b**)

**Figure 16.** Waveforms of dynamic performances with 50% load condition. (**a**) $v_{gs1}$, $v_{gs2}$, $v_{gs3}$, $v_{gs,Aux1}$, $v_{gs,Aux2}$, $i_{L1}$, $i_{L2}$, transfer signal. (**b**) Change from the step-up mode to the step-down mode.

### 4.2. Step-Down Mode to Step-Up Mode (CCM Condition)

The experimental waveforms with 50% load condition are shown in Figure 17. The gate driving signals $v_{gs1}$ and $v_{gs2}$ are complementary to the gate driving signal $v_{gs3}$. Moreover, the gate driving signal $v_{gs,Aux1}$ is complementary to the gate driving signal $v_{gs,Aux2}$. When the transfer signal changes from 1 to 0, the inductor currents $i_{L1}$ and $i_{L2}$ linearly demagnetize to zero, and the direction of the inductor currents immediately reverses to the opposite direction. The experimental results presented in Figure 17 validate that the proposed control strategy can effectively switch from the step-down mode to the step-up mode.

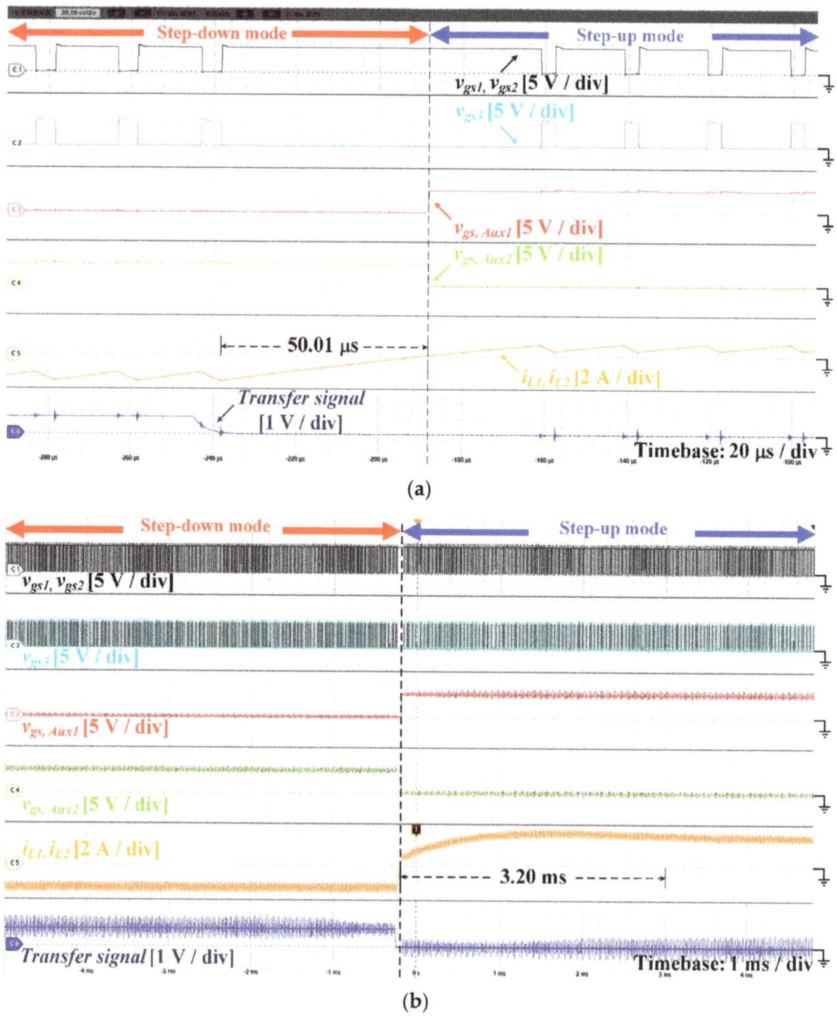


**Figure 17.** Waveforms of dynamic performances with 50% load condition. (**a**) $v_{gs1}$, $v_{gs2}$, $v_{gs3}$, $v_{gs,Aux1}$, $v_{gs,Aux2}$, $i_{L1}$, $i_{L2}$, transfer signal. (**b**) Change from the step-down mode to the step-up mode.

### 4.3. Step-Up Mode to Step-Down Mode (BCM Condition)

The experimental waveforms with BCM load conditions are shown in Figure 18. The gate driving signals $v_{gs1}$ and $v_{gs2}$ are complementary to the gate driving signal $v_{gs3}$. Moreover, the gate driving signal $v_{gs,Aux1}$ is complementary to the gate driving signal $v_{gs,Aux2}$. When the transfer signal changes from 0 to 1, the inductor currents $i_{L1}$ and

$i_{L2}$ linearly demagnetize to zero, and the direction of the inductor currents immediately reverses to the opposite direction. The experimental results presented in Figure 18 validate that the proposed control strategy can effectively switch from the step-up mode to the step-down mode within one switching cycle.

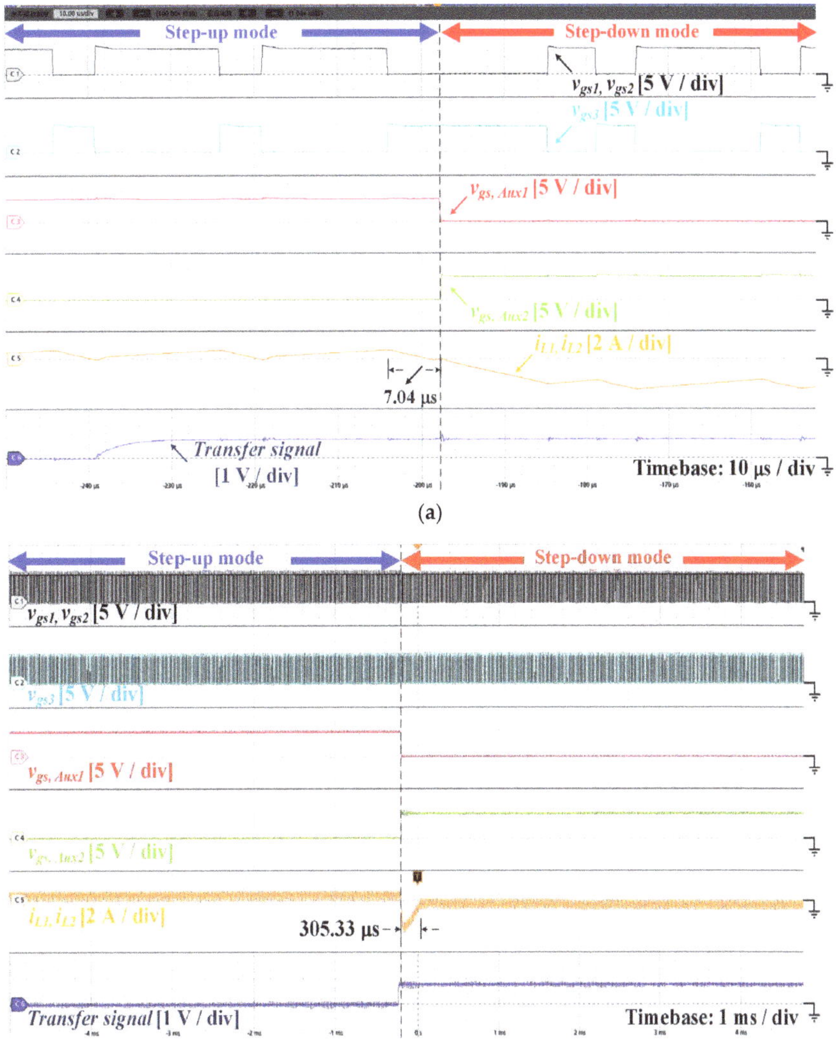


(a)

(b)

**Figure 18.** Experimental results of dynamic performances with BCM load condition. (**a**) $v_{gs1}$, $v_{gs2}$, $v_{gs3}$, $v_{gs,Aux1}$, $v_{gs,Aux2}$, $i_{L1}$, $i_{L2}$, transfer signal. (**b**) Change from the step-up mode to the step-down mode.

## 4.4. Step-Up Mode to Step-Down Mode (DCM Condition)

The experimental waveforms with DCM load conditions are shown in Figure 19. The gate driving signals $v_{gs1}$ and $v_{gs2}$ are complementary to the gate driving signal $v_{gs3}$. Moreover, the gate driving signal $v_{gs,Aux1}$ is complementary to the gate driving signal $v_{gs,Aux2}$. When the transfer signal changes from 0 to 1, the inductor currents $i_{L1}$ and $i_{L2}$ linearly demagnetize to zero and then commutate immediately. The experimental results in Figure 19 validate that the proposed control strategy can effectively switch from the step-up mode to the step-down mode within one switching cycle.

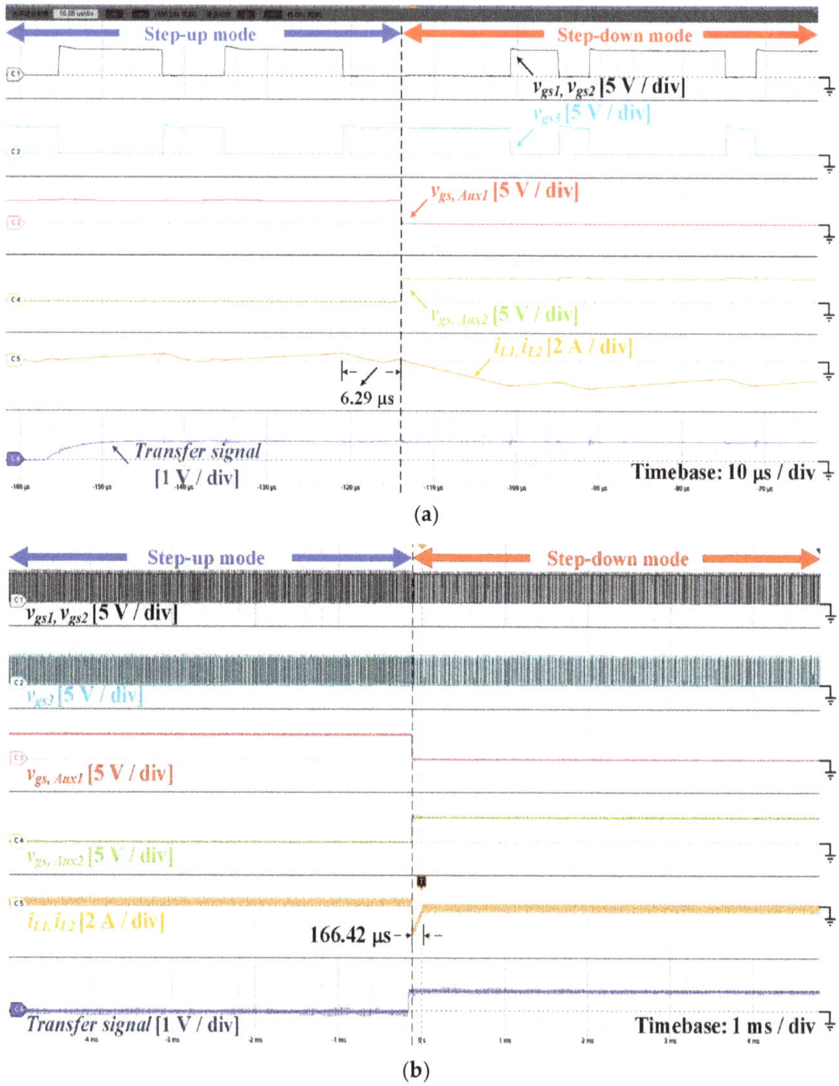


**Figure 19.** Experimental results of dynamic performances with DCM load condition. (**a**) $v_{gs1}$, $v_{gs2}$, $v_{gs3}$, $v_{gs,Aux1}$, $v_{gs,Aux2}$, $i_{L1}$, $i_{L2}$, transfer signal. (**b**) Change from the step-up mode to the step-down mode.

When the converter operates in DCM or BCM, because the inductor current decreases to zero within one switching cycle, the energy conversion can be completed in one switching cycle. The controller combines analog and digital methods for the transfer state. On the other hand, when the converter operates in CCM, the path for the freewheeling of the inductor current must be provided to avoid the occurrence of voltage spikes on the power switches. Therefore, the inductor current of the converter in the CCM condition requires longer duration for reaching the steady state. The comparison of the transition times of the converter operating in different modes is shown in Table 3.

Table 3. Comparisons of transition time.

| Mode | DCM | BCM | CCM |
|---|---|---|---|
| Step-up to Step-down | 6.29 μs | 7.04 μs | 15.8 μs |
| Step-down to Step-up | <20 μs | <20 μs | 50.01 μs |

### 4.5. Efficiency under Step-Up Mode

The efficiency curve of the converter at step-up mode with different load conditions is illustrated in Figure 20. The result shows that the maximum efficiency is 96.37% at 40% load and the full load efficiency is 93.66%.

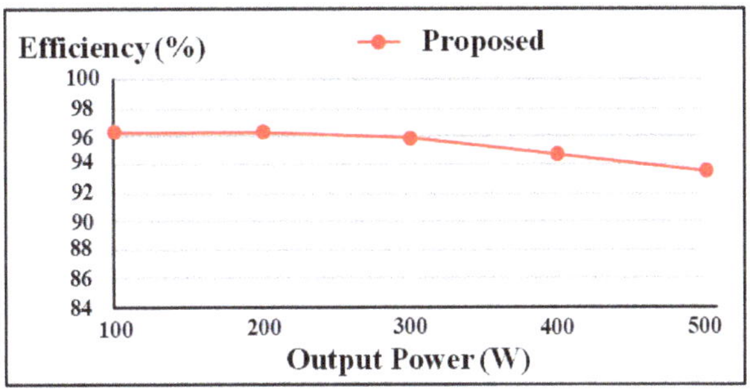


Figure 20. Efficiency curve under step-up mode.

### 4.6. Efficiency under Step-Down Mode

The efficiency curve of the converter at step-down mode for different load conditions is illustrated in Figure 21. The result shows that the maximum efficiency is 94.02% at 60% load and the full load efficiency is 93.08%.

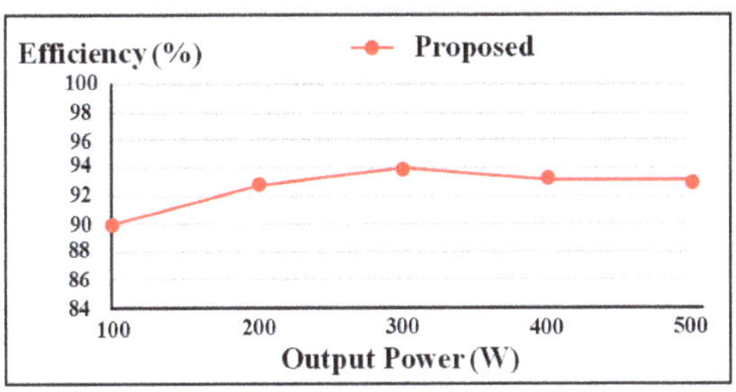


Figure 21. Efficiency curve under step-down mode.

## 5. Conclusions

The DC-DC converters have been widely used for connecting RES and ESS systems due to the need for energy storage for power generation with renewable energy. This paper proposes a bidirectional DC-DC converter with rapid energy conversion as the main circuit architecture for the energy conversion between the DC bus and the battery. The proposed method can improve the time required for energy transfer and provide smooth power flow under the same specification of the main components. In addition, the operational principles and the transferring states are analyzed in this paper.

A prototype converter for a 24 V battery, 200 V DC bus, and 500 W output power is constructed to confirm the feasibility of theoretical analyses. When the converter is operating in the step-up mode, the maximum efficiency of 96.37% is obtained at 40% load. In addition, the maximum efficiency of 94.02% is obtained at 60% load in the step-down mode. The experimental results presented indicated that the minimum transfer period is about 6.29 μs on the DCM stage. The resulting transition time for the experiment circuit can be completed within one switching cycle on DCM and BCM, and a minimum conversion time of 15.8 μs was obtained for the converter operating on CCM condition when step-up mode changed to step-down mode. Finally, the experimental results validate the effectiveness and precision of the proposed control strategies and rapid energy conversion.

While this study proposed a simplified idea for the feasibility of the theoretical proposal, many concerns can be discussed for future works. This includes constructing a complete system environment to control the converter, deriving a small-signal model of the converter, and improving the digital control design for a better conversion period.

**Author Contributions:** Conceptualization, J.-F.C. and B.-Z.C.; methodology, B.-Z.C.; validation, B.-Z.C. and H.L.; original draft preparation, H.L.; writing—Review and editing, L.C.; supervision, J.-F.C. All authors have read and agreed to the published version of the manuscript.

**Funding:** This study is supported by the Delta Electronics Foundation, and it was financially supported by the Hierarchical Green-Energy Materials (Hi-GEM) Research Center, from The Featured Areas Research Center Program within the framework of the Higher Education Sprout Project by the Ministry of Education (MOE) in Taiwan, and the Ministry of Science and Technology under Project MOST 110-2221-E-006-125, MOST 110-2634-F-006-017.

**Institutional Review Board Statement:** Not applicable.

**Informed Consent Statement:** Not applicable.

**Conflicts of Interest:** The authors declare no conflict of interest.

## References

1. Uang, Y.; Javanroodi, K.; Nik, V.M. Climate Change and Renewable Energy Generation in Europ—Long-Term Impact Assessment on Solar and Wind Energy High-Resilution Future Climate Data and Considering Climate Uncertainties. *Energies* **2022**, *15*, 302. [CrossRef]
2. Arraño-Vargas, F.; Shen, Z.; Jiang, S.; Fletcher, J.; Konstantinou, G. Challenges and measures in Power Systems with High Share of Renewables—The Australian Expperience. *Energies* **2022**, *15*, 429. [CrossRef]
3. Cardenas, B.; Styles, L.S.; Rouse, J.; Garvey, S.D. Short-, Medium- and Long-Duration Energy Storage in a 100% Reneable Electricity Grid: A UK Case Study. *Energies* **2022**, *14*, 8524. [CrossRef]
4. Chen, S.; Yang, B.; Pu, T.; Chang, C.; Lin, R. Active Current Sharing of a Parallel DC-DC Converters System Using Bat Algorithm Optimized Two-DOF PID Control. *IEEE Access* **2019**, *7*, 84757–84769. [CrossRef]
5. Bora, B.; Gudy, S.K. Performance and Analysis of Adaptable PI and SMC for a Hybrid PV-Battery System. In Proceedings of the 2020 IEEE International Women in Engineering (WIE) Conference on Electrical and Computer Engineering (WIECON-ECE), Bhubaneswar, India, 26–27 December 2020; pp. 84–89. [CrossRef]
6. Han, Y.; Ning, X.; Yang, P.; Xu, L. Review of Power Sharing, Voltage Restoration and Stabilization Techniques in Hierarchical Controlled DC Microgrids. *IEEE Access* **2019**, *7*, 149202–149223. [CrossRef]
7. Wu, H.; Sun, K.; Chen, L.; Zhu, L.; Xing, Y. High step-up/step-down soft-switching bidirectional DC-DC converter with coupled-inductor and voltage matching control for energy storage systems. *IEEE Trans. Ind. Electron.* **2016**, *63*, 2892–2903. [CrossRef]
8. Kwon, O.; Kim, J.; Kwon, J.; Kwon, B. Bidirectional Grid-Connected Single-Power-Conversion Converter with Low-Input Battery Voltage. *IEEE Trans. Ind. Electron.* **2018**, *65*, 3136–3144. [CrossRef]
9. Liang, T.J.; Lee, J. Novel high-conversion-ratio high-efficiency isolated bidirectional DC-DC converter. *IEEE Trans. Ind. Electron.* **2015**, *62*, 4492–4503. [CrossRef]
10. Bai, S.; Lukic, S.M. Unified active filter and energy storage system for an MW electric vehicle charging station. *IEEE Trans. Power Electron.* **2013**, *28*, 5793–5803. [CrossRef]
11. Bryden, T.S.; Hilton, G.; Dimitrov, B.; de León, C.P.; Cruden, A. Rating a Stationary Energy Storage System Within a Fast Electric Vehicle Charging Station Considering User Waiting Times. *IEEE Trans. Transp. Electrif.* **2019**, *5*, 879–889. [CrossRef]
12. Venkatramanan, D.; John, V. A Reconfigurable Solar Photovoltaic Grid-Tied Inverter Architecture for Enhanced Energy Access in Backup Power Applications. *IEEE Trans. Ind. Electron.* **2020**, *67*, 10531–10541. [CrossRef]
13. IEEE Guide for Batteries for Uninterruptible Power Supply Systems (Revision of IEEE 1184-1994). In *IEEE Std 1184-2006(R2011)*; IEEE: New York, NY, USA, 2006; pp. 1–73. [CrossRef]
14. Mozos, A.B.; Mouli, G.R.C.; Bauer, P. Evaluation of topologies for a solar powered bidirectional electric vehicle charger. *IET Power Electron.* **2019**, *12*, 3675–3687. [CrossRef]

15. Zheng, Y.; Li, S.; Dey, S.; Smedley, K. Family of isolated bidirectional resonant converters with duty-cycle control and automatic power flow transition. *IET Power Electron.* **2018**, *11*, 1582–1590. [CrossRef]
16. Sharma, A.; Nag, S.S.; Bhuvaneswari, G.; Veerachary, M. An improved mode transition technique for a non-isolated bidirectional DC-DC converter. *IEEE Trans. Circuits Syst. II* **2020**, *67*, 3093–3097. [CrossRef]
17. Cornea, O.; Andreescu, G.; Muntean, N.; Hulea, D. Bidirectional power flow control in a DC microgrid through a switched-capacitor cell hybrid DC-DC converter. *IEEE Trans. Ind. Electron.* **2017**, *64*, 3012–3022. [CrossRef]
18. Vuyyuru, U.; Maiti, S.; Chakraborty, C. Active power flow control between DC Microgrids. *IEEE Trans. Smart Grid* **2019**, *10*, 5712–5723. [CrossRef]
19. Bayati, M.; Farahmandrad, M.; Tehrani, K. More and Faster Energy Transfer Capability for Battery Chargers of Electric Vehicles. In Proceedings of the 2021 World Automation Congress (WAC), Paris, France, 24–26 August 2021; pp. 228–232. [CrossRef]
20. Khan, I.A.; Pi, D.; Yue, P.; Li, B.; Khan, Z.U.; Hussain, Y.; Nawaz, A. Efficient behavior specification and bidirectional gated recurrent units-based intrusion detection method for industrial control systems. *IET Electron. Lett.* **2020**, *56*, 27–30. [CrossRef]
21. Callegaro, L.; Ciobotaru, M.; Pagano, D.J.; Turano, E.; Fletcher, J.E. A simple smooth transition technique for the noninverting buck-boost converter. *IEEE Trans. Power Electron.* **2018**, *33*, 4906–4915. [CrossRef]
22. Tang, Y.; Chen, Y.; Madawala, U.K.; Thrimawithana, D.J.; Ma, H. A new controller for bidirectional wireless power transfer systems. *IEEE Trans. Power Electron.* **2018**, *33*, 9076–9087. [CrossRef]
23. Xu, J. PWM modulation and control strategy for LLC-DCX converter to achieve bidirectional power flow in facing with resonant parameters variation. *IEEE Access* **2019**, *7*, 54693–54704. [CrossRef]
24. Waghmare, T.; Chaturvedi, P. Study of Bidirectional DC-DC Converter: Control Schemes and Switching Techniques. In Proceedings of the 2020 IEEE First International Conference on Smart Technologies for Power, Energy and Control (STPEC), Nagpur, India, 25–26 September 2020; pp. 1–6. [CrossRef]
25. Bhaskar, M.S.; Meraj, M.; Iqbal, A.; Padmanaban, S. Nonisolated Symmetrical Interleaved Multilevel Boost Converter with Reduction in Voltage Rating of Capacitors for High-Voltage Microgrid Applications. *IEEE Trans. Ind. Appl.* **2019**, *55*, 7410–7424. [CrossRef]
26. Zhu, R.; Hoffmann, F.; Vázquez, N.; Wang, K.; Liserre, M. Asymmetrical bidirectional DC-DC converter with limited reverse power rating in smart transformer. *IEEE Trans. Power Electron.* **2020**, *35*, 6895–6905. [CrossRef]

energies

MDPI

Article

# Design and Control of Modular Multilevel Converter for Voltage Sag Mitigation

Fazal Muhammad [1,*], Haroon Rasheed [1], Ihsan Ali [2], Roobaea Alroobaea [3] and Ahmed Binmahfoudh [4]

1 Electrical Engineering Department, Bahria University Karachi Campus, Karachi 75260, Pakistan; haroonrasheed.bukc@bahria.edu.pk
2 Department of Computer System and Technology, Faculty of Computer Science and Information Technology, Universiti Malaya, Kuala Lampur 50603, Malaysia; ihsanalichd@siswa.um.edu.my
3 Department of Computer Science, College of Computers and Information Technology, Taif University, P.O. Box 11099, Taif 21944, Saudi Arabia; r.robai@tu.edu.sa
4 Department of Computer Engineering, College of Computers and Information Technology, Taif University, P.O. Box 11099, Taif 21944, Saudi Arabia; A.binmahfoudh@tu.edu.sa
* Correspondence: fazalkhan00@gmail.com; Tel.:+92-332-8068744

**Abstract:** Voltage sag in a power system is an unavoidable power quality issue, and it is also an urgent concern of sensitive industrial users. To ensure the power quality demand and economical operation of the power system, voltage sag management has always drawn great attention from researchers around the world. The latest research that realizes the power quality conditioning has used dynamic voltage restorers (DVRs), static VAR compensator (SVCs), adaptive neuro-fuzzy inference systems (ANFISs), and fuzzy logic controllers based on DVR to mitigate voltage sag. These devices, methods, and control strategies that have been recently used for voltage sag mitigation have some limitations, including high cost, increased complexity, and lower performance. This article proposes a novel, efficient, reliable, and cost-effective voltage sag mitigation scheme based on a modular multilevel converter (MMC) that ensures effective power delivery at nominal power under transient voltage conditions. The proposed method, the MMC, compensates for the energy loss caused by voltage sags using its internal energy storage of the submodules, and ensures reliable power delivery to the load distribution system. Furthermore, control strategies are developed for the MMC to control DC voltage, AC voltage, active power, and circulating current. Detailed system mathematical models of controllers are developed in the dual synchronous reference frame (DSRF). Validation of the results of back-to-back MMC for dynamic load distribution system is analyzed which proves the effectiveness of the proposed scheme for voltage sag mitigation.

**Keywords:** dynamic load distribution; voltage sag mitigation (VSM); modular multilevel converter (MMC); circulating current suppression control (CCSC); dual synchronous reference frame (DSRF); proportional resonance controller (PRC); inner current controller (ICC)

**Citation:** Muhammad, F.; Rasheed, H.; Ali, I.; Alroobaea, R.; Binmahfoudh, A. Design and Control of Modular Multilevel Converter for Voltage Sag Mitigation. *Energies* **2022**, *15*, 1681. https://doi.org/10.3390/en15051681

Academic Editor: Andrea Mariscotti

Received: 8 January 2022
Accepted: 18 February 2022
Published: 24 February 2022



## 1. Introduction

Voltage sag is one of the major concerns of the modern industry, as it can interrupt sensitive electrical loads and in the worst case cause production problems. According to the official definition of voltage sag, "It is the phenomenon in which the magnitude of voltage is reduced below 10% of nominal RMS value over a time ranging from one half-cycle to one minute" [1]. A voltage sag is defined by IEEE standard 1159–1995 as "A decrease in the RMS voltage from 10% to 90% of nominal value for the duration of 0.5 cycles to 1 min" [2]. Time duration and voltage drop for voltage sag differ from one grid code to another. According to different countries' grid codes and regulations, different limits are allowed for voltage drop and time duration in case of voltage sag events, as shown in Table 1 [3].

**Table 1.** The voltage drop and time duration allowed differently based on grid codes in different countries.

| S. No | Country | Voltage Drop | Time |
|---|---|---|---|
| 1 | USA | 15% | 0.6 s |
| 2 | UK | 15% | 0.14 s |
| 3 | China | 20% | 0.625 s |
| 4 | Italy | 0% | 0.2 s |
| 5 | Japan | 15% | 1.0 s |
| 6 | Germany | 0% | 0.15 s |
| 7 | South Africa | 0% | 0.15 s |
| 8 | Spain | 0% | 1.5 s |
| 9 | Australia | 15% | 0.45 s |
| 10 | Denmark | 20% | 0.5 s |

It is one of the serious power quality challenges that are caused by faults, energization of heavy distribution feeders, and abrupt rise of heavy load. Voltage sag can damage sensitive industrial load equipment and create high power losses in power systems [4]. Different devices, methods, and control strategies were used, i.e., dynamic voltage restorer (DVR) in [5], static VAR compensator (SVC) in [6], ANFIS-based UPQC in [7], and fuzzy logic controller based DVR in [8], to mitigate voltage sag and to address power quality conditioning of load distribution system.

### *1.1. Voltage Sag Mitigation Based on Various Technologies*

Power system quality has gained a lot of interest over the last decade. There are a variety of devices and control methods used to mitigate the risks associated with power quality in the power system. Different technologies and methods are developed based on power system quality to control the voltage sag of the AC grid system.

#### 1.1.1. Dynamic Voltage Restore (DVR)

Dynamic voltage restorer is the device that ensures optimal power system performance while minimizing the adverse effects of voltage sags, harmonics, unbalanced voltages, and voltage swells. A DVR includes four components, i.e., voltage source inverter, energy storage system, passive filters, and injection transformer (IT), as illustrated in Figure 1. In ref. [9], a DVR is used to inject three-phase voltages in series with grid voltage to compensate for voltage disturbances caused by asymmetrical faults.

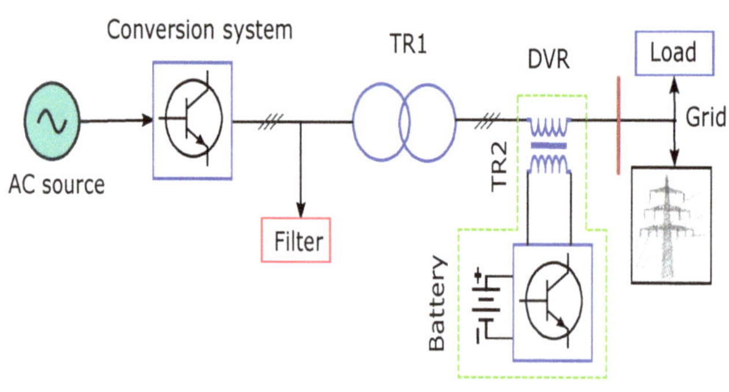


**Figure 1.** Dynamic voltage restorer–integrated AC grid system to mitigate voltage sag.

It maintains nominal voltages at the point of connection and compensates voltage sag, harmonics, and voltage unbalance. Nevertheless, some limitations in terms of low-voltage ride-through still exist. Ref. [5] mitigated voltage sag effectively during different types of grid faults using an efficient dynamic voltage restorer (DVR) since its cost is high and it has high complexity. In ref. [10], a fuzzy logic controller-based dynamic voltage restorer was designed that mitigates voltage sag up to 50%, but the system has the limitation of the frequency fluctuation. The DVR in ref. [11] mitigates voltage sag, voltage unbalance factor to less than one percent, and voltage total harmonic distortion to less than five percent.

### 1.1.2. Static VAR Compensator (SVC)

In an electrical power system, a static VAR compensator controls parameters such as bus voltage by utilizing static VAR generators or absorbers. It is a flexible AC transmission system device as shown in Figure 2, which has been extensively used in recent years to rectify various power quality problems caused by faults causing voltage sags in the system. In ref. [12], the authors compared static VAR compensators' effectiveness for managing voltage sag events and concluded that SVC contributes the least to transient margin (TM). In ref. [6], the static VAR compensator is used in the AC grid system to enhance power quality and ensure power efficiency, but its performance was lower than that of STATCOM and DVR.

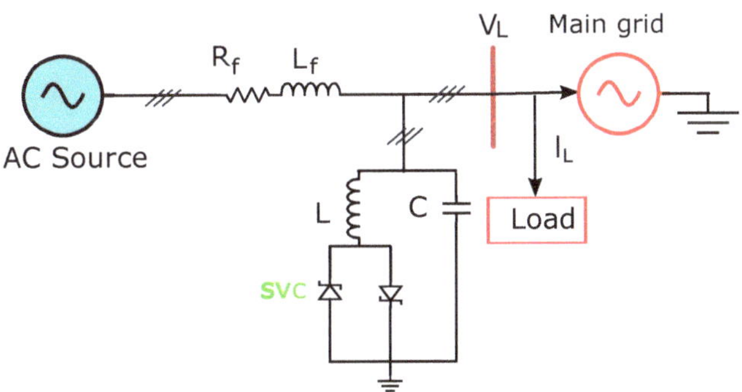


**Figure 2.** Typical configuration of SVC used for mitigation of voltage sag in AC grid system.

### 1.1.3. Unified Power Quality Conditioner

Similar to a DVR, UPQC has the capability of dealing with power quality issues of the power system, such as harmonics and voltage fluctuations. It is designed to enhance power system performance while minimizing disturbances that adversely affect the performance of critical loads. In addition to regulating the flow of power, UPQCs compensate for harmonics, reactive power, and voltage disturbances. A UPQC contains two voltage source converters linked with a common DC link in a single-phase/three-phase configuration. In Figure 3, the UPQC combines shunt and series controllers into one DC bus. Shunt controllers can produce or absorb reactive power at the common point of coupling.

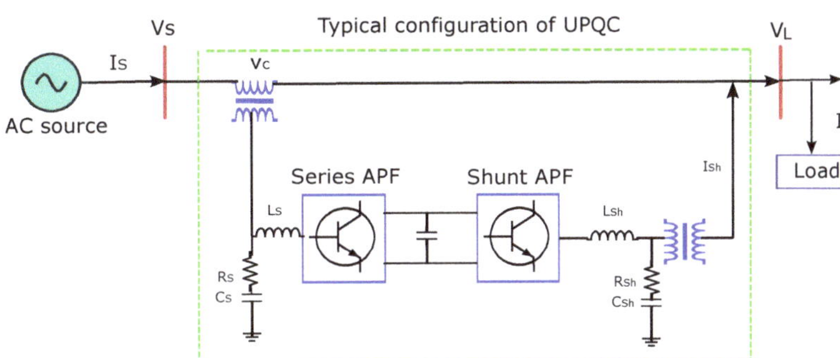


**Figure 3.** General configuration of UPQC for voltage sag mitigation in AC grid system.

Accordingly, the series controller is connected to the AC grid systems and controls the parameters of the lines, as it is described in detail in [13]. Ref. [14] used UPQC with the fuzzy logic controller to minimize voltage sag and current harmonics. Its results indicate a reduction in a harmonic distortion from 8.93% to 3.34%. According to [15], UPQC is used in a hybrid PV and wind system microgrid to reduce voltage sag and swell by using reactive current injection or absorption. Ref. [7] apply an adaptive neuro-fuzzy interference system (ANFIS) based on a unified power quality conditioner (UPQC) to address voltage sag mitigation and THD reduction since it is costly and quite complex. A power quality controller will additionally compensate for voltage sags, surges, and harmonics, and will limit voltage imbalance in microgrids that are linked to the main grids in the ref. [16]. Table 2 presents a comparison between the various popular technologies used for voltage sag mitigation.

**Table 2.** Comparison of various technologies used for voltage sag mitigation.

| Factors | DVR | SVC | UQPC |
|---|---|---|---|
| Rating | High rating | Low rating | High rating |
| Speed of operation | Fast | Less than DVR | Faster |
| Compensation method | Series compensation | Shunt compensation | Both series and shunt |
| Active & Reactive power | Reactive/active | Reactive | Both active and reactive |
| Harmonics | Much less | Less | Least |
| Problem addressed | Voltage sag/Swell | Voltage sag/Swell | Sag/Swell/Transients |
| Complexity | High | High | Higher |
| Cost | High | High | Higher |

The comparison of different voltage sag mitigation technologies includes cost, rating, and various aspects of performance. In conclusion, a DVR provides superior stability to that of SVC in terms of voltage sag, voltage swell, and voltage fluctuation mitigation, while UPQC provides superior protection against poor-quality sources for sensitive loads. The different control methods and devices used to mitigate voltage sag, mentioned in Table 2, has some limitations, i.e., increased complexity, high cost, and lower performance.

Modular multilevel converter (MMC) is an emerging and state-of-the-art power electronic-based technology first introduced in ref. [17]. It has the potential of scalability to meet voltage level requirements, low level of power losses, high reliability, enhanced efficiency, and better harmonic suppression quality [18–20]. It requires the grid side filter to be smaller in size since it has lower harmonics in its output voltage [21]. A modular multilevel converter is an ideal converter with the flexibility of hundreds of output voltage levels for high-voltage three-phase motor drives [22], electric railways, and high-voltage transmission [23]. To address the limitations of the above methods and control strategies mentioned for voltage sag mitigation technologies in Table 2, the MMC can be designed to operate continuously in islanded mode with satisfactory power quality. The voltage sag in the upstream AC grid system caused by asymmetrical faults can be successfully mitigated by using the MMC to ensure satisfactory power delivery to the load distribution system. Using modern MMC technology, this paper presents a novel approach to efficiently mitigate voltage sag sustained in upstream AC grid systems.

### 1.2. Contributions

The following contributions are summarized in this paper:

- An analysis of an MMC component dimensioning and control system is provided, followed by a consideration of whether an MMC can be specifically designed for dynamic load distribution ensuring power quality.
- The article explores the feasibility and potential of the MMC to mitigate voltage sag due to asymmetrical fault in the upstream AC grid side and to deliver constant power to load distribution.
- The paper examines the effect of voltage sag of the magnitude of up to −60% for 75 ms in the upstream AC grid when using the internal energy storage of back-to-back modular multilevel converters.

### 1.3. Organization

The remainder of this article is organized as follows: Section 2 of this article explains the design principles of MMC, selection of characteristic parameters, component sizing, and all the relevant involved procedures in detail. Section 3 describes the proposed system in detail. Section 4 explains the mathematical modeling of different control systems of the MMC and the control strategies developed for the regulation of circulating current, active and reactive power, and AC and DC voltages of the MMC-rectifier and the MMC-inverter. Section 5 of this research article explains, in detail, the simulation results for the voltage sag mitigation strategy of the MMC and gives a comparison of the commonly employed and proposed control strategy for voltage sag mitigation. Section 6 concludes the research article.

## 2. MMC Component Design Principles

It requires extensive planning to design a modular multilevel converter, including sizing, and dimensioning of its components. The design procedure of the MMC includes determining the number of submodules (SMs) per arm, DC link voltage, submodule switching devices as per industrial standards and ratings, submodule capacitor's capacitance, and arm inductance.

The flow chart of the MMC design is shown in Figure 4. Initially, it is required to determine the apparent power of MMC, which can be calculated by

$$S = \sqrt{3} V_{LL} I_{RMS} \tag{1}$$

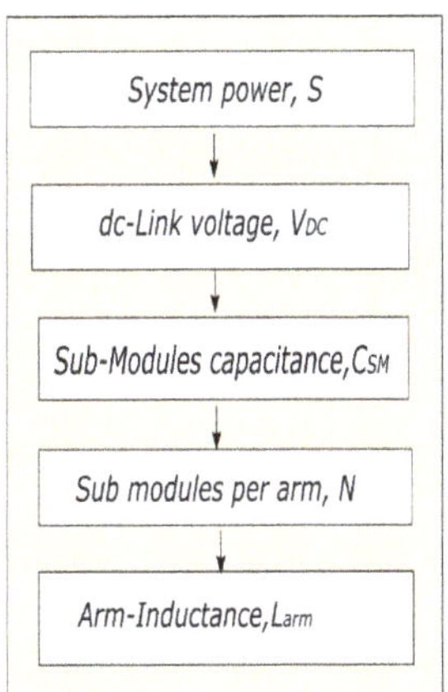


**Figure 4.** Component design flow chart of modular multilevel converter.

The DC-link voltage of a modular multilevel converter determines the average voltage of a submodule capacitor. The ratings of semiconductor-based IGBT technology limit the average voltage $v_{SM}$ of a submodule of the MMC. The DC-link voltage of the MMC can be determined by multiplying the average voltage of the capacitor with the total number of submodules per arm, taking into account standard current and voltage ratings of semiconductors. The submodule capacitor acts as an energy buffer, and its capacitance dimensioning is necessary for energy storage calculations. The capacitance of the capacitor of SM should be chosen with a low level of voltage ripples. The submodule capacitor must be dimensioned properly so that it allows the maximum arm current of the converter. Arm inductance of the MMC is required to limit the circulating current in the arms, short-circuit current, and to isolate the lower and upper arms of the MMC.

### 2.1. DC-Link Voltage of the MMC

According to the ideal relationship between the AC and DC side of the modular multilevel converter, without taking account of internal losses, this equation must hold.

$$P_{dc} = P_{ac} \tag{2}$$

$$V_{dc} I_{dc} = \sqrt{3} V_{LL} I_{Phase}$$

where $I_{Phase}$ is

$$I_{Phase} = \frac{2\sqrt{2}}{3m_a} I_{dc}$$

$$V_{dc} = \frac{2\sqrt{2} V_{LL,rms}}{\sqrt{3} m_a} \tag{3}$$

$V_{dc}$ can be easily calculated from

$$V_{dc} = 1.6330 \frac{V_{LL,rms}}{m_a} \tag{4}$$

The DC-link voltage of the modular multilevel converter has a direct relation with the RMS value of the AC grid line-to-line voltage and inverse proportion with the modulation index of the converter.

### 2.2. Submodule Capacitance, $C_{SM}$

The capacitor of the submodule (Figure 5) is the main component of the converter which acts as an energy buffer. Its sizing and dimensioning play a vital role while designing the submodule of the converter. The capacitor of the submodule is floating in nature owing to its charging and discharging by converter arm currents. The arm current causes inevitable voltage ripples, which have an inverse relation with the capacitance of the capacitor of the submodule.

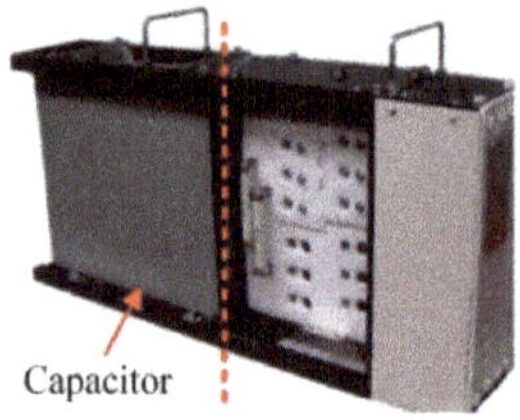


**Figure 5.** Submodule of MMC.

Choosing a reasonable capacitance of capacitor for submodule can reduce the magnitude of voltage ripple. Capacitor voltage ripple can also be controlled by eliminating the harmonics in circulating currents. Selection of the arbitrary high value of capacitance of capacitor for submodule may be exorbitant and unnecessary. The value of $C_{SM}$ can be calculated as proposed in [24].

$$C_{SM} = \frac{S}{3Nm_a V_c^2 \varepsilon \omega} (1 - (\frac{m cos\theta}{2})^{\frac{3}{2}}) \tag{5}$$

where $S$ is the apparent power of converter that is $S = \frac{P}{cos\theta}$, $N$ is the total number of submodules per arm, $m_a$ is voltage modulation index, $V_c$ is time average of submodule capacitor voltage, $\omega$ is the fundamental frequency, $\varepsilon$ is ripple voltage of the capacitor (peak to average), and $\theta$ is the power factor of AC upstream grid side power factor. The above equation offers to derive the capacitance of capacitor $C_{SM}$; only the voltage ripple of capacitor needs consideration in designing and dimensioning of the capacitor of submodule.

In ref. [25], 10% voltage ripple of the capacitor of SM is considered suitable. Another approach in reference [25] that calculates the capacitance of the capacitor of SM by using the stored energy $E_c$ in all submodules is given by

$$E_c = 6N\frac{1}{2}CV_c^2 \tag{6}$$

$$E_c = 3NC(\frac{V_{dc}}{N})^2$$

$$E_c = \frac{3CV_{dc}^2}{N} \tag{7}$$

To bring some abstraction and to ease in the comparison of different systems, energy–power ratio, E.P., is defined by

$$Energy\ power\ ratio = \frac{E_c}{S} \tag{8}$$

from the above equation of $C_{SM}$ and energy to power ratio, the capacitance of the capacitor of the submodule can be written in terms of energy to power ratio, total number of submodules, apparent power, and DC-link voltage. Using energy–power ratio, the capacitance of the capacitor can be adjusted as

$$C_{arm} = \frac{EP.N.S}{3V_{dc}^2} \tag{9}$$

$$C_{SM} = C_{arm}N \tag{10}$$

In ref. [25], 30–40 kJ/MVA energy is needed per converter station HVDC-based MMC, and 60–80 kJ/MVA for back to back converter system.

### 2.3. Determining of Submodules per Arm of MMC

The potential advantage of an MMC over a conventional VSC is its adaptability and scalability to different power and voltage levels. While designing the MMC for different power and voltage levels, the number of SMs per arm is one of the basic parameters to be determined. The selection of the number of submodules per arm plays an important role to handle the high voltage and power of the system. DC-link voltage is equally distributed to the "$N$" number of submodules of an arm, and the average voltage $V_{SM}$ of a submodule capacitor is $V_{dc}/N$. The mathematical expression for determining the number of submodules is

$$N = \frac{V_{dc}}{V_{SM}} \tag{11}$$

To determine the number of submodules also needs a deep consideration of the selection of semiconductor ratings because semiconductors limit the voltage of submodules. Current and voltage ratings are the key parameters of the integrated gate bipolar transistor (IGBT) module. Today, commercially available semiconductor-based IGBT technology has limited kV ratings. The limitation of the kV ratings should be considered while determining the number of submodules. The semiconductor-based high-voltage IGBT module with the modules package available in the market is shown in Table 3.

**Table 3.** Market availability of high-voltage IGBT modules.

| Blocking Voltage | Current Rating (Ampere) | | | | |
|---|---|---|---|---|---|
| 1.7 kV | 1000 | 1200 | 1600 | 3600 | - |
| 3.3 kV | 450 | 600 | 800 | 1200 | - |
| 4.5 kV | 900 | 1000 | 1350 | 1500 | - |
| 6.5 kV | 225 | 300 | 600 | 900 | 1000 |

In case of failure or damage of any submodule in either arm, redundant submodules should be provided to improve the reliability and stability of the system.

### 2.4. Arm's Inductance, $L_{arm}$

The arm inductor $L_{arm}$ of the MMC is also called outdoor air-core reactor, which has multiple functions in a modular multilevel converter. It allows the converter to control reactive power and second harmonic circulating current, and limits the rise of the current in short circuit conditions. The concept in ref. [26] is of an arm inductor forming a resonance frequency $\omega_r$ with submodule capacitance

$$\omega_r = \sqrt{\frac{N}{L_{arm}C_{SM}}} \cdot \sqrt{\frac{2(n^2-1)+m^2n^2}{8n^2(n^2-1)}} \tag{12}$$

In the above equation, $n$-th harmonic currents can only be limited by arm resistance. Modular multilevel converter should operate above the high value of resonant frequency (RF) which occurs at $m = 1$ and $n = 2$. The harmonic number n is given by the equation $n = 3k \pm 1$. The above equation can be written as

$$f_r = \frac{1}{2\pi}\sqrt{\frac{5N}{48L_{arm}C_{SM}}} \tag{13}$$

The product of arm inductance and SM capacitance must have minimum value depending on operating frequency $\omega_r$ and the number of SM per arm:

$$L_{arm}C_{SM} > \frac{5N}{\omega^2} \tag{14}$$

Arm inductance also limits fault current in the case of DC-link short-circuit condition. It is affected by several factors, including the submodule capacitor voltage $V_{dc}/n$, the modulation technique, the switching frequency, and a controller optionally used for suppression of circulating current. Based on the literature, the typical value for arm inductance is 0.15 p.u. of base AC input in [27].

## 3. Proposed System Description

A system of back-to-back MMC that ensures the power quality demand of load distribution is shown in Figure 6. It consists of an upstream AC grid system, back-to-back MMC, DC grid system, and dynamic load distribution system. In the proposed system, we intend to explore the feasibility of the MMC for voltage sag mitigation and to achieve reliable power for load distribution. The concept and feasibility of the proposed system are to be proven through simulation results studies, showing the functionality of sag mitigation, effective control system, and quick response of the MMC converter. Characteristic parameters, component sizing, and dimensioning are performed as per design participles of MMCs described in Section 2 of this article.

A 10 MW/11 kV dynamic load distribution system is connected with 220 kV AC and 135 kV DC grid systems linked through the MMC-rectifier and the MMC-inverter. The MVA rating of the MMC-rectifier and the MMC-inverter is 220 MVA. The structure of the MMC is illustrated in Figure 7. The structure of the MMC contains six arms, each of them consisting of 60 submodules with a capacitance of 0.0105 p.u. and a series arm reactor with an inductance of 0.0164 p.u. The upstream 220 kV AC grid voltage is stepped down by core type (Y-D) power transformer to 66 kV and fed to the MMC-rectifier, which is linked with 135 kV DC grid system. The (Y-D) power transformers have the capability of limiting the zero-sequence current. Since the MMC will not experience zero-sequence current, the zero-sequence current controller is not necessary. A core type (Y-D) power transformer is used to step down MMC-inverter voltage from 66 kV to 11 kV, feeding the 10 MW dynamic load distribution system. The proposed system is built up to mitigate voltage sag occurring in upstream AC grid systems due to asymmetrical faults using the internal energy stored in the submodules of back-to-back MMC to compensate the voltage sag and to ensure satisfactory power delivery for the operation of dynamic load distribution.

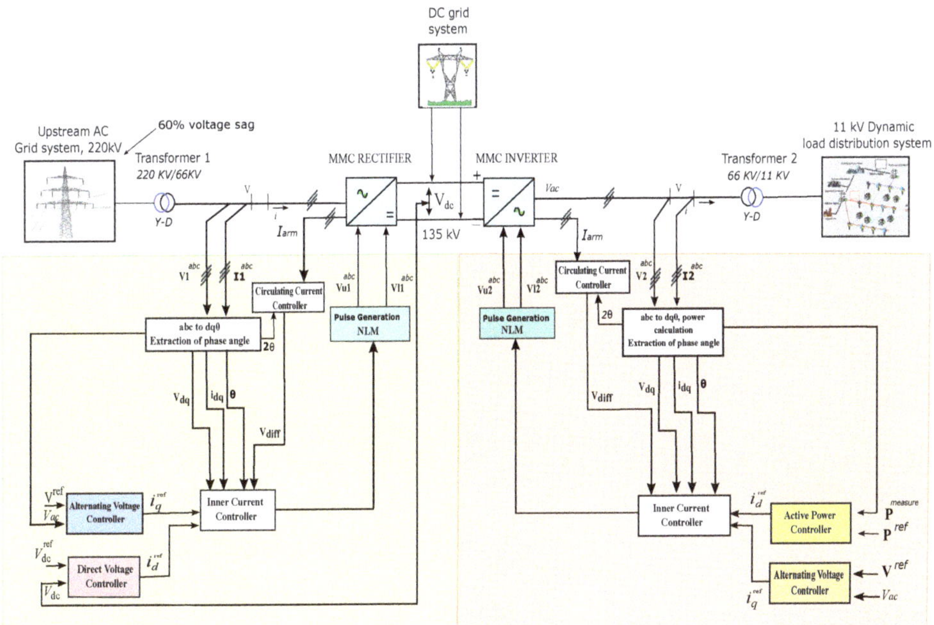

**Figure 6.** A detailed control system model of the MMC-rectifier and the MMC-inverter.

Therefore, to accomplish the principal objective of the proposed scheme, it requires an effective control system for quick response to the system. A dual synchronous reference frame (DSRF)-based current controller is developed for the current control mode of the MMC-rectifier and the MMC-inverter. The outer control loop for the MMC-rectifier consists of AC and DC voltage control, whereas AC voltage and active power control for the MMC-inverter. Circulating current suppression controller is designed to control the differential current in between the phases of modular multilevel converter which distorts the arm currents and increases the peak value of arm current causing system power losses.

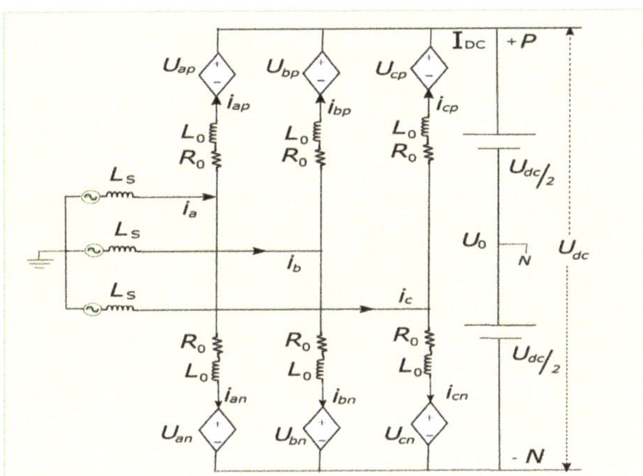

**Figure 7.** Equivalent circuit of MMC.

During the unbalanced grid conditions, positive-sequence and negative-sequence current components exist which are fully controllable for an MMC. The controller for positive- and negative-sequence current components is illustrated in Figure 6.

Moreover, for efficiency and accuracy in reproducing dynamic behavior, analytical and simulation reasons, and increasing performance for the analysis of converter response, we use the aggregate model of a modular multilevel converter. This MMC model relies on the assumption of perfect capacitor balancing. The equivalent circuit of the MMC for the mathematical modeling of system-level control design is illustrated in Figure 7. In an equivalent circuit of the MMC, $u_{ap}$, $u_{bp}$, and $u_{cp}$ are the upper arm voltages, whereas $u_{an}$, $u_{bn}$, and $u_{cn}$ are lower arm voltages. $i_{ap}$, $i_{bp}$, $i_{cp}$ and $i_{an}$, $i_{bn}$, $i_{cn}$ are the upper and lower arm currents in three-phase legs of the MMC. The sum of currents in the upper and lower arm of each phase of the MMC yields phase currents $i_a$, $i_b$, and $i_c$.

$$[i_a \quad i_b \quad i_c]^T = [i_{ap} \quad i_{bp} \quad i_{cp}]^T + [i_{an} \quad i_{bn} \quad i_{cn} \quad ]^T \tag{15}$$

The differential current in the arm of each phase is $i_{ad}$, $i_{bd}$, and $i_{cd}$ given by

$$[i_{ad} \quad i_{bd} \quad i_{cd}]^T = \frac{[i_{ap} \quad i_{bp} \quad i_{cp}]^T - [i_{an} \quad i_{bn} \quad i_{cn}]^T}{2} \tag{16}$$

Currents in the upper arm and lower arm in terms of their difference are illustrated in the following equations:

$$[i_{ap} \quad i_{bp} \quad i_{cp}]^T = \frac{1}{2}[i_a \quad i_b \quad i_c]^T + [i_{ad} \quad i_{bd} \quad i_{cd}]^T \tag{17}$$

$$[_{an} \quad i_{bn} \quad i_{cn}]^T = \frac{1}{2}[i_a \quad i_b \quad i_c]^T - [i_{ad} \quad i_{bd} \quad i_{cd}]^T \tag{18}$$

Using Kirchhoff's voltage law, the upper and lower arm phase voltages are

$$\begin{bmatrix} u_a \\ u_b \\ u_c \end{bmatrix} - \begin{bmatrix} u_0 + \frac{u_{dc}}{2} - u_{ap} \\ u_0 + \frac{u_{dc}}{2} - u_{bp} \\ u_0 + \frac{u_{dc}}{2} - u_{cp} \end{bmatrix} = 2L\frac{d}{dt}\begin{bmatrix} i_{ap} \\ i_{bp} \\ i_{cp} \end{bmatrix} + 2R\begin{bmatrix} i_{ap} \\ i_{bp} \\ i_{cp} \end{bmatrix} \tag{19}$$

$$\begin{bmatrix} u_a \\ u_b \\ u_c \end{bmatrix} - \begin{bmatrix} u_0 - \frac{u_{dc}}{2} + u_{an} \\ u_0 - \frac{u_{dc}}{2} + u_{bn} \\ u_0 - \frac{u_{dc}}{2} + u_{cn} \end{bmatrix} = 2L\frac{d}{dt}\begin{bmatrix} i_{an} \\ i_{bn} \\ i_{cn} \end{bmatrix} + 2R\begin{bmatrix} i_{an} \\ i_{bn} \\ i_{cn} \end{bmatrix} \tag{20}$$

where $u_a$, $u_b$, and $u_c$ are phase voltages of the AC side of the converter, and $u_{dc}$ is the DC voltage between the two poles. The voltage between two neutral points and ground is $u_0$. Adding Equations (19) to (20) gives

$$\begin{bmatrix} u_a \\ u_b \\ u_c \end{bmatrix} - \begin{bmatrix} u_0 + \frac{u_{an} - u_{ap}}{2} \\ u_0 + \frac{u_{bn} - u_{bp}}{2} \\ u_0 + \frac{u_{cn} - u_{cp}}{2} \end{bmatrix} = L\frac{d}{dt}\begin{bmatrix} i_a \\ i_b \\ i_c \end{bmatrix} + R\begin{bmatrix} i_a \\ i_b \\ i_c \end{bmatrix} \tag{21}$$

Let $u'_a$, $u'_b$ and $u'_c$ are average of the difference between upper and lower voltages

$$\begin{gathered} \begin{bmatrix} u'_a \\ u'_b \\ u'_c \end{bmatrix} = \begin{bmatrix} \frac{u_{an} - u_{ap}}{2} \\ \frac{u_{bn} - u_{bp}}{2} \\ \frac{u_{cn} - u_{cp}}{2} \end{bmatrix} \\ \begin{bmatrix} u_a \\ u_b \\ u_c \end{bmatrix} - \begin{bmatrix} u_0 + u'_a \\ u_0 + u'_b \\ u_0 + u'_c \end{bmatrix} = L\frac{d}{dt}\begin{bmatrix} i_a \\ i_b \\ i_c \end{bmatrix} + R\begin{bmatrix} i_a \\ i_b \\ i_c \end{bmatrix} \end{gathered} \tag{22}$$

Equation (22) mathematically presents the AC side of the MMC. Controllers are designed with this model in the following sections.

## 4. Control System Modeling

This section explains, in detail, the inner current loop and outer voltage loops designed for the MMC-rectifier and the MMC-inverter.

### 4.1. Inner Current Controller under Balanced Grid Conditions

The inner current controller is the most fundamental part of the MMC control system, as shown in Figure 8. The controller is a built-in SRF with PLL to generate frequency and phase angle information for the measurement of AC voltages. A dq vector control method is used for generating the output voltage required for the inner current controller. Applying Park and Laplace transformation to Equation (22) yields

$$\begin{bmatrix} u_d \\ u_q \end{bmatrix} = \begin{bmatrix} sL+R & -wL \\ wL & sL+R \end{bmatrix} \begin{bmatrix} i_d \\ i_q \end{bmatrix} + \begin{bmatrix} u_d \\ u_q \end{bmatrix} \tag{23}$$

Equation (23) shows that the dq components $u_d$ and $u_q$ of the AC voltage in rotating dq reference frame. The output voltage of the MMC $u_d$ and $u_q$ can be determined by following decoupled feedforward PI controller in ref. [28].

$$\begin{cases} u_d & = -(Kp + \frac{Ki}{s})(i_d^{ref} - i_d) + wLi_q + u_d \\ u_q & = -(Kp + \frac{Ki}{s})(i_q^{ref} - i_q) + wLi_d + u_q \end{cases} \tag{24}$$

Substituting Equation (24) into Equation (23)

$$\begin{bmatrix} si_d \\ si_q \end{bmatrix} = -\frac{1}{L} \begin{bmatrix} R - (Kp + \frac{Ki}{s}) & 0 \\ 0 & R - (Kp + \frac{Ki}{s}) \end{bmatrix} \begin{bmatrix} i_d \\ i_q \end{bmatrix} + \begin{bmatrix} i_d \\ i_q \end{bmatrix} - \frac{1}{L}(Kp + \frac{Ki}{s}) \begin{bmatrix} i_d^* \\ i_q^* \end{bmatrix} \tag{25}$$

It is clear from Equation (25) that $i_d$ and $i_q$ are decoupled. The inner current control by Equation (24) is illustrated in the block diagram of the inner current control loop.

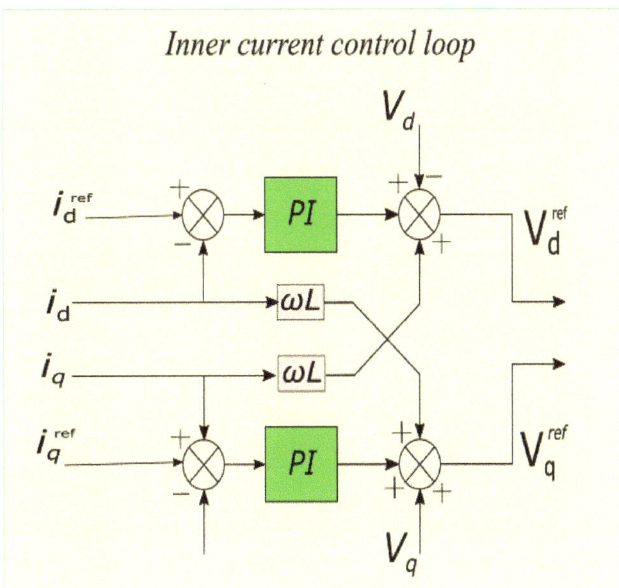


**Figure 8.** MMC inner current control loop.

### 4.2. Inner Current Controller under Unbalanced Grid Conditions

When the grid is under unbalanced conditions, positive-, negative-, and zero-sequence components exist. These component needs to be controlled independently. These components are derived from Equation (22) and can be separated into three independent systems.

$$v^{+}_{abc}(t) - v^{+}_{abc}(t) = L\frac{d}{dt}i^{+}_{abc} + Ri^{+}_{abc} \tag{26}$$

$$v^{-}_{abc}(t) - v^{-}_{abc}(t) = L\frac{d}{dt}i^{-}_{abc} + Ri^{-}_{abc} \tag{27}$$

$$v^{0}(t) - (v^{0}(t) + v_{0}(t)) = L\frac{d}{dt}i^{0}_{abc} + Ri^{0}_{abc} \tag{28}$$

Equations (26) and (27) can be expressed in $dq^{+}$ and $dq^{-}$ reference frame as

$$\frac{d}{dt}\begin{bmatrix} i_d^+ \\ i_q^+ \end{bmatrix} = \begin{bmatrix} -\frac{R}{L} & w \\ -w & -\frac{R}{L} \end{bmatrix}\begin{bmatrix} i_d^+ \\ i_q^+ \end{bmatrix} - \frac{1}{L}\begin{bmatrix} u_d^+ \\ u_q^+ \end{bmatrix} + \frac{1}{L}\begin{bmatrix} u_d^+ \\ u_q^+ \end{bmatrix} \tag{29}$$

$$\frac{d}{dt}\begin{bmatrix} i_d^- \\ i_q^- \end{bmatrix} = \begin{bmatrix} -\frac{R}{L} & -w \\ w & -\frac{R}{L} \end{bmatrix}\begin{bmatrix} i_d^- \\ i_q^- \end{bmatrix} - \frac{1}{L}\begin{bmatrix} u_d^- \\ u_q^- \end{bmatrix} + \frac{1}{L}\begin{bmatrix} u_d^- \\ u_q^- \end{bmatrix} \tag{30}$$

### 4.3. Positive- and Negative-Sequence Current Control

Applying Laplace transformation to $dq^{+}$ and $dq^{-}$, the positive-sequence voltage of the MMC $u_d^+$ and $u_q^+$ and negative-sequence voltage $u_d^-$ and $u_q^-$ can be determined by following the proportional integral feedforward controller.

Positive- and negative-sequence current control is provided by using Equations (31) and (32). The outer loop provides $i_d^+$ & $i_q^+$ as a command reference for positive-sequence current control, whereas $i_d^-$ and $i_q^-$ are set to zero to eliminate negative-sequence current components, as illustrated in Figure 9.

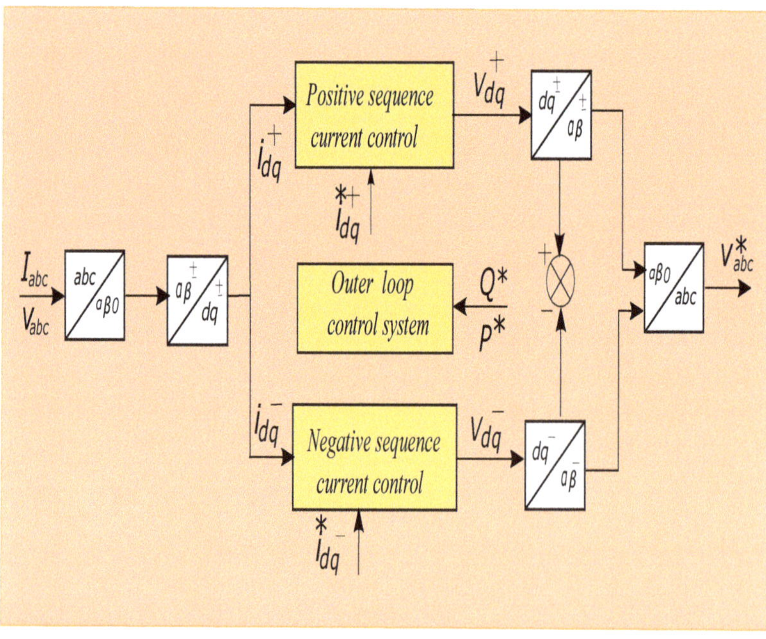


**Figure 9.** Control system.

$$\begin{cases} u_d^+ &= -(Kp + \frac{Ki}{s})(i_d^* - i_d^+) + wLi_q^+ + u_d^+ \\ u_q^+ &= -(Kp + \frac{Ki}{s})(i_q^* - i_q^+) - wLi_d^+ + u_q^+ \end{cases} \tag{31}$$

$$\begin{cases} u_d^- &= -(Kp + \frac{Ki}{s})(i_d^* - i_d^-) + wLi_q^- + u_d^- \\ u_q^- &= -(Kp + \frac{Ki}{s})(i_q^* - i_q^-) - wLi_d^- + u_q^- \end{cases} \tag{32}$$

The inner current control from Equations (31) and (32) is illustrated in the block diagram of the DSRF inner current control loop shown in Figure 10.

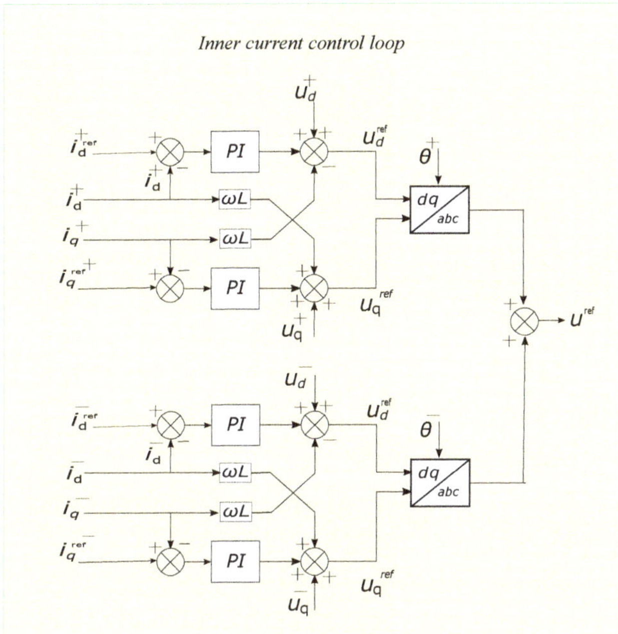


**Figure 10.** MMC DSRF inner current control loop.

### 4.4. Circulating Current Suppression Control

An active power transform is directly responsible for circulating current in the modular multilevel converter. The transformation of energy from the AC side of the modular multilevel converter to its DC side reflects with DC circulating current. There is some low-frequency fluctuation in the submodule voltage because of the floating nature of the submodule's capacitor. This fluctuation appears at the end, at the arm voltage $\sum v$ in the form of dominated second harmonic voltage. Circulating current in between the phases of modular multilevel converter distorts the arm currents, increases the peak value of arm currents, and increases system power losses.

The equivalent circuit of the MMC for circulating current analysis is shown in Figure 11. The dynamics for circulating current can be described by the following equation:

$$V_{dc} = \sum v + 2i_{cc}R_{arm} + 2L_{arm}\frac{di_{cc}}{dt} \tag{33}$$

$$V_{dc} = v_u + v_L + 2i_{cc}R_{arm} + 2L_{arm}\frac{di_{cc}}{dt} \tag{34}$$

The above equation implies that $(v_u + v_L)$ are only variables for controlling the circulating current in the arm of MMC. Splitting $\sum V$ into DC-link voltage and circulating current,

$$V_{dc} = v_u + v_L + 2V_{cc} \tag{35}$$

$$V_{dc} - 2V_{cc} = v_u + v_L$$

$$V_{dc} = V_{dc} - 2V_{cc} + 2R_{arm}i_{cc} + 2L_{arm}\frac{di_{cc}}{dt} \tag{36}$$

$$V_{cc} = i_{cc}R_{arm} + L_{arm}\frac{di_{cc}}{dt} \tag{37}$$

Time domain dynamics for three-phase MMC is as follows:

$$V_{cc_a} = i_{cc_a}R_{arm} + L_{arm}\frac{di_{cc_a}}{dt} \tag{38}$$

$$V_{cc_b} = i_{cc_b}R_{arm} + L_{arm}\frac{di_{cc_b}}{dt} \tag{39}$$

$$V_{cc_c} = i_{cc_c}R_{arm} + L_{arm}\frac{di_{cc_c}}{dt} \tag{40}$$

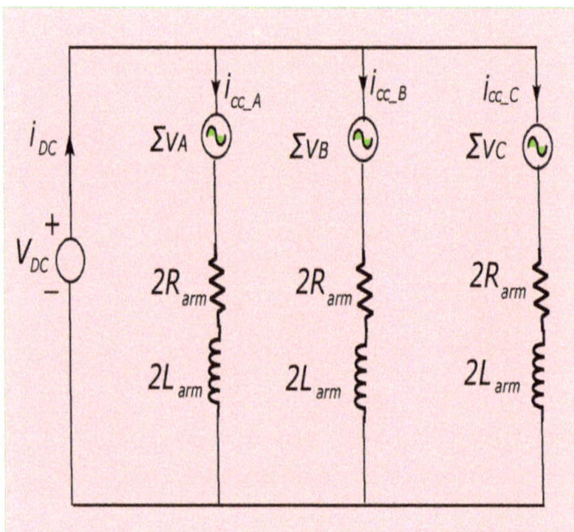


**Figure 11.** Equivalent circuit of the MMC for circulating current.

Circulating currents are formalized in Equations (43)–(45) below. Three-phase circulating currents consist of negative-sequence double-line frequency component.

$$i_{cc_a} = \frac{1}{3}I_{dc} + I_{cc}cos(2\omega t + \theta) \tag{41}$$

$$i_{cc_B} = \frac{1}{3}I_{dc} + I_{cc}cos(2\omega t + \theta - \frac{2\pi}{3}) \tag{42}$$

$$i_{cc_c} = \frac{1}{3}I_{dc} + I_{cc}cos(2\omega t + \theta - \frac{4\pi}{3}) \tag{43}$$

$$i_{cc_a} + i_{cc_b} + i_{cc_c} = 0 \tag{44}$$

To control the second harmonic circulating current, DC terms are neglected in the analysis.

$$\overrightarrow{I}_{cc} = I_{cc}e^{J(-2\omega t+\theta)} \tag{45}$$

Control of the second harmonic current can be accomplished by utilizing a vector method. In a dq rotating frame, the current is controlled with $2\omega$ due to its negative-sequence double-line frequency component. The equations for $V_{cc_a}$, $V_{cc_b}$, and $V_{cc_c}$ are transformed into $\alpha\beta$-frame by rotating vectors:

$$\vec{V}_{cc} = I_{cc} R_{arm} e^{J(-2\omega t+\theta)} + L_{arm} \frac{d(I_{cc} e^{J(-2\omega t+\theta)})}{dt} \quad (46)$$

Converting $\alpha\beta$-frame to dq-frame, we have

$$e^{-j\theta} \vec{V}_{cc} = e^{-} j\theta I_{cc} R_{arm} e^{J(-2\omega t+\theta)} + \quad (47)$$

$$e^{-j\theta} L_{arm} \frac{d(I_{cc} e^{J(-2\omega t+\theta)})}{dt} \quad (48)$$

$$\vec{V}_{cc,dq} = I_{cc} R_{arm} e^{j\theta} + L_{arm} \frac{d(I_{cc} e^{j\theta})}{dt} - 2j\omega L_{arm} e^{j\theta} \quad (49)$$

$$\begin{bmatrix} V_{cc_{-d}} \\ V_{cc_{-q}} \end{bmatrix} = R_{arm} \begin{bmatrix} I_{cc_{-d}} \\ I_{cc_{-q}} \end{bmatrix} + L_{arm} \frac{d}{dt} \begin{bmatrix} I_{cc_{-d}} \\ I_{cc_{-q}} \end{bmatrix} + \begin{bmatrix} 2L_{arm}\omega i_{cc_{-q}} \\ -2L_{arm}\omega i_{cc_{-p}} \end{bmatrix} \quad (50)$$

Circulating current suppression controller (CCSC) is built up with the help of Equation (52), as shown in Figure 12. The references are set to zero for current to see zero-control voltages at a steady state.

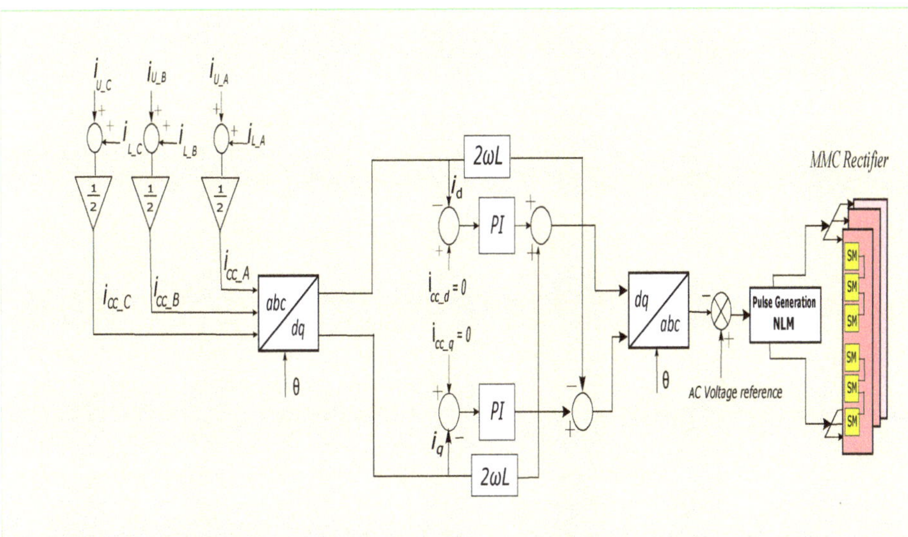


**Figure 12.** Circulating current controller of the converter.

### 4.5. Description of Outer Loop Controllers of Proposed System

The principal objective of this research work is to operate back-to-back MMC in islanded mode continuously with satisfactory power quality conditions. In the proposed system, the MMC-rectifier controls the DC voltage and AC voltage, whereas the MMC-inverter controls AC voltage and active power.

The inner current controller is designed in a dual synchronous reference frame (DSRF) to allow negative-sequence current injection to upstream AC grid and nonlinear load system under unbalanced fault conditions. DC voltage outer control loop is designed for the MMC-rectifier to control the DC voltage, as shown in Figure 13.

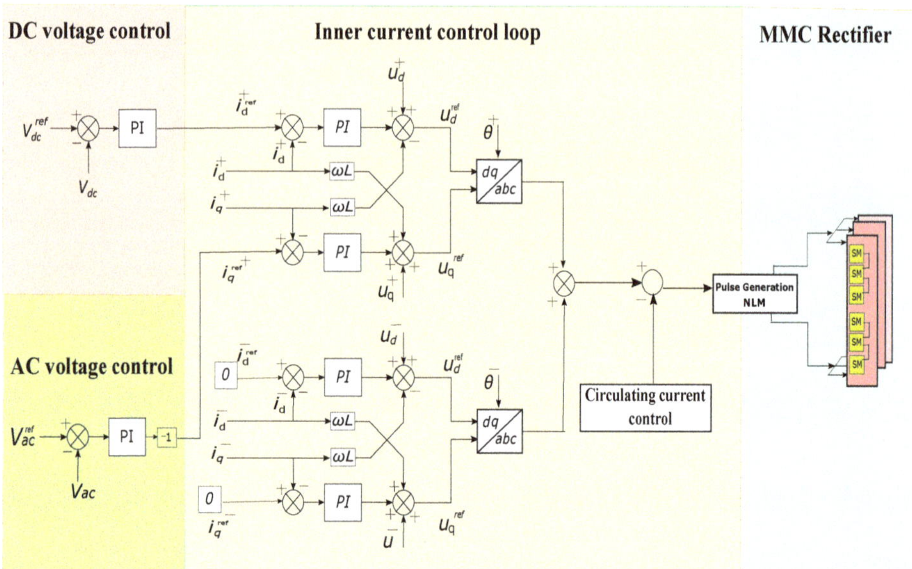


**Figure 13.** Outer and inner control loop for the MMC-rectifier.

This is achieved by creating a *d*-axis reference current for the inner current loop (ICL), comparing reference DC voltage to DC voltage measured, with an error signal $I_d$ feeding the PI controller. Output reference $I_d^{+ref}$ is fed to the inner current control loop.

The control loop can be represented by the following equation:

$$I_d^{+ref} = (V_{dc}^{ref} - V_{dc}).(K_p + \frac{K_i}{s}) \quad (51)$$

The PI controller is tuned at $k_p = 8$ and $k_i = 150$. Simulation results for DC voltage and power control of the MMC-rectifier is shown in Figure 14. Outer loop AC voltage controller for both the MMC-rectifier and the MMC-inverter are shown in Figures 13 and 15. The main objective of this control is the shaping of AC voltage. It is carried out by creating a *d*-axis reference current for the inner current loop (ICL), comparing reference AC voltage to AC voltage measured, with an error signal iq feeding PI controller. Output reference $I_q^{+ref}$ is used for the inner current control loop. The control loop can be represented by the following equation:

$$I_q^{+ref} = (V_{ac}^{ref} - V_{ac})(K_p + \frac{K_i}{s}) \quad (52)$$

The PI controller is tuned at $k_p = 2$ and $k_i = 260$. Outer loop active power controller is designed for the MMC-inverter shown in Figure 15. The output reference $I_d^{+ref}$ is used for inner current control loop. The control loop can be represented by the following equation:

$$P_s = \frac{3}{2}[v_d.i_d] \quad (53)$$

$$I_d^{+ref} = \frac{3}{2}(P_s - P_{ref})(K_p + \frac{K_i}{s}) \quad (54)$$

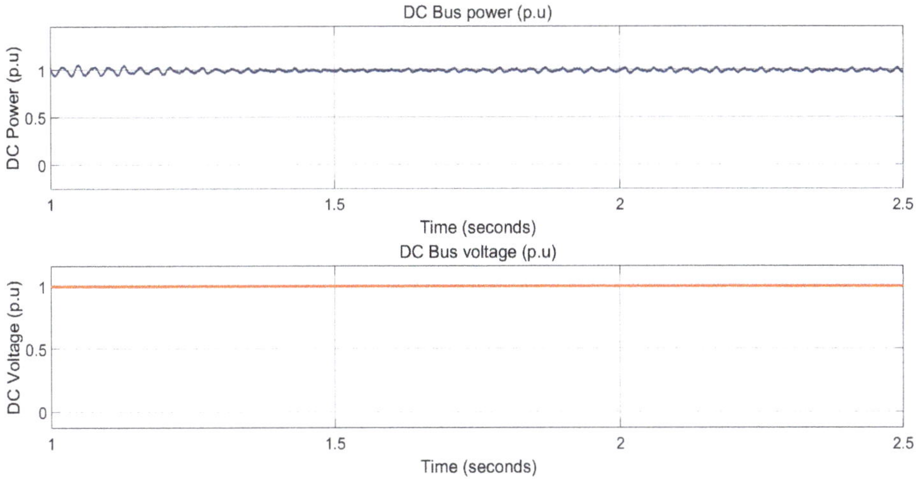


**Figure 14.** DC voltage control of the MMC-rectifier.

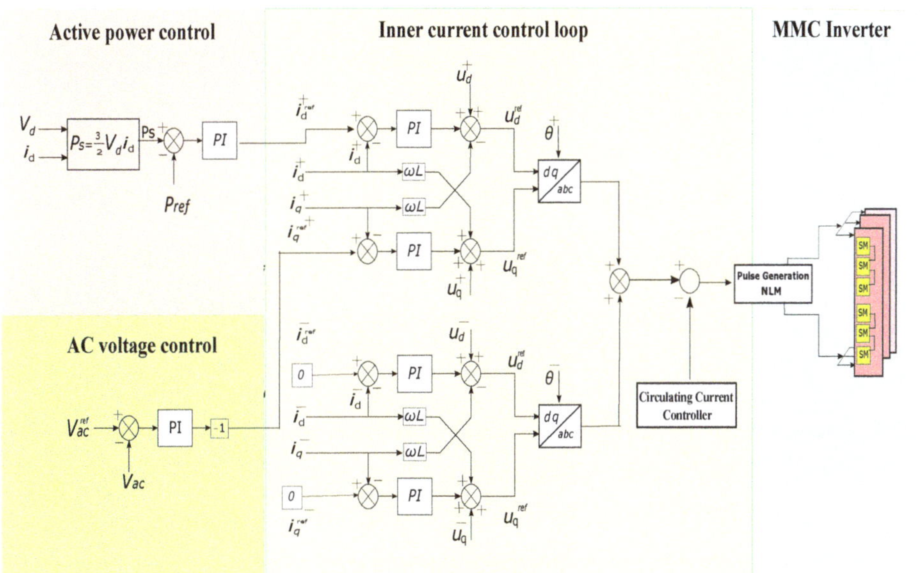


**Figure 15.** Outer and inner control loop for MMC-inverter.

The PI controller is tuned at $k_p = 0.1193$ and $k_i = 55.891$ to control the AC side active power of MMC-inverter. Simulation results for active power control of MMC-inverter are shown in Figure 16. This shows that the load is receiving healthy nominal power. Circulating current suppression control is an outer loop control that effectively suppresses the problematic AC component of circulating differential current in modular multilevel converter arms. The controller designed for circulating current suppression control of the MMC-rectifier and the MMC-inverter is the same as that shown in Figure 12. The circulating current controller feeds its reference signals to the inner current loop (ICL). The PI controller is tuned at $k_p = 1$ and $k_i = 5$. A simulation for the circulating current suppression controller is shown in Figure 17.

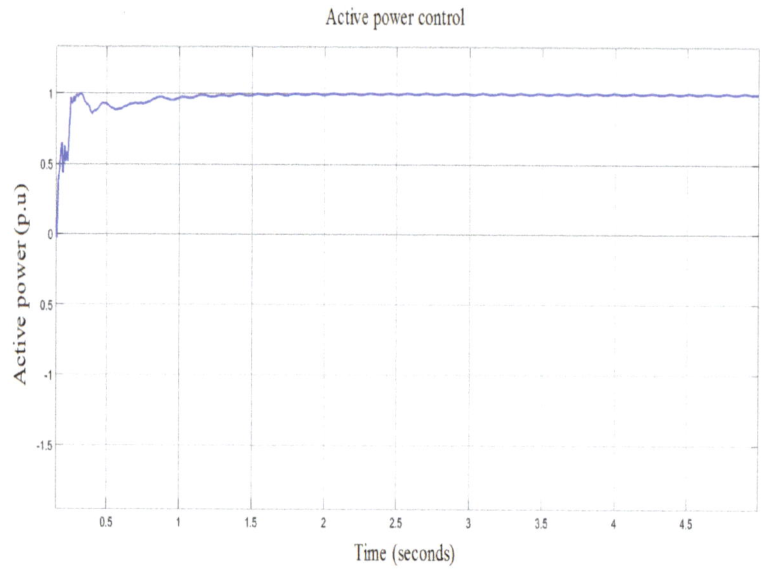


**Figure 16.** Active power control of inverter.

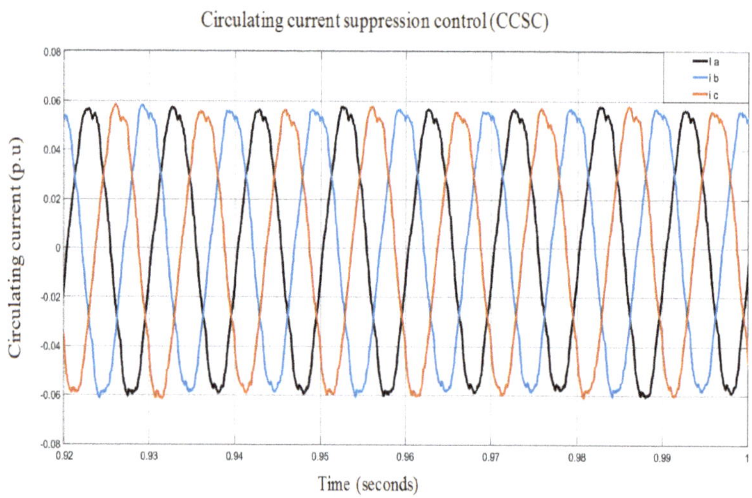


**Figure 17.** Circulating current suppression control of the MMC-rectifier.

*4.6. Modulation Strategy*

The modulation strategy adopted for the MMC-rectifier and the MMC-inverter is the nearest level modulation technique. This modulation technique allows for the operation of the switching frequency equal to fundamental frequency, reducing the associated conduction losses of the converter effectively. It works by translating the modulating signal into discrete stair waveform, directing the pulse generator how many modules to insert. It is therefore adopted for the system because it is the most common modulation strategy for high-power modular multilevel converter applications. Figure 18 shows the NLM-generated reference signals for MMC. The $y$-axis shows the number of SMs and $x$-axis is the time in seconds.

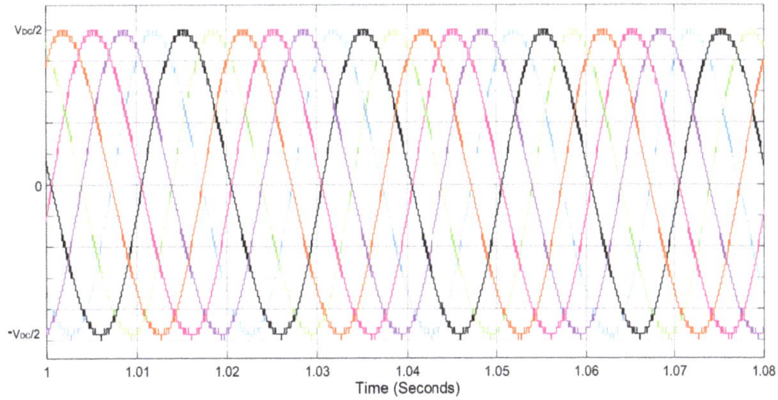


**Figure 18.** Nearest level modulation signals for MMC.

## 5. Results of the Simulation and Discussion

The simulation for the proposed scheme is conducted in a MATLAB-based Simulink system. The simulation environment and system parameters are illustrated in Table 4.

**Table 4.** Parameters of upstream AC, DC grid and modular multilevel converter.

| S. No | Parameter | Values |
|---|---|---|
| 1 | Upstream AC grid voltage | 220 kV |
| 2 | Transformer rating/AC grid side | 220/66 kV (Core type) |
| 3 | No of SMs in MMC | 60 |
| 4 | Arm inductance $L_{arm}$ | 0.0164 (p.u.) |
| 5 | Arm resistance $R_{arm}$ | 0.0300 (p.u.) |
| 6 | Submodule capacitance | 0.0105 (p.u.) |
| 7 | DC grid voltage | 135 kV |
| 8 | AC grid voltage | 66 kV |
| 9 | Transformer rating/load side | 66/11 kV (Core type) |
| 10 | Dynamic load active power | 10 MW |
| 11 | Dynamic load reactive power | 25 MVAR |
| 12 | Sampling time | 20 μs |

Steady-state operation of both the MMC-rectifier and the MMC-inverter with submodule voltage, arm current, and capacitors voltage is shown in Figure 19.

The voltage of the submodule of MMC is calculated in p.u. where base voltage is set to 2.25 kV, dividing DC voltage by the number of submodules "$N$". Figure 19b,g shows that the arm current flows in one phase of each of the MMC-rectifier and MMC-inverter. Arm current is calculated in p.u. as the base current is set as 1.924 kA. Voltage ripple of submodules is approximately 11% which is composed of both AC and DC quantity. Capacitors of submodules can handle this voltage efficiently. IGBT module package of 4.8 kV is used and derated to 2.25 kV having 11% ripples.

A single-phase voltage sag of up to −60% magnitude, asymmetrical fault, is modeled through the 220 kV source of the system model on the overlaying AC grid side and is the case to be simulated in this research article. A single-phase fault is applied to the simulation model, as shown in Figure 20. When the simulation time is 1.6 s, the effect of the voltage sag is −60% magnitude for 75 ms. AC grid with voltage sag of −60% magnitude feeds the MMC-rectifier.

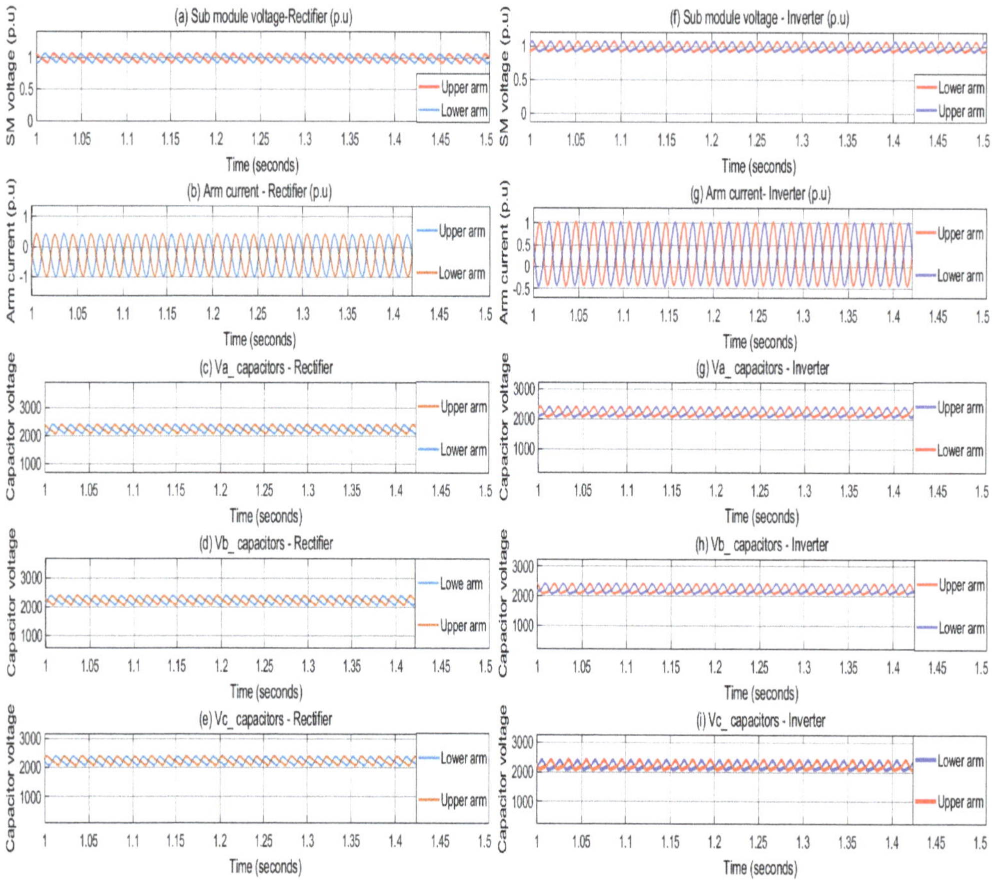


**Figure 19.** Steady state operation of the MMC-rectifier and the MMC-inverter with SMs voltage, current, and capacitors voltage.

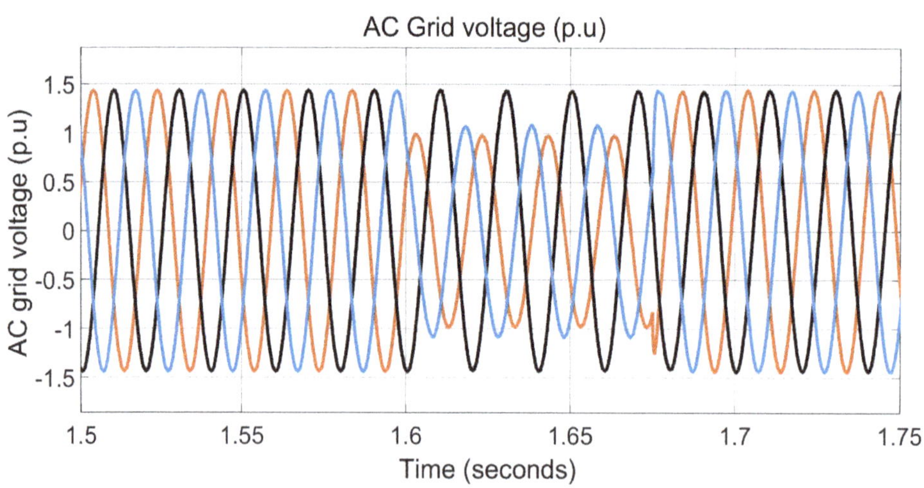


**Figure 20.** Upstream AC grid voltage sag of −60% magnitude.

*Voltage Sag Mitigation Strategy*

The main objective of this research is to explore the feasibility and potential of the MMC to mitigate voltage sag that is on the overlaying grid due to asymmetrical fault and to deliver a constant power to load distribution. During severe asymmetrical fault conditions, e.g., single-phase fault, in the overlaying grid, voltage sag is for up to −60% within 75 ms, the MMC should provide support to ensure the constant power flow for load distribution. This can be realized by supplying energy from integrated energy in the submodules of the MMC-rectifier and the MMC-inverter to compensate for the missing energy of overlaying grid due to asymmetrical fault.

Energy Storage of Converter

The capacitor of the submodule acts as an energy buffer. Energy stored in the capacitor of the submodule is given by the following equation:

$$E_c = \frac{1}{2} C V_c^2 \tag{55}$$

The capacitance of the capacitor is 10.48 mF and the submodule's voltage $V_{SM}$ is 2.25 kV. The installed energy of a submodule as per the above equation is 26.53 kJ. Energy stored in one arm and phase leg of the converter is 1.591 MJ and 3.18 MJ, respectively. The installed energy of the three-phase MMC converter with 60 submodules is 9.5 MJ. Energy stored in back-to-back MMC is 19 MJ.

During asymmetrical fault in the overlaying grid side, the capacitor of submodules should safely discharge its stored energy to compensate for energy loss made by voltage sag during asymmetrical fault. The modulation index of the MMC is 0.8, and the threshold of capacitor discharging is decided by the steady-state modulation index of MMC. Therefore, the MMC-rectifier allows for DC voltage drop from 1.0 p.u. to 0.8 p.u. without affecting its operation. During the fault conditions, the internal energy storage of the MMC-rectifier and the MMC-inverter should have supplied 10.4% and 12.4% of its installed energy, respectively, as their energy is discharged to approximately 0.896 p.u. and 0.876 p.u. The instantaneous energy of the MMC-rectifier and the MMC-inverter is shown in Figure 21.

Energy stored in the MMC-rectifier is discharged to approximately 0.896 p.u. to compensate for the energy loss caused by voltage sag in the upstream AC grid. The installed energy of the MMC-rectifier is 9.5 MJ. Only 10.4% of installed energy, 0.99 MJ, is discharged to compensate for the effect of the transient voltage sag without affecting the operation of the MMC-rectifier.

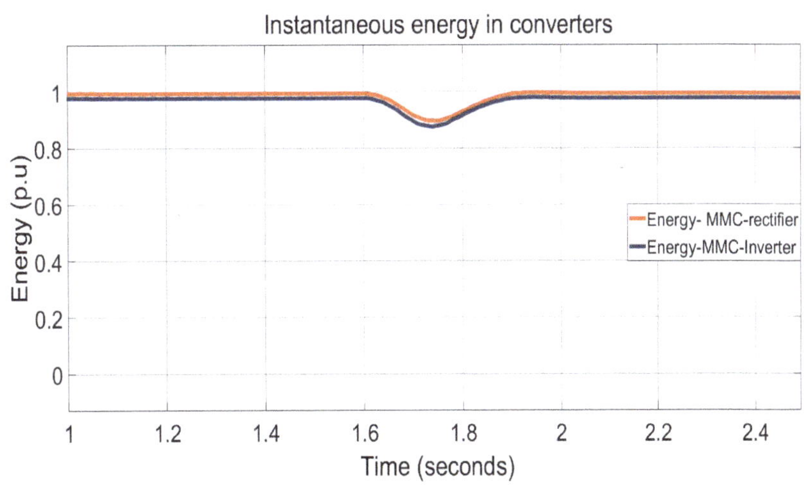


**Figure 21.** Instantaneous energy in the MMC-rectifier and the MMC-inverter.

The output DC power of the MMC-rectifier becomes oscillating, as seen in Figure 22, because of the energy interaction between the MMC-rectifier and the MMC-inverter. It can be seen that the DC power peak has reached 1.18 p.u., since it is not harmful, because the DC power equally flows in the upper and lower arms of the both the MMC-rectifier and the MMC-inverter. In this scenario, IGBT bears safely the overcurrent, which is around 25%, that lasts for a one-half cycle.

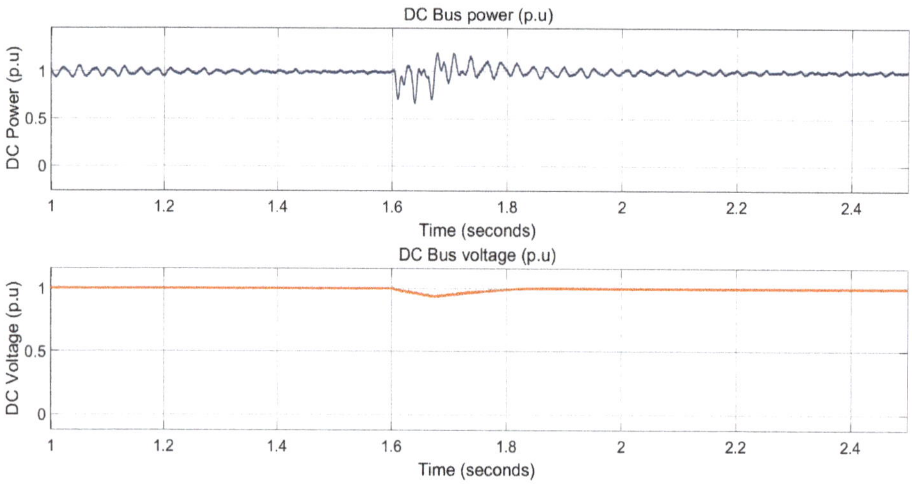


**Figure 22.** DC bus power.

The power consumed by the MMC-rectifier is shown in Figure 23. This shows that when −60% voltage sag is present, active power drawn is reduced to 0.82 p.u., while only 10.4% of installed energy, 0.99 MJ, is discharged to compensate for the effect of the transient voltage sag without affecting the operation of the MMC-rectifier. The MMC-inverter has compensated for the remaining oscillations after the MMC-rectifier using its integrated energy of the submodules.

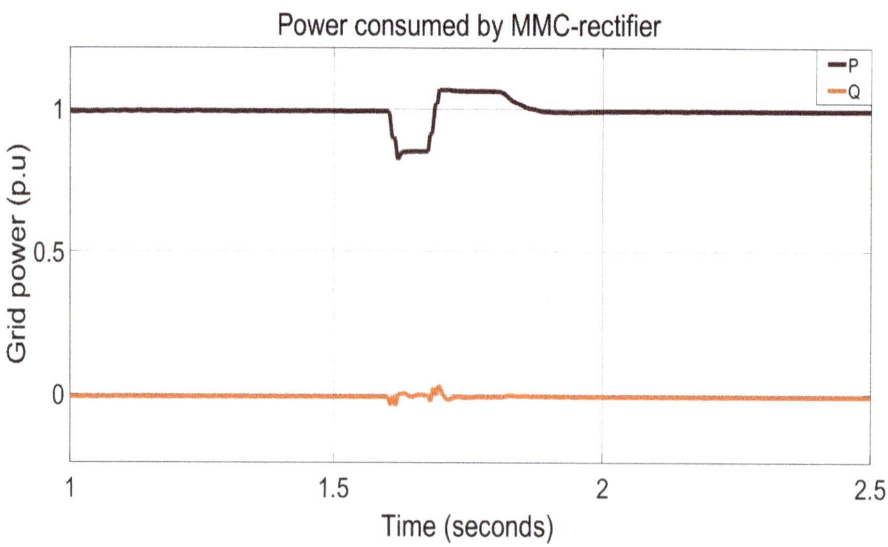


**Figure 23.** Power drawn from grid to feed rectifier.

The installed energy of the MMC-inverter is 9.5 MJ. Only 12.4% of installed energy, 1.178 MJ, is discharged for mitigation of the transient voltage sag completely. Figure 24 shows the power delivered from MMC-inverter to the dynamic load distribution. This shows that the load distribution system is healthy and receives AC voltage at a nominal value. These results confirm that the MMC is feasible for an option to mitigate voltage sag and to ensure satisfactory power delivery to the dynamic load distribution. In Figure 25, the output voltage and current of MMC-inverter are normal and healthy with no evidence of voltage sag or energy loss.

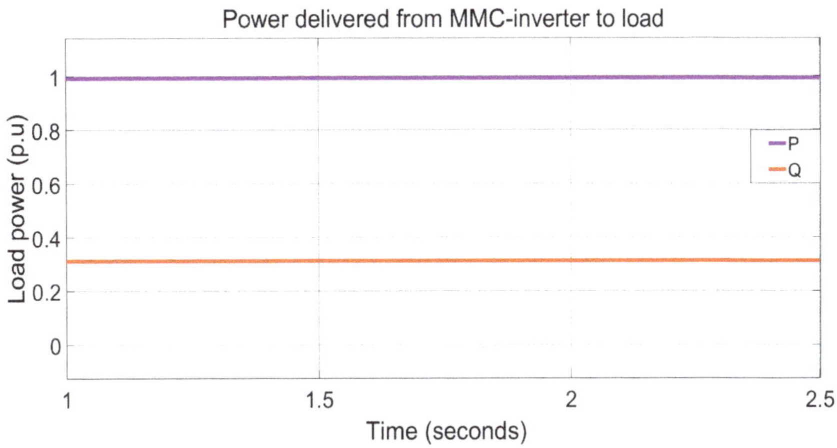


**Figure 24.** Power delivered from MMC-inverter to load distribution system.

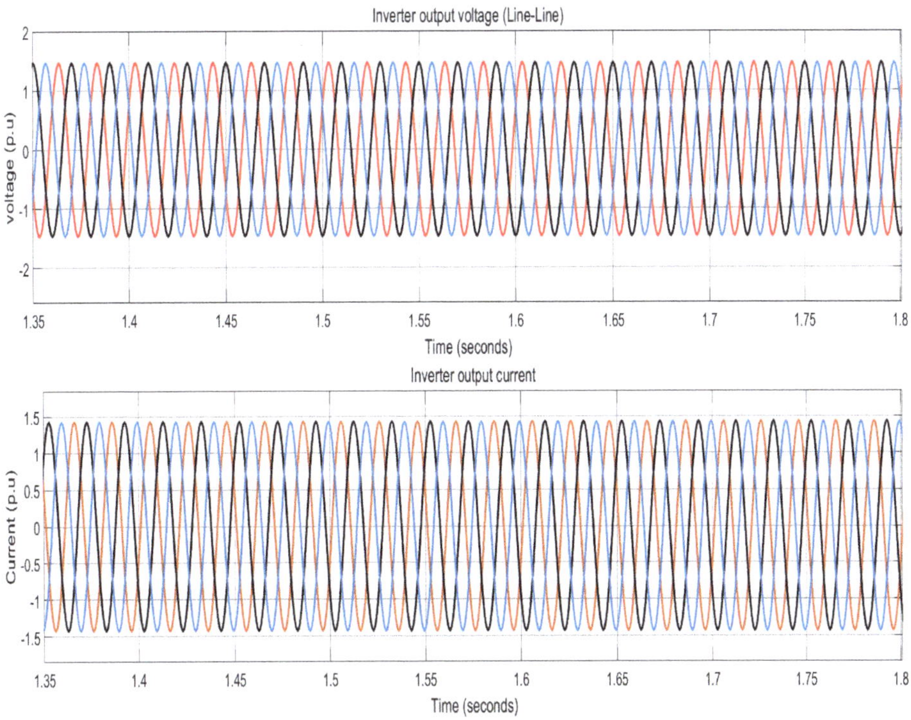


**Figure 25.** Inverter output voltage and current.

It is concluded from Figures 24 and 25 that transient voltage sag of about −60% for the period of 75 ms, the worst sag scenario, in upstream AC grid side is controlled effectively by back-to-back MMC-rectifier and MMC-inverter.

Table 5 shows the comparison of the commonly employed and proposed control strategies for voltage sag mitigation.

**Table 5.** Comparison of the commonly employed and proposed control strategies for voltage sag mitigation.

| Method | Issue Mitigated | Outcomes | Limitations | Reference |
|---|---|---|---|---|
| DVR | Voltage sag | -Voltage sag is mitigated during different types of grid faults. | -High cost.<br>-Quite complex. | [5] |
| Fuzzy logic controller-based DVR | Voltage sag | -Mitigated voltage sag up to 50%.<br>-Rapid detection rate compared to conventional methods. | -All types of sags are not tested.<br>-Frequency fluctuation. | [8] |
| SVC | Voltage sag | -Voltage sag of all kinds addressed.<br>-During sag events, inject reactive power to maintain voltage. | -Lower performance than DVR. | [6] |
| Adapted SVC control strategy | Voltage sag | -Voltage sag is addressed with reactive current injection.<br>-Voltage stability is fully supported during faults scenario. | -No grid code standards are followed. | [29] |
| ANFIS-Based UPQC | Voltage sag | -Voltage sag is addressed with reactive current injection.<br>-Support voltage stability during faults. | -Increased cost.<br>-High complexity. | [7] |
| MMC-based voltage sag mitigation (Proposed method) | Voltage sag | -Compensate for missing energy of upstream AC grid due to fault using integrated energy in the SMs of MMC.<br>-Mitigated voltage sag up to 60%.<br>-Response time is very fast (<1 ms).<br>-Support voltage stability during faults.<br>-Simple design and control mechanism.<br>-It has high redundancy possibilities.<br>-It has smaller footprint.<br>-Low cost (no need of custom PQ regulator).<br>-Low harmonics and low conduction power losses. | -When a semiconductor fails in a sub-module, there is a need for protection. The capacitor will release stored energy causing an explosion. | Proposed in this research. |

## 6. Conclusions

In this article, we focus primarily on providing satisfactory power delivery to the load distribution system and meeting nominal power demand under transient voltage sag conditions. A back-to-back modular multilevel converter is designed to mitigate transient voltage sags. Using the internal energy storage of submodules of back-to-back modular multilevel converters, transient voltage sag due to asymmetrical fault with a magnitude of −60% for 75 ms in the upstream AC grid is mitigated effectively. The ability of the improved voltage sag mitigation method is proved through simulation results studies. Furthermore, effective control strategies are developed for inner current control (ICC) in the dual synchronous reference frame (DSRF) to control undesirable components, i.e., double-line positive- and negative-sequence components of the circulating current. The effectiveness and performance of the proposed method are well-validated by the results of the simulation.

*Future Work and Limitations*

Based on this research work, future studies can explore the following aspects:

- An investigation into AC grid stabilization and reactive power injection into upstream AC grid with the back-to-back MMCs.
- Fault handling and detection studies for internal submodules of MMC. Examining and simulating internal submodule faults and detection methods, as well as exploring arm–arm short circuits in the converter.
- Techno-economical studies and reliability analysis of the back-to-back modular multilevel converters.

The limitation of this research scheme is as follows:

- Short circuits in the individual module, as well as flash-over faults between two arms (between the converter arms), cannot be simulated using this model, since the aggregated modeling strategy is used, which considers only the entire converter arm.

**Author Contributions:** Conceptualization, F.M. and H.R.; methodology, F.M.; formal analysis, F.M.; investigation, H.R.; writing—review and editing, I.A., R.A. and A.B.; supervision, H.R., I.A., R.A. and A.B.; funding acquisition, I.A., R.A. and A.B. All authors have read and agreed to the published version of the manuscript.

**Funding:** This research was funded by Taif University researchers supporting Project (number TURSP-2020/36), Taif University, Taif, Saudi Arabia.This research work was also partially supported by the faculty of computer science and information technology, University of Malaya, under postgraduate research grant PG035-2016A.

**Institutional Review Board Statement:** Not applicable.

**Informed Consent Statement:** Not applicable.

**Acknowledgments:** The authors are grateful to the Taif University researchers supporting Project (number TURSP-2020/36), Taif University, Taif, Saudi Arabia.This research work was also partially supported by the faculty of computer science and information technology, University of Malaya, under postgraduate research grant PG035-2016A.

**Conflicts of Interest:** The authors declare no conflict of interest.

## References

1. Ye, X.; Liu, H.; Hao, S.; Zhang, K.; Lü, G. Comprehensive mitigation strategy of voltage sag based on sensitive load clustering. *Int. Trans. Electr. Energy Syst.* **2021**, *31*, e12914. [CrossRef]
2. Huang, A.; Xiao, X.; Wang, Y. Evaluation Scheme of Voltage Sag Immunity in Sensitive Industrial Process. *IEEE Access* **2021**, *9*, 66398–66407. [CrossRef]
3. Al-Shetwi, A.Q.; Hannan, M.; Jern, K.P.; Mansur, M.; Mahlia, T. Grid-connected renewable energy sources: Review of the recent integration requirements and control methods. *J. Clean. Prod.* **2020**, *253*, 119831. [CrossRef]
4. Alkahtani, A.A.; Alfalahi, S.T.; Athamneh, A.A.; Al-Shetwi, A.Q.; Mansor, M.B.; Hannan, M.; Agelidis, V.G. Power quality in microgrids including supraharmonics: Issues, standards, and mitigations. *IEEE Access* **2020**, *8*, 127104–127122. [CrossRef]
5. Srivatchan, N.; Rangarajan, P. Half cycle discrete transformation for voltage sag improvement in an islanded microgrid using dynamic voltage restorer. *Int. J. Power Electron. Drive Syst.* **2018**, *9*, 25–32. [CrossRef]
6. Fardad, N.; Soleymani, S.; Faghihi, F. Voltage Sag Investigation of Microgrid in the presence of SMES and SVC. *Signal Process. Renew. Energy* **2019**, *3*, 23–34.
7. Kumar, A.S.; Rajasekar, S.; Raj, P.A.D.V. Power quality profile enhancement of utility connected microgrid system using ANFIS-UPQC. *Procedia Technol.* **2015**, *21*, 112–119. [CrossRef]
8. Farsadi, M.; Shahrak, A.G.; Tabrizi, S.D. DVR with fuzzy logic controller and photovoltaic for improving the operation of wind farm. In Proceedings of the 2013 8th International Conference on Electrical and Electronics Engineering (ELECO), Bursa, Turkey, 28–30 November 2013; pp. 191–195.
9. Omar, R.; Rahim, N.; Sulaiman, M. Dynamic voltage restorer application for power quality improvement in electrical distribution system: An overview. *Aust. J. Basic Appl. Sci.* **2011**, *5*, 379–396.
10. He, J.; Liang, B.; Li, Y.W.; Wang, C. Simultaneous microgrid voltage and current harmonics compensation using coordinated control of dual-interfacing converters. *IEEE Trans. Power Electron.* **2016**, *32*, 2647–2660. [CrossRef]
11. Al-Shetwi, A.Q.; Hannan, M.; Jern, K.P.; Alkahtani, A.A.; PG Abas, A. Power quality assessment of grid-connected PV system in compliance with the recent integration requirements. *Electronics* **2020**, *9*, 366. [CrossRef]

12. Molinas, M.; Suul, J.A.; Undeland, T. Low voltage ride through of wind farms with cage generators: STATCOM versus SVC. *IEEE Trans. Power Electron.* **2008**, *23*, 1104–1117. [CrossRef]
13. Khadem, S.K.; Basu, M.; Conlon, M.F. Integration of UPQC for Power Quality improvement in distributed generation network-a review. In Proceedings of the 2020 2nd IEEE PES International Conference and Exhibition on Innovative Smart Grid Technologies, Delft, The Netherlands, 26–28 October 2020; pp. 1–5.
14. Rasheed, A.; Rao, G.K. Improvement of power quality for microgrid using fuzzy based UPQC controller. *Indian J. Sci. Technol.* **2020**, *8*, 1.
15. Gayatri, M.; Parimi, A.M.; Kumar, A.P. Utilization of Unified Power Quality Conditioner for voltage sag/swell mitigation in microgrid. In Proceedings of the 2016 Biennial International Conference on Power and Energy Systems: Towards Sustainable Energy (PESTSE), Bengaluru, India, 21–23 January 2016; pp. 1–6.
16. Hoseinnia, S.; Akhbari, M.; Hamzeh, M.; Guerrero, J. A control scheme for voltage unbalance compensation in an islanded microgrid. *Electr. Power Syst. Res.* **2019**, *177*, 106016. [CrossRef]
17. Marquardt, R. Stromrichterschaltung mit Verteilten Energiespeichern und Verfahren zur Steuerung Einer Derartigen Stromrichterschaltung. German Patent DE10,103,031, 24 January 2001.
18. Ji, M.; Wang, Y.; Tan, K.; Wang, C. A voltage-balanced hybrid MMC topology for DC fault ride-through. In Proceedings of the 2019 IEEE Innovative Smart Grid Technologies-Asia (ISGT Asia), Chengdu, China, 21–24 May 2019; pp. 2282–2286.
19. Mahmoud, A.A.; Hafez, A.A.; Yousef, A.M. Modular multilevel converters for renewable energies interfacing: Comparative review. In Proceedings of the 2019 IEEE Conference on Power Electronics and Renewable Energy (CPERE), Aswan, Egypt, 23–25 October 2019; pp. 397–406.
20. Pereira, H.A.; Mendes, V.F.; Harnefors, L.; Teodorescu, R. Comparison of 2L-VSC and MMC-based HVDC Converters: Grid Frequency Support Considering Reduced Wind Power Plants Models. *Electr. Power Compon. Syst.* **2020**, *45*, 2007–2016. [CrossRef]
21. Franquelo, L.G.; Rodriguez, J.; Leon, J.I.; Kouro, S.; Portillo, R.; Prats, M.A. The age of multilevel converters arrives. *IEEE Ind. Electron. Mag.* **2018**, *2*, 28–39. [CrossRef]
22. Rohner, S.; Hiller, M.; Sommer, R. A new highly modular medium voltage converter topology for industrial drive application. In Proceedings of the Power Electronics and Applications, EPE 13th European Conference, Barcelona, Spain, 8–9 September 2009.
23. Saeedifard, M.; Iravani, R. Dynamic performance of a modular multilevel back-to-back HVDC system. *IEEE Trans. Power Deliv.* **2010**, *25*, 2903–2912. [CrossRef]
24. Tu, Q.; Xu, Z.; Xu, L. Reduced switching-frequency modulation and circulating current suppression for modular multilevel converters. *IEEE Trans. Power Deliv.* **2011**, *26*, 2009–2017.
25. Jacobson, B.; Karlsson, P.; Asplund, G.; Harnefors, L.; Jonsson, T. VSC-HVDC transmission with cascaded two-level converters. In *Cigré Session*; Cigre: Paris, France, 2010; pp. B4–B110.
26. Ilves, K.; Antonopoulos, A.; Norrga, S.; Nee, H.P. Steady-state analysis of interaction between harmonic components of arm and line quantities of modular multilevel converters. *IEEE Trans. Power Electron.* **2011**, *27*, 57–68. [CrossRef]
27. Sharifabadi, K.; Harnefors, L.; Nee, H.P.; Norrga, S.; Teodorescu, R. *Design, Control, and Application of Modular Multilevel Converters for HVDC Transmission Systems*; John Wiley & Sons: Hoboken, NJ, USA, 2016.
28. Allebrod, S.; Hamerski, R.; Marquardt, R. New transformerless, scalable modular multilevel converters for HVDC-transmission. In Proceedings of the 2008 IEEE Power Electronics Specialists Conference, Rhodes, Greece, 15–19 June 2008; pp. 174–179.
29. Chhor, J.; Sourkounis, C. Optimal voltage control strategy for grid-feeding power converters in AC microgrids. *Electr. Power Syst. Res.* **2019**, *176*, 105945. [CrossRef]

electronics

Article

# A Low-Power Hiccup-Mode Short-Circuit Protection Technique for DC-DC Boost Converter

Hao Wang [1], Tianmei Shen [2], Nairiga Wu [1] and Fang Tang [1,*]

[1] Chongqing Engineering Laboratory of High Performance Integrated Circuits, School of Microelectronics and Communication Engineering, Chongqing University (CQU), Chongqing 400044, China; 20163990@cqu.edu.cn (H.W.); nairiga@cqu.edu.cn (N.W.)
[2] Kunming Branch of the 705 Research Institute, China State Shipbuilding Corporation Limited, Kunming 650106, China; 20134551@cqu.edu.cn
* Correspondence: frankfangtang@gmail.com

**Abstract:** An improved low-power hiccup-mode technology for short-circuit protection is proposed in this paper, which can effectively suppress short-circuit currents and greatly minimize the power dissipation of hiccup mode. The circuit can start normally after the short circuit is recovered, and there is no voltage overshoot. At the same time, the proposed pre-charge circuit can effectively suppress the large initial inrush current in the pre-charge stage. These technologies are used in a Peak-Current-Mode-Control (PCMC) Pulse-Width-Modulation (PWM) DC-DC boost converter designed with a 0.35 standard CMOS process. Compared with the conventional structure, the post-simulation results show that the initial inrush current during the start-up phase in the proposed structure is reduced by about 90%. When the output short circuit occurs, the inductor current drops to approximately zero and the power dissipation of the converter is very low at this time. The converter repeatedly detects the state of the output load after a period of about 24 ms. Eventually, the converter will restart after the short circuit is recovered and there is no voltage overshoot.

**Keywords:** hiccup mode; short-circuit protection; sleep mode; low power; inrush current; voltage overshoot

**Citation:** Wang, H.; Shen, T.; Wu, N.; Tang, F. A Low-Power Hiccup-Mode Short-Circuit Protection Technique for DC-DC Boost Converter. *Electronics* **2022**, *11*, 870. https://doi.org/10.3390/electronics11060870

Academic Editor: Carlos Andrés García-Vázquez

Received: 25 January 2022
Accepted: 3 March 2022
Published: 10 March 2022



## 1. Introduction

DC-DC converters are widely used in various electronic systems because of their high conversion efficiency, low power dissipation, large drive capability, and so on [1,2]. The trend of electronic systems developing towards high reliability and low power puts forward higher requirements for power supply systems. Nowadays, the reliable design of converters faces many challenges. The large initial current [3] and the voltage overshoot [4] of the DC-DC boost converter during the start-up phase will have a negative impact on the subsequent circuit. At the same time, both the large current caused by the output short circuit and the voltage overshoot caused by the short-circuit recovery are also important factors that cannot be ignored regarding the system failure and large power dissipation [5]. Therefore, how to reduce inrush current and voltage overshoot is important to converters. Meanwhile, the short-circuit protection technology of the converter is also one of the recent research hot spots.

There have been many studies on how to reduce the inrush current during the start-up phase of the converter. Aiming at the energy transfer characteristics of the boost converter, there was a soft-start structure in [6], which reduces the inrush current to a certain extent, but in the stage of $V_{OUT} < V_{CC}$, the inrush current still exists. In [7], they proposed a two-stage circuit startup structure, which further reduces the inrush current. Some other methods are also used to reduce the initial inrush current. For example, in the start-up phase, an adaptive resistor is connected in series with the load capacitor to suppress the charging current at the load, thereby limiting the inductor current [8], but this method requires the design of a complex control circuit to control the resistance. There are some other methods for reducing the inrush current, but there are few methods that can effectively suppress the initial inrush current of the inductor without adding additional pins and while being easy to implement.

There are also many studies on short-circuit protection technology. The over-current limiting technology proposed by studies [9,10] have certain protective effects on the system. However, the output current during the short circuit is relatively large, which leads to high power dissipation. The methods proposed in [11] can maintain a low output current, but it ignores the voltage overshoot caused by short-circuit recovery. In addition, in [12,13], they proposed an improved short-circuit protection technology based on hiccup mode, which effectively reduces the voltage overshoot after short-circuit recovery, but the power dissipation of converters during short circuit is still a problem. In [14,15], although the power dissipation during short circuit is reduced by sleep time, the output current is still large during the detection of whether the output returns to normal.

The pre-charge circuit proposed in this paper can suppress the initial inrush current well in the circuit startup stage. At the same time, a short-circuit protection technique based on hiccup mode can reduce the inductor current to nearly zero during the short-circuit phase. The converter only consumes a small amount of current in the process of detecting whether the short circuit is recovered, which greatly reduces the power dissipation during the short-circuit period. In addition, the circuit can restart normally after the short circuit is recovered, and there is no voltage overshoot. In Section 2, the causes of both initial inrush current and voltage overshoot are briefly introduced, and the working principle of the conventional short-circuit protection technology based on hiccup mode is introduced at the same time. In Section 3, the pre-charge circuit is given, and the working principle of the proposed hiccup mode is analyzed. The simulation results are given in Section 4. Finally, the conclusion is given in Section 5.

## 2. Analysis of Inrush Current and Overshoot and Introducing the Conventional Hiccup-Mode Technique

The conventional structure of Peak-Current-Mode-Control (PCMC) Pulse-Width-Modulation (PWM) boost converter is shown in Figure 1a. Energy is transformed from supply to output by charging and discharging the inductor L. The error amplifier EA amplifies the difference between the output feedback voltage $V_{FB}$ and the reference voltage $V_{ref}$, and the output result is compared with the slope-compensated current signal to obtain the clock signal, which is driven by logic to control the on and off of the power transistors $M_N$ and $M_P$. In the soft-start phase of the converter, the linearly increasing $V_{SS}$ replaces $V_{ref}$ to guide $V_{OUT}$ to increase slowly, which is a popular way to reduce the output voltage overshoot.

### 2.1. Principle of Inrush Current and Output Voltage Overshoot

The large initial inrush current has always been a problem for boost converters, and this phenomenon has its inherent reasons. At the beginning of the converter power up, $V_{CC} > V_{OUT}$, as shown in Figure 1b, the converter operates in the $T_{ON}$ phase, and the slope of $i_L$ can be expressed as follows:

$$K_{ON} = \frac{V_{CC}}{L} \tag{1}$$

In the $T_{OFF}$ phase, as shown in Figure 1c, it can be expressed as follows:

$$K_{OFF} = \frac{V_{CC} - V_{OUT}}{L} \tag{2}$$

This means that the inductor will always be charged. In other words, the inductor current $i_L$ will rise continuously, and the result is a large initial current, as shown in Figure 1d. In order to solve this problem, pre-charge has been proposed, and it has a good effect in reducing inrush current. In the pre-charge phase, the oscillator of the converter does not operate, and the on and off of the power transistors are controlled by the pre-charge. At this time, the pre-charge provides a low-level signal to control $M_N$ off and $M_P$ on. The conventional pre-charge method keeps $M_N$ off and $M_P$ on to allow $V_{OUT}$ to be charged until it is greater than $V_{CC}$. The converter always operates in the $T_{OFF}$ phase and the function of $M_P$ is equivalent to that of a resistor. This does reduce the inrush current, but $i_L$ still maintains an upward trend during the process, and the inrush current still exists, as shown in Figure 1e.

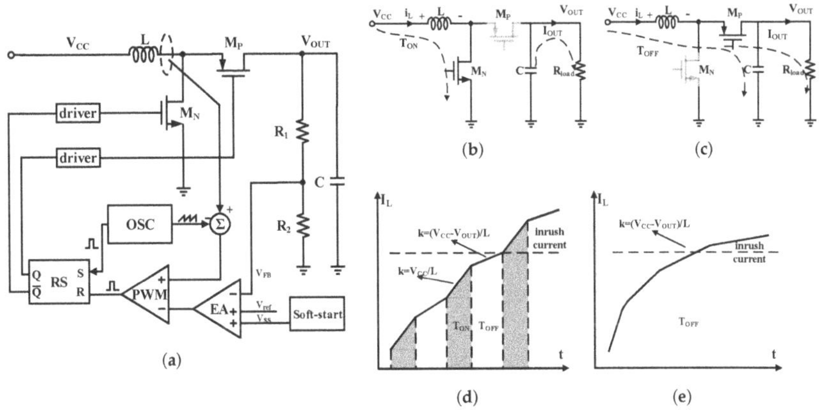


**Figure 1.** (**a**) The conventional structure of the PCMC PWM boost converter. (**b**) The $T_{ON}$ phase ($M_N$ is on). (**c**) The $T_{OFF}$ phase ($M_N$ is off). (**d**) The generation process of inrush current. (**e**) Inrush current of the conventional pre-charging.

Output voltage overshoot is also a problem during short-circuit recovery, especially. When the circuit is short-circuited, the output current becomes very large, and a large amount of energy is stored in the inductor. When the short circuit is recovered, the energy in the inductor is transferred to the load in the form of current, which results in a large output voltage overshoot inevitably. The inductance value, load capacitance and short-circuit current all affect the magnitude of the overshoot voltage.

### 2.2. The Conventional Hiccup-Mode Short-Circuit Protection Technique

In the normal operation of the converter, a mechanism is required to protect the circuit when the circuit is short-circuited, such as shutting down the converter and restarting the converter after the short circuit is recovered. Hiccup mode is a commonly used short-circuit protection technique.

The conventional mode is shown in Figure 2. $V_{ref1}$, $V_{ref2}$ and $I_{REF}$ are generated by the reference circuit. $C_1$ is the charging capacitor used to generate a linearly increasing soft-start voltage $V_{SS}$. Transistor $M_{N1}$ is used as a discharge switch to discharge capacitor $C_1$ to restart soft-start. When the converter is short-circuited, the feedback voltage $V_{FB}$ drops sharply. When it is lower than the reference voltage $V_{ref1}$, the comparator outputs the identification voltage to allow the hiccup-mode controller to operate. The controller generates a pulse to briefly turn on $M_{N1}$ to discharge $C_1$. Due to the existence of the reference current $I_{REF}$ and the shutdown of $M_{N1}$, $V_{SS}$ rises slowly from zero again in an attempt to restart the circuit. When the short circuit is not recovered, this process will be repeated, so it is also called hiccup mode. Some methods start hiccup mode by detecting the fall of $V_{FB}$ or $V_{EA}$, as is shown in Figure 2a,b, respectively.

However, the power dissipation of hiccup mode is a problem that cannot be ignored. When the hiccup-mode controller tries to detect whether the short circuit is recovered, it needs to open the circuit. In this process, the controller attempts to restart the circuit. Although there is a peak current limit, the current at this time is still very large, as is shown in Figure 2c. At the same time, as previously analyzed, a larger current is also the main cause of voltage overshoot during short-circuit recovery. Therefore, how to reduce the current in hiccup mode has great significance.

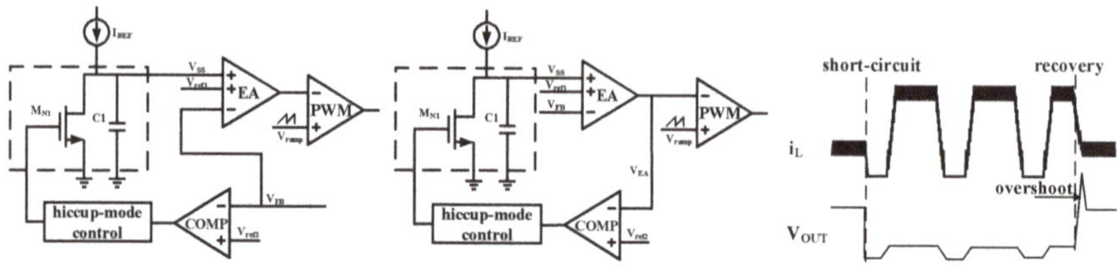


(a) (b) (c)

**Figure 2.** The conventional hiccup mode (**a**) The hiccup mode is activated by $V_{FB}$. (**b**) The hiccup mode is activated by $V_{EA}$. (**c**) The waveform during short-circuit recovery in the conventional hiccup mode with current limiting.

## 3. Proposed Boost DC-DC Converter with Hiccup-Mode Short-Circuit Protection Technique

The top level of the proposed boost converter is shown in Figure 3a. Compared with Figure 1a, it adds hiccup-mode short-circuit protection technique in the start-up module. The converter is composed of a power stage and a feedback control system. The power stage consists of a main PMOS transistor $M_P$ and a synchronous rectification NMOS transistor $M_N$. In addition, an inductor L and a capacitor C form a filter network. Particularly, $V_{INT}$ is the supply voltage of the internal module, such as EA (error amplifier), CS (current sensing), oscillator and so on.

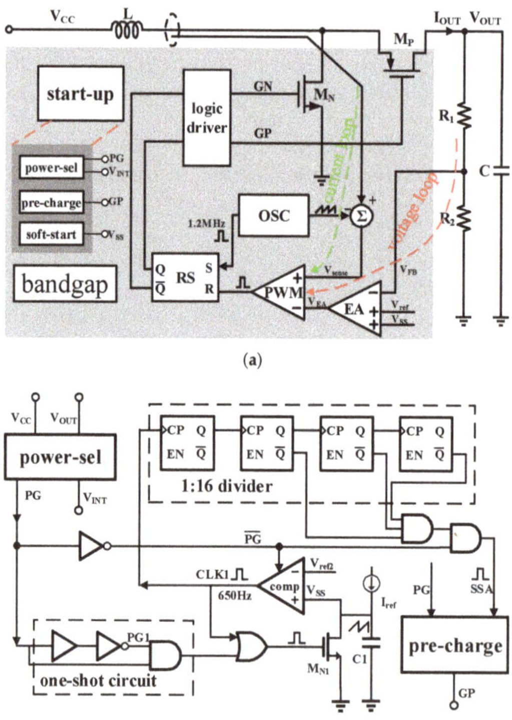


(a)

(b)

**Figure 3.** (**a**) Top level of the proposed boost DC-DC converter. (**b**) The proposed start-up circuit with hiccup mode.

The power-on phase of the converter can be divided into three stages. Stage 1: when $V_{OUT} < V_{CC}$, PG is low and $V_{INT} = V_{CC}$. Pre-charge module starts to operate and provides a stable charging current for the load capacitor, and $V_{OUT}$ increases. Stage 2: when $V_{CC} < V_{OUT} < 5$ V (the output of the converter during normal operation is 5 V), PG is high and $V_{INT} = V_{OUT}$. Pre-charge module closes and soft-start module starts to operate. Linearly increasing voltage $V_{SS}$ produced by soft-start module is used to avoid overshoot of output voltage. Meanwhile, the bandgap starts to operate and the oscillator begins providing the converter with a clock frequency. Stage 3: $V_{OUT} = V_{INT} = 5$ V, the difference between the feedback voltage $V_{FB}$ and the reference voltage $V_{ref}$ is amplified by the error amplifier, and the result $V_{EA}$ is compared with the current sensing signal $V_{sense}$, and finally, a control signal for controlling the power transistors on and off is obtained.

The slope compensation signal is added in Figure 3a to avoid sub-harmonic oscillation, which is well-known when the duty cycle is larger than 0.5. The start-up circuit integrates the short-circuit protection technique. When the converter is short-circuited, $V_{OUT}$ decreases sharply, PG becomes low, the the converter enters hiccup mode to implement short-circuit protection and can restart the converter after short-circuit recovery.

### 3.1. *Proposed Start-Up Circuit with Hiccup-Mode Circuit-Short Protection Technique*

The proposed start-up circuit is shown in Figure 3b. The start-up circuit consists of power-sel module, pre-charge module, a comparator, a 1:16 divider, a charge capacitor $C_1$, a discharge NMOS $M_{N1}$ and some other logic gates. The power-sel module realizes the function of selecting the internal power supply voltage $V_{INT}$ and giving the identification signal PG. The details are as follows: when $V_{OUT} < V_{CC}$, $V_{INT} = V_{CC}$ and PG = 0; when $V_{OUT} > V_{CC}$, $V_{INT} = V_{OUT}$ and PG = 1. When the converter is powered on, $V_{OUT} < V_{CC}$, so PG is low and $V_{INT}$ is equal to $V_{CC}$, respectively. The pre-charge module starts to operate and provides a stable charge current for the load capacitor by controlling GP. This stable charge current avoids larger initial inrush current in the pre-charge phase. Meanwhile, the comparator is enabled and CLK1 is low, because $V_{SS}$ is smaller than $V_{ref2}$ ($V_{ref2}$ is generated by the reference circuit to determine the peak of $V_{SS}$ in hiccup mode). As the load capacitor is charged, $V_{OUT}$ increases linearly. When $V_{OUT} > V_{CC}$, PG is high and $V_{INT}$ is equal to $V_{OUT}$. At this time, both the comparator and pre-charge module are off, then CLK1 is low and GP is controlled by logic controller, respectively. There is a logic circuit named one-shot circuit, whose function is that $M_{N1}$ will open for a short time to discharge C1 while PG changes from low to high. Then, $V_{SS}$ increases linearly, and the converter operates in soft-start state normally. $V_{FB}$ increases by following $V_{SS}$ until that $V_{FB} = V_{ref}$, as shown in Figure 3a. Finally, $V_{OUT}$ keeps stable and $V_{SS} = V_{OUT}$.

When the output is shorted, $V_{OUT}$ falls quickly and PG becomes low. Meanwhile, the pre-charge module operates to avoid large load current. CLK1 is high and $C_1$ discharged, because $V_{SS}$ is larger than $V_{ref2}$. In order to ensure that the capacitor can fully discharge, enough delay time is designed. Although pre-charge operates, $V_{OUT}$ remains low because of short circuit. So, PG remains low and the comparator keeps operating. $V_{SS}$ increases until it is equal to $V_{ref2}$, then CLK1 becomes high and C1 discharges again. If the short circuit persists, this process is repeated. The result is that CLK1 becomes a clock signal and the converter waits for staring-up again, and the converter operates in hiccup mode. Now the load current is also large, and if short circuit lasts for a long period of time, power dissipation is high. In order to minimize power dissipation, a 1:16 divider is designed to make sure that the converter operates in sleep mode for most of the time when output is shorted. When SSA (Sleep-Switch Auto, enables signal for entering sleep mode) is low, the pre-charge module stops operating and makes GP high to ensure PMOS power transistor closes. At this time, load current approaches zero. The corresponding waveform is shown in Figure 4. We can know from the figure that $I_{OUT}$ drops to zero when the output is short-circuited. This achieves short-circuit protection. The converter does not restart until both short-circuit recovery is established and SSA is high. When $V_{OUT} > V_{CC}$, PG is high, $V_{SS}$ drops to zero and re-increases to ensure that the converter can soft-start smoothly.

The restart process is the same as the power-on start-up process, and $V_{OUT}$ can rise to the normal voltage.

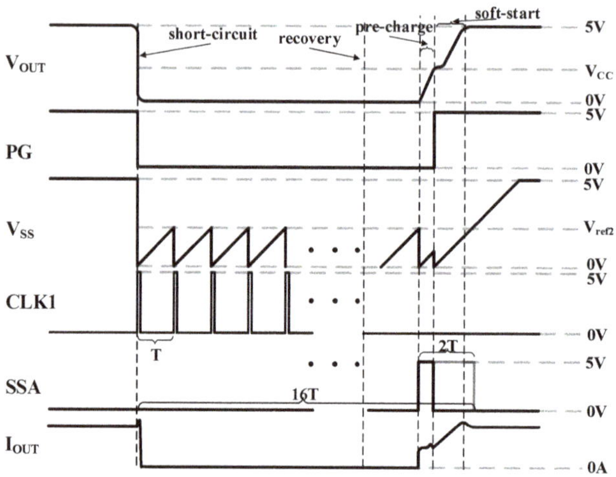


**Figure 4.** The corresponding waveform of the start-up circuit.

### *3.2. The Schematic of Proposed Pre-Charge*

The proposed pre-charge circuit is shown in Figure 5. The circuit in the dotted box belongs to the external circuit and $M_P$ is PMOS power transistor.

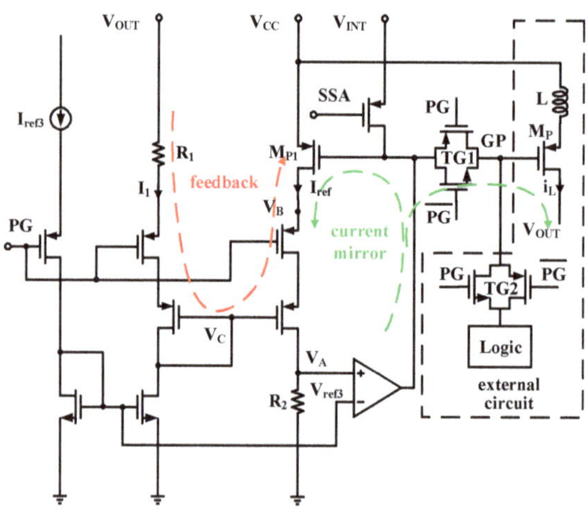


**Figure 5.** The proposed pre-charge circuit.

When PG is low, transmission gate TG1 is on and TG2 is off. At this time, GP is controlled by pre-charge module instead of logic. The operational amplifier operates properly to stabilize GP, and the virtual short of the op amp makes $V_A = V_{ref3}$, so $I_{ref1}$ can be expressed as:

$$I_{ref1} = \frac{V_A}{R_2} = \frac{V_{ref3}}{R_2} \tag{3}$$

At first, $V_{OUT}$ is small and feedback does not work. At this time, both $M_P$ and $M_{P1}$ work in the saturation region. They form a current mirror and $i_L$ is as follows:

$$i_L = I_{ref1} \frac{(W/L)_{MP}}{(W/L)_{MP1}} \tag{4}$$

W represents the channel width of MOS and L represents the effective channel length of MOS, respectively. Then, $V_{OUT}$ increases and $M_P$ works in linear region. In order to ensure that the current is as stable as possible, the feedback path is designed. When $V_{OUT}$ increases, $I_1$ increases and $V_C$ increases, and the result is that $V_B$ increases as $V_{OUT}$ increases. $I_{ref1}$ declines and $V_A$ declines, then GP declines to make sure $i_L$ is stable. Finally, the circuit can provide the load capacitor with a stable current. There is only a small drop in current when $V_{OUT}$ is close to $V_{CC}$. When SSA is low, GP is high and $M_P$ is off. Meanwhile, the converter works in the sleep mode and output current falls to zero. In particular, when PG is high, transmission gate TG1 is off and TG2 is on. At this time, GP is controlled by logic instead of the pre-charge module.

## 4. Simulation Results

The proposed hiccup-mode short-circuit protection technique is designed with the 0.35 standard CMOS process. The layout of the proposed boost converter is shown in Figure 6, and the die size is about 2 × 1.6 mm$^2$. The external inductance is 4.7 uH and the filter capacitor is 10 uF during the simulation. The converter output voltage is 5 V, and the input voltage range is 2 V to 3.5 V. The total feedback resistance is 100 kΩ, and the built-in reference is set to 0.5 V.

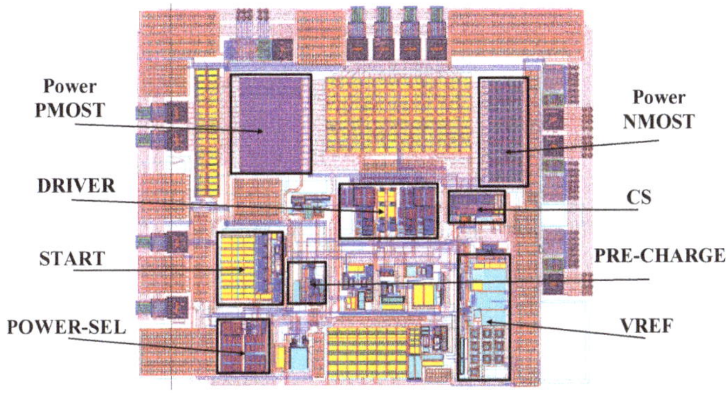


**Figure 6.** Layout of the proposed boost converter.

Figure 7 presents initial inrush current of both the conventional pre-charge and the proposed pre-charge. It can be seen from Figure 7 that the conventional pre-charge has a peak current of 2 A due to the inductance being continuously charged, while the proposed pre-charge has a peak current of only about 200 mA. Under the same conditions, thanks to the controllable gate voltage of the power transistor, the proposed pre-charge circuit can effectively suppress the initial inrush current, which is reduced by 90%.

Figure 8 presents the simulation results of proposed hiccup-mode short-circuit protection technique. The results of Figure 8a show that when short circuit occurs, $V_{SS}$ drops to 0 and repeatedly works in the form of a triangular wave, and is compared with a reference voltage to generate a clock CLK1 with a frequency of about 650 Hz. During the short-circuit period, the converter has a sleep period of about 21 ms, named sleep mode. At this time, the N and P power transistors are in the off state. After the sleep mode, there is about 3 ms to turn on the pre-charge module to detect whether the short circuit is recovered, named detection mode. When short-circuit recovery occurs in detection mode, the converter restarts directly. In particular, if the short-circuit recovery occurs during sleep mode, the converter will restart at the beginning of the detection mode. Figure 8b presents the curve of $V_{OUT}$ and inductor current $i_L$ during short circuit. The working frequency of the converter is 1.2 MHz during normal operation, and the working frequency is only 650 Hz during the short-circuit period. At the same time, the inductor current is close to 0, which is only 200 mA in the detection mode, which is also far smaller than the inductor current during normal operation. Therefore, the power dissipation during the short circuit is very low.

Meanwhile, when the short circuit is recovered, the converter restarts normally and there is no voltage overshoot. This is because the inductor current is very small before restarting, and there is almost no stored energy in the inductor, so there is no voltage overshoot. In fact, the restart process of the converter is similar to the process of power-on startup.

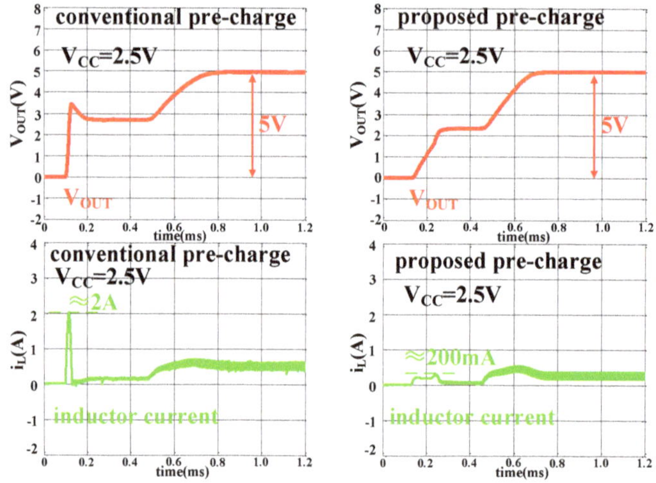


**Figure 7.** Initial inrush current of the conventional pre-charge and the proposed pre-charge.

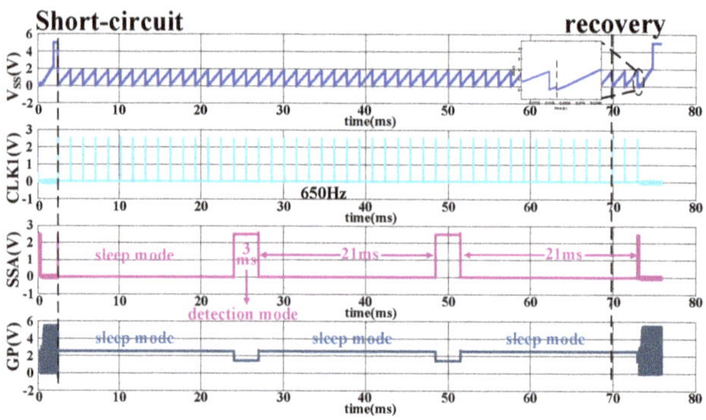


(**a**)

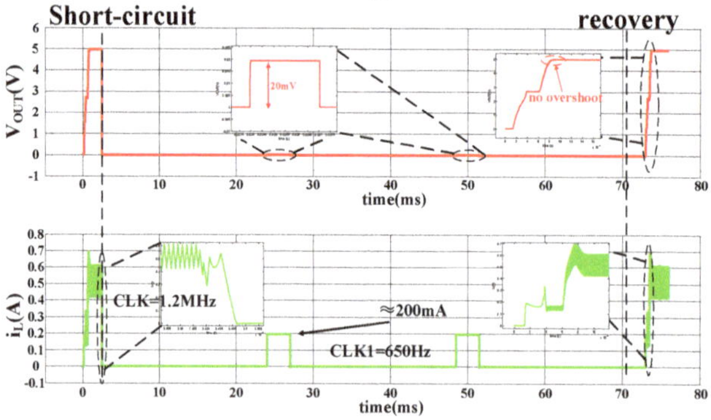


(**b**)

**Figure 8.** Simulation results of the proposed hiccup-mode short-circuit protection technique. (**a**) The waveform of $V_{SS}$, CLK1, SSA and GP. (**b**) The waveform of $V_{OUT}$ and inductor current.

Figure 8 shows the corresponding simulation results when the short-circuit recovery occurs during sleep mode, while Figure 9 shows the relevant simulation results when the short-circuit recovery occurs during detection mode. It can be seen from Figure 9 that when the converter is in the detection mode, once the short circuit recovers, the circuit can be restarted immediately. The inductor current at this time charges the load through a continuous current of about 200 mA. Whenever the short circuit recovers, there is no voltage overshoot when the circuit restarts. This is because $V_{SS}$ will decrease to 0 immediately before the restart, which ensures the soft start works normally and $V_{OUT}$ can always rise slowly.

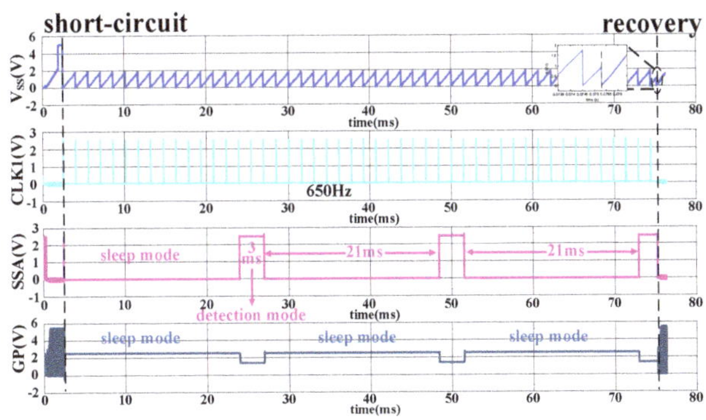


(a)

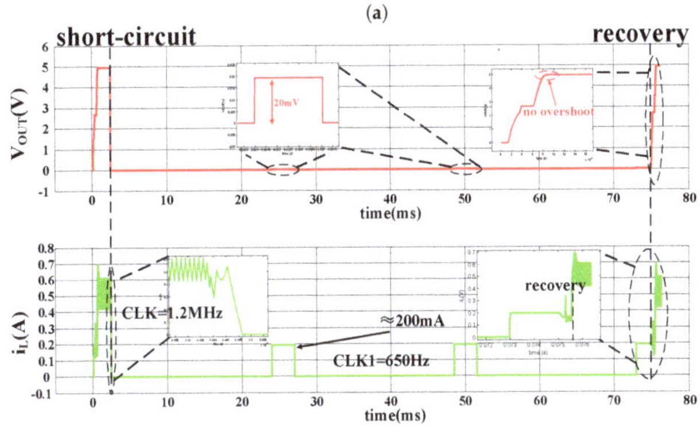


(b)

**Figure 9.** Simulation results in another case of the proposed hiccup-mode short-circuit protection technique. (**a**) The waveform of $V_{SS}$, CLK1, SSA and GP. (**b**) The waveform of $V_{OUT}$ and inductor current.

Table 1 summaries the performance of the proposed boost converter. The effective power during the short-circuit period is 0, so the power consumed by the converter can be expressed as:

$$P_{diss} = V_{in}I_{in} \quad (5)$$

where $V_{in}$ is equal to 2.5 V and $V_{OUT}$ is equal to 5 V. When the converter is in detection mode, $I_{in}$ is approximately equal to 200 mA. This is approximately equal to the inductor current, and the current consumed by the internal module is much less than 200 mA, which is almost negligible when calculating the power dissipation. The power dissipation at this time is as follows:

$$P_{diss(detection)} \approx V_{in}I_{in} = 2.5\text{ V} \times 200\text{ mA} = 500\text{ mW} \quad (6)$$

**Table 1.** Performance of the proposed boost converter.

| Parameters | Value |
| --- | --- |
| Input voltage range | 2 V to 3.5 V |
| $V_{OUT}$ | 5 V |
| External inductor | 4.7 uH |
| External capacitor | 10 uF |
| Total feedback resistance | 100 kΩ |
| Feedback reference voltage | 0.5 V |
| Switch frequency(normal) | 1.2 MHz |
| Working frequency(hiccup-mode) | 650 Hz |
| Maximum inrush current | 200 mA |
| Output voltage overshoot | no |
| Short-circuit power consumption | 62.5 mW |
| Power dissipation in detection mode | 500 mW |
| Power dissipation in sleep mode | 0.7 mW |

When the converter is in sleep mode, the converter has only a few modules that consume current, $I_{in}$ is approximately equal to 280 uA, and the power dissipation at this time is as follows:

$$P_{diss(sleep-mode)} \approx V_{in}I_{in} = 2.5\text{ V} \times 280\text{ uA } = 0.7\text{ mW} \quad (7)$$

The time ratio between sleep mode and detection mode is 7:1, so the average power dissipation during short circuit is as follows:

$$P_{diss(short-circuit)} \approx P_{diss(detection)}/8 = 500\text{ mW}/8 = 62.5\text{ mW} \quad (8)$$

It can be seen that the short-circuit power dissipation is very low.

Table 2 gives the comparison of the proposed converter with state-of-the-art converters. The inrush current of [8] is greater than that of its normal current, while inrush current of this work is only 40% of normal current. When the converter is short-circuited, other works reduce the short-circuit current to a reasonable range. Among them, ref. [12] reduces it to about 55% of the normal current, while this work reduces it to 40% of the normal current. This greatly reduces the short-circuit power dissipation. At the same time, the working frequency of the proposed converter during short circuit is only 0.55% of the normal frequency, which also reduces the short-circuit power dissipation.

**Table 2.** Comparison of the proposed converter with other state-of-the-art converters.

| | [14] 2012 | [15] 2013 | [12] 2014 | [8] 2016 | This Work |
| --- | --- | --- | --- | --- | --- |
| Process (nm) | 500 | 600 | 600 | N/A | 350 |
| Area ($mm^2$) | N/A | N/A | 3.456 | N/A | 3.2 |
| Input voltage range | 4.5–18 V | 4.5–18 V | 4.5–30 V | 5 V | 2–3.5 V |
| $V_{OUT}$ | 3.3 V | 1.2 V | 3–12 V | 18 V | 5 V |
| Inrush current | N/A | N/A | N/A | 1.25 A | 200 mA |
| $i_L$(normal) | 5 A | 4 A | 5 A | 1 A | 500 mA |
| $i_L$(short-circuit) | 4.5–7 A | 3.8–7 A | 2.78 A | N/A | 200 mA |
| Voltage overshoot | N/A | no | no | no | no |
| Switch frequency (normal) | 500 kHz | 500 kHz | 370 kHz | 1 MHz | 1.2 MHz |
| Working frequency (hiccup-mode) | 110 kHz | 110 kHz | 115 kHz | N/A | 650 Hz |
| Short-circuit power dissipation | medium | medium | low | N/A | low |

N/A: Not mentioned.

## 5. Conclusions

In this paper, a low-power hiccup-mode short-circuit protection technique with overshoot and inrush current suppression for a DC-DC boost converter is analyzed and its process is verified. Compared with the conventional pre-charge circuit, the proposed pre-charge circuit greatly minimizes the initial inrush current, and inrush current is reduced by about 90%. This current is also smaller than the inductor current during normal operation. When the circuit is short-circuited, the converter operates with very low power dissipation, it can restart normally after the short circuit is recovered, and there is no voltage overshoot.

**Author Contributions:** Formal analysis, H.W.; Funding acquisition, F.T.; Investigation, N.W.; Methodology, H.W.; Software, H.W.; Supervision, T.S.; Visualization, T.S.; Writing—review & editing, H.W. All authors have read and agreed to the published version of the manuscript.

**Funding:** This work was sponsored by Natural Science Foundation of Chongqing, China, No. cstc2019jcyj-zdxmX0014.

**Conflicts of Interest:** The authors declare no conflict of interest.

## References

1. Liu, H.; Hu, H.; Wu, Y. Overview of High-Step-Up Coupled-Inductor Boost Converters. *IEEE J. Emerg. Sel. Top. Power Electron.* **2016**, *42*, 689–704. [CrossRef]
2. Tofoli, F.L.; Pereira, D.d.C.; Josias de Paula, W.; Oliveira Júnior, D.d.S. Survey on non-isolated high-voltage step-up dc–dc topologies based on the boost converter. *IET Power Electron.* **2014**, *8*, 2044–2057. [CrossRef]
3. Martínez-Salamero, L.; García, G.; Orellana, M.; Lahore, C.; Estibals, B. Start-Up Control and Voltage Regulation in a Boost Converter Under Sliding-Mode Operation. *IEEE Trans. Ind. Electron.* **2013**, *10*, 4637–4649. [CrossRef]
4. Li, S.Z.; Zou, X.C.; Chen, X.F.; Gan, Q. Designing a compact soft-start scheme for voltage-mode DC–DC switching converters. *Microelectron. J.* **2010**, *41*, 430–439. [CrossRef]
5. Xiao, Z. An Instantaneously Triggered Short-Circuit Protection Architecture for Boost Switching DC/DC Converters. *IEEE Trans. Power Electron.* **2018**, *33*, 5677–5685. [CrossRef]
6. Bian, Z.P.; Yao, R.H.; Luo, F. Soft-Start Circuit Based on Switched-Capacitor for DC-DC Switching Regulator. *IEICE Trans. Electron.* **2012**, *10*, 1692–1694. [CrossRef]
7. Lou, F.; Ma, D.S. Design of digital tri-mode adaptive-output buck-boost power converter for power-efficient integrated systems. *IEEE Trans. Ind. Electron.* **2010**, *57*, 2151–2160.
8. Fan, S.; Xue, Z.; Guo, Z.; Wang, Y.; Geng, L. VRSPV Soft-Start Strategy and AICS Technique for Boost Converters to Improve the Start-Up Performance. *IEEE Trans. Power Electron.* **2016**, *31*, 3663–3672. [CrossRef]
9. Fang, Y.; Wei, T.; He, L.; Xie, Y.; Xing, Y. A novel over-current protection technique applied to peak-current type DC–DC converter. *J. Electromagn. Anal. Appl.* **2010**, *2*, 352–356. [CrossRef]
10. Narula, S.; Singh, B.; Bhuvaneshwari, G. Interleaved CSC converter-based power factor corrected switched mode power supply for arc welding. *IET Power Electron.* **2016**, *9*, 2404–2415. [CrossRef]
11. Xiao, J.C.; Xin, W. A high-performance foldback current limiting circuit for improving regulators latch-up effect. In Proceedings of the 2006 International Conference on Communications, Circuits and Systems, Guilin, China, 25–28 June 2006; pp. 2248–2250.
12. Li, Y.; Lai, X.; Ye, Q.; Yuan, B. Novel short-circuit protection technique for DC–DC buck converters. *IET Circuits Devices Syst.* **2014**, *57*, 90–99. [CrossRef]
13. Yan, Y.; Li, Q.; Gu, E. Synchronized Hiccup Mode Overcurrent Protection Scheme for Decentralized Multiphase Current Mode DC-DC Converters. In Proceedings of the 2021 IEEE Applied Power Electronics Conference and Exposition (APEC), Phoenix, AZ, USA, 14–17 June 2021; pp. 771–777.
14. Shi, L.F.; Mao, X.Y.; He, H.S.; Liu, F.B.; Lai, X.Q. Design of a hiccup mode over-current protection circuit for DC–DC switching converters. *Microelectron. J.* **2012**, *43*, 57–62. [CrossRef]
15. Li, Y.; Mao, X.; Wang, H.; Wen, C.; Wen, L. An Improved Hiccup Mode Short-Circuit Protection Technique With Effective Overshoot Suppression for DC–DC Converters. *IEEE Trans. Power Electron.* **2013**, *28*, 877–885. [CrossRef]

*electronics*

MDPI

*Article*

# Real-Time Fault Location Using the Retardation Method

**Moneer Nabwani [1,2], Michael Suleymanov [1], Yosef Pinhasi [1] and Asher Yahalom [1,3,*]**

1 Department of Electrical & Electronic Engineering, Ariel University, Kiriat Hamada, P.O. Box 3, Ariel 4077625, Israel; moneer.nabwani@iec.co.il (M.N.); michaels@ariel.ac.il (M.S.); yosip@ariel.ac.il (Y.P.)
2 Israel Electric Company (IEC), Netiv Haor 1, P.O. Box 10, Haifa 3100001, Israel
3 Princeton Plasma Physics Laboratory, Princeton University, Princeton, NJ 08543, USA
* Correspondence: asya@ariel.ac.il

**Abstract:** A new method for short-circuit fault location is proposed. The method is based on instantaneous signal measurement and its first and second derivatives, which are the novel elements of the current approach. The derivatives allow associating a precise time stamp to the occurrence of the fault. Due to retardation phenomena, the difference between the times in which a signal is registered in two detectors can be used to locate the fault. We offer several mathematical models to describe the fault. Although a description of faults in terms of a lumped circuit is useful for elucidating the methods for detecting the fault, this description will not suffice to describe the fault signal propagation; hence, a distributed models is needed, which is given in terms of the telegraph equations. Those equations were used to derive a transmission line transfer function, and an exact analytical description of the fault signal propagating in the transmission line was obtained. The analytical solution was verified both by numerical simulations and experimentally.

**Keywords:** fault location; real-time

**Citation:** Nabwani, M.; Suleymanov, M.; Pinhasi, Y.; Yahalom, A. Real-Time Fault Location Using the Retardation Method. *Electronics* **2022**, *11*, 980. https://doi.org/10.3390/electronics11070980

Academic Editor: Ahmed Abu-Siada

Received: 27 December 2021
Accepted: 11 March 2022
Published: 22 March 2022



## 1. Introduction

One of the most important results of special relativity is the fact that no signal can travel faster than the speed of light in a vacuum $c$ [1]. The same is true for a signal generated from a fault occurring in a power transmission line such as a short or a disconnection. As the signal due to the fault reaches detectors along the line at different times (due to a finite propagation speed), one can use the differences in the time of arrival to locate the fault along the line. This location technique is passive in the sense that one does not need to inject any signal to the power line in order to locate the fault; rather, the fault itself is the source of the signal. Another advantage is that the detection and location are performed in real-time. Still, several issues are raised regarding the proposed technique:

- What is the signal velocity, and what is the needed sampling rate?
- What should be measured (voltage and/or current)?
- How many detectors are needed?
- What are the dispersion effects on the prorogation of the signal, and how do they affect accuracy?
- What are the best practices for signal processing in order to obtain an accurate time of arrival?
- How does the current technique compare in terms of accuracy with previous works?

We try to answer those questions in the current paper. The discussion was limited to low-voltage transmission lines, while the discussion of high-voltage transmission lines will be the subject of a future paper.

Power transmission lines have a broad range of faults. These fault classifications appeared in various previous articles [2–9]. There are several approaches to fault location algorithms, including various approaches regarding measurements and data processing and their proposed applications. The bridge circuit method [10] employs an adjustable impedance to calculate the location of the fault. Mustari et al. [11] employed a neural network approach. The method of de Morais Pereira and Zanetta [12] was based on steady-state measured phasors in local terminals. M. N. Alam et al. [13] presented a method based on surface electromagnetic waves propagating along a transmission line (see also [14]). In M. Aldeen and F. Crusca's study [15], the faults were modeled as unknown inputs, decoupled from the state and output measurements through coordinate transformations, and then estimated via the use of the observer theory. The article by Qais Alsafasfeh et al. [16] presented a method that integrates the symmetrical components technique with principal component analysis (PCA) for fault classification and detection. In another research work [17], Petri nets were used to obtain the modeling and location detection of faults in power systems. Another widely used method is that of wavelet transform analysis [18–21]. We compared the accuracy of the current method to the accuracy of previous work in the concluding section.

The plan of the paper is as follows: First, we present the basic idea of the method. Then, we provide a description of the fault in terms of a lumped circuit, which is useful for elucidating the methods for detecting the fault; here, we shall demonstrate the ability to determine the signal arrival time using derivatives. This description will not suffice, however, to describe the fault signal propagation; hence, a distributed model is needed, which is given in terms of the telegraph equations [22–25]. After introducing the main formalism and the telegraph equations of the distributed system, we give a specific example of a two-wire power line. We present several models to describe the development of a short and the signal generated in a possible detector due to that short. These include exponential, Gaussian, and step function forms. For the step function model, an inverse Laplace transform allowed us to determine the time-dependent signal at the sensor position analytically. At this stage, we compared the analytical and numerical solutions. Next, the experimental setup is described. We show the high level of conformity of the theoretical and experimental measurements using the appropriate data processing. Finally, we determined the system accuracy and compared it with previous methods.

## 2. Methodology: Fault Detection by Retardation

Consider a fault of unknown location that causes a signal to propagate to both sided of the transmission line (see Figure 1). The signal is registered by a detector, which determines its time of arrival, $t_1$ for Detector 1 and $t_2$ for Detector 2.

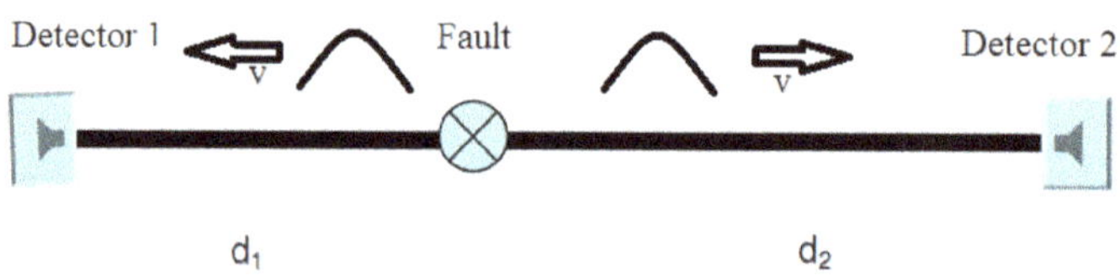


**Figure 1.** Transmission line with a short and detectors.

If the unknown time at which the fault occurred is $t_0$ and the constant velocity of the signal propagation is $v$, the following relations follow:

$$v(t_1 - t_0) = d_1; \qquad v(t_2 - t_0) = d_2 \tag{1}$$

in which $d_1$ and $d_2$ are the unknown distance from the fault to Detectors 1 and 2, respectively. The total distance between the two detectors is known as:

$$d = d_1 + d_2 \tag{2}$$

Defining the time difference $t_{di} \equiv t_2 - t_1$, the distance from Detector 1 to the short may be written as follows:

$$d_1 = \frac{1}{2}(d - v\, t_{di}) \tag{3}$$

Hence, if $t_{di}$ is a random variable of the standard deviation $\sigma_{t_{di}}$, the corresponding standard deviation of $d_1$ is:

$$\sigma_{d_1} = \frac{v}{2}\sigma_{t_{di}} \tag{4}$$

provided we assume that the velocity $v$ is known. Now, $\sigma_{t_{di}}$ can be evaluated as:

$$\sigma_{t_{di}} = \sqrt{\sigma_{t_1}^2 + \sigma_{t_2}^2 - 2C_{t_1 t_2}}. \tag{5}$$

in which the covariance $C_{t_1 t_2}$ is given in terms of the following expectation value:

$$C_{t_1 t_2} = E[(t_1 - \bar{t_1})(t_2 - \bar{t_2})], \qquad \bar{t} \equiv E[t] \tag{6}$$

If the uncertainty in $t_1$ is uncorrelated with the uncertainty in $t_2$, the covariance is null, and we have a simplified expression:

$$\sigma_{t_{di}} = \sqrt{\sigma_{t_1}^2 + \sigma_{t_2}^2}. \tag{7}$$

If $\sigma_{t_1} = \sigma_{t_2}$, this will result in:

$$\sigma_{t_{di}} = \sqrt{2}\sigma_{t_1}. \tag{8}$$

Thus, we can rewrite $\sigma_{d_1}$ in terms of $\sigma_{t_1}$ as follows:

$$\sigma_{d_1} = \frac{v}{\sqrt{2}}\sigma_{t_1} \tag{9}$$

Now, suppose that the data are sampled at intervals of $T$; if the signal is detected initially at time $t_D$, this means that the signal arrived at any time between $t_D$ and $t_D - T$. Since we do not have any information about what time the signal really arrived, we assumed that $t_1$ is a random variable distributed uniformly in the interval $[t_D - T, t_D]$. The probability density function of $t_1$ is depicted in Figure 2.

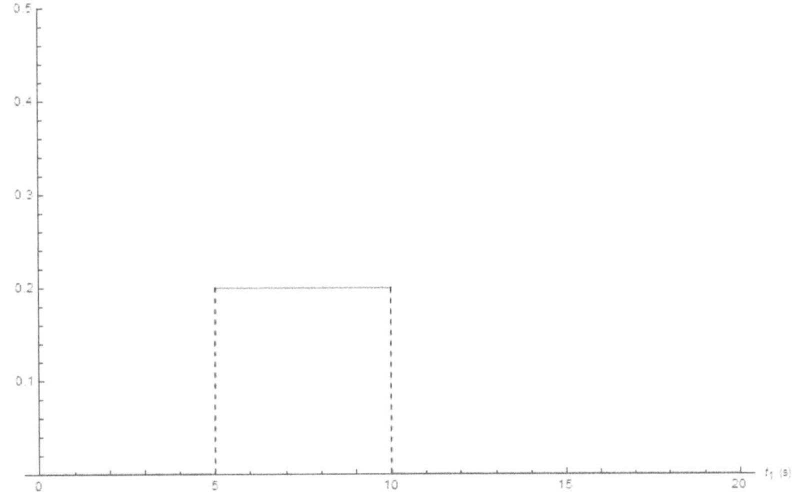


**Figure 2.** Uniform distribution probability density function of $t_1$; in the illustration, we assumed $T = 5$ and $t_D = 10$.

Since the moments of a uniform distribution function are known, we can easily evaluate the expectation and standard deviation of $t_1$ as follows:

$$\bar{t_1} = t_D - \frac{1}{2}T, \qquad \sigma_{t_1} = \frac{T}{2\sqrt{3}} \tag{10}$$

Inserting the result of Equation (10) into Equation (9) leads to:

$$\sigma_{d_1} = \frac{vT}{2\sqrt{6}} \tag{11}$$

Thus, the accuracy at which we need to know the location of the fault will determine the sampling rate $f_S$ according to the following formula:

$$f_S = \frac{1}{T} = \frac{v}{2\sqrt{6}\,\sigma_{d_1}} \tag{12}$$

As we shall show later, for a power line, the typical propagation of the signal is close to light's velocity in a vacuum:

$$v \simeq c \simeq 3 \times 10^8 \text{ m/s}. \tag{13}$$

Thus, if the required distance precision is about 1 m, then the time measurement sampling rate should be:

$$f_S \simeq 6.12 \times 10^7 \text{Hz} \tag{14}$$

This is much lower than the clock rate of current computer processors. We do not deal here with additional sources of uncertainty, such as the noise level of transmission lines, and leave that for future work.

## 3. Methodology: Signal Detection

What kind of signal should we measure for the fault location, and should it be the voltage or current? How should it be processed in order to avoid the accumulation of a large amount of unnecessary data due to the high sampling rate dictated by Equation (14)? We shall try to answer those question using a lumped circuit model, as described in Appendix A [26–28], in which we show that a fault signal can be obtained by measuring either the current or voltage.

The high sampling rate that is needed for accurate location as described in Equation (12) imposes a storage challenge as the amount of data accumulated may be prohibitive. However, taking the derivative of the signal, which may be the voltage or current, allowed us to overcome this obstacle (see Appendix A). By choosing a high detection threshold, one avoids false positives, and this allowed us to store a relatively small amount of data that was sufficient for the detection and location of the short. In some cases, a second derivative was required.

Generally, the first voltage derivative is enough (except for the load voltage), while in the case of current measurement, the second derivative is necessary.

We further noticed an additional restriction on the required resolution $T$, in order to avoid the case that the pulse goes undetected, that is between sampling points, we needed to have a resolution smaller than the pulse duration, which is the same duration as the time it takes the short to form; hence:

$$T < \tau_S \tag{15}$$

Now, if $\tau_S \simeq 10^{-8}$ s, this means that:

$$T < 10^{-8} \text{ s} \tag{16}$$

This limitation is even more restrictive than the one appearing in Equation (14), leading to:

$$T_S = \frac{1}{f_S} \simeq 1.63 \times 10^{-8}\ \text{s} \tag{17}$$

Obviously, a lumped model neglects the effect of spatial distances, and hence the effect of signal propagation. To describe the effect of signal propagation properly, a distributed model is needed, and with such a model we could study the short signal propagation and related phenomena including dispersion; this is discussed in the following section.

## 4. Methodology: Distributed Model

Until now, we have ignored the signal propagation in the circuit and assumed that the changes in the voltage and the current occur immediately and simultaneously everywhere. This assumption is not compatible with the theory of special relativity, which states that any signal must propagate with finite velocity, smaller than the speed of light in a vacuum [1]. To describe this behavior, we used the transmission line propagation model [29], a section of which is depicted in Figure 3.

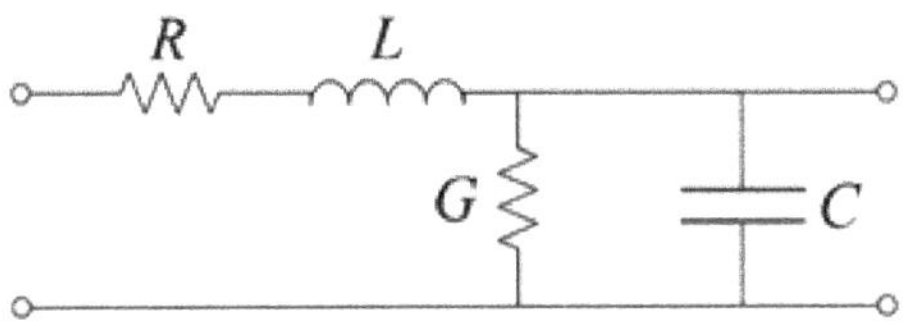


**Figure 3.** Transmission line structure for a single unit length.

This approach leads to the telegraph equations. We describe this model in the time and frequency domains and draw the relevant conclusion from each presentation.

### 4.1. The Time Domain

The equations that describe the voltage and the current dependence in the time domain are the telegraph equations given below [29]:

$$\frac{\partial V(x,t)}{\partial x} = -RI(x,t) - L\frac{\partial I(x,t)}{\partial t}, \qquad \frac{\partial I(x,t)}{\partial x} = -GV(x,t) - C\frac{\partial V(x,t)}{\partial t} \tag{18}$$

where $R$ is the resistance, $L$ is the inductance, $G$ is the conductance, and $C$ is the capacitance, per unit length each (see Figure 3). This pattern is repeated indefinitely (see Figure 4).

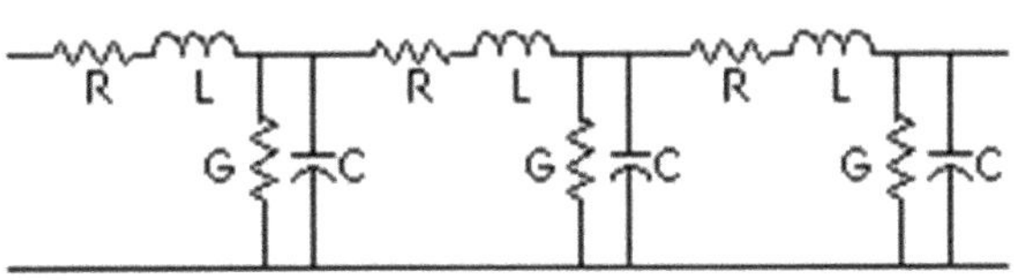


**Figure 4.** Transmission line structure repeated.

The above equations can be solved for an infinite transmission line excited at $x = 0$ by an excitation $V_{in}(t)$ representing the effect of the short on the voltage. The solution for the voltage is:

$$V(x,t) \approx V_{in}\left(t - \sqrt{LC}x\right) \exp\left(-\frac{\sqrt{LC}}{2}\left(\frac{R}{L} + \frac{G}{C}\right)x\right) \tag{19}$$

The above solution describes a voltage signal propagating at a velocity:

$$v = \frac{1}{\sqrt{LC}} \tag{20}$$

and an exponential decay with a decay factor of:

$$\alpha = \frac{\sqrt{LC}}{2}\left(\frac{R}{L} + \frac{G}{C}\right). \tag{21}$$

Assuming the transmission line to be a two-wire cable as described in Appendix A, the following parameters are obtained [29]:

$$R = \frac{2R_S}{\pi d}, \quad L = \frac{\mu}{\pi \cosh^{-1}\left(\frac{D}{d}\right)}, \quad G = \frac{\pi\sigma}{\cosh^{-1}\left(\frac{D}{d}\right)}, \quad C = \frac{\pi\varepsilon}{\cosh^{-1}\left(\frac{D}{d}\right)} \tag{22}$$

where each wire has a diameter $d$, and the distance between the wires is $D$. The material between the wires has a permittivity $\varepsilon$, permeability $\mu$, and (a very small) conductivity $\sigma$. The wire resistance is calculated, as in Equation (A6), using the surface resistance given in Equation (A5). The short propagation velocity (20) can be now calculated using the parameters of Equation (22), to yield:

$$v = \frac{1}{\sqrt{\varepsilon\mu}} = \frac{c}{n} \tag{23}$$

where $n$ is the index of refraction around the transmission line. It should be noted that the electromagnetic wave is propagating in the region between the conductors and not in the conductors themselves, where the propagation is much slower and the decay is very strong. We noticed that if the two wires are surrounded by air $n \simeq 1$, we would recover the velocity of Equation (13).

Notice that since according to Equations (A5) and (A6), the resistance is frequency dependent as dictated by the skin effect, the resistance term in Equation (18) is not a simple multiplication and should be replaced by a convolution. Thus, a frequency domain formalism is more adequate for this type of problem, as will be described next.

### 4.2. The Frequency Domain

In the frequency domain, the telegraph equation takes the form [29]:

$$\frac{\partial V(x,\omega)}{\partial x} = -(R + i\omega L)I(x,\omega), \quad \frac{\partial I(x,\omega)}{\partial x} = -(G + i\omega C)V(x,\omega), \quad i \equiv \sqrt{-1} \tag{24}$$

Combining these two equations, we can separate the voltage and current variables, in terms of the following two equations:

$$\frac{\partial^2 V(x,\omega)}{\partial x^2} = \gamma^2 V(x,\omega), \qquad \frac{\partial^2 I(x,\omega)}{\partial x^2} = \gamma^2 I(x,\omega) \tag{25}$$

where we define:

$$\gamma \equiv \sqrt{(R + i\omega L)(G + i\omega C)} \tag{26}$$

These equations have a solution of the form:

$$\begin{aligned} V(x,\omega) &= V^{(+)}(\omega)e^{-\gamma x} + V^{(-)}(\omega)e^{\gamma x} \\ I(x,\omega) &= \tfrac{1}{Z_0}\left(V^{(+)}(\omega)e^{-\gamma x} - V^{(-)}(\omega)e^{\gamma x}\right) \end{aligned} \tag{27}$$

The functions $V^{(\pm)}(\omega)$ are derived from the initial conditions. The impedance $Z_0$ is defined as follows:

$$Z_0 \equiv \sqrt{\frac{R + i\omega L}{G + i\omega C}} \tag{28}$$

In the case where the resistivity and the leakage admittance are small enough, such that:

$$R << \omega L, \qquad G << \omega C \tag{29}$$

we can approximate the impedance:

$$Z_0 \approx \sqrt{\frac{L}{C}} \tag{30}$$

and the real and imaginary parts of $\gamma$ take the form:

$$\mathrm{Re}(\gamma) \approx \frac{\sqrt{LC}}{2}\left(\frac{R}{L} + \frac{G}{C}\right) = \alpha, \qquad \mathrm{Im}(\gamma) \approx \omega\sqrt{LC} \tag{31}$$

As the frequency rises, the approximation becomes more accurate. Hence, for the higher-frequency Fourier components associated with the short formation, this approximation is more effective. Notice that while $\mathrm{Re}(\gamma)$ describes absorption and coincides with the same expression for absorption obtained in Equation (19), $\mathrm{Im}(\gamma)$ describes propagation.

### 4.3. Time-Dependent vs. Stationary Shorts

Steady-state shorts are easily analyzed in transmission line theory; here, we shall try to elucidate the connection between the transient phenomena of the short's appearance and its asymptotic behavior as a steady- state phenomenon. Our model is depicted in Figure 5.

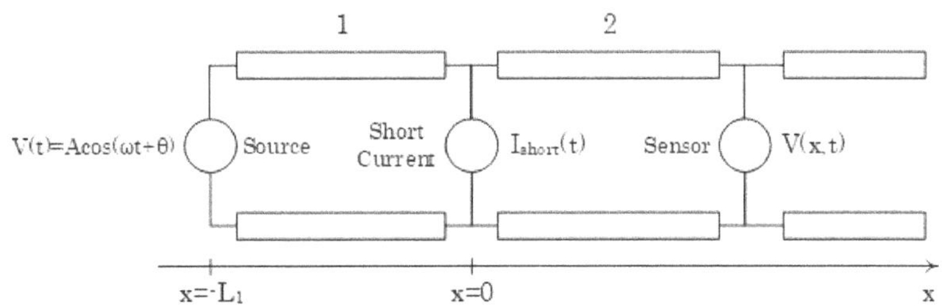


**Figure 5.** Detailed short in the transmission line model.

We assumed that the short appears at some point ($x = 0$) in the transmission line, in which the continuity of the voltage and current dictates:

$$V_1(0,\omega) = V_2(0,\omega), \qquad I_1(0,\omega) = I_2(0,\omega) + I_{short}(\omega). \tag{32}$$

We assumed that the short current does not exist at $t = 0$; likewise, the short current after a long time can be calculated using the fact that the voltage on the short vanishes at $t \to \infty$. The second assumption can be formulated as:

$$\lim_{t\to\infty} I_{short}(t) = I_{short\ asymptotic}(t) \tag{33}$$

To obtain the asymptotic short current, the following calculation was performed. First, we used the fact that after a considerable duration, the voltage on the short vanishes to obtain the following:

$$V_{short\ asymptotic}(\omega) = V^{(+)}(\omega) + V^{(-)}(\omega) = 0 \tag{34}$$

This leads to the result:

$$V^{(-)}(\omega) = -V^{(+)}(\omega) \tag{35}$$

Inserting this result into Equation (27) and taking into account that $V_{asymptotic}(\omega, -L_1)$ is the source voltage lead to the following result:

$$V_{in}(\omega) = V_{asymptotic}(\omega, -L_1) = V^{(+)}(\omega)\left(e^{\gamma L_1} - e^{-\gamma L_1}\right) \tag{36}$$

Hence:

$$V^{(+)}(\omega) = \frac{V_{in}(\omega)}{2\sinh(\gamma L_1)} \tag{37}$$

The asymptotic short current in the frequency domain can be calculated using Equation (27) at $x = 0$ and taking into account Equations (35) and (37):

$$I_{short\ asymptotic}(\omega) = I(\omega, 0) = \frac{2V^{(+)}}{Z_0} = \frac{V_{in}(\omega)}{Z_0 \sinh(\gamma L_1)} \tag{38}$$

Taking into account $V_{in}(t)$ given in Equation (A8) (which is equivalent to a sum of delta functions in the frequency domain), the asymptotic short current in the time domain is:

$$I_{short\ asymptotic}(t) = \frac{A_0}{Z_0}\mathrm{Re}\frac{e^{j\omega_0 t}}{\sinh(\gamma(\omega_0)L_1)} \tag{39}$$

The second assumption was that the short current vanishes at $t = 0$:

$$I_{short}(t = 0) = 0 \tag{40}$$

In the current model, this requirement is fulfilled by multiplying the asymptotic expression with some reasonable function that vanishes at $t \to 0$ and approaches unity at $t \to \infty$, for example:

$$I_{short}(t) = I_{short\ asymptotic}(t)u(t)\left(1 - e^{-\frac{t}{\Delta}}\right) \tag{41}$$

in which $u(t)$ is a step function. The calculation results for the short current on the load side are as follows:

$$I_2(x,t) = \frac{e^{-\alpha(x+L_1)}}{Z_0}V_{in}(t - t_{d1}) - \frac{e^{-\alpha x}}{2}I_{short}(t - t_d) - \frac{e^{-\alpha(x+2L_1)}}{2}I_{short}(t - t_{d2}) \tag{42}$$

where $t_d = \frac{x}{v}$, $t_{d1} = \frac{x+L_1}{v}$, and $t_{d2} = \frac{x+2L_1}{v}$. $v$ is the signal propagation velocity. The current at the load side is affected by three terms, each with unique retardation. The source voltage is retarded by a time $t_{d1}$, while the short current is retarded by a time $t_d$ for the direct signal and by $t_{d2}$ for the same signal reflected by the circuit source. Each retardation time is proportional to the distance it needs to travel from its source and inversely proportional to the velocity of propagation. The model also shows that the signals suffer attenuation proportional to the distance they travel.

### 4.4. Signal Dispersion

We now consider the problem of dispersion. This problem is interesting since we would like to know in what ways does the line distort signals propagating on it and, in particular, how the signal produced by the short is effected. We start from the definition of $\gamma$ given in Equation (26), and we assume that $G = 0$. Consequently, the propagation index is:

$$\gamma(\omega) = i\omega\sqrt{LC}\sqrt{1 + \frac{R}{i\omega L}}, \tag{43}$$

Now, let us investigate the condition:

$$Ra \equiv \frac{R(\omega)}{i\omega L} \ll 1 \tag{44}$$

The resistance $R(\omega)$ is frequency dependent due to the skin effect and can be calculated from the surface resistance using Equations (A5) and (A6) as follows:

$$R = \frac{2}{\sqrt{\pi}d}\sqrt{\frac{\mu_c}{\sigma_c}f} \tag{45}$$

Thus, we obtain the ratio:

$$\frac{R}{\omega L} = \frac{1}{\sqrt{\pi^3}dL}\sqrt{\frac{\mu_c}{\sigma_c}}\frac{1}{\sqrt{f}} \tag{46}$$

Thus, the approximation given in Equation (44) is even better for higher-frequency components. Introducing the inductance impedance:

$$Z_L = i\omega L = i2\pi f L \tag{47}$$

we can cast Equation (46) in the form:

$$\frac{R}{Z_L} \cong -\frac{2.72i}{\sqrt{f}} \tag{48}$$

in which the numerical values for the parameters are taken from Tables A1, A2, and A5. Equation (48) is depicted in Figure 6, which shows that the condition given in Equation (44) holds even better for a high frequency.

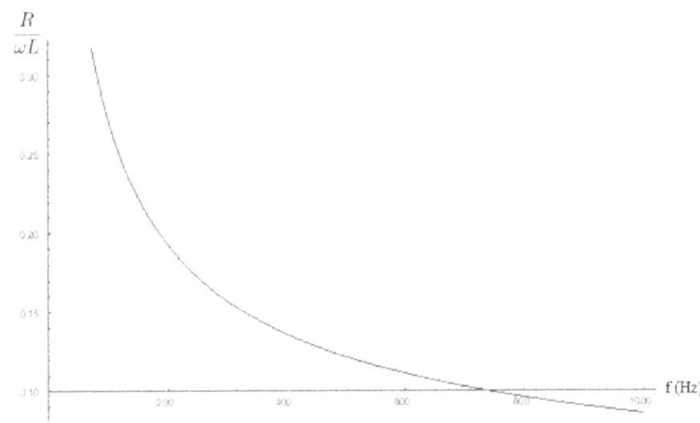


**Figure 6.** The ratio of resistance to inductance impedance as a function of frequency according to Equation (48).

For a small parameter $|x| \ll 1$, we can write:

$$\sqrt{1+x} = 1 + \frac{x}{2} - \frac{x^2}{8} + \frac{x^3}{16} + O[x]^4. \tag{49}$$

Hence, taking into account that $\frac{R}{\omega L} \ll 1$, $\gamma(\omega)$ of Equation (43) has up to third order the form:

$$\gamma(\omega) \cong i\frac{\omega}{\mathrm{v}}\left[1 - j\frac{R}{2\omega L} + \frac{1}{8}\left(\frac{R}{\omega L}\right)^2 + \frac{j}{16}\left(\frac{R}{\omega L}\right)^3\right] \tag{50}$$

This expression can be separated into an imaginary and real parts as follows. The imaginary part of $\gamma$ takes the form:

$$\mathrm{Im}\gamma \cong \frac{\omega}{\mathrm{v}} + \frac{\mu_c}{4\pi^2 d^2 L^2 \sigma_c \mathrm{v}} \tag{51}$$

where we have taken into account Equation (46). As Equation (51) is linear in $\omega$, we concluded that there is no dispersion during the signal propagation that requires non-linear phase terms. To appreciate the linearity of $\text{Im}\gamma$, we depict it as a function of the frequency in both Figure 7 for low frequencies and in Figure 8 for high frequencies without making an expansion approximation; the linearity and, hence, the lack of dispersion are apparent for a wide frequency range.

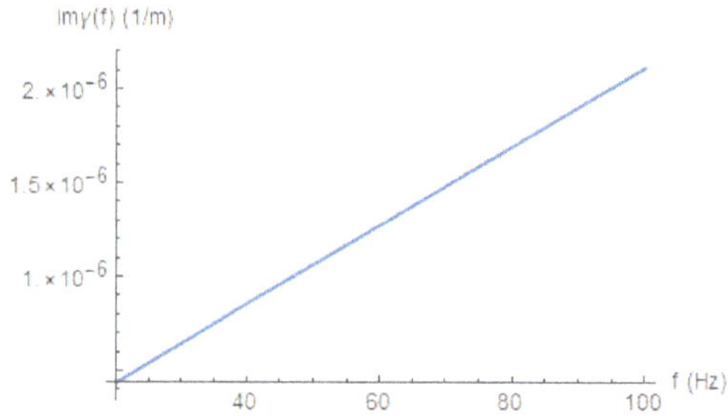


**Figure 7.** $\text{Im}\gamma$ low frequency dependence.

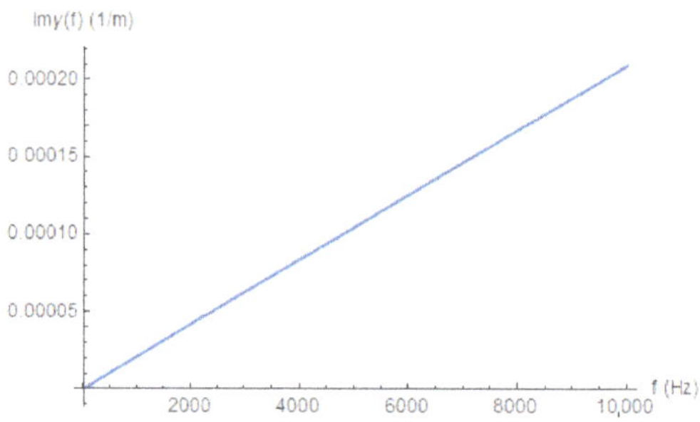


**Figure 8.** $\text{Im}\gamma$ high frequency dependence.

For the real part of $\gamma$, we obtain:

$$\text{Re}\gamma \cong \frac{\sqrt{\mu_c}}{2^{\frac{3}{2}} L \sqrt{\sigma_c} \text{v}} \sqrt{\omega} \tag{52}$$

The real part depends on the frequency, but this part is relatively small, compared to the imaginary part. To appreciate the frequency dependence of $\text{Re}\gamma$, we depict its behavior for low frequencies (Figure 9), high frequencies (Figure 10), and extremely high frequencies (Figure 11).

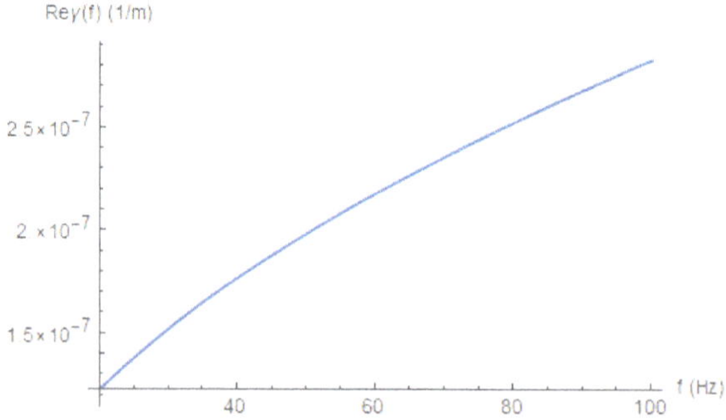


**Figure 9.** $\text{Re}\gamma$ for low frequencies.

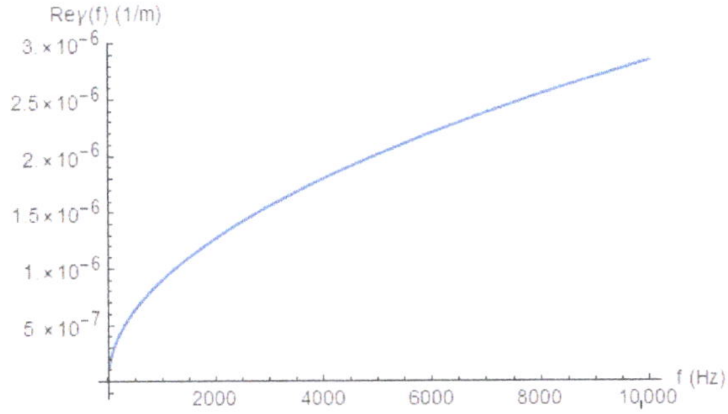


**Figure 10.** Re$\gamma$ for high frequencies.

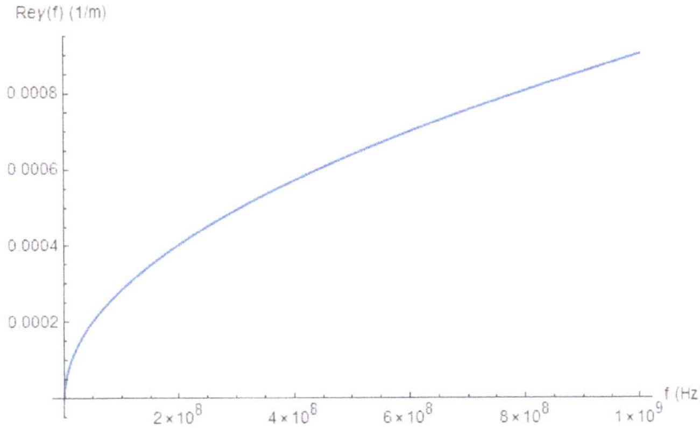


**Figure 11.** Re$\gamma$ for very high frequencies.

To appreciate how small Re$\gamma$ is with respect to Im$\gamma$, we plot the ratio of those quantities in Figure 12.

Another indication of the dominance of Im$\gamma$ over Re$\gamma$ is the frequency dependence of $|\gamma|$, which follows quite closely the linear behavior of Im$\gamma$, as depicted in Figure 13.

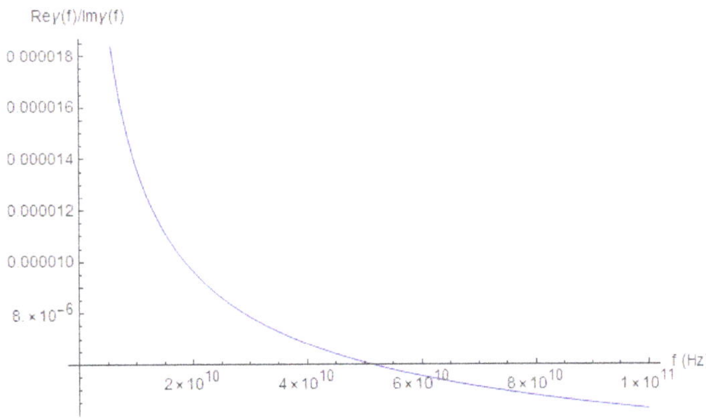


**Figure 12.** The ratio of the real part to the imaginary part of $\gamma$.

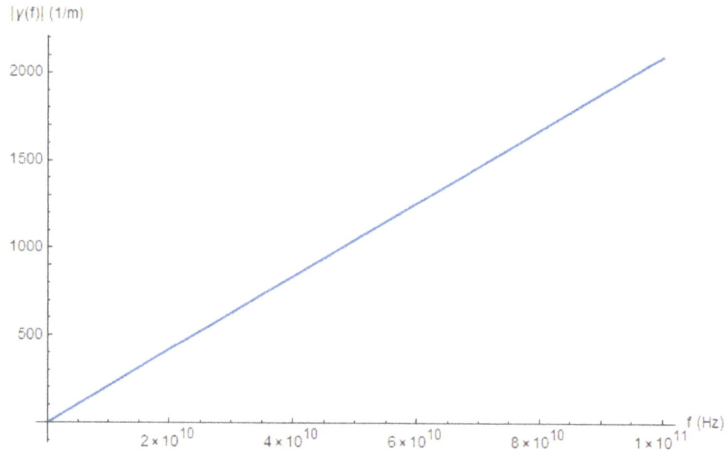


**Figure 13.** $|\gamma|$ as a function of frequency.

Finally, we remind ourselves that we assumed zero admittance $G = 0$ in our calculations. Practically, this means that we neglected the air admittance with respect to the capacitive admittance:

$$Y_C = i\omega C = i2\pi f C. \tag{53}$$

To check if this assumption is justified, we calculated the ratio of the air admittance to capacitive admittance:

$$\frac{G}{Y_C} \simeq -\frac{1.38 \times 10^{-6} i}{f} \tag{54}$$

where we used the parameters of Table A3 for the air admittance and Equation (22) for the capacitance. The ratio, which is quite small, becomes even smaller for higher frequencies, as depicted in Figure 14, and thus justifies our initial assumption.

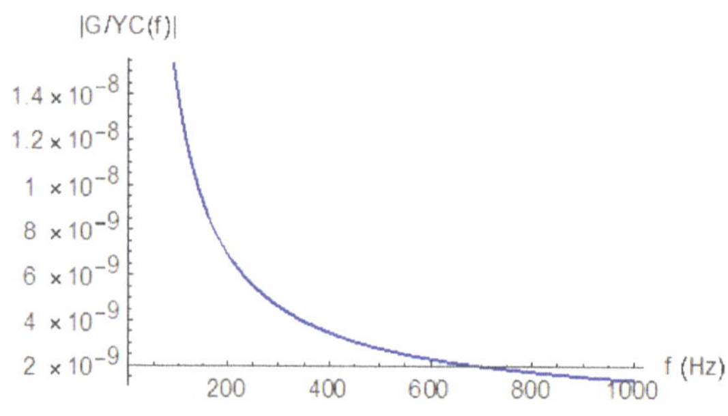


**Figure 14.** The frequency dependence of $\left|\frac{G}{Y_C}\right|$.

The conclusion of this subsection is that dispersion is not significant in the media in which the pulse generated by the short propagates. This will be further elaborated in the next subsection, as we study the prorogation of a signal along the line.

*4.5. Signal Propagation*

We assumed that at the entrance to a transmission line, we inject a short signal of the form:

$$V(0,t) = 2e^{-\frac{t^2}{2\sigma_0^2}} \cos(\omega_0 t) \tag{55}$$

Thus, we assumed that the short generates a pulse signal, while the standard voltage is a periodic trigonometric function of frequency $\omega_0$, which is modulated by the pulse.

The pulse was assumed to be Gaussian with a width $\sigma_0$. The same signal in the frequency domain takes the form:

$$V(0,\omega) = \sqrt{2\pi}\sigma_0\left(e^{-\frac{\sigma_0^2}{2}(\omega+\omega_0)^2} + e^{-\frac{\sigma_0^2}{2}(\omega-\omega_0)^2}\right) \tag{56}$$

This signal is depicted in Figure 15.

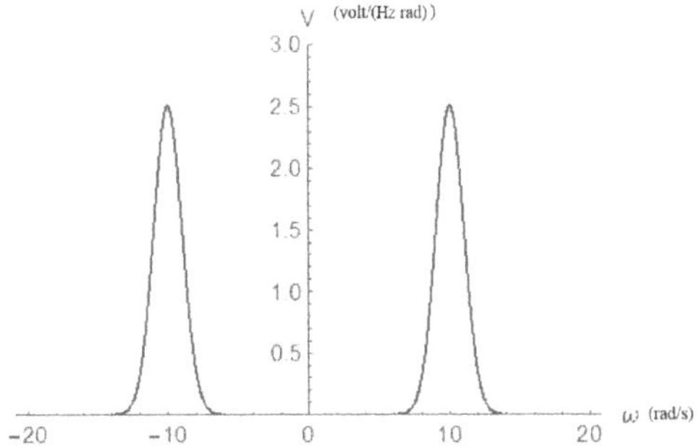


**Figure 15.** The entrance voltage signal in the frequency domain.

Now, since that signal is injected at the entrance, it can propagate in only one direction; hence, Equation (27) takes the form:

$$V(x,\omega) = V(0,\omega)e^{-\gamma(\omega)x} \tag{57}$$

We now expand $\gamma$ around $\omega_0$ till the second order. This results in:

$$\gamma(\omega_0 + \omega') \cong \gamma(\omega_0) + \gamma'(\omega_0)\omega' + \frac{1}{2}\gamma''(\omega_0)\omega'^2. \tag{58}$$

Taking into account Equations (56)–(58) and performing an inverse Fourier transform, we arrive at the following expression for a propagating signal:

$$V(x,t) = 2\sigma_0 \mathrm{Re}\left[\frac{e^{i\omega_0 t - \gamma_0 x} e^{-\frac{(t-t_d)^2}{2\sigma^2(x)}}}{\sigma(x)}\right] \tag{59}$$

where the delay time $t_d$ is:

$$t_d = -i\gamma_0' x = \frac{x}{v}\frac{1+\frac{3}{4}Ra}{\sqrt{1+Ra}} \cong \frac{x}{v}; \tag{60}$$

hence, this is approximately the distance divided by the velocity, as expected. We noticed that the expression:

$$v_g = \left|Im\gamma_0'\right|^{-1} = v\left(\frac{\sqrt{1+Ra}}{1+\frac{3}{4}Ra}\right), \tag{61}$$

is the group velocity, which is the velocity of a wave packet. The width of the signal is:

$$\sigma(x) = \sqrt{\sigma_0{}^2 + \gamma''_0 x} \tag{62}$$

where:

$$\gamma''_0 = \frac{-iRa(1+\frac{3}{2}Ra)}{16v\omega_0\sqrt{1+Ra}^3} \cong -\frac{R(\omega_0)}{16v\omega_0^2 L} \tag{63}$$

The term $\gamma''_0 x$ signifies the pulse broadening as it propagates along the line. Let us assume that $\sigma_0 \approx 10^{-6}$ s. For a frequency of 1 MHz and a distance of 1 km:

$$\gamma''_0 x \cong -6.39 \times 10^{-19} - 2.85 \times 10^{-33} \mathrm{i} << \sigma_0^2 = 10^{-12}; \tag{64}$$

hence, the dispersion is negligible. However, for a frequency of 1 kHz and a distance of 10 km:

$$\gamma''_0 x \cong -6.39 \times 10^{-12} - 2.85 \times 10^{-17} \mathrm{i} \approx \sigma_0^2 = 10^{-12} \tag{65}$$

the widening is comparable to the initial width. Hence, despite the fact that dispersion seems small, it accumulates over long distances.

*4.6. Bifurcations*

A power transmission line often bifurcates, as depicted in Figure 16.

In the bifurcation junction, the signal is transmitted to the bifurcating channels and is also reflected to the original channel in which the short was originally formed. The amount of signal reflected or transmitted is quantified by the reflection $R$ and transmission $T$ coefficients. Those coefficients can in turn be calculated as follows:

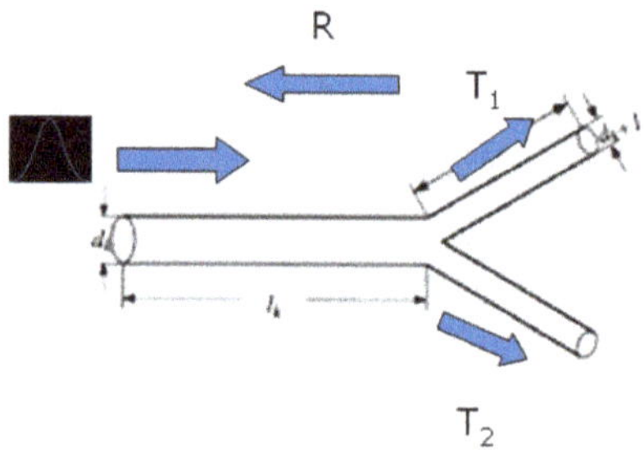


**Figure 16.** Schematic bifurcation in a power line.

$$T_{1\to 2} = \frac{2Z_2}{Z_1 + Z_2}, \qquad T_{2\to 1} = \frac{2Z_1}{Z_1 + Z_2} \tag{66}$$

$$R_{2\to 1} = \frac{Z_2 - Z_1}{Z_1 + Z_2} = -R_{1\to 2} \tag{67}$$

in which $Z_1$ is the impedance of the line before the bifurcation junction and $Z_2$ is the impedance of the line after the bifurcation junction. In case the signal meets multiple bifurcation junctions along its path, multiple reflection occur, as depicted schematically in Figure 17.

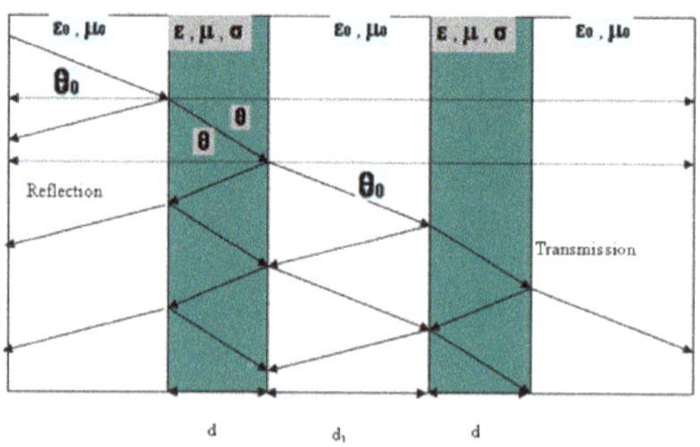


**Figure 17.** Multiple reflections from a changing propagation media.

Multiple reflections will result in multiple signals arriving at the detector, as depicted schematically in Figure 18.

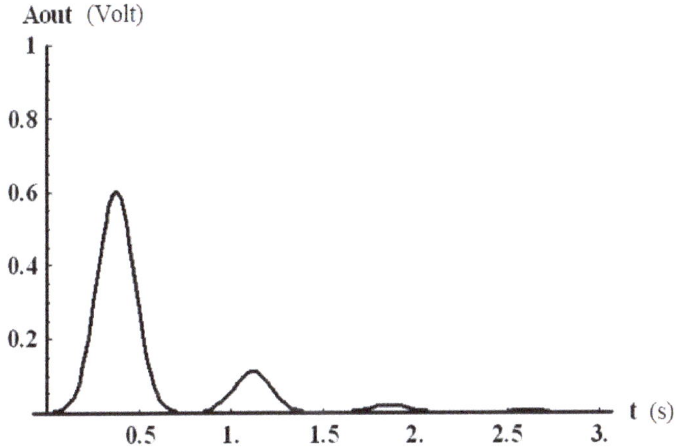


**Figure 18.** Multiple reflections arriving at the detector.

We note that reflected signals arrive at the detector later and in reduced amplitude due to the longer path they need to travel and the additional attenuation the signal suffers during propagation (see Equation (59)) and reflection. We note that if a line bifurcates into multiple identical lines as in Figure 19, the total impedance at the channel after the bifurcation will be equal to the original impedance $Z_0$ divided by the number of transmission lines $N$; hence:

$$Z_1 = Z_0, \qquad Z_2 = \frac{Z_0}{N} \tag{68}$$

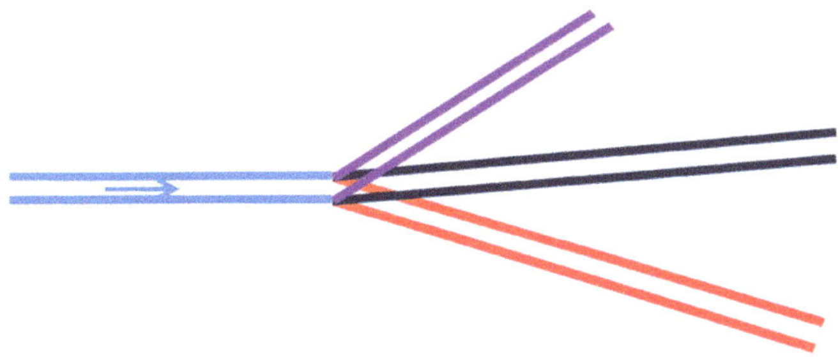

**Figure 19.** A transmission line bifurcating into multiple channels.

This implies, according to Equation (66), a transmission coefficient of:

$$T_{1\to2} = \frac{2Z_2}{Z_1 + Z_2} = \frac{2\frac{Z_0}{N}}{Z_0 + \frac{Z_0}{N}} = \frac{2}{N+1} \tag{69}$$

$$R_{1\to2} = \frac{Z_1 - Z_2}{Z_1 + Z_2} = \frac{Z_0 - \frac{Z_0}{N}}{Z_0 + \frac{Z_0}{N}} = \frac{N-1}{N+1}; \tag{70}$$

hence, for a bifurcation to a large number of channels, the reflection coefficient will tend to one, while for a continuation into a single identical channel, there will not be obviously a reflection.

We now examine the reflection effect and see if one can use the reflected signal instead of a second sensor, thus reducing the amount of hardware needed in order to implement the method described. First, let us look at Figure 20: The short occurs between the sensor and the bifurcation point. Thus, a signal is propagating from the short to the sensor, and an additional signal propagates to the bifurcation point where it is reflected. Provided that the short will not introduce an impenetrable obstacle, the signal will eventually reach the detector at a later time. The direct signal arrival time will be satisfied according to Equation (1):

$$t_1 - t_0 = \frac{L_1}{v} \tag{71}$$

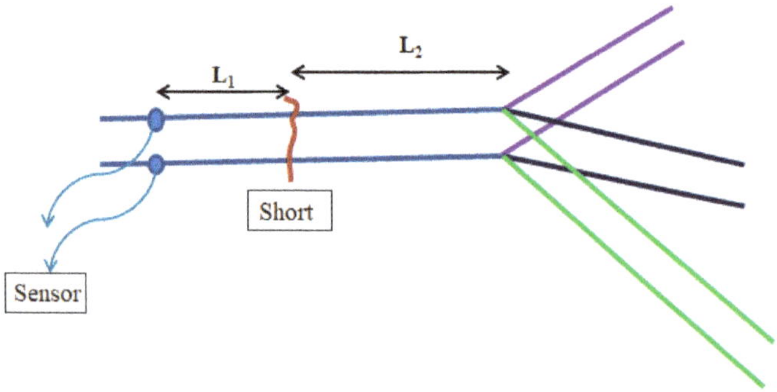


**Figure 20.** A short occurs between the sensor and the bifurcation point.

The reflected signal will arrive at a later time such that:

$$t_2 - t_0 = \frac{L_1 + 2L_2}{v} \tag{72}$$

Hence, the time difference between the direct and reflected signal allowed us to calculate the distance between the short and bifurcation point as:

$$L_2 = \frac{1}{2}v(t_2 - t_1) \tag{73}$$

Now, since the distance $L$ between the sensor and the bifurcation point is known in advance, we may calculate the distance between the short and the bifurcation point as follows:

$$L_1 = L - L_2 = L - \frac{1}{2}v(t_2 - t_1) \tag{74}$$

Thus, in this case, a single detector will suffice, and we will not need two detectors, as described in Section 2. This reduces the cost of the system and removes redundant issues such as sensor synchronization. The last advantageous scenario is depicted in Figure 21.

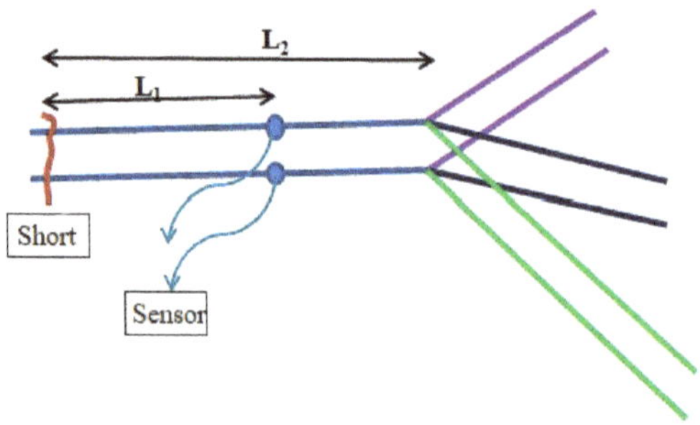


**Figure 21.** The sensor is between the short and the bifurcation point.

The direct signal arrival time will be satisfied according to Equation (1):

$$t_1 - t_0 = \frac{L_1}{v} \tag{75}$$

The reflected signal will arrive at a later time such that:

$$t_2 - t_0 = \frac{2L_2 - L_1}{v} \tag{76}$$

The time difference in this case will yield:

$$L_2 - L_1 = \frac{1}{2}v(t_2 - t_1) \tag{77}$$

which does not reveal any information about the short location, but rather, some trivial information about the distance between the sensor and the bifurcation point, which is already known. Of course, if there exist additional bifurcation points on the signal path (for example, left to the short), then we are in the previous case again, and one sensor will suffice. We may deduce that putting sensors on bifurcation points will reduce the amount of sensors needed. Finally, we looked at the case in which the signal arrives at a sensor located after the branching point of the net, as in Figure 22.

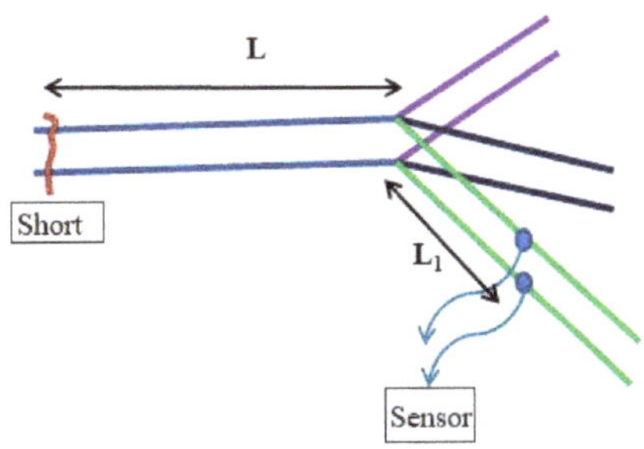


**Figure 22.** The sensor is on a bifurcation of the transmission line.

The sensor will receive a signal at:

$$t_1 - t_0 = \frac{L + L_1}{v} \tag{78}$$

This will not suffice to locate the short unless a reflected signal is received from another point along the network. To conclude, we deduced that putting sensors on bifurcation points will reduce the amount of sensors needed along the network. Moreover, relying on reflections may solve the problem of sensor synchronization.

### 4.7. Laplace Analysis

We now address the transmission line pulse propagation problem using the technique of Laplace analysis [30]. The simplified power system of our interest is schematically presented in Figure 23.

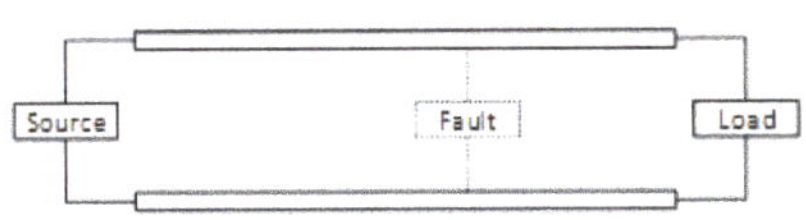


**Figure 23.** Schematic of the fault scenario.

The fault produces two separate signals propagating towards the source and the load, where the sensors are located. In each section, as a result of the fault, the voltage perturbation is perceived as an input signal, propagating towards a sensor, as shown in Figure 24.

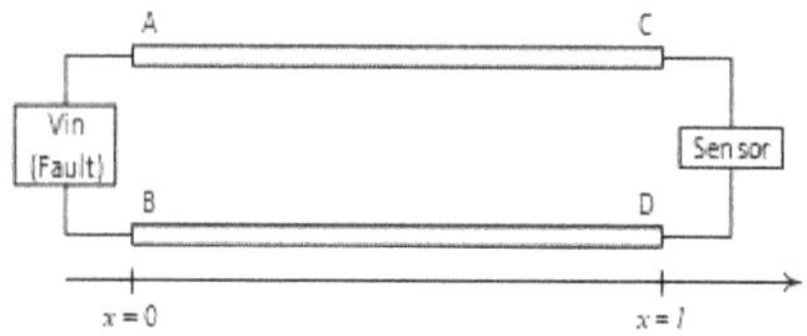


**Figure 24.** Schematic of the fault modeling.

In the transmission line, we assumed the validity of the telegraph Equation (24) and replaced $i\omega \to s$. The Laplace form of the telegraph equations admits the solution:

$$V(x,s) = V_{(-)}(s)e^{-x\gamma(s)} + V_{(+)}(s)e^{x\gamma(s)} \tag{79}$$

$$I(x,s) = \frac{V_{(-)}(s)}{Z_0(s)}e^{-x\gamma(s)} - \frac{V_{(+)}(s)}{Z_0(s)}e^{x\gamma(s)} \tag{80}$$

where,

$$Z(s) = R + sL,\ Y(s) = G + sC \tag{81}$$

$$\gamma(s) = \sqrt{Z(s)Y(s)} \tag{82}$$

$$Z_0(s) = \sqrt{\frac{Z(s)}{Y(s)}} \tag{83}$$

It is realistic to presume that the conductance $G$ of the separating dielectric material between the wires is insignificant. The series resistivity is calculated taking into account the skin effect:

$$R(s) = \frac{1}{\pi d}\sqrt{\frac{2\mu_c}{\sigma_c}}\sqrt{s} \equiv \xi\sqrt{s} \tag{84}$$

where $d$ is the conductor wire diameter and:

$$\xi = \frac{1}{\pi d}\sqrt{\frac{2\mu_c}{\sigma_c}}. \tag{85}$$

Consequently, the expression (82) can be written as follows:

$$\begin{aligned} \gamma(s) &= s\sqrt{LC}\sqrt{1+\tfrac{R(s)}{sL}} = s\sqrt{LC}\sqrt{1+\tfrac{\xi}{\sqrt{s}L}} \approx \\ &\approx s\sqrt{LC}\left(1+\tfrac{\xi}{2\sqrt{s}L}\right) = s\sqrt{LC} + \tfrac{\xi}{2}\sqrt{\tfrac{C}{L}}\sqrt{s} \end{aligned} \tag{86}$$

The sensor impedance $Z_L$ was designed to be infinite so as not to influence the measurement results. Substituting boundary conditions,

$$V_{fault}(s) = V(0,s) = V_{(-)}(s) + V_{(+)}(s) \tag{87}$$

and:

$$Z_L = \infty \tag{88}$$

which implies zero current at $x = l$:

$$I(l,s) = \frac{V_{(-)}(s)}{Z_0(s)}e^{-l\gamma(s)} - \frac{V_{(+)}(s)}{Z_0(s)}e^{l\gamma(s)} = 0 \tag{89}$$

or:

$$V_{(-)}(s) = V_{(+)}(s)e^{2l\gamma(s)} \tag{90}$$

Combining the above equation with Equation (87) yields:

$$V_{(+)} = \frac{V_{fault}(s)}{1+e^{2l\gamma(s)}} \tag{91}$$

and:

$$V_{(-)} = \frac{V_{fault}(s)}{1+e^{-2l\gamma(s)}} \tag{92}$$

The voltage signal at the sensor due to the fault is thus:

$$V_{out}(s) \equiv V(l,s) = \frac{2e^{-l\gamma(s)}}{1+e^{-2l\gamma(s)}}V_{fault}(s) \tag{93}$$

Identifying the last expression as the summation of a geometric series, it can be re-written as the sum:

$$V_{out}(s) = 2V_{fault}(s)\sum_{n=0}^{\infty}(-1)^n e^{-(1+2n)l\gamma(s)} \tag{94}$$

Defining:

$$\tau_n \equiv (1+2n)l\sqrt{LC} \tag{95}$$

and:

$$B_n \equiv (1+2n)l\frac{\xi}{2}\sqrt{\frac{C}{L}} \tag{96}$$

the expression for the voltage at the sensor, Equation (94), takes the form:

$$V_{out}(s) = 2V_{fault}(s)\sum_{n=0}^{\infty}(-1)^n e^{-\tau_n s+B_n\sqrt{s}} \tag{97}$$

This sum can be interpreted as the sum of multiple reflected waves each with its unique delay time. The system's transfer function may be now calculated as follows:

$$H(s) \equiv \frac{V_{out}(s)}{V_{fault}(s)} = 2\sum_{n=0}^{\infty}(-1)^n e^{-\tau_n s+B_n\sqrt{s}} \tag{98}$$

This allows for obtaining the signal measured at the sensor, due to any fault waveform. For example, if the fault is a sudden short circuit at $x=0$ and $t=0$, the voltage at the fault location is:

$$V_{in}(t) \equiv V(0,t) = V_0 - V_0 u(t). \tag{99}$$

Using the superposition principle, $V_{out}(t)$ may be expressed as the sum of a DC input and a step function input response.

$$V_{out}(t) = V_{out(V_{in}=V_0)}(t) - V_{out(V_{in}=V_0 u(t))}(t) \tag{100}$$

Since the impedance at the edge is infinite, the voltage along the line due to the DC input is simply $V_0$. Moreover, the Laplace transform of a step function satisfies:

$$\mathcal{L}\{u(t)\} = \frac{1}{s}, \tag{101}$$

hence:

$$V_{out(V_{in}=V_0 u(t))}(s) = \frac{V_0 H(s)}{s} \tag{102}$$

Hence, using the transfer function definition, Equation (98), and performing an inverse transform back to the time domain, we obtain:

$$V_{out}(t) = V_0 - 2\mathcal{L}^{-1}\left\{\frac{V_0}{s}\sum_{n=0}^{\infty}(-1)^n e^{-\tau_n s + B_n\sqrt{s}}\right\} \quad (103)$$

Taking a known inverse Laplace transform [22]:

$$\mathcal{L}^{-1}\left\{e^{-\tau s}\frac{e^{-B\sqrt{s}}}{s}\right\} = erfc\left(\frac{B}{2\sqrt{t-\tau}}\right)u(t-\tau) \quad (104)$$

The voltage at the sensor, as a result of the fault, is:

$$V_{out}(t) = V_0 - 2V_0\sum_{n=0}^{\infty}(-1)^n erfc\left(\frac{B_n}{2\sqrt{t-\tau_n}}\right) u(t-\tau_n) \quad (105)$$

The above result will suffice if the rise time of the short is fast enough and can be ignored.

## 5. Methodology: Experimental Setup

### *5.1. Experiment Hardware*

The experimental demonstration of the fault location technique involved the setup shown schematically in Figure 25, in which a 10V DC voltage was supplied from a TTi QL355TP power supply source, shown in Figure 26, to a 50 m two-wire cable and switches. The sensor was a MS09404A Mixed Signal Oscilloscope with a 4 GHz bandwidth, which has two channels, as shown in Figure 27.

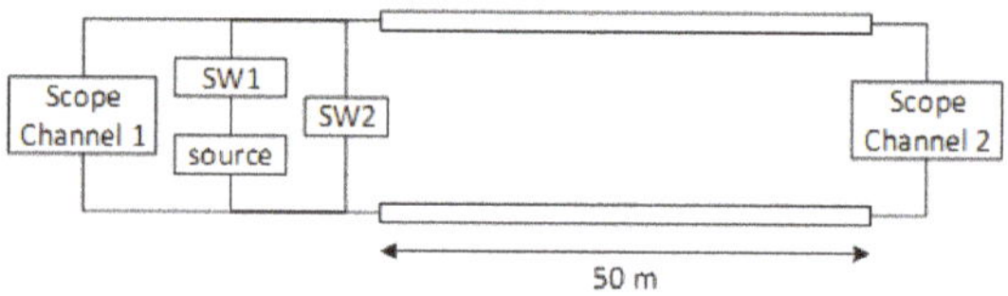


**Figure 25.** Schematic of the experimental setup.

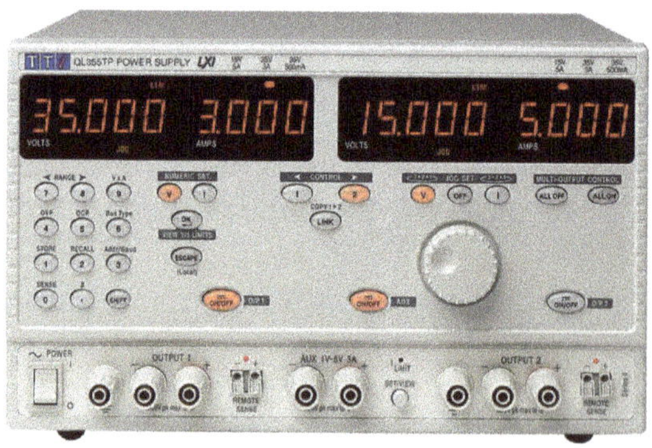


**Figure 26.** TTi QL355TP power supply.

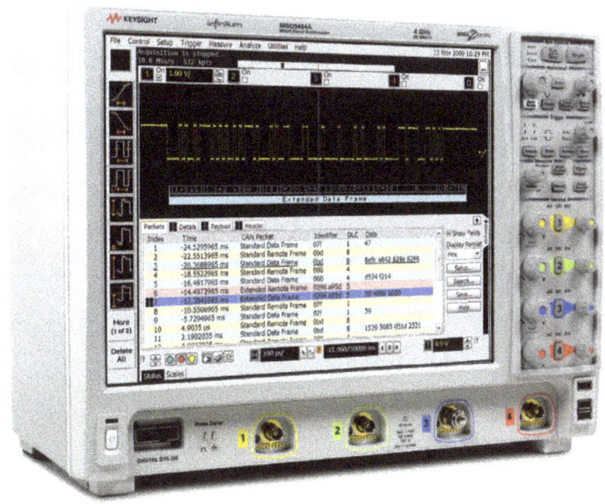

**Figure 27.** MS09404A Mixed Signal Oscilloscope of a 4 GHz bandwidth having two channels.

Figure 28 illustrates the dimensions of the transmission line.

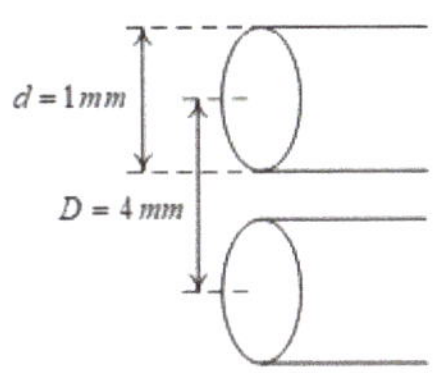


**Figure 28.** Dimensions of the transmission line.

The transmission line was comprised of a two-wire power cord. Each wire was made of copper and had a $d = 1$ mm diameter, while the distance between the wires was $D = 4$ mm. These parameters allowed us to calculate the inductance L and capacitance C from the dielectric constant, magnetic permeability, and conductivity as follows (see Equation (22)):

$$\begin{aligned}
&\varepsilon_{r(insulator)} = 3.45, \mu_{r(insulator)} = 1, \\
&\sigma_{copper} = 3 \times 10^6 \frac{S}{m}, n = \sqrt{\varepsilon_r \mu_r} \simeq 1.86 \\
&C = \frac{2\pi\varepsilon}{\cosh^{-1}\left(\frac{D^2 - 2d^2}{2d^2}\right)} = 7.22 \times 10^{-11} \frac{F}{m} \\
&L = \frac{\varepsilon\mu}{C} = 5.27 \times 10^{-7} \frac{H}{m}
\end{aligned} \tag{106}$$

### *5.2. Experimental Process*

On one side of the transmission line, we connected the DC source, 2 switches, and the scope channel; another channel was connected to the other side. Closing the first switch when the second switch is open, we obtained a steady-state DC voltage along the line. The short circuit fault was achieved by closing the second switch.

## 6. Results: Data Processing

### *6.1. Location Accuracy*

Measuring the voltage of both sides of the line, we can see clearly the delay in signal propagation along the line. Additional reflected waves (Figures 29 and 30) are also seen. The scope channel data were exported and processed using the MATLAB software.

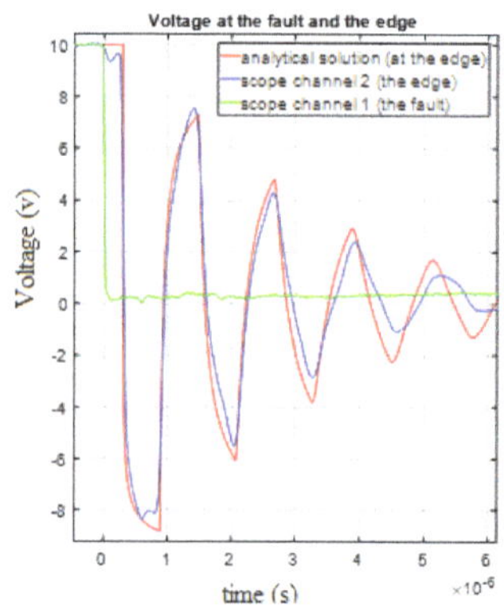


**Figure 29.** Comparison of the experimental results and theoretical predictions of Equation (105) for the measured voltage.

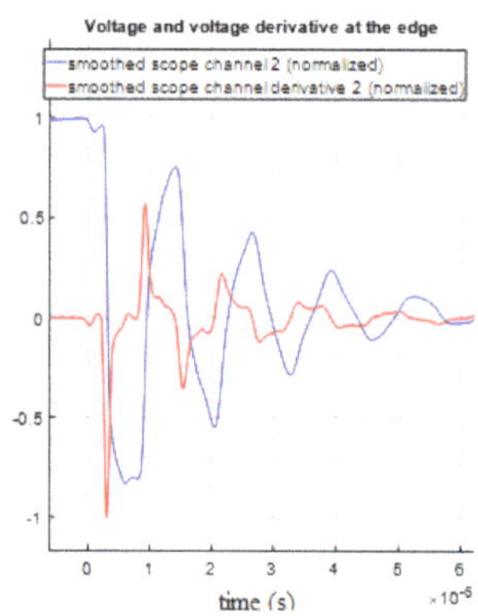


**Figure 30.** Comparison of the sensor voltage and its derivative.

Our aim was of course to discover how long it takes the signal to arrive from the fault to the measuring point. Because the waves arriving at the edge are reflected in the opposite direction (towards the fault), where they are repeatedly reflected, one fault may supply several data points that we can process to our advantage, depending on how fast the transient signals decay. In this study, it was possible to measure five signals (peaks) each time, and so, we chose the optimal data processing technique.

To obtain the best accuracy, we performed numerous voltage smoothing and voltage derivative smoothing, changing the time window and thus the number of samples. Obviously, the wider the window, the less noise we had to distort our results. However, a wide window may distort the reflection signal as well, compromising our accuracy. We used five different averaging windows and thus obtained five different errors for each reflection. The data are given in the Appendix B; in the table, each line represents a different value of the voltage smoothing window and voltage derivative smoothing window and the errors that were deduced for each reflection and processing method.

Figure 30 shows that the voltage derivative peaks were much sharper; therefore, the short circuit fault location was calculated by finding the time intervals between the local extremum points in the voltage derivative at the transmission line edge. In addition, the smoothing filter window may slightly change the results, and the optimum configuration was 200 points in a window. In this situation, the accuracy in the short circuit fault location detection was ±0.005% using the second peak.

This can be compared to the theoretical accuracy predicted by Equation (11). In the current case, the sampling rate was 4 GHz, and hence, the time between samples was $T = 2.5 \times 10^{-10}$ s and the velocity $v = \frac{c}{n} = 1.62 \times 10^8$ m/s. Thus, the accuracy may be as high as 0.008 m, which is ±0.008%, quite close to the best experimental result. This indicates that the theoretical limit is achievable provided the data are processed correctly.

## 6.2. Comparison with Theory

We now compare the results of the theory given in Equation (105) to the (smoothed) experimental results. The signal reflection scheme is shown in Figure 31. The results of the comparison are depicted in Figure 32.

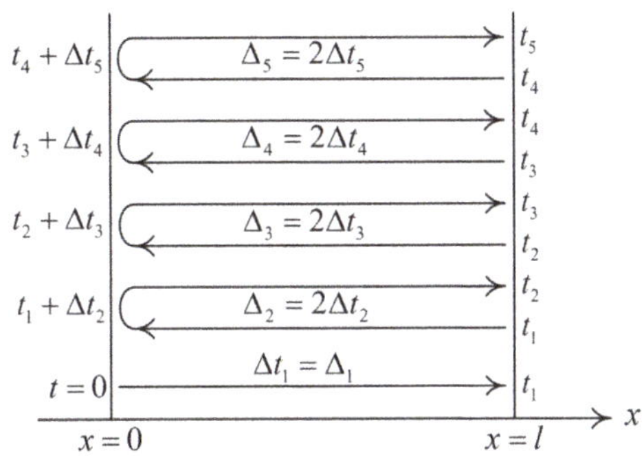


**Figure 31.** The propagation times for the pulse reflections.

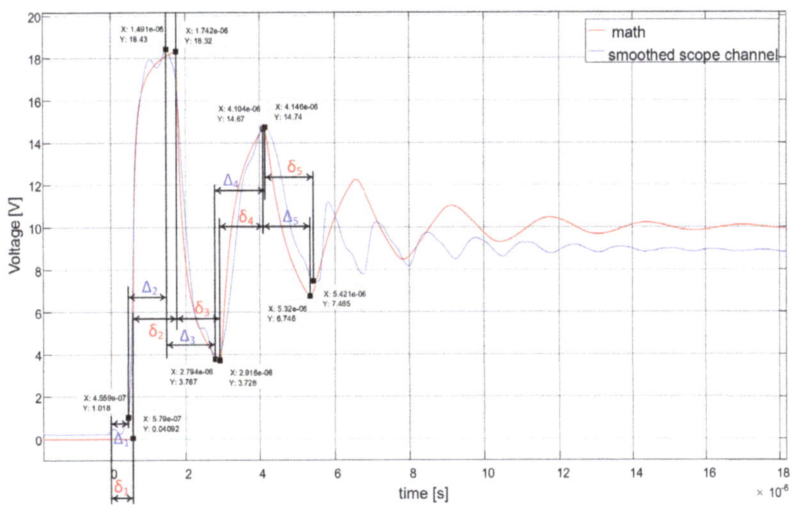


**Figure 32.** Comparison results between Equation (105) and the smoothed scope channel.

The unprocessed timings of various peaks were correlated with signal reflections and are given in Table 1.

**Table 1.** Distance calculations based on raw data without applying any processing technique.

| $t_i$ (s) | $\Delta_i$ (s) | $\Delta t_i$ (s) | $\Delta L$ | *Error* (%) |
|---|---|---|---|---|
| $4.5 \times 10^{-7}$ | | $4.5 \times 10^{-7}$ | 79 | −21% |
| $1.5 \times 10^{-6}$ | $1.0 \times 10^{-6}$ | $5.2 \times 10^{-7}$ | 90 | −10% |
| $2.8 \times 10^{-6}$ | $1.3 \times 10^{-6}$ | $6.5 \times 10^{-7}$ | 113 | 13% |
| $4.1 \times 10^{-6}$ | $1.3 \times 10^{-6}$ | $6.5 \times 10^{-7}$ | 113 | 13% |
| $5.4 \times 10^{-6}$ | $1.3 \times 10^{-6}$ | $6.6 \times 10^{-7}$ | 114 | 14% |

The same results derived from the theoretical calculation are given in Table 2.

**Table 2.** Distance calculations from the simulation results.

| $t_j$ (s) | $\delta_j$ (s) | $\Delta t'_I$ (s) | $L'$ |
|---|---|---|---|
| $5.8 \times 10^{-7}$ | | $5.8 \times 10^{-7}$ | 100 |
| $1.7 \times 10^{-6}$ | $1.2 \times 10^{-6}$ | $5.8 \times 10^{-7}$ | 101 |
| $2.9 \times 10^{-6}$ | $1.2 \times 10^{-6}$ | $5.9 \times 10^{-7}$ | 102 |
| $4.2 \times 10^{-6}$ | $1.2 \times 10^{-6}$ | $6.2 \times 10^{-7}$ | 107 |
| $5.3 \times 10^{-6}$ | $1.2 \times 10^{-6}$ | $5.9 \times 10^{-7}$ | 102 |

Obviously, the theoretical distance evaluation was better than the experimental one; however, processing dramatically improved the situation, as we saw in the previous section. The distances can be derived from the time differences since $\Delta L = v\,\Delta t = 1.62 \times 10^8\ \Delta t$ m. In the above, we denote $\Delta t_i$ and $\Delta t'_i$ for one-direction experimental and theoretical propagation times, respectively. As we can see, the propagation times were sufficiently stable and consistent with the theoretical model. Hence, the pulse location information can also be obtained from the voltage signal (without the signal derivative if precision is not required).

However, there was a difference between the theoretical and the experimental graphs. In the experimental data, there were additional peaks in the first reflected waves; after the third (main) peak, the "oscillation" frequency was twice as great as in the theoretical model. This was because of the reflected waves from the middle of the transmission line: in the experimental setup, the line was not homogeneous, as can be deduced from the additional peaks we received.

Finally, we present the comparison of the theoretical and empirical derivatives given in Table 3 and Figure 33. The timing of the peak derivatives seemed to fit much better the theoretical curve, and the location errors were much smaller. As noted previously, processing may improve the accuracy drastically.

**Table 3.** Distance calculations from the signal derivatives and respective location errors.

| | $t_i$ | $\Delta t_i$ | $\Delta L$ | *Error* (%) |
|---|---|---|---|---|
| 1 | $3.11 \times 10^{-7}$ | | 52.1 | 4.28% |
| 2 | $9.16 \times 10^{-7}$ | $6.06 \times 10^{-7}$ | 101.5 | 1.55% |
| 3 | $1.52 \times 10^{-6}$ | $6.08 \times 10^{-7}$ | 101.9 | 1.90% |
| 4 | $2.16 \times 10^{-6}$ | $6.31 \times 10^{-7}$ | 105.8 | 5.82% |
| 5 | $2.77 \times 10^{-6}$ | $6.15 \times 10^{-7}$ | 103.1 | 3.14% |
| 6 | $3.40 \times 10^{-6}$ | $6.28 \times 10^{-7}$ | 105.3 | 5.32% |

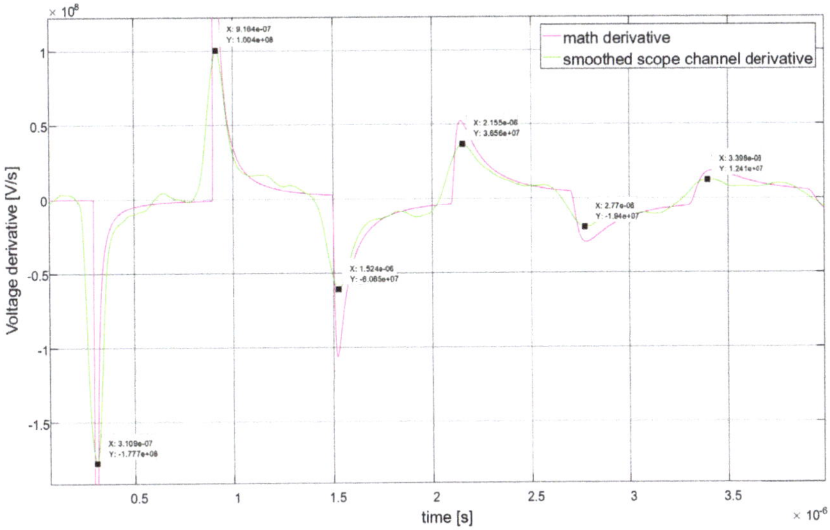


**Figure 33.** Comparison between the theoretical derivative and smoothed scope channel derivative.

## 7. Synchronization between Sensors

In order to take full advantage of the accuracy of the current method (see Table 4 and [2–4,31–33]) proper synchronization between sensors is required. Indeed, before measuring a transient phenomenon [34], an accurate method for measuring it must be chosen [35]. According to a broad literary review of a shelf product that will be used in the final device for synchronization between the measurements of the sensors, it was found that the White Rabbit Precision Time Protocol (PTP) was suitable for robust sub-nanosecond synchronization [36]. For a description of the characteristics of this technology, see [37–39].

**Table 4.** Comparison of fault location accuracy methods. DSE—distribution state estimation algorithm. SMT—synchronized measurement technology.

| Method | Accuracy (%) | Reference |
|---|---|---|
| The developed algorithm | 0.005 | Here |
| Both ends of a single high-voltage line | 0.01 | [31] |
| H-matrix and the measurements of V and C | 3.226 | [32] |
| Varying the fault resistance fault locations algorithm | 0.81 | [33] |
| Proposed Algor. | 12.58 | [2] |
| Simple Ohm's Law | 16.22 | [2] |
| Absol. Value Imped. | 16.17 | [2] |
| Loop React. | 21.98 | [2] |
| Takagi | 17.45 | [2] |
| Santoso et al.'s algorithm | 14.4–17.5 | [2] |
| Impedance-based method | 3.91 | [3] |
| DSE | large | [4] |

**Table 4.** *Cont.*

| Method | Accuracy (%) | Reference |
|---|---|---|
| Fuzzy inference | 5 | [5] |
| HV travel waves | 0.78 | [6] |
| SMT | 3 | [7] |
| Steady-state measured phasors (three phase) | 0.313 | [12] |
| Steady-state measured phasors (phase to ground) | 0.410 | [12] |
| ANFIS | 0.0774 | [11] |

## 8. Conclusions and Summary

In the framework of the current research, a propagation of a short signal in a two-wire cable was investigated. It was shown that the short can be detected either by voltage or current measurements using the voltage/current first or second derivatives. The short can be detected either at the source or the load side. It was also shown that using bifurcations and reflection, the amount of detectors may be reduced. In order to achieve a high level of accuracy regarding the short's location, the inter-sampling duration should be of the nanosecond order (GHz sampling frequency). Of course, if there is no significant slope over a predetermined number of samples (say 1000 samples), there is no need to save them. Hence, the samples may be written as a moving window to some buffer register and deleted if the gradient is not significant. We also studied the phenomena of the dispersion of the short signal and determined in what cases it was significant.

Our method was based on the idea that in any system information even in the form of an electromagnetic wave will advance from one side of the system to the other in a fast, but finite velocity $v$ stratifying $v < c$, where $c$ is the speed of light in a vacuum. When considering a steady-state, we may neglect the propagation phenomena and employ lumped models; however, this is not applicable for short durations, as considered here.

The accuracy was significantly better in the retardation approach than in various methods mentioned in previous work. In Table 4, the accuracy of the different methods is compared.

It is expected that in the future, better wave sampling techniques and devices will be developed, as well as better processing methodologies, thus improving accuracy.

As the temporal difference between the fault signal arrivals and the width of the signal (see for example Figure A37) are much less by many orders of magnitude than the period of the AC voltage, one can assume that for all practical purposes, the AC signal is constant. Hence, we expect an AC experiment to yield similar results.

The above system description was derived for an electrical transmission line as the wave behavior is well known and there is considerable expertise in making the calculations; however, the same holds true for optical fibers and optical reflections, as well as for pipes, such as water, gas, or oil pipes, for which one can use acoustic echoes. The calculations for the fault's location are the same, although the propagation rates are different in the acoustic case. In the optical case, the equation $v = \frac{c}{n}$ still holds true. In both cases, underground pipes and underground fiber optic cables can be damaged at hard-to-reach locations, and an accurate determination of the fault's location is helpful in knowing where to start digging.

**Author Contributions:** Conceptualization, A.Y. and Y.P.; methodology, A.Y. and Y.P.; software, A.Y. and M.S.; validation, A.Y. and Y.P.; formal analysis, A.Y., Y.P. and M.S.; investigation, A.Y., Y.P. and M.S.; resources, A.Y. and Y.P.; data curation, M.S.; writing—original draft preparation, M.N.; writing—review and editing, A.Y.; visualization, A.Y., Y.P. and M.S.; supervision, A.Y. and Y.P.; project administration, A.Y. and Y.P. All authors have read and agreed to the published version of the manuscript.

**Funding:** This research received no external funding.

**Institutional Review Board Statement:** Not applicable.

**Informed Consent Statement:** Not applicable.

**Data Availability Statement:** Not applicable.

**Conflicts of Interest:** The authors declare no conflict of interest.

## Appendix A. The Lumped Circuit Model

The schematic description of the short circuit is presented in Figure A1.

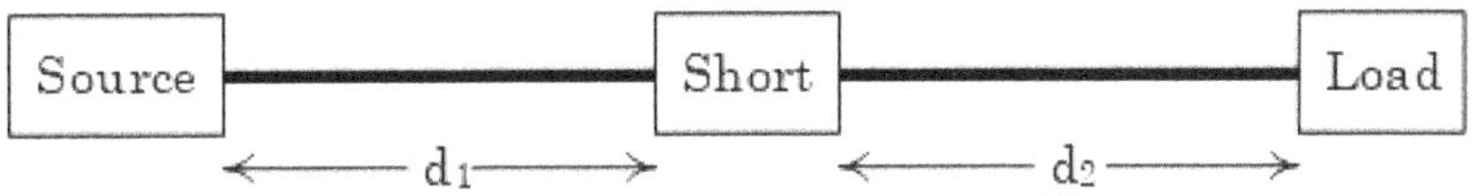


**Figure A1.** Transmission line with a short.

This is modeled in Figure A2.

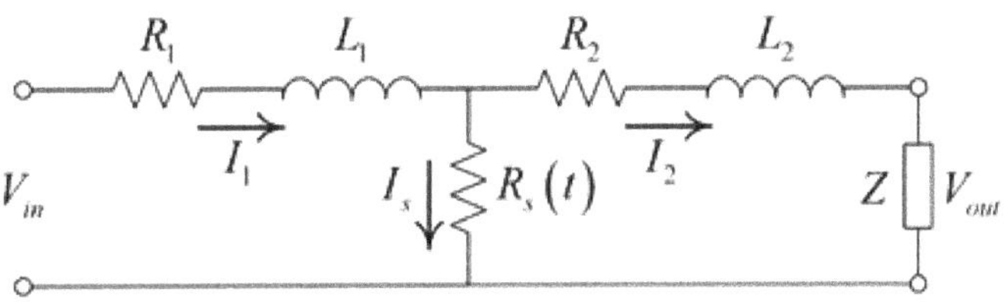


**Figure A2.** Short circuit example.

The parts of the transmission line before and after the short circuit are considered as a resistor and inductor connected in series. In order to mathematically analyze this, we used the Kirchhoff voltage law and Ohm's law:

$$\begin{aligned} V_{in}(t) &= R_1 I_1(t) + L_1 \tfrac{dI_1(t)}{dt} + V_S(t) \\ V_S(t) &= R_2 I_2(t) + L_2 \tfrac{dI_2(t)}{dt} + V_{out}(t) \\ V_{out}(t) &= Z I_2(t) \end{aligned} \tag{A1}$$

in which $V_{in}$ is the source voltage, $I_1$ is the source current, $V_S$ is the short voltage, $I_2$ is the load current, and $V_{out}$ is the load voltage. $R_1$ and $L_1$ are the resistance and inductance before the short, and $R_2$ and $L_2$ are the resistance and inductance after the short, while $Z$ is the load impedance.

### *Appendix A.1. The Case of Resistive Impedance*

In the first step, we ignored the inductance for mathematical simplicity. We also noticed that by Ohm's law:

$$V_S(t) = R_S(t) I_S(t) \tag{A2}$$

where $R_S(t)$ is the time-dependent short resistance and $I_S(t)$ is the current flowing through the short. According to Kirchhoff's current law:

$$I_S(t) = I_1(t) - I_2(t) \tag{A3}$$

Thus, the currents in this case can be calculated algebraically as follows:

$$\begin{aligned} I_1(t) &= \tfrac{Z+R_2+R_S(t)}{R_1\cdot[Z+R_2+R_S(t)]+R_S(t)\cdot(Z+R_2)} \cdot V_{in}(t) \\ I_2(t) &= \tfrac{R_S(t)}{R_1\cdot[Z+R_2+R_S(t)]+R_S(t)\cdot(Z+R_2)} \cdot V_{in}(t) \\ I_S(t) &= \tfrac{Z+R_2}{R_1\cdot[Z+R_2+R_S(t)]+R_S(t)\cdot(Z+R_2)} \cdot V_{in}(t) \end{aligned} \tag{A4}$$

In the current model, the transmission line was a two-wire copper cable; each wire has a diameter $d$, and the distance between them is $D$; the cable is depicted in Figure A3. The total cable length is $l$. The values used for the demonstration are given in Table A1.

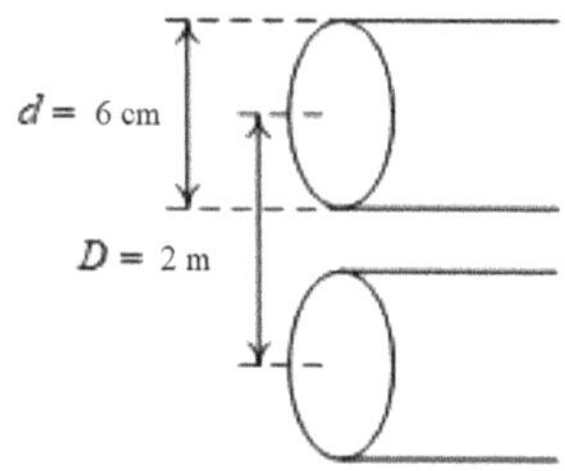


**Figure A3.** Cable geometry for the demonstration.

**Table A1.** Two-wire cable parameters.

| Parameter | Value | Unit |
|---|---|---|
| $\sigma_c$(Copper) | $5.96 \times 10^7$ | S/m |
| d | 0.06 | m |
| D | 2 | m |
| $l$ | 1000 | m |

The surface resistance in ohms ($\Omega$) may be written as follows [29]:

$$R_{Surface} = \sqrt{\frac{\omega \mu_c}{2\sigma_c}} = \sqrt{\frac{\pi f \mu_c}{\sigma_c}} \tag{A5}$$

In the above, $\omega$ is the angular frequency, $f$ is the frequency, $\mu_c$ is the magnetic permeability of the material, and $\sigma_c$ is the conductivity. Hence, the resistance per unit length $[\Omega/\mathrm{m}]$ can be written as:

$$R = 2\frac{R_{surface}}{2\pi\frac{d}{2}} = 2\frac{R_{surface}}{\pi d} \tag{A6}$$

The values used for the cable description appear in Table A2.

**Table A2.** Two-wire cable resistance.

| Parameter | Value | Unit |
|---|---|---|
| $\mu_c$ | $4\pi \times 10^{-7}$ | H/m |
| f | 50 | Hz |
| $R_{surface} = \sqrt{\frac{\pi f \mu_c}{\sigma_c}}$ | $1.82 \times 10^{-6}$ | $\Omega$ |
| $R = 2\frac{R_{surface}}{\pi d}$ | $1.93 \times 10^{-5}$ | $\frac{\Omega}{m}$ |

We noticed that the value of $f = 50$ Hz used above is typical for many power lines. However, as the typical duration of the short is miniscule, the signal generated by the short will include a broad band of frequencies, each suffering a different impedance. This is dealt with in a later part of this paper. However, here, we assumed for simplicity that the resistance is constant.

The short circuit appears at distance $l_1$ from the input. The short-circuit $R_S(t)$ resistance is shown in Figure A3. The short is described by a time-dependent resistance, which was assumed to behave exponentially:

$$R_S(t) = R_{S0}e^{-\frac{t}{\tau_S}}. \tag{A7}$$

The initial resistance $R_S(0) = R_{S0}$ was assumed to be very large and represents the region's air resistance. However, at time $t = 0$, a short is initiated, causing an exponential decrease in the short resistance at a typical duration of $\tau_S$ = 10–100 ns, which depends on the conditions and geometry of the short-circuit region. The short-circuit parameters are concentrated in Table A3.

**Table A3.** Short-circuit parameters.

| Parameter | Value | Unit |
|---|---|---|
| $l_1$ | 300 | m |
| $l_2 = l - l_1$ | 700 | m |
| $\tau_S$ | 100 | ns |
| $\rho_{S0}$ (air) | $1.3 \times 10^{16}$ | $\Omega$ m |
| $l_S$ | 0.1 | m |
| $A_S$ | 0.0004 | $m^2$ |
| $R_{S0} = \frac{\rho_{S0} l_S}{A_S}$ | $3.25 \times 10^{18}$ | $\Omega$ |

Thus, the current flows through the short instead of the load, causing a decrease in the load current. Moreover, since the overall impedance of the circuit decreases and is now dominated by the impedance of the short, the current at the source becomes much higher.

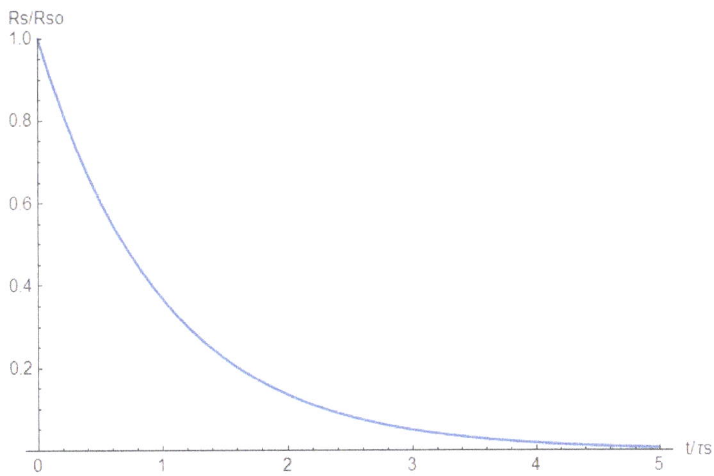


**Figure A4.** Short-circuit resistance.

The transmission line resistances and the load impedance are described in Table A4.

**Table A4.** Transmission line resistance and load impedance.

| Parameter | Value | Unit |
|---|---|---|
| $R_1 = R \cdot l_1$ | 0.00579 | Ω |
| $R_2 = R \cdot l_2$ | 0.0135 | Ω |
| $R_T = R_1 + R_2$ | 0.0193095 | Ω |
| Z | 15.625 | Ω |

The input voltage was assumed to be of the form:

$$V_{in}(t) = A_0 \cos(\omega_0 t), \tag{A8}$$

where $A_0$ = 220 [V], $\omega_0 = 2\pi f_0 = 100\pi$ rad/s. We shall now present the calculation results.

Appendix A.1.1. Current

The source current just before the short is depicted in Figure A5 and after the short is depicted in Figure A6.

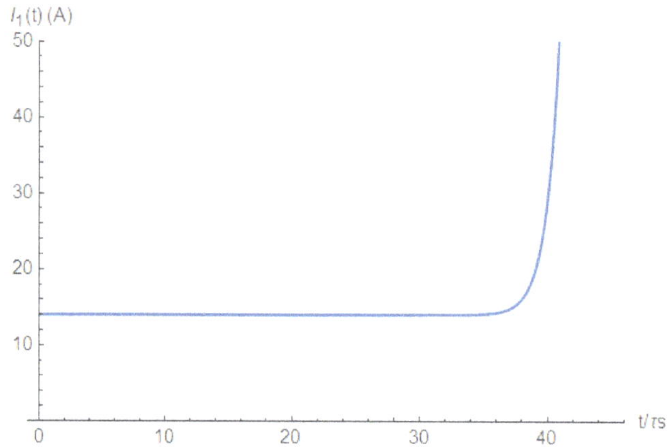


**Figure A5.** Source current before the short.

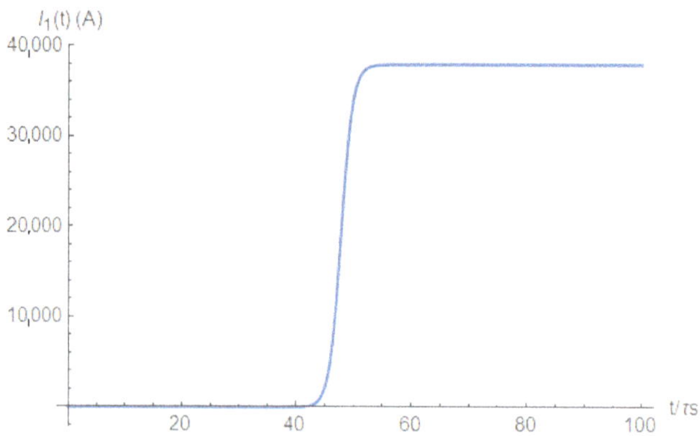


**Figure A6.** Source current after the short.

Hence, the current at the source is a highly sensitive indicator of the occurrence of a short. To avoid storing unnecessary data and for precise timing of the short's occurrence, one can look at the source current derivative (Figure A7); this has a distinctive pulse shape. Hence, by taking the derivative of the signal and by fixing a high detection threshold, one can avoid recording unnecessary data.

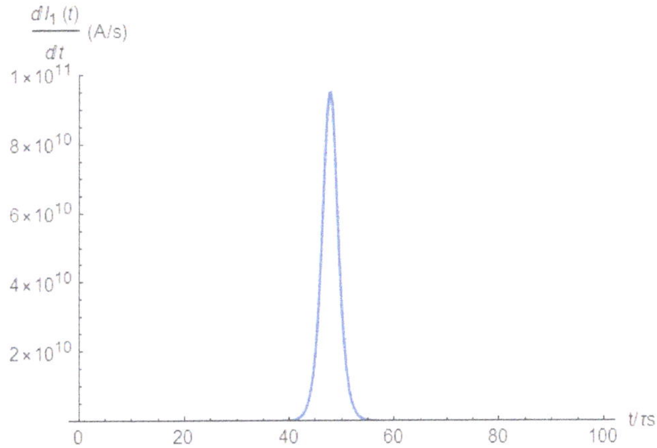


**Figure A7.** Source current derivative.

On the load side, the current vanishes after the short occurs (Figure A8); hence, the load current is also an excellent indicator of the short's occurrence.

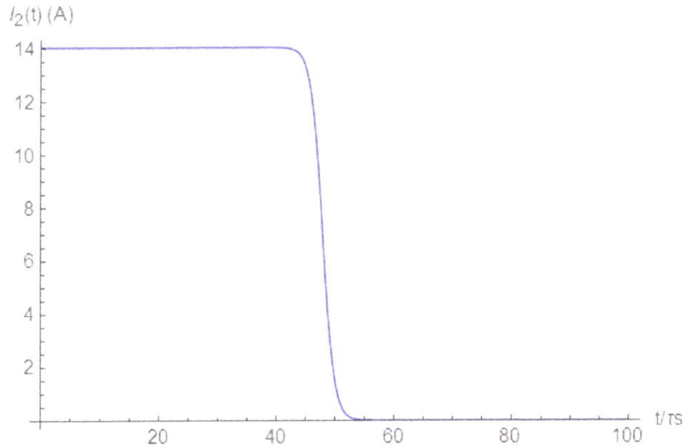


**Figure A8.** Load current.

Again, we see that on the load side, the current behavior allows for identification of the short by taking the current derivative (Figure A9). This method allows precise timing with the need to store a small amount of data, as indicated above.

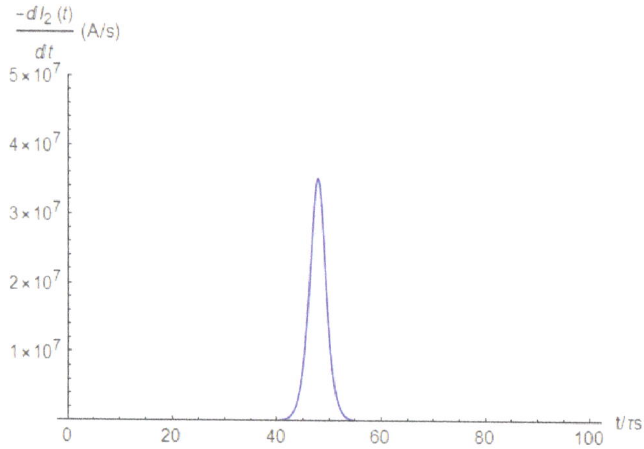


**Figure A9.** Load current derivative.

Appendix A.1.2. Voltage

The short circuit pulse may also be detected by the voltage measurement. For example, the voltage measured at half the distance between the source and the short will yield a voltage $V_1$ as follows:

$$V_1(t) = V_{in}(t) - \frac{1}{2}R_1 I_1(t) \tag{A9}$$

This voltage is depicted in Figure A10.

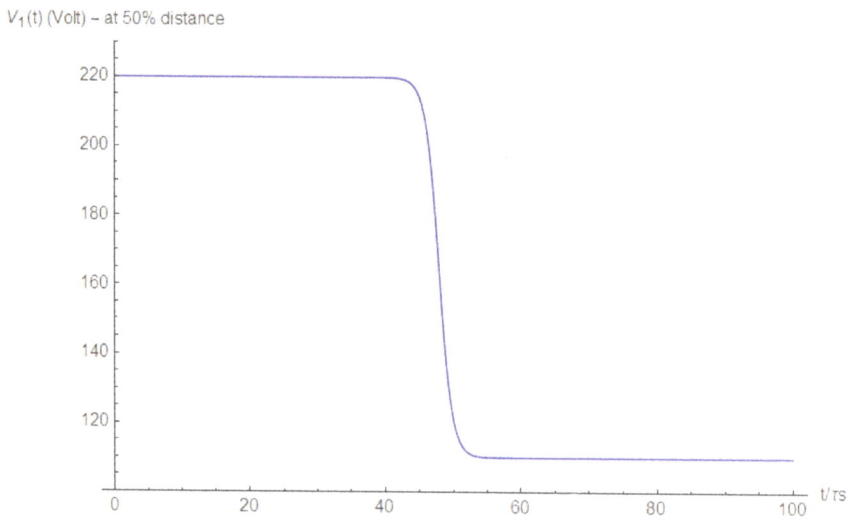


**Figure A10.** Voltage at half the distance between the source and the short.

Again, the distinctive pulse shape of the voltage derivative (Figure A11) is apparent with the same advantage mentioned before.

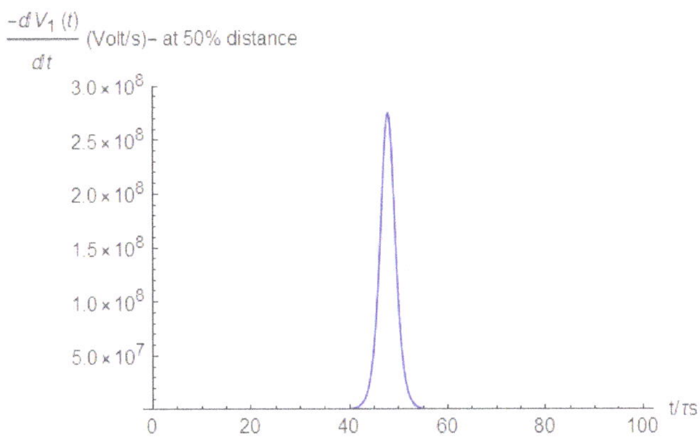


**Figure A11.** Voltage derivative at half the distance between the source and the short.

The voltage at the short vanishes (Figure A12), since the resistance approaches zero during the short's creation, providing the same behavior, which allows the pulse form in the voltage derivative (Figure A13).

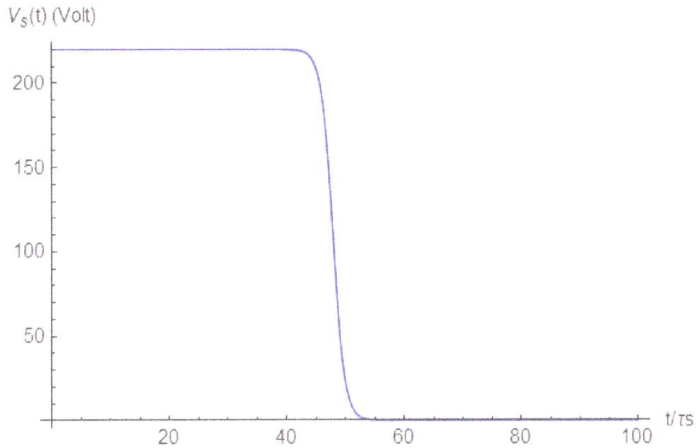


**Figure A12.** Voltage at the short.

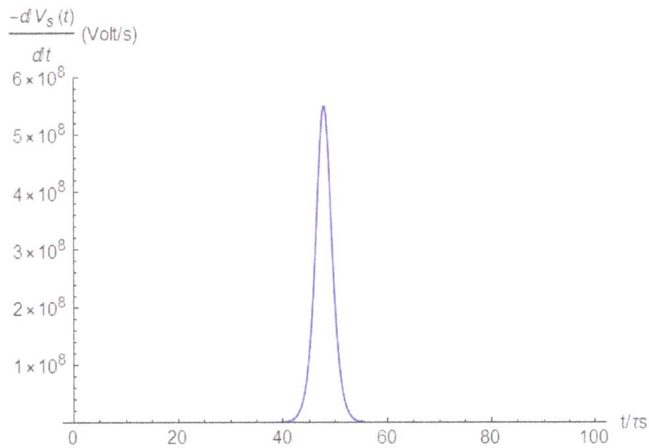


**Figure A13.** Voltage derivative at the short.

Similarly, the voltage $V_2$ at half the distance between the short and the load may be measured (Figure A14), and the pulse may be detected using the voltage derivative (Figure A15).

$$V_2(t) = V_s(t) - \frac{1}{2}R_2 I_2(t) \tag{A10}$$

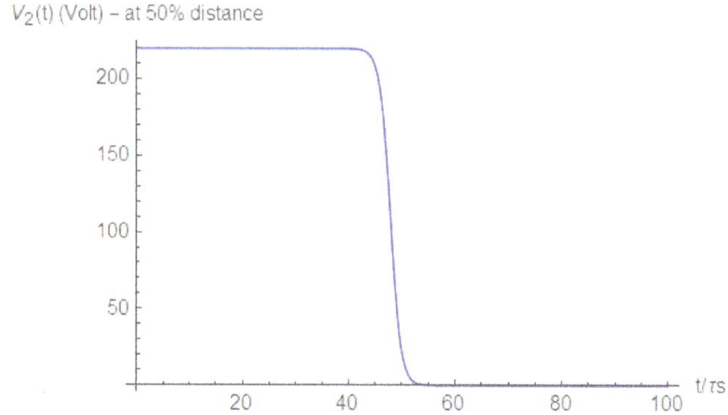


**Figure A14.** Voltage at half the distance between the short and the load.

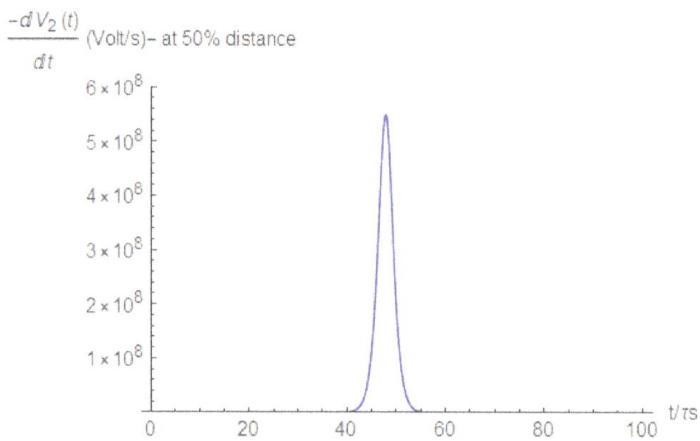


**Figure A15.** Voltage derivative at half the distance between the short and the load.

Summarizing the results of our first model, we saw that the current and voltage measurements enabled the short's pulse detection, indicating the short's occurrence. Likewise, it was shown that detection was possible both at the source and the load side. In order to detect the pulse, a resolution of the $\tau_S$ order is needed. In the current model, the inductance was neglected, leading to a simplified description; in the next section, we will look at the case were induction is taken into account, leading to a somewhat more complex mathematical analysis. Moreover, in a lumped model, there is no pulse propagation along the transmission line, and for this purpose, we introduce a distributed model later in this paper.

*Appendix A.2. The Case of Resistive and Inductive Impedance*

In this model, the transmission line inductance is no longer neglected as in the previous section. Therefore, the equations given in (A1) become coupled differential equations that can only be solved numerically. The two-wire cable induction can be calculated and is shown in Table A5.

**Table A5.** Transmission line inductance.

| Parameter | Value | Unit |
|---|---|---|
| $L = \frac{\mu}{\pi} ar\cosh\left(\frac{D}{d}\right)$ | $1.68 \times 10^{-6}$ | $\frac{H}{m}$ |
| $L_1 = L \cdot l_1$ | 0.000503938 | $H$ |
| $L_2 = L \cdot l_2$ | 0.00117585 | $H$ |
| $L_T = L_1 + L_2$ | 0.00167979 | $H$ |

The current in the circuit before the short occurs can be calculated analytically and is given as follows:

$$I_{10}(t) = I_{20}(t) = \frac{A_0 \cdot A_t}{Z + R_T} \cdot \cos[\omega \cdot t + \varphi_t] \quad \text{(A11)}$$

where: $A_t \equiv \frac{1}{\sqrt{1+(\omega \cdot \tau_p)^2}}$, $\varphi_t \equiv -\arctan[\omega \cdot \tau_p]$, $\tau_p \equiv \frac{L_T}{Z}$. This current is depicted in Figure A16.

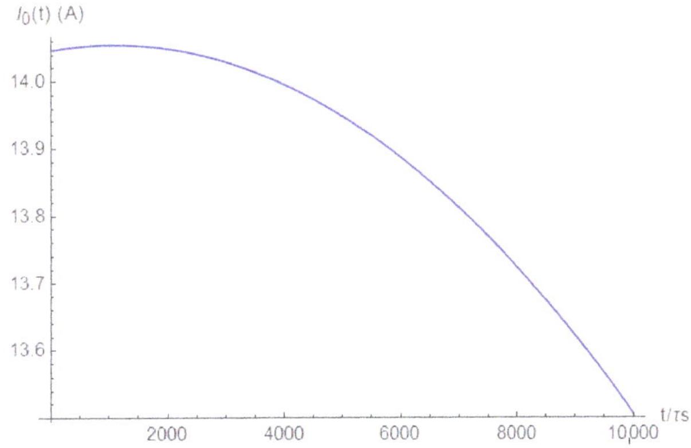


**Figure A16.** The initial current.

We now study the numerical solutions of Equation (A1) given the above initial form.

Appendix A.2.1. The Current

The source current is demonstrated in Figures A17 (right after the short) and A18 (a long time after the short).

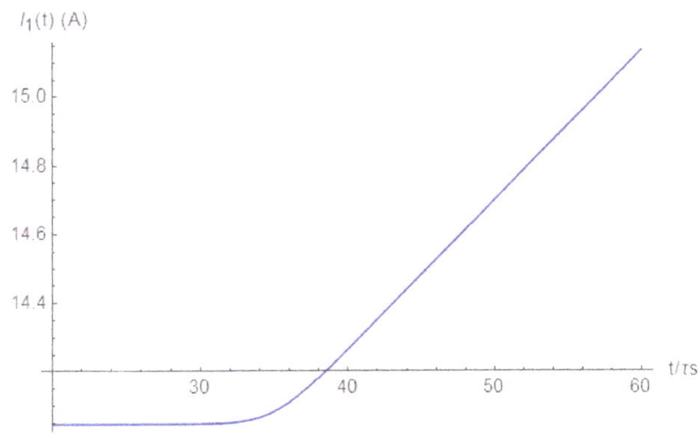


**Figure A17.** Source current right after the short.

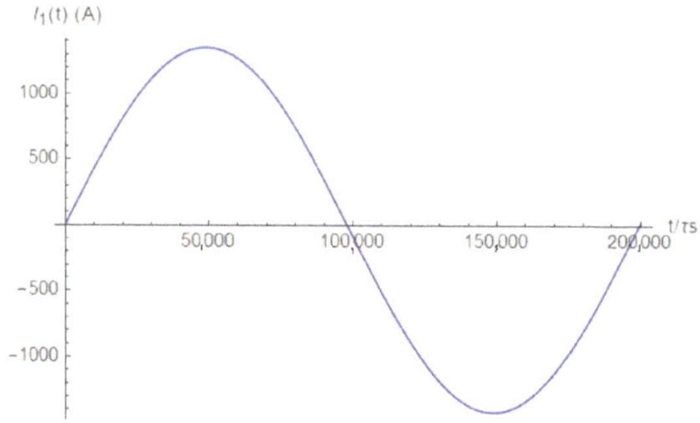


**Figure A18.** Source current a long time after the short.

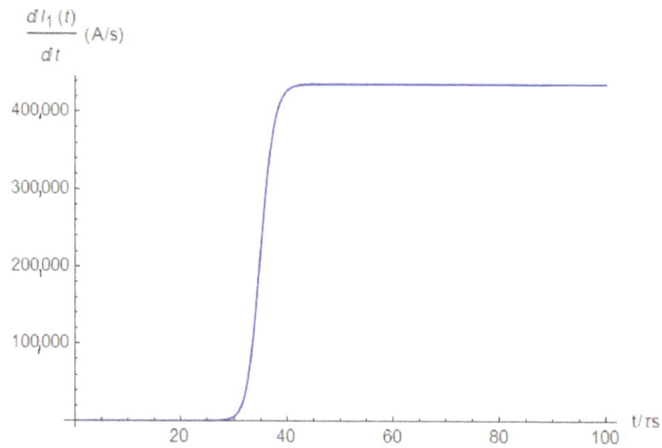


**Figure A19.** The source current derivative.

In the previous model, the derivative had a pulse shape. In the current model, the first derivative did not exhibit a pulse shape (Figure A19); however, the second derivative (Figure A20) did exhibit a pulse shape with all the benefits mentioned previously.

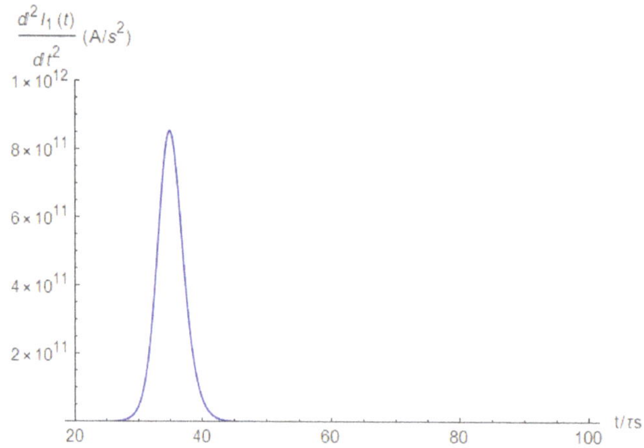


**Figure A20.** The source current second derivative.

Current measurements on the load side also provide the short pulse detection ability. The load current is shown in Figures A21 (right after the short) and A22 (a long time after the short). Here, the current decay is evident.

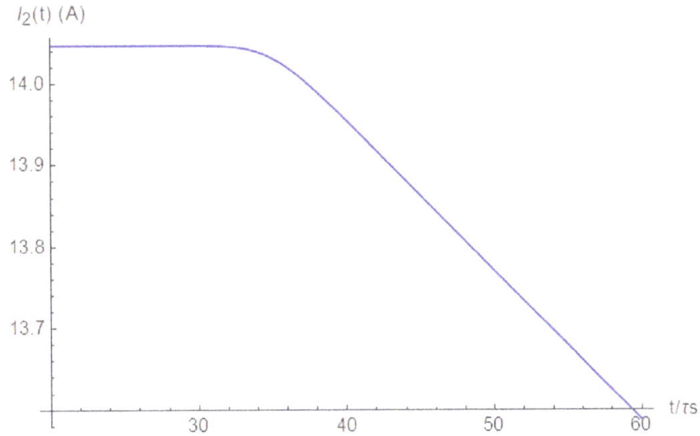


**Figure A21.** Load current (right after the short).

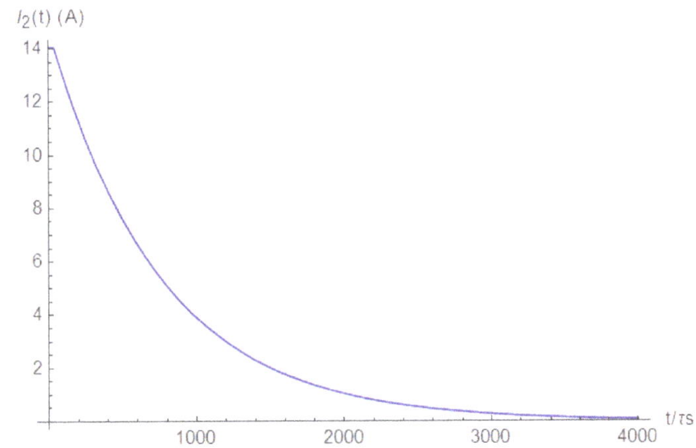


**Figure A22.** Load current (after some time).

The current's first and second derivatives are shown in Figures A23 and A24, respectively.

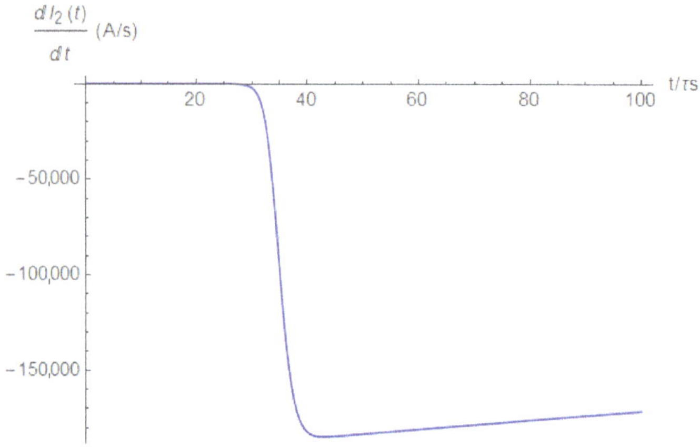


**Figure A23.** The load current derivative.

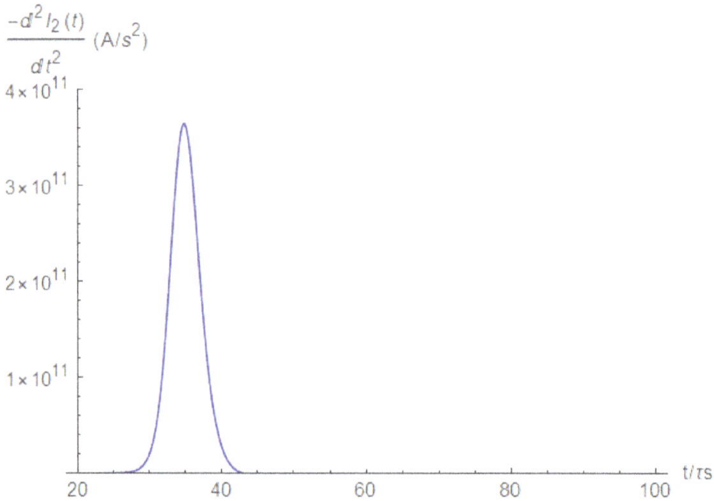


**Figure A24.** The load current second derivative.

We see that the short may be detected and accurately timed by the current's second derivative, measured at either the source or the load side. Next, we investigated the current at the short itself. It is obvious that it is zero before the short happens, after which it grows linearly (Figure A25). After some time, the short current reaches the characteristic source current values (Figure A26).

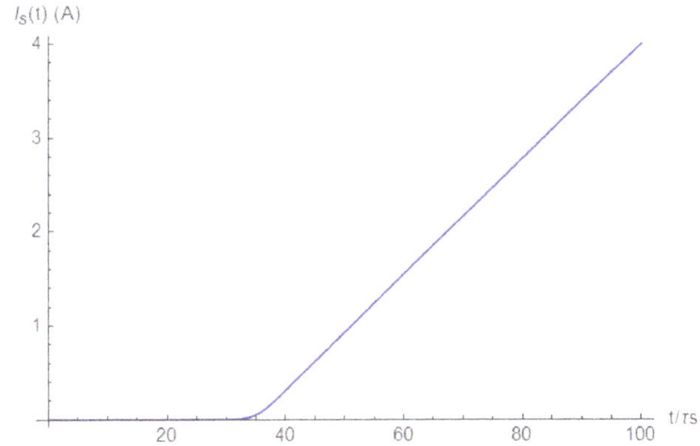


**Figure A25.** Short current (right after the short).

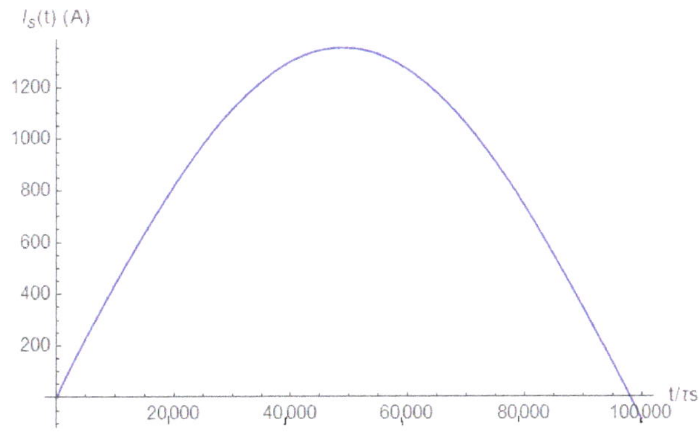


**Figure A26.** Steady-state short current.

Appendix A.2.2. The Voltage

The short-circuit pulse may also be detected by the voltage measurement between the source and the short. If the voltage is measured at half a distance, due to a high short current, the voltage will be:

$$V_1(t) = V_{in}(t) - \frac{1}{2}R_1 I_1(t) - \frac{1}{2}L_1 \frac{dI_1(t)}{dt} \tag{A12}$$

The voltage and its derivative are shown in Figures A27 and A28, respectively. We deduced that for the voltage case, a first derivative will suffice for the short location even when inductance is not neglected.

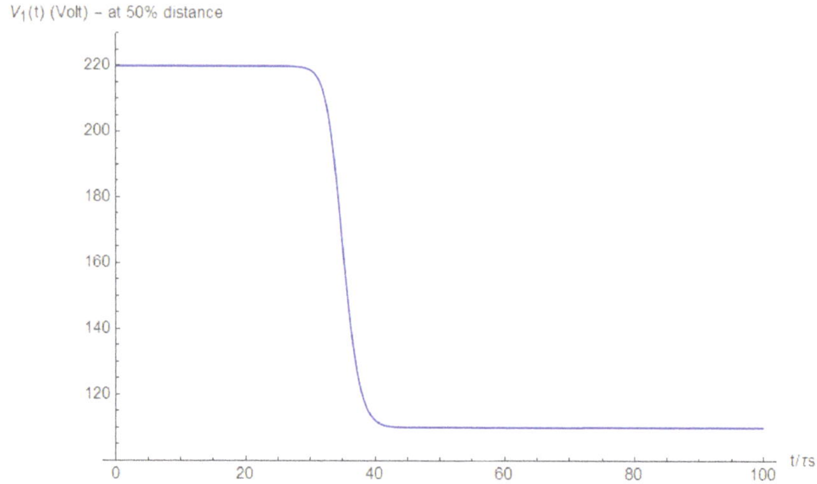


**Figure A27.** Voltage at half the distance between the source and the short.

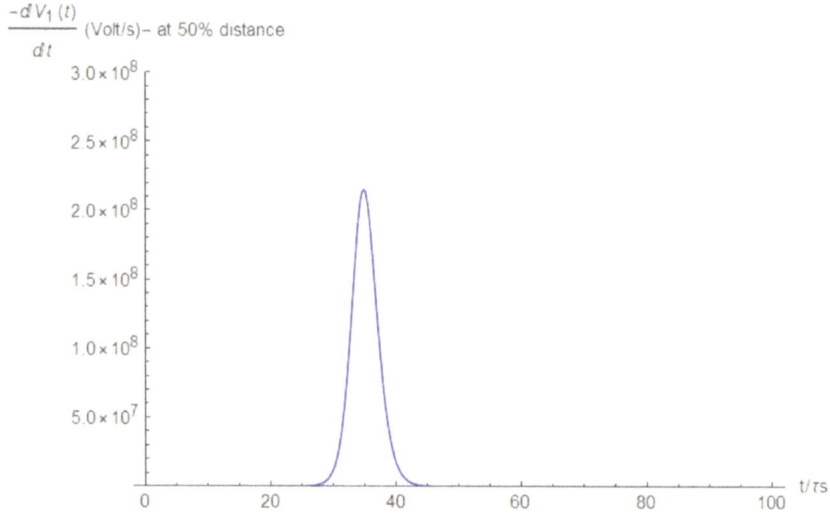


**Figure A28.** Voltage derivative at half the distance between the source and the short.

Figures A29 and A30 describe the voltage and its derivative at the short itself. We see that the voltage difference at the short when its resistivity goes to zero is also zero, and the pulse behavior of the voltage derivative is depicted.

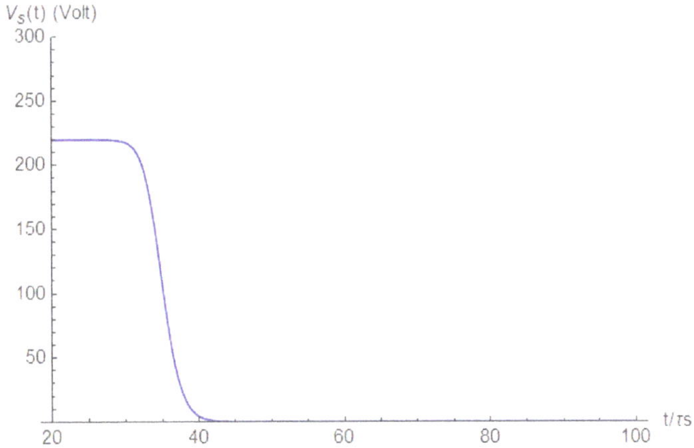


**Figure A29.** The short voltage.

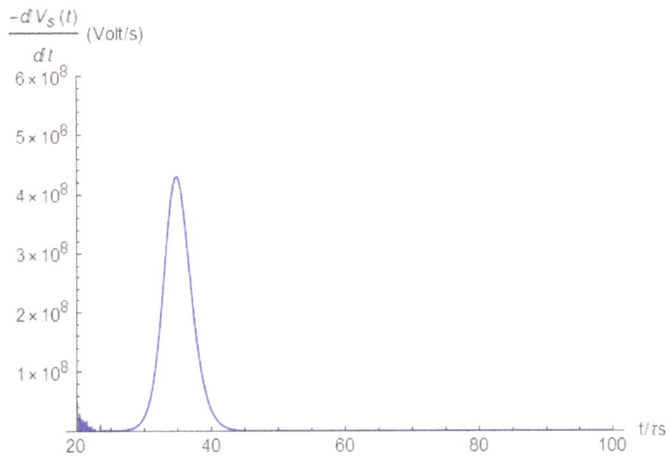


**Figure A30.** The short voltage derivative.

Analogous results may be obtained if the voltage is measured at the load side, even if the measurement is not at the load itself. For example, for half the distance voltage measurement, one obtains:

$$V_2(t) = V_s(t) - \frac{1}{2}R_2 I_2(t) - \frac{1}{2}L_2 \frac{dI_2(t)}{dt} \quad \text{(A13)}$$

Figures A31 and A32 show the voltage at half the distance between the short and the load. The voltage a brief duration after the short has formed is depicted in Figure A31, and a longer duration of the same is depicted in Figure A32.

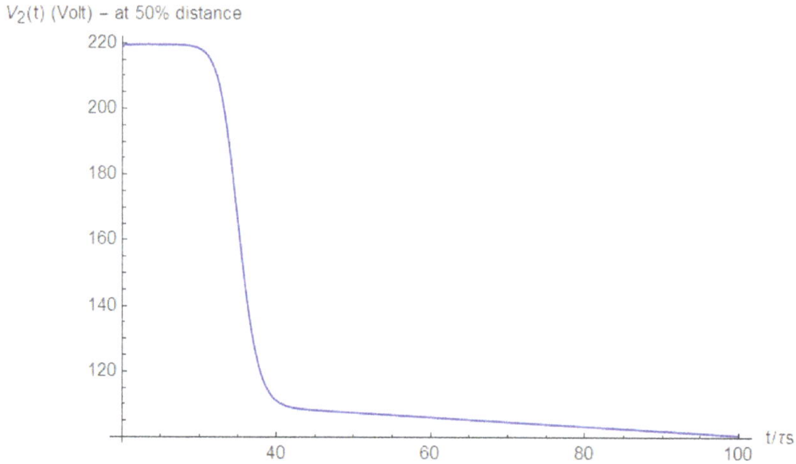


**Figure A31.** Voltage at half the distance between the short and the load (right after).

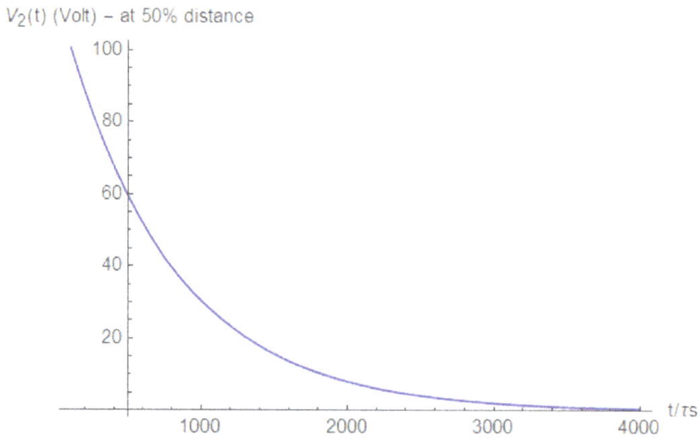


**Figure A32.** Voltage at half the distance between the short and the load (after some time).

The voltage derivative displays a pulse behavior (Figure A33), and thus, in this case, a first derivative will suffice and a second derivative is not needed.

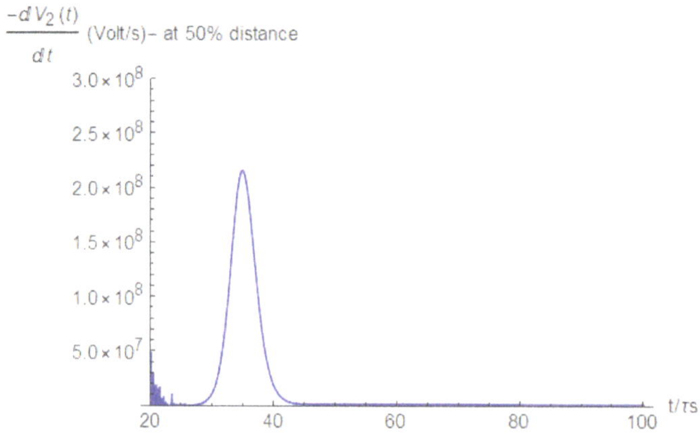


**Figure A33.** Voltage derivative at half the distance between the short and the load.

Finally, the load voltage is described, which is proportional to the load current (see also Equation (A1)):

$$V_{out}(t) = Z \cdot I_2(t) \tag{A14}$$

The load voltages, after a brief duration since the short's occurrence and later, are shown in Figures A34 and A35, respectively.

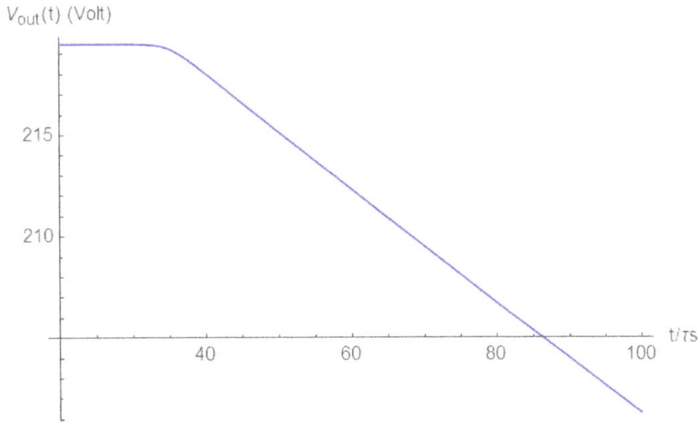


**Figure A34.** Load voltage (right after the short).

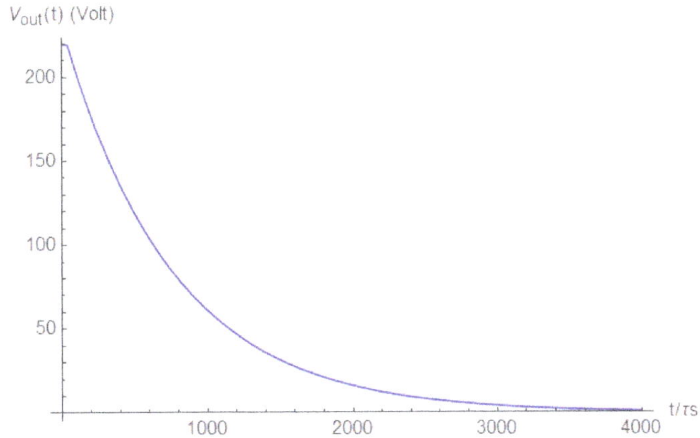


**Figure A35.** Load voltage (after some time).

The first and the second load voltage derivatives are shown in Figures A36 and A37. In this case, the desirable pulse shape was obtained for the second derivative.

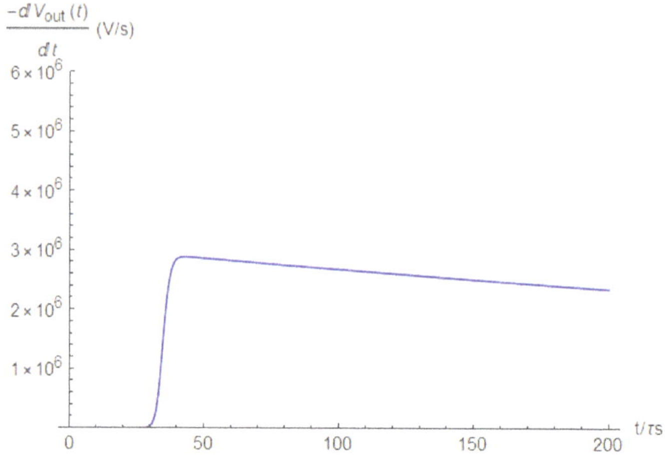


**Figure A36.** Load voltage derivative.

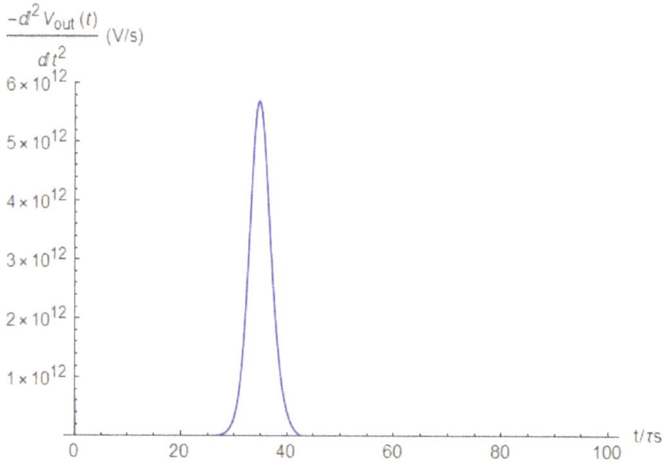


**Figure A37.** Load voltage second derivative.

## Appendix B. Errors Received from the Smoothing Voltage and Derivative Smoothing Voltage Window

| Voltage smoothing window | Voltage derivative smoothing window | $Error_1$ [%] | $Error_2$ [%] | $Error_3$ [%] | $Error_4$ [%] | $Error_5$ [%] | max(abs(error)) | mean(abs(error)) | max(error)-min(error) |
|---|---|---|---|---|---|---|---|---|---|
| 50 | 50 | -1.80 | -2.06 | 2.98 | 1.56 | -5.03 | 5.03 | 2.69 | 8.01 |
| 50 | 100 | -2.06 | -1.93 | 1.95 | 1.04 | -1.41 | 2.06 | 1.68 | 4.01 |
| 50 | 150 | -3.48 | -2.32 | 1.04 | 0.91 | 1.56 | 3.48 | 1.86 | 5.04 |
| 50 | 200 | -4.12 | -3.35 | 2.34 | 0.01 | -0.89 | 4.12 | 2.14 | 6.46 |
| 50 | 250 | -5.16 | -2.19 | 2.46 | -0.51 | -0.51 | 5.16 | 2.16 | 7.62 |
| 50 | 300 | -2.96 | -2.45 | 1.17 | 2.46 | -0.64 | 2.96 | 1.94 | 5.43 |
| 50 | 350 | -4.25 | -4.64 | 2.98 | 2.59 | -0.89 | 4.64 | 3.07 | 7.62 |
| 50 | 400 | -2.06 | -4.51 | 1.56 | 2.98 | 0.53 | 4.51 | 2.33 | 7.49 |
| 50 | 450 | -3.22 | -3.22 | 1.43 | 2.46 | 1.43 | 3.22 | 2.35 | 5.69 |
| 50 | 500 | 0.01 | -2.83 | 1.82 | 3.76 | 4.66 | 4.66 | 2.62 | 7.49 |
| 50 | 550 | 5.70 | -2.06 | 0.91 | 3.89 | 2.85 | 5.70 | 3.08 | 7.75 |
| 50 | 600 | 11.51 | -2.32 | 0.66 | 2.85 | 3.89 | 11.51 | 4.24 | 13.83 |
| 100 | 50 | -2.06 | -1.93 | 1.95 | 1.04 | -1.41 | 2.06 | 1.68 | 4.01 |
| 100 | 100 | -1.28 | -1.15 | 1.43 | 1.43 | 1.56 | 1.56 | 1.37 | 2.84 |
| 100 | 150 | -2.06 | -1.02 | 1.17 | 1.56 | 0.01 | 2.06 | 1.16 | 3.62 |
| 100 | 200 | -2.83 | -2.06 | 2.08 | 1.43 | -1.02 | 2.83 | 1.88 | 4.91 |
| 100 | 250 | -3.74 | -2.32 | 2.21 | 1.56 | -1.28 | 3.74 | 2.22 | 5.94 |
| 100 | 300 | -3.87 | -2.83 | 1.69 | 2.34 | 0.53 | 3.87 | 2.25 | 6.20 |
| 100 | 350 | -3.48 | -3.74 | 1.82 | 3.63 | -0.38 | 3.74 | 2.61 | 7.37 |
| 100 | 400 | -3.48 | -3.61 | 1.43 | 3.50 | 0.14 | 3.61 | 2.43 | 7.11 |
| 100 | 450 | -0.89 | -3.48 | 0.79 | 3.24 | 1.69 | 3.48 | 2.02 | 6.72 |
| 100 | 500 | 0.53 | -2.83 | 0.27 | 5.18 | 2.21 | 5.18 | 2.20 | 8.01 |
| 100 | 550 | 3.89 | -2.83 | 0.66 | 4.79 | 1.95 | 4.79 | 2.82 | 7.62 |
| 100 | 600 | 9.44 | -1.28 | -0.12 | 3.50 | 2.21 | 9.44 | 3.31 | 10.72 |
| 150 | 50 | -3.48 | -2.32 | 1.04 | 0.91 | 1.56 | 3.48 | 1.86 | 5.04 |
| 150 | 100 | -2.06 | -1.02 | 1.17 | 1.56 | 0.01 | 2.06 | 1.16 | 3.62 |
| 150 | 150 | -0.77 | -0.38 | 2.08 | 0.79 | 0.01 | 2.08 | 0.80 | 2.84 |
| 150 | 200 | -1.67 | -0.64 | 2.98 | 1.56 | 0.14 | 2.98 | 1.40 | 4.65 |
| 150 | 250 | -2.70 | -1.02 | 2.59 | 2.85 | -0.89 | 2.85 | 2.01 | 5.56 |
| 150 | 300 | -2.83 | -1.80 | 2.85 | 3.24 | 0.01 | 3.24 | 2.15 | 6.07 |
| 150 | 350 | -2.83 | -2.45 | 2.72 | 4.02 | -1.28 | 4.02 | 2.66 | 6.85 |
| 150 | 400 | -2.83 | -2.70 | 2.72 | 3.50 | -0.38 | 3.50 | 2.43 | 6.33 |
| 150 | 450 | -2.06 | -2.96 | 2.72 | 3.11 | 1.30 | 3.11 | 2.43 | 6.07 |
| 150 | 500 | 1.17 | -3.22 | 1.95 | 4.40 | 3.89 | 4.40 | 2.93 | 7.62 |
| 150 | 550 | 1.43 | -2.45 | 1.82 | 4.27 | 3.89 | 4.27 | 2.77 | 6.72 |
| 150 | 600 | 4.40 | -1.28 | 1.04 | 2.98 | 4.66 | 4.66 | 2.87 | 5.94 |
| 200 | 50 | -4.12 | -3.35 | 2.34 | 0.01 | -0.89 | 4.12 | 2.14 | 6.46 |
| 200 | 100 | -2.83 | -2.06 | 2.08 | 1.43 | -1.02 | 2.83 | 1.88 | 4.91 |
| 200 | 150 | -1.67 | -0.64 | 2.98 | 1.56 | 0.14 | 2.98 | 1.40 | 4.65 |
| 200 | 200 | -0.51 | 0.01 | 2.72 | 2.21 | 0.79 | 2.72 | 1.25 | 3.23 |
| 200 | 250 | -1.28 | 0.01 | 4.02 | 2.46 | -0.38 | 4.02 | 1.63 | 5.30 |
| 200 | 300 | -2.19 | -0.64 | 4.53 | 3.50 | -0.77 | 4.53 | 2.32 | 6.72 |
| 200 | 350 | -2.06 | -1.54 | 4.53 | 2.72 | 1.43 | 4.53 | 2.46 | 6.59 |
| 200 | 400 | -2.32 | -2.32 | 4.14 | 4.02 | 2.34 | 4.14 | 3.03 | 6.46 |
| 200 | 450 | -2.45 | -2.32 | 3.50 | 5.18 | 3.50 | 5.18 | 3.39 | 7.62 |
| 200 | 500 | -1.80 | -2.19 | 3.63 | 4.92 | 3.63 | 4.92 | 3.23 | 7.11 |
| 200 | 550 | 0.27 | -1.80 | 2.72 | 5.18 | 4.53 | 5.18 | 2.90 | 6.98 |
| 200 | 600 | 1.04 | -1.02 | 2.98 | 3.50 | 3.50 | 3.50 | 2.41 | 4.52 |
| 250 | 50 | -5.16 | -2.19 | 2.46 | -0.51 | -0.51 | 5.16 | 2.16 | 7.62 |
| 250 | 100 | -3.74 | -2.32 | 2.21 | 1.56 | -1.28 | 3.74 | 2.22 | 5.94 |
| 250 | 150 | -2.70 | -1.02 | 2.59 | 2.85 | -0.89 | 2.85 | 2.01 | 5.56 |
| 250 | 200 | -1.28 | 0.01 | 4.02 | 2.46 | -0.38 | 4.02 | 1.63 | 5.30 |
| 250 | 250 | -0.12 | 0.53 | 4.79 | 2.46 | 1.04 | 4.79 | 1.79 | 4.91 |
| 250 | 300 | -0.89 | 0.27 | 4.92 | 4.40 | -0.38 | 4.92 | 2.17 | 5.81 |
| 250 | 350 | -1.54 | 0.01 | 4.53 | 5.18 | 0.66 | 5.18 | 2.38 | 6.72 |
| 250 | 400 | -1.80 | -0.64 | 4.27 | 6.73 | 2.46 | 6.73 | 3.18 | 8.53 |
| 250 | 450 | -1.93 | -1.15 | 4.53 | 6.73 | 4.02 | 6.73 | 3.67 | 8.66 |
| 250 | 500 | -1.93 | -1.28 | 4.66 | 5.70 | 5.31 | 5.70 | 3.77 | 7.62 |
| 250 | 550 | -2.19 | -0.77 | 4.02 | 5.57 | 4.92 | 5.57 | 3.49 | 7.75 |
| 250 | 600 | 1.17 | -0.25 | 4.02 | 4.14 | 4.27 | 4.27 | 2.77 | 4.52 |

**Figure A38.** Load voltage second derivative.

## References


1. Einstein, A. On the Electrodynamics of Moving Bodies. *Ann. Der Phys.* **1905**, *17*, 891–921. [CrossRef]
2. Kulkarni, S.S. Fault Location and Characterization in AC and DC Power Systems. Ph.D. Thesis, University of Texas at Austin, Austin, TX, USA, 2012.
3. Ghimire, S. *Analysis of Fault Location Methods on Transmission Lines*; University Of New Orelans: New Orelans, LA, USA, 2014.
4. Xu, Z. *Fault Location and Incipient Fault Detection in Distribution Cables*; Western University: London, ON, Canada, 2011.
5. Chen, K.; Huang, C.; He, J. Fault detection, classification and location for transmission lines and distribution systems: A review on the methods. *High Voltage* **2016**, *1*, 25–33. [CrossRef]
6. Glik, K.; Rasolomampionona, D.D.; Kowalik, R. Detection, classification and fault location in HV lines using travelling waves. *Prz. Elektrotechniczny (Electr. Rev.)* **2012**, *88*, 269–275.
7. Radojević, Z.M.; Kim, C.H.; Popov, M.; Preston, G.; Terzija, V. New approach for fault location on transmission lines not requiring line parameters. In Proceedings of the International Conference on Power System Transients Proceedings, Kyoto, Japan, 2–6 June 2009.
8. Patel, M.; Patel, R.N. Fault analysis in transmission lines using neural network and wavelets. In Proceedings of the 2015 2nd International Conference on Signal Processing and Integrated Networks (SPIN), Noida, India, 19–20 February 2015; pp. 719–724. [CrossRef]
9. Bunnoon, P. Fault detection approaches to power system: State-of-the-art article reviews for searching a new approach in the future. *Int. J. Electr. Comput. Eng.* **2013**, *3*, 553. [CrossRef]
10. Dong, A.; Huo, L.; Liang, L.; Wang, Q. Research on the practical detection for a power cable fault point. In Proceedings of the 2010 International Conference on Computer and Communication Technologies in Agriculture Engineering, Chengdu, China, 12–13 June 2010; Volume 2, pp. 80–84. [CrossRef]
11. Mustari, M.R.; Hashim, M.N.; Osman, M.K.; Ahmad, A.R.; Ahmad, F.; Ibrahim, M.N. Fault Location Estimation on Transmission Lines using Neuro-Fuzzy System. *Procedia Comput. Sci.* **2019**, *163*, 591–602. [CrossRef]
12. de Morais Pereira, C.E.; Zanetta, L.C. Fault location in transmission lines using one-terminal postfault voltage data. *IEEE Trans. Power Deliv.* **2004**, *19*, 570–575. [CrossRef]
13. Alam, M.N.; Bhuiyan, R.H.; Dougal, R.; Ali, M. Novel surface wave exciters for power line fault detection and communications. In Proceedings of the 2011 IEEE International Symposium on Antennas and Propagation (APSURSI), Pokane, WA, USA, 3–8 July 2011; pp. 1139–1142. [CrossRef]
14. Srinivasarao, Y.; Pavani, S.; Sudharmi, G. Detection of Fault Location in Transmission Lines. *Int. J. Appl. Eng. Res.* **2017**, *12*, 1.
15. Aldeen, M.; Crusca, F. IEE Proceedings-Generation, Transmission and Distribution. In *Observer-Based Fault Detection and Identification Scheme for Power Systems*; The Institution of Engineering and Technology: London, UK, 2016; Volume 153, pp. 71–79. ISSN 1350-2360. [CrossRef]
16. Alsafasfeh, Q.; Abdel-Qader, I.; Harb, A. Symmetrical pattern and PCA based framework for fault detection and classification in power systems. In Proceedings of the 2010 IEEE International Conference on Electro/Information Technolgy, Normal, IL, USA, 20–22 May 2010; pp. 1–5. [CrossRef]
17. Ashouri, A.; Jalilv, A.; Noroozian, R.; Bagheri, A. A new approach for fault detection in digital relays-based power system using Petri nets. In Proceedings of the 2010 Joint International Conference on Power Electronics, Drives and Energy Systems 2010 Power India, New Delhi, India, 20–23 December 2010; pp. 1–8. [CrossRef]
18. Bhoomaiah, A.; Reddy, P.K.; Murthy, K.L.; Naidu, P.A.; Singh, B.P. Measurement of neutral currents in a power transformer and fault detection using wavelet techniques. In Proceedings of the 17th Annual Meeting of the IEEE Lasers and Electro-Optics Society, 2004, LEOS 2004, Boulder, CO, USA, 20 October 2004; pp. 170–173. [CrossRef]
19. Dubey, R.; Samantaray, S.R.; Tripathy, A.; Babu, B.C.; Ehtesham, M. Wavelet based energy function for symmetrical fault detection during power swing. In Proceedings of the 2012 Students Conference on Engineering and Systems, Allahabad, India, 16–18 March 2012; pp. 1–6. [CrossRef]
20. Shu, H.; Wang, X.; Tian, X.; Wu, Q.; Peng, S. On the use of S-transform for fault feeder detection based on two phase currents in distribution power systems. In Proceedings of the 2010 2nd International Conference on Industrial and Information Systems, Dalian, China, 10–11 July 2010; Volume 2, pp. 282–287. [CrossRef]
21. Long, Z.; Younan, N.H.; Bialek, T.O. Underground power cable fault detection using complex wavelet analysis. In Proceedings of the 2012 International Conference on High Voltage Engineering and Application, Shanghai, China, 17–20 September 2012; pp. 59–62. [CrossRef]
22. Magnusson, P.C.; Alexander, G.C.; Tripathi, V.K.; Weisshaar, A. *Transmission Lines and Wave Propagation*; CRC Press: Boca Raton, FL, USA, 2000.
23. Farahmand, F. *Introduction to Transmission Lines*; A Primer on Electromagnetic Fields; International Publishers Association: Geneva, Switzerland, 2012.
24. Hui, H.T. *Transmission Lines–Basic Theories*; (NUS/ECE Lecture Notes); WordPress.com: San Francisco, CA, USA, 2011.
25. Matzner, H.; Levy, S. *Experiment 3–Transmission Lines, Part 1*; HIT: Holon, Israel, 2008.
26. Nabwani, M.; Suleymanov, M.; Pinhasi, Y.; Yahalom, A. Retardation in the Service of Real Time Fault Detection and the Difference Between Distributed and Lumped Fault Models. In Proceedings of the Material Technologies and Modeling the Tenth International Conference, Ariel, Israel, 20–24 August 2018.

27. Yosef, P.; Asher, Y. Fault Location in a Transmission Line. U.S. Patent US11079422B2, 3 August 2021.
28. Yosef, P.; Asher, Y. Fault Location in a Transmission Line. IL Patent 261763, 1 February 2022.
29. David, M. Pozar. In *Microwave Engineering*, 4th ed.; John Wiley & Sons: Hoboken, NJ, USA, 2012.
30. Michael, S.; Asher, Y.; Yosef, P. Real Time Fault Location in High Voltage Network Power Lines. In Proceedings of the 2016 ICSEE International Conference on the Science of Electrical Engineering, Hilton Queen of Sheba, Eilat, Israel, 16–18 November 2016. [CrossRef]
31. Ankamma, R.J.; Bizuayehu, B.; Jemal, M.A.; Tsegaye, N.G. Fault Location Calculation Based on Two Terminal Data High Voltage Transmission Line. *Int. J. Res. Eng. Technol.* **2016**, *4*, 1–18.
32. Josef, T.; Jaroslav, D.; Jan, K.; Radek, H. *Fault Location in Hv Systems*; Research, Development and Innovation Information System: Prague, Czech Republic, 2022.
33. Li, J.; Yang, Q.; Mu, H.; Le Blond, S.; He, H. A new fault detection and fault location method for multi-terminal high voltage direct current of offshore wind farm. *Appl. Energy* **2018**, *220*, 13–20. [CrossRef]
34. Arieh, S. *Transient Analysis of Electric Power Circuits Handbook*; Springer: Heidelberg, Germany, 2005; p. 466.
35. *Guide for Measurement of Radio Frequency Interference From Hv And Mv Substations*; Draft 6a, CIRED/CIGRÉ JWG C4.202; CIGRE: Paris, France, 2008.
36. Lipiński, M.; Włostowski, T.; Serrano, J.; Alvarez, P. White Rabbit: A PTP application for robust sub-nanosecond synchronization. In Proceedings of the ISPCS2011, Munich, Germany, 12–16 September 2011.
37. Moreira, P.; Alvarez, P.; Serrano, J.; Darwezeh, I.; Wlostowski, T. Digital Dual Mixer Time Difference for Sub-Nanosecond Time Synchronization in Ethernet. In Proceedings of the 2010 IEEE International Frequency Control Symposium, Newport Beach, CA, USA, 1–4 June 2010.
38. Rizzi, M.; Lipinski, M.; Serrano, T.W.J.; Ferrari, S.; Rinaldi, S. White Rabbit clock characteristics. In Proceedings of the ISPCS2016, Stockholm, Sweden, 4–9 September 2016.
39. Wlostowski, T. Precise Time and Frequency Transfer in a White Rabbit Network. Ph.D. Thesis, Warsaw University of Technology, Warsaw, Poland, 2010.

electronics

Article

# Application of Novel AC–AC Matrix VFD for Power Factor Improvement in Conventional AC–DC–AC VFD-Loaded Power Distribution Lines

Gytis Petrauskas [1,*], Gytis Svinkunas [2], Audrius Jonaitis [2] and Andreas Giannakis [3]

1 Department of Automation, Faculty of Electrical and Electronics Engineering, Kaunas University of Technology, LT-51394 Kaunas, Lithuania
2 Department of Electric Power Systems, Faculty of Electrical and Electronics Engineering, Kaunas University of Technology, LT-51394 Kaunas, Lithuania; gytis.svinkunas@ktu.lt (G.S.); audrius.jonaitis@ktu.lt (A.J.)
3 Department of Electric Power Engineering, Norwegian University of Science and Technology, 7491 Trondheim, Norway; andreas.giannakis@ntnu.no
* Correspondence: gytis.petrauskas@ktu.lt; Tel.: +370-687-17526

**Abstract:** In this study, an innovative approach to matrix-converter-based AC—AC variable frequency drives (VFDs) is introduced. The possibility of using AC–AC matrix VFDs for reactive power compensation in conventional AC–DC–AC VFD-loaded power distribution lines is investigated. It is found that the interaction of a large number of conventional AC–DC–AC VFDs with a conventional capacitor-based local compensation device leads to overcompensation in 0.4 kV power distribution lines. This is due to the fact that the conventional compensation device is designed to compensate the lagging reactive power produced by inductive loads, such as AC motors. This highlights the demand for the compensation of leading reactive power that is not predicted by the designer. To solve this problem, the modification of a certain number of previously installed VFDs by replacing their conventional AC–DC–AC converters with AC–AC matrix converters is proposed. This can lead to improvements in the power factor in 0.4 kV power distribution lines. In this study, the range of reactive power produced by conventional AC–DC–AC VFDs was determined mathematically, by simulation, and experimentally. The range of reactive power produced by the novel AC–AC matrix VFD was also determined. On that basis, the number of VFDs to be modified is defined to keep the power factor close to unity.

**Keywords:** matrix converter; space vector modulation; power factor; variable frequency drive

**Citation:** Petrauskas, G.; Svinkunas, G.; Jonaitis, A.; Giannakis, A. Application of Novel AC–AC Matrix VFD for Power Factor Improvement in Conventional AC–DC–AC VFD-Loaded Power Distribution Lines. *Electronics* **2022**, *11*, 997. https://doi.org/10.3390/electronics11070997

Academic Editor: Giampaolo Buticchi

Received: 14 February 2022
Accepted: 22 March 2022
Published: 23 March 2022



## 1. Introduction

Power generation and transmission are complex processes that require the interconnected operation of the components of the power system. The reactive power is one of the main components in the system. Loads such as electric motors require reactive power for their operation. To improve the performance of the AC power systems, this reactive power has to be managed. This is known as reactive power compensation [1].

Typically, the current in old-fashioned power distribution lines (PDL) lags behind voltage, because of inductive loads such as motors. Local reactive power compensation devices are usually designed to compensate the lagging reactive power [2].

The evolution of variable frequency drive (VFD) technology has changed the type of reactive power in 0.4 kV PDL. VFDs are finding increasing applications in various industrial and infrastructure sectors. The global VFD market size is expected to be 31.3 billion USD by 2025, from an estimated 19.2 billion USD in 2020, rising at a market growth of 6% for the mentioned forecast period.

A typical induction motor connected directly to a power distribution line is a lagging reactive power source. When connecting this motor via an AC–DC–AC frequency converter, a conventional VFD is formed. This VFD, unlike the induction motor, produces the leading reactive power. Data on the reactive power produced by VFDs obtained experimentally are given below. Reactive power compensators, traditionally designed to compensate the inductive (lagging) reactive power, are not capable of compensating the leading reactive power produced by the VFD. It carries financial losses. Companies that are unable to compensate the VFD-produced reactive power pay fines. Some studies have suggested solutions to this problem. The authors of [3] proposed to compensate the reactive power using reactive elements. The authors of [4] defined power quality coefficients that determine the distortion caused by variable-speed asynchronous electric drives. The authors of [5] suggested the use of dynamic reactive power compensation equipment.

Recently, VFDs based on a novel matrix converter (MC) are becoming more widespread in buildings and industry. With the development of semiconductor technologies, the popularity of MC is constantly increasing. In the AC–AC matrix VFD, indirect space vector modulation (ISVM) is the most common way to control converter switches. Using this control method, the MC is virtually divided into two parts. The first part is the current source rectifier (CSR). The second part is the voltage source inverter (VSI). This division makes it possible to control the current displacement angle independently of the voltage on the motor side. Such a VFD connected to the PDL can control the reactive power at the connection point. With regard to the method of control mentioned above, the MC-based induction motor VFD has the ability to exchange both leading and lagging reactive power with the connected power distribution line. This means that the AC–AC matrix VFD can be used for conventional AC–AC–AC VFD-produced reactive power compensation. It is, therefore, proposed to replace the electronic equipment of one of the conventional VFDs installed in a commercial or industrial building. Instead of three stages (diode supply-side rectifier, intermediate DC circuit, and inverter), nine bidirectional electronic switches are installed, leaving the VFD induction motor the same. In this way, a PDL power factor close to unity can be achieved.

Some authors of the reviewed publications suggested using ISVM-controlled MC for reactive power compensation. The authors of [6] suggested the use of a permanent magnet machine powered by a MC as the compensation device. The authors of [7] proposed to connect the AC–AC matrix VFD to the PDL, thus compensating for the reactive power produced by light-emitting diode light sources. The authors of [8] proposed using an MC in combination with a transformer for power factor control. The authors of [9,10] presented a modulation strategy allowing the expansion of the MC reactive power range. The authors of [11] suggested the use of AC–AC matrix VFDs for voltage sag impact reduction. The only study that considered the reactive power in the induction motor MC VFD was [12]. However, the authors of this publication only proposed a control strategy that provides current control and a unity power factor. They did not measure the ability to compensate for the reactive power produced by other sources, such as conventional AC–DC–AC VFDs. Therefore, a comprehensive study of MC-based induction motor VFDs is needed to determine the range of reactive power compensation they provide before they can be put into practical application. The number of conventional AC–DC–AC VFDs whose reactive power can be compensated for by a single AC–AC matrix VFD must be determined. Furthermore, the effect of motor load torque and rotor speed on the reactive power of conventional AC–DC–AC VFDs must be investigated.

## 2. Overcompensation Problem in Current Power Distribution Lines

Typically, the current in conventional PDLs lags behind voltage, because of inductive loads such as AC motors (Figure 1a).

However, currently, as shown in Figure 1b, more and more AC motors are powered by AC–DC–AC frequency converters. One component of the conventional VFD directly connected to the PDL is the diode rectifier. As shown in Figure 2, another large and sensitive element is the electrolytic capacitor. The value of this capacitor is usually designed as 85 μF/kW. The range of voltage variation in such a case is 5–10% of the nominal voltage.

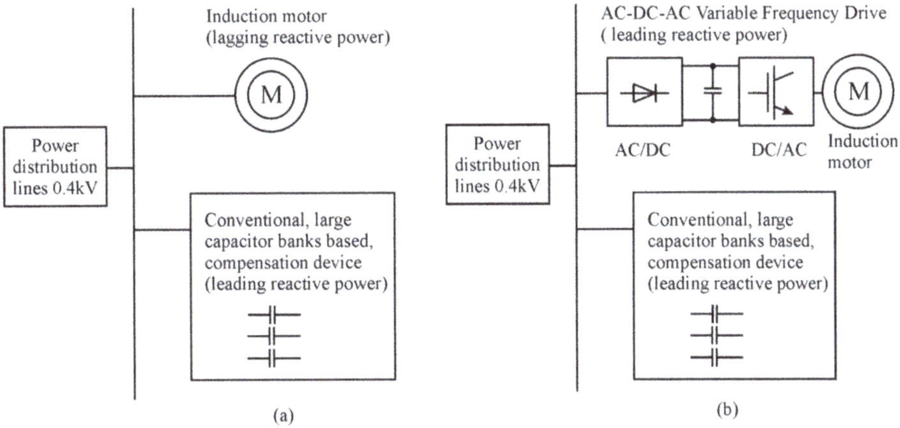


**Figure 1.** The loads and compensation devices of a 0.4 kV PDL: (**a**) conventional load equipment; (**b**) current load equipment (causes overcompensation).

*2.1. Theoratical Analysis of the Reactive Power Generated by Diode Rectifier Supply-Side Variable Frequency Drive*

The reactive power of such equipment directly depends on the voltage at the power distribution line connection point and the current of the diode rectifier (Figure 2).

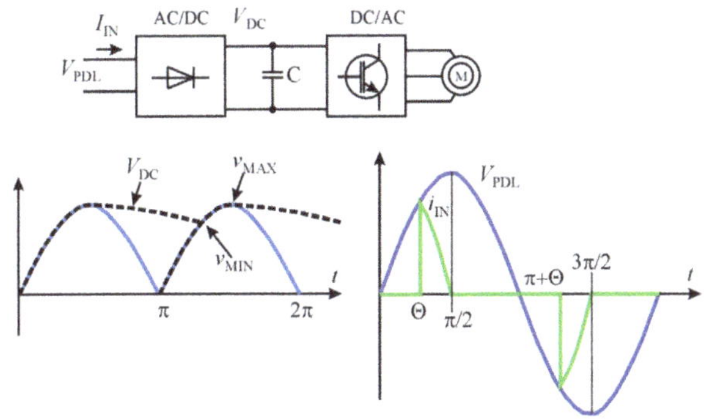


**Figure 2.** Time diagrams of the voltage and current of single-phase diode rectifier supply-side VFD.

The input current of the AC–DC device, due to the large number of harmonics, is quite complex and can be described by the following Equation (1) [7]:

$$I_{IN}(t) = I_0 + I_1\, sin(\omega t + \varphi_1) + I_2\, sin(\omega t + \varphi_2) \cdots \quad (1)$$

The consumption or generation of reactive power depends on $I_1 sin(\omega t + \varphi_1)$. Because of this, it must be expressed in a Fourier series, and $\varphi_1$ must be determined as follows:

$$\tan\varphi_1 = \frac{I_1''}{I_1'} = \frac{(\pi - 2\theta\ -\ sin2\theta)}{(1 + cos2\theta)}. \quad (2)$$

In the case of $\theta_1 = 64°$, $\varphi_1 = 19.1°$ when $k_V = 10\%$, tan $\varphi_1 = 0.34$. In the case of $\theta_2 = 72°$, $\varphi_2 = 16.1°$ when $k_V = 5\%$, tan $\varphi_2 = 0.28$. These are the minimum and maximum limits of the current displacement angle. Since $\varphi_1 > 0$ and $\varphi_2 > 0$, it can be stated that the conventional AC–DC–AC VFD injects some amount of the leading reactive power into the PDL. It injects 0.28–0.34 kVar in the case of powering a 1 kW load.

For more powerful VFDs, a three-phase diode rectifier supply is used (Figure 3). An equation for the expression of the reactive power on the forward-conducting angle of the diodes can be derived for this rectifier.

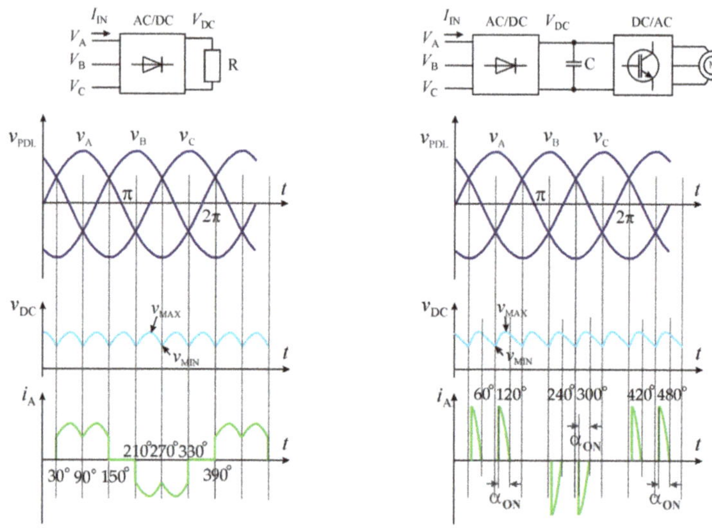


**Figure 3.** Time diagrams of the voltage and current of three-phase diode rectifier supply-side VFD: (**a**) pure resistor load; (**b**) conventional VFD.

The voltage pulsations and current of diode rectifier-based devices can be described by the following common equations:

$$V_A = V_m sin\omega t, \quad (3)$$

$$i_{IN} \approx C\frac{dV_A}{dt} = C\ V_m\ \omega\ cos\omega t. \quad (4)$$

$$\Delta V_{PULS} = V_{MAX} - V_{MIN}, \quad (5)$$

$$k_V = \frac{\Delta V_{PULS}}{V_{MAX}}. \quad (6)$$

According to the technical requirements, the voltage pulsation $k_V$ on the DC capacitor is in the range of 5–10%. According to this, angle $\alpha_{ON}$ determines the moment forward-conduction of the VFD's supply-side diode rectifier.

$$\begin{aligned} V_{DC} = \ & V_A - V_B = V_m sin\omega t - V_m sin(\omega t - 120^\circ) = 2cos\frac{\omega t + (\omega t - 120^\circ)}{2} \times sin\frac{120^\circ}{2} = \\ & 2cos(\omega t - 60^\circ) \times \left(\frac{\sqrt{3}}{2}\right) = \sqrt{3}cos(\omega t - 60^\circ) = \sqrt{3}sin(\omega t + 30^\circ). \end{aligned} \quad (7)$$

The maximum value is obtained when $\omega t = 60^\circ$ or $120^\circ$, where $sin\ (60^\circ + 30^\circ - \alpha_{ON}) =$ 0.9–0.95. Therefore,

$$\alpha_{ON} = 26^\circ,\ at\ k_V = 10\%;$$

$$\alpha_{ON} = 18^\circ,\ at\ k_V = 5\%;$$

$$\alpha_{ON} = 13^\circ,\ at\ k_V = 2,5\%.$$

The type of reactive power (leading or lagging) is determined by the component $i_1 = I_{1m} sin(\omega t + \varphi_1)$ in the Fourier series (Equation (1)).

$$\begin{aligned} I_1' &= \frac{1}{\pi}\int_{60^\circ-\alpha_{ON}}^{60^\circ} cos(\omega t + 30^\circ) sinxdt\omega t + \frac{1}{\pi}\int_{120^\circ-\alpha_{ON}}^{120^\circ} cos(\omega t - 30^\circ) sinxdt\omega t + \\ &\frac{1}{\pi}\int_{240^\circ-\alpha_{ON}}^{240^\circ} cos(\omega t + 30^\circ) sinxdt\omega t + \frac{1}{\pi}\int_{300^\circ-\alpha_{ON}}^{300^\circ} cos(\omega t - 30^\circ) sinxdt\omega t. \end{aligned} \tag{8}$$

$$\begin{aligned} I_1'' &= \frac{1}{\pi}\int_{60^\circ-\alpha_{ON}}^{60^\circ} cos(\omega t + 30^\circ) cosxdt\omega t + \frac{1}{\pi}\int_{120^\circ-\alpha_{ON}}^{120^\circ} cos(\omega t + 30^\circ) cosxdt\omega t + \\ &\frac{1}{\pi}\int_{240^\circ-\alpha_{ON}}^{240^\circ} cos(\omega t + 30^\circ) cosxdt\omega t + \frac{1}{\pi}\int_{300^\circ-\alpha_{ON}}^{300^\circ} cos(\omega t - 30^\circ) cosxdt\omega t. \end{aligned} \tag{9}$$

$$I_1' = \frac{1}{2\pi}\left[\sqrt{3} + cos(150^\circ - 2\alpha_{ON}) - cos(30^\circ - 2\alpha_{ON})\right]. \tag{10}$$

$$I_1'' = \frac{1}{2\pi}\left[\frac{4\sqrt{3}}{2} - sin(150^\circ - 2\alpha_{ON}) - sin(210^\circ - 2\alpha_{ON})\right] \tag{11}$$

$$\tan\varphi_1 = \frac{\frac{1}{2\pi}\left(2\sqrt{3}\alpha_{ON} - sin(150^\circ - 2\alpha_{ON}) + sin(30^\circ - 2\alpha_{ON})\right)}{\frac{1}{2\pi}\left(\sqrt{3} - cos(150^\circ - 2\alpha_{ON}) - cos(30^\circ - 2\alpha_{ON})\right)}. \tag{12}$$

According to Equation (12),

$$\begin{aligned} \alpha_{ON} &= 13^\circ,\ k_V = 2,5\%,\ \varphi_1 = 8.86^\circ,\ \tan\varphi_1 = 0.152, \\ \alpha_{ON} &= 18^\circ,\ k_V = 5\%, \varphi_1 = 11.98^\circ,\ \tan\varphi_1 = 0.22, \\ \alpha_{ON} &= 26^\circ,\ k_V = 10\%,\ \varphi_1 = 17.27^\circ,\ \tan\varphi_1 = 0.311. \end{aligned}$$

The results of the calculation show that, at the same pulsation level, the three-phase rectifier generates less reactive power than the single-phase rectifier.

### 2.2. Experiemntal Analysis of the Reactive Power Generated by Diode Rectifier Supply-Side Ariable Frequency Drive

The experimental analysis of the reactive power in the VFD loaded PDL was performed with a METREL MI 2892 power quality analyzer. That instrument records all measurement (Classic and Modern), regardless of selected method. This device complies with the measurement methods defined in the IEEE 1459 standard. The old measurement methods were applicable when currents and voltages were close to sinusoidal ones. However, they are not suitable for modern measurements when we have power electronics equipment such as VFD representing a nonlinear load for industrial and commercial PDL. The methodology used in this study measures the fundamental and nonfundamental components separately, as shown in Figure 4.

The experimentally obtained results of power analysis are presented in Table 1 (single-phase supply AC-DC-AC VFD) and in Table 2 (three-phase supply AC-DC-AC VFD). The following power distribution line voltage parameters were investigated:

- $P_C$—active power (combined) of all voltage and current harmonic;
- $N_C$—reactive power (combined) of all voltage and current harmonic
- $S_C$—apparent power (combined) of all harmonic
- $PF$—total effective power factor (combined) of all harmonics;
- $P_{Fund}$—active power of fundamental harmonics;
- $Q_{VFund}$—reactive power of fundamental harmonics;
- $S_{VFund}$—apparent power vector of fundamental harmonics;
- $PF_{VFund}$—displacement factor vector of fundamental harmonics.

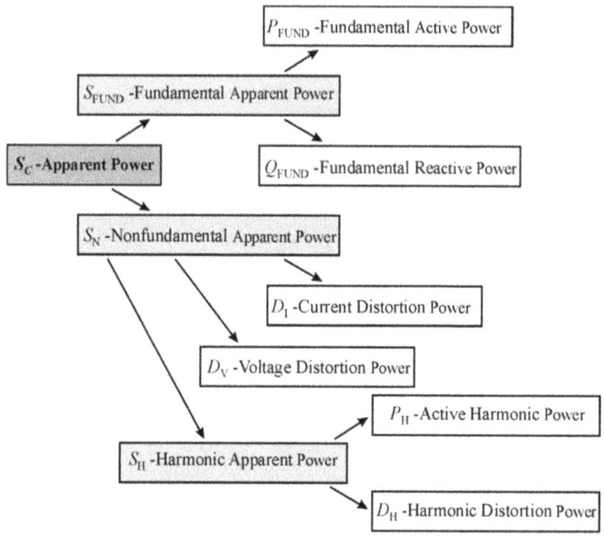


**Figure 4.** The power management organization according to the IEEE1459.

**Table 1.** Power components at the point of connection of Siemens Micromaster 420 single-phase diode rectifier supply-side VFD to the PDL, keeping motor load torque constant.

| $f_{OUT}$, Hz | Combined | | | | Fundamental | | | | |
|---|---|---|---|---|---|---|---|---|---|
| | $P_C$, W | $N_C$, var | $S_C$, VA | $PF_C$ | $P_{Fund}$, W | $Q_{VFund}$, var | $S_{VFund}$, VA | $PF_{VFund}$ | $tan(\varphi)$ |
| 10 | 32.8 | −58.01 | 58.1 | 0.565 | 32.6 | −16.5 | 36.9 | 0.894 | −0.5 |
| 20 | 40.9 | −71.4 | 71.7 | 0.571 | 40.7 | −16.7 | 44.5 | 0.927 | −0.405 |
| 30 | 50.1 | −82.5 | 84.1 | 0.596 | 49.9 | −16.6 | 53.3 | 0.950 | −0.328 |
| 40 | 60.4 | −95.5 | 98.4 | 0.614 | 60.2 | −17.0 | 63.6 | 0.963 | −0.278 |
| 50 | 67.6 | −106 | 109 | 0.618 | 67.4 | −17.6 | 70.8 | 0.968 | −0.256 |

**Table 2.** Power components at the point of connection of Rockwell Automation Power Flex 70 three-phase diode rectifier supply-side VFD to the PDL, keeping motor load torque constant.

| $f_{OUT}$, Hz | Combined | | | | Fundamental | | | | |
|---|---|---|---|---|---|---|---|---|---|
| | $P_C$, W | $N_C$, var | $S_C$, VA | $PF_C$ | $P_{Fund}$, W | $Q_{VFund}$, var | $S_{VFund}$, VA | $PF_{VFund}$ | $tan(\varphi)$ |
| 10 | 175 | −438 | 472 | 0.3705 | 171 | −102 | 200 | 0,858 | −0.596 |
| 20 | 257 | −585 | 640 | 0.4021 | 253 | −103 | 273 | 0.925 | −0.407 |
| 30 | 342 | −730 | 807 | 0.424 | 336 | −104 | 352 | 0.954 | −0.311 |
| 40 | 433 | −857 | 961 | 0.450 | 427 | −100 | 439 | 0.973 | −0.235 |
| 50 | 530 | −997 | 1131 | 0.469 | 522 | −108 | 534 | 0.978 | −0.208 |

The forward-conducting and reverse-blocking angles of the VFD's supply-side diode were determined from the experimental voltage and current time diagrams presented in Figure 5. These data are consistent with the theoretical assumptions (Figure 2) and with the data from the mathematical calculations performed according to Equation (2). According to the experimental data, the forward-conducting angle was 72.6°, leading to the calculation of $\tan(\varphi) = 0.28$. A similar result can be obtained from the ratio of the reactive power of fundamental harmonics to the active power of fundamental harmonics measured experimentally using a power quality analyzer: $\tan(\varphi) = Q_{VFund}/P_{Fund} = 17.6/67.4 = 0.26$.

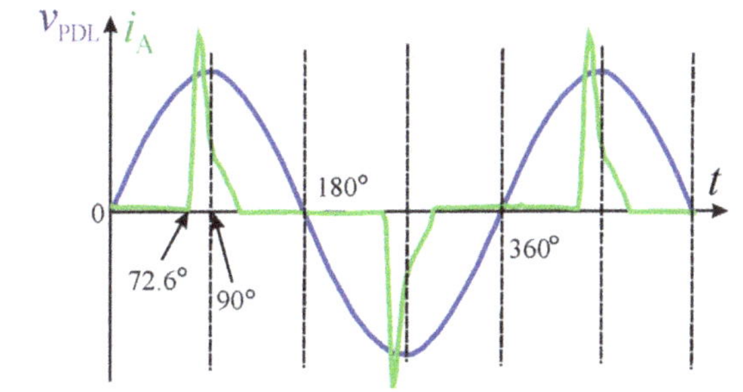


**Figure 5.** Current and voltage time diagrams at the point of connection of Siemens Micromaster 420 single-phase diode rectifier supply-side VFD to the PDL, with a motor supply voltage frequency of 50 Hz.

The forward-conducting and reverse-blocking angles of the VFD's supply-side diode were determined from the experimental voltage and current time diagrams presented in Figure 6. These data confirm the theoretical assumptions presented in Figure 3. These data were also compared with the data of the mathematical calculations performed according to Equation (12). A similar result can be obtained from the ratio of the reactive power of fundamental harmonics to the active power of fundamental harmonics measured experimentally using a power quality analyzer: tan(φ) = $Q_{VFund}/P_{Fund}$ = 108/522 = 0.20. Thus, the ratio is higher than the theoretical one.

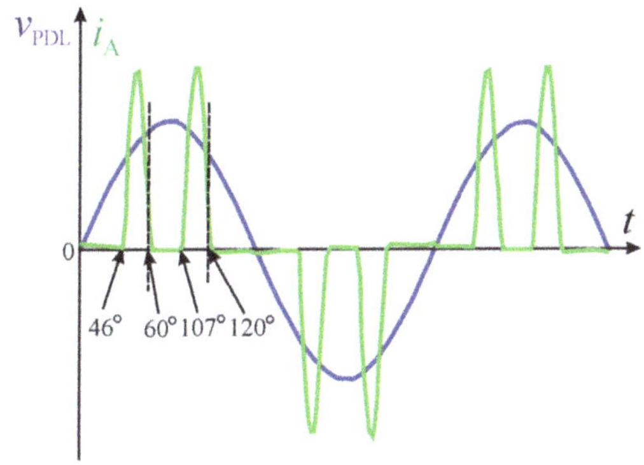


**Figure 6.** Current and voltage time diagrams at the point of connection of Rockwell Automation PowerFlex 70 three-phase diode rectifier supply-side VFD to the PDL, with a motor supply voltage frequency of 50 Hz.

The reactive power produced by AC–DC–AC VFDs cannot be compensated by using a conventional capacitor-based local compensation device. In this case, a universal compensation device is required, which has the ability to compensate for both types of reactive power: leading and lagging.

Inductive reactors designed to reduce harmonic distortion can perform the function of compensating the reactive power produced by VFDs. However, this method of compensation has significant drawbacks. The first disadvantage is the cost of compensation equipment due to the high cost of ferrous metals. Another drawback is the high energy loss. A third disadvantage is the high weight and dimensions of the compensation equipment. These drawbacks are evident in Table 3, which presents the data for mass-produced reactive power compensation reactors.

The use of an induction motor powered by an MC as the universal compensation device was researched. Depending on the structure of the converter and the specific controls, it can be said that AC–AC matrix VFDs may be suitable for both leading and lagging reactive power compensation.

**Table 3.** Inductive reactors of Schneider electric designed for 400 V, 50 Hz PDL.

| Relative Impedance, % | Capacitor Power, kVar | Inductance, mH | IMP, A | Reactor Power, kVar | Max Losses, kW | Power Losses in Percent of Reactor Power, % | Weight, kg | Reference Number |
|---|---|---|---|---|---|---|---|---|
| 4% | 6.5 | 11.439 | 10 | 1.07 | 0.1 | 9.3 | 10 | LVR14065A40T |
| | 12.5 | 6.489 | 20 | 2.44 | 0.15 | 6.14 | 15 | LVR14125A40T |
| | 25 | 3.195 | 40 | 4.80 | 0.2 | 4.16 | 22 | LVR14250A40T |
| | 50 | 1.598 | 80 | 9.58 | 0.4 | 4.17 | 33 | LVR14500A40T |
| | 100 | 0.799 | 160 | 19.09 | 0.6 | 3.14 | 55 | LVR14X00A40T |

## 3. Special Features of the AC–AC Matrix VFD

The researched compensation device consisted of an MC, whose input was connected to the PDL and powered the induction motor (Figure 7). The MC used the forced switching of nine bidirectional switches. These nine switches generated the supply voltage to the motor stator windings of the required frequency. Unlike conventional AC–DC–AC VFDs, the AC–AC matrix VFD contains no capacitors. These nine bidirectional switches created combinations of 27 active, passive, and zero vectors, as shown in Table 4.

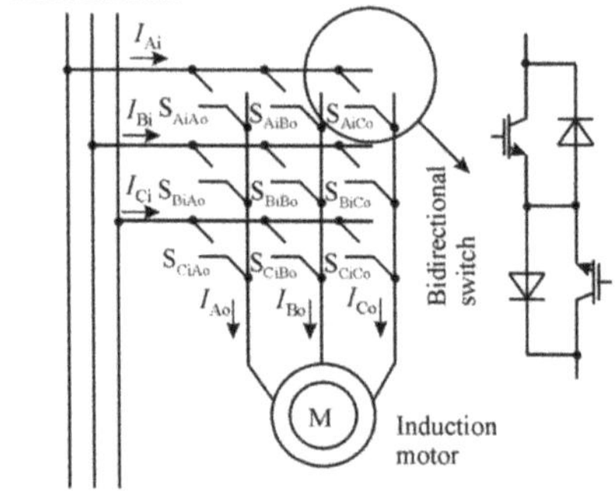


**Figure 7.** Functional diagram of MC-based induction motor VFD.

**Table 4.** Active combinations created by nine bidirectional switches.

| Vector. | Ao | Bo | Co | $S_{AiAo}$ | $S_{BiAo}$ | $S_{CiAo}$ | $S_{AiBo}$ | $S_{BiBo}$ | $S_{CiBo}$ | $S_{AiCo}$ | $S_{BiCo}$ | $S_{CiCo}$ |
|---|---|---|---|---|---|---|---|---|---|---|---|---|
| −3 | Ai | Ci | Ci | 1 | 0 | 0 | 0 | 1 | 0 | 0 | 0 | 1 |
| +2 | Bi | Ci | Ci | 1 | 0 | 0 | 0 | 0 | 1 | 0 | 0 | 1 |
| −1 | Bi | Ai | Ai | 0 | 1 | 0 | 1 | 0 | 0 | 1 | 0 | 0 |
| +3 | Ci | Ai | Ai | 0 | 0 | 1 | 1 | 0 | 0 | 1 | 0 | 0 |
| −2 | Ci | Bi | Bi | 0 | 0 | 1 | 0 | 1 | 0 | 0 | 1 | 0 |
| +1 | Ai | Bi | Bi | 1 | 0 | 0 | 0 | 1 | 0 | 0 | 1 | 0 |
| −6 | Ci | Ai | Ci | 0 | 0 | 1 | 1 | 0 | 0 | 0 | 0 | 1 |
| +5 | Ci | Bi | Ci | 0 | 0 | 1 | 0 | 1 | 0 | 0 | 0 | 1 |
| −4 | Ai | Bi | Ai | 1 | 0 | 0 | 0 | 1 | 0 | 1 | 0 | 0 |
| +6 | Ai | Ci | Ai | 1 | 0 | 0 | 0 | 0 | 1 | 1 | 0 | 0 |
| −5 | Bi | Ci | Bi | 0 | 1 | 0 | 0 | 0 | 1 | 0 | 1 | 0 |
| +4 | Bi | Ai | Bi | 0 | 1 | 0 | 1 | 0 | 0 | 0 | 1 | 0 |
| −9 | Ci | Ci | Ai | 0 | 0 | 1 | 0 | 0 | 1 | 1 | 0 | 0 |
| +8 | Ci | Ci | Bi | 0 | 0 | 1 | 0 | 0 | 1 | 0 | 1 | 0 |
| −7 | Ai | Ai | Bi | 1 | 0 | 0 | 1 | 0 | 0 | 0 | 1 | 0 |
| +9 | Ai | Ai | Ci | 1 | 0 | 0 | 1 | 0 | 0 | 0 | 0 | 1 |
| −8 | Bi | Bi | Ci | 0 | 1 | 0 | 0 | 1 | 0 | 0 | 0 | 1 |
| +7 | Bi | Bi | Ai | 0 | 1 | 0 | 0 | 1 | 0 | 1 | 0 | 0 |

Using the ISVM control method, the required stator winding voltage and PDL current were generated from the active and zero vectors. Figure 8 shows the active vectors produced from the previously presented combinations.

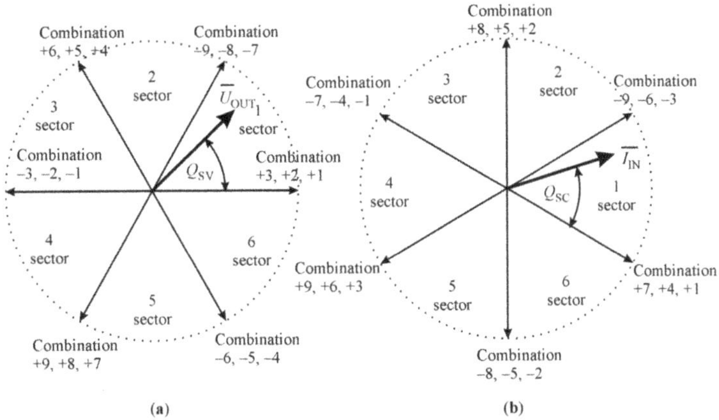


**Figure 8.** Graphics of the active vectors presented in Table 4: (**a**) graphic of the motor stator windings voltage vectors; (**b**) graphic of the PDL current vectors.

The motor stator winding voltage and PDL current are formed from two certain sector vectors. The frequency of MC motor supply voltage is determined by the angle $Q_{SC}$, and the current displacement at the connection to the PDL point is determined by the angle $Q_{SV}$ [7].

$$Q_{SC} = (\omega_i t - \varphi_{in}) + 30^\circ, \tag{13}$$

$$-30^\circ \leq \omega_i t - \varphi_{in} \leq +30^\circ, \tag{14}$$

where $t$ is the time for synchronization with the PDL frequency, and $\varphi_{in}$ is the current displacement angle.

## 4. Application of the AC–AC Matrix VFD as the Reactive Power Compensation Device

To improve the power factor in current 0.4 kV PDLs, it is proposed to replace some of the conventional AC–DC–AC VFDs with the novel AC–AC matrix VFD, as presented in Figure 9.

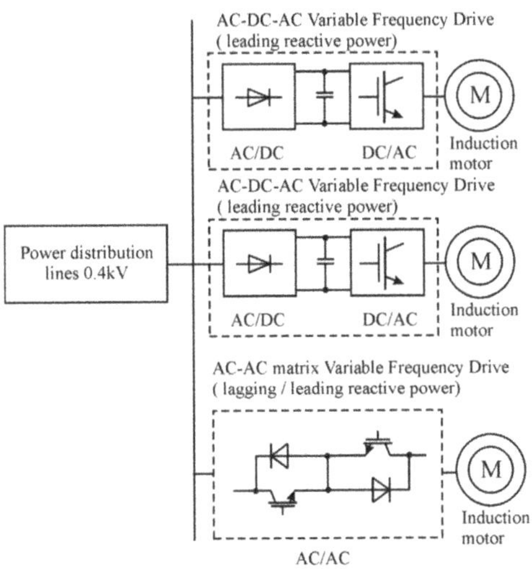


**Figure 9.** The structure of researched system.

As described above, the main difference between conventional AC–DC–AC VFDs and the AC–AC matrix VFD is related to electronic power circuits and control. The conventional AC–DC–AC VFD power circuits consist of a diode supply-side rectifier, an intermediate DC circuit, and an inverter. The AC–AC matrix VFD power circuits consist of nine bidirectional electronic switches based on IGBT. As shown by the theoretical calculations, initial experimental study, and initial simulation in Matlab/Simulink, the conventional AC–DC–AC VFD produces leading reactive power at the point of connection to the PDL. Moreover, theoretical calculations and initial simulations in Matlab/Simulink showed that the AC–AC matrix VFD has the ability to directly control the reactive power produced at the point of connection to the PDL. When these VFDs are connected to the same PDL, the matrix VFD must compensate for the reactive power produced by the conventional AC–DC–AC VFD. It is, therefore, proposed to replace the electronic equipment of one of the conventional VFDs installed in a commercial or industrial building. Instead of three stages (diode supply-side rectifier, intermediate DC circuit, and inverter), nine bidirectional electronic switches are installed, leaving the VFD induction motor the same. In this way, a PDL power factor close to unity can be achieved. By using the AC–AC matrix VFD, the transfer of active power to the motor and the reactive power compensation are performed simultaneously. A Matlab/Simulink simulation study was performed to determine the reactive power compensation range of the system shown in Figure 9.

The input current and motor supply voltage of the MC can be described by the following equations [13]:

$$I_{IN} = \frac{\sqrt{3}}{2} I_{OUT}\, m cos\varphi_{OUT}, \quad (15)$$

$$V_{OUT} = \frac{\sqrt{3}}{2} V_{IN}\, m cos\varphi_{IN}, \quad (16)$$

where $I_{IN}$ is the MC input current, $V_{OUT}$ is the MC motor supply voltage, $m$ is the modulation index, $\varphi_{IN}$ is the input displacement angle, and $\varphi_{OUT}$ is the motor supply displacement angle.

The AC–AC matrix VFD electronic switches were simulated using standard lossless Matlab/Simulink blocks. On the basis of Equations (15) and (16), the power equation of the AC–AC matrix VFD was derived as follows:

$$P_{IN} = \frac{3}{2} V_{PDL} I_{IN} cos\varphi_{IN} = P_{OUT} = \frac{3}{2} V_{OUT} I_{OUT} cos\varphi_{OUT}, \quad (17)$$

where $V_{PDL}$ is the voltage of the PDL, and $f_{PDL}$ is the frequency of the PDL.

From Equation (15), it can be seen that an increase in the induction motor winding current displacement angle $\varphi_{OUT}$ causes a decrease in PDL branch current $I_{IN}$. From Equation (16), it can be seen that an increase in the PDL current displacement angle $\varphi_{IN}$ causes a decrease in induction motor winding voltage $V_{OUT}$.

According to Equation (17), it can be seen that the active power of the AC–AC matrix VFD must be a constant value, thereby ensuring the operation of the powered induction motor. To ensure constant $P_{OUT}$ in the case of variable cos $\varphi_{IN}$, the motor supply voltage $V_{OUT}$ must be regulated within a certain range. To achieve this, constant $P_{OUT}$ can be maintained during the control of the modulation index $m$ (0–1).

For application of the AC–AC matrix VFD as the reactive power compensation device, the lower point of $V_{OUT}$ at induction motor stator windings was determined.

$$V_{OUT} = \frac{V_{IN}}{\sqrt{3}}. \quad (18)$$

The voltage was reduced to extend the modulation limits. To reduce voltage, the stator windings have to be connected in a delta arrangement (Δ).

The voltage ratio of the induction motor to ensure reactive power compensation is as follows:

$$\frac{V_{OUT}}{V_{PDL}} = \frac{1}{\sqrt{3}} = 0.58. \tag{19}$$

According to Equation (16),

$$\frac{\sqrt{3}}{2} mcos\varphi_{IN} = 0.58, \tag{20}$$

$$mcos\varphi_{IN} = 0.67. \tag{21}$$

The following equations determine the limits within which m and $\varphi_{IN}$ vary:

$$0.67 < m < 1, \tag{22}$$

$$0 < \varphi_{IN} < 48^{\circ}. \tag{23}$$

$$0 < \tan\varphi_{IN} < 1.1. \tag{24}$$

Equation (21) is valid only if the AC–AC matrix VFD motor supply voltage frequency corresponds to the PDL voltage frequency. In the event that the MC motor supply voltage frequency is not constant, one possible condition is

$$\frac{V_{OUTf}}{f_{OUT}} = const. \tag{25}$$

In the case of a VFD load with constant torque variation, the following equations apply:

$$V_{outf} = \frac{f_{OUT} \cdot V_{OUT}}{f_{PDL}}, \tag{26}$$

$$\frac{f_{OUT} \cdot V_{OUT}}{f_{PDL}} = \frac{\sqrt{3}}{2} V_{IN}\, mcos\varphi_{IN}, \tag{27}$$

$$mcos\varphi_{IN} = 0.67 \times \frac{f_{OUT}}{f_{PDL}}. \tag{28}$$

Equation (28) shows that the ability to adjust the angle increases if the motor operating frequency $f_{OUT}$ is lower than the frequency of the power distribution line. In the case of an AC–AC matrix VFD load with squared torque variation, the following equations apply:

$$V_{outf} = V_{OUT} \left(\frac{f_{OUT}}{f_{PDL}}\right)^2, \tag{29}$$

$$mcos\varphi_{IN} = 0.67 \times \left(\frac{f_{OUT}}{f_{PDL}}\right)^2. \tag{30}$$

As shown in Equation (30), it is possible to further adjust the angle $\varphi_{IN}$ under squared torque load. However, the reactive power is determined not only by the angle $\varphi_{IN}$, but also by the active power $P_{OUT}$ of the AC–AC matrix VFD. As the active power decreases, the reactive power also decreases, unless this is compensated for by an increase in allowable compensation angle $\varphi_{IN}$. Assuming that the AC–AC matrix VFD motor is always loaded at the maximum allowable torque, when the AC–AC matrix VFD operating frequency is varied, the motor current remains at the maximum permissible $I_{OUT}$.

$$P_{OUT} \sim \frac{f_{OUT}}{f_{PDL}} V_{OUT} I_{OUT}. \tag{31}$$

Keeping the active power at a nominal speed of 100%, the reactive power of the AC–AC matrix VFD at variable speed was calculated. On the basis of Equations (28) and (30), the compensation angles $\varphi_{IN}$ were calculated. The tan$\varphi_{IN}$ and the reactive power $Q$ required for compensation were calculated on the basis of the compensation angles.

The calculation results presented in Figure 10 show that, as the frequency of the AC–AC matrix VFD decreases, the reactive power increases. Thus, a decrease in the operating frequency of the VFD increases the possibility of reactive power compensation. Reactive power compensation is available over the entire operating frequency range of the AC–AC matrix VFD.

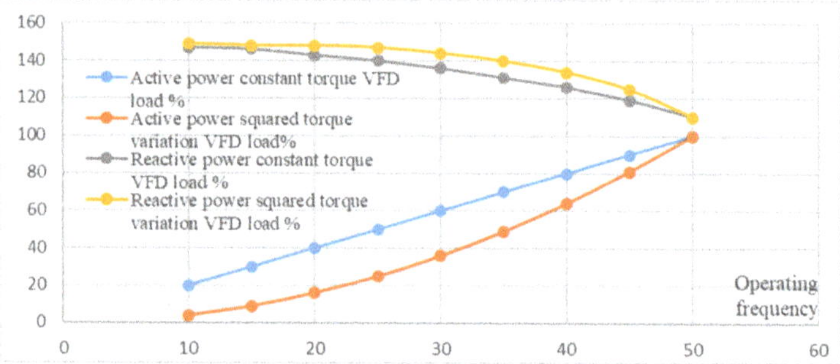


**Figure 10.** Relative active and reactive power at the point of AC–AC matrix VFD connection to the PDL as a function of the VFD operating frequency in the cases of VFD load with constant torque and squared torque variation.

### 4.1. Structure of the Matlab/Simulink Model for Simulating the AC–AC Matrix VFD

The Matlab/Simulink models of the reactive power compensation systems consisted of three pieces (Figure 11).

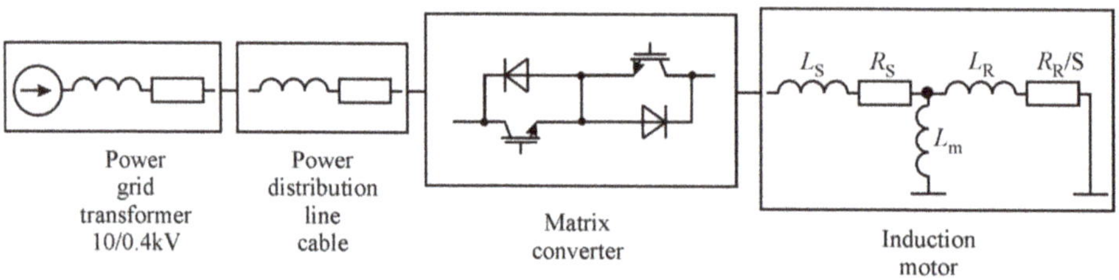


**Figure 11.** Structure of the AC–AC matrix VFD Matlab/Simulink model.

The first part consisted of PDL simulation blocks that simulated 10/0.4 kV transformer and PDL cables using the following standard Matlab/Simulink blocks: ideal sinusoidal AC voltage source and three-phase series RLC branch.

The second part consisted of Matlab/Simulink blocks for simulating MC power circuits (Figure 12) using standard blocks simulating an IGBT device in parallel with a series RC snubber. In the second part, space vector modulation control modules were also attached [14]. Subsystems running Park and Clarke transformations were used for this purpose. The sector separation subsystem was implemented using standard Matlab/Simulink logic and comparison blocks.

The third part consisted of Matlab/Simulink blocks for simulating the MC-powered induction motor according to an equivalent diagram of the induction motor [15] using standard blocks simulating a three-phase series RL branch.

For recording the active and reactive power data, the PLL-driven, positive-sequence block was used to record the positive-sequence active power P (W) and reactive power Q (var) of a periodic set of voltages and currents.

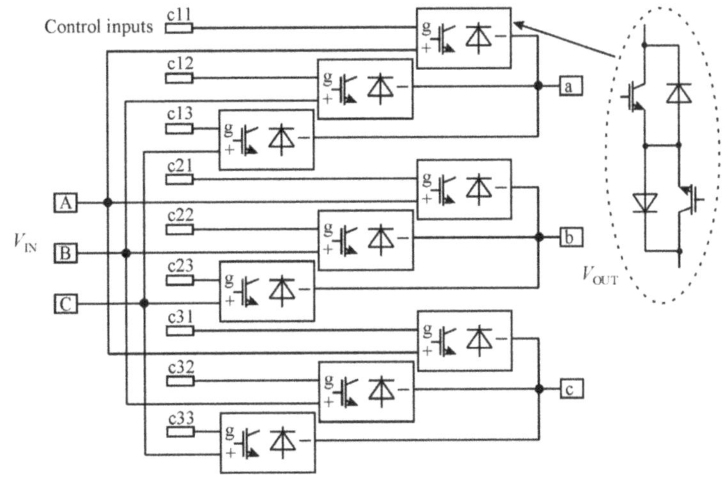


**Figure 12.** Structural diagram of the nine ideal bidirectional switches in Matlab/Simulink model.

### 4.2. Structure of the Matlab/Simulink Model for Simulating the Conventional AC–DC–AC VFD

The Matlab/Simulink model of the conventional AC-DC-AC VFD consisted of four parts. The first part of this model used a standard Matlab/Simulink block to simulate the full-bridge three-phase diode rectifier connected to the PDL using a universal bridge mask Matlab/Simulink block, in which the "power electronic device—diode" settings were applied. The second part implemented the capacitor bank of the DC circuit using a parallel RLC branch Matlab/Simulink block, in which the "branch type—C" settings were applied. The third part consisted of a Matlab/Simulink block implementing an IGBT full-bridge three-phase inverter using a universal bridge mask Matlab/Simulink block, in which the "power electronic device—IGBT/diode" settings were applied. The "PWM generator 2 level" block was also used to control the inverter. The fourth part consisted of Matlab/Simulink blocks for simulating the AC–DC–AC powered induction motor according to an equivalent diagram of the induction motor using standard blocks simulating a three-phase series RL branch.

### 4.3. Reactive Power Compensation Limits of the AC–AC Matrix VFD

The reactive power produced by the MC-based VFD depends directly on the current displacement angle at the point of connection to the PDL. This compensation range was determined by way of Matlab/Simulink simulations. The values of reactive power were determined by keeping the real power transferred to the induction motor constant. As a result, changes in real power caused by current displacement angle changes (rectifier stage) were compensated for by voltage modulation index correction (inverter stage).

These ranges of reactive power compensation are quite wide because an extension of these limits was applied. While the VFD's MC was connected to a 400 V line-to-line PDL, the induction motor windings were connected in a "delta" arrangement and powered by 230 V line-to-line voltage. This enabled changing the ISVM modulation index within a wider range. Consequently, at a displacement angle of zero, the modulation index was significantly reduced. However, when the displacement angle increased, there was a much wider voltage correction range due to the delta connection.

The simulation results presented in Figure 13 clearly show that, using the MC-based VFD, the reactive power can be compensated for in both directions (leading and lagging), depending on the current displacement angle.

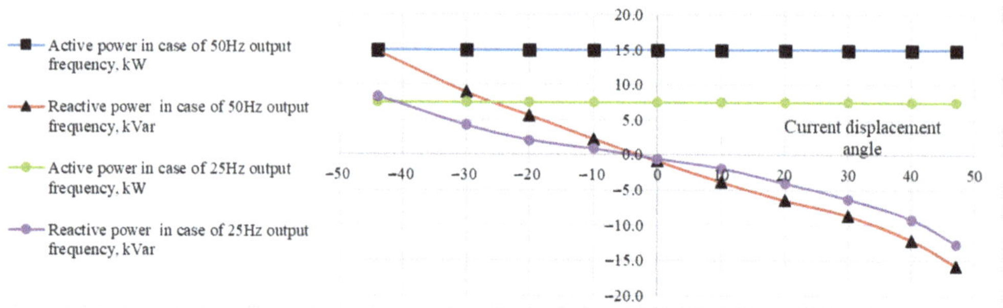


**Figure 13.** The reactive power compensation ranges of the 15 kW MC-based VFD connected to 0.4 kV PDL.

The simulation results presented in Figure 13 show that the possible displacement angles ranged from −44° to +47°. MC-based VFD control in this angle range allowed the real power transferred to the motor to be kept constant. For the 15 kW induction motor, the reactive power produced at these displacement angles varied from the leading 14.7 kVar to the lagging 15.7 kVar. Increasing the displacement angle above this specified range led to a decrease in the voltage at the induction motor stator windings and a decrease in the real power transferred to the induction motor.

As mentioned previously, the voltage drop due to the displacement angle change has to be recovered by increasing the value of the modulation index. Upon reaching the limits determined by the displacement angle specified range, the modulation index reaches its maximum value (shown in Table 5) and cannot recover a voltage drop caused by a further increase in the current displacement angle.

**Table 5.** Reactive power at the point of the VFD connection to the PDL.

| Number of Novel AC–AC Matrix VFDs | Total Power of Novel AC–AC Matrix VFDs, kW | Number of Conventional AC–DC–AC VFDs | Total Power of Conventional AC–DC–AC VFDs, kW | Reactive Power Produced by the Number Conventional AC–DC–AC VFDs, kVar | Reactive Power at the Point of VFD Connection to PDL, kVar | Displacement Angle in the AC–AC Matrix VFD, ° | Modulation Index in the AC–AC Matrix VFD |
|---|---|---|---|---|---|---|---|
| 1 | 15 | 1 | 15 | −3.37 | 0 | −13 | 0.72 |
| 1 | 15 | 2 | 30 | −6.21 | 0 | −25 | 0.79 |
| 1 | 15 | 3 | 45 | −8.56 | 0 | −30 | 0.81 |
| 1 | 15 | 4 | 60 | −10.55 | 0 | −34 | 0.85 |
| 1 | 15 | 5 | 75 | −12.11 | 0 | −38 | 0.92 |
| 1 | 15 | 6 | 90 | −13.36 | −0 | −40 | 0.96 |
| 1 | 15 | 7 | 105 | −14.40 | −0.34 | −42 | 1.00 max value |

These simulation results were confirmed by mathematical calculation according to Equation (21). The influence of the indirect space vector modulation control method was mathematically evaluated. The modulation index calculated to compensate for the voltage drop across the delta-connected induction motor windings was calculated using Equation (21) as presented in Figure 14.

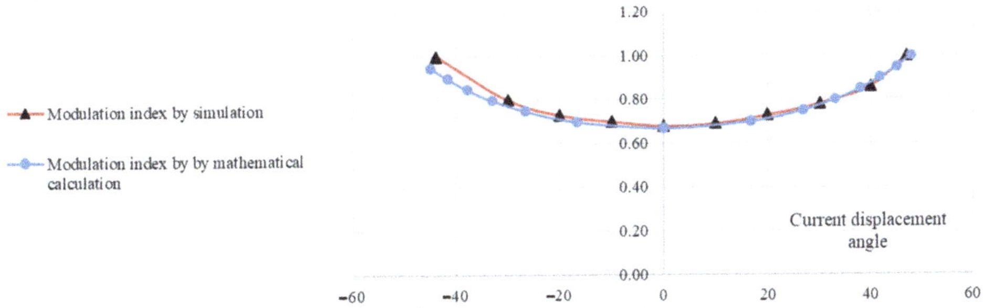


**Figure 14.** ISVM modulation index for voltage drop recovery.

*4.4. Compensation of Reactive Power Produced by Conventional AC–DC–AC Variable Frequency Drives*

In order to evaluate the ability of the AC–AC matrix VFD to compensate the reactive power produced by conventional AC–DC–AC VFDs and keep the power factor close to unity, a Matlab/Simulink model was used. As shown in Figure 15, the model included a 0.4 kV PDL branch to which one AC–AC matrix VFD and a number of conventional AC–DC–AC VFDs were connected. The 15 kW induction motors were powered by the MC converter and by conventional AC–DC–AC frequency converters.

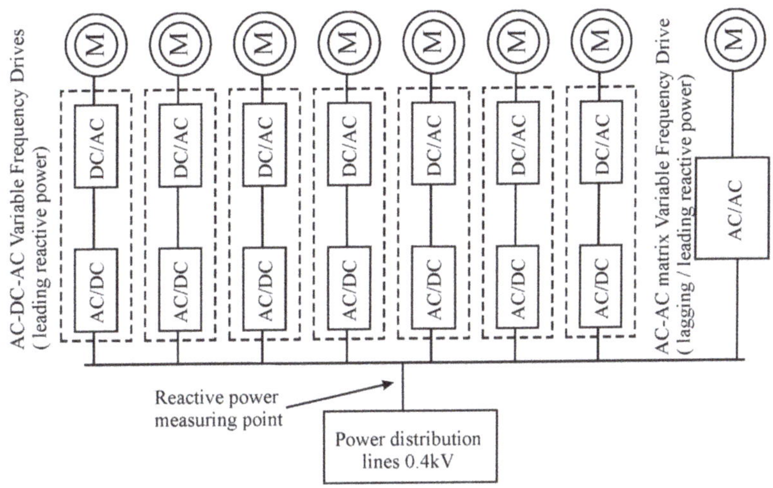


**Figure 15.** The researched system for conventional AC–DC–AC VFD reactive power compensation.

In the course of the study, the number of conventional AC–DC–AC VFDs was varied from one to seven. To compensate the leading reactive power in the PDL produced by the conventional AC–DC–AC VFDs, the current displacement angle of the AC–AC matrix VFD was regulated within the permissible limits.

As mentioned above, an increase in the current displacement angle caused a decrease in the voltage of the MC-powered induction motor stator winding and a decrease in transferred active power. This is directly related to ISVM control, which is applied as the main control strategy for AC–AC matrix VFDs. To recover this decrease, the modulation index was increased. This allowed the active power that was transferred to the induction motor by the MC constant to be maintained. However, the Matlab/Simulink simulation shows that the limits for the MC motor supply voltage recovery by the modulation index were not wide enough. As shown in Table 5, the current displacement angle and modulation index of the ISVM had to be significantly increased as the number of conventional AC–DC–AC

VFDs increased. When connecting more than six conventional AC–DC–AC VFDs, the modulation index had to be increased to the maximum allowed value. The unchanged active power transferred to the induction motor was maintained. However, under such conditions, the AC–AC matrix VFD could no longer fully compensate for the reactive power and maintain the unity power factor. Therefore, it can be argued that an ISVM-controlled AC–AC matrix VFD can only compensate for the reactive power produced by up to six conventional AC–DC–AC VFDs.

The relationship between the reactive power produced by the AC–DC–AC VFD at the point of VFD connection to the PDL and the motor load torque was investigated using the Matlab/Simulink simulation. This relationship was formed at different motor stator winding voltage frequencies and different rotor speeds. The simulation results are shown in Figure 16. Accordingly, it can be stated that VFD produces leading reactive power with varying motor load torque and rotor speed. The value of this reactive power depends directly on the motor load torque and rotor speed. This assumes that the proportions between the number of novel AC–AC matrix VFDs connected to the PDL and the number of conventional AC–DC–AC VFDs will remain similar to those described in Table 5 to maintain the power factor close to unity with varying motor load torque and rotor speed.

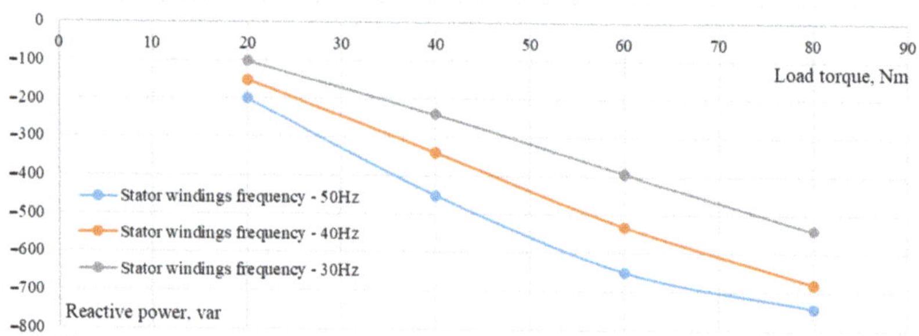


**Figure 16.** The reactive power at the point of single AC–DC–AC VFD connection to the PDL as a function of motor load torque.

## 5. Discussion

The idea to conduct this study arose from the observation of reactive power compensation problems in large building PDLs. It was observed that the interaction of a number of conventional AC–DC–AC VFDs with a conventional capacitor-based local compensation device leads to overcompensation in the 0.4 kV PDLs of the Kaunas Arena building. Capacitive reactive power compensators are installed in this building. This is due to the fact that conventional compensation devices are designed to compensate for the lagging reactive power produced by inductive loads, such as AC motors. This highlighted the demand for the compensation of capacitive reactive power that is not predicted by the designer.

Theoretical calculations show that a conventional AC–DC–AC VFD generates leading reactive power equal to about 30% of real power. Experimental studies confirmed this. This leading reactive power is produced by both single-phase and three-phase supply AC–DC–AC VFDs. Thus, this study is relevant and timely, as confirmed by the example provided above.

Energy companies have proposed installing large quantities of inductive reactors for power factor improvement. First of all, this is not ecologically acceptable due to the high consumption of nonferrous metals. Another shortcoming would be a significant financial cost for the company operating the Kaunas Arena building. The use of AC–AC matrix VFDs would avoid both of these disadvantages.

An AC–AC matrix VFD operating as the reactive power compensation device has another obvious advantage over capacitor-based compensation devices, namely there is no need to use large capacitors. This is an advantage in terms of volume and reliability.

Another important application of the AC–AC matrix VFD is the voltage regulation of the distribution grid. Due to the rapidly evolving solar energy, the distribution grid faces the problem of voltage fluctuations, which can be solved by using the AC–AC matrix VFD as a static reactive power compensator. A particularly important feature of this compensator is that it can generate both lagging and leading power. Future studies will investigate the application possibilities of AC–AC matrix VFD bidirectional reactive power in detail.

## 6. Conclusions

Due to its specific properties, a conventional AC–DC–AC VFD generates leading reactive power. From a technical point of view, this reactive power can be compensated for by induction reactors. However, the high cost, weight, and dimensions of these reactors make this method of compensation unusable. As a result, it was proposed to replace a certain number of conventional AC–DC–AC VFDs with AC–AC matrix VFDs. Studies have shown that, in this way, it is possible to bring the power factor closer to unity.

The application of indirect space vector modulation provides an opportunity to control the current displacement angle of the MC independently from the voltage of the PDL. As a result, the AC–AC matrix VFD can be applied for reactive power compensation. It was determined that the AC–AC matrix VFD can provide the appropriate amount of both leading and lagging reactive power. This feature assumes the application of this AC–AC matrix VFD as the reactive power compensation device.

An increase in the current displacement angle causes a decrease in the voltage of the stator windings and active power transferred to the induction motor. To recover this voltage decrease, the modulation index of the ISVM has to be increased. The determined applicable displacement angle ranges from $-44°$ to $+47°$. To maintain constant power transfer to the motor, the range of the current displacement must not exceed the determined range.

The AC–AC matrix VFD has the ability to compensate for the reactive power produced by conventional AC–DC–AC VFDs that capacitor-based local reactive power compensation devices cannot compensate for. A single AC–AC matrix VFD has the ability to compensate for the reactive power produced by six conventional AC–DC–AC VFDs with the same nominal power.

The compensation power value of the innovative AC–AC matrix VFD is independent from the drive load and frequency. This AC–AC matrix VFD can perform reactive power compensation, even when the VFD operation mode changes.

**Author Contributions:** Conceptualization, G.P.; methodology, G.S.; software, G.P.; validation, G.S. and G.P.; formal analysis, G.S.; investigation, G.S. and G.P.; resources, G.P., G.S. and A.J.; data curation, G.S.; writing—original draft preparation, G.P. and G.S.; writing—review and editing, A.J. and A.G.; visualization, G.P.; project administration, A.J.; funding acquisition, A.J. All authors have read and agreed to the published version of the manuscript.

**Funding:** This research was supported by the European Economic Area (EEA) and Norway Financial Mechanism 2014–2021 under Grant EMP474.

**Conflicts of Interest:** The authors declare no conflict of interest.

## References

1. Wang, X.; Dai, K.; Chen, X.; Zhang, X.; Wu, Q.; Dai, Z. Reactive Power Compensation and Imbalance Suppression by Star-Connected Buck-Type D-CAP. *Energies* **2019**, *12*, 1914. [CrossRef]
2. Vlahinić, S.; Franković, D.; Komen, V.; Antonić, A. Reactive Power Compensation with PV Inverters for System Loss Reduction. *Energies* **2019**, *12*, 4062. [CrossRef]
3. Blasco, P.A.; Montoya-Mira, R.; Diez, J.M.; Montoya, R.; Reig, M.J. Compensation of Reactive Power and Unbalanced Power in Three-Phase Three-Wire Systems Connected to an Infinite Power Network. *Appl. Sci.* **2020**, *10*, 113. [CrossRef]
4. Petrushyn, V.; Horoshko, V.; Plotkin, J.; Almuratova, N.; Toigozhinova, Z. Power Balance and Power Factors of Distorted Electrical Systems and Variable Speed Asynchronous Electric Drives. *Electronics* **2021**, *10*, 1676. [CrossRef]
5. Zhang, Y.; Tang, F.; Qin, F.; Li, Y.; Gao, X.; Du, N. Research on Dynamic Reactive Power Compensation Scheme for Inhibiting Subsequent Commutation Failure of MIDC. *Sustainability* **2021**, *13*, 7829. [CrossRef]

6. Holtsmark, N.; Molinas, M. Reactive power compensation capability of a matrix converter-based FACTS device. In Proceedings of the 2011 IEEE Trondheim PowerTech, Trondheim, Norway, 19–23 June 2011; pp. 1–6. [CrossRef]
7. Petrauskas, G.; Svinkunas, G. Application of Matrix VFD for Power Factor Improvement in LED Lighting Sources Loaded Power Distribution Lines. *Energies* **2021**, *14*, 3546. [CrossRef]
8. Szczesniak, P.; Kaniewski, J. A Voltage Regulator/Conditioner Based on a Hybrid Transformer with Matrix Converter. In Proceedings of the IECON 2014—40th Annual Conference of the IEEE Industrial Electronics Society, Dallas, TX, USA, 29 October–1 November 2014; IEEE: New York, NY, USA, 2014; pp. 3292–3297.
9. Schafmeister, F.; Kolar, J.W. Novel Hybrid Modulation Schemes Significantly Extending the Reactive Power Control Range of All Matrix Converter Topologies with Low Computational Effort. *IEEE Trans. Ind. Electron.* **2012**, *59*, 194–210. [CrossRef]
10. Roostaee, M.; Eskandari, B.; Azizi, M.R. Predictive current control with modification of instantaneous reactive power minimization for direct matrix converter. In Proceedings of the 2018 9th Annual Power Electronics, Drives Systems and Technologies Conference (PEDSTC), Tehran, Iran, 13–15 February 2018; pp. 199–205. [CrossRef]
11. Petrauskas, G.; Svınkunas, G. Modified control of the matrix converter-based drive for voltage sag impact reduction. *Turk. J. Electr. Eng. Comput. Sci.* **2017**, *25*, 2243–2260. [CrossRef]
12. Vijayagopal, M.; Empringham, L.; de Lillo, L.; Tarisciotti, L.; Zanchetta, P.; Wheeler, P. Current control and reactive power minimization of a direct matrix converter induction motor drive with Modulated Model Predictive Control. In Proceedings of the 2015 IEEE International Symposium on Predictive Control of Electrical Drives and Power Electronics (PRECEDE), Valparaiso, Chile, 5–6 October 2015; pp. 103–108. [CrossRef]
13. Svinkunas, G.; Petrauskas, G. Switching Ripple Harmonics Attenuation in DFIG and Matrix Converter-Based WECS. *Electronics* **2021**, *10*, 2589. [CrossRef]
14. Petrauskas, G.; Svinkunas, G. Innovative Filter Topology for Power Grid Protection from Switching Ripple Harmonics Produced by Matrix Converter. *Iran. J. Sci. Technol. Trans. Electr. Eng.* **2019**, *43*, 495–505. [CrossRef]
15. Petrauskas, G.; Svınkunas, G.; Juraskıene, R. Simplification of harmonics filter for a matrix converter-based drive. *Turk. J. Electr. Eng. Comput. Sci.* **2018**, *26*, 1011–1026. [CrossRef]

*electronics*

MDPI

*Article*

# A Fault Detection Method of IGBT Bond Wire Fatigue Based on the Reduction of Measured Heatsink Thermal Resistance

Dan Luo [1], Minyou Chen [1], Wei Lai [1,*], Hongjian Xia [1], Zhenyu Deng [2], Zhi Wang [1] and Kai Yu [3]

1 State Key Lab of Power Transmission Equipment & System Security and New Technology, School of Electrical Engineering, Chongqing University, Chongqing 400044, China; luodancqu@cqu.edu.cn (D.L.); minyouchen@cqu.edu.cn (M.C.); hongjian_xia@cqu.edu.cn (H.X.); 20191102066t@cqu.edu.cn (Z.W.)
2 NXP (Chongqing) Semiconductors Co., Ltd., Chongqing 400044, China; zhenyu.deng@nxp.com
3 CRRC Yong Ji Electric Co., Ltd., Yongji 044502, China; 5829700167@126.com
* Correspondence: laiweicqu@cqu.edu.cn

**Abstract:** Bond wire lift-off is one of the major failure mechanisms in the insulated gate bipolar transistor (IGBT) modules. Detecting the fault of bond wires is important to avoid the open-circuit fault of IGBT to ensure the reliable operation of power converters. In this paper, we propose a novel bond wire fatigue detection method for IGBT, which could be used in normal working conditions. Firstly, we investigated the dependence of bond wire fatigue on heatsink thermal resistance. An aging rate K was proposed to compare the measured thermal resistance with the initial value, which could indicate the bond wire fatigue. Then, this proposed method was verified by simulation and experimental results under different current levels. Finally, a power cycling test was used to show the aging process of the IGBT module, which shows the feasibility of proposed method.

**Keywords:** IGBT; bond wire fatigue; Heatsink thermal resistance; fault detection; heat flow

**Citation:** Luo, D.; Chen, M.; Lai, W.; Xia, H.; Deng, Z.; Wang, Z.; Yu, K. A Fault Detection Method of IGBT Bond Wire Fatigue Based on the Reduction of Measured Heatsink Thermal Resistance. *Electronics* **2022**, *11*, 1021. https://doi.org/10.3390/electronics11071021

Academic Editor: Hamid Reza Karimi

Received: 9 March 2022
Accepted: 23 March 2022
Published: 24 March 2022



## 1. Introduction

Power converters are widely used with the rapid development of renewable energy [1]. As the key component of power converters, insulated gate bipolar transistors (IGBTs) are the most adopted power devices [2] according to the demand for operational reliability, energy efficiency, and cost competitiveness.

Figure 1 shows the structure of wire-bonded IGBT, which consists of several layers of different materials. IGBT die is soldered on the direct-bonded copper (DBC) plate, and the DBC plate is soldered on the baseplate. The silicon IGBT die is connected to the substrate by several paralleled aluminum bond wires. As a result of physical limits, a coefficient of thermal expansion (CTE) mismatch occurs between silicon die and aluminum bond wires, causing swelling stress and shrinkage stress during power cycling and finally leading to the bond wire lift-off [3]. Bond wire lift-off is one of the domain failure mechanisms of the IGBT module [4], and it usually appears in the last 10% of the remaining life stage of the module. High-junction temperature fluctuations [5] and high-frequency operation [6] will also accelerate the aging of bond wires. The failure of bond wires in IGBT will easily lead to open-circuit faults and cause damage to the whole system [7]. Consequently, the reliability of bond wires is important for the IGBT module.

The fault detection method could obtain the state of the device, which is important to ensure the safe operation of power converters. Many bond wire failure detection methods have been proposed in recent studies. J. Lehmann [8] presented a modified DCB layout in an IGBT module to measure the fatigue of bond wires. This method needs to change the structure of the IGBT module to set the auxiliary circuit inside. H. Shiratsuchi [9] and H. Tomonaga [10] proposed to monitor IGBT current distribution by the magnetic field, which could be used to show the status of bond wires [11]. These sensors need to be integrated into the module to measure the magnetic field. Such methods are capable of detecting the bond wire fatigue; however, the invasive modifications or sensors would bring unpredictable reliability issues to the power module itself. To overcome such challenges, non-invasive detection methods are also proposed. The fatigue of bond wires changes the electrical gate charge circuit, which finally affects the switching process. Hence, these parameters have been used by several researchers to indicate the fault of bond wires, such as gate current [12], charging time [13], transconductance [14], and stray inductance [15]. Although those methods are non-invasive and only gate voltage needs to be measured, the sensitivity and accuracy of the measured results could be easily affected by the high sampling frequency and resolution of gate signal [16]. The on-state voltage of IGBT is independent of the junction temperature at the inflection point, which has a zero-temperature characteristic. Hence, U. Choi [17], A. Singh [18], and M. Du [19] used the conduct voltage at the inflected point with special gate voltage to measure bond wire lift-off, while P. Sun [20] measured the short-current at special gate voltage. Although these methods showed adequate sensitivity to detect the bond wire fault irrespective of the junction temperature, they could only be used at abnormal working conditions. In general, the weaknesses of those detection methods lead to an increase in costs and complexity in normal operations. Therefore, a more convenient detection method is still needed to reflect the health status of bond wires in IGBT modules in normal working conditions.

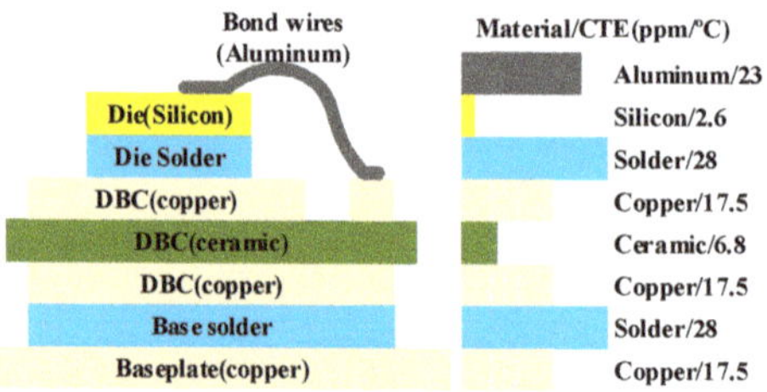


**Figure 1.** Structure of wire-bonded IGBT module.

To address this challenge, a novel method is proposed in this paper to detect the bond wire fatigue based on the heatsink thermal resistance decrement, which could be used in normal conditions. Two aspects are demonstrated as follows. (1) The effect of IGBT bond wire lift-off on heatsink thermal resistance measurement is presented. The measured results decrease due to the heat flow of bond wires. (2) A fault detection method based on the reduction in measured heatsink thermal resistance is proposed. In this method, the fault of bond wires is indicated after a decrease in measured results of heatsink thermal resistance, which is independent of the working conditions. By these means, the fault of bond wires could be conveniently detected to avoid open-circuit fault in normal conditions.

The rest of this paper is organized as follows. Our detection method is proposed in Section 2. A multi-physics field FEM model is established to analyze the effect of bond wire lift-off in Section 3. The experimental results verify this method in Section 4. A power cycling test is used to show the aging process in Section 5. Section 6 concludes this paper.

## 2. Failure Mechanism and Detection Method

### 2.1. Effect of Bond Wire Lift-Off on Heatsink Thermal Resistance Measurement

Power loss of die $P_D$ is far greater than bond wires $P_{WB}$ in a healthy module [17]; hence, IGBT die is the major heat source in the IGBT module $P_{loss}$, while the power loss of bond wires is nearly negligible (Equation (1)). The heat mainly concentrates under die and transfers through each layer to the baseplate, and dissipates to the ambient temperature by the heatsink, as shown in Figure 2a.

$$P_{loss} = P_D + P_{WB} \overset{P_D \gg P_{WB}}{\approx} P_D \tag{1}$$

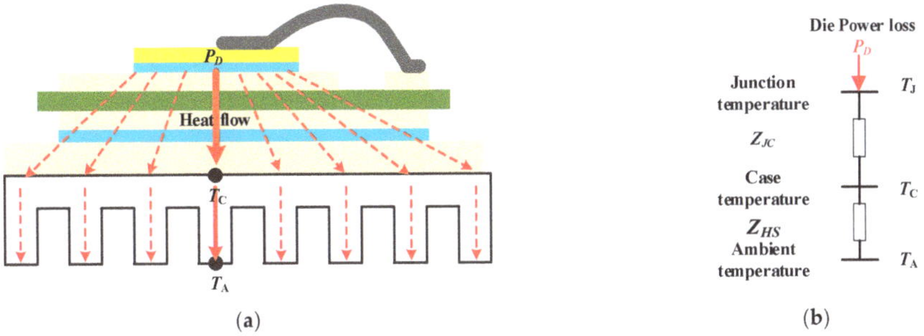


**Figure 2.** (**a**) Heat flow in healthy module. (**b**) Thermal network of healthy module.

The case temperature $T_C$ rises when the power loss $P_{loss}$ flows through the heatsink to ambient $T_A$, which is related to the equivalent thermal resistance of heatsink $Z_{E\_HS}$, as in Equation (2).

$$T_C = P_{loss} \cdot Z_{E\_HS} + T_A = P_D \cdot Z_{E\_HS} + T_A \quad (2)$$

The thermal network in the healthy module is shown in Figure 2b. The power loss of die transfers through the IGBT module $Z_{JC}$ and heatsink $Z_{HS}$ to the ambient. Hence, the measured thermal resistance of heatsink could be obtained based on (3), which equals the theoretical value of equivalent thermal resistance of heatsink $Z_{E\_HS}$ on the case point.

$$Z_{E\_HS} = \frac{T_C - T_A}{P_{loss}} = Z_{HS} \quad (3)$$

However, the total resistance of parallel bond wires increases after bond wire lift-off [21] which leads to a power loss increase in the remaining bond wires $P_{WB}$ [20], as in Equation (4), where $R_{WB}$ is the resistance of a single bond wire, $I_C$ is the conduction current, $N_{wire}$ is the number of total bond wires, and $N_{fail}$ is the number of broken wires.

$$P_{WB} = I_C^2 \cdot R_{WB} / (N_{wires} - N_{fail}) \quad (4)$$

Our previous study [22,23] proposes that the heat flow of bond wires is different from the die, which transfers to both the die side and the emitter side at the same time, as in Figure 3a. Hence, the power loss of bond wires could be divided into two parts based on the direction of heat flow, which transfers through the die side $P_{WB\_D}$ and emitter side $P_{WB\_E}$, as in Equation (5).

$$P_{WB} = P_{WB\_D} + P_{WB\_E} \quad (5)$$

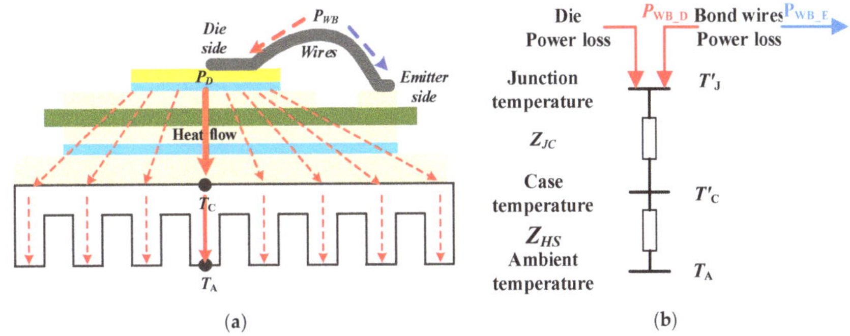


**Figure 3.** (**a**) Heat flow after bond wire lift-off. (**b**) Thermal network of IGBT module after wire lift-off.

Only $P_{WB\_D}$ transfers through die to the case point and dissipates to ambient temperature by heatsink with die power loss. This power loss $P'_{loss}$ on case point finally causes a temperature rise on the heatsink $T'_C$, as in Equations (6) and (7).

$$P'_{loss} = P_D + P_{WB\_D} \tag{6}$$

$$T'_C = P'_{loss} \cdot Z_{E\_HS} + T_A = (P_D + P_{WB\_D}) \cdot Z_{E\_HS} + T_A \tag{7}$$

The thermal network after bond wire lift-off is shown in Figure 3b. The thermal resistance of the module and equivalent thermal resistance of heatsink stay the same after bond wire lift-off; only part of the bond wire power loss transfers through the die to the heatsink.

Previous research [24,25] proposed that the heat flow in the IGBT module is only affected by the structure after solder fatigue. Although the power loss of bond wires increases, the heat flow stays the same after lift-off due to the same structure. Hence, the equivalent thermal resistance of heatsink on the case point could be calculated based on the same Equation as (3) as in Equation (8).

$$Z_{E\_HS} = \frac{(T'_C - T_A)}{P_D + P_{WB\_D}} \tag{8}$$

Although the power loss of die $P_D$ and part of the bond wire power loss $P_{WB\_D}$ transfer through the die to the heatsink, the power loss of the whole IGBT module $P_{IGBT}$ is measured as the calculated power loss, which is different from the theoretical power loss $P'_{loss}$ on the case point, as shown in Figure 4. Hence, using the power loss of the whole module will overestimate the power loss on the case point, which is higher than the theoretical power loss $P'_{loss}$, as in Equation (9).

$$P_{IGBT} = P_D + P_{WB} = P_D + P_{WB\_D} + P_{WB\_E} = P'_{loss} + P_{WB\_E} \tag{9}$$

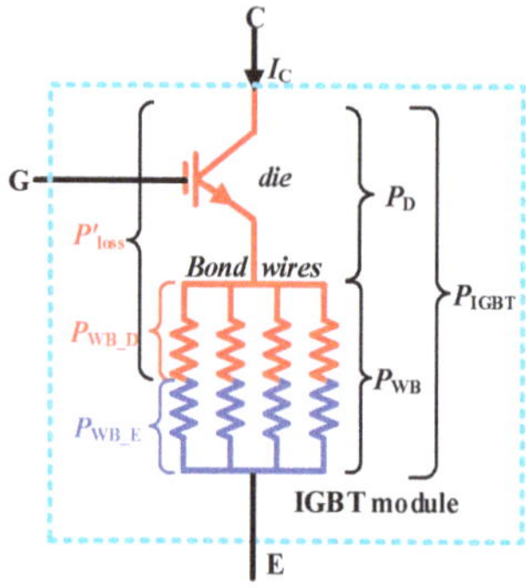


**Figure 4.** Power loss of IGBT module.

Using the overestimated power loss to calculate the thermal resistance of heatsink will underestimate the measured thermal resistance of heatsink $Z_{M\_HS}$ on the case point after bond wire lift-off, as in Equation (10).

$$Z_{M\_HS} = \frac{T'_C - T_A}{P_D + P_{WB}} = \frac{(P_D + P_{WB\_D}) \cdot Z_{E\_HS}}{P_D + P_{WB\_D} + P_{WB\_E}} < \frac{(P_D + P_{WB\_D}) \cdot Z_{E\_HS}}{P_D + P_{WB\_D}} = Z_{E\_HS} \tag{10}$$

### *2.2. Fault Detection Method of IGBT Bond Wire Fatigue Based on Heatsink Thermal Resistance*

Both solder fatigue and bond wire lift-off will lead to increases in on-state voltage $V_{CE}$, power loss $P_{loss}$, junction temperature $T_J$, and case temperature $T_C$. The thermal impedance of the heatsink increases slightly [25] or remains the same [26] after solder fatigue. However, we propose that, unlike the healthy and solder fatigue modules, the measured heatsink thermal resistance would decrease after bond wire lift-off, as shown in Table 1. Hence, using the decrement in heatsink thermal resistance can easily distinguish the bond wire lift-off from the healthy module and solder fatigue module. To determine the status of bond wires, a method based on the reduction in heatsink thermal resistance is presented.

**Table 1.** Different performance of healthy and fatigue modules.

| Parameters | Healthy | Solder Fatigue | Bond Wire Lift-Off |
|---|---|---|---|
| $V_{CE}$ | stay | Increase ↑ | Increase ↑ |
| $P_{loss}$ | stay | Increase ↑ | Increase ↑ |
| $T_C$ | stay | Increase ↑ | Increase ↑ |
| $T_J$ | stay | Increase ↑ | Increase ↑ |
| $Z_{M\_HS}$ | stay | Increase ↑ or unchanged- | Decrease ↓ |

The initial thermal resistance of heatsink $Z_{HS\_initial}$ is measured through the power loss of the IGBT module $P_{IGBT}$ and the temperature difference $\Delta T_{CA}$ between case and ambient temperature on the case point, as in Equation (11). The $Z_{HS\_initial}$ is set as the reference value of the healthy module.

$$Z_{HS\_initial} = \frac{\Delta T_{CA}}{P_{IGBT}} \quad (11)$$

The compared heatsink thermal resistance $Z_{HS\_K}$ is measured at the same current level with new power loss $P'_{IGBT}$ and temperature difference $\Delta T'_{CA}$, as in Equation (12).

$$Z_{HS\_K} = \frac{\Delta T'_{CA}}{P'_{IGBT}} \quad (12)$$

The aging ratio $K$ is defined to detect bond wire fatigue as in Equation (13), which is solely related to the difference between power loss of the whole module and theoretical power loss on the case point.

$$K = \frac{Z_{HS\_K} - Z_{HS\_inital}}{Z_{HS\_inital}} \cdot 100\% \quad (13)$$

The value of $K$ decreases after bond wire lift-off. The number of remaining bond wires will be updated after $K$ decreases. The aged IGBT should be replaced to avoid open-circuit failure when the critical failure is reached. The flowchart of the proposed method is shown in Figure 5.

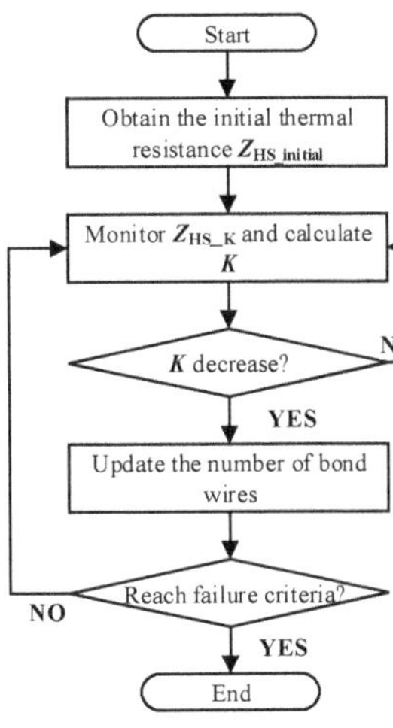


**Figure 5.** Flowchart of the fault detection method of bond wires.

The proposed method is capable of monitoring the bond wire lift-off by measuring the decrement in heatsink thermal resistance. This decrement occurs in the IGBT operation due to the different heat spreading paths of bond wires. The key advantage of this method is that the thermal resistance of the heatsink is measured independent of working conditions, and does not depend on the current, die temperature, or solder fatigue.

## 3. Simulation Validation

### *3.1. Electro-Thermal Modeling for IGBT Module*

A multi-physics field finite element method (FEM) model was established based on a 1200 V/50 silicon IGBT module (STARPOWER, Jiaxing, China, GD50HFL120C1). The unpacked power module free of gel is shown in Figure 6a. The geometry of the IGBT layout was supplied by the power module manufacture STARPOWER, shown in Figure 6b. The FEM model is tested with and without bond wires removed. The simulation module is used to show the mechanisms of heat transfer after bond wire lift-off.

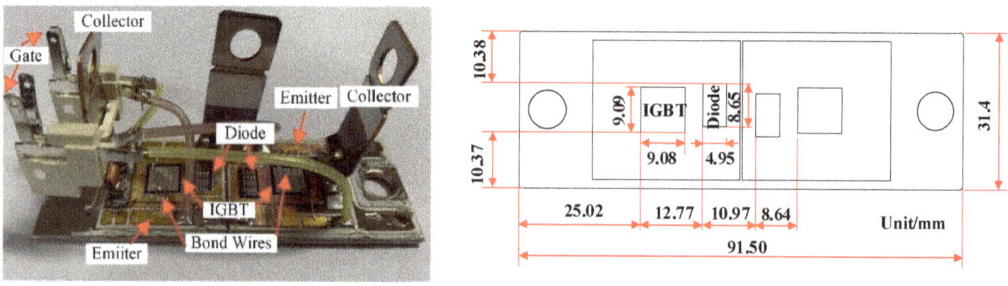


(**a**) (**b**)

**Figure 6.** (**a**) IGBT module GD50HFL120C1. (**b**) Dimensions of GD50HFL120C1.

The bond wires of diode and gate have been removed in the FEM model. The structure and boundaries of the FEM model are shown in Figure 7a. The boundary conditions are set as follows. The ambient temperature is set at 20 °C. All of the boundaries' conditions are set as electric insulation in the physical field of electric currents, except a DC source with a high potential on the collector side and ground potential on the emitter side. A convection coefficient of 3000 W/m$^2$·K is defined on the bottom surface of the baseplate to simulate a wind-cooling heatsink; the rest of the surfaces are set as 12.5 W/m$^2$·K for free convection air in the physical field of heat transfer in solids. The physical fields of heat transfer in solids and electric currents are defined as the source and destination in the multi-physics field of temperature coupling, respectively. All domains are used in electromagnetic heat source and boundary electromagnetic heat source, where the electric currents are set as the electromagnetic source and heat transfer in solids is set as the heat transfer source.

(**a**) (**b**) (**c**)

**Figure 7.** (**a**) FEM model; (**b**) meshed IGBT module; (**c**) mesh of baseplate.

The meshed model is shown in Figure 7b, which consists of 1,152,643 domain elements, 216,192 boundary elements, and 7236 edge elements. To study the effect of bond wire lift-off on the thermal resistance measurement, the bottom of the baseplate layer meshed with more free triangular distribution than other areas as in Figure 7c, which consist of 164,632 boundary elements and 1619 edge elements.

The material properties of the FEM model are shown in Table 2. The conduct characteristic of IGBT die $r_{die}$ is temperature-dependent in different current levels. Hence, the electrical conductivity of die $\sigma_{die}$ could be obtained based on the die temperature [27], where $l$ and $A$ are the height and conduct area of IGBT die as in Equations (14) and (15).

$$r_{die} = \frac{V_{CE\_25}}{I_C} + \frac{V_{CE\_150} - V_{CE\_25}}{I_C \cdot (150 - 25)}(T_J - 25) \quad (14)$$

$$\sigma_{die} = \frac{1}{r_{die}} \cdot \frac{l}{A} \quad (15)$$

**Table 2.** Material properties of the FEM model.

| Part | Material | Density kg/m$^3$ | Electrical Conductivity S/m | Thermal Conductivity W/(m·K) |
|---|---|---|---|---|
| Bond wires | Aluminum | 2700 | $2.8 \times 10^7$ | 237.2 |
| IGBT/Diode | Silicon | 2329 | $\sigma(T_J, I_C)$ | 124 |
| Solder | 96.5Sn3.5Ag | 7400 | $9.1 \times 10^6$ | 54 |
| Baseplate | Copper | 8960 | $6 \times 10^7$ | 380 |
| Insulation | Ceramic | 3780 | $1 \times 10^{-6}$ | 30 |

This IGBT die shows different temperature characteristics under each current condition, with a negative temperature coefficient (NTC) at 10 A, a zero temperature coefficient (ZTC) at 20 A, and a positive temperature coefficient (PTC) at 30 A and 40 A, as shown in Figure 8a. The decrement in the measured heatsink thermal resistance is determined by the fatigue of bond wires, which is not affected by the conducted current. Hence, the simulation will be tested under each current level. The case temperature is measured by the point below the center of the die on the baseplate, as in Figure 8b.

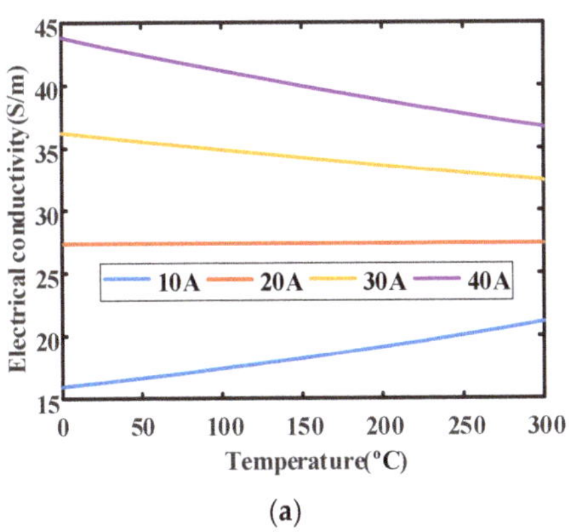


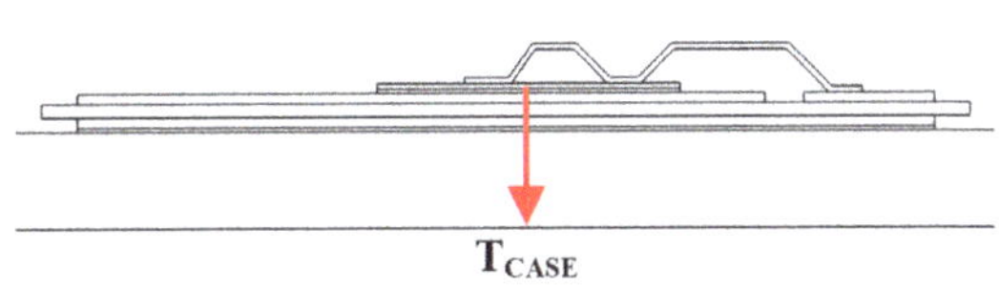


**Figure 8.** (**a**) Temperature characteristics of IGBT electrical conductivity under different current levels; (**b**) case temperature measure point.

### 3.2. FEM Results

Figure 9a shows the power loss after bond wires are removed under different current conditions, where the dashed columns are die power loss and solid columns are bond wire power loss. The power loss of the whole module increases due to the bond wire power loss after lift-off. The power loss of the die stays almost the same, while the power loss of the bond wires significantly increases after lift-off. The power loss of bond wires rises from 0.19 W to 0.79 W at 10 A, while the die power loss stays at 11.16 W. The higher current shows a higher power loss increase in wires, which rises from 3.1 W to 12.6 W at 40 A, while the die power loss remains at 69 W. The case temperature variation is shown in Figure 9b, which shows a slight rise after bond wires are removed due to the increased power loss.

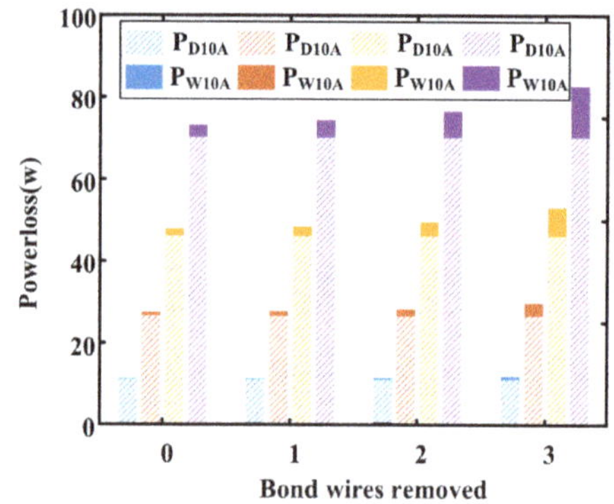


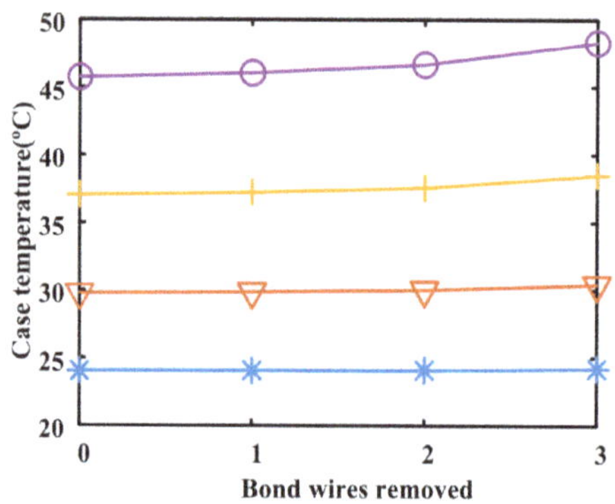


**Figure 9.** FEM results. (**a**) Power loss of IGBT module; (**b**) case temperature at case point.

Not only the power loss but also the proportion of bond wires to total power loss increase after bond wire lift-off, as shown in Table 3. The proportion of bond wires in total power loss rises after lift-off, increasing from 1.7% to 6.59% at 10 A, 2.83% to 10.59% at 20 A, 3.65% to 13.34% at 30 A, and 4.3% to 15.42% at 40 A. The proportion of bond wire power loss to whole power loss of the IGBT module changed obviously after lift-off at a larger current due to the lower on-state resistance of IGBT die.

**Table 3.** Proportion of bond wire power loss in total loss.

| Conduct Current | 0 Wire Removed | 1 Wire Removed | 2 Wires Removed | 3 Wires Removed |
|---|---|---|---|---|
| 10 A | 1.71% | 2.31% | 3.49% | 6.59% |
| 20 A | 2.83% | 3.82% | 5.72% | 10.59% |
| 30 A | 3.65% | 4.91% | 7.31% | 13.34% |
| 40 A | 4.30% | 5.77% | 8.55% | 15.42% |

The measured thermal resistances of heatsink after bond wires were removed are shown in Figure 10a. Although the thermal resistances show different initial values at each current level due to the spreading of thermal resistance [28], they all decrease after bond wires are removed. The thermal resistance decreases from 0.357, 7 K/W to 0.353, 2 K/W at 10 A, 0.357, 3 K/W to 0.350, 1 K/W at 20 A, 0.357, 1 K/W to 0.348, 0 K/W at 30 A, and 0.356, 9 K/W to 0.346, 3 K/W at 40 A after bond wires are removed.

The aging ratio $K$ of simulation results are shown in Figure 10b. Although the aging ratio shows similar values in each current level, the decrement is more significant at a higher current. The value of $K$ reaches 1.27%, 2.02% 2.56%, and 2.96% at each current level with one bond wire remaining. According to the results, the value of $K$ decreases after bond wire lift-off. Hence, the decrement in $K$ could indicate the fatigue of bond wires to avoid an open circuit.

The heat flow in the IGBT module before and after bond wires are removed is shown in Figure 11. The distribution of heat flow is mainly concentrated under the die before bond wires are removed, while the heat flow of bond wires is nearly negligible. Although the heat flow under the chip stays nearly the same after bond wire lift-off (blue lines), the heat flow of bond wires changes. Bond wire fatigue leads to a rise in bond wire power loss, and heat mainly concentrates in the die side and emitter side. This increase in the heat flow in the emitter side shows that part of the bond wire power loss did not transfer through the die side (red lines). Hence, using the power loss of the whole module to calculate the heatsink thermal resistance will underestimate the result.

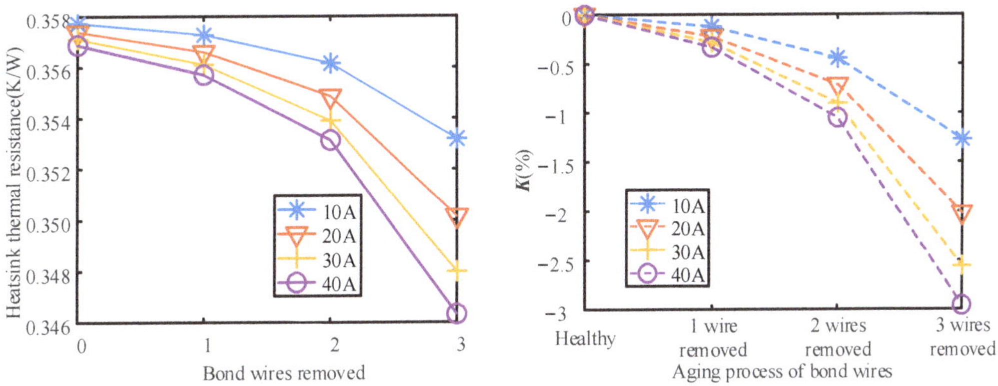


**(a)** **(b)**

**Figure 10.** FEM results. (**a**) Heatsink thermal resistance $Z_{M-HS}$; (**b**) aging ratio $K$.

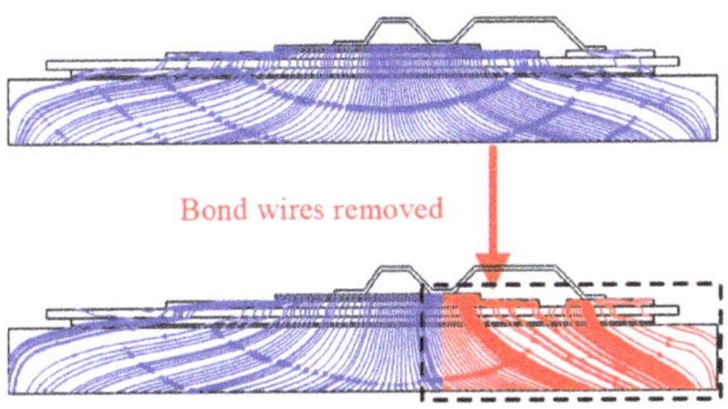


**Figure 11.** Heat flow in IGBT module after bond wires are removed.

## 4. Experimental Validation

### *4.1. Test Bench Setup*

The main electrical circuit for the test is shown in Figure 12a. The gate voltage was set at 15 V to control the IGBT. A DC source, which varied from 10 A to 40 A with 10 A intervals, was used to emulate the different operating conditions. The collector–emitter voltage was measured by oscilloscopes.

The test environment is shown in Figure 12b. The case temperature was measured from the thermocouple data logger (PICO, Cambridgeshire, UK, TC08). A fan-cooled heatsink (GD Rectifiers, West Sussex, UK, PS260/xxF(6)) was set as the air-cooling heatsink to be tested. Two graves were incised on top of the heatsink to place the thermal couples. The case temperature was measured by the thermal couple, as shown in Figure 12c. A thermal pad was attached to the module baseplate to reduce the air gap between the heatsink and IGBT. To emulate the aging of bond wires, the thermal resistance of heatsink was measured with different bond wires removed. The temperature curve of the case point on heatsink at 30 A is shown in Figure 12d, where the blue solid line is the case temperature, the black dashed line is the ambient temperature, and the red dashed line is the steady temperature of the case point. The forced air-cooling heatsink reaches its steady state after 500 s. Hence, the power loss and case temperature are measured after 10 min until reaching thermal steady state in each test.

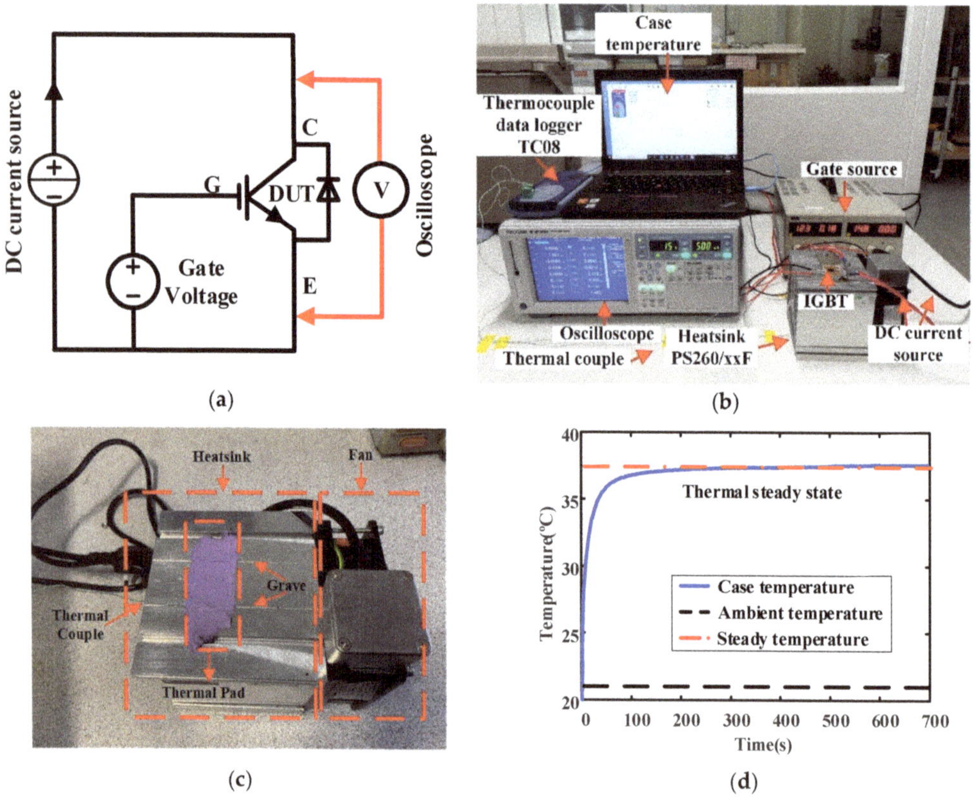


**Figure 12.** Test bench: (**a**) main electrical circuit; (**b**) test environment; (**c**) thermal couples on heatsink; (**d**) temperature curve of case point on heatsink at 30 A.

## 4.2. Experimental Results Analysis

The experiment results show that the aging of bond wires finally leads to a rise in the power loss of the IGBT module; the module fuses at 40 A with one bond wire remaining, as in Figure 13a. The power loss shows a small rise after bond wire lift-off at the smaller current, which increases more significantly at a higher current level. These increased power losses finally lead to a rise in the temperature at case point, as shown in Figure 13b. The experimental results show that the increase in case temperature is more significant at a higher current, while the variation is small at 10 A due to its low power loss.

The measured heatsink thermal resistances are shown in Figure 13c. Although the initial measured thermal resistance is different for each current level due to the thermal spreading resistance, they all show a decreasing trend after bond wires are removed. The measured thermal resistance decreases from 0.4427 K/W to 0.3960 K/W at 10 A, 0.3803 K/W to 0.3657 K/W at 20 A, and 0.3729 K/W to 0.3548 K/W at 30 A after three bond wires are removed. Although this module fused at 40 A with one wire remaining, it still decreases from 0.3609 K/W to 0.3568 K/W with two wires removed.

Figure 13d illustrates that the value of *K* shows similar decreasing trends at higher currents, and shows a large decrement at 10 A. The aging rate *K* is only dependent on the heat flow of bond wires, which are free of operating conditions. Hence, *K* can be used to detect the fault of bond wires to avoid the open circuit.

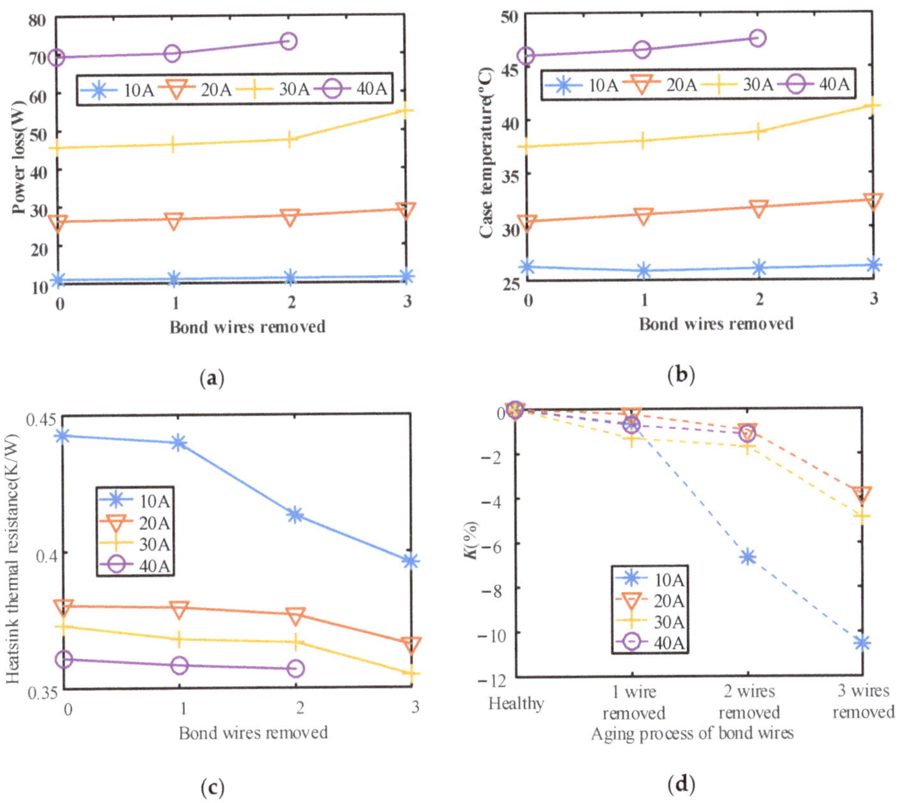


**Figure 13.** Experimental results: (**a**) power loss of IGBT module; (**b**) case temperature; (**c**) heatsink thermal resistance $Z_{M\_HS}$; (**d**) aging ratio *K*.

## 4.3. Discussion

The comparison of bond wire fault detection method is discussed in Table 4. As a non-invasive detection method, the proposed method does not need to change the structure of the IGBT module, which could be used in normal conditions. The case temperature and power loss are easier to measure than the gate signal during operation. Hence, these advantages could help maintain the power converter to avoid an open circuit. However, there are still some problems during measurement.

As previously mentioned, the thermal resistance of heatsink decreases after bond wires are removed. Although the ***K*** shows a similar reduction at all current levels in the simulation, it shows an anomalous reduction at 10 A in the experimental results. This strange performance is caused by the measurement of the case temperature.

JESD51-1 [29] proposed to measure an accurate thermal resistance with at least a 20 °C temperature rise on-die, because the high power loss of die could make the detection of the case temperature more accurate. The increased power loss after bond wires are removed is shown in Figure 14. The experimental results show that power loss significantly increases at higher current levels after bond wires are removed, while the increased power loss at 10 A is far smaller than other current levels.

The small increase in power loss at 10 A makes it difficult to detect the variation in case temperature after lift-off, which causes a small change in the case temperature of less than 0.1 °C. These small changes in the case temperature lead to inaccurate measured results, shown in Figure 13d. The low power loss makes it difficult to detect the case temperature variation after bond wires are removed at a low current, while this is easier to detect at a higher current. Thus, it is important to ensure the sensitivity of the case temperature detection method to provide accurate measurement results, which would be better to use in higher current working conditions.

**Table 4.** Comparisons of bond wire fault detection.

| Detection Type | | Measured Parameter | Working Conditions | Reference |
|---|---|---|---|---|
| Invasive Detection Method | Added layout | Voltage | Normal condition | [8] |
| | Current sensor | Current | Normal condition | [9–11] |
| Non-Invasive Detection Method | Gate charging current | Gate signal | Normal condition | [12] |
| | Gate charging time | Gate signal | Special gate driver | [13] |
| | Transconductance | Gate signal | Normal condition | [14] |
| | Stray inductance | Gate signal | Normal condition | [15] |
| | Conduct voltage at the inflected point with special gate voltage | Conduct voltage | Special conduct current at a special gate voltage | [17–19] |
| | Short-circuit current with special gate voltage | Current | Short-circuit at a special gate voltage | [20] |
| | Measured thermal resistance of heatsink | Thermal resistance | Normal condition | This paper |

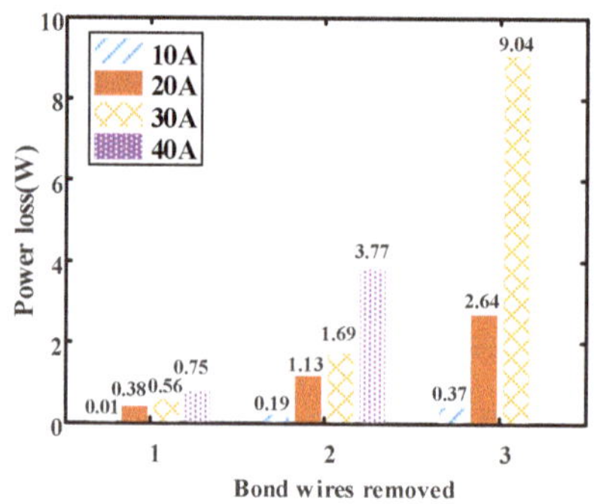


**Figure 14.** The growth of the IGBT module's power loss after bond wires are removed.

## 5. Thermal Resistance Measurement of Heatsink in Power Cycling

To verify the detection method proposed in this paper, a power cycling test is used to simulate the power converter. The power cycling circuit is shown in Figure 15. A single-phase full-bridge is built based on Ref. [30], which could generate a high-temperature swing with low power loss due to the inductive load. The power cycling circuit is shown in Figure 16. The full-bridge is composed of IGBT1, IGBT2, IGBT3, and the device under test (DUT). The $V_{CE}$ is measured by the $V_{CE}$ measurement circuit. A small current source (100 mA) is used to measure the junction temperature of IGBT by the $V_{CE}$ measurement circuit with the Temperature-Sensitive Parameter (TSEP) method. The auxiliary IGBT is used to block the branch of DUT during junction temperature measurement.

The test bench is shown in Figure 16. To show the feasibility of the proposed method, a different IGBT module (Semikron, Zhuhai, China, SKM50GB12T4) was used for this experiment. The DC source was set at 50 V and a 0.02 mH inductor was used as the load. The gate source and clamp source supply the gate driver and $V_{CE}$ measurement circuit. The IGBT was controlled by DSP. All of the data were collected by the NI DAQ. The control signal and acquisition signal were processed by LabVIEW on computer. The IGBT was placed on the cold plate and cooled by the chiller. The DUT was tested with a temperature variation of 83.2 °C and mean die temperature of 93.3 °C. The working frequency and switching frequency were 0.03 Hz and 6 k Hz, respectively.

The DUT failed after 56,879 cycles. The open package of the failed IGBT module is shown in Figure 17a. There are five failed wires on the chip with one wire remaining in Figure 17b.

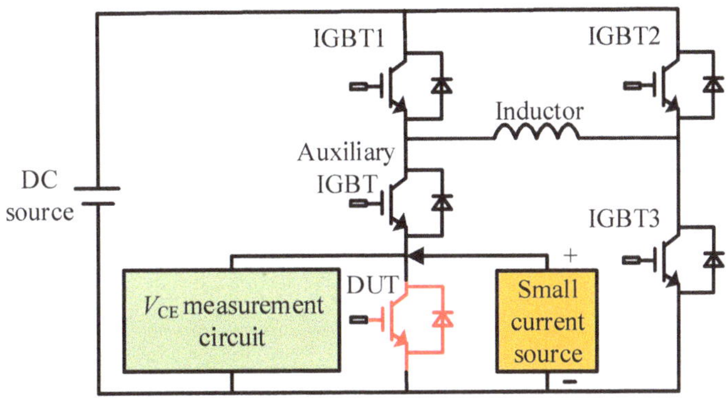


**Figure 15.** Power cycling circuit.

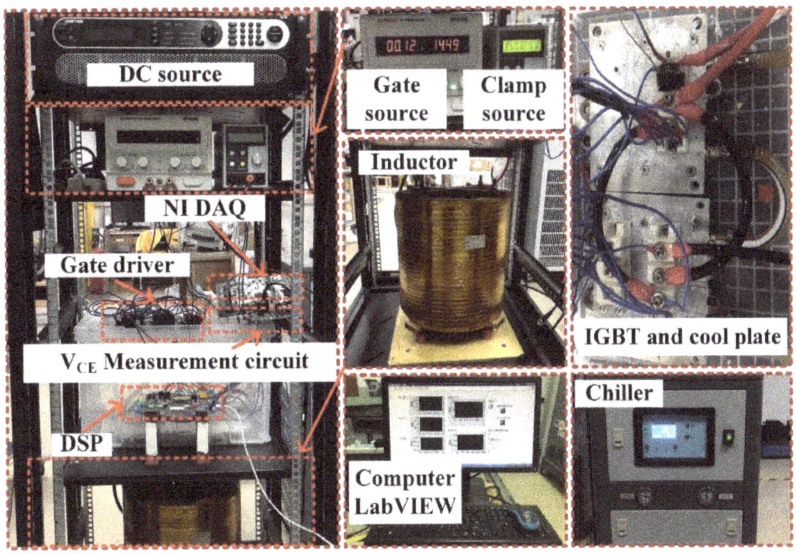


**Figure 16.** Test bench.

Previous research proposed that the thermal resistance will decrease after bond wires failed [23], which is shown in Figure 18a, where the brown wave is normalized conduct voltage and the blue wave is normalized measured thermal resistance of the IGBT module. The measured thermal resistance of the IGBT module could be measured in this test due to the special custom circuit test circuit; however, it is nearly impossible to measure the thermal resistance of the module during operation in a normal power converter. Although the measured thermal resistance of heatsink seems to show a notable change during cycling, it shows a significant decrease after bond wires fault, which is shown in Figure 18b, where the brown wave is normalized conduct voltage and the blue wave is normalized measured thermal resistance of heatsink.

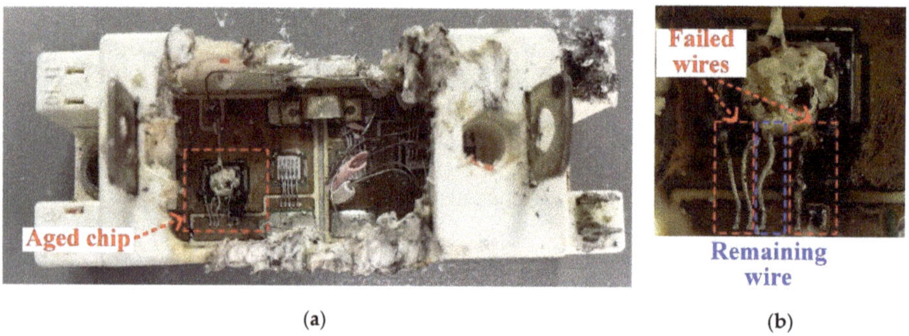


**Figure 17.** Cycling results of IGBT module SKM50GB12T4: (**a**) whole module; (**b**) failed wires.

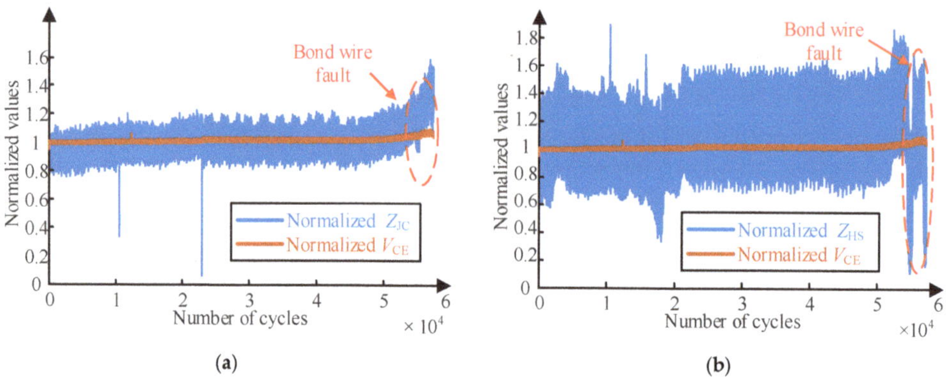


**Figure 18.** Normalized measured results of (**a**) IGBT module and (**b**) heatsink.

All in all, these results show that the measured thermal resistance of heatsink is easy to obtain to indicate the bond wire fault of IGBT modules, which might be helpful for the condition monitoring of power converters.

## 6. Conclusions

We proposed a novel method to detect bond wire fatigue by the decrement in measured heatsink thermal resistance. The advantage of this method is that it is not necessary to detect junction temperature, and thus, it could be used in normal working conditions. A multi-physics field FEM model was established to validate this method. The simulation results show that the heat flow changes after bond wire lift-off, which overestimates the calculated power loss and underestimates the measured heatsink thermal. Hence, the decrement in measured thermal resistance of heatsink could be used to detect the fatigue of bond wires. This method was verified by our experimental results. A power cycling test was used to show the variation in measured thermal resistance of heatsink during the aging process of the IGBT module, which verified the fault detection method.

The scope of this paper is monitoring the fatigue of bond wires under normal operation conditions, which could provide information for health management to avoid open-circuit faults. However, it is still difficult to measure the state of IGBT in the power converter. In future work, this method will be further studied to be integrated into a real power converter, which could help to ensure the reliable operation of the converter.

**Author Contributions:** Conceptualization, methodology, validation, and writing—original draft preparation, D.L.; writing—review and editing, W.L., H.X. and Z.D.; supervision, W.L. and M.C.; funding acquisition, K.Y.; validation, Z.W. All authors have read and agreed to the published version of the manuscript.

**Funding:** This research was supported by the Shanxi Provincial Department of Science and Technology unveiled the bidding project (20201101017); the National Key Research and Development Program of China (grant numbers 2018YFB0905800); the National Natural Science Foundation of China (grant numbers 51707024); the Fundamental Research Funds for the Central Universities Project (no. 2020CDJQY-A026); and the National "111" Project (B08036).

**Data Availability Statement:** The data presented in this study are available on request from the corresponding author.

**Acknowledgments:** The authors would like to thank Li Ran, Ruizhu Wu, Borong Hu, and Xuan Guo at the University of Warwick for the help and advice.

**Conflicts of Interest:** The authors declare no conflict of interest.

## References

1. Li, B.; Ma, Z.; Hidalgo-Gonzalez, P.; Lathem, A.; Fedorova, N.; He, G.; Zhong, H.; Chen, M.; Kammen, D.M. Modeling the impact of EVs in the Chinese power system: Pathways for implementing emissions reduction commitments in the power and transportation sectors. *Energy Policy* **2021**, *149*, 111962. [CrossRef]
2. Yang, S.; Bryant, A.; Mawby, P.; Xiang, D.; Ran, L.; Tavner, P. An Industry-Based Survey of Reliability in Power Electronic Converters. *IEEE Trans. Ind. Appl.* **2011**, *47*, 1441–1451. [CrossRef]
3. Lai, W.; Chen, M.; Ran, L.; Xu, S.; Jiang, N.; Wang, X.; Alatise, O.; Mawby, P. Experimental investigation on the effects of narrow junction temperature cycles on die-attach solder layer in an IGBT module. *IEEE Trans. Power Electron.* **2016**, *32*, 1431–1441. [CrossRef]
4. Baker, N.; Liserre, M.; Dupont, L.; Avenas, Y. Improved reliability of power modules: A review of online junction temperature measurement methods. *IEEE Ind. Electron. Mag.* **2014**, *8*, 17–27. [CrossRef]
5. Pedersen, K.B.; Kristensen, P.K.; Pedersen, K.; Uhrenfeldt, C.; Munk-Nielsen, S. Vce as early indicator of IGBT module failure mode. In Proceedings of the 2017 IEEE International Reliability Physics Symposium (IRPS), Monterey, CA, USA, 2–6 April 2017; pp. FA-1.1–FA-1.6. [CrossRef]
6. Morozumi, A.; Yamada, K.; Miyasaka, T.; Sumi, S.; Seki, Y. Reliability of power cycling for IGBT power semiconductor modules. *IEEE Trans. Ind. Appl.* **2003**, *39*, 665–671. [CrossRef]
7. An, Q.T.; Sun, L.Z.; Zhao, K.; Sun, L. Switching function model-based fast-diagnostic method of open-switch faults in inverters without sensors. *IEEE Trans. Power Electron.* **2010**, *26*, 119–126. [CrossRef]
8. Lehmann, J.; Netzel, M.; Herzer, R.; Pawel, S. Method for electrical detection of bond wire lift-off for power semiconductors. In Proceedings of the ISPSD'03, 2003 IEEE 15th International Symposium on Power Semiconductor Devices and ICs, 2003 Proceedings, Cambridge, UK, 14–17 April 2003. [CrossRef]
9. Shiratsuchi, H.; Matsushita, K.; Omura, I. IGBT die current imaging system by scanning local magnetic field. *Microelectron. Reliab.* **2013**, *53*, 1409–1412. [CrossRef]
10. Tomonaga, H.; Tsukuda, M.; Okoda, S.; Noda, R.; Tashiro, K.; Omura, I. 16-channel micro magnetic flux sensor array for IGBT current distribution measurement. *Microelectron. Reliab.* **2015**, *55*, 1357–1362. [CrossRef]
11. Sheng, M.; Alvi, M.H.; Lorenz, R.D. Real-time Bond Wire Lift-off Monitoring via Module Integrated Current Sensors. In Proceedings of the 2019 IEEE Energy Conversion Congress and Exposition (ECCE), Baltimore, MD, USA, 29 September–3 October 2019. [CrossRef]
12. Zhou, S.; Zhou, L.; Sun, P. Monitoring potential defects in an IGBT module based on dynamic changes of the gate current. *IEEE Trans. Power Electron.* **2013**, *28*, 1479–1487. [CrossRef]
13. Wang, K.; Zhou, L.; Sun, P.; Du, X. Monitoring bond wires fatigue of multichip IGBT module using time duration of the gate charge. *IEEE Trans. Power Electron.* **2020**, *36*, 888–897. [CrossRef]

14. Wang, K.; Zhou, L.; Sun, P.; Du, X. Monitoring Bond Wire Defects of IGBT Module Using Module Transconductance. *IEEE J. Emerg. Sel. Top. Power Electron.* **2021**, *9*, 2201–2211. [CrossRef]
15. Yang, Y.; Zhang, P. A novel bond wire fault detection method for IGBT modules based on turn-on gate voltage overshoot. *IEEE Trans. Power Electron.* **2021**, *36*, 7501–7512. [CrossRef]
16. Zhang, Q.; Yang, Y.; Zhang, P. A Novel Method for Monitoring the Junction Temperature of SiC MOSFET On-line Based on On-state Resistance. In Proceedings of the 2019 22nd International Conference on Electrical Machines and Systems (ICEMS), Harbin, China, 11–14 August 2019; pp. 1–5. [CrossRef]
17. Choi, U.M.; Blaabjerg, F. Separation of wear-out failure modes of IGBT modules in grid-connected inverter systems. *IEEE Trans. Power Electron.* **2017**, *33*, 6217–6223. [CrossRef]
18. Singh, A.; Anurag, A.; Anand, S. Evaluation of Vce at inflection point for monitoring bond wire degradation in discrete packaged IGBTs. *IEEE Trans. Power Electron.* **2017**, *32*, 2481–2484. [CrossRef]
19. Du, M.; Kong, Q.; Ouyang, Z.; Wei, K.; Hurley, W.G. Strategy for diagnosing the aging of an IGBT module by on-state voltage separation. *IEEE Trans. Electron Devices* **2019**, *66*, 4858–4864. [CrossRef]
20. Sun, P.; Gong, C.; Du, X.; Peng, Y.; Wang, B.; Zhou, L. Condition monitoring IGBT module bond wires fatigue using short-circuit current identification. *IEEE Trans. Power Electron.* **2016**, *32*, 3777–3786. [CrossRef]
21. Held, M.; Jacob, P.; Nicoletti, G.; Scacco, P.; Poech, M.H. Fast power cycling test of IGBT modules in traction application. In Proceedings of the Second International Conference on Power Electronics and Drive Systems, Singapore, 26–29 May 1997; Volume 1, pp. 425–430. [CrossRef]
22. Luo, D.; Lai, W.; Chen, M.; Xu, S.; Xiao, Y. A Fault Detection Method for IGBT Bond Wires Partial Lift off Based on Thermal Resistance Assessment. In Proceedings of the 2018 IEEE Region Ten Symposium (Tensymp), Sydney, NSW, Australia, 4–6 July 2018; pp. 135–139. [CrossRef]
23. Luo, D.; Chen, M.; Lai, W.; Xia, H.; Deng, Z. A study on the effect of bond wires lift-off on IGBT thermal resistance measurement. *Electronics* **2021**, *10*, 194. [CrossRef]
24. Hu, Z.; Du, M.; Wei, K. Online calculation of the increase in thermal resistance caused by solder fatigue for IGBT modules. *IEEE Trans. Device Mater. Reliab.* **2017**, *17*, 785–794. [CrossRef]
25. Wang, Z.; Bo, T.; Wei, Q.; Qu, L. Real-time aging monitoring for IGBT modules using case temperature. *IEEE Trans. Ind. Electron.* **2016**, *63*, 1168–1178. [CrossRef]
26. Xiang, D.; Ran, L.; Tavner, P.; Bryant, A.; Yang, S.; Mawby, P. Monitoring solder fatigue in a power module using case-above-ambient temperature rise. *IEEE Trans. Ind. Appl.* **2011**, *47*, 2578–2591. [CrossRef]
27. Gao, B.; Yang, F.; Chen, M.; Ran, L.; Ullah, I.; Xu, S.; Mawby, P. A Temperature Gradient-Based Potential Defects Identification Method for IGBT Module. *IEEE Trans. Power Electron.* **2017**, *32*, 2227–2242. [CrossRef]
28. Feng, T.Q.; Xu, J.L. An analytical solution of thermal resistance of cubic heat spreaders for electronic cooling. *Appl. Therm. Eng.* **2004**, *24*, 323–337. [CrossRef]
29. *EIA/JEDEC Standard JESD51-1*; Integrated Circuit Thermal Measurement Method—Electrical Test Method. Electronic Industries Association: Arlington, VA, USA, 1995. Available online: www.jedec.org(accessed on 8 March 2022).
30. Choi, U.; Jørgensen, S.; Blaabjerg, F. Advanced Accelerated Power Cycling Test for Reliability Investigation of Power Device Modules. *IEEE Trans. Power Electron.* **2016**, *31*, 8371–8386. [CrossRef]

electronics

MDPI

Article

# Fault Diagnosis Method of Six-Phase Permanent Magnet Synchronous Motor Based on Vector Space Decoupling

**Hanying Gao *, Jie Guo, Zengquan Hou, Bangping Zhang and Yao Dong**

School of Electrical and Electronic Engineering, Harbin University of Science and Technology, Harbin 150080, China; 1603011307@stu.hrbust.edu.cn (J.G.); 1920310189@stu.hrbust.edu.cn (Z.H.); zbp123456@163.com (B.Z.); a1090897186@163.com (Y.D.)
* Correspondence: ghy@hrbust.edu.cn; Tel.: +86-137-9666-8369

**Abstract:** Compared with the three-phase motor, the six-phase motor has lower torque ripple and higher fault tolerance performance, which makes it widely used in aviation, ships, industrial manufacturing, and other application fields. However, the probability of failure of the polyphase motor system increases with the increase in the number of phases. In order to deal with the open phase fault and power switch fault of the six-phase inverter, a fault diagnosis method for the six-phase inverter based on vector space decoupling (VSD) is proposed. The open phase fault index is first determined according to the VSD decoupling inverse transform and the current constraints. The fault index is then optimized from the perspective of preventing misdiagnosis and improving reliability, and the open phase fault can be diagnosed in one fundamental cycle. In addition, the current trajectory of harmonic plane after switch fault is analyzed, and the back propagation (BP) neural network is used to identify the harmonic plane current trajectory of different types of switch fault. Finally, the correctness and feasibility of the proposed fault diagnosis method are verified by simulations and experiments. The obtained results demonstrate that the proposed method can quickly and accurately locate the open phase fault and switch fault without additional hardware. The proposed method is simple, efficient, and robust.

**Keywords:** six-phase permanent magnet synchronous motor (PMSM); inverter fault diagnosis; neural network; open circuit fault; power switch failure

**Citation:** Gao, H.; Guo, J.; Hou, Z.; Zhang, B.; Dong, Y. Fault Diagnosis Method of Six-Phase Permanent Magnet Synchronous Motor Based on Vector Space Decoupling. *Electronics* **2022**, *11*, 1229. https://doi.org/10.3390/electronics11081229

Academic Editor: Ahmed Abu-Siada

Received: 21 March 2022
Accepted: 7 April 2022
Published: 13 April 2022



## 1. Introduction

The polyphase permanent magnet synchronous motor has its unique advantages in improving power, restraining torque ripple, and fault-tolerant control. It is suitable for ships [1], aerospace [2], energy utilization [3], rail transit [4], and other application domains with high-reliability requirements. However, compared with the three-phase motor, the number of power devices and sensors in the driving system of the polyphase motor is also doubled. In the case of failure, it will threaten personal safety or lead to system collapse. Therefore, it is necessary to detect and diagnose the different operating parameters in the early stage of system fault.

In the polyphase motor drive system, the inverter switch failure and phase failure rate are the highest [5]. The fault diagnosis technology can instantly and accurately locate the fault location of the motor drive system and provide key information for maintenance and repair. In addition, it provides an accurate basis for the selection of the subsequent fault-tolerant control technology of the polyphase motor. Therefore, a fast, stable, and accurate fault diagnosis method for polyphase motor drive systems is crucial.

Three main fault diagnosis methods for motor drive systems exist: model-based [6–8], knowledge-based [9,10], and signal-based [11]. In [12], the related terms of DC side voltage and current of inverter are modeled and analyzed, and the residual evaluation function of DC side voltage is constructed to diagnose the fault of switch. A phase current observer is designed to convert the phase current error into coordinates to obtain the fault vector angle, so as to locate the switch fault [13]. A current interval sliding mode observer is designed to improve the prediction process of the phase current [14]. The performance of the model-based fault diagnosis method highly depends on the model's accuracy. The changes in the motor parameters and operating conditions will also have a certain impact on the diagnosis accuracy.

In the knowledge-based diagnosis method, the main neural network algorithm and auxiliary neural network are used to analyze the amplitude-frequency characteristics of DC bus voltage harmonics and diagnose the fault of the three-level inverter power device [15]. In [16], the intermittent faults of power equipment are studied, and the fault levels under different fault conditions are classified by using fuzzy theory. In [17], a new method is mentioned to diagnose the fault of a wind turbine by developing a knowledge base to simulate the thinking mode of experts. This method has a high computational load, and low real-time performance, and it is difficult to implement.

The fault diagnosis based on signals can be divided into voltage and current methods. The typical shape of the fault track is matched by calculating the distance from the harmonic plane current track to the centroid [18]. This algorithm is complex and requires a lot of calculations. In [19], a new method is mentioned to decompose and reconstruct the voltage signal by using wavelet packets. It also uses the characteristic frequency related to the fault signal in its power spectrum to judge the open circuit fault type of the switch. In [20], by analyzing the frequency and angle of each current vector, whether the power switch of the three-phase inverter fails can be judged. In [21], the sampled current is decomposed using the wavelet transform and transformed by coordinate, and the fault trajectory is diagnosed using the support vector machine.

In this paper, a fault diagnosis method based on VSD is proposed for phase failure and power switch failure. The phase failure index is first determined according to the VSD transformation and phase failure constraints. A series of fault indexes are then optimized in order to improve the diagnosis reliability. Afterward, the switch fault is diagnosed as the characteristic signal by analyzing the current trajectories of harmonic planes after VSD transformation. Finally, considering the 10 kW double Y-phase shifted 30° six-phase PMSM, the correctness of the diagnosis algorithm is verified by simulations and experiments.

## 2. Mathematical Model of the Six-Phase Permanent Magnet Synchronous Motor

The six-phase PMSM drive system is mainly composed of a six-phase voltage source inverter (VSI). Its topology is shown in Figure 1. Note that N1 and N2 are the neutral points of the two windings.

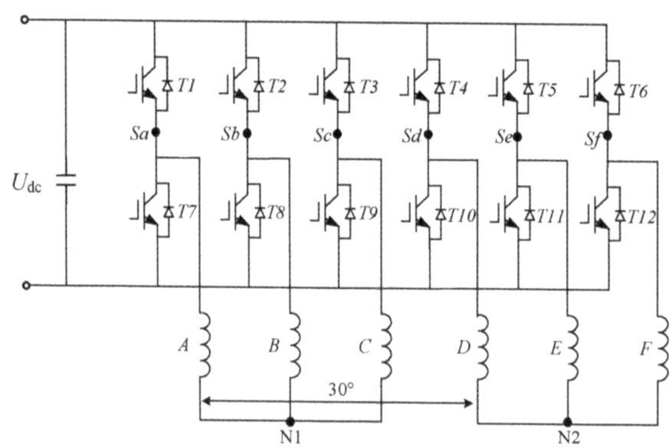


**Figure 1.** Topology of the six-phase inverter with neutral isolation.

The six-phase PMSM is a nonlinear and strongly coupled system. Therefore, an appropriate decoupling transformation is required to simplify the analysis. In addition, the VSD decoupling method, which is more suitable for polyphase motors, can control the harmonic component. The vector space decoupling matrix is given by:

$$C_{6s} = \frac{1}{\sqrt{3}} \begin{bmatrix} 1 & -\frac{1}{2} & -\frac{1}{2} & \frac{\sqrt{3}}{2} & -\frac{\sqrt{3}}{2} & 0 \\ 0 & \frac{\sqrt{3}}{2} & -\frac{\sqrt{3}}{2} & \frac{1}{2} & \frac{1}{2} & -1 \\ 1 & -\frac{1}{2} & -\frac{1}{2} & -\frac{\sqrt{3}}{2} & \frac{\sqrt{3}}{2} & 0 \\ 0 & -\frac{\sqrt{3}}{2} & \frac{\sqrt{3}}{2} & \frac{1}{2} & \frac{1}{2} & -1 \\ 1 & 1 & 1 & 0 & 0 & 0 \\ 0 & 0 & 0 & 1 & 1 & 1 \end{bmatrix} \tag{1}$$

In this matrix, the first two rows of variables correspond to the $\alpha$-$\beta$ subspace, the middle two rows correspond to the $z_1$-$z_2$ subspace, and the last two rows correspond to the $o_1$-$o_2$ zero sequence. The conversion of electromechanical energy is only related to $\alpha$-$\beta$.

The rotation coordinate transformation is expressed as:

$$C_{6s/6r} = \begin{bmatrix} \cos\theta & \sin\theta & 0 \\ -\sin\theta & \cos\theta & 0 \\ 0 & 0 & I_4 \end{bmatrix} \tag{2}$$

where $I_4$ is a fourth-order unit matrix. The motor mathematical model after VSD transformation is detailed in the sequel.

The magnetic flux equation is:

$$\begin{bmatrix} \psi_d \\ \psi_q \end{bmatrix} = \begin{bmatrix} L_d & 0 \\ 0 & L_q \end{bmatrix} \begin{bmatrix} i_d \\ i_q \end{bmatrix} + \psi_f \begin{bmatrix} 1 \\ 0 \end{bmatrix} \tag{3}$$

where $L_d$ and $L_q$, $i_d$ and $i_q$, $\psi_d$ and $\psi_q$ are the inductance, current, and stator flux of the direct axis and quadrature axis, respectively.

The voltage equation is:

$$\begin{bmatrix} u_d \\ u_q \end{bmatrix} = \begin{bmatrix} R & 0 \\ 0 & R \end{bmatrix} \begin{bmatrix} i_d \\ i_q \end{bmatrix} + \frac{\mathrm{d}}{\mathrm{d}t} \begin{bmatrix} \psi_d \\ \psi_q \end{bmatrix} + \omega_e \begin{bmatrix} -\psi_q \\ \psi_d \end{bmatrix} \tag{4}$$

where $u_d$ and $u_q$ are the stator voltages of the direct axis and quadrature axis, respectively.

The electromagnetic torque equation is:

$$T_e = 3p[\psi_f i_q + (L_d - L_q) i_d i_q] \tag{5}$$

where $T_e$ and $\psi_f$ are the electromagnetic torque and permanent magnet flux linkage, respectively.

The equation of motion is:

$$T_e - T_L - B\omega_m = J\frac{d\omega_m}{dt} \tag{6}$$

where $T_L$, $B$, $\omega_m$, and $J$ are load torque, damping coefficient, mechanical angular velocity, and moment of inertia, respectively.

## 3. Open Phase Fault Diagnosis Strategy

After VSD transformation, the property of the $\alpha$-$\beta$ plane variable is similar to that of the three-phase motor, while the $z_1$-$z_2$ plane is a unique property of the six-phase PMSM [22]. Simultaneously, the change of the $z_1$-$z_2$ component during inverter fault is explored, which can be used for inverter fault diagnosis.

### 3.1. Calculation of the Phase Failure Index

In the normal state, the fundamental subspace current $i_\alpha$, $i_\beta$ in the VSD static coordinate system is obtained by decoupling transformation of the six-phase stator current:

$$i_\alpha = i_a - \frac{1}{2}i_b - \frac{1}{2}i_c + \frac{\sqrt{3}}{2}i_d - \frac{\sqrt{3}}{2}i_e \quad (7)$$

$$i_\beta = \frac{\sqrt{3}}{2}i_b - \frac{\sqrt{3}}{2}i_c + \frac{1}{2}i_d + \frac{1}{2}i_e - i_f \quad (8)$$

Substituting the six-phase stator current of the motor during normal operation into Equations (7) and (8), the following is obtained:

$$\begin{cases} i_\alpha = I_m\sin(\theta) - \frac{1}{2}I_m\sin(\theta - \frac{2\pi}{3}) - \frac{1}{2}I_m\sin(\theta + \frac{2\pi}{3}) + \frac{\sqrt{3}}{2}I_m\sin(\theta - \frac{\pi}{6}) - \frac{\sqrt{3}}{2}I_m\sin(\theta - \frac{5\pi}{6}) = 3I_m\sin(\theta) \\ i_\beta = \frac{\sqrt{3}}{2}I_m\sin(\theta - \frac{2\pi}{3}) - \frac{\sqrt{3}}{2}I_m\sin(\theta + \frac{2\pi}{3}) + \frac{1}{2}I_m\sin(\theta - \frac{\pi}{6}) + \frac{1}{2}I_m\sin(\theta - \frac{5\pi}{6}) - I_m\sin(\theta + \frac{\pi}{6}) = 3I_m\cos(\theta) \end{cases} \quad (9)$$

Similarly, the current relationship in the harmonic subplane can be obtained:

$$\begin{cases} i_x = i_a - \frac{1}{2}i_b - \frac{1}{2}i_c - \frac{\sqrt{3}}{2}i_d + \frac{\sqrt{3}}{2}i_e \\ i_y = -\frac{\sqrt{3}}{2}i_b + \frac{\sqrt{3}}{2}i_c + \frac{1}{2}i_d + \frac{1}{2}i_e - i_f \end{cases} \quad (10)$$

Substituting the six-phase stator current of the motor during normal operation into Equation (10) results in:

$$\begin{cases} i_x = 0 \\ i_y = 0 \end{cases} \quad (11)$$

It can be deduced from this analysis that the output electromagnetic torque of the motor is mainly provided by the fundamental subspace current components $i_\alpha$ and $i_\beta$. In the normal state, the current $i_{z1}$, $i_{z2}$ of the harmonic subplane is null. The currents $i_\alpha$,$i_\beta$ and $i_{z1}$,$i_{z2}$ after VSD decoupling transformation are shown in Figure 2. The fundamental subspace is circular in the steady-state current $i_\alpha$,$i_\beta$. The harmonic subspace current $i_{z1}$,$i_{z2}$ is the point at the origin, which is coherent with the theoretical calculation results.

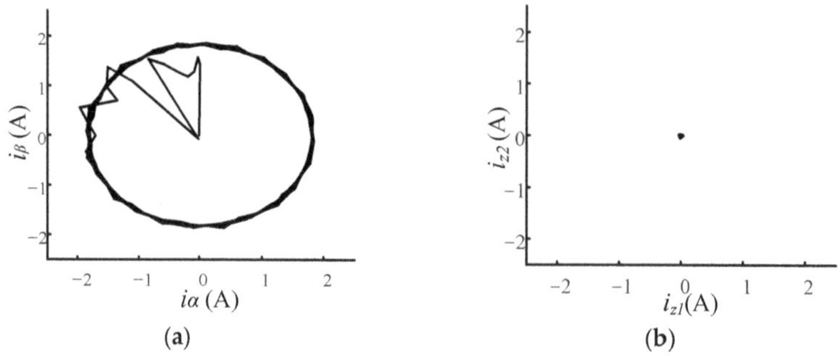


**Figure 2.** $\alpha$-$\beta$ and $z_1$-$z_2$ plane current waveform during normal operation: (**a**) $\alpha$-$\beta$ plane current waveform in case of normal operation; (**b**) $z_1$-$z_2$ plane current waveform in case of normal operation.

By inverting the normal VSD transformation, the following is obtained:

$$\begin{bmatrix} i_a \\ i_b \\ i_c \\ i_d \\ i_e \\ i_f \end{bmatrix} = \frac{1}{\sqrt{3}} \begin{bmatrix} 1 & 0 & 1 & 0 & 1 & 0 \\ -\frac{1}{2} & \frac{\sqrt{3}}{2} & -\frac{1}{2} & -\frac{\sqrt{3}}{2} & 1 & 0 \\ -\frac{1}{2} & -\frac{\sqrt{3}}{2} & -\frac{1}{2} & \frac{\sqrt{3}}{2} & 1 & 0 \\ \frac{\sqrt{3}}{2} & \frac{1}{2} & -\frac{\sqrt{3}}{2} & \frac{1}{2} & 0 & 1 \\ -\frac{\sqrt{3}}{2} & \frac{1}{2} & \frac{\sqrt{3}}{2} & \frac{1}{2} & 0 & 1 \\ 0 & -1 & 0 & -1 & 0 & 1 \end{bmatrix} \begin{bmatrix} i_\alpha \\ i_\beta \\ i_{z1} \\ i_{z2} \\ i_{o1} \\ i_{o2} \end{bmatrix} \quad (12)$$

When a phase failure occurs, it can be seen from Equation (12) that the fundamental and harmonic plane currents have a certain relationship. It is assumed that the phase failure of phase *a* winding occurs. That is, when the constraint condition is $i_a = 0$, the first row of the VSD inverse matrix (cf. Equation (12)) is multiplied by the plane current in order to obtain $i_\alpha + i_{z1} = 0$. When the motor is in normal operation, the harmonic current is $i_{z1} = 0$. In summary, the open circuit fault index $F_a$ can be computed as:

$$F_a = -\frac{i_{z1}}{i_\alpha} \tag{13}$$

Similarly, the open-circuit fault index of six-phases (cf. Equation (14)) can be obtained:

$$\begin{cases} F_a = -\frac{i_{z1}}{i_\alpha} \\ F_b = \frac{i_{z1}}{-i_a+\sqrt{3}i_\beta-\sqrt{3}i_{z2}} \\ F_c = \frac{i_{z1}}{-i_a-\sqrt{3}i_\beta+\sqrt{3}i_{z2}} \\ F_d = \frac{i_{z1}}{i_a+1/\sqrt{3}i_\beta+1/\sqrt{3}i_{z2}} \\ F_e = \frac{i_{z1}}{i_a-1/\sqrt{3}i_\beta-1/\sqrt{3}i_{z2}} \\ F_f = -\frac{i_{z2}}{i_\beta} \end{cases} \tag{14}$$

The numerator of this fault index is the harmonic plane current. Due to the fact that the harmonic plane current component is null in the normal state, the fault index is always kept null during the normal operation of the motor. When a phase fault occurs, the fault index of the phase rises 1, so as to distinguish the normal operation and inverter phase fault.

### 3.2. Optimization Scheme of the Winding Open Circuit Fault Index

In the case where only the fault index $F_n$ ($n = a,b,c,d,e,f$) is used, because the denominator of the fault indicator is too small near the zero-crossing area, a spike pulse will appear, which may lead to misdiagnosis. Therefore, the fault index $F_n$ should be improved and optimized. The optimization process is illustrated in Figure 3.

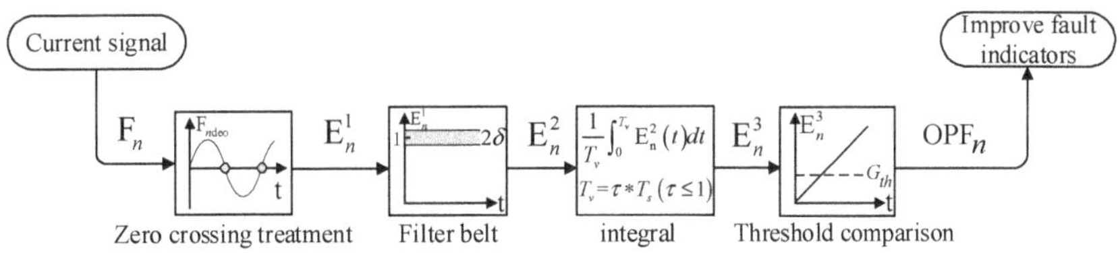


**Figure 3.** Fault index optimization process of disconnected phase.

The fault index $F_n$ is first subjected to peak pulse processing using Equation (15), in order to obtain the first fault index $E_n^1$ :

$$\begin{cases} E_n^1 = 0\ ,\ |F_{ndeo}| \le \Delta \\ E_n^1 = F_n,\ |F_{ndeo}| > \Delta \end{cases} \tag{15}$$

where $F_{ndeo}$ is the denominator value of fault index $F_n$ and $\Delta$ is the threshold value of the zero-crossing area width. This value cannot be too large, in order to ensure the diagnosis speed. The specific value is determined according to the experiment. In this paper, $\Delta$ is set to 0.1.

The fault index after peak pulse processing is then filtered using Equation (16) in order to obtain the second fault index $E_n^2$:

$$\begin{cases} E_n^2 = E_n^1,\ (1-\delta \le E_n^1 \le 1+\delta) \\ E_n^2 = 0\ ,\ (1+\delta \le E_n^1 \& E_n^1 \le 1-\delta) \end{cases} \tag{16}$$

where $\delta$ is the width of the filter band, which is used to filter out the unnecessary interference while not affecting the subsequent processing. The recommended value range of $\delta$ is [0.1–0.3]. In this paper, it is set to 0.1.

The processed fault index will be set to 1 after the fault occurs. However, the fault index will be set to zero in the zero-crossing area, due to the processing of the zero-crossing spike pulse. Therefore, it is necessary to integrate the fault index according to Equation (17) in order to obtain the third fault index $E_n^3$:

$$E_n^3 = \frac{1}{T_v}\int_0^{T_v} E_n^2(t)dt \tag{17}$$

$$T_v = \tau * T_s(\tau \leq 1) \tag{18}$$

where $T_s$ is the periodic value of the current fundamental wave. The larger the value of $\tau$, the slower the diagnosis speed, but the higher the diagnosis accuracy. On the contrary, the smaller the value of $\tau$, the faster the diagnosis speed, but the smaller the diagnosis accuracy. In this paper, $\tau$ is set to 0.6, in order to ensure the diagnosis accuracy and rapidity. The traditional integration consists in integrating the fault indicators of the whole cycle, which will prolong the fault diagnosis time and reduce the diagnosis efficiency. Selecting $T_v$ Equation (18) can speed up the detection speed.

Finally, the third fault index $E_n^3$ is compared with the threshold in order to obtain the improved fault index $OPF_n$:

$$\begin{cases} OPF_n = 1, \ E_n^3 > G_{th} \\ OPF_n = 0, \ E_n^3 < G_{th} \end{cases} \tag{19}$$

where $G_{th}$ is the judgment threshold selected between 0 and 1. If the value of $G_{th}$ is too large, the diagnosis speed will slow down. Because these steps have a negative integration link to suppress the fluctuation of fault indicators, the value can be appropriately reduced, and $G_{th}$ can have a value between 0.1 and 0.3.

It can be deduced that using the constraint condition that the fault phase current is null after the phase failure, combined with the vector space decoupled VSD inverse matrix, the fault index of the phase failure is derived. This can diagnose and locate 15 types of phase faults of the inverter, including 6 types of single-phase faults and 9 types of two-phase faults of different windings.

## 4. Power Switch Fault Diagnosis Strategy

Several types of power switch faults exist. They mainly include the short circuit and open circuit faults. The short circuit fault has a short duration and is difficult to detect. For further treatment, the short circuit fault is changed into open an circuit fault by a fuse device. The open-circuit faults are mostly single-switch and two-switch faults. The fault of two switches on the same bridge arm can be treated as open phase fault. Therefore, this paper only tackles the open circuit fault of a single switch and two switches on different bridge arms.

### *4.1. Harmonic Plane Current Fault Trace Analysis*

Figure 4 shows the current waveform of the $z_1$-$z_2$ plane when different switches fail after VSD conversion of the six-phase stator current. When T1 fails, the current trajectory of the $z_1$-$z_2$ plane points to the negative half of the $z_1$ axis. When T1 and T2 fail, the current trajectories of $i_{z1}$ and $i_{z2}$ are fan-shaped. Similarly, when other switches fail, they also have these trajectory characteristics.

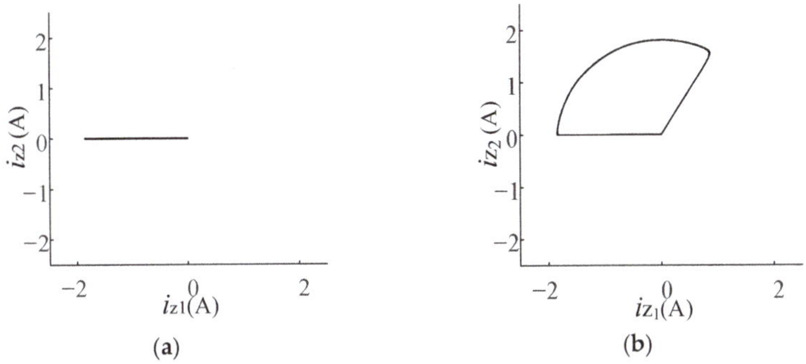


**Figure 4.** $z_1$-$z_2$ plane current in case of different power switch faults; (**a**) $z_1$-$z_2$ plane current waveform in case of T1 fault; (**b**) $z_1$-$z_2$ plane current waveform in case of T1 and T2 faults.

The reference direction of the current vector during each phase fault can be expressed as:

$$\mathrm{dn-ref} = \arctan(\frac{A_{z2}}{A_{z1}})\begin{cases} \pi - (k-1)\frac{\pi}{6}, f_n \leq 1 \\ -(k-1)\frac{\pi}{6}, f_n \geq 0 \end{cases} \tag{20}$$

where $A_{z1}$ and $A_{z2}$ represent the amplitude of harmonic currents $i_{z1}$ and $i_{z2}$, respectively. $n$ represents the *a,b,c,d,e,f* phase, while the fault phase corresponding to $k = 1, 5, -3, -4, 6, 4$ is *a,b,c,d,e,f*, respectively. The fault current vector turns counterclockwise by angle dn-ref from the $x$ axis. $f_n(t) = \int_0^T i_n dt$ is the average value of fault current in a current cycle.

It can be deduced that in the case of a single-switch or two-switch open-circuit fault of the inverter, the current trajectory in the harmonic $i_{z1}$ and $i_{z2}$ planes regularly changes. The current vector trajectory profile under single-switch and multi-switch faults can be developed using the current values in the $i_{z1}$ and $i_{z2}$ harmonic planes, as shown in Figures 5 and 6.

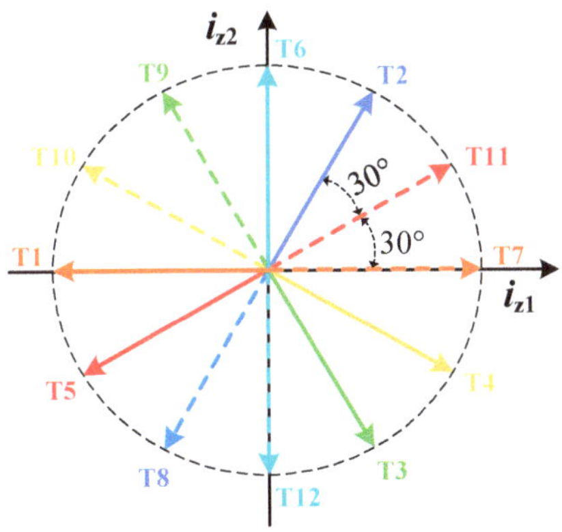


**Figure 5.** Reference direction of $i_{z1}$ and $i_{z2}$ current trace in the case of a single switch failure.

According to the different reference directions of the current track under the $z_1$-$z_2$ plane fault, the intelligent algorithm has the characteristics of strong analysis ability and several diagnosis types. Therefore, the neural network algorithm is used for fault diagnosis based on the characteristics of fault current trajectory.

In order to distinguish the fault types and improve the diagnosis efficiency, different fault types should be classified and coded. The 12 bit binary number $F_1F_2F_3F_4F_5F_6F_7F_8F_9F_{10}F_{11}F_{12}$ is used to encode the single-switch and two-switch faults of the six-phase inverter. When all the bits are equal to 1, the switch has an open-circuit fault. On the contrary, when all the bits are 0, the switch has no fault. Note that there are 36 fault conditions. The specific fault types are shown in Table 1.

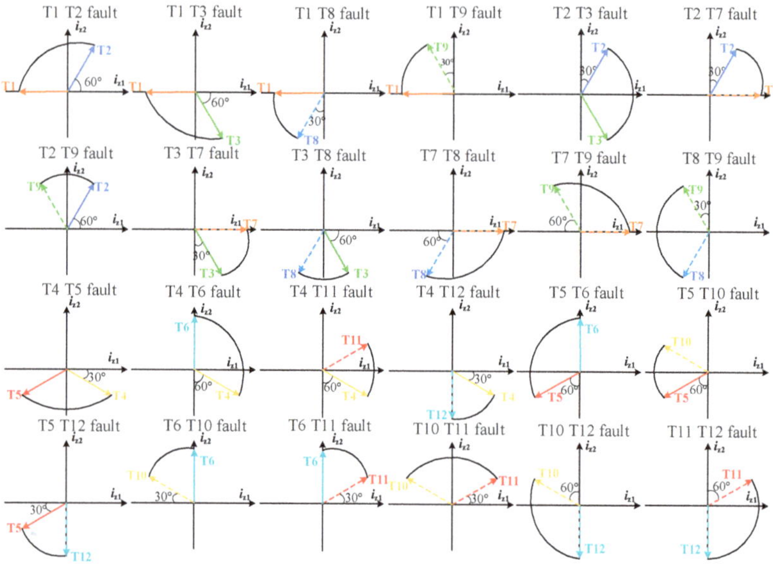


**Figure 6.** Reference direction of $i_{z1}$ and $i_{z2}$ current trace, in the case of two-power-switch faults.

**Table 1.** Fault type code.

| Fault Switch | Fault Type | Fault Code | Fault Switch | Fault Type | Fault Code |
|---|---|---|---|---|---|
| Nothing | Normal | 000000000000 | | | |
| T1 | Single-switch fault of the first set of winding | 100000000000 | T4 | Single-switch fault of the second set of winding | 000100000000 |
| T2 | | 010000000000 | T5 | | 000010000000 |
| T3 | | 001000000000 | T6 | | 000001000000 |
| T7 | | 000000100000 | T10 | | 000000000100 |
| T8 | | 000000010000 | T11 | | 000000000010 |
| T9 | | 000000001000 | T12 | | 000000000001 |
| T1T2 | Failure of two switches of the first set of winding | 110000000000 | T4T5 | Failure of two switches of the second set of winding | 000110000000 |
| T1T3 | | 101000000000 | T4T6 | | 000101000000 |
| T1T8 | | 100000010000 | T4T11 | | 000100000010 |
| T1T9 | | 100000001000 | T4T12 | | 000100000001 |
| T2T3 | | 011000000000 | T5T6 | | 000011000000 |
| T2T7 | | 010000100000 | T5T10 | | 000010000100 |
| T2T9 | | 010000001000 | T5T12 | | 000010000001 |
| T3T7 | | 001000100000 | T6T10 | | 000001000100 |
| T3T8 | | 001000010000 | T6T11 | | 000001000010 |
| T7T8 | | 000000110000 | T10T11 | | 000000000110 |
| T7T9 | | 000000101000 | T10T12 | | 000000000101 |
| T8T9 | | 000000011000 | T11T12 | | 000000000011 |

### 4.2. Principle and Algorithm of BP Neural Network

The neural network is a nonlinear system. Each neuron can output a signal according to the input layer signal and activation function. In addition, a large number of neurons can jointly form a highly nonlinear system [23]. Figure 7 presents a representative neuron structure model.

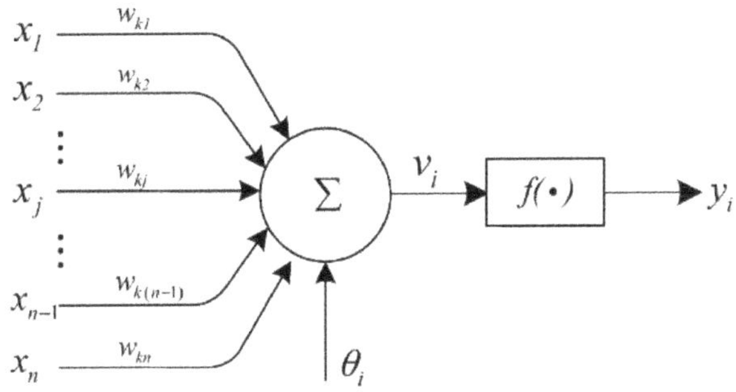


**Figure 7.** Single neuron structure.

In the latter, $x_1$,$x_2$,$x_3$ ... $x_n$ is the input parameter, $W_{k1}$, $W_{k2}$, $W_{k3}$ ... $W_{k(n-1)}$, $W_{kn}$ represents the weight of each parameter, $\theta_i$ denotes the threshold, $f(\cdot)$ s the activation function, and $y_i$ represents the neuron output. The relationship between each variable is given by:

$$\begin{cases} u_i = \sum_{j=1}^{n} w_{ij}x_j \\ v_i = u_i + \theta \end{cases} \tag{21}$$

The output $y_i$ of the neuron is expressed as:

$$y_i = f\left(\sum_{j=1}^{n} w_{ij}x_j + \theta_i\right) \tag{22}$$

The structure of the BP neural network is shown in Figure 8. Due to the gradient descent method, the training can fall into the local minimum, which leads to the inability of effective training [24]. Therefore, this paper uses the Levenbrg–Marquardt algorithm based on the combination of the gradient method and Newton method, in order to optimize the BP neural network. This method has less iterations and a fast iteration speed. Therefore, it can approach the optimal weight faster in order to complete the training.

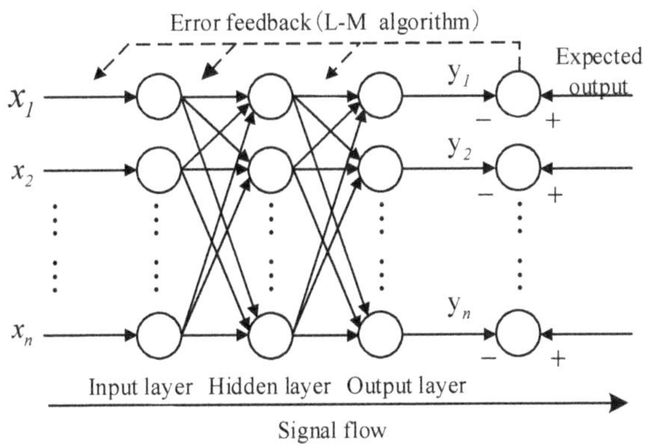


**Figure 8.** Neural network structure model.

Considering that the residual of the output layer is $\gamma(x)$, which has a nonlinear relationship with $x$, the iterative formula of the Gauss Newton method is given by:

$$\begin{cases} x_{k+1} = x_k + \Delta x_{gn} \\ x_{k+1} = x_k - H^{-1}\nabla f \end{cases} \tag{23}$$

where $\nabla f$ and $H$ are respectively the gradient and Hessian matrix of $\gamma(x)$ given by:

$$\begin{gathered}\nabla f = 2J_r^T r = 2\sum_{i=1}^{m} r_i \frac{\delta r_i}{\delta x_i} \\ H_{jk} = 2J_r^T J_r = 2\sum_{i=1}^{m}\left(\frac{\delta r_i}{\delta x_j}\frac{\delta r_i}{\delta x_k} + r_i \frac{\partial^2 r_i}{\partial x_j \partial x_k}\right)\end{gathered} \tag{24}$$

and therefore:

$$\Delta x_{gn} = -\left(J_r^T J_r\right)^{-1} J_r^T r \tag{25}$$

The Levenbrg–Marquardt algorithm improves the iteration step size:

$$y_i = f\left(\sum_{j=1}^{n} w_{ij} x_j + \theta_i\right) \tag{26}$$

where I is the identity matrix and $\mu$ represents the damping factor usually having a value range of $[10^{-8}, 1]$, which ensures that $J_r^T J_r + \mu I$ is a positive definite matrix and that the iteration is in the downward direction.

Before using the neural network diagnosis, the relevant parameters of the neural network should be set according to the needs, the Sigmoid function should be selected as the activation function, the three-layer neural network should be used for training, and the other parameters should be set by default.

## 5. Simulation Analysis

### 5.1. Simulation Analysis of Winding Open-Circuit Fault

In order to verify the proposed fault diagnosis algorithm, a six-phase motor simulation model is built, which is mainly composed of a rectifier, controller, inverter, and motor. The motor parameters are: rated output power P = 10 kW, rated voltage U = 311 V, rated current I = 20 A, rated speed N = 1500 r/min, rated torque T = 60 N·m, pole pair p = 5, quadrature axis and direct axis inductance $L_d = L_q$ = 8.5 mH, and power factor = 0.9.

The current waveform obtained by increasing and decreasing load and disturbance under normal conditions of the motor is shown in Figure 9. After the motor speed is increased, the load is set to 30 N·m and runs stably at 500 r/min. The six-phase stator current waveform is balanced and the sinusoidal degree is satisfactory. When the load of the motor suddenly changes to 45 N·m at 0.2 s, the current amplitude rapidly increases and runs in a stable state. When the load decreases to 40 N·m at 0.4 s, the current decreases. When the interference is added at 0.6 s, the motor current slightly increases and then decreases to the stable state.

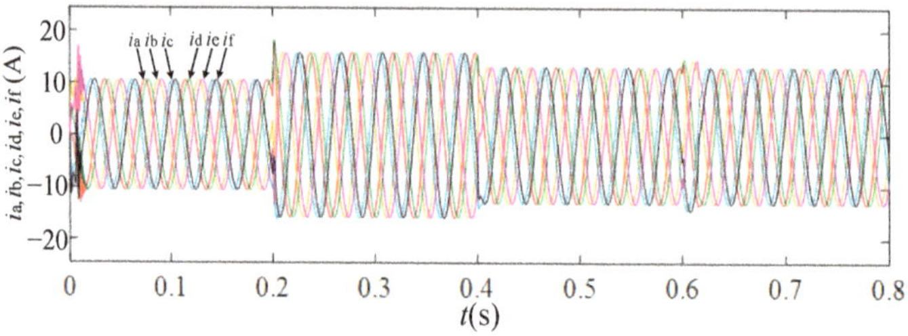


**Figure 9.** Current waveform of load and disturbance during normal operation.

The fault index waveform corresponding to this case is shown in Figure 10. The fault index of the six-phase open circuit is null. In addition, there is no rise when the motor stably operates, suddenly increases or decreases load, and suddenly adds disturbance. The fault index has no false diagnosis under different conditions without fault, which proves that the diagnosis method has a good robustness.

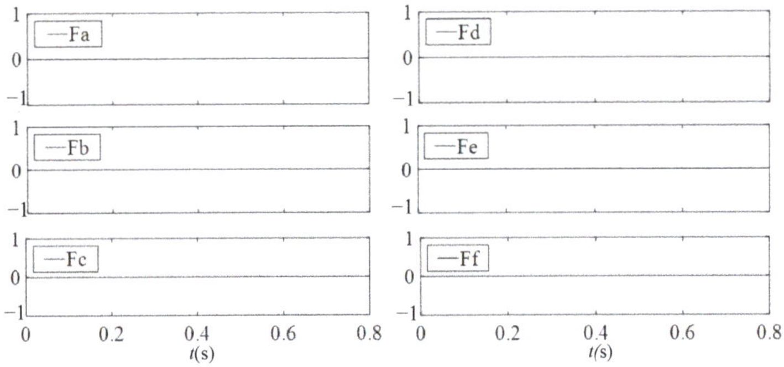


**Figure 10.** Fault indicators under normal operation load and disturbance.

The current waveform of the open-circuit fault of phase *a* at 0.3 s is shown in Figure 11. The load torque is 30 N·m, and the phase failure of phase *a* occurs at 0.3 s. The phase *a* current becomes zero after 0.3 s. Among the remaining four-phase currents, phase *b* and phase *c* have the same amplitude and complementary waveforms, and the waveform amplitude and phase of phase *d*, *e* and *f* change to varying degrees.

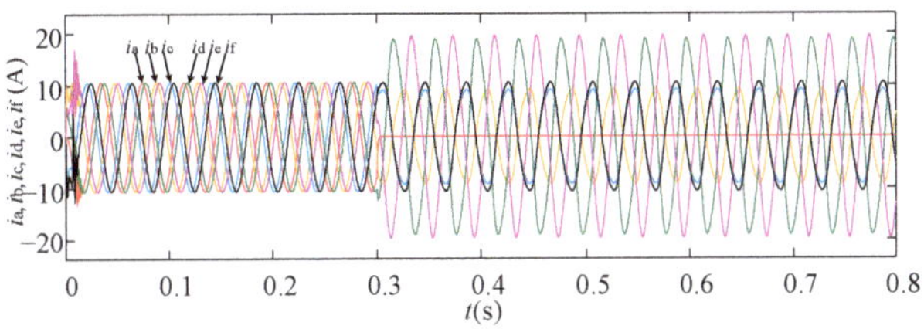


**Figure 11.** Phase *a* current waveform of open-circuit fault in 0.3 s.

The fault index waveform under these conditions is shown in Figure 12. After 0.3 s, the phase *a* fault index rises to 1 and the fault indexes of other phases do not change.

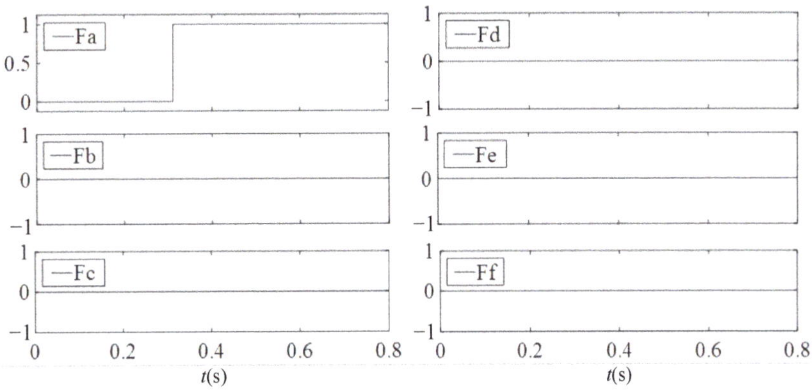


**Figure 12.** Phase *a* fault indicator waveform when open fault occurs within 0.3 s.

The current waveform of the open-circuit fault of phase *a* and phase *d* at 0.3 s and 0.5 s is shown in Figure 13. The load torque is 30 N·m, the phase failure of phase *a* and phase *d* occurs at 0.3 s and 0.5 s, respectively. The phase *a* current becomes zero after 0.3 s and the phase *d* current also becomes zero after 0.5 s.

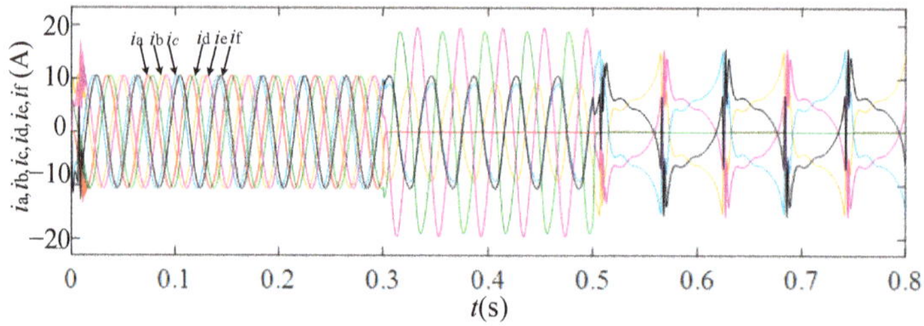

**Figure 13.** Current waveforms of phase *a* and phase *d* at 0.3 s and 0.5 s open-circuit fault, respectively.

The fault index waveform under these conditions is shown in Figure 14. After 0.3 s, the phase *a* fault index rises to 1, and the phase *d* fault index rapidly rises to 1 within the second half cycle of 0.5 s, so as to accurately locate the two-phase open-circuit fault.

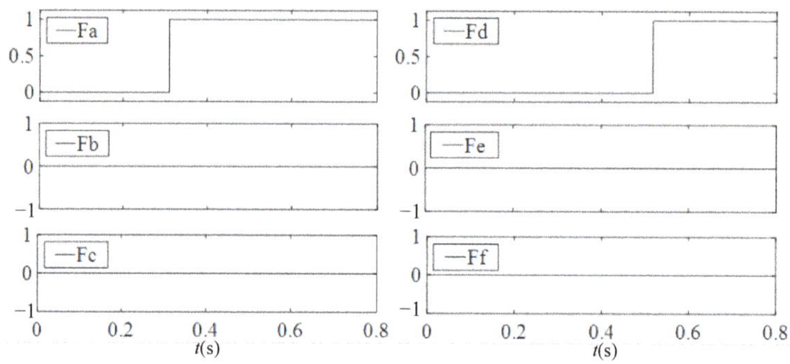

**Figure 14.** Fault index waveforms of phase *a* and phase *d* at 0.3 s and 0.5 s open-circuit fault, respectively.

The current waveform of the open-circuit fault of phase *a* and phase *f* at 0.3 s and 0.5 s is shown in Figure 15. The load torque is 30 N·m, the phase failure of phase *a* and phase *f* occurs at 0.3 s and 0.5 s, respectively. The phase *a* current becomes zero after 0.3 s and the phase *f* current also becomes zero after 0.5 s. Among the remaining four-phase currents, phase *b* and phase *c* have the same amplitude and opposite phase. Similarly, phase *d* and phase *e* have the same amplitude and opposite phase.

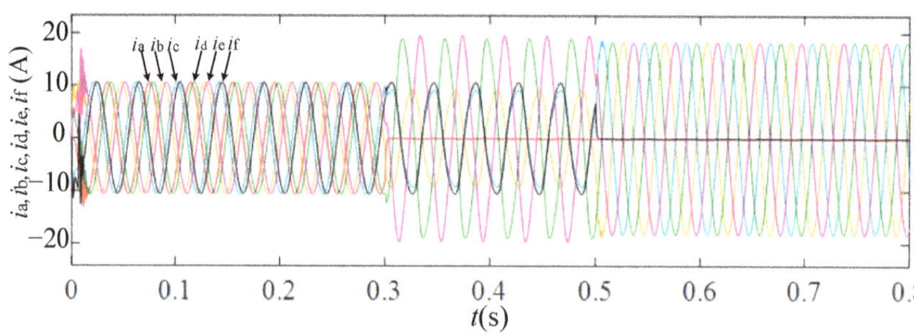

**Figure 15.** Current waveforms of phase *a* and phase *f* at 0.3 s and 0.5 s open-circuit fault, respectively.

The fault index waveform under these conditions is shown in Figure 16. After 0.3 s, the phase *a* fault index rises to 1, and the phase *f* fault index rapidly rises to 1 within the second half cycle of 0.5 s, so as to accurately locate the two-phase open-circuit fault.

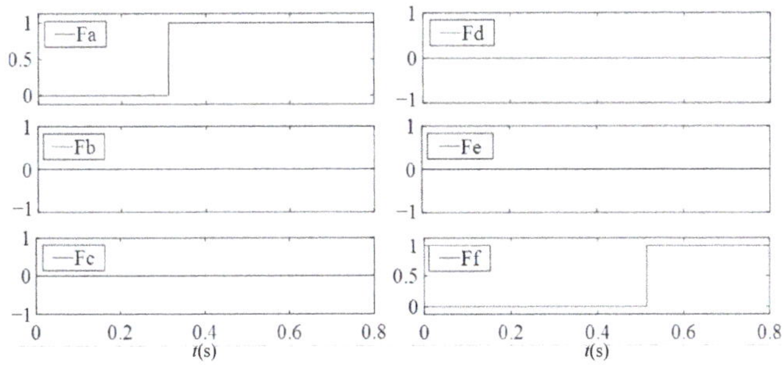


**Figure 16.** Fault index waveforms of phase *a* and phase *f* at 0.3 s and 0.5 s open-circuit fault, respectively.

### 5.2. Switch Fault Simulation

Figure 17 shows the harmonic plane current trajectories obtained by simulation in the case of 12 single-switch faults. It can be observed that the fault trajectories of the 12 switches of the six-phase VSI inverter are consistent with the theory.

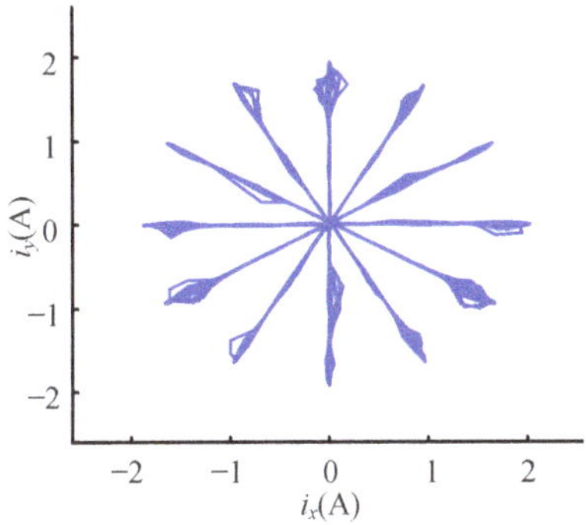


**Figure 17.** $z_1$-$z_2$ harmonic plane current trace in case of single-switch fault.

Figure 18 illustrates the harmonic plane current trajectories obtained by simulation, in the case of 24 types of two-switch faults. The harmonic plane current trajectories are fan-shaped patterns with different center angles in the case of two-switch faults, that are used as characteristic signals for fault diagnosis.

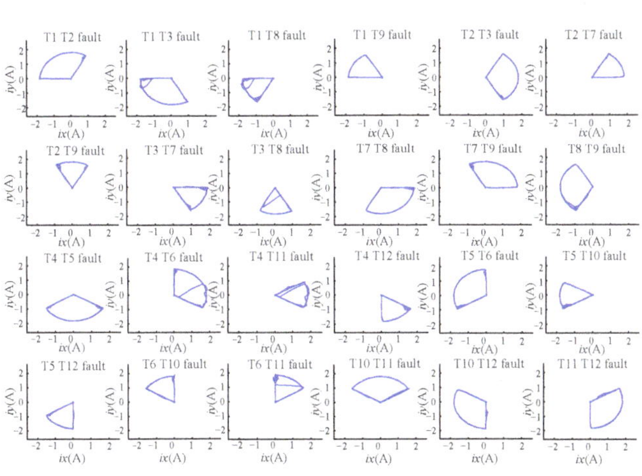


**Figure 18.** $z_1$-$z_2$ harmonic plane current trace in case of two-switch fault.

Figure 19 presents the training model of the neural network. The current vector characteristic trajectories of single-switch fault and two-switch fault, are used as samples for training. The hidden layer neuron is 9, the expected error is 0.001, and the maximum number of iterations is 1000. The Newff function is used for training and the Sigmoid function is used as the activation function. The feedback process of the optimization error is performed using the L–M algorithm.

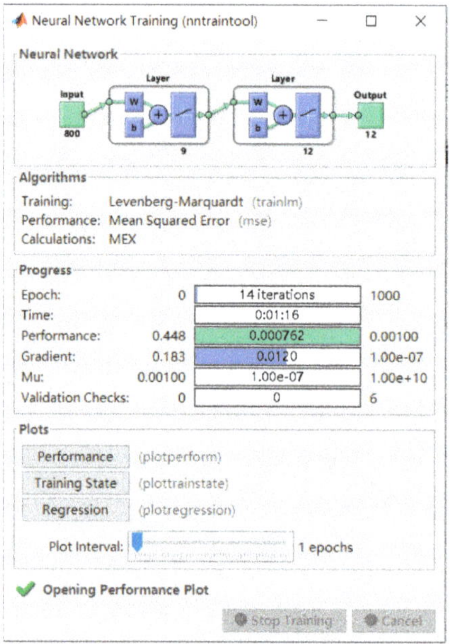


**Figure 19.** Neural network training model.

Figure 20 shows the training error convergence curve of the neural network. The neural network optimized using the L–M algorithm has a high convergence speed. When the training is completed 14 times, the mean square error is $7.6208 \times 10^{-4}$, which meets the error requirements.

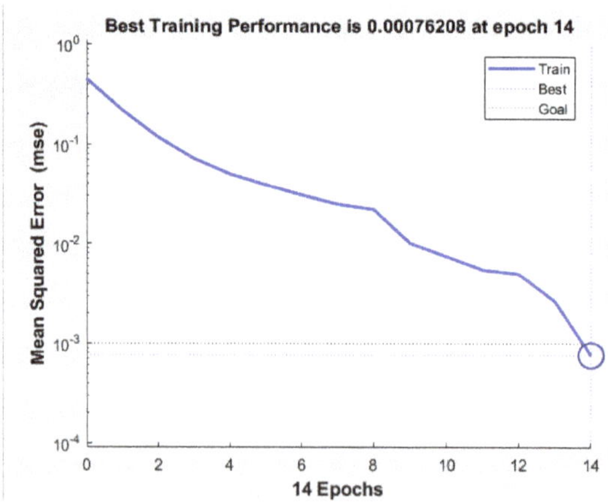


**Figure 20.** Error convergence curve of neural network.

The coincidence degree between the error regression curve and the expected curve shown in Figure 21 is high. This proves that the neural network has good fitting and classification ability for different fault current trajectories.

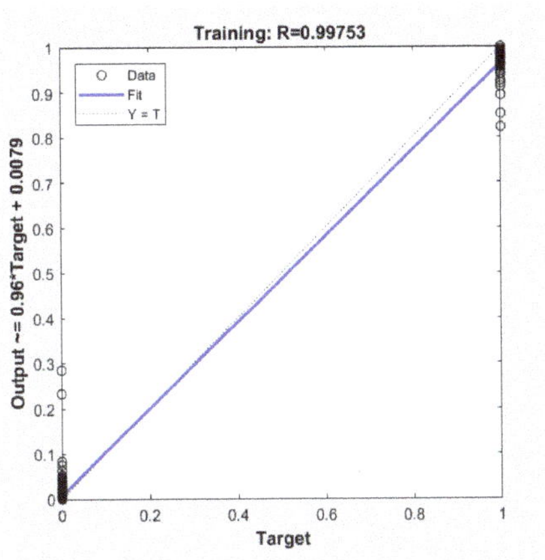


**Figure 21.** Regression curve of neural network training error.

The diagnosis results of the neural network are shown in Table 2. It can be seen that the trained network can identify the single-switch and two-switch faults. The error between the output value of 12 neurons in the output layer and the coding value of the fault type is very small, which leads to an accurate diagnosis.

**Table 2.** Neural network diagnosis results.

| Serial Number | Actual Output | Ideal Output | Fault Type |
|---|---|---|---|
| 1 | 0.9574 0.0003<br>0.0923 0.0119<br>0.0492 0.0087<br>0.0029 0.0565<br>0.0984 0.0278<br>0.0106 0.0012 | 100000000000 | T1 |
| 2 | 0.0094 0.9869<br>0.0147 0.0131<br>0.0106 0.0236<br>0.0293 0.0001<br>0.0882 0.0017<br>0.0453 0.0031 | 010000000000 | T2 |
| 3 | 0.0895 0.0282<br>0.9365 0.0013<br>0.0003 0.0002<br>0.0718 0.1507<br>0.0004 0.0077<br>0.0002 0.0221 | 001000000000 | T3 |
| 4 | 0.0016 0.0000<br>0.0071 0.9399<br>0.0615 0.0297<br>0.0001 0.0578<br>0.0000 0.0313<br>0.0001 0.0373 | 000100000000 | T4 |

Table 2. *Cont.*

| Serial Number | Actual Output | Ideal Output | Fault Type |
|---|---|---|---|
| 5 | 0.0002 0.0181<br>0.0001 0.0198<br>0.9987 0.0001<br>0.0003 0.0001<br>0.0000 0.0001<br>0.2961 0.0052 | 000010000000 | T5 |
| 6 | 0.0068 0.0009<br>0.0001 0.0384<br>0.0000 0.9631<br>0.0064 0.0002<br>0.0103 0.0000<br>0.0200 0.0000 | 000001000000 | T6 |
| 7 | 0.0010 0.0060<br>0.0217 0.0580<br>0.1198 0.0036<br>0.9840 0.0385<br>0.0447 0.0691<br>0.0316 0.0005 | 000000100000 | T7 |
| 8 | 0.1813 0.0000<br>0.0052 0.0019<br>0.0398 0.0241<br>0.0077 0.9211<br>0.0005 0.0055<br>0.0000 0.0904 | 000000010000 | T8 |
| 9 | 0.0049 0.0096<br>0.0003 0.0111<br>0.0893 0.0558<br>0.0040 0.0764<br>0.9695 0.1351<br>0.0002 0.0058 | 000000001000 | T9 |
| 10 | 0.0156 0.0001<br>0.0099 0.0003<br>0.085 0.0324<br>0.0000 0.0054<br>0.0009 0.9973<br>0.0001 0.0237 | 000000000100 | T10 |
| 11 | 0.0000 0.0019<br>0.0021 0.0021<br>0.0055 0.0009<br>0.0001 0.0000<br>0.0014 0.6796<br>0.9954 0.0058 | 000000000010 | T11 |
| 12 | 0.0137 0.0000<br>0.0233 0.0171<br>0.0014 0.0001<br>0.0244 0.0005<br>0.0000 0.0001<br>0.0227 0.9101 | 000000000001 | T12 |
| 13 | 0.9455 0.9961<br>0.0199 0.0014<br>0.0007 0.0059<br>0.0699 0.0720<br>0.0533 0.0006<br>0.0000 0.0012 | 110000000000 | T1T2 |

**Table 2.** *Cont.*

| Serial Number | Actual Output | Ideal Output | Fault Type |
|---|---|---|---|
| 14 | 0.9635 0.0585<br>0.9787 0.0003<br>0.0013 0.0002<br>0.1210 0.0322<br>0.0418 0.1409<br>0.0006 0.0002 | 101000000000 | T1T3 |
| 15 | 0.9465 0.0002<br>0.0262 0.0004<br>0.0414 0.0618<br>0.0002 0.9202<br>0.0022 0.0302<br>0.0000 0.0214 | 100000010000 | T1T8 |
| 16 | 0.9799 0.1623<br>0.0730 0.0002<br>0.0022 0.0116<br>0.0092 0.0692<br>0.9608 0.1598<br>0.0000 0.0007 | 100000001000 | T1T9 |
| 17 | 0.0247 0.9807<br>0.9339 0.0015<br>0.0016 0.0001<br>0.0754 0.0216<br>0.0840 0.0010<br>0.0391 0.0007 | 011000000000 | T2T3 |
| 18 | 0.0001 0.9602<br>0.0222 0.0063<br>0.0264 0.0007<br>0.9491 0.0050<br>0.0870 0.0007<br>0.0175 0.0411 | 010000100000 | T2T7 |
| 19 | 0.0004 0.9792<br>0.0095 0.0001<br>0.0016 0.0549<br>0.0009 0.0012<br>0.9561 0.0095<br>0.0141 0.0937 | 010000001000 | T2T9 |
| 20 | 0.0479 0.0051<br>0.9673 0.0080<br>0.0000 0.0008<br>0.9701 0.0603<br>0.0000 0.0512<br>0.0000 0.0003 | 001000100000 | T3T7 |
| 21 | 0.0076 0.0000<br>0.9344 0.0029<br>0.0003 0.0068<br>0.0003 0.9902<br>0.0001 0.0724<br>0.0958 0.0069 | 001000010000 | T3T8 |
| 22 | 0.0000 0.0000<br>0.0195 0.9876<br>0.8665 0.0032<br>0.0001 0.0001<br>0.0000 0.0051<br>0.0507 0.0106 | 000110000000 | T4T5 |

**Table 2.** *Cont.*

| Serial Number | Actual Output | Ideal Output | Fault Type |
|---|---|---|---|
| 23 | 0.0002 0.0186<br>0.0001 0.9968<br>0.0311 0.9535<br>0.0064 0.0002<br>0.0000 0.0000<br>0.0366 0.0000 | 000101000000 | T4T6 |
| 24 | 0.0000 0.0092<br>0.0005 0.9422<br>0.0242 0.0008<br>0.0000 0.0008<br>0.0000 0.0036<br>0.9658 0.0414 | 000100000010 | T4T11 |
| 25 | 0.0000 0.0000<br>0.0051 0.9606<br>0.0115 0.0004<br>0.0428 0.0034<br>0.0000 0.0009<br>0.0274 0.9119 | 000100000001 | T4T12 |
| 26 | 0.0002 0.0494<br>0.0000 0. 0011<br>0.8858 0.9982<br>0.0017 0.0001<br>0.0032 0.0001<br>0.0969 0.0000 | 000011000000 | T5T6 |
| 27 | 0.1324 0.0013<br>0.0000 0.0000<br>0.9892 0.0006<br>0.0000 0.0357<br>0.0004 0.8010<br>0.0008 0.0048 | 000010000100 | T5T10 |
| 28 | 0.0001 0.0001<br>0.0122 0.1004<br>0.9647 0.0000<br>0.0021 0.0003<br>0.0000 0.0363<br>0.0163 0.7030 | 000010000001 | T5T12 |
| 29 | 0.0315 0.0002<br>0.0368 0.0010<br>0.0001 0.9669<br>0.0163 0.0004<br>0.0328 0.8387<br>0.0001 0.0033 | 000001000100 | T6T10 |
| 30 | 0.0000 0.0015<br>0.0001 0.1827<br>0.0001 0.6872<br>0.0673 0.0000<br>0.0026 0.0000<br>0.9825 0.0000 | 000001000010 | T6T11 |
| 31 | 0.0050 0.0000<br>0.1506 0.0274<br>0.0174 0.0003<br>0.9406 0.9372<br>0.0022 0.0685<br>0.0001 0.0021 | 000000110000 | T7T8 |

**Table 2.** *Cont.*

| Serial Number | Actual Output | Ideal Output | Fault Type |
|---|---|---|---|
| 32 | 0.0724 0.0930<br>0.0004 0.0007<br>0.0014 0.0508<br>0.9778 0.1332<br>0.9280 0.0191<br>0.0000 0.0002 | 000000101000 | T7T9 |
| 33 | 0.0837 0.0270<br>0.0092 0.0007<br>0.0446 0.0027<br>0.0275 0.9727<br>0.9892 0.1027<br>0.0000 0.0011 | 000000011000 | T8T9 |
| 34 | 0.0000 0.0000<br>0.0401 0.0074<br>0.0001 0.0153<br>0.0005 0.0002<br>0.0005 0.9980<br>0.6568 0.1243 | 000000000110 | T10T11 |
| 35 | 0.0002 0.0001<br>0.0033 0.1092<br>0.1596 0.0033<br>0.0033 0.0407<br>0.0000 0.9995<br>0.0083 0.9203 | 000000000101 | T10T12 |
| 36 | 0.0008 0.0044<br>0.0451 0.0283<br>0.0005 0.0000<br>0.0014 0.0002<br>0.0000 0.0009<br>0.9609 0.9795 | 000000000011 | T11T12 |

## 6. Experimental Verification

Figure 22 shows the overall block diagram of the six-phase PMSM control system, which provides the required DC bus voltage for the inverter after rectification by a three-phase uncontrolled rectifier bridge. The DSP-28335 chip outputs the IGBT driving waveform. The resolver and current hall sensor detect the position and phase current of the motor rotor. The Labview host computer platform can receive the motor running state and fault index waveform in DSP in real time.

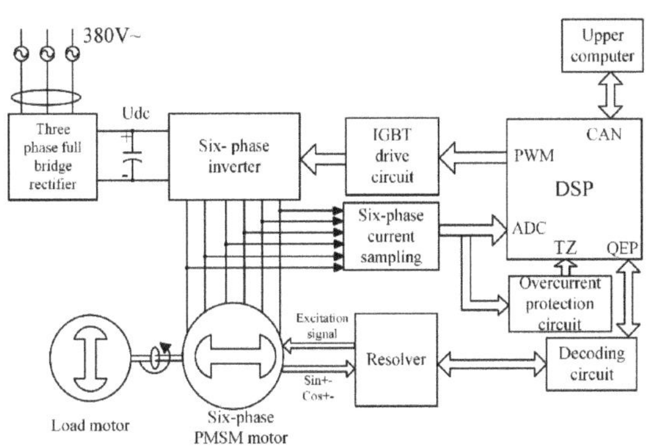


**Figure 22.** Overall block diagram of the six-phase PMSM control system.

In order to verify the proposed fault diagnosis algorithm, a six-phase motor experimental platform based on DSP is built. The experimental platform is shown in Figure 23. It is mainly composed of a rectifier, controller, inverter, and motor. The motor parameters are: rated output power P = 10 kW, rated voltage U = 311 V, rated current I = 20 A, rated speed N = 1500 r/min, rated torque T = 60 N·m, pole pair p = 5, quadrature axis and straight axis inductor $L_d = L_q$ = 8.5 mH, and power factor = 0.94.

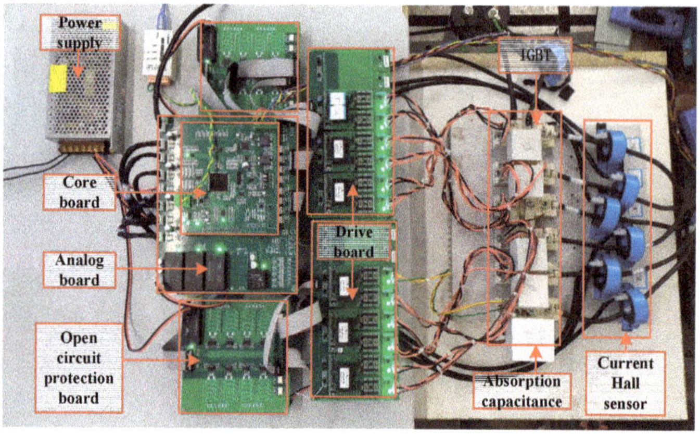


**Figure 23.** Experimental platform of the six-phase motor control system.

Figure 24 shows the phase current waveform when the motor is running at 500 r/min. The current is observed by collecting the voltage at the two ends of the sampling resistance. The voltage value is 100 mV, which corresponds to a current value of 2 A. The amplitudes of the phase currents are equal with a high sinusoidal degree. The current waveforms of phase *a*, *b*, and *d* under the same conditions are shown in Figure 24a. The phase difference between phase *d* current and phase *a* current is 30 degrees. The waveform of the motor is smooth and sinusoidal during normal operation.

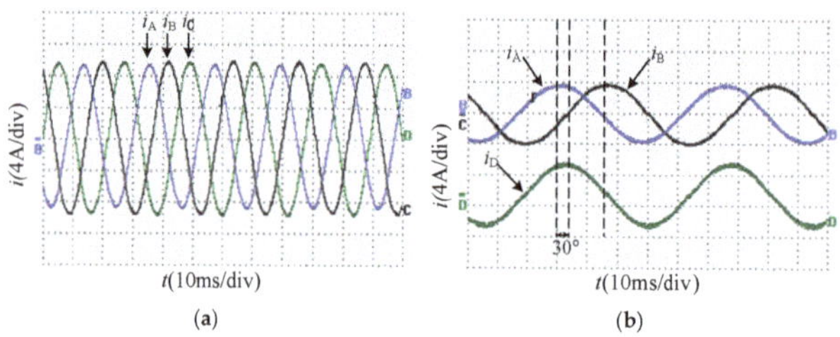


**Figure 24.** Waveform of motor under stable operation at 500 r/min: (**a**) current waveform of phase *a*, *b*, and *c*; (**b**) current waveform of phase *a*, *b*, and *d*.

Figure 25 shows the motor speed and six-phase current waveform under sudden load. When the motor speed increases from 400 r/min to 500 r/min in Figure 25a, the actual motor speed gradually increases in the form of the slope with the given speed; the transition is gentle and the overshoot is small in the process of speed increase. In Figure 25b, the amplitude of the motor phase current gradually increases during loading, and the dynamic adjustment process is smooth and stable.

Figure 26 shows the motor speed and the six-phase current waveform during sudden load reduction. When the motor decelerates from 500 r/min to 400 r/min in Figure 26a, the actual motor speed gradually decreases in the form of the slope with the given speed. The transition is smooth and the overshoot is small during deceleration. In Figure 26b, the amplitude of the motor phase current gradually decreases during load reduction, and the dynamic regulation process is smooth and stable.

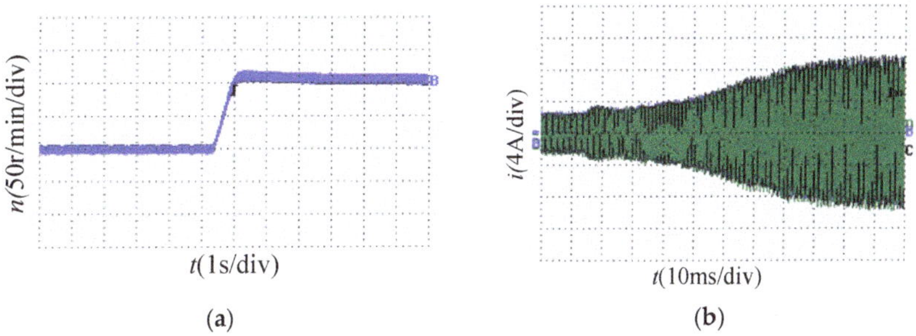


**Figure 25.** Waveform of the motor under sudden load: (**a**) speed waveform; (**b**) current waveform of the six-phase motor.

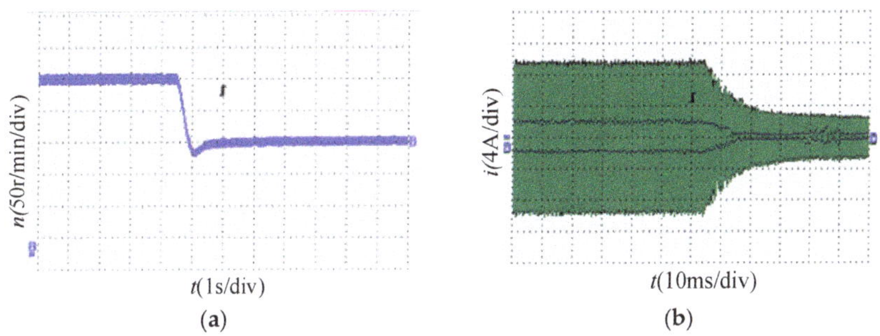


**Figure 26.** Waveform of the motor under sudden load: (**a**) speed waveform; (**b**) current waveform of the six-phase motor.

Figure 27 shows the fault index waveform of the motor during stable operation and sudden load increase and decrease. During stable operation and sudden load increase and decrease in the motor, the values of the six-phase open phase fault index are zero, and there is no change. The fault index has no misdiagnosis under fault-free operation, which proves that it has good robustness.

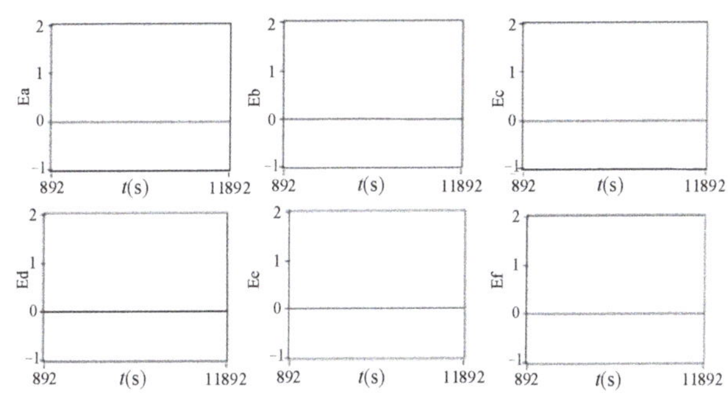


**Figure 27.** Fault index waveform under stable operation at 500 r/min.

Figure 28 presents the phase current and fault index waveform of motor phase $a$ in the case of phase failure. Figure 28a presents the current waveform before and after the fault, in which the blue, black, and green lines respectively represent the current waveform of phase $a$, $b$, and $c$ of the motor. After the phase failure of phase $a$, the current value of this phase drops to zero, the current amplitude of phase *b* and *c* becomes almost twice the original one, and the phase positions are complementary. Figure 28b illustrates the waveform of the fault phase current and fault index. After the phase failure of phase *a*, the phase failure index Ea of phase *a* increases to a high level after 32 ms, and the optimized fault index has no obvious burr and peak pulse. Therefore, the fault diagnosis and early warning can be performed in a short time.

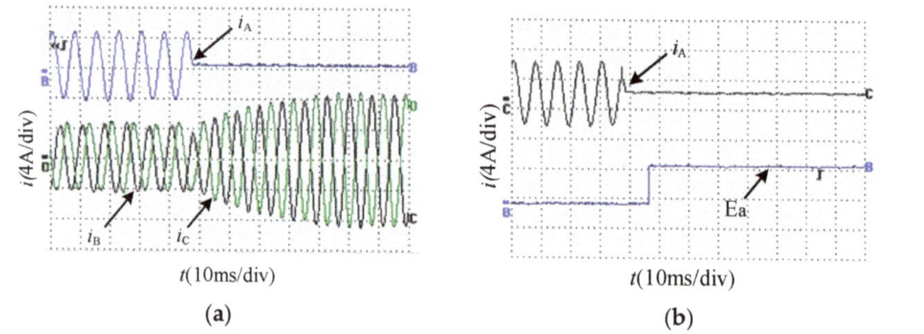


**Figure 28.** Phase current and fault index waveform in case of phase failure of phase *a*: (**a**) fault current waveform; (**b**) fault phase current and fault index waveform.

Figure 29 shows the fault index waveform of the six-phase disconnection when the *a* phase disconnection fault occurs. After the fault occurs, only the fault index Ea of phase *a* rises to 1, while the fault indexes of the other phases do not change, and there is no misdiagnosis.

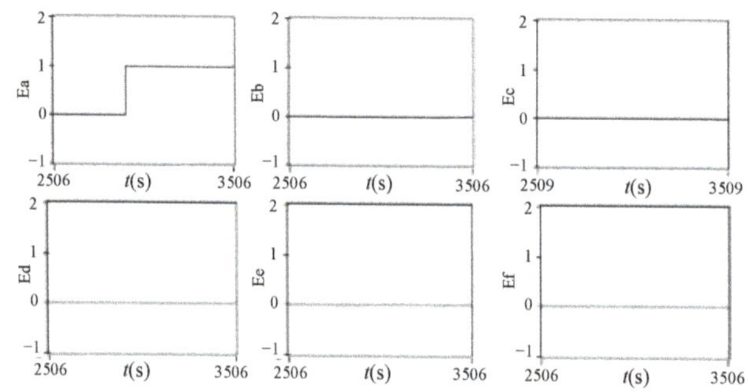


**Figure 29.** Fault index waveform in case of phase failure of phase *a*.

The phase current and fault index waveform of phase *a* and phase *d* of the motor with phase failure, are shown in Figure 30. The blue, black, and green lines shown in Figure 30a represent the current waveforms of motor phases *a*, *b*, and *d*, respectively. After the phase failure of phase *a* and *d*, the corresponding phase current value drops to zero, and the amplitude of the phase *b* current increases. The six-phase failure index waveform of phase *a* and phase *d* is shown in Figure 31. Before the phase failure occurs, the six-phase failure index remains null. After the failure occurs, the failure indexes Ea and Ed of phase *a* and phase *d* rise to 1, while the other four-phase failure indexes remain null without misdiagnosis.

Figure 32 shows the phase current and fault index waveform of phase failure of phase *a* and phase *f*. The blue, black, and green lines shown in Figure 32a represent the current waveforms of motor phases *a*, *b*, and *f*, respectively. A phase failure occurs in phase *a* and phase *f*, and the current value drops to zero. After the failure of phase *a* and phase *f*, their failure indexes Ea and Ef jump to the high level after 22 ms and 24 ms, respectively. These situations show that the proposed open phase fault diagnosis method has a high diagnosis speed, and can accurately locate the fault phase position.

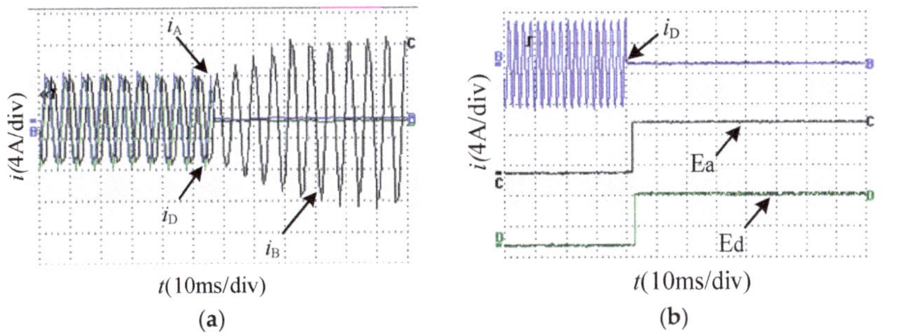


**Figure 30.** Phase current and fault index waveform in case of phase failure of phase *a* and phase *d*: (**a**) fault current waveform; (**b**) fault phase current and fault index waveform.

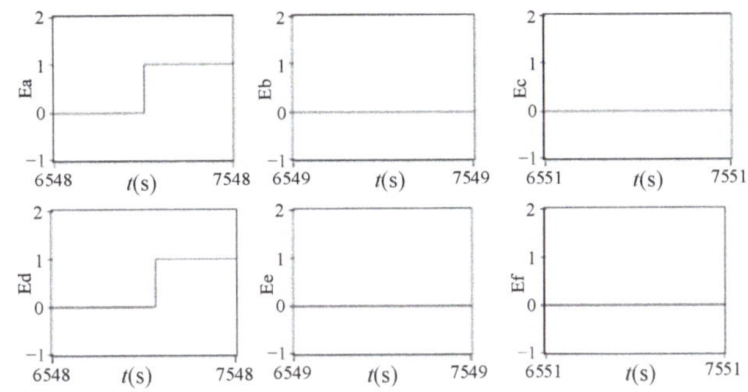


**Figure 31.** Fault index waveform in case of phase failure of phase *a* and phase *d*.

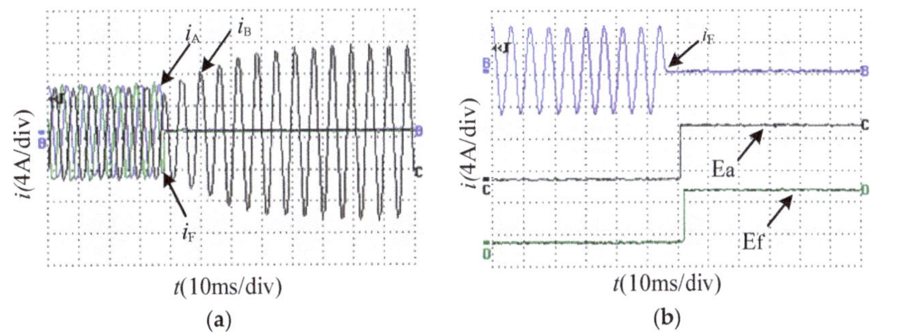


**Figure 32.** Phase current and fault index waveform in case of phase failure of phase *a* and phase *f*: (**a**) fault current waveform; (**b**) fault phase current and fault index waveform.

The six-phase failure index waveform of phase *a* and phase *f* is shown in Figure 33. During normal operation, the failure indexes of the two phases are 0. After phase *a* and phase *f* fail, their failure indexes Ea and Ef rise to 1, and the failure indexes of the other phases do not change. It can then be deduced that the fault index of each phase can accurately locate the position of the fault phase. In addition, the diagnosis results are accurate, and there will be no false diagnosis. Finally, the experimental results are coherent with the theoretical analysis, which verifies the correctness and efficiency of the proposed diagnosis algorithm.

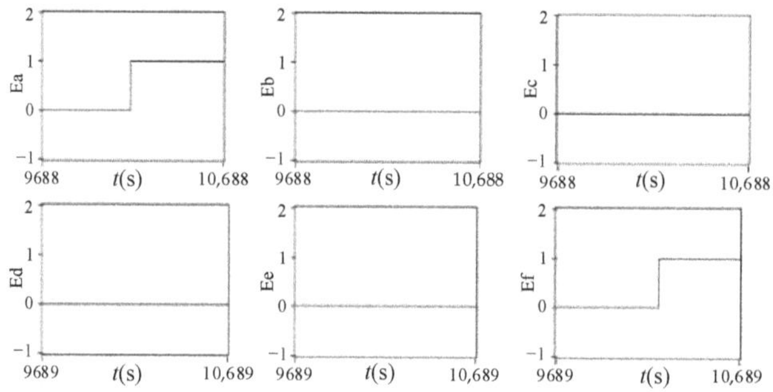


**Figure 33.** Fault index waveform in case of phase failure of phase *a* and phase *f*.

## 7. Conclusions

This paper proposed a fault diagnosis method based on vector space decoupling (VSD). The open phase fault index is derived according to the VSD inverse transformation and the current constraints after the fault. The fault index is optimized in order to prevent misdiagnosis and improve reliability. The diagnosis of a single-phase and double-phase fault can be performed in one fundamental cycle. In addition, by considering the unique harmonic plane current track after VSD transformation as the characteristic signal, a neural network is used to identify different fault current tracks, in order to diagnose the switch fault. The simulation and experimental results show that the proposed scheme has a high detection speed, ability to deal with several fault types, and good robustness. Finally, the proposed approach can accurately locate the winding open-circuit fault and power switch fault, which can be used for inverter open-circuit fault diagnosis and provides a basis for maintenance personnel.

**Author Contributions:** Conceptualization, H.G.; formal analysis, H.G. and J.G.; data curation, Z.H. and J.G.; writing—review and editing, B.Z. and Y.D. All authors have read and agreed to the published version of the manuscript.

**Funding:** This research work was funded by the National Natural Science Foundation of China grant number 51177031.

**Data Availability Statement:** The data presented in this study are available on request from the corresponding author.

**Conflicts of Interest:** The authors declare no conflict of interest.

## References

1. Boglietti, A.; Cavagnino, A.; Tenconi, A.; Vaschetto, S. The Safety Critical Electric Machines and Drives in the more Electric Aircraft: A survey. In Proceedings of the 2009 35th Annual Conference of IEEE Industrial Electronics, Porto, Portugal, 3–5 November 2009; pp. 2587–2594.
2. Ma, W.M.; Wang, D.; Cheng, S.W.; Chen, J.Q. Common basic scientific problems and technological development frontier of high performance motor system. *Chin. J. Electr. Eng.* **2016**, *36*, 2025–2035.
3. Yan, M.; Ma, W.M.; Ou, Y.B.; Mang, M.Z.; Lu, L. Study on performance of dual nine phase energy storage motor system. *Chin. J. Electr. Eng.* **2015**, *35*, 6.
4. Jeong, Y.S.; Sul, S.K.; Schulz, S.E.; Patel, N.R. Fault Detection and Dault Tolerant Control of Interior Permanent Magnet Motor Drive System for Electric Vehicle. *IEEE Trans. Ind. Appl.* **2003**, *41*, 1458–1463.
5. IEEE Committee Report. Report of Large Motor Reliability Survey of Industrial and Commercial Installations. *IEEE Trans. Ind. Appl.* **1985**, *23*, 153–158.
6. Gao, H.W.; Su, J.Y.; Yang, G.J. Open circuit fault diagnosis method of five phase inverter based on pole voltage error standardization. *Chin. J. Electr. Eng.* **2016**, *36*, 10.
7. Zhang, Z.; Luo, G.A. Hybrid Diagnosis Method for Inverter Open-Circuit Faults in PMSM Drives. *China Electrotech. Soc. Trans. Electr. Mach. Syst.* **2020**, *4*, 180–189. [CrossRef]
8. Diallo, D.; Benbouzid, M.E.H.; Hamad, D.; Pierre, X. Fault Detection and Diagnosis in an Induction Machine Drive: A Pattern Recognition Approach Based on Concordia Stator Mean Current Vector. *IEEE Trans. Energy Convers.* **2005**, *20*, 512–519. [CrossRef]

9. Yang, Z.L.; Wu, Z.G.; Li, H. Open circuit fault diagnosis method of inverter based on DC side current detection. *Chin. J. Electr. Eng.* **2008**, *28*, 18–22.
10. Dabour, S.M.; Rashad, E.; Hammad, R. Open-Circuit Fault Detection of Asymmetrical Six-phase Induction Motor Fed from Z-Source Inverter. In Proceedings of the 2019 21st International Middle East Power Systems Conference (MEPCON), Cairo, Egypt, 17–19 December 2019; pp. 1216–1222.
11. Zhang, Z.; Qi, S.; Liu, X. A Diagnosis Method for Open Circuit Fault of Dual Three-Phase Permanent Magnet Synchronous Motor Drive System. In Proceedings of the 2018 21st International Conference on Electrical Machines and Systems (ICEMS), Jeju, Korea, 7–10 October 2018; pp. 1425–1430.
12. Yang, C.; Gui, W.; Chen, Z.; Zhang, J.; Peng, T.; Yang, C. Voltage Difference Residual-Based Open-Circuit Fault Diagnosis Approach for Three-Level Converters in Electric Traction Systems. *IEEE Trans. Power Electron.* **2019**, *35*, 3012–3028. [CrossRef]
13. Chen, Y.; Zhang, J.J.; Chen, Z.Y. On line diagnosis of open circuit fault of three-phase inverter circuit based on current observer. *J. Electrotech.* **2019**, *34*, 609–617.
14. Chen, C.B.; Wang, X.X.; Gao, S.; Zhang, Y.M. Open circuit fault diagnosis method of inverter based on interval sliding mode observer. *Chin. J. Electr. Eng.* **2020**, *40*, 4569–4579.
15. Danjiang, C.; Yinzhong, Y. Open circuit fault diagnosis method of three-level inverter based on multi neural network. *J. Electrotech.* **2013**, *28*, 120–126.
16. Cai, F.Y. *Research on Fault Diagnosis and Fault Tolerant Operation of Underwater Permanent Magnet Synchronous Motor Driver*; Harbin Institute of Technology: Harbin, China, 2016.
17. Zhang, X.D.; Xu, B.J.; Wu, G.X.; Zuo, Y.B. Research on wind turbine fault diagnosis expert system. *Manuf. Autom.* **2014**, *36*, 21–25.
18. Trabelsi, M.; Nguyen, N.K.; Semail, E. Real-Time Switches Fault Diagnosis Based on Typical Operating Characteristics of Five-Phase Permanent Magnetic Synchronous Machines. *IEEE Trans. Ind. Electron.* **2016**, *63*, 4683–4694. [CrossRef]
19. Zhou, C.Y.; Shen, Y.X. Open circuit fault diagnosis of double three-phase voltage source inverter based on wavelet analysis. *J. Mot. Control.* **2020**, *24*, 65–75.
20. Chen, Y.; Liu, Z.L.; Chen, Z.Y. Fast diagnosis and location method of inverter open circuit fault based on current vector feature analysis. *J. Electrotech.* **2018**, *33*, 883–891.
21. Cui, J.; Wang, Q.; Gong, C.Y. Research on inverter power tube fault diagnosis technology combined with wavelet and Concordia transform. *Chin. J. Electr. Eng.* **2015**, *35*, 3110–3116.
22. Hu, Y.; Li, Y.; Ma, X.; Li, X.; Huang, S. Flux-Weakening Control of Dual Three-Phase PMSM Based on Vector Space Decomposition Control. *IEEE Trans. Power Electron.* **2021**, *36*, 8428–8438. [CrossRef]
23. Xie, R.; Wang, X.M.; Li, Y.; Zhao, K.R. Research and application on improved BP neural network algorithm. In Proceedings of the 2010 5th IEEE Conference on Industrial Electronics and Applications, Taichung, Taiwan, 15–17 June 2010; pp. 1462–1466.
24. Cheng, H.; Cui, L.; Li, J. Application of improved BP neural network based on LM algorithm in desulfurization system of thermal power plant. In Proceedings of the 2017 Chinese Automation Congress (CAC), Jinan, China, 20–22 October 2017; pp. 5917–5920.

energies

MDPI

*Article*

# Analysis and Operation of a High DC-AC Gain 3-$\phi$ Capacitor Clamped Boost Inverter

**Dogga Raveendhra [1,*], Poojitha Rajana [2], Beeramangalla Lakshminarasaiah Narasimharaju [3], Yaramasu Suri Babu [4], Eugen Rusu [5,*] and Hady Habib Fayek [6]**

1 EEE Department, Gokaraju Rangaraju Institute of Engineering and Technology, Hyderabad 500090, India
2 Zunik Energies Pvt. Ltd., I-2, TIDES Business Incubator, IIT Roorkee, Roorkee 247667, India; ceo@zunikenergies.com
3 Electrical Engineering Department, National Institute of Technology, Warangal 506004, India; blnraju@nitw.ac.in
4 Electrical & Electronics Engineering Department, R.V.R. & J.C. College of Engineering, Guntur 522019, India; ysuribabu@gmail.com
5 Department of Mechanical Engineering, Faculty of Engineering, 'Dunarea de Jos' University of Galati, Domneasca Street, 800008 Galati, Romania
6 Electromechanics Engineering Department, Faculty of Engineering, Heliopolis University, Cairo 11785, Egypt; hady.habib@hu.edu.eg
* Correspondence: doggaravi19@gmail.com (D.R.); eugen.rusu@ugal.ro (E.R.)

**Citation:** Raveendhra, D.; Rajana, P.; Narasimharaju, B.L.; Babu, Y.S.; Rusu, E.; Fayek, H.H. Analysis and Operation of a High DC-AC Gain 3-$\phi$ Capacitor Clamped Boost Inverter. *Energies* **2022**, *15*, 2955. https://doi.org/10.3390/en15082955

Academic Editors: Nicu Bizon and Adolfo Dannier

Received: 21 February 2022
Accepted: 13 April 2022
Published: 18 April 2022



**Abstract:** This article introduces a three-phase capacitor clamped inverter with inherent boost capability by relocating the filter components from the AC side to the configuration's midpoint. This topology has several distinguishing characteristics, including: (a) low component count; (b) high DC-AC gain; (c) decreased capacitor voltage stresses; (d) improved power quality (extremely low voltage and current THDs) without the use of an AC-side filter; and (e) decreased voltage stresses on power semiconductor devices. Simulations were carried out on the MATLAB Simulink platform, and results under steady-state conditions, load and reference change conditions, and phase sequence change conditions, along with THD profiles, are presented. This inverter's performance was compared to that of similar converters with intrinsic gain. A 1200 W experimental prototype was built to demonstrate the system's feasibility and benefits. When compared to existing topologies, simulation and experimental results indicate that the proposed inverter provides superior high gain, smooth control, low stress, and a long life time.

**Keywords:** step up inverter; single stage inverter; capacitor clamped inverter; high-gain converter; DC–AC power converter

## 1. Introduction

To overcome the disadvantages of the traditional inverters, such as voltage sources and current source inverters, the ZSI/qZSI [1,2] is widely accepted for various applications. The ZSI/qZSI offers a voltage step-up/down function in single-stage power conversion without requiring additional power processing stage, as shown in Figure 1a, and its application in an electric vehicle is depicted in Figure 1b. Moreover, ZSI/qZSI's reliability is high because shoot-through is an integral part of the operation. The study of these converters has mainly focused on control techniques [3–5], applications [6], and PWM schemes [7,8].

In addition, various power electronics topologies have been proposed in the literature to meet the different objectives, such as reducing the number of switches, reducing the passive component count, and increasing the voltage gain and voltage stresses on the switches/capacitors. The SBI [9] is one such topology that was introduced to decrease the number of passive components (one inductor and one capacitor) in comparison to the ZSI/qZSI by having one additional switch. Although the SBI can perform voltage step-up or step-down in a single stage, its voltage gain is significantly less than that of impedance source converters such as the ZSI/qZSI (1-D). To enhance the source current profile and voltage gain, a group of qSBIs was reported in [10,11]. These included dc-link and embedded-type qSBIs, in addition to current-fed SBIs. However, the embedded-type qZSI necessitates the use of two distinct DC sources, which is undesirable. The qSBI has similar characteristics to the qZSI, except that the shoot-through mode is used for voltage boosting. A comprehensive comparison of the qSBI and qZSI was described [12]. However, the shoot-through duty cycle in these topologies cannot exceed (1–M), where M is the modulation index, thereby limiting the voltage gain in all of the above-mentioned topologies. To achieve the desired output voltage with a high voltage gain and good power quality, a high-duty cycle must be used, lowering M. The lower the value of M, the lower the overall DC-AC conversion gain and the higher the output harmonics. Either the M or the B.F. can be increased to increase the overall DC-AC gain. Numerous PWM techniques have been proposed for modifying the modulating waves, and it has been demonstrated that PWM schemes can only slightly increase M [7]. Numerous high-gain inverter circuits with and without a galvanic isolation transformer have been proposed [13–25] to increase the boost factor in impedance source converters. Transformer-based ZSIs have been introduced [13,14]. However, the transformer's leakage inductance results in voltage spikes at the DC-bus. To achieve a high voltage gain, transformerless ZSIs with additional passive components, such as inductors, capacitors, and diodes, have been proposed [15–25]. They have been labelled L-ZSI [15], SL-ZSI [16], SL-qZSI [17], EB-ZSI [18], DA-qZSI [19], CA-qZS [20], and EB-qZSI [21], and by incorporating switched-inductor, switched-capacitor, and hybrid switched-capacitor/switched-inductor designs, high boosting factors can be achieved. The addition of passive elements and power electronic components, on contrast, increases the converter's cost, size, volume, losses, and weight [24,25].

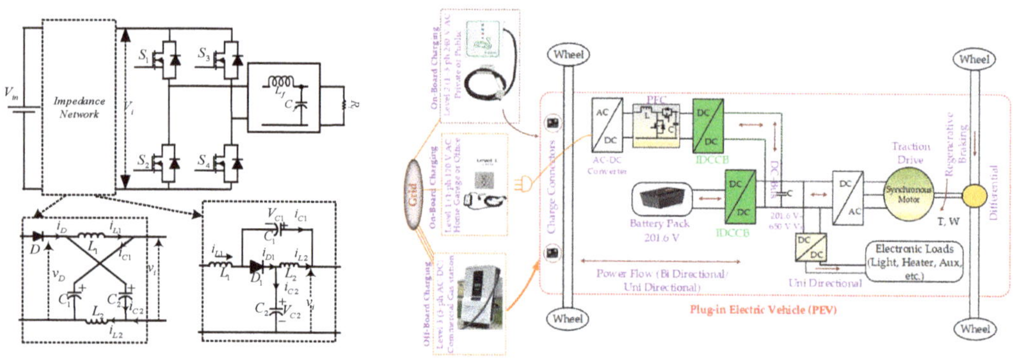


(a) (b)

**Figure 1.** (**a**) Classical impedance source converters and (**b**) their applications in EV.

For single-phase and three-phase applications, the aforementioned topologies have been proposed. A simple single-phase HB ZSI with reduced capacitor voltage stress was described in [26]. Although this topology is straightforward and compact, it has a low boost factor. In [27], an HB-SBI with discontinuous input current was introduced to increase the gain of the inverter. To address the shortcomings of the HB-SBI, [28] proposed the HB-qSBI. The overall voltage gain is low in all of these half-bridge topologies [26–28]. By comparison, in the SL-ZSI [16], SL-qZSI [17], EB-ZSI [18], DA-qZSI [19], CA-qZSI [20], and EB-qZSI [21] topologies, the voltage stresses on the capacitors are greater, and these topologies employ a greater number of capacitors. The capacitor voltage is typically greater than the input voltage in order to perform the impedance-source stage's voltage boost function. As a result, high-voltage Z capacitors must be used, potentially adding volume and cost to the system. Because of the strong probability of capacitor failure during the field operation of power electronic converters [29], and the stringent reliability restrictions imposed by the aerospace, automotive, defense, space, and energy industries, stresses and use of capacitors must be reduced to improve inverter reliability [30].

Overall, all of the mentioned topologies have common issues, such as the requirement for capacitors having a high voltage rating (greater than supply), common mode voltages, PWM-natured voltages after the inverter switching leg, and a higher component count when achieving a boosted DC-AC gain in a single stage manner. To address these issues, the capacitor clamped boost inverter with high voltage gain was introduced [31], which disperses the X-shaped passive components of the impedance source inverter rather than concentrating them in one location (between the input and inverter switching network in impedance source inverters). The passive components are distributed evenly between the input and output ports on each leg. However, ref. [31] does not deal with detailed steady-state analysis. Hence, in this study, a detailed analysis in various modes based on the inductor current $i_L$ bands (upper ($i_{Lmax}$) and lower bands ($i_{Lmin}$)) was conducted. Based on the upper and lower band values, the operation was divided into three zones in this study. Based on the aforementioned zones, the operation of the converter was further divided into two cases: *case-I* (*zone-1* and *zone-3*) and *case-II* (*zone-2*). In addition, this manuscript includes capacitor voltage profiles and capacitor life time calculations. This paper also discusses the design of a sliding mode controller for the CCBI to track the required voltages and currents to fulfil the specified load characteristics. Also shown are the CCBI's performance under load, reference, and phase sequence change conditions, and its THD profiles. In addition, the performance of this inverter for non-linear loads was also examined. The operating principles, steady-state analysis of various cases, differential modulation technique, capacitor voltage profiles, capacitor life time calculations, and sliding mode controller for the CCBI are presented in Section II. Simulation results and a discussion of the results under steady-state conditions, load and reference change, and phase sequence change conditions, along with THD profiles, are discussed in Section III. A performance investigation of this inverter for non-linear loads is also presented. In the same section, comparative analysis of the proposed inverter with existing similar converters is also presented, along with experimental verification. Section IV presents the conclusions.

## 2. Operation of the CCBI

The proposed inverter, depicted in Figure 2a, includes six switches, three small inductors, and three capacitors. Because of the time-varying duty cycle, the intrinsic boost feature of this proposed inverter provides flexibility for grid-connected and stand-alone applications, for a large range of AC output voltages, which are even higher than the DC voltage. This capability is not accessible in standard VSIs, where the DC input voltage is always greater than the AC output voltage [11]. The following offers an analysis of the converter in various modes, a differential modulation scheme, capacitor voltage profiles, a life time analysis, and a sliding mode controller.

### 2.1. Analysis of Converter

The operation of the converter shown in Figure 2a is explained with the help of a single-phase equivalent circuit, which is shown in Figure 2b. The equivalent circuit contains one leg (phase-A) along with considerations of the effect of other phases. Boost inverter upper switches are represented with odd numbers, whereas lower switches are represented with even numbers, and these switches operate in a complementary manner. Every inverter leg contains one inductor, one capacitor, and two switches. Analysis of the boost inverter is explained via mathematical modeling.

For one cycle ($0 < t < T$) of load current ($i_{Load}$), the average value of the inductor current ($i_L$) is positive and negative for the periods ($0 < t < T/2$) and ($T/2 < t < T$) respectively, which is shown in Figure 3, where $T = 1/f$. It can be observed that $i_L$ is oscillating at the switching frequency of $f_s$ between two bands, namely, the upper band ($i_{Lmax}$) and the lower band ($i_{Lmin}$). During the positive half cycle, the upper band is always positive. However, the lower band current value is negative in *zone-1* ($0 < t < T_a$) and *zone-3* ($T/2 - T_a < t < T/2$), whereas it is positive in *zone-2* ($T_a < t < T/2 - T_a$). Here, $T_a$ is the time when *zone-1* comes to an end.

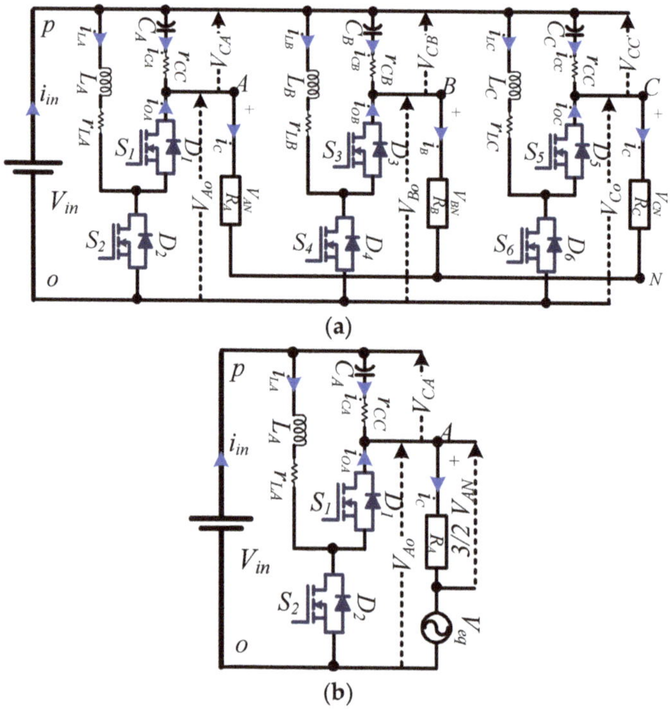


**Figure 2.** (**a**) Proposed converter; (**b**) single-phase equivalent circuit.

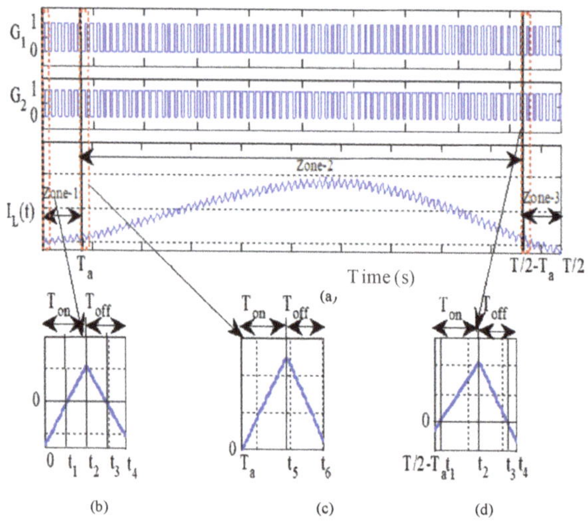


**Figure 3.** (**a**) Gate signals of $S_1$ and $S_2$ and inductor current for positive half cycle; inductor current in one switching cycle of (**b**) *zone-1*, (**c**) *zone-2*, and (**d**) *zone-3*.

Based on the aforementioned zones, the operation of converter is divided into two cases, *case-I* (*zone-1* and *zone-3*) and *case-II* (*zone-2*) as shown in Figure 3. To simplify the analysis further, only one switching sample is considered of N (=$f_s/f$) samples, and the detailed description of both cases is given below.

*case-I (zone-1 and zone-3):* It is interesting to note that, in this case, all the semiconductor devices sequentially ($D_2$, $S_2$, $D_1$, and $S_1$) participate in one switching sample of $G_2$, which is shown in Figure 4. Based on the conduction of these switching devices, the operation of the circuit in this case is further classified into four modes, and its equivalent circuit in each mode is shown in Figure 4. The operation of the inverter in various modes for *case-I* is explained below.

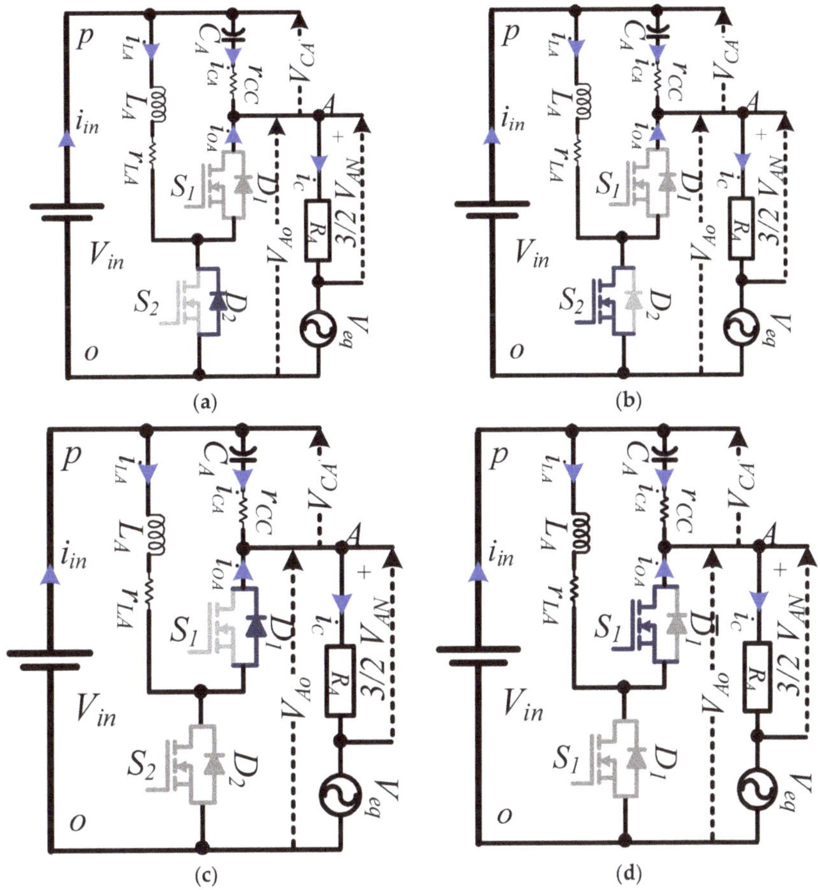


**Figure 4.** Equivalent circuit in various modes: (**a**) *Mode-1*; (**b**) *Mode-2*; (**c**) *Mode-3*; (**d**) *Mode-4*;

*Mode-1* ($0 < t < t_1$)*:* This mode starts when the gate signal is applied to the lower switch $S_2$. However, $S_2$ cannot be turned on instantly due to the fact that the inductor does not allow the sudden change in current, as it is has a negative value in the previous mode. This leads the diode $D_1$ to be turned on and provide a path for a negative inductor current, as shown in Figure 4a.

This $i_L$ increases linearly in the presence of a positive supply voltage, as shown in Figure 5, and its current equation can be written as:

$$i_L(t) = \frac{V_{in}}{L}(t - T_o) + i_L(T_o) \tag{1}$$

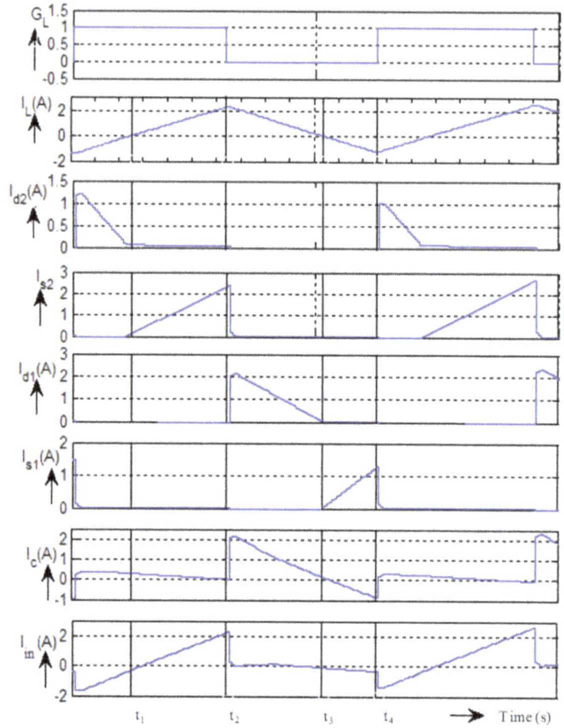

**Figure 5.** Waveforms during *case-I*.

This mode ends at $t = t_1$, where $i_L$ becomes zero and diode $D_2$ turns off. The time duration of this *Mode-1* can be calculated as:

$$t_1 = L\frac{i_L(T_o)}{V_{in}} + T_o \tag{2}$$

*Mode-2* ($t_1 < t < t_2$): This mode starts at $t = t_1$ when $S_2$ comes into conduction. In the presence of a positive supply voltage ($V_{in}$) across the inductor, its current is increases linearly from 0 to $I_{Lmax}$, as shown in Figure 6, and its equation can be written as:

$$i_L(t) = \frac{V_{in}}{L}(t - t_1) \tag{3}$$

This mode ends at $t = t_2$, when the switching pulse for $S_2$ is removed and the time duration of this mode can be evaluated as:

$$t_2 - t_1 = DT - t_1 \tag{4}$$

*Mode-3* ($T_{on} < t < t_3$): This mode starts when the gate signal is given to switch $S_1$. From the previous state, it is clear that the initial inductor current is positive ($I_{Lmax}$), which brings diode $D_1$ to conduction, even in the presence of firing pulses at $G_1$, as shown in Figure 6. During this period, a negative voltage ($V_{in} - V_{AO}$) appears across the inductor, which leads to the decrement of the inductor current with a negative slope of $(V_{in} - V_{AO})/L$, which can be expressed as:

$$i_L(t) = \frac{V_{in} - V_{AO}}{L}(t - t_2) + i_L(t_2) \tag{5}$$

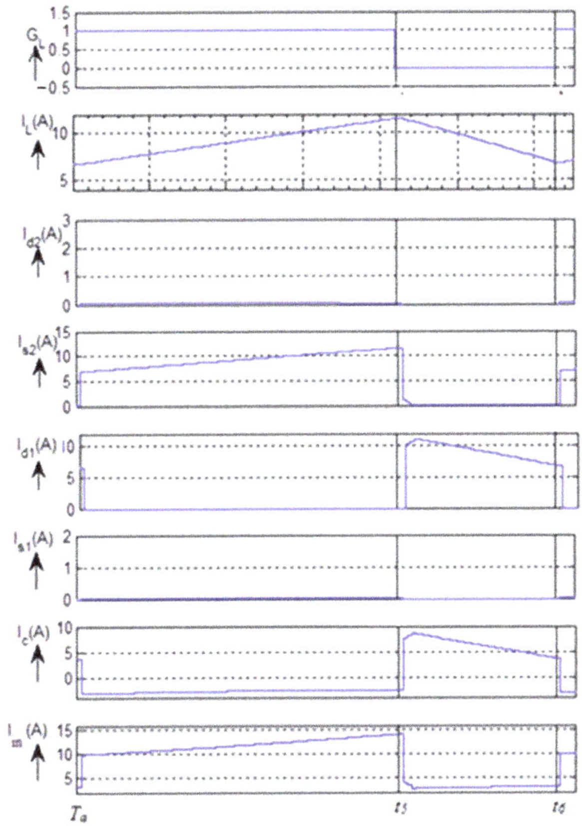


**Figure 6.** Waveforms during *case-II*.

This mode ends at $t = t_3$, when $i_L$ reaches zero and forces the diode to be turned off. The duration of this mode can be determined as:

$$t_3 - DT = \frac{Li_L(t_2)}{V_{AO} - V_{in}} \tag{6}$$

*Mode-4 ($t_3 < t < T$):* The zero-initial current of the inductor and the presence of the switching pulse brings $S_1$ to conduction mode. Now, the inductor discharges in the presence of a negative voltage ($V_{in} - V_{AO}$) and, hence, the inductor current is decreased to a specific negative value ($I_{Lmin}$). This negative inductor current is the initial current for *Mode-1*. The expression for the inductor current is given as:

$$i_L(t) = \frac{V_{in} - V_{AO}}{L}(t - t_3) \tag{7}$$

This mode ends with the removal of the gate pulse of $S_1$ and the duration of this mode can be calculated as $t_4 - t_3 = T - t_3$.

The same cycle of operation repeats until $t = T_a$. The inductor voltage balance equation during *case-I* can be written as:

$$\begin{aligned} &V_{in}d_{Ak}T_s = (V_{AO} - V_{in})(1 - d_{Ak})T_s \\ &\Rightarrow V_{AO} = \frac{V_{in}}{1-d_{Ak}} \end{aligned} \tag{8}$$

where $d_{Ak}$ is the duty cycle at the '$k^{th}$' switching sample of switch $S_2$ of phase A. Waveforms during this case shown in Figure 5. As $d_A$ is a time-varying value, and when $d > d_a$ at $t = T_a$, *case-I* ends.

*case-II* ($T_a < t < T/2 - T_a$): This is the special case where *Mode-2* and *Mode-3* operations of *case-I* only take place as there is no negative value of $i_{Lmin}$ and the remaining two cases are absent. In this case, only two semiconductor switches take active participation in conversion, namely, $S_2$ and $D_1$, whereas the remaining two are in an idle state. Hence, two modes are sufficient to explain the operation; waveforms during this case are shown in Figure 6. It can be observed from the characteristics presented in Figures 5 and 6 that *case-II* is a special case of *case-I*, where only two modes (*Mode-2* and *Mode-3*) are presented during operation. Waveforms of lower switch gate pulses, inductor current, upper and lower diode and switch currents, coupled capacitor current, and input current are presented in Figure 6 respectively. A detailed explanation is presented below.

*Mode-1*($T_a < t < t_5$): This mode starts at $t = T_a$, at which instant firing pulses ($G_2$) are given to $S_2$. Due to a positive initial inductor current and the presence of $G_2$, switch $S_2$ is turned on. In the presence of a positive voltage across the inductor, it is charged to a specific value, which can be expressed as:

$$i_L(t) = \frac{V_{in}}{L}(t - T_p) + i_L(T_p) \tag{9}$$

This mode ends at $t = t_5$ when firing pulses are removed from $G_2$.

*Mode-1* ($t_5 < t < t_6$): This mode of operation starts when pulses are given to $S_1$. Although pulses are presented at $S_1$, it cannot be turned on due to the positive initial current in the inductor. This turns on diode $D_1$. Now, the negative voltage across the inductor causes a decrement in the current with the negative slope of $(V_{in} - V_{AO})/L$, which can be expressed as:

$$i_L(t) = \frac{V_{in} - V_{AO}}{L}(t - T_{onb}) + i_L(T_{onb}) \tag{10}$$

This mode ends at $t = t_6$ when firing pulses are removed from $S_1$. A similar operation (*Mode-1* and *Mode-2*) continues for several switching cycles until $(T/2 - T_a)$. The voltage balance equation of the inductor obeys Equation (8), as discussed in *case-I*. Furthermore, the similarly boosted inverter operates in a negative half cycle, with the major role of $S_1$, $D_2$ in $T/2$ to $(T/2 + T_a)$, and $(T - T_a)$ to $T$, in addition to $D_1$, $S_1$, $D_2$, and $S_2$ during $(T/2 + T_a)$ to $(T - T_a)$.

### 2.2. Differential Modulation Technique for Three-Phase Boost Inverter

As shown in (8), once the duty cycle becomes zero, $V_{AO} = V_{in}$; this shows that this converter outputs a dc bias voltage in relation to the negative supply terminal. The primary goal of this study was to generate three-phase sinusoidal voltages across the load terminals of Figure 2a. Based on the gain of this boost inverter, we assume these voltages are modulated with the following duty cycles ($d_A$, $d_B$, and $d_C$) as follows [13]:

$$\begin{aligned} V_{AO} &= \tfrac{V_{dc}}{1-d_A} = V_{in} + A + A\sin\omega t \\ V_{BO} &= \tfrac{V_{dc}}{1-d_B} = V_{dc} + A + A\sin(\omega t - 120^o) \\ V_{CO} &= \tfrac{V_{dc}}{1-d_C} = V_{dc} + A + A\sin(\omega t - 240^o) \end{aligned} \tag{11}$$

where $d_A$, $d_B$, and $d_C$ are the duty cycles of the $A$, $B$, and $C$ phases, respectively, $V_{in}$ is the bias DC voltage of the boost inverter (i.e., supply voltage), and A is the sinusoidal voltage amplitude. It should be noted that the CCBI one-leg voltage with regard to the negative terminal of a source ($V_{AO}$, $V_{BO}$, $V_{CO}$) always has the same sign as $V_{in}$; thus, $(V_{AO})_{DC}$ must be added to maintain this necessary condition:

$$\left.\begin{array}{l} V_{AO}(t) = (V_{AO})_{DC} + (V_{AO})_{AC} \sin wt \\ V_{BO}(t) = (V_{BO})_{DC} + (V_{BO})_{AC} \sin wt \\ V_{CO}(t) = (V_{CO})_{DC} + (V_{CO})_{AC} \sin wt \end{array}\right\} \quad (12)$$

Here

$$V_{dc} + \mathrm{A} = (V_{AO})_{DC} = (V_{BO})_{DC} = (V_{CO})_{DC} \& |V_{AO}|_{AC} = |V_{BO}|_{AC} = |V_{CO}|_{AC} \quad (13)$$

From the $1-\phi$ equivalent model of the $3-\phi$ boost inverter, as shown in Figure 2b, the phase voltage $V_{AN}$ applied to $R_L$ of the $3-\phi$ system can be obtained as [2]:

$$V_{AN} = \frac{2}{3}\left[V_{A0} - \frac{1}{2}(V_{BO} + V_{CO})\right] \quad (14)$$

The first term in (14) is the voltage $V_{AO}$ produced by the same phase of the boost converter, whereas the second and third terms account for the influence of the other two phases. It can be understood that the phase voltage of the load is a function of all of the three phases' leg voltages ($V_{AO}$, $V_{BO}$, and $V_{CO}$), and for the particular load phase voltage, the other phases' leg voltages' combined effort can be grouped as:

$$V_{eq} = \frac{1}{2}(V_{BO} + V_{CO}) \quad (15)$$

The difference in $V_{AN} V_{eq}$ should also cause the same phase current in the $1-\phi$ model, so an equivalent resistance can be introduced, as mentioned below:

$$R_{eq} = \frac{3}{2} R_L \quad (16)$$

From Equations (12) and (14):

$$\begin{array}{ll} V_{AN} & = \frac{2}{3}\Big\{[(V_{AO})_{DC} + (V_{AO})_{DC} \sin \omega t] \\ & -\frac{1}{2}[(V_{BO})_{DC} + (V_{BO})_{DC} \sin(\omega t - 120) \\ & +(V_{CO})_{DC} + (V_{CO})_{DC} \sin(\omega t - 240)]\Big\} \\ \Rightarrow & V_{AN} = (V_{AO})_{AC} \sin \omega t \end{array} \quad (17)$$

In a similar way, $V_{BN}$ and $V_{CN}$ have a 120° phase shift at the load terminals. Therefore, the phase-to-neutral voltages at the load are:

$$\begin{array}{l} V_{AN} = (V_{AO})_{AC} \sin \omega t \\ V_{BN} = (V_{AO})_{AC} \sin(\omega t - 120^o) \\ V_{CN} = (V_{AO})_{AC} \sin(\omega t - 240^o) \end{array} \quad (18)$$

These ideal outcomes can be achieved by calculating the three-phase duty cycles using:

$$\left.\begin{array}{l} D_A(t) = 1 - \frac{V_{in}}{(V_{AO})_{DC} + (V_{AO})_{AC} \sin wt} \\ D_B(t) = 1 - \frac{V_{in}}{(V_{BO})_{DC} + (V_{BO})_{AC} \sin(wt - 120^o)} \\ D_C(t) = 1 - \frac{V_{in}}{(V_{CO})_{DC} + (V_{CO})_{AC} \sin(wt - 240^o)} \end{array}\right\} \quad (19)$$

## 2.3. Capacitor Voltage Profile

One of the main objectives of this study was to reduce the capacitor peak voltages, which can be calculated for the CCBI as follows:

$$\begin{aligned} V_{CA} &= (V_{AO})_{DC} + (V_{AO})_{AC} \sin \omega t - V_{in} \\ \Rightarrow V_{CA} &= (V_{CO})_{DC} + (V_{CO})_{AC} \sin \omega t \end{aligned} \tag{20}$$

Here $(V_{CO})_{DC} = (V_{AO})_{DC} - V_{in}$, and can be calculated as:

$$\begin{aligned} (V_{CO})_{DC} &= (V_{AO})_{DC} - V_{in} \\ \Rightarrow (V_{CO})_{DC} &= \left(\frac{D}{1-D}\right) V_{in} \end{aligned} \tag{21}$$

Whereas in case of other topologies, the capacitor voltages are higher due to the requirement of a higher dc link voltage for the required DC-AC conversion. Capacitor voltage profiles for the DC-AC conversion of 1 to 1.8 in the proposed case and other similar impedance source inverters are shown in Figure 7, which depicts the reduction in voltage stress on the capacitor.

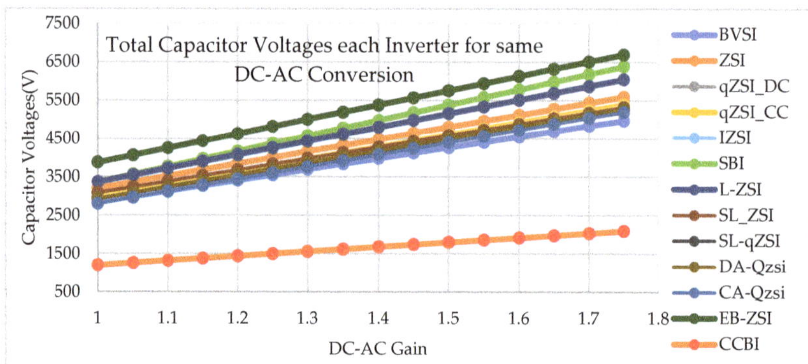


**Figure 7.** Capacitor voltage profiles.

## 2.4. Life Time Calculation for Capacitor

To examine the life time benchmarks of different capacitor solutions and online condition monitoring, life models are used. Generally, the life time of the capacitors is greatly influenced by two factors, namely, voltage stress and temperature. The most extensively accepted empirical model for capacitor life is:

$$\tau = \tau_o \times \left(\frac{V}{V_o}\right)^{-n} \times \exp\left[\left(\frac{E_a}{K_B}\right)\left(\frac{1}{\theta} - \frac{1}{\theta_0}\right)\right] \tag{22}$$

where $\tau$ is the life time under use conditions, $\tau_o$ is the life time under test conditions, $V$ is the voltage at use conditions, and $V_0$ is the voltage at test conditions. $\theta$ and $\theta_0$ are the temperature (Kelvin) at use and test conditions, respectively. $E_a$ is the activation energy, $K_B$ is Boltzmann's constant ($8.62 \times 10^{-5}$ eV/K), and n is the voltage stress exponent.

From (22), it is clear that $E_a$ and n are the key parameters to determine the life time; its values were found to be 1.19 and 2.46 for high dielectric constant ceramic, and 1.3–1.5 and 1.5–7 for MLC-Caps.

For Al-Caps and film capacitors, a simplified model from (22) is popularly applied as follows [14]:

$$\tau = \tau_o \times \left(\frac{V}{V_o}\right)^{-n} \times 2^{\frac{\theta_0 - \theta}{10}} \tag{23}$$

### 2.5. Sliding Mode Controller

When a sliding mode controller is adopted, the system performs effectively in both steady-state and dynamic operations. Although more complicated control approaches, such as THD, can increase system performance, the observed results look satisfactory in many circumstances of practical importance, while the basic controller lowers system cost. An experimental prototype was created, and the experimental findings show that the converter is capable of step-up [2].

The following reasonable assumptions must be considered when designing the sliding mode controller for the proposed converter: power switches that are ideal, converters that operate at high switching frequencies, and power supplies that are free of sinusoidal ripple. Each phase of the proposed converter has two state variables. The sliding surface equation of state space in a three-phase system is expressed as:

$$\left.\begin{array}{l} s_1(i_{La}, v_a) = K_{a1}\varepsilon_{a1} + K_{a2}\varepsilon_{a2} = 0 \\ s_2(i_{Lb}, v_b) = K_{b1}\varepsilon_{b1} + K_{b2}\varepsilon_{b2} = 0 \\ s_3(i_{Lc}, v_c) = K_{c1}\varepsilon_{c1} + K_{c2}\varepsilon_{c2} = 0 \end{array}\right\} \quad (24)$$

where:

$$\begin{bmatrix} \varepsilon_{a1} \\ \varepsilon_{b1} \\ \varepsilon_{b1} \end{bmatrix} = \begin{bmatrix} i_{la} \\ i_{lb} \\ i_{lc} \end{bmatrix} - \begin{bmatrix} i_{la_{ref}} \\ i_{lb_{ref}} \\ i_{lc_{ref}} \end{bmatrix} \& \begin{bmatrix} \varepsilon_{a2} \\ \varepsilon_{b2} \\ \varepsilon_{c2} \end{bmatrix} = \begin{bmatrix} v_a \\ v_b \\ v_c \end{bmatrix} - \begin{bmatrix} v_{a_{ref}} \\ v_{b_{ref}} \\ v_{c_{ref}} \end{bmatrix} \quad (25)$$

In sliding mode control theory, sensing of all state variables is required to generate the proper control signals and obtain the required AC supply. The generation of the inductor current reference is difficult to assess because it is dependent on several factors, such as supply voltage, load demand, and load voltage. As a result, $i_L$-$i_{Lref}$ can be generated directly from the high frequency component of the inductor current feedback signal, which must be removed due to the control strategy by designing a suitable high pass filter. The addition of a high pass filter increases system order and has the potential to change system dynamics. To overcome this issue, the selected values of the CCBI's switching frequency were higher than the filter cut-off frequency. The trajectory of the sliding surface for this design is shown in Figure 8.

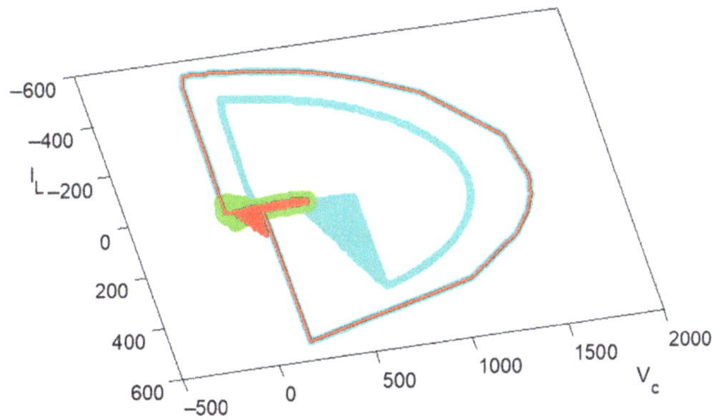


**Figure 8.** Trajectory of the sliding surface.

## 3. Results and Discussions

The proposed $3-\phi$ CCBI, as shown in Figure 9, was successfully assessed by means of both simulations and prototype-based hardware results. Simulations were carried out using the MATLAB Simulink environment, and the parameters considered for the simulations are summarized in Table 1, as shown below.

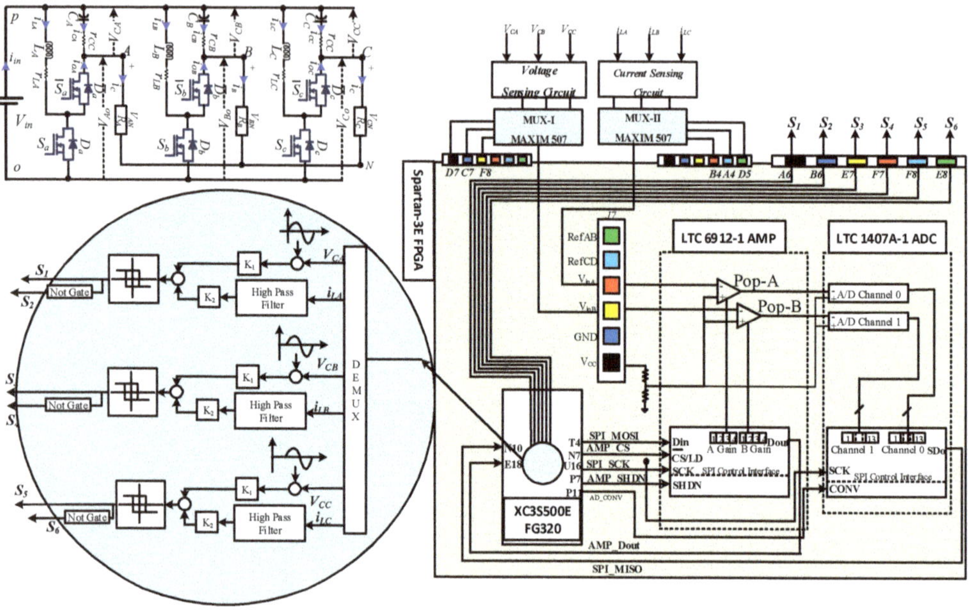

**Figure 9.** Complete system diagram of the hardware setup.

**Table 1.** Electrical parameters of the system.

| | |
|---|---|
| Input voltage ($V_{in}$) | 200 V |
| Output voltage ($V_{orms}$) | 400 V |
| Output power ($P_o$) | 1.2 KW |
| Maximum switching frequency ($F_{smax}$) | 20 KHz |
| Frequency ($F_o$) | 50 Hz |

The following results were acquired at the average switching frequency ($F_s$) equal to 10 kHz. A sliding mode controller was used to achieve good dynamic response, high robustness, and noise-free response while tracking the required $3-\phi$ AC from DC supply. System state variables were continuously monitored and controlled near to a zero error response with the hysteresis band = 0.3, filter constant = 0.01, $K_1$ = 0.304, and $K_2$ = 0.2. System performance was evaluated in both a steady state and transient states while feeding power to different types of loads (linear and nonlinear) under different test conditions. $3-\phi$ Phase voltages, line voltages, and load currents obtained from this inverter are shown in Figures 10–12, respectively. From these results, it can be clearly seen that the input low-level DC supply was successfully converted to ideal sinusoidal three-phase AC power.

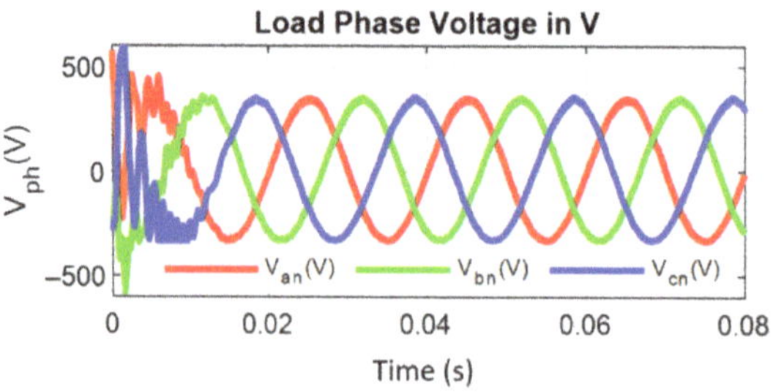


**Figure 10.** Phase voltage of the inverter during steady-state conditions.

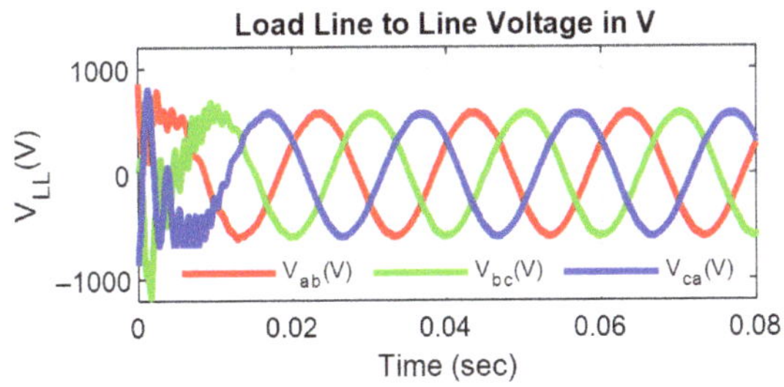


**Figure 11.** Line voltages of the inverter during steady-state conditions.

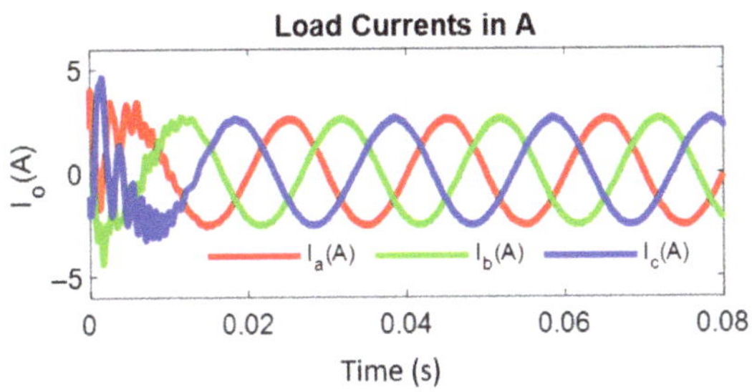


**Figure 12.** Load currents of the inverter during steady-state conditions.

In the boost inverter topology, at least one capacitor is placed in every leg of the respective phase of the converter for the boosting operation. The negative terminals of each of the three capacitors of the three-phase inverter are connected to a common point in this topology, and these are also shown in Figure 13. With reference to this point, the common mode capacitor voltage (*CMMCV*) is defined as the average of all of the three capacitor voltages ($V_{AO}$, $V_{BO}$, and $V_{CO}$) and is shown in Figure 14. In Figure 14, the conventional CMMCV is calculated for the topology proposed by Cecati and compared with the proposed topology. Figure 14 shows that the CMMCV across all of the three capacitors is greatly reduced by the proposed scheme, due to the fact that the individual capacitor voltage is also lower than that of the conventional topology. Figures 15 and 16 were captured for the critical evaluation of the harmonic content contained at the output. These results show the THD waveforms of phase voltage, line voltage, and load current, respectively; from these results, it can be clearly understood that this inverter offers good quality of AC output without any lowest order harmonics (<3% of fundamental) for resistive load. All of the harmonic quantities are lower than 1.5% of the fundamental.

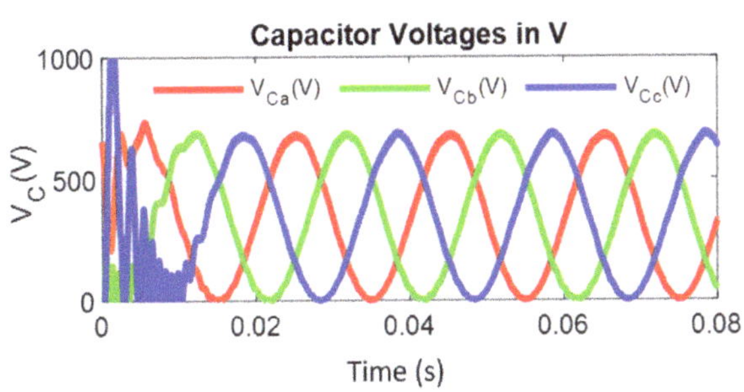


**Figure 13.** Capacitor voltages of the inverter during steady-state conditions.

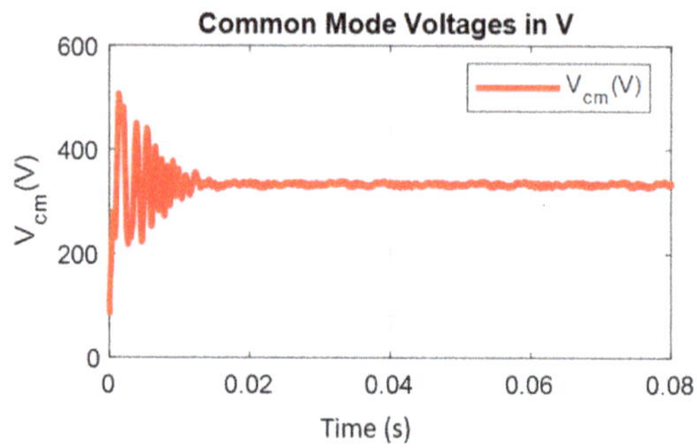


**Figure 14.** Common mode capacitor voltages.

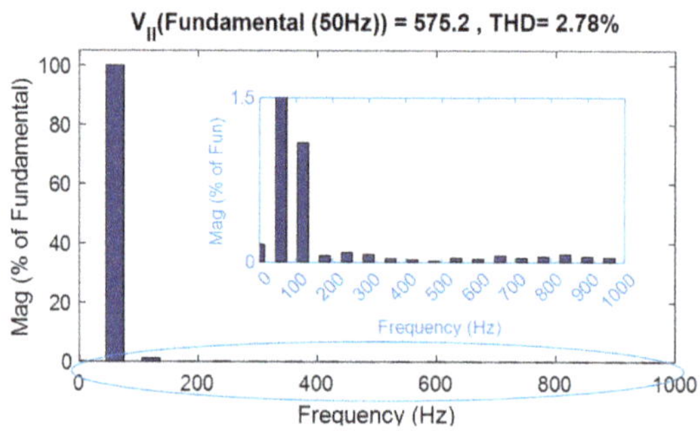


**Figure 15.** THD waveform of line voltage.

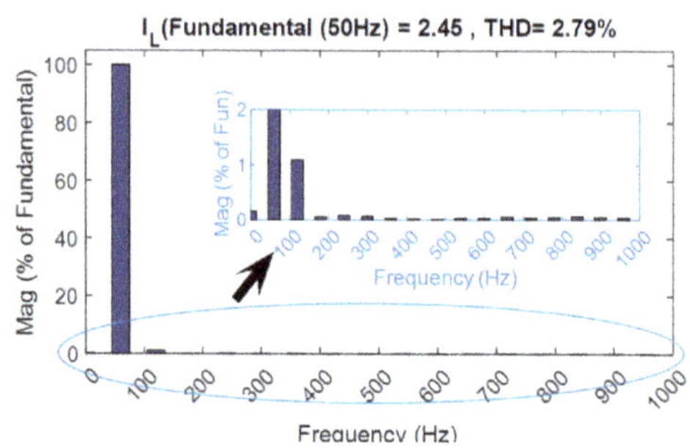


**Figure 16.** THD waveform of load current.

In order to assess the dynamic performance of the converter, sudden changes were incorporated during the operation of the converter at load (increased by 50% and decreased by 50%), and in the reference voltage (decreased by 50% and changed the phase by 180°), and results were captured in each case for analysis. Figure 17 shows the current drawn by the 100% load (1.2 kW) from (0 to 0.04 s) and 50% load (0.6 kW) from (0.04 to 0.08 s), whereas Figure 18 depicts the opposite case of loading, i.e., 50% loading from (0 to 0.04 s) and 100% load from (0.04 to 0.08 s). Figure 19 was captured when the mode of operation is suddenly changed from active mode to regeneration mode at 0.04 s. Although the mode is changed from the active mode to the regeneration mode, the voltage amplitude remains constant, and its harmonics also remain the same. Figures 20–23 show the phase voltages, line voltages, load currents, and capacitor voltages observed when the reference voltage is suddenly changed from 100% to 50% at 0.04 s. Whenever the reference voltage is changed, the output voltage changes, and the current also changes accordingly for the resistive load. THD waveforms in the case in which the reference voltage is changed were captured after the disturbance was settled, and are shown in Figures 23–25. These results (Figures 20–23) reveal that the CCBI with a sliding mode offers good dynamic response in stable operation, even for all kinds of disturbances, as discussed earlier.

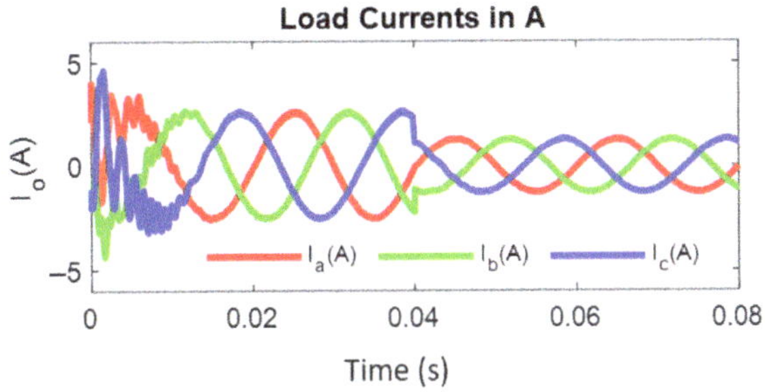


**Figure 17.** Load current for a linear load with a step change in the load from 100 to 50%.

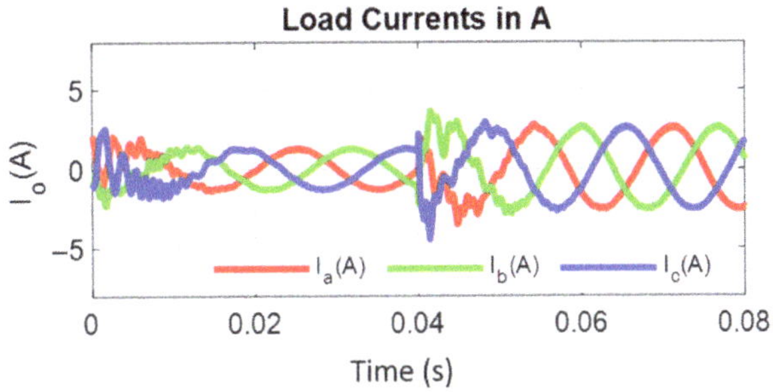


**Figure 18.** Load current for a linear load and a step change in the load from 100 to 50%.

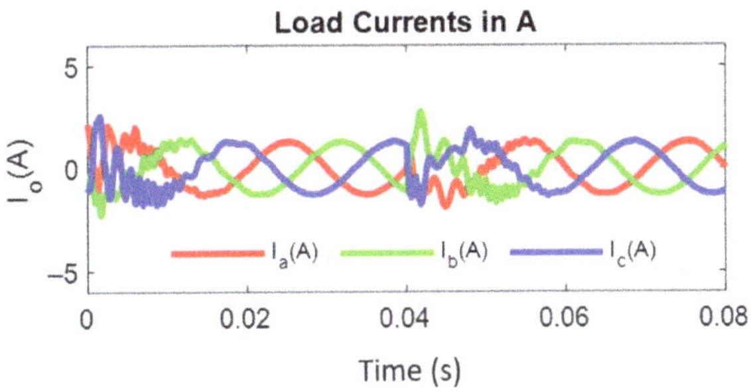


**Figure 19.** Load current for the inversion mode of the load.

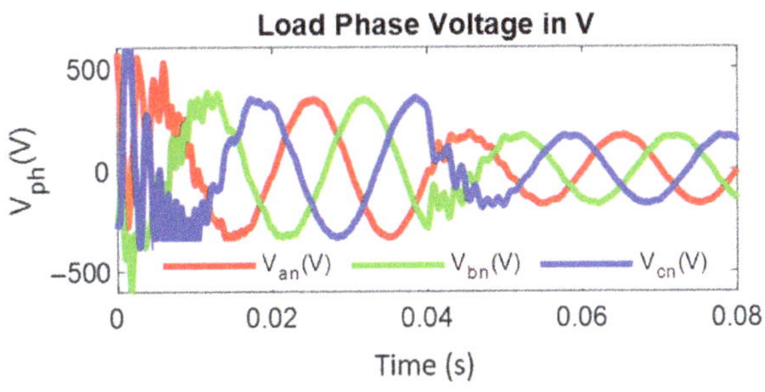


**Figure 20.** Phase voltages for a linear load and a step change in the reference load voltage from 100 to 50%.

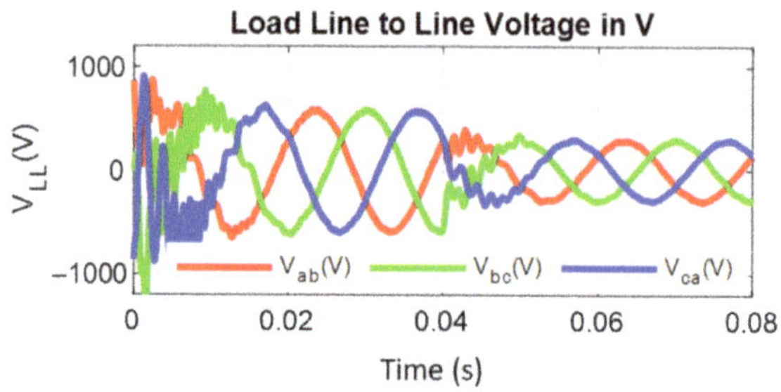


**Figure 21.** Line voltages for a linear load and a step change in the reference load voltage from 100 to 50%.

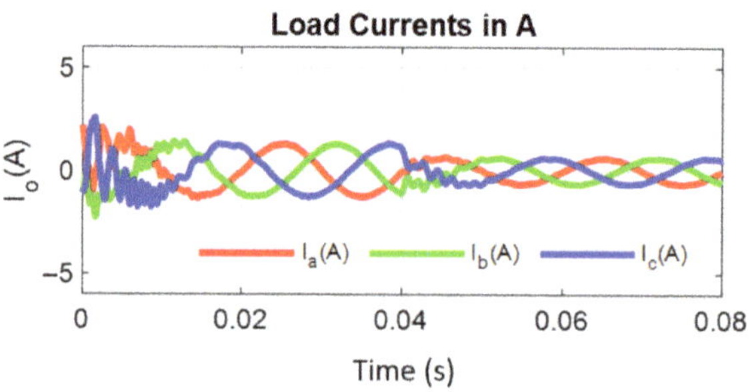


**Figure 22.** Load currents for a linear load and a step change in the reference load voltage from 100 to 50%.

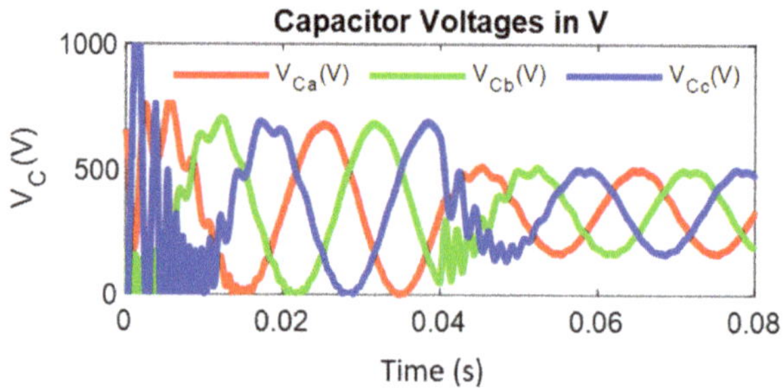


**Figure 23.** Capacitor voltages for a linear load and a step change in the reference load voltage from 100 to 50%.

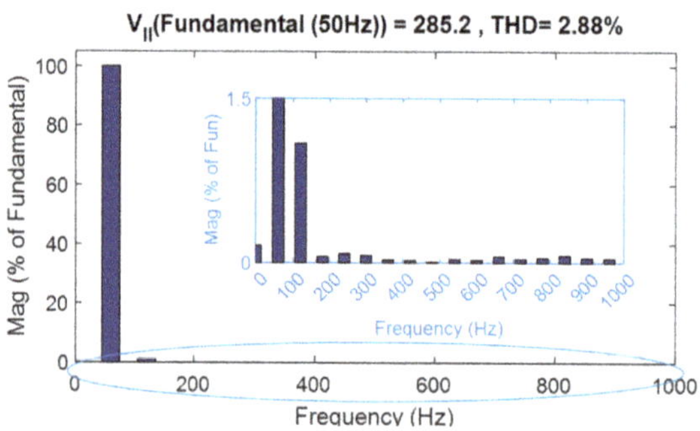


**Figure 24.** Load voltage and load current for a linear load and a step change in the reference load voltage from 50%.

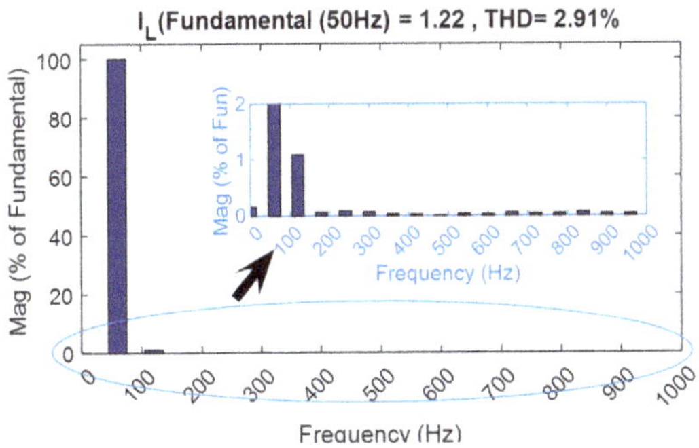


**Figure 25.** Load voltage and load current for a linear load and a step change in the reference load voltage from 100 to 33%.

For critical evaluation of the converter, the inverter output is fed to a nonlinear load (three-phase diode bridge rectifier with R Load of 255 Ω), and Figures 26–33 show the CCB inverter-fed diode bridge output currents and voltage, the diode bridge input line voltage and currents, the capacitor voltages of CCB and CMMCV, and the harmonic spectra of diode bridge input voltage and currents, respectively. Under steady-state mode, the diode bridge rectifier-fed resistive load absorbs the highly distorted current of 31.39% THD and voltage of 7.25% THD, as shown in Figures 32 and 33, respectively. All of the foregoing data show that the CCB inverter has good behavior, and particularly superior dynamic behavior, which is mostly due to the lower values of the boost capacitances and voltage across the capacitor. This performance is especially notable when compared to that of a current source inverter (CSI); whereas the proposed system employs three independent small inductors, the CSI employs only one large inductor, resulting in much poorer dynamic performance.

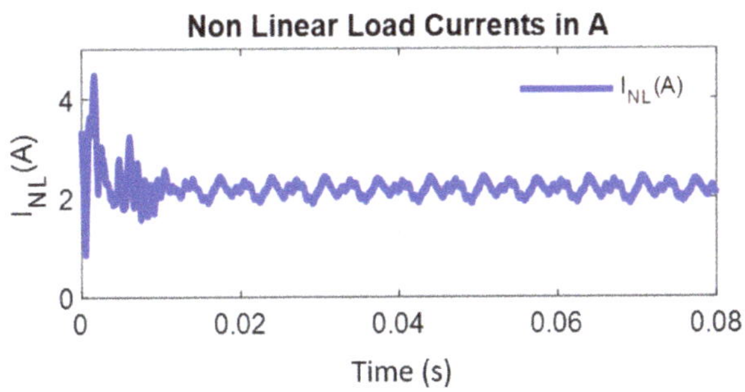


**Figure 26.** Non-linear load (diode bridge) output currents.

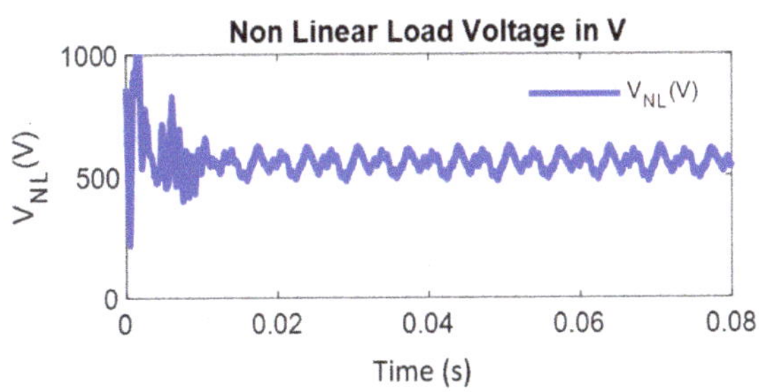


**Figure 27.** Non-linear load (diode bridge) output voltages.

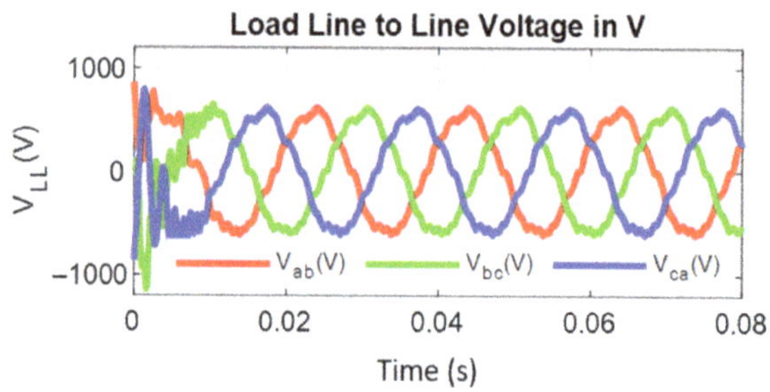


**Figure 28.** Non-linear load (diode bridge) input voltages.

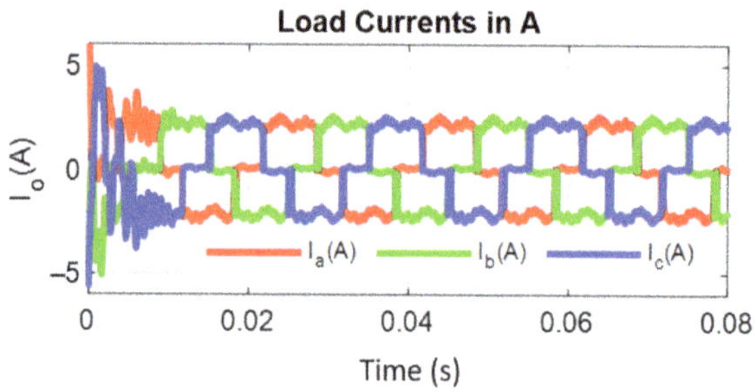


**Figure 29.** Non-linear load (diode bridge) input currents.

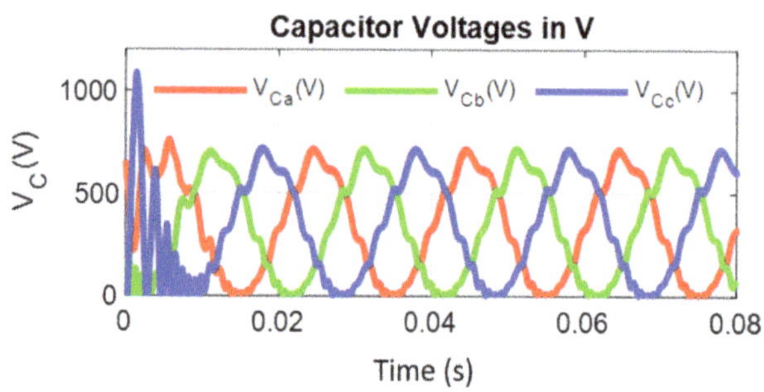


**Figure 30.** Capacitor voltages of the CCBI.

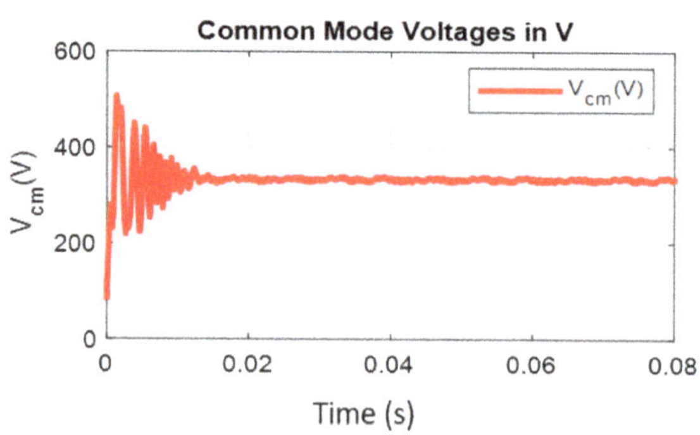


**Figure 31.** Common mode voltages of the CCBI.

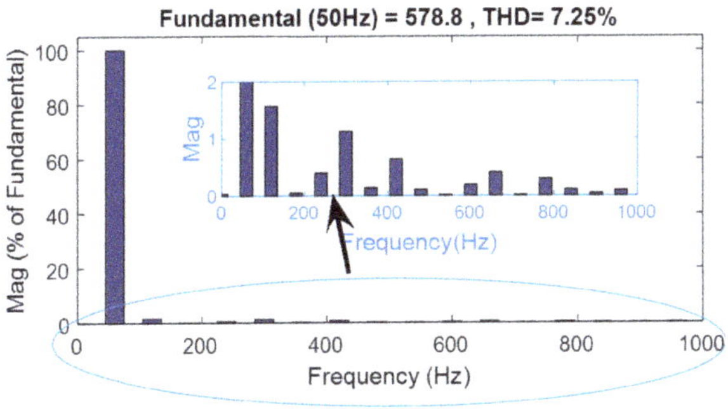


**Figure 32.** CCBI voltage THD for the nonlinear (diode bridge) load.

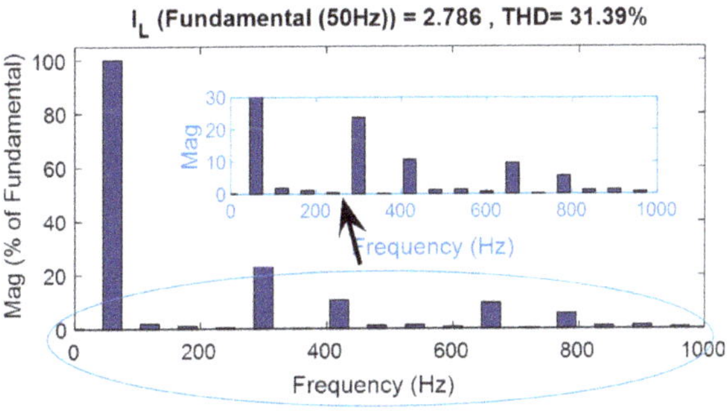


**Figure 33.** CCBI current THD for the nonlinear (diode bridge) load.

*Comparison of the CCBI with ZSI:* For the same source and load, and required gain, ZSI is implemented with the shoot-through duty of 0.4091 and AC-side filter components of inductor $L_f$ = 0.25 mH and capacitor $C_f$ = 44 μH. Its load parameters, z-source capacitor voltages, and CMMV are presented in Figures 34–36.

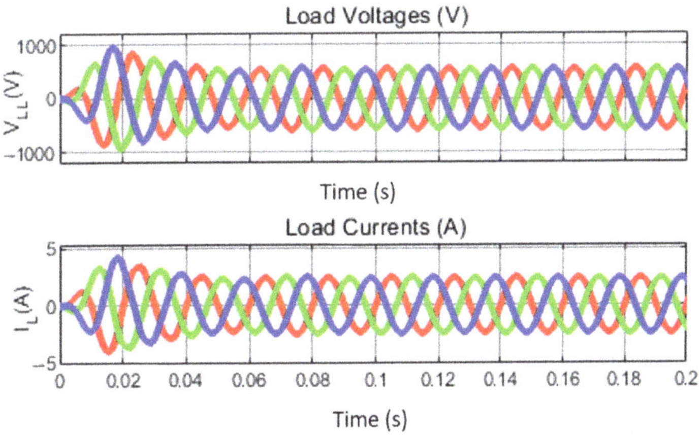


**Figure 34.** Load voltage and current of ZSI.

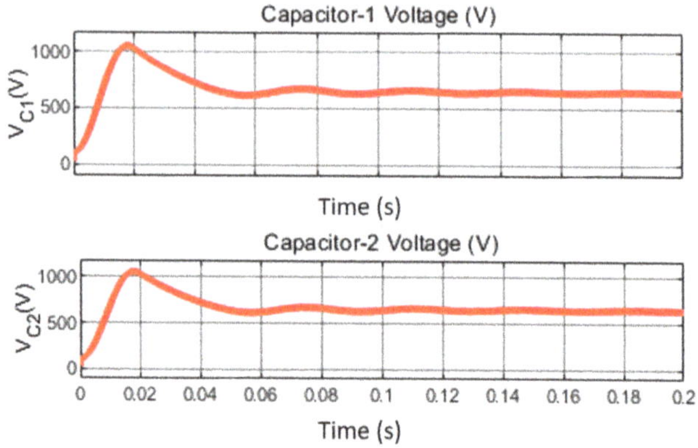


**Figure 35.** Z-source capacitor voltages of ZSI.

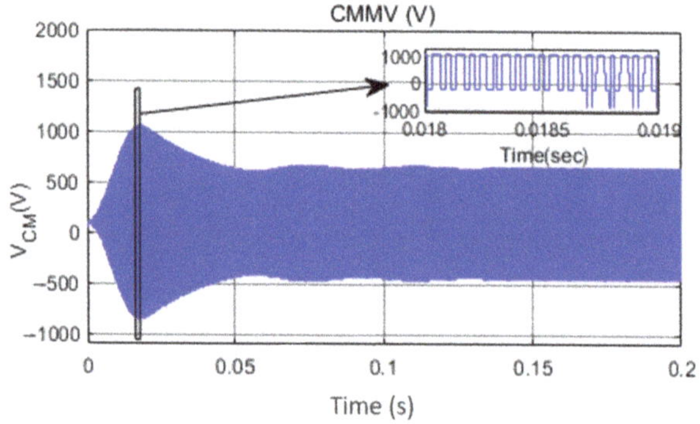


**Figure 36.** Common mode voltages of ZSI.

From these results (Figures 21 and 34), it can be understood that peak voltages (956 V in the impedance source inverter and 910 V in the CCBI) and settling time to reach the steady state (0.065 s for the impedance source inverter and 0.021 s for the CCBI) are higher in the case of the impedance source inverter. The peak capacitor voltage is 1124 V in the case of the impedance source inverter, whereas it is 1056 V in the case of the CCBI. It can also be seen that CMMV in the case of the impedance source inverter has a PWM nature, as depicted in Figure 36, whereas it has a steady nature in the case of the CCBI, as depicted in Figure 14. Hence, it can be understood that the CCBI offers better performance for the single-stage power conversion.

In addition, in terms of the number of components, voltage and current THDs, capacitor voltage stresses, and boost factors, the performance of this inverter was compared to that of existing inverter topologies. Figures 37–40 provide these comparative characteristics. In comparison to other topologies, the implementation of the CCBI requires fewer components, as seen in Figure 37. As a result, the converter's cost, size, and volume are reduced. THD (both voltage and current) profiles of the proposed inverter, and the studied inverter topologies, were captured for comparative analysis. It is worth noting that, with the exception of the boost and CCBI topologies, AC filters are utilized on the AC side in all other topologies. The CCBI is able to deliver greater performance in terms of THDs even under these conditions, as seen in Figure 38.

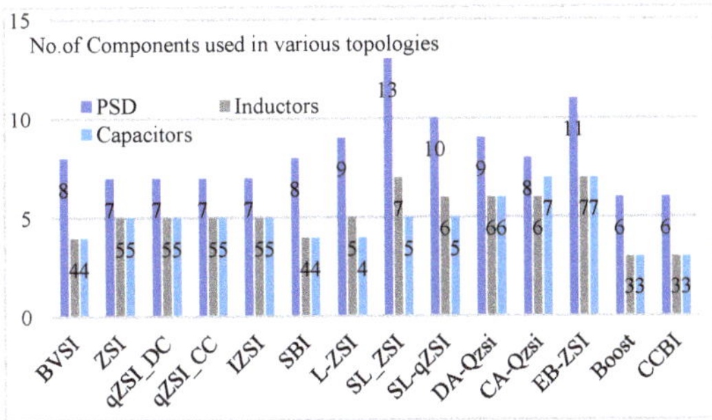


Figure 37. Bar chart of no. of components used in different topologies.

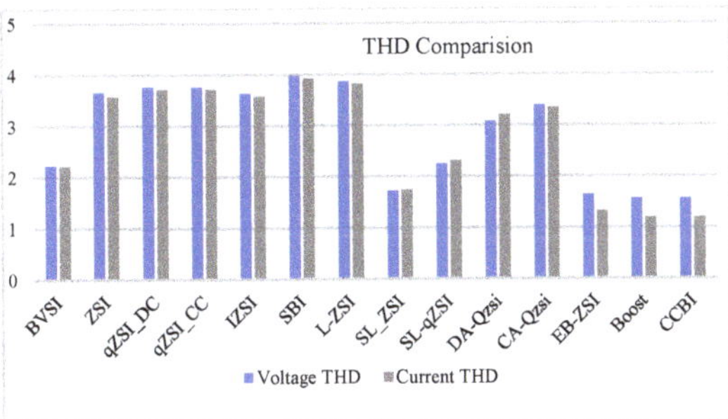


Figure 38. THD values in different topologies.

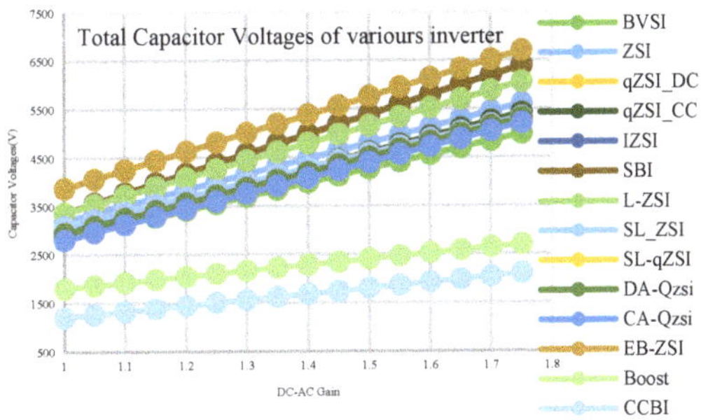


Figure 39. Total capacitor stress in different topologies.

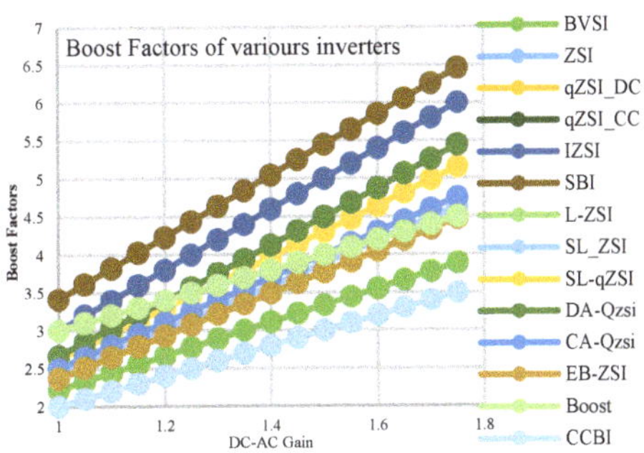


Figure 40. Boost factors in different topologies.

The total capacitor stresses in the CCBI are quite low as compared to other topologies, as shown in Figure 39. Because the capacitor is the most vulnerable component in an inverter in terms of reliability, reducing voltage stresses on the capacitor improves its reliability. As shown in Figure 40, the proposed converter is capable of providing superior gain than the existing topologies despite having fewer boosting factors. This function aids in the reduction in stresses on the inverter's capacitors and switches. Overall, the proposed inverters provide higher performance in terms of number of components, voltage and current THDs, capacitor voltage stresses, and boost factors, as evidenced by these comparative data.

*Experimentation Results:* For the experimental verifications, a laboratory-made test bench was developed, as illustrated in Figure 41. It consists mainly of six IRF460 MOSFETs (500 V, 16 A) driven by a TLP25-based optically isolated driver circuit, three EZPE50506MTA capacitors (15 μF), and three inductors (0.6 mH). Inductor currents and capacitor voltages are sensed by a TELCON-25 and AD202JN-based signal measurement and conditioning circuit.

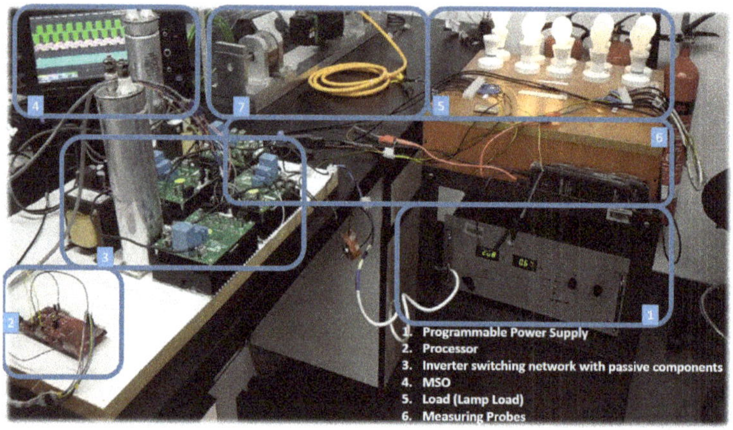


**Figure 41.** Prototype of the boost inverter.

The quantities sensed by these sensor-based resistor networks are then applied to the corresponding multiplexer (HEF4052B) input terminals via filtering, amplifying, and biasing circuits. All of the sensed parameters are sent to the FPGA Spartan-3E kit via a multiplexer circuit. Two 2-channel multiplexers are used in time division multiplexing to independently process inductor currents and capacitor voltages. Inductor currents and capacitor voltages are time division multiplexed and processed on the FPGA kit's on-board ADC (LTC1407A). Internally, these signals are demultiplexed using VHDL code. Using demultiplexed inductor currents and capacitor voltages, a VHDL-programmed sliding mode controller generates gating pulses to the inverter.

This prototype was tested with a 150 V DC supply to demonstrate the proposed inverter's step-up capability, and the results were monitored in a closed-loop manner, with the control logic developed in an FPGA Sparta-3e XC3S500e board. The CCBI converts 150 V DC to three-phase AC with a peak voltage of 282 volts, 163.29 V (peak) phase voltage, and 4.89 A (peak) phase current. These conversion pole voltages are shown in Figure 42. Load currents are depicted in Figure 43. These findings show that the converter's performance is consistent with the simulation results. Simulation and hardware tests confirm that the inverter is performing proper DC-AC conversion.

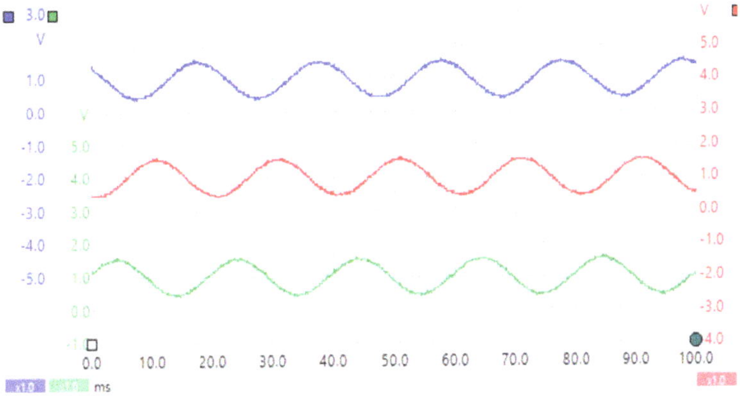


**Figure 42.** Pole voltages [300 V/div].

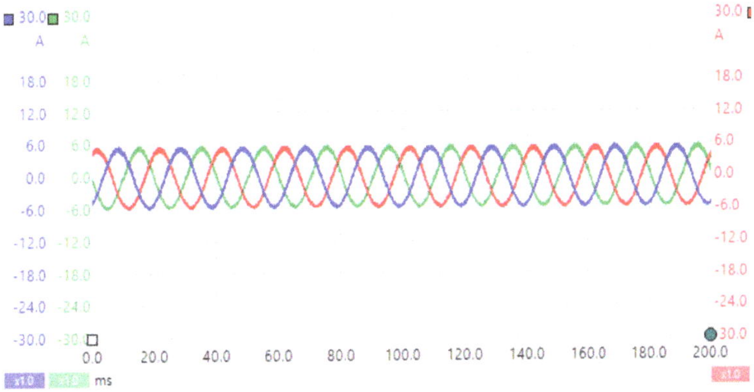


**Figure 43.** Load current [1 A/div].

## 4. Conclusions

This research suggested and successfully validated a unique three-phase, step-up DC-AC converter for distributed power generation using both simulation and experimental data. This converter successfully demonstrated single-stage operation in the same way as any other impedance source converter. Both simulation and experimental results verified that the operating voltages across the capacitors are reduced, resulting in increased capacitor and converter reliability and longevity. In addition to the technology, this inverter offers a lower boosting factor for the necessary DC-AC conversion, thus requiring a lower dc link voltage. When compared to other impedance source converters for the same DC-AC conversion, this feature has a high side gate isolation voltage requirement. In this paper, detailed operations in various modes are presented, along with differential pulse width modulation and a sliding mode controller.

**Future work:** A performance investigation of the CCBI in electrical vehicle loads with different drive cycles, and in distributed power generation with different environmental conditions, can comprise the future scope of work.

**Author Contributions:** Writing—original draft and resources, D.R. and P.R.; conceptualisation, methodology, D.R. and P.R.; investigation, D.R., B.L.N. and Y.S.B.; review and editing, D.R., H.H.F. and E.R. All authors have read and agreed to the published version of the manuscript.

**Funding:** Unitatea Executiva Pentru Finantarea Invatamantului Superior a Cercetarii Dezvoltarii si Inovarii: PN-III-P4-ID-PCE-2020-0008.

**Institutional Review Board Statement:** Not applicable.

**Informed Consent Statement:** Not applicable.

**Data Availability Statement:** Not applicable.

**Conflicts of Interest:** The authors declare no conflict of interest.

## Nomenclature

| | |
|---|---|
| Z-source inverter | ZSI |
| Quasi Z-source inverter | q-ZSI |
| Continuous input current quasi Z-source inverter | CICq-ZSI |
| Discontinuous input current quasi Z-source inverter | DICq-ZSI |
| Switched boost inverter | SBI |
| Current-fed switched boost inverter | CF-SBI |
| Quasi SBI | qSBI |
| Improved ZSI | IZSI |
| Pulse width modulation | PWM |
| Total harmonic distortion | THD |
| Capacitor clamped boost inverter | CCBI |
| Boost factor | BF |
| Full-bridge | FB |
| Half-bridge | HB |
| Enhanced boost ZSI | EB-ZSI |
| Diode assisted qZSI | DA-qZSI |
| Capacitor assisted qZSI | CA-qZSI |
| Enhanced boost quasi ZSI | EB-qZSI |

## References


1. Anderson, J.; Peng, F. A class of quasi-Z-source inverters. In Proceedings of the IEEE IAS'08 Industry Applications Society Annual Meeting, Edmonton, AB, Canada, 5–9 October 2008; pp. 1–7.
2. Loh, P.C.; Liu, Y.; Abu-Rub, H.; Ge, B.; Blaabjerg, F.; Ellabban, O. *Impedance Source Power Electronic Converters*; Wiley-IEEE Press: Trenton, NJ, USA, 2016; p. 113. ISBN 978111903710.
3. Raveendhra, D.; Rajana, P.; Kumar, K.S.R.; Jugge, P.; Devarapalli, R.; Rusu, E.; Fayek, H.H. A High-Gain Multiphase Interleaved Differential Capacitor Clamped Boost Converter. *Electronics* **2022**, *11*, 264. [CrossRef]
4. Shawky, A.; Ahmed, M.; Orabi, M.; El Aroudi, A. Classification of Three-Phase Grid-Tied Microinverters in Photovoltaic Applications. *Energies* **2020**, *13*, 2929. [CrossRef]
5. Hu, X.; Liang, W.; Gao, B.; Ma, P.; Zhang, Y. Integrated step-up non-isolated inverter with leakage current elimination for grid-tied photovoltaic system. *IET Power Electron.* **2019**, *12*, 3749–3757. [CrossRef]
6. Law, K.H. Mathematics Modelling and Simulation of Batteries Charging Capability in Quasi Z Source Impedance Network. In Proceedings of the 2019 7th International Conference on Smart Computing & Communications (ICSCC), Sarawak, Malaysia, 28–30 June 2019; pp. 1–5.
7. Shen, M.; Wang, J.; Joseph, A.; Peng, F.Z.; Tolbert, L.M.; Adams, D.J. Constant boost control of the Z-source inverter to minimize current ripple and voltage stress. *IEEE Trans. Ind. Appl.* **2006**, *42*, 770–778. [CrossRef]
8. Tang, Y.; Xie, S.; Ding, J. Pulse width modulation of Z-source inverters with minimum inductor current ripple. *IEEE Trans. Ind. Electron.* **2014**, *61*, 98–106. [CrossRef]
9. Mishra, R.S.; Joshi, A. Analysis and PWM control of switched boost inverter. *IEEE Trans. Ind. Electron.* **2013**, *60*, 5593–5602.
10. Nguyen, M.K.; Le, T.V.; Park, S.J.; Lim, Y.C. A class of quasi-switched Boost inverters. *IEEE Trans. Ind. Electron.* **2015**, *62*, 1526–1536. [CrossRef]
11. Nag, S.S.; Mishra, S. Current–Fed Switched Inverter. *IEEE Trans. Ind. Electron.* **2014**, *61*, 4680–4690. [CrossRef]
12. Reddivari, R.; Jena, D. A Correlative Investigation of Impedance Source Networks: A Comprehensive Review. *IETE Tech. Rev.* **2021**, *1*, 256–268. [CrossRef]
13. Qian, W.; Peng, F.Z.; Cha, H. Trans-Z-source inverters. *IEEE Trans. Power Electron.* **2011**, *26*, 3453–3463. [CrossRef]
14. Nguyen, M.K.; Lim, Y.C.; Choi, J.H.; Choi, Y.O. Trans-switched boost inverters. *IET Power Electron.* **2016**, *9*, 1065–1073. [CrossRef]
15. Pan, L. L-Z-Source Inverter. *IEEE Trans. Power Electron.* **2014**, *29*, 6534–6543. [CrossRef]
16. Zhu, M.; Yu, K.; Luo, F.L. Switched inductor Z-source inverter. *IEEE Trans. Power Electron.* **2010**, *25*, 2150–2158.
17. Nguyen, M.K.; Lim, Y.C.; Cho, G.B. Switched-inductor quasi-Z-source inverter. *IEEE Trans. Power Electron.* **2011**, *26*, 3183–3191. [CrossRef]
18. Fathi, H.; Madadi, H. Enhanced-Boost Z-Source Inverters with Switched Z-Impedance. *IEEE Trans. Ind. Electron.* **2016**, *63*, 691–703. [CrossRef]

19. Vinnikov, D.; Roasto, I.; Jalakas, T.; Strzelecki, R.; Adamowicz, M. Analytical comparison between capacitor assisted and diode assisted cascaded quasi-Z-source inverters. *Electr. Rev.* **2012**, *88*, 212–217.
20. Gajanayake, C.J.; Luo, F.L.; Gooi, H.B.; So, P.L.; Siow, L.K. Extended boost Z-source inverters. *IEEE Trans. Power Electron.* **2010**, *25*, 2642–2652. [CrossRef]
21. Jagan, V.; Kotturu, J.; Das, S. Enhanced-Boost Quasi-Z-Source Inverters with Two-Switched Impedance Networks. *IEEE Trans. Ind. Electron.* **2017**, *64*, 6885–6897. [CrossRef]
22. Nguyen, M.; Tran, T.; Lim, Y. A Family of PWM Control Strategies for Single-Phase Quasi-Switched-Boost Inverter. *IEEE Trans. Power Electron.* **2019**, *34*, 1458–1469. [CrossRef]
23. Xiaoquan Zhu, Bo Zhang, Dongyuan Qiu, Enhanced boost quasi-Z-source inverters with active switched-inductor boost network. *Power Electron. IET* **2018**, *11*, 1774–1787. [CrossRef]
24. Ho, V.; Hyun, J.S.; Chun, T.W.; Lee, H.H. Embedded quasi-Z-source inverters based on active switched-capacitor structure. In Proceedings of the IEEE IECON, 42nd Annual Conference of the IEEE Industrial Electronics Society, Florence, Italy, 23–26 October 2016; pp. 3384–3389.
25. Ho, V.; Yang, S.G.; Chun, T.W.; Lee, H.H. Topology of modified switched-capacitor Z-source inverters with improved boost capability. In Proceedings of the IEEE Applied Power Electronics Conference and Exposition (APEC), Tampa, FL, USA, 26–30 March 2017; pp. 685–689.
26. Babaei, E.; Asl, E.S. High voltage gain half-bridge Z-source inverter with low voltage stress on capacitors. *IEEE Trans. Ind. Electron.* **2017**, *64*, 191–197. [CrossRef]
27. Babaei, E.; Asl, E.S.; Babayi, M.H. Steady-state and small-signal analysis of high-voltage gain half-bridge switched boost inverter. *IEEE Trans. Ind. Electron.* **2016**, *63*, 3546–3553. [CrossRef]
28. Asl, E.S.; Babaei, E.; Sabahi, M. High voltage gain half-bridge quasi-switched boost inverter with reduced voltage stress on capacitors. *IET Power Electron.* **2017**, *10*, 1095–1108. [CrossRef]
29. Wang, H.; Liserre, M.; Blaabjerg, F. Toward reliable power electronics—Challenges, design tools and opportunities. *IEEE Ind. Electron. Mag.* **2013**, *7*, 17–26. [CrossRef]
30. Cupertino, A.F.; Lenz, J.M.; Brito, E.M.S.; Pereira, H.A.; Pinheiro, J.R.; Seleme, S.I. Impact of the mission profile length on lifetime prediction of PV inverters. *Microelectron. Reliab.* **2019**, *100–101*, 113427. [CrossRef]
31. Raveendhra, D.; Pathak, M.K. Three-Phase Capacitor Clamped Boost Inverter. *IEEE J. Emerg. Sel. Top. Power Electron.* **2019**, *7*, 1999–2011. [CrossRef]

energies

MDPI

Article

# An Accurate Switching Transient Analytical Model for GaN HEMT under the Influence of Nonlinear Parameters

**Dong Yan, Lijun Hang *, Yuanbin He, Zhen He and Pingliang Zeng**

College of Automation, Hangzhou Dianzi University, Hangzhou 310018, China; ayyzyd@163.com (D.Y.); yuanbinhe@hdu.edu.cn (Y.H.); hezhen_hz@hdu.edu.cn (Z.H.); plzeng@hotmail.com (P.Z.)
* Correspondence: ljhang@hdu.edu.cn

**Abstract:** The Gallium Nitride high electron mobility transistor (GaN HEMT) has been considered as a potential power semiconductor device for high switching speed and high power density application since its commercialization. Compared with the traditional Si transistors, GaN HEMT has faster switching speed and lower on-off loss. As a result, it is more sensitive to the nonlinear parameters due to the fast switching speed. The subsequent voltage and current overshooting will affect the efficiency and safety of the GaN HEMT and power electronic systems. In this paper, an accurate switching transient analytical model for GaN HEMT is proposed, which considers the effects of parasitic inductances, nonlinear junction capacitances and nonlinear transconductance. The model characteristic of turn-ON process and turn-OFF process is illustrated in detail, and the equivalent circuits are derived for each switching transition. The accuracy of the proposed model can be verified by comparing the predicted switching waveform and switching loss with that of the experimental results based on the double pulse test (DPT) circuit. Compared with the conventional model, the proposed model is more accurate and matches better with the experimental results than the conventional model. Finally, this model can be used for analyzing the influences of gate resistance, nonlinear junction capacitances, and parasitic inductances on switching transient waveform and refining calculation switching loss.

**Keywords:** GaN HEMT; switching transient analytical model; parasitic inductance; nonlinear junction capacitance; nonlinear transconductance; double pulse test

**Citation:** Yan, D.; Hang, L.; He, Y.; He, Z.; Zeng, P. An Accurate Switching Transient Analytical Model for GaN HEMT under the Influence of Nonlinear Parameters. *Energies* **2022**, *15*, 2966. https://doi.org/10.3390/en15082966

Academic Editor: Denis N. Sidorov

Received: 2 March 2022
Accepted: 14 April 2022
Published: 18 April 2022



## 1. Introduction

GALLIUM NITRIDE (GaN) is a typical representative of the third generation wide band gap semiconductor materials and has a broad application prospect. In recent years, Gallium Nitride high electron mobility transistor (GaN HEMT) has been successfully applied to high frequency and high power density converters [1]. GaN HEMT can not only attain small conduction resistance, but also push the switching frequency to several megahertz when comparing with the state-of-the-art Si MOSFETs. Nevertheless, high frequency switching transient still brings some extra problems, such as voltage and current overshooting, and even instability. As a result, it is more sensitive to the nonlinear parameters because of high dv/dt and di/dt. Meanwhile, the subsequent switching process ringing will affect the efficiency and reliability of the GaN HEMT devices and power electronic systems [2]. In order to fully utilize the GaN HEMT, it is necessary to analyze the parameters' effects upon switching loss and the switching transient characteristic.

Generally, the switching transient of devices can be analyzed by experimental testing [3] or software simulation method [4]. However, these methods cannot give a definitive or analytical explanation of the influence mechanism of the nonlinear parameters in devices. Therefore, an accurate switching transient analytical model for estimating the switching loss and influence of nonlinear parameters is necessary.

The popular switching transient analytical model is the piecewise linear model [5]. However, the nonlinear parameters are not taken into account so that the calculated results do not agree with the experimental results very well, especially in high frequency applications, since the switching transient and switching loss are not well evaluated. In study [6], a second-order analytical model is proposed. More comprehensive analytical models of Si MOSFETs are presented in [7–9]. These models consider the parasitic inductances and the nonlinear junction capacitances properly. On this basis, the authors of [10] apply the analysis model to develop the SiC MOSFET model, whereas, in some of the derivation, the parameter setting is idealistic. The nonlinear parameters' modeling results in [11] is too complex due to the numerical method or iterative calculations. In studies [12–14], the impact of nonlinear capacitance and nonlinear transconductance in SiC MOSFET is discussed, and a semi-physical semi-behavioral analytical model for switching transient of SiC MOSFET power module is proposed. However, these models are still not suitable for GaN HEMT devices, because GaN HEMT have no reverse recovery characteristics, and have smaller charge capacitance and parasitic inductance in contrast to Si MOSFET and SiC MOSFET [15].

In study [16], a switching loss model for GaN HEMT is proposed. The model considers the nonlinear junction capacitances as piecewise linear structure, so it still cannot precisely represent the nonlinear characteristics. In studies [17–19], the adopted interpolation method is used to fit nonlinear junction capacitances of GaN HEMT. However, this method has no specific mathematical expression and the results are too complex. In study [20], the package and PCB parasitic inductances, as well as the nonlinear capacitances are considered in the GaN HEMT model, but this model only includes the turn-ON process without switching loss calculation and parameter analysis. Some analytical models are proposed for cascode GaN HEMT [21,22]. These models cannot be directly applied to the GaN HEMT because of the different structure. Therefore, a more accurate switching transient analytical model for GaN HEMT is needed to explain the influence of nonlinear parameters and calculate the loss accurately. The main contributions of this paper are:

(1) An accurate model for nonlinear junction capacitances of GaN HEMT is given and an extraction method for nonlinear transconductance is proposed. The proposed approximation method for capacitances and transconductance have better performance than conventional approaches.
(2) An accurately switching transient analytical model for GaN HEMT considering the influence of nonlinear parameters is proposed in this paper. The analytical derivation of turn-ON process and turn-OFF process is illustrated in detail, and the equivalent circuits are derived for each switching transition. The accuracy of the proposed model is verified by DPT circuit. Experiment results verify that the proposed model is more precise compared with the traditional model. The experimental results show that the maximum relative error of the switching loss can be kept within 10% by the proposed model.
(3) The effects of gate resistance, gate source capacitance, gate drain capacitance, drain source capacitance, and parasitic inductances on switching characteristic and switching losses are systematically investigated based on the proposed model.

## 2. Novel Analytical Model of GaN HEMT Device

In order to test and evaluate the switching characteristics of GaN HEMT, a double pulse test (DPT) circuit with inductive load can be adopted as illustrated in Figure 1, where $Q$ and $Q_{\_H}$ represent the lower switch and upper GaN HEMT device of the half bridge. In this paper, the GaN HEMT device of GS61004B produced by GaN systems company will be illustrated as an example. $C_{gs}$, $C_{gd}$, and $C_{ds}$ represent the capacitors of gate source, gate drain, and drain source, respectively. To simplify the circuit analysis, the parasitic inductance and package stray inductance are accounted into $L_d$, $L_s$, $L_g$. $L_{loop}$ lumps all of the parasitic inductances and package stray inductances along the power loop. The GaN HEMT device does not contain a p-n diode, but the way they conduct in the reverse direction is similar to diodes because this channel is formed by turning on the two-dimensional electron gas (2DEG) in the device.

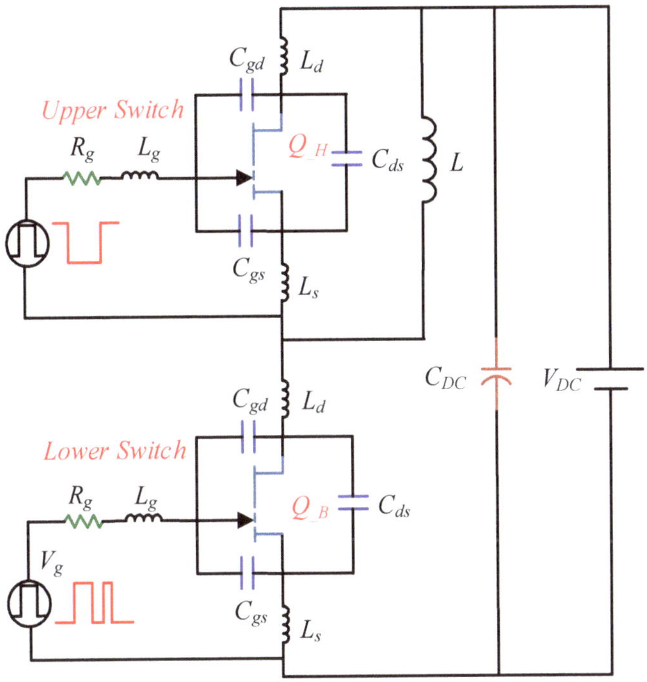


**Figure 1.** Circuit schematic of double pulse test bench.

In Figure 1, it is very important to determine the parasitic inductances and nonlinear parameters to obtain an accurate switching transient analytical model. Generally, the parasitic inductances can be acquired by Maxwell 3-D software and the nonlinear parameters can be obtained by the modeling method, such as the input capacitance $C_{iss} = C_{gs} + C_{gd}$, the output capacitance $C_{oss} = C_{ds} + C_{gd}$, and the reverse capacitance $C_{rss} = C_{gd}$. $C_{iss}$ and $C_{rss}$ can be fitted by simple piecewise linear equation due to their low degree of nonlinearity. However, $C_{oss}$ is a parameter that changes dramatically with the variation of $v_{ds}$ at low voltage level, which is with high degree of nonlinearity, and is more difficult to be accurately modeled by the above methods. A nonlinear function tanh($x$) will be introduced to represent $C_{oss}$ in this paper, as shown in Equation (1). This function can be used to approximate the $C_{oss}$ during switching process.

$$\tanh(x) = \frac{e^x - e^{-x}}{e^x + e^{-x}} \tag{1}$$

Based on tanh($x$), a nonlinear function shown in Equation (2) is proposed to fitting the nonlinear characteristic of the $C_{oss}$ of GS61004B, which can model different changing rate of $C_{oss}$ when $v_{ds}$ began to change in low voltage.

$$C_{oss} = C_{omax}(1 + v_{ds}(1 + k_1(1 + \tanh(k_2 \cdot v_{ds} + k_3))))^{k_4} \tag{2}$$

where $C_{omax}$, $k_1$, $k_2$, $k_3$, and $k_4$ are the fitting parameters and are given in Table 1. This method has more accuracy modeling results. The modeling results $C_{oss}$ of GaN HEMT is verified in Figure 2a, and it can be concluded that the maximum error between the fitted data and extracted data from the datasheet is less than 3%. Furthermore, this model can also be applied to other different types of devices by adjusting the parameters.

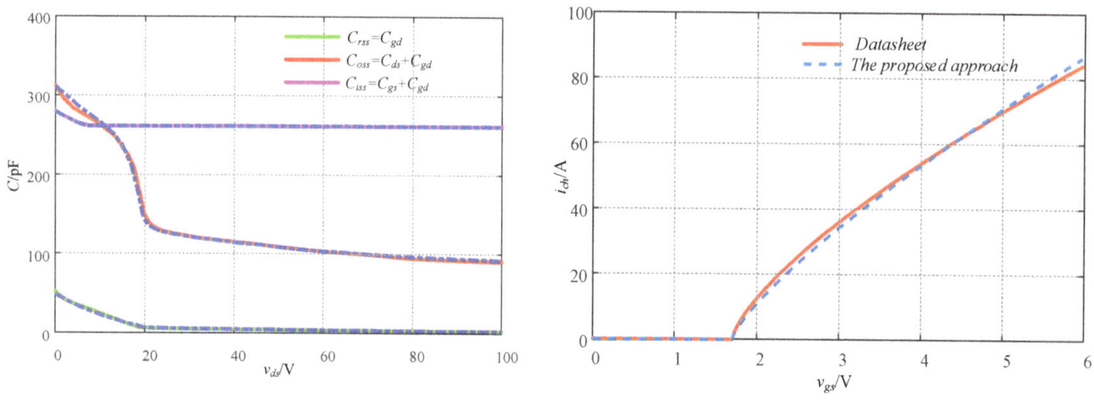


**Figure 2.** Modeling results and datasheet: (**a**) Nonlinear capacitances; (**b**) Transfer characteristic curves.

Transconductance $g_{fs}$ describes the dynamic behavioral feature of $i_{ch}$ when $v_{gs}$ changes. Using a single value of transconductance has a large opportunity to cause great errors. In this paper, the slope of the tangent line in the transfer characteristic curve is used to approximate the transconductance at the corresponding operation point. Thus, the equation of $g_{fs}$ can be expressed as Equation (3).

$$\begin{cases} g_{fs} = 0 & v_{gs} < V_{th} \\ g_{fs} = \frac{di_{ch}}{dv_{gs}} = k_5(v_{gs} - V_{th})^{k_6 - 1} & v_{gs} \geq V_{th} \end{cases} \quad (3)$$

where $k_5$ and $k_6$ are the fitting parameters which are given in Table 1 as well. $V_{th}$ is the threshold voltage of gate to source for the device. Figure 2b shows the curve of $i_{ch}$ versus $v_{gs}$. It is obvious that the modeling result makes an accurate estimation based on the proposed model.

**Table 1.** Fitting parameters.

| Symbol | Value |
|---|---|
| $C_{omax}$ | 311 pF |
| $k_1$ | −0.456 |
| $k_2$ | −0.51 |
| $k_3$ | 9.5 |
| $k_4$ | −0.256 |
| $k_5$ | 30 |
| $k_6$ | 0.75 |

## 3. Transient Characteristic Analysis of Turn-On Process Based on the Proposed Model

A typical waveform during turn-ON process of the GaN HEMT half bridge is shown in Figure 3. There are four stages during the switching process and each stage has different features according to the change of the circuit structure. At each stage, the equations can be obtained from the corresponding equivalent circuit which is illustrated in Figure 4. By solving the equations, we can obtain five variables, namely, the gate current $i_g$, the gate source voltage $v_{gs}$, the drain current $i_d$, the drain source voltage $v_{ds}$, and the drain source voltage $v_{ds\_H}$. The nonlinear junction capacitance and nonlinear transconductance modeling results proposed in Section 2 will be applied to the turn-ON process in the following analysis. Thus, the analytical model of each variable during turn-ON process can be derived. Furthermore, the switching loss during the turn-ON process can be obtained.

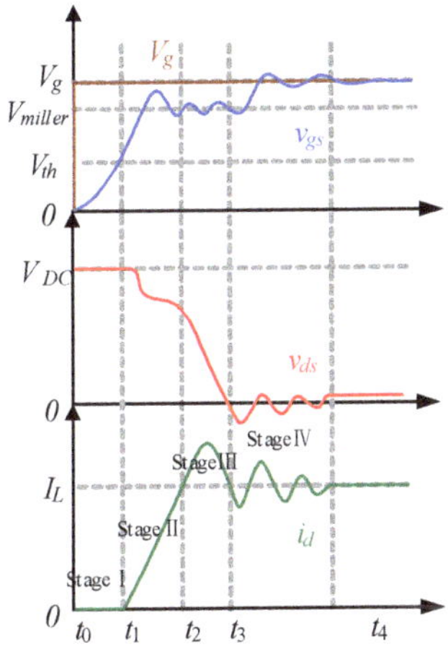


**Figure 3.** Typical waveforms during turn-ON process.

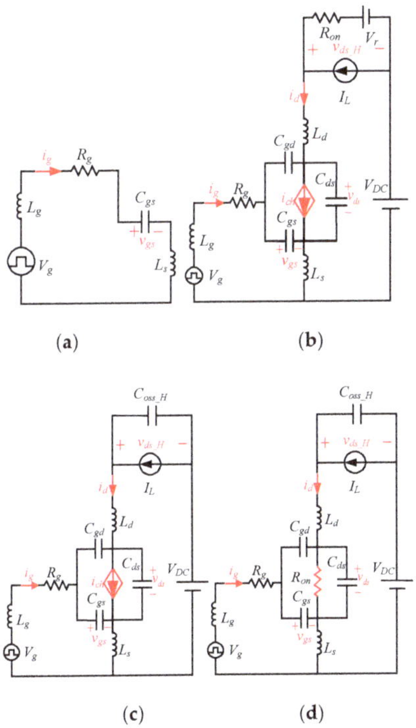


**Figure 4.** Equivalent circuits for different stages during turn-ON period: (**a**) Stage I; (**b**) Stage II; (**c**) Stage III; (**d**) Stage IV.

### 3.1. Stage I ($t_0$–$t_1$): Turn-On Delay Period

As shown in Figure 3, at time $t_0$, the gate drive voltage $V_g$ is applied. In this stage, the gate current charges the capacitance $C_{gs}$ which causes $v_{gs}$ to rise. The equivalent circuit of this stage is shown in Figure 4a. According to the equivalent circuit, the following equations can be obtained:

$$V_g = (L_g + L_s)\frac{di_g}{dt} + R_g i_g + v_{gs} \tag{4}$$

$$i_g = C_{gs}\frac{dv_{gs}}{dt} \tag{5}$$

This stage ends at time $t_1$ when $v_{gs}$ reaches the threshold voltage $V_{th}$. During this stage, $Q$ is still OFF, the drain source voltage $v_{ds}$, and the drain current $i_d$ are constant, so there is no switching loss.

### 3.2. *Stage II ($t_1$–$t_2$): Current Rise Period*

As shown in Figure 3, at time $t_1$, $v_{gs}$ reaches the threshold voltage $V_{th}$, and the channel current $i_{ch}$ begins to rise which can be expressed by Equation (3). At the same time, the capacitances $C_{oss}$ begins to discharge through the channel of $Q$. The drain current id is clamped to the sum of the channel current $i_{ch}$ and the $C_{oss}$ discharging current. However, $Q_{\_H}$ is still in reverse conduction condition, where $R_{on}$ is the on-state resistance and $V_r$ represents the reverse conducting voltage drop of $Q_{\_H}$. There will be a voltage drop $\Delta V$ between $v_{ds}$ and its initial voltage $V_{DC}$ + $V_r$ on the parasitic inductance of the power loop. As shown in the equivalent circuit of Figure 4b, the equations of this stage are as follows:

$$V_g = v_{gs} + R_g i_g + L_g\frac{di_g}{dt} + L_s(\frac{di_d}{dt} + \frac{di_g}{dt}) \tag{6}$$

$$i_g = C_{gs}\frac{dv_{gs}}{dt} + C_{gd}(\frac{dv_{gs}}{dt} - \frac{dv_{ds}}{dt}) \tag{7}$$

$$i_d = g_{fs}(v_{gs} - V_{th}) + C_{ds}\frac{dv_{ds}}{dt} - C_{gd}\frac{d(v_{gs} - v_{ds})}{dt} \tag{8}$$

$$V_{DC} = v_{ds} + v_{ds\_H} + (L_d + L_s)\frac{di_d}{dt} + L_s\frac{di_g}{dt} \tag{9}$$

$$v_{ds\_H} = R_{on}(I_L - i_d) - V_r \tag{10}$$

This stage ends when the drain current $i_d$ reaches the load current $I_L$. Meanwhile, $v_{gs}$ reaches Miller plateau voltage. It should be especially noted that the voltage drop $\Delta V$ is shown in the following equation:

$$\Delta V = (L_d + L_s)\frac{di_d}{dt} \tag{11}$$

In case of $L_d$ and $L_s$ being larger, $\Delta V$ will drop faster, so $i_d$ would change even slower than $v_{ds}$. The transient will skip the voltage drop stage III into ringing stage IV directly. While for small $L_d$ and $L_s$ of properly designed PCB and reasonable current rising speed, stage III will occur as expected.

During this stage, the energy loss can be expressed as:

$$E_{on} = \int_{t2}^{t1} v_{ds} \cdot i_{ch} dt \tag{12}$$

The power loss in this stage can be determined by:

$$\begin{aligned} P_{on} &= f_s \cdot E_{on} \\ &= f_s \cdot \textstyle\int_{t2}^{t1} v_{ds} \cdot i_{ch} dt \end{aligned} \tag{13}$$

where $f_s$ is the switching frequency. However, only id can be measured experimentally, ich cannot. During the turn-ON process, a part of the energy stored in $C_{oss}$ is dissipated through the channel, which is not included in the loss calculated by $i_d$. Thus, the relationship can be expressed as:

$$P_{on} = f_s \cdot \int_{t2}^{t1} v_{ds} \cdot i_d dt + P_{Coss} \tag{14}$$

where $P_{Coss}$ represents the energy stored in the output capacitance when the applied voltage is $V_{DC} + V_r$ and can be expressed as:

$$P_{Coss} = f_s \cdot \int_0^{V_{DC}+V_r} v_{ds} \cdot C_{oss}(v_{ds}) dv_{ds} \tag{15}$$

$C_{oss}(v_{ds})$ can be determined by Equation (2), which is plotted in Figure 2a.

### 3.3. Stage III ($t_2$–$t_3$): Voltage Decline Period

As shown in Figure 3, at time $t_2$, the upper switch GaN HEMT device $Q_{\_H}$ stops conducting, and the channel current $i_{ch}$ surpasses the load current $I_L$. This surpasses current charge of the output capacitance $C_{oss\_H}$ of $Q_{\_H}$. The upper switch can be equivalent to the output capacitor $C_{oss\_H}$, and $v_{ds\_H}$ is increased. Correspondingly, $v_{ds}$ decreases and $C_{oss}$ continues to discharge through the channel of $Q$. $i_d$ may have a ringing because of the effect of parasitic $LC$ elements on the device. The equivalent circuit of this stage is illustrated in Figure 4c. The circuit equations can be expressed as:

$$i_d = g_{fs}(v_{gs} - V_{th}) + C_{ds}\frac{dv_{ds}}{dt} - C_{gd}\frac{d(v_{gs} - v_{ds})}{dt} \tag{16}$$

$$V_g = v_{gs} + R_g i_g + L_g\frac{di_g}{dt} + L_s(\frac{di_d}{dt} + \frac{di_g}{dt}) \tag{17}$$

$$i_g = C_{gs}\frac{dv_{gs}}{dt} + C_{gd}(\frac{dv_{gs}}{dt} - \frac{dv_{ds}}{dt}) \tag{18}$$

$$V_{DC} = v_{ds} + v_{ds\_H} + (L_d + L_s)\frac{di_d}{dt} + L_s\frac{di_g}{dt} \tag{19}$$

$$i_d = I_L + C_{oss\_H}\frac{dv_{ds\_H}}{dt} \tag{20}$$

This period ends when $v_{ds}$ decreases to zero. A current overshoot is caused by the rapid *dv/dt* action on the output capacitance:

$$\Delta i_d = C_{oss\_H}\frac{dv_{ds\_H}}{dt} - C_{oss}\frac{dv_{ds}}{dt} \tag{21}$$

In addition, the Miller plateau time on $v_{gs}$ for GaN HEMT is less than Si MOSFET because of small transfer capacitance $C_{gd}$. This characteristic can help us to understand why GaN HEMT can turn-ON at a high speed. The turn-ON switching loss calculation method is the same as that of Equation (15).

### 3.4. Stage IV ($t_3$–$t_4$): Ringing Period

As shown in Figure 3, at time $t_3$, $Q$ will work in the linear region and behave equivalent to a loop resistor, as illustrated in Figure 4d. Usually, the ringing is serious due to small damping resistance with large parasitic inductances. The key equations can be expressed as:

$$V_g = v_{gs} + R_g i_g + L_g\frac{di_g}{dt} + L_s(\frac{di_d}{dt} + \frac{di_g}{dt}) \tag{22}$$

$$i_g = C_{gs}\frac{dv_{gs}}{dt} + C_{gd}(\frac{dv_{gs}}{dt} - \frac{dv_{ds}}{dt}) \tag{23}$$

$$V_{DC} = v_{ds} + v_{ds\_H} + (L_d + L_s)\frac{di_d}{dt} + L_s\frac{di_g}{dt} \tag{24}$$

$$i_d = \frac{v_{ds}}{R_{on}} + C_{ds}\frac{dv_{ds}}{dt} - C_{gd}\frac{d(v_{gs} - v_{ds})}{dt} \tag{25}$$

$$i_d = I_L + C_{oss\_H}\frac{dv_{ds\_H}}{dt} \tag{26}$$

where $R_{on}$ represents the on-state resistance of $Q$. During this stage, the turn-ON switching loss can be expressed as:

$$P_{on} = f_s \cdot \int_{t4}^{t3} (i_d - I_L)^2 \cdot R_{on} dt \quad (27)$$

This period ends when the ringing is fully damped.

## 4. Transient Characteristic Analysis of Turn-Off Process Based on the Proposed Model

There are still four stages and the typical waveforms during turn-OFF process. At each stage, the equations can be obtained from each equivalent circuit. Similarly, the nonlinear junction capacitance and nonlinear transconductance modeling results proposed in Section 2 are applied to the turn-OFF process and five state variables can be calculated according to the equations obtained from the equivalent circuit of each stage. Furthermore, the switching loss during the turn-OFF process can be obtained. The calculation of each stage will be discussed as follows.

### *4.1. Stage I ($t_5$–$t_6$): Turn-Off Delay Period*

As shown in Figure 5, at time $t_5$, the gate drive voltage $V_g$ declines to 0 V. In this stage, $C_{gs}$ and $C_{gd}$ are discharged, so $v_{gs}$ begins to reduce. However, $Q$ is still in the turn-ON state, $I_L$ keeps flowing through the channel of $Q$. The equivalent circuit of this stage is shown in Figure 6a. The following equations can be obtained:

$$0 = (L_g + L_s)\frac{di_g}{dt} + R_g i_g + v_{gs} \quad (28)$$

$$i_g = (C_{gs} + C_{gd})\frac{dv_{gs}}{dt} \quad (29)$$

This period ends at time $t_6$ when $v_{gs}$ reaches the Miller plateau voltage $v_{miller}$.

$$v_{miller} = V_{th} + \frac{I_L}{g_{fs}} \quad (30)$$

$v_{ds}$ and $i_d$ are constants, so there is no turn-OFF switching loss in this stage.

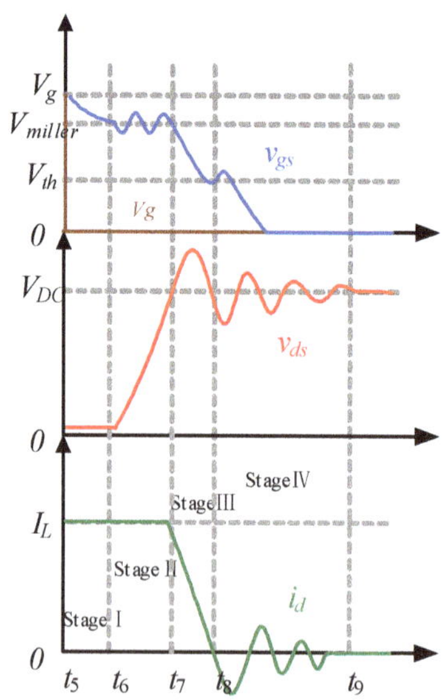


**Figure 5.** Typical waveforms during turn-OFF process.

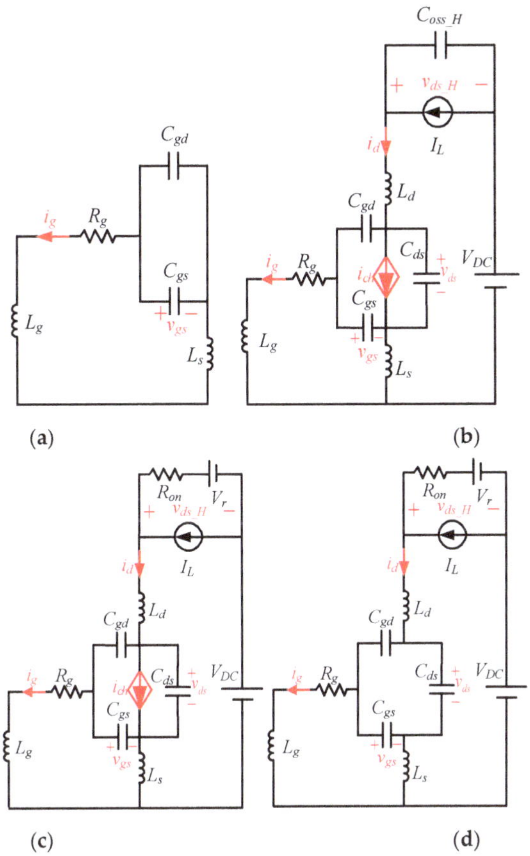


**Figure 6.** Equivalent circuits for different stages during turn-OFF process: (**a**) Stage I; (**b**) Stage II; (**c**) Stage III; (**d**) Stage IV.

### *4.2. Stage II ($t_6$–$t_7$): Voltage Rise Period*

As shown in Figure 5, at time $t_6$, the channel of $Q$ starts to turn-OFF, so the channel current $i_{ch}$ decreases and begins to charge the capacitances $C_{ds}$ and $C_{gd}$, then $v_{ds}$ is rapidly increased. At the same time, $C_{oss\_H}$ begins to discharge, resulting in the decline of $v_{ds\_H}$. However, $Q_{\_H}$ is still not reverse conducting because $i_d$ is limited by the power loop parasitic inductances. Therefore, the decrease rate of $i_d$ is smaller than that of $i_{ch}$. The equivalent circuit is shown in Figure 6b, and the equations of this stage are as follows:

$$0 = v_{gs} + R_g i_g + L_g \frac{di_g}{dt} + L_s\left(\frac{di_d}{dt} + \frac{di_g}{dt}\right) \tag{31}$$

$$i_g = C_{gs}\frac{dv_{gs}}{dt} + C_{gd}\left(\frac{dv_{gs}}{dt} - \frac{dv_{ds}}{dt}\right) \tag{32}$$

$$i_d = g_{fs}(v_{gs} - V_{th}) + C_{ds}\frac{dv_{ds}}{dt} - C_{gd}\frac{d(v_{gs} - v_{ds})}{dt} \tag{33}$$

$$V_{DC} = v_{ds} + v_{ds\_H} + (L_d + L_s)\frac{di_d}{dt} \tag{34}$$

$$i_d = I_L + C_{oss\_H}\frac{dv_{ds\_H}}{dt} \tag{35}$$

This period ends when $v_{ds\_H}$ reaches $-V_r$. During this stage, the energy loss can be calculated as:

$$E_{off} = \int_{t7}^{t6} v_{ds} \cdot i_{ch} dt \tag{36}$$

The power loss in this stage can be determined by:

$$\begin{aligned} P_{off} &= f_s \cdot E_{off} \\ &= f_s \cdot \int_{t7}^{t6} v_{ds} \cdot i_{ch} dt \end{aligned} \tag{37}$$

Usually, the energy stored in $C_{oss}$ is quite large due to the high peak $v_{ds}$. The power loss is different from turn-ON process because $i_d$ contains the charge current from $C_{oss}$. Thus, the relationship can be modified as:

$$P_{off} = f_s \cdot \int_{t7}^{t6} v_{ds} \cdot i_d dt - P_{Coss} \tag{38}$$

where $P_{Coss}$ is the energy stored in $C_{oss}$ when the applied voltage is $v_{\text{Peak}}$ and can be calculated as:

$$P_{Coss} = f_s \cdot \int_{0}^{v_{peak}} v_{ds} \cdot C_{oss}(v_{ds}) dv_{ds} \tag{39}$$

$C_{oss}(v_{ds})$ can be determined by Equation (2) and is plotted in Figure 2a.

### 4.3. Stage III ($t_7$–$t_8$): Current Decline Period

As shown in Figure 5, at time $t_7$, $Q\_H$ starts to reverse conduct. During this stage, $i_d$ and $v_{gs}$ decreases, $v_{ds}$ continues to increase, and $C_{ds}$ and $C_{gd}$ are charged. The equivalent circuit is shown in Figure 6c, and the circuit equations can be expressed as:

$$0 = v_{gs} + R_g i_g + L_g \frac{di_g}{dt} + L_s\left(\frac{di_d}{dt} + \frac{di_g}{dt}\right) \tag{40}$$

$$i_g = C_{gs} \frac{dv_{gs}}{dt} + C_{gd}\left(\frac{dv_{gs}}{dt} - \frac{dv_{ds}}{dt}\right) \tag{41}$$

$$i_d = g_{fs}(v_{gs} - V_{th}) + C_{ds} \frac{dv_{ds}}{dt} - C_{gd} \frac{d(v_{gs} - v_{ds})}{dt} \tag{42}$$

$$V_{DC} = v_{ds} + v_{ds\_H} + (L_d + L_s) \frac{di_d}{dt} + L_s \frac{di_g}{dt} \tag{43}$$

$$v_{ds\_H} = R_{on}(I_L - i_d) - V_r \tag{44}$$

This stage ends when $i_d$ drops to zero. The turn-OFF power loss calculation method is the same as Stage II.

### 4.4. Stage IV ($t_8$–$t_9$): Ringing Period

As shown in Figure 5, at time $t_8$, when $v_{gs}$ drops below $V_{th}$, the channel of $Q$ is totally shut down. Then $v_{gs}$ continues to decrease until it reaches zero. During this stage, $v_{ds}$ and $i_d$ both have a ringing due to the oscillation between the parasitic inductances and $C_{oss}$. The equivalent circuit is shown in Figure 6d, and the circuit equations can be expressed as:

$$0 = v_{gs} + R_g i_g + L_g \frac{di_g}{dt} + L_s\left(\frac{di_d}{dt} + \frac{di_g}{dt}\right) \tag{45}$$

$$i_g = C_{gs} \frac{dv_{gs}}{dt} + C_{gd}\left(\frac{dv_{gs}}{dt} - \frac{dv_{ds}}{dt}\right) \tag{46}$$

$$i_d = C_{ds} \frac{dv_{ds}}{dt} - C_{gd} \frac{d(v_{gs} - v_{ds})}{dt} \tag{47}$$

$$V_{DC} + V_r = v_{ds} + (L_d + L_s) \frac{di_d}{dt} \tag{48}$$

During this stage, the turn-OFF power loss calculation method can be expressed as:

$$P_{off} = f_s \cdot \int_{t9}^{t8} i_d \cdot v_{ds} dt \tag{49}$$

This period ends when the ringing is fully damped.

## 5. Experimental Verifications

In order to verify the accuracy of the proposed switching transient analytical model, a DPT experiment prototype was built, as shown in Figure 7. The double pulses will be applied to the gate of the bottom GaN HEMT by TMS320F28335. The device under test is the GS61004B 100 V enhanced GaN HEMT and the driver is Si8271. The oscilloscope is MDPO3054 by Tektronix Inc. The switching current is measured with 2000 MHz bandwidth current shunt SSDN-10 by T&M Research Products Inc. The parasitic inductance and package stray inductance can be acquired by Maxwell 3-D. The specifications and the main circuit parameters are shown in Table 2. The comparisons among the switching waveforms provided by the experiment, the proposed model, and the traditional mode in [16] are shown in Figure 8. The calculated results from the models for the variables $v_{ds}$ and $i_d$ with fitting value from Figure 2 will be used to evaluate the model accuracy. The loss calculation results are shown in Figure 9.

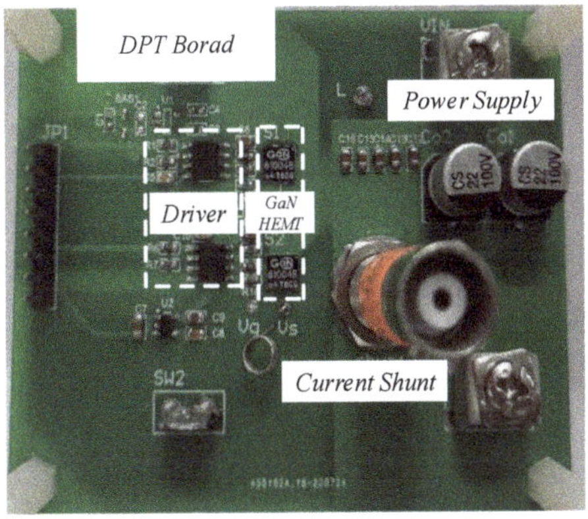


**Figure 7.** Experiment circuit prototype.

**Table 2.** Specifications and main circuit parameters.

| Parameters | Value |
|---|---|
| $V_{DC}$ | 40 V |
| $I_L$ | 7 A |
| $R_{on}$ | 0.57 Ω |
| $L_s$ | 5.6 nH |
| $L_d$ | 3.7 nH |
| $L_g$ | 2.8 nH |
| $R_g$ | 10 Ω |
| $V_g$ | 6 V |

It can be concluded that the proposed model matches more accurately with the experimental results than the traditional model according to the slope and spike of $i_d$ and $v_{ds}$ from Figure 8. The oscillation frequency and amplitude of the waveforms from the experimental results deviate from the traditional model while the proposed model results agree with that accurately. The reason is that the traditional model treats the nonlinear parameters in the device as fixed values.

Subsequently, the energy loss of $E_{on}$ and $E_{off}$ is calculated by the proposed loss calculation method as shown in Figure 9. The results show that the energy loss increases when the load current changes from 5 A to 10 A. It is obvious that the loss calculation results of the proposed model in this paper are more precise compared with the traditional model. Experiment results verify the accuracy of the proposed loss calculation method with a maximum relative error of less than 10%. Finally, it can be concluded that the proposed switching transient analytical model is quite accurate and the loss analysis results based on the model is believable.

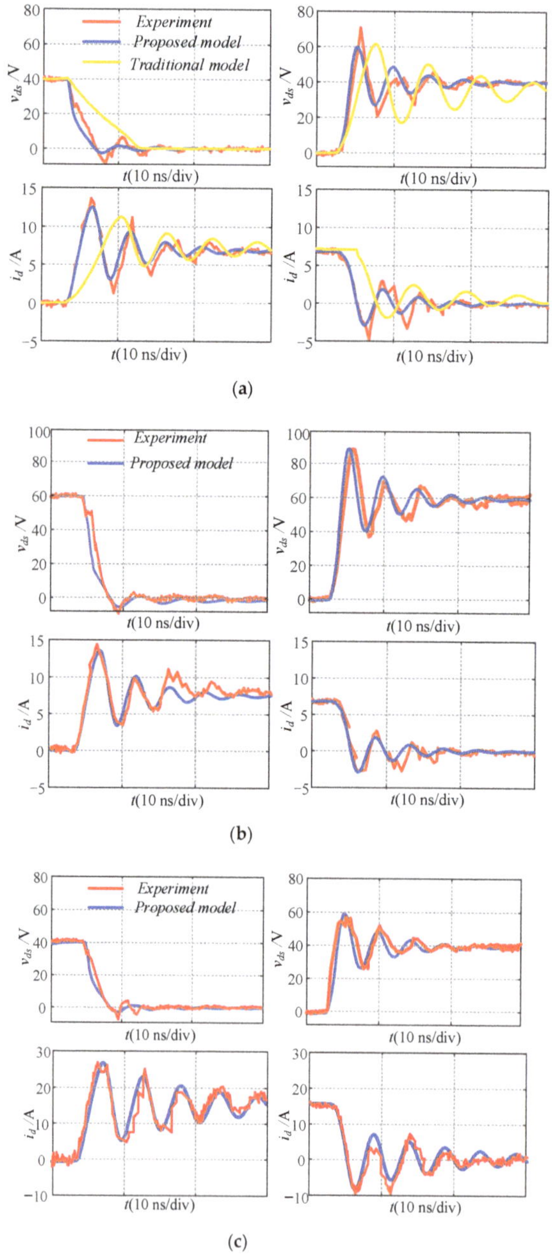


**Figure 8.** Switching waveforms of the experiment, the proposed model, and the traditional model: (**a**) $V_{DC}$ = 40 V, $I_L$ = 7 A; (**b**) $V_{DC}$ = 60 V, $I_L$ = 7 A; (**c**) $V_{DC}$ = 40 V, $I_L$ = 15 A.

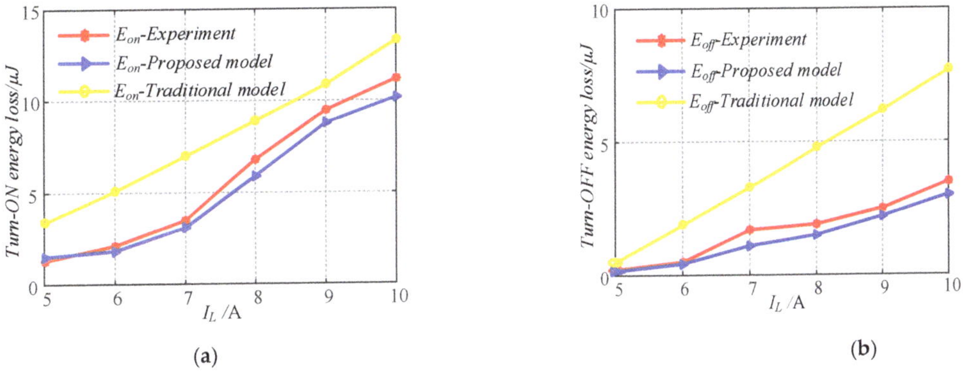


**Figure 9.** Switching loss of the experiment, the proposed model, and the traditional model: (**a**) turn-ON energy loss; (**b**) turn-OFF energy loss.

## 6. Effect of Parameter Variation on the Analytical Model

The effectiveness of the proposed switching transient analytical model is verified in the previous section. Moreover, it is important to evaluate how the gate resistance $R_g$, the nonlinear junction capacitances $C_{gs}$, $C_{gd}$, $C_{ds}$, and the parasitic inductance $L_{loop}$ affect the switching transient and switching loss, thus the further improvement of packaging and application of the device can be inspired.

Figure 10 shows the effect of different driving resistors $R_g$ on the switching transition behavior. The waveforms of $v_{ds}$ and $i_d$ are plotted. It can be seen that the oscillation of $v_{ds}$ and $i_d$ can be effectively suppressed by increasing the $R_g$. However, the larger $R_g$ limits the gate charging current and reduces the charging speed, thus resulting the switching time increase of the device. In addition, Figure 10 also calculates the influence of different $R_g$ on switching loss. With the increase of $R_g$, the turn-ON and turn-OFF switching loss is also increased. However, with the increase of $R_g$, the current overshoot for turn-ON transient and the voltage overshoot for turn-OFF transient decreased. Therefore, considering the impact of switching loss and overshoot in practical application, it is necessary to select an appropriate $R_g$ value.

In order to simplify the analysis, $L_{loop}$ lumps all of the parasitic inductances and package stray inductances along the power loop. As the package stray inductances are accounted into $L_d$ and $L_s$, so $L_{loop}$ is the sum of $L_d$ and $L_s$. The effect of different $L_{loop}$ on the switching transition behavior is shown in Figure 11. During the turn-ON process, the change rate of $i_d$ and $v_{ds}$ become slower with the increase of $L_{loop}$. Meanwhile, the current overshoot decreases but the voltage overshoot increases. In addition, the ringing period of $i_d$ and $v_{ds}$ becomes larger. These factors cause the turn-ON switching loss to become larger due to overlap area increases. During the turn-OFF process, the increase of $L_{loop}$ has an obvious influence on $i_d$ overshoot, which also leads to the increase of loss. It can be concluded that the total loss also increases, therefore, the value of $L_{loop}$ should be minimized as much as possible on the PCB design and optimization process.

For the same type of GaN HEMT, the value of the nonlinear junction capacitances should be the same. However, due to different production processes or different production lines, the capacitance values may change. In order to explore the effect of the nonlinear junction capacitance on the switching transient, we make corresponding changes for each nonlinear junction capacitance value within a reasonable range.

The influence of $C_{gs}$ is shown in Figure 12. Set the variation value of $C_{gs}$ as 200 pF, 400 pF, 600 pF. It is observed that the increasing of $C_{gs}$ caused the charge time to become longer, thus resulting in the turn-ON delay time notably increasing. Meanwhile, the current overshoot and the voltage overshoot during the turn-OFF process can be suppressed by increasing $C_{gs}$. In general, $C_{gs}$ can effectively reduce the waveform oscillation, but will also lead to an increase in switching loss. Therefore, an extra capacitor paralleled with $C_{gs}$ can be used to reduce the oscillation when the oscillation is too large. However, the value of the parallel capacitor should be selected properly, otherwise the loss will be too large.

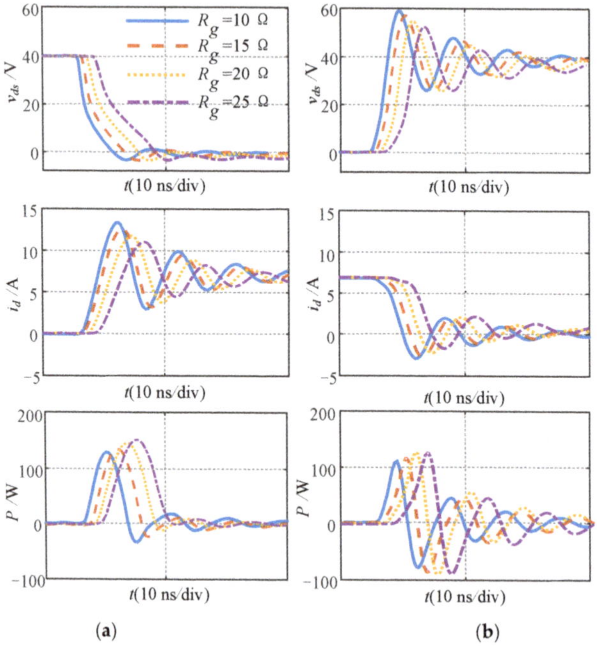


**Figure 10.** Influence of different $R_g$ on the switching transient: (**a**) turn-ON process; (**b**) turn-OFF process.

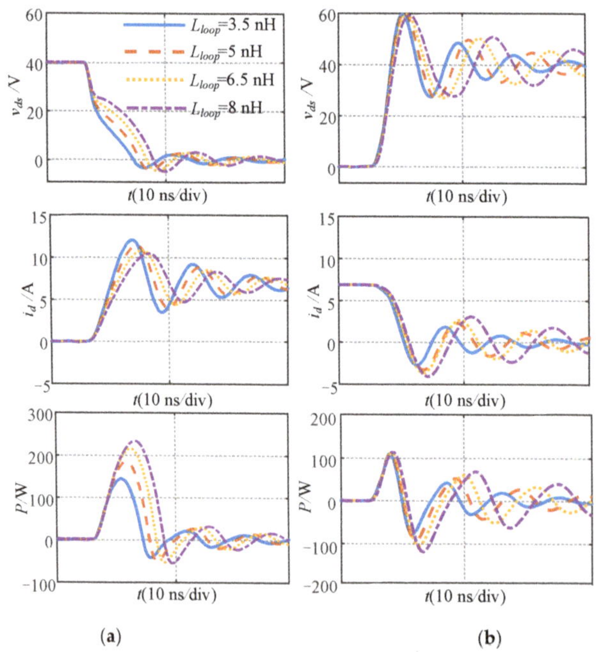


**Figure 11.** Influence of different $L_{loop}$ on the switching transient: (**a**) turn-ON process; (**b**) turn-OFF process.

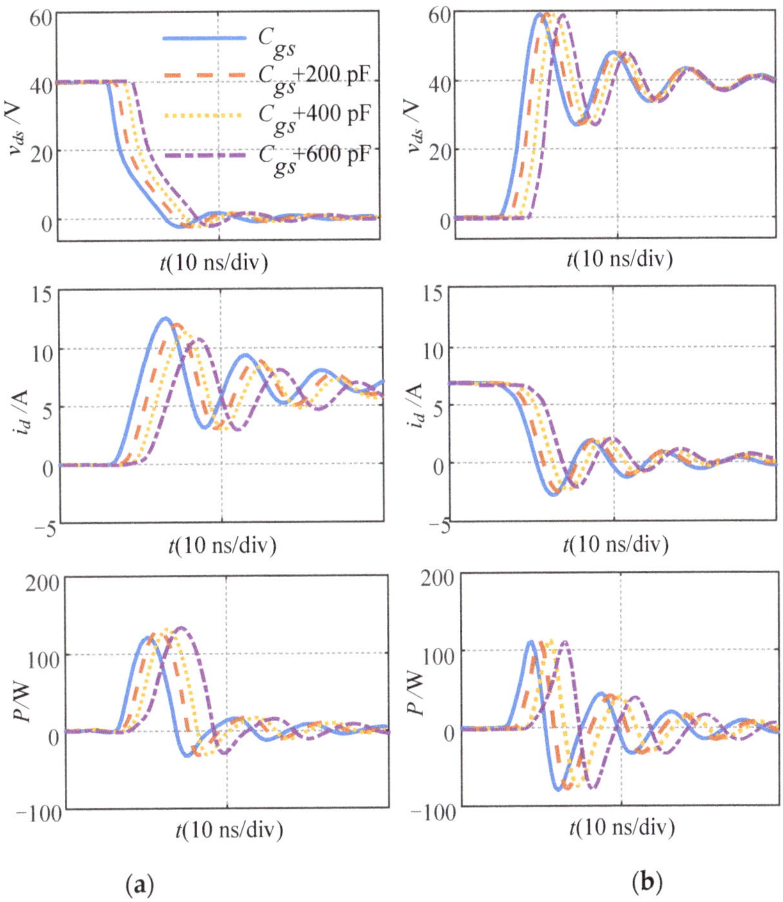


**Figure 12.** Influence of different $C_{gs}$ on the switching transient: (**a**) turn-ON process; (**b**) turn-OFF process.

The effect of $C_{gd}$ is shown in Figure 13. As stated previously, the increase of $C_{gd}$ has a large effect on the switching loss model. It can be concluded that the magnitude of $C_{gd}$ mainly affects the turn-ON process, i.e., the change rate of $i_d$ and $v_{ds}$ decreases and the conducting time increases. Although the increase of $C_{gd}$ is only 5 pF–15 pF, it has a great impact on the on-off time and switching loss due to the capacitor which affects the length of Miller platform time. Therefore, $C_{gd}$ should be designed as small as possible in device production. Compared with the Si devices, GaN HEMT is able to achieve much faster switching speeds and lower switching losses due to its smaller $C_{gd}$.

The influence of $C_{ds}$ is shown in Figure 14. As we know, $C_{oss}$ is a nonlinear junction capacitance of $v_{ds}$. It can be concluded that the increase of $C_{ds}$ mainly affects the change rate of $i_d$. Since $C_{oss}$ increases with the increase of $C_{ds}$, according to Equation (17), the energy stored in $C_{oss}$ will increase, which will result in an increase in the switching loss. In addition, the switching speed is increased and the ringing period becomes larger. For the convenience of overall consideration, the influencing parameter of switching characteristics and their specific impact analysis are listed in Table 3.

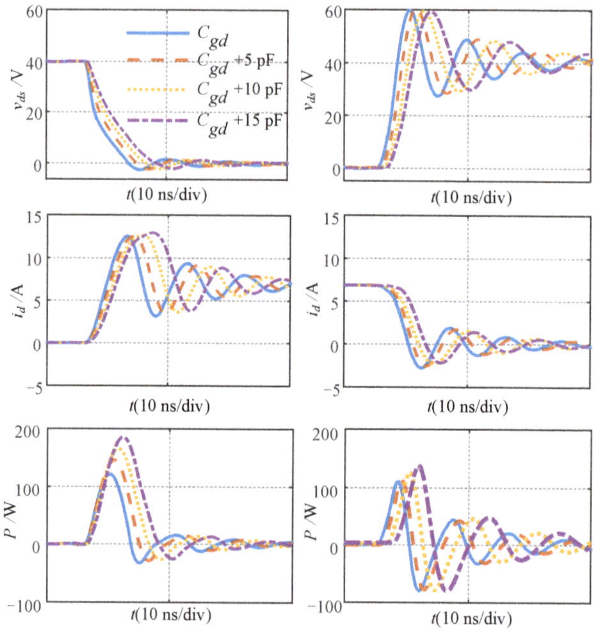


(a) (b)

**Figure 13.** Influence of different $C_{gd}$ on the switching transient: (**a**) turn-ON process; (**b**) turn-OFF process.

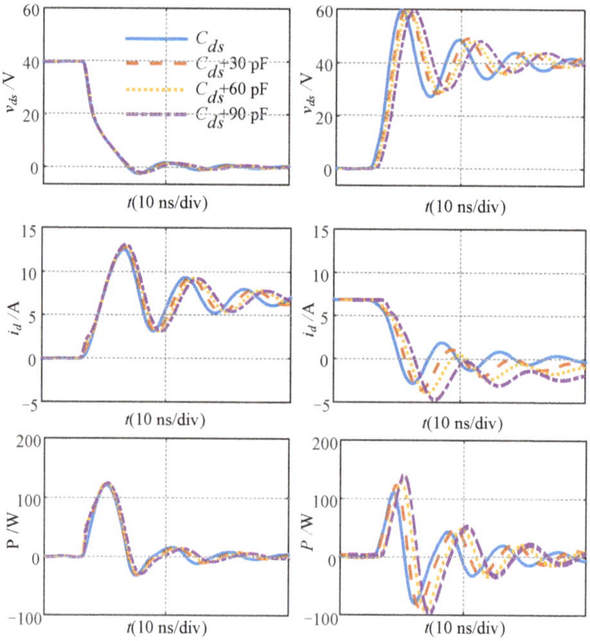


(a) (b)

**Figure 14.** Influence of different $C_{ds}$ on the switching transient: (**a**) turn-ON process; (**b**) turn-OFF process.

**Table 3.** Analysis of The Influencing Parameter of Switching Characteristics.

| Parameters | Turn-On Switching Speed | Turn-Off Switching Speed | Turn-On Current Overshoot | Turn-Off Voltage Overshoot | Turn-Off Current Overshoot | Turn-On Loss | Turn-Off Loss |
|---|---|---|---|---|---|---|---|
| $R_g$ | Increase | Increase | Decrease | Decrease | Decrease | Increase | Increase |
| $L_{loop}$ | Increase | Increase | Decrease | Invariant | Increase | Increase | Increase |
| $C_{gs}$ | Increase | Increase | Decrease | Invariant | Decrease | Increase | Increase |
| $C_{gd}$ | Increase | Increase | Increase | Invariant | Invariant | Increase | Increase |
| $C_{ds}$ | Invariant | Increase | Invariant | Invariant | Increase | Invariant | Increase |

## 7. Conclusions

In this paper, an accurate switching transient analytical model for GaN HEMT under the influence of nonlinear parameters was proposed. The proposed model covers the process of turn-ON and turn-OFF, which is of great significance in understanding the dynamic characteristics during the switching process. In addition, this model can be applied to evaluate the influence of parameter changes on the switching characteristics of devices, and calculate the switching loss. Finally, the following conclusions can be drawn:

(1) Switching speed. The switching process of the GaN HENT are essentially the charging process of its nonlinear junction capacitances. With the increases of $C_{gs}$, $C_{gd}$, $C_{ds}$, $L_{loop}$ and $R_g$, the switching speed is slowed down. Since the channel current is gate-source voltage dependent in the saturation region, it is out of the question that $C_{gs}$ can affect the current slew rate. As for the voltage slew rate, it is clarified that the charging of $C_{gd}$ leads to the change in the drain source voltage. The loop parasitic inductances $L_{loop}$ and gate resistance $R_g$ decrease the change rate of $i_d$ and $v_{ds}$ with the increase of $L_{loop}$ and $R_g$.

(2) Current and voltage overshoot. The current and voltage overshoot in switching transient can be suppressed by increasing $R_g$. The current overshoot can be suppressed by increasing $C_{gs}$. The current and voltage overshoot can be worsened by increasing $L_{loop}$. The current overshoot can be worsened by increasing $C_{ds}$. $C_{gd}$ cannot affect the current and voltage overshoot.

(3) Switching loss. Both $C_{gs}$, $C_{gd}$, $C_{ds}$, $L_{loop}$ and $R_g$ increases switching loss by slowing down the switching speed and prolonging the switching process.

(4) Design guidelines. It can be concluded that parameter changes of the GaN HEMT have a profound influence on the switching performance; hence, it calls for special attention to the selection of components and circuit design. The voltage change rate is most sensitive to $C_{gd}$, $L_{loop}$ and $R_g$; for high-speed high-voltage applications, these values should not be large. The current change rate is most sensitive to $C_{gs}$, $C_{ds}$, $L_{loop}$, and $R_g$. Therefore, the PCB traces should be carefully routed to reduce the stray inductance in the circuit. The other parasitic elements should be minimized to achieve fast switching and low loss.

**Author Contributions:** All authors contributed to the writing and revisions; writing—review and editing, L.H.; writing—original draft, D.Y.; review, Y.H.; review, Z.H.; supervision, P.Z. All authors have read and agreed to the published version of the manuscript.

**Funding:** This research was funded by the National Natural Science Foundation of China, grant numbers 51777049, 51707051.

**Institutional Review Board Statement:** Not applicable.

**Informed Consent Statement:** Not applicable.

**Data Availability Statement:** Data available on request due to privacy or ethical restrictions.

**Conflicts of Interest:** The authors declare no conflict of interest.

## References


1. Afonso, J.L.; Tanta, M.; Pinto, J.G.O.; Monteiro, L.F.C.; Machado, L.; Sousa, T.J.C.; Monteiro, V. A Review on Power Electronics Technologies for Power Quality Improvement. *Energies* **2021**, *14*, 8585. [CrossRef]
2. Maddi, H.L.R.; Yu, S.; Zhu, S.; Liu, T.; Shi, L.; Kang, M.; Xing, D.; Nayak, S.; White, M.H.; Agarwal, A.K. The Road to a Robust and Affordable SiC Power MOSFET Technology. *Energies* **2021**, *14*, 8283. [CrossRef]
3. Ramachandran, R.; Nymand, M. Experimental demonstration of a 98.8% efficient isolated DC–DC GaN converter. *IEEE Trans. Ind. Electron.* **2017**, *64*, 9104–9113. [CrossRef]
4. Ren, Y.C.; Xu, M.; Zhou, J.H.; Lee, F.C. Analytical loss model of power MOSFET. *IEEE Trans. Power Electron.* **2006**, *21*, 310–319.
5. Xiong, Y.; Sun, S.; Jia, H.; Shea, P. New physical insights on power MOSFET switching losses. *IEEE Trans. Power Electron.* **2009**, *24*, 525–531. [CrossRef]
6. Wang, J.; Chung, H.S.; Li, R.T. Characterization and experimental assessment of the effects of parasitic elements on the MOSFET switching performance. *IEEE Trans. Power Electron.* **2013**, *28*, 573–590. [CrossRef]
7. Turzynski, M.; Kulesza, W.J. A simplified behavioral MOSFET model based on parameters extraction for circuit simulations. *IEEE Trans. Power Electron.* **2016**, *31*, 3096–3105. [CrossRef]
8. Christen, D.; Biela, J. Analytical switching loss modeling based on datasheet parameters for MOSFET in a Half-bridge. *IEEE Trans. Power Electron.* **2019**, *34*, 3700–3710. [CrossRef]
9. Costinett, D.; Maksimovic, D.; Zane, R. Circuit-oriented treatment of nonlinear capacitances in switched mode power supplies. *IEEE Trans. Power Electron.* **2015**, *30*, 985–995. [CrossRef]
10. Sakairi, H.; Yanagi, T.; Otake, H.; Kuroda, N.; Tanigawa, H. Measurement methodology for accurate modeling of SiC MOSFET switching behavior over wide voltage and current ranges. *IEEE Trans. Power Electron.* **2018**, *33*, 7314–7325. [CrossRef]
11. Roy, S.K.; Basu, K. Analytical estimation of turn ON switching loss of SiC MOSFET and schottky diode pair from datasheet parameters. *IEEE Trans. Power Electron.* **2019**, *34*, 9118–9130. [CrossRef]
12. Sun, J.; Yuan, L.; Duan, R.; Lu, Z.; Zhao, Z. A semi-physical semi-behavioral analytical model for switching transient of SiC MOSFET module. *IEEE J. Emerg. Sel. Top. Power Electron.* **2021**, *9*, 2258–2270. [CrossRef]
13. Chen, K.; Zhao, Z.; Yuan, L.; Lu, T.; He, F. The impact of nonlinear junction capacitance on switching transient and its modeling for SiC MOSFET. *IEEE Trans. Electron Device* **2014**, *62*, 333–338. [CrossRef]
14. Wang, X.; Zhao, Z.; Li, K.; Zhu, Y.; Chen, K. Analytical methodology for loss calculation of SiC MOSFETs. *IEEE J. Emerg. Sel. Top. Power Electron.* **2018**, *7*, 71–83. [CrossRef]
15. Xie, R.; Wang, H.; Tang, G.; Yang, X.; Chen, K.J. An analytical model for false turn-on evaluation of high-voltage enhancement-mode GaN transistor in bridge-leg configuration. *IEEE Trans. Power Electron.* **2017**, *32*, 6416–6433. [CrossRef]
16. Wang, K.; Yang, X.; Li, H.; Ma, H.; Zeng, X.; Chen, W. An Analytical Switching Process Model of Low-Voltage eGaN HEMTs for Loss Calculation. *IEEE Trans. Power Electron.* **2016**, *31*, 635–647. [CrossRef]
17. Chen, J.; Luo, Q.; Huang, J.; He, Q.; Du, X. A Complete Switching Analytical Model of Low-Voltage eGaN HEMTs and Its Application in Loss Analysis. *IEEE Trans. Ind. Electron.* **2020**, *67*, 1615–1625. [CrossRef]
18. Kozak, J.P.; Barchowsky, A.; Hontz, M.R.; Koganti, N.B.; Stachina, W.H.; Reed, G.F.; Mao, Z.; Khana, R. An Analytical Model for Predicting Turn-ON Overshoot in Normally-OFF GaN HEMTs. *IEEE J. Emerg. Sel. Top. Power Electron.* **2020**, *8*, 99–110. [CrossRef]
19. Xin, Y.; Chen, W.; Sun, R.; Shi, Y.; Zhang, B. Analytical Switching Loss Model for GaN-Based Control Switch and Synchronous Rectifier in Low-Voltage Buck Converters. *IEEE J. Emerg. Sel. Top. Power Electron.* **2019**, *7*, 1485–1495. [CrossRef]
20. Long, X.; Liang, W.; Jun, Z.; Chen, G. A Normalized Quantitative Method for GaN HEMT Turn-ON Overvoltage Modeling and Suppressing. *IEEE Trans. Power Electron.* **2019**, *66*, 2766–2775. [CrossRef]
21. Huang, X.; Li, Q.; Liu, Z.; Lee, F.C. Analytical Loss Model of High Voltage GaN HEMT in Cascode Configuration. *IEEE Trans. Power Electron.* **2014**, *29*, 2208–2219. [CrossRef]
22. Li, Q.; Liu, B.; Duan, S. Simplified analytical model for estimation of switching loss of cascode GaN HEMTs in totem-pole PFC converters. *Chin. J. Electr. Eng.* **2019**, *5*, 1–9. [CrossRef]

electronics

MDPI

Article

# Three-Level Active Power Filter Based on Model Predictive Control

**Hanying Gao [1,*], Weihao Zhang [1], Mingjie Ren [2] and Xiangnan Liu [1]**

1 School of Electrical and Electronic Engineering, Harbin University of Science and Technology, Harbin 150080, China; hrbzwh123@163.com (W.Z.); 1920300072@stu.hrbust.edu.cn (X.L.)
2 School of Automation, Harbin University of Science and Technology, Harbin 150080, China; 1920510064@stu.hrbust.edu.cn
* Correspondence: ghy@hrbust.edu.cn; Tel.: +86-137-9666-8369

**Abstract:** The model predictive control (MPC) algorithm is used in the harmonic compensation of active power filter (APF), which has a fast dynamic response and does not require a PWM modulation model. However, this method has some shortcomings, such as the mass computing and difficult selections of weight factors. To solve these problems, this paper proposes a single objective function improved MPC algorithm based on sector judgment, which only takes the reference current and feedback current as the objective function, omits the weight coefficient setting process, and reduces the number of rolling optimization from 27 to 7, thus reducing the computing time and control complexity. The improved model predictive control is applied to APF. Finally, the simulations and experiments show that the improved MPC algorithm is accurate and efficient.

**Keywords:** model predictive control; objective function; sector judgment; active power filter; small vectors

**Citation:** Gao, H.; Zhang, W.; Ren, M.; Liu, X. Three-Level Active Power Filter Based on Model Predictive Control. *Electronics* **2022**, *11*, 1291. https://doi.org/10.3390/electronics11091291

Academic Editor: Maciej Ławryńczuk

Received: 3 March 2022
Accepted: 8 April 2022
Published: 19 April 2022



## 1. Introduction

With the advancement of power electronics technology and the development of manufacturing industry, more and more nonlinear devices are put into use. The input of these nonlinear loads brings a huge harmonic interference and a reactive power impact on the power grid. If not treated, it brings a huge pollution to the power grid, and causes major economic losses and safety accidents. The active power filter has excellent dynamic response, real-time performance, and controllability. It is an ideal device for compensating harmonics and improving power quality [1–4]. The three-level APF has the advantages of large capacity, more output levels and small filter inductance. It is widely used in medium and low voltage occasions [5–7]. At present, the studies on the three-level APF mainly focus on the optimization of control algorithm.

The studies on the optimization of the three-level control algorithm mainly focus on the harmonic current tracking strategy. The mature tracking strategies include the proportional integral control, hysteresis control, proportional resonance control, and model predictive control [8–11]. The PI control algorithm is not suitable for ac tracking [12]. More precisely, the response speed of hysteresis controller is affected by the loop width, while the loop width of different systems is difficult to adjust, and it is also difficult to balance the selection between switching frequency and tracking effect, which limits the use of this algorithm [13]. In general, in order to take the computational load and compensation performance into consideration, the proportional resonance control algorithm is only used for tracking specific harmonic frequency segments [14]. As a novel current tracking control algorithm, the model prediction has the advantages of fast dynamic response, flexible control and good stability. It has been widely used in the control of power converters [15,16].

The model predictive control mainly samples the device output current at the current time, and substitutes it into the established system mathematical model in order to predict the future device output current information according to the current or past current information [17]. Compared with other control algorithms, the model predictive control has great advantages in nonlinear systems, and can better deal with problems such as multi-input, multi variable, and multi constraint conditions [18]. However, the model prediction includes a rolling optimization process, which requires lot of processing time, thus reducing the dynamic performance of the model predictive control [19]. The weight factor reflects the importance of the objective function. It plays a crucial role in the control effect of the target. Therefore, when controlling the multi-objective function, the desired control effect can be obtained by adjusting the weight factor [20,21]. Although the model prediction can unify multiple constraints by adjusting the weight factor, it is difficult to adjust the weight coefficient in engineering applications. In [22], the model predictive control algorithm is applied to the three-phase four leg APF in order to predict the reference current and compensation current, and therefore perform harmonic compensation. However, this method has many rolling times, and the weight factor should be adjusted, which increases the complexity of the system. The authors in [23] apply model predictive control to the current tracking control of the three-level APF system. Although the values of the required parameters in the system are analyzed, a specific setting method of the parameters does not exist. The authors in [24] first classify different types of the objective function and weight factors, and then adjust the weight factors. This method has an excellent control effect. However, it is complex and it has a large computational load, which limits its use in practical application. Another approach [25] combines the model predictive control algorithm with a neural network algorithm. Compared with the traditional MPC algorithm, this method does not require an accurate mathematical model, and significantly improves the dynamic response ability of the APF system. However, the implementation process of this method is complex, and therefore it is difficult to be used in several application domains. The method in [26] applies the model predictive control algorithm to the T-type three-level neutral point clamped (NPC) inverter. It uses the predictive control for the grid current and capacitor voltage. Although the maximum power output and DC side capacitor voltage are stable, a specific setting method for the weight factor does not exist. The method in [27] introduces the idea of mixed logic dynamic (MLD) modeling into MPC control. Although the accuracy of the prediction model is improved, the modeling of this method is complex and difficult to implement. The authors in [28] select the optimal voltage vector according to the inverter output power and DC side voltage. They then optimize multiple objective functions by a constraint term, omitting the setting process of weighting factors. However, this method has more rolling times and a large computational load.

In this paper, a single objective function improved model predictive control algorithm based on sector judgment is proposed. It only takes the given current and feedback current as the objective function, determines the optimal voltage vector according to the influence of small vector and the change of DC voltage, omits the setting process of weight coefficient, and, finally, reduces the rolling optimization times to 7 times by judging the sector. The complexity of the algorithm is reduced and the dynamic response ability of APF system is improved. Finally, the improved MPC algorithm is applied to NPC three-level APF, and the correctness and feasibility of the algorithm are proved by system simulation and experiments.

## 2. NPC Topological Structure Analysis

### 2.1. NPC Topological Structure and Working Principle

Figure 1 shows the NPC three-level topology, which is composed of 6 diodes, 2 capacitors, and 12 switches. The midpoint of the 3 pairs of diodes is connected with point O. By controlling the switch, each phase of the topology can output three voltage states. The working process of the topology is analyzed as follows.

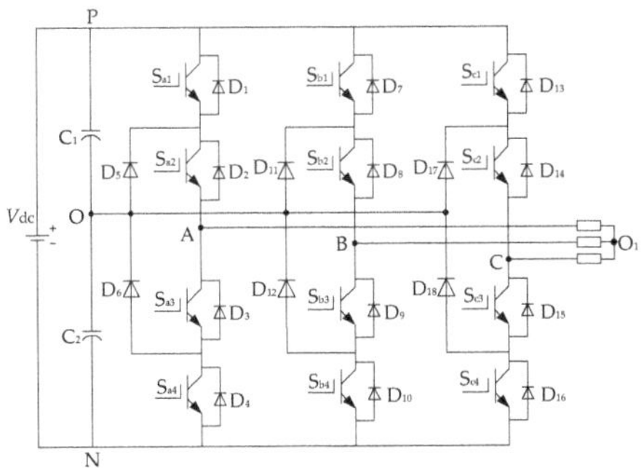


**Figure 1.** NPC three-level topology.

Considering point O as the zero potential reference point, in order to maintain the balance of two capacitors, that is, the voltage of each capacitor is $+V_{dc}/2$, three levels can be output by changing the on and off states of the switch according to the modulation strategy: $+V_{dc}/2$, 0, and $-V_{dc}/2$.

For the convenience of analysis, the topological phase voltage output $+V_{dc}/2$ is defined as P state, output $-V_{dc}/2$ is N state and output 0 is O state. The NPC topology line voltage output has five states: $+V_{dc}$, $+V_{dc}/2$, 0, $-V_{dc}/2$, and $-V_{dc}$. Table 1 can be obtained by summarizing the three working states of NPC three-level topology.

**Table 1.** Relationship between switch status and output voltage.

| Output Status | Switch Status | | | | Output Voltage |
|---|---|---|---|---|---|
| | $S_1$ | $S_2$ | $S_3$ | $S_4$ | |
| P | 1 | 1 | 0 | 0 | $V_{dc}/2$ |
| O | 0 | 1 | 1 | 0 | 0 |
| N | 0 | 0 | 1 | 1 | $-V_{dc}/2$ |

Where: 1 represents the switch on, and 0 represents the switch off.

### 2.2. Mathematical Model of Three-Level NPC APF

According to the analysis, the output level function is given by:

$$S_x \begin{cases} 1(\mathrm{P})\ S_{x1}, S_{x2} \text{ on } S_{x3}, S_{x4} \text{ off} \\ 0(\mathrm{O})\ S_{x2}, S_{x3} \text{ on } S_{x1}, S_{x4} \text{ off} \\ -1(\mathrm{N})\ S_{x3}, S_{x4} \text{ on } S_{x1}, S_{x2} \text{ off} \end{cases} \tag{1}$$

where: $S_x$ is the output level function, and $x$ has the value a, b, or c.

Block diagram of NPC three-level APF, as shown in Figure 2. Assuming that the grid is symmetrical without distortion and only contains the fundamental component, the three-phase load symmetrically ignores the line parasitic inductance and only considers the AC side output filter inductance. It also ignores the system line and device loss, and only considers the inductance internal resistance. Moreover, the three-phase inductance internal resistances are equal ($R = R_a = R_b = R_c$), and the DC side capacitances are all equal to C.

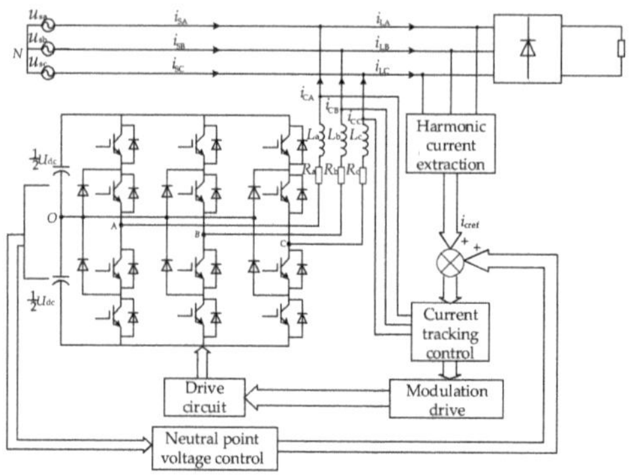


**Figure 2.** Schematic diagram of NPC three-level APF.

According to these assumptions and to the Kirchhoff voltage law (KVL), the mathematical model of the APF system is expressed as:

$$\begin{cases} (u_{AO} + u_{ON}) = Ri_{CA} + L\frac{di_{CA}}{dt} + u_{sa} \\ (u_{BO} + u_{ON}) = Ri_{CB} + L\frac{di_{CB}}{dt} + u_{sb} \\ (u_{CO} + u_{ON}) = Ri_{CC} + L\frac{di_{CC}}{dt} + u_{sc} \end{cases} \quad (2)$$

where: $u_{sa}$, $u_{sb}$, and $u_{sc}$ are the three-phase voltage, $i_{CA}$, $i_{CB}$, and $i_{CC}$ are the system compensation currents, $u_{AO}$, $u_{BO}$, and $u_{CO}$ are the potential voltages between the output point of each phase and the reference point O of the NPC topology, $u_{ON}$ is the voltage between the DC side neutral point and grid neutral point. $U_{AO}$, $u_{BO}$, and $u_{CO}$ can be expressed as (3).

$$u_{xo} = \frac{1}{2}S_x U_{dc},\ x = \text{a, b, c}, \quad (3)$$

and therefore:

$$\begin{cases} u_{sa} + u_{sb} + u_{sc} = 0 \\ i_{CA} + i_{CB} + i_{CC} = 0 \end{cases} \quad (4)$$

By combining Equations (2)–(4), the difference between the point O potential and the point N potential is given by:

$$u_{ON} = \frac{1}{6}(S_a + S_b + S_c)U_{dc} \quad (5)$$

The three-level NPC APF output voltage and switching function then meet the following relationships:

$$\begin{cases} u_{AN} = u_{AO} + u_{ON} = \frac{1}{6}(2S_a - S_b - S_c)U_{dc} \\ u_{BN} = u_{BO} + u_{ON} = \frac{1}{6}(-S_a + 2S_b - S_c)U_{dc} \\ u_{CN} = u_{CO} + u_{ON} = \frac{1}{6}(-S_a - S_b + 2S_c)U_{dc} \end{cases} \quad (6)$$

### *2.3. Analysis of Neutral Point Potential of NPC APF*

Two main reasons exist for the capacitor voltage fluctuation: the various losses in APF system, including the loss of switch tube and line, as well as the load fluctuation and the power grid fluctuation. The load fluctuation and grid fluctuation belong to the external disturbance and cannot be controlled. Therefore, in order to suppress the capacitor voltage fluctuation, maintain the stability of the APF system and achieve a better tracking effect on the harmonic current, the relationship between the NPC topology output voltage vector and neutral point potential fluctuation is analyzed as follows.

As shown in Figure 3, there are 27 voltage vectors in NPC topology switching state vector, including 6 medium vectors, 6 large vectors, 12 small vectors, and 3 zero vectors. Among them, the medium vectors include: PNO(V1), PON(V3), OPN(V5), NPO(V7), NOP(V9), and ONP(V11). The large vector includes: PNN(V2), PPN(V4), NPN(V6), NPP(V8), NNP(V10), and PNP(V12). The small vectors includes: ONO(V13), POP(V14), ONN(V15), POO(V16), OON(V17), PPO(V18), NON(V19), OPO(V20), NOO(21), OPP(V22), NNO(V23), OOP(V24). Zero vectors include: NNN(V25), OOO(V26), and PPP(V27).

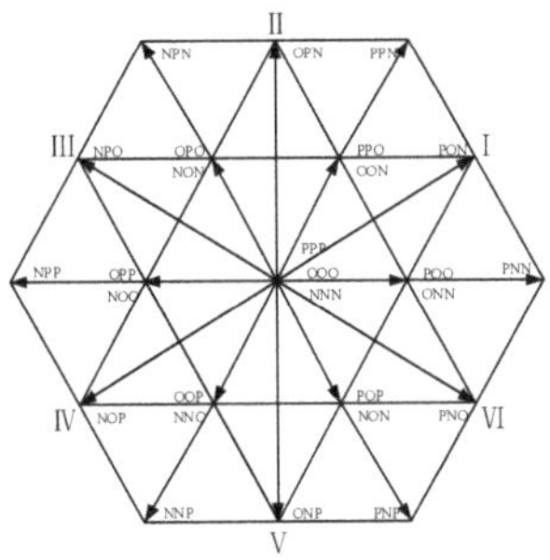


**Figure 3.** Voltage vector diagram of NPC three-level topology.

In order to analyze the influence of vectors on the neutral point potential, the equivalent circuits corresponding to these five vectors are analyzed, (cf. Figure 4).

It can be seen from Figure 4a, that the large vector (PNN) acts on the entire bus voltage and does not constitute a loop with the point O. The same applies for the other five large vectors. Therefore, when the large vector acts, it only affects the entire bus voltage on the DC side. When the output voltage vector is a medium vector, one phase is connected to point O, and the other phases are, respectively, connected to the bus voltage in order to form a loop with point O, and therefore affect its potential. However, this effect is uncontrollable. That is, the point O potential cannot be controlled by the medium vectors. The small vectors include positive and negative small vectors. It can be seen from Figure 4c,d that no matter which small vector acts, an electrical connection with point O which constitutes a circuit exists, so the small vectors will affect the neutral point potential balance. When the output state of the switch is a zero vector, it can be seen from Figure 4e that although the three bridge arms are connected to the point O, they do not form a current loop and do not affect the voltage balance. The other two zero vectors of the bridge arms are not connected to the point O, and will not affect the voltage balance. The influences of these four voltage vectors are summarized in Table 2.

**Table 2.** Influence of voltage vectors on neutral point potential.

| Vector Types | | Switch Status | Effect on Neutral Point Potential |
|---|---|---|---|
| Zero vectors | | −1−1−1, 111, 000 | without influence |
| Large vectors | | 1−1−1, 11−1, −11−1, −111, −1−11, 1−11 | without influence |
| Medium vectors | | 10−1, 01−1, −110, −101, 0−11, 1−10 | Influential but uncontrollable |
| Small vectors | Positive small vectors | 110, 100, 101, 010, 011, 001 | Influential and controllable |
| | Negative small vectors | −1−10, 00−1, 0−10, 0−1−1, −100, −10−1 | |

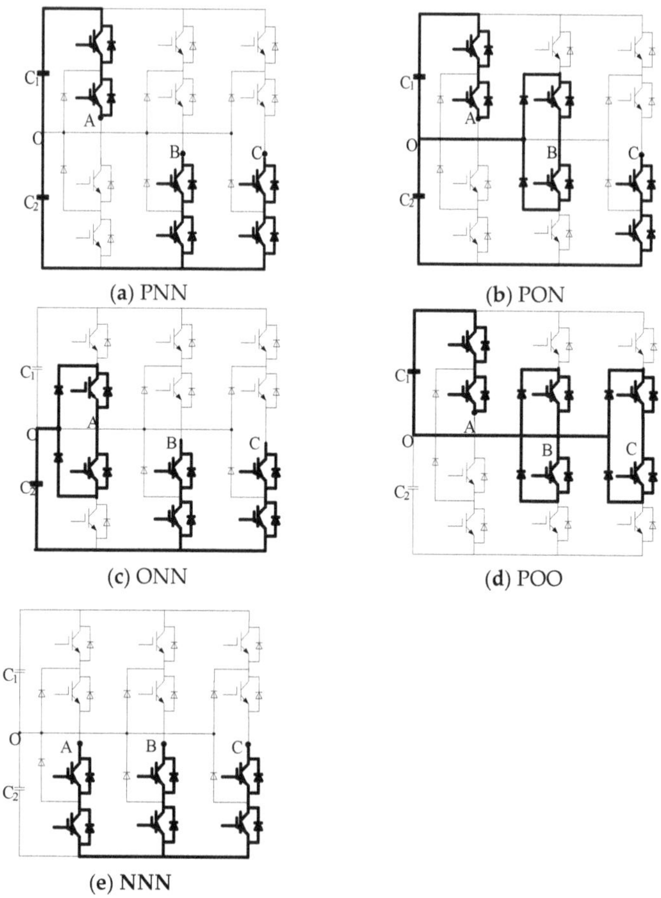


**Figure 4.** Equivalent circuit diagram of different vectors: (**a**) large vector (PNN); (**b**) middle vector (PON); (**c**) small vector (ONN); (**d**) small vector (POO); (**e**) zero vector (NNN).

Assuming the neutral current is $i_o$, the load currents are $i_a$, $i_b$, and $i_c$, the output direction of the inverter is positive direction, the corresponding relations among $i_o$, the redundant small vectors, and load currents can be obtained, (cf. Table 3).

**Table 3.** Corresponding relationship between $i_o$ and redundant small vectors and output currents.

| Positive Small Vectors | $i_o$ | Negative Small Vectors | $i_o$ |
|---|---|---|---|
| 001 | $-i_c$ | 0 0 −1 | $-i_c$ |
| 110 | $i_c$ | −1 −1 0 | $i_c$ |
| 010 | $-i_b$ | 0 −1 0 | $-i_b$ |
| 101 | $i_b$ | −1 0 −1 | $i_b$ |
| 100 | $-i_a$ | −1 0 0 | $-i_a$ |
| 011 | $i_a$ | 0 −1 −1 | $i_a$ |

According to Table 3, the mathematical relationship between $i_o$ and the APF system output compensation current is expressed as:

$$i_o = -(i_{ca}|S_a| + i_{cb}|S_b| + i_{cc}|S_c|) \tag{7}$$

The relationship between $i_o$ and the DC side capacitance is given by:

$$\frac{d(V_{dc1} - V_{dc2})}{dt} = \frac{1}{C} i_o \tag{8}$$

where: $V_{dc1}$ and $V_{dc2}$ are the voltage values of capacitors $C_1$ and $C_2$, respectively.

By setting the voltage deviation to $V_O = V_{dc1} - V_{dc2}$, according to Equations (7) and (8), the following is obtained:

$$\frac{dV_o}{dt} = -\frac{1}{C} \Sigma i_{cx} |S_x| \tag{9}$$

where: $x$ = a, b, c.

In summary, the reason of NPC topology DC side neutral point potential fluctuations is that $i_o$ is not null. Regardless of the magnitude and direction of $i_o$, as long as a current flow exists between the neutral point O and the load circuit, the current will affect the point O voltage. Under the action of different vectors, different currents have different effects on the capacitance voltage. However, only the small vectors really affect the capacitance voltage and can be controlled in the system. Therefore, the small vectors can be used to solve the problem of neutral point fluctuation.

## 3. Model Predictive Control

### *3.1. Fundamental Theory*

Figure 5 presents the overall control structure block diagram of the three-level APF system, including the harmonic current detection unit and model predictive control unit. The harmonic current control unit uses the $i_p$-$i_q$ harmonic current detection method in order to obtain the three-phase harmonic reference current $i^*_{fa}$, $i^*_{fb}$, and $i^*_{fc}$. The model predictive control samples the compensation current, capacitor voltage, and grid voltage, and predicts the compensation current and reference current in the future. It also selects the optimal switching vector output combined with the current capacitor voltage error.

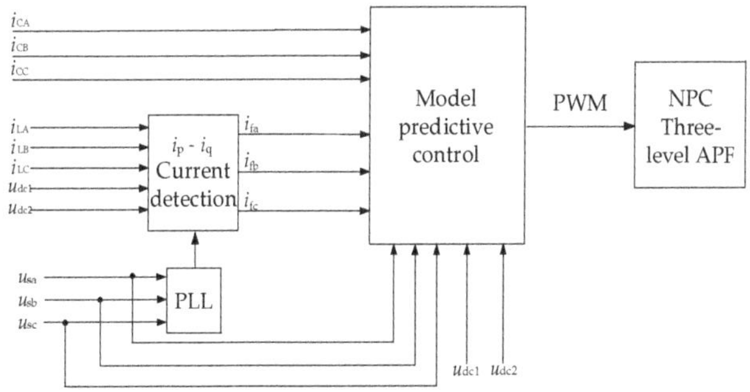


**Figure 5.** Overall control scheme of three-level APF system.

Due to the limited switching state of the main circuit, MPC predicts the output of each switching state using the prediction model. By calculating the function values of all the switching states, the switch state corresponding to the minimum function is selected and applied to the next cycle. The detailed process is as follows. The output prediction model $f_{mpc}$ of the system is first developed according to the switching state $S_{sta}$ of the APF and the system control variable $x$. The value function $f_{val}$ is then established. At time $t_k$, the output $x(t_k)$ is calculated using the system prediction model $f_{mpc}$ in order to calculate the predicted output $x_i^{mpc}(t_{k+1})$ corresponding to n different switching states. Afterwards, the value function of the n switching states can be calculated using the output value $x^{ref}(t_{k+1})$ and predicted output value $x_i^{mpc}(t_{k+1})$. Finally, the optimal switching state $S_{sta}(t_k)$ corresponding to the minimum function value is obtained, and the switching state is output. The diagram of the system is illustrated in Figure 6.

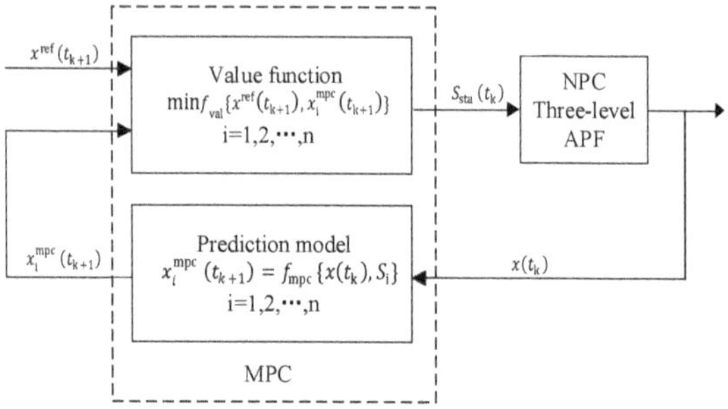


**Figure 6.** APF MPC block diagram.

## 3.2. Traditional MPC

Model predictive control (MPC) belongs to optimal control in essence. It can predict the future output information of the system according to the current and past sampling information of the system. The switching states of the converter are limited, for example, there are 27 switching states in NPC three-level topology. Therefore, the traditional model is also called finite control set model prediction.

The model predictive control consists of the predictive model, rolling optimization and feedback correction. The overall control block diagram is shown in Figure 7. Establishing a systematic prediction model is the first step of MPC, that is, building a prediction model of future information according to the previous and current information, and then rolling optimization. Through rolling optimization, the size of the value function corresponding to all the cases can be judged, and the switch state output corresponding to the minimum value function can be selected. In the process of rolling optimization, the initial positioning link of the current information to the optimization process includes a feedback correction. Therefore, an additional feedback correction link does not exist in practical applications.

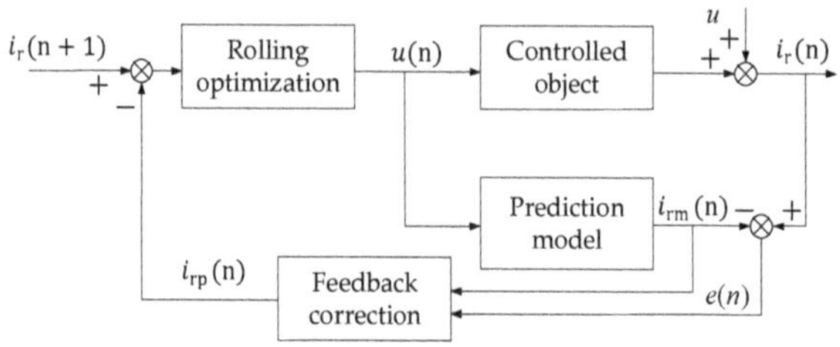


**Figure 7.** Model predictive control block diagram.

In order to reduce the control complexity and improve the real-time performance, the Clarke transform is performed on the APF mathematical model of the traditional MPC in order to obtain the following:

$$\begin{cases} u_\alpha = R_L i_\alpha + L\frac{di_\alpha}{dt} + e_\alpha \\ u_\beta = R_L i_\beta + L\frac{di_\beta}{dt} + e_\beta \end{cases} \quad (10)$$

where: $\begin{bmatrix} u_\alpha \\ u_\beta \end{bmatrix} = C_{abc-\alpha\beta}\begin{bmatrix} u_{AN} \\ u_{BN} \\ u_{CN} \end{bmatrix}$, $\begin{bmatrix} i_\alpha \\ i_\beta \end{bmatrix} = C_{abc-\alpha\beta}\begin{bmatrix} i_{CA} \\ i_{CB} \\ i_{CC} \end{bmatrix}$, $\begin{bmatrix} e_\alpha \\ e_\beta \end{bmatrix} = C_{abc-\alpha\beta}\begin{bmatrix} u_{sa} \\ u_{sb} \\ u_{sc} \end{bmatrix}$, $C_{abc-\alpha\beta} = \frac{2}{3}\begin{bmatrix} 1 & -\frac{1}{2} & -\frac{1}{2} \\ 0 & \frac{\sqrt{3}}{2} & -\frac{\sqrt{3}}{2} \end{bmatrix}$.

By applying the Clarke transformation on the pair of Equation (6), the following is obtained:

$$\begin{bmatrix} u_\alpha \\ u_\beta \end{bmatrix} = C_{abc-\alpha\beta} \begin{bmatrix} u_{AN} \\ u_{BN} \\ u_{CN} \end{bmatrix} = \frac{1}{6} \begin{bmatrix} 2S_a - S_b - S_c \\ \sqrt{3}(S_b - S_c) \end{bmatrix} u_{dc} \tag{11}$$

Equation (10) is then discretized to obtain:

$$\frac{di_\alpha}{dt} \approx \frac{i_\alpha(k+1) - i_\alpha(k)}{T_s} \tag{12}$$

$$\frac{di_\beta}{dt} \approx \frac{i_\beta(k+1) - i_\beta(k)}{T_s} \tag{13}$$

where: $i_\alpha(k)$ and $i_\beta(k)$ are the actual compensation currents of APF at time k, $i_\alpha(k+1)$ and $i_\beta(k+1)$ represents the predicted compensation currents at *k*+1 time, and $T_s$ represents the sampling period of the system.

Using Equations (10), (12), and (13), the following can be obtained:

$$\begin{cases} i_\alpha\left(k+1\right) = \frac{T_s}{L}[u_\alpha(k) - e_\alpha(k) - Ri_\alpha(k)] + i_\alpha(k) \\ i_\beta\left(k+1\right) = \frac{T_s}{L}\left[u_\beta(k) - e_\beta(k) - Ri_\beta(k)\right] + i_\beta(k) \end{cases} \tag{14}$$

where $u_{\alpha\beta}(k)$ is the APF output voltage at time *k*, and $e_{\alpha\beta}(k)$ is the APF grid voltage at time k.

In order to maintain a good current tracking effect and take into account the balance of medium electric potential, Equation (9) is discretized:

$$\frac{dV_o}{dt} \approx \frac{V_o(k+1) - V_o(k)}{T_s} \tag{15}$$

where: $V_o(k+1)$ and $V_o(k)$ are the deviation prediction value and deviation sampling value of capacitor $C_1$ and capacitor $C_2$ at *k* + 1 time and *k* time, respectively.

Equation (16) can be obtained from Equations (9) and (15):

$$V_o(k+1) = V_o(k) - \frac{T_s}{C}\Sigma i_{cx}(k)|S_x(k)|, (x = a, b, c) \tag{16}$$

According to the objective function of the NPC three-level APF system, the square error function is selected as the value function model:

$$g_1 = |i_\alpha^*(k+1) - i_\alpha(k+1)|^2 + \left|i_\beta^*(k+1) - i_\beta(k+1)\right|^2 \tag{17}$$

Considering the current tracking effect as the objective function, while taking into account the balance of neutral point potential, the value function can be expressed as:

$$g_2 = |i_\alpha^*(k+1) - i_\alpha(k+1)|^2 + \left|i_\beta^*(k+1) - i_\beta(k+1)\right|^2 + |V_o(k+1)| \tag{18}$$

Considering the weights of $g_1$ and $g_2$, the final expression of the value function is given by:

$$g = \lambda_1|i_\alpha^*(k+1) - i_\alpha(k+1)|^2 + \lambda_1\left|i_\beta^*(k+1) - i_\beta(k+1)\right|^2 + \lambda_2|V_o(k+1)| \tag{19}$$

where: $\lambda_1$ and $\lambda_2$ are weight coefficients.

The current prediction value given in Equation (19) can be obtained by Lagrange interpolation:

$$i^*(k+1) = \sum_{l=0}^{n} (-1)^{n-1} \frac{(n+1)!}{l!(n+1-l)!} i^*(k+l-n) \tag{20}$$

For the sake of accuracy and computational complexity, $n = 2$ is selected, and then:

$$i_x^*(k+1) = 3i_x^*(k) - 3i_x^*(k-1) + i_x^*(k-2) \; (x = \alpha, \beta) \tag{21}$$

### 3.3. *Improved MPC*

#### 3.3.1. Improved MPC Objective Function

The problem of setting the weight coefficient of the objective function can be solved by outputting the optimal small vector to equalize the DC side voltage, and omitting the DC side capacitor voltage error objective function in the cost function.

The realization of DC sides capacitor voltage stability is the premise of maintaining neutral points potential balances. The diagram of the DC sides capacitor voltage stabilizing control is illustrated in Figure 8. Where $e_a$, $e_b$, and $e_c$ are the grid voltage, $i_a$, $i_b$, and $i_c$ are the three-phase grid current. Firstly, the active current and reactive current components $i_p$ and $i_q$ in p-q coordinate system are obtained by Clarke transformation and Park transformation. Then, the fundamental component in the p-q coordinate system can be obtained by filtering the high-order harmonic component through the low-pass filter LPF. The difference between the DC side total voltage reference value and the DC side total voltage feedback can be processed by the PI controller to obtain the active power reference current $\Delta i_p$. The active current component $\Delta i_p$ is superimposed with the active fundamental current component in the p-q coordinate system to obtain the final active fundamental current component. The three-phase fundamental current components $i_{af}$, $i_{bf}$, and $i_{cf}$ are obtained from the final active fundamental component and reactive fundamental component through the inverse transformation of Clarke and Park. Finally, by subtracting the three-phase fundamental current from the three-phase total current, the three-phase harmonic components $i_{ah}$, $i_{bh}$, and $i_{ch}$ can be obtained.

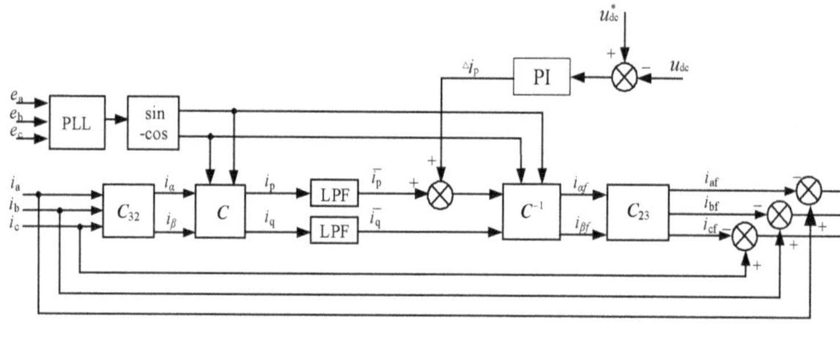


**Figure 8.** Schematic diagram of the overall voltage control.

The improved model predictive control algorithm can control the voltage of two capacitors on the DC side, by outputting a specific small vector. This method can omit the DC side capacitor voltage error value function, so as to solve the problem of setting the weight coefficient of the objective function. The specific form of the improved value function is given by:

$$g = \left| i_{\alpha\beta}^*(k+1) - i_{\alpha\beta}(k+1) \right| \tag{22}$$

Compared with the traditional model prediction value function, this function only needs to consider an objective function of current tracking error, which reduces the computational load and does not introduce a weight coefficient. Therefore, there is no problem of weight coefficient setting.

#### 3.3.2. Single Objective Function Improved Model Predictive Control Strategy Based on Sector Judgment

Inspired by the three-level SVPWM vector control, the sector judgment is introduced into the model predictive control. Figure 9 presents the three-level NPC voltage vector diagram.

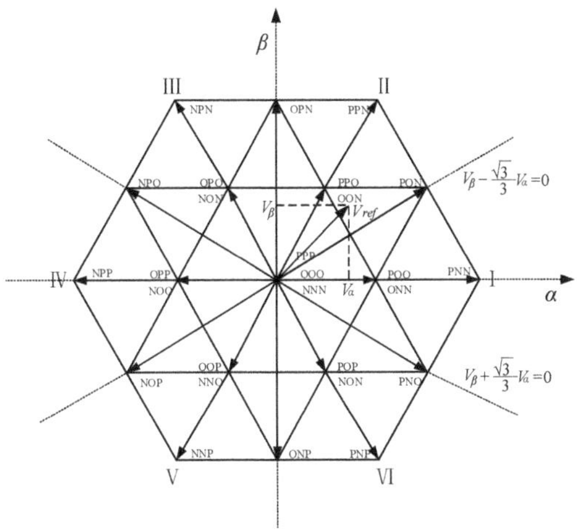


**Figure 9.** Three-level voltage vector diagram.

It can be seen that the linear $V_\beta + \left(\sqrt{3}/3\right) V_\alpha = 0, V_\beta - \left(\sqrt{3}/3\right) V_\alpha = 0$ and $\beta$ coordinate axes divide the large hexagon into six regions, quadrilateral areas of I, II, III, IV, V, and VI, while each quadrilateral contains 8 vectors. It is first assumed that the projection of the reference voltage vector $V_{\text{ref}}$ on the $\alpha$ and $\beta$ axes is $V_\alpha$ and $V_\beta$. The six quadrangles are divided according to three lines and voltage vector $V_{\text{ref}}$. Specific regional divisions are shown in Table 4.

**Table 4.** Regional divisions.

| Region of $V_{ref}$ | $V_\alpha$ and $V_\beta$ |
|---|---|
| I | $V_\alpha > 0,\ (\frac{\sqrt{3}}{3})V_\alpha > V_\beta \geq \left(-\frac{\sqrt{3}}{3}\right)V_\alpha$ |
| II | $V_\alpha > 0,\ V_\beta > 0,\ V_\beta \geq \left(\frac{\sqrt{3}}{3}\right)V_\alpha$ |
| III | $V_\alpha \leq 0,\ V_\beta > 0,\ V_\beta > \left(-\frac{\sqrt{3}}{3}\right)V_\alpha$ |
| IV | $V_\alpha < 0,\ (-\frac{\sqrt{3}}{3})V_\alpha \geq V_\beta > \left(\frac{\sqrt{3}}{3}\right)V_\alpha$ |
| V | $V_\alpha < 0,\ V_\beta < 0,\ (\frac{\sqrt{3}}{3})V_\alpha \geq V_\beta$ |
| VI | $V_\alpha > 0,\ V_\beta < 0,\ (-\frac{\sqrt{3}}{3})V_\alpha \geq V_\beta$ |

The quadrilateral region of the reference voltage vector can be determined by assessing the mathematical relationship of the projection of the reference voltage vector on the $\alpha$ and $\beta$ axes. Therefore, the mathematical model of the reference voltage is developed:

$$\begin{cases} u_\alpha(k) = e_\alpha(k) + Ri_\alpha(k) - \frac{L}{T_s}[i_\alpha(\text{k}+1) - i_\alpha(k)] \\ u_\beta(k) = e_\beta(k) + Ri_\beta(k) - \frac{L}{T_s}[i_\beta(\text{k}+1) - i_\beta(k)] \end{cases} \tag{23}$$

In order to achieve a better current tracking effect of the system, it is necessary to predict that the output current is equal to the predicted harmonic given current:

$$i^*_{\alpha\beta}(k+1) = i_{\alpha\beta}(k+1) \tag{24}$$

By simultaneously using Equations (23) and (24), the following can be obtained:

$$\begin{cases} u_\alpha(k) = e_\alpha(k) + Ri_\alpha(k) - \frac{L}{T_s}[i_\alpha^*(k+1) - i_\alpha(k)] \\ u_\beta(k) = e_\beta(k) + Ri_\beta(k) - \frac{L}{T_s}\left[i_\beta^*(k+1) - i_\beta(k)\right] \end{cases} \tag{25}$$

where: $u_\alpha = V_\alpha$ and $u_\beta = V_\beta$.

After this evaluation, the region of the reference voltage vector can be determined. In addition, 8 vectors exist in each region, and there are 2 small vectors among the 8 vectors. Consequently, a redundant vector can be removed. Therefore, the proposed model predictive control algorithm has only 7 rolling optimized vectors, which greatly reduces the rolling times and system calculation, compared with the traditional prediction model.

### 3.3.3. Flow and Analysis of Improved Model Predictive Control

The control flowchart is shown in Figure 10. Based on the traditional model predictive control, firstly, each parameter is sampled, the system parameters are initialized, $u_\alpha$ and $u_\beta$ are calculated, the position of the reference voltage vector is then judged according to the size of $u_\alpha$ and $u_\beta$, and, finally, its quadrilateral region is determined. Afterwards, the seven voltage vectors in this area are subject to rolling optimization. The compensation current is predicted and the value functions containing only current tracking error, are calculated. After seven rolling optimizations, the switching state corresponding to the value function that minimizes the current tracking error is output. Finally, the output voltage vector is judged. In the case of a small vector, a small vector output that is conducive to the capacitor voltage balance on the DC side is selected. Otherwise, it is directly output. Consequently, the optimal switching state output of single objective function improved model predictive control based on sector judgment is completed.

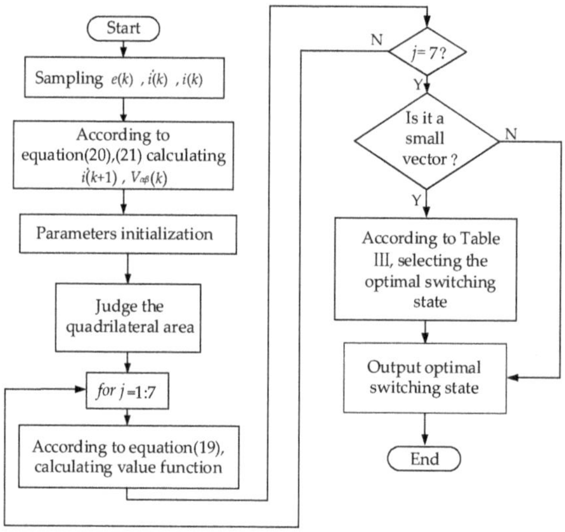


**Figure 10.** Flow chart of single objective function improved model predictive control based on sector judgment.

## 4. System Simulation and Result Analysis

The feasibility and efficiency of the proposed algorithm are verified by developing the simulation model of the NPC three-level APF. The model predictive control is written using the S-function module of MATLAB/Simulink. The two model predictions are simulated and compared. Table 5 presents the system simulation parameters.

**Table 5.** Specific parameters of APF simulation model.

| Simulation Parameters | Value |
|---|---|
| Effective value of grid Phase voltage/V | 220 |
| Grid frequency/Hz | 50 |
| Sampling frequency/kHz | 20 |
| Capacitance $C_1$, $C_2$/μF | 2000 |
| Capacitor $C_1$ voltage/V | 400 |
| Capacitor $C_2$ voltage/V | 400 |
| Filter inductance $L_a$, $L_b$, $L_c$/mH | 5 |
| Load resistance/Ω | 20 |

### 4.1. Simulation and Analysis of Traditional Model Predictive Control

Figure 11 shows the current waveform of A-phase power grid before compensation. It can be seen that when APF is not working, the current waveform is saddle shaped and the waveform distortion is serious because it contains a large number of high-order harmonics. Figure 12 shows the FFT analysis diagram of A-phase grid current before compensation. It can be seen that THD is 27.08%, which is much higher than the national standard of 5%, which seriously reduces the power quality of the grid.

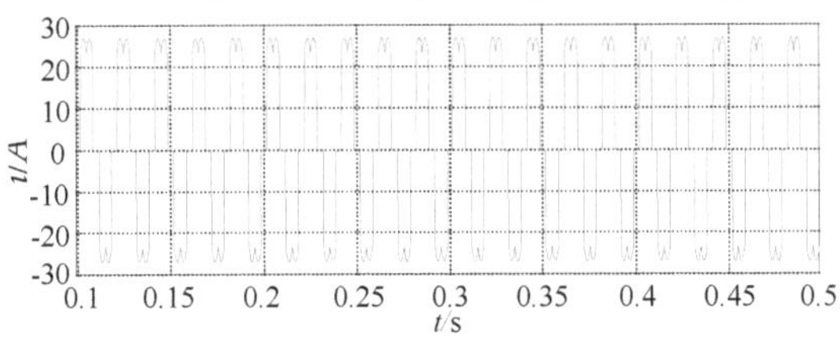


**Figure 11.** Current waveform of A-phase power grid before compensation.

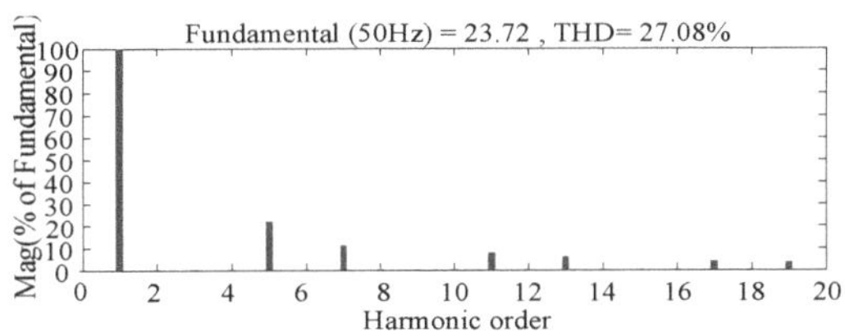


**Figure 12.** FFT analysis diagram of A-phase current before compensation.

Figures 13 and 14 present the current waveform and THD diagram of A-phase power grid under the traditional model predictive control algorithm after APF system works. Compared with Figure 10, it can be seen that the current waveform after compensation changes from saddle shape to sinusoidal shape, and the sinusoidal degree is high. Its THD is 4.51% and less than 5%, which meets the national standard and realizes harmonic compensation.

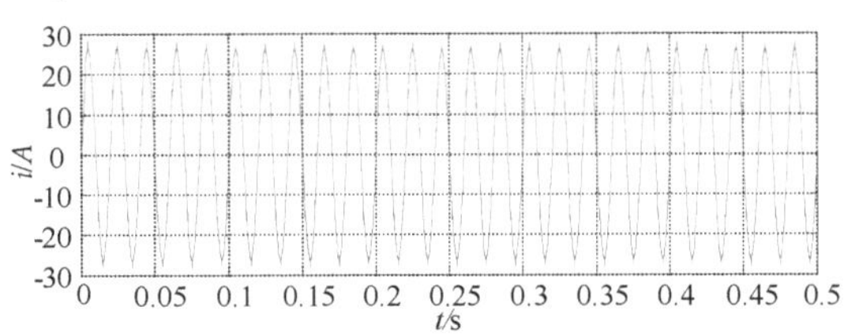


**Figure 13.** A-phase current waveform after compensation of traditional model predictive control.

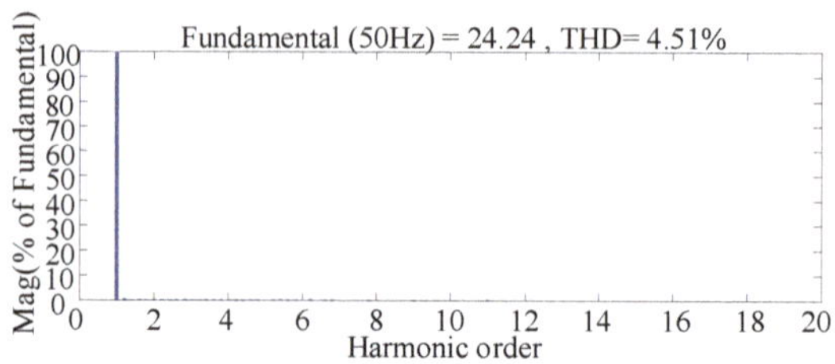


**Figure 14.** A-phase current THD diagram after compensation of traditional model predictive control.

Figure 15 is the reference current tracking effect diagram under traditional model predictive control. Under the traditional model predictive control, the feedback current can better track the reference current, but there are some deviations in the peak and trough. Figure 16 presents the error diagram of reference current and feedback current, the error range fluctuates within ±2.5 A, and the maximum error occurs at the peak and trough positions.

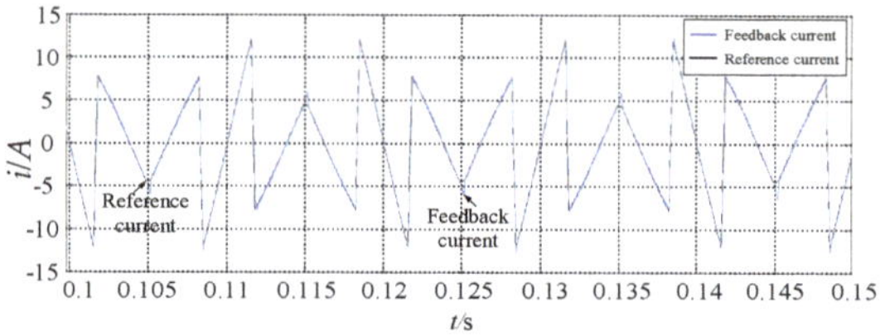


**Figure 15.** Reference current tracking effect diagram of traditional model predictive control.

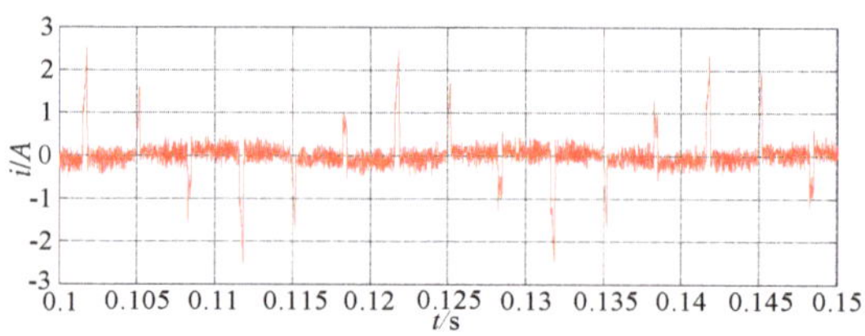


**Figure 16.** Error diagram of reference current and feedback current of traditional model predictive control.

Figure 17 presents the overall voltage waveform of the capacitor at the DC side, and its voltage amplitude fluctuates around 801 V, which can maintain stable fluctuation. Figure 18 presents the waveform of neutral point potential at DC side, with a fluctuation range of ±1.4 V and stable.

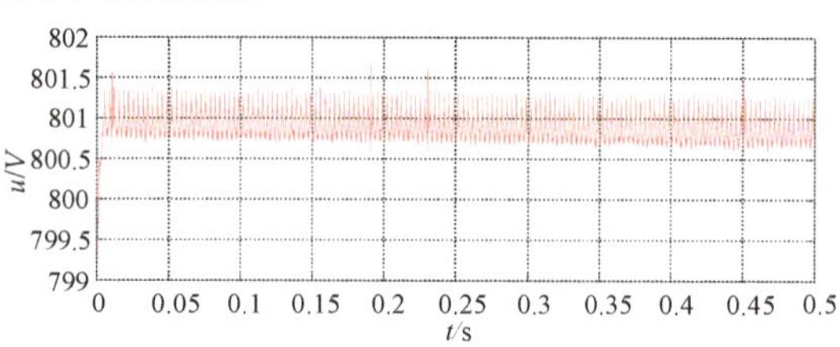


**Figure 17.** Overall voltage waveform of DC side capacitor of traditional model predictive control.

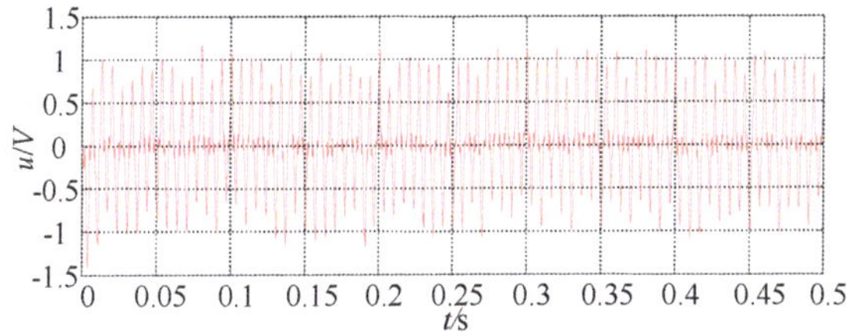


**Figure 18.** DC neutral point potential waveform of traditional model predictive control.

In order to observe the MPC the selection of switching state vector, the switching state flag is set to observe the process of finding the optimal switching vector during rolling optimization. Figure 19 is a value diagram of the switching states flag of the traditional model predictive control algorithm. The value of flag is between 1 and 27, which verifies that the rolling optimization times of the traditional model predictive control algorithm is 27.

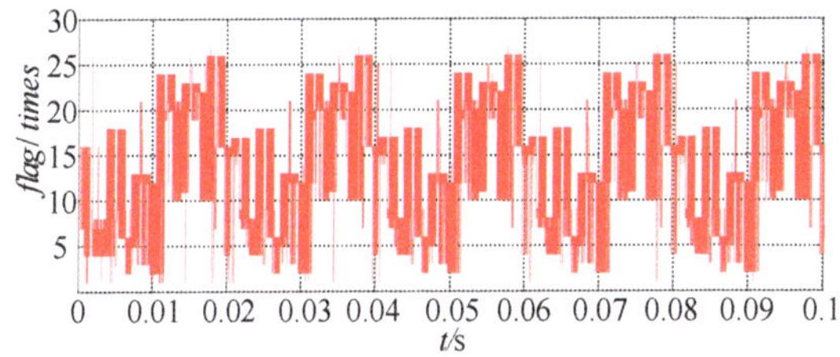


**Figure 19.** Switching state flag value diagram of traditional model predictive control.

### 4.2. Single Objective Function Improved Model Predictive Control Based on Sector Judgment

On the basis of keeping the previously mentioned simulation parameters unchanged, this section uses the single objective function improved model predictive control algorithm based on sector judgment to simulate the APF system.

Figures 20 and 21 present the current waveform and THD diagram of APF system after adding the improved model prediction algorithm. It can be seen that the sinusoidal degree of the current waveform is high. It can be deduced using the Fourier analysis that the THD is only 1.35%, and the APF system achieves an excellent compensation effect.

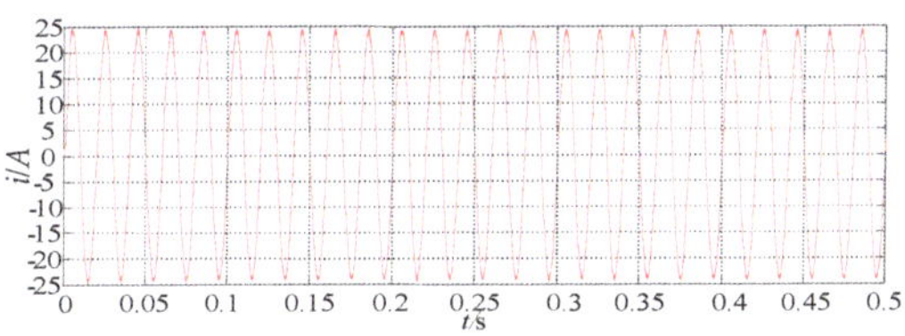


**Figure 20.** A-phase power grid current waveform after compensation based on improved model predictive control.

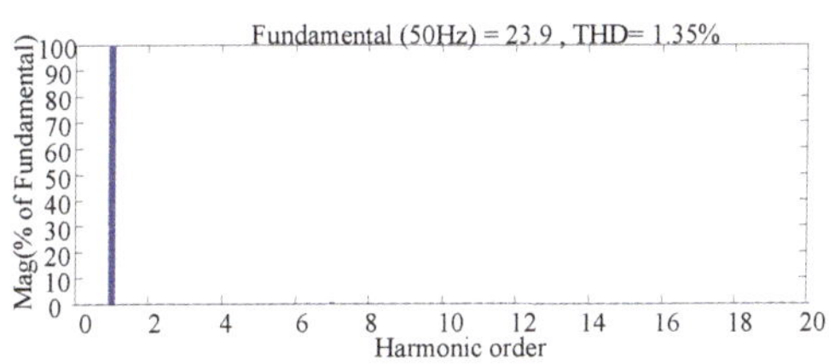


**Figure 21.** A-phase grid current THD diagram after compensation of improved model predictive control.

Figure 22 presents the effect diagram of the harmonic current tracking, while Figure 23 shows the error result diagram of the reference current and feedback current. It can be seen that the APF output current and reference current mainly coincide, and the tracking effect is efficient. The error between the reference current and feedback current is ±0.7 A, which is smaller than that of the traditional MPC algorithm.

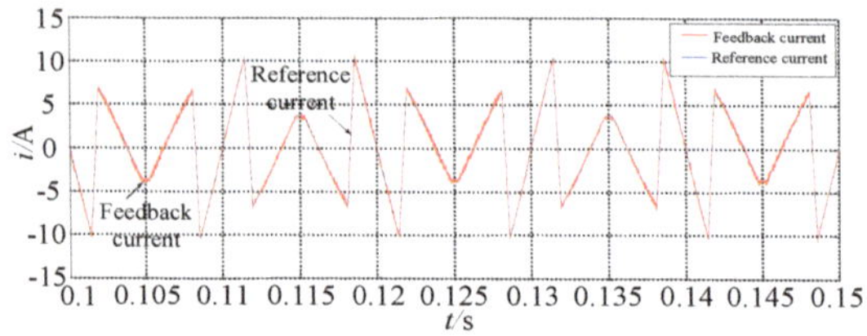


**Figure 22.** Effect of harmonic current tracking of improved model predictive control.

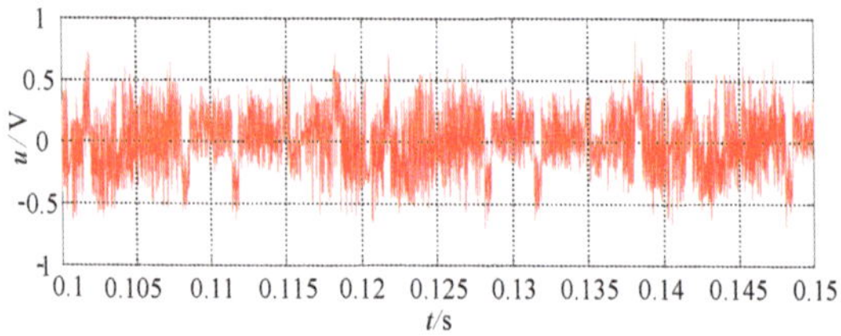


**Figure 23.** Error of reference current and feedback current of improved model predictive control.

Figure 24 shows the overall voltage waveform of the DC side capacitor, which can be stable around 800 V with a small fluctuation. Figure 25 illustrates the fluctuation diagram of the neutral point potential, with a fluctuation range of ±0.37 V, performing a good voltage stabilizing effect.

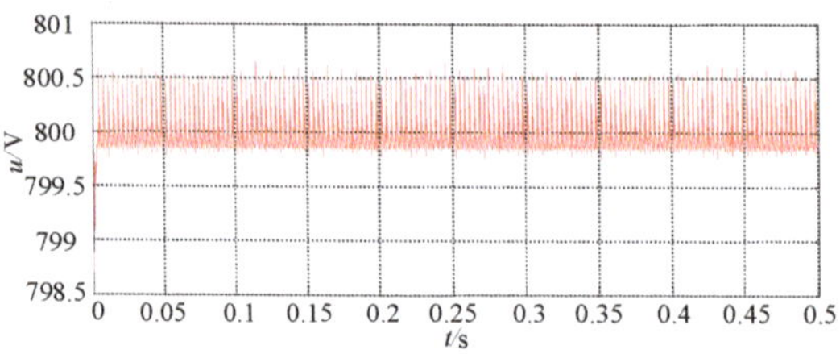


**Figure 24.** Overall voltage waveform of DC side capacitor based on improved model predictive control.

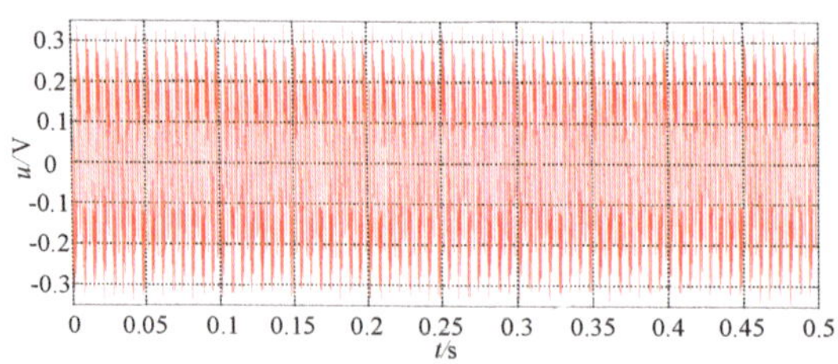


**Figure 25.** Waveform of neutral point potential at DC side of improved model predictive control.

Figure 26 presents the value of the switch state flag, which value ranges between 1 and 7. In order to further observe the sector judgment, flag1 is defined as the sector judgment flag. Figure 27 shows the result values of flag1. It can be seen that the value of flag1 ranges between 1 and 6, which verifies the efficiency of the method of judging the sector position by the voltage vectors.

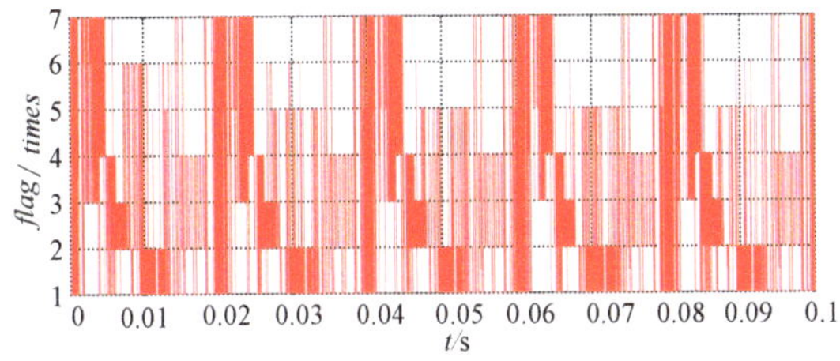


**Figure 26.** Result diagram of switch state flag value of improved model predictive control.

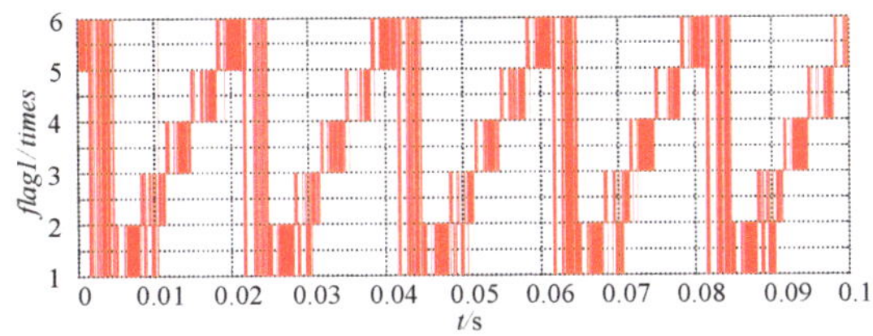


**Figure 27.** Value result of sector judgment flag1 of improved model predictive control.

In this section, the two MPC algorithms are simulated. Compared with the traditional MPC, the improved model prediction algorithm can judge the output voltage vector and select the switching state conducive to the neutral point potential balance of the capacitor on the DC side for output, so as to realize the voltage sharing control of the capacitor. Using the improved model prediction algorithm, the overall capacitor voltage is closer to 800 V, the neutral point potential fluctuation is smaller, and the harmonic current tracking effect is better. According to the diagram of the switch state flag, the rolling times are reduced from 27 to 7, which reduces the rolling optimization times. In conclusion, the control effect of the improved model prediction algorithm is significantly enhanced.

### 4.3. System Performance Simulation and Analysis

In practical applications, the operating conditions are often complex. A better compensation effect can be achieved only when the APF system has a good adaptability. Therefore, it is crucial to analyze the dynamic response capacity of the APF system. In order to study the dynamic performance of APF system based on improved model predictive control based on sector judgment, it is simulated and analyzed under different working conditions.

#### 4.3.1. System Simulation Analysis under Sudden Load Change

Based on the simulation model presented in Section 4.2, a same nonlinear load is input at 0.2 s and cut off at 0.4 s. Figure 28 shows the three-phase power grid current waveforms when the load suddenly changes. At 0.2 s, the current amplitude suddenly increases due to the load increase, which results in large spikes. However, the APF system can quickly compensate it and then eliminate the spikes. Figure 29 shows the THD diagram of the A-phase current before load mutation, while Figure 30 presents the THD diagram of the A-phase after load mutation. It can be seen that the THD content is 3.41%, which is higher than 1.94%. However, it is still less than 5%, which indicates that the system has a good dynamic compensation performance.

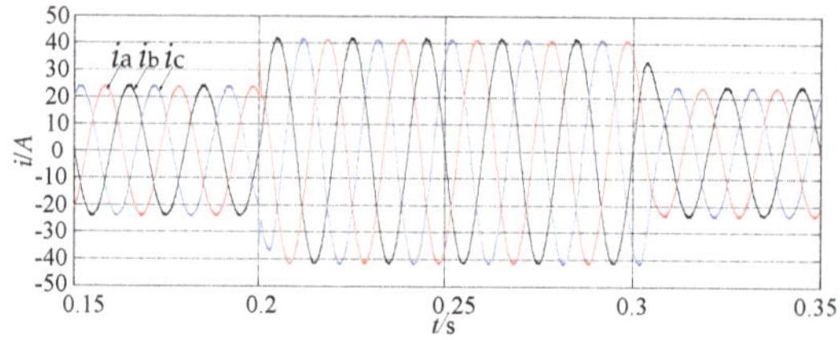


**Figure 28.** Power grid current waveform before and after load sudden change of traditional MPC.

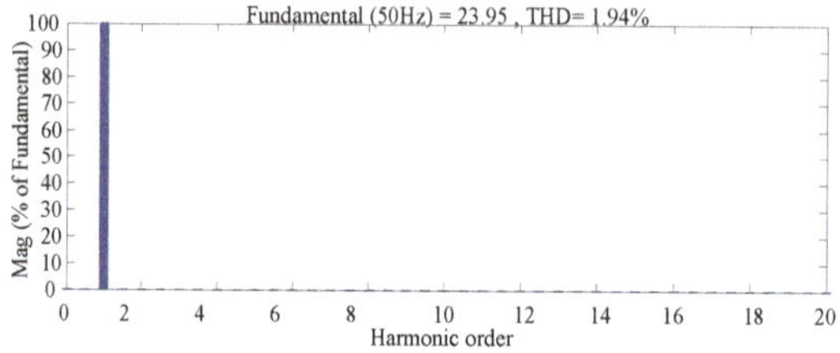


**Figure 29.** THD diagram of A-phase current before load sudden change of traditional MPC.

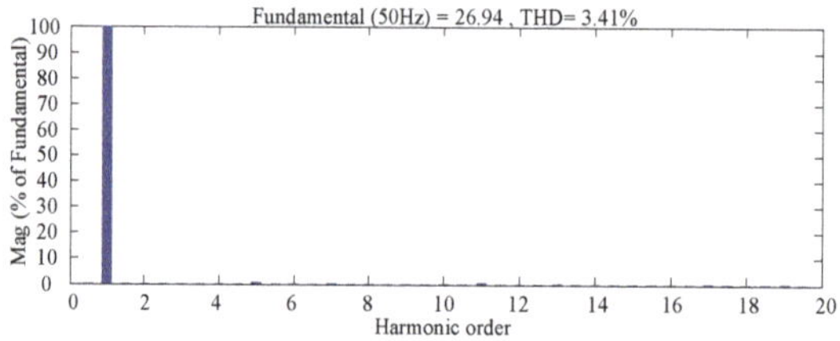


**Figure 30.** THD diagram of A-phase current after load sudden change of traditional MPC.

Figures 31 and 32 are the overall voltage waveform and neutral point potential waveform of DC side capacitor under the traditional model predictive control algorithm respectively. The DC side voltage fluctuates greatly when the load is input, and then it stabilizes at about 800.5 V, about 0.5 V deviates from the reference voltage. The neutral point potential fluctuates greatly when the load is removed, and the maximum fluctuation is about 4.9 V.

Figures 33 and 34 present the overall voltage waveform and midpoint potential waveform of the DC side capacitor under the improved model predictive control algorithm. It can be seen that their fluctuation is smaller than that in Figures 31 and 32.

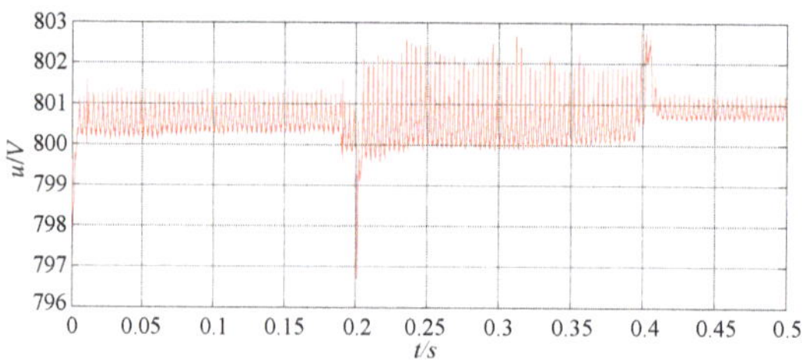


**Figure 31.** Overall voltage waveform of DC side capacitor before load and after sudden change of traditional MPC.

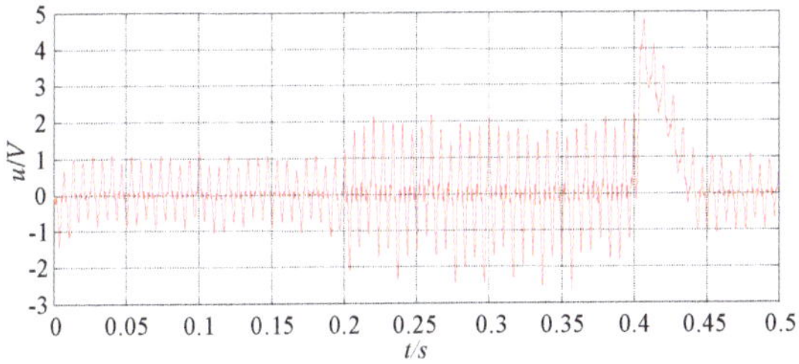


**Figure 32.** Waveform of neutral point potential before and after load sudden change of traditional MPC.

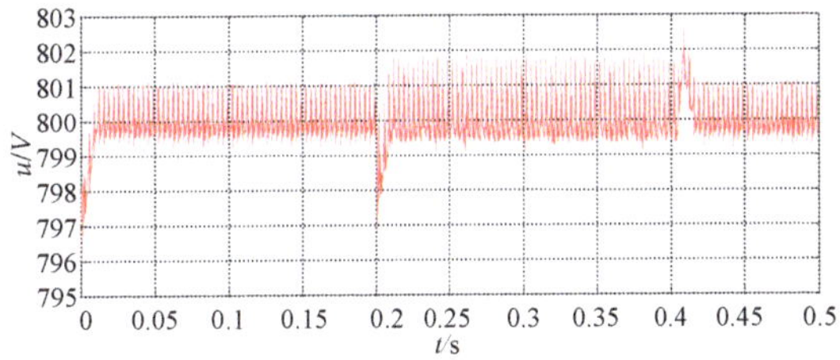


**Figure 33.** Overall voltage waveform of DC side capacitor before load and after sudden change of improved MPC.

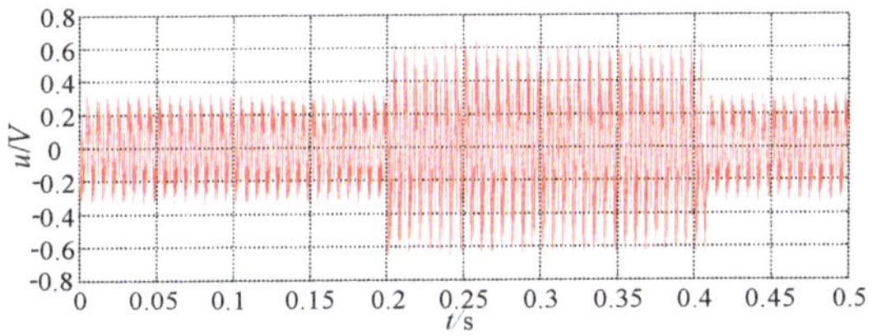


**Figure 34.** Waveform of neutral point potential before and after load mutation of improved MPC.

Compared with the simulation waveform of traditional MPC algorithm, the algorithm based on improved model predictive control has better control effect on DC side capacitor voltage and neutral point potential voltage, and it can better maintain the stability of the DC side voltage and reduce the voltage fluctuation range.

#### 4.3.2. System Simulation Analysis under Unbalanced Load

This section studies the compensation effect of the APF under load imbalance, where a 10 Ω unbalanced resistance is connected to the A-phase circuit in order to simulate the system. Figure 35 shows the waveform diagram of the power grid current under unbalanced load. It can be seen that the current distortion is significant, the amplitude of the B-phase and C-phase current is high, the amplitude of the A-phase is low, and the power grid current is unbalanced. Figure 36 presents the THD diagram of the A-phase current before compensation, while Figures 37 and 38 show the waveform diagram of the three-phase grid current and A-phase current THD diagram after compensation.

As can be seen from Figures 36 and 38 that under the action of the improved MPC control algorithm based on sector judgment, the grid current THD after compensation is reduced from 27.54% to 2.19%, performing an excellent compensation effect. It can be seen from Figure 37 that the APF system can make up the unbalanced three-phase current into a balanced state and it has a strong ability to suppress the three-phase imbalance.

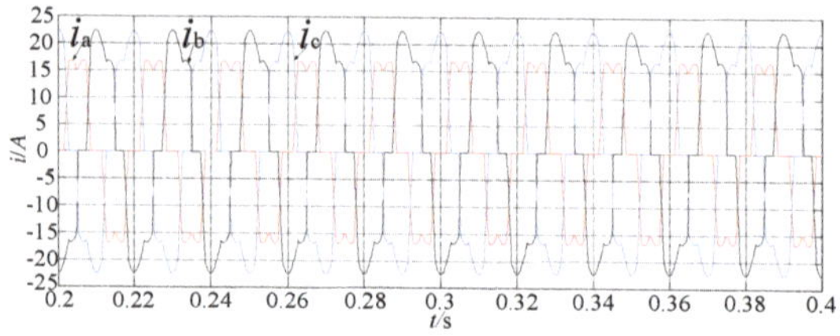


**Figure 35.** Power grid current waveforms under unbalanced load.

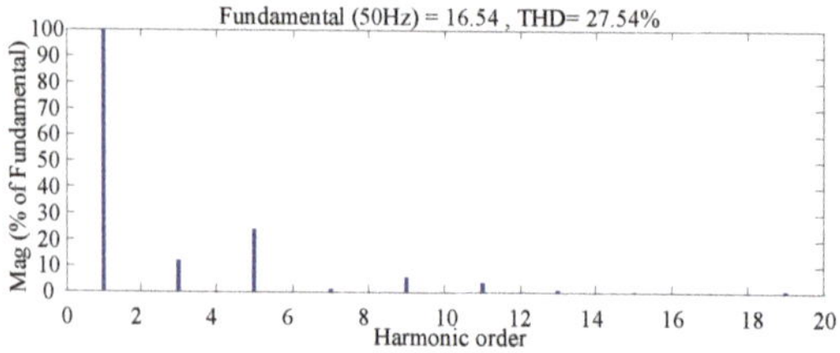


**Figure 36.** THD diagram of A-phase grid current under unbalanced load.

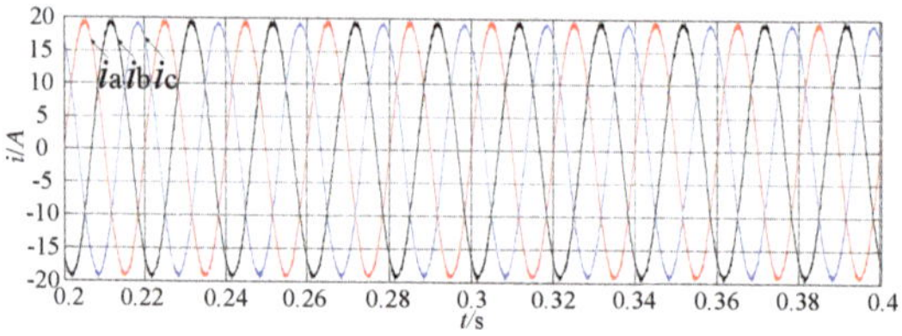


**Figure 37.** Power grid current waveforms after compensation under unbalanced load of improved MPC.

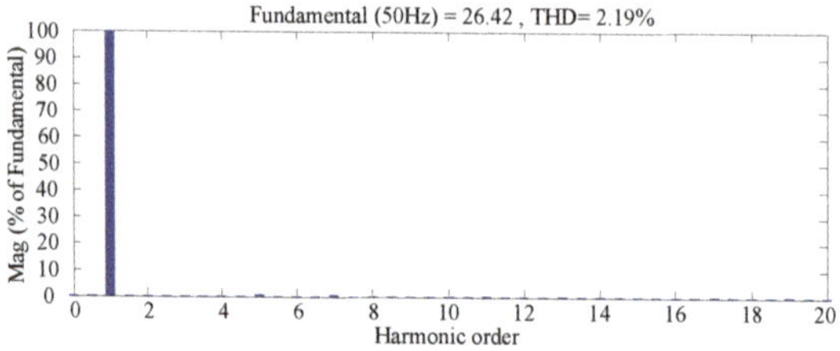


**Figure 38.** THD diagram of A-phase grid current after compensation under unbalanced load of improved MPC.

## 5. Experiment and Result Analysis

Based on the simulation, the following experiments are performed in order to verify the feasibility and correctness of the proposed control algorithm in the APF system. The experimental parameters are shown in Table 6.

According to these system parameters, an experimental platform is built. It mainly includes a topology unit, main control unit, sampling unit, and driving unit. The topology unit is composed of a NPC three-level topology and a filter inductance. The main control unit is composed of DSP and FPGA to complete the data operation and drive the signal output, respectively. The sampling unit is mainly responsible for sampling grid voltage and current, DC side capacitor voltage and APF system output current. The driving unit is responsible for amplifying the PWM wave output by the controller in order to drive the power switching devices. The experimental system platform is shown in Figure 39.

**Table 6.** Key parameters of the system.

| System Parameter | Parameter Value |
|---|---|
| Voltage of capacitor $C_1$, $C_2$ | 400 V |
| The capacitor $C_1$, $C_2$ | 1000 μF |
| The sampling frequency $f_s$ | 20 kHz |
| The grid frequency $f$ | 50 Hz |
| The inductor $L_A$, $L_B$, $L_C$ | 10 mH |
| The balanced grid phase voltages $U_A$, $U_B$, $U_C$ | 220 V |
| The balanced resistance load $R$ | 40 Ω |
| The unbalanced resistance load $R$ | 20 Ω |

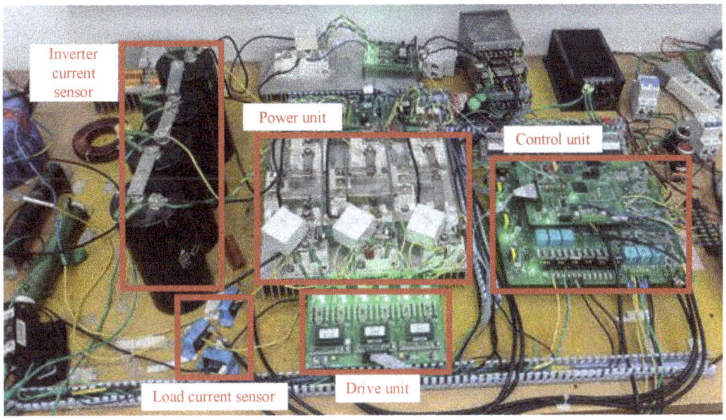


**Figure 39.** Experimental platform.

*5.1. DC Capacitor Voltage Waveform*

The waveform of the capacitor voltage is presented in Figure 40. The APF device detects the change of capacitor voltage by the Hall voltage sensor. The experimental waveform shows that the DC side voltage is stable at 800 V during operation. The results show that the capacitor voltage can remain stable with a small fluctuation, which meets the requirements of voltage stability. Therefore, the correctness of the proposed voltage stabilizing algorithm is proved.

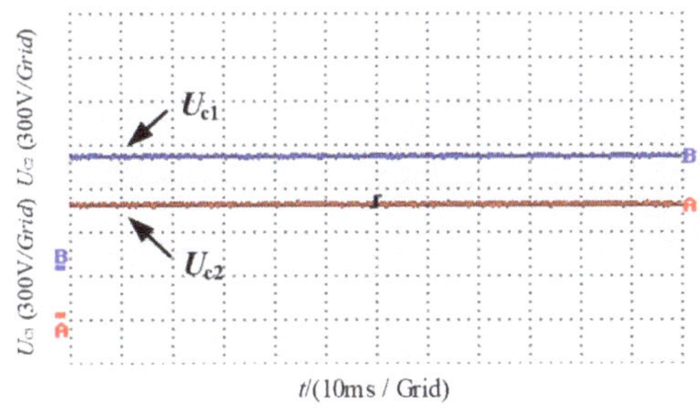


**Figure 40.** Waveforms of capacitance voltage at DC side.

*5.2. Harmonic Current Extraction and Tracking Waveform*

Figure 41 channel A shows the harmonic current waveform in the A-phase load current, while Figure 41 channel B presents the compensation current waveform by the APF system. The diagram shows that the harmonic current wave is coherent with the simulation waveform. The variation trend of the harmonic current and output current is mainly the same, which indicates that the system can better perform the extraction and tracking of harmonic current. This demonstrates that the proposed improved MPC algorithm has a high current tracking performance.

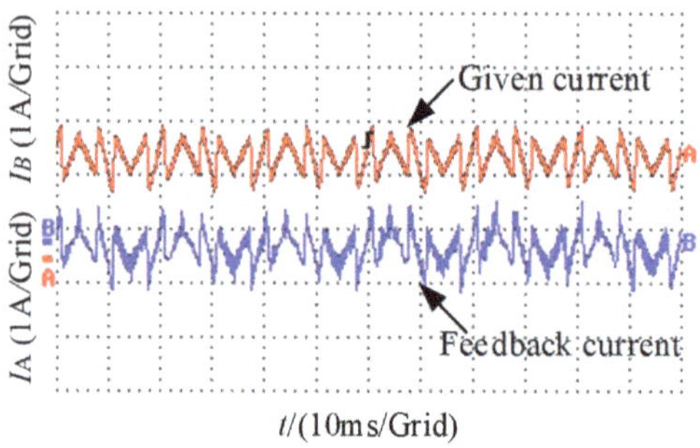


**Figure 41.** Waveforms of A-phase harmonic tracking current.

### 5.3. Algorithm Time-Consuming Comparison

In the experiment, mark points are set at the beginning and end of the algorithm program, and the running time of the algorithm program segment is obtained by the timer, so as to compare the time consumption between the traditional MPC and the proposed improved MPC.

The program running time measured by this method when using the traditional MPC algorithm for compensation is 59.2 μs, while the running time of the MPC algorithm program section is 39 μs, as shown in Figure 42. When the improved MPC strategy is used for harmonic compensation, the running time of the whole program is 38.4 μs. Figure 43 shows that the running time of the improved MPC algorithm is 18 μs.

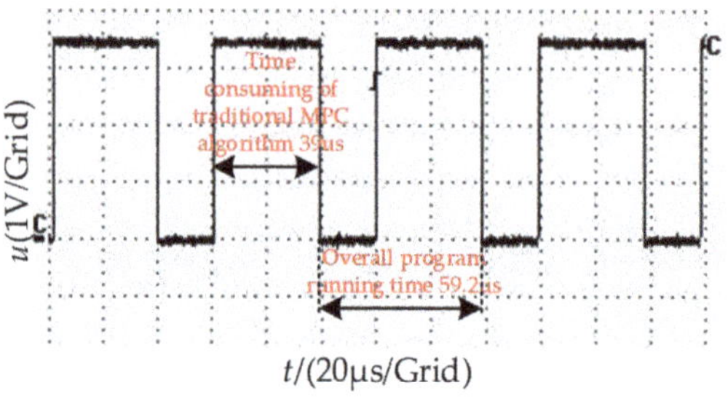


**Figure 42.** The time consumption of traditional MPC algorithm.

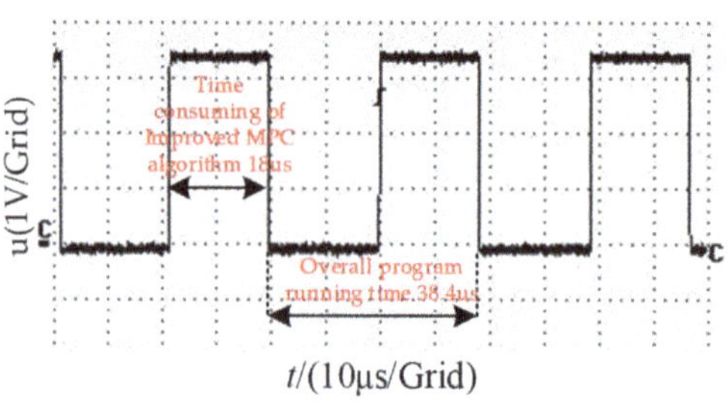


**Figure 43.** The time consumption of improved MPC algorithm.

The traditional algorithm requires 27 comparison operations when selecting the optimal vector each time, while the improved MPC algorithm only requires 7 comparison operations. Therefore, the running time of the MPC algorithm in Figure 43 is shorter than that in Figure 42.

### 5.4. Experiment of Power Grid Balance and Load Balance

The three-phase gird voltages $U_A$, $U_B$, and $U_C$ are 220 V, 220 V, and 220 V, respectively. The resistance is 40 Ω after three-phase rectification.

Figure 44a is the waveform of three-phase grid current before compensation, and Figure 44b is the FFT analysis diagram of A-phase current before compensation. Because the load contains nonlinear devices, the grid current is seriously distorted and the sinusoidal degree is low. Taking A-phase as an example, the THD of A-phase grid current is 24.86%, which reduces the power quality of the grid.

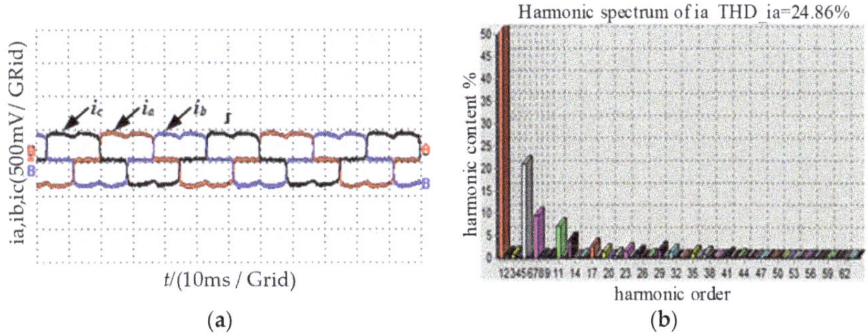


**Figure 44.** Experimental waveform before compensation: (**a**) three-phase grid current and (**b**) A-phase current THD.

Figure 45a shows the current waveform after system compensation using the proposed improved algorithm. The FFT analysis of the A-phase current is shown in Figure 45b. It can be seen that the compensated current recovers the sinusoidal shape, and the sinusoidal degree is satisfactory. In the latter, the THD of the A-phase current is 2.86%, which improves the power quality. The experimental waveform is similar to the simulation results. The compensation effect is mainly coherent with the simulation results, and the compensation effect is clear. When the power grid and load are balanced, the system dynamic compensation performance is high, which verifies the efficiency of the single objective function improved model predictive control algorithm.

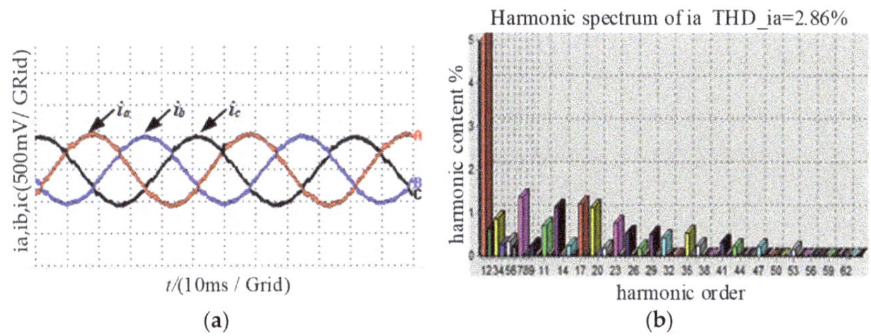


**Figure 45.** Experimental waveform after compensation: (**a**) three-phase grid current and (**b**) A-phase current THD.

### 5.5. Experiment of Power Grid Balance and Load Imbalance

Figure 46a shows the waveform of the three-phase power grid when the system is connected to unbalanced load. It can be seen that the three-phase power grid is asymmetric, and the current amplitudes are not equal. It can be observed from Figure 46b that the current THD of A-phase power grid is 28.32%.

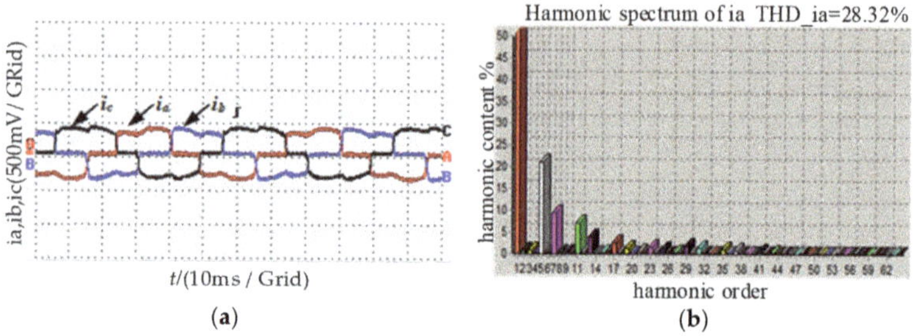


**Figure 46.** Experimental waveform before compensation when load is unbalanced: (**a**) three-phase grid current abd (**b**) A-phase current THD.

Figure 47a presents the current waveform after compensation. It can be seen that the distorted and asymmetric power grid current waveform is compensated into a three-phase symmetrical sinusoidal shape. The A-phase current is analyzed by the FFT (cf. Figure 47b), the THD is 3.80%, which is close to the simulation results. Using the improved MPC strategy, the system compensates the harmonic current, and, also, restrains the three-phase imbalance.

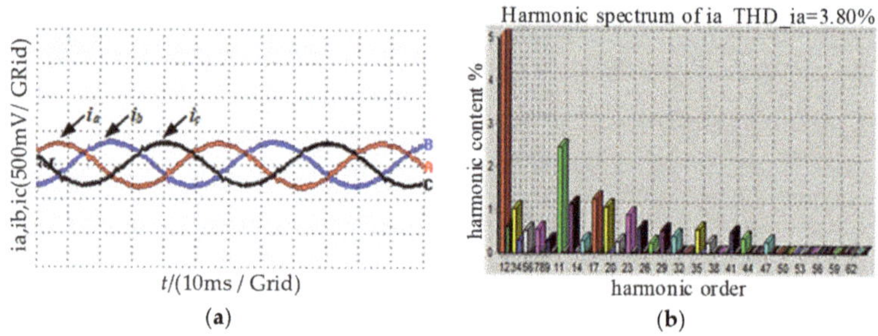


**Figure 47.** Experimental waveform after compensation when load is unbalanced: (**a**) three-phase grid current and (**b**) A-phase current THD.

*5.6. Load Sudden Change Experiment*

Figure 48a shows the waveform of the three-phase power grid after compensation in case of sudden load change. It can be observed that the current suddenly increases due to the load increase. A FFT analysis is conducted on the A-phase current of power grid after sudden change, (cf. Figure 48b), It can be seen that the THD is 4.48%, which still meets the national standard. This indicates that although the load current significantly changes in case of sudden load change, the APF system based on the single objective function improved MPC algorithm can still quickly respond, and perform the rapid tracking of sudden harmonic current with a high dynamic performance.

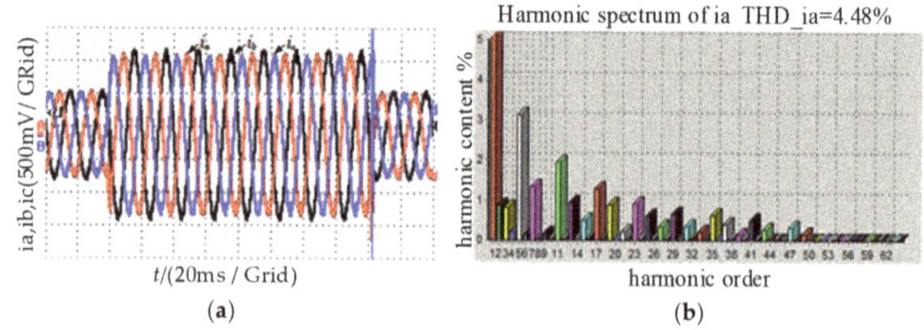


**Figure 48.** Experimental waveform of sudden load change after compensation: (**a**) three-phase grid current and (**b**) A-phase current THD.

## 6. Conclusions

This paper tackles the NPC three-level active power filter and deeply studies the principle and shortcomings of the traditional MPC algorithm, in order to develop a novel model predictive control current tracking algorithm. Dealing with the problems of many rolling times, high computational load, and difficulty to adjust the weight factor of the multi-objective function, an improved MPC algorithm of single objective function based on sector judgment, is proposed. The sector judgment concept is introduced into the model predictive control. By judging the mathematical relationship between the three lines and the voltage reference vector, the location of the area where the reference vector is located is determined, and the number of rolling optimizations is reduced from 27 to 8. After removing the redundant small vector, the rolling number can be reduced to 7, which reduces the complexity of the algorithm and the computational load. The current tracking error, neutral point potential fluctuation on DC side, compensation effect and rolling times of the two MPC algorithms, are compared by simulations and engineering implementations. The obtained results demonstrate that the proposed single objective function improved MPC algorithm based on sector judgment, has a better current tracking effect and voltage stabilizing effect. In addition, it has a good dynamic response ability and compensation performance, which proved that the proposed algorithm is feasible and efficient.

**Author Contributions:** Conceptualization, H.G.; formal analysis, H.G. and W.Z.; data curation, M.R. and W.Z.; writing—review and editing, X.L. All authors have read and agreed to the published version of the manuscript.

**Funding:** This research work was funded by the National Natural Science Foundation of China grant number 51177031.

**Data Availability Statement:** The data presented in this study are available on request from the corresponding author.

**Conflicts of Interest:** The authors declare no conflict of interest.

## References

1. Arulmurugan, R.; Chandramouli, A. Design and implementation of a Nine-Level inverter with shunt active filter for distributed system. In Proceedings of the 2018 4th International Conference on Electrical Energy Systems (ICEES), Chennai, India, 7–9 February 2018; pp. 224–228.
2. Fu, J.; Chen, L.; Zhao, H.; Zhang, P. High and low frequency control strategy for APF DC-Link ripple voltage under unbalanced load. In Proceedings of the 2019 IEEE 2nd International Conference on Electronics and Communication Engineering (ICECE), Xi'an, China, 9–11 December 2019; pp. 326–330.
3. Sun, X.F.; Gu, L.; Li, Z.C. Damping strategy research of background harmonic in novel power distribution system. *Acta Energ. Sol. Sin.* **2016**, *37*, 1359–1366.
4. Wang, X.; Gao, Y.; Lin, L. Research status and prospect of active power filter. *Power Syst. Prot. Control* **2019**, *47*, 177–186.
5. Gadgune, S.Y.; Jadhav, P.T.; Chaudhary, L.R. Implementation of shunt apf based on diode clamped and cascaded H-bridge multilevel inverter. In Proceedings of the 2015 IEEE International Conference on Electrical. Computer and Communication Technologies (ICECCT), Coimbatore, India, 5–7 March 2015; pp. 1–7.
6. Yang, Z.; Li, S.; Zhao, X. A source-type harmonic energy unbalance suppression method based on carrier frequency optimization for cascaded multilevel APF. In Proceedings of the 2016 IEEE Energy Conversion Congress and Exposition (ECCE), Milwaukee, WI, USA, 18–22 September 2016; pp. 1–8.
7. Banaei, M.R.; Jannati Oskuee, M.R.; Khounjahan, H. Reconfiguration of semi-cascaded multilevel inverter to improve systems performance parameters. *IET Power Electron.* **2014**, *7*, 1106. [CrossRef]
8. Wei, T.S. An improved compensation current detection method of static synchronous compensator. *J. Phys. Conf. Ser.* **2021**, *1871*, 012066.
9. Sen, L.; Yan, Z. Low losses hysteresis current control method for active power filter. In Proceedings of the 2010 International Conference on Electrical and Control Engineering, Wuhan, China, 25–27 June 2010; pp. 3860–3865.
10. Gao, H.Y.; Liu, X.N.; Ren, M.J. A novel multilevel controller. *Electronics* **2021**, *10*, 1222. [CrossRef]
11. Foster, J.G.; Pereira, L.R.; Gonzatti, R.B. A review of FCS-MPC in multilevel converters applied to active power filters. In Proceedings of the 2019 IEEE 15th Brazilian Power Electronics Conference and 5th IEEE Southern Power Electronics Conference (COBEP/SPEC), Santos, Brazil, 1–4 December 2019; pp. 1–6.

12. Zeng, Z.; Yang, J.; Yu, N. Research on PI and repetitive control strategy for shunt active power filter with LCL-filter. In Proceedings of the 7th International Power Electronics and Motion Control Conference, Harbin, China, 2–5 June 2012; pp. 2833–2837.
13. Narimani, M.; Wu, B.; Yaramasu, V.; Zargari, N.R. Finite Control-Set Model Predictive Control (FCS-MPC) of Nested Neutral Point-Clamped (NNPC) Converter. *IEEE Trans. Power Electron.* **2015**, *30*, 7262–7269. [CrossRef]
14. Zhang, X.; Wang, Y.; Yu, C.; Guo, L.; Cao, R. Hysteresis Model Predictive Control for High-Power Grid-Connected Inverters with Output LCL Filter. *IEEE Trans. Ind. Electron.* **2016**, *63*, 246–256. [CrossRef]
15. Ben-Brahim, L.; Gastli, A.; Trabelsi, M.; Ghazi, K.A.; Houchati, M.; Abu-Rub, H. Modular Multilevel Converter Circulating Current Reduction Using Model Predictive Control. *IEEE Trans. Ind. Electron.* **2016**, *63*, 3857–3866. [CrossRef]
16. Dekka, B.; Wu., V.; Yaramasu., N.; Zargari, R. Integrated model predictive control with reduced switching frequency for modular multilevel converters. *IET Electr. Power Appl.* **2017**, *11*, 857–863. [CrossRef]
17. Preindl, M.; Schaltz, E.; Thogersen, P. Switching Frequency Reduction Using Model Predictive Direct Current Control for HighPower Voltage Source Inverters. *IEEE Trans. Ind. Electron.* **2011**, *58*, 2826–2835. [CrossRef]
18. Li, Y.; Zhang, Z.; Zhang, Z.; Wang, J.; Kennel, R. Model Predictive Control of a Shunt Active Power Filter with Improved Dynamics Under Distorted Grid Conditions. In Proceedings of the 2020 IEEE 9th International Power Electronics and Motion Control Conference (IPEMC2020-ECCE Asia), Nanjing, China, 29 November–2 December 2020; pp. 1033–1037.
19. Iturra, R.G.; Cruse, M.; Mütze, K.; Dresel, C.; Soleimani, I.; Thiemann, P. Model Predictive Control for Shunt Active Power Filter with Harmonic Power Recycling Capability. In Proceedings of the 2018 International Conference on Smart Energy Systems and Technologies (SEST), Seville, Spain, 10–12 September 2018; pp. 1–6.
20. Tomlinson, M.; Mouton, H.d.T.; Kennel, R.; Stolze, P. A Fixed Switching Frequency Scheme for Finite-Control-Set Model Predictive Control—Concept and Algorithm. *IEEE Trans. Ind. Electron.* **2016**, *63*, 7662–7670. [CrossRef]
21. Davari, S.A.; Khaburi, D.A.; Kennel, R. An Improved FCS–MPC Algorithm for an Induction Motor with an Imposed Optimized Weighting Factor. *IEEE Trans. Power Electron.* **2012**, *27*, 1540–1551. [CrossRef]
22. Acuña, P.; Morán, L.; Rivera, M.; Dixon, J.; Rodriguez, J. Improved Active Power Filter Performance for Renewable Power Generation Systems. *IEEE Trans. Power Electron.* **2014**, *29*, 687–694. [CrossRef]
23. Hu, M.; Shen, Y.; Wang, X. A Harmonic Compensation Method for Three-level Active Power Filter Based on Predictive Control. In Proceedings of the 2018 37th Chinese Control Conference (CCC), Wuhan, China, 25–27 July 2018; pp. 3634–3639.
24. Cortes, P.; Kouro, S.; Rocca, B.L.; Vargas, R.; Rodriguez, J.; Leon, J.I.; Vazquez, S.; Franquelo, L.G. Guidelines for weighting factors design in Model Predictive Control of power converters and drives. In Proceedings of the IEEE International Conference on Industrial Technology, Churchill, Australia, 10–13 February 2009; pp. 1–7.
25. Fei, J.; Liu, L. Real-Time Nonlinear Model Predictive Control of Active Power Filter Using Self-Feedback Recurrent Fuzzy Neural Network Estimator. *IEEE Trans. Ind. Electron. August* **2022**, *69*, 8366–8376. [CrossRef]
26. Rahim, A.; Takeuchi, M.; Funato, H.; Junnosuke, H. T-type three-level neutral point clamped inverter with model predictive control for grid connected photovoltaic applications. In Proceedings of the 19th International Conference on Electrical Machines and Systems (ICEMS), Chiba, Japan, 13–16 November 2016; pp. 1–5.
27. Almér, S.; Mariéthoz, S.; Morari, M. Sampled Data Model Predictive Control of a Voltage Source Inverter for Reduced Harmonic Distortion. *IEEE Trans. Control. Syst. Technol.* **2013**, *21*, 1907–1915. [CrossRef]
28. Acuña, P.; Morán, L.; Rivera, M.; Aguilera, R.; Burgos, R.; Agelidis, V.G. A Single-Objective Predictive Control Method for a Multivariable Single-Phase Three-Level NPC Converter-Based Active Power Filter. *IEEE Trans. Ind. Electron.* **2015**, *62*, 4598–4607. [CrossRef]

*electronics* MDPI

*Article*

# A Dual-Mode Control Scheme to Improve Light Load Efficiency for Active-Clamp Flyback Converter

**Thanh Nhat Trung Tran [1], Hung-Chia Wang [2] and Jian-Min Wang [2,*]**

1 Department of Power Mechanical Engineering, National Formosa University, No. 64, Wunhua Rd., Huwei Township, Yunlin County 632, Taiwan; trungtranspk@gmail.com
2 Department of Vehicle Engineering, National Formosa University, No. 64, Wunhua Rd., Huwei Township, Yunlin County 632, Taiwan; 10758103@gm.nfu.edu.tw
* Correspondence: jmw@nfu.edu.tw; Tel.: +886-5-631-5692

**Abstract:** A novel dual-mode control scheme is proposed in this paper that permits the active-clamp flyback (ACF) converter to operate in both the quasi-resonant (QR) mode under light load and the active-clamp mode under medium or heavy load. The mode transition is performed based on the external dual-mode control circuit. In addition, the proposed converter incorporates a new QR mode valley switching (VS) control circuit that reduces switching loss in the main switch by achieving VS. Under medium to full load, the proposed converter becomes an ACF converter designed to achieve zero-voltage switching (ZVS), which reduces switching losses in both power switches. The proposed dual-mode control ACF converter has the following advantages: (1) compared with conventional ACF converters, the proposed ACF converters minimize switching losses by combining VS and ZVS; (2) under light load conditions, the frequency-limiting QR control mechanism is used to avoid disadvantageous switching losses caused by high switching frequencies. The 65 W ACF converter prototype with a DC 155 V input and a DC 19 V/3.42 A output under 65 kHz switching frequency was implemented. The experimental results demonstrate the feasibility of the proposed control scheme. The efficiency of the proposed converter reached 79% at a load of 3.5 W, which is 11% higher than the conventional ACF converter.

**Keywords:** active-clamp flyback; dual-mode control; light load; valley switching; zero-voltage switching

**Citation:** Tran, T.N.T.; Wang, H.-C.; Wang, J.-M. A Dual-Mode Control Scheme to Improve Light Load Efficiency for Active-Clamp Flyback Converter. *Electronics* **2022**, *11*, 1308. https://doi.org/10.3390/electronics11091308

Academic Editor: Pablo García Triviño

Received: 8 March 2022
Accepted: 19 April 2022
Published: 20 April 2022



## 1. Introduction

The flyback converter architecture, as shown in Figure 1, extended from the buck-boost converter, is one of the most popular converters in the power electronics industry. This type of converter is characterized by its simple structure, few components, and easy compliance with electrical isolation safety specifications. Due to the fact of these advantages, flyback architecture is widely used in converters with output power levels less than 100 W such as light-emitting diode (LED) indoor lighting and portable chargers. However, in the transformer of the flyback converter, energy is stored in an air gap in the core. To store enough energy for the transformer, the size of the air gap must be increased with a corresponding increase in the leakage inductance of the transformer, resulting in a voltage spike on the switching device when it is turned off. The snubber circuit is a popular solution for suppressing this voltage spike; however, it lowers converter efficiency.

In recent years, numerous researchers have proposed efficiency improvements for flyback converters. Generally, these improvements can be classified into two major categories: constant-frequency control and variable-frequency control. In constant-frequency control, the active-clamp technique has been widely used in flyback converters. The ACF converter structure and key waveforms are presented in Figure 2a. In this technique, a circuit loop, comprising an auxiliary switch ($Q_2$) and a capacitor ($C_{clamp}$), is added at the two terminals on the primary side of the transformer. This loop is used to perform ZVS at the main switch, improving converter efficiency and avoiding the high voltage spikes generated when the main switch is turned off [1–7]. However, conduction loss caused by resonant circulating current and additional switching in the $Q_2$ switch induces high power loss under light load [8]. Numerous studies [8,9] have described methods for addressing this drawback. Reference [8] employs a secondary-side post-regulator for minimizing power consumption. Consequently, this method requires an external post-regulator, resulting in a circuit with increased cost and size. In [9], a dual-mode control method was adopted in which the operation mode of the ACF converter was divided into two modes based on the load: the conventional flyback mode and the active-clamp mode. Briefly, the proposed converter works as a conventional flyback converter with the auxiliary switch turning off to decrease the converter's switching and conduction losses under no-load conditions but turns into an ACF converter under normal-load/full-load conditions. However, this method is only effective for reducing the power consumption of the ACF converter under no-load conditions. In categories of variable-frequency control, quasi-resonant (QR) control is the mainstream technique [10–14]. The quasi-resonant flyback (QRF) converter structure and key waveforms are presented in Figure 2b. In the QR control technique, the resonance signal generated by the parasitic capacitance of the switch, $C_{OSS}$, and the magnetizing inductance of the transformer, $L_m$, are used to turn on the main switch when its drain-source voltage reaches a minimum valley point. Hence, switching loss is significantly reduced. However, this technique cannot achieve ZVS.

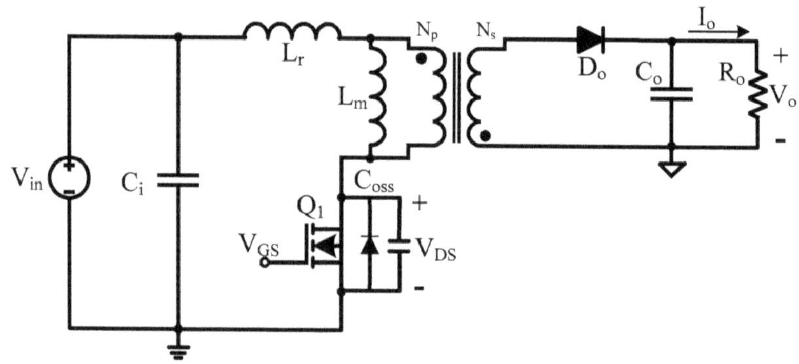


**Figure 1.** Flyback converter topology.

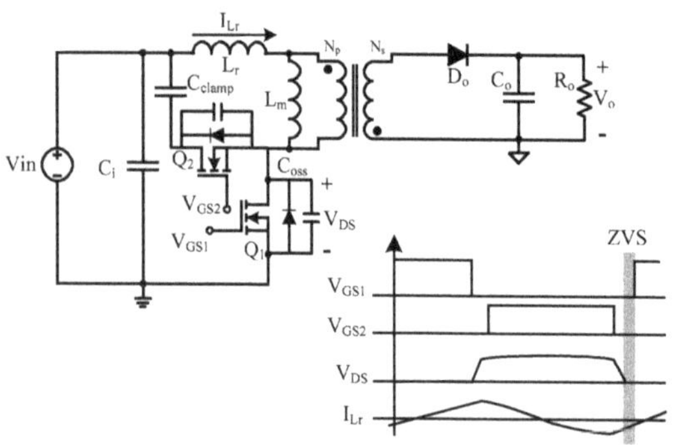


(a)

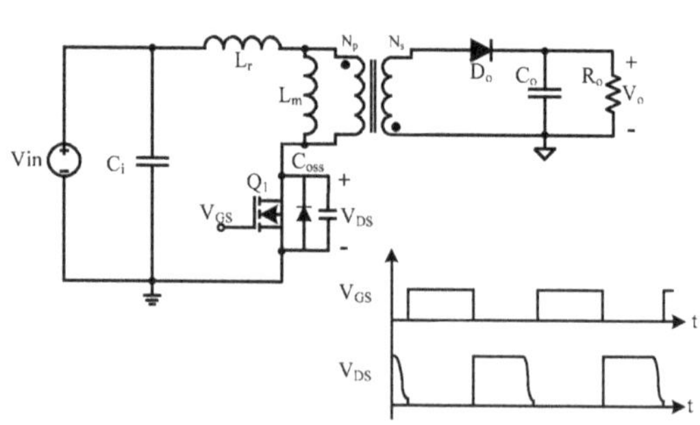


(b)

**Figure 2.** (**a**) Active-clamp flyback converter structure and key waveforms; (**b**) quasi-resonant flyback converter structure and key waveforms.

Thus, as summarized above, constant-frequency and variable-frequency control have advantages and disadvantages. By combining the advantages of ACF and QRF, a novel dual-mode control scheme combining ZVS and VS is proposed in this paper.

Based on the structure of the ACF converter, in this dual-mode control scheme, an external logic circuit, called a dual-mode switching circuit, is integrated for switching the converter to the QR mode under light load adopting VS for the main switch to substantially reduce switching surges. However, the converter operating in the QR mode may cause high switching frequency and increased switching losses. Thus, the frequency-limiting QR control technology is combined to avoid the disadvantage of the increased switching frequency. This technique limits the maximum switching frequency during light load operation to allow the converter to achieve the frequency-limiting function. Under heavy load, the dual-mode switching circuit switches the converter back to the conventional ACF converter (active-clamp mode) to enable both power switches to achieve ZVS to obtain high efficiency.

## 2. Proposed Novel Dual-Mode ACF Converter Structure

### *2.1. The Proposed Circuit Structure and Principle of Operation*

Figure 3 presents the block diagram of the proposed dual-mode control scheme and the related control sequence diagram under different load conditions. The control circuit consists of the current-mode controller UC3843 produced by STMicroelectronics (Geneva, Switzerland), a QR control circuit, and a dual-mode switching circuit. Among these components, the current mode controller UC3843 plays a critical role in stabilizing the output voltage. The QR control circuit and dual-mode switching circuit are the core circuits of the active-clamp and QR modes, respectively, as shown in Figure 3a. The working principle of the control circuit is explained in terms of two states: light load and heavy load.

#### 2.1.1. Light Load

Under light load, the controller proposed in this paper controls the ACF converter operating in the QR mode, and the related control sequence diagram is illustrated in Figure 3b. In this case, the output signal ($V_C$) of the hysteresis comparator is low; thus, the high-side drive circuit is disabled, turning off the auxiliary switch ($Q_2$) of the converter. At this moment, the feedback voltage signal ($V_{FB}$) at the converter output is then compared with the internal reference voltage of the feedback circuit to generate the error signal ($V_{Comp}$), which is sent to the current mode controller. The controller processes error signals and generates the control driving signal ($V_{PWM}$). Notably, the QR mode controller used in this paper has a mechanism to limit the maximum switching frequency, which is different from the conventional first VS method (as indicated by the red dashed circle in Figure 3b); this mechanism avoids disadvantageous switching losses due to the high switching frequencies.

#### 2.1.2. Heavy Load

When the converter transitions from light to heavy load, the operation mode of the converter is switched from the QR mode to the active-clamp mode by the controller. Figure 3c depicts the related control sequence diagram. At this moment, the magnitude of the error signal ($V_{Comp}$) generated by the feedback circuit increases, simultaneously increasing the duty cycle of the drive signal ($V_{PWM}$) generated by the current mode controller. Due to the hysteresis comparator output signal ($V_C$) in this state being high, the high-side drive circuit is enabled. Consequently, the auxiliary switch of the ACF $Q_2$ is activated in accordance with the drive signal of $V_{Q2}$. In other words, the driving signals $V_{Q1}$ and $V_{Q2}$ are complementary operation signals.

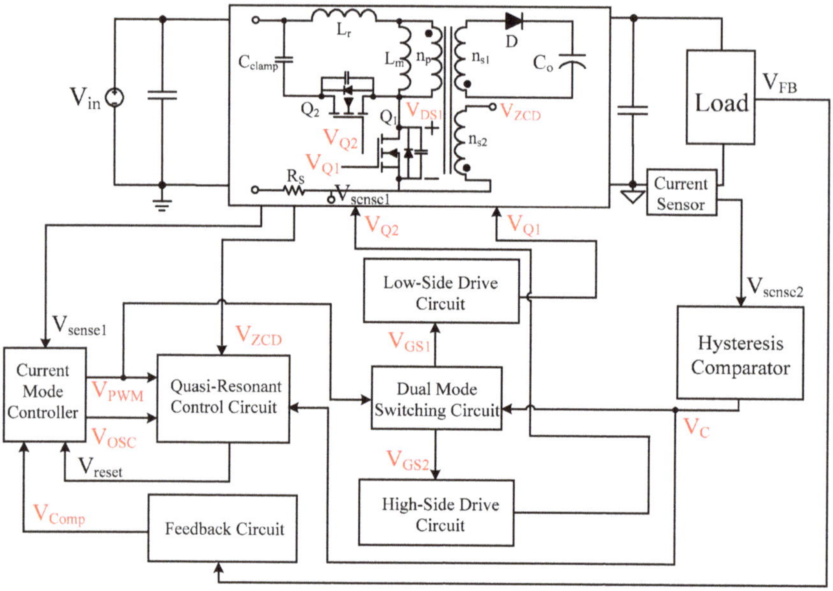


(**a**)

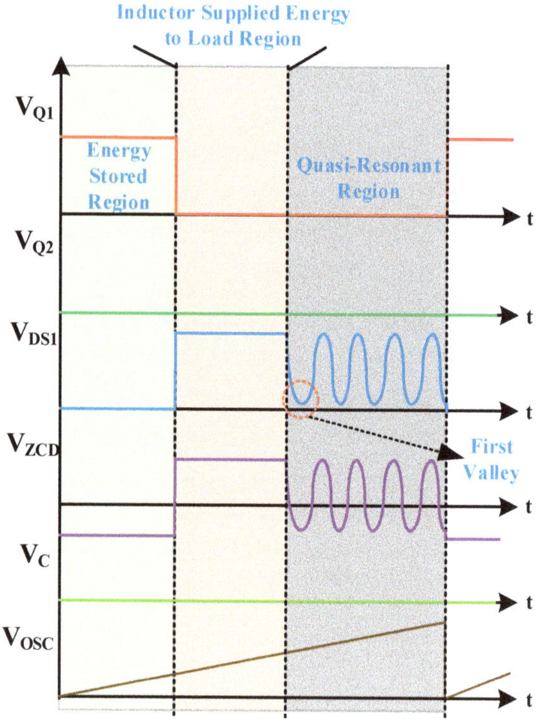


(**b**)

**Figure 3.** *Cont.*

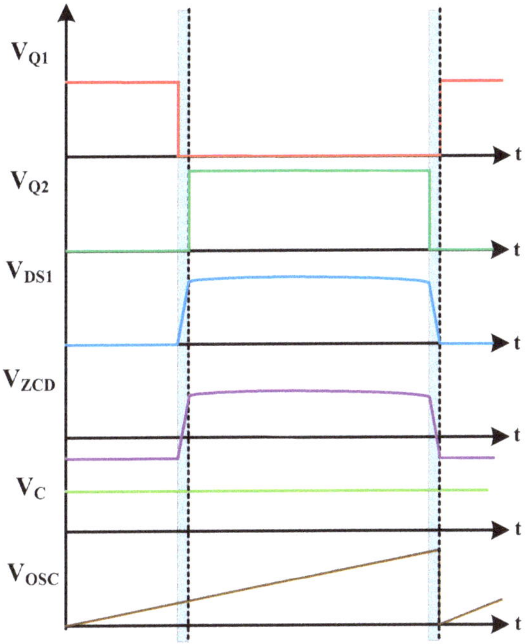


(c)

**Figure 3.** (**a**) Dual-mode control strategy proposed in this paper; (**b**) sequence diagram of the key waveform under light load; (**c**) sequence diagram of the key waveform under heavy load.

### *2.2. The Control Circuit of the Proposed Novel Dual-Mode ACF Converter*

This section describes the control circuit of the dual-mode ACF proposed in this paper.

#### 2.2.1. Dual-Mode Switching Circuit

The gray block in Figure 4 is a dual-mode switching circuit. When the converter is operating under light load, the hysteresis comparator output ($V_C$) is low, since the output voltage of the current sensor ($V_{sense2}$) is less than the reference voltage ($V_{REF}$). Therefore, the drive signal ($V_{Q2}$) of the high-side auxiliary switch remains low. The input signal ($V_{pulse}$) of the second AND gate (AND2) determines the state of the valley trigger signal ($V_{reset}$) because the voltage signal ($V_C$) becomes high after the NOT gate. The valley trigger signal ($V_{reset}$) resets the main controller, which can precisely control the low-side main switch ($Q_1$) to achieve VS and to enable the converter to operate in the QR mode.

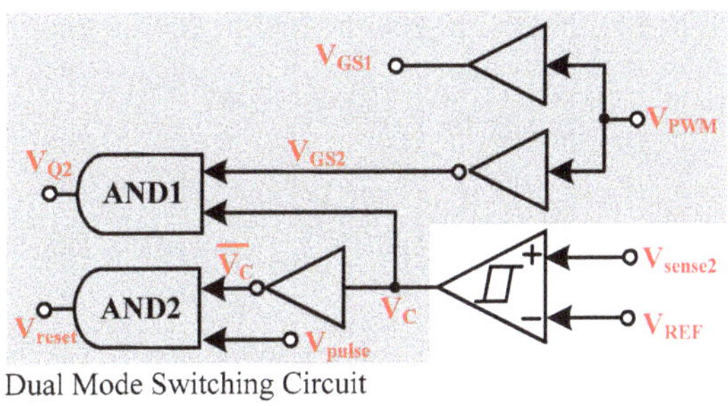


**Figure 4.** Dual-mode switching circuit structure.

When the converter is operating under heavy load, the hysteresis comparator output ($V_C$) is high; thus, the output state of the first AND gate (AND1) is determined by the voltage signal, $V_{GS2}$. At this moment, the high-side auxiliary switch begins to operate. For the second AND gate (AND2), the output voltage signal ($V_{reset}$) is low since the signal ($V_C$) is high, and the main controller is not affected by the reset signal. In this stage, the QR control circuit is disabled, enabling the transition of the converter to the active-clamp mode.

#### 2.2.2. Proposed QR Control Circuit

Before introducing the QR control technique proposed in this paper, the disadvantages of the conventional QRF converter are discussed. Conventional QR control methods rely on the phenomenon of resonance generated by the parasitic capacitance of the main switch and the magnetizing inductance of the transformer to turn on the main switch at the first resonance valley, which can reduce switching loss. However, as the output load increases, the switching frequency also increases, which leads to canceling out the advantages of VS as illustrated in Figure 5.

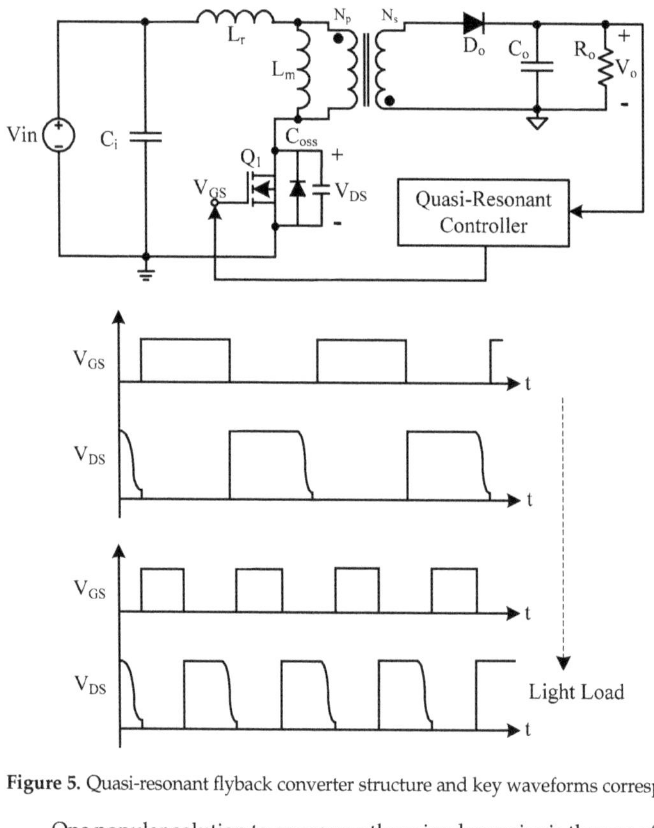


**Figure 5.** Quasi-resonant flyback converter structure and key waveforms correspond to load conditions.

One popular solution to overcome these inadequacies is the use of a frequency-limiting mechanism. In this paper, a control technology similar to that in [13] is adopted in which only pulse signals are used to adjust the maximum frequency of the oscillator. The operating principle of the proposed converter is divided into two parts: the operating principle of the ACF converter operating in the QR mode and the operating principle of the QR control circuit that limits the maximum switching frequency.

- The operating principle of the ACF converter operating in the QR mode:

The operating principle of the ACF converter in the QR mode is described as follows. This operation mode is divided into three states as shown in Figure 6, and the complete key waveforms are presented in Figure 7. The following assumptions are used to simplify the analysis:

1. The main switch ($Q_1$) and the auxiliary switch ($Q_2$) have no leakage current and forward resistance;
2. The leakage inductance ($L_r$) is negligible;
3. The parasitic capacitances, $C_{oss1}$ and $C_{oss2}$, are much smaller than $C_{clamp}$;
4. Non-ideal characteristics, such as the forward voltage drop and the resistance effect of the secondary-side rectifier diode, are ignored;
5. The output capacitance is large enough to be considered as a constant voltage source;
6. The turn ratio $n = N_p/N_s$.

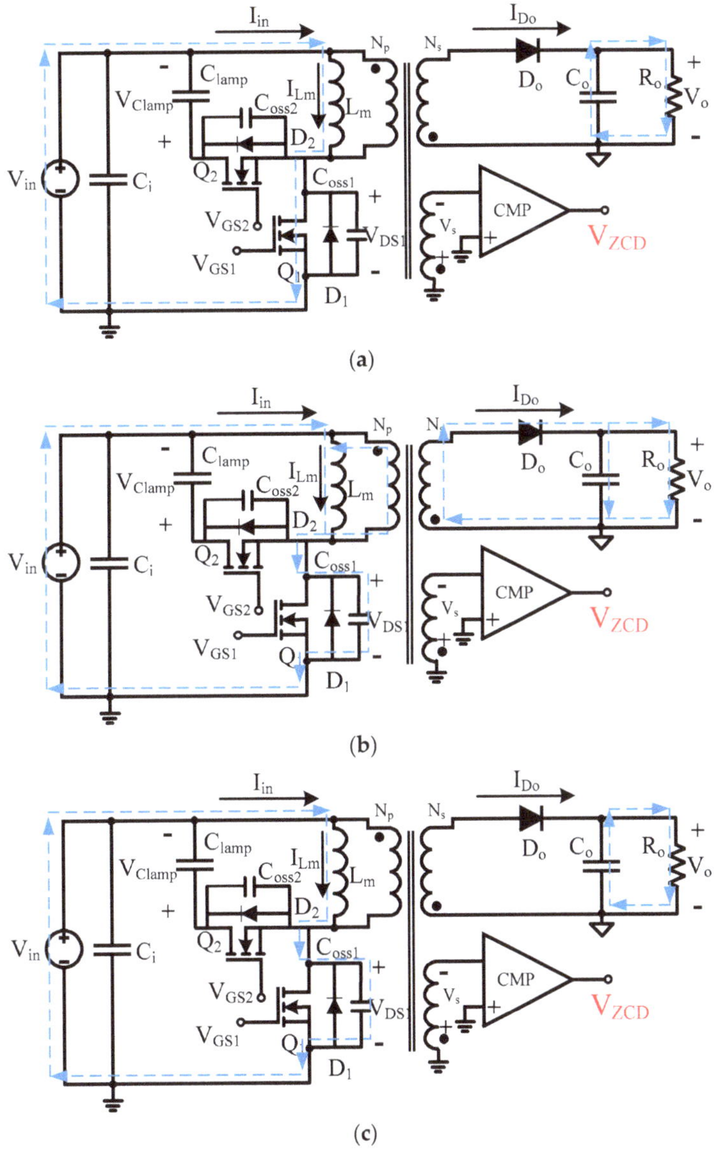


**Figure 6.** (**a**–**c**) Three operating states of the ACF converter in the QR mode.

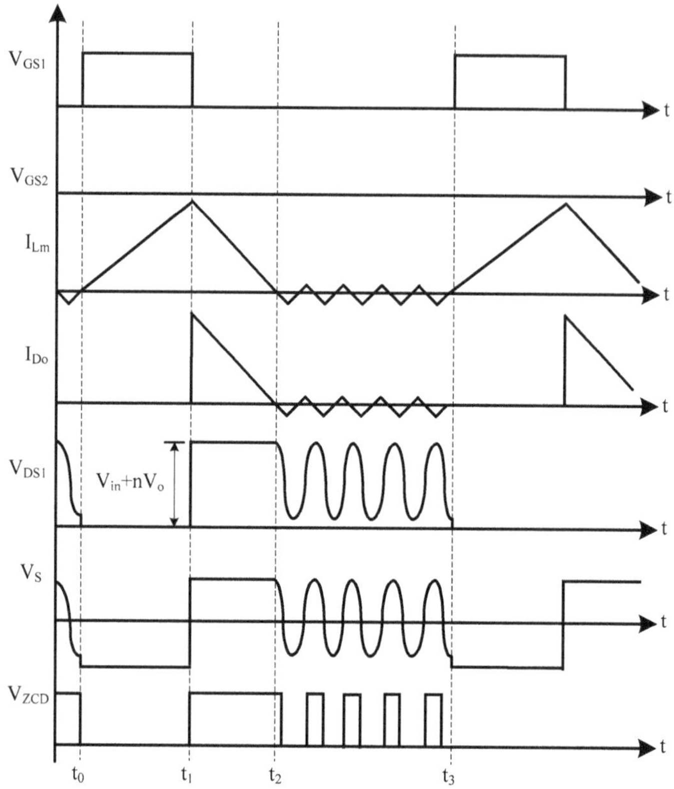


**Figure 7.** Key waveforms of the ACF converter operating in the QR Mode.

State 1 ($t_0 \leq t < t_1$):

When $t = t_0$, the main switch ($Q_1$) is turned on, and the auxiliary switch ($Q_2$) is turned off. The input source begins to supply energy to the transformer, and the magnetizing inductor current ($I_{Lm}$) increases linearly. Since the primary and secondary sides of the transformer have opposite polarity, the output rectifier diode ($D_o$) is reverse-biased. The output energy is supplied by the output filter capacitor ($C_o$). The detailed current path is presented in Figure 6a. For the auxiliary winding on the secondary side of the transformer, the voltage $V_{ZCD}$ remains low because the voltage VS is less than zero.

State 2 ($t_1 \leq t < t_2$):

When $t = t_1$, the main switch ($Q_1$) and the auxiliary switch ($Q_2$) are turned off. At this time, the magnetizing inductance ($L_m$) then begins to charge the parasitic capacitor ($C_{oss1}$) of the main switch ($Q_1$). As a result, the drain-source voltage ($V_{DS}$) of $Q_1$ on the primary side of the transformer increases rapidly. When $V_{DS1} \geq V_{in} + nV_o$, the diode, $D_o$, conducts. The energy stored in the transformer is then transferred to the output capacitor and the output load. The detailed current path is shown in Figure 6b. For the auxiliary winding on the secondary side of the transformer, the voltage $V_{ZCD}$ changes from low to high due to the transient transition of $V_s$.

State 3 ($t_2 \leq t < t_3$):

When $t = t_2$, the energy stored in the transformer has been completely transferred to the secondary side, and the current $I_{Do}$ decreases to zero. The converter is thus operating in the discontinuous conduction mode (DCM). The magnetizing inductor ($L_m$) and the parasitic capacitor ($C_{oss1}$) begin to resonate. The detailed current path is shown in Figure 6c. In the QR mode, when the voltage drops to the valley, the main switch ($Q_1$) turns on to reduce switching loss. At this time, the output energy is provided to the load by the output filter capacitor ($C_o$). The auxiliary winding on the secondary side of the transformer is in the resonant state due to the voltage $V_s$, resulting in a square wave signal at the output of the comparator with the same resonant frequency.

- The operating principle of the QR control circuit with the maximum switching frequency-limiting function:

  This section explains the operating principle of the QR control circuit with the maximum switching frequency-limiting function. Figure 8 depicts the QR control block with the maximum switching frequency-limiting function and related key waveforms. Notably, the circuit only operates when the ACF converter is in the resonant state. When voltage $V_{DS1}$ resonates through point 0 at $t_0$, the voltage $V_{ZCD}$ transiently changes from high to low. After being processed by the RC delay circuit, the voltage $V_{ZCD1}$ will be delayed for a period $\Delta t$. The falling edge of this square wave may coincide with the valley of the switching voltage $V_{DS}$. Therefore, the pulse generator and the gain controller can generate an adjustable pulse signal $V_{pulse1}$ at time $t_1$. To control the maximum switching frequency limit, the original oscillator signal ($V_{OSC}$) must be obtained from the current mode controller. After processing by the squarer, this signal becomes a triangular signal at the same frequency, $V_{OSC1}$. This method is used to accentuate the peak voltage of $V_{OSC}$. The multiplier is the main focus of this control strategy. The input voltages of the multiplier are $V_{OSC1}$ and $V_{pulse1}$; thus, the peak value of the output voltage $V_{pulse2}$ gradually increases. Finally, the voltage $V_{sum}$ is generated by adding $V_{OSC1}$ to $V_{pulse2}$. As the number of resonances increases, the peak value of $V_{sum}$ also increases. When the voltage value of $V_{sum}$ is higher than the reference voltage $V_{TH}$ of comparator CMP1 at $t_2$, the output signal $V_{reset}$ of CMP1 triggers the current mode controller to reset the oscillator. This technique enables the converter to achieve maximum switching frequency limiting. Thus, controlling the magnitude of the reference voltage $V_{TH}$ determines the valley point of the converter.

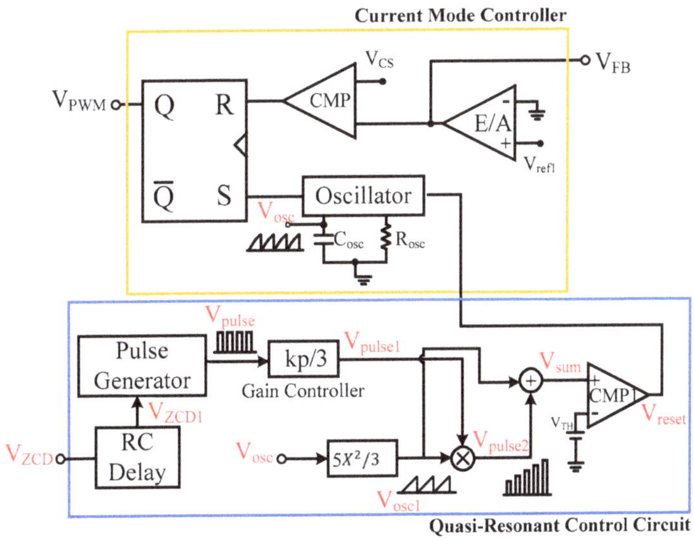


(a)

**Figure 8.** *Cont.*

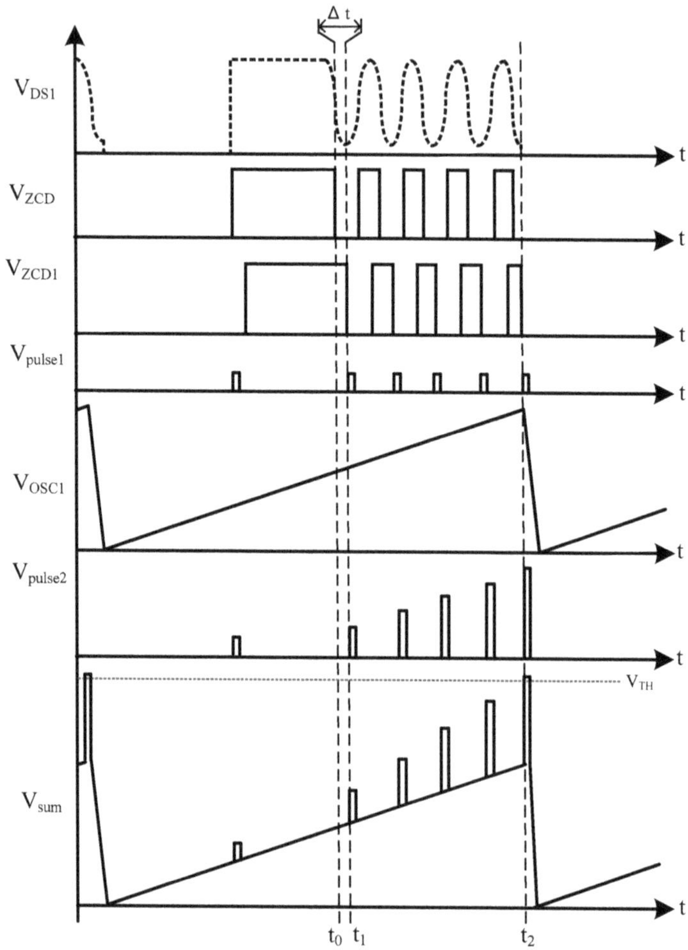


(b)

**Figure 8.** (a) QR control block with the maximum switching frequency-limiting function and (b) key waveform sequence diagram.

## 3. Design of the Proposed Novel Dual-Mode ACF Converter

The converter structure and operating principles have been described in detail in the previous section. This section describes the design of the proposed dual-mode ACF converter. A 65 W power supply is used in the following example calculations and the description of the component design of the converter. The specifications of the prototype converter are listed in Table 1.

In the following, the design procedure of the converter is briefly presented.

Step 1:

Given that the maximum output power $P_{o,max}$ = 65 W and the efficiency $\eta$ = 0.9, the magnetizing inductance ($L_m$) of the transformer can be determined [15]:

$$L_m = \frac{\eta \times (V_{in,min} \times D_{max})^2}{2 \times f_{sw} \times P_{o,max}} = 409\ \mu\text{H} \quad (1)$$

**Table 1.** Specifications of the proposed ACF converter.

| Parameters | Value |
|---|---|
| Input voltage, $V_{in}$ | 155 V |
| Maximum output power, $P_{o,max}$ | 65 W |
| Output voltage, $V_{out}$ | 19 V |
| Output current, $I_{out}$ | 3.42 A |
| Switching frequency, $F_{sw}$ | 65 kHz |
| Maximum duty cycle, $D_{max}$ | 0.4 |
| Efficiency, $\eta$ | 90% |
| Maximum magnetic flux density, $B_{max}$ | 2000 G |

Step 2:

After the magnetizing inductance ($L_m$) of the transformer is known, the peak current of the transformer primary side winding can be calculated:

$$I_{pp} = \frac{P_{o,max}}{\eta \times V_{in} \times D_{max}} + \frac{V_{in,min} \times D_{max}}{2 \times L_m \times f_{sw}} \cong 2.357 \text{ A} \tag{2}$$

Step 3:

The maximum output power ($P_{o,max}$) of the converter is 65 W, and the switching frequency is 65 kHz. In this paper, the PQ26/20 iron core produced by TDK (Tokyo, Japan) was used to calculate the primary winding turns of the transformer. Based on the actual design considerations, the magnetizing inductance ($L_m$) is 400 μH. From the core datasheet, the effective cross-sectional area of the core ($A_e$) is 1.19 cm, and the maximum magnetic flux density ($B_{max}$) is 2000 G. The primary winding turns of the transformer ($N_p$) can be calculated as follows:

$$N_p = \frac{L_m \times I_{pp}}{A_e \times B_{max}} \times 10^8 \cong 39 \tag{3}$$

Step 4:

The secondary winding turns ($N_s$) can also be calculated:

$$N_s = \frac{V_o \times (1 - D_{max}) \times N_p}{V_{in} \times D_{max}} \cong 7 \tag{4}$$

Step 5:

The output diode must be selected considering the maximum output current and the withstand voltage. The withstand voltage and withstand current of the diode are calculated as follows:

$$V_{D,max} = \frac{V_{in,max}}{\frac{N_p}{N_s}} + V_o \cong 46.82 \text{ V} \tag{5}$$

$$I_{sec,peak} = I_{L,peak} \times n \cong 13.22 \text{ A} \tag{6}$$

$$I_{sec,rms} = I_{sec,peak}\sqrt{\frac{1 - D_{max}}{3}} \cong 5.912 \text{ A} \tag{7}$$

In accordance with the results of Equations (5)–(7), the ultrafast rectifier diode BYV32E-150 produced by NXP Semiconductors (Eindhoven, Netherlands) was selected as the output rectifier diode with a peak reverse voltage of 150 V and a peak forward current of 20 A.

Step 6:

To achieve ZVS for the ACF converter power switch, enough energy must be stored in the resonant inductance ($L_r$) to completely discharge the parasitic capacitance ($C_r$) of the power switch; the calculations of this energy are as follows:

$$E_{Lr} \geq E_{Cr} \tag{8}$$

$$(I_{pp})^2 \times L_r \geq C_r \times (V_{in} + n \times V_o)^2 \tag{9}$$

where $E_{Lr}$ and $E_{Cr}$ are resonant inductance energy and parasitic capacitance energy, respectively.

Step 7:

After the resonant inductance has been determined, the appropriate value of the clamping capacitor ($C_{clamp}$) can be calculated as follows:

$$C_{clamp} = \frac{(1 - D_{max})^2}{\pi^2 \times L_r \times f_{sw}} \tag{10}$$

## 4. Experimental Results

In this experiment, the control IC UC3843 was used as the main controller. The proposed dual-mode ACF converter had the following specifications: the input voltage was 155 VDC, the output voltage was 19 VDC, and the maximum output power was 65 W. The experimental data included the measured critical switching waveforms, control signal waveforms, dual-mode switching waveforms, ZVS waveforms, and other key parameters. Finally, the comparison of efficiency between the conventional ACF converter and the proposed dual-mode ACF converter is presented in this section. The slope shape of the oscillator waveform is critical for the control technique proposed in this paper, because more precise slope changes result in more accurate VS. Generally, the traditional oscillator is an RC circuit; thus, the slightly curved slope causes the failure of frequency-limited VS. To overcome this challenge, a squarer was used to convert the oscillator waveform into a triangular waveform with a linear slope, improving the accuracy of VS. The relevant waveform is presented in Figure 9.

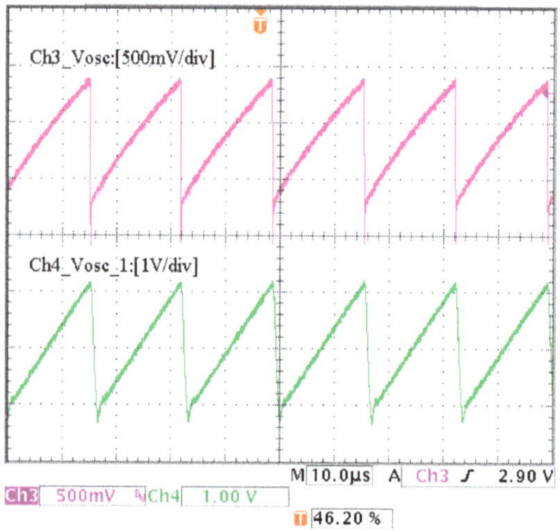


**Figure 9.** Oscillator waveform with (Ch4) and without (Ch3) the frequency-limiting control mechanism.

Figure 10a shows that when the voltage $V_{DS1}$ voltage starts to resonate generating a valley, a voltage pulse signal is generated. Through the multiplier, a linearly rising voltage pulse signal $V_{pulse2}$ is then generated. Next, the oscillating voltage signal $V_{OSC1}$ and the voltage signal $V_{pulse2}$ are added by the adder to generate the voltage signal $V_{sum}$. Finally, the voltage signal $V_{sum}$ is compared with the reference voltage $V_{TH}$ to generate a reset voltage signal $V_{reset}$, which restarts the control IC UC3843 to complete the frequency-limiting VS as shown in Figure 10b. In other words, the voltage level of the reference voltage $V_{TH}$ determines the VS point, and higher $V_{TH}$ values cause a slower system switching frequency. Conversely, lower $V_{TH}$ causes a faster system switching frequency.

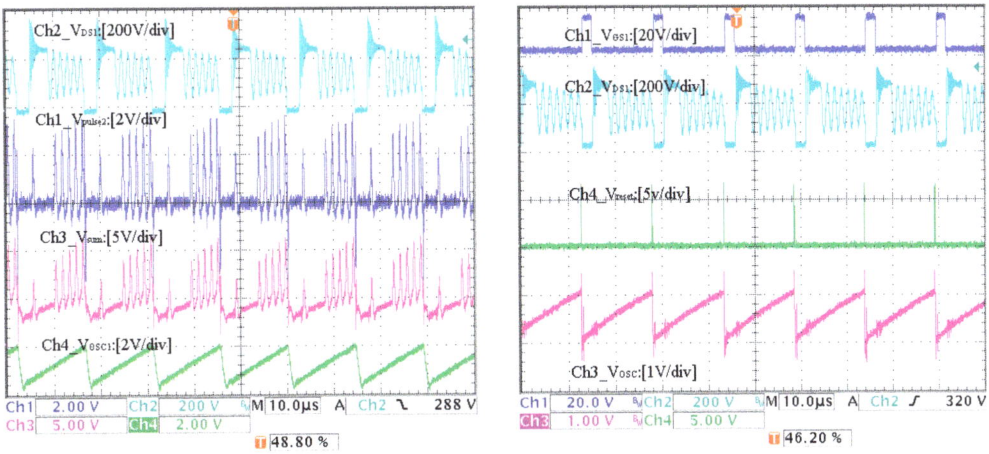


(**a**) (**b**)

**Figure 10.** (**a**) Power switch voltage waveform and its generated signal waveforms; (**b**) key control signal waveforms of the proposed ACF converter in QR mode.

Figure 11 presents the key voltage waveforms of the converter operating at different output powers. If the converter operates was at an output power of 6.5 W, the switching frequency of the converter was approximately 70 kHz as presented in Figure 11a. For an output power of 19.5 W, the switching frequency of the converter decreased to approximately 68 kHz as presented in Figure 11b. Although the output power of the converter varied, the switching frequency of the circuit can be limited within a fixed range by the frequency-limiting QR control function of the QR control circuit proposed in this paper. If the output power increased above 19.5 W, the converter changed the operation mode from the QR to the active-clamp mode as presented in Figure 12. Figure 13 shows the ZVS waveforms of the power switches of the proposed converter. The figure shows that before the main switch ($Q_1$) was turned on, the drain-source voltage of the main switch $V_{DS1}$ decreased to zero. The same principle was applied to the auxiliary switch ($Q_2$); thus, both power switches achieved ZVS.

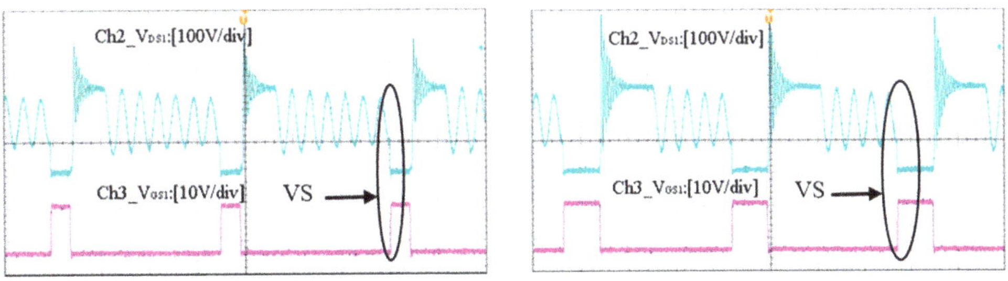


(**a**) (**b**)

**Figure 11.** Key voltage waveforms of the proposed ACF converter operating at (**a**) 6.5 W and (**b**) 19.5 W. The converter is in the QR mode.

Figure 14 presents the key waveforms for switching between the two modes of the converter according to the change in output power when the input voltage $V_{in}$ = 155 VDC. If the output power was 6.5 W, the driving signal $V_{GS2}$ of the auxiliary switch ($Q_2$) was off, and the driving signal $V_{GS1}$ of the main switch ($Q_1$) adjusted the pulse width modulation in accordance with the load. At this time, the converter operated in the QR mode. If the output power increased from 6.5 to 65 W, the drive signal $V_{GS2}$ of the auxiliary switch ($Q_2$) entered a working state, and the main switch ($Q_1$) and the auxiliary switch ($Q_2$) operated in a complementary manner to enable the converter to operate in the active-clamp mode as presented in Figure 14a. By contrast, if the output power decreased from 65 to 6.5 W, the driving signal $V_{GS2}$ of the auxiliary switch ($Q_2$) changed from the working state to the off state, and the operation mode of the converter changed from the active-clamp mode to the QR mode as shown in Figure 14b. These experimental data demonstrated that the converter proposed in this paper can switch between the two modes.

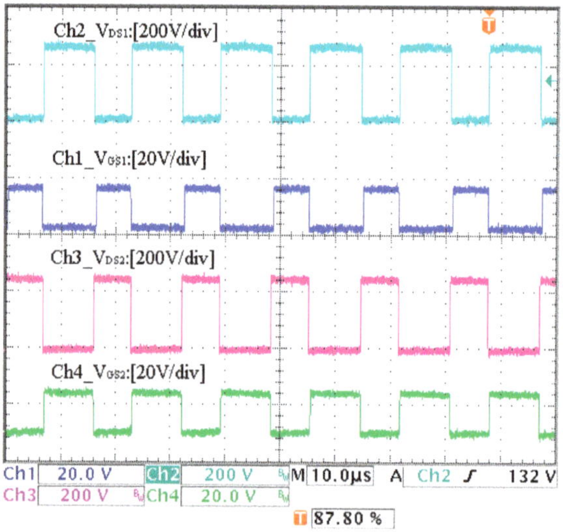


**Figure 12.** Waveforms of the main switch and auxiliary switch of the proposed ACF converter under medium- to full-load conditions (above 19.5 W). The converter is in the active-clamp mode.

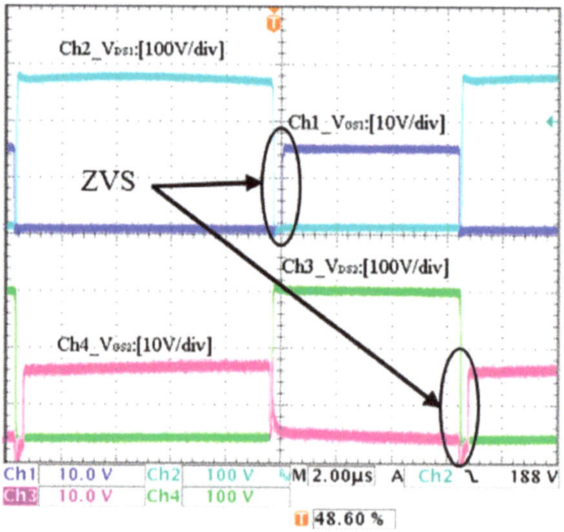


**Figure 13.** ZVS waveforms of the power switches of the proposed ACF converter in the active-clamp mode.

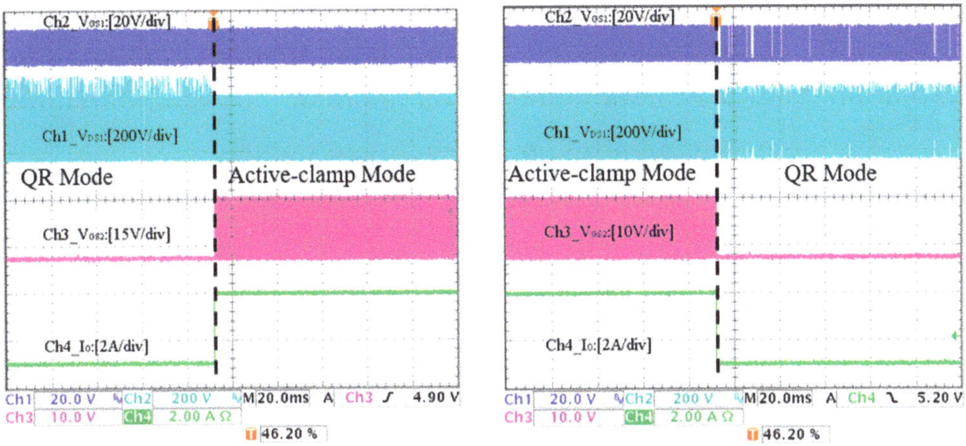


(a) (b)

**Figure 14.** Key waveforms of dual-mode switching: (**a**) QR mode to the active-clamp mode; (**b**) active-clamp mode to the QR mode.

Table 2 presents the efficiency measurement data for the dual-mode ACF converter. In the table, input current, output current, output voltage, and efficiency are presented under different load conditions with a voltage of 155 VDC as the input power supply. Figure 15 provides a comparison of the efficiency curves of the conventional ACF converter and the dual-mode ACF converter at different output power levels. The curve comparison diagram reveals that for an output power of 3.5 W, the efficiency difference between the two converters is at a maximum of 11%. If the output load increases, conduction losses gradually increase; thus, the efficiency difference between the two converters decreases. Table 2 and Figure 15 show that the control technique proposed in this paper has good performance under light load conditions. In addition, Table 3 presents the load regulation measurement data. Based on the measurement results, it can be seen that the minimum load regulation value is 1.09% at the output power of 65 W (i.e., full load).

**Table 2.** Efficiency of the proposed dual-mode ACF converter.

| Input Voltage, $V_{in}$ | Input Current, $I_{in}$ | Output Voltage, $V_o$ | Output Current, $I_o$ | Efficiency, $\eta$ |
|---|---|---|---|---|
| 155 V | 0.030 A | 19.3 V | 0.19 A | 79% |
| 155 V | 0.052 A | 19.3 V | 0.34 A | 82.5% |
| 155 V | 0.075 A | 19.3 V | 0.51 A | 84.6% |
| 155 V | 0.098 A | 19.3 V | 0.68 A | 85.7% |
| 155 V | 0.119 A | 19.3 V | 0.84 A | 87.5% |
| 155 V | 0.229 A | 19.2 V | 1.67 A | 90.3% |
| 155 V | 0.349 A | 19.2 V | 2.55 A | 90.6% |
| 155 V | 0.464 A | 19.2 V | 3.41 A | 90.7% |

**Table 3.** Load regulation measurement data of the proposed dual-mode ACF converter.

| Output Power, $P_o$ | Output Voltage, $V_o$ | Load Regulation, $LR$ |
|---|---|---|
| 3.5 W | 19.289 V | 1.52% |
| 6.5 W | 19.303 V | 1.59% |
| 13.0 W | 19.304 V | 1.60% |
| 19.5 W | 19.305 V | 1.61% |
| 26.0 W | 19.293 V | 1.54% |
| 32.5 W | 19.258 V | 1.36% |

**Table 3.** *Cont.*

| Output Power, $P_o$ | Output Voltage, $V_o$ | Load Regulation, *LR* |
|---|---|---|
| 39.0 W | 19.236 V | 1.24% |
| 45.5 W | 19.226 V | 1.19% |
| 52.0 W | 19.227 V | 1.19% |
| 58.5 W | 19.222 V | 1.17% |
| 65.0 W | 19.207 V | 1.09% |

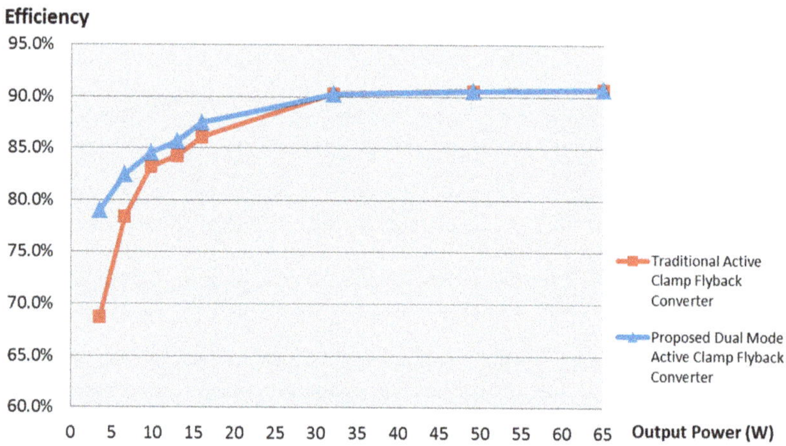


**Figure 15.** Efficiency curves of the conventional ACF converter and dual-mode ACF converters at different output powers.

## 5. Conclusions

This paper presented a new dual-mode switching control strategy for the ACF converter working in DCM, which can switch to the QR mode at light loads to improve overall efficiency. This paper proposed a new dual-mode switching control strategy for an ACF converter operating in DCM to enable operation in the QR mode under light load to optimize the overall operation. The control techniques and operating principles were described in detail. The operating principles and the results of an analysis of the control techniques were carefully discussed. A prototype of the 65 W dual-mode ACF converter was developed, and experiments provided satisfactory results. Experimental results revealed that the auxiliary switch of the resonant tank closed when the dual-mode ACF converter was under light load, which enabled the main switch to perform VS and achieve QR mode control. On the contrary, the converter operated in the active-clamp mode under medium and heavy load conditions, thereby achieving ZVS for both power switches to maintain the high efficiency of conventional ACF. In other words, the proposed dual-mode ACF converter uses QR mode VS characteristics to improve the poor efficiency of conventional ACF under light load. However, the switching frequency of the QR mode is critical to determining the light load efficiency. In general, it is recommended to design the QR mode switching frequency as close to the active-clamp mode switching frequency as possible, because excessively high switching frequencies increase switching losses at light loads. This proposed dual-mode control ACF topology is suitable for applications with relatively low power consumption (under 200 W).

**Author Contributions:** Conceptualization, J.-M.W. and T.N.T.T.; methodology, H.-C.W. and J.-M.W.; validation, T.N.T.T., H.-C.W. and J.-M.W.; resources, J.-M.W.; writing—original draft preparation, J.-M.W. and T.N.T.T.; writing—review and editing, J.-M.W. and T.N.T.T. All authors have read and agreed to the published version of the manuscript.

**Funding:** This research was funded by the Ministry of Science and Technology, R.O.C. (grant number: MOST 110-2622-E-150-002).

**Data Availability Statement:** Not applicable.

**Conflicts of Interest:** The authors declare no conflict of interest.

## References

1. Watson, R.; Hua, G.C.; Lee, F.C. Characterization of an active clamp flyback topology for power factor correction applications. *IEEE Trans. Power Electron.* **1996**, *11*, 191–198. [CrossRef]
2. Watson, R.; Lee, F.C.; Hua, G.C. Utilization of an active-clamp circuit to achieve soft switching in flyback converters. *IEEE Trans. Power Electron.* **1996**, *11*, 162–169. [CrossRef]
3. Papanikolaou, N.P.; Tatakis, E.C. Active voltage clamp in flyback converters operating in CCM mode under wide load variation. *IEEE Trans. Power Electron.* **2004**, *51*, 632–640. [CrossRef]
4. Liu, P.-H. Design consideration of active clamp flyback converter with highly nonlinear junction capacitance. In Proceedings of the 2018 IEEE Applied Power Electronics Conference and Exposition (APEC), San Antonio, TX, USA, 4–8 March 2018; pp. 783–790.
5. Wu, S.-T.; Cheng, Y.-T. Design and Implementation of a Single-Stage PFC Active-Clamp Flyback Converter with Dual Transformers. *Electronics* **2021**, *10*, 2588. [CrossRef]
6. Leng, C.-M.; Chiu, H.J. Three-Output Flyback Converter with Synchronous Rectification for Improving Cross-Regulation and Efficiency. *Electronics* **2021**, *10*, 430. [CrossRef]
7. Medina-Garcia, A.; Schlenk, M.; Morales, D.P.; Rodríguez, N. Resonant Hybrid Flyback, a New Topology for High Density Power Adaptors. *Electronics* **2018**, *7*, 363. [CrossRef]
8. Jung, J.; Ahmed, S. Flyback converter with novel active clamp control and secondary side post regulator for low standby power consumption under high-efficiency operation. *IET Power Electron.* **2011**, *4*, 1058–1067. [CrossRef]
9. Kim, J.h.; Ryu, M.-H.; Min, B.d.; Song, E.-H. A Method to Reduce Power Consumption of Active-Clamped Flyback Converter at No-Load Condition. In Proceedings of the IECON 2006—32nd Annual Conference on IEEE Industrial Electronics, Paris, France, 6–10 November 2006; pp. 2811–2814.
10. Wang, C.; Xu, S.; Shen, W.D.; Lu, S.L.; Sun, W.F. A Single-Switched High-Switching-Frequency Quasi-Resonant Flyback Converter. *IEEE Trans. Power Electron.* **2019**, *34*, 8775–8786. [CrossRef]
11. Li, J.T.; van Horck, F.B.M.; Daniel, B.J.; Bergveld, H.J. A High-Switching-Frequency Flyback Converter in Resonant Mode. *IEEE Trans. Power Electron.* **2017**, *32*, 8582–8592. [CrossRef]
12. Park, H.P.; Jung, J.H. Design Methodology of Quasi-Resonant Flyback Converter With a Divided Resonant Capacitor. *IEEE Trans. Power Electron.* **2021**, *68*, 10796–10805. [CrossRef]
13. Wang, J.; Lin, C.-W.; Huang, K.-Y.; Wong, J.-S. The Novel Quasi-Resonant Flyback Converter With Autoregulated Structure for Parallel/Serial Input. *IEEE Trans. Power Electron.* **2020**, *67*, 992–1004. [CrossRef]
14. Jeng, S.-L.; Peng, M.T.L.; Hsu, C.Y.; Chieng, W.-H.; Shu, J.P.H. Quasi-resonant flyback DC/DC converter using GaN power transistors. *World Electr. Veh. J.* **2012**, *5*, 567–573. [CrossRef]
15. Pressman, A.I. *Switching Power Supply Design*; McGraw-Hill: New York, NY, USA, 1998.

*electronics*

*Article*

# Motor Dynamic Loading and Comprehensive Test System Based on FPGA and MCU

**Chunxiang Zhu [1], Linxin Bao [1], Bowen Zheng [2], Jiacheng Qian [1], Yongdong Cai [1] and Binrui Wang [3,*]**

1 Engineering Training Center, China Jiliang University, Hangzhou 310018, China; 16a2900069@cjlu.edu.cn (C.Z.); 1900201509@cjlu.edu.cn (L.B.); 2000303206@cjlu.edu.cn (J.Q.); 1800304230@cjlu.edu.cn (Y.C.)
2 School of Electronic Information, Hangzhou Dianzi University, Hangzhou 310018, China; bwzheng@hdu.edu.cn
3 School of Mechanical and Electrical Engineering, Hangzhou Dianzi University, Hangzhou 310018, China
* Correspondence: wangbrpaper@163.com

**Abstract:** In view of the problem that the traditional motor test system cannot directly test the transient parameters of the motor and the dynamic arbitrary load loading requirements during motor loading, as well as the high cost of implementation, this research uses STM32+FPGA as the core to form the main control of the motor test system unit, combining the superior control performance of the ARM processor and the high-speed data processing advantages of FPGA. FPGA and STM32 are controlled by the FSMC bus communication and data ping-pong algorithm. Using this method, a small-size control core board in the motor test system is manufactured. It can be embedded in the existing traditional dynamometer system to improve the dynamometer transient parameter test and the dynamic motor loading performance. The experimental results show that the system can basically meet the requirements of the motor transient test and dynamic loading, and can achieve the fastest data refresh rate of 1 ms when measuring the motor's speed and torque, as well as arbitrary waveform loading within a 100 M sampling frequency, with a loading error of 0.8%. It satisfies the motor transient test and dynamic loading requirements.

**Keywords:** motor transient test; dynamic loading; STM32+FPGA; FSMC; ping-pong algorithm

**Citation:** Zhu, C.; Bao, L.; Zheng, B.; Qian, J.; Cai, Y.; Wang, B. Motor Dynamic Loading and Comprehensive Test System Based on FPGA and MCU. *Electronics* **2022**, *11*, 1317. https://doi.org/10.3390/electronics11091317

Academic Editor: Ching-Ming Lai

Received: 22 March 2022
Accepted: 17 April 2022
Published: 21 April 2022



## 1. Introduction

In recent years, the application scope of motors has been expanding, and the universality of motor operating conditions determines the complexity of the motor experiment. Traditional testing devices and methods are time-consuming, use a wide range of instruments, and have a low precision, resulting in low intelligence and low efficiency, which greatly affect the accuracy and quality of testing. Most motor tests use a dynamometer as a loading instrument, connected to the motor being tested, and then simultaneously interpret different motors as required. The dynamometer is controlled by the central control computer, and then different test items are analyzed by the power analyzer. A dynamometer is one of the important devices for performance and equipment testing in motor manufacturing and product R&D. It can be used to simulate and control the load of the tested motor so as to measure the torque, speed, current, voltage, power, efficiency, and other parameters of the motor, as well as for other special dynamic test items, such as safety test, dynamic balance test, NVH test (vibration/noise/life), etc. [1–6]. With the development of modern technology, in addition to conventional items such as motor speed, torque, voltage, and current, the requirements for the motor transient performance test are also increased [7–10]. In the traditional dynamometer system, the loads can only be loaded one by one, but the motor performance parameters are measured for a certain load point, which cannot meet the transient performance test of the motor. The loading accuracy and testing accuracy cannot meet the modern requirements. Therefore, enterprises have put forward the demand for high-performance loading and testing equipment. The equipment needs to meet the high bandwidth and high-precision loading and measurement of a variety of different waveforms, including step signal, square wave, trapezoidal wave, and sine wave. The measurement error should not be higher than 0.5%, and the loading error should not be higher than 0.2% [11–13]. According to the above analysis, the key technical points of the transient performance test of the motor and driver mainly include the following two aspects: First, the loading function of any load curve, that is, it can provide step, sine wave, square wave, sawtooth wave, and even arbitrary waveform loading of torque or speed [14–17]. The second is parameter transient waveform measurement, including the parameter transient waveform measurement of torque, speed, voltage, current, efficiency, and speed torque curve. According to this technical point, in order to meet the high-speed loading and high-speed processing at the same time, this paper innovatively designs a dual core dynamometer loading and testing integrated control system with FPGA and MCU communicating through FSMC. Regarding the industrial application requirements of the system, the corresponding software and hardware functions are designed to meet the requirements of various waveform loadings and measurements of different motor platforms, as well as the requirements of high-speed communication and responses with industrial computers. Therefore, this paper implements a motor dynamic loading and transient parameter testing device. In the process of motor testing, by loading any load curve and testing the transient performance of the motor, it can meet the use requirements of dynamometer equipment embedded in motor testing enterprises [18–20].

## 2. Composition Principle of the Motor Test System

High-speed measurements and dynamic loadings of the motor transient parameters are realized, as shown in the structural block diagram shown in Figure 1. The dynamometer is coaxially connected to the tested motor. Various loads are loaded on the tested motor through the dynamometer control system so as to simulate the torque, speed, current, voltage, power, efficiency, and other parameters of the motor in actual operation, as well as other special dynamic test items, such as the safety test, dynamic balance test, and NVH (vibration and noise life) test. When loading, the upper computer sends instructions to control the free loading engine to output the control signal, which is input to the dynamometer drive controller. After the power of the controller is amplified, the control signal is loaded on the load motor so as to drive the load motor in order to load the tested motor. At the same time, the torque−speed transducer feeds back the speed and torque of the measured motor to the free loading engine. The free loading engine adjusts the output of the load in real time according to the feedback value to realize constant speed loading or constant torque loading. The torque calculation formula is as follows:

$$T_P = N(f - f_0)/(f_P - f_0) \tag{1}$$

$$T_r = N(f_0 - f)/(f_0 - f_r) \tag{2}$$

where $T_p$ is the forward torque, $T_r$ is the reverse torque, $N$ is the torque full scale, $f_0$ is the torque zero output frequency value (KHz), $f_p$ is the forward full-scale output frequency value (KHz), $f_r$ is the reverse full scale output frequency value (KHz), and $f$ is the measured torque output frequency value (KHz). The speed calculation formula is as follows:

$$S = 60 \times f/Z \tag{3}$$

where $S$ is the rotational speed (rpm), $f$ is the measured rotational speed output frequency (Hz), and $Z$ is the number of teeth (lines) of the speed-measuring code disk of the sensor.

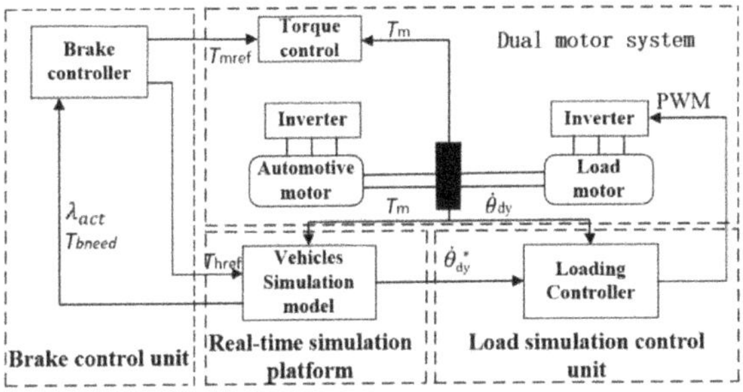


**Figure 1.** System structure block diagram.

## 3. System Hardware Design

### *3.1. Overall Structure*

The circuit structure of the motor test system based on STM32 and FPGA is shown in Figure 2. The system is composed of a user-defined signal output and a speed and torque acquisition circuit. Firstly, STM32 receives the control signals, such as the loading waveform type and waveform parameters, sent by the PC ethernet port, which generates the corresponding analog output or digital output, and obtains the output control signal through FMSC communication, FPGA analysis, magnetic coupling isolation circuit, and then the conversion of the corresponding functional modules. This signal is input to the tested motor or load motor driver to drive the corresponding motor. Secondly, when the real-time data waveform needs to be uploaded, STM32 receives the acquisition data command sent by the PC ethernet. After the speed torque feedback signal is sampled by the magnetic coupling isolation, level conversion circuit, and ADC, it is stored in the on-chip dual port RAM of FPGA, and then STM32 reads the data in the way of a ping-pong operation through the FMSC interface and sends it to the upper computer through the ethernet. The composition and functions of each part of the circuit are introduced below. The components are described below.

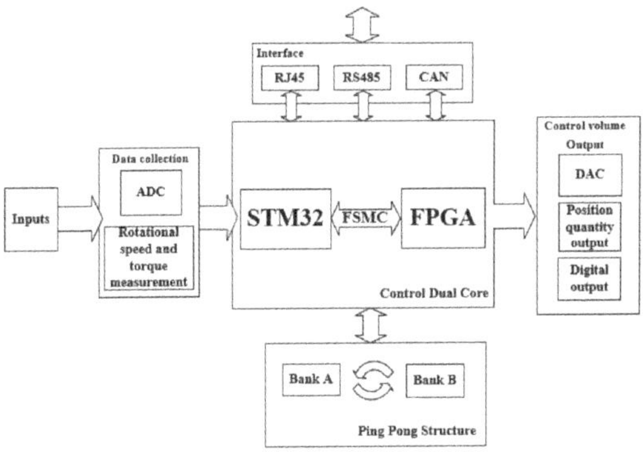


**Figure 2.** Hardware block diagram.

### *3.2. Circuit Design of STM32 Module*

The minimum system circuit structure of the STM32 module is shown in Figure 3. The STM32 system selects STM32H743ZI2, with ARM-Cortex M7 as the core as the main control chip. The STM32H743ZI2 chip is a 32-bit microprocessor. It not only has rich peripheral resources, but also has a fast data processing speed and low power consumption. It is very suitable for occasions with high data speed and high system response requirements. According to the functional requirements of the project, circuits including the ethernet communication circuit, digital output circuit (DO), low-speed ADC circuit, display key circuit, SDRAM data cache, and so on, are designed. The ethernet communication circuit is mainly responsible for the data transmission between the upper computer and the lower computer. The digital output circuit mainly realizes the digital output control relay to pull in or open. The low speed ADC is mainly used to collect the working environment, such as the external temperature/humidity. The display key circuit realizes the human−computer interaction function. The SDRAM memory circuit mainly realizes the high-speed data cache.

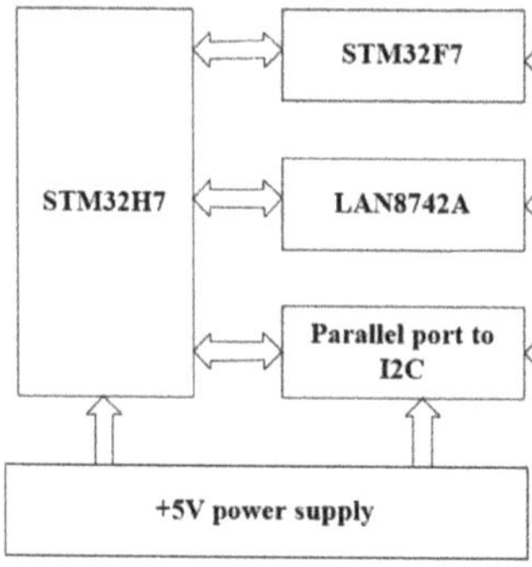


**Figure 3.** STM32 module minimum system circuit block diagram.

### 3.3. FPGA Module Circuit Design

The main control chip used in FPGA is the fourth generation EP4CE40F23C8N, which has rich logic resources and about 40,000 gates. There are about 532 user IO ports, so it is very suitable for multi-channel data acquisition and synchronous processing. The FPGA main control core circuit mainly includes a 10M magnetic coupling isolation circuit, pulse output circuit (CO), speed/torque input circuit (CI), high-speed DAC analog output circuit, two-stage operational amplifier circuit, high-speed ADC analog acquisition circuit, low-pass filter circuit, and DDR3 data buffer circuit. It mainly completes the data ping-pong operation, analog quantity acquisition and output, speed and torque measurements, receiving control commands, outputting pulse signals, receiving encoder signals, and other functions of motor testing.

The structural diagram is shown in Figure 4.

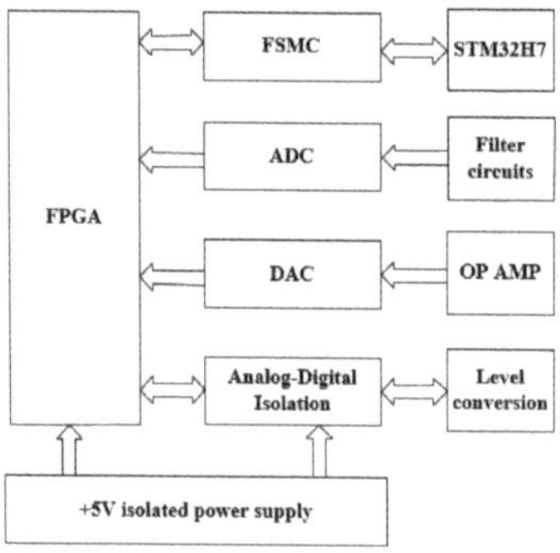


**Figure 4.** FPGA core circuit block diagram.

### 3.4. FSMC Communication Design between STM32 and FPGA

FSMC, a flexible static storage controller, can be connected with synchronous or asynchronous memory and a 16-bit PC memory card. The FMSC interface of STM32H7 supports memory such as SRAM, NAND FLASH, NOR FLASH, and PSRAM. The communication between STM32H7 and FPGA adopts a parallel FSMC bus. The collected data are saved by constructing a dual port RAM built in FPGA, and STM32H7 accesses it through the FMSC bus. According to the requirements of the collected data volume, the FMSC bus adopts a 24-bit address line and 16-bit data line, and contains six control signals at the same time. The STM32 chip selects FPGA to read and write data. The structural block diagram is shown in Figure 5.

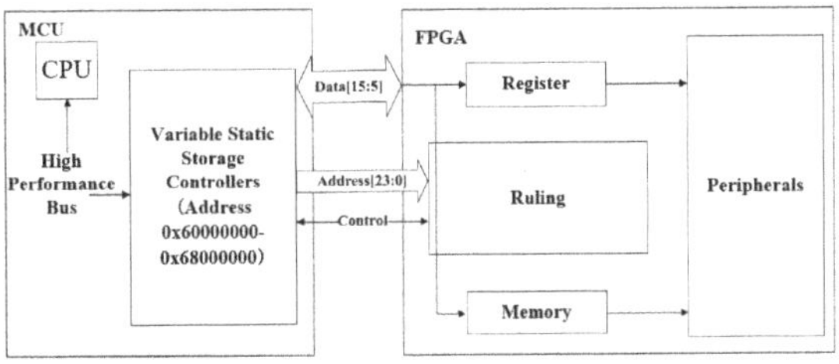


**Figure 5.** FMSC communication structure block diagram.

## 4. System Software Architecture

### *4.1. Overall Software Structure*

The software design of the motor test system can be divided into the upper computer and lower computer. As shown in Figure 6, the upper computer completes the functions of the man−machine interface, speed and torque loading, speed and torque transient data acquisition, control and display, ethernet connection between upper and lower computers, etc. The lower computer parses and executes the instructions sent by the upper computer, collects data, stores data, and uploads data in real time. The following section introduces the software design of each part seperately.

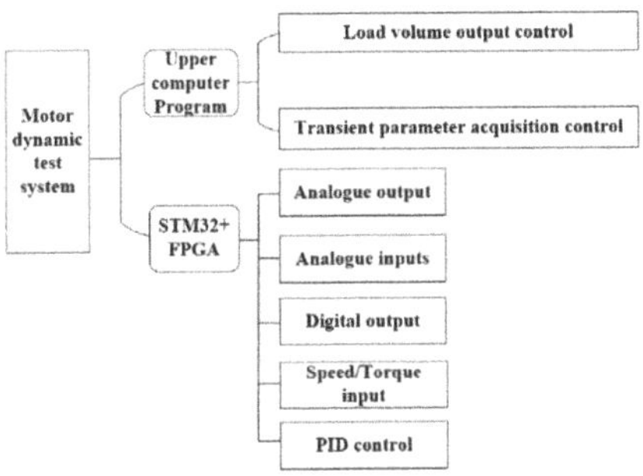


**Figure 6.** Software overall structure.

### *4.2. STM32 Software Design*

The software architecture of the lower computer mainly includes the STM32H7 end and FPGA end. The main tasks of the STM32H7 terminal include analysis of the host computer instructions, real-time data transmission, TCP/IP connection, and low-speed ADC data acquisition. There are many tasks and complex functions, so it is necessary to reasonably plan the scheduling relationship between tasks. In order to improve the real-time performance of the software and reduce the coupling between tasks, the software is transplanted with the real-time embedded operating system—Ferrets. The operating system provides task management, time management, semaphore, message queue, memory management, and other functions, and the kernel supports both the priority scheduling algorithm and polling scheduling algorithm. The system is open source, tailorable, and free, which can effectively save the cost of instrument development and is suitable for industrial application. The software flow chart of the master operating system with the embedded real-time operating system FreeRTOS as the core is shown in Figure 7. After the system is initialized, the FreeRTOS operating system is initialized, and then the FPGA AO control thread is started at the same time so as to control the output of the analog quantity. The FPGA AI control thread is used to control the acquisition of the analog quantity. The

FPGA CI control thread is used to collect speed and torque. The FPGA PID control thread is used to control the speed and torque of the measured motor and load motor. The TCP server thread is used for data and instruction transmission, and has the function of TCP disconnection and reconnection.

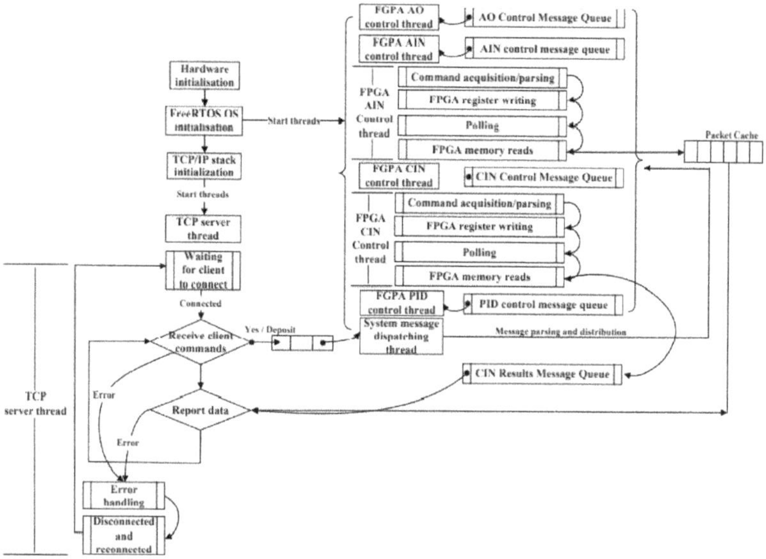


**Figure 7.** STM32H7 software flow chart.

## 4.3. FPGA Software Design

FPGA mainly completes external analog input acquisition, motor encoder signal input acquisition, data storage, receiving control instructions from ARM, controlling the analog output, controlling the position output, and controlling the digital output. The software architecture of FPGA is shown in Figure 8.

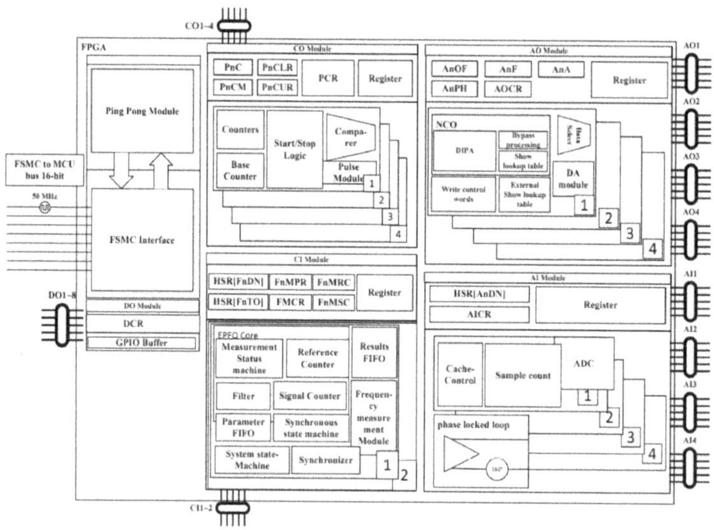


**Figure 8.** FPGA software flow chart.

### 4.4. Ping-Pong Algorithm for Data

Ping-pong operation is a common data flow control processing technique that can realize seamless transmission of data flow. The specific algorithm is shown in Algorithom 1 below. During the first buffering cycle, the input data are cached to Data Buffer Module 1. In the second buffering cycle, the input data are cached to Data Buffer Module 2, and the data in Data Buffer Module 1 are output to the data stream operation processing unit through the selection of "Output Data Stream Selection Unit". In the third buffering cycle, the data are cached into Data Buffer Module 1 by the input data stream selection unit, and Data Buffer Module 2 is selected by the output data stream selection unit to the data stream processing entity. In this design, the ADC data are driven by the ADC and enter the sampling module. After oversampling 200 times, the ADC data are sent to the data ping-pong module at a sampling rate of 1 KHz, so as to achieve a 1 ms data update rate. The ADC data acquisition process is shown in Figure 9.

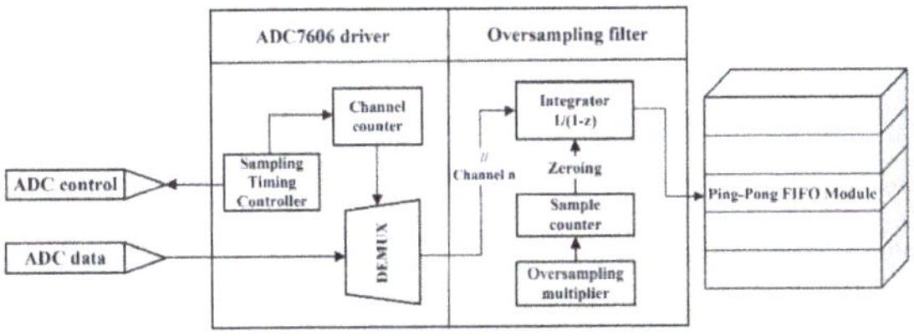


**Figure 9.** ADC data acquisition process.

**Algorithom 1** ping-pong operation

```
Input: DATA_Ti(i = 1,2,3, ... n)
  if(buffer_1 ! = Full )
    buffer_1 = DATA_Ti
  else if (buffer_1 = Full)
    buffer_2 = DATA_Ti
  End if
```

## 5. Experimental Verification and Result Analysis

The system test mainly focuses on the measurement of the motor load and motor transient parameters. The test system is built with an SFT series electric dynamometer as the core, the model of the speed−torque sensor is the WSP speed−torque power sensor, and the output signals of the torque−speed are pulses with an amplitude of 5 V. The output range of the torque frequency signal is 5 KHz to 15 KHz, and the center frequency is 10 KHz. The output range of the speed frequency signal is 0 to 6 KHz, and the calculation formula for the speed and torque is given by Equations (1)–(3). The connection mode between the control system and the upper computer is 100 M ethernet. Figures 10 and 11 show the hardware platform of the project and PC control interface, respectively.

**Figure 10.** Motor test platform.

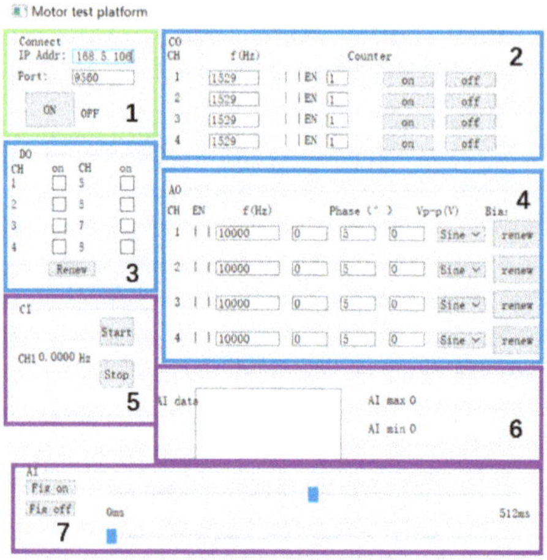


**Figure 11.** PC control interface.

## 5.1. Motor Load Test

The loading quantity includes the analog quantity output, pulse output, and digital quantity output. The analog quantity output also includes regular sine wave output, square wave output, triangular wave, and DC quantity output. The frequency range of the output of the three AC signals is 1 Hz to 10 MHz. The relative error of the frequency output is shown in Figure 12. It can be seen from the figure that the relative error of the frequency output is within 2%, and there is almost no error in the low frequency band.

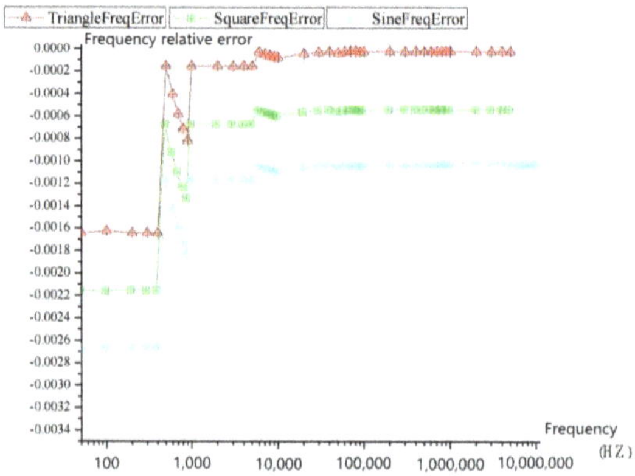


**Figure 12.** Relative error of the load output frequency.

The output range of the analog DC bias output voltage is −5 V to +5 V. The relative error of the DC bias output is shown in Figure 13. The abscissa represents the peak value and the ordinate represents the output relative error. As can be seen from the figure, the maximum relative error of the output peak value should not exceed 0.4% in order to meet the needs of practical application. The continuous pulse output mode can control the output frequency and phase, so as to control the forward and reverse rotation of the motor. The square wave loading amplitude error, sine wave loading amplitude error, and triangular wave loading amplitude error are shown in Figures 14–16, respectively. With the increase in frequency, the relative error increases slightly, and the relative error is within 0.8%.

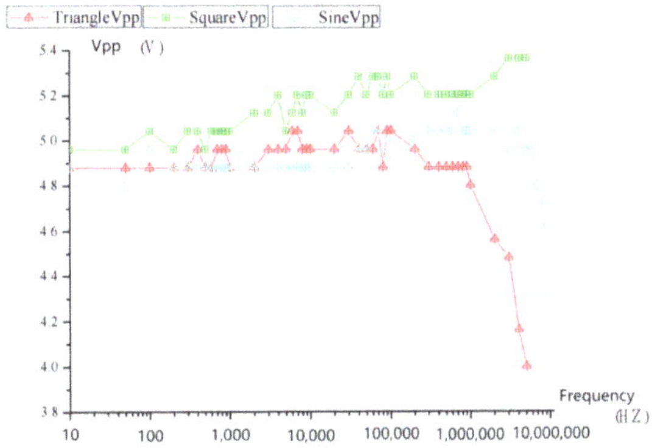


**Figure 13.** Real measurement when setting the 5 V bias output.

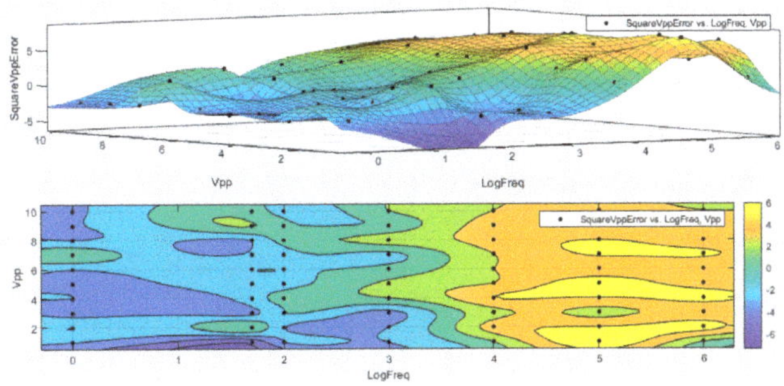


**Figure 14.** Square wave loading amplitude error.

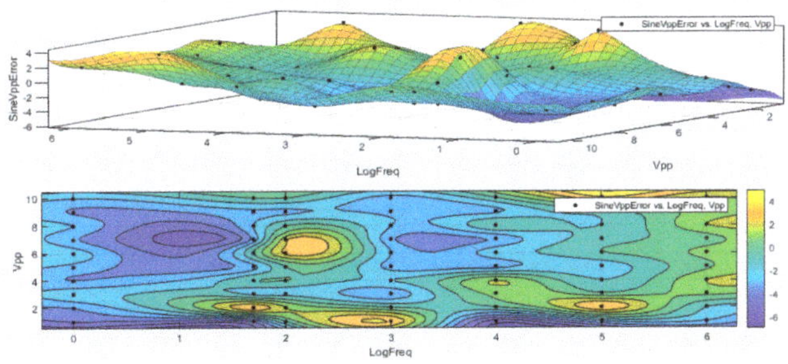


**Figure 15.** Sine wave loading amplitude error.

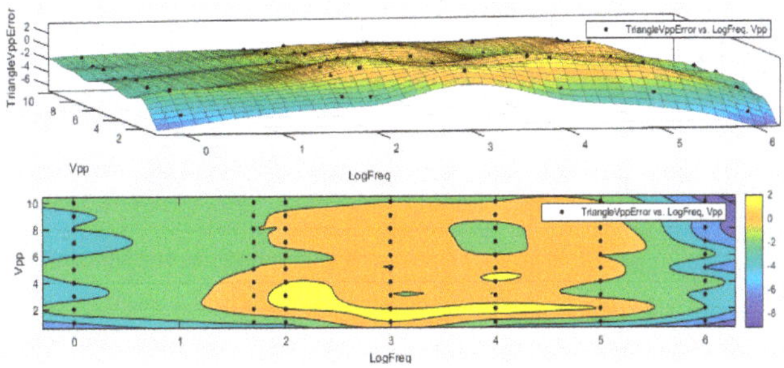


**Figure 16.** Triangular wave loading amplitude error.

### 5.2. Motor Transient Parameter Test

According to the system requirements, the main concerns of the motor transient test include the following: the dynamic test of the motor torque and the dynamic input test of the motor speed signal. Because the feedback signal of the speed torque sensor used in this research is the frequency pulse, in which the speed input frequency range is 0–6 KHz, the torque input frequency range is 5 KHz–10 KHz, and the maximum data update rate is 1 ms. The measured system can measure a frequency ranging from 1 Hz to 1 MHz, and the fastest dynamic refresh rate is 1 ms. Figure 17 shows the speed−step response curve, in which the abscissa is time and the ordinate is motor speed. It can be seen from the curve that the system responds within a millisecond level and reaches a stable state after 1 s. The time to reach stability is also related to the lag effect of the servo motor. The test results show that the system has a fast response speed and preliminarily meets the expected design requirements. Figure 18 shows frequency measurement and analog measurement, which would be used to measure the speed and torque signal of the motor.

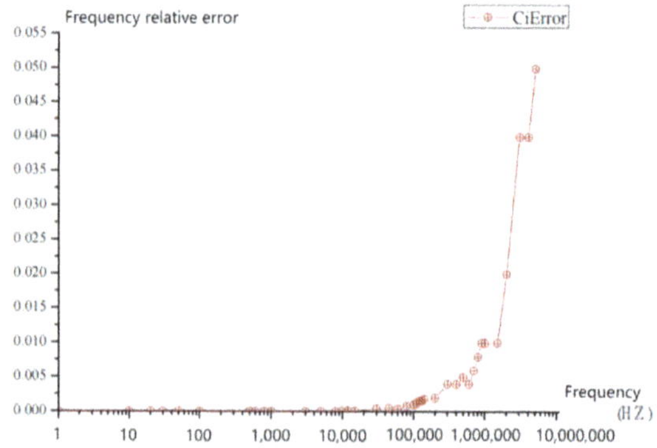


**Figure 17.** Speed and torque frequency accuracy.

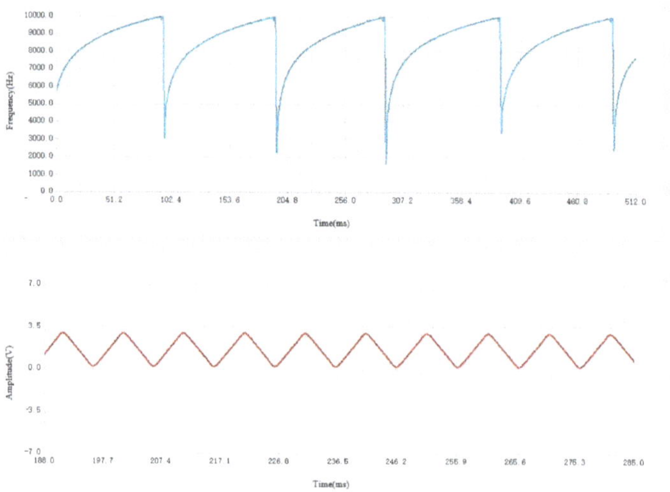


**Figure 18.** The host computer displays the speed frequency and AD sampling waveform in real time.

## 6. Conclusions

Considering the problem that the traditional motor test system cannot directly test the transient parameters of the motor, the requirements of loading a dynamic arbitrary load when the motor is loaded, as well as the high cost, this paper realizes the main control unit of the motor test system with STM32+FPGA as the core, and FSMC bus communication and the data ping-pong processing algorithm are used to control the dual core and transmit the data stream continuously. The feasibility of the system is verified through its physical design and production. The results show that the project has the following characteristics:

1. It can be embedded into the traditional dynamometer system to upgrade the motor transient test and dynamic loading performance of the original equipment;
2. The overall function of the system can be tailored, and the dynamometer control core based on modular design can independently realize multi-channel or single channel transient test and dynamic loading functions;
3. The uninterrupted transmission of the MS level data stream is realized, and the uninterrupted transmission of the data stream based on the data ping-pong operation algorithm ensures the real-time upload and dynamic refreshing of data.

The core of the dynamometer control system based on STM32+FPGA constructed in this way effectively improves the dynamic free loading of the dynamometer system and the measurement of motor transient parameters. It can be further customized and updated according to customer needs. It can provide a valuable experience for the development of dynamometer technology, especially in the measurement of motor dynamic parameters.

**Author Contributions:** Conceptualization, C.Z.; methodology, C.Z. and B.W.; software, C.Z.; validation, C.Z. and B.W.; formal analysis, C.Z.; investigation, B.W.; resources, C.Z.; data curation, L.B. and B.Z.; writing—original draft preparation, C.Z. and B.W.; visualization, J.Q.; supervision, Y.C.; project administration, C.Z.; funding acquisition, C.Z. All authors have read and agreed to the published version of the manuscript.

**Funding:** This research was funded by The National Natural Science Foundation of China with grant number 62001149 and The Key R&D Program of Zhejiang Province with grant number 2021C01069.

**Conflicts of Interest:** The authors declare no conflict of interest.

## References

1. Bunker, B.; Franchek, M.A.; Thomason, B.E. Robust multivariable control of an engine-dynamometer system. *IEEE Trans. Control. Syst. Technol.* **1997**, *5*, 189–199. [CrossRef]
2. Flieh, H.M.; Totoki, E.; Lorenz, R.D. Dynamic Shaft Torque Observer Structure Enabling Accurate Dynamometer Transient Loss Measurements. *IEEE Trans. Ind. Appl.* **2018**, *54*, 6121–6132. [CrossRef]
3. Sun, X.; Shi, Z.; Cai, Y.; Lei, G.; Guo, Y.; Zhu, J. Driving-Cycle-Oriented Design Optimization of a Permanent Magnet Hub Motor Drive System for a Four-Wheel-Drive Electric Vehicle. *IEEE Trans. Transp. Electrif.* **2020**, *6*, 1115–1125. [CrossRef]
4. Ghahderijani, M.M.; Dehkordi, B.M. Comprehensive Robust and Fast Control of $Z$-Source-Inverter-Based Interior Permanent Magnet Synchronous Motor Drive System. *IEEE Trans. Ind. Electron.* **2021**, *68*, 11783–11793. [CrossRef]
5. Langbauer, C.; Diengsleder-Lambauer, K.; Lieschnegg, M. Downhole Dynamometer Sensors for Sucker Rod Pumps. *IEEE Sens. J.* **2021**, *21*, 8543–8552. [CrossRef]
6. Udovichenko, A.V.; Kaluzhskij, D.; Uvarov, N.; Mekhtiyev, A. Electric Motors for Variable-Speed Drive of Lock Valves. *Electronics* **2021**, *10*, 2727. [CrossRef]
7. Hong, D.K.; Choi, J.H.; Kim, D.J.; Chun, Y.D.; Woo, B.C.; Koo, D.H. Development of a High Speed Induction Motor for Spindle Systems. *IEEE Trans. Magn.* **2013**, *49*, 4088–4091. [CrossRef]
8. Zhao, J.; Zhao, J.; Wang, H.; Song, J.; Dong, F. Precision Position Measurement of Linear Motors Mover Based on Temporal Image Correlation. *IEEE Trans. Instrum. Meas.* **2019**, *68*, 3268–3277. [CrossRef]
9. Xu, J.; Zhang, B.; Fang, H.; Guo, H. Guaranteeing the fault transient performance of aerospace multiphase permanent magnet motor system: An adaptive robust speed control approach. *CES Trans. Electr. Mach. Syst.* **2020**, *4*, 114–122. [CrossRef]
10. Sarani, E.; Vaez-Zadeh, S. Design Procedure and Optimal Guidelines for Overall Enhancement of Steady-State and Transient Performances of Line Start Permanent Magnet Motors. *IEEE Trans. Energy Convers.* **2017**, *32*, 885–894. [CrossRef]
11. Zhang, Z.; Wang, L.; Zhang, J.; Ma, R. Study on Requirements for Load Emulation of the Vehicle With an Electric Braking System. *IEEE Trans. Veh. Technol.* **2017**, *66*, 9638–9653. [CrossRef]
12. Song-Manguelle, J.; Ekemb, G.; Mon-Nzongo, D.L.; Jin, T.; Doumbia, M.L. A Theoretical Analysis of Pulsating Torque Components in AC Machines With Variable Frequency Drives and Dynamic Mechanical Loads. *IEEE Trans. Ind. Electron.* **2018**, *65*, 9311–9324. [CrossRef]
13. Kim, N.; Rousseau, A.; Rask, E. Parameter Estimation for a Lithium-Ion Battery From Chassis Dynamometer Tests. *IEEE Trans. Veh. Technol.* **2016**, *65*, 4393–4400. [CrossRef]

14. Al-Badri, M.; Pillay, P.; Angers, P. A Novel Algorithm for Estimating Refurbished Three-Phase Induction Motors Efficiency Using Only No-Load Tests. *IEEE Trans. Energy Convers.* **2015**, *30*, 615–625. [CrossRef]
15. Korkmaz, E.; Bediz, B.; Gozen, B.A.; Ozdoganlar, O.B. Dynamic characterization of multi-axis dynamometers. *Precis. Eng.* **2014**, *38*, 148–161. [CrossRef]
16. Kang, T.J.; Kim, J.; Lee, S.B.; Yung, C. Experimental Evaluation of Low-Voltage Offline Testing for Induction Motor Rotor Fault Diagnostics. *IEEE Trans. Ind. Appl.* **2015**, *51*, 1375–1384. [CrossRef]
17. Durantay, L.; Velly, N.; Pradurat, J.F.; Chisholm, M. New Testing Method for Large High-Speed Induction Motors. *IEEE Trans. Ind. Appl.* **2017**, *53*, 660–666. [CrossRef]
18. Lee, S.B.; Naeini, A.; Jayaram, S.; Stone, G.C.; Šašić, M. Surge Test-Based Identification of Stator Insulation Component With Partial Discharge Activity for Low Voltage AC Motors. *IEEE Trans. Ind. Appl.* **2020**, *56*, 2541–2549. [CrossRef]
19. de Sousa, L.B.; Sass, F.; Fortes, M.Z. Comparison of Three-Phase Induction Motors Testing Methods using ABNT 17094-3:2018 and IEC 60034-2-1:2014. *IEEE Lat. Am. Trans.* **2021**, *19*, 746–754. [CrossRef]
20. Reddy, P.B.; El-Refaie, A.M.; Zou, M.; Pan, D.; Alexander, J.P.; Tapadia, N.; Grace, K.; Huh, K.-K.; Johnson, F. Performance Testing and Analysis of Synchronous Reluctance Motor Utilizing Dual-Phase Magnetic Material. *IEEE Trans. Ind. Appl.* **2018**, *54*, 2193–2201. [CrossRef]

*energies*

*Article*

# Energy Balance Control for Improving Transient Performance of DC Bus Voltage in Power Electronic Transformer for the Voltage-Sensitive Loads

**Gaohui Feng [1,2,3], Pengsheng Bu [1,2,3] and Liqiang Yuan [4,*]**

1 China Coal Technology & Engineering Group Taiyuan Research Institute Co., Ltd., Taiyuan 030000, China; fgh1980fgh@163.com (G.F.); bupengsheng@126.com (P.B.)
2 China National Engineering Laboratory for Coal Mining Machinery, Taiyuan 030000, China
3 Shanxi Tiandi Coal Mining Machinery Co., Ltd., Taiyuan 030000, China
4 Department of Electrical Engineering, Tsinghua University, Beijing 100084, China
* Correspondence: ylq@tsinghua.edu.cn

**Abstract:** The power electronic transformer (PET), as a main topology for the energy router in the energy internet, consists of the rectifiers, the dual active bridge (DAB), and the inverter, and these three parts are connected by two dc buses. So, the performance of the dc bus voltages is very important because it can totally affect the output waveforms of the dc and ac voltage, especially for the voltage-sensitive loads. Compared with the proportion integration (PI) control scheme, the energy control method utilizes the energy as the control variable, and the control strategy derived from the energy relationship, including the passive elements and all the interfaces, is more direct and explicit. In this paper, considering the energy between the dc bus capacitors and the input inductor and the load and the source in the PET topology, the energy balance control (EBC) strategy is proposed. For the two dc bus voltages, the energy balance relationship of the different time scales is used to decouple the interaction in the control scheme. The EBC strategy can obviously reduce the fluctuation and the transient time of the two dc bus voltages when the load power or voltage reference is changed. Thus, under the limited voltage fluctuation, the EBC strategy can reduce the dc bus capacitance in order to reduce the volume and weight of the converter and enhance the reliability. The simulation and experimental results verify the effectiveness of the proposed control strategy.

**Keywords:** dc bus voltage; energy balance control; power electronic transformer; different time scales

**Citation:** Feng, G.; Bu, P.; Yuan, L. Energy Balance Control for Improving Transient Performance of DC Bus Voltage in Power Electronic Transformer for the Voltage-Sensitive Loads. *Energies* **2022**, *15*, 3129. https://doi.org/10.3390/en15093129

Academic Editor: Adolfo Dannier

Received: 23 March 2022
Accepted: 22 April 2022
Published: 25 April 2022



## 1. Introduction

With the gradual implementation of China's 'double carbon' goal in the distribution network, the power electronic transformer (PET), as the main topology of the energy router, has received wide attention [1–3]. In general, for a high-voltage, high-power converter in a high-power system, because of the limited voltage capacity of the power semiconductor, the topology of the module combination is usually used.

At present, the PET is connected to the distribution network through the high-voltage interface [4–7]. There are two main topologies of the high-voltage converter, which are a cascaded H-bridge topology [8,9] and a modular multilevel converter topology [10,11].

Using the three-stage structure with two dc buses, consisting of the rectifier, the dual active bridge (DAB), and the inverter, the load current feed-forward control strategy was utilized in [12] to improve the transient performance of the two dc bus voltages, and additional current sensors were required. The power synchronization control strategy was used to reduce the second harmonic voltage ripple by transmitting the second harmonic power from the primary dc bus to the secondary dc bus in [13], and the output power was set to be same phase with the input power at the same time to reduce the second voltage ripples. In this reference, the control strategy used the power balance among the three stages to reduce the dc bus voltage ripples, but the transient performance was not shown. In the cascaded topology shown in [14], in considering the differences of the driving circuits in the PWM rectifiers, the differences of the switching characteristics of the IGBT, and the differences of the leakage inductances of the high-frequency transformer (HFT) in the DAB [15], four different modulation technologies and the control strategies for the cascaded dc bus voltage and power balance were introduced. The conclusion was that the rectifiers with the same duty cycle and the DAB used to realize the cascaded dc bus (CDB) voltage balance represent the optimal control scheme after comprehensive consideration of the control performance and the hardware cost. The energy stored in the dc-link capacitor of the Static Synchronous Compensator (STATCOM), especially the cascaded inverter-based STATCOM, whose dc-link capacitance is relatively large, and the virtual inertial control strategy using the energy are presented, and their effectiveness is also validated by simulation [16]. The energy-balance model was analyzed in [17] for the two-stage, single-phase, grid-connected photovoltaic inverters, and the dc voltage control scheme was proposed. In the feed-forward control scheme, through introducing the energy changes of the inverter inductors to the feed-forward variable, the control method has a better dynamic performance in the dc bus voltage compared with the conventional feed-forward control scheme [18].

For the second harmonic of the dc bus voltage in the single-phase converter, the PI controller does not have the capability to reduce the voltage ripple accurately, and it also could cause a third harmonic in the grid current. To avoid the odd harmonic current and enlarge the grid current total harmonic distortion (THD), corresponding solutions were proposed in many papers. In [19], the moving average filter instead of the low-pass filter was used for filtering the high and low frequency ripples of the dc bus voltage, and the feed-forward component was added to the shunt controller to regulate the grid current at a constant level in the presence of grid disturbances. In another method, to reduce the output current distortion, a Finite Impulse Response (FIR) filter notch filter is employed in the bus control system, and, to improve the dynamic performances, the input power feed-forward is used [20]. In the single-phase grid inverter, the LLCL filters are used to make zero impedance, and the grid-side inductor filter can be reduced. For the resonance phenomenon in the system, an active damping method based on proportional resonance controllers was proposed to suppress the resonance [21]. In the single-phase, grid-connected inverter with LCL filters, the PI inner loop is stabilized by using an inherent one-beat delay, achieved by a digital controller. Based on the inner loop system, a detailed design scheme of a repetitive controller is presented, through which direct control of the grid current is realized; the reference is tracked perfectly to a zero-phase shift, and high-attenuation gain is achieved in the high frequency range. In particular, the grid-voltage feed-forward control and the current reference feed-forward control are adopted to suppress grid-voltage disturbance and increase the dynamic tracking performance [22]. In [23], a repetitive dc link voltage predictor was proposed to improve the compensation performance. An area-equalization-based algorithm was used to calculate the second harmonic and generated the required pulse width. The minimum ripple-energy requirement was derived in [24], and a bidirectional buck-boost converter was used as the ripple energy storage circuit, which can effectively reduce the energy storage capacitance.

Based on the previous research, a novel control strategy is proposed in this paper, named the energy balance control strategy. According to the energy relationship among the sources, the passive elements, the equivalent load, and the EBC can improve the transient performance of the voltage and power balance demands of the cascaded H-bridge. The first section in the paper introduces the application background and the device research status of PET and the current situation of the dc bus voltage control method and deduces the basic principle of the energy balance control method. It is the background introduction of the research. Section 2 introduces the equivalent circuit model of the single-phase, two-stage PET and derives the energy relationship in the device, which is the basis of the following research in the paper. Based on the topology in Section 2, the concrete research points are carried out in Sections 3 and 4. The two parts are parallel. The energy balance control method and the stability of the two-stage bus voltages (the low bus voltage and the cascade bus voltage) are deduced and verified, respectively. In addition, the influence of the parameter range of the passive components in the model is analyzed for the control law performance in the last, small part of Section 4. On the basis of the previous research contents, the voltage balance control strategy, implemented by the energy balance, is studied in Section 5 in order to form the overall energy balance control strategy in the cascaded PET. In Section 6, the main performance of the proposed energy balance control method is verified by simulation and experimental results. Section 7 is the conclusion of the full paper. The contents of each section and the relationship between them are shown in Figure 1.

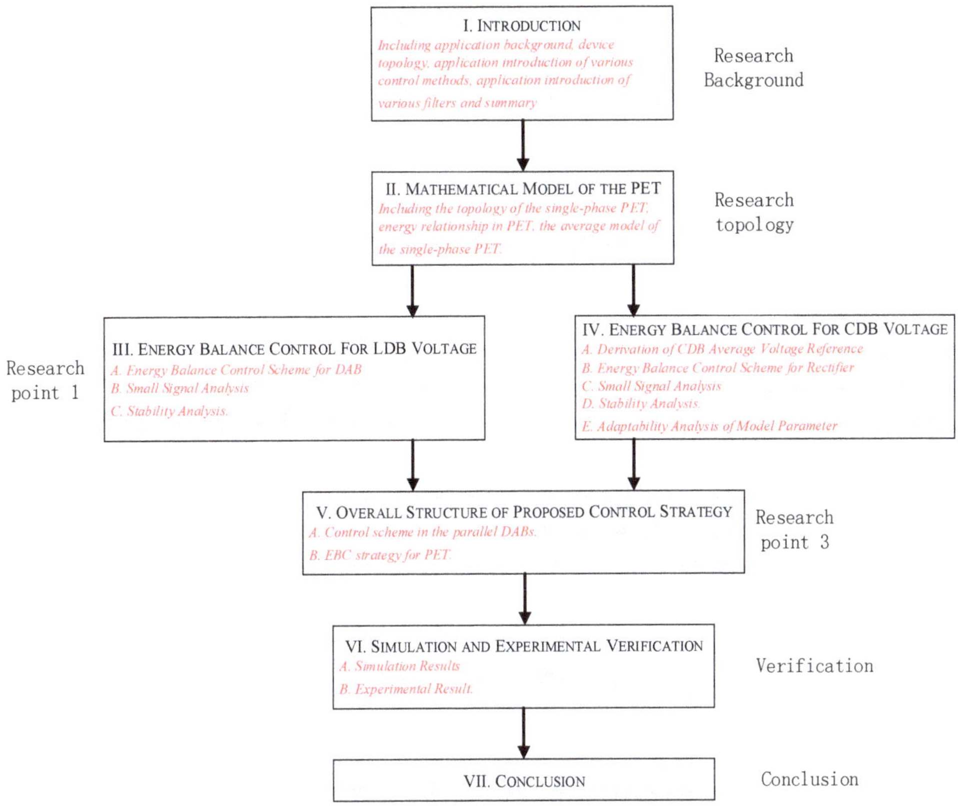


**Figure 1.** The contents of each section and their relationship.

## 2. Mathematical Model of the PET

The topology of the single-phase PET is shown in Figure 2a, which contains a single-phase rectifier composed of four cascaded H-bridges, four dual active bridges (DAB) in parallel, and a three-phase inverter. The instantaneous power relationship in PET is shown in Figure 2b. The energy flow in PET is shown in Figure 2c.

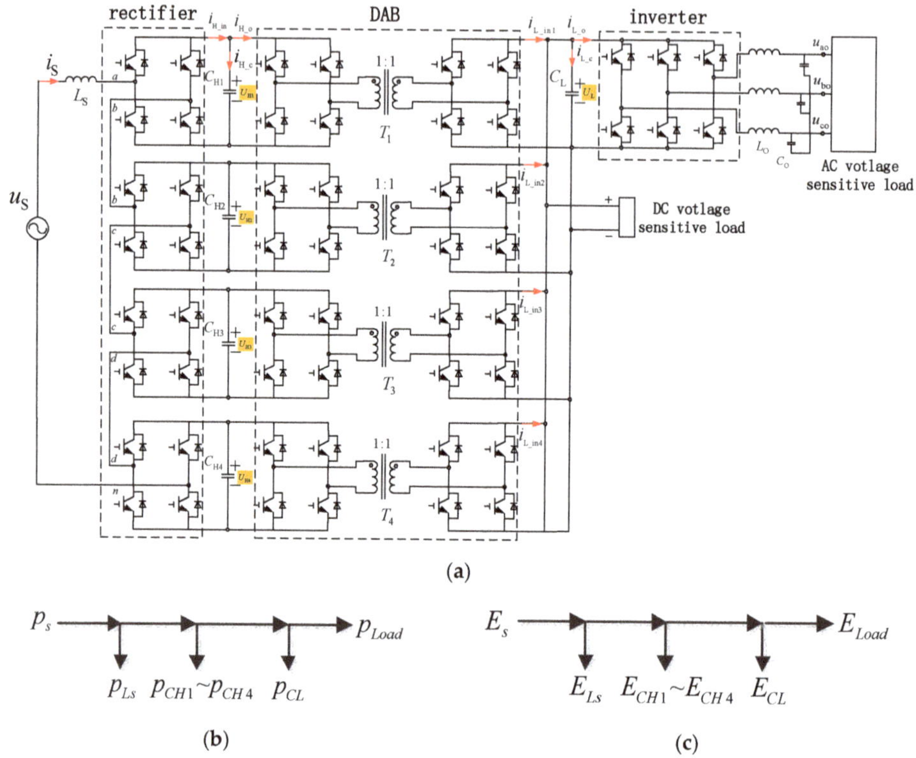


**Figure 2.** The single-phase PET: (**a**) structure, (**b**) instantaneous power relationship in PET, (**c**) energy relationship in PET.

The single-phase rectifier consists of an input inductor $L_S$ and four H-bridge modules. Each module uses the same modulation method. The $d_{rj}$ is the duty cycle of the module $j$ and the $U_{Hj}$ is the CDB voltage of the module $j$ ($j$ = 1, 2, 3, 4). Ignoring the internal impedances of the inductor and the converters, the cascaded rectifier model can be described as

$$\begin{cases} L_S \frac{di_S}{dt} = u_S - d_{r1}U_{H1} - d_{r2}U_{H2} - d_{r3}U_{H3} - d_{r4}U_{H4} \\ C_{H1}\frac{dU_{H1}}{dt} = d_{r1}i_S - \frac{U_{H1}}{Z/4} \\ C_{H2}\frac{dU_{H2}}{dt} = d_{r2}i_S - \frac{U_{H2}}{Z/4} \\ C_{H3}\frac{dU_{H3}}{dt} = d_{r3}i_S - \frac{U_{H3}}{Z/4} \\ C_{H4}\frac{dU_{H4}}{dt} = d_{r4}i_S - \frac{U_{H4}}{Z/4} \end{cases} \tag{1}$$

where $Z$ is the equivalent impedance of the cascaded rectifier.

Ignoring the DAB internal impedances, the transmitted power of the DAB $P_{DAB}$ is

$$P_{DAB} = \frac{nU_H U_L d(1-d)}{2f_s L_s} = kU_H U_L \tag{2}$$

where $L_s$ is the HFT leakage inductance, $U_H$ is the primary dc voltage, $U_L$ is the second dc voltage, $n$ is the transformer turns ratio, $f_s$ is the switching frequency, $d$ is the duty cycles, and $k = nd(1 - d)/2f_sL_s$ is the phase shift ration. When $d$ equals 0.5, the $P_{DAB}$ is the maximum, and when $d$ equals 1 or 0, the $P_{DAB}$ is zero.

The three-phase inverter model is not the focus of this paper. Here, the load $Z$ is used for replacing the equivalent impedance of the ac load and the dc load. So, the average model of the single-phase PET is given in Figure 3, where $k_1$~$k_4$ are the phase shift ratios of the four DAB modules.

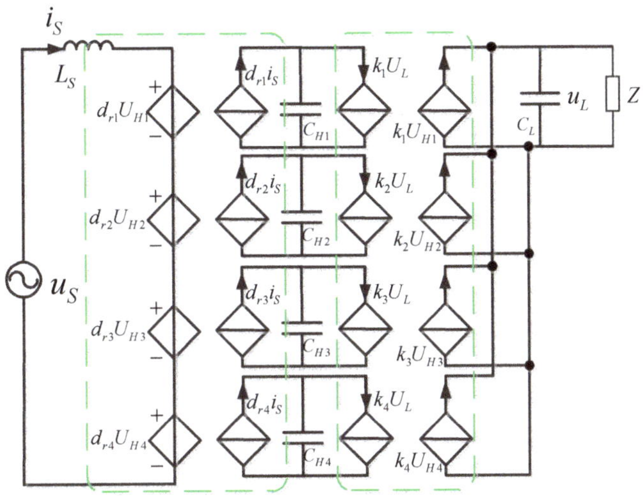


**Figure 3.** The average model of the single-phase PET.

For improving the transient performance of the PET, the CDB voltage and the LDB voltage in the PET should be controlled. In Figure 2a, the four CDB voltages ($U_{H1}$, $U_{H2}$, $U_{H3}$, and $U_{H4}$) and the LDB voltage ($U_L$) should reach their reference as quickly as possible during the transient process. At the same time, in the real system, every H-bridge model is designed to be identical in order to reduce the production consumption and the control complexity. However, the leakage inductance variation of the HFTs can reach 10–20% [14], and the transmitted power variation will also be 10–20%. In extreme cases, if the voltage balance control among the four cascaded modules is not adopted, the dc bus voltage of the DAB module with the largest leakage inductance will be four times the normal dc bus voltage, while the dc bus voltages of the other modules will reduce to nearly 0, which is absolutely unacceptable.

So, the stability of two-stage bus voltages in the topology is very important, especially for the voltage-sensitive loads, such as the data server, the navigation and positioning instrument, and so on. In this paper, Sections 3 and 4 have a parallel relationship. In Section 3, the energy balance relationship around the low-voltage bus is deduced, and the control law and its stability analysis are also introduced in detail. Moreover, in Section 4, the energy balance relationship with the cascade buses is deduced, the stability and the control law for the cascaded voltage balance are analyzed, and the influence of the parameter range of the passive components in the model is analyzed for the control law performance.

## 3. Energy Balance Control for LDB Voltage

### *3.1. Energy Balance Control Scheme for DAB*

Based on the equivalent circuit in Figure 3 and the energy relationship in Figure 2c, the energy relationship between the DAB, the LDB capacitor, and the equivalent load can be described as

$$E_{DAB} = E_{CL} + E_{Load} \tag{3}$$

where the four DAB transmitted energies are regarded as the same under the voltage balance control. Considering the energy relationship in one switching period, it can be written as

$$P_{DAB} = \frac{1}{2}C_L(U_L^{*2} - U_L{}^2)/T_s + P_{Load} \tag{4}$$

where $T_s$ is the switching period, $f_s = 1/T_s$.

Then, the total energy relationship becomes

$$4 \times \frac{nU_{H\_s}U_{L\_s}d(1-d)}{2f_sL_s} = \frac{1}{2}C_L(U_L^{*2} - U_L{}^2)/T_s + P_{Load} \tag{5}$$

where $U_{H\_s}$ and $U_{L\_s}$ are the expectation values of the CDB and the LDB voltage. After simplification and approximation, the duty cycle $d$ for the DAB is

$$d = \frac{\left(\frac{1}{2}C_L(U_L^{*2} - U_L{}^2)f_s + P_{Load}\right) \times 2f_sL_s}{4nU_{H\_s}U_{L\_s}} \tag{6}$$

Considering the control delay of the DAB stage as a first-order inertial element with a constant time $T_d$, the control scheme is depicted in Figure 4, and the DAB model is in the green block, where $k_{d1} = nU_{H1}/2f_sL_{s1}$, $k_i = 1/U_L$.

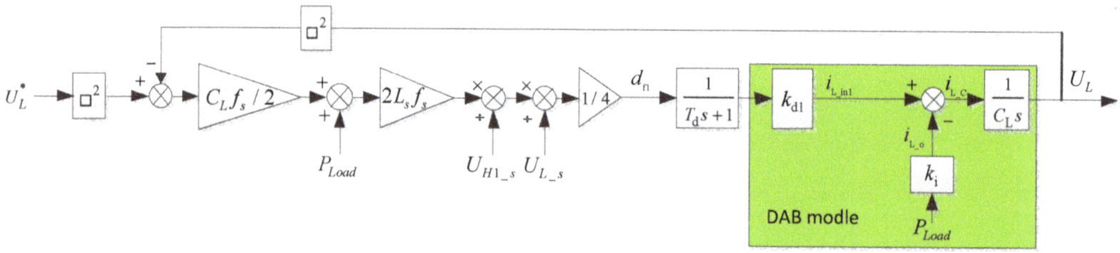


**Figure 4.** EBC Control model for the LDB voltage.

## 3.2. Small-Signal Analysis

Performing small-signal decomposition of the system variables in Figure 4, we have

$$\begin{cases} U_L^* = \overline{U}_L^* + \widetilde{u}_L^* \\ U_L = \overline{U}_L + \widetilde{u}_L \\ P_{Load} = \overline{P}_{Load} + \widetilde{P}_{Load} \end{cases} \tag{7}$$

where $\overline{U}_L^*$ is the reference of the LDB voltage, $\overline{U}_L$ is the real value of the LDB voltage, and $\overline{P}_{Load}$ is the real value of the load power. These values are the large-signal component. $\widetilde{u}_L^*$,$\widetilde{u}_L$, and $\widetilde{P}_{Load}$ are the small-signal values of the counterparts. The square values of the small signals are small enough to be ignored. The small-signal expression of the duty cycle is

$$\widetilde{d} = \frac{(\frac{1}{2}C_L(\overline{U}_L^*\widetilde{u}_L^* - \overline{U}_L\widetilde{u}_L)f_s + \widetilde{P}_{Load}) \times 2f_sL_s}{4nU_{H\_s}U_{L\_s}} \tag{8}$$

The corresponding small-signal control model is given in Figure 5.

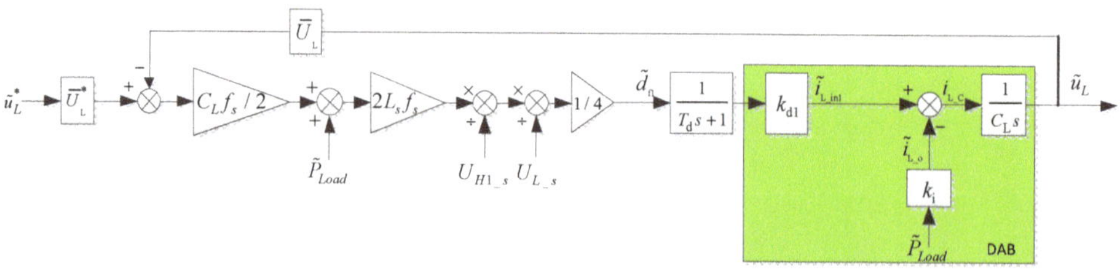


**Figure 5.** Small-signal model for the LDB voltage based on EBC controller.

The closed-loop transfer function of the LDB voltage control system based on EBC is

$$G_{dc\_L}(s) = \frac{\widetilde{u}_L}{\widetilde{u}_L^*} = \frac{f_s\overline{U}_L^* U_{H1\_s}}{8\pi(T_d s+1)s + U_{H1\_s} f_s \overline{U}_L} \tag{9}$$

From (9), it can be seen that the LDB voltage is irrelevant to the dc bus capacitance $C_L$. Considering that the expectation value of the CDB voltage $U_{H1\_s}$ and the LDB voltage $U_{L\_s}$ remains constant, the load power $\widetilde{P}_{Load}$ fluctuation in the DAB model is counteracted by the load power fluctuation, including in the feed-forward terms in the EBC. So, the load power fluctuation has no impact on the dc bus voltage, as is shown in Figure 5.

When the system adopts the low-bandwidth PI controller, the small-signal control model is as depicted in Figure 6.

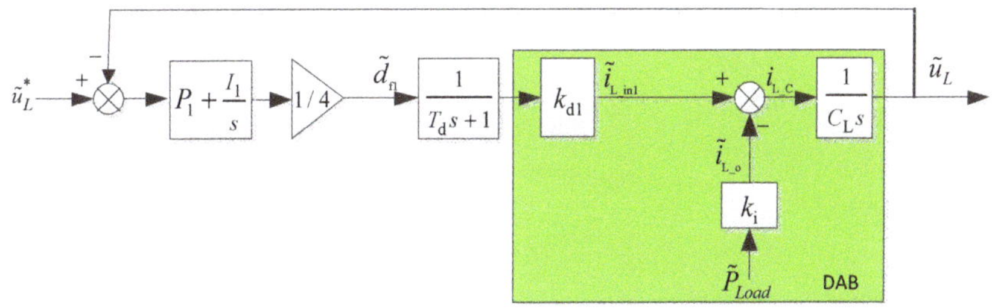


**Figure 6.** Small-signal model for the LDB voltage based on PI controller.

The closed-loop transfer function of the LDB voltage control system based on the PI controller is

$$G_{dc\_L}(s) = \frac{\widetilde{u}_L}{\widetilde{u}_L^*} = \frac{(P_1 s + I_1)U_{H1\_s}}{8\pi f_s L_s C_L s^2(T_d s+1) + (P_1 s + I_1)U_{H1\_s}} \tag{10}$$

From (10), it can be seen that the LDB voltage with the PI controller is affected by the capacitance $C_L$. When $U_{H1\_s}$ remains constant, from Figure 6, the dc bus voltage is affected by the load power $\widetilde{P}_{Load}$ fluctuation. The closed-loop transfer functions are

$$G_{dp\_L}(s) = \frac{\widetilde{u}_L}{\widetilde{P}_{Load}} = \frac{8\pi f_s L_s (T_d s+1)s}{8\pi f_s L_s C_L s^2(T_d s+1)U_{L\_s} + (P_1 s + I_1)U_{H1\_s}U_{L\_s}} \tag{11}$$

It is shown that compared with the PI controller, the EBC controller can eliminate the impacts of the LDB capacitor and the fluctuations of the load power by importing the load power into the control model. The robustness of the EBC is increased.

### *3.3. Stability Analysis*

The stability analysis of the EBC closed-loop transfer function for the LDB voltage was conducted. The switching period is set as $T_s = 5 \times 10^{-5}$ s. Considering the sampling and the control delay, $T_d = 2T_s$, the LDB voltage reference $\overline{U}_L^*$ is 700 V and the CDB voltage expectation value $U_{H1\_s}$ is 700 V. When the LDB voltage $\overline{U}_L$ changes from 400 V to 1000 V, the denominator of Equation (9) has two poles. The two poles are far away from the origin, with the increasing of the voltage value, as shown in Figure 7. This figure shows that all the poles are located in the left half plane, and so, the system is stable.

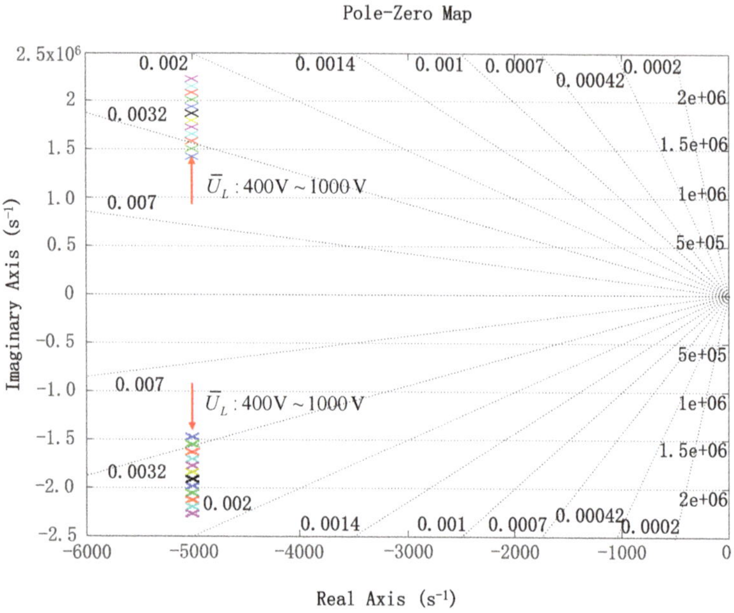


**Figure 7.** Pole distribution of the LDB voltage control system based on EBC controller.

## 4. Energy Balance Control for CDB Voltage

### *4.1. Derivation of CDB Average Voltage Reference*

Based on the equivalent model in Figure 3 and the instantaneous power relationship in Figure 2b, the instantaneous power relationship of the PET is as follows:

$$p_s = p_{Ls} + p_{CH1} + p_{CH2} + p_{CH3} + p_{CH4} + p_{CL} + p_{Load} \tag{12}$$

In the steady state, the voltages in the cascade bus capacitors are equal, and the instantaneous powers of the cascaded capacitors are also same, namely $P_{HC1} = P_{CH2} = P_{CH3} = P_{CH4} = P_{CH}$; Equation (12) can be simplified as

$$p_s = p_{Ls} + 4p_{CH} + p_{CL} + p_{Load} \tag{13}$$

Assume that the grid voltage and current are

$$u_s = \sqrt{2}U_s \sin(\omega t) \tag{14}$$

$$i_s = \sqrt{2}I_s \sin(\omega t + \theta) \tag{15}$$

where $U_s$ and $I_s$ are the RMS values of the grid voltage and current, $\omega$ is the grid angular frequency, and $\theta$ is the initial phase angle of grid current.

Thus, the instantaneous power of the grid is

$$p_s = u_s \cdot i_s = U_s I_s[\cos\theta - \cos(2\omega t + \theta)] = P_s - U_s I_s \cos(2\omega t + \theta) \tag{16}$$

where the dc component is the constant power of the grid, and it is expended by the load. The second harmonic power of the grid will cause the second harmonic voltage ripple in the CDB voltages.

The instantaneous power of the input inductors is

$$p_{Ls} = L_s \frac{di_s}{dt} i_s = L_s I_s^2 \sin(2\omega t + 2\theta) \cdot \omega \tag{17}$$

The whole instantaneous power of the four cascaded capacitors is

$$4p_{CH} = 4C_H \frac{dU_H}{dt} U_H \tag{18}$$

The instantaneous power of the capacitors in the LDB is

$$p_{CL} = C_L \frac{dU_L}{dt} U_L \tag{19}$$

In the steady state, the LDB voltage $U_L$ is constant, then $P_{CL} = 0$.
The load power equals the grid constant power.

$$P_s = p_{Load} \tag{20}$$

The phase of the grid voltage and current are set to be the same value. Substituting (14)~(20) into (13), it becomes

$$4C_H \frac{dU_H}{dt} U_H + A\cos(2\omega t - \varphi) = 0 \tag{21}$$

where

$$\begin{array}{l} A = I_s\sqrt{U_s^2 + (L_{ac} I_s \omega)^2} \\ \varphi = \arctan \frac{L_{ac} I_s \omega}{U_s} \end{array} \tag{22}$$

Then, the average reference of the CDB voltage is

$$\begin{array}{l} U_H^* = \sqrt{{U_{Have}^*}^2 - B} \\ B = \frac{A}{4C_H \cdot \omega} \sin(2\omega t - \varphi) \end{array} \tag{23}$$

where $B$ is the square value of the second harmonic voltage amplitude of the CDB voltage.

## 4.2. Energy Balance Control Scheme for Rectifier

Based on the equivalent model in Figure 3 and the energy relationship in Figure 2c, the energy relationship is as follows:

$$E_s = E_{Ls} + 4E_{CH} + E_{CL} + E_{Load} \tag{24}$$

According to the analysis of the two-stage, single-phase photovoltaic grid-connected inverters [17], for the single PWM rectifier, when the system is in steady state, the absolute value of the grid current remains constant after half line cycle $T_g/2$; then, the stored energy of the filter inductor also remains constant after $T_g/2$, namely $E_{Ls} = 0$. When ignoring the internal impedances of the PET, the energy relationship of the PET in $T_g/2$ is

$$\frac{T_g}{2} P_{Load} + \frac{1}{2} \cdot 4C_H({U_H^*}^2 - {U_H}^2) + \frac{1}{2} C_L({U_L^*}^2 - {U_L}^2) = U_s I_s^* \frac{T_g}{2} \tag{25}$$

Thus, the grid current reference is

$$I_s^* = \frac{P_{Load} + 4C_H f_g({U_H^*}^2 - {U_H}^2) + C_L f_g({U_L^*}^2 - {U_L}^2)}{U_s} \tag{26}$$

Considering the control delay of the rectifier stage as a first-order inertial element with a constant time $T_R$, the control scheme is depicted in Figure 8, and the rectifier model is in the green block, where $k_g = 4C_H f_g$, $k_T = 1/U_S$, $P_R = L_{ac}/T_R$, $u_r = 4dU_H$, and $k_R = U_S/4U_H$.

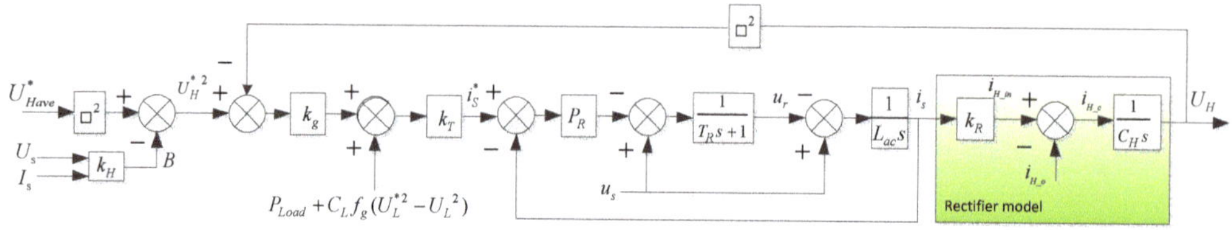


**Figure 8.** EBC control model for the CDB voltage.

### 4.3. Small-Signal Analysis

In the condition that the calculated value $B$ is consistent with the actual second harmonic voltage ripple, and the equivalent load power in the CDB side is replaced by the load power in the LDB side, the EBC control model can be depicted as shown in Figure 9.

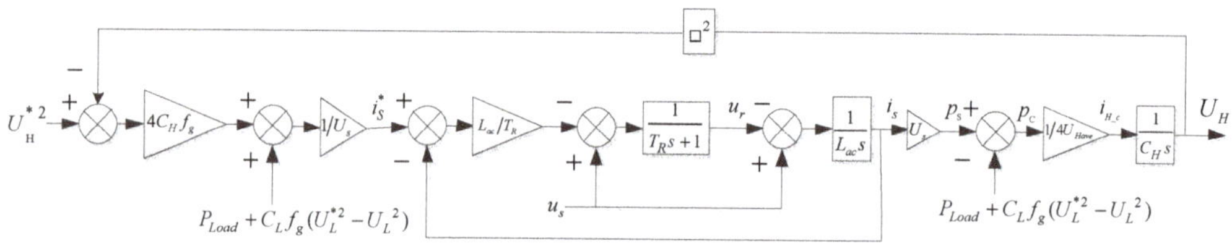

**Figure 9.** Simplified EBC control model for the CDB voltage.

Performing small-signal decomposition of the system variables in Figure 9, we have

$$\begin{cases} I_S^* = \overline{I}_S^* + \widetilde{i}_S^* \\ U_H^* = \overline{U}_H^* + \widetilde{u}_H^* \\ U_H = \overline{U}_H + \widetilde{u}_H \\ P_{Load} = \overline{P}_{Load} + \widetilde{P}_{Load} \\ U_L^* = \overline{U}_L^* + \widetilde{u}_L^* \\ U_L = \overline{U}_L + \widetilde{u}_L \end{cases} \tag{27}$$

where $\overline{I}_s^*$ is the grid current, $\overline{U}_H^*$ is the reference of the CDB voltage, $\overline{U}_H$ is the real value of the CDB voltage, $\overline{U}_L^*$ is the reference of the LDB voltage, $\overline{U}_L$ is the real value of the LDB voltage, and $\overline{P}_{Load}$ is the real load power. These values are the large-signal component. $\widetilde{i}_s^*$, $\widetilde{u}_H^*$, $\widetilde{u}_H$, $\widetilde{u}_L^*$, $\widetilde{u}_L$, and $\widetilde{P}_{Load}$ are the variation in the transient process; these variations are the small-signal values of the counterparts. The square values of the small signals are small enough to be ignored. The small-signal expression of the current reference is

$$\widetilde{i}_s^* = \frac{\widetilde{P}_{Load} + 4C_H f_g(\overline{U}_H^* \widetilde{u}_H^* - \overline{U}_H \widetilde{u}_H) + C_L f_g(\overline{U}_L^* \widetilde{u}_L^* - \overline{U}_L \widetilde{u}_L)}{U_s} \tag{28}$$

The corresponding small-signal model is shown in Figure 10.

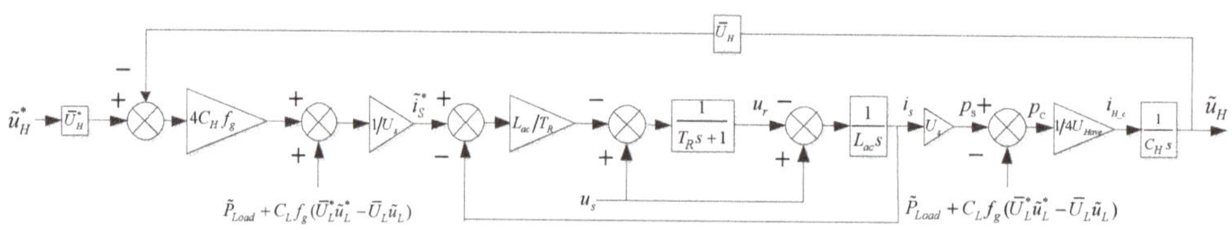

**Figure 10.** Small-signal model for the CDB voltage based on EBC controller.

The closed-loop transfer function of the CDB voltage control system based on the EBC controller is

$$G_{dc\_H}(s) = \frac{\widetilde{u}_H}{\widetilde{u}_H^*} = \frac{f_g \overline{U}_H^*}{[T_R s(T_R s+1)+1]U_{Have}s + f_g \overline{U}_H} \tag{29}$$

It can be seen that the CDB voltage is irrelevant to the dc bus capacitance $C_H$. Considering that the grid voltage $U_s$ remains constant, the load power $\widetilde{P}_{Load}$ fluctuation in the rectifier model and the stored energy fluctuation of the LDB capacitor are counteracted by the same power fluctuation, including in the feed-forward terms in the EBC. So, the fluctuation of the load power and the LDB capacitor power have no impact on the dc bus voltage, as shown in Figure 10.

When the system adopts the low-bandwidth PI controller, the small-signal model is as depicted in Figure 11.

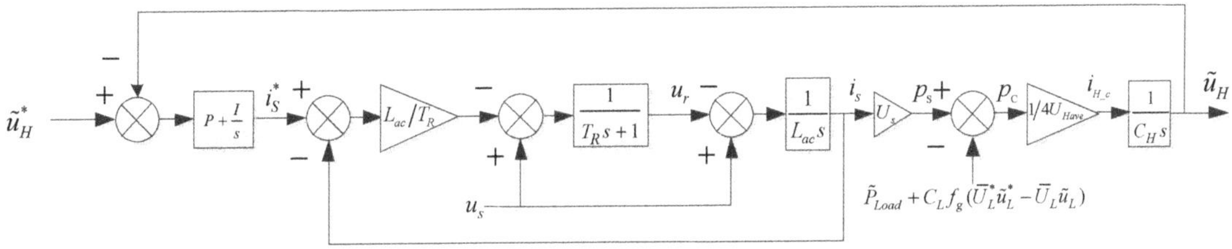

**Figure 11.** Small-signal model for the CDB voltage based on PI controller.

The closed-loop transfer function of the CDB voltage control system based on the PI controller is

$$G_{dc\_H}(s) = \frac{\widetilde{u}_H}{\widetilde{u}_H^*} = \frac{(Ps+I)U_S}{4s^2 U_{Have} C_H[T_R s(T_R s+1)+1] + (Ps+I)U_S} \tag{30}$$

It can be seen that the CDB voltage with the PI controller is affected by the capacitance $C_H$. When the grid voltage $U_S$ remains constant, the dc bus voltage is affected by the load power $\widetilde{P}_{Load}$ fluctuation, the capacitance on the LDB side $C_L$, and the LDB voltage fluctuation $\widetilde{u}_L^*$. The closed-loop transfer functions are

$$G_{dp\_H}(s) = \frac{\widetilde{u}_H}{\widetilde{P}_{Load}} = \frac{[T_R s(T_R s+1)+1]s}{4s^2 U_{Have} C_H[T_R s(T_R s+1)+1] + (Ps+I)U_S} \tag{31}$$

$$G_{di\_H}(s) = \frac{\widetilde{u}_H}{\widetilde{u}_L^*} = \frac{C_L f_g \overline{U}_L^*[T_R s(T_R s+1)+1]s}{4s^2 U_{Have} C_H[T_R s(T_R s+1)+1] + (Ps+I)U_S} \tag{32}$$

It can be concluded that compared with the PI controller, the EBC controller can eliminate the impacts of the CDB capacitors and the fluctuations of the load power and the power of the LDB capacitor by importing their power into the control structure. The robustness of the EBC for the CDB voltage is increased.

### *4.4. Stability Analysis*

The stability analysis of the EBC closed-loop transfer function for the CDB voltage was conducted. The switching period is set as $T_s = 1 \times 10^{-4}$ s. Considering the sampling and the control delay, $T_R = 2T_s$, the grid frequency $f_g = 50$ Hz, and the dc bus voltage reference $U_{Have} = 700$ V. When the dc bus voltage $\overline{U}_H$ changes from 400 V to 1000 V, the denominator of Equation (29) has three poles. Two poles are far away from the origin and the other one is the dominant pole, moving away from the origin with the increasing of the voltage value, as is shown in Figure 12. This figure shows that all the poles are located in the left half plane, and the system is stable.

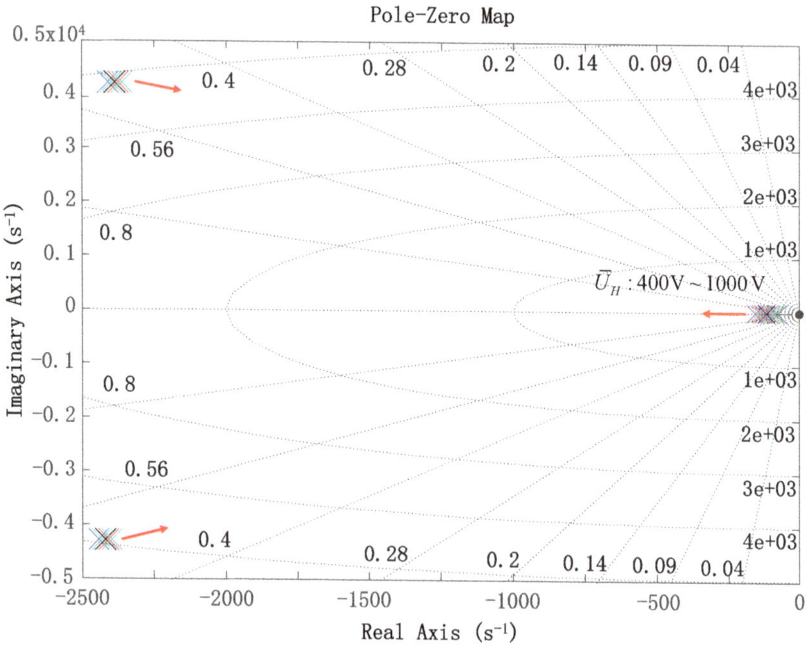


**Figure 12.** Pole distribution of the CDB voltage control system based on EBC controller.

### *4.5. Adaptability Analysis of Model Parameter*

When EBC control mode is adopted, the parameter deviation of the passive components in the model has limited influence on the control effect. In the simulation shown in Figure 13a, the load power increases rapidly from 1.4 kw to 28 kw at 0.7 s, when the capacitance value of two-stage bus has the same direction deviation of ±30%. From the simulation waveform, it can be seen that the voltage of the cascade bus basically does not overshoot in the dynamic process, and the maximum difference of the intermediate bus voltage in the steady-state process is within 2 V, which is consistent with the relationship between the capacitance value of the intermediate bus and the ripple amplitude of the secondary voltage in Formula (23). At the same time, the difference of the capacitance value will affect the average value of the bus voltage. There is almost no overshoot of low-voltage bus voltage in the dynamic process. In the steady state, the bus voltage fluctuates slightly with the increase in power, but the fluctuation value is very small and can be ignored.

As shown in Figure 13b, when the capacitance value of the two-stage bus has a reverse deviation of ±30%, the simulation waveform of the bus voltage is similar to that with the same deviation. It shows that EBC has good adaptability to capacitance parameter deviation, but the ripple of the steady-state process is affected, and the dynamic performance is not affected.

As shown in Figure 13c, when the grid side inductance value deviates by ±30%, the CDB voltage and the LDB voltage have almost no overshoot in the dynamic process. It is verified that EBC has good adaptability to the inductance parameters, and the change of inductance parameters has no impact on the steady-state and dynamic control effects.

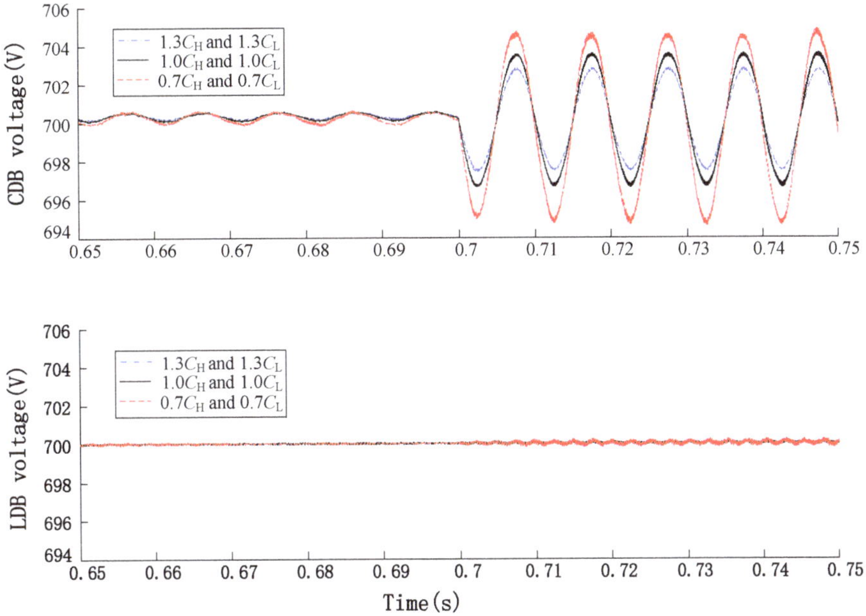


(a)

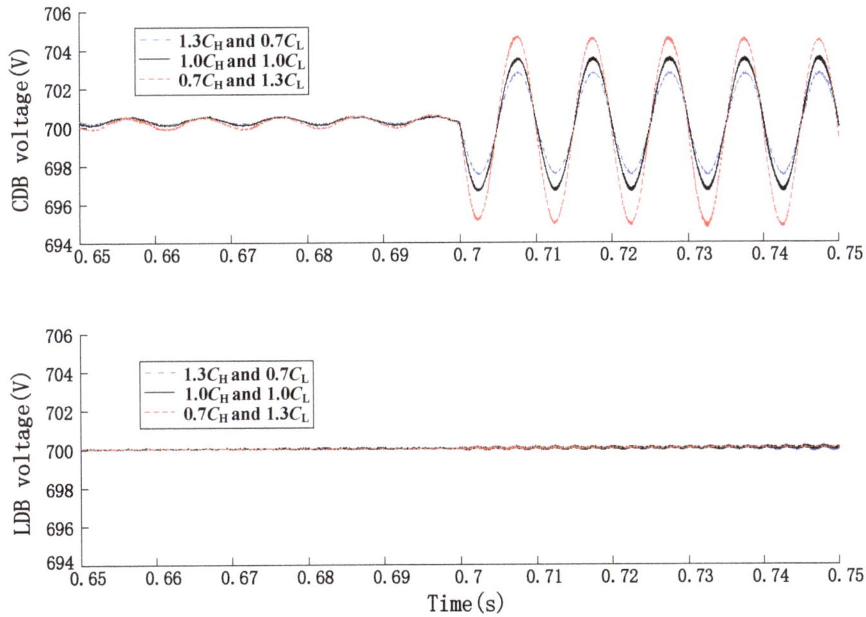


(b)

**Figure 13.** *Cont.*

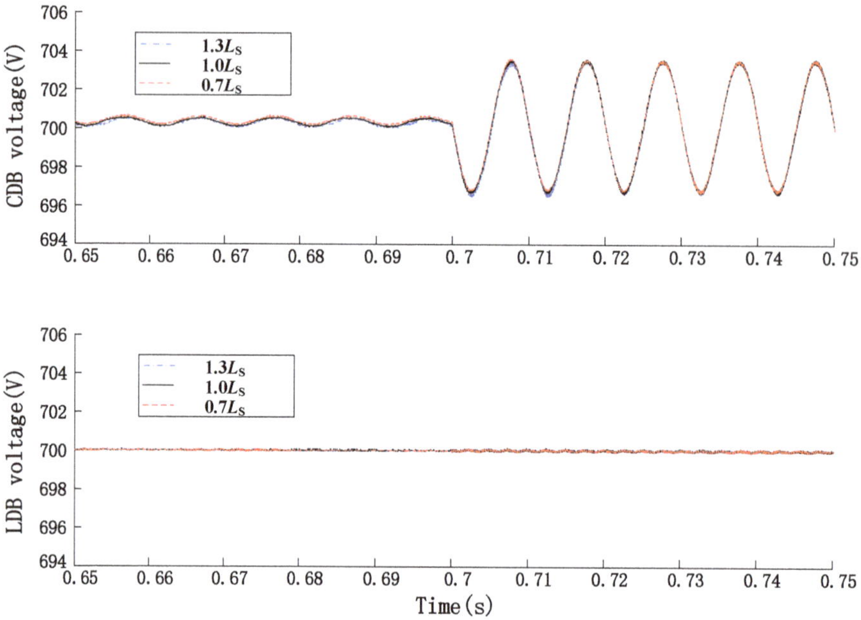


(**c**)

**Figure 13.** Influence of capacitance and inductance deviation on bus voltage: (**a**) ±30% same-direction deviation in two-stage bus capacitance, (**b**) ±30% reverse-direction deviation in two-stage bus capacitance, (**c**) ±30% deviation in the grid side inductance.

## 5. Overall Structure of Proposed Control Strategy

When the CDB voltages of each cascaded module are equal under the voltage balance control, and the duty cycles transmitted to each of the rectifier modules are the same, the received powers of each module are also equal. Therefore, when the CDB voltages are equal, the same duty cycle transmitted to each of the rectifier modules can conveniently ensure the power balance with a small error.

In the DAB controller, the first DAB module controls the LDB voltage, and the other three DAB modules control their respective primary side voltages to follow that of the first DAB module. The control scheme of the parallel DABs is shown in Figure 14; the four DAB modules adjust their respective duty cycle ($d_{f1}$, $d_{f2}$, $d_{f3}$, $d_{f4}$) and use the EBC to change the transmitted power and guarantee that the other three CDB voltages follow the first CDB voltage. The $d_{f1}$ is the phase shift control duty cycle of the first DAB module, and $d_{f2}$, $d_{f3}$, $d_{f4}$ are the phase shift control duty cycle of the second, third, and fourth DAB modules, respectively.

The cascaded rectifier modules adopt the same duty cycle for achieving the transmitted power balance. At the same time, the carrier phase shift is utilized to increase the equivalent switching frequency.

The energy balance control diagram of PET is revealed in Figure 15. In the rectifier, the control scheme adopts a double-loop control structure. The automatic energy regulator (AER) is used in the outer loop for controlling the CDB voltage, and the automatic current regulator (ACR) is used in the inner loop for controlling the grid current. In the DAB, the AER controls the LDB voltage and the CDB voltage balance, and the phase shift (PS) modulation is used in the DAB. For the inverter, the ordinary double-loop d-q decoupling control scheme is adopted. For the DC/DC converter, the ACR is used to control the output current or the output power when the PET is on the grid. In the off-grid state, the DC/DC converter is used to control the LDB voltage.

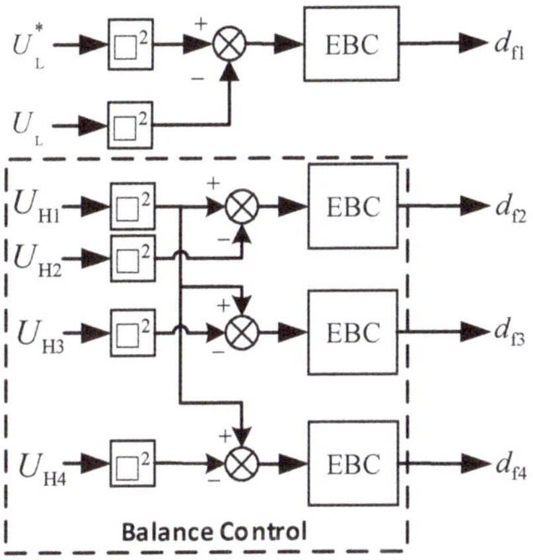


**Figure 14.** Control scheme in the parallel DABs.

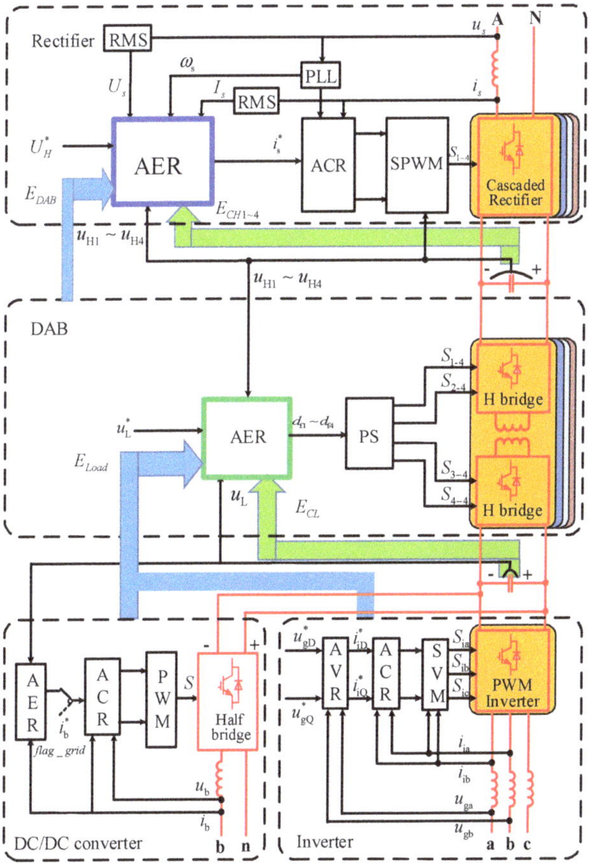


**Figure 15.** Proposed EBC strategy for PET.

## 6. Simulation and Experimental Verification

### 6.1. Simulation Results

As a comparison, the PI control scheme is used in the PET. In the CDB control loop, the voltage sampling values are filtered with a low-pass filter, whose bandwidth is 30 Hz. The simulation parameters are listed in Table 1.

**Table 1.** Parameters of Simulation Circuit.

| Parameters | Values | Parameters | Values |
|---|---|---|---|
| Input ac voltage $u_s$ | 1732 V | CDB capacitor $C_H$ | 4700 uF |
| CDB voltage $U_H$ | 700 V | LDB capacitor $C_L$ | 19,000 uF |
| LDB voltage $U_L$ | 700 V | Transformer turns ratio $n$ | 1:1 |
| Input inductor $L_s$ | 5 mH | HFT leakage inductor $L_s$ | 328 uH |

When the CDB voltage is controlled by the EBC controller and the LDB voltage is controlled by the PI controller, the waveforms of the variables are as presented in Figure 16, in which the red curves $U_{Hx\_ave}$ represent the average values of the real voltage $U_{Hx}$. It is shown that the four CDB voltages are stable and balanced in the steady state, and this proves the effectiveness of the EBC scheme.

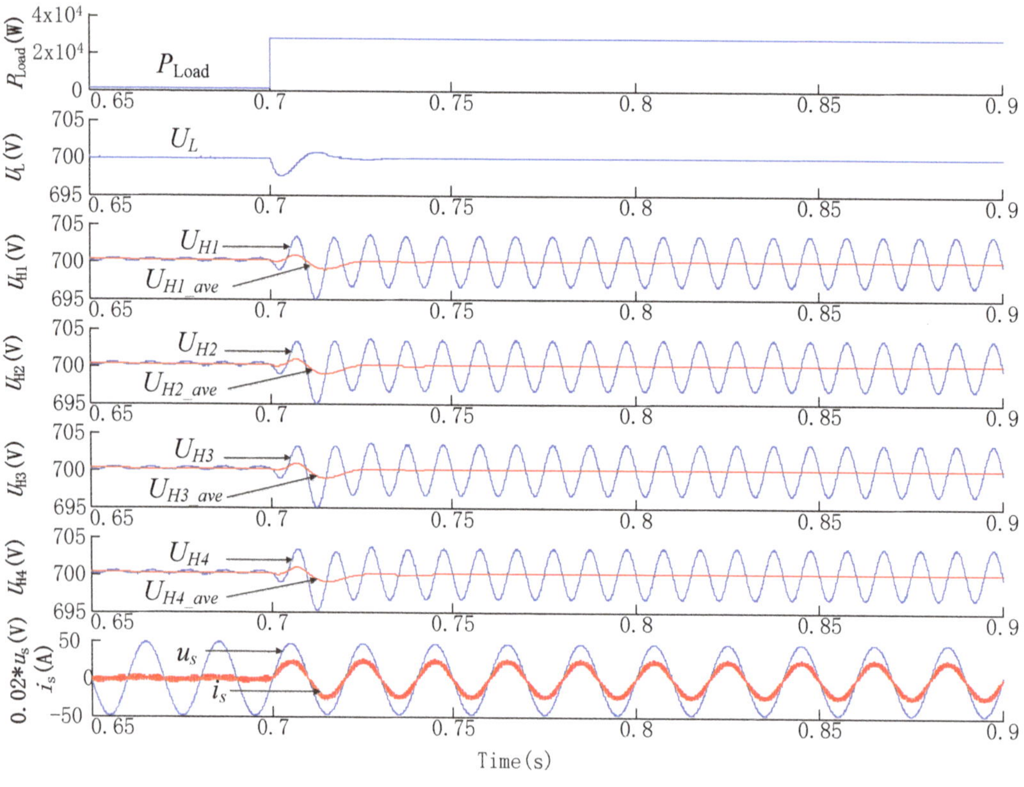


**Figure 16.** Comparison of the four CDB voltages for the load power steps (from 1.4 kW to 28 kW).

For the transient performance of the CDB voltages and the LDB voltage, Figure 17 demonstrates the waveforms of the $U_{H1}$, $U_{H1\_ave}$, and the $U_L$ when the load power increases from 1.4 kW to 28 kW at 0.7 s, in which the PI controller and the EBC controller are used, respectively. The EBC control strategy can achieve a better transient performance than the PI controller, both in the CDB voltage and the LDB voltage, as is shown in Figure 17b,c. Two different time-scale EBC controllers used in two dc bus voltages simultaneously can realize the optimal transient performance, which has the minimum transient time and a very small overshoot, as is shown in Figure 17d.

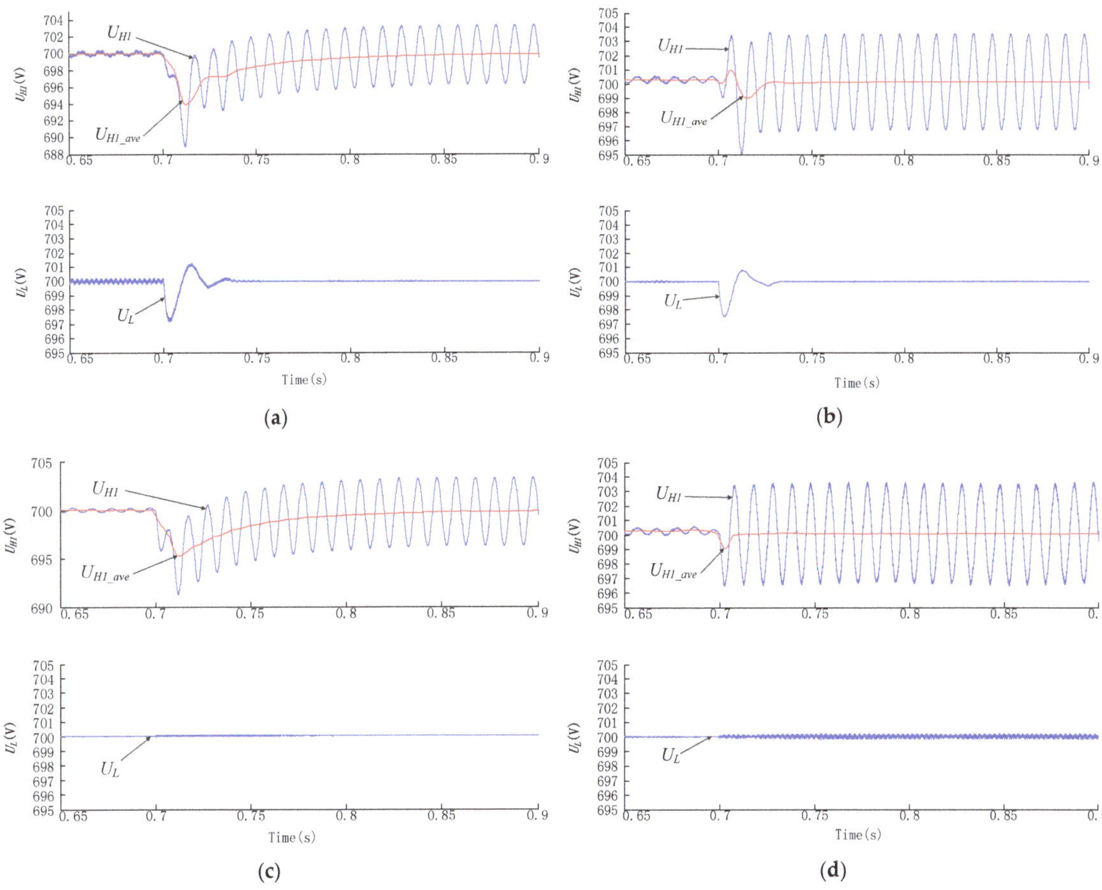


**Figure 17.** Comparison of the CDB voltage and the LDB voltage for load power steps (from 1.4 kW to 28 kW): (**a**) with PI controller in CDB voltage and LDB voltage, (**b**) with EBC controller in CDB voltage and with PI controller in LDB voltage, (**c**) with PI controller in CDB voltage and with EBC controller in LDB voltage, (**d**) with EBC controller in CDB voltage and LDB voltage.

The corresponding track of the time-domain waveform in the phase plane is shown in Figure 18, which is composed of the average of CDB voltage $U_{H1\_ave}$ in the horizontal axis and the LDB voltage $U_L$ in the vertical axis. The smaller the area that is surrounded by the track in the figure means the smaller the change of the energy storage of the two-stage bus capacitance in the transient process, which means the smaller the voltage fluctuation. It can be found from the comparison that the transient performance of the two-stage dc bus voltage can be effectively improved by using EBC.

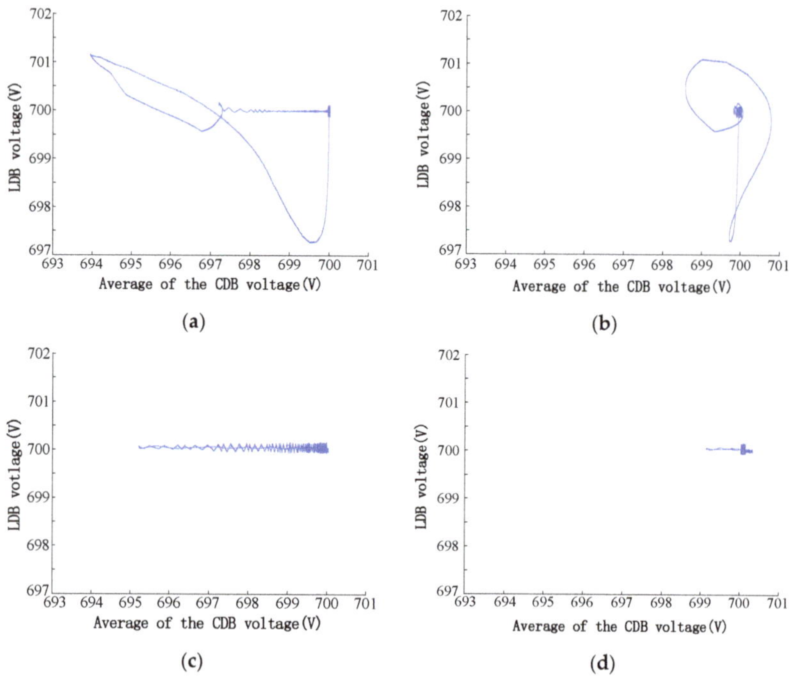


**Figure 18.** Phase-plan diagrams of the average of the CDB voltage and the LDB voltage: (**a**) With PI controller in CDB voltage and LDB voltage, (**b**) with EBC controller in CDB voltage and with PI controller in LDB voltage, (**c**) with PI controller in CDB voltage and with EBC controller in LDB voltage, (**d**) with EBC controller in CDB voltage and LDB voltage.

When the voltage reference changes from 680 V to 700 V at 0.7 s, the waveforms of the variables with the EBC controller are shown in Figure 19. It is shown that four CDB voltages are kept in good consistence whether in the steady state or the transient state.

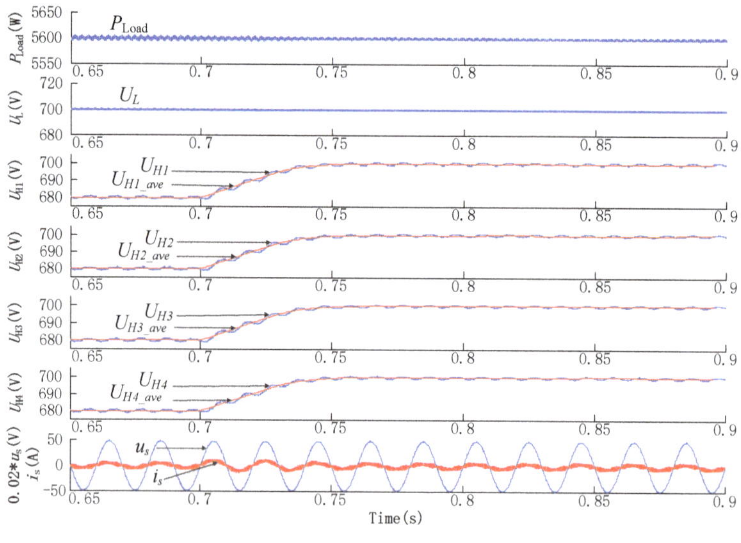


**Figure 19.** Comparison of the four CDB voltages for voltage reference rises (from 680 to 700 V).

To verify the voltage balance control effect in the different parameters of the DAB modules, the leakage inductance of the HFTs in the 4th DAB module is set to be 10% larger than the others in the simulation setup ($L_{s1}$ = $L_{s2}$ = $L_{s3}$ = 328 uH, $L_{s4}$ = 360 uH). The four CDB voltages and the parallel transmitted power are shown in Figure 20. It can be seen that when the voltage balance control is not adopted, the $U_{H4}$ is out of control when the transmission power increases. When the PI controller is used for voltage balance control, the voltage and power balance of the four modules can be realized after a period of time. After adopting the EBC controller, the voltage balance speed of the four modules is accelerated, and the power in the four DAB modules can be quickly balanced.

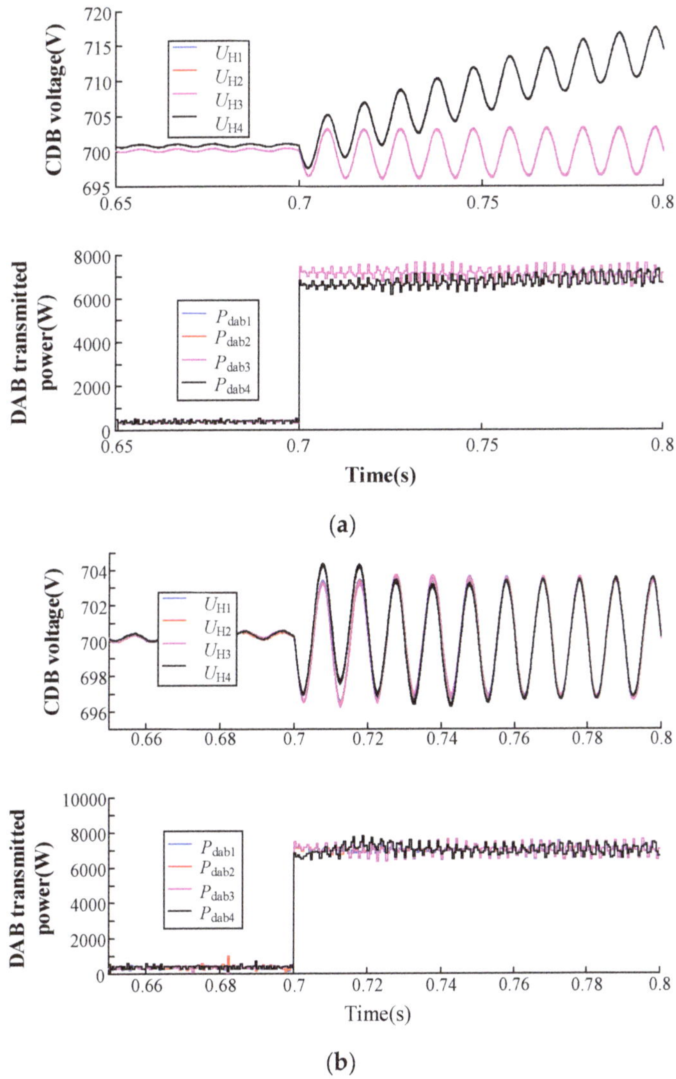


**Figure 20.** *Cont.*

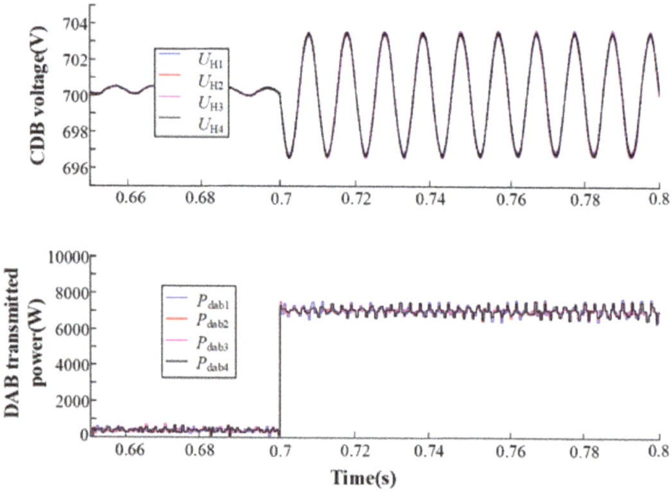


(c)

**Figure 20.** Four CDB voltages and parallel transmitted power waveforms: (**a**) without voltage balance control, (**b**) with voltage balance control using PI controller, (**c**) with voltage balance control using EBC controller.

## 6.2. Experimental Results

The photograph of the experimental prototype is displayed in Figure 21. Considering the load power limitation in the lab, the load power in the experiment changes from 5.20 kW to 9.68 kW, and the other parameters are the same as those used in the simulations.

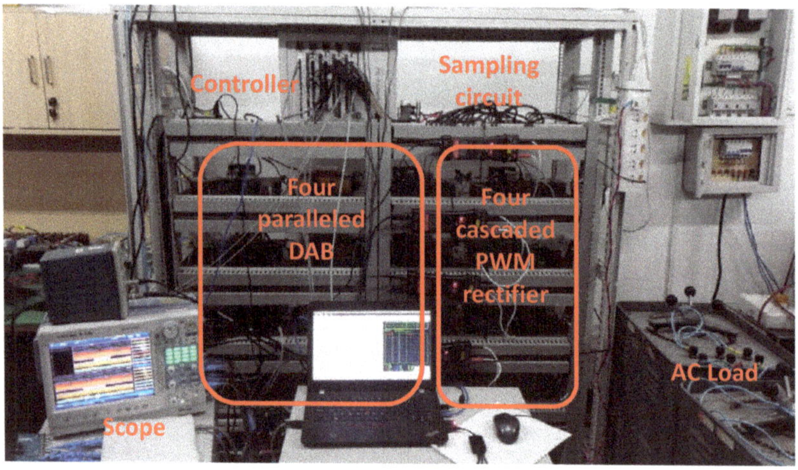


**Figure 21.** Photograph of the prototype.

Figure 22 demonstrates the experimental waveforms of the four CDB voltages ($U_{H4}$, $U_{H3}$, $U_{H2}$, and $U_{H1}$), the LDB voltage ($U_L$), the input current ($i_s$), the input voltage ($u_s$), and output ac line voltage ($u_{ab}$), and the output ac currents ($i_a$, $i_b$, and $i_c$) used by the PI controller and the EBC controller, respectively, in the rectifier and the DAB. For displaying clearly the deviation of the CDB voltage during the transient process, the second CDB voltage $U_{H2}$ is shown in an ac coupling method, and the others are shown in a dc coupling method. With the load power steps, the recovery time of the CDB voltage with the EBC controller is smaller than that in the PI controller, as is shown in Figure 22a,b. The CDB voltage deviation with the EBC controller is similar than that in the PI controller, as is shown in Figure 22b,c. The experiment results have a little difference from the simulation waveforms; the reason is that the changed load power is not obvious for the previous load, and the calculation deviation for the second harmonics voltage ripple has been influenced by measurement error.

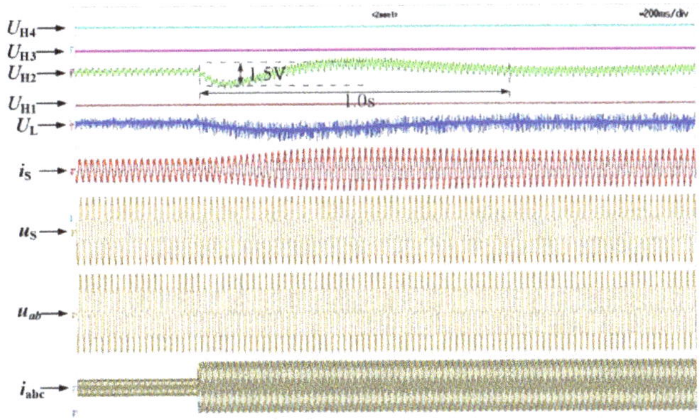


(**a**)

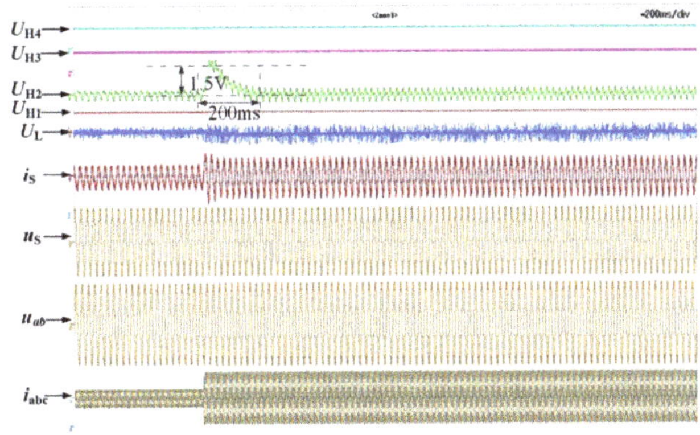


(**b**)

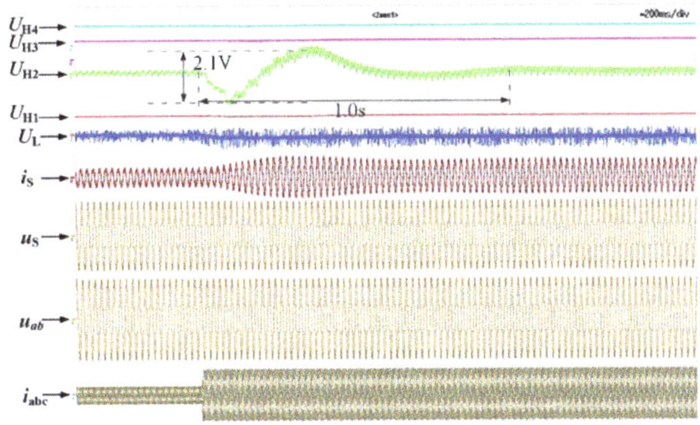


(**c**)

**Figure 22.** *Cont.*

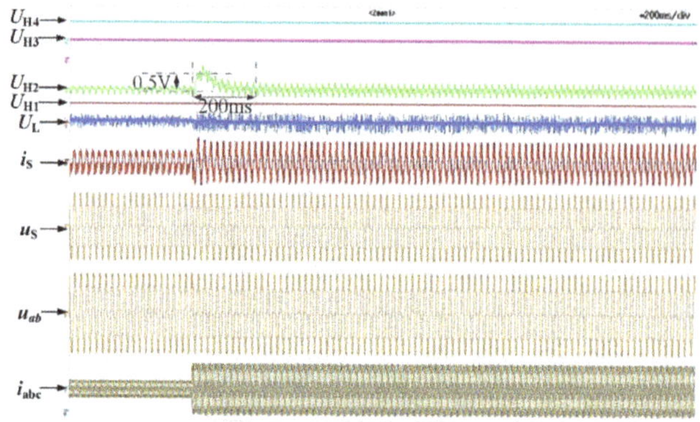


(**d**)

**Figure 22.** The transient experiment waveforms for load power steps (from 5.20 kW to 9.68 kW) (from top to bottom: $U_{H4}$ (500 V/div), $U_{H3}$ (500 V/div), $U_{H2}$ (2 V/div, ac coupling), $U_{H1}$ (500 V/div), $U_L$ (200 V/div), $i_s$ (5 A/div), $u_s$ (1 kV/div), $u_{ab}$ (400 V/div), $i_{abc}$ (20 A/div)): (**a**) with the PI controller in the CDB and LDB voltage, (**b**) with the EBC controller in the CDB voltage, with the PI controller in the LDB voltage, (**c**) with the PI controller in the CDB voltage, with the EBC controller in the LDB voltage, (**d**) with the EBC controller in the CDB and LDB voltage.

The different time scales of the EBC controllers are used in the rectifier and the DAB simultaneously; the transient performance of the two dc bus voltages is more improved than those with the PI controller, as is shown in Figure 22d. The reason is that the interaction between the energy of the CDB capacitors and the LDB capacitors has been considered in the EBC control strategy.

The trajectories of the above four experimental waveforms in the phase plane are shown in Figure 23. The area surrounded by the trajectories obtained by the EBC controller is the smallest, which verifies that EBC has good transient performance. In the figure, the horizontal axis is the voltage of LDB $U_L$, and the vertical axis is the ac coupling value of the 2nd CDB voltage $U_{H2}$.

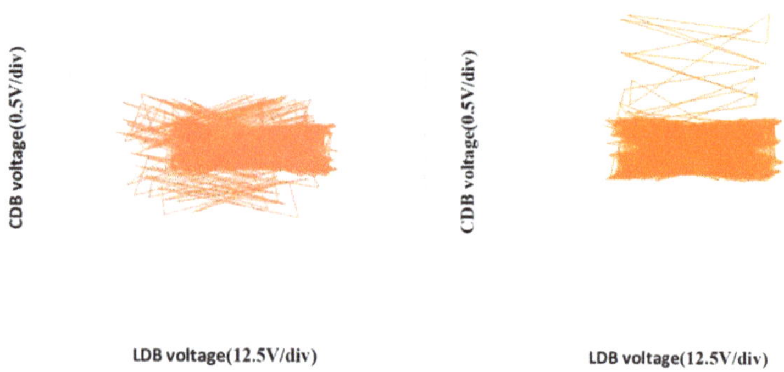


(**a**) (**b**)

**Figure 23.** *Cont.*

(**c**) (**d**)

**Figure 23.** Phase-plane diagrams of the average of the CDB voltage and the LDB voltage: (**a**) with the PI controller in the CDB and LDB voltage, (**b**) with the EBC controller in the CDB voltage, with the PI controller in the LDB voltage, (**c**) with the PI controller in the CDB voltage, with the EBC controller in the LDB voltage, (**d**) with the EBC controller in the CDB and LDB voltage.

The experimental results also verify the control performance of the EBC controller when the CDB voltage reference changes from 680 V to 700 V. Figure 24 demonstrates the experimental waveforms of the signal of the voltage reference changes, the four CDB voltages ($U_{H4}$, $U_{H3}$, $U_{H2}$, and $U_{H1}$), the LDB voltage ($U_L$), the input current ($i_s$), the input voltage ($u_s$), and the output ac line voltage ($u_{ab}$); the output ac currents ($i_a$, $i_b$, and $i_c$) used the PI controller and the EBC controller, respectively, in the different current-limited references. In Figure 24a, the current-limited reference is set up to 10 A; the input current rises to 10 A and then decreases slowly for the charging of the dc bus voltages under the PI controller. In Figure 24b, the current-limited reference is also set up to 10 A; the input current rises to 10 A and then keeps within the limited value until the dc bus voltage reaches the reference value. If the current-limited reference can increase further, for example from 10 A to 30 A, the dc bus voltage can realize the faster transient performance in Figure 24c. Therefore, by utilizing the reasonable current-limit value, the EBC controller can achieve a better transient performance than the PI controller.

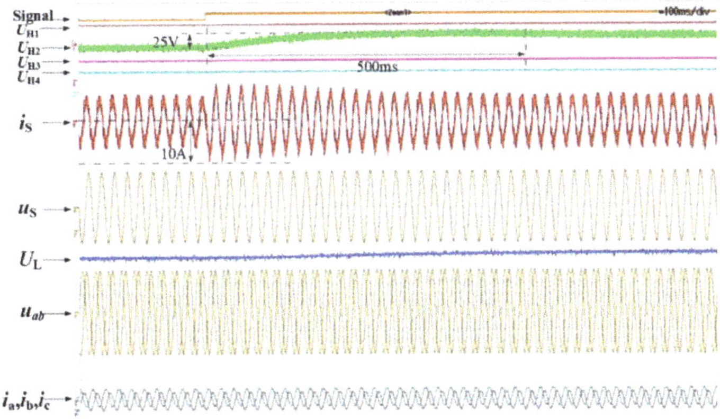


(**a**)

**Figure 24.** *Cont.*

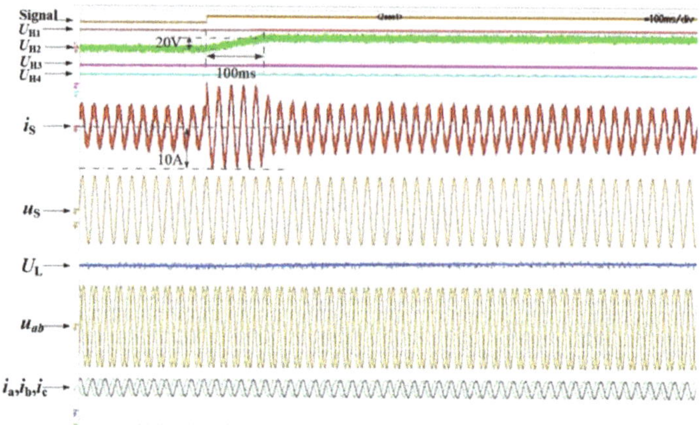


(**b**)

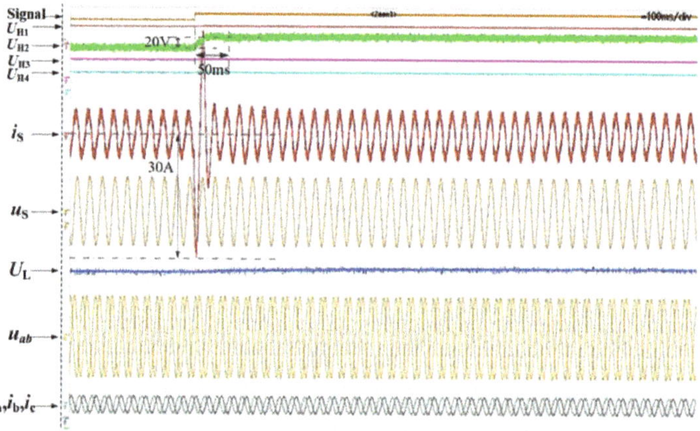


(**c**)

**Figure 24.** The transient experiment waveforms for voltage reference rises (from 680 to 700 V) (from top to bottom: the signal for the voltage reference changes, $U_{H1}$ (500 V/div), $U_{H2}$ (25 V/div), $U_{H3}$ (500 V/div), $U_{H4}$ (500 V/div), $i_s$ (5 A/div), $u_s$ (1 kV/div), $U_L$ (200 V/div), $u_{ab}$ (400 V/div), $i_{abc}$ (20 A/div)): (**a**) with the PI controller in current-limited reference, (**b**) with the EBC controller in current-limited reference, (**c**) with the EBC controller in increased current-limited reference.

## 7. Conclusions

For the voltage-sensitive loads connected with the PET, the EBC strategy can achieve a better transient performance of the dc bus voltages than the conventional PI control strategy. The essential reason is that the interaction between the energy of the CDB capacitors and the LDB capacitors has been considered in the EBC control strategy with different time scales. Through the small-signal analysis of the control strategy in the CDB voltage and LDB voltage, the EBC control strategy can eliminate the impacts of the dc bus capacitors and the fluctuations of the load power compared with the PI control scheme. The stability of the EBC strategy is proved by the stability analysis. The expression of the second harmonic in the CDB voltage is deduced through the instantaneous power-balance relationship in the single-phase PET, and it is helpful in improving the transient performance. Combined with the voltage balance control scheme, the proposed EBC strategy can realize the voltage balance in the cascaded modules. The EBC strategy can be extended to PET, which consists of $n$-H-bridge rectifier modules and $n$-DAB converter modules. The simulation and the experimental results prove the effectiveness of the proposed control strategy.

**Author Contributions:** Conceptualization, G.F.; data curation, P.B.; formal analysis, G.F.; methodology, G.F.; project administration, L.Y.; software, P.B.; supervision, L.Y. All authors have read and agreed to the published version of the manuscript.

**Funding:** This research was funded by the National Key Research and Development Program of China grant number 2020YFB1314002 And The APC was funded by 2020YFB1314002.

**Institutional Review Board Statement:** Not applicable.

**Informed Consent Statement:** Not applicable.

**Data Availability Statement:** The data presented in this study are available on request from the corresponding author.

**Conflicts of Interest:** The authors declare that there are no conflicts of interest regarding the publication of this paper.

## References

1. Mishra, D.K.; Ghadi, M.J.; Li, L.; Hossain, M.J.; Zhang, J.; Ray, P.K.; Mohanty, A. A review on solid-state transformer: A breakthrough technology for future smart distribution grids. *Int. J. Electr. Power Energy Syst.* **2021**, *133*, 107255. [CrossRef]
2. Chen, M.; Chou, S.-F.; Blaabjerg, F.; Davari, P. Overview of Power Electronic Converter Topologies Enabling Large-Scale Hydrogen Production via Water Electrolysis. *Appl. Sci.* **2022**, *12*, 1906. [CrossRef]
3. Sha, G.L.; Duan, Q.; Sheng, W.X.; Pan, A.Q.; Zheng, Y. A Multiport Electric Energy Routing Scheme Applied to Battery Energy Storage System. *J. Eng.* **2021**, *2021*, 6637926. [CrossRef]
4. Zheng, L.; Kandula, R.P.; Divan, D. Current-Source Solid-State DC Transformer Integrating LVDC Microgrid, Energy Storage, and Renewable Energy Into MVDC Grid. *IEEE Trans. Power Electron.* **2022**, *37*, 1044–1058. [CrossRef]
5. Wang, D.; Guan, Z.; Tian, J.; Mao, C.X.; Yang, Y.; Wang, Z. A hybrid redundancy scheme for medium-voltage three-phase cascaded H-bridge electronic power transformer. *IET Power Electron.* **2022**, 1–14. [CrossRef]
6. Xu, X.; Huang, W.; Hu, Y.; Tai, N.; Ji, Y.; Xie, N. Power Management of AC/DC Hybrid Distribution Network With Multi-Port PET Considering Reliability of Power Supply System. *Front. Energy Res.* **2021**, *9*, 721385. [CrossRef]
7. Li, K.; Wen, W.; Zhao, Z.; Yuan, L.; Cai, W.; Mo, X.; Gao, C. Design and Implementation of Four-Port Megawatt-Level High-Frequency-Bus Based Power Electronic Transformer. *IEEE Trans. Power Electron.* **2021**, *36*, 915–927. [CrossRef]
8. Teng, J.; Sun, X.; Zhang, Y.; Fu, H.; Liu, X.; Zhao, W.; Li, X. Two Types of Common-Mode Voltage Suppression in Medium Voltage Motor Speed Regulation System Based on Solid State Transformer With Dual DC Bus. *IEEE Trans. Power Electron.* **2022**, *37*, 7082–7099. [CrossRef]
9. Nair, A.C.; Fernandes, B.G. A Solid State Transformer based Fast Charging Station for Various Categories of Electric Vehicles with Batteries of Vastly Different Ratings. *IEEE Trans. Ind. Electron.* **2022**, *in press*. [CrossRef]
10. Perez, M.A.; Bernet, S.; Rodriguez, J.; Kouro, S.; Lizana, R. Circuit Topologies, Modeling, Control Schemes, and Applications of Modular Multilevel Converters. *IEEE Trans. Power Electron.* **2015**, *30*, 4–17. [CrossRef]
11. Liu, W.; Zhang, K.; Chen, X.; Xiong, J. Simplified model and submodule capacitor voltage balancing of single-phase AC/AC modular multilevel converter for railway traction purpose. *IET Power Electron.* **2016**, *9*, 951–959. [CrossRef]
12. Ma, D.; Lin, S.; Cheng, Q.; Sun, Q. The DC Bus Voltage Control Based on Virtual Inertia for SST. In Proceedings of the 2018 IEEE 7th Data Driven Control and Learning Systems Conference (DDCLS), Enshi, China, 25–27 May 2018; pp. 575–580.
13. Zhao, T.; She, X.; Bhattacharya, S.; Wang, G.; Wang, F.; Huang, A. Power synchronization control for capacitor minimization in Solid State Transformers (PET). In Proceedings of the 2011 IEEE Energy Conversion Congress and Exposition, Phoenix, AZ, USA, 17–22 September 2011; pp. 2812–2818.
14. Wang, G.; She, X.; Wang, F.; Kadavelugu, A.; Zhao, T.; Huang, A.; Yao, W. Comparisons of different control strategies for 20 kVA solid state transformer. In Proceedings of the 2011 IEEE Energy Conversion Congress and Exposition, Phoenix, AZ, USA, 17–22 September 2011; pp. 3173–3178.
15. Zheng, L.; Kandula, R.P.; Kandula, R.P.; Divan, D. New Modulation and Impact of Transformer Leakage Inductance on Current-Source Solid-State Transformer. *IEEE Trans. Power Electron.* **2022**, *37*, 562–576. [CrossRef]
16. Liu, Y.; Yang, S.; Zhang, S.; Peng, F.Z. Comparison of Synchronous Condenser and STATCOM for Inertial Response Support. In Proceedings of the 2014 IEEE Energy Conversion Congress and Exposition (ECCE), Pittsburgh, PA, USA, 14–18 September 2014; pp. 2684–2690.
17. He, F.; Zhao, Z.; Yuan, L.; Lu, S. A DC-link voltage control scheme for single-phase grid-connected PV inverters. In Proceedings of the 2011 IEEE Energy Conversion Congress and Exposition, Phoenix, AZ, USA, 17–22 September 2011; pp. 3941–3945.
18. Ge, J.; Zhao, Z.; Yuan, L.; Lu, T. Energy Feed-Forward and Direct Feed-Forward Control for Solid-State Transformer. *IEEE Trans. Power Electron.* **2015**, *30*, 4042–4047. [CrossRef]
19. Dash, S.K.; Ray, P.K. A New PV-Open-UPQC Configuration for Voltage Sensitive Loads utilizing novel Adaptive Controllers. *IEEE Trans. Ind. Electron.* **2022**, *in press*. [CrossRef]

20. Bahraini, F.; Abrishamifar, A.; Ayatollahi, A. Fast DC Bus Voltage Regulation for a Low Cost Single-Phase Grid-Connected PV Microinverter With a Small DC Bus Capacitor. In Proceedings of the 2020 11th Power Electronics, Drive Systems, and Technologies Conference (PEDSTC), Tehran, Iran, 4–6 February 2020.
21. Alemi, P.; Bae, C.J.; Lee, D.C. Resonance Suppression Based on PR Control for Single-Phase Grid-Connected Inverters with *LLCL* Filters. *IEEE J. Emerg. Sel. Top. Power Electron.* **2016**, *4*, 459–467. [CrossRef]
22. Gao, Y.G.; Jiang, F.Y.; Song, J.C.; Zheng, L.J.; Tian, F.Y.; Geng, P.L. A novel dual closed-loop control scheme based on repetitive control for grid-connected inverters with an LCL filter. *ISA Trans.* **2018**, *74*, 194–208. [CrossRef] [PubMed]
23. Ouyang, H.; Zhang, K.; Zhang, P.; Kang, Y.; Xiong, J. Repetitive Compensation of Fluctuating DC Link Voltage for Railway Traction Drives. *IEEE Trans. Power Electron.* **2011**, *26*, 2160–2171. [CrossRef]
24. Wang, R.; Wang, F.; Boroyevich, D.; Burgos, R.; Lai, R.; Ning, P.; Rajashekara, K. A High Power Density Single-Phase PWM Rectifier With Active Ripple Energy Storage. *IEEE Trans. Power Electron.* **2011**, *26*, 1430–1443. [CrossRef]

*electronics*

MDPI

*Article*

# Evaluation of GaN HEMTs in $H^3$TRB Reliability Testing

**Jose A. Rodriguez *, Tsz Tsoi, David Graves and Stephen B. Bayne**

Center for Pulsed Power and Power Electronics, Department of Electrical and Computer Engineering, Texas Tech University, Lubbock, TX 79409, USA; tsz.tsoi@ttu.edu (T.T.); david.graves@ttu.edu (D.G.); stephen.bayne@ttu.edu (S.B.B.)

* Correspondence: jose.a.rodriguez@ttu.edu

**Abstract:** Gallium Nitride (GaN) power devices can offer better switching performance and higher efficiency than Silicon Carbide (SiC) and Silicon (Si) devices in power electronics applications. GaN has extensively been incorporated in electric vehicle charging stations and power supplies, subjected to harsh environmental conditions. Many reliability studies evaluate GaN power devices through thermal stresses during current conduction or pulsing, with a few focusing on high blocking voltage and high humidity. This paper compares GaN-on-Si High-Electron-Mobility Transistors (HEMT) device characteristics under a High Humidity, High Temperature, Reverse Bias ($H^3$TRB) Test. Twenty-one devices from three manufacturers were subjected to 85 °C and 85% relative humidity while blocking 80% of their voltage rating. Devices from two manufacturers utilize a cascade configuration with a silicon metal-oxide-semiconductor field-effect transistor (MOSFET), while the devices from the third manufacturer are lateral p-GaN HEMTs. Through characterization, three sample devices have exhibited degraded blocking voltage capability. The results of the $H^3$TRB test and potential causes of the failure mode are discussed.

**Keywords:** power electronics; reliability; wide-bandgap

**Citation:** Rodriguez, J.A.; Tsoi, T.; Graves, D.; Bayne, S.B. Evaluation of GaN HEMTs in $H^3$TRB Reliability Testing. *Electronics* **2022**, *11*, 1532. https://doi.org/10.3390/electronics11101532

Academic Editors: Ching-Ming Lai, Yitao Liu and Elias Stathatos

Received: 18 March 2022
Accepted: 9 May 2022
Published: 11 May 2022



## 1. Introduction

Gallium Nitride (GaN) High-Electron-Mobility Transistors (HEMT) have become favorable devices for power electronics applications due to their high bandgap, mobility, and critical field [1]. GaN devices have lower thermal conductivity than silicon (Si) devices but offer lower overall power dissipation and improved efficiency [2]. Compared to Silicon Carbide (SiC) power devices, commercially available GaN HEMTs have lower blocking voltage capability due to their lateral design. However, they have lower parasitic capacitance, making them suitable for high-frequency operation [3]. The lower blocking voltage for GaN HEMTs can be overcome using modular power converter designs by stacking several low voltage devices for high voltage operation with high efficiency [2]. These characteristics have allowed the development of high-power density and high-efficiency power electronics systems, resulting in GaN HEMTs being heavily integrated into power supplies and fast chargers for EV applications [4]. These applications require high-power density systems that limit the available cooling capacity and thus operate the GaN HEMT at high operating temperatures. Implementing GaN HEMTs in EV charging stations means operating in potentially humid environments. Thus there is great interest in evaluating the reliability of GaN power HEMTs under thermal stresses.

Reliability testing of GaN HEMTs has primarily focused on thermal stresses. A known reliability issue with GaN HEMTs is its dynamic on-state resistance due to hot-carrier injection and trap buffers [5], resulting in higher conduction losses and reduced current conduction. This effect is temperature-dependent and can be suppressed by using a substrate with high thermal conductivity, such as a SiC substrate [6]. However, most GaN HEMTs for power electronics applications are grown on a Si substrate due to cost [4], so reliability tests have focused on thermal stresses induced by a current bias. Several works in reliability testing of GaN HEMTs evaluated its current capabilities in single or repetitive pulsing [7,8], where the change in its current conduction determines the degradation. Q. Song et al. have evaluated GaN cascode HEMTs under surge current events and transient overvoltage conditions [9]. There have also been works focused on accelerated aging due to thermal stresses from thermal power cycling between 25 and 125 °C [10]. S. Song et al. observed electrical cracks in the degraded devices, corresponding to increased leakage current, while the threshold voltage and on-state resistance have been unaffected [11]. These works have not included a combination of humidity, high temperature, and blocking voltage in their testing, as they can accelerate power devices' failure mechanisms [12].

The High Humidity, High Temperature, Reverse Bias ($H^3TRB$) accelerated lifetime test evaluates the robustness of the semiconductor device. Humidity can penetrate the device's packaging, causing accelerated cracking and corrosion [12]. The presence of high ambient temperature can further accelerate corrosion once moisture is present in the package [13]. Several works have evaluated Si IGBT modules and discrete devices using the $H^3TRB$ test [14–16] to identify characteristics that help estimate the remaining lifetime. Electrochemical corrosion and electrochemical migration were the primary failure mechanisms of these Si IGBTs [16]. These works have implemented current sensing to monitor the leakage current of the devices during testing. Changes in leakage current are potential signs of accelerated aging of the device or device degradation [17]. An end-of-life $H^3TRB$ reliability study was conducted for 600/650 V enhancement-mode GaN HEMTs [18]. These devices were tested at 85 °C with relative humidity (RH) of 85% and blocking voltage of 480 V, with pauses for device characterization every 500 h. A. Brunko et al. found no device failure unit after 2300 h of testing. The PCB housing for their devices had degraded after 1200 h of testing, highlighting the challenge of conducting reliability testing under humidity. Although different enhancement-mode GaN HEMTs were tested, the discussion about the observed failure modes was limited. The proposed work establishes a testbed to subject commercial GaN HEMTs to an $H^3TRB$ test to evaluate their reliability and to investigate any failure mechanism that arises due to humidity.

This paper presents a comparative study of GaN HEMTs evaluated under the $H^3TRB$ test, following the JEDEC standard JESD22-A101. GaN power devices from three manufacturers were utilized based on blocking voltage and current rating. Seven devices from each manufacturer were selected for a total of twenty-one devices. These devices were placed in an environmental chamber with an ambient temperature of 85 °C and 85% RH while blocking 80% of their voltage rating as part of the JEDEC standard. Their leakage current was monitored for each device under test (DUT) during testing to detect when device failure has occurred. Testing was conducted for one thousand hours. After testing, the characteristics of the devices were measured to identify degradation. The result of the test and potential failure mechanisms are discussed.

## 2. Methodology

This section presents the methodology for performing the $H^3TRB$ test on the GaN HEMTs. A sample of twenty-one devices was used for testing, sharing a blocking voltage rating of 650 V and a forward current rating between 30 A and 42 A. Device group D1 consists of TP65H035G4WS from Transphorm, device group D2 consists of GAN063-65WSAQ from Nexperia, and device group D3 consists of GS-065-030 from GaN Systems. Devices from groups "D1" and "D2" utilize a HEMT cascode configuration, where a GaN depletion-mode HEMT is in series with an internal low voltage Si metal–oxide–semiconductor field-effect transistor (MOSFET) to achieve a normally-off transistor device. This topology is the most common among GaN HEMTs for power electronics applications [4]. Devices from group "D3" are p-GaN HEMTs that use a p-type gate to achieve a normally-off device. The main advantage of p-GaN HEMTs over the cascode configuration is direct control of the GaN device and a less complex device structure. However, it has a lower threshold voltage that is prone to instability [19]. Table 1 lists the electrical parameters of the DUTs [20–22].

**Table 1.** Characteristics of devices under test.

| Device Parameter | Group "D1" | Group "D2" | Group "D3" | Unit |
|---|---|---|---|---|
| Part # | TP65H035WS | GAN063-650WSA | GS-065-030-2-L | |
| Blocking Voltage | 650 | 650 | 650 | V |
| Forward Current | 46.5 | 34.5 | 30 | A |
| On-state Resistance | 35 | 50 | 50 | mΩ |
| Threshold Voltage | 4 | 3.9 | 1.7 | V |
| Forward Current @ $V_{th}$ | 1 | 1 | 10 | mA |

Each DUT has its biasing circuit consisting of a fuse and a current sense resistor (CSR) for low-side sensing. Figure 1 shows the biasing circuit for a single DUT. A Sorensen XHR 600-1.7 high voltage power supply provides the high voltage bias of 520 V across the drain and source of the DUT, following JEDEC standards. The 200 mA fuse allows the $H^3$TRB test to run uninterrupted from device catastrophic failures. A 100 Ω CSR is utilized to measure the leakage current of each device during testing. The DUTs are mounted onto a daughterboard which allows several devices to be tested simultaneously. An acrylic coating is applied to the daughterboard to withstand operation inside the environmental chamber. High voltage putty was also applied around unused and exposed terminals to minimize the likelihood of voltage breakdown not caused by device degradation. The DUTs are subjected to 85 °C and 85% RH inside a T2RC-A-F4T Environmental Chamber for a thousand hours. Figure 2 shows the overall $H^3$TRB testbed.

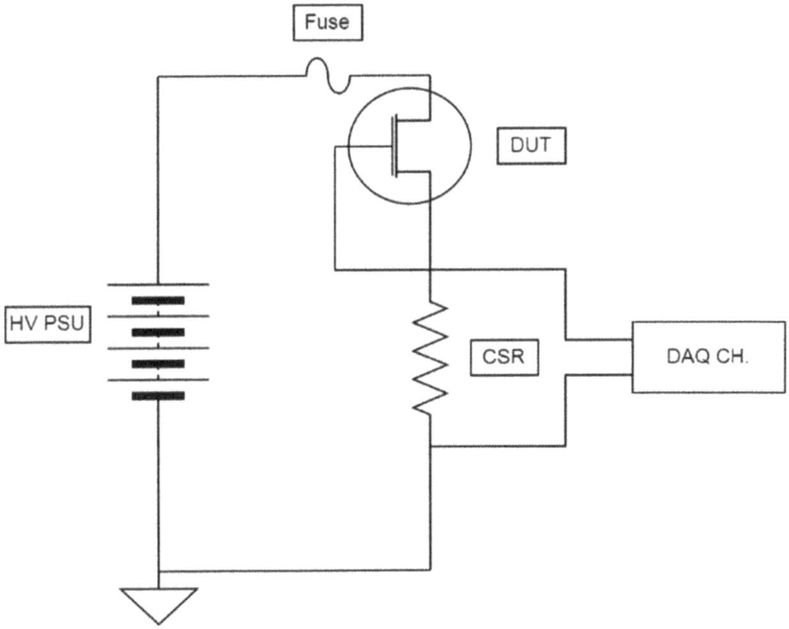


**Figure 1.** Schematic of $H^3$TRB testbed.

An NI data acquisition system (USB-6255 myDAQ) measured the voltage drop across the CSR with a sampling rate of 1 sample every 10 s. Fast sampling rates are unnecessary as the devices are DC-biased, and the leakage current is expected to increase gradually. The average leakage current per hour was calculated, and the averaged data are presented in Figure 3.

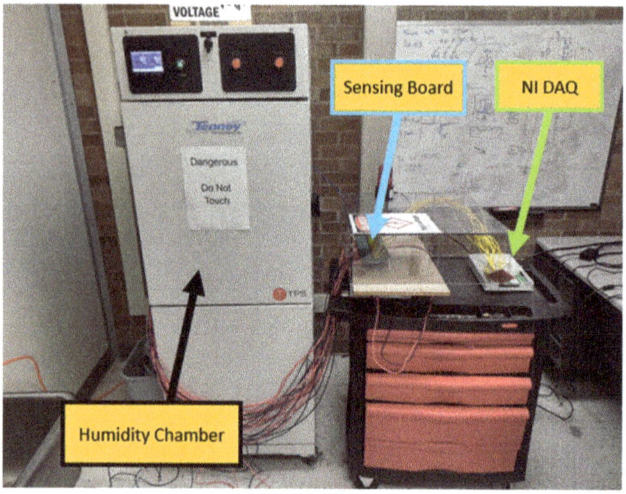


**Figure 2.** $H^3$TRB testbed.

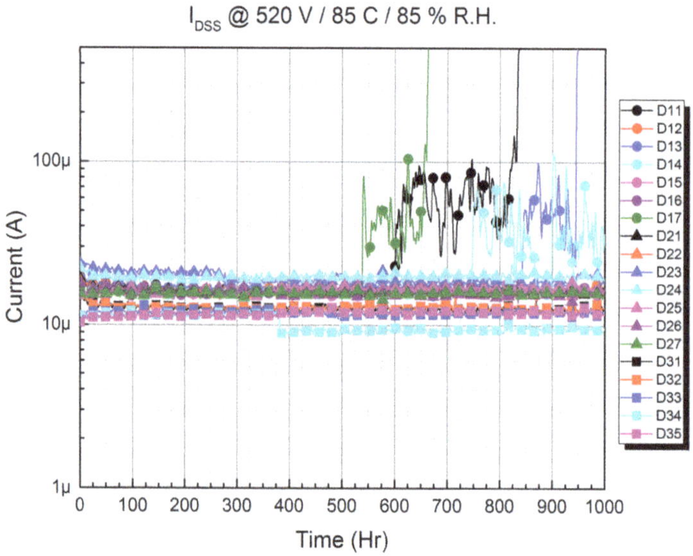


**Figure 3.** Leakage current of DUT during testing.

The average leakage current for all devices was between 10 μA and 20 μA. Devices "D11", "D13", and "D17" exhibit an increase in leakage current before sharply increasing above the leakage current threshold of 200 μA. Table 2 shows a summary of the results from the $H^3$TRB test. The fuses for these three DUTs were blown, suggesting that the devices had shorted during testing. From Figure 3, device "D11" is expected to have shorted 825 h into the test, device "D13" has shorted after 950 h, and device "D17" has shorted after 650 h. Device "D14" has also shown increased leakage current but has not sharply risen above the threshold like the other three devices. Leakage current data are missing for devices "D36" and "D37" due to a software error but were otherwise tested under the same test conditions as the other devices. Out of the sample of devices tested, devices "D11", "D13", and "D17" are considered degraded.

**Table 2.** Summary of $H^3$TRB test.

| Device Group | Failure Present | Affected Devices | Issue |
|---|---|---|---|
| D1 | Yes | D11, D13, D17 | Opened fuse |
| D2 | No | None | None |
| D3 | No | None | None |

## 3. Results

The characterization results of the DUTs are presented. The DUTs were characterized using a B1505A curve tracer before and after the test to identify degraded electrical characteristics. The blocking voltage curve, transfer characteristic curve, gate-source leakage curve, and output characteristic curve was measured for all DUTs.

Figure 4 shows the blocking voltage curve for device group "D1". For all DUTs except devices "D11", "D13", and "D17", the leakage current is consistent between testing. The breakdown voltage curve for devices "D11", "D13", and "D17" quickly increases to the leakage current threshold of 100 µA, confirming that the devices are unable to hold off voltage in the off state. Device group "D1" utilizes the GaN cascode topology, where the Si MOSFET initially holds off the voltage in the off state. As the blocking voltage increases, the gate-source of the internal GaN HEMT is biased negatively, pinching off its channel and thus allowing the internal GaN HEMT to block voltage. The degraded DUTs' inability to hold off voltage suggests that either a short is present between the drain and source of the DUT or the DUT's internal Si MOSFET is damaged. Figure 5 shows the gate-source leakage current for device group "D1". The gate current is below 40 pA for all devices and between testing. The internal Si MOSFET oxide layer of the degraded devices is intact.

Figure 6 shows the transfer characteristics for device group "D1". The threshold voltage was measured with a compliance limit of 10 mA on the SMU of the B1505A. The threshold voltage curve for the non-degraded devices is consistent between testing. However, the curve tracer cannot measure the threshold voltage curve for the degraded devices due to its drain current limit.

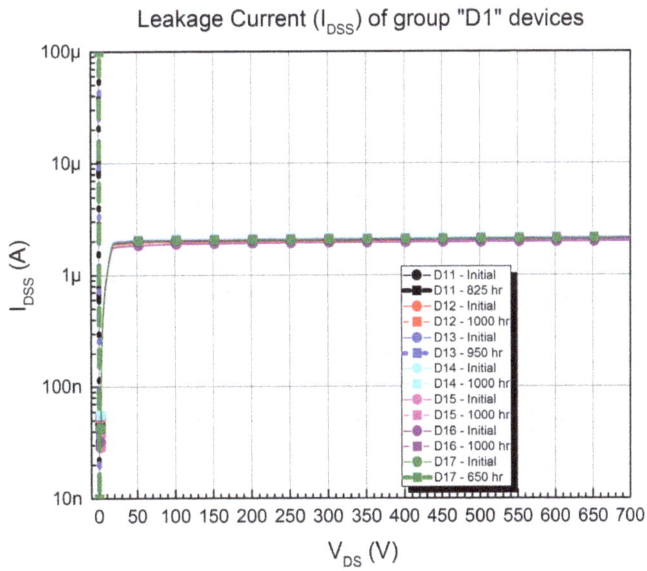


**Figure 4.** Blocking voltage curve for device group "D1".

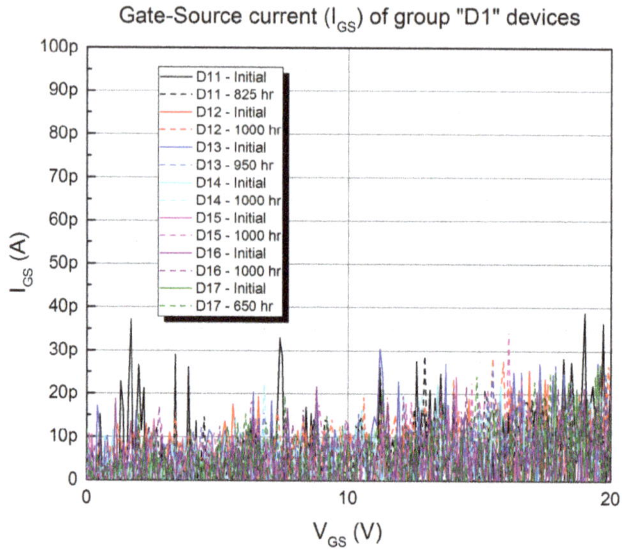


**Figure 5.** Gate leakage current curve for device group "D1".

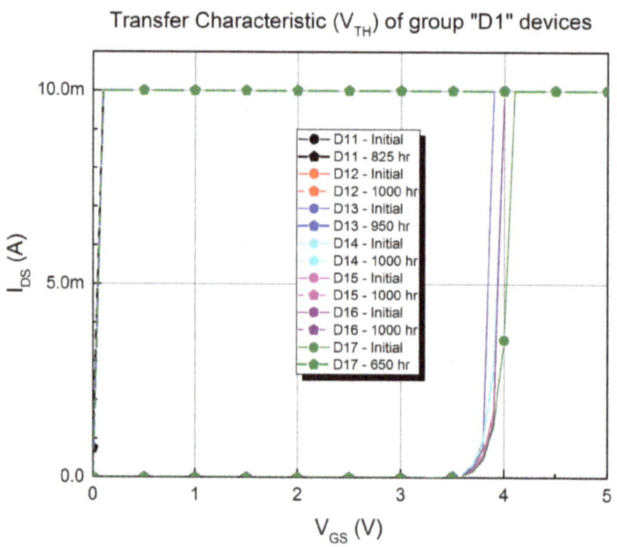


**Figure 6.** Transfer characteristic curve for device group "D1".

Figures 7–9 show the output characteristics between testing of the degraded devices "D11", "D13", and "D17", respectively. For all three devices, significant current flow is measured at 0 VGS and 4 VGS, where it was in the off state in the initial characteristics. The forward current reached 6 A, 1 A, and 2 A with a 3 VDS bias at 0 VGS for devices "D11", "D13", and "D17", respectively. However, the forward current of the degraded devices increases approximately to the same level as their initial characteristics at 8 VGS and above. Table 1 and Figure 6 show that the degraded devices' initial threshold voltage is 4 V. The output characteristic curves show that the internal Si MOSFET channel resistance can be controlled with the gate–source voltage.

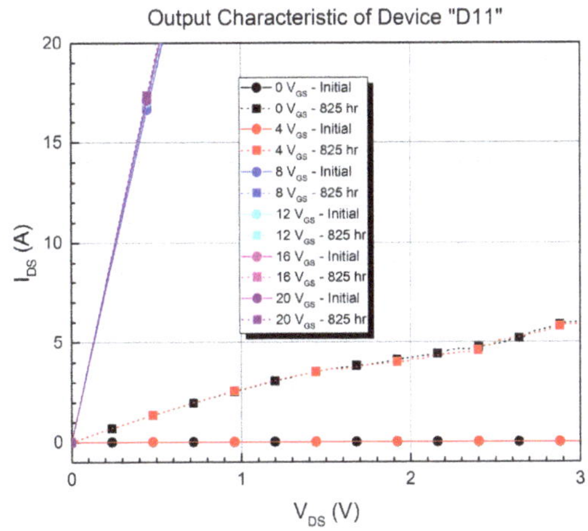


**Figure 7.** Output characteristic curve of device "D11".

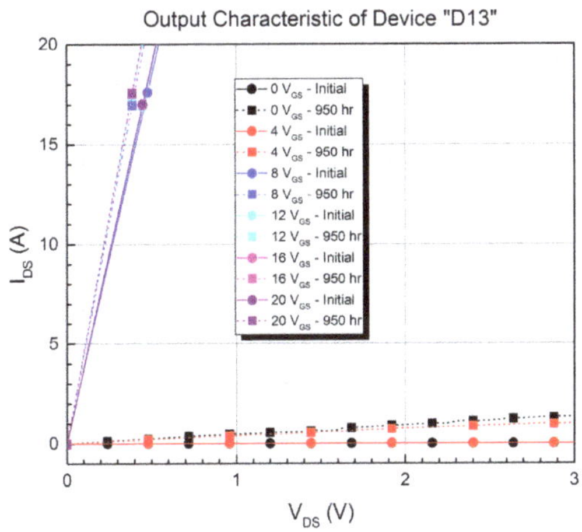


**Figure 8.** Output characteristic curve of device "D13".

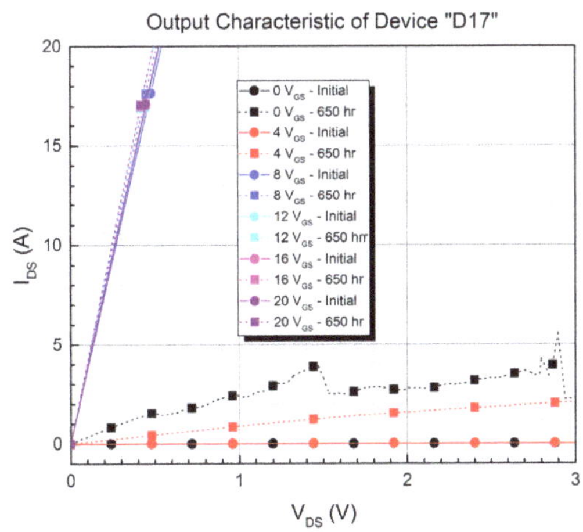


**Figure 9.** Output characteristic curve of device "D17".

The drain–source resistance was calculated using the same data from the previous output characteristic figures. Figures 10–12 show the drain–source resistance of the degraded devices "D11", "D13", and "D17", respectively. The gate–source voltage was swept from 0 to 20 VGS. The measured drain–source resistance varies between 0.4 and 2.0 Ω from 0 to 4 VGS, whereas at 5 VGS and above, the drain–source resistance drops towards its initial on-state resistance value. The resistance difference between the initial and final on-state resistance was 0.3 mΩ for device "D11". However, devices "D13" and "D17" have a much more significant drop of 4.0 mΩ and 2.0 mΩ. The overall current-conducting capabilities of the degraded devices were not affected, but their ability to block voltage was.

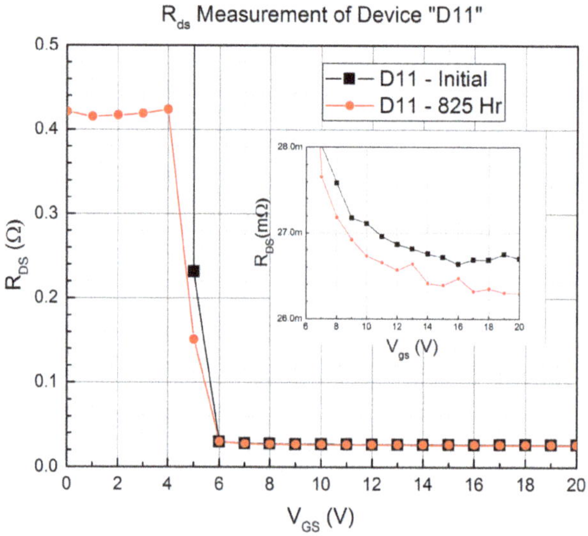


**Figure 10.** Drain–source resistance curve of device "D11".

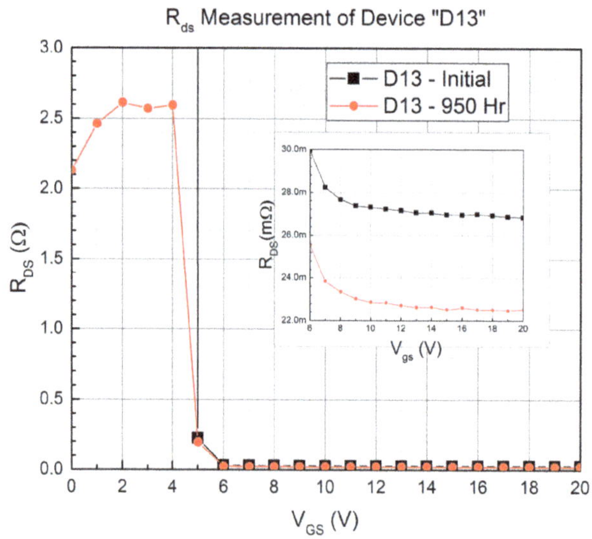


**Figure 11.** Drain–source resistance curve of device "D13".

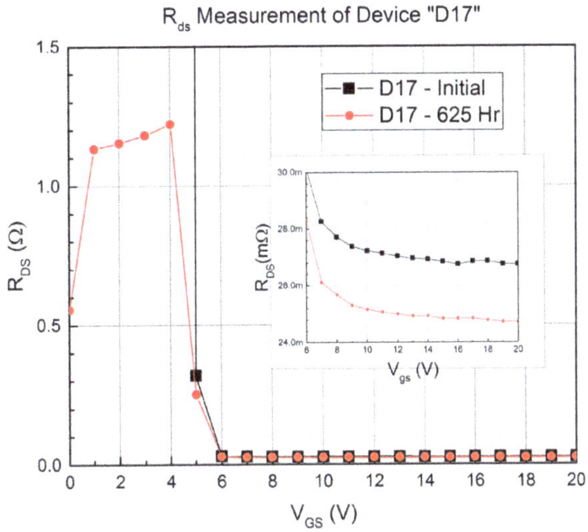


**Figure 12.** Drain–source resistance curve of device "D17".

The remaining devices from groups "D2" and "D3" were characterized. Figures 13 and 14 show the blocking voltage curve for device groups "D2" and "D3", respectively. The leakage current in the figures is consistent between testing. As shown in Figure 3, the leakage current from device groups "D2" and "D3" were constant throughout testing.

Figures 15 and 16 show the transfer characteristic curve for device groups "D2" and "D3", respectively. The threshold voltage was measured with a compliance limit of 100 mA for device group "D3". The high compliance limit was set to the threshold voltage at 10 mA following the datasheet specifications of device group D3, while the other two groups specified threshold voltage at 1 mA. While the threshold voltage curve for device group "D2" is consistent between testing, there was an increase in the rate of drain current for devices "D31", "D32", "D36", and "D37". This increase in the current drain rate suggests that the transconductance has shifted. All devices from device group "D3" have minor threshold voltage shifts up to 0.2 V, and both positive and negative shifts are observed.

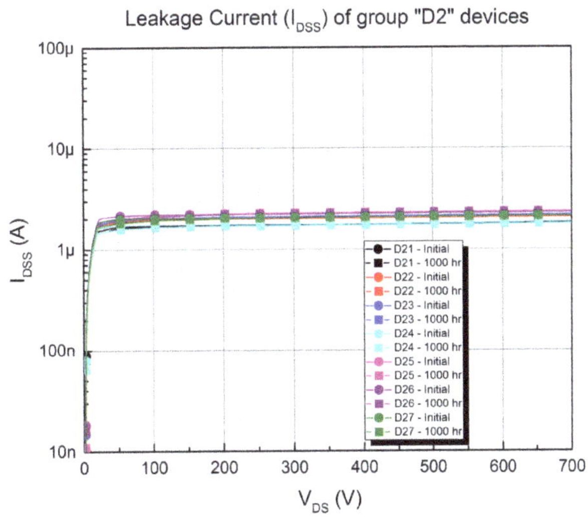


**Figure 13.** Blocking voltage curve for device group "D2".

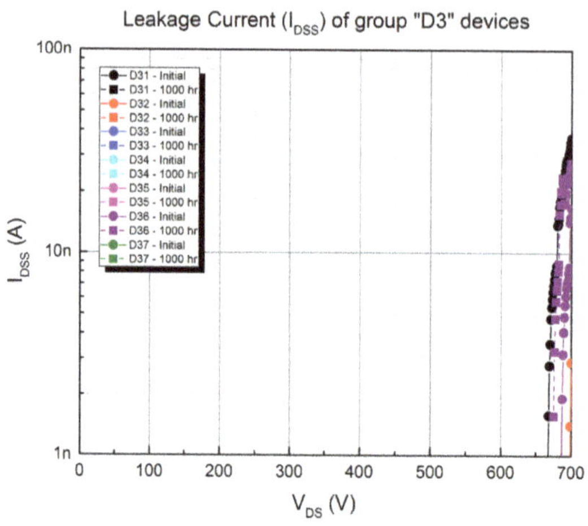


**Figure 14.** Blocking voltage curve for device group "D3".

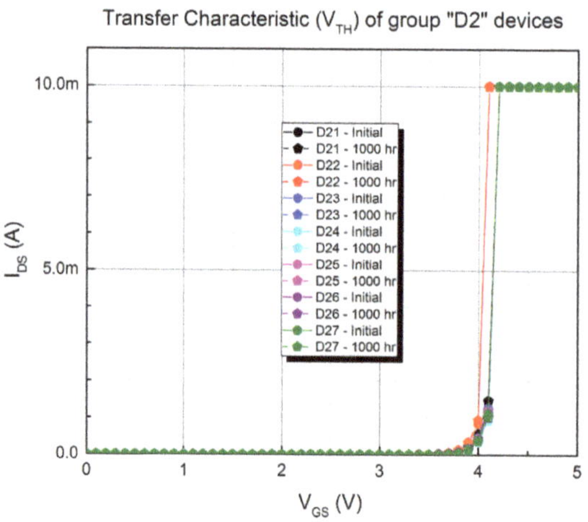


**Figure 15.** Transfer characteristic curve for device group "D2".

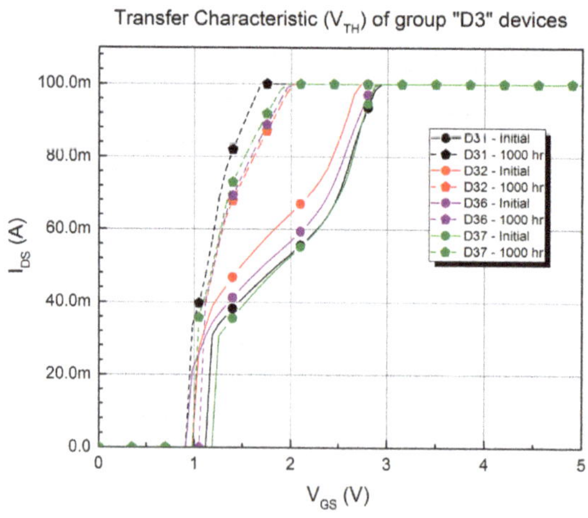


**Figure 16.** Transfer characteristic curve for device group "D3".

Figures 17 and 18 shows the output characteristic curve of device "D27" and "D37". Since the previous characterization curves for device group "D2" have not shown any changes, it is expected that all the devices in that group have a similar forward current profile, as shown in Figure 17. An increased current flow is observed for device group "D3" once the gate–source voltage is above the threshold. However, the forward current reaches the same level as the initial characteristics at 3 VGS and above. This behavior is observed for all devices in device group "D3". The devices' forward current and blocking voltage capability from device groups "D2" and "D3" have not been affected by the $H^3TRB$ test.

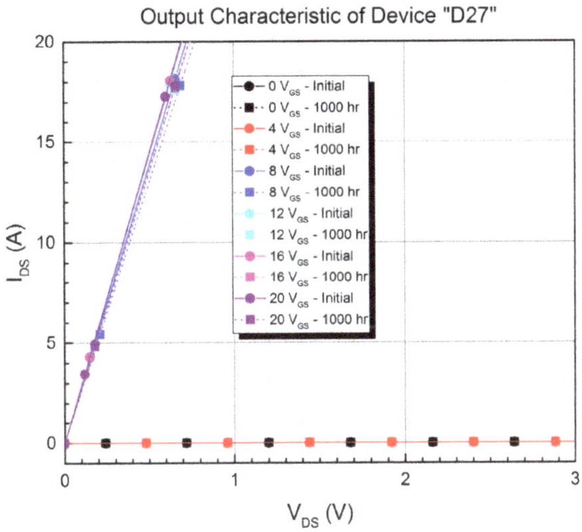


**Figure 17.** Output characteristic curve of device "D27".

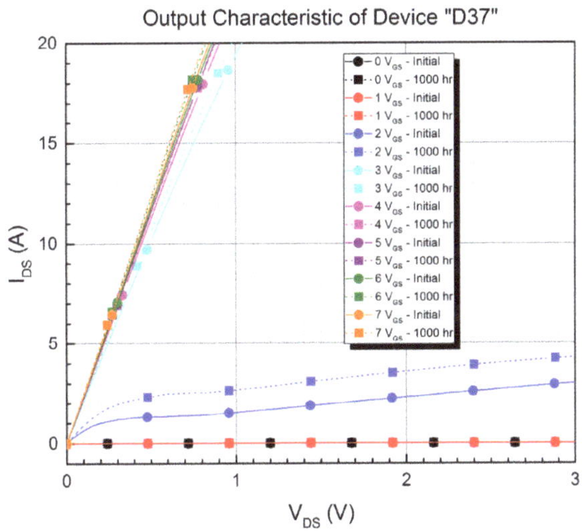


**Figure 18.** Output characteristic curve of device "D37".

## 4. Discussion

Utilizing the output characteristics and drain–source resistance curves of the degraded devices "D11", "D13", and "D17", some inferences can be made about the likely cause of degradation. For each degraded device, the drain–source resistance is between 0.41 Ω and 2.6 Ω when the gate bias is below the initially tested threshold voltage value. However, the degraded drain–source resistance falls below the initial values once the gate bias surpasses the threshold voltage. This change in drain–source resistance indicates that the internal Si MOSFET is intact and that there is a secondary path between the gate and drain of the internal GaN HEMT that allows current flow when the device is in the off state. Transphorm and Q. Song et al. [9,23] observed and documented this short in their reliability testing and caused dielectric failure in the device when subject to a high electric field. The test conditions of the previous works are different from the presented work. However, all three cases subject the DUT to heat, either localized or from the environment, and a high electric field due to a transient or static voltage bias. Based on the characteristic curves of the degraded devices, it is believed that the failure mode observed in the degraded devices are like the failure modes seen in [9,23].

Although the underlying cause for this failure is known, the results from the presented work still provide novel information on the long-term reliability of GaN HEMTs operating in high humidity, high temperature environments. Similar failure modes have been observed in three different testing environments. The test conditions and time to failure reported by Transphorm are of interest for comparison due to using the same devices. Transphorm provides reliability data under High Temperature Reverse Bias (HTRB) testing following the JEDEC standard JESD22-A108. The devices were biased at 520 V between drain and source with an operating temperature of 150 °C. Transphorm tested 231 devices for qualification, with all devices passing after one thousand hours of testing [24]. When comparing the HTRB results presented by Transphorm to the presented work results, there is a clear distinction between the number of device failures seen. The lower operating temperature of the $H^3$TRB test should result in a significantly longer lifetime. However, more devices failed well before the one thousand hour mark indicating a different factor is a dominant cause for equiree. With both tests being held to their respective JEDEC standard, the additional humidity component is the only substantial environmental factor. Thus, it is evident that the presence of high humidity drastically reduces the time to failure of the devices in group "D1". The high humidity in the environment likely exacerbated this failure mode, as it increases the presence of mobile charges, such as Na+ and K+, for charge trapping [25,26]. As humidity penetrates the packaging, positive mobile charges attach to the passivation layer on the edge termination, enhancing the local electric field above the critical field of the insulator to create a short [9,27].

Through analyzing the electrical characteristics of the DUTs, some inferences can be made as to possible reasons why the high humidity did not cause any failures in device group "D2". The difference between the processing techniques of each manufacturer is speculated to create a passivation layer that is less susceptible to defects caused by charge trapping in high humidity environments. As the passivation layer degrades due to charge trapping, these processing techniques influence the enhancements on localized electric fields and whether dielectric breakdown occurs. Such differences include passivation material and implementation of field rings. Simulation work was conducted by R. Natarajan et al. on different passivation materials for GaN HEMT devices and found differences in the electric field distribution, thus influencing the breakdown voltage [28]. From a design perspective, increasing the number of field rings between the gate and drain of the GaN HEMT would better shield the passivation from a high localized electric field [29]. Further investigation into the device structure for device groups "D1" and "D2" is equireed.

Humidity is also expected to have changed the transfer characteristic curve for some devices in device group "D3". Unlike GaN cascode HEMTs, p-GaN are more prone to threshold voltage instability issues [19,30–32] due to their Magnesium acceptors [19,31]. Much of the reliability assessment for threshold voltage for wide-bandgap has been under a high temperature, gate bias (HTGB) test [33,34], or gate bias stress under characterization [30,32] for p-GaN HEMTs. Ultimately, performance degradation was not observed in the device group "D3".

## 5. Conclusions

This paper evaluates commercial GaN devices under high humidity, high temperature, and high voltage accelerated tests. Twenty-one GaN HEMTs from three manufacturers were tested at 85 °C, 85%, while blocking voltage at 520 V for one thousand hours. Degradation was observed in three devices. Although two manufacturers used similar device topology, all degraded devices were from one manufacturer. The degraded devices could not block voltage, but their forward current capabilities were unaffected. It is suspected that the humidity has accelerated charge trapping in the degraded devices, resulting in higher localized electric fields. It is speculated that the differences seen between device groups "D1" to "D2" are due to either a processing difference in the passivation layer or a design difference in the number of field rings. Future work includes failure analysis of the degraded devices through decapsulation to verify the failure mechanism and an investigation into the device structure of the devices.

**Author Contributions:** Conceptualization, S.B.B.; methodology, J.A.R.; data collection, J.A.R. and T.T.; formal analysis, J.A.R., T.T. and D.G.; writing—review and editing, J.A.R., T.T. and D.G.; supervision, S.B.B.; project administration, S.B.B.; funding acquisition, S.B.B. All authors have read and agreed to the published version of the manuscript.

**Funding:** This research received no external funding.

**Conflicts of Interest:** The authors declare no conflict of interest.

## References

1. Lidow, A.; Strydom, J.; de Rooij, M.; Reusch, D. *GaN Transistors For Efficient Power Conversion*; John Wiley and Sons: West Sussex, UK, 2015.
2. Guacci, M.; Anderson, J.A.; Pally, K.L.; Bortis, D.; Kolar, J.W.; Kasper, M.J.; Deboy, G. Experimental Characterization of Silicon and Gallium Nitride 200 V Power Semiconductors for Modular/Multi-Level Converters Using Advanced Measurement Techniques. *IEEE J. Emerg. Sel. Top. Power Electron.* **2020**, *8*, 2238–2254. [CrossRef]
3. Jafari, A.; Nikoo, M.S.; Perera, N.; Yildirim, H.K.; Karakaya, F.; Soleimanzadeh, R.; Matioli, E. Comparison of Wide-Band-Gap Technologies for Soft-Switching Losses at High Frequencies. *IEEE Trans. Power Electron.* **2020**, *35*, 12595–12600. [CrossRef]
4. Pushpakaran, B.N.; Subburaj, A.S.; Bayne, S.B. Commercial GaN-Based Power Electronic Systems: A Review. *J. Electron. Mater.* **2020**, *49*, 6247–6262. [CrossRef]
5. Yang, W.; Yuan, J.-S.; Krishnan, B.; Shea, P. Characterization of Deep and Shallow Traps in GaN HEMT Using Multi-Frequency C-V Measurement and Pulse-Mode Voltage Stress. *IEEE Trans. Device Mater. Reliab.* **2019**, *19*, 350–357. [CrossRef]
6. Hu, M.; Liu, G.; Du, E.; Mu, F. Thermal effect on dynamic Ron degradation of p-GaN AlGaN/GaN HEMTs on SiC substrates. *IEICE Electron. Express* **2020**, *17*, 20200255. [CrossRef]
7. Ray, W.B.; Schrock, J.A.; Bilbao, A.V.; Kelley, M.; Lacouture, S.; Hirsch, E.; Bayne, S.B. Analysis of GaN power MOSFET exporsure to pulsed overcurrents. In Proceedings of the IEEE Pulsed Power Conference (PPC), Austin, TX, USA, 31 May–4 June 2015; pp. 1–5. [CrossRef]
8. Li, H.; Li, X.; Wang, X.; Wang, J.; Alsmadi, Y.; Liu, L.; Bala, S. E-mode GaN HEMT short circuit robustness and degradation. In Proceedings of the IEEE Energy Conversion Congress and Exposition (ECCE), Cincinnati, OH, USA, 1–5 October 2017; pp. 1995–2002. [CrossRef]
9. Song, Q.; Zhang, R.; Kozak, J.P.; Liu, J.; Li, Q.; Zhang, Y. Failure Mechanisms of Cascode GaN HEMTs Under Overvoltage and Surge Energy Events. In Proceedings of the IEEE International Reliability Physics Symposium (IRPS), Monterey, CA, USA, 21–25 March 2021; pp. 1–7. [CrossRef]
10. Song, S.; Munk-Nielsen, S.; Uhrenfeldt, C. How Can a Cutting-Edge Gallium Nitride High-Electron-Mobility Transistor Encounter Catastrophic Failure within the Acceptable Temperature Range? *IEEE Trans. Power Electron.* **2020**, *35*, 6711–6718. [CrossRef]
11. Cimmino, D.; Ferrero, S. High-Voltage Temperature Humidity Bias Test (HV-THB): Overview of Current Test Methodologies and Reliability Performances. *Electronics* **2020**, *9*, 1884. [CrossRef]
12. Wang, N.; Cotton, I.; Evans, K. Impact of Thermal Cycling in Humid Environments on Power Electronic Modules. *IEEE Trans. Compon. Packag. Manuf. Technol.* **2012**, *2*, 1085–1091. [CrossRef]
13. Qiu, L.; Guzonas, D.A.; Qian, J. Corrosion of silicon nitride in high temperature alkaline solutions. *J. Nucl. Mater.* **2016**, *476*, 293–301. [CrossRef]
14. Deng, H.; Meng, J.; Wang, D.-B.; Zhang, W. Breakdown Voltage Impact on Lifetime of 1200V IGBT modules under H3TRB-HVDC Testing. In Proceedings of the IEEE 26th International Symposium on Physical and Failure Analysis of Integrated Circuits (IPFA), Hangzhou, China, 2–5 July 2019; pp. 1–4. [CrossRef]

15. Hoffmann, F.; Kaminski, N.; Schmitt, S. Investigation on the Impact of Environmental Stress on the Thermo- Mechanical Reliability of IGBTs by Means of Consecutive H3TRB and PCT Testing. In Proceedings of the 2021 33rd International Symposium on Power Semiconductor Devices and ICs (ISPSD), Nagoya, Japan, 30 May–3 June 2021; pp. 371–374. [CrossRef]
16. Zorn, C.; Kaminski, N. Temperature–humidity–bias testing on insulated-gate bipolar transistor modules—Failure modes and acceleration due to high voltage. *IET Power Electron.* **2015**, *8*, 2329–2335. [CrossRef]
17. Dusmez, S.; Ali, S.H.; Heydarzadeh, M.; Kamath, A.S.; Duran, H.; Akin, B. Aging Precursor Identification and Lifetime Estimation for Thermally Aged Discrete Package Silicon Power Switches. *IEEE Trans. Ind. Appl.* **2017**, *53*, 251–260. [CrossRef]
18. Brunko, A.; Gloth, M.; Kaminski, N. Humidity Capability of Enhancement Mode GaN High Electron Mobility Transistors. In Proceedings of the IEEE 8th Workshop on Wide Bandgap Power Devices and Applications (WiPDA), Redondo Beach, CA, USA, 7–11 November 2021; pp. 242–245. [CrossRef]
19. Meneghini, M.; Hilt, O.; Wuerfl, J.; Meneghesso, G. Technology and Reliability of Normally-Off GaN HEMTs with p-Type Gate. *Energies* **2017**, *10*, 153. [CrossRef]
20. Transphorm. TP65H035WS Datasheet. Available online: https://www.transphormusa.com/en/document/datasheet-tp65h035ws-650v-gan-fet/ (accessed on 22 April 2022).
21. Nexpedia. GAN063-650WSA Datasheet. Available online: https://assets.nexperia.com/documents/data-sheet/GAN063-650WSA.pdf (accessed on 22 April 2022).
22. GaN Systems. GS-065-030-2-L Datasheet. Available online: https://gansystems.com/wp-content/uploads/2021/08/GS-065-030-2-L-DS-Rev-210630.pdf (accessed on 22 April 2022).
23. Transphorm. GaN for Automotive Applications. Available online: https://www.transphormusa.com/ja/document/autoreliabilityfaq-transphorm/ (accessed on 22 April 2022).
24. Barr, R. Qualification Report TPH65H035WS. Available online: https://www.transphormusa.com/en/document/qualification-report-tp65h035ws/ (accessed on 22 April 2022).
25. Ohring, M.; Kasprzak, L. *Reliability and Failure of Electronic Materials and Devices Second Edition*; Elsevier: Amsterdam, The Netherlands, 2014.
26. Sigel, A.; Sigel, H.; Sigel, R.K.O. *The Alkali Metal Ions: Their Role for Life*; Springer: Berlin, Germany, 2016.
27. Matsushima, H.; Yamada, R.; Shima, A. Two Mechanisms of Charge Accumulation in Edge Termination of 4H-SiC Diodes Caused by High-Temperature Bias Stress and High-Temperature and High-Humidity Bias Stress. *IEEE Trans. Electron Devices* **2018**, *65*, 3318–3325. [CrossRef]
28. Natarajan, R.R.; Parthasarathy, E.; Murugapandiyan, P. Influence of High-k Passivation Layer on Gate Field Plate AlGaN/GaN/AlGaN Double Heterojunction HEMT. *Silicon* **2022**, *14*, 1–9. [CrossRef]
29. Nezar, A.; Salama, C.A.T. Breakdown voltage in LDMOS transistors using internal field rings. *IEEE Trans. Electron Devices* **1991**, *38*, 1676–1680. [CrossRef]
30. Efthymiou, L.; Murukesan, K.; Longobardi, G.; Udrea, F.; Shibib, A.; Terrill, K. Understanding the Threshold Voltage Instability During OFF-State Stress in p-GaN HEMTs. *IEEE Electron Device Lett.* **2019**, *40*, 1253–1256. [CrossRef]
31. Tallarico, A.N.; Stoffels, S.; Posthuma, N.; Decoutere, S.; Sangiorgi, E.; Fiegna, C. Threshold Voltage Instability in GaN HEMTs with p-Type Gate: Mg Doping Compensation. *IEEE Electron Device Lett.* **2019**, *40*, 518–521. [CrossRef]
32. Guo, H.; Gong, H.; Shao, P.; Yu, X.; Wang, J.; Wang, R.; Zheng, Y. Over 1200 V Normally-OFF p-NiO Gated AlGaN/GaN HEMTs on Si with a Small Threshold Voltage Shift. *IEEE Electron Device Lett.* **2022**, *43*, 268–271. [CrossRef]
33. Habersat, D.B.; Lelis, A.J.; Green, R. Towards a Robust Approach to Threshold Voltage Characterization and High Temperature Gate Bias Qualification. In Proceedings of the IEEE International Reliability Physics Symposium (IRPS), Dallas, TX, USA, 28 April–30 May 2020; pp. 1–4. [CrossRef]
34. Meneghini, M.; Rossetto, I.; Bisi, D.; Ruzzarin, M.; Van Hove, M.; Stoffels, S.; Zanoni, E. Negative Bias-Induced Threshold Voltage Instability in GaN-on-Si Power HEMTs. *IEEE Electron Device Lett.* **2016**, *37*, 474–477. [CrossRef]

energies

MDPI

*Article*

# An Approach to the Design and the Interactions of a Fully Superconducting Synchronous Generator and Its Power Converter

**Sebastian Lengsfeld [1,*,†], Sebastian Sprunck [1,*,†], Simon Robin Frank [2,*,†], Marco Jung [1,3], Marc Hiller [2], Bernd Ponick [4] and Stefan Mersche [2]**

1 Converters and Drive Technology Department, Fraunhofer Institute for Energy Economics and Energy System Technology IEE, 34119 Kassel, Germany; marco.jung@iee.fraunhofer.de
2 Institute of Electrical Engineering (ETI), Karlsruhe Institute of Technology (KIT), 76131 Karlsruhe, Germany; marc.hiller@kit.edu (M.H.); stefan.mersche@kit.edu (S.M.)
3 Institute of Technology, Resource and Energy-Efficient Engineering, Bonn-Rhein-Sieg University of Applied Sciences, 53757 Sankt Augustin , Germany
4 Institute for Drive Systems and Power Electronics, Leibniz University Hannover, 30167 Hannover, Germany; ponick@ial.uni-hannover.de
* Correspondence: sebastian.lengsfeld@iee.fraunhofer.de (S.L.); sebastian.sprunck@iee.fraunhofer.de (S.S.); s.frank@kit.edu (S.R.F.)
† These authors contributed equally to this work.

**Abstract:** The design of a fully superconducting wind power generator is influenced by several factors. Among them, a low number of pole pairs is desirable to achieve low AC losses in the superconducting stator winding, which greatly influences the cooling system design and, consecutively, the efficiency of the entire wind power plant. However, it has been identified that a low number of pole pairs in a superconducting generator tends to greatly increase its output voltage, which in turn creates challenging conditions for the necessary power electronic converter. This study highlights the interdependencies between the design of a fully superconducting 10 MW wind power generator and the corresponding design of its power electronic converter.

**Keywords:** fully superconducting generator; synchronous generator; HTS; power converter; wind energy

**Citation:** Lengsfeld, S.; Sprunck, S.; Frank, S.R.; Jung, M.; Hiller, M.; Ponick, B.; Mersche, S. An Approach to the Design and the Interactions of a Fully Superconducting Synchronous Generator and its Power Converter. *Energies* **2022**, *15*, 3751. https://doi.org/10.3390/en15103751

Received: 26 April 2022
Accepted: 17 May 2022
Published: 19 May 2022



## 1. Introduction

For several years, wind power plants have shown a trend toward higher power classes [1,2]. To enable technically feasible solutions for Megawatt-scale systems, the generator needs to be designed for a low rated rotor speed and a high rated input torque. However, this design approach drastically increases the mass and the volume of the generator [3]. Because a larger generator also leads to higher investment costs, lightweight concepts such as those proposed in [4–6] have been investigated. Besides these conventional lightweight concepts, the use of superconductors is very promising, as shown in [7], because superconducting generators offer a much better power to mass ratio. To exploit the full potential of the superconductor, the generator should be fully superconducting [8], i.e., both the DC and the AC winding should be designed and manufactured with superconducting materials. However, the AC current leads to AC losses in the superconductor, which have to be handled by the cryogenic cooling system, reducing the overall efficiency of the system. Because each loss contribution at cryogenic temperatures has to be extracted through a proportionally larger power consumption of the cooling system, the generator has to be optimised for low AC losses. One of the key parameters to achieve low AC losses is the number of pole pairs, as described in Section 2.

The influence of the number of pole pairs on the cost of a generator is demonstrated in [9]. Additionally, the number of pole pairs is also coupled to the output voltage and the electric frequency of the generator, which in turn influence the design of the power electronic converter (PEC) . Its weight, size, cost and efficiency have an impact on the overall system as well. However, the consequences for the design of a PEC for fully superconducting generators with minimised AC losses have not yet been discussed in literature.

This study investigates the interdependencies between the design of a fully superconducting synchronous generator (FSSG) and its PEC in the context of the project SupraGenSys. First, the influence of the number of pole pairs on the output voltage of the generator will be investigated analytically in Section 2. An exemplary design of a fully superconducting generator is then presented in Section 3. This design is investigated regarding the influence of the number of pole pairs and of the pole pitch on the overall design feasibility in Section 4. The influence of these design parameters on the output voltage of the generator is provided for each studied parameter. Based on the feasible generator design options, Section 5 highlights available choices for the rectifier–machine interface. Section 6 then investigates possible design concepts of PECs and their influence on the overall FSSG system design.

This study is not intended to provide a complete design method for FSSGs but to enable an understanding of the challenges and available options for the interface between an FSSG and its PEC.

## 2. Dependencies and System Design Conflicts

One of the key aspects regarding the design of superconducting generators is the minimisation of the AC losses occurring inside their superconducting AC winding. Together with the conductive heat flow from the environment through the electrical connections and into the superconducting winding they form the major part of the heat that has to be extracted by the cryogenic cooling system. The higher these losses are, the larger (more powerful) the cooling system has to be designed and the higher is its power consumption, lowering the overall efficiency of the system. A rough estimation for the AC losses in High Temperature Superconductor (HTS) tapes $P_{\mathrm{AC}}$ was introduced in ([10], p. 21) and [11] and can be expressed with

$$P_{\mathrm{AC}} = B_{\mathrm{t}} \cdot f_1 \cdot d \cdot l_{\mathrm{tape}} \cdot I_{\mathrm{c}} \quad , \tag{1}$$

where $B_{\mathrm{t}}$ is the tangential flux density that the HTS tape is exposed to, $f_1$ is the frequency of the AC current of the generator, $d$ is the width of the HTS tape, $l_{\mathrm{tape}}$ is the effective length of the HTS tape and $I_{\mathrm{c}}$ is the critical current of the HTS tape. With a lower operating temperature, the critical current of the HTS tape increases, but at the same time the electrical power required for the cryogenic cooling system increases as well. Therefore, an operating temperature of 65 K for the armature winding is selected as a tradeoff between a high critical current and a low cryogenic cooling power. The field winding is only driven by DC current; hence, a lower operating temperature of 30 K becomes feasible.

This equation was developed in ([10], p. 21) as an estimate of the AC losses and is only valid for coil currents close to the critical current. This condition is observed for all investigations presented in this study.

In other applications, e.g., distribution grids, superconductors are often used to increase the current handling capability compared to conventional copper or aluminium wires in limited spaces. Based on these applications, it might therefore be expected to use high coil currents in FSSGs as well. However, one possibility to minimise AC losses in HTS tapes that follows (1) is the reduction in the width of the tape $d$. Because HTS tapes greatly impact the overall investment costs for a FSSG, a tape width of $d = 2\,\mathrm{mm}$ was selected, providing a critical current of $I_{\mathrm{c}} = 52\,\mathrm{A}$. This tape width is currently the most narrow one available.

A second approach to minimise AC losses is the electric frequency $f_1$, which—for a synchronous generator—is directly coupled to the rotor speed $n$ through the number of pole pairs $p$:

$$f_1 = n \cdot p \quad . \tag{2}$$

Combining (1) and (2) to (3) clearly shows the dependency of the HTS tape AC losses $P_{AC}$ on the number of pole pairs $p$:

$$P_{AC} = B_t \cdot n \cdot p \cdot d \cdot l_{tape} \cdot I_c \ . \tag{3}$$

This equation also shows the dependency of the AC losses on the tangential flux density $B_t$ and the effective HTS tape length $l_{tape}$, which in turn are coupled to the power density of the generator. Apart from the AC losses, the number of coils of the armature winding is coupled to the number of pole pairs as well. Each coil has two coil sides which need to be placed in slots. A common description for this physical placement is the number of slots per pole and per phase

$$q = \frac{Z_1}{2 \cdot p \cdot m} \ , \tag{4}$$

which is coupled to the chosen winding scheme. $Z_1$ is the number of slots and $m$ is the number of phases. The armature coils per phase can be connected in series, in parallel or in a series–parallel combination, as long as all coils are used. For a fixed number of turns per coil, a series connection increases the voltage between the phase contacts whereas a parallel connection increases the phase current. These properties have to be handled by the PEC and need to be known before it is designed. They can be identified through the apparent power $S_1$ of the generator according to

$$S_1 = 3 \cdot I_1 \cdot U_1 \ , \tag{5}$$

where $I_1$ and $U_1$ are the RMS phase current and the RMS phase voltage at the terminals of the generator, respectively, assuming a star connection of $m = 3$ phases. The internal connection of the coils inside the armature winding can be described through the number of parallel branches per phase $a$. Through this number, the phase current for the PEC is given as

$$I_1 = I_{coil} \cdot a \ . \tag{6}$$

Expressing the apparent power $S_1$ of the generator in (5) through the real power $P_1$ and the power factor $\cos\varphi$ allows the calculation of the phase voltage according to

$$U_1 = \frac{S_1}{3 \cdot I_{coil} \cdot a} = \frac{P_1}{3 \cdot \cos\varphi \cdot I_{coil} \cdot a} \ . \tag{7}$$

Equation (7) demonstrates the strong dependency of the phase voltage $U_1$ on the number of parallel branches $a$, because $U_1 \propto \frac{1}{a}$. For a power rating in the MW range and a coil current of $I_{coil} \approx I_c = 52$ A, large phase voltages $U_1$ have to be expected even without a series connection of the coils inside the armature winding. Although a suitable PEC could theoretically be designed, practical issues such as component properties and device costs usually aim for a more balanced distribution of voltage and current load for a PEC, which in this case requires the parallel connection of the FSSG coils.

However, the possibility to connect coils in parallel is limited by the number of pole pairs. By selecting $q = 1$ in (4) and choosing a single-layer winding, the number of parallel branches $a$ cannot to be higher then the number of pole pairs $p$, which in this case defines the number of connectable coils per phase. Combined with (3), this behaviour presents a conflict between the design of the PEC and the minimisation of AC losses.

After presenting an exemplary FSSG design in Section 3, the following sections will therefore explore design options both for the FSSG and the PEC to find suitable compromises between these design goals.

## 3. Design of a Fully Superconducting Synchronous Generator (FSSG)

This section will introduce the basic design of an FSSG to provide an example for armature and field winding made from HTS tapes, which is used to demonstrate the design conflict mentioned above.

### 3.1. *Basic Design Constraints of a 10 MW FSSG for Wind Power Plants*

The target application requires a rated output power of $P_N = 10\,\text{MW}$, a rated input torque approximately of $T_N = 10\,\text{MN m}$ and a rated rotor speed of $n_N = 10\,\text{min}^{-1}$. The system design has to enable a high efficiency, a low mass to power ratio and feasible capital expenditures (CAPEX). A sectional model of the generator design is shown in Figure 1.

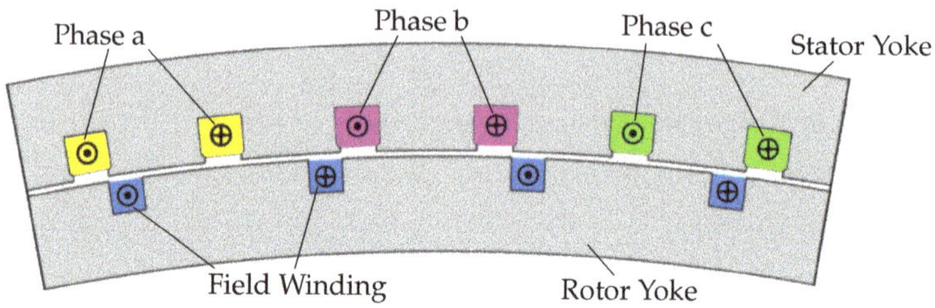


**Figure 1.** Generator model with $q = \frac{1}{2}$ for FEM calculations.

### 3.2. *Armature Winding*

The armature winding carries AC current and, therefore, its design requires different measures to reduce the AC losses. It uses a star configuration presented in [12], which distributes six coils into each slot with the field lines, where each coil is composed of 50 turns of HTS tape. This concept uses narrow HTS tapes with a width of $d = 2\,\text{mm}$, enabling low AC losses according to (1) but having a low critical current of $I_c = 52\,\text{A}$ at a temperature of 65 K. However, the individual coils of the star configuration are connected in series, because otherwise unsymmetrical current distributions must be expected.

In the presented example, the star configuration will be used in a concentrated winding (see Figure 1). This approach enables small winding heads, which will simplify the manufacturing process and reduce the danger of accidental damage to the narrow, sensitive HTS tape during manufacturing. In common FSSG designs, an air gap winding is often preferred [2,13], but by using a slot-based winding, the field penetration of the coils can be significantly reduced and much better operation conditions of the coils and lower losses can be achieved [12]. Due to the selected winding scheme, the number of parallel branches $a$ cannot be higher than half the number of pole pairs $p$ (8). The winding is covered by a 25 mm thick, multi-layered cryostat.

$$a_{\max} = \frac{p}{2} \tag{8}$$

### 3.3. *Field Winding*

The field winding is distributed in small slots and its coils are designed as racetrack coils. Similar to the armature winding, it uses a cryostat with 25 mm thickness as well, but it is operated at a lower temperature of $T = 30\,\text{K}$. To achieve an air gap flux density of $B \approx 2\,\text{T}$, the rotor coils have 170 turns, each using an HTS tape with a width of $d = 4\,\text{mm}$.

## 4. Parameter Studies for an FSSG

As described in Section 2, the number of pole pairs $p$ influences both the AC losses inside the HTS tape and the phase voltage at the terminals of the generator. To find a feasible range for the number of pole pairs, a parameter study was performed (Section 4.1), investigating its impact on the design of the FSSG.

For this investigation, a generator design based on the constraints and design rules of Section 3 is varied. These variants are each parameterised analytically and then modelled and simulated in FEMAG-DC. The simulation is used to validate the targeted output torque and to acquire the phase inductance, power factor, output voltages and the flux density distribution for the HTS coils. These results, especially the flux density, are then used to estimate the AC losses according to (1).

Beside the number of pole pairs, the power density is a major design parameter with an impact to the AC losses and the PEC. Due to the fixed number of turns per coil and coil current, it is mainly defined by the pole pitch. Hence, a second parameter study investigates the effects of varying pole pitches (Section 4.2).

All parameter studies consider the target values and design principles described in Section 3. The study regarding the influence of varying pole pair numbers assumed a fixed pole pitch of 350 mm whereas the following pole pitch investigation assumed a fixed number of pole pairs of $p = 35$.

### *4.1. Parameter Study 1: Variation of the Number of Pole Pairs*

Figure 2 shows a linear, positively coupled relation between the inner diameter of the stator and the number of pole pairs, whereas the axial stack length of the iron sheets decreases with higher number of pole pairs. These changes in turn influence the iron mass and the required HTS tape length.

Even though a lower number of pole pairs reduces the diameter of the generator, the corresponding increase in its axial length has a significant impact on its mass and on the required amount of HTS tape. In contrast, a design with a higher number of pole pairs will result in a shorter, more lightweight generator with much lower demand of HTS tape length. On the other hand, the AC losses are increasing with the number of pole pairs, which increases the energy demand of the cryogenic cooling system significantly.

Figure 3 shows the dependency of the phase voltage, the phase inductance and the power factor with a changing number of pole pairs. The number of pole pairs has no impact on the power factor, because the winding and thus the field distribution are not affected. A higher number of pole pairs enables a higher number of parallel branches and, therefore, a lower output voltage according to (7) and (8). The number of parallel branches together with the change of the HTS tape length results in the shown behaviour of the inductance, which is also the reason for the small differences between the interpolated curve and the calculated values of the inductances.

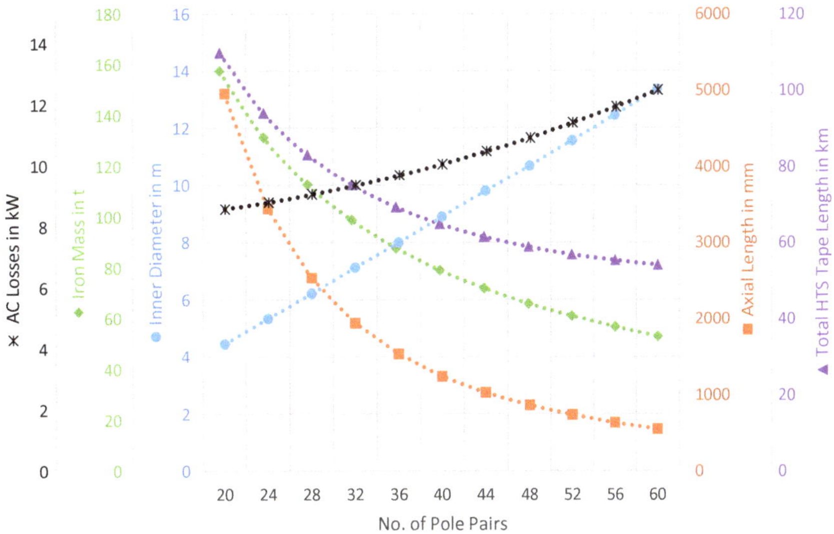


**Figure 2.** Dependency of the AC losses, the iron mass, the inner diameter, the axial length and the total HTS tape length on the number of pole pairs.

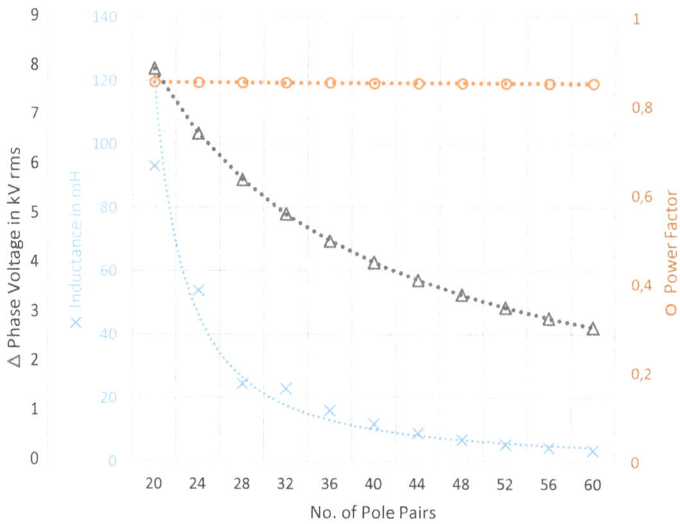


**Figure 3.** Dependency of the phase voltage, the power factor and the inductance on the number of pole pairs.

The results of this parameter study can be summarised to the basic design rule: A smaller number of pole pairs has a positive impact on the AC losses in an FSSG but in turn requires an increase in the total iron core mass and HTS tape length. Therefore, a decreasing number of pole pairs results in a higher material cost for the generator.

### 4.2. *Parameter Study 2: Variation of the Pole Pitch*

The previous investigation showed that the axial length of the generator greatly influences its mass, and, consecutively, its material cost. It was therefore interesting to investigate design methods that could counter this mass increase, e.g., by choosing a higher bore diameter, which in turn changes the pole pitch. A larger pole pitch will have a negative impact on the iron losses but can enable more desirable machine lengths to reduce the mass and the HTS tape length.

Figure 4 shows the properties of the generator for varying pole pitches. Because the number of pole pairs is fixed for this part of the parameter study, a reduced pole pitch results in a smaller diameter and a larger active length. Hence, a smaller pole pitch enables lower total iron losses, but the larger active length of such a machine requires more HTS tape and increases the AC losses inside the superconductor.

The generator output voltage, the phase inductance and the power factor are shown in Figure 5. For larger pole pitches, neither the iron mass, the output voltage, the power factor nor the phase inductance decrease monotonously but exhibit extrema between pole pitches of 350 to 400 mm. The output voltage is inversely coupled to the power factor as described in (7), which in turn is influenced by the phase inductance that is affected by the saturation of the iron sheets.

### 4.3. *Summary of Parameter Studies*

Overall, the study regarding the influence of the pole pitch suggests an optimal value between 300 and 400 mm for the investigated FSSG concept. However, choosing the number of pole pairs is a tradeoff between CAPEX, generator mass, size and efficiency.

A smaller number of pole pair numbers reduces the total losses of the FSSG but results in heavier, longer machine designs that require more HTS tape. At the same time, lower pole pairs create higher generator output voltages of several kV with relatively low phase currents of 52 A. Within the framework of the SupraGenSys project, it was found that the decision for a specific number of pole pairs cannot be made without investigating the power rectifier as well. Therefore, the FSSG design has to go hand in hand with the design of the PEC.

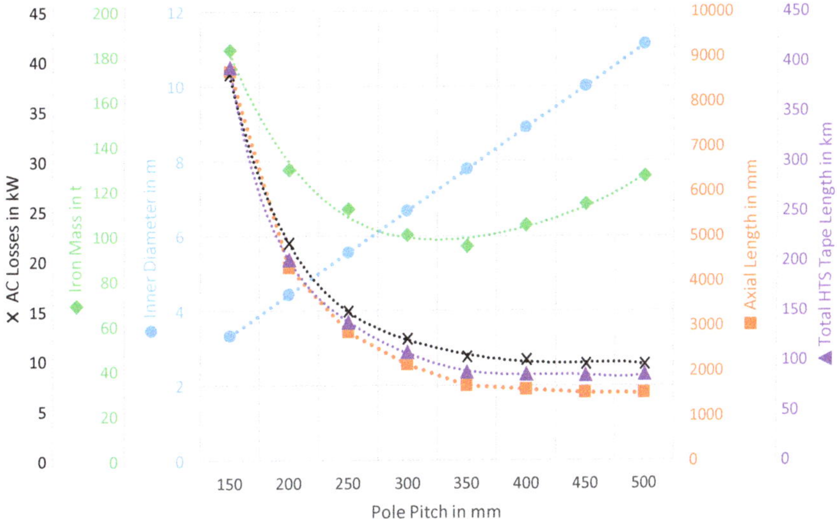


**Figure 4.** Dependency of the AC losses, the iron mass, the inner diameter, the axial length and the total HTS tape length on the pole pitch.

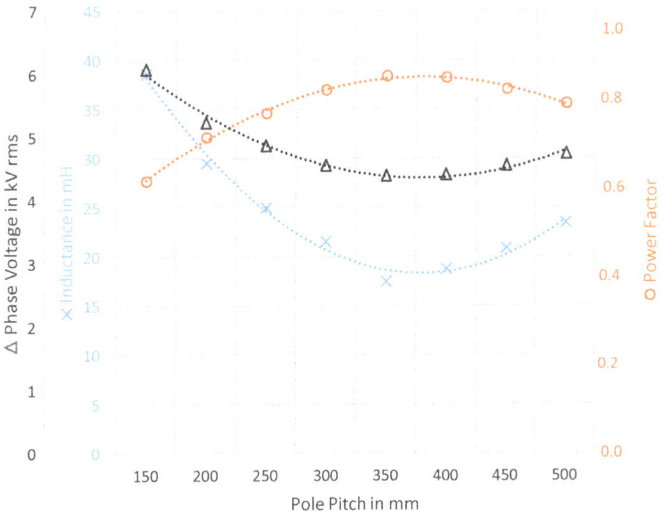


**Figure 5.** Dependency of the phase voltage, the power factor and the inductance on the pole pitches.

## 5. Design Options for the Rectifier–Machine Interface

As described in Section 2, the RMS phase voltage $U_1$ of a synchronous generator can be calculated according to (7). Depending on the topology of the PEC, either the peak phase voltage or the peak phase-to-phase voltage define the maximum voltage that power semiconductors inside a power rectifier are exposed to. Ignoring overvoltage events and observing (8) for the investigated FSSG concept, these peak values calculate according to (9) and (10), respectively.

$$\hat{u}_1 = \sqrt{2} \cdot U_1 = \frac{\sqrt{2}}{3} \cdot \frac{P_1}{\cos\varphi \cdot I_{\text{coil}} \cdot a} = \frac{2 \cdot \sqrt{2}}{3} \cdot \frac{P_1}{\cos\varphi \cdot I_{\text{coil}} \cdot p} \tag{9}$$

$$\hat{u}_{\text{phase-phase}} = \sqrt{3} \cdot \hat{u}_1 = \frac{2 \cdot \sqrt{2}}{\sqrt{3}} \cdot \frac{P_1}{\cos\varphi \cdot I_{\text{coil}} \cdot p} \tag{10}$$

By applying a sweep of the number of pole pairs $p$ to (10) with an assumed power factor of $\cos\varphi = 0.85$, the respective voltages displayed in the bottom-left graph of Figure 6 can be calculated. These calculated peak values are in good agreement with the simulated RMS values presented in Figure 3.

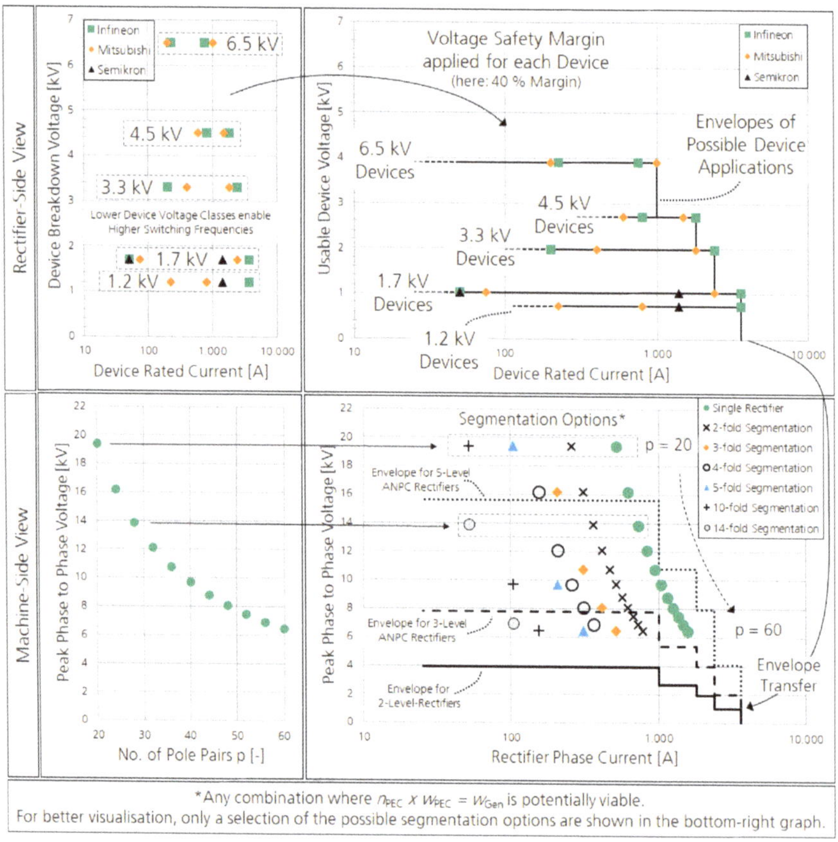


**Figure 6.** Interdependencies between the power electronic rectifier and the generator design. **Top**: Non-complete survey of available power IGBT modules (**left**) and their working voltage capabilities (**right**) with an assumed safety margin of 40%. **Bottom**: Peak phase-to-phase voltages for the investigated generator concepts (**left**) and the possible segmentation options for each concept (**right**). A combination of these investigations, marked by the solid, dashed and dotted lines in the bottom-right graph allows the identification of technically and economically feasible generator/rectifier combination candidates.

These peak voltages define the maximum voltage stress that the PEC has to endure when connected to the generator. The voltage stress for individual power semiconductor devices inside the PEC can be reduced, which will be addressed in Section 6. A non-complete market survey of available IGBT power semiconductors that focuses on the available *range* of devices for different breakdown voltage classes is presented in the top-left graph of Figure 6. This survey does not include SiC devices, as commercial products of that technology are not available above breakdown voltages of 3.3 kV. Test setups with experimental SiC devices with breakdown voltages of 6.5 kV, 10 kV and 15 kV have been reported in [14–18] but are not yet commercially available.

From the perspective of the PEC, it is preferable to use lower-voltage power semiconductors, as these devices enable the use of higher switching frequencies and therefore the design of more compact and lightweight passive devices. However, these lower-voltage devices require lower phase-to-phase voltages at the terminals of the FSSG, which is only possible if the generator is constructed with a larger number of pole pairs $p$ (see (9)). This in turn would increase the electrical frequency (2) and hence the AC losses inside the HTS tape (3). It is therefore necessary to find a compromise between a lower-voltage, compact, suitably dimensioned rectifier and a higher-voltage, low pole pair, low AC loss FSSG concept.

The following subsections will highlight the respective design possibilities and their consequences for the machine–rectifier interface.

### 5.1. *Rectifier-Side Possibilities*

The breakdown voltage ratings shown in the top-left graph of Figure 6 must not be confused with the actual working voltage of these power semiconductor devices. Influences such as turn-off overvoltages, background radiation ruggedness and lifetime optimisation require a safety margin applied to these values, often in the range of 30% to 50%. In turn, these devices can only be operated at about 50% to 70% of their rated voltage. For example, operation of a 1.7 kV device at 60% of its breakdown voltage results in a working voltage of 1020 V, which is a common application, for e.g., contemporary PV inverters [19–21]. By applying this safety margin to each surveyed device, we receive the top-right graph of Figure 6.

This enables the outlining of possible application areas for single power semiconductors, i.e., topologies without a series or parallel connection of identical semiconductors (see Section 6), as indicated by black solid lines. The right-hand and upper envelopes for each device are "hard" borders in the sense that an application outside of these boundaries would exceed the rated current or voltage handling capabilities of these devices, respectively. The lower envelopes for each voltage class are defined by the upper envelope of the next lower voltage class devices. These lower envelopes are "soft" borders, as it is very possible to use, e.g., a 6.5 kV device at a working voltage of 2 kV or even below. However, it is unusual to do so because devices of lower voltage classes often offer a better performance at such relatively low voltages and are usually less expensive than higher-voltage devices. No left-hand side envelope is shown in the top-right graph of Figure 6, as it is also possible to operate each device well below its rated current, though devices with lower current ratings are preferable in this case.

### 5.2. *Machine-Side Possibilities*

Each generator concept in the bottom-left graph of Figure 6 provides a number of three-phase systems $s_{\mathrm{Gen}}$ that is—for the investigated FSSG concept presented in Section 3—dependent on half the number of pole pairs $p$ of the generator:

$$s_{\mathrm{Gen}} = \frac{p}{2} \tag{11}$$

Because a series connection of the corresponding phases of these branches would further increase the voltage at the terminals of the generator, the following approach will not cover this possibility.

To use all three-phase systems for power conversion, each of them has to be connected to a PEC. However, from the PEC point of view, the combination of rather high voltages of several kV (see Figure 3) and a rather low phase current of 52 A is quite an unusual combination which can certainly be managed with available power semiconductors, but those high-voltage devices are usually designed for much higher currents so that their utilisation factor is very poor. Through parallelisation of the corresponding phases of the available three-phase systems, it is possible to combine the individual phase currents so that the power semiconductors inside the PEC are utilised better. However, not all of the branches have to be connected to the *same* PEC: as long as all branches are used, it is possible to connect a subset of the available branches $s_{\mathrm{Gen}}$ to one PEC and the remaining subset(s) to other, identical PECs. If $n_{\mathrm{PEC}}$ is the number of individual PECs and $s_{\mathrm{PEC}}$ is the number of parallel three-phase systems per PEC, then this requirement can be expressed through

$$n_{\mathrm{PEC}} \cdot s_{\mathrm{PEC}} \stackrel{!}{=} s_{\mathrm{Gen}} = \frac{p}{2} \; . \tag{12}$$

At the same time, the PEC phase current $I_{\mathrm{PEC}}$ is given by

$$I_{\mathrm{PEC}} = s_{\mathrm{PEC}} \cdot I_{\mathrm{coil}} \; . \tag{13}$$

The bottom-right graph in Figure 6 displays these possible machine/rectifier segmentation options, where each "row" corresponds to the pole pair concept with the same voltage in the bottom-left graph. The possible options per concept row are marked through different symbols. For example, a machine with $p = 20$ pole pairs provides $s_{\mathrm{Gen}} = 10$ available three-phase systems according to (12). With $I_{\mathrm{coil}} = 52$ A and (13), the electrical power of such a generator can be segmented in four different ways, as shown in Table 1. For better visualisation, not all available options for all investigated pole pairs are shown in the bottom-right graph of Figure 6.

**Table 1.** Exemplary options for the electric segmentation of a machine with $p = 20$, $s_{\mathrm{Gen}} = 10$ and $I_{\mathrm{coil}} = 52$ A.

| No. of PECs $n_{\mathrm{PEC}}$ | Parallel Three-Phase Systems per PEC $s_{\mathrm{PEC}}$ | PEC Phase Current $I_{\mathrm{PEC}}$ |
|---|---|---|
| 1 | 10 | 520 A |
| 2 | 5 | 260 A |
| 5 | 2 | 104 A |
| 10 | 1 | 52 A |

With these segmentation options, it is much easier to design a technically and economically feasible rectifier, as its phase current $I_{\mathrm{PEC}}$ can be matched to the current handling capabilities of available power semiconductors more easily, as can be seen when comparing the top-right and the bottom-right graphs of Figure 6. Additionally, a segmentation of the output power of the generator to several rectifiers increases the system reliability by reducing the impact of single device failures and introduces the possibility to control the torque ripple of the generator within tighter limits.

## 6. Rectifier Concepts and Topologies for FSSGs

State-of-the-art PECs can be categorised into two main groups:

- Conventional converters that are designed for a specific conversion task and usually try to minimise the number of functional components;
- Modular or cell-based converters that combine multiple copies of a basic converter "cell" to address various conversion tasks and/or to enable a more flexible system.

Conventional converters tend to have a comparably low degree of complexity and fewer components than modular converters. Two examples of this group that will be discussed in Section 6.1 are the B6 topology, consisting of six power semiconductors with a single DC capacitor, and the ANPC topology, consisting of 18 power semiconductors and a split DC capacitor.

Cell-based converters, which are discussed in Section 6.2, tend to be more flexible and often offer additional benefits such as redundancy or improved power quality, but usually require more components and a more complex control system. They present the possibility to use lower-voltage semiconductors with current ratings that fit better to the relatively low FSSG phase currents, produce lower losses and can be operated at higher switching frequencies [22]. Moreover, the increased number of voltage levels results in a better approximation of the sinusoidal AC voltages and therefore lower harmonics in the generator phase currents. Depending on the ratio of load current and harmonic content, this in turn can lead to lower AC losses in the HTS tape as well and opens possibilities to reduce the size of filters or to omit them entirely. Furthermore, assuming the same semiconductor technology, the resulting voltage slopes $du/dt$ can be reduced, resulting in lower insulation stress.

### 6.1. Conventional Converter Concepts

Suitable conventional converter concepts for the above-mentioned compromise for an FSSG can be identified directly through the bottom-right graph of Figure 6. By transferring the envelopes of the top-right graph of Figure 6 to its bottom-right graph, we can visually select the machine concepts that can be addressed with the envelope that corresponds to a given rectifier concept. The steps in the envelopes correspond to the different semiconductor voltage class applications surveyed in the upper graphs of Figure 6. The marked machine options that fall below and to the left of the respective envelopes are candidates that make further investigations viable.

#### 6.1.1. Two-Level Active Rectifiers

A standard B6 bridge rectifier (Figure 7a) requires that each individual semiconductor can endure the maximum phase-to-phase voltage on its own. The solid black envelope in the bottom-right graph of Figure 6 corresponds to this requirement, leading to the conclusion that this approach would not provide any viable options even for a generator with $p = 60$ pole pairs, as the minimum possible voltage for this design is well above the operational voltage of individual 6.5 kV power semiconductor modules.

However, power semiconductors could be connected in series to distribute the voltage load of the generator to multiple devices, yet this distribution also has to be ensured during each switching process. The additional effort required to secure this voltage balancing both in the blocking state as well as during switching transients cannot be neglected and is rarely used in practical applications, because multi-level topologies in this case often offer a better alternative. It is therefore not advisable to use a B6 topology, even a series-connected one, for the investigated FSSG concepts.

#### 6.1.2. Three-Level Active Rectifiers

Several multi-level topologies, such as an ANPC shown in Figure 7b, offer an even voltage load between the individual semiconductors both during static and switching transient operation. The third voltage level also allows a better approximation of ideal sine waves, reducing the need for filter equipment. Rectifier concepts using multi-level topologies could enable the use of lower-voltage devices and/or the use of machine concepts with a lower number of pole pairs.

The dashed black envelope in the bottom-right graph of Figure 6 shows the application limit of 3-level ANPC rectifiers for FSSG wind turbines. Due to the high voltages of the generator, 6.5 kV devices are required for this topology, enabling a match with concepts with at least $p \geq 50 \ldots 52$ pole pairs. According to (3), such designs are subjected to relatively large AC losses. Additionally, the high-voltage devices are limited to switching frequencies of a few hundred Hz to a few kHz at best. This topology provides a technical possibility to solve the interface problem, but its performance will certainly not be satisfactory—neither from the perspective of the FSSG nor from the perspective of the PEC.

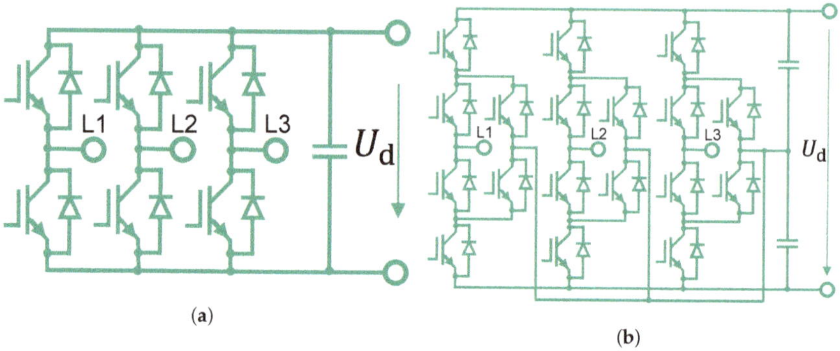


**Figure 7.** Example topologies for active rectifiers. (**a**) Standard B6, (**b**) 3-Level ANPC as an example of multi-level topologies.

Higher-level topologies can extend the voltage load capabilities of a PEC, but at the same time increase the complexity with little benefit regarding size, weight and efficiency of the PEC. For example, a 5-Level ANPC can in theory bear twice the voltage load of a 3-Level ANPC, but this approach would still require 6.5 kV devices to enable FSSG concepts with phase voltages below approximately 16 kV (see the dotted black line in the bottom-right graph of Figure 6), corresponding to $p > 24$ pole pairs. Devices with voltage ratings of 4.5 kV and 3.3 kV can also be used but only for FSSGs with higher numbers of pole pairs.

Additionally, higher-level topologies require non-negligible effort to balance the additional capacitors, which increases with the number of levels. It is therefore questionable if these conventional PEC design approaches can provide desirable effort to performance ratios for the investigated application.

*6.2. Cell-Based Multi-Level Converters*

Due to the specific requirements of the medium voltage FSSG and the above described difficulties resulting for a B6 or ANPC converter, multi-level topologies can present an interesting alternative. In this section, several multi-cell converters are therefore discussed in terms of their suitability for use with FSSGs.

Two scenarios are considered: one in which the generator is connected to a medium voltage (MV) AC grid through an inverter and a second one in which it is connected to a MV DC grid via a galvanically isolated DC/DC converter. For both scenarios, the three-phase AC voltage of the generator is first rectified using $n_{PEC}$ rectifiers. These two scenarios are pictured in Figure 8.

The common DC voltage link at the rectifier output provides a certain flexibility, because the same rectifier unit can be used for both scenarios. Therefore, only topologies which have a DC voltage link are considered here. These topologies can be constructed with single IGBTs or SiC-MOSFETs and provide a lower complexity and/or component count compared to topologies that require bidirectional blocking capabilities, e.g., current source inverters that require an additional diode [23,24].

Possible topologies for the rectifier unit are presented in Sections 6.2.1–6.2.4. A converter concept based on cascaded power cells is then investigated for the use with the FSSG in Section 6.2.5, followed by proposals for an AC side converter in Section 6.2.6.

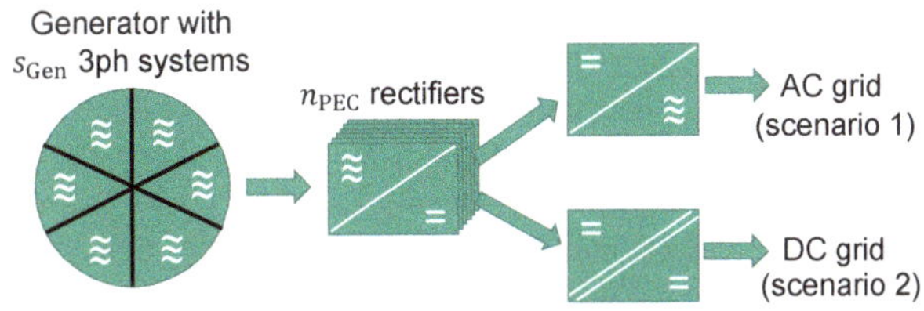


**Figure 8.** Proposed converter concept with a common DC voltage link for a superconducting generator with $s_{Gen}$ three-phase systems. For the AC grid connection, a galvanic isolation is also to be ensured, which can be realised, e.g., with a grid-side transformer on system-level or, depending on the topology, by several medium frequency (MF) transformers on cell level.

### 6.2.1. Flying Capacitor Converter

The flying capacitor (FC) converter was first introduced in [25]. An example of its cell-like structure is pictured for one phase of a 5-Level FC in Figure 9a. For $n$ levels it consists of $n-2$ flying capacitors per phase and one common DC link capacitor. The flying capacitors have to be pre-charged before operation and their voltages have to be held at nominal voltage during operation, which results in a much higher modulation and balancing effort compared to a B6 converter and even in a higher effort compared to NPC converters. On the other hand, this topology limits the maximum voltage across each power semiconductor to the $(n-1)$th part of the DC link voltage. Thus, in comparison to topologies with a lower number of levels, semiconductors with a significantly lower blocking voltage can be used, which in turn enables higher switching frequencies.

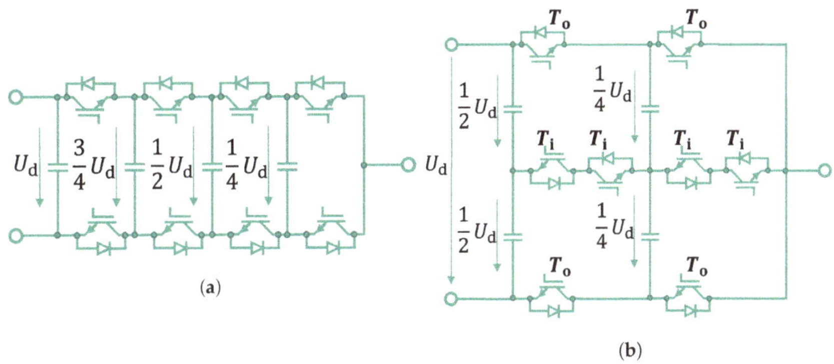


**Figure 9.** (**a**) 5-Level Flying Capacitor (FC) Converter. (**b**) 5-Level Stacked-Multi-Cell Converter (SMC).

The main disadvantage of this topology is the requirement of relatively expensive and bulky capacitors. This is especially problematic when the DC voltage is increased, because the stored energy in one capacitor grows with the square of the applied voltage, which then translates to larger capacitors. By using fast switching semiconductors and new operation principles, e.g., Quasi-Two-Level operation, the size of the capacitors can be reduced significantly, making this topology interesting for future MV applications [26].

### 6.2.2. Stacked-Multi-Cell Converter

The stacked-multi-cell converter (SMC), introduced in [27], is a further development of the FC topology. A number of $N$ FC converters are stacked on top of each other, resulting in a lower maximum voltage for each FC and therefore smaller and less expensive capacitors. Figure 9b shows a 5-Level-SMC ($N=2$) and the resulting relative voltages of the FCs.

However, compared to the FC topology, the same number of semiconductors is required for the same number of voltage levels and the blocking voltage of the outer transistors $T_o$ has to be twice as high (compare Figure 9). Because the required blocking voltage of the inner transistors $T_i$ is only half the required blocking voltage of the outer transistors $T_o$, each outer transistor can be theoretically replaced by a series connection of two transistors, so that the blocking voltage of all semiconductors is the same.

Several proposals can be found in literature to further improve this kind of converter, e.g., in [28] by decreasing the number of required voltage sources or in [29] by adding four low-frequency switches that are switched only twice per fundamental period to double the RMS value of the AC voltage and the number of voltage levels for the same DC link voltage.

6.2.3. Modular Multi-Level Converter

The modular multi-level converter (MMC), pictured in Figure 10, is known from high-voltage direct current (HVDC) electric power conversion applications. However, it can also be used for MV applications. The basic topology that converts a DC voltage to a three-phase AC voltage is introduced in [30] and consists of two converter arms for each of the three phases. One arm itself is realised by a series connection of several cells. Each cell consists of a capacitor which can either be switched into the respective arm voltage or be bypassed using a half-bridge circuit. Alternatively, a full bridge circuit can be used, allowing for the insertion of a negative capacitor voltage as well. Due to the identical structure of the cells, the whole concept is highly modular and high voltages as well as a high number of voltage levels can be reached. Therefore, this topology enables low harmonics at the cost of a larger number of semiconductors. Hence, compared to the previously presented topologies, it could theoretically further reduce the HTS AC losses.

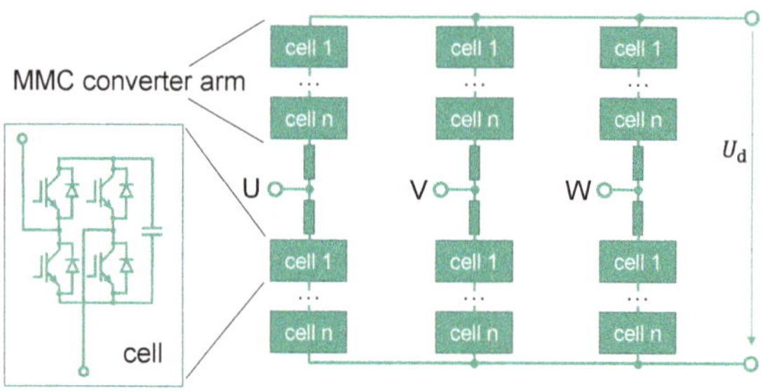


**Figure 10.** Topology of the Modular Multi-Level Converter (MMC) and the exemplary circuit diagram of one cell, configured as full bridge circuit.

In future scenarios where the wind generator could be either directly connected to an MV AC or to an MV DC offshore grid, the converter has to ensure a galvanic isolation between the generator and the grid. Without such an isolation, the FSSG winding would require extensive isolation strength. Assuming a grounded stator and a symmetrical DC grid, this isolation would need to be designed for voltages in the range of the amplitude of the AC grid phase voltage or half of the DC grid voltage, respectively. All topologies mentioned above present no intrinsic galvanic isolation between their input and output. When using these topologies, the isolation has to be guaranteed on system level, e.g., with a transformer operated at grid frequency.

6.2.4. Cascaded Power Cells

Using a topology based on the cascaded H-bridge (CHB) converter introduced in [31] allows for galvanic isolation at cell level. On the generator AC side, $n_{\mathrm{Cell,ph}}$ H-bridges are connected in series to address high generator phase voltages even with low-voltage semiconductors. At the output of each H-bridge, a highly efficient DC/DC converter stage with a very high galvanic isolation is connected. The DC outputs can then again be connected in series to reach the required DC link or DC grid voltage.

Its high modularity makes this topology very interesting for a superconducting wind generator converter. It offers the possibility to use the same converter concept for different generator designs by simply matching the number of cells to the respective voltage requirements. Theoretically, a very high generator phase voltage can be realised, limited only by practical issues such as isolation clearances. Figure 11 shows the topology for one three-phase system. The combination of an H-bridge and a DC/DC converter is called "power cell" in the following investigation.

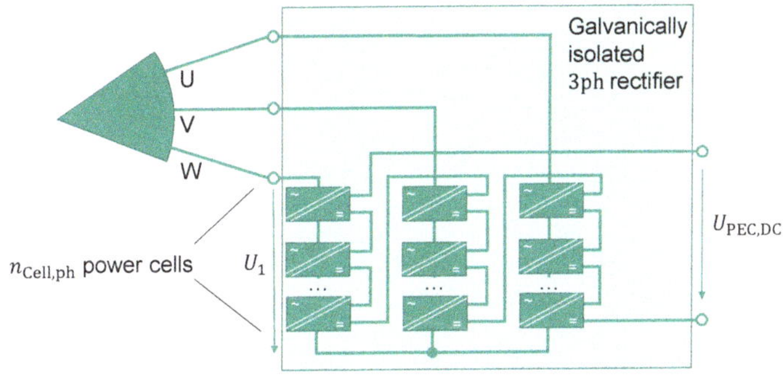


**Figure 11.** Cascaded power cell topology for one generator three-phase system.

In this example, all DC terminals are connected in series to enable the direct connection to a future MVDC grid. A proposal for the design of a single power cell is pictured in Figure 12. On the AC side, the H-bridge rectifies the single-phase AC voltage of the generator. Because very low electric frequencies in the range of a few Hz (see (2)) and rather large phase inductances in the range of several mH are expected for the FSSG (see Figure 3), it is possible to realise this input rectifier stage with commercial IGBTs.

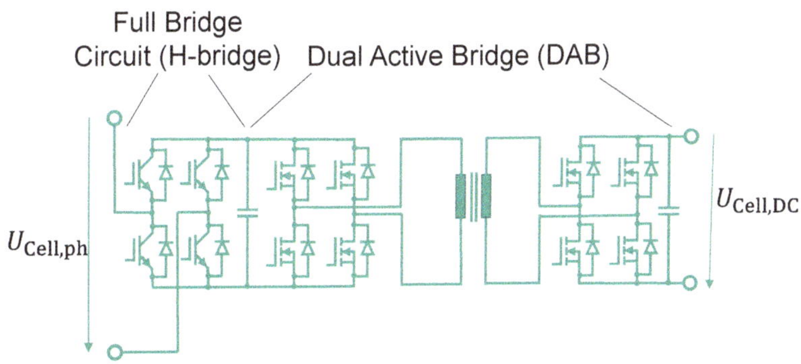


**Figure 12.** Possible design for a single power cell, consisting of an H-bridge rectifier and an isolated DAB.

The DC/DC converter can be realised with different topologies, e.g., with a resonant converter such as an LLC or a CLLC converter or non-resonant options such as a dual active bridge (DAB). A single-phase and a three-phase DAB are proposed in [32]. The single-phase DAB is compared with an LLC converter for a fixed, high conversion ratio in [33], whereas [34] compares a single-phase DAB with an LLC and a CLLC converter for a 1:1 conversion ratio. Hereinafter, a single-phase DAB is assumed for reasons of simplification. However, other topologies might show advantages compared to the DAB for that specific application and are worth investigating further.

The isolation in a DAB is provided using a medium frequency (MF) transformer and the semiconductors are usually switched with several tens of kHz. Due to lower switching losses compared to Si-IGBTs, SiC-MOSFETs are well suited for this task.

One challenge regarding this topology is the non-constant power input of one power cell, because it is a single-phase converter. If a constant power flow through the DAB is desired, the input DC link capacitor has to buffer the difference between the AC generator power and the constant power through the DAB, resulting in a reactive power at the input with double the electrical generator frequency $f_1$. It is possible to use the capacitors on the primary and secondary side of the power cell to buffer the AC power ripple with a suitable control of the DAB if they are dimensioned accordingly. Alternatively, the power transfer through the DAB could be modulated to follow the single-phase AC power of the generator. A constant total DC power flow at the converter output is then realised by paralleling the DC outputs of the converters of all three generator phases.

Further challenges for this approach are the high isolation strength required for the MF transformers as well as the demand of high reliability in terms of isolation failure. With $n_{\mathrm{Cell,tot}}$ MF transformers, each transformer must provide a failure rate that is lower by a factor of $1/n_{\mathrm{Cell,tot}}$ compared to a topology with one central transformer to reach the same overall system failure rate. Despite these challenges, the proposed topology concept should be investigated further due to its high degree of modularity and its flexibility. One possible adaption to the FSSG application is presented in the next subsection.

#### 6.2.5. Cascaded H-Bridge Concept for the FSSG

Due to their high availability even at lower rated currents, the following investigations of the Cascaded H-bridge concept are performed for 1200 V and 1700 V semiconductor modules. According to Section 5.1 these semiconductors ensure working voltages of 720 V or 1020 V, respectively, which define the maximum DC link voltage $U_{\mathrm{Cell,DC}}$ of each power cell. For sine-triangle modulation without overmodulation, this in turn is identical to the maximum possible AC voltage amplitude $\hat{u}_{\mathrm{Cell,ph,max}}$ of each power cell [35].

Because the power cells of one phase are connected in series and the phases themselves are star connected, the sum of the individual power cell AC voltage amplitudes must be equal to or higher than the amplitude value of the generator phase voltage $\hat{u}_1$, that is calculated with (9) and (11). The minimum number of series connected power cells can then be determined using

$$n_{\mathrm{Cell,ph,min}} = \left\lceil \frac{\hat{u}_1}{\hat{u}_{\mathrm{Cell,ph,max}}} \right\rceil . \tag{14}$$

Assuming the exemplary values used in Section 5 ($\cos\varphi = 0.85$, $s_{\mathrm{Gen}} = 10$, $I_{\mathrm{coil}} = 52\,\mathrm{A}$ and $P_1 = 10\,\mathrm{MW}$), an FSSG phase voltage amplitude of $\hat{u}_1 \approx 10.67\,\mathrm{kV}$ is calculated. Therefore, at least 15 power cells with 1200 V semiconductors or eleven power cells with 1700 V semiconductors are to be series connected per phase. A higher amount of cells is possible and will increase tolerance against failures. In case of a failure inside a single power cell—except for isolation failure of the transformer—the faulty cell could then be short circuited at the input and output, allowing the system to still operate at the nominal power. The minimum total number of power cells can be calculated using

$$n_{\mathrm{Cell,tot}} = 3 \cdot n_{\mathrm{Cell,ph}} \cdot \frac{s_{\mathrm{Gen}}}{s_{\mathrm{PEC}}} = 3 \cdot n_{\mathrm{Cell,ph}} \cdot n_{\mathrm{PEC}} . \tag{15}$$

As displayed in Table 2 for a three-phase machine with ten branches and no direct branch parallelisation, this results in 450 or 330 cells for 1200 V or 1700 V semiconductors, respectively.

To reduce the number of power cells, several three-phase systems can be parallelised directly as illustrated in Figure 13. This approach reduces the possibilities in the independent branch control but leads to a less complex converter design with higher currents per converter phase.

**Table 2.** Total number of required power cells and corresponding maximum reachable DC link voltages for a 10 MW FSSG depending on the used semiconductor voltage class and the number of parallel connected branches. The maximum reachable voltages for the options without parallel branches, marked with *, are purely theoretical. In practice, several DC cells or DC three-phase rectifier outputs would be paralleled for these options.

| Semiconductor Voltage Class $U_{CES}$ | 1200 V | 1700 V | 1200 V | 1700 V |
|---|---|---|---|---|
| Parallel branches $s_{PEC}$ | 1 | 1 | 5 | 5 |
| Number of 3ph-rectifier $n_{PEC}$ | 10 | 10 | 2 | 2 |
| Converter RMS phase current $I_{PEC}$ | 52 A | 52 A | 260 A | 260 A |
| Min. no. of required power cells $n_{Cell,tot}$ | 450 | 330 | 90 | 66 |
| Max. total DC voltage $U_{Grid,DC,max}$ | 324 kV * | 336.6 kV * | 64.8 kV | 67.32 kV |

For a machine with ten three-phase systems, five parallel systems result in a converter phase current of 260 A, which is a common current rating for several commercial 1200 V and 1700 V SiC-MOSFET and Si-IGBT semiconductor modules. The minimum number of cells can then be reduced down to 90 or 66 cells for 1200 V or 1700 V semiconductors, respectively (See Table 2).

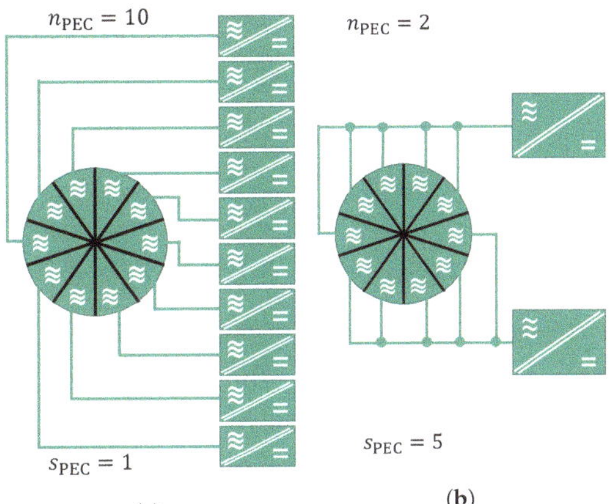


**Figure 13.** Two possible rectifier concepts. (**a**) Each branch is connected to one three-phase rectifier; therefore, 50 A modules can be used. (**b**) Direct parallelisation of five generator branches leads to a lower number of three-phase rectifiers but reduces the possibilities in the independent branch control. Here, 250 A modules can be used.

Assuming a 1:1 transfer ratio of the DAB and a series connection of all power cells on the DC side, this concept can provide a maximum total DC voltage $U_{Grid,DC,max}$ according to

$$U_{Grid,DC,max} = 3 \cdot U_{Cell,DC,max} \cdot n_{Cell,ph} \cdot n_{PEC} \quad . \tag{16}$$

For the above mentioned exemplary FSSG, this would lead to the maximum total DC voltages displayed in Table 2. Lower total DC voltages can be reached by parallelising certain power cell outputs $U_{Cell,DC}$, three-phase converter outputs $U_{PEC,DC}$ or by adapting the transfer ratio of the DAB. The latter can be achieved by changing the winding turns ratio of the MF transformer or by modifying the modulation of the DAB [32]. Higher voltages could be reached by increasing the number of cells per phase $n_{Cell,ph}$. However, for all types of power cells a direct connection on the DC side to a MV DC grid with, e.g., $\pm 25$ kV is theoretically possible for the exemplary generator concept with ten branches.

The concept can be easily adapted to generators with different pole pair numbers or number of branches, respectively. Using (14), (15) and the respective cell voltage maximum possible AC amplitudes $\hat{u}_{\text{Cell,ph,max}}$, the necessary minimum total number of power cells for the 10 MW FSSG can be calculated. The resulting values are illustrated in Figure 14.

It is evident that the number of power cells is hardly dependent on the number of pole pairs or three-phase systems. The generator phase voltage $\hat{u}_1$ decreases with $1/s_{\text{Gen}}$ according to (9) and (11). At the same time, the number of three-phase converters $n_{\text{PEC}}$ increases with $s_{\text{Gen}}$ ((12) with constant $s_{\text{PEC}}$) so that the total number of power cells stays nearly constant. The deviations result from the ceiling function in (14). For some FSSG concepts (corresponding to a certain numbers of pole pairs), not all power cell concepts are possible due to the fact that the ratio of the number of three-phase systems $s_{\text{Gen}}$ and parallel three-phase systems per converter $s_{\text{PEC}}$ would not be an integer. However, with an appropriate choice of pole pairs, this topology offers the possibility to use standard 1200 V or 1700 V semiconductors over a wide range of number of pole pairs and three-phase systems, even if an unusual combination of large phase voltage and low phase current is present.

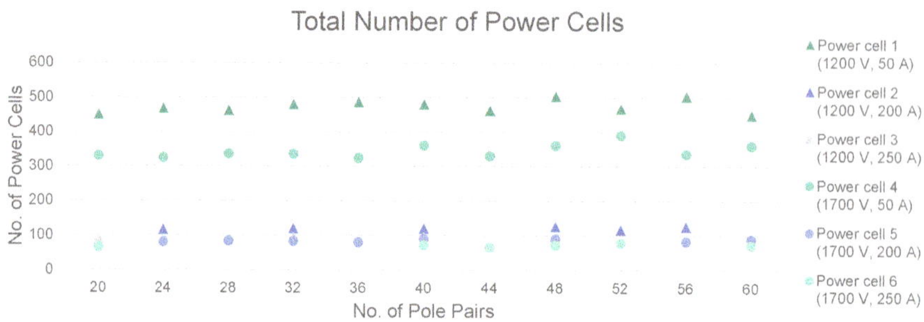


**Figure 14.** Total number of power cells $n_{\text{Cell,tot}}$ necessary for the 10 MW FSSG depending on the number of pole pairs $p$ and the semiconductor voltage and current rating. A current rating of 50 A corresponds to $s_{\text{PEC}} = 1$ (one three-phase rectifier for each branch), 200 A to $s_{\text{PEC}} = 4$ (one three-phase rectifier for four parallelised branches) and 250 A to $s_{\text{PEC}} = 5$ (one three-phase rectifier for five parallelised branches). A power factor of $cos(\varphi) = 0.85$ and a HTS current of $I_{\text{coil}} = 52$ A are assumed as well as a semiconductor voltage utilisation of 60% and a current utilisation of 105% to handle the HTS current.

### 6.2.6. Realising an AC Grid Connection

Besides the direct connection to a MV DC grid, the generator could also be connected to an MV AC grid. To achieve this, the power cells could be extended by an additional single-phase AC inverter at the output, which can be implemented with 1200 V or 1700 V components. The AC outputs are then connected in series to reach the grid phase voltage. Figure 15 shows an exemplary circuit that pictures the basic idea. As for the DC grid connection, not all outputs per phase need to be connected in series. It is also conceivable to parallelise several three-phase system outputs, provided that all sets are used and all subsystems are identical.

If a constant power flow over the DABs is required, this solution shows a high demand of reactive power on both sides of the converter: on the machine side with double the electrical generator frequency $f_1$ (2); on the grid side with double the grid frequency $f_{\text{grid}}$. Accordingly, the required energy storage can be disadvantageous compared to other concepts or operations.

Another possibility is the use of a central DC link at the output of all DABs, combined with a single common grid inverter. Using this configuration, the output H-bridges from above are not necessary and with an appropriate series or parallel connection of the power cell DC outputs, the voltage of the central DC link can be designed to be in the same range as the AC grid voltage. One possible topology for the grid inverter could be the above mentioned MMC, but other topologies can be viable as well, depending on the grid voltage and the grid-side requirements for the inverter, e.g., fault ride through capability. Those investigations are not the subject of this study.

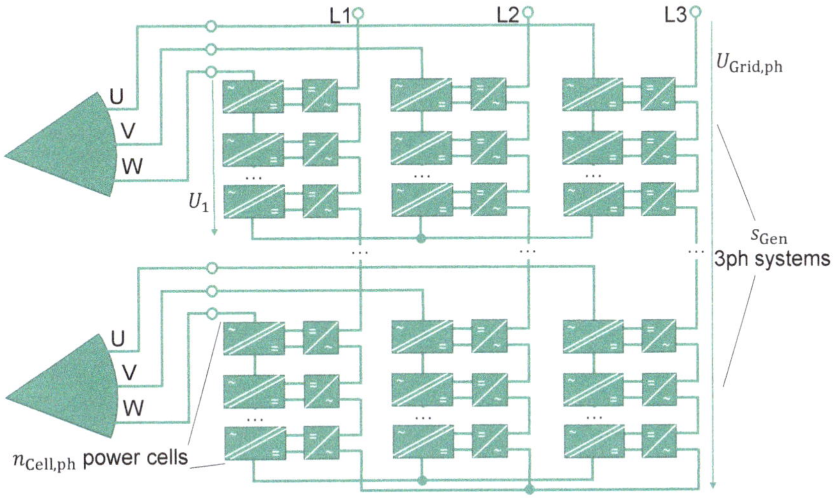


**Figure 15.** Proposed topology for a direct MV AC grid connection. Each power cell is extended by another H-bridge single-phase AC inverter.

In summary, the proposed multi-cell topology based on series connected power cells, each consisting of an H-bridge and a DAB, is well-suited as a rectifier and galvanic isolation unit for the FSSG application. Because it consists of low-voltage, low-current power semiconductors, it can be optimised for the low phase currents of the FSSG. Despite the challenges of the double-frequency reactive power and the high isolation and failure rate requirements on the MF transformer, this topology is worth investigating further, especially due to its high modularity, which allows for the easy adaptation to various generator designs.

## 7. Conclusions

To make use of the potential of superconductors in wind power generators, the design process of an FSSG has to consider the AC losses of the HTS winding. Only a few design measures enable the reduction in these losses, but they can cause conflicts with the design of corresponding PECs. This aspect has not yet been discussed in literature so far. Therefore, this study highlights the interactions between the design of an FSSG and its PEC.

The first sections discuss the general dependencies of the AC losses on the generator design. The AC loss reduction can be achieved through the use of narrow HTS tapes, which operate at very low coil currents, and by a reduction in the electric frequency and the perpendicular field strength. The frequency is influenced by the number of pole pairs whereas the perpendicular field strength is coupled to the pole pitch. Both of these design options are investigated by their respective parameter studies. The parameter optimisation leads to FSSG designs with high output voltages, low output currents and low electric frequencies which have to be handled by their PEC.

After these generator-related investigations, the following sections focused on the influence of such unusual parameter combinations on standard PEC designs. It was found that available power semiconductor modules can supply the necessary blocking voltage if multi-level topologies, e.g., an ANPC, are used. However, the current utilisation factor of these semiconductor modules would be poor due to the low phase currents. Additionally, the corresponding converter concepts tend to require higher numbers of pole pairs for the generator, which would lead to higher AC losses inside the HTS tape. Therefore, multi-cell designs are discussed in the last section as an alternative to standard PEC topologies. Among several concepts, a topology based on series connected power cells is presented, which offers practicable solutions with available power semiconductors at the cost of a larger component count and a higher system complexity.

Overall, the design of PECs for FSSGs is challenging, but there are technical solutions available. This study provides insight on how to consider the requirements of both an FSSG and its PEC and presents guidance for future designs.

**Author Contributions:** Conceptualization S.L., S.S. and S.R.F.; Data curation S.L., S.S. and S.R.F.; Formal analysis S.L., S.S. and S.R.F.; Funding acquisition B.P., M.J. and M.H.; Investigation S.L., S.S. and S.R.F.; Methodology S.L., S.S. and S.R.F.; Project administration S.L.; Resources S.L., S.S. and S.R.F.; Software S.L., S.S. and S.R.F.; Supervision M.J. and M.H.; Validation B.P., M.J., S.M. and M.H.; Visualization S.L., S.S. and S.R.F.; Writing—original draft S.L., S.S. and S.R.F.; Writing—review and editing B.P., M.J., S.M. and M.H. All authors have read and agreed to the published version of the manuscript.

**Funding:** The underlying work of this article was funded by the German Federal Ministry for Economic Affairs and Climate Action (project name "SupraGenSys", funding reference numbers 03EE3010A and 03EE3010E). The responsibility for the content of this article lies with the authors and does not necessarily reflect the opinion of the SupraGenSys project consortium.

**Conflicts of Interest:** The authors declare no conflict of interest. The funders had no role in the design of the study; in the collection, analyses, or interpretation of data; in the writing of the manuscript, or in the decision to publish the results.

## Abbreviations

The following abbreviations are used in this manuscript:

| | | | |
|---|---|---|---|
| CHB | Cascaded H-bridge | MF | Medium Frequency |
| DAB | Dual Active Bridge | MMC | Modular Multilevel Converter |
| FC | Flying Capacitor | MV | Medium Voltage |
| HTS | High Temperature Superconductor | PEC | Power Electronic Converter |
| HVDC | High-Voltage Direct Current | SMC | Stacked-Multicell Converter |

## Nomenclature

| | | | |
|---|---|---|---|
| $a$ | Number of parallel branches | $P_1$ | Real power of the generator |
| $a_{max}$ | Maximum number of parallel branches | $P_{AC}$ | AC losses in HTS tapes |
| $B_t$ | Tangential flux density | $P_N$ | Rated output power of the generator |
| $d$ | Width of the HTS tape | $q$ | Number of slots per pole and per phase |
| $f_1$ | Electric frequency in the stator winding | $S_1$ | Apparent power of the generator |
| $f_{grid}$ | AC grid frequency | $s_{Gen}$ | Number of independent three-phase systems |
| $I_1$ | RMS phase current of the generator | $s_{PEC}$ | Number of parallel three-phase systems per PEC |
| $I_{coil}$ | Coil current | $T_N$ | Rated input torque of the generator |
| $I_c$ | Critical current of the HTS tape | $U_{Cell,DC,max}$ | Maximum DC voltage of one power cell |
| $I_{PEC}$ | Phase current of the PEC | $U_{Cell,DC}$ | DC voltage of one power cell |
| $l_{tape}$ | Effective length of the HTS tape | $U_{Cell,ph}$ | RMS phase voltage of one power cell |
| $m$ | Number of phases | $U_{Grid,DC,max}$ | Maximum total DC voltage with series power cell concept |
| $n$ | Rotational speed | $U_{PEC,DC}$ | DC voltage of one three-phase rectifier |
| $n_{Cell,ph,min}$ | Minimum number of power cells per phase | $U_1$ | RMS phase voltage of the generator |
| $n_{Cell,ph}$ | Number of power cells per phase | $Z_1$ | Number of stator slots |
| $n_{Cell,tot}$ | Total number of power cells | $\cos\varphi$ | Power factor |
| $n_N$ | Rated rotor speed of the generator | $\hat{u}_1$ | Amplitude of the generator phase voltage |
| $n_{PEC}$ | Number of PEC | $\hat{u}_{phase\text{-}phase}$ | Amplitude of the generator phase-to-phase voltage |
| $p$ | Number of pole pairs | $\hat{u}_{Cell,ph,max}$ | Maximum possible AC voltage amplitude of one power cell |

## References

1. Polinder, H.; Ferreira, J.A.; Jensen, B.B.; Abrahamsen, A.B.; Atallah, K.; McMahon, R.A. Trends in Wind Turbine Generator Systems. *IEEE J. Emerg. Sel. Top. Power Electron.* **2013**, *1*, 174–185. [CrossRef]
2. Kalsi, S.S. Superconducting Wind Turbine Generator Employing $M_gB_2$ Windings Both on Rotor and Stator. *IEEE Trans. Appl. Supercond.* **2014**, *24*, 47–53. doi: [CrossRef]

3. Shrestha, G.; Polinder, H.; Ferreira, J.A. Scaling laws for direct drive generators in wind turbines. In Proceedings of the 2009 IEEE International Electric Machines and Drives Conference, Miami, FL, USA, 3–6 May 2009; pp. 797–803. [CrossRef]
4. Stuebig, C.; Haberjan, L.; Seibel, A.; Steffen, J.; Thalemann, F.; Wecker, M.; Ponick, B. Segmented permanent magnet ring generator with active damping of axial oscillations. In Proceedings of the 2016 XXII International Conference on Electrical Machines (ICEM), Lausanne, Switzerland, 4–7 September 2016; IEEE: Piscataway, NJ, USA, 2016; pp. 564–570. doi: [CrossRef]
5. Steffen, J.; Lengsfeld, S.; Jung, M.; Ponick, B.; Herranz Gracia, M.; Spagnolo, A.; Klöpzig, M.; Schleicher, K.; Schäfer, K. Design of a Medium Voltage Generator with DC-Cascade for High Power Wind Energy Conversion Systems. *Energies* **2021**, *14*, 3106. [CrossRef]
6. Alexandrova, Y.; Semken, S.; Polikarpova, M.; Pyrhonen, J. Defining proper initial geometry of an 8 MW liquid-cooled direct-drive permanent magnet synchronous generator for wind turbine applications based on minimizing mass. In Proceedings of the 2012 XXth International Conference on Electrical Machines, Marseille, France, 2–5 September 2012; pp. 1250–1255. doi: [CrossRef]
7. Haran, K.S.; Kalsi, S.; Arndt, T.; Karmaker, H.; Badcock, R.; Buckley, B.; Haugan, T.; Izumi, M.; Loder, D.; Bray, J.W.; et al. High power density superconducting rotating machines—Development status and technology roadmap. *Supercond. Sci. Technol.* **2017**, *30*, 123002. [CrossRef]
8. Maples, B.; Hand, M.; Musial, W. *Comparative Assessment of Direct Drive High Temperature Superconducting Generators in Multi-Megawatt Class Wind Turbines*; National Renewable Energy Lab.(NREL): Golden, CO, USA, 2010. [CrossRef]
9. Guan, Y.; Zhu, Z.Q.; Li, G.; Azar, Z.; Thomas, A.S.; Vedreño-Santos, F.; Odavic, M. Influence of Pole Number and Stator Outer Diameter on Volume, Weight, and Cost of Superconducting Generators with Iron-Cored Rotor Topology for Wind Turbines. *IEEE Trans. Appl. Supercond.* **2017**, *27*, 5204009. doi: [CrossRef]
10. Oomen, M. AC Loss in Superconducting Tapes and Cables. Ph.D. Thesis, Universiteit Twente, Enschede, The Netherlands, 2000.
11. Oomen, M.P.; Rieger, J.; Hussennether, V.; Leghissa, M. AC loss in high-temperature superconducting conductors, cables and windings for power devices. *Supercond. Sci. Technol.* **2004**, *17*, S394–S399. [CrossRef]
12. Vargas-Llanos, C.R.; Lengsfeld, S.; Noe, M.; Arndt, T.; Grilli, F. Influence of Coil Position on AC Losses of Stator Superconducting Windings of a Synchronous Machine for a 10 MW Wind Turbine. *IEEE Trans. Appl. Supercond.* **2021**, *31*, 5206509. [CrossRef]
13. Terao, Y.; Sekino, M.; Ohsaki, H. Electromagnetic Design of 10 MW Class Fully Superconducting Wind Turbine Generators. *IEEE Trans. Appl. Supercond.* **2012**, *22*, 5201904. [CrossRef]
14. Thoma, J.; Volzer, B.; Kranzer, D.; Derix, D.; Hensel, A. Design and Commissioning of a 10 kV Three-Phase Transformerless Inverter with 15 kV Silicon Carbide MOSFETs. In Proceedings of the 2018 20th European Conference on Power Electronics and Applications, EPE 2018 ECCE Europe, Riga, Latvia, 17–21 September 2018.
15. Kranzer, D.; Thoma, J.; Volzer, B.; Derix, D.; Hensel, A.; Ise, F. Development of a 10 kV Three-Phase Transformerless Inverter with 15 kV Silicon Carbide MOSFETs for Grid Stabilization and Active Filtering of Harmonics. In Proceedings of the 2017 19th European Conference on Power Electronics and Applications (EPE'17 ECCE Europe), Warsaw, Poland, 11–14 September 2017; pp. 1–8.
16. Thoma, J.; Kolb, S.; Salzmann, C.; Kranzer, D. Characterization of high-voltage-sic-devices with 15 kV blocking voltage. In Proceedings of the 2016 IEEE International Power Electronics and Motion Control Conference, PEMC 2016, Varna, Bulgaria, 25–28 September 2016; pp. 946–951. [CrossRef]
17. Sabri, S.; Van Brunt, E.; Barkley, A.; Hull, B.; O'Loughlin, M.; Burk, A.; Allen, S.; Palmour, J. New generation 6.5 kV SiC power MOSFET. In Proceedings of the 2017 IEEE 5th Workshop on Wide Bandgap Power Devices and Applications, WiPDA 2017, Santa Ana Pueblo, NM, USA, 30 November–1 December 2017; pp. 246–250. [CrossRef]
18. Pala, V.; Brunt, E.V.; Cheng, L.; O'Loughlin, M.; Richmond, J.; Burk, A.; Allen, S.T.; Grider, D.; Palmour, J.W.; Scozzie, C.J. 10 kV and 15 kV silicon carbide power MOSFETs for next-generation energy conversion and transmission systems. In Proceedings of the 2014 IEEE Energy Conversion Congress and Exposition, ECCE 2014, Pittsburgh, PA, USA, 14–18 September 2014; pp. 449–454. [CrossRef]
19. Fronius Schweiz, A.G. Fronius Primo. Available online: https://www.fronius.com/~/downloads/Solar%20Energy/Datasheets/SE_DS_Fronius_Primo_DE.pdf (accessed on 13 April 2022).
20. KACO New Energy GmbH. Blueplanet 125-150 TL3. Available online: https://kaco-newenergy.com/index.php?eID=dumpFile&t=f&f=8178&token=fd499b0bbd7a44b2bc9b5738a801782be60d53bb (accessed on 13 April 2022).
21. SMA Solar Technology AG. SUNNY TRIPOWER 8.0/10.0 mit SMA SMART CONNECTED. Available online: https://files.sma.de/downloads/STP8-10-3AV-40-DS-en-32.pdf (accessed on 13 April 2022).
22. Huber, J.E.; Kolar, J.W. Optimum Number of Cascaded Cells for High-Power Medium-Voltage AC-DC Converters. *IEEE J. Emerg. Sel. Top. Power Electron.* **2017**, *5*, 213–232. [CrossRef]
23. Lorenzani, E.; Immovilli, F.; Migliazza, G.; Frigieri, M.; Bianchini, C.; Davoli, M. CSI7: A Modified Three-Phase Current-Source Inverter for Modular Photovoltaic Applications. *IEEE Trans. Ind. Electron.* **2017**, *64*, 5449–5459. [CrossRef]
24. Sahan, B.; Araújo, S.V.; Nöding, C.; Zacharias, P. Comparative evaluation of three-phase current source inverters for grid interfacing of distributed and renewable energy systems. *IEEE Trans. Power Electron.* **2011**, *26*, 2304–2318. [CrossRef]
25. Meynard, T.A.; Foch, H. Multilevel choppers for high voltage applications. *EPE J.* **1992**, *2*, 45–50. [CrossRef]
26. Mersche, S.; Bernet, D.; Hiller, M. Quasi-two-level flying-capacitor-converter for medium voltage grid applications. In Proceedings of the 2019 IEEE Energy Conversion Congress and Exposition, ECCE 2019, Baltimore, MD, USA, 29 September–3 October 2019; pp. 3666–3673. [CrossRef]

27. Gateau, G.; Meynard, T.A.; Foch, H. Stacked multicell converter (SMC): Properties and design. In Proceedings of the PESC Record–IEEE Annual Power Electronics Specialists Conference, Vancouver, BC, Canada, 17–21 June 2001; Volume 3, pp. 1583–1588. [CrossRef]
28. Hosseini, S.H.; Khoshkbar Sadigh, A.; Sabahi, M. New configuration of stacked multicell converter with reduced number of DC voltage sources. *IET Conf. Publ.* **2010**, *2010*, 2–7. [CrossRef]
29. Sadeghi, M.; Hosseini, S.H.; Alizadeh, R.; Banaei, M.R. New mixed stacked multicell converter with interesting advantages. *IET Power Electron.* **2012**, *5*, 1298–1304. [CrossRef]
30. Lesnicar, A.; Marquardt, R. An innovative modular multilevel converter topology suitable for a wide power range. In Proceedings of the 2003 IEEE Bologna PowerTech—Conference Proceedings, Bologna, Italy, 23–26 June 2003; Volume 3, pp. 6–11. [CrossRef]
31. Baker, I.R.H.; Bannister, L.H. Electric Power Converter. U.S. Patent 3,867,643, 18 February 1975.
32. Doncker, W.A.A.D.; Divan, D.M.; Kheraluwala, M.H. A Three-phase Soft-Switched High-Power-Density dc/dc Converter for High-Power Applications. *IEEE Trans. Ind. Appl.* **1991**, *27*, 63–73. [CrossRef]
33. Mukherjee, S.; Kumar, A.; Chakraborty, S. Comparison of DAB and LLC DC-DC Converters in High-Step-Down Fixed-Conversion-Ratio (DCX) Applications. *IEEE Trans. Power Electron.* **2021**, *36*, 4383–4398. [CrossRef]
34. Yuan, L.; Lu, Z.; Sun, J.; Duan, R.; Zhao, Z. Comparative Evaluation of Isolated Bidirectional DC/DC Converter in High-Power High-Frequency Occasions. In Proceedings of the ICEMS 2018—2018 21st International Conference on Electrical Machines and Systems, Jeju, Korea, 7–10 October 2018; pp. 2285–2290. [CrossRef]
35. Bernet, S.; Brückner, T.; Malinowski, M.; Weber, J.; Weitendorf, N. *Selbstgeführte Stromrichter am Gleichspannungszwischenkreis*; Springer: Berlin, Germany, 2012; p. 283.

*energies*

MDPI

*Article*

# Swiveling Magnetization for Anisotropic Magnets for Variable Flux Spoke-Type Permanent Magnet Motor Applied to Electric Vehicles

**Yin-Hui Lee and Min-Fu Hsieh ***

Department of Electrical Engineering, National Cheng Kung University, Tainan 701, Taiwan; yhlee.work@gmail.com
* Correspondence: mfhsieh@mail.ncku.edu.tw

**Abstract:** This paper investigates the application of anisotropic low-coercive force (LCF) magnets to a novel variable-flux spoke-type permanent magnet synchronous motor (VFS-PMSM) for electrical vehicles with a wide speed range. In the VFS-PMSM, flux is regulated by swiveling the magnetization of the anisotropic LCF magnets instead of directly magnetizing or demagnetizing them. The previously proposed VFS-PMSM uses only isotropic LCF magnets for easily swiveling the magnetic pole direction, resulting in lower torque density. The challenge thus lies in the feasibility to swivel the magnetic pole direction of the anisotropic LCF magnet, and the impact of the different magnetization strengths of the anisotropic magnets on the motor performance. This paper first studies the feasibility to swivel the magnetization direction of anisotropic LCF magnets through experiments. It is confirmed that the magnetization direction can be successfully swiveled by 90 degrees with a reduced external magnetizing field. Then, two VFS-PMSM topologies and various rotor configurations are compared in terms of key performance indices to determine critical sizing factors for performance enhancement. Finite element analysis is used for simulations. In comparison with the VFS-PMSM equipped with isotropic LCF magnets, the maximum torque of the proposed topology can be improved for the same flux adjustment ability. Alternatively, the flux adjustment ability can also be enhanced by 37.43% for the same maximum torque.

**Keywords:** electric vehicles; magnetization; memory machine; variable-flux motor; permanent magnet synchronous motors

**Citation:** Lee, Y.-H.; Hsieh, M.-F. Swiveling Magnetization for Anisotropic Magnets for Variable Flux Spoke-Type Permanent Magnet Motor Applied to Electric Vehicles. *Energies* **2022**, *15*, 3825. https://doi.org/10.3390/en15103825

Academic Editor: Calin Iclodean

Received: 14 April 2022
Accepted: 19 May 2022
Published: 23 May 2022



## 1. Introduction

Electric vehicles (EVs) have been rapidly developed in recent years due to the growing concerns about environmental protection and the demand for green transportation. Traction motors are treated as key to electric vehicle performance. Interior Permanent Magnet Synchronous Motors (IPMSM) are a mature candidate for EV tractions with their merits of high torque density and high efficiency [1,2], and therefore, they have been widely studied. Huynh et al. analyzed and compared the performances of IPMSMs and permanent-magnet-assisted synchronous reluctance machines (PMa-SynRMs) on constant power speed range (CPSR) for EV applications [3]. Kim et al. studied the shape parameters of IPMSMs and improved their efficiency by about 1% [4]. However, common EVs usually face various road conditions, such as high torque for hill climbing or high speed for long-distance cruises, and the constant flux linkage contributed by permanent magnets (PMs) may limit the speed of IPMSMs [5]. Flux-weakening control is usually used to weaken the field and increase the speed range [6,7]. Inoue et al. analyzed the performance of current control and direct torque control (DTC) with flux-weakening applied and summarized their advantages and disadvantages [8]. However, weakening the flux would sacrifice part of the stator current to suppress flux and lead to extra copper loss. Additionally, motor mathematical models are often needed to calculate the current phase advance angle, and the errors are likely to occur under dynamic operations [9]. The motor speed can also be expanded with an auxiliary coil to accurately control rotor magnetic flux [10,11]. Some designs, such as the double excited synchronous machine [12] and the synchronous/permanent magnet hybrid AC machine [13], place the auxiliary windings and permanent magnets in series or parallel. Nevertheless, the presence of the slip rings is the main drawback of this type of motor. To mitigate the problem of slip rings, a doubly salient PM motor was proposed [14], which placed the auxiliary windings, stator windings, and PMs in the stator, resulting in the stator being prone to saturation. A motor having an auxiliary winding with a hybrid radial and axial magnetic circuit was proposed [15,16]; however, the torque and power density may likely reduce, and the power consumption and copper loss are still inevitable for weakening the field [17]. The appearance of memory motors, also known as variable flux motors (VFMs), effectively resolves the above problems [18]. In the memory motors, an impulse current can be generated by stator windings to change the magnetization state (MS) of a low coercive force (LCF) magnet to achieve the purpose of adjusting air gap flux density. The loss caused by the impulse current to change the MS of the LCF magnet is almost negligible for motors in operation [19].

Low coercive force (LCF) magnets (e.g., AlNiCo) are often used in the VFMs to reduce the impulse current needed for magnetization or demagnetization of the magnets. However, using only LCF magnets is prone to unexpected demagnetization [20]. To prevent this situation, Kato et al. [21] proposed a variable-flux flux-intensifying interior permanent magnet machine (VFI-IPMM), where demagnetization can be avoided because the d-axis current is positive at heavy load. To further enhance the torque density of the VFMs, various topologies of hybrid variable flux motors (H-VFM) combining LCF magnets and high coercive force (HCF) magnets were proposed [22,23]. The H-VFM rotor can be divided into series and parallel configurations according to the arrangements of permanent magnets. The series type of H-VFM [24] possesses a higher torque density at low speed since the flux of the HCF and LCF magnets can be combined by the series arrangement. In the series type of H-VFM, a much smaller impulse current is needed to switch the MS of the LCF magnet from flux weakening to enhancing due to the influence of the HCF magnet. Conversely, it requires a larger impulse current to switch from flux enhancing to weakening. For the parallel type of H-VFM [25], an active flux leakage bypass is established which is advantageous for flux regulation. Hua et al. compared the series and parallel types of H-VFMs and revealed that the operating point of the LCF magnet in the parallel type is unstable [26]. To have a stable operation of the LCF magnet and excellent flux regulation ability, a variable-flux permanent-magnet synchronous machine with a quasi-series magnet configuration and passive flux barrier was proposed [27,28]; however, the problem of a large magnetizing current to switch from flux enhancing to weakening is yet to be resolved. Yang et al. proposed a hybrid-magnet-circuit VFM [29–31] that combines the advantages of both types of H-VFMs mentioned above to achieve high torque density, good flux adjustment ability, and reduce magnetizing currents. However, the rotor configuration seems complex and requires more expensive HCF rare-earth magnets.

Lee et al. proposed a novel variable flux spoke-type permanent magnet motor (VFS-PMSM) [32], which combines the advantages of the series and parallel types of H-VFMs. The flux regulation can be achieved by swiveling the magnetic pole of the isotropic LCF magnet. By so doing, the flux produced by the LCF magnets is ensured to align with that produced by the HCF magnet, and therefore accidental demagnetization can be avoided. The magnetizing impulse current is also greatly reduced for switching the MS and regulating air gap flux density. However, there is still room for improvement in torque density and flux regulation ability due to the use of the weaker isotropic LCF magnets.

To simultaneously achieve the above-mentioned advantages, i.e., high torque density, excellent flux regulation ability, low demagnetization risk, low magnetizing impulse current demand, high efficiency at high speed, and expanded motor operating range, this paper proposes a novel VFS-PMSM topology combining HCF magnets and anisotropic LCF magnets, instead of the isotropic LCF magnet used in a previous study [32]. The method of swiveling the magnetic pole of the LCF magnet [32] is adopted, which differs from the previously mentioned methods that magnetize/demagnetize the magnets in the same axis [18–31]. It should be noted that there is a significant difference in flux generated and magnetizing field required between the easy and hard axes in the anisotropic magnets. Therefore, it is generally believed that the magnetic pole of an anisotropic magnet cannot be swiveled. This is the key challenge and the major difference from the isotropic design previously reported [32]. In this paper, analyses and experiments are conducted to study the feasibility of swiveling magnetization for the anisotropic magnet applied to the VFS-PMSM. Then the torque density and flux regulation ability of the motor are investigated. Several configurations of the VFS-PMSM are compared to highlight a suitable arrangement to reduce magnetizing current while enhancing the maximum torque and flux regulation ability.

This paper is organized as follows. In Section 2, the characteristics of the proposed variable flux PM motor topologies and the previously developed VFS-PMSM are analyzed and compared. Their features are explained in detail. The feasibility of applying the anisotropic AlNiCo in the VFS-PMSM is experimentally verified in Section 3. In Section 4, the key characteristics of the proposed anisotropic VFS-PMSM with different rotor types are compared by adjusting the sizing parameters of the LCF magnets and HCF magnets. The electromagnetic characteristics of the proposed motor topologies are simulated and compared with other types of motors in Section 5, followed by Section 6 to conclude this paper.

## 2. Variable Flux PM Motor Topologies and Proposed VFS-PMSM

The VFS-PMSM [32], like the general series and parallel types of H-VFMs, uses both LCF magnets and HCF magnets in the rotor; however, the principal difference of this design is to use the magnetic field excited by the stator to "swivel" the magnetization vector of the LCF magnet to achieve flux regulation. When the flux density of the air gap needs to be enhanced, the impulse current produced by the stator winding generates a magnetizing field on the +d axis (phase advance is $-90^\circ$) to magnetize the LCF magnet radially. The flux of the HCF magnet is guided to the air gap and the stator through the LCF magnet. To weaken the flux density of the air gap, the impulse current generates a magnetizing field on the -d axis (phase advance is $+90^\circ$) to swivel the magnetization direction of the LCF magnet to align in the tangential direction. The flux of the HCF magnet is guided to the flux leakage bypass via the LCF magnet. The details for the magnetization of the LCF magnets can be referred to in a previous study [32]. Theoretically, the impulse current applied on the +d or $-$d axis will not result in an undesirable torque surge. To successfully adjust the air gap flux density of the VFS-PMSM by swiveling the magnetization vector of the LCF magnet (called the "swiveling magnetization technique" in what follows), isotropic AlNiCo was used in the previously developed VFS-PMSM [32] (called the "isotropic VFS-PMSM" in what follows). The magnetic characteristics of isotropic magnets should be the same regardless of the direction of magnetization.

To investigate the feasibility of anisotropic AlNiCo magnets for swiveling magnetization, the demagnetization curves of the isotropic and anisotropic AlNiCo magnets in different directions, as shown in Figure 1, are first measured by a B-H tracer. Both magnets are 10 mm $\times$ 10 mm $\times$ 10 mm cubes. As shown in Figure 1, the "$0^\circ$" indicates one side of the cube, and the "$90^\circ$" is the other. As can be seen, the isotropic magnet has almost identical demagnetization curves in the two directions and its remanence $Br$ and maximum energy product are lower than those in the easy axis of the anisotropic magnets. It should be noted that the swiveling magnetization technique involves the magnetization of an anisotropic LCF magnet in two directions separated by $90^\circ$ (i.e., either on the easy axis or hard axis), the arrangements of the anisotropic LCF magnet should be considered for magnetizations to regulate the flux and torque production.

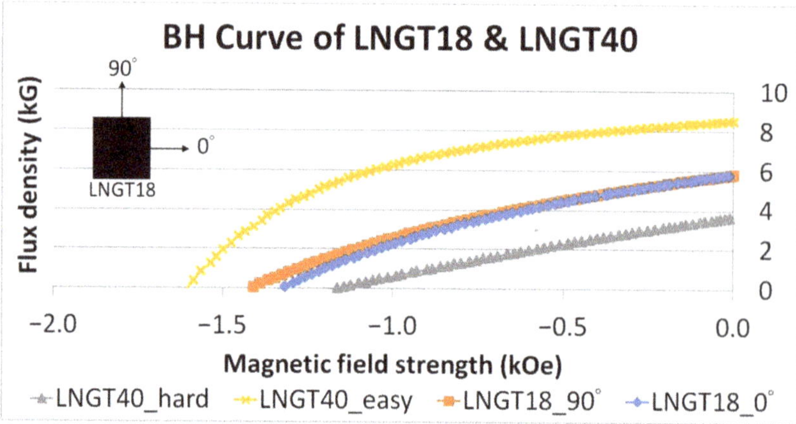


**Figure 1.** B-H curve of LNGT18 (isotropic) and LNGT40 (anisotropic).

One merit of the VFMs is their air gap flux density regulation ability to improve the constant power and high-efficiency operating range. Therefore, the key index used to evaluate a VFM is the highest torque at flux enhancing state and the back electromotive force (back EMF) difference between the flux enhancing and weakening cases at a certain speed. In the enhancing state, the motor air gap flux density is maximized and in the weakening state, it is minimized. This paper proposes two topologies that apply anisotropic AlNiCo to the VFS-PMSM, as shown in Figure 2 (called the "anisotropic VFS-PMSM" in what follows). In Topology I, the easy axis of the anisotropic magnet is arranged to be aligned with the radial direction of the rotor, as shown in Figure 2b while the hard axis of the anisotropic magnet in Topology II is aligned radially (Figure 2c).

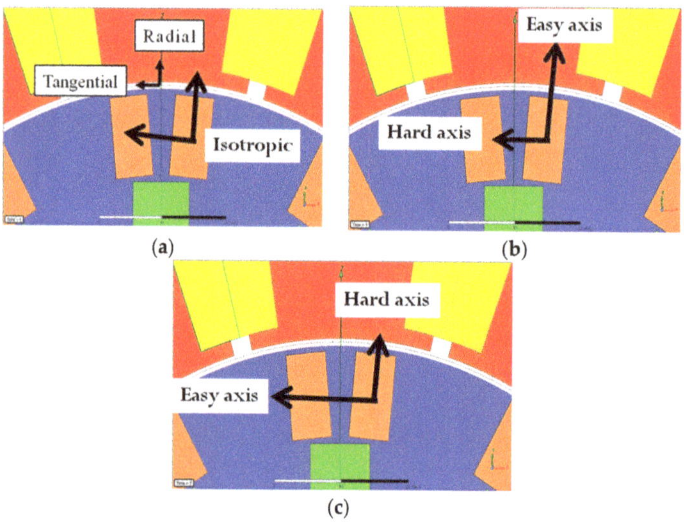


**Figure 2.** (**a**) Isotropic variable-flux spoke-type permanent magnet synchronous motor (VFS-PMSM), (**b**) Anisotropic Topology I, and (**c**) Anisotropic Topology II.

*2.1. Torque Generation*

The material properties and dimensions of the motor are given in Table 1. As can be seen, the difference between Topologies I, II, and the isotropic VFS-PMSM only lie in the LCF magnets used. The AlNiCo magnet (model: LNGT 18) is used for the isotropic

VFS-PMSM, and the anisotropic AlNiCo magnet (LNGT 40) is for Topologies I and II. The performance of the three models in Figure 2 is compared by simulations using finite element analysis (FEA) of transient electromagnetic fields (FEA, ANSYS Maxwell). The material properties of the LCF magnets (e.g., the hysteresis loops) used in the simulations are obtained from the measurements conducted in Section 3. The material properties of the HCF magnets can be found in the database of the FEA software. Note that the two anisotropic models are only preliminary designs by directly replacing the isotropic magnets with the anisotropic magnets. As shown in Figure 3, the maximum torque produced at flux enhancing is compared between the isotropic VFS-PMSM and Topologies I and II of the anisotropic VFS-PMSM. As can be seen, the isotropic VFS-PMSM produces larger torque than the other two, and Topology II has slightly higher torque than Topology I. This seems that the application of the anisotropic magnets is not effective; however, as previously mentioned, it would not be appropriate to directly replace the isotropic magnets with anisotropic ones due to their different strengths in different directions, as indicated in Figure 1. Therefore, with slightly lower peak torque than that of the isotropic VFS-PMSM, the topology with anisotropic LCF magnets will be optimized in a later section.

**Table 1.** Motor parameters and material properties.

| Item | Value |
|---|---|
| DC-link voltage Udc (V) | 48 |
| Phase Current (peak) (A) | 60 |
| Power (kW) | 2 |
| Rated Speed (rpm) | ~1500 |
| Outer diameter of stator (mm) | 160 |
| Outer diameter of rotor (mm) | 112 |
| Air gap (mm) | 1 |
| Stack length (mm) | 36 |
| Winding turns per phase (turns) | 36 |
| HCF PM grade | N32H |
| LCF PM grade (isotropic VFS-PMSM/Topology I & II (T_I&II)) | LNGT18/LNGT40 |
| HCF PM coercivity (kA/m) | 886 |
| LCF PM (isotropic VFS-PMSM/T_I&II) coercivity (kA/m) | 100.5/115.7 |
| HCF PM remanence (T) | 1.146 |
| LCF PM (isotropic VFS-PMSM/T_I&II) remanence (T) | 0.58/0.866 |
| Stator & rotor steel | 25CS1500HF |

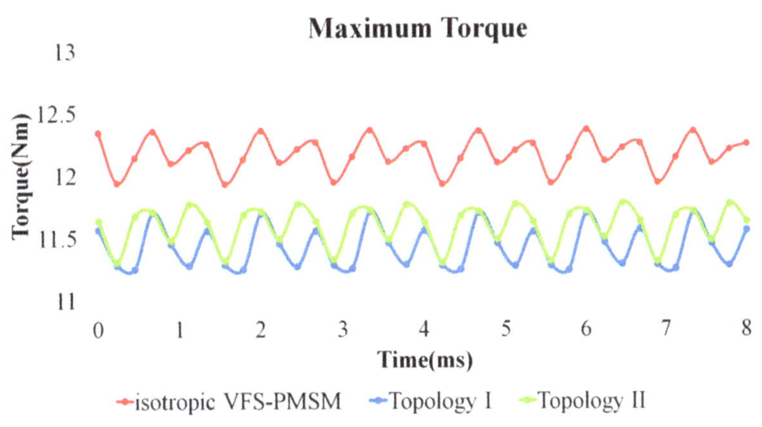


**Figure 3.** Maximum torque of isotropic VFS-PMSM, Topologies I and II in enhancing state obtained through Maximum Torque per Ampere (isotropic VFS-PMSM = 12.12 Nm, Topology I = 11.42 Nm, Topology II = 11.6 Nm).

### 2.2. Flux Regulation Ability

The flux regulation ability can be indirectly evaluated through the difference in the back EMF between the flux enhancing and weakening states. The back EMF difference $E_r$ is defined as:

$$E_r = E_+ - E_- \quad (1)$$

where $E_+$ is the back EMF at flux enhancing and $E_-$ is that at flux weakening. Note that the air gap flux density is considered proportional to the back EMF at the same speed (assuming no saturation), and thus the back EMFs at flux enhancing and weakening represent the maximum and minimum air gap flux densities, respectively. The larger the back EMF difference is, the better the flux regulation ability is. Figure 4 shows the simulated back EMF of the isotropic VFS-PMSM, Topologies I and II at flux enhancing and weakening, and $E_r$ of the three cases can be obtained and expressed by $E_{r\_iso\_VFS}$, $E_{r\_T-I}$ and $E_{r\_T-II}$, respectively. As shown in Figure 4, $E_{r\_T-II}$ is the largest, and $E_{r\_T-I}$ is the smallest.

### 2.3. Discussion

The correlation between the LCF magnet features and the back EMF difference $E_r$ should be investigated. The remanence of the LCF magnets toward the radial and tangential directions (indicated in Figure 2) are considered separately here for their individual contributions to air gap flux. As shown in Table 2, the LCF magnet in Topology I has the largest radial remanence, followed by that of the isotropic VFS-PMSM and then Topology II. It seems that the remanence of the LCF magnet in the radial direction does not directly correlate to the maximum torque production. For the tangential remanence of the LCF magnet, Topology II has the highest, the isotropic VFS-PMSM is the second, and Topology I the least. From these discussions, it can be seen that $E_r$ and the flux regulation ability are strongly related to the tangential remanence of the LCF magnet rather than the radial one. This is a key to the arrangement of anisotropic LCF magnets for improving flux regulation ability (placing the easy axis of the LCF magnet tangentially).

**Table 2.** Radial and tangential remanence of isotropic VFS-PMSM, Topology I, and Topology II.

| Remanence of LCF Magnets | Isotropic VFS-PMSM | Topology I | Topology II |
|---|---|---|---|
| Radial | 0.58 T | 0.895 T | 0.364 T |
| Tangential | 0.58 T | 0.364 T | 0.895 T |

For torque production, from the study previously mentioned, the remanence of the LCF magnet should not be the main cause that affects the maximum torque (neither radial nor tangential). It is presumed that the HCF magnet is key to the maximum torque of the motor, and when enhanced, as mentioned, optimization will be conducted.

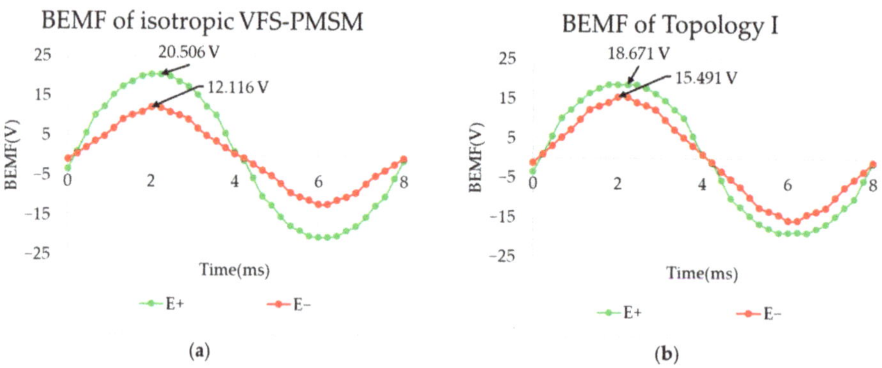


**Figure 4.** *Cont.*

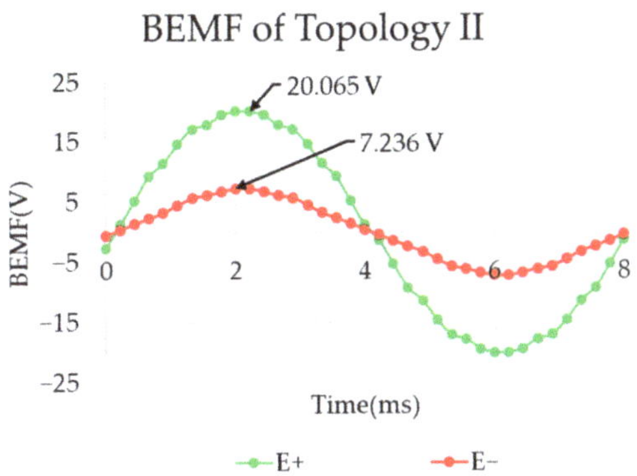


(c)

**Figure 4.** Back EMF of (**a**) isotropic VFS-PMSM, (**b**) Topology I, and (**c**) Topology II at enhancing and weakening ($E_{r\ iso\_VFS}$ = 8.39, $E_{r\ T-I}$ = 3.18 V, $E_{r\ T-II}$ = 12.83 V).

## 3. Feasibility Verification of Anisotropic AlNiCo Application in VFS-PMSM

The aforementioned VFS-PMSM topologies are studied using anisotropic AlNiCo as the LCF magnet to adjust the air gap flux density. It is generally believed that the easy axis of an anisotropic magnet should only be used for applications as a stronger field can be produced. Furthrmore, for the swiveling magnetization technique, the MS of the anisotropic LCF magnet is required to be switched between the hard axis and easy axis by an external magnetizing field. Therefore, whether the magnetization of an anisotropic magnet can be successfully swiveled should be investigated, and experiments can be conducted for validation. Moreover, the magnetizing field strength required to swivel the magnetization direction should also be tested. Then, the demagnetization curves obtained by the "swiveling magnetization technique" and the "general magnetizing method" (i.e., magnetization in the same axis) can be compared. Note that commercial finite element simulation software is not capable of simulating the swiveling magnetization technique, and thus the following experiments are planned, and the results are obtained.

The magnets to be measured are cube anisotropic AlNiCo magnets LNGT40. The measurements include:

I. The initial magnetization curve and demagnetization curve in the easy axis and hard axis when they are originally unmagnetized (general magnetization method);
II. The initial magnetization curve and demagnetization curve in the hard axis after the magnet is already magnetized in the easy axis (swiveling magnetization technique);
III. The initial magnetization curve and demagnetization curve in the easy axis after the magnet is already magnetized in the hard axis (swiveling magnetization technique).

The experimental setup is shown in Figure 5, where item "A" is the B-H tracer, item "B" is the magnetizing coil, item "C" is the magnet holder, item "D" is the magnet to be measured, and item "E" is the pickup secondary coil. The experimental steps are explained as follows.

(I) The unmagnetized magnet is placed in the B-H tracer with its easy axis aligned with the pickup coil. The external magnetizing field is slowly increased until the measurement of the initial magnetization curve in the easy axis is completed (general magnetizing method). Then, the coil current is released, and then a sufficiently large external field opposite to the previous magnetizing direction is applied and gradually reduced until the current in the magnetizing coil drops to zero to complete the measurement of the demagnetization curve in the easy axis (general magnetizing method). Then, these are all repeated for the hard axis of another piece of magnet.

Moreover, from all these processes there will be two pieces of anisotropic magnets magnetized in different axes.

(II) The hard axis of the magnet previously magnetized in the easy axis is then placed aligning with the magnetizing and pickup coils. The magnetizing field is gradually increased until saturation to complete the measurement of the initial magnetization curve for the swiveling magnetization technique (turning from the easy axis to the hard axis). Then, after the magnetizing current is released, an opposite magnetizing field being sufficiently large is applied and gradually reduced until the current drops to zero to complete the demagnetization curve measurement for the swiveling magnetization technique.

(III) Adopting the same process as described in Step 2 but using another piece of magnet previously magnetized in the hard axis, the initial magnetization curve and demagnetization curve with swiveling magnetization turning from the hard axis to the easy axis can be obtained.

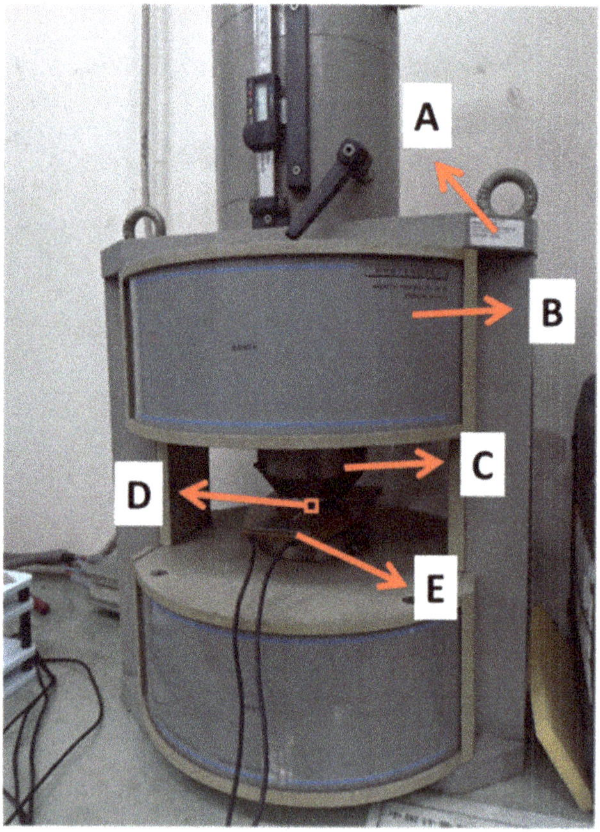


**Figure 5.** Experimental setup. Item "A" is the B-H tracer, item "B" is the magnetizing coil, item "C" is the magnet holder, item "D" is the magnet to be measured, and item "E" is the pickup secondary coil.

From the results obtained via the above steps, the initial magnetization curves in the easy and hard axis by the swiveling magnetization technique and the general method are presented in Figure 6. As can be observed, the swiveling magnetization technique can fully magnetize the magnet with an external magnetizing field smaller than that of the general method in the easy axis. On the hard axis, the swiveling magnetization technique can also magnetize the magnet with a slightly smaller field, as shown in Figure 6b. To show the difference between the two methods clearly, the results in Figure 6 are summarized in

Table 3. The MS of the magnet can be defined as 100% when it is fully magnetized, and thus the MS can be expressed by the percentage of magnetization. As can be seen in Table 3, for the MS to achieve 100% in the easy axis, the general magnetization method requires a magnetizing field strength of 4.55 kOe, and for the swiveling method, only 3.32 kOe, which is 27% lower. The comparison in Table 3 shows the magnetizing field strengths required by the two magnetization methods for the MS to achieve 100%, 75%, and 50%. When the magnet is magnetized to 50% MS in the hard axis, the swiveling method can also save 10.4% magnetizing current compared to the general method. Note that the impulse current for magnetization and changing the MS in a VFS-PMSM is generated by its stator winding. Therefore, the smaller the magnetizing field needed, the smaller the inverter capacity required. From Figure 6 and Table 3, the swiveling magnetization technique can effectively reduce the magnetizing field required for producing an equivalent demagnetization curve to that with the general method, as shown in Figure 7. This indicates that the swiveling method can produce the same characteristics of the magnets as the general method does.

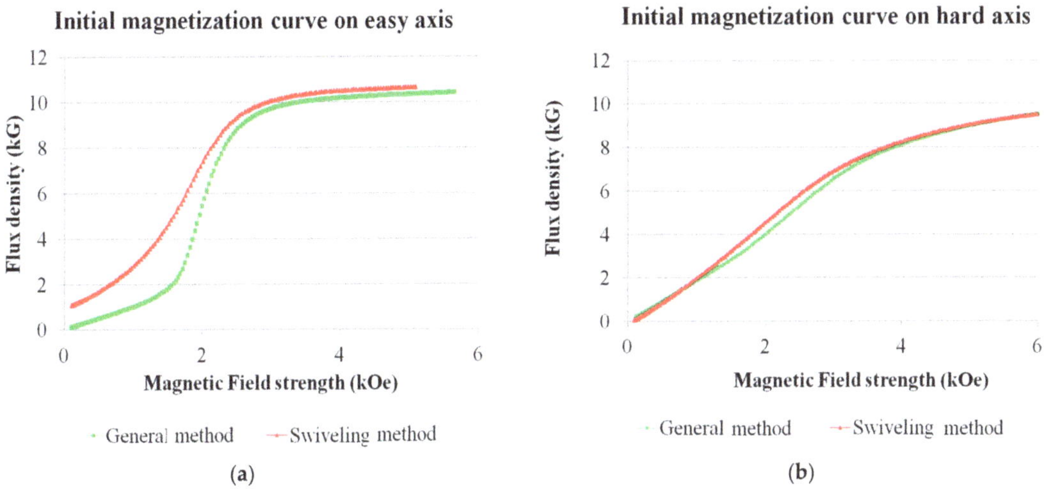


**Figure 6.** (**a**) Initial magnetization curve on the easy axis, and (**b**) initial magnetization curve on the hard axis.

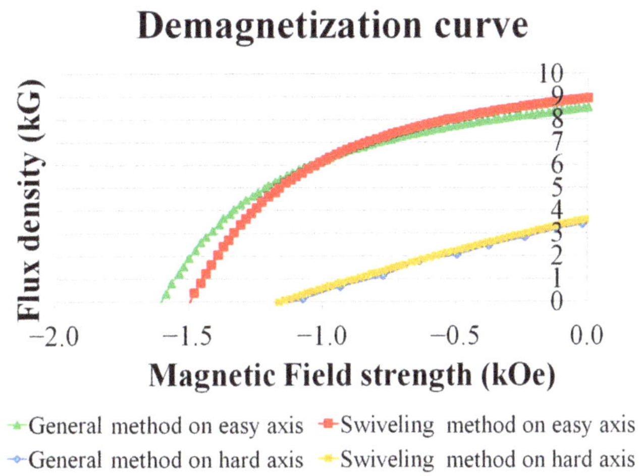


**Figure 7.** Comparison of demagnetization curves by the swiveling method and the general method in easy axis and hard axis.

Note that from the previous study, the anisotropic LCF magnets with the easy axis aligning tangentially are adopted in the following analyses.

**Table 3.** Comparison of external magnetizing field strength required by the swiveling method and the general method in the easy axis and hard axis.

| | **Magnetizing Field Strength Needed in Easy Axis** | | |
|---|---|---|---|
| | **Swiveling (A)** | **General (B)** | **Decrease Rate (B − A)/B** |
| MS = 100% | 3.32 | 4.55 | 27.0% |
| MS = 75% | 2.07 | 2.27 | 8.8% |
| MS = 50% | 1.58 | 1.95 | 19.0% |
| | **Magnetizing Field Strength Needed in Hard Axis** | | |
| | **Swiveling (A)** | **General (B)** | **Decrease Rate (B − A)/B** |
| MS = 100% | 4.42 | 4.55 | 2.9% |
| MS = 75% | 2.76 | 2.95 | 6.4% |
| MS = 50% | 1.9 | 2.12 | 10.4% |

## 4. Performance Comparison of Anisotropic VFS-PMSM with Different Rotor Configurations

From the discussion in Section 2, the main factor that affects the peak torque of the VFS-PMSM at flux enhancing would be the HCF magnet. Therefore, various HCF magnet configurations in the rotor are studied, covering the flat-type, V-type, and spoke-type magnet arrangements. Their performances are compared based on the same amount of LCF magnets used in terms of maximum torque at flux enhancing, flux regulation ability, and torque ripple at maximum torque. This aims at choosing the most suitable rotor configuration for the VFS-PMSM that would produce high peak torque. Note that as previously mentioned, the LCF magnets here are anisotropic with the easy axis being aligned tangentially.

### *4.1. Scenario I*

The first condition for comparison includes (referring to Figure 8):

- The size and position of the LCF magnet remain the same;
- The thickness of the HCF magnet $W_h$, is fixed;
- The angle $\theta_h$ between the two HCF magnets for one pole is set as a variable.

The angle $\theta_h$ for the spoke-type rotor is 36 degrees. For the V-type rotor, three angles are considered: 72, 108, and 144 degrees. The angle is 180° for the flat-type rotor. These rotor configurations are shown in Figure 8.

It can be seen in Figure 9 that as $\theta_h$ increases, the current phase advance to achieve maximum torque per ampere (MTPA) also increases, which means that the proportion of reluctance torque increases. A larger $\theta_h$ brings the HCF magnets closer to the air gap and can provide more torque. However, the maximum torque at flux enhancing decreases with the increase of $\theta_h$. This is likely because as $\theta_h$ increases and the thickness of the HCF magnet remains, a larger $\theta_h$ leads to a smaller HCF magnet length $L_h$, which greatly reduces the air gap flux density. As shown in Figure 10, the torque ripple gently rises, and the flux regulation ability decreases as $\theta_h$ increases. The flux concentration factor decreases as $\theta_h$ increases, and the amount of flux leakage will be lower accordingly even if the flux leakage bypass is opened. Therefore, it is believed that the flux adjustment ability will be affected.

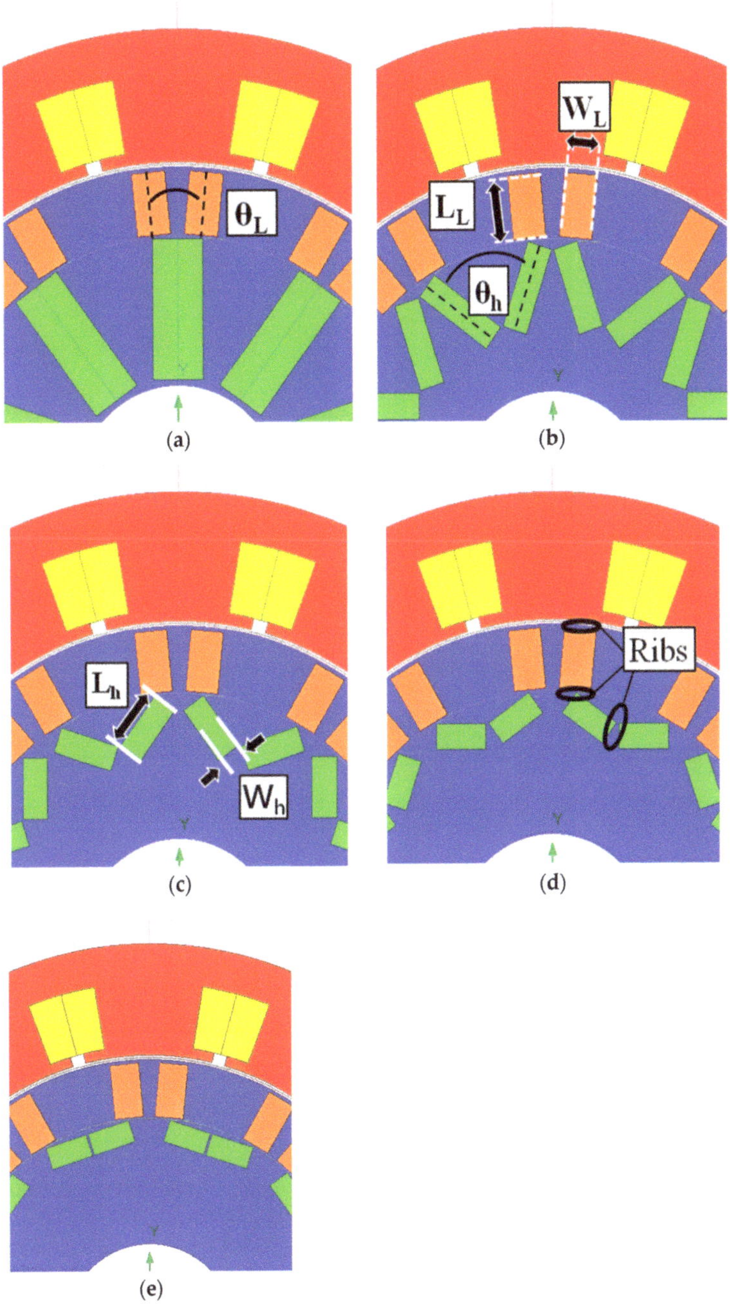


**Figure 8.** Rotor configurations with different $\theta_h$: (**a**) $\theta_h$ = 36° (spoke-type), (**b**) $\theta_h$ = 72° (V-type), (**c**) $\theta_h$ = 108° (V-type), (**d**) $\theta_h$ = 144° (V-type), and (**e**) $\theta_h$ = 180° (flat type).

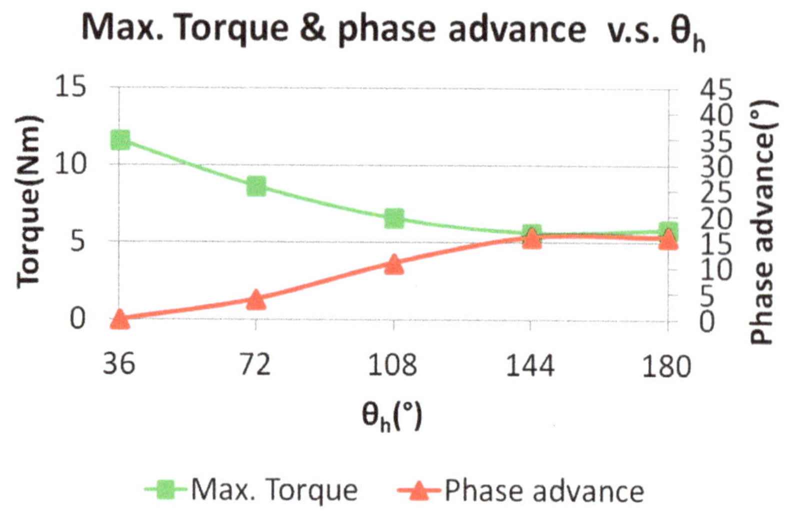


**Figure 9.** Maximum torque and phase advance change with $\theta_h$.

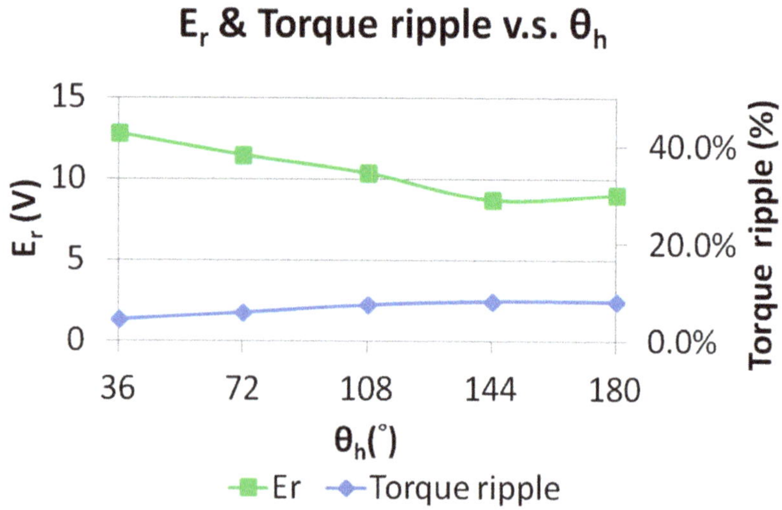


**Figure 10.** $E_r$ and torque ripple variation with $\theta_h$.

*4.2. Scenario II*

From the discussion for Scenario I, it can be concluded that as $\theta_h$ increases, the maximum torque at flux enhancing decreases. The main reason is that the amount of HCF magnets is reduced as $\theta_h$ increases. To enhance the maximum torque, the angle $\theta_L$ of the LCF magnet is adjusted. In the rotor of the anisotropic VFS-PMSM, the LCF magnet should be placed in a location related to the HCF magnet, as indicated in Figure 11. The reason for this arrangement is to allow the MS of the LCF magnet to be easily changed by the swiveling magnetization technique.

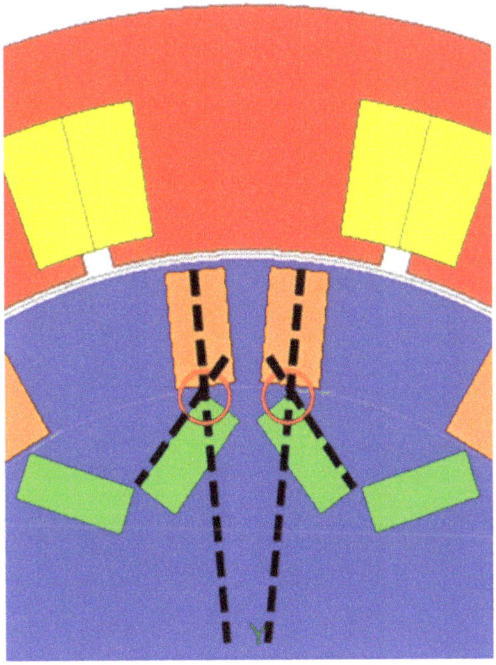

**Figure 11.** Schematic diagram of the low coercive force (LCF) magnet arrangement to align with the high coercive force (HCF) magnet.

In addition, the minimum width of the ribs is 0.5 mm, as shown in Figure 8d, by considering the manufacturing capability, and thus the thickness $W_h$ and length $L_h$ of the HCF magnet will be limited and vary with $\theta_L$ due to the limited rotor space, rib sizes and the position constraint given in Figure 11. The simulated maximum torque at different $\theta_h$ varying with $\theta_L$ is shown in Figure 12, where under the limited magnet size, the torque of the V-type and flat-type increases as $\theta_L$ decreases while that of the spoke-type hardly changes. The reason is that $L_h$ of the V-type and flat-type increases as $\theta_L$ decreases. Although this also decreases $W_h$, the HCF magnet length $L_h$ appears to have a greater impact on torque production, and torque increases as $\theta_L$ decreases. On the other hand, no matter how $\theta_L$ changes, it has only a very minor effect on $L_h$ for the spoke-type rotor, but $W_h$ is greatly reduced. Therefore, the effect of the two changes (i.e., $L_h$ and $W_h$) cancels each other out. Meanwhile, it can be seen from Figure 12 that even if the amount of HCF magnets for the V-type and flat-type are increased to the maximum usage limited by the ribs size between two pieces of HCF magnets, the maximum torque is still lower than that of the spoke-type rotor, whose amount of HCF magnets are the largest for the space constraint.

Torque ripple increases as $\theta_L$ increases, as shown in Figure 13, but the maximum value does not exceed 7.5%. The spoke-type rotor has the best performance on torque ripple, which is only 4.1%. As shown in Figure 14, the flux regulation ability increases with $\theta_L$ when $\theta_h$ is 36°, 72°, and 108°. This is because as $\theta_L$ increases, $W_h$ will increase, which will then increase the magnetic field strength of the HCF magnet, resulting in more flux passing through the LCF magnet at flux weakening. However, as shown in Figure 14, when $\theta_h$ is 144° and 180°, the flux regulation ability decreases. The reason is that the flux adjustment ability is enough to reduce the flux of the HCF magnet at flux weakening to almost zero, which makes $E_-$ almost zero, as shown in Figure 15. Therefore, as $E_+$ decreases, the flux regulation ability decreases.

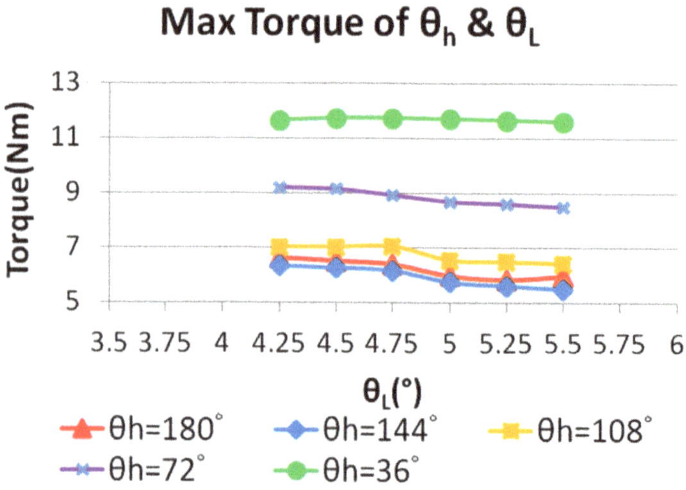


**Figure 12.** Maximum torque variation with $\theta_h$ and $\theta_L$.

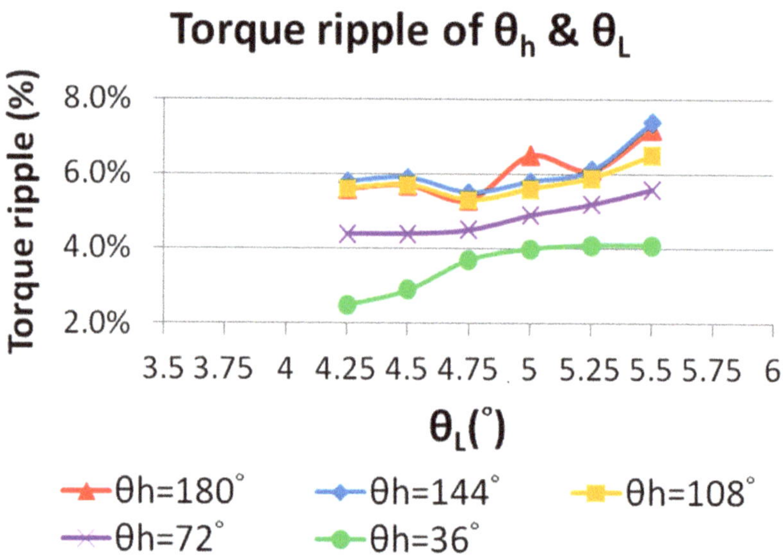


**Figure 13.** Torque ripple variations with $\theta_h$ and $\theta_L$.

It can be concluded from Figure 9 in Scenario I that the reluctance torque generated by changing $\theta_h$ is not the main factor that affects the maximum torque of the anisotropic VFS-PMSM. Instead, the HCF magnet length, $L_h$, which decreases with the increase of $\theta_h$, is the main factor affecting the maximum torque. For Scenario II, $W_h$ and $L_h$ are increased by changing $\theta_L$ due to the manufacturing limitations of the ribs. Although the maximum torque of the V-shaped and the flat-type increases with the decrease of $\theta_L$, it still cannot be larger than that of the spoke-type. As shown in Figure 14, the flux regulation ability decreases as $\theta_h$ increases. Under the same $\theta_h$, if $\theta_L$ becomes smaller, the flux regulation ability will also decrease, but the torque will increase instead. Subsequent designs of the anisotropic VFS-PMSM will be based on the above results.

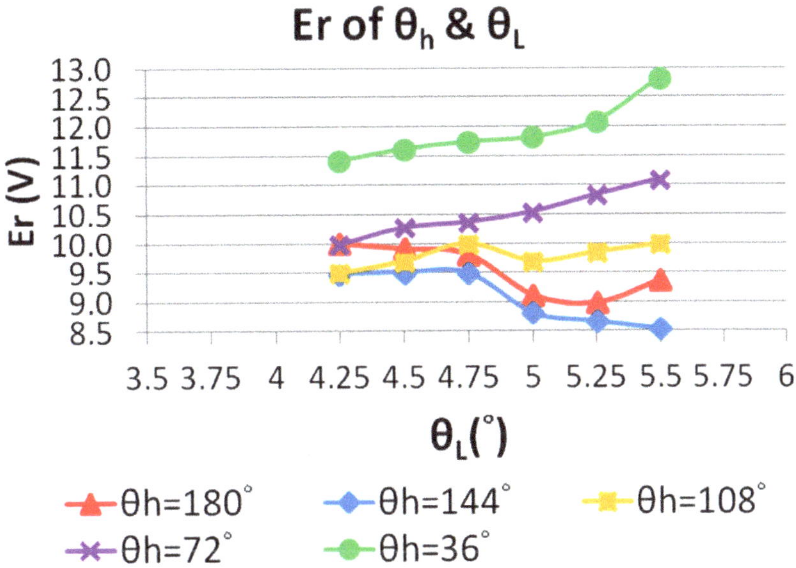


**Figure 14.** $E_r$ variation with $\theta_h$ and $\theta_L$.

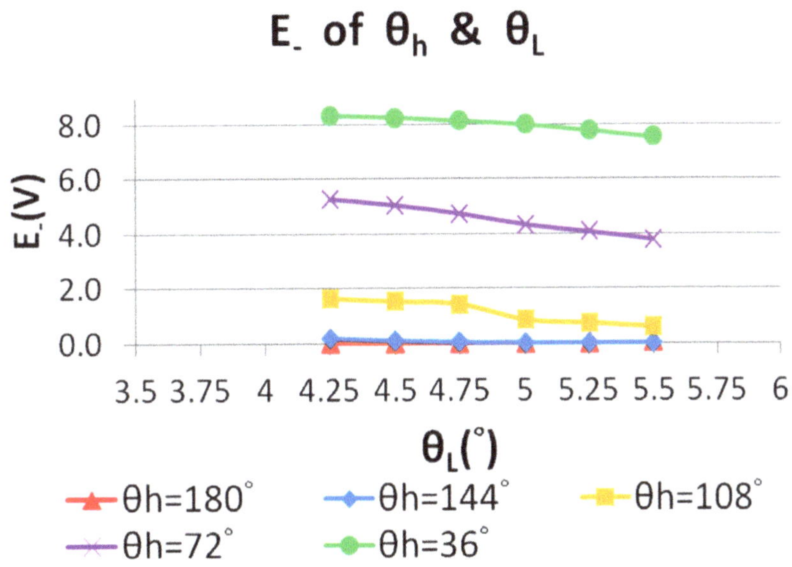


**Figure 15.** $E_{-}$ variation with $\theta_h$ and $\theta_L$.

To sum up, as previously mentioned, the maximum torque and flux regulation ability (i.e., $E_r$) are the two key performance indices to consider, and, therefore, $\theta_h$ and $\theta_L$ can be determined from Figures 12–14 for maximizing the torque and flux regulation. Note that due to the constraints explained in Figure 11, the other dimensions, i.e., $W_h$ and $W_L$ will also be determined simultaneously. These four parameters are listed in Table 4. However, the parameters, $L_h$ and $L_L$ have not been decided and will be discussed in Section 5.

**Table 4.** Rotor parameters of the final type of design.

| Parameters (Unit) | Value |
|---|---|
| $\theta_h$ (°) | 36 |
| $\theta_L$ (°) | 5.5 |
| $W_h$ (mm) | 8 |
| $W_L$ (mm) | 6 |

## 5. Comparison of Anisotropic and Isotropic VFS-PMSMS by Simulation

From the comparison presented in Section 2, it is confirmed that the radial magnetic flux of the LCF magnet has a slight effect on the maximum torque of the VFS-PMSM, while the tangential magnetic flux of the LCF magnet has a great influence on the flux regulation ability; therefore, Topology II has more advantages than Topology I. From the comparison given in Section 4, the conclusion of Section 2 regarding the fact that the tangential magnetic flux of the LCF magnet has a great impact on flux regulation ability is reconfirmed; meanwhile, it is also shown that under the limited size, the spoke-type motor has better key performance such as maximum torque at flux enhancing, flux regulation ability, and torque ripple than other rotor types. Therefore, from the previous comparison, Topology II with the spoke-type rotor is chosen as the final type of design, with the rotor parameters shown in Table 4.

The reason for $L_h$ and $L_L$ not being able to be determined in the previous stage discussed in Section 4 is that $L_h$ has a divergent effect on the maximum torque (proportional) and the flux regulation $E_r$ (inversely proportional), as shown in Figure 16. Note that $L_L$ will vary with $L_h$ and will not be discussed here. To highlight the strengths of the proposed design, two comparisons are made: (a) with the isotropic VFS-PMSM, and (b) with the general types of VFMs, i.e., the series and parallel types. For a fair comparison, the following criteria are considered:

(I) The comparison under the same maximum torque production;
(II) The comparison under the same flux regulation ability $E_r$.

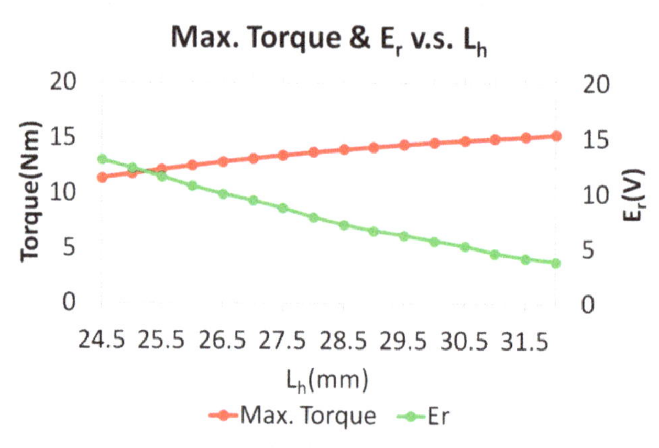


**Figure 16.** Maximum torque and $E_r$ of the target type rotor versus $L_h$.

Based on the two principles given above, the parameters, $L_h$ and $L_L$, can be determined accordingly referring to the results shown in Figure 16. For example, when comparing with the isotropic VFS-PMSM, a certain $L_h$ (and thus $L_L$) can be chosen from Figure 16 to achieve the same maximum torque as the isotropic VFS-PMSM (Criterion I), and, therefore, the $E_r$ of these two motors can be directly compared. For the basis in Criterion II, it is the maximum torque to be compared.

As summarized in Table 5, with a similar $E_r$, the maximum torque of the isotropic VFS-PMSM is 12.12 Nm, and that for the final type of design is 13.47, improved by 11.14%. In this condition, the rotor parameter $L_h$ is 27.5 mm. For a similar maximum torque, the flux regulation ability $E_r$ of the final type of design is enhanced by 37.43% when $L_h$ is 25.5 mm. This clearly shows the advantages of the proposed design.

**Table 5.** Comparison of maximum torque and $E_r$ between final type and isotropic VFS-PMSM.

| | VFS-PMSM (A) | Final Type (B) | Difference (%) ((B − A)/A) |
|---|---|---|---|
| Max. Torque (Nm) | 12.12 | 13.47 | +11.14% |
| $E_r$ (V) | 8.39 | 8.68 | +3.46% |
| | **VFS-PMSM(A)** | **Final Type (B)** | **Difference (%) ((B − A)/A)** |
| Max. Torque (Nm) | 12.12 | 12.16 | +0.33% |
| $E_r$ (V) | 8.39 | 11.53 | +37.43% |

In Figure 16, it can also be observed that the decrease of $E_r$ with respect to $L_h$ is faster than the increase of the maximum torque against $L_h$. Thus, $L_h$ should be carefully chosen for a target $E_r$ while gaining the torque as much as possible.

To make a complete comparison, the general series and parallel types of variable flux motors [25,26] are designed for the same HCF magnet usage. The stator and rotor dimensions also remain the same as that for the final type of design. From the simulation results shown in Table 6, when the flux regulation ability is similar, the maximum torque of the final type of design improves by about 7.25% compared to that of the general series type VFM. When the maximum torque is similar, the flux regulation $E_r$ of the final type of design is 26.84% higher than that of the general series type VFM. The general parallel type does not consider the potential risk of demagnetization, and thus the simulated maximum torque and $E_r$ in Table 7 are better than that of the final type of design. However, if the risk of demagnetization is considered, it will be unstable [26].

**Table 6.** Comparison of maximum torque and $E_r$ between final type and series type.

| | Series (A) | Final Type (B) | Difference (%) ((B − A)/A) |
|---|---|---|---|
| Max. Torque (Nm) | 12.83 | 13.76 | +7.25% |
| $E_r$ (V) | 7.86 | 7.84 | −0.25% |
| | **Series (A)** | **Final Type (B)** | **Difference (%) ((B − A)/A)** |
| Max. Torque (Nm) | 12.83 | 12.86 | +0.23% |
| $E_r$ (V) | 7.86 | 9.97 | +26.84% |

**Table 7.** Comparison of maximum torque and $E_r$ between final type and parallel type.

| | Parallel (A) | Final Type (B) | Difference (%) ((B − A)/A) |
|---|---|---|---|
| Max. Torque (Nm) | 12.3 | 11.4 | −7.32% |
| $E_r$ (V) | 12.74 | 13.01 | +2.11% [1] |
| | **Parallel (A)** | **Final Type (B)** | **Difference (%) ((B − A)/A)** |
| Max. Torque (Nm) | 12.3 | 12.53 | +1.87% |
| $E_r$ (V) | 12.74 | 10.66 | −16.33% [1] |

[1] Note demagnetization risk is not considered.

The proposed motor is a variable flux motor, and thus it has different torque-speed curves at different air gap flux densities between the two extreme conditions represented by $E_+$ and $E_-$, as discussed in Section 2.2. The torque-speed and the power-speed curves of the final type of design with the flux density condition (flux enhancing) to produce the maximum torque are shown in Figure 17a. After switching to a weaker air gap flux density to produce the maximum speed (20,358 rpm), the torque-speed and output power-speed curves are shown in Figure 17b. Finally, the torque-speed curves at various air-gap flux densities can be merged to produce the maximized operating range, as shown in Figure 18. It can be seen that the final type of design has both high torque and high speed with a CPSR of 3.99 (CPSR is defined as the ratio of the maximum speed maintaining the same power as the based speed to the based speed).

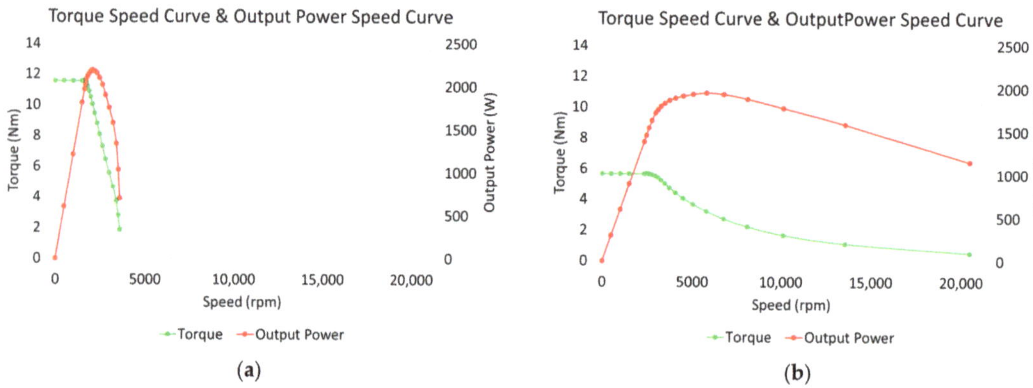


**Figure 17.** Torque-speed and output power-speed curve of the final type of design that produces (**a**) maximum torque; (**b**) maximum speed.

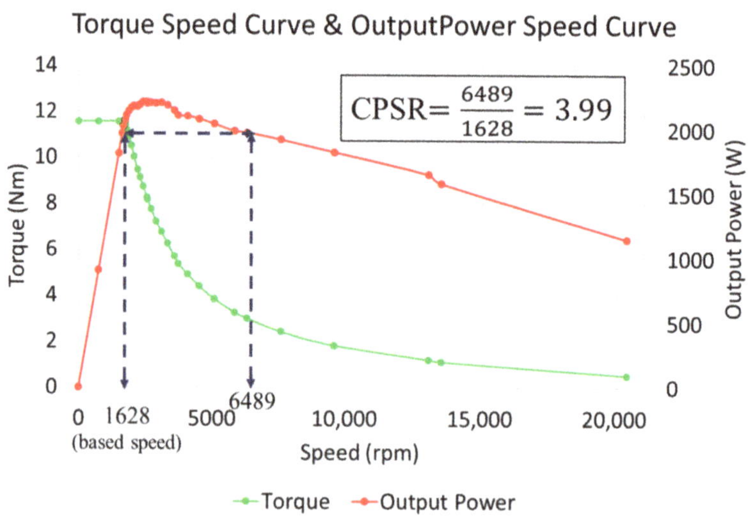


**Figure 18.** Complete torque-speed and output power-speed curve of the final type of design.

From the studies shown above, the proposed anisotropic VFS-PMSM appears to possess more advantages in terms of maximum torque, flux adjustment ability, and the stable LCF magnet operating point compared with the isotropic VFS-PMSM and the conventional series type of VFM. Therefore, these features make the proposed design suitable for EV traction applications.

## 6. Conclusions

This paper has proposed a variable flux spoke-type permanent magnet synchronous motor using anisotropic AlNiCo as the LCF magnets based on the previously developed isotropic VFS-PMSM. It is confirmed through the experiments that the magnetization direction of the anisotropic AlNiCo can be successfully swiveled with a reduced external magnetizing field/current (a reduction of up to 27% to achieve full magnetization). The spoke-type rotor has been found to be the most suitable configuration for the proposed anisotropic VFS-PMSM. Finally, the comparison with other conventional variable flux motors has shown clear advantages of the proposed design in terms of flux adjustment ability, maximum torque and magnetizing current required. Compared with the general series VFM, the maximum torque of the final type of design improves by about 7.25% and the flux regulation is 26.84% higher. Therefore, it can be concluded that the proposed design is suitable to be applied as an EV traction motor. Further research will be conducted in the future to improve the maximum torque and consider the influence of rotor temperature on motor performance.

**Author Contributions:** Conceptualization, Y.-H.L. and M.-F.H.; methodology, Y.-H.L.; software, Y.-H.L.; validation, Y.-H.L. and M.-F.H.; formal analysis, Y.-H.L. and M.-F.H.; investigation, Y.-H.L. and M.-F.H.; resources, M.-F.H.; data curation, Y.-H.L.; writing—original draft preparation, Y.-H.L. and M.-F.H.; writing—review and editing, Y.-H.L. and M.-F.H.; visualization, Y.-H.L.; supervision, M.-F.H.; project administration, M.-F.H.; funding acquisition, M.-F.H. All authors have read and agreed to the published version of the manuscript.

**Funding:** This research was funded by the Ministry of Science and Technology of Taiwan under contract 110-2221-E-006-184-MY2 and the National Space Organization under contract NSPO-S-109343.

**Institutional Review Board Statement:** Not applicable.

**Informed Consent Statement:** Not applicable.

**Data Availability Statement:** Not applicable.

**Acknowledgments:** The authors would like to thank Y.Z. Weng and W.D. Chen, for their assistance in the experiments. ANSYS is acknowledged for providing Maxwell, the finite element package.

**Conflicts of Interest:** The authors declare no conflict of interest.

## References

1. Mun, J.; Park, G.; Seo, S.; Kim, D.; Kim, Y.; Jung, S. Design Characteristics of IPMSM With Wide Constant Power Speed Range for EV Traction. *IEEE Trans. Magn.* **2017**, *53*, 8105104. [CrossRef]
2. Hwang, M.-H.; Han, J.-H.; Kim, D.-H.; Cha, H.-R. Design and Analysis of Rotor Shapes for IPM Motors in EV Power Traction Platforms. *Energies* **2018**, *11*, 2601. [CrossRef]
3. Huynh, T.A.; Hsieh, M.F. Comparative Study of PM-Assisted SynRM and IPMSM on Constant Power Speed Range for EV Applications. *IEEE Trans. Magn.* **2017**, *53*, 8211006. [CrossRef]
4. Kim, H.J.; Lee, C.S. Shape Parameters Design for Improving Energy Efficiency of IPM Traction Motor for EV. *IEEE Trans. Veh. Technol.* **2021**, *70*, 6662–6673. [CrossRef]
5. Zhang, Y.; Cao, W.; McLoone, S.; Morrow, J. Design and flux-weakening control of an interior permanent magnet synchronous motor for electric vehicles. *IEEE Trans. Appl. Supercond.* **2016**, *26*, 0606906. [CrossRef]
6. Fadel, M.; Sepulchre, L.; Pietrzak-David, M. Deep Flux-Weakening Strategy with MTPV for High-Speed IPMSM for Vehicle Application. *IFAC-PapersOnLine* **2018**, *51*, 616–621. [CrossRef]
7. Liu, X.; Chen, H.; Zhao, J.; Belahcen, A. Research on the performances and parameters of interior PMSM used for electric vehicles. *IEEE Trans. Ind. Electron.* **2016**, *63*, 3533–3545. [CrossRef]
8. Inoue, Y.; Morimoto, S.; Sanada, M. Comparative Study of PMSM Drive Systems Based on Current Control and Direct Torque Control in Flux-Weakening Control Region. *IEEE Ind. Appl.* **2012**, *6*, 2382–2389. [CrossRef]
9. Liu, J.; Gong, C.; Han, Z.; Yu, H. IPMSM Model Predictive Control in Flux-Weakening Operation Using an Improved Algorithm. *IEEE Trans. Ind. Electron.* **2018**, *65*, 9378–9387. [CrossRef]
10. Leonardi, F.; Matsuo, T.; Li, Y.; Lipo, T.A.; McCleer, P. Design considerations and testresults for a doubly salient PM motor with flux control. In Proceedings of the IAS '96. Conference Record of the 1996 IEEE Industry Applications Conference Thirty-First IAS Annual Meeting, San Diego, CA, USA, 6–10 October 1996.

11. Finken, T.; Hameyer, K. Study of Hybrid Excited Synchronous Alternators for Automotive Applications Using Coupled FE and Circuit Simulations. *IEEE Trans. Magn.* **2008**, *44*, 1598–1601. [CrossRef]
12. Fodorean, D.; Djerdir, A.; Viorel, I.; Miraoui, A. A Double Excited Synchronous Machine for Direct Drive Application—Design and Prototype Tests. *IEEE Trans. Energy Convers.* **2007**, *22*, 656–665. [CrossRef]
13. Luo, X.G.; Lipo, T.A. A synchronous/permanent magnet hybrid AC machine. *IEEE Trans. Energy Convers.* **2000**, *15*, 203–210.
14. Li, Y.; Lipo, T.A. A doubly salient permanent magnet motor capable of field weakening. In Proceedings of the Power Electronics Specialists Conference, Atlanta, GA, USA, 18–22 June 1995.
15. Kosaka, T.; Kano, Y.; Matsui, N.; Pollock, C. A novel multi-pole permanent magnet synchronous machine with SMC bypass core for magnet flux and SMC field-pole core with toroidal coil for independent field strengthening/weakening. In Proceedings of the 2005 European Conference on Power Electronics and Applications, Dresden, Germany, 11–14 September 2005.
16. Ozawa, I.; Kosaka, T.; Matsui, N. Less rare—Earth Magnet-high power density hybrid excitation motor designed for hybrid electric vehicle. In Proceedings of the 2009 13th European Conference on Power Electronics and Applications, Barcelona, Spain, 8–10 September 2009.
17. Naoe, N.; Fukami, T. Trial production of a hybrid excitation type synchronous machine. In Proceedings of the IEEE Electric Machines and Drives Conference, Cambridge, MA, USA, 17–20 June 2001.
18. Ostovic, V. Memory motors. *IEEE Ind. Appl. Mag.* **2003**, *9*, 52–61. [CrossRef]
19. Ostovic, V. Memory motors-a new class of controllable flux PM machines for a true wide speed operation. In Proceedings of the IEEE Industry Applications Conference, Chicago, IL, USA, 30 September–4 October 2001.
20. Limsuwan, N.; Kato, T.; Akatsu, K.; Lorenz, R.D. Design and Evaluation of a Variable-Flux Flux-Intensifying Interior Permanent-Magnet Machine. *IEEE Trans. Ind. Appl.* **2014**, *50*, 1015–1024. [CrossRef]
21. Kato, T.; Limsuwan, N.; Yu, C.Y.; Akatsu, K.; Lorenz, R.D. Rare earth reduction using a novel variable magnetomotive force flux-intensified IPM machine. *IEEE Trans. Ind. Appl.* **2014**, *50*, 1748–1756. [CrossRef]
22. Zhu, Z.Q.; Hua, H.; Pride, A.; Deodhar, R.; Sasaki, T. Analysis and Reduction of Unipolar Leakage Flux in Series Hybrid Permanent-Magnet Variable Flux Memory Machines. *IEEE Trans. Magn.* **2017**, *53*, 2500604. [CrossRef]
23. Zhou, Y.; Chen, Y.; Shen, J. Analysis and improvement of a hybrid permanent-magnet memory motor. *IEEE Trans. Energy Convers.* **2016**, *31*, 915–923. [CrossRef]
24. Liu, F.L.; Cheng, L.M.; Wang, M.Q.; Qiao, G.Y.; Zheng, P.; Yang, H. Comparative study of hybrid-PM variable-flux machines with different series PM configurations. *AIP Adv.* **2019**, *9*, 125241. [CrossRef]
25. Wang, M.Q.; Yu, B.; Tong, C.D.; Oiao, G.Y.; Liu, F.L.; Yang, S.J.; Zheng, P. Optimization on Magnetization-Regulation Performance of a Variable-Flux Machine with Parallel Permanent Magnets. In Proceedings of the 2020 IEEE 19th Biennial Conference on Electromagnetic Field Computation (CEFC), Pisa, Italy, 16–18 November 2020.
26. Hua, H.; Zhu, Z.Q.; Pride, A.; Deodhar, R.; Sasaki, T. Comparative study of variable flux memory machines with parallel and series hybrid magnets. *IEEE Trans. Ind. Appl.* **2019**, *55*, 1408–1419. [CrossRef]
27. Zhang, S.; Zheng, P.; Jahns, T.M.; Cheng, L.; Wang, M.; Sui, Y. A Novel Variable-Flux Permanent-Magnet Synchronous Machine with Quasi-Series Magnet Configuration and Passive Flux Barrier. *IEEE Trans. Magn.* **2018**, *54*, 8109705. [CrossRef]
28. Yang, H.; Zheng, H.; Lin, H.; Zhu, Z.Q.; Lyu, S. A Novel Variable Flux Dual-Layer Hybrid Magnet Memory Machine with Bypass Airspace Barriers. In Proceedings of the 2019 IEEE International Electric Machines & Drives Conference (IEMDC), San Diego, CA, USA, 12–15 May 2019.
29. Yang, H.; Zheng, H.; Lyu, S.; Lin, H.; Zhu, Z.Q.; Fu, W. Analysis of Flux Regulation Principle in a Novel Hybrid-Magnet-Circuit Variable Flux Memory Machine. In Proceedings of the 2019 22nd International Conference on Electrical Machines and Systems (ICEMS), Harbin, China, 11–14 August 2019.
30. Yang, H.; Lyu, S.; Lin, H.; Zhu, Z.; Zheng, H.; Wang, T. A Novel Hybrid-Magnetic-Circuit Variable Flux Memory Machine. *IEEE Trans. Ind. Electron.* **2020**, *67*, 5258–5268. [CrossRef]
31. Yang, H.; Zheng, H.; Lin, H.Y.; Zhu, Z.Q.; Fu, W.N.; Liu, W.; Lei, J.X.; Lyu, S.K. Investigation of Hybrid-Magnet-Circuit Variable Flux Memory Machines with Different Hybrid Magnet Configurations. *IEEE Trans. Ind. Appl.* **2020**, *57*, 340–351. [CrossRef]
32. Lee, Y.H.; Hsieh, M.F.; Chen, P.H. Novel Swiveling Variable Flux Motor with Spoke Type Interior Permanent Magnet. *IEEE Access* **2022**, *submitted*.

energies

MDPI

*Article*

# A Literature Review of the Control Challenges of Distributed Energy Resources Based on Microgrids (MGs): Past, Present and Future

**Darioush Razmi and Tianguang Lu** *

Department of Electrical Engineering, Shandong University, Jinan 250061, China; darush.razmi@yahoo.com
* Correspondence: tlu@sdu.edu.cn

**Abstract:** Different types of distributed generation (DG) units based on renewable and non-renewable energy sources can create a local energy system in microgrids. The widespread penetration of distributed energy resources (DERs) has affected many power system issues, such as the control and operation of these networks. For the optimal operation of microgrids, optimal energy planning and management in the new space governing the distribution system requires extensive research and analysis. Getting acquainted with the latest research about the evaluation of the problems and challenges in the design of control systems plays an important role in providing a guidance map for researchers to find the recent challenges and propose new solutions. This paper tried to list the challenges of distributed generation sources for MG applications, opportunities, and solutions. These challenges are reported in hierarchical control strategies and power-sharing categories. Therefore, Model Predictive Control (MPC)-based approaches are reviewed for different recent control levels and power sharing strategies in a comprehensive and simple point of view. The performance comparison of MPC methods together and different allocated fitness functions and implementation algorithms are dedicated. Another hand, the potential of MPC methods to control inverters for increasing the reliability of the grid, which this feature could not be achieved by using conventional strategies, while has not been investigated by researchers widely, is introduced in a short review. Therefore, this paper shows an intersection guidance map for readers to facilitate future research works in these exciting and undiscovered fields.

**Keywords:** model predictive control; distributed generation sources; power sharing; reliability

**Citation:** Razmi, D.; Lu, T. A Literature Review of the Control Challenges of Distributed Energy Resources Based on Microgrids (MGs): Past, Present and Future. *Energies* **2022**, *15*, 4676. https://doi.org/10.3390/en15134676

Academic Editor: Abu-Siada Ahmed

Received: 8 April 2022
Accepted: 28 May 2022
Published: 26 June 2022



## 1. Introduction

Energy sources that use fossil fuels to generate electricity always have a devastating effect on the environment, so the attention of distribution network planners has been attracted to the use of renewable energy sources [1]. On the other hand, the existing advances in technologies related to the connection of renewable energy sources to the grid have caused the distributed generation resources connected to the grid to be considered more than before. These distributed generation resources, if installed in the right place and in the right size, can include many economic and technical benefits, such as reducing power losses, improving power quality, improving reliability, eliminating distribution density and causing economic benefits for the power grid [2].

The fundamental difficulty in integrating renewable energy sources is their intermittent nature and unpredictability. Intermittent power means that distributed renewable generations cannot produce power continuously and change dramatically in the short term. Uncertainty means that it is very difficult to predict renewable energy sources [3]. Therefore, this situation may cause changes in energy losses and an unexpected droop or increase in voltage. Increasing the influence of these resources along with the random nature of their output power has created many challenges in the operation of distribution networks [4].

The injected power by distributed generations changes size and direction of power flow in distribution networks. In recent years, the integration of distributed renewable generations in distribution systems has received much attention. With penetration of distributed generation in distribution networks, the concept of smart grids was formed [5].

Distributed energy sources are a key element in the formation of smart grids. The large distribution systems can be subdivided into a set of micrograds to build control and operational infrastructure capabilities in future distribution systems [6]. These sources are crucial in the development of microgrids. One approach to maximizing the potential benefits of distributed renewable generations and energy storage batteries is to optimize their allocation and sizing in a distribution system [7]. In this situation, the integration of energy storage devices such as energy storage batteries in distribution networks is one of the possible solutions to facilitate the penetration of high levels of distributed renewable generations in power system [8].

According to their applicability in microgrids, distributed generation sources are divided into two major types [9]:

- Traditional sources, which consist of rotating units in which the connection interface to the network consists of electric machines;
- Resources that inject power into the network by power electronic devices.

The control and operation of resources, in which the interface with the microgrid consists of power electronic devices, are different from the sources of energy generation in which electric machines act as an interface. Thus, the dynamic behavior and control of a microgrid with a traditional grid will be different [10]. The existence of distributed generation sources in the distribution network has an influence on the network's current and voltage, which can have a positive or negative impact on system performance [11]. Its positive aspects include increasing reliability, improving power quality, improving voltage profile and reducing losses. On the other hand, the most important problem created by installing distributed generation sources in distribution networks is increasing the level of the short circuit [12].

The task of the power grid is not only to supply electricity to consumers with the lowest number of blackouts. The quality of power delivered to consumers along with providing high reliability, generating power from clean sources and reducing costs at various levels are of great importance [13]. Therefore, the idea of a microgrid to generate electrical energy from various types of energy sources was proposed. A microgrid can be considered as a part of a power system that includes one or more distributed generation that will not lose its functionality if separated from the system [14]. A grid-connected device for power storage can also be classified as a DER system, often referred to as a distributed energy storage system (DESS) [15]. Using an interface, DER systems can be managed and synchronized within an intelligent network [16].

The microgrid can be disconnected from the centralized network and operate independently, which makes the network more flexible and helps reduce network disruptions [17]. Microgrids are increasingly being utilized to combine distributed energy sources, such as hybrid solar energy systems, greatly lowering carbon emissions [18]. The distributed power supply systems are small-scale power generation or storage technologies (usually in the range of 1 to 10,000 kilowatts) that are used as an alternative or upgrading a traditional power system. The following are the benefits of employing distributed generation sources [19]:

- Decentralization of the generation system and transmission to consumption centers using small-scale generation units with more appropriate technical conditions, which generally depends on the level of penetration in the network;
- On-site load energy supply reduces transmission and distribution losses, leads to capacity liberalization and reduces transmission capacity in the development planning phase.

Despite all the advantages of distributed generation, one of its possible negative effects on the distribution network is voltage level and protection coordination issues. The DGs develops a network protection system, and as a result, the operation and control of the network will be difficult. With connecting the DGs to the network, harmonics are created in the network and short-circuit impedance is reduced as well. In the rest of this paper, the features and challenges of distributed energy resources in microgrids are reviewed in Section 2, then the hierarchical control methods are discussed in Section 3. In Section 4, the power sharing control strategies are presented. Section 5 is related to the presented solution in the literature based on MPC control strategies. Section 6 first provides an overview of microgrid reliability, then provides predictive control approaches to increase microgrid reliability. Challenges and future perspectives are followed in Section 7; finally, the conclusion section is presented in Section 8.

## 2. Distributed Energy Resources in Microgrids

Any low-capacity energy generator of any type can be used in microgrids, which are divided into two main types [20]:

- Fossil fuel-based generators;
- Renewable energy-based generators.

Due to the small size of the network, diesel generators and gas microturbines are used in the microgrid, which are in the category of fossil fuel generators [21]. These units are of special importance in microgrids due to their high controllability in energy production, although they are considered as units for the production of environmental pollutants. The renewable generators, which are increasing their penetration in the power grids, are another type of generators that are of special importance due to their low or cheap fuel and their lack of environmental pollution. The most important of these units are wind turbines, solar cells and fuel cells [22]. Due to the unpredictability of the behavior of distributed generation sources, including wind turbines, solar cells and fuel cells, the frequency and power produced by these sources are variable and will not be able to connect directly to the main grid [23]. Therefore, in order to connect these sources to the power grid, power electronic converters with the ability to convert DC or AC voltage to desirable AC voltage will be required [24]. A power electronic converter can respond to rapid changes to control active and reactive powers and frequency control [25]. However, the use of DG units complicates the structure and poses challenges in the operation, control, stability, protection and security of power distribution networks [26]. Hence, new strategies have been proposed to connect these resources to the grid. According to these standards, whenever a fault occurs in the grid, the DG units are disconnected from the main grid and operate as an island [27]. However, in order to achieve the desired reliability and power quality, it is necessary to review these standards.

### 2.1. *Wind Turbines*

Nowadays, demand for renewable energy sources has increased significantly. Among renewable energies, wind energy is one of the most economical methods of generating electricity that does not pollute the environment. Wind generators are used in both stand-alone systems for powering remote loads and in grid-connected applications. The wind energy conversion system receives the wind energy from the wind turbine and converts it into electrical energy through a generator, which is available in two forms, fixed speed and variable speed. Due to the low energy production, stress in mechanical parts and low power quality, the fixed speed wind energy conversion systems have been replaced by variable speed energy conversion systems that reduce mechanical stress and aerodynamic noises. These systems can be controlled in such a way as to enable the turbine to operate at its maximum power factor under different wind conditions and receive maximum energy from the wind. Block diagrams of different types of wind turbines are shown in Figure 1 [28].

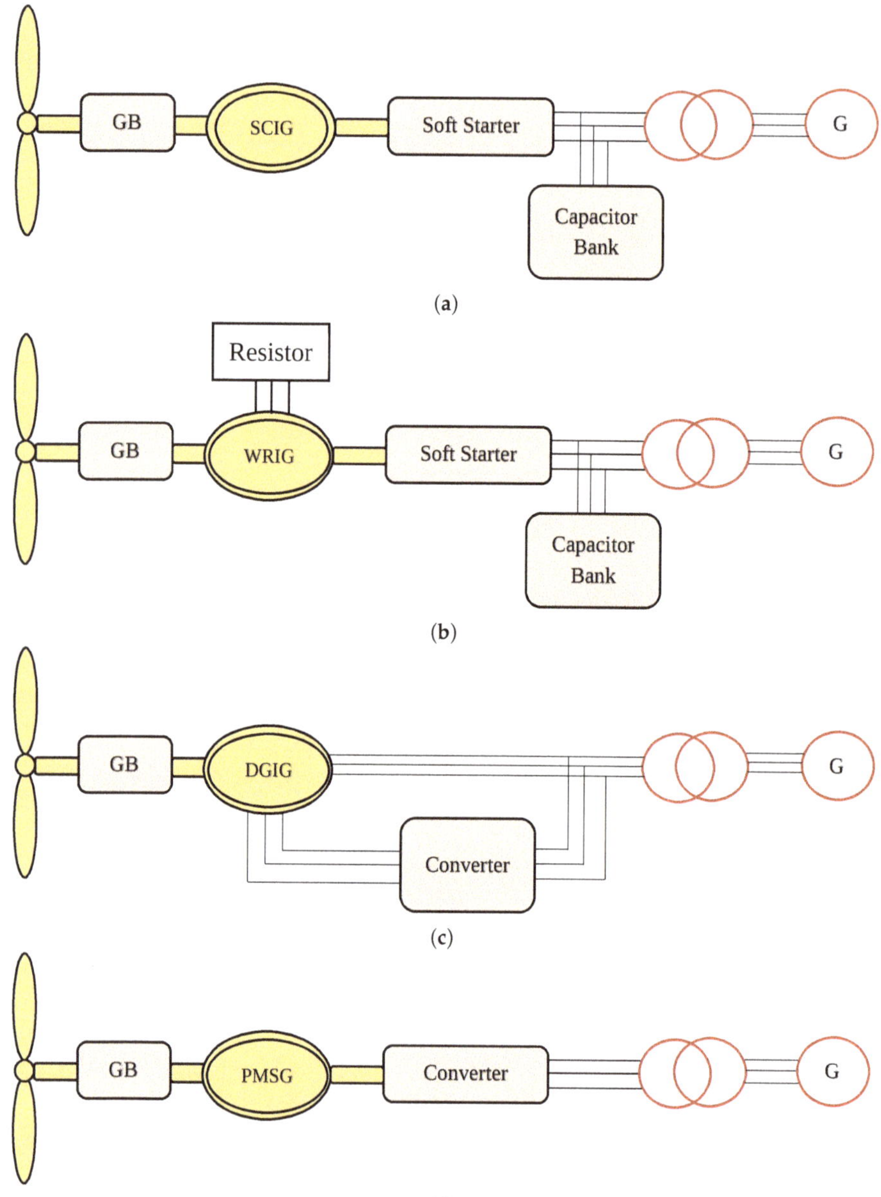


**Figure 1.** Block diagram of wind turbines. (**a**) Block diagram of squirrel cage induction generator. (**b**) Block diagram of wound rotor induction generator. (**c**) Block diagram of doubly-fed induction generator. (**d**) Block diagram of permanent magnetic synchronous generator.

### *2.2. Photovoltaic (PV) Generation Units*

Photovoltaic systems for general and agricultural use operate in both connected and island operating modes. The advantage of moving systems is the ability to track the sun and increase the energy of the sun during the day. In grid-connected mode, the electrical energy from the photovoltaic system is injected into the main grid using the inverters while changing the shape and matching the voltage level and frequency of the electrical energy from the photovoltaic system. The use of photovoltaic power plants connected to the main grid in a centralized or decentralized manner (while amplifying the current energy in the distribution network), due to the injection of voltage and current, prevents the voltage drop of the distribution network. Block diagrams of photovoltaic solar cells are shown in Figure 2 [29,30].

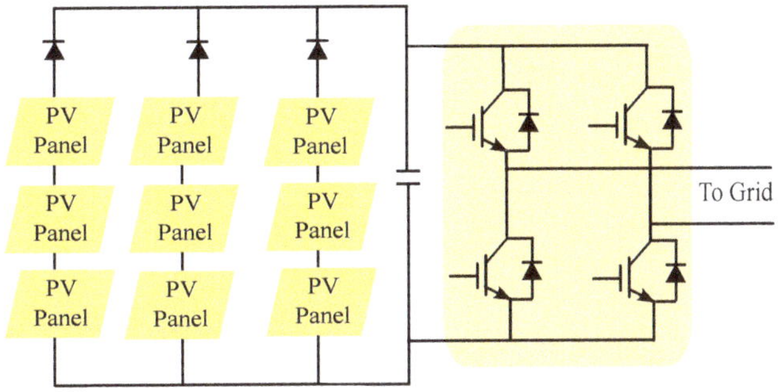


**Figure 2.** Basic diagram of photovoltaic solar cells.

### 2.3. Microturbines

Microturbines are small turbines that generate electricity using wind, water or various fuels, such as natural gas. Microturbines have more advantages than other small-scale technologies, some of which are fewer moving parts, compact size, light weight, higher efficiency, lower emission, lower power costs and use of wasted fuel [31].

### 2.4. Fuel Cells

Among the distributed energy sources, fuel cells have been highly regarded due to their high efficiency. In a fuel cell, the method of generating electricity is from the chemical reaction of hydrogen fuel and air oxygen. One of the characteristics of the fuel cell is its slow dynamics due to the chemical process [32].

### 2.5. Diesel Generators

A diesel generator is a combination of a diesel engine, a generator and various accessories such as chassis, control systems, emergency circuit breakers, heat generating system, automatic start system, etc. that are used to generate electricity. Diesel generators have lower initial costs, easier maintenance and shorter turning times than small turbines. The production capacity of a diesel generator depends only on the amount of fuel. The relationship between fuel content and diesel generator output power can be modeled as a linear relationship [33].

### 2.6. Batteries

Rechargeable batteries belong to the group of electrochemical cells that store electrical energy. These types of cells are called secondary cells, because electrochemical reactions occur reversibly in them. Rechargeable batteries exist in a range of sizes and kinds, ranging from small button batteries to grid-connected megawatt batteries. Lead-acid, nickel-cadmium, nickel-metal, lithium-ion and lithium-ion-polymer are some of the most popular chemical compounds used in rechargeable batteries. Rechargeable batteries are less expensive than other disposable batteries and have less environmental impact. Some rechargeable batteries are produced and supplied in the same sizes as other batteries, except that they can be charged and reused. In grid energy storage applications, rechargeable batteries are used to balance the grid load when they want to inject electrical energy into the grid for use at peak times. New power grids, such as solar power, also use rechargeable batteries to store energy during the day and consume it at night [34,35].

### 2.7. Challenges and Limitations of Distributed Generation Resources

With penetration of distributed generation sources in the microgrid, the dynamics of the system are affected [36]. In this case, the analysis of the connection of distributed generation resources to the network becomes complicated, especially when it is necessary to analyze the types of distributed generation sources in the distribution network, which usually passes power in one direction [37]. Researchers and operators of microgrids face these challenges when distributed generation resources are widely used. Some of these challenges are [38,39]:

- Reverse power flow: Connecting distributed generation sources to the distribution network can cause power flow in the reverse direction, which causes faults in the detection of protection systems;
- Reactive power: Many types of distributed generation sources use asynchronous generators that cannot inject reactive power into the grid;
- System frequency: Deviation from the nominal frequency of the system occurs with an imbalance between production and consumption. Increasing distributed generation resources affect system frequency and complicate the control process;
- Voltage levels: Distributed generation sources change the voltage level of feeders due to changes in the direction of the load distribution.
- Protection: The structure of most distribution networks is radial. This causes the load to be distributed unilaterally and the corresponding protection systems to be designed accordingly;
- Harmonic injection into the grid by distributed generation sources that are connected to the grid by an inverter. Short-circuit faults increase with respect to the location of distributed generation sources;

## 3. Hierarchical Control of Microgrid

Hierarchical control of microgrids has been proposed in several studies. The overall microgrid control structures are in three levels that each have specific objectives, and different methods for their operation have been proposed and controller of each level must be calculated and designed. Figure 3 shows the hierarchical control of the microgrid [40,41].

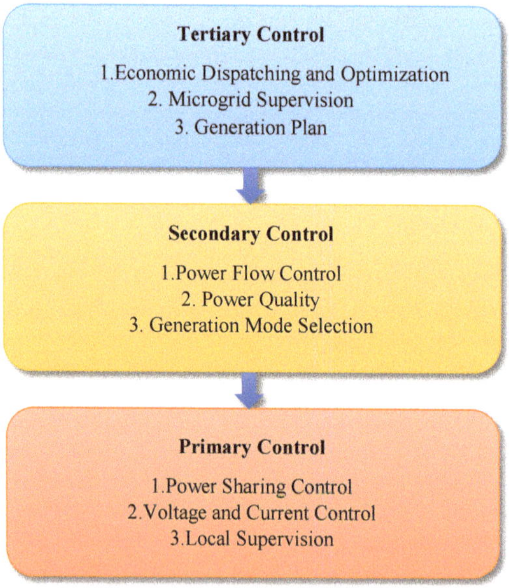


**Figure 3.** The hierarchical control strategy.

### 3.1. Microgrid Primary Control

The main controller is in charge of ensuring the reliability of the microgrid system as well as increasing the performance and stability of each converter's local voltage control system. This control level also regulates the reference voltage required for the internal voltage and current loops so that the distribution of active and reactive powers between the distributed generation sources is done optimally. The most common method for the primary controller is droop control, which aims to compensate for the mismatch between the generated power and demand. Based on this requirement, the droop control generates a reference voltage signal for the source and then the internal control loop (voltage and current) ensures that the actual voltage is equal to its reference value. In reality, in grid-connected mode, the primary controller of a microgrid regulates the active and reactive powers of each DG unit, as well as the voltage and frequency in island mode [42–44]. Various approaches to primary control have been proposed, especially for the inverter voltage source interface of distributed energy sources [45]. Most of these approaches use an inner/outer control loop based on the PI controller [46]. In [47], an intelligent load strategy based on a three-phase-single-phase AC-DC converter is proposed as a virtual synchronous machine to reduce frequency fluctuations in the island microgrid. This approach can also carry out grid support operations. By increasing the sources of distributed generation in a microgrid, the inertia of the system decreases, In case of load changes or generation loss, the primary frequency controller faces the problem of high-frequency change rate. A new way to increase system inertia is to use solid-state transformers as virtual synchronous machines in island operation mode. The solid state transformers can play an effective role in improving the island microgrid frequency [48]. Several studies have evaluated the role of electric vehicles in regulating the primary frequency of microgrids; with the penetration of electric vehicles in microgrids, the load profile will change significantly. The authors in [49] illustrated that the high penetration of electric vehicles can have a significant effect on reducing the primary frequency fluctuations in microgrids. In [50], an adaptive droop controller for optimal power distribution in the DC microgrid is presented. The problem of economic distribution of microgrids has been proposed by distributed hierarchical control. The droop controller receives the reference values from the economic regulator and provides the desired output power while maintaining system stability. The conventional droop control method is a decentralized method that is mostly used to control parallel-connected converters in the DC microgrid. One of the disadvantages of the conventional droop control method is incorrect power sharing and voltage deviation in the DC bus [51]. The small signal stability in a hybrid microgrid has been analyzed by considering changes in renewable energy sources. Due to the fact that the wind speed fluctuates, the active output power of the DGs will change significantly, and as a result, power sharing will also change. A suitable power sharing strategy is provided to manage the fluctuations of energy sources that regulate the voltage and frequency of each DG [52].

### 3.2. Microgrid Secondary Control

A centralized control strategy is proposed for distributed multipurpose inverters that allows the detection of adverse current components. It also allows the inverter to monitor and measure the desired values through the Internet protocol [53]. With the increase of renewable energy sources and due to its oscillating nature, the separation of secondary and tertiary control levels has negative effects on the overall efficiency of the microgrid. An optimal and robust control strategy for the secondary control level and the distributed MPC approach are proposed to solve the optimization problem [54]. The secondary controller, also known as a microgrid energy management system, is used to ensure the reliability, economics and safety operation of microgrids in both grid-connected and island modes. This level of control is important when the microgrid with the presence of distributed energy sources enters the island mode. This level of control can include a synchronization control loop. It also intervenes when the primary control is not able to return the frequency to the desired values and returns the system frequency values to the reference values [55]. In general, the secondary controller will be responsible for recovering the frequency deviation and voltage range. Furthermore, this control level provides network connection conditions for the microgrid when synchronizing. Many control systems have been proposed for this level of control, which may be grouped into three categories: distributed, centralized and decentralized [56]. The microgrid controllers in the centralized method are based on the same internal loop controllers. In this method, there is a central microgrid controller called MGCC (Microgrid Cenral Control), which allows each DG to communicate with the distribution management system. This type of control is used for small microgrids that are manually controlled. In fact, the definition of centralized and decentralized control for secondary control is based on the status of the MGCC [57]. In centralized control, the frequency and voltage amplitude of DGs are compared with the reference values received from the main grid (in grid-connected mode) [58]. The advantage of this control structure is that due to the fact that the message transmission path is one-way, the telecommunication system does not suffer from high traffic, but the disadvantage of this method is the existence of a central controller, which if it malfunctions can cause the system to also be affected. In order to connect the microgrid to the main grid, the voltage and frequency of the network must be measured [59]. These values are used as a secondary control reference. Furthermore, the phase angle between the microgrid and the main grid must be synchronized using the synchronization control loop. The consensus-based secondary frequency control is provided for islanded microgrids under weak communication conditions. In this research, to reduce the effects of weak communication conditions, a new approach with time-varying control gain is proposed [60]. A distributed secondary control approach is proposed to distribute power and voltage restoration in island mode, which, unlike other secondary controller methods, is based on the idea of feedback, and only by knowing the voltage of each bus, can the secondary controller be designed for each DG [61]. The authors of [62] proposed a method for island mode of inverter-based microgrids. The proposed method is able to predict system dynamics at high and low frequencies. The wireless power sharing approach is proposed for DGs and converts the output impedance of the inverter to the pure impedance by adjusting the virtual impedance. Table 1 summarizes the advantages and disadvantages of control strategies.

**Table 1.** Advantages and disadvantages of control strategies.

| Control Structure | Control Approach | Advantages | Disadvantages | Application | Ref |
|---|---|---|---|---|---|
| Primary Control | Adaptive droop control | Reduction of rotational current between parallel inverters | Resistances between converter terminals must be specified | DC/DC converter | [50] |
| | Conventional droop | Simple and easy to implement, quick response | Unsuitable power sharing and stability | DC system | [51] |
| | Dynamic droop control | Suitable power sharing and stability | Slow response, complex to implement | AC system | [52] |
| Secondary Control | Centralized control | System controllability | Occurs in the current sharing bus, the output voltage will drop | Cluster of MGs | [53] |
| | Distributed control | Suitable for short transmission lines, accurate load sharing | Unsuitable for long transmission lines, security | DC/DC converter | [54] |

### 3.3. *Microgrid Tertiary Control*

The purpose of this control layer is to manage the microgrid power flow in the grid-connected mode by adjusting the voltage and frequency, which is done by the secondary controller. By measuring the P/Q ratio at the Point of Common Coupling (PCC), the active and reactive powers of the grid can be compared with the desired reference values. This level of control is the last level and the slowest in terms of response. In fact, tertiary control consists of the optimal operation of the microgrid in the economic and technical sectors.

Technically, if there is an error or unplanned islanding in the microgrid, the third-party control tries to absorb the active power from the grid, so that if the main grid is not available, the frequency will be reduced. It is noteworthy that depending on the allocation of active and reactive powers values, the distribution of active and reactive powers can be from the microgrid to the main grid or vice versa [63]. Table 2 evaluates the performance modes of microgrids [64].

**Table 2.** Comparison of microgrid performance modes [64].

| Parameters | Sensitivity to Load Fluctuations | The Complexity of Control | Objective of Control | Control Strategy |
|---|---|---|---|---|
| Grid-Connected Mode | Low sensitivity due to better performance as a fixed source | Easy | PQ | P-Q Control |
| Island Mode | High sensitivity to load fluctuation due to low inertia | Complex | VF | V-F Control |

## 4. Power Sharing Theory

In order to investigate the distribution of active and reactive powers by distributed generation sources and power distribution equations, the equivalent circuit of a distributed generation source is considered according to Figure 4. In this equivalent circuit, the distributed generation source is modeled with an alternating voltage source that is connected to the ac bus via a Z-impedance supply line and supplies the load. The power transferred from the distributed generation source to the ac bus is as follows:

$$S = P + jQ \tag{1}$$

$$S = \bar{E}I^* = E\angle\left[\frac{E}{Z}\angle(\theta-\delta) - \frac{V}{Z}\angle\theta\right] = \frac{E^2}{Z}\angle\theta - \frac{EV}{Z}\angle(\theta+\delta) \tag{2}$$

$$P = \frac{E^2}{Z}Cos\theta - \frac{EV}{Z}Cos(\theta+\delta) = \frac{(E^2 - EVCos\delta)Cos\theta}{Z} + \frac{EVsin\theta sin\delta}{Z} \tag{3}$$

$$Q = \frac{E^2}{Z}Sin\theta - \frac{EV}{Z}Sin(\theta+\delta) = \frac{(E^2 - EVCos\delta)Sin\theta}{Z} - \frac{EVsin\theta sin\delta}{Z} \tag{4}$$

In which $S$, $P$ and $Q$ are apparent power, active and reactive powers transmitted to the load, respectively, $E$ and $V$ are inverter output voltage amplitude and AC bus and $\delta$, $Z$ and $\theta$ are load angle, line impedance amplitude and phase, respectively. Table 3 shows the general equations of droop and power sharing based on different impedance angles according to Equations (1)–(4).

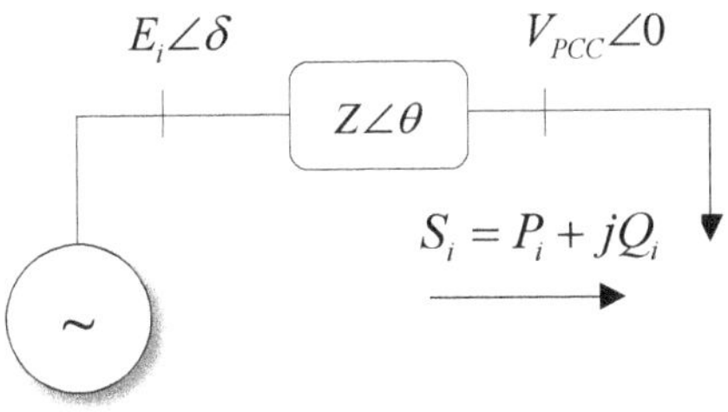


**Figure 4.** Single line diagram of power system [65].

According to the equations of active and reactive powers distribution (3) and (4), the line impedance is proportional with the inverse power distribution, i.e., $P \propto \frac{1}{Z}, Q \propto \frac{1}{Z}$. This means that in the high line impedance, the amount of power transmitted to the load will be lower [66]:

$$P = \frac{(E^2 - EVCos\delta)Cos\theta}{Z} + \frac{EVsin\theta sin\delta}{Z} \tag{5}$$

$$Q = \frac{(E^2 - EVCos\delta)Sin\theta}{Z} - \frac{EVsin\theta sin\delta}{Z} \quad (6)$$

Therefore, the output power of each unit can be controlled by controlling the line impedance.

**Table 3.** Power sharing and droop equations [65].

| Line Impedance Type | Line Impedance Angle | Power Equation | Droop Equation |
|---|---|---|---|
| L | $\theta_i = 90°$ | $P_i = \frac{E_i V_{PCC} Sin\delta_i}{X_i}$ <br> $Q_i = \frac{E_i^2 - E_i V_{PCC} Cos\delta_i}{X_i}$ | $f_i^* = f_0 - k_{pi}P_i$ <br> $E_i^* = E_0 - k_{qi}Q_i$ |
| RL | $0° < \theta_i < 90°$ | $P_i = \frac{E_i^2 Cos\theta_i - E_i V_{PCC} Cos(\theta_i+\delta_i)}{Z_i}$ <br> $Q_i = \frac{E_i^2 Sin\theta_i - E_i V_{PCC} Cos(\theta_i+\delta_i)}{Z_i}$ | $f_i^* = f_0 - k_{pi}P_i + k_{qi}Q_i$ <br> $E_i^* = E_0 - k_{pi}P_i - k_{qi}Q_i$ |
| R | $\theta_i = 0°$ | $P_i = \frac{E_i^2 - E_i V_{PCC} Cos\delta_i}{R_i}$ <br> $Q_i = -\frac{E_i V_{PCC} Sin\delta_i}{R_i}$ | $f_i^* = f_0 + k_{qi}Q_i$ <br> $E_i^* = E_0 - k_{pi}P_i$ |
| RC | $90° < \theta_i < 0°$ | $P_i = \frac{E_i^2 Cos\theta_i - E_i V_{PCC} Cos(\theta_i+\delta_i)}{Z_i}$ <br> $Q_i = \frac{E_i^2 Sin\theta_i - E_i V_{PCC} Sin(\theta_i+\delta_i)}{Z_i}$ | $f_i^* = f_0 + k_{pi}P_i + k_{qi}Q_i$ <br> $E_i^* = E_0 - k_{pi}P_i + k_{qi}Q_i$ |
| C | $\theta_i = -90°$ | $P_i = -\frac{E_i V_{PCC} Sin\delta_i}{X_i}$ <br> $Q_i = -\frac{E_i^2 - E_i V_{PCC} Cos\delta_i}{X_i}$ | $f_i^* = f_0 + k_{pi}P_i$ <br> $E_i^* = E_0 + k_{qi}Q_i$ |

*Control Strategies for Power Sharing*

One of the most important and challenging issues in the control of island microgrids is the control of power sharing among distributed generation sources, so that each unit must feed the load in proportion to its nominal value. The main purpose of power sharing is to increase the network capacity to meet power demand. Power sharing among distributed generation units is one of the most important challenges; lack of power sharing among them leads to unequal load sharing, which may damage the load and often leads to instability. Various methods have been proposed to control the power sharing in microgrids, one of the most famous of which is the droop control method. The droop control method is a decentralized control method that uses local control signals such as frequency and voltage to control power sharing in microgrids. In the droop control method, active and reactive powers, frequency and voltage amplitude of distributed generation sources are the control variables that are used to share power [67,68].

In microgrids with a large amount of reactance to resistance ratio (X/R), due to the relationship between active power and frequency, as well as the relationship between reactive power and voltage, the conventional droop control method is used. However, in low voltage microgrids, due to the small ratio of reactance to line resistance, there is a resistance connection between the inverters, and therefore, the active power depends on the voltage and the reactive power depends on the frequency. In island microgrids with inverter-based distributed generation sources, mismatching the impedance of the supply lines leads to a voltage difference between the inverter terminals, which can cause a rotational current between inverters and over-current in the distributed generation sources. In order to solve the challenges in microgrids, including the dependence of active and reactive powers on low voltage microgrids by the distributed generation sources in the microgrids, as well as the rotational current between these sources, various methods, such as the virtual power method, voltage method and virtual frequency and virtual impedance methods, are proposed. Figure 5 shows the challenges and proposed solutions [69–78].

A network-based power sharing strategy under unknown impedance is presented [79–81]. This improved strategy can provide acceptable power sharing while keeping the steady-state frequency constant. In addition, it improves microgrid dynamic performance and power sharing errors under unknown impedance. Network-based active power sharing strategies have been proposed in previous research [82–84]. However, there are two main drawbacks: (1) Due to the presence of voltage and frequency droop loops, the frequency droop cannot be eliminated. (2) Communication delays under the uncertainties parameters increase system sensitivity. In microgrids, the virtual impedance method was used to eliminate the dependence of active and reactive powers between power supplies that were connected directly to the PCC in microgrids through resistance lines [85,86]. Virtual impedance has been proposed to stabilize the line impedance of distributed generation sources to improve the power sharing performance by droop control method [87], which can be placed in the inverter output without a physical connection. By implementing virtual impedance in the controller of the distributed generation unit, the line-equivalent impedance can be divided into two parts: physical impedance and virtual impedance. When virtual impedance is considered, the distributed output source can be considered as a voltage droop source controlled by the droop method in series with the virtual impedance [88]. Table 4 shows the challenges of active power sharing for different types of control strategies.

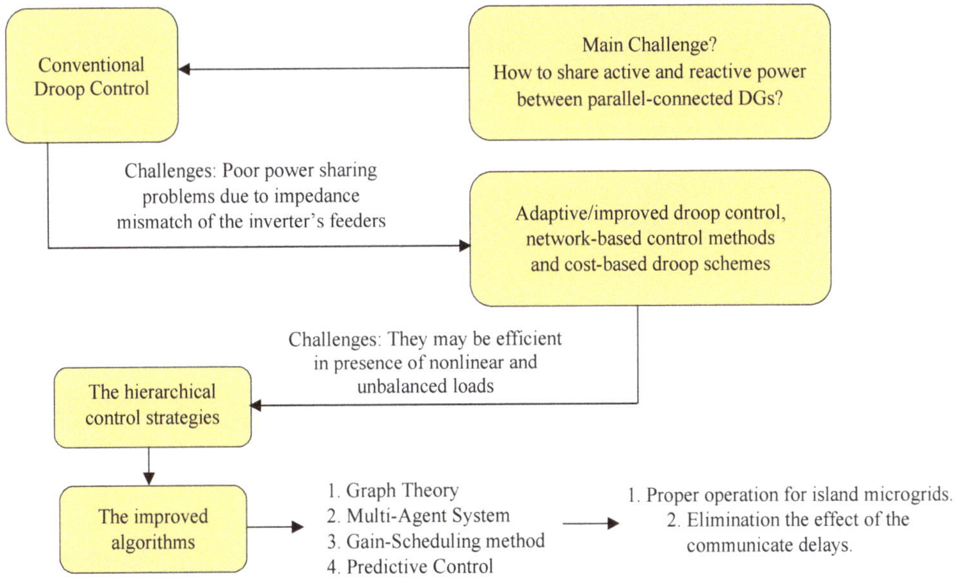


**Figure 5.** Challenges and proposed solutions.

**Table 4.** Active power sharing strategies.

| Methods | Challenges | Ref. |
|---|---|---|
| Under an inductive/resistive feeder impedance condition | 1. Sensitivity to communication delays.<br>2. Failure of recovery of frequency drop in the presence of complex control loop. | [69–78] |
| Under an unknown impedance condition | 1. No proportional active power sharing.<br>2. Not considering total cost of generation. | [79–81] |
| Proportional active power sharing strategies | 1. Sensitivity to communication delays.<br>2. Not suitable for complex MG. | [82–84] |

Virtual impedance is used for various purposes, such as independent control of active and reactive powers, the realization of proper power sharing, increasing power transmission capacity, etc. [89]. Therefore, if the virtual impedance is properly designed and controlled, it can have a significant impact on improving the performance of control methods by reducing the dependence of power distribution on line impedance. In order to facilitate the use of the conventional droop control method in microgrids and to solve the problem of line impedance mismatch, a virtual and physical impedance interface inductor is considered to reduce the dependence of active and reactive power sharing as well as to stabilize the output impedance [90,91].

## 5. Control Approaches Based on MPC

The basic idea of model predictive control or Receding Horizon Control (RHC) was introduced in the 1960s and then in the 1970s in industrial control applications, especially in the chemical industry. Then, this control approach entered the academic and research space and was welcomed by researchers. Various papers have described and compared the types of predictive controllers well. MPC is an intuitive method for systems that have constraints. This fact demonstrates why predictive controllers are so popular among other controller types. In general, the predictive controller sends the appropriate control signal to the converter by minimizing the cost function that provides the proper behavior of the system. For each sampling time, MPC calculates a series of control signals that minimize the cost function, with this description that the first signal is applied to the converter [92]. Table 5 shows some cost functions for power electronics applications.

**Table 5.** Example of cost functions for power electronic applications.

| Strategy | Cost Function | Ref |
|---|---|---|
| FCS-MPC | $g = \lvert i_\alpha^*(k+1) - i_\alpha^p(k+1)\rvert + \lvert i_\beta^*(k+1) - i_\beta^p(k+1)\rvert + \lambda n_{sw}$ | [93] |
| CSC-AFE | $g = (q)^2 + \lambda(\hat{i}_L - i_L^*)^2$ | [94] |
| VSC-AFE | $g = \lvert \hat{i}_k - i_k^* \rvert$ | [95] |
| Motor drive | $g = (T - T^*)^2 + \lambda(\hat{\phi} - \phi^*)^2$ | [96] |
| MPDPC | $g = \lvert P^* - P^{k+1}\rvert^2 + \lvert Q^* - Q^{k+1}\rvert^2$ | [97] |

In recent years, widespread research has been done on microgrid control systems. In small-scale microgrids, the most important issue is to attain reliable performance, balancing supply and demand, while in other structures, optimal performance is required. Several types of research have been performed using the hysteresis band control method to energy management due to the simplicity of the structure and its implementation. On the other hand, predictive controllers have found a wide range of applications in the field of power electronics and drives, also minimizing the cost function by solving optimization problems while considering technical constraints [98].

The predictive controller is very suitable for electronic power converters and drives due to its fast dynamic response. Some of the features that make predictive control a suitable method to microgrids are as follows [99]:

- It is equipped with a feedback system that enables it to cope with uncertainty and disturbances;
- It may take into account prediction for both consumption and production;
- Its basic principles are based on the futuristic behavior of the system that rely on the demand and production of renewable energy;
- The multivariate structure of the predictive controller can manage the performance of microgrid units in a coordinated manner since coordination between diverse sources of renewable energy or energy storage systems in a microgrid is particularly challenging;
- When a disconnection or failure of one of the generation units occurs in a microgrid, the predictive controller can adapt to the current situation and provide safe operation for the entire system.

In modern industrial control, the model predictive control method is used for many industrial applications. These methods can be divided into two categories: 1. Finite control set MPC (FCS-MPC), which uses the advantage of limiting the number of switching possibilities to optimize the problem; 2. Continuous control set MPC (CCS-MPC), which requires a modulator to generate switching pulses according to the control system settings [100].

As we know, in a converter, the switches can only be switched on and off in two states, and their combination creates a limited number of different states. Using this inherent feature, the converter switching model can be easily presented and the prediction can be summarized only in the limited situations mentioned. The main elements of this control schemes are the mathematical model of the system and the predefined cost function. One of the most important advantages of FCS-MPC is its simplicity. The operation of this controller is that first, the system variables are measured or estimated, then the system model is extracted and discretized according to the controlled variable, which can be current, voltage and power. By discretizing the system model and knowing the value of current variables, controlled variables can be predicted in future times [101]. Table 6 provides a general classification of MPC approaches.

**Table 6.** General classification of the MPC approaches.

| Microgrid Based on MPC | |
|---|---|
| Predictive model | Dynamic of converters and RLC circuits, topology features, forecast, dynamic of DGs, ESSs, local demand |
| Control scheme | Centralized<br>Decentralized<br>Distributed/hierarchical |
| Objectives in cost function | Converter outputs: voltage, current, active power reactive power, circulating currents, Compensation of frequency and voltage |
| Control parameters | Voltage, frequency, active/reactive power |
| Solving algorithm | Exhaustive search (FCS-MPC, CCS-MPC)<br>Specific toolbox and solver |

All predictive control strategies are based on solving an optimization problem so that it can select the appropriate control signal and apply to power electronic converter. Cost functions also define the appropriate behavior for the system. Thus, cost functions can be complex depending on what type of variable is considered.

### *5.1. MPC-Based Primary Control*

In [102], the FCS-MPC approach is proposed as the primary control layer to control the output power of each DGs (grid-connected mode) and the voltage regulate in PCC (in island mode). In the grid-connected mode, the Direct Power Model Predictive Control approach (DPMPC) is applied to manage the power sharing between each DG and the grid, and in island mode, Voltage Model Predictive Control (VMPC) approach is provided as a droop control and secondary to adjust the output voltage of DGs. In [103], a research on the design of predictive controller based on discrete time model for voltage control in island microgrids is presented. In this research, several distributed generation units are connected in parallel to build a microgrid. The purpose of the controller design is to control the mains voltage on different loads. The design of this controller is done using a discrete-time function that provides extensive microgrid tracking function; the controller is for a single-phase microgrid design. The performance of this controller has been tested in various ways. This controller offers acceptable performance in microgrids against the dynamics of different loads and shows better tracking performance.

In the reference [104], a study on the method of controlling multiple photovoltaic systems in a DC microgrid using a predictive controller is presented. Direct current microgrid sys- tems are beneficial due to higher efficiency, reliability and easier connection of renewable energy sources compared to AC microgrid systems. The finite set of predictive control models is a robust control technique that predicts the future behavior of the system in a number of arbitrary sampling steps over the time horizon, based on a set of possible control operators [105]. The proposed predictive controller for DC microgrids includes maximum power point tracking and bidirectional DC-DC converter control to evacuate the battery power storage system. Using predictive controller features with the ability to add design constraints, the proposed system ensures maximum output power from the PV system due to the battery SoC. The controller increases the speed of the control loop because it increases the prediction and correction of the error before the switching signal is applied to the DC-DC converters. In [106], a MPC-based VSG control structure is proposed for the primary controller for voltage control and power sharing, i.e., firstly this predictive controller is applied to the inner loop, which provides a faster dynamic response as well as increased bandwidth and stability; then, a VSG has been proposed to proper power sharing in an outer loop, and a comprehensive comparison between the proposed control structure and the conventional structures is presented in Table 7.

**Table 7.** Comparison of control strategies.

| Control Type | Advantages | Ref |
|---|---|---|
| DMPC | It has the benefits of MPC.<br>Compatibility of constraints and good optimization performance.<br>Flexibility distribution framework.<br>There is also fault tolerance, connection and execution in the network. | [102] |
| MPC | Ensures maximum output power from the PV system.<br>The controller increases the speed of the control loop.<br>Increases error prediction and correction before the switching signal is applied to DC/DC converters. | [103] |
| MPC | Has the ability to predict the future behavior of the system.<br>The proposed technique is able to cope with the limitations that often occur in practice.<br>The predictive controller uses current system measurement, current system dynamics, target process variables and future event calculations that are rarely considered in other control techniques.<br>This control uses the proposed cost function, which ensures high tracking performance of the microgrid system. | [104] |
| Droop | Frequency return under optimal active power control.<br>The voltage restoration and power sharing reactive simultaneously.<br>Select a critical path for voltage restoration or accurate reactive power sharing with AC side voltage limitation. | [105] |
| Droop | The voltage controller is fully centralized at the initial level and no digital communication is required.<br>The design method is scalable.<br>The controller ensures the stability of the overall microgrid system.<br>The stable performance of the microgrid system is in accordance with IEEE standards.<br>The controller provides control power according to load changes and microgrid structure. | [106] |

### *5.2. MPC-Based Secondary Control*

In [107], a unified model predictive voltage and current control (UMPVIC) strategy is presented for both grid-connected mode and island mode, which can be flexibly applied in the primary control layer for proper load sharing and in the tertiary layer for power flow. A fuzzy control algorithm is proposed to reduce the voltage and frequency deviations caused by the primary droop layer. This fuzzy controller can optimize secondary layer coefficients to improve voltage quality. The dynamic response of converters is much faster than the frequency load. Usually, the droop control method is used to share the active and reactive powers of microgrids. The disadvantages of this method are the deviation of voltage and frequency from its nominal value in the steady state. A distributed secondary controller based on the prediction model with a state space approach is proposed to return the voltage and frequency of the microgrids. Therefore, a secondary controller is provided to eliminate voltage and frequency deviations, and the performance of the proposed method is evaluated by evaluating it under a feeder impedance that is inconsistent with balanced and unbalanced linear load conditions [108]. A distributed model predictive control strategy based on voltage observer for multiple energy storage systems is proposed. In order to reduce the effects of communication delay on the voltage observer, an improved distributed predictive control algorithm is proposed. This proposed scheme strengthens the delayed system. A small-signal dynamic model with a predictive controller is used to analyze the dynamic performance [109].

## 6. Fault Tolerant of Microgrids

Assessing the reliability of the distribution system is one of the most important studies for power system operators and planners. Existing distribution networks are designed to be powered on one side, but with the presence of distributed energy resources, it is possible to transfer power from both sides. In other words, power may be transferred from the distribution network to the transmission network. For this reason, conventional risk management methods will not be used to calculate the reliability of this type of network, and researchers are looking for new solutions to calculate the reliability of this type of network. Reliability is predicted based on expected performance; in other words, performance without failure. Increasing the use of power generation units using renewable energy in microgrids has led to attention to the reliability of these systems. Since such distributed generation sources have a lot of uncertainty, necessary measures should be considered to provide sustainable power to the consumer [110,111].

Recently, the predictive control method of the model has found increasing interest among researchers in the field of reliability. In [112], a combined method of dynamic planning based on predictive control is presented. In [113], a predictive-based approach is proposed for the management of grid-connected photovoltaic power generation. In [114], a microgrid scheduling method based on MPC is proposed to increase the flexibility of distribution networks. A robust MPC-based two-layer model has been developed considering the worst case scenario for microgrids. It is shown that the proposed solution can ensure the reliability and feasibility of microgrids in the presence of uncertainties. A microgrid power management framework that is connected to the grid via a transformer and contains a local consumer, a wind turbine and an energy storage system is provided to minimize costs based on optimal battery planning. The model predictive control approach has been used to improve power, cost and production in the presence of uncertainties [115]. In [116], a predictive control algorithm is proposed to solve the problem of economic optimization in the presence of distributed generation sources.

## 7. Challenges and Future Perspectives

Among different types of microgrids, DC microgrids have problems in the field of power quality and one of the important obstacles to the development of this type of microgrids is limited DC loads. Therefore, with the development of technologies, it is expected that DC loads will be more compatible. On the other hand, due to the high cost of energy storage systems in AC microgrids, such advances in technology can make these systems more cost-effective. Since the issue of power quality and grid stability are the most important issues in the field of microgrids, with the widespread penetration of renewable energy sources in microgrids, more efforts should be made in this area [117].

An integrated distributed planning model can provide an effective solution for balancing the efficiency and allocation of costs of DERs. Among the items that can play an important role include [118]:

- Improving connection cost estimates can develop an integrated planning model for energy resources to make photovoltaic system equipment installation more efficient;

- Estimating and predicting more accurate, robust economic distribution of distributed generation resources can help reduce the risk of recovering potential future costs due to uncertainty in the future development of distributed generation resources.

Accurate forecasts of distributed energy sources are of particular importance in the long-term planning of distribution and transmission. In both over and under forecasting, the transmission line in turn can have special consequences for the system. In other words, if this forecast is done too much, the reliability and flexibility of the system will be affected. In case of under forecasting, unnecessary generation distribution resources can be added.

The two forecast methodologies for future distributed energy sources are shown in Table 8. Top-down approaches provide a significant improvement in data and methodological complexity. Time series, econometric and bass diffusion models are among the top-down models that have been employed to anticipate DER adoption. Time-series models extrapolate historical, cyclic data to future outcomes. Econometric models are statistical models that are used to explain observable data and can be expressed in a variety of ways. The most often utilized strategy for forecasting DER adoption is the use of bass diffusion models.

**Table 8.** Methods used for DER forecasting [118].

| | **Top-Down** | | | **Bottom-Up** |
|---|---|---|---|---|
| | **Time Series** | **Econometric** | **Bass Diffusion** | |
| Prospective | Simple, easy to estimate and validate | High familiarity and use in other domains | Easy to specify | Modeling unique attributes of consumers |
| Consequence | Lack of expression of technical restrictions | Prediction aggregate adoption than feeder level is better | Sensitive to transient market effects | Requires computational investment resources |

## 8. Conclusions

Microgrids are a modern concept for the future of energy systems that have made it possible to use renewable energy and require electrical and control structures and equipment for optimal performance. One of the most important and fundamental issues in microgrids are how to control distributed generation sources to supply local load. Since most distributed generation sources are connected to the grid by electronic power converters, appropriate methods must be used to control electronic power converters to ensure the microgrid's economic performance. The purpose of this research is to review what has been done so far about energy source-based microgrids in the literature. Hierarchical control structures are expressed and its control method is evaluated. The MPC play an important role in microgrid performance and system reliability. Therefore, a review of MPC-based control structures and following that a comparison between its control strategies are performed.

**Author Contributions:** Conceptualization, D.R. and T.L.; Investigation, D.R. and T.L.; Writing—original draft preparation, D.R.; Writing—review and editing, D.R. and T.L.; Supervision, T.L.; Project administration, T.L.; Funding acquisition, T.L. All authors have read and agreed to the published version of the manuscript.

**Funding:** This work is financially supported by the General Program of National Natural Science Foundation of China (No.52107110); Young Elite Scientists Sponsorship Program (No. YESS20200239); The Shandong Natural Science Foundation for Excellent Young Scientists Program of China (2022HWYQ039); The Shandong Natural Science Foundation of China (ZR2021QE018).

**Institutional Review Board Statement:** Not applicable.

**Informed Consent Statement:** Informed consent was obtained from all subjects involved in the study.

**Conflicts of Interest:** The authors declare no conflict of interest.

## References


1. Mozina, C.J. Impact of green power distributed generation. *IEEE Ind. Appl. Mag.* **2010**, *16*, 55–62. [CrossRef]
2. Wang, C.; Nehrir, M.H. Analytical approaches for optimal placement of distributed generation sources in power systems. *IEEE Trans. Power Syst.* **2004**, *19*, 2068–2076. [CrossRef]
3. Ochoa, L.F.; Padilha-Feltrin, A.; Harrison, G.P. Evaluating distributed generation impacts with a multiobjective index. *IEEE Trans. Power Deliv.* **2006**, *21*, 1452–1458. [CrossRef]
4. Ochoa, L.F.; Padilha-Feltrin, A.; Harrison, G.P. Evaluating distributed time-varying generation through a multiobjective index. *IEEE Trans. Power Deliv.* **2008**, *23*, 1132–1138. [CrossRef]
5. Abu-Mouti, F.S.; El-Hawary, M. Optimal distributed generation allocation and sizing in distribution systems via artificial bee colony algorithm. *IEEE Trans. Power Deliv.* **2011**, *26*, 2090–2101. [CrossRef]
6. Dent, C.J.; Ochoa, L.F.; Harrison, G.P. Network distributed generation capacity analysis using OPF with voltage step constraints. *IEEE Trans. Power Syst.* **2009**, *25*, 296–304. [CrossRef]
7. Atwa, Y.; El-Saadany, E.; Salama, M.; Seethapathy, R. Optimal renewable resources mix for distribution system energy loss minimization. *IEEE Trans. Power Syst.* **2009**, *25*, 360–370. [CrossRef]
8. Celli, G.; Mocci, S.; Pilo, F.; Soma, G.G. A multi-objective approach for the optimal distributed generation allocation with environmental constraints. In Proceedings of the 10th International Conference on Probablistic Methods Applied to Power Systems, Rincon, PR, USA, 25–29 May 2008; pp. 1–8.
9. Haddad, R.J.; Guha, B.; Kalaani, Y.; El-Shahat, A. Smart distributed generation systems using artificial neural network-based event classification. *IEEE Power Energy Technol. Syst. J.* **2018**, *5*, 18–26. [CrossRef]
10. Krishan, O.; Suhag, S. An updated review of energy storage systems: Classification and applications in distributed generation power systems incorporating renewable energy resources. *Int. J. Energy Res.* **2019**, *43*, 6171–6210. [CrossRef]
11. Bahramara, S.; Mazza, A.; Chicco, G.; Shafie-khah, M.; Catalão, J.P. Comprehensive review on the decision-making frameworks referring to the distribution network operation problem in the presence of distributed energy resources and microgrids. *Int. J. Electr. Power Energy Syst.* **2020**, *115*, 105466. [CrossRef]
12. Sarangi, S.; Sahu, B.K.; Rout, P.K. Distributed generation hybrid AC/DC microgrid protection: A critical review on issues, strategies, and future directions. *Int. J. Energy Res.* **2020**, *44*, 3347–3364. [CrossRef]
13. Bajaj, M.; Singh, A.K. An analytic hierarchy process-based novel approach for benchmarking the power quality performance of grid-integrated renewable energy systems. *Electr. Eng.* **2020**, *102*, 1153–1173. [CrossRef]
14. Razmjoo, A.; Kaigutha, L.G.; Rad, M.V.; Marzband, M.; Davarpanah, A.; Denai, M. A Technical analysis investigating energy sustainability utilizing reliable renewable energy sources to reduce $CO_2$ emissions in a high potential area. *Renew. Energy* **2021**, *164*, 46–57. [CrossRef]
15. Guerrero, J.; Gebbran, D.; Mhanna, S.; Chapman, A.C.; Verbič, G. Towards a transactive energy system for integration of distributed energy resources: Home energy management, distributed optimal power flow, and peer-to-peer energy trading. *Renew. Sustain. Energy Rev.* **2020**, *132*, 110000. [CrossRef]
16. Liu, Z.; Guo, J.; Wu, D.; Fan, G.; Zhang, S.; Yang, X.; Ge, H. Two-phase collaborative optimization and operation strategy for a new distributed energy system that combines multi-energy storage for a nearly zero energy community. *Energy Convers. Manag.* **2021**, *230*, 113800. [CrossRef]
17. Rayudua, K. Design of Controller for Transition of Grid Connected Microgrid to Island Mode. *Turk. J. Comput. Math. Educ. (TURCOMAT)* **2021**, *12*, 845–854.
18. Zhao, W.; Xu, M.; Ning, N.; Kang, Y.; Cai, Z. A Microgrid Capacity Optimization Method Considering Carbon Emission Cost. In Proceedings of the 2020 IEEE Sustainable Power and Energy Conference (iSPEC), Chengdu, China, 23–25 November 2020; pp. 2219–2224.
19. Shahgholian, G. A brief review on microgrids: Operation, applications, modeling, and control. *Int. Trans. Electr. Energy Syst.* **2021**, *31*, e12885. [CrossRef]
20. Sarangi, S.; Sahu, B.K.; Rout, P.K. A comprehensive review of distribution generation integrated DC microgrid protection: Issues, strategies, and future direction. *Int. J. Energy Res.* **2021**, *45*, 5006–5031. [CrossRef]
21. Ghenai, C.; Bettayeb, M. Modelling and performance analysis of a stand-alone hybrid solar PV/Fuel Cell/Diesel Generator power system for university building. *Energy* **2019**, *171*, 180–189. [CrossRef]
22. Abdmouleh, Z.; Gastli, A.; Ben-Brahim, L.; Haouari, M.; Al-Emadi, N.A. Review of optimization techniques applied for the integration of distributed generation from renewable energy sources. *Renew. Energy* **2017**, *113*, 266–280. [CrossRef]
23. Hossain, E.; Tür, M.R.; Padmanaban, S.; Ay, S.; Khan, I. Analysis and mitigation of power quality issues in distributed generation systems using custom power devices. *IEEE Access* **2018**, *6*, 16816–16833. [CrossRef]
24. Twaha, S.; Ramli, M.A. A review of optimization approaches for hybrid distributed energy generation systems: Off-grid and grid-connected systems. *Sustain. Cities Soc.* **2018**, *41*, 320–331. [CrossRef]
25. Adnani, K.; Vaez-Zadeh, S.; Jabbarnejad, A. Flexible active/reactive power ripple control of grid-connected voltage source converters under unbalance conditions. In Proceedings of the 2020 IEEE 29th International Symposium on Industrial Electronics (ISIE), Delft, The Netherlands, 17–19 June 2020; pp. 1022–1027.
26. Das, C.K.; Bass, O.; Kothapalli, G.; Mahmoud, T.S.; Habibi, D. Overview of energy storage systems in distribution networks: Placement, sizing, operation, and power quality. *Renew. Sustain. Energy Rev.* **2018**, *91*, 1205–1230. [CrossRef]

27. Krishan, O.; Suhag, S. Power management control strategy for hybrid energy storage system in a grid-independent hybrid renewable energy system: A hardware-in-loop real-time verification. *IET Renew. Power Gener.* **2020**, *14*, 454–465. [CrossRef]
28. Igwemezie, V.; Mehmanparast, A.; Kolios, A. Current trend in offshore wind energy sector and material requirements for fatigue resistance improvement in large wind turbine support structures—A review. *Renew. Sustain. Energy Rev.* **2019**, *101*, 181–196. [CrossRef]
29. Ranjbaran, P.; Yousefi, H.; Gharehpetian, G.; Astaraei, F.R. A review on floating photovoltaic (FPV) power generation units. *Renew. Sustain. Energy Rev.* **2019**, *110*, 332–347. [CrossRef]
30. Lee, J.; Shepley, M.M. Benefits of solar photovoltaic systems for low-income families in social housing of Korea: Renewable energy applications as solutions to energy poverty. *J. Build. Eng.* **2020**, *28*, 101016. [CrossRef]
31. Fershalov, A.Y.; Fershalov, Y.Y.; Fershalov, M.Y. Principles of designing gas microturbine stages. *Energy* **2021**, *218*, 119488. [CrossRef]
32. Fathy, A.; Ferahtia, S.; Rezk, H.; Yousri, D.; Abdelkareem, M.A.; Olabi, A. Optimal adaptive fuzzy management strategy for fuel cell-based DC microgrid. *Energy* **2022**, *247*, 123447. [CrossRef]
33. Alzahrani, A.M.; Zohdy, M.; Yan, B. An overview of optimization approaches for operation of hybrid distributed energy systems with photovoltaic and diesel turbine generator. *Electr. Power Syst. Res.* **2021**, *191*, 106877. [CrossRef]
34. Al-Salloomee, A.G.S.; Khosroabadi, S.; Albukariat, A.A.A. Study of power management of standalone DC microgrids with battery supercapacitor hybrid energy storage system. *Int. J. Electr. Comput. Eng.* **2022**, *12*, 114–121. [CrossRef]
35. Brahmendra Kumar, G.; Palanisamy, K. Review of energy storage system for microgrid. In *Microgrid Technologies*; Wiley: Hoboken, NJ, USA, 2021; pp. 57–90.
36. Sandhya, K.; Chatterjee, K. A review on the state of the art of proliferating abilities of distributed generation deployment for achieving resilient distribution system. *J. Clean. Prod.* **2021**, *287*, 125023. [CrossRef]
37. Deshmukh, A.N.; Chandrakar, V.K. Power quality issues and their mitigation techniques in grid tied Solar Photovoltaic Systems—A review. In Proceedings of the 2021 International Conference on Computer Communication and Informatics (ICCCI), Coimbatore, India, 27–29 January 2021; pp. 1–6.
38. Dsouza, K.; Halbe, S.; Thomas, M.; Baran, M.; Chowdhury, B.; Schwarz, P.; Proudlove, A. The challenges of valuing distributed generation. In Proceedings of the 2020 IEEE Power & Energy Society Innovative Smart Grid Technologies Conference (ISGT), Washington, DC, USA, 17–20 February 2020; pp. 1–5.
39. Mohammed, A.A.A.; Ahamed, A.M.M.M.; Kenish, S.S.E.M.E.; Mansour, M.A.O.A.; Superviser, S.E.M. Design and Integration of Distributed Generation Systems to Dongola Distribution Grid. Ph.D. Thesis, Sudan University of Science & Technology, Khartoum, Sudan, 2020.
40. Yamashita, D.Y.; Vechiu, I.; Gaubert, J.P. A review of hierarchical control for building microgrids. *Renew. Sustain. Energy Rev.* **2020**, *118*, 109523. [CrossRef]
41. Feng, X.; Shekhar, A.; Yang, F.; Hebner, R.E.; Bauer, P. Comparison of hierarchical control and distributed control for microgrid. *Electr. Power Compon. Syst.* **2017**, *45*, 1043–1056. [CrossRef]
42. Rokrok, E.; Shafie-Khah, M.; Catalão, J.P. Review of primary voltage and frequency control methods for inverter-based islanded microgrids with distributed generation. *Renew. Sustain. Energy Rev.* **2018**, *82*, 3225–3235. [CrossRef]
43. Bendib, A.; Kherbachi, A.; Kara, K.; Chouder, A. Droop controller based primary control scheme for parallel-connected single-phase inverters in islanded AC microgrid. In Proceedings of the 2017 5th International Conference on Electrical Engineering-Boumerdes (ICEE-B), Boumerdes, Algeria, 29–31 October 2017; pp. 1–6.
44. Unamuno, E.; Barrena, J.A. Equivalence of primary control strategies for AC and DC microgrids. *Energies* **2017**, *10*, 91. [CrossRef]
45. Fusero, M.; Tuckey, A.; Rosini, A.; Serra, P.; Procopio, R.; Bonfiglio, A. A comprehensive inverter-BESS primary control for AC microgrids. *Energies* **2019**, *12*, 3810. [CrossRef]
46. Alfergani, A.; Alfaitori, K.A.; Khalil, A.; Buaossa, N. Control strategies in AC microgrid: A brief review. In Proceedings of the 2018 9th International Renewable Energy Congress (IREC), Hammamet, Tunisia, 20–22 March 2018; pp. 1–6.
47. Khodabakhsh, J.; Moschopoulos, G. Primary Frequency Control in Islanded Microgrids Using Solid-State Transformers as Virtual Synchronous Machines. In Proceedings of the 2020 IEEE Energy Conversion Congress and Exposition (ECCE), Detroit, MI, USA, 11–15 October 2020; pp. 5380–5385.
48. Datta, U.; Kalam, A.; Shi, J.; Li, J. Electric vehicle charging station for providing primary frequency control in microgrid. In Proceedings of the 2019 14th IEEE Conference on Industrial Electronics and Applications (ICIEA), Xi'an, China, 19–21 June 2019; pp. 2440–2444.
49. Kariminejad, N.; Taher, S.A.; Shahidehpour, M.; Khateri, K. A hierarchical governor/turbine and electric vehicles optimal control framework for primary frequency support in power systems. *IEEE Trans. Smart Grid* **2017**, *9*, 6702–6712. [CrossRef]
50. Hu, J.; Duan, J.; Ma, H.; Chow, M.Y. Distributed adaptive droop control for optimal power dispatch in DC microgrid. *IEEE Trans. Ind. Electron.* **2017**, *65*, 778–789. [CrossRef]
51. Dragičević, T.; Lu, X.; Vasquez, J.C.; Guerrero, J.M. DC microgrids—Part II: A review of power architectures, applications, and standardization issues. *IEEE Trans. Power Electron.* **2015**, *31*, 3528–3549. [CrossRef]
52. Krismanto, A.U.; Mithulananthan, N.; Lomi, A. Dynamic droop control in microgrid for stability enhancement considering RES variation. In Proceedings of the 2017 IEEE PES Innovative Smart Grid Technologies Conference Europe (ISGT-Europe), Turin, Italy, 26–29 September 2017; pp. 1–6.

53. Morales-Paredes, H.K.; Bonaldo, J.P.; Pomilio, J.A. Centralized control center implementation for synergistic operation of distributed multifunctional single-phase grid-tie inverters in a microgrid. *IEEE Trans. Ind. Electron.* **2018**, *65*, 8018–8029. [CrossRef]
54. Guo, Z.; Jiang, H.; Zheng, Y.; Li, S. Distributed model predictive control for efficient operation of islanded microgrid. In Proceedings of the 2017 Chinese Automation Congress (CAC), Jinan, China, 20–22 October 2017; pp. 6253–6258.
55. Khayat, Y.; Shafiee, Q.; Heydari, R.; Naderi, M.; Dragičević, T.; Simpson-Porco, J.W.; Dörfler, F.; Fathi, M.; Blaabjerg, F.; Guerrero, J.M.; et al. On the secondary control architectures of AC microgrids: An overview. *IEEE Trans. Power Electron.* **2019**, *35*, 6482–6500. [CrossRef]
56. Tayab, U.B.; Roslan, M.A.B.; Hwai, L.J.; Kashif, M. A review of droop control techniques for microgrid. *Renew. Sustain. Energy Rev.* **2017**, *76*, 717–727. [CrossRef]
57. Liu, B.; Wu, T.; Liu, Z.; Liu, J. A small-AC-signal injection-based decentralized secondary frequency control for droop-controlled islanded microgrids. *IEEE Trans. Power Electron.* **2020**, *35*, 11634–11651. [CrossRef]
58. Han, Y.; Zhang, K.; Li, H.; Coelho, E.A.A.; Guerrero, J.M. MAS-based distributed coordinated control and optimization in microgrid and microgrid clusters: A comprehensive overview. *IEEE Trans. Power Electron.* **2017**, *33*, 6488–6508. [CrossRef]
59. Bharath, K.; Krishnan, M.M.; Kanakasabapathy, P. A review on DC microgrid control techniques, applications and trends. *Int. J. Renew. Energy Res.* **2019**, *9*, 1328–1338.
60. Deng, S.; Chen, L.; Zheng, T.; Guo, Y.; Mei, S. Distributed Secondary Frequency Control for Microgrids Under Weak Communication Conditions. In Proceedings of the 2020 IEEE Power & Energy Society General Meeting (PESGM), Montreal, QC, Canada, 2–6 August 2020; pp. 1–5.
61. Hong, Y.; Xie, J.; Fang, F. Distributed Secondary Control of Islanded Microgrids for Fast Convergence Considering Arbitrary Sampling. *Processes* **2021**, *9*, 971. [CrossRef]
62. Pan, R.; Sun, P. Wireless fundamental frequency-based variable mode power sharing strategy for autonomous microgrid. *Electr. Eng.* **2022**, *104*, 1473–1486. [CrossRef]
63. Qureshi, T.; Alvi, R. Review on secondary and tertiary control structures for Microgrid. *Int. J. Comput. Sci. Eng.* **2018**, *6*, 251–255. [CrossRef]
64. Hirsch, A.; Parag, Y.; Guerrero, J. Microgrids: A review of technologies, key drivers, and outstanding issues. *Renew. Sustain. Energy Rev.* **2018**, *90*, 402–411. [CrossRef]
65. Chen, Y.; Guerrero, J.M.; Shuai, Z.; Chen, Z.; Zhou, L.; Luo, A. Fast reactive power sharing, circulating current and resonance suppression for parallel inverters using resistive-capacitive output impedance. *IEEE Trans. Power Electron.* **2015**, *31*, 5524–5537. [CrossRef]
66. Li, Z.; Wang, P.; Chu, Z.; Zhu, H.; Sun, Z.; Li, Y. A three-phase 10 kVAC-750 VDC power electronic transformer for smart distribution grid. In Proceedings of the 2013 15th European Conference on Power Electronics and Applications (EPE), Lille, France, 2–6 September 2013; pp. 1–9.
67. Mohamed, Y.A.R.I.; Zeineldin, H.H.; Salama, M.; Seethapathy, R. Seamless formation and robust control of distributed generation microgrids via direct voltage control and optimized dynamic power sharing. *IEEE Trans. Power Electron.* **2011**, *27*, 1283–1294. [CrossRef]
68. Mohamed, Y.A.R.I.; El-Saadany, E.F. Adaptive decentralized droop controller to preserve power sharing stability of paralleled inverters in distributed generation microgrids. *IEEE Trans. Power Electron.* **2008**, *23*, 2806–2816. [CrossRef]
69. Mohamed, Y.A.R.I.; Radwan, A.A. Hierarchical control system for robust microgrid operation and seamless mode transfer in active distribution systems. *IEEE Trans. Smart Grid* **2011**, *2*, 352–362. [CrossRef]
70. Blaabjerg, F.; Teodorescu, R.; Liserre, M.; Timbus, A.V. Overview of control and grid synchronization for distributed power generation systems. *IEEE Trans. Ind. Electron.* **2006**, *53*, 1398–1409. [CrossRef]
71. Abhishek, A.; Ranjan, A.; Devassy, S.; Verma, B.K.; Ram, S.K.; Dhakar, A.K. Review of hierarchical control strategies for DC microgrid. *IET Renew. Power Gener.* **2020**, *14*, 1631–1640. [CrossRef]
72. Eltawil, M.A.; Zhao, Z. Grid-connected photovoltaic power systems: Technical and potential problems—A review. *Renew. Sustain. Energy Rev.* **2010**, *14*, 112–129. [CrossRef]
73. Yu, X.; Khambadkone, A.M.; Wang, H.; Terence, S.T.S. Control of parallel-connected power converters for low-voltage microgrid—Part I: A hybrid control architecture. *IEEE Trans. Power Electron.* **2010**, *25*, 2962–2970. [CrossRef]
74. Sao, C.K.; Lehn, P.W. Control and power management of converter fed microgrids. *IEEE Trans. Power Syst.* **2008**, *23*, 1088–1098. [CrossRef]
75. Tuladhar, A.; Jin, H.; Unger, T.; Mauch, K. Parallel operation of single phase inverter modules with no control interconnections. In Proceedings of the APEC 97-Applied Power Electronics Conference, Atlanta, GA, USA, 23–27 February 1997; Volume 1, pp. 94–100.
76. Vasquez, J.C.; Guerrero, J.M.; Luna, A.; Rodríguez, P.; Teodorescu, R. Adaptive droop control applied to voltage-source inverters operating in grid-connected and islanded modes. *IEEE Trans. Ind. Electron.* **2009**, *56*, 4088–4096. [CrossRef]
77. Guerrero, J.M.; Matas, J.; De Vicuna, L.G.D.V.; Castilla, M.; Miret, J. Wireless-control strategy for parallel operation of distributed-generation inverters. *IEEE Trans. Ind. Electron.* **2006**, *53*, 1461–1470. [CrossRef]
78. Guerrero, J.M.; Vasquez, J.C.; Matas, J.; Castilla, M.; de Vicuña, L.G. Control strategy for flexible microgrid based on parallel line-interactive UPS systems. *IEEE Trans. Ind. Electron.* **2008**, *56*, 726–736. [CrossRef]

79. Chen, J.; Wang, L.; Diao, L.; Du, H.; Liu, Z. Distributed auxiliary inverter of urban rail train—Load sharing control strategy under complicated operation condition. *IEEE Trans. Power Electron.* **2015**, *31*, 2518–2529. [CrossRef]
80. Shafiee, Q.; Guerrero, J.M.; Vasquez, J.C. Distributed secondary control for islanded microgrids—A novel approach. *IEEE Trans. Power Electron.* **2013**, *29*, 1018–1031. [CrossRef]
81. Zhong, Q.C. Robust droop controller for accurate proportional load sharing among inverters operated in parallel. *IEEE Trans. Ind. Electron.* **2011**, *60*, 1281–1290. [CrossRef]
82. Nutkani, I.U.; Loh, P.C.; Wang, P.; Blaabjerg, F. Autonomous droop scheme with reduced generation cost. *IEEE Trans. Ind. Electron.* **2014**, *61*, 6803–6811. [CrossRef]
83. Nutkani, I.U.; Loh, P.C.; Blaabjerg, F. Droop scheme with consideration of operating costs. *IEEE Trans. Power Electron.* **2013**, *29*, 1047–1052. [CrossRef]
84. Nutkani, I.U.; Loh, P.C.; Wang, P.; Blaabjerg, F. Cost-prioritized droop schemes for autonomous AC microgrids. *IEEE Trans. Power Electron.* **2014**, *30*, 1109–1119. [CrossRef]
85. Micallef, A. Review of the current challenges and methods to mitigate power quality issues in single-phase microgrids. *IET Gener. Transm. Distrib.* **2019**, *13*, 2044–2054. [CrossRef]
86. Zhang, J.; Shu, J.; Ning, J.; Huang, L.; Wang, H. Enhanced proportional power sharing strategy based on adaptive virtual impedance in low-voltage networked microgrid. *IET Gener. Transm. Distrib.* **2018**, *12*, 2566–2576. [CrossRef]
87. Dou, C.; Zhang, Z.; Yue, D.; Song, M. Improved droop control based on virtual impedance and virtual power source in low-voltage microgrid. *IET Gener. Transm. Distrib.* **2017**, *11*, 1046–1054. [CrossRef]
88. Liu, B.; Liu, Z.; Liu, J.; An, R.; Zheng, H.; Shi, Y. An adaptive virtual impedance control scheme based on small-AC-signal injection for unbalanced and harmonic power sharing in islanded microgrids. *IEEE Trans. Power Electron.* **2019**, *34*, 12333–12355. [CrossRef]
89. Keddar, M.; Doumbia, M.; Belmokhtar, K.; Della Krachai, M. A new energy management strategy of an autonomous microgrid based on virtual impedance in multilevel droop controlled inverters. In Proceedings of the 2019 International Conference on Advanced Electrical Engineering (ICAEE), Algiers, Algeria, 19–21 November 2019; pp. 1–7.
90. Mesbah, A. Stochastic model predictive control: An overview and perspectives for future research. *IEEE Control Syst. Mag.* **2016**, *36*, 30–44.
91. Kouro, S.; Cortés, P.; Vargas, R.; Ammann, U.; Rodríguez, J. Model predictive control—A simple and powerful method to control power converters. *IEEE Trans. Ind. Electron.* **2008**, *56*, 1826–1838. [CrossRef]
92. Rodriguez, J.; Kazmierkowski, M.P.; Espinoza, J.R.; Zanchetta, P.; Abu-Rub, H.; Young, H.A.; Rojas, C.A. State of the art of finite control set model predictive control in power electronics. *IEEE Trans. Ind. Inform.* **2012**, *9*, 1003–1016. [CrossRef]
93. Michalík, J.; Peroutka, Z.; Šmídl, V. Finite control set MPC of active current-source rectifier with full state space model. In Proceedings of the IECON 2015-41st Annual Conference of the IEEE Industrial Electronics Society, Yokohama, Japan, 9–12 November 2015; pp. 004121–004126.
94. Rodriguez, J.; Pontt, J.; Silva, C.A.; Correa, P.; Lezana, P.; Cortés, P.; Ammann, U. Predictive current control of a voltage source inverter. *IEEE Trans. Ind. Electron.* **2007**, *54*, 495–503. [CrossRef]
95. Karamanakos, P.; Stolze, P.; Kennel, R.M.; Manias, S.; du Toit Mouton, H. Variable switching point predictive torque control of induction machines. *IEEE J. Emerg. Sel. Top. Power Electron.* **2014**, *2*, 285–295. [CrossRef]
96. Babqi, A.J.; Etemadi, A.H. MPC-based microgrid control with supplementary fault current limitation and smooth transition mechanisms. *IET Gener. Transm. Distrib.* **2017**, *11*, 2164–2172. [CrossRef]
97. Sun, D.; Wang, X. Low-complexity model predictive direct power control for DFIG under both balanced and unbalanced grid conditions. *IEEE Trans. Ind. Electron.* **2016**, *63*, 5186–5196. [CrossRef]
98. Qin, S.J.; Badgwell, T.A. A survey of industrial model predictive control technology. *Control Eng. Pract.* **2003**, *11*, 733–764. [CrossRef]
99. Vazquez, S.; Rodriguez, J.; Rivera, M.; Franquelo, L.G.; Norambuena, M. Model predictive control for power converters and drives: Advances and trends. *IEEE Trans. Ind. Electron.* **2016**, *64*, 935–947. [CrossRef]
100. Karamanakos, P.; Geyer, T.; Kennel, R. Reformulation of the long-horizon direct model predictive control problem to reduce the computational effort. In Proceedings of the 2014 IEEE Energy Conversion Congress and Exposition (ECCE), Pittsburgh, PA, USA, 14–18 September 2014; pp. 3512–3519.
101. Bella, S.; Houari, A.; Djerioui, A.; Chouder, A.; Machmoum, M.; Benkhoris, M.F.; Ghedamsi, K. Robust Model Predictive Control (MPC) for large-scale PV plant based on paralleled three-phase inverters. *Sol. Energy* **2020**, *202*, 409–419. [CrossRef]
102. Xiao, S.; Shadmand, M.B.; Balog, R.S. Model predictive control of multi-string PV systems with battery back-up in a community dc microgrid. In Proceedings of the 2017 IEEE Applied Power Electronics Conference and Exposition (APEC), Tampa, FL, USA, 26–30 March 2017; pp. 1284–1290.
103. Sarkar, S.K.; Badal, F.R.; Das, S.K.; Miao, Y. Discrete time model predictive controller design for voltage control of an islanded microgrid. In Proceedings of the 2017 3rd International Conference on Electrical Information and Communication Technology (EICT), Khulna, Bangladesh, 7–9 December 2017; pp. 1–6.
104. Sadabadi, M.S.; Shafiee, Q.; Karimi, A. Plug-and-play robust voltage control of DC microgrids. *IEEE Trans. Smart Grid* **2017**, *9*, 6886–6896. [CrossRef]
105. Wu, X.; Shen, C.; Iravani, R. A distributed, cooperative frequency and voltage control for microgrids. *IEEE Trans. Smart Grid* **2016**, *9*, 2764–2776. [CrossRef]

106. Zheng, C.; Dragičević, T.; Blaabjerg, F. Model predictive control-based virtual inertia emulator for an islanded alternating current microgrid. *IEEE Trans. Ind. Electron.* **2020**, *68*, 7167–7177. [CrossRef]
107. Yang, Q.; Zhou, J.; Chen, X.; Wen, J. Distributed MPC-Based Secondary Control for Energy Storage Systems in a DC Microgrid. *IEEE Trans. Power Syst.* **2021**, *36*, 5633–5644. [CrossRef]
108. Morattab, A.; Akhrif, O.; Saad, M. Decentralised coordinated secondary voltage control of multi-area power grids using model predictive control. *IET Gener. Transm. Distrib.* **2017**, *11*, 4546–4555. [CrossRef]
109. Farzin, H.; Fotuhi-Firuzabad, M.; Moeini-Aghtaie, M. Role of outage management strategy in reliability performance of multi-microgrid distribution systems. *IEEE Trans. Power Syst.* **2017**, *33*, 2359–2369. [CrossRef]
110. Sandelic, M.; Peyghami, S.; Sangwongwanich, A.; Blaabjerg, F. Reliability aspects in microgrid design and planning: Status and power electronics-induced challenges. *Renew. Sustain. Energy Rev.* **2022**, *159*, 112127. [CrossRef]
111. Muhtadi, A.; Pandit, D.; Nguyen, N.; Mitra, J. Distributed energy resources based microgrid: Review of architecture, control, and reliability. *IEEE Trans. Ind. Appl.* **2021**, *57*, 2223–2235. [CrossRef]
112. Polimeni, S.; Meraldi, L.; Moretti, L.; Leva, S.; Manzolini, G. Development and experimental validation of hierarchical energy management system based on stochastic model predictive control for Off-grid Microgrids. *Adv. Appl. Energy* **2021**, *2*, 100028. [CrossRef]
113. Tarragona, J.; Pisello, A.L.; Fernández, C.; de Gracia, A.; Cabeza, L.F. Systematic review on model predictive control strategies applied to active thermal energy storage systems. *Renew. Sustain. Energy Rev.* **2021**, *149*, 111385. [CrossRef]
114. Cai, S.; Xie, Y.; Wu, Q.; Xiang, Z. Robust MPC-based microgrid scheduling for resilience enhancement of distribution system. *Int. J. Electr. Power Energy Syst.* **2020**, *121*, 106068. [CrossRef]
115. Prodan, I.; Zio, E. A model predictive control framework for reliable microgrid energy management. *Int. J. Electr. Power Energy Syst.* **2014**, *61*, 399–409. [CrossRef]
116. Xie, L.; Ilic, M.D. Model predictive economic/environmental dispatch of power systems with intermittent resources. In Proceedings of the 2009 IEEE Power & Energy Society General Meeting, Calgary, AB, Canada, 26–30 July 2009; pp. 1–6.
117. Mariam, L.; Basu, M.; Conlon, M.F. Microgrid: Architecture, policy and future trends. *Renew. Sustain. Energy Rev.* **2016**, *64*, 477–489. [CrossRef]
118. Horowitz, K.A.; Peterson, Z.; Coddington, M.H.; Ding, F.; Sigrin, B.O.; Saleem, D.; Baldwin, S.E.; Lydic, B.; Stanfield, S.C.; Enbar, N.; et al. *An Overview of Distributed Energy Resource (DER) Interconnection: Current Practices and Emerging Solutions*; National Renewable Energy Lab. (NREL): Golden, CO, USA, 2019.

MDPI
St. Alban-Anlage 66
4052 Basel
Switzerland
www.mdpi.com

MDPI Books Editorial Office
E-mail: books@mdpi.com
www.mdpi.com/books

www.ingramcontent.com/pod-product-compliance
Lightning Source LLC
Chambersburg PA
CBHW061444220926
51850CB00004B/11
*9783036596341*